COMPREHENSIVE ORGANIC TRANSFORMATIONS

A Guide to Functional Group Preparations

COMPREHENSIVE ORGANIC TRANSFORMATIONS

A Guide to
Functional Group Preparations

By
Richard C. Larock

Richard C. Larock
Department of Chemistry
Iowa State University
Ames, Iowa 50011

Library of Congress Cataloging-in-Publication Data

Larock, Richard C., 1944–
Comprehensive organic transformations: a guide to functional
   group preparations/by Richard Larock; foreword by H. C. Brown

     p.   cm.
   Includes index.
   ISBN 0-89573-710-8
   1. Chemistry, Organic—Synthesis. I. Title.
QD262.L355   1989
547′.2—dc19   89-30333
                                                          CIP

British Library Cataloguing in Publication Data

Larock, Richard C.
   Comprehensive organic transformations:
   a guide to functional group preparations.
   1. Organic compounds. Synthesis
   I. Title
   547′.2

   ISBN 0-89573-710-8

Printed in the United States of America.
ISBN 0-89573-710-8 VCH Publishers

Printing History:
10 9 8

Published jointly by:

VCH Publishers, Inc.   VCH Verlagsgesellschaft mbH   VCH Publishers (UK) Ltd.
220 East 23rd Street   P.O. Box 10 11 61   8 Wellington Court
Suite 909   D-6940 Weinheim   Cambridge CB1 1HW
New York, New York 10010   Federal Republic of Germany   United Kingdom

The text and chemical structures in this book were set by
Science Typographers, Inc., 15 Industrial Boulevard,
Medford, L.I., New York 11763

To all the students, past, present, and future,
with whom I have had or will have the pleasure of working
on the development of new organic synthetic methodology.

FOREWORD

At one time organic chemists looked askance at those who elected to devote their research effort to the development of new, selective methods for achieving chemical transformations of organic compounds. Early practitioners in this area received many critical comments—it was felt that the proper objective of organic chemists was the synthesis of natural products, and that one should only develop a new method when such a method was required to overcome a hurdle in the synthetic procedure. But we persisted. Now there appears to be a general acceptance of the value of such research.

Now a new problem has appeared. We have brought forth so many new, highly selective synthetic methods that it is difficult for the chemist to know all of them and to select the method that would be most effective for the synthesis in hand. One approach is to publish monographs that review specialized areas. For example, my own field of borane reagents was reviewed 15 years ago (*Organic Syntheses via Boranes*, by H. C. Brown, Wiley, 1975). A more recent book of this sort is *Borane Reagents*, by A. Pelter, K. Smith, and H. C. Brown, Academic Press, 1988.

But such specialized books solve only part of the problem. In the present work, Richard C. Larock has set himself the goal of organizing the entire panoply of synthetic methods. But there are now so many methods that even brief descriptions of each procedure would have produced a work as large as the new Oxford English Dictionary (16 volumes, Oxford University Press, 1986). Instead, he adopted the unique solution of indicating the reagent and transformation, with pertinent references. This provides a concise summary that should be of enormous assistance to those searching for a selective reaction to achieve a desired transformation in the presence of difficult substituents. This problem is frequently faced today by those who undertake the involved syntheses that are often an important objective of current research.

One must admire Richard Larock's courage in undertaking this monumental task.

Herbert C. Brown
H. C. Brown and R. B. Wetherill
Laboratories of Chemistry
Purdue University
West Lafayette, Indiana

PREFACE

Organic synthesis is one of the most rapidly developing areas of chemistry. Every day useful new reagents and reactions are reported worldwide in the chemical literature. It is increasingly difficult for the organic chemist to keep up with the latest in synthetic organic methodology without spending an inordinate amount of time reading a wide variety of chemical journals, including those whose focus is not strictly synthetic organic chemistry.

In recent years a variety of books and reviews have appeared to aid the organic chemist interested in synthetic methodology, but even the best of these have now ballooned to inconvenient multivolume sets whose cost is prohibitive to those just entering the field. The intent of the present volume is to provide a comprehensive, highly condensed, systematic collection of useful synthetic methodology that both the beginning student and the long-time practitioner of organic synthesis will find useful.

This book began in 1973 as a series of course handouts designed to cover the key reactions of the major organic functional groups. Like the aforementioned publications, this reference work has grown rapidly over the years to a major treatise covering a vast amount of synthetic organic methodology. It was felt that the synthetic organic community might find this compilation useful, so a serious effort has been made in the last two years to thoroughly update and organize this material for publication.

The author takes full responsibility (and credit?) for the choice of reactions and references. Obviously not every reaction or reference could be included. In choosing material for this text the author has observed the following guidelines. All reactions to be included should be general in scope or else so unique that the methodology will find real synthetic utility. Yields should generally be at least 50%. Reagents should be readily available or easily prepared and handled in the laboratory. As much as possible, similar transformations should appear together in as concise a format as possible. Significant limitations in methodology shall be noted. No effort has been made to cover the use of protecting groups since excellent reviews on this subject are already available. Likewise, heterocyclic chemistry has consciously been omitted, except where heterocycles have been employed to effect simple functional group manipulations. Multiple group transformations have been covered, although they present certain organizational problems. To those chemists whose contributions to synthetic organic chemistry may have been slighted or altogether ignored, I apologize. It would be appreciated if major errors or omissions be brought to the author's attention so that future printings or subsequent editions may be corrected.

All reactions have been systematically organized according to the functional group being synthesized, with no attempt to cover the less important functional groups. Within each section the methodology is subdivided into major processes, such as oxidation, reduction, alkylation, etc. It is hoped that the reader will easily find the desired transformations by skimming the detailed Table of Contents, although an extensive Transformation Index is available in time of need.

Literature coverage is complete through 1987. Some 160 or more primary chemical journals and a number of books and reviews have been abstracted. Obscure journals not readily available to most synthetic organic chemists have been avoided. The names of authors have been omitted to save space. Original publications have not always been cited if they do not necessarily describe the best reaction conditions for running the reaction or purvey little of the scope of the reaction. References containing full experimental procedures, though they may be buried in an experimental section, have been favored over communications lacking such details. An attempt has been made to highlight reviews and significant publications. One immediately encounters problems in deciding where to draw the line on references. Initial reports of a useful new reaction have received complete coverage. However, the time soon comes when a truly significant reaction, such as the use of ester enolates in synthesis, appears routinely in publication after publication and no reviews have appeared. In such situations, the author has tended to include all new material and may not have had the time to omit the more inconsequential earlier references.

It is hoped that the reader finds this effort worthwhile and will not hesitate to make suggestions on ways this material may be improved.

Richard C. Larock
Ames, Iowa U.S.A.
May, 1988

ACKNOWLEDGMENTS

The preparation of a book of this magnitude requires the assistance of a number of people. The author is indebted to Iowa State University for providing the time and assistance necessary for the preparation of much of this book. The Department of Chemistry at the University of Hawaii at Manoa is gratefully acknowledged for having provided a visiting professorship which allowed the author to push this manuscript through to publication.

To those around me who have had to "endure" this book, your patience and perseverance are appreciated. To my students who have sometimes had to take a back seat to this project, I thank them for waiting. To those who volunteered to help in proofing the final text, I extend my gratitude.

I must also acknowledge a core of dedicated secretaries over the years who have continually updated this material for classroom use. Most important of these secretaries is Mrs. Nancy Qvale, who is responsible for the preparation of a major portion of this work and bore the burden of putting this manuscript in final form. Without her outstanding technical assistance and dedication, this book might never have materialized.

CONTENTS

LITERATURE ABBREVIATIONS

Acct Chem Res	*Accounts of Chemical Research*
Acta Chem Scand	*Acta Chemica Scandinavica*
Acta Chem Scand B	*Acta Chemica Scandinavica. Series B: Organic Chemistry and Biochemistry*
Adv Alicyclic Chem	*Advances in Alicyclic Chemistry*
Adv Carbohydr Chem	*Advances in Carbohydrate Chemistry*
Adv Catalysis	*Advances in Catalysis*
Adv Chem Ser	*Advances in Chemistry Series*
Adv Heterocyclic Chem	*Advances in Heterocyclic Chemistry*
Adv Org Chem	*Advances in Organic Chemistry: Methods and Results*
Adv Organometal Chem	*Advances in Organometallic Chemistry*
Adv Photochem	*Advances in Photochemistry*
Adv Phys Org Chem	*Advances in Physical Organic Chemistry*
Agric Biol Chem	*Agricultural and Biological Chemistry*
Anal Chem	*Analytical Chemistry*
Anal de Quim	*Anales de Quimica*
Angew	*Angewandte Chemie*
Angew Int	*Angewandte Chemie, International Edition in English*
Ann	*Justus Liebig's Annalen der Chemie*
Ann Chim	*Annales de Chimie*
Ann NY Acad Sci	*Annals of the New York Academy of Sciences*
Ann Rep Med Chem	*Annual Reports in Medicinal Chemistry*
Appl Microbiol	*Applied Microbiology*
Appl Microbiol Biotechnol	*Applied Microbiology and Biotechnology*
Arch Pharm	*Archiv der Pharmazie*
Arkiv Kemi	*Arkiv for Kemi*
Austral J Chem	*Australian Journal of Chemistry*
BCSJ	*Bulletin of the Chemical Society of Japan*
Ber	*Berichte der Deutschen Chemischen Gesellschaft*
Biochem	*Biochemistry*
Biochem Biophy Res Commun	*Biochemical and Biophysical Research Communications*
Biochem J	*Biochemical Journal*
Biochim Biophys Acta	*Biochimica et Biophysica Acta*
Bioorg Chem	*Bioorganic Chemistry*
BSCF	*Bulletin de la Societe Chimique de France*
Bull Acad Polon Sci, Ser Sci Chem	*Bulletin de l'Academie Polonaise des Sciences, Serie des Sciences Chimiques*
Bull Acad Sci USSR, Div Chem Sci	*Bulletin of the Academy of Sciences of the USSR. Division of Chemical Science*
Bull Korean Chem Soc	*Bulletin of the Korean Chemical Society*
Bull Soc Chim Belg	*Bulletin des Societes Chimiques Belges*
CA	*Chemical Abstracts*
Can J Chem	*Canadian Journal of Chemistry*
Cancer Lett	*Cancer Letters*

Carbohydr Res	*Carbohydrate Research*
Catal Rev	*Catalysis Reviews*
CC	*Journal of the Chemical Society: Chemical Communications*
Chem Eng News	*Chemical and Engineering News*
Chem in Britain	*Chemistry in Britain*
Chem Ind	*Chemistry and Industry*
Chem Listy	*Chemicke Listy*
Chem Pharm Bull	*Chemical and Pharmaceutical Bulletin*
Chem Phys Lipids	*Chemistry and Physics of Lipids*
Chem Rev	*Chemical Reviews*
Chem Scripta	*Chemica Scripta*
Chem Soc Rev	*Chemical Society Reviews*
Chem Weekb	*Chemisch Weekblad*
Chem Zeitung	*Chemiker Zeitung*
Chem Zentr	*Chemisches Zentralblatt*
CL	*Chemistry Letters*
Coll Czech Chem Commun	*Collection of Czechoslovak Chemical Communications*
Compt Rend	*Comptes Rendus Hebdomadaires des Seances de l'Academie des Sciences*
Compt Rend C	*Comptes Rendus Hebdomadaires des Seances de l'Academie des Sciences. Serie C: Sciences Chimiques*
Curr Sci	*Current Science*
Discuss Faraday Soc	*Discussions of the Faraday Society*
Fortschr Chem Forsch	*Fortschritte der Chemischen Forschung*
Fund Res Homogeneous Catal	*Fundamental Research in Homogeneous Catalysis*
Gazz Chim Ital	*Gazzetta Chimica Italiana*
Helv	*Helvetica Chimica Acta*
Ind Eng Chem	*Industrial and Engineering Chemistry*
Ind J Chem	*Indian Journal of Chemistry*
Ind J Chem B	*Indian Journal of Chemistry. Section B: Organic Chemistry and Medicinal Chemistry*
Inorg	*Inorganic Chemistry*
Int J Sulfur Chem	*International Journal of Sulfur Chemistry*
Intra-Science Chem Reports	*Intra-Science Chemistry Reports*
Israel J Chem	*Israel Journal of Chemistry*
Izv Akad Nauk SSSR, Ser Khim	*Izvestiia Akademii Nauk SSSR. Seriia Khimicheskaia*
J Am Oil Chem Soc	*Journal of the American Oil Chemists' Society*
J Antibiotics	*Journal of Antibiotics*
J Biol Chem	*Journal of Biological Chemistry*
J Catalysis	*Journal of Catalysis*
J Chem Ed	*Journal of Chemical Education*
J Chem Eng Data	*Journal of Chemical and Engineering Data*
J Chem Res (S)	*Journal of Chemical Research. Synopses*
J Fluorine Chem	*Journal of Fluorine Chemistry*
J Gen Chem USSR	*Journal of General Chemistry of the USSR*
J Heterocyclic Chem	*Journal of Heterocyclic Chemistry*
J Ind Chem Soc	*Journal of the Indian Chemical Society*

J Label Compds	*Journal of Labelled Compounds*
J Lipid Res	*Journal of Lipid Research*
J Med Chem	*Journal of Medicinal Chemistry*
J Mol Catal	*Journal of Molecular Catalysis*
J Nat Prod	*Journal of Natural Products*
J Pharm Sci	*Journal of Pharmaceutical Sciences*
J Photochem	*Journal of Photochemistry*
J Polym Sci, Polym Chem Ed	*Journal of Polymer Science: Polymer Chemistry Edition*
J Prakt Chem	*Journal für Praktische Chemie*
J Russ Phys Chem Soc	*Journal of the Russian Physical Chemical Society*
J Sci Ind Res B	*Journal of Scientific and Industrial Research. Part B: Physical Sciences*
J Vitaminol (Osaka)	*Journal of Vitaminology*
JACS	*Journal of the American Chemical Society*
JCS	*Journal of the Chemical Society*
JCS A	*Journal of the Chemical Society. Section A: Inorganic, Physical and Theoretical*
JCS B	*Journal of the Chemical Society. Section B: Physical Organic*
JCS C	*Journal of the Chemical Society. Section C: Organic*
JCS D	*Journal of the Chemical Society. Section D: Chemical Communications*
JCS Dalton	*Journal of the Chemical Society: Dalton Transactions*
JCS Japan	*Journal of the Chemical Society of Japan*
JCS Perkin I	*Journal of the Chemical Society: Perkin Transactions I*
JCS Perkin II	*Journal of the Chemical Society: Perkin Transactions II*
JOC	*Journal of Organic Chemistry*
JOC USSR	*Journal of Organic Chemistry of the USSR*
JOMC	*Journal of Organometallic Chemistry*
Methods Carbohydr Chem	*Methods in Carbohydrate Chemistry*
Monatsh	*Monatshefte für Chemie*
Natl Prod Repts	*Natural Product Reports*
Naturwiss	*Naturwissenschaften*
Newer Methods Prep Org Chem	*Newer Methods of Preparative Organic Chemistry*
Nouv J Chim	*Nouveau Journal de Chimie*
Org Mag Res	*Organic Magnetic Resonance*
Org Photochem	*Organic Photochemistry*
Org Prep Proc Int	*Organic Preparations and Procedures International*
Org Rxs	*Organic Reactions*
Org Syn	*Organic Syntheses*
Org Syn Coll Vol	*Organic Syntheses. Collective Volume*
Organomet	*Organometallics*
Organomet Chem Rev A	*Organometallic Chemistry Reviews. Section A: Subject Reviews*
Organomet Chem Syn	*Organometallics in Chemical Synthesis*
Phosphorus	*Phosphorus and the Heavier Group Va Elements*
Phosphorus and Sulfur	*Phosphorus and Sulfur and the Related Elements*
Photochem Photobiol	*Photochemistry and Photobiology*
Pol J Chem	*Polish Journal of Chemistry*
Polym J	*Polymer Journal*

Proc Acad Sci USSR, Chem Sec	*Proceedings of the Academy of Sciences of the USSR. Chemistry Section*
Proc Chem Soc	*Proceedings of the Chemical Society (London)*
Proc Ind Acad Sci A	*Proceedings — Indian Academy of Sciences. Section A, Part 1: Chemical Sciences*
Proc Natl Acad Sci USA	*Proceedings of the National Academy of Sciences of the United States of America*
Pure Appl Chem	*Pure and Applied Chemistry*
Quart Rev	*Quarterly Reviews — Chemical Society, London*
Rec Chem Prog	*Record of Chemical Progress*
Rec Trav Chim	*Recueil des Travaux Chimiques des Pays-Bas*
Recl J R Neth Chem Soc	*Recueil: Journal of the Royal Netherlands Chemical Society*
Rev Chem Intermed	*Reviews of Chemical Intermediates*
Rev Pure Appl Chem	*Reviews of Pure and Applied Chemistry*
Rocz	*Roczniki Chemii*
Russ Chem Rev	*Russian Chemical Reviews*
Soc Chem Ind	*Society of Chemical Industry, London Chemical Engineering Group, Proceedings*
Syn	*Synthesis*
Syn Commun	*Synthetic Communications*
Tetr	*Tetrahedron*
TL	*Tetrahedron Letters*
Topics Curr Chem	*Topics in Current Chemistry*
Topics Stereochem	*Topics in Stereochemistry*
Trans Faraday Soc	*Transactions of the Faraday Society*
Transition Met Chem	*Transition Metal Chemistry (New York)*
Z Chem	*Zeitschrift für Chemie*
Z Naturforsch B	*Zeitschrift für Naturforschung. Tiel B: Anorganische Chemie, Organische Chemie, Biochemie, Biophysik, Biologie*
Zh Obshch Khim	*Zhurnal Obshchei Khimii*

CHEMICAL ABBREVIATIONS

Ac	Acetyl
acac	acetylacetonate [$CH_3COCHCOCH_3$] (with line over)
acaen	N,N'-bis(1-methyl-3-oxobutylidene)ethylenediamine
AIBN	2,2'-azobisisobutyronitrile [$Me_2C(CN)N{=}NC(CN)Me_2$]
Am	amyl
aq	aqueous
Ar	aryl

acac — acetylacetonate [$CH_3\overset{|}{C}OCHCOCH_3$]

B⟩ 9-borabicyclo[3.3.1]nonyl

9-BBN 9-borabicyclo[3.3.1]nonane [HB⟩]

BINAP 2,2'-bis(diphenylphosphino)-1,1'-binaphthyl

bipy 2,2'-bipyridyl

Bu butyl

c cyclo

cat catalytic

COD **cis,cis**-1,5-cyclooctadiene

Cp Cyclopentadienyl

Cy cyclohexyl

DABCO 1,4-diazabicyclo[2.2.2]octane

DBA dibenzylideneacetone [$PhCH{=}CHCOCH{=}CHPh$]

DBN 1,5-diazabicyclo[4.3.0]non-5-ene

DBU 1,8-diazabicyclo[5.4.0]undec-7-ene

DDQ 2,3-dichloro-5,6-dicyano-1,4-benzoquinone

diop (2,3)-O-isopropylidene-2,3-dihydroxy-1,4-bis(diphenylphosphino)butane

DMAP 4-dimethylaminopyridine

DME 1,2-dimethoxyethane

DMF N,N-dimethylformamide

DMSO dimethylsulfoxide

dppe 1,2-bis(diphenylphosphino)ethane [$Ph_2PCH_2CH_2PPh_2$]

E^+ electrophile

EDA ethylenediamine [$H_2NCH_2CH_2NH_2$]

EDTA ethylenediaminetetraacetate

Et ethyl

fod 6,6,7,7,8,8,8-heptafluoro-2,2-dimethyl-3,5-octanedionato
[$CF_3CF_2CF_2CO\overset{|}{C}HCOC(CH_3)_3$]

Fp dicarbonyl(η^5-cyclopentadienyl)iron(I)
[$Fe(CO)_2$(cyclopentadienyl)]

Het heterocycle

HMPA = HMPT hexamethylphosphoramide

i	iso
L	ligand
LDA	lithium diisopropylamide [LiN(i-C$_3$H$_7$)$_2$]
m	meta
Me	methyl
Mes	mesityl
mesal	N-methylsalicylaldimine
Ms	methanesulfonyl
n	normal
NBA	N-bromoacetamide
NBD	norbornadiene
NBS	N-bromosuccinimide
NCS	N-chlorosuccinimide
NIS	N-iodosuccinimide
Nuc	nucleophile
o	ortho
p	para
PCC	pyridinium chlorochromate
PDC	pyridinium dichromate
PEG-400	poly(ethylene glycol)-400
Ph	phenyl
phen	1,10-phenanthroline
PPA	polyphosphoric acid
Pr	propyl
py	pyridine
R	an organic group
R$_f$	perfluoroalkyl
Salen	N,N'-ethylenebis(salicylideneiminato)
salophen	o-phenylenebis(salicylideneiminato)
sec	secondary
Sia	1,2-dimethylpropyl [(CH$_3$)$_2$CHCHCH$_3$]
S,S-chiraphos	(**S,S**)-2,3-bis(diphenylphosphino)butane [(**S,S**)-Ph$_2$PCH(CH$_3$)CH(CH$_3$)PPh$_2$]
t	tertiary
Tf	trifluoromethanesulfonyl
THF	tetrahydrofuran
THP	2-tetrahydropyranyl
TMEDA	N,N,N',N'-tetramethylethylenediamine [Me$_2$NCH$_2$CH$_2$NMe$_2$]
Tol	tolyl
tolbinap	2,2'-bis(di-p-tolylphosphino)-1,1'-binaphthyl
Ts	p-toluenesulfonyl

ALKANES AND ARENES

GENERAL REFERENCES

Houben-Weyl, "Methoden der Organischen Chemie," 4th ed, Vol V/1a (alkanes), G. Thieme, Stuttgart (1970)

Houben-Weyl, "Methoden der Organischen Chemie," 4th ed, Vol IV/3 (cyclopropanes), G. Thieme, Stuttgart (1971)

Houben-Weyl, "Methoden der Organischen Chemie," 4th ed, Vol IV/4 (cyclobutanes), G. Thieme, Stuttgart (1971)

Houben-Weyl, "Methoden der Organischen Chemie," 4th ed, Vol V/2b (arenes), G. Thieme, Stuttgart–New York (1981)

1. REDUCTION

1. Cyclic Alkanes

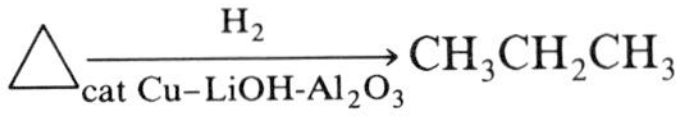

$$\triangle \xrightarrow[\text{cat Cu–LiOH-Al}_2\text{O}_3]{\text{H}_2} \text{CH}_3\text{CH}_2\text{CH}_3$$

JOC USSR *18* 585 (1982)

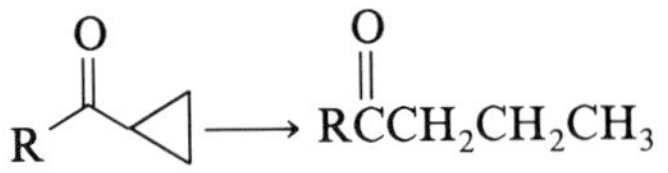

$$\underset{R}{\overset{O}{\parallel}}\text{C}\triangleright \longrightarrow \text{R}\overset{O}{\overset{\parallel}{\text{C}}}\text{CH}_2\text{CH}_2\text{CH}_3$$

Li, NH₃

Acta Chem Scand *19* 1289 (1965)
JOC *31* 3794 (1966); *35* 374 (1970)
BSCF 4449 (1968)
TL 2879 (1974)
Angew Int *18* 809 (1979) (review)

Zn

TL *22* 695 (1981)

Zn, HCl

JOC *35* 2986 (1970)

Zn, ZnCl₂

TL 2489 (1975); *22* 695 (1981)

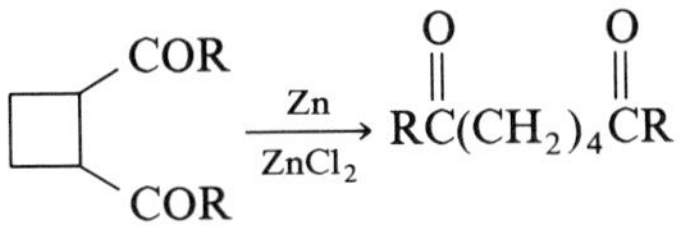

$$\square\!\!<\!\!\begin{smallmatrix}\text{COR}\\ \\ \text{COR}\end{smallmatrix} \xrightarrow[\text{ZnCl}_2]{\text{Zn}} \text{R}\overset{O}{\overset{\parallel}{\text{C}}}(\text{CH}_2)_4\overset{O}{\overset{\parallel}{\text{C}}}\text{R}$$

TL 2489 (1975)

2. Arenes

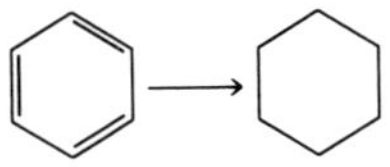

Review: Acct Chem Res *12* 324 (1979)

H₂, cat π-C₃H₅CoL₃

Inorg *15* 2379 (1976)

H_2, cat $RhCl_3$, cat $[(n\text{-}C_8H_{17})_3NCH_3]Cl$	TL *24* 4139 (1983)
	J Mol Catal *34* 221 (1986)
	(naphthalenes); *39* 185 (1987)
	(polycyclic arenes)
	JOC *52* 2804 (1987)
H_2, cat $[ClRh(C_7H_8)]_2$-polymer	CL 603 (1982)
H_2, cat Rh(I)-polymer	JOC *44* 239 (1979)
	JACS *103* 5096 (1981)
H_2, cat $[Rh(C_5Me_5)Cl_2]_2$	CC 427 (1977)
	Acct Chem Res *11* 301 (1978)
H_2, cat [Si]-ORh(allyl)H	JACS *103* 5253 (1981)

3. Alkenes

3.1. Catalytic Hydrogenation

Reviews:

R. L. Augustine, "Catalytic Hydrogenation," Marcel Dekker, Inc., New York (1965)
P. N. Rylander, "Catalytic Hydrogenation over Platinum Metals," Academic Press, New York (1967)
M. Freifelder, "Practical Catalytic Hydrogenation," Wiley-Interscience, New York (1971), Chpt 9
B. R. James, "Homogeneous Hydrogenation," J. Wiley, New York (1973)
F. J. McQuillin, "Homogeneous Hydrogenation in Organic Chemistry," D. Reidel, Boston (1976)
A. P. G. Kieboom and F. van Rantwijk, "Hydrogenation and Hydrogenolysis in Synthetic Organic Chemistry," Delft University Press, Delft (1977)
M. Freifelder, "Catalytic Hydrogenation in Organic Synthesis: Procedures and Commentary," J. Wiley and Sons, New York (1978), Chpt 4
Syn 85 (1981) (homogeneous asymmetric hydrogenation)
Chem Rev *85* 129 (1985) (heterogeneous catalytic transfer hydrogenation)
"Asymmetric Synthesis," Ed. J. D. Morrison, Academic Press, New York (1985), Vol 5, Chpts 2, 3, 10
P. N. Rylander, "Hydrogenation Methods," Academic Press, New York (1985), Chpt 2

3.1.1. Selective hydrogenation catalysts

5% Ru on Norit	JOC *24* 708 (1959)
$RuCl_2(PPh_3)_3$	CL 1083 (1977) (polyenes to monoenes)
$NaBH_4$, $CoCl_2$ (mono > di > tri > tetrasubstituted)	JOC *44* 1014 (1979)
	JACS *108* 67 (1986)
$ClRh(PPh_3)_3$ (Wilkinson's catalyst)	Discuss Faraday Soc *46* 60 (1968)
	Chem Rev *21* 73 (1973)
	Syn 329 (1978)
$[Rh(NBD)Ph_2P(CH_2)_4PPh_2]BF_4$	CC 348 (1982) (syn to allylic, homoallylic OH)
	JACS *106* 3866 (1984) (syn to OH); *107* 4339
	(1985) (syn to OH); *108* 2476 (1986)
	TL *28* 3659 (1987)

[Ir(COD)py(PCy$_3$)]PF$_6$	TL *22* 303 (1981) (α face of steroids)
	JACS *105* 1072 (1983) (syn to OH); *106* 3866
	(1984) (syn to OH); *109* 6493 (1987)
	(syn to CONR$_2$)
	Organomet *2* 681 (1983) (syn to OH)
	JOC *50* 5905 (1985) (syn to CO$_2$R, CONR$_2$); *51*
	2655 (1986) (syn to OH, CO$_2$R, C=O, OMe)
nickel boride (P-2)	JOC *38* 2226 (1973); *46* 1263 (1981)
LaNi$_5$H$_6$	JOC *52* 5695 (1987) (least substituted double
	bond)
NaH, NaO-*t*-Am, Ni(OAc)$_2$	JOC *45* 1937, 1946 (1980)

3.1.2. Enantioselective hydrogenation catalysts

Catalyst	Substrate	
Ru$_2$Cl$_4$(BINAP)$_2$(NEt$_3$)	α-amidoacrylic acids	CC 922 (1985)
	unsaturated	TL *28* 1905 (1987)
	dicarboxylic acids	
RuHCl(BINAP)$_2$	α-amidoacrylic acids	CC 922 (1985)
	unsaturated	TL *28* 1905 (1987)
	dicarboxylic acids	
Ru(OAc)$_2$(BINAP)	enamides	JACS *108* 7117 (1986)
	allylic and homoallylic	JACS *109* 1596 (1987)
	alcohols	
	α,β- and β,γ-unsaturated	JOC *52* 3174 (1987)
	carboxylic acids	
[Rh(NBD)(S,S-chiraphos)]ClO$_4$	α-amidoacrylic acid	JOC *52* 5143 (1987)
	and ester	
[Rh(NBD)(BINAP)]ClO$_4$	α-amidoacrylic acids	JACS *102* 7932 (1980)
	α-amidoacrylic acids	Tetr *40* 1245 (1984)
	and esters	
[Rh(NBD)(BINAP)]BF$_4$	α,β-unsaturated ester	JACS *108* 2476 (1986)
Rh(O$_2$CCF$_3$)$_2$(tolbinap)	enamides	TL *28* 4829 (1987)
chiral (aminoalkyl)-	trisubstituted acrylic	JACS *109* 7876 (1987)
ferrocenylphosphine-Rh	acids	
chiral pyrrolidinodi-	itaconic acid	CL 567 (1978)
phosphines-Rh		
	α-amidoacrylic acids	Ber *119* 3326 (1986)
Ph$_2$PCH(CH$_2$OR)CH$_2$PPh$_2$-Rh	α-amidoacrylic acids	J Chem Res (S) 117
	and esters	(1982)
{R,R-1,2-bis((phenyl-O-	α-amidoacrylate	JACS *109* 1746 (1987)
anisoyl)phosphino)-ethane}-		
Rh(I)		

3.2. Diimide Reduction (HN=NH)

JOC *30* 3965 (1965)
J Chem Ed *42* 254 (1965) (review)
Angew Int *4* 271 (1965) (review)
Syn Commun *12* 287 (1982) (HONH$_2$, EtOAc)
JOC *52* 4665 (1987) (TsNHNH$_2$, NaOAc)

3.3. Hydroboration–Protonolysis

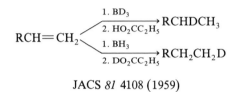

JACS *81* 4108 (1959)

3.4. Transition Metal Salts and Metal Hydrides

LiH, VCl$_3$ (RCH=CH$_2$ only)	JOC *45* 1041 (1980)
NaH, NaO-*t*-Bu, FeCl$_3$	TL 3947 (1977)
NaH, NaO-*t*-Am, Ni(OAc)$_2$	TL 1069 (1977)
NaBH$_4$, CoCl$_2$	JACS *108* 67 (1986)
LiAlH$_4$, TiCl$_4$ or ZrCl$_4$	TL 15 (1976) JOMC *142* 71 (1977)
LiAlH$_4$, CoCl$_2$	JOC *43* 2567 (1978)
LiAlH$_4$, NiCl$_2$	TL 4481 (1977)
LaNi$_5$H$_6$	CC 163 (1984)

3.5. Miscellaneous Reagents

Na, *t*-BuOH, amide	JOC *44* 2369 (1979)
NaH$_2$PO$_2$·H$_2$O, Na$_2$CO$_3$, cat Pd-C, H$_2$O	JOC *50* 3408 (1985)
R$_3$SiH, CF$_3$CO$_2$H	Tetr *23* 2235 (1967) Syn 633 (1974) Org Prep Proc Int *12* 13 (1980) JACS *108* 1239 (1986)

3.6. Conjugate Reduction

$$RCH=CHX \longrightarrow RCH_2CH_2X$$

X = CHO

H$_2$, CO, cat Co$_2$(CO)$_8$	Compt Rend C *281* 877 (1975)
H$_2$, cat Rh-Al$_2$O$_3$	Compt Rend C *284* 577 (1977)

H$_2$, cat ClRh(PPh$_3$)$_3$	JCS C 270 (1967) JOC *34* 3684 (1969)
H$_2$; cat HRhCO(PPh$_3$)$_3$, Rh$_4$(CO)$_{12}$ or Rh$_6$(CO)$_{16}$; cat diphosphine	J Mol Catal *16* 51 (1982) (chiral)
H$_2$, NaBH$_4$-cat Ni(OAc)$_2$ or PdCl$_2$	JOC *42* 551 (1977)
(Et$_3$NH)O$_2$CH, cat Pd-C	JOC *43* 3985 (1978)
i-Bu$_2$AlH, HMPA	JOC *51* 537 (1986)
Et$_3$SiH, cat ClRh(PPh$_3$)$_3$/K$_2$CO$_3$	TL 5035 (1972) Organomet *1* 1391 (1982)
Ph$_2$SiH$_2$, cat Pd(PPh$_3$)$_4$, cat ZnCl$_2$	JACS *108* 7314 (1986)
polymethylhydrosiloxane, cat RhCl$_3$, cat [(*n*-C$_8$H$_{17}$)$_3$NCH$_3$]Cl	J Mol Catal *37* 359 (1986)
n-Bu$_3$SnH, cat Pd(PPh$_3$)$_4$	TL *23* 477 (1982)
n-Bu$_3$SnH, cat Pd(PPh$_3$)$_4$, HOAc	TL *23* 1825 (1982)
NaHFe$_2$(CO)$_8$	JACS *100* 1119 (1978)
HCo(CO)$_4$	JOC *27* 3698 (1962) JACS *85* 2782 (1963)
LaNi$_5$H$_6$	CC 163 (1984)
LiHCuC≡C(CH$_2$)$_2$CH$_3$	JACS *96* 1623 (1974)
Fe(CO)$_5$, NaOH	JOC *37* 1542 (1972) Syn 596 (1976)
CO, H$_2$O, cat Rh$_6$(CO)$_{16}$	CL 379 (1973)
Li, NH$_3$	Org Rxs *23* 1 (1976) JOC *46* 5371 (1981)
Na$_2$S$_2$O$_4$, NaHCO$_3$, cat Aliquat®	TL *26* 831 (1985)
NaTeH	CL 847 (1980)

X = COR

electrolysis	JOC *52* 276 (1987)
Li, NH$_3$ (Birch reduction)	JCS 3045 (1954) Org Rxs *23* 1 (1976) (review) JACS *108* 4561 (1986)
Li·4 NH$_3$	JOC *43* 4647 (1978)
Na or Li, HMPA or NH$_3$	JACS *92* 2800 (1970)
Li, NH$_3$, EtOH	JACS *86* 1761 (1964)
Li, NH$_3$, *t*-BuOH	JOC *32* 2851 (1967); *42* 183 (1977); *50* 2607 (1985); *52* 2263, 3346 (1987) JACS *106* 3539 (1984)
Li, EtNH$_2$, *t*-BuOH	Syn 400 (1980)

Na, HMPA, (ROH)	JACS *92* 2783 (1970)
Yb, NH_3	JOC *43* 4555 (1978)
C_8K, $HN(SiMe_3)_2$	Syn 30 (1979)
Mg, $TiCl_4$, *t*-BuOH	TL *27* 4153 (1986)
$TiCl_3$, MeOH, NH_3	Helv *13* 1308 (1930)
Zn, HCl	Quart Rev *23* 522 (1969)
Zn, cob(I)alamin, HOAc	Helv *62* 2361 (1979)
Zn, $NiCl_2$, H_2O, $CH_3OCH_2CH_2OH$, ultrasound	TL *28* 2347 (1987)
Zn, $NiCl_2$, H_2O, $CH_3OCH_2CH_2OH$, NH_3-NH_4Cl or Et_3N, ultrasound	TL *28* 2351 (1987)
H_2, cat $HRuClL_4$ or H_2RuL_4 (chiral ligands)	TL *27* 5497 (1986) (chiral)
H_2, cat $H_2Ru(PPh_3)_4$ or $RuCl_2(PPh_3)_3$, glycoside	TL 4083 (1976) (chiral)
H_2, CO, cat $Co_2(CO)_8$	Compt Rend *281* 877 (1975)
H_2, cat $Co_2(CO)_6(PR_3)$	JOMC *284* 101 (1985) (chiral)
H_2, cat $K_3Co(CN)_5H$	TL 115 (1979) JOC *45* 3860 (1980)
H_2, cat bis(dimethylglyoximato)-cobalt(II)-quinine	CL 265 (1973) (chiral)
H_2, cat Rh-C, Na_2CO_3	JOC *50* 2438 (1985)
H_2, cat Rh-Al_2O_3	Compt Rend C *284* 577 (1977)
H_2, cat $ClRh(PPh_3)_3$	JCS C 1894 (1966) JOC *34* 3684 (1969) JACS *107* 8066 (1985)
H_2, cat $[Rh(diolefin)L_2]BF_4$ (L = chiral phosphine)	JOC *43* 1787 (1978) (chiral)
H_2, cat $RhCl_3 \cdot 3 H_2O$, cat $[(n\text{-}C_8H_{17})_3NCH_3]Cl$	J Mol Catal *34* 229 (1986) TL *28* 1321 (1987)
H_2, cat $HRh(PPh_3)_4$, $PhCHOHCH_3$	Can J Chem *57* 218 (1979)
H_2, cat $[Ir(COD)py(PCy_3)]PF_6$	TL *22* 303 (1981)
H_2, cat NaH-NaO-*t*-Am-$Ni(OAc)_2$	JOC *45* 1946 (1980)
H_2, cat Pd-C	JOC *23* 1853 (1958); *52* 2875, 5594 (1987) JACS *107* 7967 (1985); *108* 4561 (1986)
H_2, cat PtO_2	JOC *23* 1853 (1958) JACS *108* 7967 (1985)
NaH, NaO-*t*-Am, $Ni(OAc)_2$, $(MgBr_2)$	JOC *44* 2203 (1979)

NaH, t-AmOH, Ni(OAc)$_2$, (Me$_3$SiCl)	TL 27 5487 (1986)
HBI$_2$	TL 3865 (1976)
NaBH$_4$	JCS C 616 (1968)
NaBH$_4$, diglyme, amine base	JCS 5280 (1965)
LiHB(sec-Bu)$_3$	JACS 109 6389 (1987)
KHB(sec-Bu)$_3$	JOC 40 146 (1975); 41 2194 (1976)
KHBPh$_3$	JOC 52 5564 (1987)
NaBH$_3$CN, ZnCl$_2$	JOC 50 1927 (1985)
i-Bu$_2$AlH, (cat MeCu), HMPA	JOC 51 537 (1986) Syn Commun 16 639 (1986)
i-Bu$_3$Al, cat Ni(mesal)$_2$	JOC 47 4640 (1982)
HAl(OR)$_2$ (R = t-Bu, i-Pr)	TL 3865 (1976)
HAl[N(i-Pr)$_2$]$_2$	TL 3865 (1976)
LiAlH$_4$, 4 CuI	TL 4453 (1975) JOC 41 1939 (1976)
LiAlH$_4$, cat CuI, HMPA	CC 1013 (1980)
LiAlH$_4$, Cu—(2,5-dimethylphenyl)	JOC 46 192 (1981)
LiHAl(SR)$_3$ (R = Me, t-Bu)	TL 2397 (1974)
2 LiHAl(OMe)$_3$, CuBr	JOC 40 3619 (1975); 42 3180 (1977)
NaH$_2$Al(OCH$_2$CH$_2$OCH$_3$)$_2$, CuBr	JOC 40 3619 (1975); 42 3180 (1977)
PhSiH$_3$, cat Mo(CO)$_6$/H$_2$O	JOC 52 2576 (1987)
Ph$_2$SiH$_2$, cat Pd(PPh$_3$)$_4$, cat ZnCl$_2$	TL 26 1353 (1985) JACS 108 7314 (1986)
R$_3$SiH (R = Me, Et), TiCl$_4$/H$_2$O	Chem Pharm Bull 25 1468 (1977)
Et$_3$SiH, cat ClRh(PPh$_3$)$_3$	TL 5035 (1972) Organomet 1 1390 (1982) JACS 105 7358 (1983); 107 3731 (1985); 108 4586 (1986)
polymethylhydrosiloxane, cat RhCl$_3$, cat [(n-C$_8$H$_{17}$)$_3$NCH$_3$]Cl	J Mol Catal 37 359 (1986)
Ph$_2$SnH$_2$	TL 2221 (1966)
Ph$_3$SnH	TL 2221 (1966) BSCF 1928 (1967) Helv 56 1062 (1973)
n-Bu$_3$SnH	TL 28 1313 (1987)
n-Bu$_3$SnH, cat Pd(PPh$_3$)$_4$	TL 23 477 (1982)

n-Bu$_3$SnH, cat Pd(PPh$_3$)$_4$, ZnCl$_2$ TL *23* 1825 (1982)
Nouv J Chim *8* 611 (1984)

n-Bu$_3$SnH, cat Cl$_2$Pd(PPh$_3$)$_2$, cat   Nouv J Chim *8* 611 (1984)
   n-Bu$_3$SnO$_3$SCF$_3$

NaTeH   CL 847 (1980)
TL *27* 3411 (1986)

MHCr$_2$(CO)$_{10}$ (M = Na, K)   Syn 596 (1976)

MHFe(CO)$_4$ (M = Na, K)   JCS Perkin I 1273 (1975)
Syn 596 (1976)

NaHFe$_2$(CO)$_8$, HOAc   JACS *100* 1119 (1978)

(Et$_4$N)HFe$_3$(CO)$_{11}$ JOMC *171* 85 (1979)

HCo(CO)$_4$ JACS *85* 2782 (1963)

LaNi$_5$H$_6$ CC 163 (1984)
JOC *52* 5695 (1987)

LiHCu(n-Bu)   JACS *96* 3686 (1974)

LiHCuC≡C(CH$_2$)$_2$CH$_3$ JACS *96* 1623 (1974)

Li$_2$CuH$_3$ JOC *43* 183 (1978)

(R$_3$NH)O$_2$CH (R = Et, n-Bu), cat Pd-C   JOC *43* 3985 (1978)

Fe(CO)$_5$, NaOH   JOC *37* 1542 (1972)

Na$_2$Fe(CO)$_4$, Fe(CO)$_5$, HOAc   Acct Chem Res *8* 342 (1975)

CO, H$_2$O, cat Rh$_6$(CO)$_{16}$ CL 379 (1973)

CO, cat Rh$_4$(CO)$_{12}$-resin   CL 203 (1975)

benzylic alcohol, cat Cl$_2$Ru(PPh$_3$)$_3$ JOC *40* 1887 (1975)

Na$_2$S$_2$O$_4$, NaHCO$_3$, Adogen 464®, H$_2$O,   Tetr *42* 4603 (1986)
   benzene

Na$_2$S$_2$O$_4$, NaHCO$_3$, cat Aliquat®   TL *26* 831 (1985)

, AlCl$_3$, CH$_3$OH   Syn 308 (1984)

PhCH$_2$NH$_2$/KO-t-Bu/H$_3$O$^+$ JACS *89* 2794 (1967)

baker's yeast   TL *27* 4737 (1986) (α-chloro enones, chiral)

Beauveria sulfurescens   JOC *47* 792 (1982) (enantiospecific); *52* 4893 (1987) (enantiospecific)

$X = CO_2H$

Li, NH$_3$ Org Rxs *23* 1 (1976)

Na, NH$_3$, t-BuOH   TL *27* 3923 (1986)

C_8K/H_3O^+	Syn 30 (1979)
$TiCl_3$, MeOH, NH_3	Helv *13* 1308 (1930)
$CrSO_4$, H_2O	JACS *88* 4964 (1966)
cob(I)alamin, Zn, HOAc	Helv *62* 2361 (1979)
SmI_2, CH_3OH	JACS *102* 2693 (1980)
H_2, cat $RuCl_2$	JACS *88* 5150 (1966)
H_2, cat Ru-BINAP	JOC *52* 3174 (1987) (chiral)
H_2, cat $ClRh(PPh_3)_3$	JOC *34* 3684 (1969)
H_2, cat Raney Ni-NaBr-tartaric acid	JOC *52* 1139 (1987) (chiral)
$PhSiH_3$, cat $Mo(CO)_6/H_2O$	JOC *52* 2576 (1987)
NaTeH	Syn 545 (1978)
$HCo(CN)_5^{3-}$	Catal Rev *1* 37 (1967) (review)
$HONH_2$, EtOAc	Syn Commun *12* 287 (1982)

$X = CO_2Na$

H_2, cat Raney Ni-NaBr-tartaric acid	JOC *52* 1139 (1987) (chiral)

$X = CO_2R$

Mg, CH_3OH	CL 633 (1983) (lactone) TL *27* 2409 (1986); *28* 5287 (1987) JOC *52* 4641 (1987)
Mg, CH_3OH, ultrasound	CC 344 (1986)
cob(I)alamin, Zn, HOAc	Helv *62* 2361 (1979)
$CrSO_4$, H_2O	JACS *88* 4964 (1966) Org Syn Coll Vol *5* 993 (1973)
Zn, $NiCl_2$, H_2O, $CH_3OCH_2CH_2OH$, ultrasound	TL *28* 2347 (1987)
SmI_2, CH_3OH	JACS *102* 2693 (1980)
H_2, cat bis(dimethylglyoximato)-cobalt(II)-quinine	CL 265 (1973) (chiral)
H_2, cat $RhCl_3 \cdot 3\,H_2O$, cat $[(n\text{-}C_8H_{17})_3NCH_3]Cl$	TL *28* 1321 (1987)
H_2, cat $ClRh(PPh_3)_3$	JACS *90* 1673 (1968) Discuss Faraday Soc *46* 60 (1968) JOC *34* 3684 (1969)
H_2, cat $[Rh(NBD)Ph_2P(CH_2)_4PPh_2]BF_4$	JACS *108* 2476 (1986)
H_2, cat $[Rh(NBD)(+)\text{-}BINAP]BF_4$	JACS *108* 2476 (1986) (chiral)
H_2, cat $PdCl_2$	JACS *91* 2579 (1969)

H$_2$, cat Pd-C	JOC *52* 4603 (1987) (lactone)
H$_2$, cat PtO$_2$	JOC *52* 4641 (1987)
NaBH$_4$	JOC *31* 620 (1966)
NaBH$_4$, CF$_3$CH$_2$OH, EtOH	JOC *50* 3948 (1985)
NaBH$_4$, cat NiCl$_2$	JACS *106* 5585 (1984) (lactone) TL *28* 4037 (1987) (lactone)
LiHB(*sec*-Bu)$_3$, *t*-BuOH	JOC *40* 2846 (1975); *41* 2194 (1976)
i-Bu$_2$AlH, cat MeCu, HMPA	JOC *51* 537 (1986)
LiAlH$_4$, CuI	CC 1013 (1980)
NaH$_2$Al(OCH$_2$CH$_2$OCH$_3$)$_2$, CuBr	JOC *40* 3619 (1975); *42* 3180 (1977) JACS *109* 5432 (1987)
LiHAl(OMe)$_3$, CuBr	JOC *42* 3180 (1977)
PhSiH$_3$, cat Mo(CO)$_6$/H$_2$O	JOC *52* 2576 (1987)
Ph$_2$SiH$_2$, cat Pd(PPh$_3$)$_4$, cat ZnCl$_2$	JACS *108* 7314 (1986)
Et$_3$SiH, cat ClRh(PPh$_3$)$_3$/H$_3$O$^+$	Chem Pharm Bull *22* 2767 (1974) Syn Commun *15* 965 (1985) JACS *109* 5432 (1987)
polymethylhydrosiloxane, cat RhCl$_3$, cat [(n-C$_8$H$_{17}$)$_3$NCH$_3$]Cl	J Mol Catal *37* 359 (1986)
NaTeH	Syn 545 (1978); 847 (1980)
KHCr$_2$(CO)$_{10}$	Syn 596 (1976)
NaHFe$_2$(CO)$_8$	JACS *100* 1119 (1978)
(Et$_4$N)HFe$_3$(CO)$_{11}$	JOMC *171* 85 (1979)
HCo(CN)$_5^{3-}$	Catal Rev *1* 37 (1967) (review)
LaNi$_5$H$_6$	CC 163 (1984) JOC *52* 5695 (1987)
LaNi$_{4.5}$Al$_{0.5}$H$_5$	JOC *52* 5695 (1987)
LiHCuC≡C(CH$_2$)$_2$CH$_3$	JACS *96* 1623 (1974)
(R$_3$NH)O$_2$CH, cat Pd-C	JOC *43* 3985 (1978)
CO, H$_2$O, cat Rh$_6$(CO)$_{16}$	CL 379 (1973)
CO, cat Rh$_4$(CO)$_{12}$-resin	CL 203 (1975)
Fe(CO)$_5$, NaOH	JOC *37* 1542 (1972)
Na$_2$Fe(CO)$_4$, Fe(CO)$_5$, HOAc	JACS *100* 1119 (1978)
baker's yeast	TL *28* 1447 (1987) (enantioselective on methyl 2,4,4-trichloro-2-butenoates)

$X = CONR_2$

Mg, CH_3OH	TL *21* 2915 (1980)
	JCS Perkin I 2912 (1981)
$TiCl_3$, CH_3OH, NH_3	Helv *13* 1308 (1930)
cob(I)alamin, Zn, HOAc	Helv *62* 2361 (1979)
H_2, cat nickel boride	JOC *37* 3552 (1972)
$NaBH_4$	JOC *31* 620 (1966)
$LiHB(sec-Bu)_3$	TL *27* 4717 (1986)
$PhSiH_3$, cat $Mo(CO)_6/H_2O$	JOC *52* 2576 (1987)
Ph_2SiH_2, cat $Pd(PPh_3)_4$, cat $ZnCl_2$	JACS *108* 7314 (1986)
NaTeH	Syn 545 (1978)
$NaHFe_2(CO)_8$	JACS *100* 1119 (1978)
CO, H_2O, cat $Rh_6(CO)_{16}$	CL 379 (1973)
CO, cat $Rh_4(CO)_{12}$-resin	CL 203 (1975)

$X = CN$

Mg, CH_3OH	JOC *40* 127 (1975); *51* 449 (1986)
$CrSO_4$, H_2O	JACS *88* 4964 (1966)
H_2, cat $ClRh(PPh_3)_3$	Discuss Faraday Soc *46* 60 (1968)
	JOC *34* 3684 (1969)
$NaBH_4$	JOC *31* 620 (1966)
	Can J Chem *56* 41 (1978)
$NaH_2Al(OCH_2CH_2OCH_3)_2$, $CuBr/sec-BuOH$	JOC *45* 167 (1980)
$PhSiH_3$, cat $Mo(CO)_6/H_2O$	JOC *52* 2576 (1987)
Ph_2SiH_2, cat $Pd(PPh_3)_4$, cat $ZnCl_2$	JACS *108* 7314 (1986)
$KHCr_2(CO)_{10}$	Syn 596 (1976)
$NaHFe(CO)_4$	JCS Perkin I 1273 (1975)
$NaHFe_2(CO)_8$	JACS *100* 1119 (1978)
$HCo(CN)_5^{3-}$	Catal Rev *1* 37 (1967) (review)
$(R_3NH)O_2CH$, cat Pd-C	JOC *43* 3985 (1978)
CO, H_2O, cat $Rh_6(CO)_{16}$	CL 379 (1973)
$Fe(CO)_5$, NaOH	JOC *37* 1542 (1972)
$Na_2Fe(CO)_4$, $Fe(CO)_5$, HOAc	JACS *100* 1119 (1978)
$NaH_2PO_2 \cdot H_2O$, Na_2CO_3, cat Pd-C, H_2O	JOC *50* 3408 (1985)

$X = NO_2$

H_2, cat ClRh(PPh$_3$)$_3$	JCS C 1894 (1966)
	JOC *34* 3684 (1969)
NaBH$_4$	JACS *108* 1039 (1986)
n-Bu$_3$SnH/H$_2$F$_2$, MeOH	TL *28* 5365 (1987)

$$\overset{\text{H}\quad\text{H}}{\underset{}{X-C=C-Y \longrightarrow X-\underset{|}{\overset{|}{C}}-\underset{|}{\overset{|}{C}}-Y}}$$

X	Y	Reagent(s)	
CHO	COR	NaI, HCl	Syn 245 (1980)
RCO	COR	NaI, HCl	Syn 245 (1980)
		SnCl$_2$, HCl, HOAc	Ber *89* 822 (1956)
			JOC *51* 4169 (1986)
		TiCl$_3$, H$_2$O	JOC *39* 258 (1974)
			Syn 956 (1982)
			JACS *109* 7230 (1987)
RCO	CO$_2$R	TiCl$_3$, H$_2$O	JOC *39* 258 (1974)
HO$_2$C	CO$_2$H	H$_2$, catalyst	M. Freifelder, "Catalytic Hydrogenation in Organic Synthesis: Procedures and Commentary," Wiley-Interscience, New York, (1978), pp 16–18
		H$_2$, cat RuCl$_2$	JACS *88* 5150 (1966)
		Ti, electrolysis	Can J Chem *56* 2269 (1978)
		TiCl$_3$, H$_2$O	JOC *39* 258 (1974)
		CrSO$_4$	JACS *88* 4964 (1966)

EtO$_2$C \ / CO$_2$Et

CH$_3$ N   CH$_3$
        |
        H

JCS 3257 (1960)

---OCOCO---

EtO$_2$C \ / CO$_2$Et

CH$_3$ N   CH$_3$
        |
        H

JCS 3257 (1960)

RO$_2$C	CO$_2$R	CrSO$_4$	JACS *88* 4964 (1966)
			Org Syn *49* 98 (1969)
		LiAlH$_4$, TiCl$_4$, Et$_3$N	TL *28* 2393 (1987)

EtO$_2$C \ / CO$_2$Et

CH$_3$ N   CH$_3$
        |
        H

JCS 3257 (1960)

CN   CN $CrSO_4$ JACS *88* 4964 (1966)

$$R_2C=C\underset{X}{\overset{CN}{<}} \xrightarrow[\text{Et}_3\text{N}]{\text{HCO}_2\text{H}} R_2CHCHXCN$$

$X = CO_2Et,\ CN,\ SO_2Ph$

Chem Pharm Bull *25* 2396 (1977)

4. Alkynes

4.1. Catalytic Hydrogenation

Reviews:

M. Freifelder, "Practical Catalytic Hydrogenation," Wiley-Interscience, New York (1971), Chpt 8
M. Freifelder, "Catalytic Hydrogenation in Organic Synthesis: Procedures and Commentary," J. Wiley and Sons, New York (1978), Chpt 3
P. N. Rylander, "Hydrogenation Methods," Academic Press, New York (1985), Chpt 3

4.1.1. Catalysts

Ni	JCS 5032 (1952)
PtO_2	JACS *74* 3636 (1952)
Ru	JOC *24* 708 (1959)
Re_2S_7	JACS *76* 1519 (1954); *81* 3587 (1959)

4.2. Hydroalumination–Protonolysis

$$RC\equiv CH \longrightarrow RCH_2CHD_2$$

i-Bu_2AlH/D_2O	Ann *618* 267 (1958)
i-$Bu_2AlH/Cp_2TiCl_2/D_2O$	CL 429 (1982)

4.3. Miscellaneous Reagents

NaH, NaO-t-Am, $Ni(OAc)_2$	TL 1069 (1977)
NaH, NaO-t-Bu, $FeCl_3$	TL 3947 (1977)
$LaNi_5H_6$	CC 163 (1984)
	JOC *52* 5695 (1987)
$LaNi_{4.5}Al_{0.5}H_5$	JOC *52* 5695 (1987)
$KO_2CN{=}NCO_2K,\ H^+$	TL *28* 5457 (1987)

5. Organic Halides

Review: Syn 425 (1980)

$$R-X \longrightarrow R-H$$

5.1. Low-Valent Metals

Li, THF, t-BuOH (alkyl, vinyl, aryl)	Chem Ind 405 (1960) JACS *109* 7230 (1987)
Li, THF, t-BuOH, NH₃ (3° alkyl)	JOC *52* 4784 (1987)
Li(Hg), EtOH (dihalocyclopropanes)	JACS *91* 1767 (1969)
Na, THF, t-BuOH (alkyl, vinyl, aryl)	Org Syn *48* 68 (1968) JACS *108* 1265 (1986)
Na-K, tris(3,6-dioxaheptyl)amine (2° alkyl)	TL *28* 2503 (1987)
K, crown ether, toluene (alkyl F and Cl)	TL *22* 2583 (1981)
Mg/D₂O (aryl)	JACS *106* 1750 (1984)
Mg, i-PrOH (alkyl, vinyl, aryl)	Proc Chem Soc 219 (1963)
Mg, TiCl₃, THF (alkyl, aryl)	CC 781 (1975)
Zn, D₂O (2° alkyl)	JOC 50 2557 (1985)
Zn, HOAc (alkyl, allylic, aryl)	Org Syn Coll Vol *2* 320 (1943) JACS *91* 1767 (1969) TL *28* 3225 (1987)
Zn, KOH, ROH (cyclopropyl)	JOC *37* 1734 (1972)
Zn, KI, CH₃OH (RCOCH=CClR)	JACS *72* 1645 (1972)
Zn-Ag, CH₃OH (RCOCH=CClR)	JOC *41* 636 (1976)
Zn, H₂O, cat NiCl₂-NaI-PPh₃ (aryl, 3° alkyl)	JOC *47* 2622 (1982)
CrSO₄, DMF, H₂O (1°, 2° alkyl; allylic; benzylic)	JACS *85* 2768 (1963)
Cr(ClO₄)₂(EDA)₂, DMF, H₂O (alkyl, aryl)	JACS *88* 4094 (1966); *92* 137 (1970) Angew Int *7* 247 (1968) Tetr *24* 3503 (1968)
Cr(ClO₄)₂, EDA, n-BuSH, (electrolysis) (1°, 2° alkyl)	Angew Int *19* 46 (1980)
SmI₂, THF (1° RBr, RI; benzylic chloride, bromide)	JACS *102* 2693 (1980)

5.2. Metal Hydrides

Review: JOC *45* 849 (1980)

NaH, NaO-*t*-Am, Ni(OAc)$_2$ or ZnCl$_2$ (1°, 2°, 3° alkyl; vinylic; allylic; benzylic; aryl)	TL 3951 (1977) JOC *46* 1270 (1981)
KH (1° alkyl)	JOC *52* 4299 (1987)
MgH$_2$, HMgOC$_6$H$_3$[CH(CH$_3$)$_2$]$_2$-2,6 (1° RI)	JOC *43* 1557 (1978)
NaBH$_4$, DMSO (3° alkyl, allylic, benzylic)	TL 3495 (1965) JACS *88* 1473 (1966) JOC *34* 3923 (1969); *51* 3502 (1986) CC 338 (1970)
NaBH$_4$, DMSO or HMPA or sulfolane (1°, 2° alkyl)	JOC *43* 2259 (1978)
NaBH$_4$, sulfolane (3° alkyl)	JOC *36* 1568 (1971)
NaBH$_4$ (phase transfer) (1°, 2° alkyl; benzylic)	JOC *46* 3909 (1981)
NaBH$_4$, H$_2$O, CH$_3$CN, hν (aryl)	JACS *95* 5085 (1973)
NaBH$_4$, (*t*-BuO)$_2$, hν (aryl)	TL *27* 109 (1986)
NaBH$_4$, cat R$_3$SnCl (R = Me, *n*-Bu), hν (1°, 2° alkyl; aryl)	JOC *40* 2554 (1975)
NaBH$_4$, PdCl$_2$ (aryl)	TL 4699 (1973) CL 1029 (1981)
Zn(BH$_4$)$_2$ (3° alkyl, benzylic)	Angew Int *22* 562 (1983)
NaBH$_3$CN (1° RCl, RBr, RI; 2° RBr, RI; allylic; benzylic)	Syn 35 (1975) JOC *42* 82 (1977)
NaBH$_3$CN-polymer (1° RBr, RI)	JOC *42* 82 (1977) CC 1088 (1979)
(*n*-Bu$_4$N)BH$_3$CN (1° RCl, RBr, RI; 2° RBr, RI; allylic; benzylic)	JOC *42* 82 (1977)
NaBH$_3$CN, ZnCl$_2$ (3° alkyl, allylic, benzylic)	TL *24* 3369 (1983)
Na or K[H(CN)B⟨⟩], HMPA (allylic, benzylic)	JOC *42* 82 (1977)
LiR$_2$B⟨⟩ (3° alkyl, allylic, benzylic)	JACS *97* 2558 (1975) Tetr *37* 2261 (1981)
LiHBEt$_3$ (1°, 2° alkyl; 1°, 2° benzylic)	JACS *95* 1669 (1973); *107* 2046 (1985) JOC *45* 849 (1980); *47* 2590 (1982)
LiHBEt$_3$, cat Pd(PPh$_3$)$_4$ (1° allylic Cl)	JOC *47* 4380 (1982)

KHB(sec-Bu)$_3$, CuI (1° alkyl, aryl)   CC 762 (1974)

KHBPh$_3$ (1° alkyl)   JOC 52 5564 (1987)

NaHB(OMe)$_3$ (1° allylic)   TL 28 5977 (1987)

LiAlH$_4$ (1°, 2°, 3° alkyl; benzylic)   JACS 70 3664, 3738 (1948); 71 1675
                                                             (1949); 94 8905 (1972)
                                                             TL 2483 (1973); 28 3883 (1987)
                                                             JOC 45 849 (1980); 47 276 (1982); 49
                                                             3545 (1984)

LiAlH$_4$ (aryl)   JOC 33 619 (1968); 34 3918 (1969)
                                                             TL 1223 (1969)

LiAlH$_4$, ultrasound (aryl)   TL 23 1643 (1982)

LiAlH$_4$, TiCl$_3$ (aryl)   Rec Trav Chim 101 112 (1982)

LiAlH$_4$, TiCl$_4$ (aryl, vinylic)   CL 291 (1973)

LiAlH$_4$, CoCl$_2$ (2° alkyl)   JACS 108 67 (1986)

LiAlH$_4$, NiCl$_2$ or CoCl$_2$ (1°, 2°, 3°   TL 4481 (1977)
    alkyl; aryl)   JOC 43 1263 (1978)

LiHAl(i-Bu)$_2$(n-Bu) (1° alkyl,   JOC 49 1717 (1984)
    benzylic, allylic)

LiHAl(OCH$_3$)$_3$, CuI (1°, 2°, 3° alkyl;   JACS 95 6452 (1973)
    aryl; allylic; vinylic)

R$_3$SiH (R = Et, n-Bu) or n-BuSiH$_3$,   JOC 41 1393 (1976)
    cat AlCl$_3$ (1°, 2°, 3° alkyl)

polymethylhydrosiloxane, cat Pd(PPh$_3$)$_4$,   JOC 51 734 (1986)
    (PhCH$_2$)$_3$N, Δ (aryl)

R$_3$SnH (R = n-Bu, Ph) (alkyl, aryl)   JOC 28 703, 2332 (1963)
                                                             Adv Organometal Chem 1 47 (1964)
                                                             JACS 87 4007 (1965)
                                                             Acct Chem Res 1 299 (1968)
                                                             CC 875 (1969)
                                                             Syn 499 (1970)
                                                             TL 28 3883 (1987)

HFe(CO)$_4^-$-polymer (2° benzylic)   JOC 43 1598 (1978)

LiHCu(n-Bu) (2°, 3° alkyl; aryl)   JACS 96 3686 (1974)

Li$_4$CuH$_5$ (1° alkyl)   TL 3695 (1977)
                                                             JOC 43 183 (1978)

5.3. Miscellaneous Reagents

H$_2$, cat Raney Ni (aryl)   JOC 52 3200 (1987)

N-benzyl-1,4-dihydronicotinamide; cat   JOC 50 3283 (1985)
    ClRh(PPh$_3$)$_3$ or Pd(OAc)$_2$ (allylic,
    benzylic, aryl, vinylic)

$NaH_2PO_2 \cdot H_2O$, Na_2CO_3, cat Pd-C, H_2O (benzylic, aryl)	JOC *50* 3408 (1985)
NaO_2CH, cat $Pd(PPh_3)_4$, Δ (aryl)	TL 1913 (1978) JOC *51* 734 (1986)
HCO_2H, cat Pd-C, DMF, Δ (aryl)	Syn 876 (1982)
$(Et_3NH)O_2CH$, cat Pd-C or $(Ar_3P)_2Pd(OAc)_2$ (aryl)	JOC *42* 3491 (1977)
$NaOCH_3$, cat $Pd(PPh_3)_4$ (aryl)	JOC *43* 1619 (1978)
$KPPh_2$ (1,1-dichlorocyclopropanes to chlorocyclopropanes)	JOC *52* 3923 (1987)
$KP(O)(OMe)_2$ (1,1-dibromocyclopropanes to bromocyclopropanes)	JOC *50* 3713 (1985)
$AlCl_3$, EtSH (polycyclic aryl)	TL *23* 689 (1982)
$h\nu$, Et_3N, anthracene (aryl)	CC 1703 (1987)
electrolysis (dihalo to monohalo cyclopropanes)	JACS *91* 1767 (1969)

$$\underset{R_2C-CR}{\overset{X \quad O}{\overset{|\qquad \|}{}}} \longrightarrow R_2CHCR\overset{O}{\overset{\|}{}}$$

Review: Org Rxs *29* 163 (1983)

H_2, cat Pd-C (X = Br)	JOC *23* 1938 (1958)
HI (X = Br, I)	JCS 4356 (1956) JACS *81* 3634 (1959)
NaI, R_3N (R = Me, Et), SO_2 (X = Cl, Br)	Syn 59 (1979)
NaI, C_5H_5N, SO_3 (X = Cl, Br)	Syn 59 (1979)
LiI, BF_3 (X = Cl, Br, I)	TL 137 (1971)
NaI, $SnCl_2$, H_2O, THF (X = Cl, Br)	Syn 570 (1986)
NaI, HOAc (X = Br)	JOC *16* 573 (1951)
NaI, aq H_2SO_4 (X = Cl, Br, I)	TL *21* 3195 (1980)
NaI, Me_3SiCl (X = Cl, Br)	JOC *45* 3531 (1980)
NaI, $MeSiCl_3$ (X = Br)	JOC *48* 3667 (1983)
Me_3SiI (X = Cl, Br)	Syn Commun *11* 101 (1981)
PI_3 or P_2I_4 (X = Br, I)	TL *22* 1431 (1981)
Zn, H^+ (X = Br)	JACS *81* 3127, 3634, 3644 (1959) JOC *36* 1153 (1971)
Zn, H_2O, DMSO, C_6H_6 (X = Cl)	JOC *52* 5280 (1987)
Zn, HOAc (X = Cl, Br)	Ann *681* 196 (1965) JOC *37* 2363 (1972); *45* 2036 (1980); *48* 2590 (1983); *50* 3957 (1985); *52* 4772 (1987) JACS *101* 4003 (1979); *105* 2435 (1983)

Zn, HOAc, KI (X = Cl)	JACS *83* 3114 (1961)
Zn, HOAc, EtOH, TMEDA (X = Cl)	TL *28* 3299 (1987) (4,4-dichloro-cyclobutenones)
Fe graphite (X = Br)	JOC *47* 876 (1982)
SnCl$_2$, HCl (X = Br)	JACS *68* 1813 (1946)
SnCl$_2$, *i*-Bu$_2$AlH (X = Br)	CL 2069 (1984)
TiCl$_3$ (X = Cl, Br)	Syn Commun *3* 237 (1973) JACS *108* 1239 (1986) TL *28* 1541 (1987)
Ti$_2$(SO$_4$)$_3$ (X = Cl, Br)	TL *28* 1541 (1987)
VCl$_2$ (X = Cl, Br)	Syn 807 (1976)
CrSO$_4$, H$_2$O, DMF (X = Cl, Br)	JACS *85* 2768 (1963)
CrCl$_2$ (X = Br)	JACS *67* 1728 (1945) JCS 3869 (1953)
CrCl$_3$, LiAlH$_4$ (X = Br)	TL 3829 (1977)
Ce$_2$(SO$_4$)$_3$, NaI (X = Cl, Br)	Syn Commun *9* 241 (1979)
SmI$_2$, MeOH (X = Cl)	JOC *51* 1135 (1986)
NaBH$_4$, Pd(OAc)$_2$ or Pd(NO$_3$)$_2$ or Hg(OAc)$_2$ or Ni(OAc)$_2$ (X = Br)	TL 513 (1961)
PhSiH$_3$, cat Mo(CO)$_6$, (cat PPh$_3$), NaHCO$_3$ (X = Cl, Br)	JOC *52* 5570 (1987)
Ph$_2$SiH$_2$, cat Pd(PPh$_3$)$_4$ (X = Br)	JOC *52* 5570 (1987)
polymethylhydrosiloxane or NaO$_2$CH, cat Pd(PPh$_3$)$_4$, Δ (X = Br)	JOC *51* 734 (1986)
R$_3$SnH or R$_2$SnH$_2$ (R = *n*-Bu, Ph) (X = Cl, Br)	JOC *28* 2165 (1963)
n-Bu$_3$SnH (X = Cl)	JOC *52* 307 (1987)
Ph$_3$SnH (X = Cl, Br)	JOC *51* 5182 (1986)
HFe(CO)$_4^-$-polymer (X = Br)	JOC *43* 1598 (1978)
(Et$_4$N)HFe$_3$(CO)$_{11}$ (X = Br)	JOMC *171* 85 (1979)
Fe(CO)$_5$ (X = Br)	JOC *44* 641 (1979)
Fe(CO)$_5$, KOH (X = Br)	TL 2257 (1975)
Fe$_2$(CO)$_9$, H$_2$O, DMF (X = Br)	JACS *94* 7202 (1972); *100* 1759 (1978)
Mo(CO)$_6$, Al$_2$O$_3$ (X = Br)	JOC *44* 2568 (1979)
cat Co$_2$(CO)$_8$, NaOH, (PhCH$_2$NEt$_3$)Cl (X = Br)	TL 2861 (1977)
PhNMe$_2$ (X = Br)	Chimia *21* 464 (1967)

(X = F, Cl, Br)   JOC *51* 5400 (1986) (α-halo aldehydes, ketones)

(X = F, Cl, Br)   JOC *52* 2142 (1987)

N-benzyl-1,4-dihydronicotinamide; cat   JOC *50* 3283 (1985)
  ClRh(PPh$_3$)$_3$ or Pd(OAc)$_2$ (X = Cl, Br)

Ph$_2$PH (X = Cl, Br)   JOC *34* 2687 (1969)

Ph$_3$P, MeOH, C$_6$H$_6$ (X = Cl, Br)   TL 471, 583 (1962)
     JACS *85* 2183 (1963)
     JCS 1379 (1965)
     JOC *31* 4031 (1966)

PhSH, iron polyphthalocyanine   CL 1241 (1975)
  (X = Cl, Br)

RS$^-$ (X = Cl, Br, I)   BCSJ *44* 828 (1971)
     JOC *46* 2596 (1981)

NaHSO$_3$ (X = I)   JACS *72* 362 (1950)

Na$_2$S$_2$O$_4$, H$_2$O, DMF (X = Cl, Br)   Syn Commun *12* 261 (1982)

C$_5$H$_5$N/Na$_2$S$_2$O$_4$ (X = Cl, Br)   JOC *39* 562 (1974)

PhSeH, K$_2$CO$_3$ (X = I)   JOC *46* 2596 (1981)

NaTeH (X = Cl, Br)   CL 119 (1983)

— TeLi(Na)   (X = Cl, Br, I)   JOC *47* 3946 (1982)

(EtO)$_2$P(O)TeNa (X = Cl, Br)   JOC *47* 1124 (1982)

$$\underset{\text{R}_2\text{C}-\text{CX}}{\overset{\text{Y}\quad\text{O}}{\vert\quad\parallel}} \longrightarrow \underset{\text{R}_2\text{CHCX}}{\overset{\text{O}}{\parallel}}$$

X	Y	Reagent(s)	
OH	Cl	Zn, HOAc	TL *28* 5339 (1987)
	Cl, Br	— TeLi(Na)	JOC *47* 3946 (1982)
			JOC *51* 5400 (1986)
OR	Cl	H$_2$, cat Pd-C	JOC *52* 3777 (1987)
		Zn, HOAc	TL *28* 5339 (1987)

$$\underset{\substack{| \\ R_2C-CX}}{\overset{\substack{Y \quad O \\ | \quad \parallel}}{}} \longrightarrow R_2CH\overset{O}{\overset{\parallel}{C}}X \quad (\textit{Continued})$$

X	Y	Reagent(s)	
		$NaH_2PO_4 \cdot H_2O$, Na_2CO_3, cat Pd-C, H_2O	JOC *50* 3408 (1985) (α,α-dihalo ester)
		NaI, $SnCl_2$, H_2O, THF	Syn 570 (1986)
		NaTeH	CL 119 (1983)
	Cl, Br	$PhSiH_3$, cat $Mo(CO)_6$, cat PPh_3, NaHCO_3	JOC *52* 5570 (1987)

			JOC *51* 5400 (1986)
	Br	SmI_2, CH_3OH	JOC *51* 1135 (1986)
		$HFe(CO)_4^-$-polymer	JOC *43* 1598 (1978)
$CONR_2$	Cl	Zn, HOAc	TL *28* 5339 (1987)
		$NaH_2PO_4 \cdot H_2O$, Na_2CO_3, cat Pd-C, H_2O	JOC *50* 3408 (1985) (α,α-dihalo amide)
	Br	NaTeH	CL 119 (1983)

JOC *47* 3946 (1982)

$$\underset{\substack{| \\ R_2C=CR}}{\overset{X}{}} \longrightarrow R_2CHCH_2R$$

H_2, cat Pd-C
                                 Can J Chem *37* 1870 (1959)
                                 JACS *104* 2198 (1982); *109* 4752 (1987)

Li, NH_3, EtOH
                                 TL *28* 3209 (1987)

6. Amines

$$RNH_2 \longrightarrow RH$$

RSO_2Cl (R = Ph, *p*-Tol, Me)/H_2NOSO_3H, NaOH
                                 JACS *82* 753 (1960); *86* 1152 (1964)

RSO_2Cl (R = CF_3, *p*-Tol, *p*-BrC_6H_4, *p*-$NO_2C_6H_4$) (disulfonimide)/$NaBH_4$, HMPA
                                 JOC *40* 2018 (1975); *43* 2259 (1978)

H_2NOSO_3H, OH^-
                                   JACS *100* 341 (1978)

MeCO$_2$CHO/POCl$_3$/n-Bu$_3$SnH, AIBN

JACS *90* 4182 (1968)
TL 2291 (1979)
CC 345 (1979)
Syn 68 (1980)

ClO$_4^-$/NaBH$_4$/Δ

TL 2689 (1976)

HNF$_2$

JACS *85* 97 (1963); *86* 2233 (1964)

NaNO$_2$, H$_3$PO$_2$, cat Cu$_2$O (R = aryl)

JACS *108* 1000 (1986)

i-AmONO, H$_2$SO$_4$, hydroquinone, dioxane
  (R = aryl)

CC 605 (1973)

n-AmONO, THF (R = aryl)

JCS Perkin I 541 (1973)

RONO (R = t-Bu, PhCH$_2$), DMF (R = aryl)

JOC *42* 3494 (1977)

$$\text{ArNH}_2 \longrightarrow (\text{ArN}_2)\text{X} \xrightarrow[\text{agent}]{\text{reducing}} \text{Ar}-\text{H}$$

Review: Org Rxs *2* 262 (1944)

H$_3$PO$_2$

JACS *71* 2137 (1949); *72* 3013 (1950);
  76 290 (1954)
Org Syn Coll Vol *3* 295 (1955); *4* 947 (1963)

H$_3$PO$_3$

JACS *74* 3074 (1952)

Ca(H$_2$PO$_2$)$_2$

Ber *35* 162 (1902)

NaBH$_4$, DMF

JACS *83* 1251 (1961)

Et$_3$SiH or n-Bu$_3$SnH

Tetr *26* 4609 (1970)

SnCl$_2$, NaOH

JACS *64* 376 (1942)

cat ClRh(PPh$_3$)$_3$ or ClRh(CO)(PPh$_3$)$_2$, DMF

JOC *36* 1725 (1971)

cat Cu$_2$O, dioxane

JOC *32* 3844 (1967)

NaOCH$_3$, CH$_3$OH

JOC *33* 1924 (1968)
CC 1469 (1971)

KO-t-Bu, t-BuOH, Et$_2$O

JOC *33* 1924 (1968)

PhSH

CL 1051 (1979)

H$_2$CO

JACS *61* 2418 (1939)

HMPA

JOC *39* 1317 (1974)

DMF

Ind J Chem B *20* 767 (1981)

Me$_2$NCONMe$_2$

JOC *28* 568 (1963)
Ind J Chem B *20* 767 (1981)

various ethers and 3° amines

Angew *70* 211 (1958)

$$\text{ArCH}_2\text{NR}_2 \xrightarrow[\text{catalyst}]{\text{H}_2} \text{ArCH}_3$$

Review: Org Rxs *7* 263 (1953)

7. Nitro Compounds

$$RNO_2 \longrightarrow RH$$

Reagent(s)	R	
$n\text{-}Bu_3SnH$, AIBN	3° alkyl, 2° α-ketoalkyl, 1° allylic or benzylic	JACS *103* 1557 (1981) TL *22* 1705 (1981); *23* 2957 (1982) CC 33 (1982) JOC *50* 3692 (1985); *51* 2832 (1986); *52* 5061 (1987) Tetr *41* 4013 (1985) Syn 693 (1986) (review)
$AlCl_3$ or $SnCl_4$, Et_3SiH	benzylic, α- or β-thiophenyl alkyl	TL *28* 2277 (1987)
NaTeH, EtOH	3° alkyl	BCSJ *58* 1067 (1985)
NaSMe, DMF or HMPA	3° alkyl	JACS *100* 289 (1978); *101* 647 (1979)
KOH, $HOCH_2CH_2OH$, Δ	2°, 3° alkyl	TL 3203 (1971); 1243 (1979)

(structure: pyridine ring with CONH$_2$ substituent and CH$_2$Ph on nitrogen), $h\nu$ — α-cyano, -keto or -ester — JACS *102* 2851 (1980); *105* 4017 (1983)

$$R_2\overset{\overset{\displaystyle NO_2}{|}}{C}CH{=}CH_2 \xrightarrow[\text{cat } Pd(PPh_3)_4]{(NH_4)O_2CH, \ PPh_3} R_2CHCH{=}CH_2$$

JOC *51* 3734 (1986)

See page 115, Section 4 for reduction with double bond transposition.

8. Ethers

$$ROR' \xrightarrow[\text{2. Zn, HOAc}]{\text{1. Me}_3\text{SiCl, NaI}} RH$$

R′ = Me, $SiMe_3$; R = 1°, 2° alkyl or benzylic

Syn 32 (1981)

$$RCH{=}CHCH_2OR' \xrightarrow[\text{cat } Pd(PPh_3)_4]{\text{LiHBEt}_3} RCH{=}CHCH_3$$

R′ = Me, Ph, $SiMe_2(t\text{-Bu})$

JOC *47* 4380 (1982)

$$RCH{=}CHCH_2OPh \ or \ R\overset{\overset{\displaystyle OPh}{|}}{C}HCH{=}CH_2 \xrightarrow[\text{cat } PdCl_2(PPh_3)_2]{(Et_3NH)O_2CH} RCH_2CH{=}CH_2$$

TL 613 (1979)

$$ArOR \longrightarrow ArH$$

K, THF	CC 1549 (1987)
Na, ROH	Ber *69* 1643 (1936)
	Can J Chem *52* 2136 (1974)
EtSH, AlCl$_3$	TL *23* 689 (1982)

$$\underset{RC}{\overset{O}{\parallel}}\!-\!\underset{CHR}{\overset{OR'}{\mid}} \xrightarrow[MeOH]{SmI_2} \underset{RCCH_2R}{\overset{O}{\parallel}}$$

R' = Me, SiMe$_3$, Ts

JOC *51* 1135 (1986)
TL *28* 3065 (1987)

9. Alcohols and Phenols

9.1. Direct Reduction

H$_2$, cat PtO$_2$, CF$_3$CO$_2$H	JOC *29* 2325 (1964) (3° alkyl)
Li, NH$_3$, NH$_4$Cl	JOC *40* 3151 (1975) (1°, 2°, 3° benzylic)
K, Fe(CO)$_5$/HCl	TL *21* 801 (1980) (2°, 3° benzylic)
BH$_3$·py, CF$_3$CO$_2$H	Chem Pharm Bull *27* 2405 (1979) (2° benzylic)
BH$_3$, BF$_3$·OEt$_2$	TL 1849 (1967) (3° benzylic cyclopropyl)
NaBH$_4$, BF$_3$·OEt$_2$	Proc Chem Soc 357 (1962) (2° benzylic)
	J Sci Ind Res B *21* 583 (1962) (2° benzylic)
NaBH$_4$, PdCl$_2$	CL 1029 (1981) (benzylic)
NaBH$_4$, CF$_3$CO$_2$H	Syn 172 (1977) (2°, 3° benzylic)
NaBH$_3$CN, ZnI$_2$	JOC *51* 3038 (1986) (1°, 2° benzylic;
	2° allylic; 3° alkyl)
LiAlH$_4$, AlCl$_3$	JOC *29* 121 (1964) (3° alkyl, benzylic)
	TL 2447 (1967) (allylic)
AlCl$_3$, cat Pd-C, ⬡	Syn 397 (1978) (1°, 2°, 3° benzylic)
AlCl$_3$, EtSH	TL *23* 689 (1982) (polycyclic ArOH)
R$_3$SiH, BF$_3$	TL 2955 (1976) (2° alkyl; 3° benzylic)
R$_3$SiH (R = Ph, Et), HOAc	JACS *90* 2578 (1968) (benzylic)
Et$_3$SiH, CF$_3$CO$_2$H	JOC *34* 4 (1969) (3° benzylic);
	36 758 (1971) (3° alkyl);
	52 2226 (1987) (2° benzylic)
	JACS *91* 2967 (1969) (3° alkyl; benzylic)
	J Gen Chem USSR *43* 2294 (1973)
	(2° ferrocenyl carbinol)
	Syn 633 (1974) (review)

Me$_3$SiCl, NaI, CH$_3$CN, hexane	TL *28* 3817 (1987) (2°, 3° benzylic)
Me$_3$SiCl, NaI/Zn, HOAc	Syn 32 (1981) (1°, 2° alkyl or benzylic)
Me$_2$SiI$_2$	TL 4941 (1979) (2°, 3° benzylic)
P$_2$I$_4$	CL 247 (1983) (1°, 2°, 3° benzylic)
PPh$_3$, I$_2$	Ber *109* 1586 (1976) (1° allylic, 2° benzylic)

$$\underset{RC-CHR}{\overset{O\ \ \ OH}{\underset{||\ \ \ \ |}{}}} \longrightarrow \underset{RCCH_2R}{\overset{O}{\underset{||}{}}}$$

HI, HOAc, H$_2$O	JACS *109* 4690 (1987)
Zn, HOAc	JCS 3045 (1954)
SmI$_2$, Ac$_2$O	JOC *51* 1135 (1986)
	JACS *109* 4424 (1987)
Me$_3$SiI/Na$_2$S$_2$O$_3$, H$_2$O	Syn Commun *9* 665 (1979)
P, I$_2$	Syn 161 (1975)
Ph$_2$PLi/HOAc, MeI	JOC *51* 2378 (1986)

9.2. Via Phosphorus Compounds

$$ROH \xrightarrow[\text{2. ClPOX}_2]{\text{1. NaH or RLi}} RO\overset{\overset{O}{||}}{P}X_2 \xrightarrow{\text{Reagent(s)}} RH$$

<u>X</u>	Reagent(s)	
OEt	electrolysis	JOC *44* 4508 (1979) (phenol)
	Li, NH$_3$, Et$_2$O	JOC *42* 344 (1977) (phenol)
	Li or Na, NH$_3$	JCS 522 (1955) (phenol)
	Li, Na or K; NH$_3$; Et$_2$O	JOC *38* 2314 (1973) (phenol)
	Ti, THF	JOC *43* 4797 (1978) (phenol)
NMe$_2$	Li, NH$_3$	JOC *45* 1172 (1980) (2° alcohol)
	Li, NH$_3$, EtOH, THF	JACS *94* 5098 (1972) (2° alcohol)
	Li, MeNH$_2$	CC 1342 (1987) (1° alcohol)
	Li, EtNH$_2$	JACS *94* 5098 (1972) (2° alcohol)
	Li, EtNH$_2$, *t*-BuOH, THF	JACS *94* 5098 (1972) (1°, 2°, 3° alcohol); *108* 3835 (1986) (3° alcohol)
		TL *28* 2863 (1987) (1° alcohol)

9.3. Via Sulfonates

MsCl/electrolysis (1°, 2° alkyl)	TL 2157 (1979)
Me$_2$NSO$_2$Cl/Na, NH$_3$ (2° alkyl)	TL 3365 (1978)
TsCl or MsCl/NaI, Zn (1°, 2° alkyl)	TL 3325 (1976)
	Coll Czech Chem Commun *44* 246 (1979)
	JACS *107* 2471 (1985)

MsCl/KO-t-Bu/H$_2$, cat Pd-C (2° alkyl)   JOC *47* 2685 (1982)

MsCl/NaBH$_4$, t-BuOH, CH$_3$OCH$_2$CH$_2$OCH$_3$ JOC *52* 1309 (1987)
  (1° alkyl)

TsCl/NaBH$_4$, DMSO or HMPA or sulfolane   JOC *43* 2259 (1978)
  (1°, 2° alkyl)

TsCl/NaBH$_3$CN (1° alkyl; benzylic)   Syn 135 (1975)
                                    JOC *42* 82 (1977)

TsCl/(n-Bu$_4$N)BH$_3$CN (1° alkyl)   JOC *42* 82 (1977)

TsCl/LiHBEt$_3$ (1°, 2° alkyl)   JOC *41* 3064 (1976); *45* 1 (1980);
                                  50 2668, 5646 (1985)
                              JOMC *156* 171 (1978)
                              Carbohydr Res *110* 19 (1982)
                              JACS *108* 468 (1986); *109* 6858,
                                8102 (1987)
                              CC 1786 (1987)

MsCl/LiHBEt$_3$ (1°, 2° alkyl)   JOC *42* 2166 (1977)
                                CC 1139 (1987)
                                JACS *109* 6858 (1987)
                                TL *28* 5161 (1987)

C$_5$H$_5$N·SO$_3$/LiAlH$_4$ (allylic, benzylic) JOC *34* 3667 (1969)
                                TL 1837 (1969); *27* 4813 (1986)
                                JACS *92* 6636, 6637 (1970)

TsCl or MsCl/LiAlH$_4$ (1°, 2° alkyl)   JACS *73* 2872 (1951); *77* 1820 (1955);
                                92 553 (1970); *107* 686 (1985)
                                JOC *47* 2685 (1982); *52* 4776 (1987)

TsCl/LiAlH$_4$, LiH (1° alkyl)   JACS *108* 468 (1986)

MsCl/2 LiHAl(OMe)$_3$,   JACS *95* 6452 (1973)
  CuI (1°, 2° alkyl)

TsCl or MsCl/LiHCu(n-Bu)   JACS *96* 3686 (1974)
  (1°, 2° alkyl; allylic)

TsCl/Li$_4$CuH$_5$ (1° alkyl)   JOC *43* 183 (1978)

(R$_f$SO$_2$)$_2$/cat (Ph$_3$P)$_2$PdCl$_2$ or   CC 1452 (1986)
  Pd(PPh$_3$)$_4$, n-Bu$_3$N, HCO$_2$H (aryl)

$$\underset{\text{RC–CHR}}{\overset{\text{O X}}{\Vert \;\;\vert}} \longrightarrow \underset{\text{RCCH}_2\text{R}}{\overset{\text{O}}{\Vert}}$$

X	Reagent(s)	
OTs	SmI$_2$, MeOH	JOC *51* 1136 (1986)
OMs	(thienyl)—TeLi(Na)	JOC *47* 3946 (1982)

JACS *109* 1564 (1987)

9.4. Via Other Derivatives

$$ROH \longrightarrow ROX \xrightarrow[\text{agent}]{\text{reducing}} RH$$

Review: Pure Appl Chem *53* 15 (1981)

$\underline{X}$	Reducing Agent	
PhC=S	n-Bu$_3$SnH	JCS Perkin I 1574 (1975) (2° alkyl); 885 (1980) (2° alkyl) JOC *46* 4300 (1981) (2° alkyl)
ClC=O	n-Pr$_3$SiH, (t-BuO)$_2$, Δ	JCS Perkin I 1207 (1980) (1°, 2° alkyl)
R'OC=S	n-Bu$_3$SnH	JCS Perkin I 1718 (1977) (2° alkyl)
PhOC=S	n-Bu$_3$SnH	JACS *103* 932 (1981); *105* 4059 (1983) (both 2° alkyl) JOC *52* 3706 (1987) (2° alkyl)
PhSC=S	n-Bu$_3$SnH	Chem Pharm Bull *26* 1786 (1978) (2° alkyl)
MeSC=S	n-Bu$_3$SnH	JCS Perkin I 1574 (1975); 1718 (1977) (both 2° alkyl) Austral J Chem *30* 1269 (1977) (2° alkyl) Chem Pharm Bull *26* 1786 (1978) (2° alkyl) JACS *101* 6116 (1979) (2° alkyl); *108* 3443 (1986) (2° alkyl) Nouv J Chim *4* 59 (1980) (2° alkyl) Org Syn *64* 57 (1985) (2° alkyl) JOC *51* 2148 (1986) (2° alkyl); *52* 1057 (2° benzylic), 3096 (2° alkyl), 4647 (2° alkyl) (1987) TL *27* 2679 (2° alkyl), 3057 (2° alkyl) (1986); *28* 3615 (2° alkyl), 3883 (2° alkyl) (1987) CC 1351 (2° alkyl), 1802 (2° alkyl) (1987)
	K, t-BuNH$_2$, 18-crown-6	CC 1175 (1979) (1°, 2° alkyl) TL *28* 2937 (1987) (2° alkyl)
R$_2$NC=S	K, t-BuNH$_2$, 18-crown-6	CC 1175 (1979) (1°, 2° alkyl) JCS Perkin I 1510 (1981) (1°, 2° alkyl)

(pyrrole)NC=S	n-Bu$_3$SnH	JCS Perkin I 1574 (1975) (1°, 2° alkyl); 1109 (1984) (2° alkyl) JOC *46* 4843 (1981) (2° alkyl) TL *24* 865 (1983) (2° alkyl); *28* 6425 (1987) (2° alkyl)
	n-Bu$_3$SnH, cat Pd(PPh$_3$)$_4$	JACS *107* 2471 (1985) (2° allylic)
R'N=CNHR'	H$_2$, cat Pd-C	Ber *107* 1353 (1974) (1°, 2°, 3° alkyl; benzylic)

See also page 41, Section 15.

$$R_2CHOH \xrightarrow[n\text{-Bu}_3P]{\left(\text{(benzothiazole)}S\right)_2} R_2CHS-\text{(benzothiazole)} \xrightarrow{n\text{-Bu}_3SnH} R_2CH_2$$

TL *27* 5385 (1986)

$$R_2\overset{\overset{\displaystyle OH}{|}}{C}C\equiv CH \xrightarrow[\substack{2.\ \text{reducing agent} \\ 3.\ Fe(NO_3)_3\ \text{or}\ (NH_4)_2Ce(NO_3)_6}]{1.\ Co_2(CO)_8} R_2CHC\equiv CH$$

BH$_3$·SMe$_2$, CF$_3$CO$_2$H	Syn Commun *16* 1535 (1986) TL *28* 1857 (1987)
NaBH$_4$, CF$_3$CO$_2$H	JACS *107* 4999 (1985)

10. Sulfur Compounds

See also pages 28–31, Sections 9.3 and 9.4, for reduction of sulfur-containing derivatives of alcohols and phenols.

Raney nickel reviews:

E. E. Reid, "Organic Chemistry of Bivalent Sulfur," Chemical Publishing Co., New York (1958), Vol 1, p 115
N. Kharasch and C. Y. Meyers, "The Chemistry of Organic Sulfur Compounds," Pergamon Press, New York (1966), Vol 2, p 35

$$RSH \longrightarrow RH$$

R$_2$CHMgX, cat NiCl$_2$-PPh$_3$ (R = aryl)	CC 840 (1982)
Mo(CO)$_6$, HOAc or SiO$_2$ (R = alkyl, aryl)	CC 169 (1980)
P(OEt)$_3$, hν (R =1° alkyl, benzyl)	JACS *78* 6414 (1956)

$$RSR \xrightarrow{\text{Raney Ni}} 2\ RH$$

TL *28* 1799 (1987)

$$ArSR \longrightarrow ArH$$

Raney Ni	TL *26* 39 (1985); *28* 1459 (1987)
R$_2$CHMgX, cat NiCl$_2$-PPh$_3$ (R = Me)	CC 840 (1982)

AlCl$_3$, EtSH (R = Et, *i*-Pr; Ar = polycyclic)   TL *23* 689 (1982)

$$R_2C \!=\! CRSR' \xrightarrow[\text{LiAlH}_4]{\text{TiCl}_4} R_2CHCH_2R$$

CL 291 (1973)

$$R_2C \!=\! CRSR' \xrightarrow[\text{Ni catalyst}]{R_2CHMgX} R_2C \!=\! CHR$$

TL *22* 3463 (1981)
CC 840 (1982)

$$RCH \!=\! CHCH_2SPh \longrightarrow RCH \!=\! CHCH_3$$

Li, NH$_3$ JACS *103* 4615 (1981)
                                                    TL *24* 5531 (1983)

Li, EtNH$_2$ TL 3707 (1969)
                                                    Tetr *27* 5861 (1971)
                                                    Proc Natl Acad Sci USA *68* 1294 (1971)
                                                    JACS *95* 4444 (1973)
                                                    Syn 129 (1974)
                                                    JOC *45* 4097 (1980)

Na, *i*-PrOH, THF   TL *28* 5665 (1987)

LiHBEt$_3$, cat Pd(PPh$_3$)$_4$ JOC *47* 4380 (1982)

$$\overset{\displaystyle SAr}{\underset{\displaystyle |}{RCHC}} \!\equiv\! CH \xrightarrow[\text{NH}_3]{\text{Li}} RCH_2C \!\equiv\! CH$$

JOC *46* 5041 (1981)

$$RR'CHSAr \longrightarrow RR'CH_2$$

Raney Ni   Tetr *37* 4027 (1981)
                                                    Syn 937 (1982)

LiAlH$_4$, TiCl$_4$ CL 291 (1973)

LiAlH$_4$, CuCl$_2$ BCSJ *44* 2285 (1971)

$$ArSAr \longrightarrow 2\ ArH$$

Mo(CO)$_6$, SiO$_2$ Pure Appl Chem *52* 607 (1980)

TiCl$_4$, LiAlH$_4$ CL 291 (1973)

$$ArSOCH_3 \xrightarrow[\text{cat NiCl}_2\text{-PPh}_3]{R_2CHMgX} ArH$$

CC 840 (1982)

$$R_2CHSOAr \longrightarrow R_2CH_2$$

Al(Hg)   JOC *46* 5244 (1981)

Raney Ni   CC 1688 (1986)

$$ArSO_2CH_3 \xrightarrow[\text{cat NiCl}_2\text{-PPh}_3]{R_2CHMgX} ArH$$

CC 840 (1982)

$$R_2CHSO_2Ar \longrightarrow R_2CH_2$$

Li, EtNH$_2$ JOC *39* 2135 (1974)
                                                BSCF 513 (1976)
                                                JCS Perkin I 761 (1981)
                                                CL 25, 1711 (1981)
                                                JACS *107* 686 (1985)

Li, H$_2$NCH$_2$CH$_2$NH$_2$ BSCF 513 (1976)
                                                Tetr *37* 1233 (1981)

Na(Hg), MeOH, (phosphate)   JCS 4881 (1952)
                                                JOC *38* 2747 (1973); *51* 5100 (1986)
                                                BSCF 3065 (1973); 1363 (1975); 513, 525 (1976)
                                                TL 3477 (1976); *28* 813 (1987)
                                                JACS *102* 853 (1980); *106* 3811 (1984); *107* 2033
                                                 (1985); *108* 1035 (1986); *109* 6205 (1987)
                                                JCS Perkin I 761 (1981)
                                                Syn 55 (1981)
                                                CC 451 (1986)

Na, EtOH   CL 2105 (1984)
                                                TL *27* 4817 (1986)

Raney Ni   BSCF 525 (1976)

$$RCH{=}CHSO_2Ar \longrightarrow RCH{=}CH_2$$

Na(Hg), Na$_2$HPO$_4$, MeOH   TL *27* 2187 (1986)

Al(Hg), H$_2$O   CC 351 (1973)

LiAlH$_4$, CuCl$_2$ CC 351 (1973)

n-BuMgCl, cat Ni(acac)$_2$ TL *28* 6273 (1987)

RMgX (R = *i*-Pr, *n*-Bu), Ni or Pd catalyst   TL *24* 4311 (1983)

Na$_2$S$_2$O$_4$, H$_2$O   TL *23* 3265 (1982)

$$RCH{=}CHCH_2SO_2Ph \longrightarrow RCH{=}CHCH_3$$

Li, EtNH$_2$ JOC *39* 2135 (1974)
                                                BSCF 513 (1976)
                                                JCS Perkin I 761 (1981)
                                                CL 25 (1981)

Li, H$_2$NCH$_2$CH$_2$NH$_2$ BSCF 513 (1976)

Na, EtOH   BCSJ *55* 1325 (1982)

Na(Hg), CH$_3$OH   BSCF 513 (1976)

Na(Hg), CH_3OH, Na_2HPO_4	JCS Perkin I 761 (1981)
	JOC 51 5100 (1986)
LiHBEt$_3$, cat Pd(PPh$_3$)$_4$	JOC 47 4380 (1982)

$$R_2C(SR)_2 \longrightarrow R_2CH_2$$

Li, EtNH$_2$	JCS 4413 (1960)
Na, NH$_3$, EtOH	JACS 80 4604 (1958)
Raney Ni	JACS 66 909 (1944); 84 2938 (1962); 109 3025 (1987)
	Can J Chem 37 1870 (1959)
	Org Rxs 12 356 (1962)
	Chem Rev 62 347 (1962)
	JOC 33 3551 (1968); 50 2359, 2607 (1985); 52 2875, 3346 (1987)
LiAlH$_4$, Δ	JACS 86 478 (1964)
LiAlH$_4$, TiCl$_4$	CL 291 (1973)
LiAlH$_4$, CuCl$_2$	BCSJ 44 2285 (1971)
LiAlH$_4$, CuCl$_2$, ZnCl$_2$	Int J Sulfur Chem 7 173 (1972)
n-Bu$_3$SnH, AIBN	JOC 45 3393 (1980)
N$_2$H$_4$, KOH	JACS 81 5834 (1959)

$$\overset{\displaystyle X}{\underset{\displaystyle YCHR}{|}} \longrightarrow YCH_2R$$

X	Y	Reagent(s)	
SR	COR	Li, NH$_3$/H$_2$O	Syn Commun 3 265 (1973)
		Na(Hg), Na$_2$HPO$_4$, MeOH	TL 3477 (1976)
		Zn, NH$_4$Cl, H$_2$O	JOC 52 2317 (1987)
		Zn, Me$_3$SiCl	Syn Commun 7 427 (1977)
		Raney Ni	Syn Commun 3 265 (1973)
			JOC 47 4384 (1982); 50 2589 (1985)
			CC 717 (1985)
		SmI$_2$, MeOH	JOC 51 1135 (1986)
		NaSR	BCSJ 44 828 (1971)
		⟨S⟩—TeLi(Na)	JOC 47 3946 (1982)
	CO$_2$R	Li, NH$_3$/H$_2$O	Syn Commun 3 265 (1973)
		Raney Ni	CC 717 (1985)
		NaSR	BCSJ 44 828 (1971)
	CN	Li, naphthalene/H$_2$O	Syn Commun 3 265 (1973)
SOPh	COR	Al(Hg), H$_2$O	JACS 103 2886 (1981); 104 4180 (1982)
			JOC 46 5244 (1981)
		Zn, NH$_4$Cl, H$_2$O	JOC 52 2317 (1987)
		SmI$_2$, MeOH	JOC 51 1135 (1986)
	CO$_2$R	Al(Hg), H$_2$O	JACS 103 2886 (1981)

SO$_2$Ph	COR	Al(Hg), H$_2$O	JOC 45 4002 (1980);
			49 1246 (1984)
		Al(Hg), Na$_2$HPO$_4$, MeOH	TL 28 5017 (1987)
		SmI$_2$, MeOH	JOC 51 1135 (1986)
	CO$_2$R	Na(Hg), Na$_2$HPO$_4$, MeOH	TL 3477 (1976)

11. Selenium Compounds

See also page 39, Section 12.7.

$$RSeR' \longrightarrow RH$$

$$R' = Me, Ph$$

Li, EtNH$_2$	TL 2643 (1976)
Raney Ni	TL 2643 (1976)
Ph$_3$SnH	CC 41 (1978)
	JACS 102 4438 (1980)

$$RCH = CHCH_2SePh \xrightarrow[\text{cat Pd(PPh}_3)_4]{\text{LiHBEt}_3} RCH = CHCH_3$$

JOC 47 4380 (1982)

$$R_2C(SeR')_2 \longrightarrow R_2CH_2$$

Li, EtNH$_2$	TL 2643 (1976)
Raney Ni	TL 2643 (1976)
Ph$_3$SnH	JACS 102 4438 (1980)
	TL 22 1623 (1981)

12. Aldehydes and Ketones

12.1. Direct Reduction

$$RCHO (R_2CO) \longrightarrow RCH_3 (R_2CH_2)$$

H$_2$, various catalysts (aryl aldehydes and ketones)	Org Rxs 7 263 (1953)
	R. L. Augustine, "Catalytic Hydrogenation", Marcel Dekker, New York (1965)
H$_2$, cat Pd-C (aryl ketones)	Can J Chem 49 2712 (1971)
	Tetr 38 3555 (1982)
	Syn 940 (1982)
D$_2$, cat Pd-C, DOAc (aryl ketones)	JOC 52 2938 (1987)
H$_2$, cat PtO$_2$ (aryl ketones)	TL 23 2415 (1982)

cyclohexene or limonene, cat Pd-C, FeCl$_3$ (aryl aldehydes or ketones)	CC 757 (1976)
Raney Ni, aq EtOH (aryl aldehydes and ketones)	TL 21 2637 (1980)
HI, P (diaryl ketone)	JOC 13 786 (1948)
HI, P, I$_2$ (aryl ketones)	Ber 92 1705 (1959) Can J Chem 52 1229 (1974)
Li, NH$_3$/NH$_4$Cl (aryl ketones)	JOC 36 2588 (1971)
Zn, HCl	CC 919 (1969)
Zn, HCl, H$_2$O	Org Rxs 1 155 (1942) JACS 76 6368 (1954) Angew 71 726 (1959) Quart Rev 23 522 (1969) CC 893 (1986) (mechanism)
Zn, HCl, Ac$_2$O	JCS C 2887 (1968)
Zn, NaOH or KOH (diaryl ketones)	JOC 51 3502 (1986); 52 3205 (1987)
BH$_3$ (diaryl ketone)	J Sci Ind Res B 21 583 (1962)
BH$_3 \cdot$py, CF$_3$CO$_2$H (aryl ketones)	Chem Pharm Bull 27 2405 (1979)
BH$_3$, BF$_3 \cdot$OEt$_2$ (cyclopropyl or aryl ketones)	TL 1849 (1967)
NaBH$_4$, BF$_3 \cdot$OEt$_2$ (aryl ketones)	Proc Chem Soc 357 (1962) J Sci Ind Res B 21 583 (1962)
NaBH$_4$, AlCl$_3$ (diaryl ketone)	J Sci Ind Res B 21 583 (1962)
NaBH$_4$, CF$_3$CO$_2$H (aryl ketones)	Syn 763 (1978) JOC 50 5451 (1985)
NaBH$_4$, PdCl$_2$ (aryl ketones)	CL 1029 (1981)
NaBH$_3$CN, ZnI$_2$ (aryl aldehydes and ketones)	JOC 51 3038 (1986)
LiAlH$_4$, AlCl$_3$ (allyl or aryl ketones)	JCS 3755 (1957); 1406 (1960); 1405 (1961); 1658 (1962) JACS 80 2896 (1958) Tetr 21 2641 (1965) TL 28 2937 (1987)
Et$_3$SiH, BF$_3$ (ketones and aryl aldehydes)	JOC 43 374 (1978) Org Syn 60 108 (1981)
Et$_3$SiH, BF$_3 \cdot$OEt$_2$ (enone)	JOC 52 1984 (1987)

Et$_3$SiH, CF$_3$CO$_2$H (aryl aldehydes and ketones)   Bull Acad Sci USSR, Div Chem Sci 1245 (1966)
Tetr *23* 2235 (1967)
JOC *38* 2675 (1973)
Syn 633 (1974) (review)

PhSiMe$_2$H, CF$_3$CO$_2$H (aryl aldehydes)   JOC *38* 2675 (1973)

12.2. Via Hydrazones

NH$_2$NH$_2$, K$_2$CO$_3$, (HOCH$_2$CH$_2$)$_2$O   JACS *105* 7352, 7358 (1983); *106* 6690, 6702 (1984)

NH$_2$NH$_2$, K$_2$CO$_3$, HO(CH$_2$CH$_2$O)$_3$H   JOC *48* 1404 (1983)

NH$_2$NH$_2$, KOH, (CH$_2$OH)$_2$ Org Rxs *4* 378 (1948)
JACS *71* 3301 (1949); *107* 7978 (1985); *108* 3385 (1986)
JCS 2056 (1955)
Angew Int *7* 120 (1968)

NH$_2$NH$_2$, HO(CH$_2$CH$_2$O)$_n$H (n = 2 or 3), NaOH or KOH   JACS *68* 2487 (1946)
Ind J Chem *2* 229 (1964)
JOC *50* 2359 (1985); *52* 3205 (1987)

NH$_2$NH$_2$, KO-t-Bu   JOC *51* 5019 (1986)

TsNHNH$_2$/BH$_3$ BCSJ *47* 2323 (1974)

TsNHNH$_2$/[catechol]BH/NaOAc·3 H$_2$O   JOC *40* 1834 (1975)
Syn 124 (1977)
JACS *107* 5732 (1985)

TsNHNH$_2$/[catechol]BH/MeOH/n-Bu$_4$NOAc   Syn Commun *9* 275 (1979)

TsNHNH$_2$/(PhCO$_2$)$_2$BH/NaOAc·3 H$_2$O   JOC *46* 1217 (1981)

TsNHNH$_2$/(PhCO$_2$)$_2$BH/NaOH   JACS *107* 1721 (1985)

TsNHNH$_2$/NaBH$_4$ Chem Ind 153, 1689 (1964)
Ber *98* 3236 (1965)
Org Syn *52* 122 (1972)
BCSJ *47* 2323 (1974)
TL *28* 4759 (1987)

TsNHNH$_2$/NaBH$_3$CN (aliphatic aldehydes and ketones only)   JACS *93* 1793 (1971); *95* 3662 (1973); *106* 2115 (1984); *109* 7270 (1987)
Syn 35 (1975)
Org Prep Proc Int *11* 201 (1979)
JOC *50* 2607 (1987)

TsNHNH$_2$/NaBH$_3$CN, ZnCl$_2$ (aliphatic aldehydes and ketones only)   JOC *50* 1927 (1985)

TsNHNH$_2$/LiAlH$_4$ Tetr *19* 1127 (1963); *22* 487 (1966)
                                           Ber *98* 3236 (1965)

2,4,6-(i-Pr)$_3$C$_6$H$_2$SO$_2$NHNH$_2$/CuBH$_4$(PPh$_3$)$_2$ TL *21* 4031 (1980)

12.3. Via Oxygen and Sulfur Derivatives

$$-\overset{\overset{\displaystyle H}{|}}{\underset{|}{C}}-\overset{\overset{\displaystyle O}{\|}}{C}-\;\longrightarrow\;\underset{/}{\overset{\backslash}{C}}=\underset{\backslash}{\overset{X}{C}}\xrightarrow{\text{reagent(s)}}-\overset{\overset{\displaystyle H}{|}}{\underset{|}{C}}-CH_2-$$

$\underline{X}$	$\underline{\text{Reagent(s)}}$	
OSO$_2$CF$_3$	H$_2$, cat Pd-C	TL *23* 117 (1982)
OPO(OEt)$_2$	H$_2$, cat Pt-C	JACS *104* 2198 (1982)
	Li, t-BuOH, EtNH$_2$	JACS *108* 5650 (1986)

$$\underset{RCR}{\overset{\overset{\displaystyle O}{\|}}{}}\longrightarrow\;\underset{RCR}{\overset{RS\backslash\;/SR}{}}\longrightarrow RCH_2R$$

See page 31, Section 10.

12.4. Via Selenium Derivatives

$$R_2C{=}O\xrightarrow{R'SeH}R_2C(SeR')_2\longrightarrow R_2CH_2$$

See page 35, Section 11.

12.5. Decarbonylation

$$RCHO\longrightarrow RH$$

Review: Syn 157 (1969)

$\underline{\text{Reagent(s)}}$	$\underline{R}$	
R'SH, hν	alkyl	JACS *85* 4010 (1963)
Fe(CO)$_5$	vinylic	TL 447 (1973)
cat ClRh(PPh$_3$)$_3$	alkyl, aryl, vinylic	TL 3969 (1965); 2173 (1967); 1899 (1968); 2145 (1969); 823 (1970); *28* 5669 (1987)
		CC 129 (1966); 856 (1974)
		JACS *89* 2338 (1967); *90* 99 (1968); *93* 5465 (1971); *95* 1229, 2038, 7862 (1973); *106* 1421, 5312, 6364 (1984)
		JCS A 348 (1968); 612 (1969)
		JOC *41* 2288 (1976); *45* 315 (1980); *52* 3303 (1987)
		JCS Perkin I 700 (1977)

cat ClRh(Ph$_2$PCH$_2$CHRPPh$_2$)$_2$ (R = H, Me)	aryl	JACS *100* 7083 (1978) Fund Res Homogeneous Catal *3* 909 (1979) JOC *49* 3195 (1984)
cat (Ph$_3$P)$_2$Rh(CO)Cl, Ph$_2$P(CH$_2$)$_3$PPh$_2$	alkyl	TL *28* 6089 (1987)
cat Pd-C	alkyl, aryl	JOC *25* 2215 (1960); *52* 1201 (1987)

12.6. Ketone Cleavage

$$\underset{RCR'}{\overset{O}{\|}} \xrightarrow{\text{base}} RH + \underset{H_2NCR'}{\overset{O}{\|}}$$

NaNH$_2$	Org Rxs *9* 1 (1957) (review)
NaNH$_2$, DABCO	Syn 395 (1975)
MNH$_2$ (M = Li, Na, K) or KO-*t*-Bu	JACS *109* 6858 (1987)

12.7. Reductive Coupling

$$\underset{RCR}{\overset{O}{\|}} \xrightarrow[\text{or ArMgX}]{\text{ArLi}} \xrightarrow[\text{NH}_4\text{Cl}]{\text{Li, NH}_3} R_2CHAr$$

JOC *40* 271, 3306 (1975); *41* 3465 (1976); *46* 4139, 5060 (1981)

$$\underset{ArCR}{\overset{O}{\|}} \xrightarrow[\text{NH}_4\text{Cl}]{\text{R'Li} \quad \text{Li, NH}_3} ArCHRR'$$

JOC *38* 1735, 1738 (1973); *41* 1494 (1976)
Org Syn *55* 7 (1976)

$$\underset{RCR}{\overset{O}{\|}} \longrightarrow R_2C(CH_3)_2$$

(CH$_3$)$_3$Al	JOC *35* 532 (1970) CC 595 (1972)
(CH$_3$)$_2$TiCl$_2$ or (CH$_3$)$_2$TiCl$_2$-TiCl$_4$ or (CH$_3$)$_2$Zn	CC 237 (1981) JOC *48* 254 (1983)
(CH$_3$)$_2$Zn, TiCl$_4$	JOC *50* 5727 (1985) (enone)

$$\underset{ArCR}{\overset{O}{\|}} \xrightarrow[\text{2. MeTiCl}_3, \text{Me}_2\text{TiCl}_2]{\text{1. R'Li}} ArCR\overset{R' \quad CH_3}{\diagdown\diagup}$$

JOC *48* 254 (1983)

$$R_2CO \longrightarrow R_2C{=}CHSO_2Ar \xrightarrow{R'_2CuLi} R_2\overset{\overset{\displaystyle R'}{|}}{C}CH_2SO_2Ar \xrightarrow{Na(Hg)} R_2CHCH_3$$

$$Ar = p\text{-}ClC_6H_4$$

JOC *38* 2747 (1973)

$$\begin{matrix} \overset{O}{\underset{\parallel}{Ar\overset{}{C}R}} \\ \overset{|}{\underset{|}{SeMe}} \\ Ar\overset{}{C}H \\ \overset{|}{SeMe} \end{matrix} \xrightarrow[\text{2. RX}]{\text{1. base}} \begin{matrix} SeMe \\ \overset{|}{Ar\overset{}{C}R} \\ \overset{|}{SeMe} \end{matrix} \xrightarrow[\text{2. } R^1X]{\text{1. } n\text{-BuLi}} \begin{matrix} SeMe \\ \overset{|}{Ar\overset{}{C}R} \\ \overset{|}{R^1} \end{matrix} \xrightarrow[\text{2. } R^2X]{\text{1. } n\text{-BuLi}} \begin{matrix} R^2 \\ \overset{|}{Ar\overset{}{C}R} \\ \overset{|}{R^1} \end{matrix}$$

TL *27* 1719, 1723 (1986)

13. Carboxylic Acids

$$ArCO_2H \xrightarrow[R_3N]{HSiCl_3} \xrightarrow{\Delta} \xrightarrow{KOH} ArCH_3$$

JACS *92* 3232 (1970)

$$\overset{O}{\underset{\parallel}{R\overset{}{C}OH}} \longrightarrow RH$$

hν (R = ArCONRCHR)	Syn 141 (1982)
R$_2$CO, Δ (RH = tryptamine)	Heterocycles *6* 1167 (1977)
Cu, quinoline, Δ (R = ArCH=CH)	Org Syn Coll Vol *4* 732 (1963)
Cu chromite, quinoline, Δ (R = heterocyclic)	TL *27* 3045 (1986)
Cu$_2$O (R = stabilized carbanion)	Tetr *40* 3229 (1984)
H$_2$, cat Ni or Pd, Δ (R =1°, 2°, 3° alkyl; aryl)	Ber *115* 808 (1982)
PhOPOCl$_2$/PhSeH, Et$_3$N/n-Bu$_3$SnH (R = tetrahydrofuranyl)	JACS *107* 3285 (1985)

14. Acid Halides

$$RCOCl \xrightarrow[(t\text{-BuO})_2]{n\text{-Pr}_3SiH} RH \;(1°, 2° > 3°\text{ alkyl})$$

JCS Perkin I 1137 (1979)

15. Esters

$$ArCO_2R \xrightarrow[n\text{-}Pr_3N]{Me_3SiI \quad HSiCl_3} \xrightarrow{KOH} ArCH_3$$

JOC *44* 2185 (1979)

$$\underset{\displaystyle ROCR'}{\overset{O}{\overset{\|}{}}} \longrightarrow RH$$

hν, *N*-methylcarbazole (R' = Ph; R = 2° alkyl)	JACS *108* 3115 (1986)
Li, NH$_3$/EtOH (R = benzylic)	JOC *52* 4879 (1987)
Li, EtNH$_2$ (R = hindered alkyl, allylic)	JCS 1969 (1957) CC 68 (1978); 1173 (1979) JACS *106* 723 (1984) TL *27* 5471 (1986)
Li, (CH$_2$NH$_2$)$_2$	Ind J Chem B *18* 179 (1979)
Na, *t*-BuOH, HMPA	CC 567 (1978)
K, *t*-BuNH$_2$, 18-crown-6	CC 1173 (1979)
Na-K, tris(3,6-dioxaheptyl)amine	TL *28* 2503 (1987)
Zn, HOAc, THF (R = 3° allylic, R' = CF$_3$)	JACS *109* 6187 (1987)
H$_2$, cat Pd-C, HClO$_4$ (R = benzylic)	Syn Commun *12* 983 (1982)
Ph$_3$SiH, (*t*-BuO)$_2$, Δ	CL 77 (1986)
HMe$_2$SiOSiMe$_2$H, cat Me$_3$SiOTf	CC 1743 (1987) (4-acetoxyazetidin-2-ones)

$$RO_2CCO_2CH_3 \xrightarrow{n\text{-}Bu_3SnH} RH$$

R = 1°, 2° alkyl

CC 1588 (1985)
JACS *109* 7534 (1987)

$$\underset{\displaystyle RCHCH=CH_2}{\overset{OAc}{\overset{|}{}}} \text{ or } RCH=CHCH_2OAc \longrightarrow RCH_2CH=CH_2$$

(NH$_4$)O$_2$CH, cat PdCl$_2$(PPh$_3$)$_2$	TL 613 (1979)
SmI$_2$, cat Pd(PPh$_3$)$_4$, *i*-PrOH	TL *27* 601 (1986)

$$\underset{\displaystyle RC=CHR}{\overset{OAc}{\overset{|}{}}} \xrightarrow{Fe(CO)_5} RCH=CHR$$

TL 447 (1973)

$$\underset{\displaystyle RC-CHR}{\overset{O \quad O_2CR'}{\overset{\| \quad |}{}}} \longrightarrow \underset{\displaystyle RCCH_2R}{\overset{O}{\overset{\|}{}}}$$

Ca, NH$_3$ JCS 4344 (1956)

Zn, HOAc   JACS *77* 4367 (1955); *79* 5540 (1957)

Zn(Hg), HOAc   JACS *97* 1101 (1975)

Fe(CO)$_5$ TL 447 (1973)

SmI$_2$, MeOH   JOC *51* 1135 (1986)

n-Bu$_3$SnH   Ber *110* 2911 (1977)
                                      CC 1290 (1987)

$$RCOCl \longrightarrow R\overset{\overset{\textstyle O}{\|}}{C}OO\text{-}t\text{-Bu} \xrightarrow[R'H]{\Delta \text{ or } h\nu} RH$$

JACS *83* 3998 (1961); *86* 3157 (1964); *102* 678 (1980)
JOC *33* 99 (1968)
Ber *108* 2156 (1975)

$$RCO_2H \text{ or } RCOCl \longrightarrow R\overset{\overset{\textstyle O}{\|}}{C}ON{=}CPh_2 \xrightarrow[t\text{-BuSH}]{h\nu} RH$$

TL *28* 6207 (1987)

$$RCO_2H \text{ or } RCOCl \longrightarrow RCO-N \xrightarrow{\text{reagent(s)}} RH$$

Reagent(s)

n-Bu$_3$SnH   Heterocycles *21* 1 (1984)
                                      Tetr *41* 3901 (1985)

t-BuSH, hν CC 1298 (1984); 1041 (1987)

$$RCO_2H \longrightarrow R\overset{\overset{\textstyle O}{\|}}{C}SePh \xrightarrow{n\text{-Bu}_3\text{SnH}} RH$$

Helv *63* 1562, 2328 (1980)
JACS *107* 3285 (1985)

16. Nitriles

$$RCN \longrightarrow RCH_3$$

H$_2$, cat Ni-Al$_2$O$_3$ Syn 802 (1980)

H$_2$, cat Pd-C   Ann *707* 26 (1967)

(NH$_4$)O$_2$CH, cat Pd-C   Syn 1036 (1982)

$$RCN \longrightarrow RH$$

electrolysis, EtNH$_2$ TL 1975 (1968)

Li, NH$_3$, EtOH   TL *28* 547 (1987) (α-amino nitrile)

Li, EtNH$_2$ JACS *89* 6794 (1967); *91* 2059 (1969)

Li naphthalenide

TL *27* 2199 (1986)

Li, Na or K; HMPA; (*t*-BuOH)

Compt Rend *274* 797 (1972)
BSCF 1174 (1973)
TL 3851 (1975)
Syn 391 (1976)

Na, NH$_3$

JACS *89* 6794 (1967); *91* 2059 (1969)
BSCF 178 (1975)
JOC *40* 1162 (1975); *42* 3309 (1977)
TL 57 (1976); 61 (1976) (α-amino nitrile); 27 (1979); *23* 3369 (1982) (α-amino nitrile)
Chem Pharm Bull *25* 2689 (1977) (α-amino nitrile)

Na, Fe(acac)$_3$

JACS *93* 7113 (1971)

K, crown ether, toluene

TL *26* 6103 (1985)

K, Al$_2$O$_3$

JOC *45* 3227 (1980)

NaOH

Syn 290 (1979) (*o*-hydroxyaryl nitrile)

KOH

Syn Commun *10* 939 (1980)

NaBH$_4$

TL 3105 (1969) (α-amino nitrile); *23* 3369 (1982) (α-amino nitrile)

2. COUPLING REACTIONS

See also page 39, Section 12.7.

1. Symmetrical or Intramolecular Coupling

$$2\,RH \xrightarrow[\text{Hg}]{h\nu} R{-}R$$

CC 970 (1987)
TL *28* 5599 (1987)

$$2\,ArH \longrightarrow Ar{-}Ar$$

Review: Tetr *36* 3327 (1980)

Pb(OAc)$_4$, BF$_3$·OEt$_2$	JACS *102* 6504 (1980)
PhI(OAc)$_2$	JOC *52* 5662 (1987) (intramolecular)
(Et$_4$N)[I(O$_2$CCX$_3$)$_2$] (X = H, F, Cl)	TL *21* 3509 (1980)
VOF$_3$, CF$_3$CO$_2$H, [(CF$_3$CO)$_2$O]	JACS *95* 6861 (1973); *97* 5622, 5623 (1975); *98* 267 (1976) (all intramolecular) JOC *41* 3772, 4047 (1976); *43* 2521, 4076 (1978); *52* 5662 (1987) (all intramolecular) TL *27* 1785 (1986) (intramolecular)
VOCl$_3$	TL *27* 1785 (1986) (intramolecular) JOC *52* 5662 (1987) (intramolecular)
FeCl$_3$, SiO$_2$ (ArOH, ArOR)	JOC *45* 749 (1980); *46* 4545 (1981)
FeCl$_3$, Ac$_2$O	Chem Pharm Bull *33* 3599 (1985)
Fe(ClO$_4$)$_3$·6 H$_2$O, CH$_3$CN	Chem Pharm Bull *33* 3599 (1985)
K$_3$Fe(CN)$_6$ (ArOH)	JOC *46* 2547 (1981)
RuO$_2$, CF$_3$CO$_2$H, (CF$_3$CO)$_2$O, BF$_3$·OEt$_2$	TL *27* 1785, 5377 (1986); *28* 543, 5161 (1987) (all intramolecular)
CoF$_3$, CF$_3$CO$_2$H	JACS *102* 6504 (1980)
Tl(O$_2$CCF$_3$)$_3$	JACS *107* 4984 (1985) (intramolecular) JOC *52* 5662 (1987) (intramolecular

$Tl(O_2CCF_3)_3$, $BF_3 \cdot OEt_2$ | JOC *42* 764 (1977) (inter- and intramolecular);
49 3220 (1984) (intramolecular)
CC 538 (1977) (intramolecular)
JACS *102* 6504, 6513 (intramolecular) (1980)
Syn Commun *10* 827 (1980) (intramolecular)
Austral J Chem *37* 1775 (1984) (intramolecular)
TL *27* 1465, 1781, 1785 (1986); *28* 543, 5161
(1987) (all intramolecular)

$Tl(O_2CCF_3)_3$, cat $Pd(OAc)_2$ TL *22* 3793 (1981)

$$2\ RCH{=\!=}CH_2 \longrightarrow (RCH_2CH_2)_2$$

$BH_3/AgNO_3$, NaOH   JACS *83* 1001, 1002 (1961)
$LiAlH_4$, $TiCl_4$ or $Cp_2TiCl_2/Cu(OAc)_2$ CL 1155, 1337 (1978)

$$2\ ArX \longrightarrow Ar{-\!-}Ar$$

Review: Tetr *36* 3327 (1980)

Cu (Ullman)   Chem Rev *38* 139 (1946); *64* 613 (1964)
(rcvicws)
Syn 9 (1974)
Ann 329 (1977)
BCSJ *54* 3522 (1981)
JACS *106* 3297 (1984)

Co   JOC *48* 4904 (1983)

Ni   TL *23* 4215 (1982)
JOC *48* 840, 4904 (1983)

$Ni(COD)_2$ JACS *93* 5908 (1971); *103* 6460 (1981)

$Ni(PPh_3)_4$ JACS *97* 3873 (1975); *103* 6460 (1981)
TL 3375 (1975); *21* 631 (1980)
JOC *52* 4665 (1987) (intramolecular)

$Ni(ClO_4)_2$, electrolysis   JCS Dalton 1074 (1981)

$Ni(acac)_2$, electrolysis   JOC *41* 719 (1976)

$NiCl_2(PPh_3)_2$, electrolysis   JOMC *202* 435 (1980)
J Chem Res (S) 26 (1980)

NaH, NaO-*t*-Am, $Ni(OAc)_2$, bipy   TL 3951 (1977)

NaH, Na-O-*t*-Bu, $Ni(OAc)_2$, PPh_3 TL *27* 5483 (1986) (heteroaromatic)

NaH, *t*-AmOH, $Ni(OAc)_2$, PPh_3 or bipy   JOMC *264* 263 (1984)

Pd(Hg), N_2H_4 BCSJ *53* 1767 (1980)

cat Pd-C, NaOH, NaO_2CH, surfactant   Syn 537 (1978)
JOC *47* 4116 (1982) (bipyridines)

cat $Pd(OAc)_2$, R_3N JCS Perkin I 121 (1975)

cat $Pd(PPh_3)_4$, electrolysis   TL *26* 1655 (1985)

cat $Pd(PPh_3)_4$, 0.5 $(Me_3Sn)_2$, LiCl   JACS *109* 5478 (1987)

Zn, NiBr$_2$, KI	CL 917 (1979)
	BCSJ *53* 3691 (1980)
Zn, NiCl$_2$, PPh$_3$	Syn 736 (1984) (bipyridines)
Zn, cat NiCl$_2$(PPh$_3$)$_2$	TL 4089 (1977)
Zn, cat NiCl$_2$(PEt$_3$)$_2$, KI	BCSJ *57* 1887 (1984)
Zn, NiCl$_2$(PPh$_3$)$_2$, *n*-Bu$_4$NI	CC 1476 (1987) (bithiophenes)
Zn, NiBr$_2$(PPh$_3$)$_2$, *n*-Bu$_4$NI	TL *26* 3829 (1985)
Zn, Mg or Mn; cat NiCl$_2$; PPh$_3$	JOC *51* 2627 (1986)

$$2\ RX \longrightarrow R\!-\!R$$

For R = allylic and vinylic, see also page 241, Section 7.

Na (Wurtz) (R = alkyl)	JACS *80* 622 (1958)
O$_2$, Et$_3$B (R = allylic, benzylic)	JACS *93* 1508 (1971)
Ti, V (review) (R = allylic, benzylic)	Org Prep Proc Int *12* 361 (1980)
TiCl$_3$ or TiCl$_4$, LiAlH$_4$ (R = allylic, benzylic)	Syn 607 (1976)
VCl$_2$(py)$_4$ (R = benzylic)	JACS *95* 4158 (1973)
VCl$_3$, LiAlH$_4$ (R = allylic, benzylic)	Syn 170 (1977)
CrCl$_2$ (R = benzylic)	Tetr *20* 1005 (1964)
CrCl$_2$, electrolysis (R = allylic, benzylic)	Syn 901 (1978)
CrCl$_3$, LiHBEt$_3$ (R = 3° alkyl, allylic, benzylic)	TL *22* 5167 (1981)
CrCl$_3$, LiAlH$_4$ (R = allylic, benzylic)	TL 3829 (1977)
	BCSJ *55* 561 (1982)
Fe$_3$(CO)$_{12}$, NO (R = allylic, benzylic)	CL 1309 (1978)
Fe(acac)$_3$, electrolysis (R = 1° alkyl)	JOC *41* 719 (1976)
CoCl(PPh$_3$)$_3$ (R = benzylic)	CL 1277 (1981)
Ni (R = benzylic)	TL *23* 4215 (1982)
	JOC *48* 4904 (1983)
Ni(CO)$_4$, DMF (R = allylic)	JACS *73* 2654 (1951); *86* 1641 (1964); *89* 2757, 2758 (1967); *90* 2416, 2417 (1968); *96* 4724 (1974)
	TL 6237 (1966)
	Org Rxs *19* 115 (1972) (review)
Ni(PPh$_3$)$_4$ (R = allylic, aryl, vinylic)	TL 3375 (1975)
	JACS *103* 6460 (1981)

Ni(acac)$_2$, electrolysis (R = benzylic)   JOC 41 719 (1976)

Ni(ClO$_4$)$_2$, electrolysis   JCS Dalton 1074 (1981)
  (R = 1°, 3° alkyl; benzylic)

cat PdCl$_2$, NaOH, N$_2$H$_4$ (R = 1° alkyl)   BCSJ 54 3599 (1981)

⬠NLi, CuI (R = allylic)   TL 783 (1979)

$$RX \longrightarrow RM \xrightarrow{\text{reagent(s)}} R{-}R$$

RM	Reagent(s)	
ArCH$_2$Li	CoCl$_2$	BSCF 1331 (1964)
ArLi	Fe(acac)$_3$	JACS 106 2160, 3286 (1984); 109 7068 (1987)
	CoCl$_2$	BSCF 1331 (1964)
	CuCl$_2$	CC 1476 (1987) (bithiophenes)
RMgX (R = 1° alkyl)	AgOTf	JOC 41 2882 (1976) (cyclobutanes and -pentanes)
	RX, Ag or AgX	Syn 303 (1971)
RMgX (R = 2° alkyl, aryl)	TlBr	JACS 90 2423 (1968) Tetr 26 4041 (1970) JOC 51 1618 (1986)
RMgX (R = 1° alkyl, benzylic, aryl)	CoCl$_2$	BSCF 1331 (1964)
RMgX (R = 1° alkyl, benzyl, vinyl, phenyl)	CuCl	Angew Int 6 85 (1967) BCSJ 44 3063 (1971)
	AgBr	BCSJ 44 3063 (1971)
ArMgX	Ni(acac)$_2$ or NiCl$_2$L$_2$	JOC 41 2252 (1976)
	ClCH$_2$CH=CHCH$_2$Cl or ClCH$_2$C≡CCH$_2$Cl	JOC 46 2194 (1981)
RCu or R$_2$CuLi (R = 1°, 2° alkyl; aryl; vinylic; alkynyl)	Δ, O$_2$, or PhNO$_2$ or CuCl$_2$·TMEDA	JACS 88 4541 (1966); 89 5302 (1967); 93 1379 (1971) Angew Int 13 291 (1974) JOC 44 2705 (1979)
(ArCu)$_n$	CuO$_3$SCF$_3$	CC 203 (1977)

$$2\ ROH \longrightarrow \text{``(RO)}_2\text{Ti''} \longrightarrow R{-}R$$

R = allylic, benzylic

JACS 87 3277 (1965); 90 209, 3284 (1968)
CC 53 (1969)
JOC 40 2687 (1975); 43 3249 (1978)
J Med Chem 23 841 (1980)
Org Prep Proc Int 12 361 (1980) (review)

$$2 \, ROH \xrightarrow[\text{NaAlH}_4]{\text{NbCl}_5} R\!-\!R$$

R = allylic, benzylic

CL 157 (1982)

$$RM \longrightarrow R\!-\!R$$

RM	Reagent	
$K_2(PhSiF_5)$	$PdCl_2$	Organomet *1* 542 (1982)
$RHgX$ (R = aryl, vinylic)	cat $[RhCl(CO)_2]_2$	JOC *42* 1680 (1977)
R_2Hg (R = aryl, alkynyl, benzylic, vinylic)	cat $[RhCl(PPh_3)_3$	BCSJ *50* 2741 (1977)

$$2 \, RCO_2H \xrightarrow{\text{electrolysis}} R\!-\!R$$

"Organic Electrochemistry," Ed. M. M. Baizer, M. Dekker, New York (1973), p 469
"Techniques of Chemistry," Ed. N. L. Weinberg, J. Wiley, New York (1974), Vol. 5, Part 1, p 793
Org Syn *60* 1 (1981)
Can J Chem *59* 945 (1981)
Angew Int *20* 911 (1981)

2. Unsymmetrical Coupling

2.1. Organolithium Reagents

See also page 39, Section 12.7.

$$ArCH_3 \xrightarrow{\text{Li base}} ArCH_2Li \xrightarrow{E^+} ArCH_2E$$

Ar	E^+	
benzene	R_2SO_4 (R = Me, Et)	JOC *47* 3949 (1982)
phenol	RX, CO_2, R_3MCl (M = Si, Sn)	JOC *51* 1432 (1986)
naphthalene	MeI, Me_3SiCl	JOC *44* 3483 (1979); *48* 903 (1983)
pyridine	RBr	TL *24* 31 (1983)
pyrazine	RCHO	JOC *52* 3971 (1987)
oxazole	MeOD, RX, RCHO, R_2CO	TL *24* 4391 (1983); *28* 3585 (1987) JACS *107* 1423 (1985)
thiazole	RX	TL *28* 3585 (1987)

$$o\text{-}X\text{—}Ar\text{—}CHR_2 \xrightarrow[\text{2. } E^+]{\text{1. base}} o\text{-}X\text{—}Ar\text{—}\overset{\overset{\displaystyle E}{|}}{C}R_2$$

$\underline{X}$	$\underline{E^+}$	
CH_2OH	D_2O, RX, RCHO, R_2CO, CO_2 Me_3SiCl	TL *24* 1233 (1983) JACS *109* 2738 (1987)
CO_2H	RX, R_2CO	JACS *92* 1396 (1970) (also *m* and *p* isomers)
	RCHO	JACS *104* 4708 (1982) CC 479 (1986)
	$(RO)_2CO$	JACS *99* 4533 (1977) Syn 245 (1977) JOC *42* 4155 (1977)
CO_2R	RX	JCS Perkin I 1043 (1984) JACS *108* 4953 (1986)
	RCHO, R_2CO	CC 764 (1983), 520 (1987) JCS Perkin I 1043 (1984)
	RCOCl, CO_2 other electrophiles	JCS Perkin I 1043 (1984) CC 406 (1978); 205, 206 (1979) JCS Perkin I 1043 (1984)
CONHMe	RCHO, R_2CO	JOC *29* 3514 (1964); *52* 5378 (1987) J Heterocyclic Chem *6* 83 (1969)
	ArCH=NR RCN	JOC *52* 5378 (1987) JOC *47* 3787 (1982)
$CONR_2$	RX, R_2CO	JOC *38* 1668 (1973) (also *m* and *p* isomers)
	ArCH=NR RCN O_2 Me_3SiCl	JOC *52* 5378 (1987) JACS *108* 7100 (1986) JOC *52* 674 (1987) JOC *51* 3325 (1986)
$CONEtCH_2CH_2NEt_2$	RX, RCHO, $HCONMe_2$	JOC *51* 3566 (1986)

| 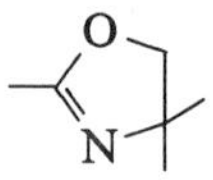 | RX | CC 388 (1986) |

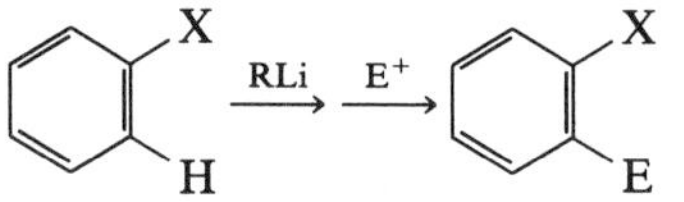

$$E^+ = RX, RCHO, R_2CO, CO_2, DMF, RCO_2R, (RCO)_2O, (RS)_2, ClSiMe_3, I_2, TsN_3$$

Reviews:

Org Rxs *8* 258 (1954); *26* 1 (1979)
Chem Rev *69* 693 (1969)
TL 3443 (1970)
Adv Chem Ser *130* 222 (1974)
JOC *41* 3653 (1976)

Z Chem *17* 1 (1977)
Tetr *39* 2009 (1983) (π-deficient heteroaromatics)
Syn 957 (1983) (synthesis of heterocycles)

X
—

C≡CH

CC 366 (1981)

F

JOC *22* 1915 (1957); *46* 203 (pyridines), 4494 (pyridines) (1981); *50* 805 (1985); *52* 713 (1987)
J Med Chem *11* 814 (1968); *29* 1982 (1986)
Compt Rend C *275* 1535 (1972) (pyridines)
JOMC *171* 273 (1979) (quinolines); *215* 139 (1981) (pyridines)
Tetr *39* 2009 (1983) (heteroaromatics)
TL *28* 4139 (1987)

Cl

TL *21* 4137 (1980) (pyridines)
JOMC *216* 139 (1981) (pyridines)

Br

TL *21* 4137 (1980) (pyridines); *24* 3291 (1983) (pyridines)
Tetr *38* 3035 (1982) (pyridines)
J Chem Res (S) 278 (1982) (pyridines)

I

TL *21* 4137 (1980) (pyridines)

CF$_3$

JACS *68* 1658 (1946)
JOMC *11* 209 (1968)

OH

JOC *19* 510 (1954)
Chem Rev *57* 583 (1957) (review)
J Chem Eng Data *14* 388 (1969)
Org Prep Proc Int *13* 426 (1981)
JACS *107* 2571 (1985)

OMe

JACS *62* 667, 987 (1940); *80* 4537 (1958); *106* 7150 (1984); *107* 2712 (1985); *109* 7068 (1987)
Ber *73* 1197 (1940); *112* 1841 (1979)
JOMC *9* 193 (1967); *11* 209, 217 (1968); *132* 321 (1977); *182* 155 (1979)
Tetr *25* 3509 (1969)
TL 3443 (1970); *27* 1971, 5125 (1986); *28* 5551 (1987)
JOC *38* 1675 (1973); *43* 3205, 3717 (1978); *46* 203, 783 (1981); *47* 2396 (1982); *50* 2690 (1985); *52* 547, 674 (1987)
J Heterocyclic Chem *17* 1333 (1980)

OR

JOC *45* 2739 (1980)
Syn 235 (1982) (pyridines)

OCH$_2$CH$_2$OMe

Chem Pharm Bull *33* 1016 (1984) (arenes, pyridines)

OCH$_2$CH$_2$NMe$_2$

Chem Pharm Bull *33* 1016 (1984) (arenes, pyridines)

OCH$_2$OMe	JACS 79 5792 (1957); 108 7100 (1986)
	JOC 44 2480 (1979); 47 2101 (1982)
	  (arenes, pyridines)
	Syn 906 (1979)
	TL 22 811, 3923 (1981); 24 3795 (1983); 28 5093
	  (1987)
	Tetr 39 2031 (1983)
	CC 1234 (1986)

OCH$_2$OR	JACS 101 257 (1979) JOC 45 2224, 5067 (1980)
OCH(CH$_3$)OEt	JOC 44 2480 (1979)
OTHP	JACS 70 4187 (1948); 79 5797 (1957); 80 4537 (1958); 107 2712 (1985)
OCONEt$_2$	JOC 48 1935 (1983); 50 5436 (1985) (pyridines) TL 24 3795 (1983); 28 5093, 5097 (1987) JACS 107 6312 (1985)
OPO(NMe$_2$)$_2$	TL 27 1391 (1986) (furans)
CH$_2$OH	Angew Int 17 521 (1978) JOC 45 1835 (1980) Ber 113 1304 (1980)
CHOHR	Syn 59 (1981)
CHOHCH$_2$NMeR	CC 968 (1974)
CR(OLi)NR$_2$	Acta Chem Scand 22 1353 (1968) (thiophene) Arkiv Kemi 32 283 (1970) (furan, thiophene) JOC 41 3651 (1976) TL 23 3979 (1982)
CH$_2$OMe	Syn Commun 5 65 (1975) TL 3973 (1975); 28 5551 (1987) JOC 48 3653 (1983); 50 2690 (1985)
CH$_2$OCH(CH$_3$)OCH$_2$CH$_3$	JOC 48 3653 (1983)
CPh$_2$OMe	JACS 74 6282 (1952)
CH(OMe)$_2$	TL 4921 (1979) JOC 48 3653 (1983); 50 805 (1985)

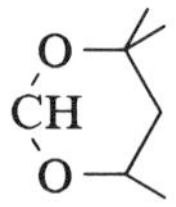

TL 27 2963 (1986)

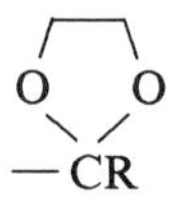

JOC 48 3653 (1983)

CH$_2$CH(OMe)$_2$ JOC 48 3653 (1983)

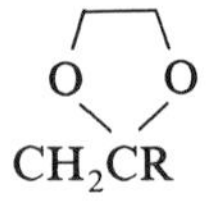

JOC 48 3653 (1983)

SR	Arkiv Kemi *13* 269 (1958) (thiophenes)
	JOMC *132* 321 (1977); *182* 155 (1979)
CR$_2$SMe	JACS *101* 257 (1979)
SO$_2$R	JACS *75* 278 (1953)
	Arkiv Kemi *13* 269 (1958) (thiophenes)
	JOC USSR *5* 313 (1969)
	Tetr *27* 433 (1971)
SO$_3$H	JOC *45* 3728 (1980)
SO$_3$R	Bull Acad Sci USSR, Div Chem Sci 129 (1980)
	JOC *51* 2833 (1986)
SO$_2$NHR	JOC *33* 900 (1968)
SO$_2$NR$_2$	Can J Chem *47* 1543 (1969)
	Angew Int *13* 270 (1974)
	TL 1499 (1975)
	Bull Acad Sci USSR, Div Chem Sci 129 (1980)
	Tetr *39* 2073 (1983)
	Syn 822 (1983) (pyridines)
NMe$_2$	Ber *75* 1491 (1942)
	JOC *31* 2047 (1966); *35* 1288 (1970); *44* 237 (1979)
	TL 3443 (1970)
	Org Syn *53* 56 (1973)
NR$_2$	JOC *50* 2690 (1985)
NMeCH$_2$CH$_2$NMe$_2$	TL *24* 5465 (1983)
NHCOC(CH$_3$)$_3$	JOC *44* 1133 (1979); *45* 4798 (1980); *46* 3564 (1981) (pyridines); *48* 3401 (1983) (pyridines)
	J Heterocyclic Chem *17* 1333 (1980)
	Chem Pharm Bull *30* 1257 (1982) (arenes, pyridines)
	Syn 499 (1982) (pyridines)
	TL *28* 5435 (1987)
	CC 1528 (1987)
NHCO$_2$C(CH$_3$)$_3$	JOC *44* 1133 (1979); *45* 4798 (1980); *51* 2781 (1986)
	Syn 499 (1982) (pyridine)
	TL *28* 5093 (1987)
—N$\diagdown$N	Helv *57* 1988 (1974)
CH$_2$NHMe	JOC *36* 1607 (1971)
CH$_2$NMe$_2$	JACS *85* 2467 (1963)
	JOC *27* 701 (1963); *28* 663, 3461 (1963); *32* 1479 (1967); *45* 5067 (1980); *52* 704 (1987)
	TL 4159 (1968); 3443 (1970); *22* 2797 (1981)
	JOMC *54* 1 (1973)
	Tetr *39* 1975 (1983)

CH=NR

JOC *41* 1564 (1976); *45* 5067 (1980)

JOC *44* 2004 (1979)

$(CH_2)_2NMe_2$

TL 603 (1966); 3443 (1970); *27* 1971 (1986)
Chem Ind 120 (1967)
Austral J Chem *21* 2319 (1968)
Acta Chem Scand B *31* 514 (1977)

CO_2H

TL *26* 1777 (1985) (furans, thiophenes)

CONHR

JOC *29* 853 (1964); *40* 1427 (1975); *44* 4463
(1979); *46* 2799 (1981); *47* 34 (1982); *50* 4362
(1985); *51* 2011 (1986)
CC 564 (1968); 1552 (1970); 1042 (1980)
TL 4159 (1968); 2559, 3965 (1978); *22* 1779
(1981); *23* 1647 (1982); *25* 2127 (1984)
(pyridines); *27* 501 (1986)
J Heterocyclic Chem *6* 475 (1969)
Tetr *27* 6171 (1971); *39* 1983 (1983)
Chem Ind 75 (1974)
Syn 797 (1975); 127 (1981) (pyridines)
Ind J Chem B *15* 512 (1977)
JACS *103* 4247 (1981)
JCS Perkin I 2227 (1982)

$CONR_2$

JOC *41* 3651 (1976); *42* 1823 (1977); *44* 4463,
4802 (1979); *47* 34, 2120, 3335, 5009 (1982); *48*
1565 (1983); *49* 318 (1984); *50* 805, 2690,
4362, 5902 (1985); *51* 271, 2011, 3325, 3566
(1986); *52* 183, 283, 674, 3181 (naphthalenes),
5668 (1987)
TL 5099, 5103, 5107 (1978); *21* 3335, 4739 (pyridines) (1980); *22* 1093, 2349 (1981);
24 2649 (pyridines), 2945, 3795, 4515, 4735
(pyridines) (1983); *26* 1149 (thiophenes),
6213 (1985); *28* 5093, 5097 (1987)
Can J Chem *57* 1598 (1979)
Heterocycles *14* 1649 (1980) (review)
JACS *102* 1457 (1980); *104* 5531 (1982); *107* 6312
(1985); *108* 7100 (1986); *109* 3402 (1987)
CC 1215 (1981)
Acct Chem Res *15* 306 (1982) (review)
Tetr *39* 1955, 1983 (1983)

CSNHR

JOC *41* 4029 (1976)

CN

Arkiv Kemi *21* 335 (1963) (thiophene)
BSCF 628 (1976) (thiophene, selenophene)
JOC *47* 2681 (1982) [m-C$_6$H$_4$(CN)$_2$]

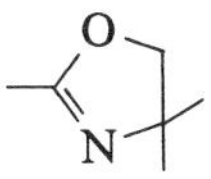

JOC *40* 2008, 3158 (1975); *42* 2649 (1977)
  (thiophenes); *43* 727 (1978); *44* 4464 (1978);
  46 3881 (1981); *47* 1585, 2633 (pyridines), 2837
  (1982); *52* 713 (1987)
Angew Int *15* 270 (1976)
Heterocycles *11* 133 (1978) (pyridines)
TL 227 (1978); *21* 3335 (1980); *23* 2091 (1982);
  24 3795 (1983); *26* 5335 (1985); *27* 3431 (1986)
JCS Perkin I 1343 (1982); 173 (1985)
Tetr *40* 2107 (1984)
JACS *104* 4015 (1982); *108* 2662, 4138 (furan)
  (1986)
CC 388 (1986)

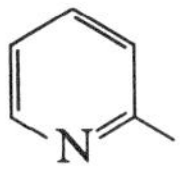

Tetr *39* 3593 (1983) (furans)

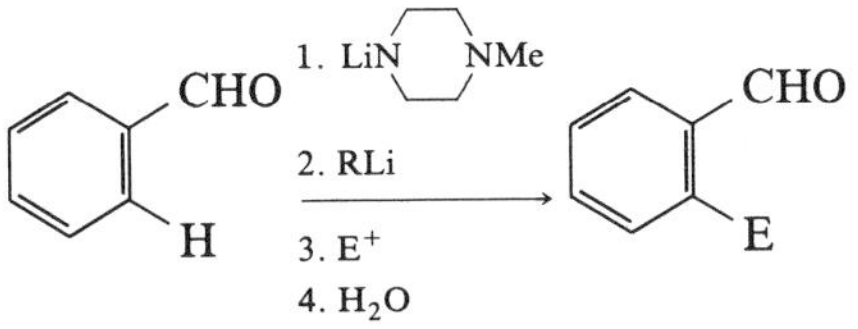

TL *23* 3979 (1982)
JOC *49* 1078 (1984); *52* 104 (1987)

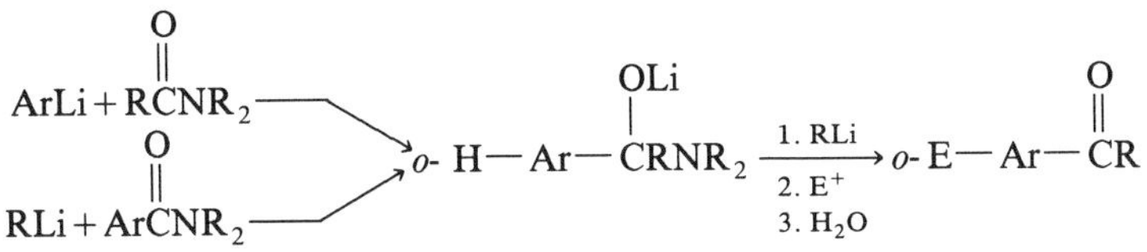

Acta Chem Scand *22* 1353 (1968) (thiophene)
Arkiv Kemi *32* 283 (1970) (furan, thiophene)
JOC *41* 3651 (1976)
TL *24* 5465 (1983); *27* 1793 (1986)

$$RLi + R'X \longrightarrow R-R'$$

R = allylic, benzylic

JACS *91* 4871 (1969)
TL 4115 (1973); 2215 (1974)
JOC *39* 1168, 3452 (1974)

$$\text{Br}\diagdown\text{Br} \quad \xrightarrow[\text{2. RX}]{\text{1. }n\text{-BuLi}} \quad \overset{\text{R}\quad\text{Br}}{\diagup\!\diagdown} \qquad \text{JACS } 97 \text{ 949 (1975)}$$

$$\xrightarrow[\text{2. MeI or E}^+]{\text{1. }n\text{-BuLi}} \quad \overset{n\text{-Bu}\quad\text{Me(E)}}{\diagup\!\diagdown} \qquad \text{BCSJ } 50 \text{ 2158 (1977)}$$

$$1° \text{ RLi} + \text{ArX} \longrightarrow \text{R—Ar}$$

JACS *91* 4871 (1969)
JOC *39* 3452 (1974)

$$o\text{-X—Ar}\overset{\text{O}}{\underset{\text{N}}{\diagup\!\!\diagdown}} \xrightarrow{\substack{\text{RLi or} \\ \text{RMgX}}} o\text{-R—Ar}\overset{\text{O}}{\underset{\text{N}}{\diagup\!\!\diagdown}} \longrightarrow o\text{-R—Ar—CO}_2\text{R}'$$

$$X = F, \text{OMe}$$

JACS *97* 7383 (1975); *104* 879, 881 (1982); *107* 682, 4238 (1985); *109* 3098, 5446 (1987)
JOC *43* 1372 (1978); *46* 783 (1981)

$$\text{X—Ar—Y} \xrightarrow[\text{2. E}^+]{\text{1. RLi}} \text{X—Ar—E}$$

Review: Acct Chem Res *15* 300 (1982)

X = H, R, Br, CH$_2$Cl, CHClCH$_3$, CCl(CH$_3$)$_2$, CH$_2$CH$_2$Br, (CH$_2$)$_3$Cl(Br), (CH$_2$)$_2$CHBrCH$_3$, OR, O(CH$_2$)$_n$Br ($n = 2,3$), OH, CR$_2$OH, CH$_2$CH$_2$OH, CH$_2$SH, CH(OLi)NR$_2'$, CN, CR$_2$CN, CH$_2$CH$_2$CN, CO$_2$R, CO$_2$H, CH$_2$CO$_2$H, (CH$_2$)$_2$CO$_2$H, NO$_2$, NH$_2$, NHCOR, SO$_2$NR$_2$, pyrrole
Y = Br, I

$$\overset{\text{O}}{E^+ = \text{H}_2\text{O, D}_2\text{O, Br}_2\text{, RX, RCH}\!\!-\!\!\text{CH}_2\text{, RCHO, R}_2\text{CO, RCO}_2\text{H, RCO}_2\text{R, RCOCl, HCONR}_2\text{, CO}_2\text{,}}$$
RCN, PhNCO, imines, Michael acceptors

Ber *71* 1903 (1938); *73* 1197 (1940); *103* 1412 (1970)
JOC *3* 108 (1938); *37* 1545 (1972); *39* 2051, 2053 (1974); *40* 2394 (1975); *41* 1184, 1187, 1268, 2628, 2704 (1976); *42* 257 (1977); *43* 1606, 3800 (1978); *45* 922 (1980); *46* 327, 1057, 1384, 2730, 2826, 4600, 4608, 4804 (1981); *47* 2608 (1982); *50* 2423, 2427 (1985); *51* 3973, 5100 (indoles) (1986); *52* 586, 704, 5668 (1987)
JACS *61* 106, 1371 (1939); *62* 344, 346, 446, 1843, 2327 (1940); *63* 1553, 2844 (1941); *64* 1007 (1942); *69* 1537 (1947); *70* 4177 (1948); *99* 4822 (1977); *106* 7150 (1984)
TL 4573 (1977); *22* 1475, 3707, 4213 (1981); *28* 1937, 2933, 4507, 6089 (1987)
Org Prep Proc Int *10* 267 (1978)
JOMC *212* 1 (1981); *215* 281 (1981)
Syn Commun *12* 49, 231 (pyrroles) (1982)

$$\text{ROH} \xrightarrow[\substack{\text{3. R'Li} \\ \text{4. (Ph}_3\text{PNMePh)I}}]{\substack{\text{1. MeLi} \\ \text{2. CuI}}} \text{R—R}'$$

R = allylic, benzylic, cyclopropylcarbinyl

R' = 1° alkyl, aryl, alkynyl

JACS *99* 2361 (1977)

2.2. Grignard Reagents

$$RMgX + R'OTs \longrightarrow R—R'$$

Org Syn Coll Vol *1* 471 (1932); *2* 47 (1943)

$$RMgX + H_2C = CHCH_2X \longrightarrow RCH_2CH = CH_2$$

Org Syn Coll Vol *1* 186 (1941)
TL 1393 (1969)

$$RMgBr + R'Br \longrightarrow R—R'$$

TL 1857 (1978)

RMgX + [structure: benzene ring with Br and I substituents] $\xrightarrow{H_2O}$ [structure: benzene ring with R substituent]

R = aryl, alkenyl, alkynyl

TL *26* 29 (1985)
JOC *50* 3104, 5524 (1985); *51* 3162 (1986); *52* 4311 (1987)

2.3. Organoboron Reagents

$$R—X + R'—B\!\! < \xrightarrow{\text{Pd catalyst}} R—R'$$

R—X	Organoborane	Catalyst	
ArBr, ArI	ArB(OH)$_2$	Pd(PPh$_3$)$_4$, Na$_2$CO$_3$	Syn Commun *11* 513 (1981) Organomet *3* 1261 (1984) JOC *49* 5237 (1984) Chem Scripta *24* 5 (1984) TL *26* 5997 (1985); *28* 5093, 5097 (1987)
		Pd(OAc)$_2$, PAr$_3$, NEt$_3$	JOC *49* 5237 (1984)
ArBr, ArI, RCH=CHBr	R$_3$B or RB⟩ (R =1° alkyl)	PdCl$_2$[bis(diphenyl-phosphino)ferrocene], NaOH or NaOCH$_3$	TL *27* 6369 (1986)

2.4. Organoaluminum Reagents

$$R_3CCl \xrightarrow{(CH_3)_3Al} R_3CCH_3$$
$$R_3COH \longrightarrow R_3CCH_3$$
$$R_2CO \longrightarrow R_2C(CH_3)_2$$
$$RCO_2H \longrightarrow RC(CH_3)_3$$

JOC *35* 532 (1970)
CC 595 (1972)

2.5. Organothallium Reagents

$$ArH \xrightarrow{Tl(O_2CCF_3)_3} ArTl(O_2CCF_3)_2 \xrightarrow[Ar'H]{h\nu} Ar—Ar'$$

JACS *92* 6088 (1970)

2.6. Organosilicon and -tin Reagents

$$\text{(structure)} \quad \text{or} \quad \text{(structure)} \xrightarrow[\text{AlBr}_3 \ (\text{CF}_3\text{CO}_2\text{H})]{\text{Me}_4\text{M (M = Si, Sn)}} \text{(structure)}$$

CC 748 (1980)

$$\text{ArCH}_2\text{SiMe}_3 \xrightarrow[\substack{\text{KF, 18-crown-6} \\ \text{or } n\text{-Bu}_4\text{NF, silica}}]{\text{PhCH}_2\text{Br}} \text{ArCH}_2\text{CH}_2\text{Ph}$$

TL 23 577 (1982)

$$\text{R}-\text{Br} + \text{R}'-\text{SnR}_3'' \xrightarrow{\text{Pd catalyst}} \text{R}-\text{R}'$$

R = benzylic, aryl; R' = Me, PhCH$_2$, Ph, CH=CH$_2$

JACS 101 4992 (1979)

$$(\text{ArN}_2)\text{X} + \text{RSnR}_3' \xrightarrow{\text{Pd catalyst}} \text{Ar}-\text{R}$$

R = Me, Ph, CH=CH$_2$

JOC 48 1333 (1983)

2.7. Organolead Reagents

$$\text{ArH or ArHgO}_2\text{CR} \xrightarrow{\text{Pb(O}_2\text{CR)}_4} \text{ArPb(O}_2\text{CR)}_3 \longrightarrow \text{ArX}$$

X	Reagent(s)	
Ar'	Ar'H, CF$_3$CO$_2$H, (AlCl$_3$)	Austral J Chem 32 1531 (1979)
CR$_2$NO$_2$	R$_2$CHNO$_2$	TL 22 783 (1981)
CH(COR)$_2$	RCOCH$_2$COR	Austral J Chem 32 1561 (1979)
CH(COR)CO$_2$R	RCOCH$_2$CO$_2$R	Austral J Chem 33 113 (1980)
CR(CO$_2$R)$_2$	$\bar{\text{C}}$R(CO$_2$R)$_2$	TL 21 965 (1980)

2.8. Sulfur, Selenium and Phosphorus Reagents

$$\text{(structure)}-\text{SPh} \xrightarrow[\text{2. 1° RX}]{\text{1. } n\text{-BuLi}} \text{(structure)} \begin{smallmatrix} \text{SPh} \\ \text{R} \end{smallmatrix}$$

TL 23 2379 (1982)

$$\text{ArCH}_2\text{SCH}_2\text{Ar} \xrightarrow[\text{P(OR)}_3]{h\nu} \text{ArCH}_2\text{CH}_2\text{Ar}$$

TL 1215 (1973)
Syn Commun 6 591 (1976)
CL 977 (1977)

$$RCH_2SAr \xrightarrow[\text{2. E}^+]{\text{1. base}} \underset{\overset{\displaystyle||}{O}}{\overset{\overset{\displaystyle E}{|}\,\overset{\displaystyle O}{||}}{RCHSAr}} \xrightarrow{\text{reduction}} RCH_2E$$

$\underline{E^+}$

RX

BSCF 3065 (1973); 1363 (1975)
Tetr *37* 1233 (1981)
JCS Perkin I 1846 (1981)
CL 1711 (1981)
JACS *106* 3811 (1984); *108* 1035 (1986)
JOC *52* 3541 (1987)
TL *28* 813, 6069 (1987)

epoxide

TL 2275 (1975)
BSCF 513 (1976)

RCO_2R

JOC *45* 4002 (1980)

$$ArCH_2SO_2CH_2Ar' \longrightarrow ArCH_2CH_2Ar'$$

hν

Angew *85* 831 (1973)
TL 861, 865 (1978); *27* 3341 (1986)
  (all intramolecular)
JACS *86* 3173 (1979); *106* 1779, 1789 (1984)
JOC *44* 1608 (1979) (intramolecular)
Org Photochem *5* 227 (1981)

hν, $P(OR)_3$

Angew *85* 831 (1973)

Δ

Angew Int *8* 274 (1969); *18* 515 (1979) (review)
Angew *85* 831 (1973)
CL 977 (1977)
JOC *50* 2939 (1985) (intramolecular);
  52 3196 (1987)

$$R_2C{=}O \xrightarrow{R'SeH} R_2C(SeR')_2 \xrightarrow[\text{2. R''X}]{\text{1. RLi}} \underset{R_2CR''}{\overset{\overset{\displaystyle SeR'}{|}}{}}\!\!R_2CR'' \xrightarrow[\substack{\text{Raney Ni,} \\ \text{or } n\text{-Bu}_3\text{SnH}}]{\text{Li-EtNH}_2,} R_2CHR''$$

TL 2643 (1976); *28* 1337 (1987)

$$\underset{\underset{\displaystyle SeMe}{\overset{\displaystyle |}{ArCH}}}{\overset{\overset{\displaystyle O}{||}}{\underset{\underset{\displaystyle SeMe}{\overset{\displaystyle |}{ArCR}}}{}}} \; \substack{\xrightarrow{\text{MeSeH}} \\ \xrightarrow{\text{ZnCl}_2} \\ \\ \xrightarrow{\text{1. base}} \\ \xrightarrow{\text{2. RX}}} \; \underset{\overset{\displaystyle |}{SeMe}}{\overset{\overset{\displaystyle SeMe}{|}}{ArCR}} \xrightarrow[\text{2. R}^1X]{\text{1. }n\text{-BuLi}} \underset{\overset{\displaystyle |}{R^1}}{\overset{\overset{\displaystyle SeMe}{|}}{ArCR}} \xrightarrow[\text{2. R}^2X]{\text{1. }n\text{-BuLi}} \underset{\overset{\displaystyle |}{R^1}}{\overset{\overset{\displaystyle R^2}{|}}{ArCR}}$$

TL *27* 1719, 1723 (1986)
CC 457 (1986)

$$Ph_3\overset{+}{P}\overset{-}{CHR} + R'X \longrightarrow Ph_3\overset{+}{P}\underset{\underset{R'}{|}}{C}HR \xrightarrow[\text{electrolysis}]{\text{NaOH or}} R'CH_2R$$

Ann *704* 109 (1967)

2.9. Organotitanium and -zinc Reagents

$$2° \text{ or } 3° \text{ RCl} \longrightarrow R—R'$$

R'_2Zn CC 1202 (1980)

$(CH_3)_2Zn$, (cat) $TiCl_4$ Angew Int *19* 900, 901 (1980)
                                                 Syn Commun *11* 261 (1981)

CH_3TiCl_3 Angew Int *19* 900, 901 (1980)

$(CH_3)_2TiCl_2$ Angew Int *19* 900, 901 (1980)

$$R_2CCl_2 \xrightarrow[\text{(CH}_3)_2\text{Zn, cat TiCl}_4]{\text{CH}_3\text{TiCl}_3 \text{ or}} R_2C(CH_3)_2$$

$$(CH_3)_3COCH_3 \xrightarrow[\text{cat TiCl}_4]{\text{(CH}_3)_2\text{Zn}} (CH_3)_3CCH_3$$

Syn Commun *11* 261 (1981)

$$R_3COH \xrightarrow[\text{or (CH}_3)_2\text{TiCl}_2]{\text{(CH}_3)_2\text{Zn, TiCl}_4} R_3CCH_3$$

CC 237 (1981)

$$\overset{\overset{\textstyle O}{\|}}{RCR} \xrightarrow[\text{(CH}_3)_2\text{TiCl}_2/\text{TiCl}_4, \text{(CH}_3)_2\text{Zn}]{\text{(CH}_3)_2\text{TiCl}_2 \text{ or}} R_2C(CH_3)_2$$

CC 237 (1981)
JOC *48* 254 (1983)

$$\overset{\overset{\textstyle O}{\|}}{ArCR} \xrightarrow[\text{2. MeTiCl}_3, \text{Me}_2\text{TiCl}_2]{\text{1. R'Li}} Ar\overset{\overset{\textstyle R' \quad CH_3}{\diagup}}{CR}$$

JOC *48* 254 (1983)

2.10. Organochromium Reagents

$$ArH \longrightarrow ArH \cdot Cr(CO)_3 \xrightarrow[\text{2. } I_2]{\text{1. RLi}} Ar\text{---}R$$

JACS *96* 7091, 7092 (1974); *97* 1247 (1975); *98* 6387 (1976); *99* 959, 1675 (1977); *101* 217, 3535
  (1979)
JOC *44* 3275 (1979)
Tetr *37* 3957 (1981)
Pure Appl Chem *53* 2379 (1981) (review)
JOMC *221* 147 (1981); *226* 183 (1982); *240* C5 (1982) (heterocycles)
CC 1361 (1982)
Organomet *2* 467 (1983)

$$ArH \longrightarrow ArH \cdot Cr(CO)_3 \xrightarrow[\text{2. } E^+]{\text{1. RLi}} E\text{---}Ar \cdot Cr(CO)_3 \xrightarrow[\text{Ce(IV)}]{I_2 \text{ or}} Ar\text{---}E$$

$$E^+ = RX, CO_2, RCHO, Me_3SiCl, RO_2CCl, PhSCl$$

JACS *101* 769 (1979)
JOC *45* 2555, 2560 (1980)
TL *21* 2069 (1980); *23* 1605 (1982); *27* 5525 (1986)
CC 1260 (1981) (indoles); 467 (1982) (indoles); 1235 (1986)

$$ArCH_2X \longrightarrow ArCH_2X \cdot Cr(CO)_3 \xrightarrow[\text{2. } E^+]{\text{1. base}} Ar\overset{\overset{E}{|}}{C}HX \cdot Cr(CO)_3 \longrightarrow Ar\overset{\overset{E}{|}}{C}HX$$

$$E^+ = D_2O, RX, RCHO, RCO_2R$$

$$X = H, R, CO_2R, OMe, SEt$$

JACS *94* 2897 (1972); *106* 2207 (1984)
CC 813 (1975); 1264 (1981); 1316 (1983)
TL 2727 (1975)
Tetr *35* 2249 (1979)
BSCF II 357 (1982)

$$ArCH_2X + RX \xrightarrow[\text{LiHBEt}_3 \text{ or LiAlH}_4]{CrCl_3} ArCH_2R$$

$$R = 3° \text{ alkyl, allylic, benzylic}$$

TL *22* 5167 (1981)
JOC *52* 511 (1987)

2.11. Organomanganese Reagents

$$H\text{---}ArMn(CO)_3^+ \xrightarrow{RM} R\text{---}ArHMn(CO)_3 \xrightarrow[\text{H}_2\text{SO}_4]{CrO_3} Ar\text{---}R$$

RM

RLi (R = Me, Ph)   JCS Dalton 1683 (1975)

RMgX (R = Me, Ph)   Organomet *1* 1053 (1982)

RCOCH$_2$Li   Organomet *1* 1053 (1982)

2.12. Organoiron Reagents

$$1°, 2°, 3° \ RMgBr + BrCH=CHR' \xrightarrow{\text{cat Fe(PhCOCHCOPh)}_3} \underset{\text{retention}}{RCH=CHR'}$$

JOC *40* 599 (1975)

2.13. Organonickel Reagents

$$H_2C=CHCH_2Br \xrightarrow[C_6H_6]{Ni(CO)_4} \left\langle\!\!\left\langle Ni \underset{Br}{\overset{Br}{<\!\!>}} Ni \right\rangle\!\!\right\rangle \xrightarrow{RX} RCH_2CH=CH_2$$

R = alkyl, vinyl, aryl

Org Rxs *19* 115 (1972) (review)

$$RM + ArX \xrightarrow{\text{Ni catalyst}} Ar\!-\!R$$

X	RM	Catalyst	
halogen	CH$_3$Li	NiCl$_2$L$_2$	JACS *97* 7262 (1975)
	RO$_2$CCH$_2$Li	NiBr$_2$ + *n*-BuLi	JACS *99* 4833 (1977)
	RMgX (R =1°,	NiCl$_2$L$_2$	JACS *94* 4374 (1972);
	2° alkyl;		*97* 7262 (1975);
	aryl; vinylic;		*106* 3286, 7150 (1984)
	benzylic; allylic)		JOMC *50* C12 (1973)
			Helv *56* 460 (1973)
			J Heterocyclic Chem *10* 243
			(1973); *12* 443 (1975)
			TL 3 (1974); *22* 4449
			(1981); *23* 4629 (1982)
			BCSJ *49* 1958 (1974)
			CL 133 (1975)
			Pure Appl Chem *52* 669
			(1980) (review)
			Tetr *38* 3347 (1982); *39*
			2699 (1983)
			JOC *47* 4319 (1982); *50*
			2086, 5370 (1985); *51*
			142, 921 (1986)
		Ni(acac)$_2$	CC 144 (1972)
			JOC *41* 2252 (1976)
			TL *22* 4449 (1981)
	ArMgX	Ni(acaen)	BCSJ *55* 845 (1982)
	n-BuZnCl	NiCl$_2$(PPh$_3$)$_2$	Bull Acad Sci USSR, Div
			Chem Sci 620 (1986)
	RZnX (R = aryl,	Ni(PPh$_3$)$_4$	JOC *42* 1821 (1977)
	benzylic)		
SH	RMgX	NiCl$_2$L$_2$	CC 637 (1979)
	(R = Me, Ph)		

SR	RMgX (R = Me, aryl, allyl)	$NiCl_2L_2$	CC 637 (1979) TL *23* 4629 (1982); *26* 39 (1985)
SOR	RMgX (R = Me, Tol)	$NiCl_2L_2$	CC 637 (1979)
SO_2R	RMgX (R = Me, Tol)	$NiCl_2L_2$	CC 637 (1979)

2.14. Organopalladium Reagents

See also page 57, Section 2.3; and page 58, Section 2.6.

$$R_f I + RX \xrightarrow[\text{Pd catalyst}]{\text{Zn}} R_f - R$$

$$RX = RCH= CHCH_2 Br \,(\text{with rearrangement}), RCH= CHBr, ArI$$

CL 137 (1982)

$$RM + R'X \xrightarrow{\text{Pd catalyst}} R - R'$$

RM	R'X	Catalyst	
—Li(ZnBr)	RBr (R = 1° alkyl, allylic, benzylic, aryl)	$Pd(PPh_3)_4$	Syn 51 (1987) TL *28* 1203 (1987)
RMgBr (R = 1°, 2° alkyl; aryl)	RI (R = 1°, 2° alkyl)		TL *27* 6013 (1986)
RMgCl(ZnCl) (R = 1°, 2° alkyl)	R'Br (R' = aryl, vinylic)		JACS *106* 158 (1984)
RMgX (R = 1°, 2° alkyl)	ArX	$PdCl_2L_2$	TL 1871 (1979) JACS *104* 180 (1982); *106* 158 (1984)
RMgX (R = 1°, 2° alkyl; PhC≡C; Ar)	ArX	$PhPdI(PPh_3)_2$	JOMC *118* 349 (1976)
ArMgX, 1° RMgX	ArX	$Pd(PPh_3)_4$ or $PdCl_2L_2$	TL *21* 845 (1980)
ArMgX	ArX	$PdCl_2[Ph_2P(CH_2)_4\text{-}PPh_2]_2$	CC 511 (1984)

$$RM + R'X \xrightarrow{\text{Pd catalyst}} R\!-\!R' \quad (\textit{Continued})$$

RM	R'X	Catalyst	
ArMgBr	ArI	cat $PdCl_2$	JOMC *125* 281 (1977)
ArMgBr(ZnCl)	ArBr	$PdCl_2[Ph_2P(CH_2)_4\text{-}PPh_2]_2$	TL *22* 5319 (1981)
n-BuZnCl	PhI	$PdCl_2L_2$	Bull Acad Sci USSR, Div Chem Sci 620 (1986)
RZnX (R = Ph, PhCH$_2$)	ArX	$Pd(PPh_3)_4$ or $PdCl_2L_2$	TL *21* 845 (1980)
RZnX (R = Ph, PhCH$_2$)	ArX	$PdCl_2(PPh_3)_2$, i-Bu$_2$AlH	JOC *42* 1821 (1977)
$EtO_2C(CH_2)_n ZnI$ ($n = 2,3$)	ArI	$PdCl_2L_2$	TL *27* 955 (1986)
$(RO_2CCHRCH_2)_2Zn$	ArX	$PdCl_2(PAr_3)_2$	JACS *109* 8056 (1987)
RZnX (R = alkynyl, benzyl, EtO$_2$CCH$_2$CH$_2$)	ArO$_3$SR$_f$	$Pd(PPh_3)_4$, LiCl	TL *28* 2387 (1987)
RZnX [R = aryl·Cr(CO)$_3$]	ArX (Ar = heterocyclic)	$Pd(PPh_3)_4$	TL *28* 2645 (1987)
RZnX (R = heterocyclic)	ArX	$Pd(PPh_3)_4$	Syn 51 (1987) TL *28* 2645, 5213 (1987)
R$_4$Sn (R = Me, Et)	PhCHBrR' (R' = H, Me)	(bipy)Pd (NCCH=CHCN)	TL *27* 5207 (1986)
R$_4$Sn (R = Me, Ph, CH=CH$_2$, PhCH$_2$)	ArX, ArCH$_2$X	PhCH$_2$PdCl(PPh$_3$)$_2$	JACS *101* 4992 (1979)
R$_4$Sn or RSnR'$_3$ (R = 1° alkyl, alkynyl, aryl, vinylic)	ArOTf	$Pd(PPh_3)_4$ or $PdCl_2(PPh_3)_2$, LiCl	JACS *109* 5478 (1987)
RSnMe$_3$ (R = Me, aryl)	ArI	ArPdI(PPh$_3$)$_2$	Proc Acad Sci USSR, Chem Sec *272* 333 (1983)
RSnMe$_3$ (R = heterocyclic)	ArX·Cr(CO)$_3$	$Pd(PPh_3)_4$	TL *28* 2645 (1987)
RSn(n-Bu)$_3$ (R = n-Bu, H$_2$C=CH)	ArCl·Cr(CO)$_3$	$Pd(PPh_3)_4$	CC 1755 (1987)
RSnR'$_3$ (R = heterocyclic)	ArX	$PdCl_2(PPh_3)_2$	TL *27* 4407 (1986)

$$RCH = CH_2 + {}^- CR'X_2 \xrightarrow[Et_3N]{PdCl_2}$$

$$\xrightarrow{25°C} \begin{array}{c} CH_2 \\ \| \\ RCCR'X_2 \end{array}$$

$$\xrightarrow{H_2} \begin{array}{c} CH_3 \\ | \\ RCHCR'X_2 \end{array}$$

$$X = CO_2R, COR$$

JACS *99* 7093 (1977); *102* 4973 (1980)
Organomet *1* 1175 (1982)
TL *23* 939 (1982)

2.15. Organocopper Reagents

Reviews:

Syn 63 (1972)
Org Rxs *22* 253 (1975)

$$ArI + Ar'I \xrightarrow{Cu} Ar — Ar'$$

Tetr *38* 2569 (1982) (intramolecular)

$$RMgX + R'X \xrightarrow[X = Br, I, OTs]{cat\ Li_2CuCl_4} R — R'$$

Syn 303 (1971)
JACS *96* 7101 (1974); *109* 7477 (1987)
Angew Int *13* 82 (1974)
JOC *41* 3505 (1976); *52* 2337, 4369 (1987)
TL 4697 (1976); *23* 3115, 3587 (1982); *27* 3903, 6193 (1986); *28* 651, 1175 (R'X = allylic acetate), 2281
 (1987)
Ber *111* 1446 (1978)
Ann 1532 (1982)
J Chem Res (S) 93 (1982)

$$1°, 2° \text{ or } 3° \text{ } RMgX + BrCH_2CH_2OR' \xrightarrow[{[P(OEt)_3]}]{cat\ CuBr} RCH_2CH_2OR'$$

$$R' = H, \text{alkyl, Ac}$$

TL 3263 (1977)

$$RMgX + Br(CH_2)_n Br \xrightarrow[HMPA]{cat\ CuBr} R(CH_2)_n Br \text{ or } R(CH_2)_n R$$

$$R = i\text{-Pr, Ar}; \ n = 3-6$$

BCSJ *59* 2035 (1986)

$$R'X \xrightarrow[R' = \text{alkyl, vinyl, aryl}]{} R — R'$$

$NaCH(CO_2Et)_2$, CuX (X = Br, I, BF_4)   CL 367 (1981)
  (R' = Ar)

NaO_2CCF_3, CuI (R' = Ar; R = CF_3)   CL 1719 (1981); 135 (1982)

RMgX, cat CuX (X = Br, CN)   JOC *52* 3847 (1987) (heterocycles)

$$R'X \xrightarrow[R' = \text{alkyl, vinyl, aryl}]{} R-R' \quad (\textit{Continued})$$

$H_2C=CHCH_2MgCl$, CuI	TL *28* 2083 (1987)
CF_3Cu	Syn 932 (1980)
ArCu	TL 3307 (1968)
	JACS *90* 2186 (1968); *102* 790 (1980)
	Org Syn *59* 122 (1980)
	JOC *52* 547 (1987)
R_2CuLi	JACS *89* 3911 (1967); *90* 5615 (1968); *91* 4871 (1969); *94* 2520 (1972); *95* 7777, 7783 (1973); *104* 4696 (1982); *48* 546 (1983)
	JOC *39* 400 (1974); *44* 2705 (1979); *50* 127 (1985); *52* 4554 (1987)
	TL 683 (1974); *23* 415 (1982); *24* 3717 (1983); *27* 4273 (1986); *28* 3135 (1987)
	Syn 752 (1982)
	JOMC *251* 133 (1983)
R_3CuLi_2	JOC *42* 2805 (1977)
Me_3Cu_2Li, Me_3CuLi_2, $Me_5Cu_3Li_2$	JOC *42* 2805 (1977)
$(RCuC\equiv CCMe_2OMe)Li$	JOC *43* 3418 (1978)
$(RCuCN)Li$	CC 88 (1973)
	JOMC *251* 133 (1983)
$(ArCuCN)Li$	JACS *107* 2712 (1985)
$(R_2CuCN)Li_2$	JACS *103* 7672 (1981); *104* 4696 (1982)
	JOC *48* 3334 (1983); *49* 3928 (1984)
$\left[\begin{array}{c}\\ \overset{}{\underset{S}{\diagdown}}\!-CuR(CN)\end{array}\right]Li_2$	JOMC *285* 437 (1985)
	TL *28* 945 (1987)
$[CH_3SOCH_2Cu(CN)R]Li_2$	JOC *52* 1885 (1987)
$(R_2CuSCN)Li_2$	JOC *48* 546 (1983)
	TL *28* 2977 (1987)
$(RCuNCy_2)Li$	JACS *104* 5824 (1982)
$(RCuPPh_2)Li$	JACS *104* 5824 (1982)
$(RCuO\text{-}t\text{-}Bu)Li$	TL 1815 (1973)

$$\overset{Br}{\underset{R}{\bowtie}} \xrightarrow[\text{2. } R'X]{\text{1. } n\text{-}Bu_2CuLi} \overset{R'}{\underset{R}{\bowtie}}$$

JACS *99* 5816 (1977)
BCSJ *52* 3632 (1979)

$$\overset{X}{\underset{X}{\bowtie}} \xrightarrow[\text{2. } R^2I]{\text{1. } R^1_2CuLi} \overset{R^1}{\underset{R^2}{\bowtie}}$$

$$X = Cl, Br \; (R^1 \neq CH_3, CH=CH_2)$$

BCSJ *50* 1600 (1977)

2.16. Organomercury Reagents

$$RHgOAc + H_2C{=}CHX \xrightarrow[R_3SnH]{NaBH_4 \text{ or}} RCH_2CH_2X$$

$$R = \text{alkyl}; \; X = \text{halogen}, CO_2R, COR, CN, CO_2COR$$

Ber *110* 2588 (1977); *112* 3759, 3766 (1979); *113* 1192, 2787 (1980); *114* 1572 (1981); *117* 859, 2132, 3160, 3175 (1984); *118* 1289, 1345, 1616 (1985); *119* 1291 (1986)
Angew Int *16* 178 (1977); *18* 154 (1979); *20* 965 (1981); *26* 479 (1987)
TL 2779, 2783 (1977); *21* 1829, 3569 (1980); *22* 2155 (1981); *24* 11, 15, 2051, 3221 (1983); *25* 2743 (1984)
Organomet *1* 675 (1982)
Tetr *41* 4025 (1985)
Ann 427 (1987)

$$ArH \longrightarrow ArHgX \xrightarrow{R_2CuLi} Ar{-}R$$

$$R = 1°, 2° \text{ alkyl; vinylic}$$

$$RCH{=}CH_2 \longrightarrow \overset{\overset{\displaystyle X}{|}}{R}CHCH_2HgCl \xrightarrow{Li_2Cu(CH_3)_3} \overset{\overset{\displaystyle X}{|}}{R}CHCH_2CH_3$$

$$X = H, OH$$

TL *22* 3435 (1981)
Organomet *1* 74 (1982)

2.17. Miscellaneous Reactions

$$R_f I(Ph)OSO_2CF_3 + ArH \longrightarrow R_f{-}Ar$$

CL 1663 (1981)

$$ArI + Ar'H \xrightarrow{h\nu} Ar{-}Ar'$$

JOC *30* 2493 (1965)

$$ArNH_2 + Ar'H \xrightarrow{t\text{-BuSNO}_2} Ar{-}Ar'$$

BCSJ *53* 2023 (1980)

$$\overset{\overset{\displaystyle O}{\|}}{R}CR \longrightarrow \overset{\overset{\displaystyle NNHCPh_2(t\text{-Bu})}{\|}}{R}CR \xrightarrow[\text{2. R'X}]{\text{1. base}} \overset{\overset{\displaystyle N{=}NCPh_2(t\text{-Bu})}{|}}{\underset{\underset{\displaystyle R'}{|}}{R}}CR \xrightarrow[\Delta]{PhSH} \overset{\overset{\displaystyle H}{|}}{\underset{\underset{\displaystyle R'}{|}}{R}}CR$$

CC 176 (1986)

$$RCO_2H + R'CO_2H \xrightarrow{\text{electrolysis}} R{-}R'$$

"Organic Electrochemistry," Ed. M. M. Baizer, M. Dekker, New York (1973), p 469
"Techniques of Chemistry," Ed. N. L. Weinberg, J. Wiley, New York (1974), Vol 5, Part 1, p 793
Angew Int *20* 911 (1981)
Ann 1532 (1982)

3. FRIEDEL-CRAFTS AND RELATED ALKYLATION REACTIONS

Reviews:

Org Rxs *3* 1 (1946)
Quart Rev *8* 355 (1954)
"Friedel-Crafts and Related Reactions," Ed. G. A. Olah, Interscience, New York (1964), Vol 2
G. A. Olah, "Friedel-Crafts Chemistry," Wiley-Interscience, New York (1973)
R. M. Roberts, A. A. Khalaf, "Friedel-Crafts Alkylation Chemistry," Marcel Dekker, New York (1984)

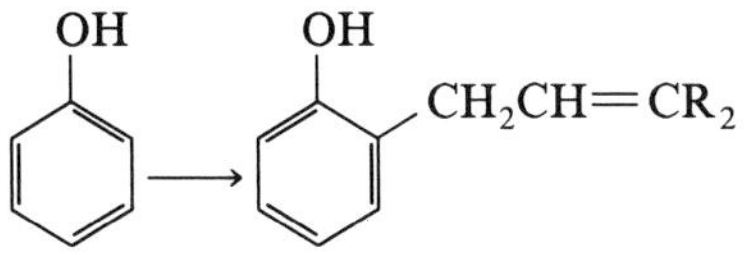

NaH/$R_2C{=}CHCH_2X$, toluene	JCS Perkin I 1601 (1982)
KOH/$R_2C{=}CHCH_2X$	TL *23* 4567 (1982) (*m*-diphenol)
NaOMe/$R_2C{=}CHCH_2X$	Ind J Chem *7* 1072 (1969)
	JOC *39* 2215 (1974)
K/$ZnCl_2$/$R_2C{=}CHCH_2X$	Syn 310 (1981)
Al_2O_3, $R_2C{=}CHCH_2X$	JOC *51* 4481 (1986)
Ag_2O, $R_2C{=}CHCH_2X$	Tetr *28* 4395 (1972) (*m*-diphenol)
$H_2C{=}CHC(CH_3){=}CH_2$, Me_3SiCl, NaI	TL *25* 5581 (1984)
$H_2C{=}CHC(CH_3){=}CH_2$, TsOH, NaI	TL *25* 5581 (1984)

$$ArH + ROH \longrightarrow Ar-R$$

HOAc	Org Syn Coll Vol *4* 47 (1963)
HF	JACS *72* 5232 (1951)
P_2O_5	JCS 2520 (1932) (intramolecular)
$AlCl_3$	JACS *60* 1421 (1938)

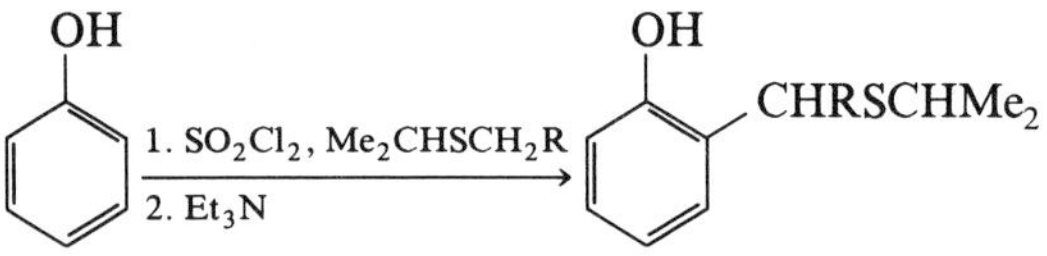

OH
$\downarrow$
(benzene ring with OH)
$\xrightarrow[\text{2. Et}_3\text{N}]{\text{1. SO}_2\text{Cl}_2,\ \text{Me}_2\text{CHSCH}_2\text{R}}$
(benzene ring with OH and CHRSCHMe$_2$)

JACS *100* 7611 (1978)
JOC *52* 5495 (1987)

$$\text{H}-\text{ArOH} \xrightarrow[\text{2. Raney Ni}]{\text{1. PhSCHClR, SnCl}_4} \text{RCH}_2\text{ArOH}$$

Syn 937 (1982)

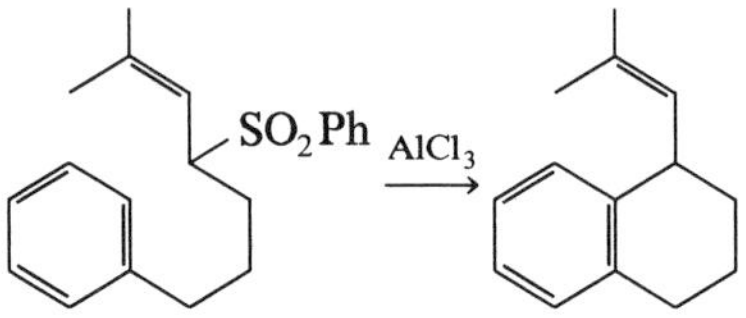

JACS *106* 7260 (1984)

$$\text{ArH} \longrightarrow \text{ArCH}_2\text{X}$$

$$\text{X} = \text{Cl, Br, I}$$

Reviews:

Org Rxs *1* 63 (1942)
"Friedel-Crafts and Related Reactions," Vol 2, pp 659–784

ClCH$_2$OCH$_2$Cl, SnCl$_4$	JOC *41* 1627 (1976)
(XCH$_2$OCH$_2$CH$_2$)$_2$ (X = Cl, Br), Lewis acid	Syn 560 (1974)
	JOC *41* 1627 (1976)
	Org Prep Proc Int *14* 3 (1982)
H$_2$CO, HCl, ZnCl$_2$, HOAc	JACS *75* 6292 (1953)
H$_2$CO, HBr, HOAc	JOC *47* 578 (1982); *52* 3200 (1987)

$$\text{ArH} + \text{H}_2\text{C}\overset{\text{O}}{\overbrace{}}\text{CHR} \xrightarrow{\text{Lewis acid}} \text{ArCHRCH}_2\text{OH}$$

JCS 5404 (1964) (intramolecular)
Tetr *25* 1807 (1969)
JOC *48* 2449, 4572 (1983); *52* 425 (1987) (all intramolecular)

4. RING-FORMING REACTIONS

See also page 653, Section 6.

1. Three-Membered Rings

Review: Houben-Weyl, Vol IV/3 (1971)

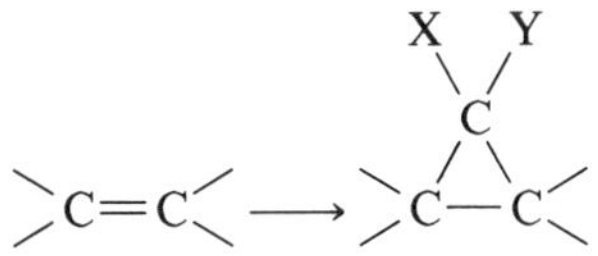

$$\underset{>}{>}C=C\underset{<}{<} \longrightarrow \underset{>}{>}C-C\underset{<}{<}$$

with the cyclopropane structure having X and Y substituents on the top carbon.

Reviews:

Syn 77 (1978) (synthesis of spiro compounds)
Tetr *36* 2531 (1980)
Acct Chem Res *13* 58 (1980); *19* 348 (1986)
Chem Rev *86* 919 (1986)

CXY	Reagent(s)	
CH_2	CH_2I_2, hν	JACS *100* 655 (1978)
		Tetr *37* 3229 (1981)
	CH_2I_2, R_3Al	JOC *50* 4412 (1985)
	(R = Me, Et, *i*-Bu)	CC 157 (1987)
	CH_2I_2, Cu	JACS *98* 2676 (1976);
		101 2139 (1979)
	CH_2I_2, Cu, ultrasound	CC 1460 (1987)
	CH_2I_2, Zn, ultrasound	TL *23* 2729 (1982)
		Org Prep Proc Int *16* 25 (1984)
	CH_2I_2, Zn-Cu	JACS *80* 5323 (1958); *81* 4256 (1959); *85* 468 (1963); *91* 6892 (1969); *107* 4984, 8256 (chiral) (1985); *108* 3443 (1986)
		JOC *24* 1825 (1959); *33* 1767, 2141 (1968); *51* 2721 (1986) (diastereoselective); *52* 3000 (1987) (diastereoselective)
		Org Rxs *20* 1 (1973) (review)
		Org Syn Coll Vol *5* 855 (1973)
		J Chem Res (S) 179 (1978)

CXY	Reagent(s)	
CH$_2$ (*continued*)	CH$_2$I$_2$, Zn-Ag	Syn 549 (1972)
		JACS *104* 4290 (1982)
		TL *28* 1865 (1987)
		JOC *52* 603 (1987)
	CH$_2$I$_2$, EtZnI	TL *23* 259 (1982)
		JACS *109* 3025 (1987)
	CH$_2$I$_2$, Et$_2$Zn	TL 3353 (1966)
		Tetr *24* 53 (1968)
		JACS *106* 3869, 6006 (1984); *107* 8254 (1985) (chiral); *108* 6343 (1986); *109* 7553 (1987)
		JOC *52* 3603 (1987)
	CH$_2$I$_2$, Sm or Sm(Hg)	JOC *52* 3942 (1987) (allylic alcohols)
	CH$_2$Br$_2$, CoCl$_2$ or NiBr$_2$, Zn, NaI	CL 395 (1981)
	CH$_2$Br$_2$, cat Ni(COD)$_2$, NaI	CL 761 (1979)
	CH$_2$Br$_2$; cat Ni(PPh$_3$)$_4$; NaI, ZnX$_2$ (X = Cl, Br or I) or AlCl$_3$	CL 761 (1979)
	CH$_2$Br$_2$, Zn, CuCl, ultrasound	JOC *50* 4640 (1985)
	CH$_2$N$_2$/Δ or hν	TL *23* 2103 (1982)
	CH$_2$N$_2$, cat Pd(OAc)$_2$	TL 1465 (1972); 629 (1975); *23* 502 (1982); *28* 4547 (1987)
		Syn 636 (1975); 714 (1981)
		JOC *51* 4836 (1986)
	CH$_2$N$_2$, CuBr, cat Cu(OTs)$_2$,·2 H$_2$O	Helv *59* 1953 (1976)
	CH$_2$N$_2$, cat (CuOTf)$_2$	JACS *95* 3300 (1973); *107* 996 (1985)
	[CpFe(CO)$_2$CH$_2$$\overset{+}{S}Me_2$]BF$_4^-$	JACS *101* 6473 (1979); *109* 3739 (1987)
		JOC *50* 5898 (1985)
	$\overset{-}{C}$H$_2$$\overset{+}{S}Me_2$ (on Michael acceptors)	TL 661 (1962)
		Ber *96* 1881 (1963)
		JOC *29* 3277 (1964); *33* 3849 (1968); *34* 3324 (1969); *37* 2354 (1972)
		Ind J Chem B *19* 563 (1980)
	$\overset{-}{C}$H$_2$$\overset{+}{S}OMe_2$ (on Michael acceptors)	JACS *84* 3822 (1962); *87* 1353 (1965); *106* 5335 (1984); *107* 4984 (1985)
		Z Naturforsch B *18* 976 (1963)
		JOC *31* 3467 (1966); *43* 2839 (1978); *50* 5898 (1985)
		JCS C 2495 (1967)
		Angew Int *12* 845 (1973)
		BSCF 888 (1974)
		TL *28* 3201 (1987)

CH_2, CHR, CR_2	$ArS\overset{+}{C}R_2$ with $\overset{O}{\overset{\|}{C}}$ above and NMe_2 below	Acct Chem Res 6 341 (1973) JACS 95 7692 (1973)
CH_2, CR_2, $CHCO_2R$	R_2CN_2, cat Cu(II)	Tetr 24 3655 (1968) Can J Chem 47 1242 (1969)
CHR (R = H, Me)	$RCHI_2$, R'_3Al	JOC 50 4412 (1985)
CHR (R = H, Me, Ph)	$RCHI_2$, Et_2Zn	Tetr 27 1799 (1971) JOC 42 3031 (1977); 47 1615, 2426 (1982)
$CHCH_3$	CH_3CHI_2, Zn, CuCl $CpFe(CO)_2CH(CH_3)SPh$, $MeSO_3F$ or $(Me_3O)BF_4$	JOC 47 1615, 2426 (1982) JACS 107 4230 (1985) JACS 103 1862 (1981) JOMC 285 231 (1985)
CHR	$RCHN_2$, CuI $CpFe(CO)_2CHRSPh$, $(Me_3O)BF_4$	CL 863 (1982) (intramolecular) JOC 50 5898 (1985) (intramolecular)
$CHCF_3$	CF_3CHN_2, hν	JCS 1881 (1964)
$CHCH(OMe)_2$	$N_2CHCH(OMe)_2$, Δ $N_2CHCH(OMe)_2$, hν	TL 23 503 (1982) TL 21 2239 (1980)
$CHCH=C(CH_3)_2$	$PhSO_2CH_2CH=C(CH_3)_2$, KO-t-Bu (on Michael acceptors) $NO_2CH(K)CH=C(CH_3)_2$ (on Michael acceptors)	BSCF 985 (1967) JOC 50 2806 (1985)
CHR (R = vinylic)	$RCH(K)NO_2$ (R = vinylic) (on Michael acceptors)	JOC 50 2806 (1985)
$C(CH_3)_2$	$Ph_2\overset{+}{S}\overset{-}{C}(CH_3)_2$ (on Michael acceptors) $(CH_3)_2CBr_2$, n-BuLi $(CH_3)_2CH(K)NO_2$, DMSO (on Michael acceptors)	JACS 89 3912 (1967) Angew Int 20 863 (1981) JOC 50 2806 (1985)
$C(CF_3)_2$	$\overset{N}{\underset{N}{\|}}\!\!>\!\!C(CF_3)_2$, Δ	JACS 88 3617 (1966)
CR_2	$R_2CH(K)NO_2$, DMSO (on Michael acceptors)	JOC 50 2806 (1985)
CHF	$CHFI_2$, hν	TL 1819 (1975)
CHCl	CH_2Cl_2, n-BuLi	JACS 82 5729 (1960) Can J Chem 59 621 (1981)
CHBr	CH_2Br_2, $NaN(SiMe_3)_2$	Syn 201 (1972)
CHX (X = Cl, Br)	$XCHN_2$	JACS 84 4350 (1962); 87 4270 (1965)

CXY	Reagent(s)	
CHX (X = F, Cl, Br, I)	HCXI$_2$, Cu	Tetr *35* 1919 (1979)
	HCX$_3$ or HCXI$_2$, Et$_2$Zn	CC 1375 (1971) (X = F, Cl, Br, I); 364 (1975) (X = Br)
		BCSJ *47* 1500 (1974) (X = I)
CF$_2$	F–C(O)C(CF$_3$)F$_2$ epoxide, Δ	JOC *35* 678 (1970)
		JCS Perkin I 2203 (1982)
	ClCF$_2$CO$_2$Na, Δ	JACS *85* 1851 (1963)
CCl$_2$	HCCl$_3$, NaOH, ultrasound	JOC *47* 1587 (1982)
	HCCl$_3$, NaOH, phase transfer	TL 4659 (1969); 1461 (1972); *27* 893 (1986)
		Ann *744* 42 (1971); 591 (1981)
		Tetr *28* 175 (1972)
		JOC *44* 447 (1979); *52* 3579 (1987)
		Syn 1004 (1981)
		J Chem Res (S) 354 (1982)
	HCCl$_3$, KO-t-Bu	JACS *76* 6162 (1954); *78* 1437 (1956); *80* 5274 (1958)
	XCCl$_3$ (X = Br, I), RLi (R = Me, n-Bu)	JACS *81* 5009 (1959)
	Cl$_3$CCO$_2$Na, Δ	JACS *85* 1851 (1963)
	Cl$_3$CCO$_2$Na, R$_4$NBr, Δ	J Chem Res (S) 72 (1977)
		Helv *65* 1191 (1982)
	Cl$_3$CCO$_2$Et, NaOR (R = Me, Et) or KO-t-Bu	JOC *24* 1733 (1959); *40* 2234 (1975)
		JACS *82* 4085 (1960)
		Org Syn Coll Vol *5* 874 (1973)
CBr$_2$	HCBr$_3$, NaOH, phase transfer	TL 1367 (1973)
		Syn 296 (1982)
		JOC *47* 3211 (1982); *52* 4732 (1987)
	HCBr$_3$, NaOH, n-Bu$_3$N	JACS *107* 8066 (1985)
	HCBr$_3$, KO-t-Bu	JOC *47* 3211 (1982)
CX$_2$ (X = Cl, Br)	HCX$_3$, NaOH	TL 1749 (1965)
CXY (X, Y = F, Cl, Br, I)	PhHgCXYZ (Z = F, Cl, Br, I)	Acct Chem Res *5* 65 (1972) (review)
		Russ Chem Rev *46* 941 (1977) (review)
		R. C. Larock, "Organomercury Compounds in Organic Synthesis," Springer, New York (1985), Chpt 10 (review)
CFCl	FCCl$_2$COCCl$_2$F, KO-t-Bu	JOC *28* 2494 (1963)
	HCCl$_2$F, NaOH	TL 1749 (1965)

CFBr	HCBr$_2$F, NaOH, phase transfer	Ann 591 (1981)
CClCH$_3$	CH$_3$CHCl$_2$, RLi	Syn 801 (1974) Can J Chem *60* 1933 (1982) JACS *108* 1251 (1986) JOC *52* 1475 (1987)
CClCH=CH$_2$	Cl$_2$CHCH=CH$_2$, (piperidyl-N-Li)	Syn 425 (1979)
CClCCl=CCl$_2$	(tetrachlorocyclopropene), Δ	TL *23* 3341 (1982)
CClPh	(N=N)C(Cl)(Ph), Δ	Org Syn *60* 53 (1981) TL *25* 901 (1984); *27* 4395 (1986)
CClCH$_2$Ph	(N=N)C(Cl)(CH$_2$Ph), hν or Δ	JOC *52* 4223 (1987)
CHMMe$_3$ (M = Si, Sn)	ClCH$_2$MMe$_3$, (piperidyl-N-Li)	TL 1677 (1978)
CHOR	ClCH$_2$OR, (piperidyl-N-Li)	TL 3779 (1976)
	ClCH$_2$OR, RLi	Angew *73* 27, 765 (1961) Ber *96* 2266 (1963); *97* 636 (1964)
	Cl$_2$CHOR, MeLi	Ber *98* 2221 (1965)
C(Ph)OMe	(N=N)C(OMe)(Ph), hν or Δ	CC 432 (1982) JACS *109* 4341 (1987)
C(CH$_3$)OSiMe$_3$	CH$_3$COSiMe$_3$, hν	JACS *103* 699 (1981)
CClOPh	(N=N)C(Cl)(OPh), Δ	JOC *47* 4177 (1982)
	HCCl$_2$OPh, NaOH, phase transfer	Angew Int *21* 916 (1982)
C(OMe)OPh	(N=N)C(OMe)(OPh), Δ	JACS *109* 3811 (1987)
CHSPh	ClCH$_2$SPh, NaOH, (PhCH$_2$NEt$_2$)Cl	TL 4247 (1975); *28* 4797 (1980)
	ClCH$_2$SPh, NaH	Compt Rend C *275* 283 (1972)

CXY	Reagent(s)	
CHSAr	ClCH$_2$SAr, KO-t-Bu	Angew 73 765 (1961)
		TL 165 (1962)
		Ber 97 1527 (1964)
		BCSJ 52 3434 (1979)
	ClCH$_2$SAr, n-BuLi	Angew 73 765 (1961)
		TL 165 (1962)
		Ber 97 1527 (1964)
CCISPh	Cl$_2$CHSPh, NaOH, (PhCH$_2$NEt$_3$)Cl	TL 4517 (1971)
	Cl$_2$CHSPh, KO-t-Bu	Ber 99 806 (1966)
CHSO$_2$C$_6$H$_4$CH$_3$-p	p-CH$_3$C$_6$H$_4$SO$_2$CHN$_2$, cat Cu(acac)$_2$	JOC 52 4760 (1987)
CHCOR	RCOCHN$_2$	Org Rxs 26 361 (1980) (intramolecular, review)
	RCOCHN$_2$, hν, PhCOPh	JACS 90 2200 (1968)
	RCOCHN$_2$, cat Mo(CO)$_6$	JOC 45 1538 (1980)
	PhCOCHN$_2$, cat Mo(CO)$_6$ or Mo$_2$(OAc)$_4$	JOC 47 4059 (1982)
	RCOCHN$_2$, cat Rh$_2$(OAc)$_4$	TL 27 2075 (1986)
		JOC 51 3878 (1986)
		JACS 109 5432 (1987) (all intramolecular)
	RCOCHN$_2$, cat Pd(OAc)$_2$	Tetr 36 3269 (1980)
		TL 23 2411 (1982)
	RCOCHN$_2$, cat Cu	JACS 103 1808, 1813 (1981) (both intramolecular)
	RCOCHN$_2$, cat Cu-bronze or CuSO$_4$	Tetr 28 4653 (1972) (intramolecular)
	RCOCHN$_2$, cat Cu, cat CuSO$_4$	Tetr 26 2815 (1970)
		JACS 109 3147 (1987)
		JOC 52 4634 (1987) (all intramolecular)
	RCOCHN$_2$, cat Cu(acac)$_2$	JOC 45 5020 (1980) (intramolecular)
	RCOCHN$_2$, cat CuSO$_4$, cat Cu(acac)$_2$	JOC 47 1522 (1982); 52 4641 (1987)
CRCOR	N$_2$CRCOR, hν, PhCOPh	JACS 90 2200 (1968)
	N$_2$CRCOR, cat CuSO$_4$	TL 1363 (1972) (intramolecular)
	N$_2$CRCOR, cat Cu(acac)$_2$	JOC 45 5020 (1980) (intramolecular)
	N$_2$CRCOR, cat Cu(II) bis(N-n-butylsalicyl-ideneaminate)	JACS 106 6006 (1984) (intramolecular)

CHCO$_2$R	Br$_2$CHCO$_2$Me, Cu	JACS 98 2676 (1976);
		101 2139 (1979)
	N$_2$CHCO$_2$R, hν or Δ	Org Rxs 18 217 (1970) (review)
		Org Syn 50 94 (1970)
		Chem Rev 75 431 (1974)
	N$_2$CHCO$_2$R, (Ar$_3$N)SbCl$_6$	JACS 108 4234 (1986)
	N$_2$CHCO$_2$Et, cat Mo(CO)$_6$	JOC 45 1538 (1980)
	N$_2$CHCO$_2$Et, cat Mo(CO)$_6$	JOC 47 4059 (1982)
	or Mo$_2$(OAc)$_4$	
	N$_2$CHCO$_2$Et, cat Rh$_6$(CO)$_{16}$	TL 22 1783 (1981);
		23 2261 (1982)
		Syn 787 (1981)
	N$_2$CHCO$_2$R, cat Rh$_2$(OAc)$_4$	Syn 787 (1981)
		TL 23 2261 (1982);
		28 833 (1987)
	N$_2$CHCO$_2$R, cat Rh$_2$(O$_2$CCF$_3$)$_4$	TL 28 833 (1987)
	N$_2$CHCO$_2$R,	TL 28 833 (1987)
	cat Rh$_2$(NHCOCH$_3$)$_4$	
	N$_2$CHCO$_2$Et, cat Rh(III)-	Tetr 38 2365 (1982)
	porphyrins	
	N$_2$CHCO$_2$R, cat Rh(III)-	TL 21 3489 (1980)
	tetraphenylporphyrin	
	N$_2$CHCO$_2$R, cat Rh$_2$(OAc)$_4$	JOC 45 695 (1980)
	or Pd(OAc)$_2$ or Cu(OTf)$_2$	
	N$_2$CHCO$_2$Et, cat Pd(OAc)$_2$	TL 1465 (1972); 629 (1975)
	N$_2$CHCO$_2$Et,	TL 23 2261 (1982)
	cat PdCl$_2$(PhCN)$_2$	
	N$_2$CHCO$_2$Et, cat π-allyl-	JOC 52 5158 (1987)
	palladium chloride	
	N$_2$CHCO$_2$Et, cat Cu	JOC 38 2221 (1973)
	N$_2$CHCO$_2$R, chiral Cu catalyst	Tetr 24 3655 (1968)
		TL 1707 (1975);
		23 685 (1982)
		JACS 104 1362 (1982);
		106 1421, 5312, 6364 (1984)
	N$_2$CHCO$_2$R, cat Cu(II)	TL 25 3559 (1984)
	bis(salicylaldehyde-t-	(intramolecular)
	butylimine)	JACS 107 5574 (1985)
	N$_2$CHCO$_2$R, cat Cu(acac)$_2$	TL 27 2139 (1986)
		(intramolecular)
	N$_2$CHCO$_2$Et,	TL 23 2261 (1982)
	cat CuCl·(i-PrO)$_3$P	
	N$_2$CHCO$_2$Et, cat CuI	JOC 50 2026 (1985)
	N$_2$CHCO$_2$Et, cat CuSO$_4$	Can J Chem 60 2383 (1982)
		JACS 106 1421 (1984);
		107 734 (1985)
		JOC 52 4898 (1987)
	N$_2$CHCO$_2$Et,	JACS 95 3300 (1973);
	cat (CuOTf)$_2$·C$_6$H$_6$	107 996 (1985)
CHCONHAr	N$_2$CHCONHAr,	JOC 51 5362 (1986)
	cat Rh$_2$(O$_2$CR)$_4$	

CXY	Reagent(s)	
$CHCONR_2$	$N_2CHCONR_2$; cat $Rh_2(OAc)_4$, $Rh_2(O_2CCF_3)_4$ or $Rh_2(NHCOCH_3)_4$	TL *28* 833 (1987)
$CRCO_2R$	$RCHXCO_2R$ (X = Cl, Br), NaH or NaOMe	JACS *80* 6568 (1958)
$C(CH{=}CH_2)CO_2Et$	$CH_3CH{=}CBrCO_2Et$, LDA (on Michael acceptors)	JOC *51* 4746 (1986); *52* 4397, 4641 (1987)
$C(COR)SO_2Ph$	$N_2C(COR)SO_2Ph$, cat $Rh_2(OAc)_4$	TL *28* 3459 (1987) (intramolecular)
$C(COR)_2$	$N_2C(COR)_2$, hν, PhCOPh	JACS *106* 6006 (1984) (intramolecular)
$C(COR)CO_2R$	$N_2C(COR)CO_2R$, hν, PhCOPh	JACS *106* 6006 (1984) (intramolecular)
	$N_2C(COR)CO_2R$, cat Cu(II)	Tetr *37* 2079 (1981) JACS *106* 6006 (1984); *109* 4717, 6187 (1987) (all intramolecular)
$C(CO_2R)_2$	$BrCH(CO_2Et)_2$, DBU, $CuBr_2$ $Br_2C(CO_2R)_2$, Cu $MeSCl/NaCH(CO_2R)_2/$ $Me_2SO_4/NaOEt$	BCSJ *55* 2687 (1982) Tetr *36* 3517 (1980) Syn 690 (1980) JACS *106* 5335 (1984)
$CPhCO_2R(CN)$	$PhCHClCO_2R(CN)$, NaOH (on enones)	Syn 34 (1980)
CXY (X, Y = CO_2Et, CN)	Br_2CXY, Cu_2Br_2, DMSO	BCSJ *54* 2539 (1981)
CHCN	$(Me_3\overset{+}{N}CH_2CN)I^-$, NaH (on enones)	Syn 301 (1982)
$C{=}C{=}C{=}CR_2$	$HC{\equiv}CC(OSO_2CF_3){=}CR_2$, KO-*t*-Bu	Acct Chem Res *15* 348 (1982)

$$\underset{/}{\overset{\backslash}{}}C{=}C\overset{\diagup X}{\underset{\diagdown}{}} + (Br)ClCHYZ \xrightarrow{\text{base}} \overset{Y\ Z}{\underset{X}{\bigtriangleup}}$$

X = CHO, COR, CO_2R, CN; Y = COR, CO_2R; Z = H, R, CO_2R, Cl

Ber *51* 907, 533 (1918)
JACS *70* 3470 (1948); *80* 6568 (1958); *82* 6416 (1960); *84* 2246 (1962)
BSCF 986 (1957); 1102 (1959); 418, 788 (1960); 200 (1961); 2462 (1964)
JOC *24* 1536 (1959); *25* 2078 (1960); *27* 4312 (1962); *29* 240 (1964)
Compt Rend *248* 887, 1465, 2840 (1959)

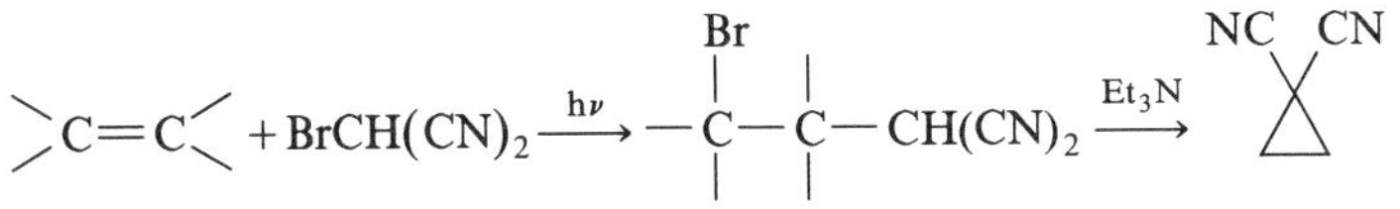

$$\text{C}=\text{C} + \text{BrCH(CN)}_2 \xrightarrow{h\nu} -\overset{\overset{\displaystyle Br}{|}}{\underset{|}{C}}-\overset{|}{\underset{|}{C}}-\text{CH(CN)}_2 \xrightarrow{Et_3N} \overset{NC\quad CN}{\triangle}$$

JACS *87* 1394 (1965)
TL 1415 (1966); 4351 (1967)
JOC *31* 2784 (1966)
Ber *100* 1281 (1967)

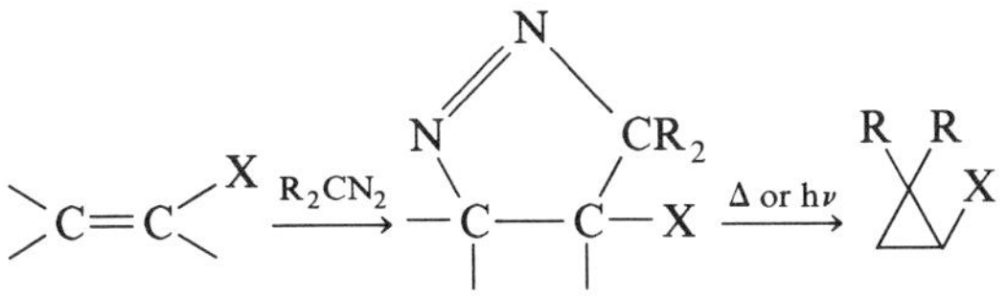

$$\text{C}=\text{C}\overset{X}{\underset{}{}} \xrightarrow{R_2CN_2} -\overset{|}{\underset{|}{C}}-\overset{|}{\underset{|}{C}}-X \xrightarrow{\Delta \text{ or } h\nu} \overset{R\quad R}{\underset{}{\triangle}}X$$

X = Ar, COR, CO₂R, CN

Reviews:

Newer Methods Prep Org Chem *1* 513 (1948)
Heterocyclic Compounds, Ed. R. Elderfield, J. Wiley, New York (1957), Vol 5
Houben-Weyl, Vol IV/3, p 43

Ber *21* 2637 (1888); *23* 701 (1890); *27* 868, 877, 879 (1894); *36* 3774, 3782 (1903); *49* 1928 (1916); *70* 1688 (1937); *71* 2673 (1938); *88* 49 (1955); *93* 883, 1710 (1960); *100* 3495 (1967)
Ann *273* 239 (1893); *496* 252 (1932); *618* 105 (1958); *678* 78 (1964); *703* 104 (1967)
J Prakt Chem [2] *133* 291 (1932) (1,3-diene); [4] *36* 73 (1967)
JACS *56* 2710 (1934); *65* 159 (1943); *72* 3815 (1950); *73* 2383 (1951); *79* 4994 (1957); *80* 6687 (1958); *81* 3776, 5153, 5472 (1959); *82* 3136, 5251 (1960); *84* 869, 3736 (1962); *86* 658 (1964); *87* 4119 (1965)
JCS 829 (1936); 5186 (1962)
Rec Trav Chim *62* 210 (1943)
JOC *25* 852 (1960); *26* 1831, 3669 (1961); *27* 213, 1030 (1962); *31* 3467 (1966); *50* 2220, 2600 (1985); *52* 120 (1987)
Can J Chem *38* 2410 (1960); *45* 691 (1967) (1,3-diene)
BSCF 550 (1960) (BF₃·OEt₂ promoted elimination); 1761 (1962)
Helv *43* 2178 (1960); *49* 1049 (1966)
TL 1719 (1964); 3947 (1965)
Angew *79* 815 (1967); *80* 42 (1968)
Z Chem *7* 421 (1967)

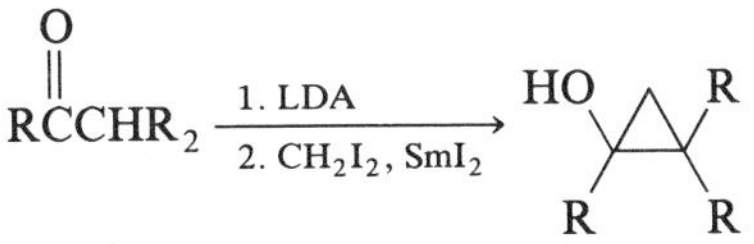

$$\overset{\overset{\displaystyle O}{\|}}{R C C H R_2} \xrightarrow[\text{2. } CH_2I_2,\, SmI_2]{\text{1. LDA}} \overset{HO\quad R}{\underset{R\quad R}{\triangle}}$$

TL *28* 1307 (1987)

$$\text{C}=\overset{|}{\underset{|}{C}}-\overset{\overset{\displaystyle O}{\|}}{C}- \xrightarrow{H_2NNH_2} HN \overset{N}{\underset{}{\diagup}} \xrightarrow[\Delta]{KOH} \triangle$$

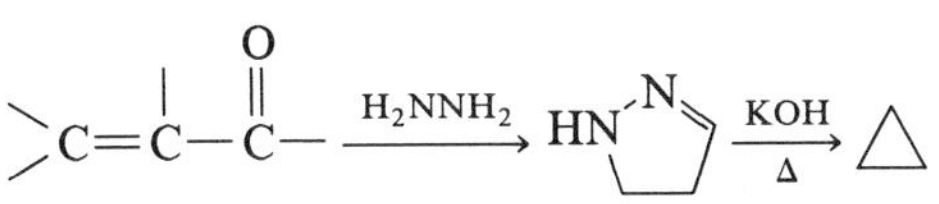

J Russ Phys Chem Soc *44* 165 [CA *6* 1431 (1912)], 849 [CA *6* 2915 (1913)] (1912); *45* 949 [CA *7* 3964 (1913)], 957 [CA *7* 3965 (1913)] (1913); *47* 1102 [CA *9* 3051 (1915)]; 1819 [CA *10* 1338 (1916)] (1915); *50* 1 (1918) [CA *18* 1485 (1924)]; *61* 781 (1929) [CA *23* 4698 (1929)]
Helv *25* 732 (1942)
JACS *66* 488 (1944); *69* 2544 (1947); *73* 3840 (1951)
JCS 4686 (1952)

$$\text{C}_6\text{H}_{11}-\text{CO}_2\text{Me} \longrightarrow \underset{\text{OMe}}{\overset{\text{OSiMe}_3}{\text{C}=\text{C}}} \xrightarrow[\text{Et}_2\text{Zn}]{\text{HCBr}_3} \text{CO}_2\text{Me}$$

JACS *106* 7283 (1984)

$$\underset{\text{R}^1\text{CH}=\text{CR}^2}{\overset{\text{SO}_2\text{Ph}}{|}} + \underset{\text{R}^3\text{CHCN}}{\overset{\text{Li}}{|}} \longrightarrow \underset{\text{R}^1\text{CH}-\text{CHR}^2}{\overset{\text{R}^3\text{CCN}}{\diagup\diagdown}}$$

JCS Perkin I 751 (1981)

$$\text{ArCH}_2\text{CH}=\text{CH}_2 \xrightarrow{\text{h}\nu} \text{Ar}-\triangleleft$$

Chimia *35* 52 (1981)

$$\underset{(\text{CH}_3)_2\text{CCH}=\text{C(Ph)CN}}{\overset{\text{R}}{|}} \xrightarrow{\text{h}\nu} \underset{\text{CH}_3}{\overset{\text{R}}{\triangle}}\underset{\text{CH}_3}{\overset{\text{Ph}}{\underset{\text{CN}}{}}}$$

JOC *43* 2839 (1978)
Can J Chem *60* 1657 (1982)
TL *27* 6225 (1986)

$$-\underset{|}{\overset{\text{O}}{\text{C}}}-\underset{|}{\text{C}}- + \text{HCClXY} \xrightarrow[\Delta]{\text{Et}_4\text{NBr}} \overset{\text{X} \quad \text{Y}}{\triangle}$$

X = Y = Cl

X = F, Y = Cl

X = Y = F

Ber *100* 1858 (1967)

$$\text{RO(CH}_2)_3\text{X} \longrightarrow \triangle$$

Mg   JACS *74* 6290 (1952)
                        JOC *20* 275 (1955)

EtMgBr, CoBr$_2$ JACS *83* 2734 (1961)

$$\text{MsO}-\text{C}-\text{C}-\text{C}-\text{OMs} \xrightarrow{\text{electrolysis}} \triangle$$

JOC *47* 3090 (1982)

$$\underset{\text{R}_3\text{SiOC}=\text{CHCH}_2\text{CH}=\text{COSiR}_3}{\overset{\text{OR} \qquad\qquad\qquad \text{OR}}{|\qquad\qquad\qquad\qquad |}} \xrightarrow{\text{TiCl}_4} \overset{\text{CO}_2\text{R}}{\triangle}\underset{\text{CO}_2\text{R}}{}$$

TL *23* 799 (1982)
Tetr *39* 847 (1983)

$$H_2C = CHCH_2Cl \xrightarrow[\text{2. base}]{\text{1. } R_2BH} \triangle$$

JACS *80* 5830 (1958); *82* 1886 (1960); *86* 1791 (1964); *91* 2149, 4306 (1969)
Israel J Chem *6* 691 (1968)

$$HC \equiv CCH_2Br \xrightarrow[\text{2. NaOH or MeLi}]{\text{1. 2 HB}} \triangleright\!-B \xrightarrow[\text{NaOAc}]{H_2O_2} \triangleright\!-OH$$

JACS *91* 4306 (1969)

$$Y - C - C - \overset{\overset{\textstyle H}{|}}{C} - X \xrightarrow{\text{base}} \triangleright\!-X$$

Review: Houben-Weyl, Vol IV/3, p 89

X	Y	
COR	OTs	BSCF 1340, 1347 (1957)
		JACS *83* 4678 (1961); *85* 41 (1963)
		JCS C 909 (1967)
		JOC *37* 2911 (1972)
	OTs, $\overset{+}{N}Me_3$	JOC *22* 1146 (1957)
	OAc	JACS *80* 1264, 5304 (1958)
	Cl	JOC *19* 1628 (1954)
		JACS *78* 112 (1956); *79* 1455 (1957)
		BSCF 1634 (1962)
		Org Syn Coll Vol *4* 597 (1963)
		JCS Perkin II 579 (1982)
	Br	JCS 1060 (1954); 2620 (1956)
		JACS *76* 4115 (1954)
		Ber *91* 768 (1958)
		JCS Perkin II 579 (1982)
	I	JACS *87* 4601 (1965)
CO$_2$R	O$_3$SC$_6$H$_4$Br	JOC *22* 1146 (1957)
	O$_2$CR	TL 2441 (1976)
	Cl	Compt Rend *245* 2304 (1957)
		BSCF 1487 (1964)
		Syn 955 (1982) (phase transfer)
	Br	BSCF 681 (1957)
CN	OTs, OMs	BSCF 1854 (1961)
	O$_2$CR	BSCF 1476 (1964)
	Cl	JACS *67* 1587 (1945); *109* 7483 (1987) (bicyclobutane)
SO$_2$Ph	OMs	TL *22* 4339 (1981)
(SO$_2$Ph)$_2$	Cl, I	CC 1374 (1983)
		JCS Perkin II 605 (1986)

For analogous opening of epoxides see page 512 Section 4.1.

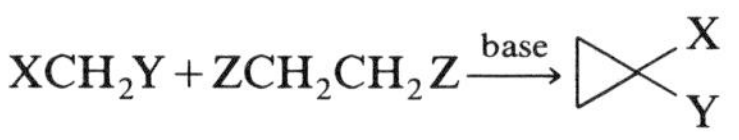

$$XCH_2Y + ZCH_2CH_2Z \xrightarrow{base} \triangleright\!\!<^X_Y$$

$\underline{X}$	$\underline{Y}$	$\underline{Z}$	
CN	Ar	Cl	JACS *67* 1249 (1945)
CO_2R, CN	Ar	Br	TL *27* 3685 (1986)
CH=CHCH=CH		Br	JCS 646 (1958)

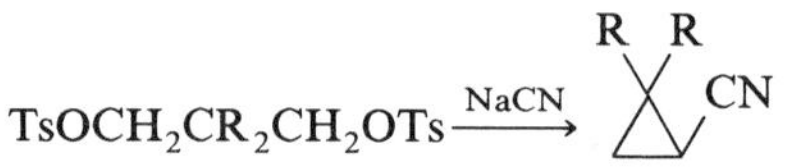

$$TsOCH_2CR_2CH_2OTs \xrightarrow{NaCN} \overset{R\ R}{\triangle}\!\!^{CN}$$

JACS *79* 3467 (1957)

$$XYCHCH_2CHXY \longrightarrow \overset{Y}{\underset{X}{\overset{X}{\Big]}}\!\!\triangleright}\overset{}{\underset{Y}{}}$$

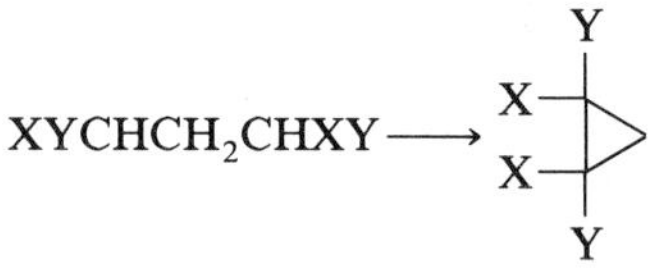

$\underline{X}$	$\underline{Y}$	Reagent(s)	
CN	CN	Br_2	JOC *22* 1130 (1957)
RCO	H	Br_2/NaI	JCS 2620 (1956)

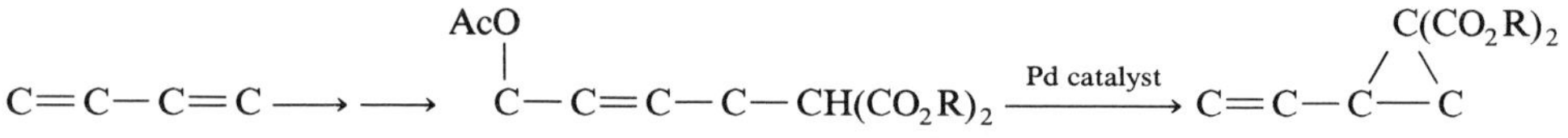

$$C{=}C{-}C{=}C \longrightarrow \longrightarrow \overset{AcO}{\underset{\;}{C}}{-}C{=}C{-}C{-}CH(CO_2R)_2 \xrightarrow{Pd\ catalyst} C{=}C{-}C{-}C \overset{C(CO_2R)_2}{\diagup\;\diagdown}$$

JOC *52* 5430 (1987)

$$n\text{-}Bu_3SnCH_2CH_2CH{=}CHCR\overset{O}{\overset{\|}{}} \xrightarrow{F_3CCO_2H} \triangleright{-}CH_2CR\overset{O}{\overset{\|}{}}$$

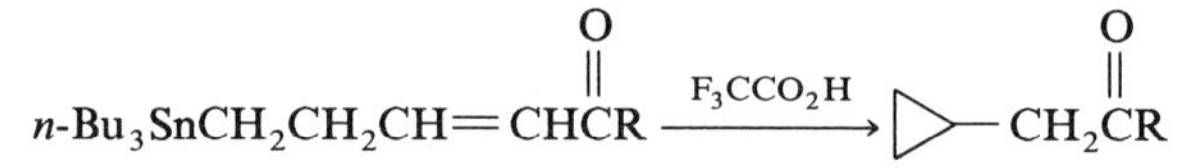

TL *23* 2577 (1982)

2. Four-Membered Rings

Review: Houben-Weyl, Vol IV/4 (1971)

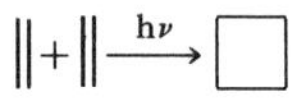

$$\| + \| \xrightarrow{h\nu} \square$$

Reviews:

   Chem Rev *66* 373 (1966) (intramolecular)
   Acct Chem Res *1* 50 (1968) (enones); *4* 41 (1971) (enone photoannelation); *15* 135 (1982)
      (intramolecular)
   Syn 287 (1969) (enones)
   Photochem Photobiol *25* 605 (1977)
   Org Photochem *5* 123 (1981)
   Rev Chem Intermed *4* 369 (1981)
   A. C. Weedon in "Synthetic Organic Photochemistry," Ed. W. M. Horspool, Plenum, New
      York (1984), Chpt 2

Recent references:

Intramolecular

 CC 235 (1979); 1195 (1980); 118 (1981); 1578 (1987)

 Helv *63* 1198 (1980)

 Org Photochem *5* 123 (1981) (enones, review)

 JOC *46* 4821 (1981); *47* 331, 829, 3121, 3297 (allene), 3597 (lactones), 3893 (1982); *50* 3155 (1985) (allene); *52* 83 [Cu(I) catalyzed], 2346, 2644, 3603 (1987)

 Tetr *37* 4543 (1981)

 JACS *103* 82 (1981); *104* 998, 6841 (1982); *105* 1292, 1299 (1983); *107* 5732 (1985); *108* 800, 1311 [Cu(I) catalyzed], 3385, 3435, 6425 (1986); *109* 2212, 2523, 2850, 6199, 7230 (1987)

 Can J Chem *60* 425 (1982)

 Acct Chem Res *15* 135 (1982) (review)

 J Chem Ed *59* 313 (1982) (review)

 TL *23* 23 (enones), 711 (1982); *24* 2961 (1983); *26* 989 (allene), 1429 (allene), 5883 (1985); *27* 2703 (allene), 5177, 6393 (1986); *28* 1295, 5083, 5087, 5497 (1987)

Intermolecular

 JACS *102* 3634 (1980); *107* 1308 (1985); *108* 306 (1986)

 Can J Chem *60* 425 (micelles), 872 (allene-lactone) (1982)

 JCS Perkin I 1697 (1984)

 JOC *50* 3026 (1985); *51* 5226 (1986); *52* 1993, 2639, 3595 (1987)

 TL *26* 4707, 6163 (1985); *27* 5975, 6049 (1986); *28* 581 (allene), 2833 (chiral), 2857, 5017 (1987)

 Tetr *42* 3547 (1986)

JOC *45* 5017 (1980); *51* 3643 (1986)
TL *22* 1953 (1981)

JOC *41* 1184 (1976)
TL 4573 (1977)
JACS *109* 7137 (1987)

X	Y	
RCO	Cl	JACS *83* 938 (1961)
		TL 2525 (1964)
		Ber *100* 720 (1967)
		JCS C 2350 (1969)
Br		Ber *91* 1616 (1958)
		Tetr *25* 613 (1969)

$$\underset{\text{C}}{\overset{\overset{\displaystyle Y}{|}}{}} - \text{C} - \text{C} - \underset{\text{C}}{\overset{\overset{\displaystyle H}{|}}{}} - \text{X} \xrightarrow{\text{base}} \square - \text{X} \quad (\textit{Continued})$$

$\underline{X}$	$\underline{Y}$	
	I	Angew Int *1* 457 (1962)
		Tetr *26* 1589 (1970)
	$O_3SC_6H_5$	JOC *27* 1883 (1962)
	OTs	Angew Int *1* 457 (1962)
		TL 1043 (1963); 2525 (1964); 2043 (1966); 3627 (1969)
		JACS *88* 4110 (1966); *89* 4133 (1967)
		BSCF 147 (1966)
		Ber *100* 720, 2978 (1967)
		Tetr *25* 5267 (1969); *26* 1589 (1970)
		JOC *37* 2911 (1972)
	$\overset{+}{N}Me_3$	Angew Int *1* 457 (1962)
		Ber *100* 720 (1967)
enone	OMs	JOC *34* 3837 (1969)
	OTs	Tetr *25* 5281 (1969)
CO_2R	Cl	BSCF 1833 (1959); 979 (1960)
$(CO_2R)_2$	Cl	CA *37* 4705[6] (1943)
		Zh Obshch Khim *22* 122 (1952) [CA *46* 11119d (1952)]
		Bull Acad Sci USSR, Div Chem Sci 1585 (1966)
	Br	JACS *71* 2941 (1949); *106* 1051 (1984)
		JOC *26* 2335 (1961)
CN	Cl	Compt Rend *231* 703 (1950)
		BSCF 1833 (1959)
	OTs	BSCF 979 (1960)
1,3-dithiane	Cl	JOC *33* 300 (1968)
$(SO_2Ph)_2$	Cl, I	CC 1374 (1983)
		JCS Perkin II 605 (1986)
$\overset{+}{P}Ph_3$	Br	JOC *30* 3215 (1965); *33* 3082 (1968)

$$X - C - C - C - X + {}^-CH(CO_2R)_2 \longrightarrow \square \overset{\displaystyle CO_2R}{\underset{}{}}{-}CO_2R$$

X = halogen, OTs

JCS *51* 1 (1887); *61* 36 (1892); *127* 2387 (1925); 1487 (1929); 1211 (1938); 3002 (1953)
Ber *40* 3883 (1907); *98* 2651 (1965); *99* 1509 (1966)

Rec Trav Chim *50* 921 (1931)
JACS *63* 1698 (1941); *84* 4982 (1962)
JOC *14* 1036 (1949); *21* 1371 (1956); *22* 1029 (1957); *26* 54 (1961); *27* 1647 (1962); *30* 1945, 3308, 4212 (1965); *31* 4069 (1966)
Org Syn Coll Vol *3* 213 (1955); *4* 288 (1963)
Ann *648* 36 (1961); *666* 81 (1963); *678* 53 (1964); *685* 74 (1965); *692* 53 (1966); *694* 1 (1966)
Naturwiss *50* 441 (1963)
Angew Int *5* 127 (1966)
Bull Acad Sci USSR, Div Chem Sci 1585 (1966)

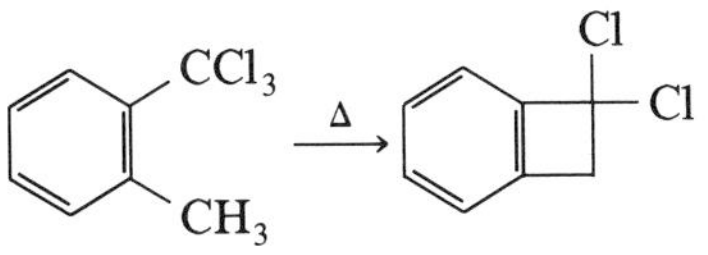

JACS *82* 749 (1960)
JOC *31* 2244 (1966); *34* 2906 (1969)

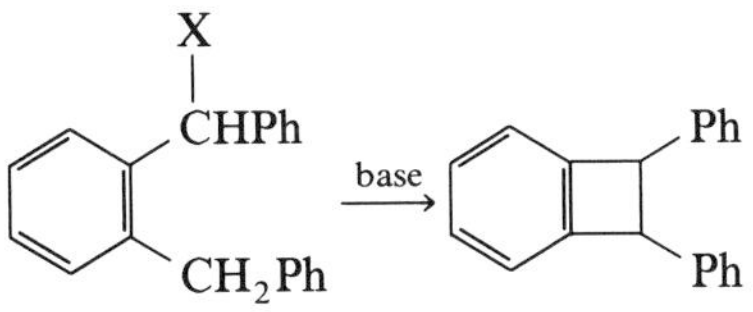

X = Cl, Br
JACS *80* 6149 (1958); *85* 4031 (1963)

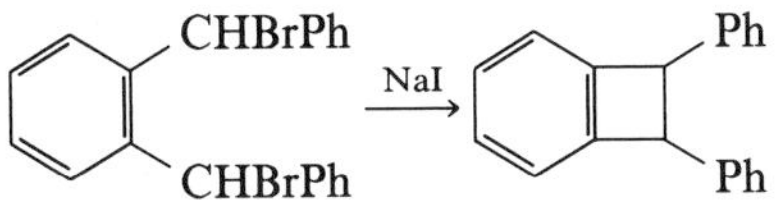

JACS *80* 6149 (1958)

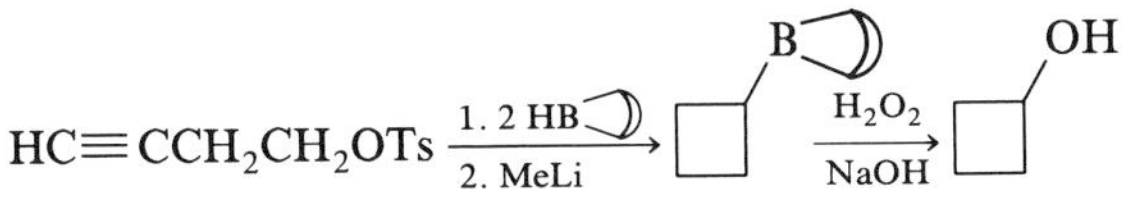

$$HC{\equiv}CCH_2CH_2OTs \xrightarrow[\text{2. MeLi}]{\text{1. 2 HB}}$$

JACS *91* 4306 (1969)

3. Five-Membered Rings

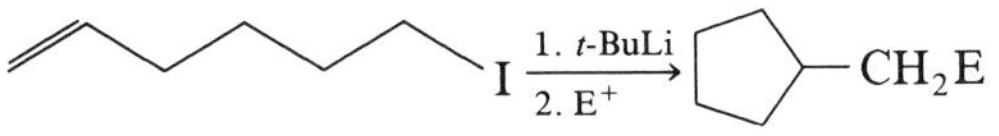

$E^+ = CH_3OH, CH_3OD, CO_2, O_2$

JACS *109* 2442 (1987)

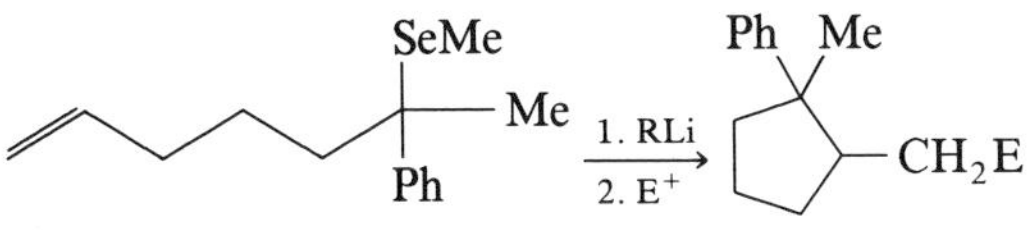

$E^+ = H_2O, PhCHO$

CC 1214 (1987)

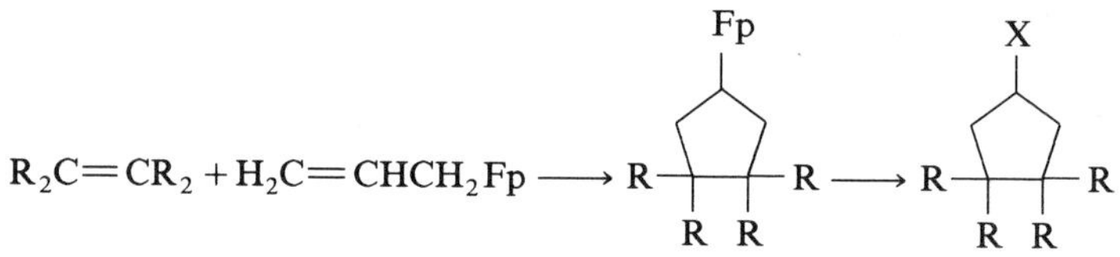
Fp
X
R2C=CR2 + H2C=CHCH2Fp
R
R R

R = H, CN, CO_2R; Fp = $CpFe(CO)_2$; X = H, Br, CO_2R'

JCS Perkin I 285, 295, 301 (1982)

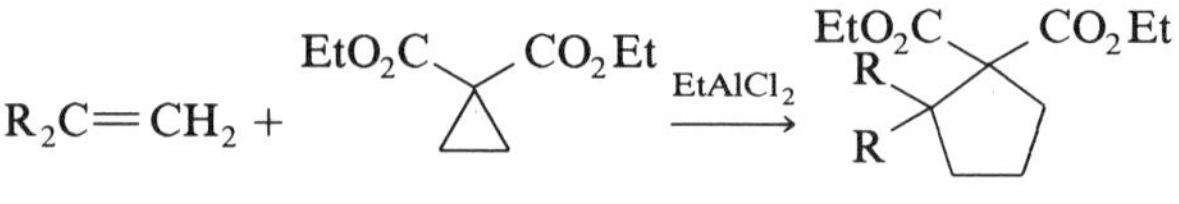
R2C=CH2 +
EtO2C
CO2Et
EtAlCl2
R

JOC *51* 4391 (1986)

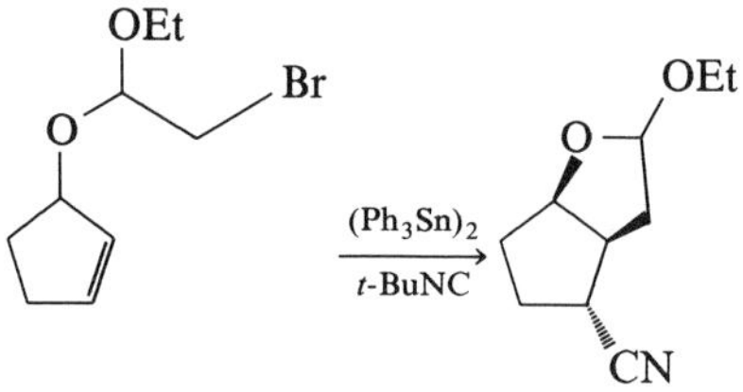
OEt
Br
O
(Ph3Sn)2
t-BuNC
CN

JACS *105* 6765 (1983)

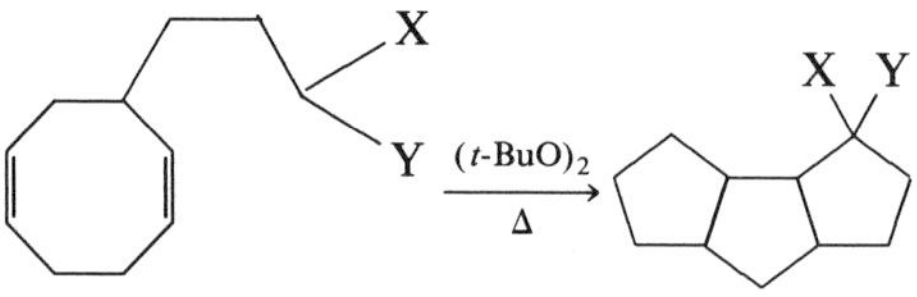
X
Y
(t-BuO)2
Δ

X, Y = CN, CO_2Et; SCH_2CH_2S

JACS *108* 1708 (1986)

4. Six-Membered Rings

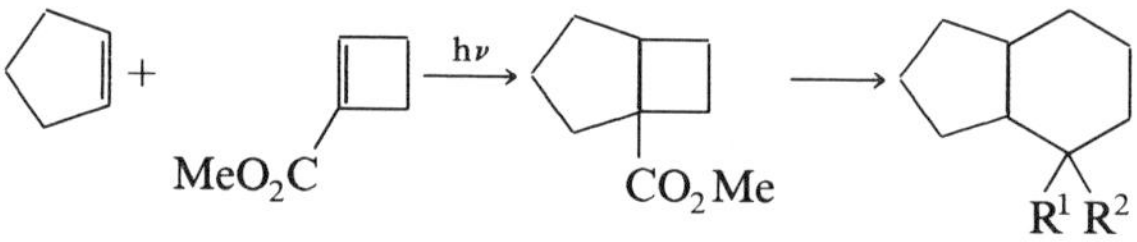
+
hν
MeO2C
CO2Me
R1 R2

R^1 = H; R^2 = CH_2OH

R^1 = H; R^2 = CO_2Me

R^1 = Me; R^2 = CO_2Me

JACS *100* 4321 (1978)

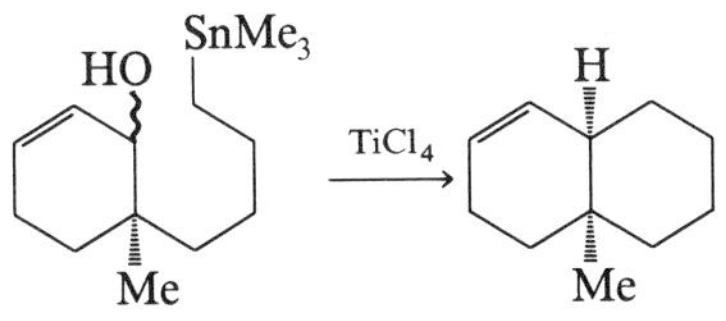

TL *22* 2077 (1981)
JACS *103* 6767 (1981)

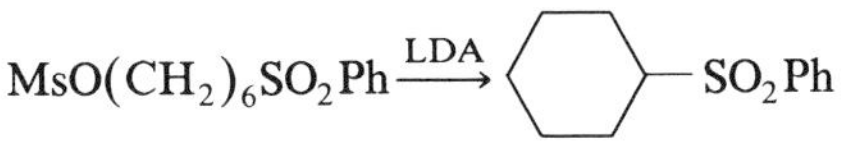

$$MsO(CH_2)_6SO_2Ph \xrightarrow{LDA} \langle \rangle - SO_2Ph$$

JOC *51* 5311 (1986)

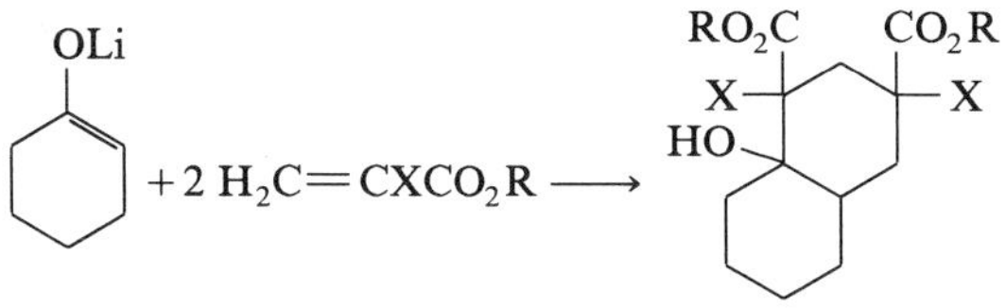

X = H, R, Br, ArS, SiMe$_3$

JACS *108* 511 (1986)

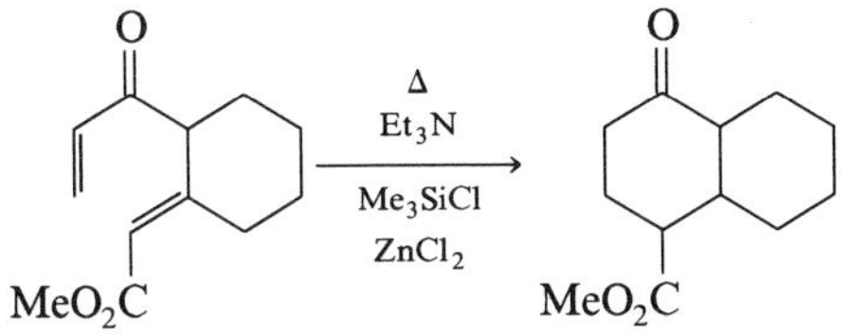

TL *25* 2167 (1984)
CC 1159 (1985); 721, 1467 (1987)
JCS Perkin I 1543 (1986)

5. Various Ring Sizes

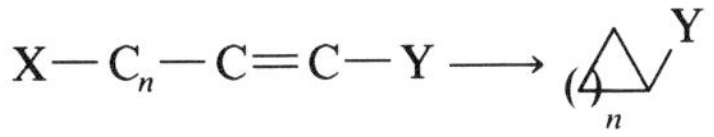

$$X - C_n - C = C - Y \longrightarrow (\triangle)_n^Y$$

X = Br, I; Y = COR, CO$_2$R, CN

See the appropriate functional group Y for conjugate addition reactions

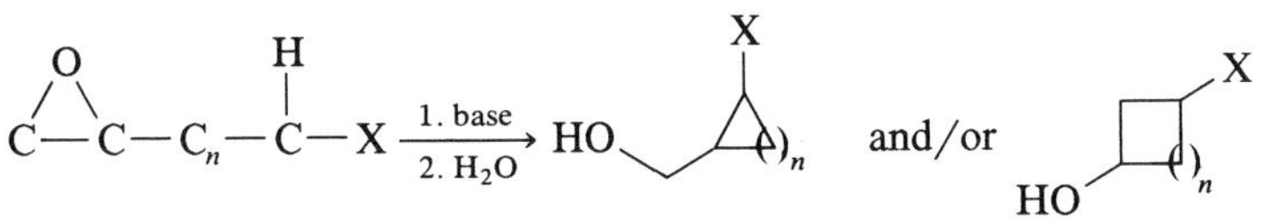

$$C \overset{O}{-} C - C_n - \overset{H}{\underset{}{C}} - X \xrightarrow[\text{2. H}_2\text{O}]{\text{1. base}} HO \diagdown \triangle)_n \quad \text{and/or} \quad \square_n^X$$

See also page 512, Section 4.1.

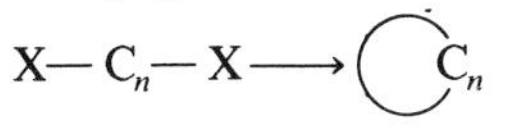

$$X - C_n - X \longrightarrow (\overset{\cdot}{C}_n$$

Reviews:

n = 3   Houben-Weyl, Vol IV/4 (1971), p 32

Russ Chem Rev *51* 368 (1982)

n = 4   Houben Weyl, Vol IV/4 (1971), p 31

Reagent(s)	n	
electrolysis	3, 4	TL 1119 (1974)
Li(Hg)	3–5	TL 4925 (1967); *28* 5263 (1987)
MeLi	3	TL *28* 5411 (1987)
t-BuLi	3–5	TL *23* 5123 (1982)
Na	3	TL 2173 (1963)
Mg	3	JACS *75* 3344 (1953)
Mg/AgOTf	4–6	JOC *41* 2882 (1976)
RMgBr, FeCl$_3$	3	JACS *83* 3232 (1961) JOC *29* 2813 (1964)
Cr(ClO$_4$)$_2$, EDA	3	JOC *33* 1027 (1968)
Ni(COD)$_2$, bipy	3–6	CL 1363 (1974)
Zn	3	J Prakt Chem *36* 300 (1887); *58* 458 (1898) JACS *66* 812 (1944); *68* 1335 (1946); *70* 946 (1948) JOC *23* 1715 (1958) BSCF 116 (1966) Syn Commun *11* 865 (1981)
Zn-Cu	3	Org Syn *44* 30 (1964)

$$X-C_n-\overset{\displaystyle H}{\underset{|}{C}}-\overset{\displaystyle O}{\overset{\|}{C}}R \xrightarrow{\text{base}} \overset{\frown}{C_n \quad C}-\overset{\displaystyle O}{\overset{\|}{C}}R$$

n	X	
4	Cl	JCS Perkin II 579 (1982)
	Br	Ber *91* 1616 (1958) JCS Perkin II 579 (1982)
5	Br	Ber *91* 1616 (1958)

$$(PhSO_2)_2CH(CH_2)_nX \xrightarrow{\text{NaOEt}} (PhSO_2)_2\overset{\frown}{C \quad (CH_2)_n}$$

$$n = 2\text{–}6$$

CC 1374 (1983)
JCS Perkin II 605 (1986)

$$(EtO_2C)_2CH(CH_2)_nBr \xrightarrow{\text{base}} (EtO_2C)_2\overset{\frown}{C \quad (CH_2)_n}$$

$$n = 3\text{–}12, 16, 20$$

JACS *106* 1051 (1984)

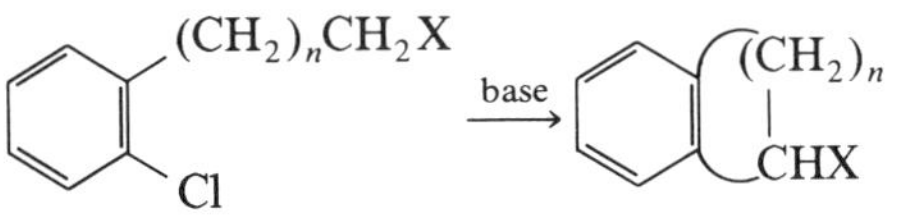

$$\text{(ortho-Cl-phenyl)-(CH}_2)_n\text{CH}_2\text{X} \xrightarrow{\text{base}} \text{(benzo-fused ring)}$$

with ring: $(CH_2)_n$ / CHX

X (ring size) = PhCO (5), CO_2Et (4–6), CN (4–7), SO_2Ph (4, 5)

JOC *27* 3836 (1962)

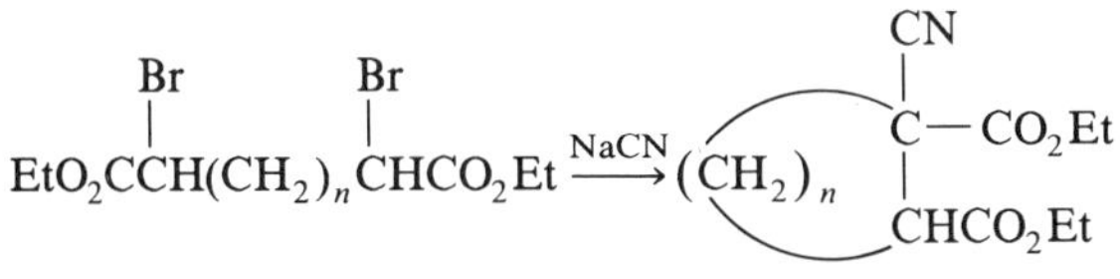

$$EtO_2CCH(CH_2)_n CHCO_2Et \xrightarrow{\text{NaCN}} (CH_2)_n$$

with Br, Br substituents; product ring bearing:

CN
|
C—CO_2Et
|
CHCO$_2$Et

n	
2	JACS *51* 1536 (1929); *64* 2696 (1942)
2–4	JOC *30* 1206 (1965)

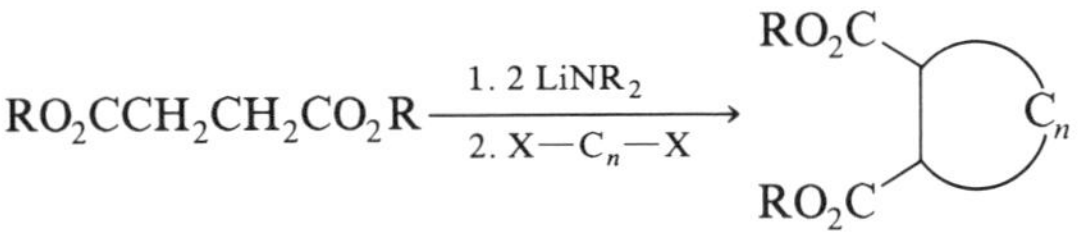

$$RO_2CCH_2CH_2CO_2R \xrightarrow[\text{2. } X-C_n-X]{\text{1. 2 LiNR}_2}$$

product: ring with RO_2C, RO_2C substituents and C_n

$$n = 1\text{–}5$$

TL 1815 (1979); *27* 5951 (1986); *28* 589 (1987)
Syn 389 (1980)
Syn Commun *14* 227 (1984)
JOC *49* 1412 (1984); *50* 5727 (1985)
JACS *107* 3343 (1985) (chiral)

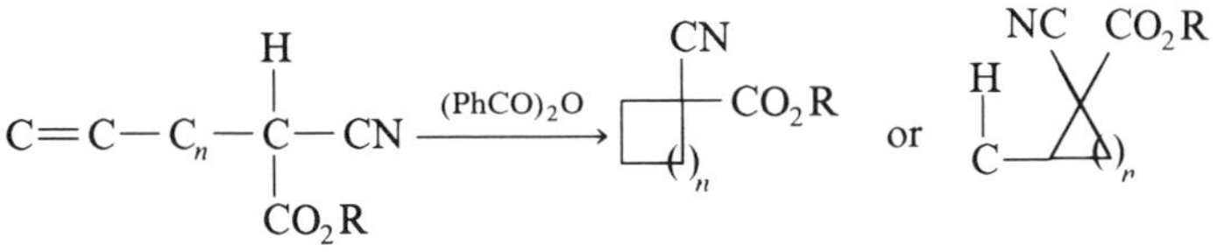

$$C{=}C-C_n-\overset{\overset{\textstyle H}{|}}{\underset{\underset{\textstyle CO_2R}{|}}{C}}-CN \xrightarrow{(PhCO)_2O}$$

(square ring) CN, CO_2R, $)_n$ $n = 3$

or

NC CO_2R
(cyclopropane) H, C, $)_n$ $n = 4$

BSCF 3070 (1970)

$$X-C_n-C{=}C \xrightarrow{R_3SnH} C_n \quad C-\overset{\overset{\textstyle H}{|}}{C} \quad \text{or} \quad C_n \overset{C}{\underset{C-H}{}}$$

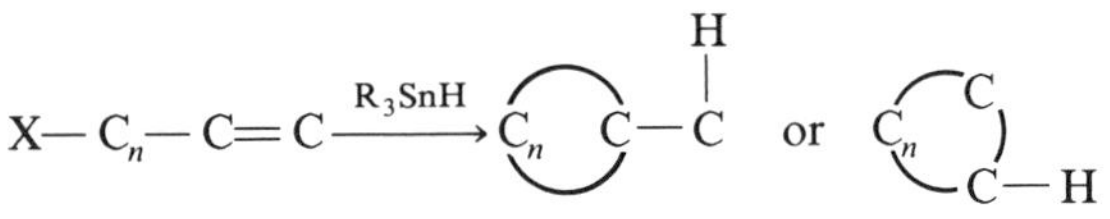

Reviews:

Acct Chem Res *4* 386 (1971)
Pure Appl Chem *40* 553 (1974) (mechanism)
A. L. J. Beckwith and K. U. Ingold in "Rearrangements in Ground and Excited States," Ed. P. de Mayo, Academic Press, New York (1980), pp 162–310
CC 482 (1980) (steric and stereoelectronic effects)
Tetr *37* 3073 (1981); *41* 3925 (1986) (theory)
J.-M. Surzur in "Reactive Intermediates," Ed. R. A. Abramovitch, Plenum, New York (1981), Vol 2, Chpt 3
Angew Int *22* 753 (1983); *24* 553 (1985)
Science *223* 883 (1984)

Tetr Symposia-in-Print No. 22 *41* 3887–4302 (1985)
B. Giese, "Radicals in Organic Synthesis—Formation of Carbon-Carbon Bonds," Pergamon
  Press, New York (1986)
Curr Sci *56* 392 (1987)

X	Ring size	
halogen	5	JACS *96* 1613 (1974); *104* 5564 (1982) (2-alkoxytetrahydrofurans); *105* 3741 (1983) (bicyclic acetals); *107* 500 (2-alkoxytetrahydrofurans), 1448 (1985) (serial cyclization); *108* 303 (2-alkoxytetrahydrofuran, subsequent radical trapping), 1106 (serial cyclization), 1708 (serial cyclization), 2116 (serial cyclization), 3102, 5890 (dihydrobenzofurans), 8098 (1986)
		J Chem Res (S) 78 (1981)
		Austral J Chem *36* 545 (1983) (rates, stereochemistry)
		CC 1445 (1983) (tetrahydroindanes); 115 (1986) (oxindoles); 1456 (1987) (bicyclic alkane)
		JOC *48* 1782 (1983); *50* 546, 5620 (pyrrolidines), 5875 (1985); *51* 4708 (1986) (tetrahydrofurans)
		TL *25* 4317 (1984) (2-alkoxytetrahydrofurans); *26* 957 (pyrrolidines); 3349 (serial cyclization, *n*-Bu$_3$GeH), 4991 (serial cyclization) (1985); *27* 3715 (1986) (tetrahydrofurans); *28* 671 (2-alkoxytetrahydrofurans), 1317 (2-alkoxytetrahydrofurans), 1623 (2-alkoxytetrahydrofurans), 2127, 3953 (pyrrolidines), 6389 (oxindole) (1987)
		CL 1725 (1985)
		JCS Perkin I 1351 (1986) (2-alkoxytetrahydrofurans)
	5, 6	JACS *94* 6059, 6064 (1972); *105* 3741 (1983) (2-alkoxytetrahydrofurans and -pyrans); *108* 5893 (1986)
		CC 472 (1974)
		JCS Perkin II 795 (1975)
		Tetr *31* 1737 (1975); *37* 3073 (1981)
		JOC *43* 6 (1978); *49* 2299 (1984) (diols); *52* 4072 (1987)
		TL *22* 2811 (1981) (rates, stereochemistry); *26* 6001 (1985) (dihydrobenzofurans and -pyrans); *27* 4525, 4529 (1986); *28* 2637 (1987) (carbapenams and carbacephams)
		CC 85 (lactones), 464 (1986); 1456 (1987) (bicyclic alkane)
	6	JOC *41* 3261 (1976)
		TL *26* 4413 (1985); *28* 1035 (1987)
		CC 1438 (1987) (lactone)
	6, 7	JOC *48* 1841 (1983) (bicyclic β-lactams)
	7	CC 472 (1974)
	10, 14, 18	JACS *108* 2787 (1986) (ketones)
$\overset{S}{\underset{OCN}{\parallel}}$—N⟩	5	JACS *109* 609 (1987)

	6	TL *26* 5675 (1985) JOC *52* 1568 (1987)
SR	5, 6	JACS *104* 1430 (1982) (pyrrolizidinones, indolizidi- nones); *106* 8201 (pyrrolizidinones, indolizidi- nones), 8209 (pyrrolizidinones) (1984)
	7	JOC *48* 1841 (1983) (bicyclic β-lactams)
SePh	5	CC 307 (1983) (butyrolactones); 1205 (1985) (butyrolactones); 588 (1986) TL *26* 6431 (1985); *28* 2637 (1987) (carbapenam)
	5, 6	JACS *106* 8201 (1984) (pyrrolizidinones, indolizidi- nones) CC 353 (1987)
	7	JOC *48* 1841 (1983) (bicyclic β-lactams)
NO$_2$	5	CL 635 (1985) (tetrahydrofurans) Tetr *41* 4013 (1985) (tetrahydrofurans)

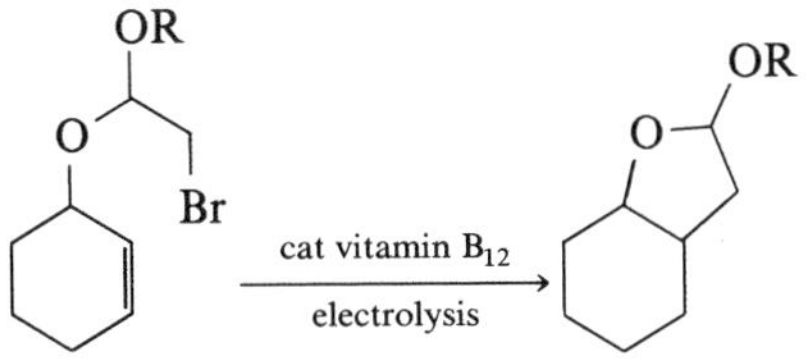

TL *28* 1313, 1317 (1987)

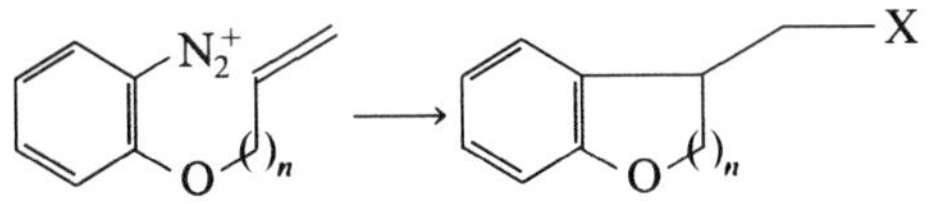

n	X	Reagent(s)	
1	Cl, Br, CN, SR	CuCl$_2$, CuBr$_2$, CuCN or RS$^-$	JACS *108* 5890 (1986)
	I	NaI	JOC *52* 1922 (1987)
2	I	NaI	JOC *52* 2568 (1987)

5.1. Carbocationic Cyclization

Reviews:

Acct Chem Res *1* 1 (1968); *8* 152 (1975)
Bioorg Chem *5* 51 (1976)
Angew *88* 33 (1976)
P. A. Bartlett in "Asymmetric Synthesis," Ed. J. D. Morrison, Academic Press, New York (1984), Vol 3, p 341

Some recent references

JACS *103* 4615 (1981); *104* 3508 (1982); *107* 522, 2712 (1985) (both Hg promoted); *109* 2517, 5852 (1987)
JOC *46* 2709 (1981); *52* 4878 (1987) (Hg promoted)
JCS Perkin I 2956 (1981)
J Chem Res (S) 20 (1981)
TL *28* 4053 (Sn promoted), 6413 (Hg promoted) (1987)

5. AROMATIZATION

1. Dehydrogenation

Reviews:

"Newer Methods of Preparative Organic Chemistry," Vol 1 (1963) (S, Se, Pt)
"The Chemistry of the Quinoid Compounds," Ed. S. Patai, J. Wiley (1974), Chpt 7
Houben-Weyl, Vol V/2b, p 107

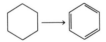

Review: Chem Rev *78* 317 (1978)

cat [Ni$_3$(PO$_4$)$_2$ + H$_2$]	CL 967 (1976)
cat Ni-Al$_2$O$_3$, C$_6$H$_6$	JACS *63* 1320 (1941)
cat Ni-Cr$_2$O$_3$, C$_6$H$_6$	JACS *63* 1320 (1941); *71* 2962 (1949)
cat Ni-Cr$_2$O$_3$, C$_6$H$_6$, PhSPh or thiophene	JACS *70* 381 (1948)
cat Pd-asbestos	JCS 583 (1957)
cat Pd-C	JCS 583 (1957) Helv *51* 1102 (1968); *52* 1023 (1969)
cat Pd, C$_6$H$_6$	JACS *71* 2962 (1949)
cat Pt-C	JCS 583 (1957)
cat Pt, C$_6$H$_6$	JACS *63* 1320 (1941); *71* 2955, 2962 (1949)
S	JCS 1832 (1932) Tetr *30* 3303 (1974)
Se	JCS 1431 (1936)

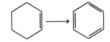

Review: Chem Rev *78* 317 (1978)

cat Pd-C	JACS *90* 6992 (1968)
cat Pt-alumina	Ann *670* 23 (1963) TL 3491 (1973)

S

JCS 1832 (1932)
JACS *60* 940 (1938)
JOC *38* 1430 (1973)

Se

Ber 1471 (1933)

quinones

Adv Org Chem *2* 239 (1960) (review)

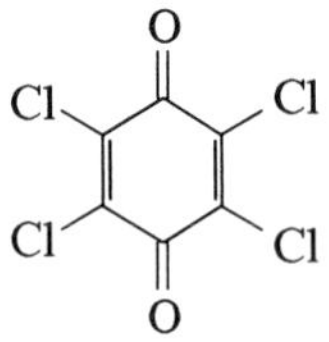

JACS *61* 1407 (1939); *62* 983 (1940)
BCSJ *44* 2841 (1971)

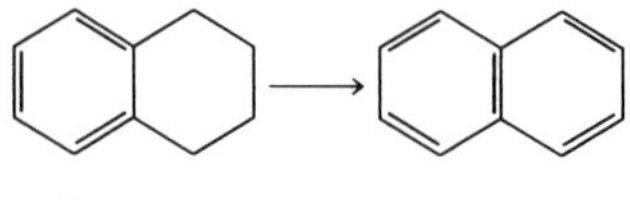

Review: Chem Rev *78* 317 (1978)

cat Ni-Al$_2$O$_3$, C$_6$H$_6$

JACS *63* 1320 (1941)

cat Pd-C

Ber *96* 707 (1963)
Syn 307 (1971)

cat Pd, C$_6$H$_6$

JACS *71* 2962 (1949)

cat Pt, C$_6$H$_6$

JACS *71* 2955 (1949)

S

Helv *10* 915 (1927)
Ber *65* 883 (1932)
JCS 1286 (1938)

Se

Ber *60* 2323 (1927)
JCS 1125 (1932); 1012 (1933); 365 (1934); 62
   (1936); 1845 (1959)

(i-AmS)$_2$

JACS *59* 2351 (1937)

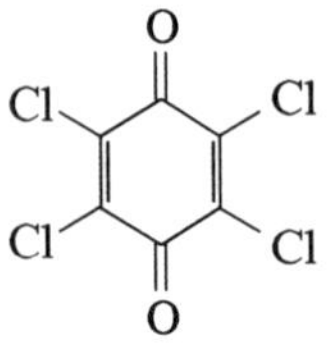

JACS *61* 1407 (1939)

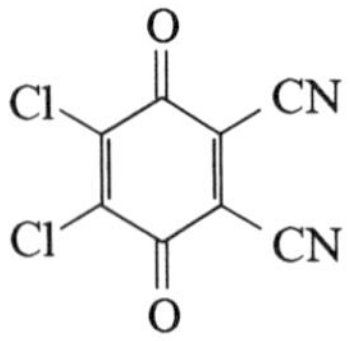

JCS 3569 (1954)
Chem Rev *67* 153 (1967) (review)

Ph$_3$COH, CF$_3$CO$_2$H

TL 3217 (1974)

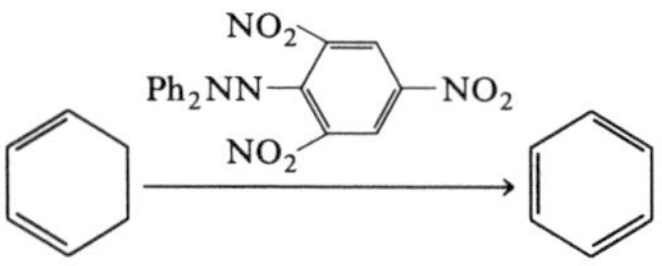

JCS 3574 (1954)

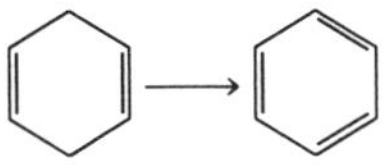

Review: Chem Rev *78* 317 (1978)

cat Pd-C	TL 191 (1969)
quinones	Adv Org Chem *2* 239 (1960) (review)

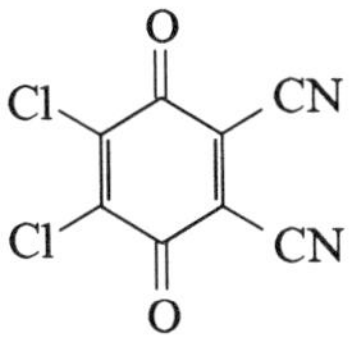

JACS *107* 5305 (1985)
JOC *51* 5100 (1986)

$(NC)_2C{=}C(CN)_2$

TL 205 (1962)
JACS *85* 3436 (1963); *87* 2751 (1965); *102* 886 (1980)

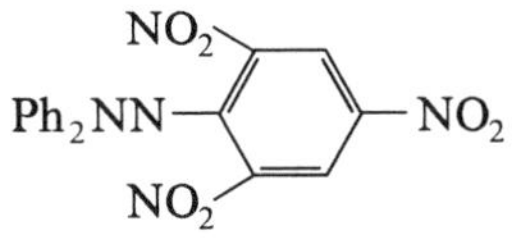

JCS 3574 (1954)

act MnO_2

Syn Commun *12* 637 (1982)

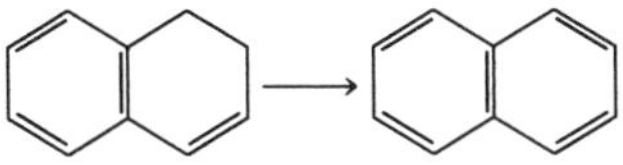

Review: Chem Rev *78* 317 (1978)

cat Pd-C	Ann *686* 40 (1965)
cat Pt	JCS 1431 (1936)
cat Pt, C_6H_6	JACS *71* 2955 (1949)
S	JOC *23* 797 (1958)
Se	JCS 62, 1431 (1936)
SeO_2	Z Chem *13* 216 (1973)

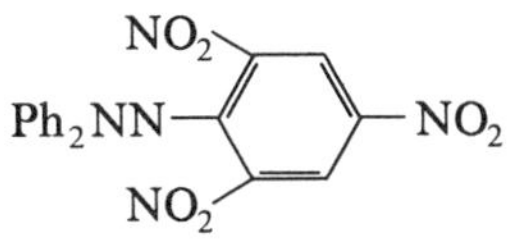

JCS 3574 (1954)

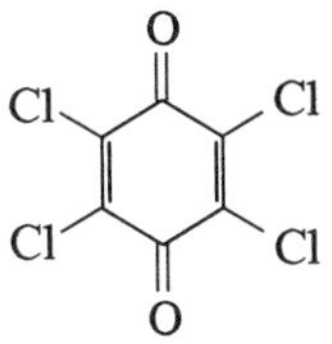

Ann *686* 40 (1965)

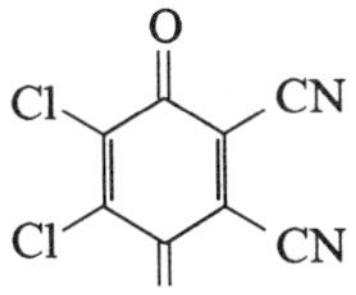

JOC *52* 2226, 3205 (1987)

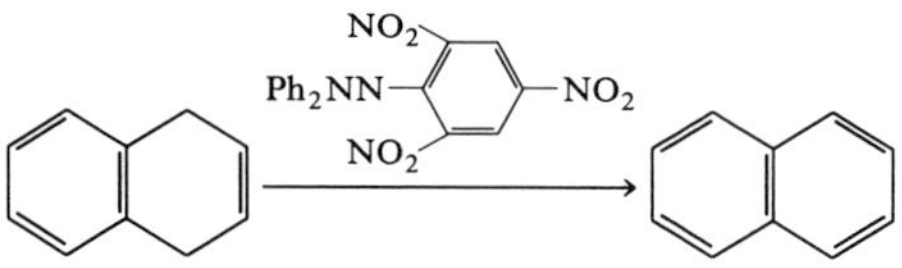

JCS 3574 (1954)

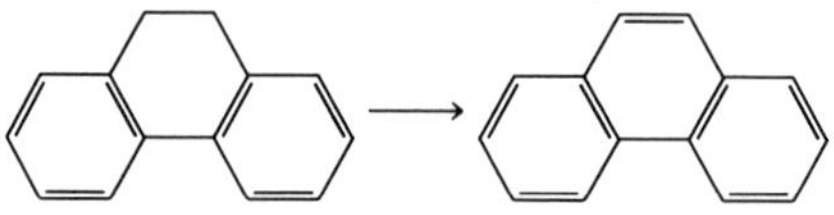

Review: Chem Rev *78* 317 (1978)

(Ph₃C)ClO₄	JCS 2773 (1959)
Ph₃COH, CF₃CO₂H	TL 3217 (1974)
n-BuLi	JACS *95* 2376 (1973); *96* 2434 (1974)

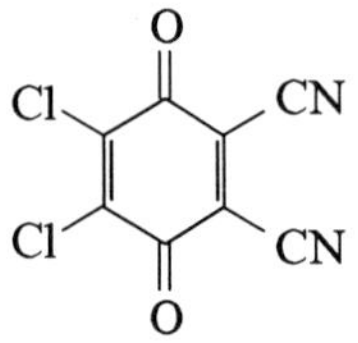

JOC *52* 3196 (1987)

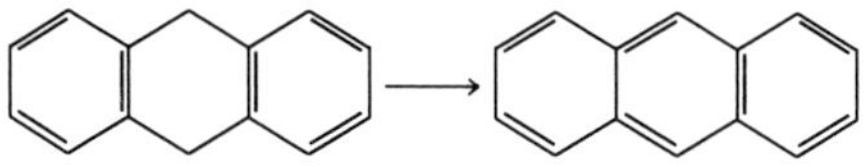

Review: Chem Rev *78* 317 (1978)

cat Pd-C	Ber *96* 707 (1963)
Cu	Ber *72* 1817 (1939); *96* 707 (1963)
SeO₂	JCS 764 (1947)

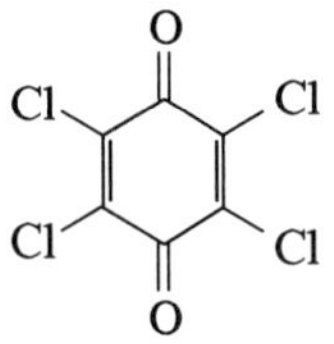

JACS *61* 1407 (1939)
Ber *96* 707 (1963)

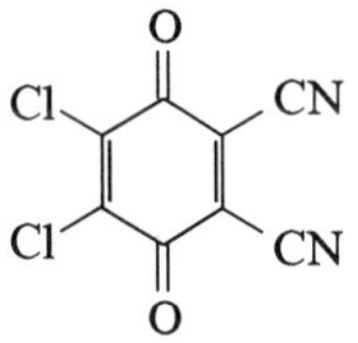

JOC *52* 3205 (1987)

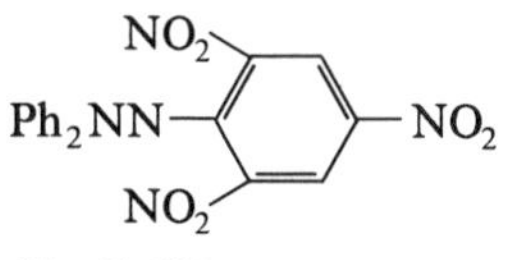

JCS 3574 (1954)

(Ph₃C)ClO₄	JCS 2773 (1959)
Ph₃COH, CF₃CO₂H	TL 3217 (1974)
PhSO₃H, PhNO₂	Ber *91* 2109 (1958)
n-BuLi/CdCl₂	JACS *95* 2376 (1973); *96* 2434 (1974)

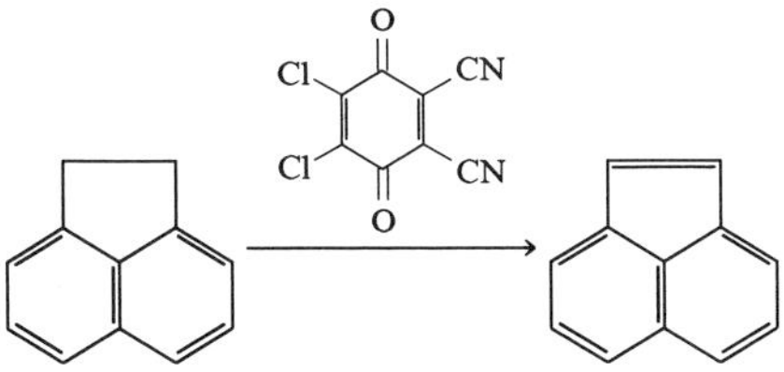

JOC *52* 3205 (1987)

2. Elimination

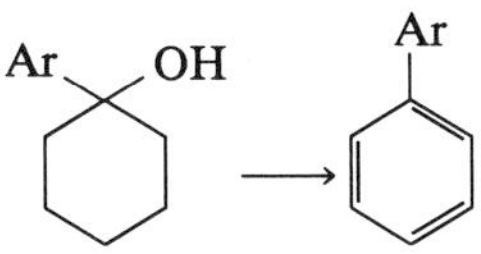

JCS 1832 (1932)

S

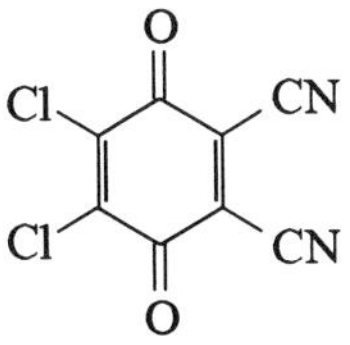

JOC *30* 2942 (1965)

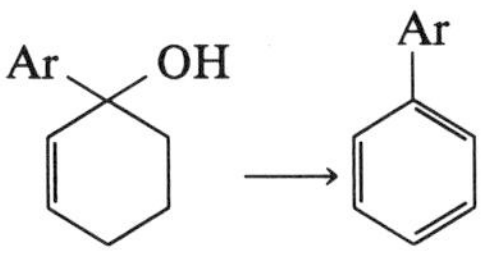

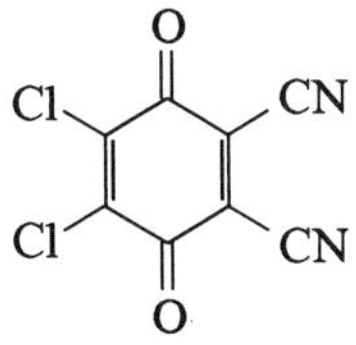

JOC *28* 2403 (1963)

Ph₃COH, CF₃CO₂H

TL 3217 (1974)

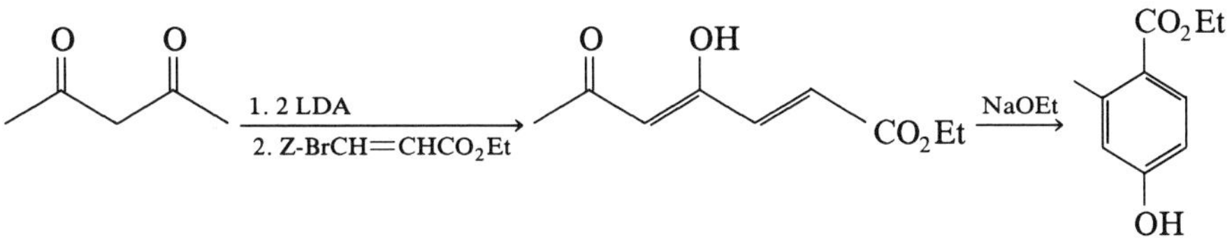

TL *26* 4419 (1985)

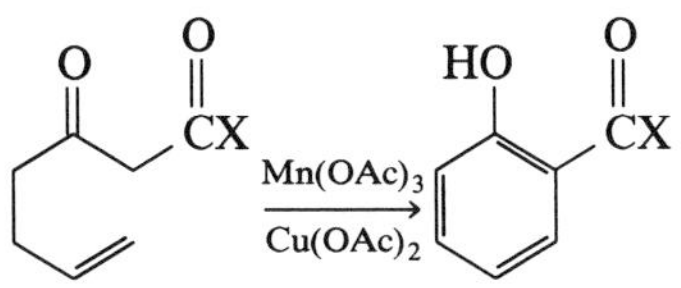

X = R, OR

TL *28* 845, 6109 (1987)

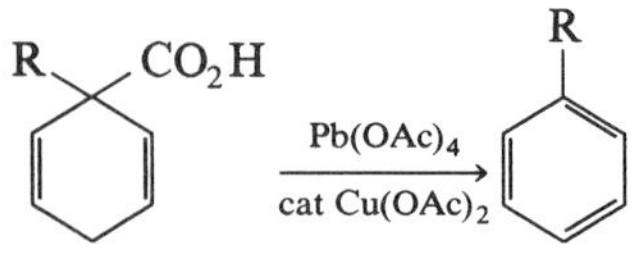

JCS 1551 (1950)
Ber *94* 2095 (1961)
TL 2079 (1976); *27* 3923 (1986)

3. Reductive Elimination

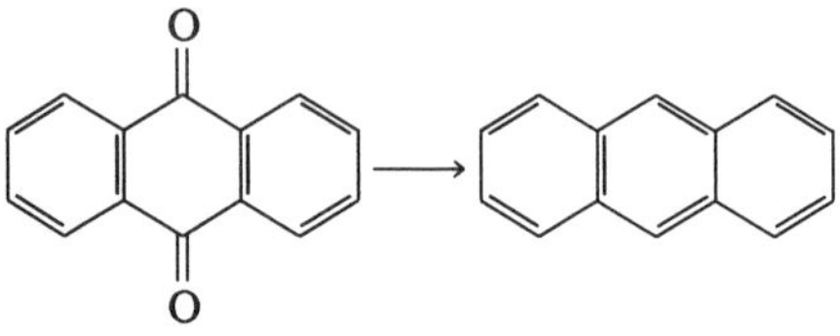

HI, HOAc   JOC *43* 3423 (1978); *44* 4813 (1979);
                                                     52 3205 (1987)

Ph_2SiH_2, Δ JOC *26* 4817 (1961)

$NaBH_4$/HCl/$NaBH_4$ JOC *39* 770 (1974)

$NaBH_4$, $BF_3 \cdot OEt_2$ or $AlCl_3$ Ind J Chem *1* 19 (1963)

4. Rearrangement

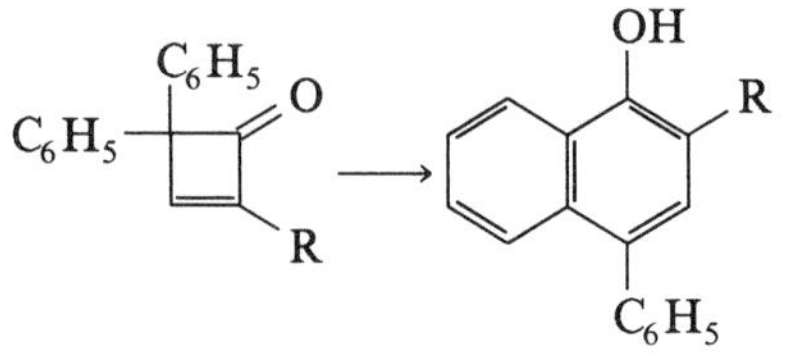

Angew Int *14* 500 (1975)

See also page 101, Section 6.

5. Cyclization and Annulation

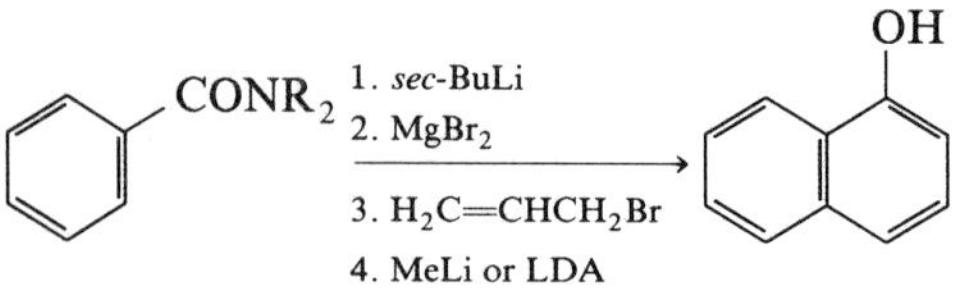

JOC *51* 271 (1986)

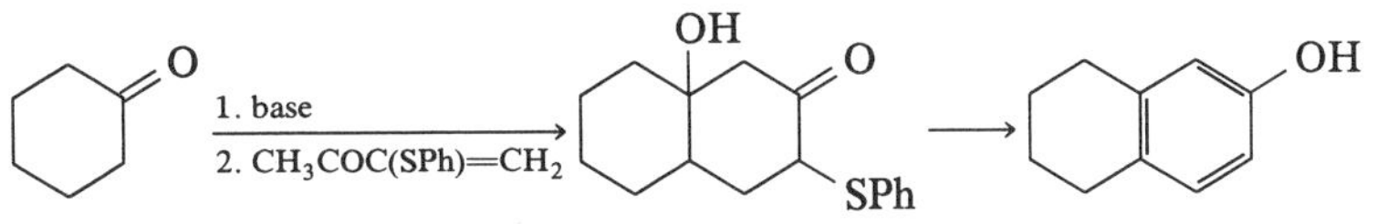
1. base
2. $CH_3COC(SPh)=CH_2$
OH
O
SPh
OH

CC 1183 (1980)
JOC *47* 1200 (1982)

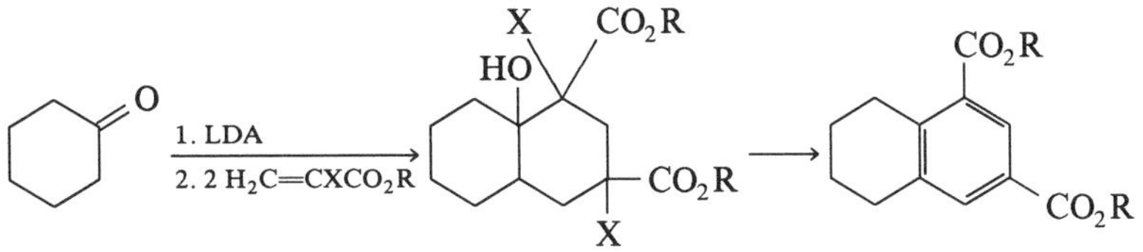
1. LDA
2. 2 $H_2C=CXCO_2R$
X
CO_2R
HO
CO_2R
X
CO_2R
CO_2R

JACS *108* 511 (1986)

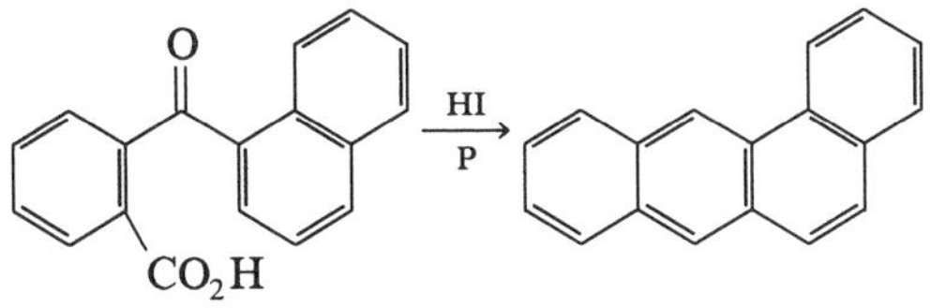
O
CO_2H
HI
P

JOC *46* 2601 (1981)

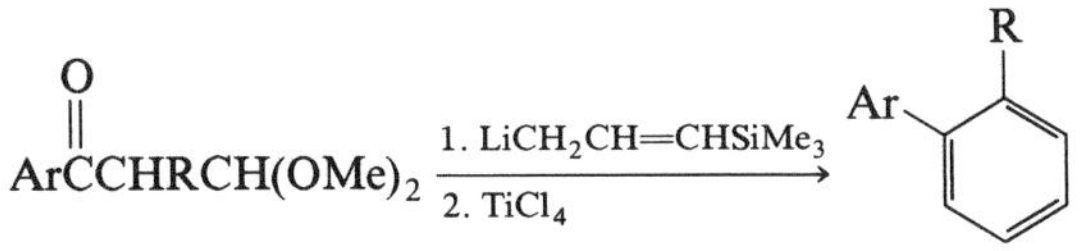
O
$ArCCHRCH(OMe)_2$
1. $LiCH_2CH=CHSiMe_3$
2. $TiCl_4$
R
Ar

TL *22* 3335 (1981)

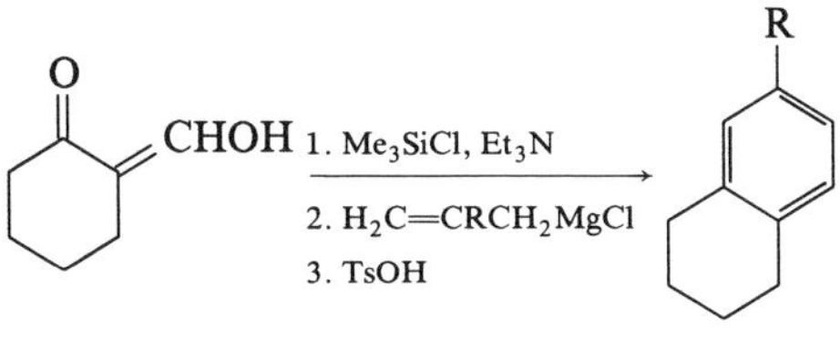
O
CHOH
1. Me_3SiCl, Et_3N
2. $H_2C=CRCH_2MgCl$
3. TsOH
R

JOC *47* 3163 (1982)
Syn 467 (1983)
TL *27* 4541 (1986)

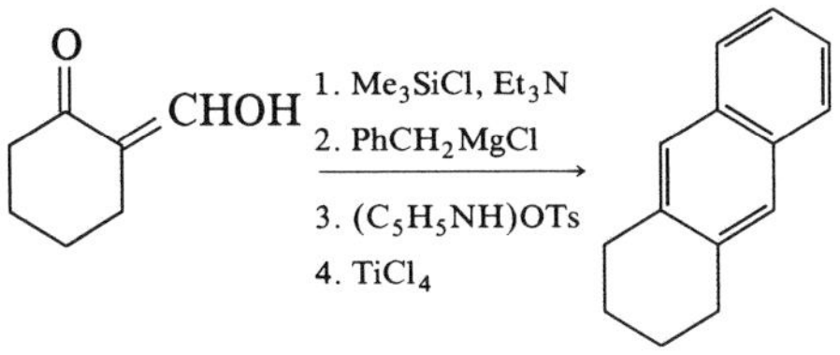
O
CHOH
1. Me_3SiCl, Et_3N
2. $PhCH_2MgCl$
3. $(C_5H_5NH)OTs$
4. $TiCl_4$

TL *27* 2571 (1986)

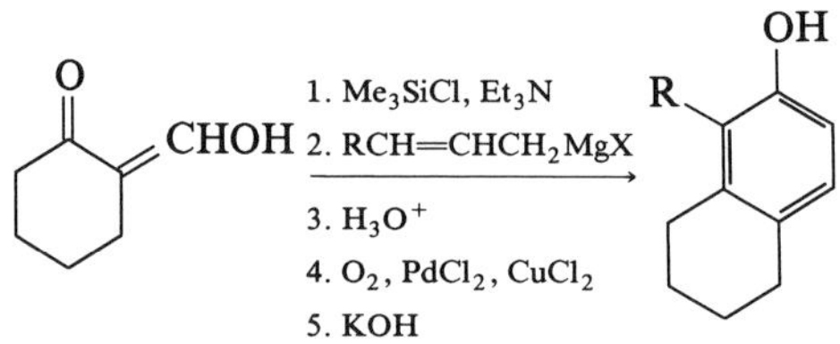

TL *23* 2823 (1982)

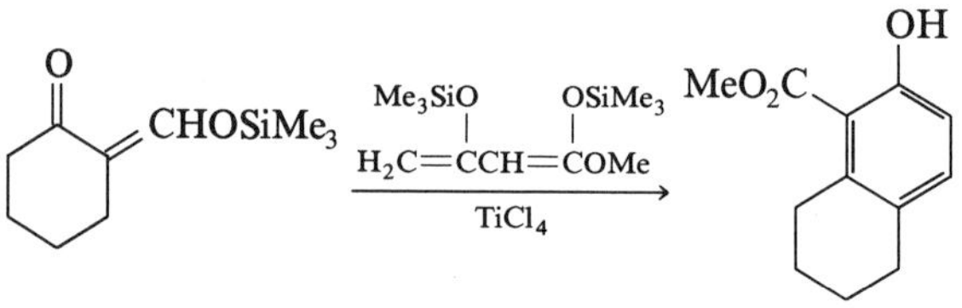

CC 578 (1979)
JACS *102* 3534 (1980)
TL *23* 2935 (1982)

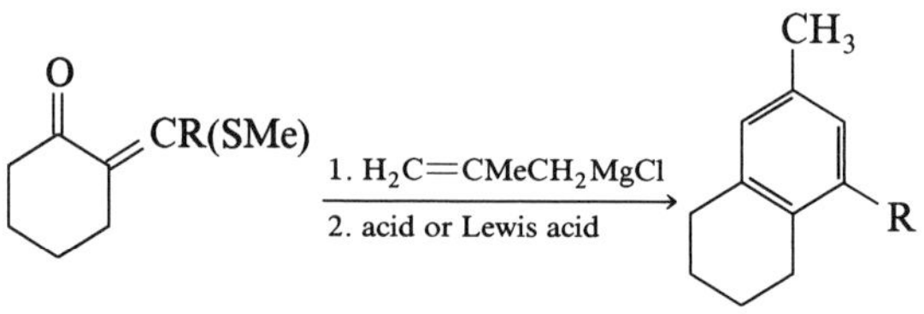

R = Me, SMe

TL *25* 5095 (1984); *26* 39 (1985)

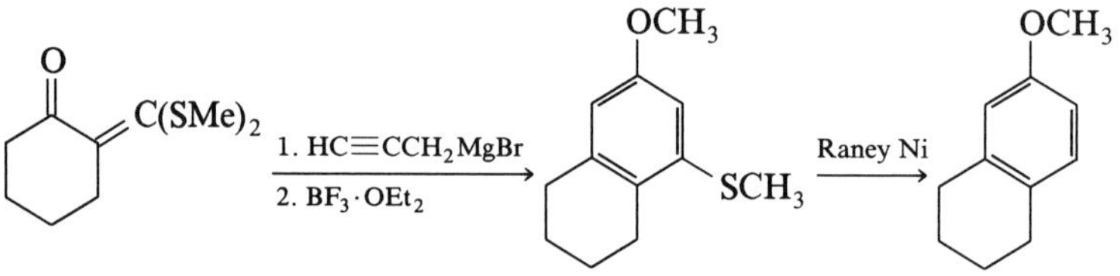

TL *28* 1459 (1987)

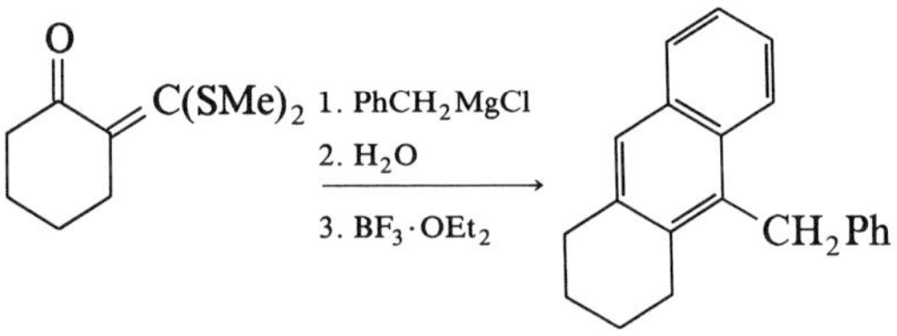

TL *27* 117 (1986)

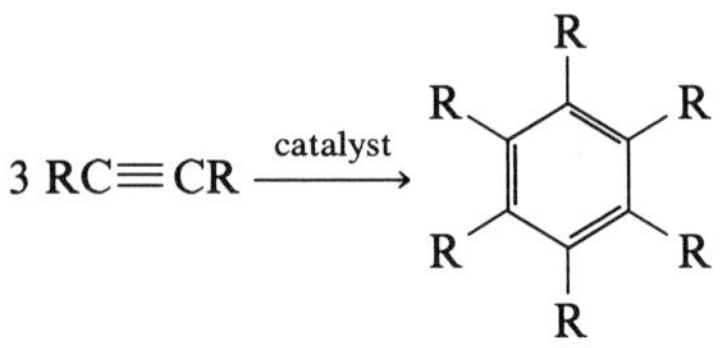

$$3\ RC\equiv CR \xrightarrow{\text{catalyst}}$$

Reviews: Russ Chem Rev *35* 510 (1966); *43* 48 (1974)
           Acct Chem Res *10* 1 (1977)
           JOC *52* 1161 (1987)

$i\text{-Bu}_2\text{AlH}, \Delta$	JACS *88* 2213 (1966)
R_3Al (R = Et, i-Bu), TiCl_4	JACS *81* 1514 (1959) JOC *26* 2234 (1961) Helv *48* 509 (1965)
Me_3SiCl, cat Pd-C	JOC *52* 1161 (1987)
$\text{M}_2\text{Cl}_6\left(\text{S}\bigcirc\right)_2$ M = Nb, Ta	Macromolecules *14* 235 (1981)
R_3Cr	JACS *81* 6090, 6203 (1959) JOMC *5* 176 (1966)
Fe	JOMC *249* 195 (1983)
$\text{Fe}_3(\text{CO})_{12}$	Ber *93* 103 (1960)
CpCo(CO)_2	Acct Chem Res *10* 1 (1977) (review) CC 953 (1982) Angew Int *23* 539 (1984) JACS *107* 1379, 5670 (1985); *108* 856, 3150 (1986)
$\text{Co}_2(\text{CO})_8$	Ber *93* 103 (1960) JACS *86* 4729 (1964)
$\text{Hg}[\text{Co(CO)}_4]_2$	Ber *93* 103 (1960); *94* 2829 (1961) Helv *48* 509 (1965)
Co carbonyl alkyne complexes	Ber *93* 103 (1960); *94* 2817, 2829 (1961)
$\text{ClRh(PPh}_3)_3$	TL *23* 2691 (1982)
$\text{Ni(CO)}_2(\text{PPh}_3)_2$	Ann *560* 104 (1948) JCS *69* (1950) JOC *26* 5155 (1961)
$\text{Ni(H}_2\text{C}{=}\text{CHCN})_2$	Ber *94* 1403 (1961)
$\text{PdCl}_2(\text{PhCN})_2$	JACS *84* 2329 (1962); *92* 2276 (1970) Can J Chem *43* 470 (1965)

$$\text{+ RC}{\equiv}\text{M(CO)}_4\text{Br} \longrightarrow$$

M = W, Cr

TL *26* 2159 (1985)

6. Diels-Alder and Related Reactions

$$\text{O + XC}{\equiv}\text{CY} \xrightarrow{\Delta}$$

JOC *52* 496 (1987)

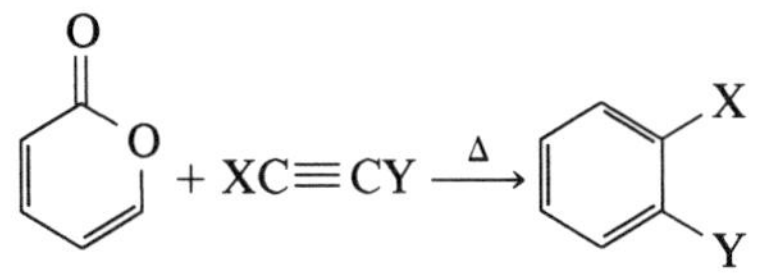

Ber *70* 1354 (1937)
JOC *29* 2534 (1964); *42* 2930 (1977); *47* 1150 (1982)
JACS *89* 952, 4793 (1967)
JOMC *34* 119 (1972)
CC 95 (1978)
TL *22* 3355 (1981)
Ann 914 (1982)

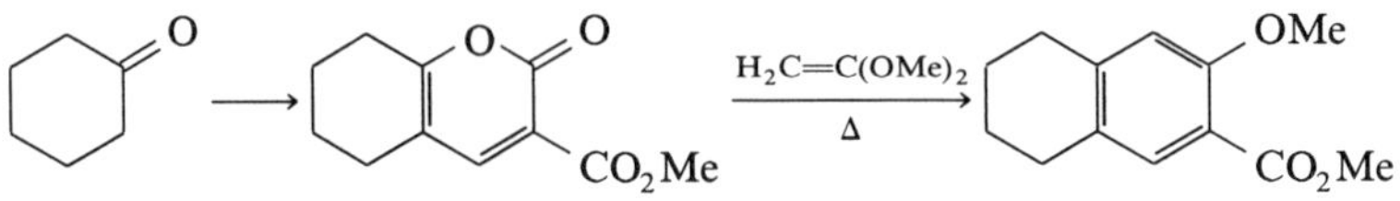

TL *23* 4551, 4555, 4559 (1982); *24* 4939 (1983)
JOC *49* 4033, 4055 (1984)
Org Syn *65* 98 (1987)

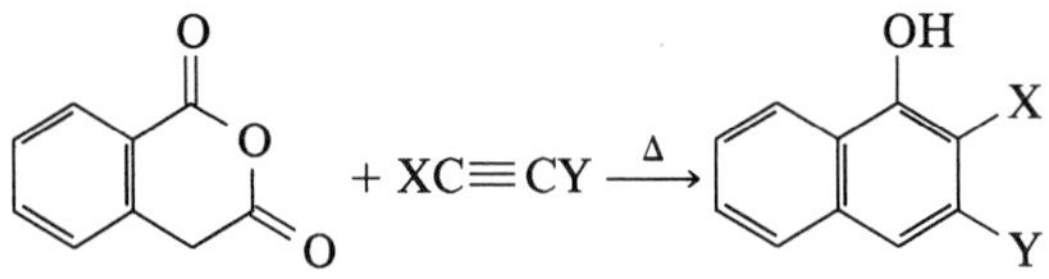

TL *22* 4283 (1981)

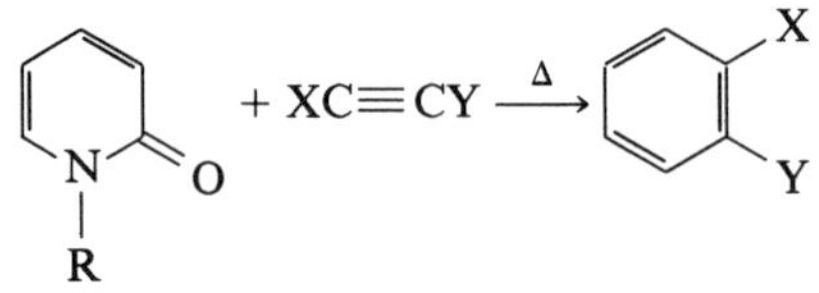

JCS Perkin I 169 (1982)

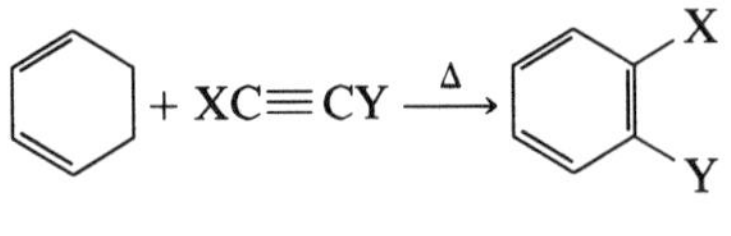

Syn 223 (1982)
JACS *107* 2712 (1985)

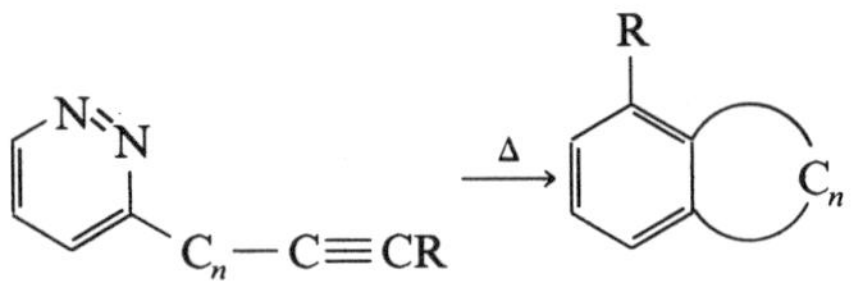

JOC *49* 2240 (1984); *51* 3250 (1986); *52* 4280 (1987) (pyridines)
Chem Zeitung *108* 331 (1984) (pyridazines)
TL *26* 2419 (pyridines), 4355 (pyridines, pyridazines) (1985); *27* 431 (pyridines), 1967 (pyridines), 2107
(pyridines), 2747 (pyridines, pyridazines) (1986); *28* 379 (1987) (pyridines)
JACS *109* 2717 (1987)

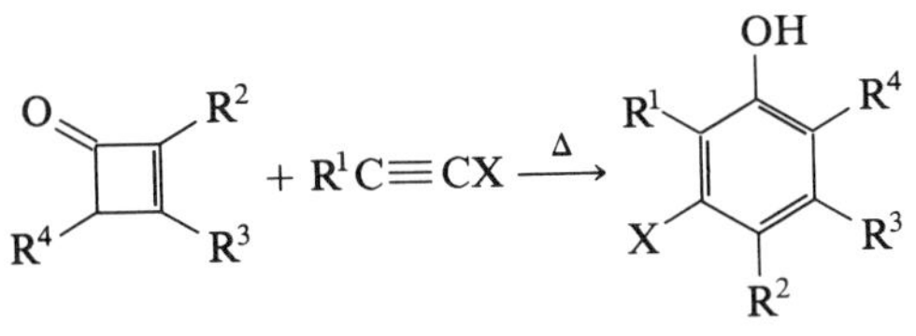

JOC *49* 1672 (1984)
JACS *108* 806 (1986)

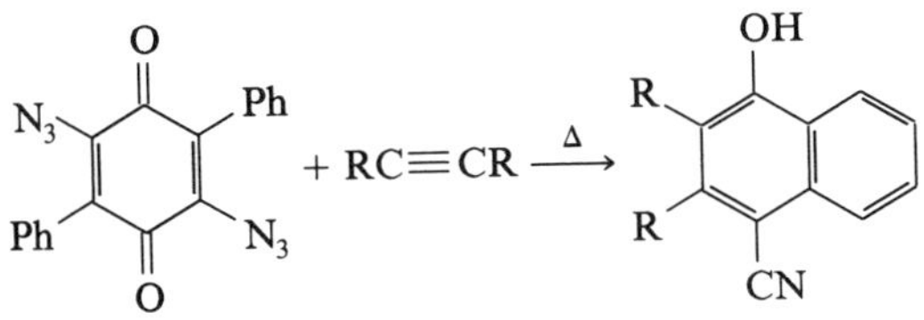

TL *28* 5013 (1987)

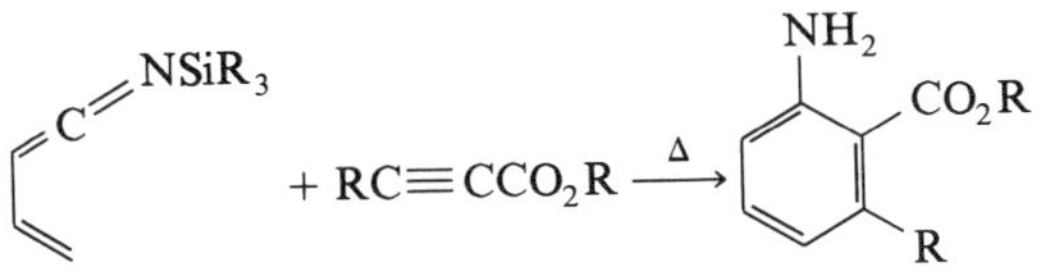

TL *28* 397 (1987)

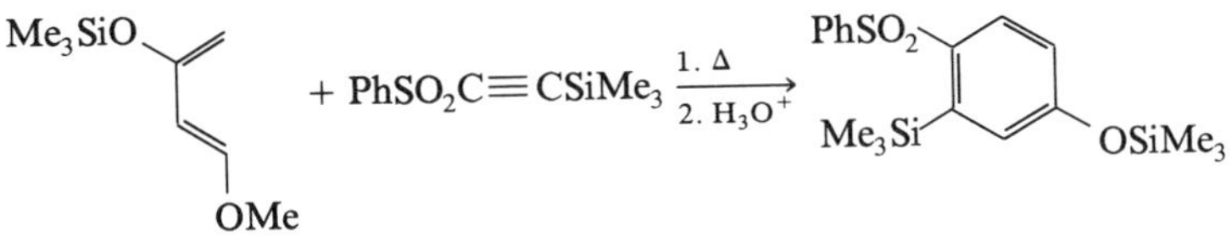

JACS *101* 7001 (1979)
JOC *46* 1951 (1981)

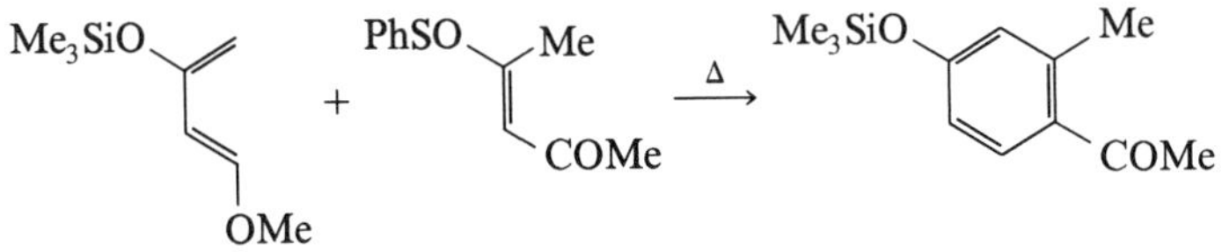

JACS *101* 7008 (1979)

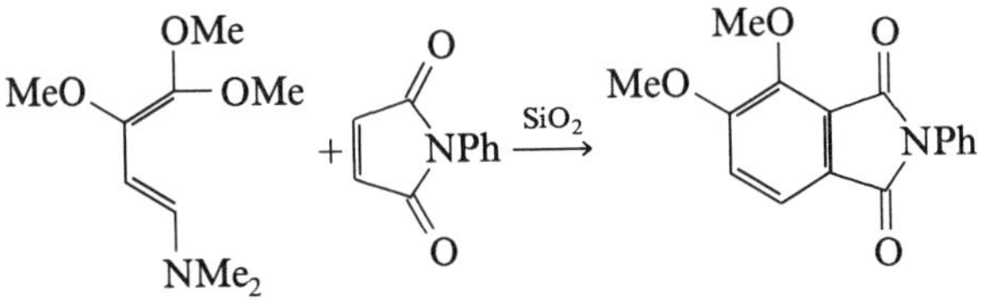

Syn 649 (1981)

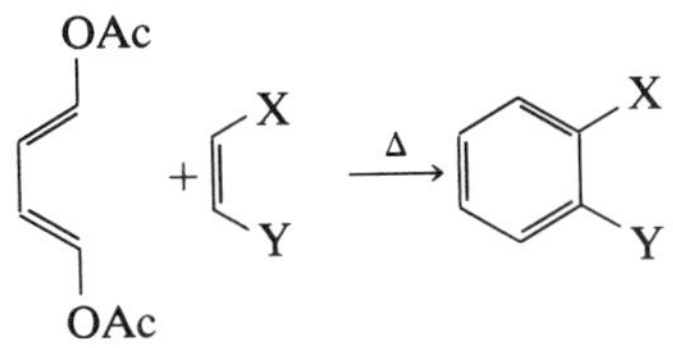

TL 1157 (1964)

ALKENES

GENERAL REFERENCES

"The Chemistry of Alkenes," Ed. S. Patai, Interscience, New York (1964)

F. Asinger, "Olefins," Pergamon Press, New York (1966)

Houben-Weyl, "Methoden der Organischen Chemie," 4th ed, Vol V/1c (dienes), G. Thieme, Stuttgart (1970)

Quart Rev *25* 135 (1971)

Syn 175 (1971)

Houben-Weyl, "Methoden der Organischen Chemie," 4th ed, Vol V/1b (alkenes), G. Thieme, Stuttgart (1972)

Houben-Weyl, "Methoden der Organischen Chemie," 4th ed, Vol V/1d (polyenes), G. Thieme, Stuttgart (1972)

1. ISOMERIZATION OF ALKENES

Review: Syn 405 (1970)

For reactions of allylic organometallics which proceed with allylic transposition see page 185, Section 5.

1. Alkene Inversion

See also page 129, Section 2, for elimination via diols, epoxides, β-dihalides, β-halo esters and bromohydrins.

Reviews:

Chem Rev 55 625 (1955)
Tetr 36 557 (1980)

hν	JACS 107 1034 (1985)
hν, (PhS)$_2$	JCS C 260 (1966)
	JACS 89 2758 (1967)
	Helv 51 548 (1968)
	Org Syn 63 192 (1984)
	TL 28 47, 6437 (1987)
PhSH, AIBN	JOC 51 260 (1986) (R = 3° alkyl); 52 4674 (1987)
cat Ph$_3$GeH, cat BEt$_3$	TL 28 3709 (1987)

Helv 38 1338 (1955)
JACS 79 2318 (1957); 86 2087 (1964); 107 1034 (1985); 108 3005 (R = Ph, R′ = CO$_2$R; added BF$_3$·OEt$_2$ or EtAlCl$_2$), 3016 (R = alkyl, R′ = CO$_2$R; added EtAlCl$_2$) (1986)
JOC 38 1247 (1973)
Tetr 31 193 (1975)

2. Simple Rearrangement

2.1. Thermal

"Molecular Rearrangements," J. Wiley and Sons, New York (1963), Vol 1, Chpt 11
Angew Int *2* 115 (1963)

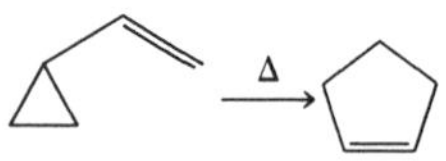

See page 273, Section 9.

2.2. Photochemical

See also page 109, Section 1 (Alkene Inversion).

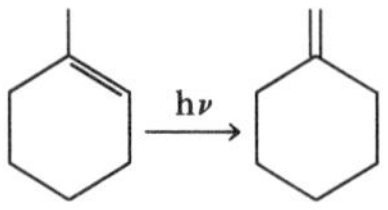

JACS *89* 5199 (1967)

$$R'CH_2CH=CHCX \xrightarrow[\text{reagent}]{h\nu} R'CH=CHCH_2CX$$

(where X bears =O)

X	Reagent	
R	—	Helv *46* 678 (1963) TL 1203 (1964) Pure Appl Chem *9* 481 (1964) BSCF 1185 (1973) Can J Chem *54* 2127 (1976) JACS *109* 2479 (1987)
OH	—	JACS *85* 1210 (1963) (dienoic acid → allenic acid) Pure Appl Chem *9* 481 (1964) JOC *33* 1671 (1968)
OR	—	Pure Appl Chem *9* 481 (1964) CC 137 (1965) JOC *32* 3222 (1967); *33* 1671 (1968); *34* 609 (1969); *38* 2558 (1973) TL 4987, 4991 (1968); 489 (1970) JACS *91* 198 (1969); *93* 5490 (1971)

J Photochem *1* 433 (1972/73)
BCSJ *46* 2181 (1973) (lactone)
Syn 1019 (1983) (lactones)

cat [structure: imidazoline ring with N=, Me, N–Me]

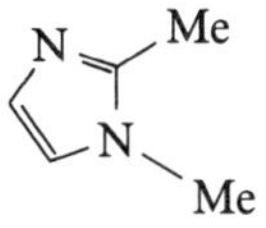

TL *24* 4299 (1983); *27*
    5555 (1986)
Can J Chem *62* 1933 (1984)
JOC *50* 873 (1985)

ephedrine

TL *26* 4945 (lactones), 6079
    (1985); *27* 2997, 3001 (1986)
    (all enantioselective)
Pure Appl Chem *58* 1257 (1986)
    (esters and lactones)

various amino alcohols

TL *27* 3001 (1986); *28* 4825
    (1987) (both enantioselective)
Pure Appl Chem *58* 1257 (1986)
    (esters and lactones)

See also page 867, Section 1; and page 873, Section 2, for reactions with strong bases followed by protonation.

2.3. Base-Promoted

Review: Syn 97 (1969)

KNH_2, NH_3

JCS 1642 (1947)
Acct Chem Res *1* 231 (1968)
Syn 97 (1969)

$KNH(CH_2)_3NH_2$, $H_2N(CH_2)_3NH_2$

Org Syn *65* 224 (1987)

LiNHEt

JOC *23* 1136 (1958)
JACS *86* 5281 (1964)

KO-*t*-Bu, DMSO

JACS *83* 3731 (1961); *84* 3164 (1962); *85* 1553
    (1963)
JCS 4234 (1963)
Ber *99* 1737 (1966)
JOC *33* 221 (1968)

[structure: cyclohexane with OK substituent]

JOC *45* 900 (1980)

$NaCH_2Ph$

TL 467 (1964)

Na, alumina

JACS *87* 4107 (1965)
JCS C 2149 (1967)
Angew Int *24* 320 (1985)

K, alumina

Angew Int *24* 320 (1985)

MgO

CL 133 (1979)

$$\underset{\underset{\text{H}}{\mid}}{\text{C}}=\text{C}-\text{C}-\text{OR} \xrightarrow[\text{DMSO}]{\text{KO-}t\text{-Bu}} \underset{\underset{\text{H}}{\mid}}{\text{C}}-\text{C}=\text{C}-\text{OR}$$

TL 1191 (1964); *28* 5887 (1987)
JCS C 82 (1966); 1903 (1968)
JCS Perkin I 1535 (1972); 1858 (1973)

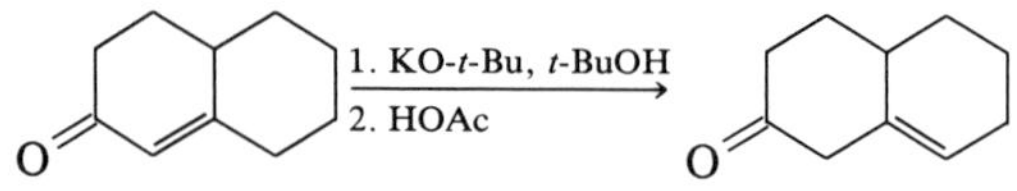

1. KO-*t*-Bu, *t*-BuOH
2. HOAc

TL 669 (1962)
JOC *30* 2513 (1965)

2.4. Acid-Catalyzed

acid alumina   JACS *87* 4107 (1965)

$(CO_2H)_2$, EtOH   JACS *75* 5421 (1953)

2.5. Sulfur Dioxide

JACS *99* 5219 (1977)

2.6. Ene Reaction–Reduction

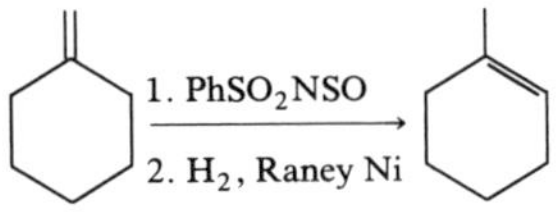

1. $PhSO_2NSO$
2. H_2, Raney Ni

TL *28* 5017 (1987)

2.7. Iodine

JCS 2760 (1957)
JCS C 260 (1966) (plus hν)
JOC *52* 2586 (1987)

2.8. Transition Metal-Catalyzed

Review:

C. W. Bird, "Transition Metal Intermediates in Organic Synthesis," Academic Press, New York (1967), Chpt 3

$$H_2C=CHCH_2R \longrightarrow CH_3CH=CHR$$

$[(o\text{-}CH_3C_6H_4O)_3P]_2Ni(C_2H_4)$, HCl   JOC *41* 3020 (1976)
$PdCl_2(PhCN)_2$ TL *28* 5473 (1987)

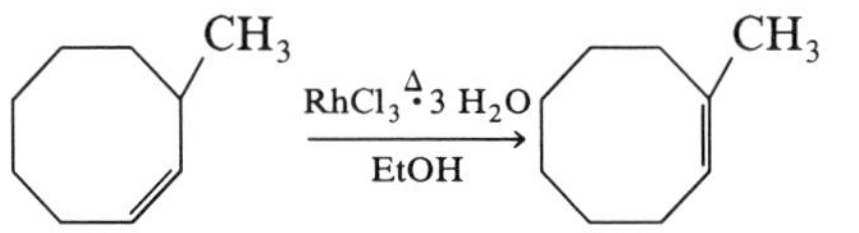

JACS *86* 1776, 2516 (diene) (1964); *107* 7352 (1985)
Helv *50* 2445 (1967) (diene)
JOC *52* 2875 (1987)

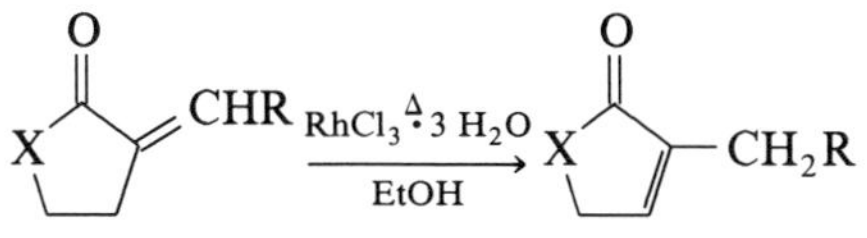

X

CH$_2$

O

JCS Perkin I 359 (1977)
JOC *45* 3017 (1980)
Syn 952 (1983)

JOC *52* 1429 (1987)

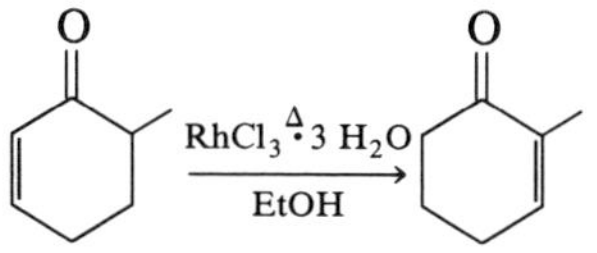

JACS *98* 7102 (1976); *109* 3025 (1987)
TL *28* 31 (1987)

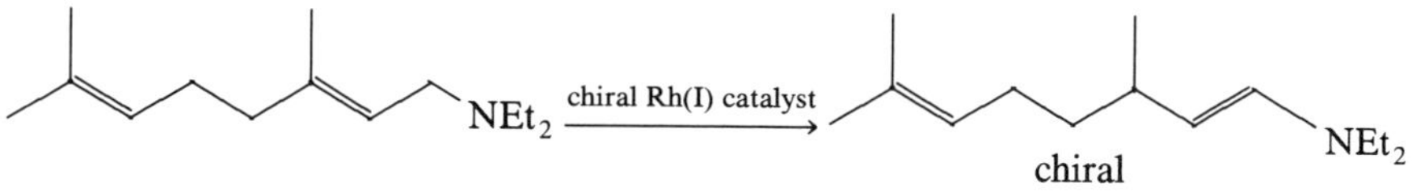

JACS *106* 5208 (1984)

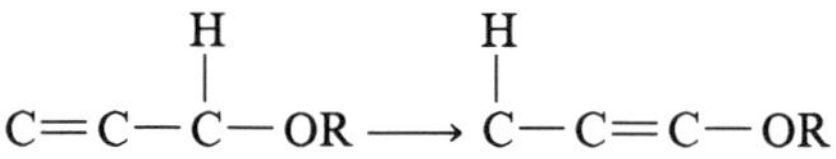

cat H$_2$Ru(PPh$_3$)$_4$

TL 1415 (1979); *21* 4927 (1980); *23* 1079 (1982); *28* 239 (1987)
CL 1435 (1981)
JACS *109* 5280 (1987)
JOC *52* 5700 (1987)

cat H$_2$Ru(CO)(PPh$_3$)$_3$

TL *21* 4927 (1980)

cat HRuCl(CO)(PPh$_3$)$_3$

TL *21* 4927 (1980)

cat ClRh(PPh$_3$)$_3$

JOC *38* 3224 (1973)
CC 277 (1974)
JACS *109* 5446 (1987)

cat PdCl$_2$(PhCN)$_2$

JCS Perkin I 2870 (1973)

2.9. Organoboranes

JACS *82* 2074 (1960); *89* 561, 567 (1967)

2.10. Addition–Elimination

CC 305 (1968)

3. Functional Group Rearrangement

NaH, CS$_2$, MeI/n-Bu$_3$SnH/m-ClC$_6$H$_4$CO$_3$H/HCl	Syn 1011 (1980)
t-BuO$_2$H, cat Ti(O-i-Pr)$_4$, diisopropyl tartrate/MeSO$_2$Cl/Li, NH$_3$	JOC *49* 1707 (1984)
t-BuO$_2$H, cat VO(acac)$_2$/MeSO$_2$Cl/Li, NH$_3$	JACS *109* 3025 (1987)
t-BuO$_2$H, cat VO(acac)$_2$/MeSO$_2$Cl/Na, NH$_3$	TL 2621 (1976) CL 465 (1977) BCSJ *52* 1757 (1979)
t-BuO$_2$H, cat VO(acac)$_2$/MeSO$_2$Cl/Na naphthalene	TL *27* 6353 (1986)
PhCMe$_2$O$_2$H; cat VO(acac)$_2$; 2,6-lutidine/CBr$_4$, PPh$_3$/Zn, HOAc, THF	JACS *109* 6187 (1987)
o- or p-NO$_2$C$_6$H$_4$SeCN, n-Bu$_3$P/H$_2$O$_2$	CC 770 (1978) JOC *50* 2981 (1985)

Hg(O$_2$CCF$_3$)$_2$/PPh$_3$	JACS *100* 4822 (1978)
PdCl$_2$(CH$_3$CN)$_2$	Can J Chem *62* 791 (1984)

$$\underset{\underset{R_2CCH=CH_2}{|}}{OAc} \xrightarrow{\text{cat PdCl}_2(\text{CH}_3\text{CN})_2} R_2C=CHCH_2OAc$$

JACS *94* 5200 (1972); *102* 7587 (1980)
TL 321 (1979); *28* 4131, 5655 (1987)
JOC *46* 5005 (1981)
Tetr *40* 1791 (1984)
Can J Chem *62* 791 (1984) (carbonates also)

$$RCH=\underset{\underset{CHCHCN}{|}}{OAc} \xrightarrow{\text{cat Pd(PPh}_3)_4} \underset{\underset{RCHCH}{|}}{OAc}=CHCN$$

Syn 687 (1982)

4. Reductive Transposition

$$RCH=CHCH_2X \longrightarrow RCH_2CH=CH_2$$

X	Reagent(s)	
$\overset{+}{P}Ph_3$	LiAlH$_4$	JOC *49* 4084 (1984)
SO$_2$Ph	LiAlH$_4$, CuCl$_2$	JACS *107* 1034 (1985)

$$\underset{\underset{R_2CCH=CH_2}{|}}{NO_2} \xrightarrow[\text{cat Pd(PPh}_3)_4,\text{ phosphine}]{\overset{\text{NaBH}_4,\text{ NaBH}_3\text{CN,}}{n\text{-BuZnCl or LiHB(}sec\text{-Bu)}_3}} R_2C=CHCH_3$$

JOC *51* 3734 (1986)

$$H_2C=CH\overset{O}{\overset{/\backslash}{CH}}{-}CH_2 \longrightarrow CH_3CH=CHCH_2OH$$

BH$_3$	Rocz *47* 771 (1973); *48* 467 (1974) Syn 62 (1979)
NaBH$_3$CN, BF$_3\cdot$OEt$_2$	JOC *46* 5214 (1981)
LiHBEt$_3$	TL 3775 (1976)
i-Bu$_2$AlH	JACS *95* 957 (1973)
LiAlH$_4$	TL 3775 (1976)
Li/NH$_3$	JACS *95* 957 (1973)
Ca, NH$_3$	JACS *95* 957 (1973) TL *28* 2021 (1987)
SmI$_2$, ROH	JOC *51* 5259 (1986)

$$RCH=\overset{\overset{\displaystyle R'}{|}}{C}CH_2OH \xrightarrow[\substack{3.\ n\text{-}Bu_3SnH \\ 4.\ H^+}]{\substack{1.\ NaH,\ CS_2 \\ 2.\ MeI,\ \Delta}} RCH_2\overset{\overset{\displaystyle R'}{|}}{C}=CH_2$$

TL *21* 1767 (1980)

$$R_2C=CHCH_2OH \xrightarrow[\substack{3.\ HBF_4 \\ 4.\ NaI}]{\substack{1.\ ClPO(O\text{-}i\text{-}Pr)_2,\ py \\ 2.\ NaFe(CO)_2Cp}} R_2CHCH=CH_2$$

JOC *51* 2126 (1986)

$$RCH=CHCH_2OAc \text{ or } R\overset{\overset{\displaystyle OAc}{|}}{C}HCH=CH_2 \xrightarrow[cat\ PdCl_2(PPh_3)_2]{NH_4(O_2CH)} RCH_2CH=CH_2$$

TL 613 (1979)

$$\overset{\overset{\displaystyle C-X}{|}}{C}=\overset{\overset{\displaystyle |}{C}}{}-CO_2R \xrightarrow{LiHBEt_3} \overset{\overset{\displaystyle C}{||}}{C}-\overset{\displaystyle C}{}-CO_2R$$

X = Br, OAc

Angew Int *22* 796 (1983)

5. Oxidative Transposition

$$R_2C=CHCH_2R \longrightarrow R_2\overset{\overset{\displaystyle OMe}{|}}{C}CH=CHR$$

(PhSe)$_2$, MgSO$_4$, MeOH, electrolysis   JACS *103* 4606 (1981)

PhSeCl, MeOH/*m*-ClC$_6$H$_4$CO$_3$H   TL *27* 6361 (1986)

$$H_2C=CHCH_2R \xrightarrow[]{^1O_2} \xrightarrow[]{reduction} HOCH_2CH=CHR$$

Angew *69* 579 (1957)
A. Schonberg, "Preparative Organische Photochemie," Springer Verlag, Berlin (1958), p 47
JACS *90* 975 (1968)
Adv Photochem *6* 1 (1968)
Acct Chem Res *1* 104 (1968); *13* 419 (1980) (mechanism)
Tetr *37* 1825 (1981)

$$RCH_2CH=CR_2 \longrightarrow RCH=\overset{\overset{\displaystyle OH}{|}}{C}HCR_2$$

"PhSeOH"/*t*-BuO$_2$H   TL 3967 (1978)
                                               JOC *43* 1688 (1978)

(PhSe)$_2$, MgSO$_4$, H$_2$O, CH$_3$CN,   JACS *103* 4606 (1981)
   electrolysis

PhSeCl, H$_2$O/m-ClC$_6$H$_4$CO$_3$H   TL 27 2391 (1986)

PhSeBr, AgO$_2$CCF$_3$/H$_2$O, MeOH,   JACS 106 1446 (1984)
  NaHCO$_3$/H$_2$O$_2$

epoxidize/PhSeNa/[O]   JACS 95 2697 (1973); 106 7854 (1984); 107 1691,
                                    1777 (1985)
                                   Tetr 34 1049 (1978)
                                   JOC 51 2148 (1986); 52 2644 (1987)

epoxidize/i-Bu$_2$AlSePh/m-ClC$_6$H$_4$CO$_3$H/   JOC 50 5897 (1985)
  Et$_3$N, Δ

$$(CH_3)_2C{=}CHR \longrightarrow (CH_3)_2C{-}CHR \xrightarrow{\text{reagent(s)}} H_2C{=}\underset{\underset{CH_3}{|}}{\overset{\overset{OH}{|}}{C}}{-}CHR$$

Reviews:

    Russ Chem Rev 41 403 (1972)
    Tetr 39 2323 (1983)
    Org Rxs 29 345 (1983)
    Syn 629 (1984)

Reagent(s)

Li, H$_2$NCH$_2$CH$_2$NH$_2$ Helv 50 153 (1967)

alkali metal amide bases   JACS 80 2849, 2855 (1958); 82 6370 (1960); 87
                                   3125 (1965); 89 4526, 4527 (1967); 92 737,
                                   2064 (1970); 95 5311 (1973); 106 2949 (1984);
                                   108 3739 (1986)
                                 JOC 29 2830 (1964); 32 435, 532 (1967); 34 3583
                                   (1969); 36 1365 (1971); 37 2060, 3919, 4250
                                   (1972); 51 1362 (1986); 52 1907 (1987)
                                   Syn 194 (1972); 602 (1975)
                                   Org Syn 53 17 (1974)
                                   TL 1929 (1974); 28 1439, 3831, 4993 (1987)
                                   Tetr 34 1541 (1978)

LiNRR' (chiral)   JOC 45 755 (1980)
                                   CL 829 (1984); 389 (1987)
                                   TL 26 5803 (1985)

⬡—N(i − Pr) MgBr   JACS 102 1433, 7986 (1980)
                                   JOC 51 793 (1986)

⬡—N(i − Pr) MgCH$_3$ TL 27 299, 303 (1986)

⬡—NAlEt$_2$ JACS 96 6513 (1974)
                                   Angew Int 17 169 (1978)
                                   BCSJ 52 1705 (1979)
                                   JOC 52 1106 (1987)

Al(O-*i*-Pr)$_3$ JOC *35* 1598 (1970); *44* 868 (1979); *50* 2948
                                           (1985)
                                          JCS Perkin I 2909 (1982)
                                          JACS *105* 5510 (1983)
                                          Helv *67* 1998 (1984)
                                          CC 418, 727 (1986)

Ti(O-*i*-Pr)$_4$ (epoxy alcohols)   JACS *103* 462 (1981)

9-BBN-OTf,   CL 1215 (1977)

Me$_3$SiOTf/DBU/HCl or HF   JACS *101* 2738 (1979)
                                          Tetr *37* 3899 (1981)

Et$_3$SiOTf/2,6-lutidine/H$_3$O$^+$ TL *28* 6417 (1987)

Me$_3$SiBr, PPh$_3$/DBU/HCl   JACS *106* 7854 (1984)

Me$_3$SiI/DBU or DBN/H$^+$ TL *21* 2329 (1980)
                                          JOC *45* 2579 (1980)

t-BuMe$_2$SiI/DBN/H$^+$ JOC *45* 924 (1980)

$$RCH{=}CHCH_2R' \xrightarrow[\text{2. H}_2\text{O}_2]{\text{1. PhSeBr, HOAc}} \overset{\overset{\textstyle OAc}{|}}{R}CHCH{=}CHR'$$

JACS *106* 1446 (1984)

$$RCH{=}CHCH_2CO_2R \xrightarrow[\text{\textit{n}-AmONO, HOAc}]{\text{KOAc, PdCl}_2} \overset{\overset{\textstyle OAc}{|}}{R}CHCH{=}CHCO_2R$$

TL *22* 131 (1981)

JACS *107* 1421 (1985)

6. Heteroatom Displacement

$$R_3Si{-}C{-}C{=}C{-}C \longrightarrow C{=}C{-}\overset{\overset{\textstyle OH}{|}}{C}{-}C$$

CH$_3$CO$_3$H, buffer   CC 679 (1976)

m-ClC$_6$H$_4$CO$_3$H/H$^+$ CC 79 (1977)
                                          TL *23* 1267 (1982)
                                          JOC *49* 4224 (1984)

m-ClC$_6$H$_4$CO$_3$H/(*n*-Bu$_4$N)F   TL *24* 4153 (1983)

$$ArSO_2CH_2CR = CR_2 \xrightarrow[\substack{\text{2. R'X} \\ \text{3. } m\text{-ClC}_6H_4CO_3H \\ \text{4. Na(Hg)}}]{\text{1. } n\text{-BuLi}} R'CH = CRCR_2OH$$

TL 441 (1979)

$$RCH = CHCH_2OH \xrightarrow{Cl_3CCN} RCH = CHCH_2O\overset{\overset{\displaystyle NH}{\|}}{C}CCl_3 \xrightarrow[\text{HgX}_2 \text{ or PdX}_2]{\Delta \text{ or}} R\overset{\overset{\displaystyle HNCOCCl_3}{|}}{C}HCH = CH_2$$

See page 937, Section 13.

7. Ene Reaction

See further examples in the following Section 8

Angew Int *8* 556 (1969); *17* 476 (1978) (intramolecular); *23* 876 (1984)
Acct Chem Res *13* 426 (1980)
Tetr *37* 3927 (1981)

8. Alkylative Transposition

See also page 521, Section 5; page 637, Section 7.1; page 775 (Carroll Reaction); page 934, Section 10; page 936, Section 11; page 937, Section 14; and page 938, Section 15.

$$n = 1, 2$$

X	Pd reagents	
Br or I	cat $Pd(OAc)_2(PAr_3)_2$	JOC *43* 2952, 4110 (1978); *44* 21 (1979) Tetr *35* 329 (1979)
N_2^+	cat $PdCl_2$-NaO_2CH or $Pd(DBA)_2$	Tetr *37* 31 (1981)
NH_2	*t*-BuONO, cat $Pd(DBA)_2$	CL 551 (1980) JOC *46* 4885 (1981)
	t-BuONO, $Pd(OAc)_2$	JOC *45* 2359 (1980)
HgX (X = OAc, Cl)	Pd(II) salts	JACS *93* 6896 (1971); *100* 287 (1978) Organomet Chem Syn *1* 455 (1972) JOC *43* 4110 (1978); *48* 399, 2870 (1983)

$$\underset{\text{HO}}{\overset{|}{R^1-C}}-C=C-R^2 \longrightarrow R^1-C=C-\underset{R}{\overset{|}{C}}-R^2$$

$CH_3Li/CuI/RLi/[n\text{-}Bu_3PNMePh]I$	JACS *99* 2361 (1977); *100* 4610 (1978) JOC *46* 2144 (1981) TL *23* 557 (1982); *27* 6353 (1986)
CH_3MgBr, cat $NiCl_2L_2$ (R = CH_3)	CC 681 (1981)
$Me_2C=CClN\bigcirc$/cat CuI, HMPA/RMgX	TL *24* 5745 (1983)

$$\left[\underset{\underset{Et}{+N}}{\overset{Me}{\underset{Me}{\bigvee}}}F \right] BF_4^-/RMgBr \qquad \text{CL 1257 (1977)}$$

$$RM + C=C-\overset{X}{\overset{|}{C}}-R' \longrightarrow R-C-C=C-R'$$

See also page 201, Section 13.

Review: Tetr *36* 1901 (1980)

RM	X	
$\overset{NNMe_2}{\overset{\|}{RCCHLiR}}$, cat CuI	O_2CR	CL 1521 (1982) (lactone)
PhMgBr, Ni catalyst	Cl, OR, $OSiR_3$, OH	CC 313 (1981)
PhMgBr, cat $PdCl_2L_2$	$OSiEt_3$	CC 313 (1981)
RMgX, cat CuCl	O_2CR	TL 633 (1979) (enyne synthesis) JOC *51* 2884, 2892 (1986)
RMgX, cat CuBr	OR	TL 3831 (1975) BSCF II 309 (1979) Tetr *35* 1517 (1979) (S_N2 or S_N2')
RMgX, cat CuBr·SMe$_2$	O_2CR	JOC *51* 1612 (1986)
RMgX, cat CuI	O_2CR	TL *23* 3583, 3587 (1982) CL 1521 (1982) (all lactones)
RMgX, cat CuCN	O_2CR	JOC *51* 2884, 2892 (1986)
RMgX, cat Li_2CuCl_4	Cl Br OAc	TL *23* 3115 (1982) Helv *65* 684 (1982) TL *23* 3115 (1982) CC 827 (1987)
RMgX, cat Cu(acac)$_2$	$PhSO_2$	TL 2393 (1979) Tetr *39* 3283, 3289 (1983)

RMgX(Li), CuBr·SMe$_2$	O$_2$CR	JOC *51* 1612 (1986) (lactone)
RMgX, CuX (X = Br, I)	O—(benzothiazol-2-yl)	Syn 885 (1979) JOMC *231* 179 (1982) JOC *46* 4482 (1982)
	S—(benzothiazol-2-yl)	CC 1085 (1978); 1252 (1986) (S$_N$2 or S$_N$2′)
RMgX (X = Cl, Br), CuBr, LiBr	OAc	Syn 469 (1982)
RCu	Br O$_2$CR	JOC *49* 1838 (1984) JOC *51* 1612 (1986) (lactone)
RC≡CCu	O—(benzothiazol-2-yl)	TL 3873 (1979)
RCu·BF$_3$	Cl, Br	JACS *99* 8068 (1977); *102* 2318 (1980) JOC *49* 1838 (1984)
	OH	JOMC *156* C9 (1978) JACS *102* 2318 (1980)
	O$_2$CR	JOC *51* 1612 (1986) (lactone)
RCu·2 *n*-Bu$_3$P	OR	TL *22* 1809 (1981)
R$_2$CuLi	Cl Br O$_2$CR	JACS *92* 737 (1970) JOC *52* 3394 (1987) CC 43 (1969) JACS *92* 735 (1970); *94* 5379 (1972); *95* 6832 (1973); *98* 7854 (1976); *101* 1035, 2493 (1979); *109* 8117 (1987) TL 4439 (1976); 1035 (1977); *23* 3093 (1982) JOMC *136* 103 (1977) Tetr *35* 1517 (1979) JCS Perkin I 2093 (1980); 1729 (1981) (both lactones) JOC *46* 2591, 5304 (1981); *48* 1531, 3986 (1983); *50* 5495 (1985) (relative rates); *51* 1612 (1986) (lactone); *52* 5452 (1987) Syn 469 (1982) CL 1521 (1982) (lactone)
	OCH$_2$OCH$_3$ OAr OCONHPh	TL *22* 119 (1981) JOC *45* 4026 (1980) JACS *101* 1035 (1979) TL *23* 3093 (1982) JOC *48* 715, 1404 (1983)

RM	X	
R$_2$CuLi (*continued*)	OTs	JACS *108* 7420 (1986)
	SOPh	CC 434 (1980)
	SO$_2$Ph	CC 434 (1980)
		JACS *109* 6396 (1987)
R$_2$CuLi·BF$_3$	OTs, OMs	JACS *108* 7420 (1986)
R$_2$CuLi·2 BF$_3$	OTs	JACS *108* 7420 (1986)
R$_2$CuLi, Me$_3$SiCl	OH	TL *28* 5521 (1987)
R$_2$CuMgX	O$_2$CR	TL 4439 (1976); *23* 3583 (1982) (lactone)
		CL 1521 (1982) (lactone)
		Can J Chem *61* 632 (1983)
(RCuCN)MgCl	Br, OAc	TL *27* 5095 (1986)
(RCuCH$_2$SOCH$_3$)Li	OAc	JOC *52* 1885 (1987)
[RCu(CN)CH$_2$SOCH$_3$]Li$_2$	OAc	JOC *52* 1885 (1987)
(RCuX)Li (X = CN, SPh)	OAc	JOMC *136* 103 (1977)
(RCuCN)Li	Br	TL *27* 5095 (1986)
	O$_2$CR	JOC *45* 4256 (1980) (lactone); *48* 1404, 3986 (1983); *49* 422 (1984); *51* 1612 (lactone), 4492 (1986)
		TL *27* 5095 (1986)
RCu(CN)Li·BF$_3$	OMs	CC 1596 (1987)
(R$_2$CuCN)Li$_2$	OMs	JACS *108* 7420 (1986)
(R$_2$CuCN)Li$_2$·BF$_3$	OTs, OMs	JACS *108* 7420 (1986)
		CC 1596 (1987)
(RCuNCy$_2$)Li	Cl	JACS *104* 5824 (1982)
(RCuPPh$_2$)Li	Cl	JACS *104* 5824 (1982)
Li$_2$Cu$_3$Me$_5$	OCONHPh	TL *23* 3093 (1982); *24* 4477 (1983)
		JOC *51* 1264 (1986); *52* 897 (1987)
(R$_3$BMe)Li, CuBr	Cl, Br	BCSJ *50* 2199 (1977)
RZnCl or R$_2$Zn, cat CuBr·SMe$_2$	Cl	JACS *109* 8056 (1987)
(RO$_2$CCHRCH$_2$)$_2$Zn, cat CuBr·SMe$_2$	Cl, Br	JACS *109* 8056 (1987)

RHgX, Li$_2$PdCl$_4$ Cl, OAc   JACS *90* 5531 (1968); *100* 287
  (R = aryl, vinylic)   (1978)
               JOC *43* 2870 (1978); *46* 1432
               (1981)
               JOMC *156* 45 (1978)

n = 1, 2; X = OAc, OSiMe$_3$; RM = RMgX-cat Cu(I), RCHLiCO$_2$R

TL *28* 5521 (1987)

X = O, NH

TL *23* 2575 (1982)

RLi (R = alkyl, aryl)   TL 4419 (1975); *22* 577 (1981)

R$_3$B, O$_2$ (R = alkyl)   JACS *93* 2792 (1971)

RCH=CHB, Ni or Pd catalyst   JOMC *233* C13 (1982)

RLi or RMgX, cat CuX   TL 2027 (1978); 2051 (1979)
  (X = Cl, Br, I; R = alkyl, allyl,   Syn 528 (1978)
  aryl, vinyl)   CL 185 (1980)
               JOC *46* 239 (1981); *50* 1607 (1985)
               JACS *105* 3360 (1983)

RMgX, CuI, SMe$_2$ JACS *105* 6515 (1983) (R = *i*-Pr$_3$SiC≡CCH$_2$);
               106 723, 6006 (1984)
               JOC *49* 1707 (1984)

RMgX, CuCN   JOC *50* 3988 (1985); *52* 1106 (1987)

R$_2$CuLi (R = alkyl, vinylic, aryl)   JACS *92* 4978, 4979 (1970); *93* 3046, 3047
               (1971); *103* 2114 (1981)
               Syn 528 (1978)
               TL 2027 (1978); 2051 (1979); *28* 4985 (1987)
               JOC *46* 122 (1981)

R$_2$CuMgBr (R = pyridyl)   J Chem Res (S) 294 (1983)

(RCuC≡CR')Li   TL 675 (1979)
  (R = alkyl, vinyl; R' = *n*-Pr, *n*-Bu)   JCS Perkin I 2084 (1980); 683 (1983)

(RCuCH$_2$SOCH$_3$)Li (R = alkyl)   JOC *52* 1885 (1987)

(RCuCN)Li (R = alkyl, aryl, vinyl)

JOC *44* 4467 (1979); *46* 4389, 5379 (1981); *52* 4898 (1987)
TL 675 (1979)
Syn 872 (1980)
JACS *103* 2907 (1981); *104* 3165 (1982)

$(R_2CuCN)Li_2$ (R = alkyl, aryl, vinyl)

JACS *104* 2305 (1982)

(RCuCN)MgX

JOC *50* 3988 (1985)

$(RO_2CCH_2CH_2)_2Zn$, cat $CuBr \cdot SMe_2$

JACS *109* 8056 (1987)

RHgCl (R = aryl, vinylic), Li_2PdCl_4

TL *27* 2211 (1986)

HCR′XY (X, Y = electron-withdrawing groups), cat $Pd(PPh_3)_4$ (R = CR′XY)

TL *22* 2575 (1981); *25* 1921 (1984); *27* 3881 (1986) (intramolecular)
JACS *103* 5969 (1981); *104* 6112 (1982) (intramolecular); *105* 147, 5940 (1983) (both intramolecular)
CC 985 (1983) (intramolecular)

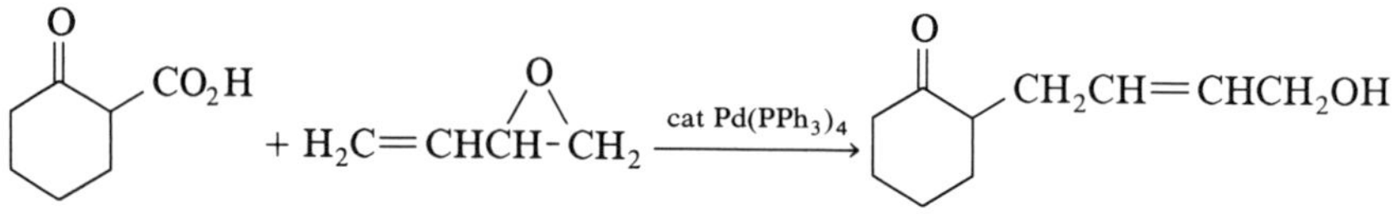

JOC *51* 5216 (1986)

$$H_2C=CHCH(OR')_2 \xrightarrow{RM} RCH_2CH=CHOR'$$

$$H_2C=CHC(OR')_3 \longrightarrow RCH_2CH=C(OR')_2$$

RM

RMgX, cat CuBr

TL 3833 (1975)
BSCF II 305 (1979)

E-$RR^1C=CHAlR^2_2$, cat $Pd(PPh_3)_4$, $(ZnCl_2)$
(R^1, R^2 = H and *i*-Bu; Me and Me)

JOC *50* 3406 (1985)

PhZnCl, (HMPA)

JOC *50* 3406 (1985)

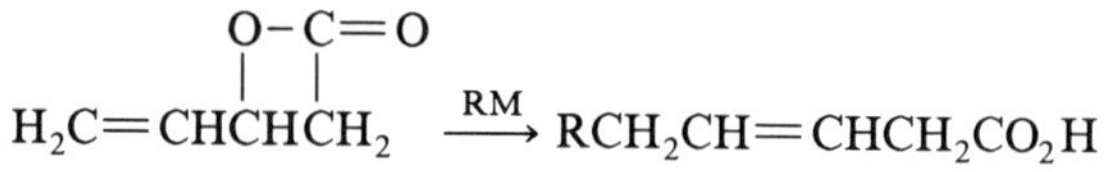

RM

$LiCR_2CS_2Me$

CL 1901 (1983)

organocopper reagents

TL *22* 1817 (1981)
CL 1307 (1981); 71, 219, 1521 (1982)

$$\underset{\substack{| \\ \text{RCHCH} = \text{CH}_2}}{\overset{\text{OH}}{}} \xrightarrow[\substack{2.\ n\text{-Bu}_3\text{SnCH}_2\text{I} \\ 3.\ n\text{-BuLi}}]{1.\ \text{KH}} \left[\underset{\substack{| \\ \text{RCHCH} = \text{CH}_2}}{\overset{\text{OCH}_2\text{Li}}{}} \right] \xrightarrow[]{2,3 \quad \text{H}_2\text{O}} \text{RCH} = \text{CHCH}_2\text{CH}_2\text{OH}$$

JACS *100* 1927 (1978); *108* 3841 (1986); *109* 3017, 6199 (1987)
TL *26* 5013, 5017 (1985); *28* 2099, 4993 (1987)
JOC *52* 2960 (1987)
See page 521, Section 5, for 2,3-Wittig rearrangement.

$$\text{RCH}_2\text{CH} = \text{CH}_2 + \text{R}'\overset{\overset{\text{O}}{\|}}{\text{CH}}(\text{R}'_2\text{CO}) \xrightarrow[\text{Lewis acid}]{\Delta,\ \text{acid or}} \underset{\substack{| \\ \text{RCH} = \text{CHCH}_2\text{CHR}'}}{\overset{\text{OH}}{}}$$

Reviews:

   Ann Chim *10* 25 (1965)
   Angew Int *8* 556 (1969)
   Syn 661 (1977)
   Acct Chem Res *13* 426 (1980)

Intermolecular:

   JCS 4111 (1953)
   JACS *79* 4972, 4976 (1957); *81* 133 (1959); *104* 555 (1982)
   Compt Rend C *263* 153 (1966)
   JOC *33* 1156 (1968); *37* 964 (1972) (R′ = CO$_2$R); *44* 3567 (1982) (R′ = CO$_2$R); *48* 464 (1983);
      50 3025 (1985) (chiral); *51* 4779 (1986) (R′ = CO$_2$R)
   CC 380, 382 (1977) (both R′ = CCl$_3$); 989 (1982) (R′ = CO$_2$R)
   TL 4867 (1979) (R′ = CCl$_3$); 1815 (1980); *28* 5755 (1987)
   Helv *64* 1682 (1981)
   Tetr *42* 2993 (1986) (R′ = CO$_2$R)

Intramolecular:

   Helv *50* 153 (1967)
   TL 1219 (1967); 3325 (1973); 3783 (1977); *23* 5111 (1982) (R′ = CO$_2$R); *26* 4167 (1985); *28*
      5945 (1987)
   JACS *89* 2748 (1967); *94* 4361 (1972); *103* 1835 (1981); *105* 7358 (1983); *106* 718 (1984); *107*
      2730 (1985); *109* 4424 (1987)
   JOC *35* 186, 858 (1970); *42* 1794 (1977); *44* 4014 (1979); *45* 4479 (1980); *47* 745, 4538 (1982);
      48 1822 (1983); *50* 4144 (1985); *52* 5419 (1987)
   CC 956 (1972)
   Syn Commun *8* 449 (1978)
   Tetr *42* 2951 (1986)

$$\text{RCH}_2\text{CH} = \text{CH}_2 \xrightarrow[\substack{\Delta\ \text{or} \\ \text{Lewis acid}}]{\text{OC(CO}_2\text{Et)}_2} \underset{\substack{| \\ \text{RCH} = \text{CHCH}_2\text{C(CO}_2\text{Et)}_2}}{\overset{\text{OH}}{}} \xrightarrow[\text{or Ce(IV)}]{\text{NaIO}_4} \text{RCH} = \text{CHCH}_2\text{CO}_2\text{H}$$

JACS *102* 2473 (1980); *106* 1092, 3797 (1984)
JOC *45* 1228 (1980); *47* 4201 (1982)

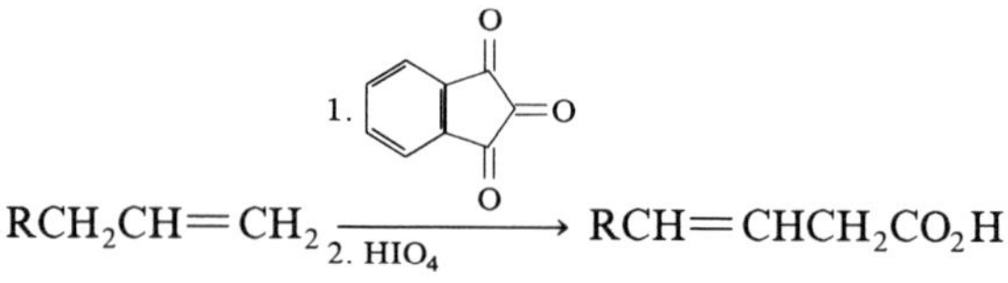

$$RCH_2CH{=}CH_2 \xrightarrow[\text{2. HIO}_4]{\text{1.}} RCH{=}CHCH_2CO_2H$$

TL *23* 1399 (1982)

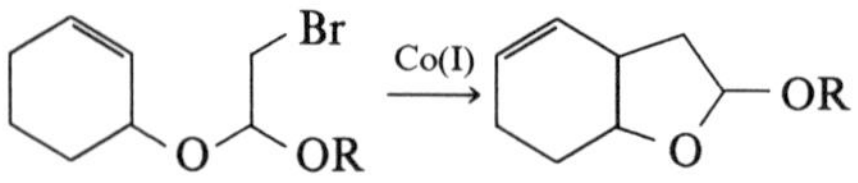

$$\text{(cyclohexenyl)} \xrightarrow{\text{Co(I)}} \text{(bicyclic)}{-}OR$$

TL *25* 4317 (1984)
JOC *50* 5875 (1985)

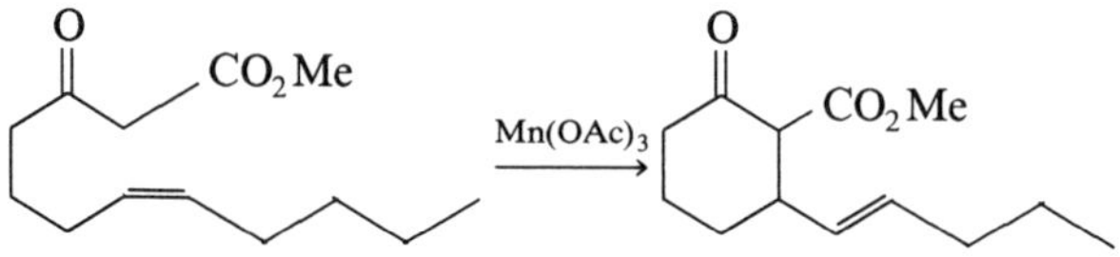

$$\xrightarrow{\text{Mn(OAc)}_3}$$

JOC *50* 3659 (1985)

$$RCH_2CH{=}CH_2 \xrightarrow[(CF_3CO)_2O]{CH_3SOCH_2CONR'_2} RCH{=}CHCH_2\overset{\overset{\textstyle SCH_3}{|}}{C}HCONR'_2$$

Syn 56 (1982)

$$CH_3CH{=}CHR \xrightarrow{TsN{=}CHCO_2R'} H_2C{=}CHCH\overset{\overset{\textstyle R}{|}}{C}H\overset{\overset{\textstyle NHTs}{|}}{C}HCO_2R'$$

JCS Perkin I 2680 (1981)
TL *23* 3015 (1982)

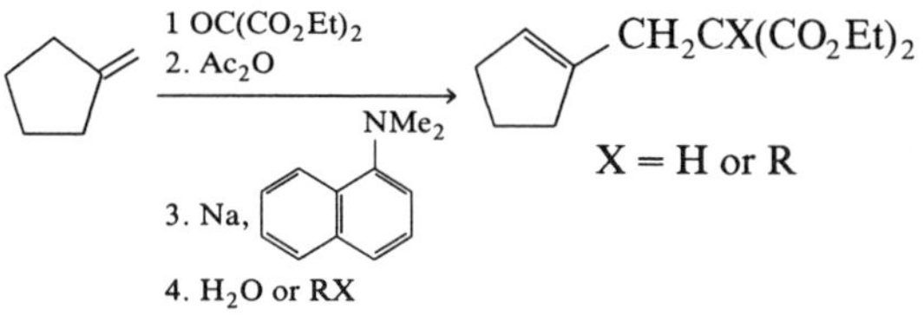

$$\text{(methylenecyclopentane)} \xrightarrow[\substack{\text{3. Na, (naphthyl-NMe}_2)\\ \text{4. H}_2\text{O or RX}}]{\substack{\text{1 OC(CO}_2\text{Et)}_2 \\ \text{2. Ac}_2\text{O}}} \text{(cyclopentenyl)}{-}CH_2CX(CO_2Et)_2$$

X = H or R

TL *22* 1885 (1981)

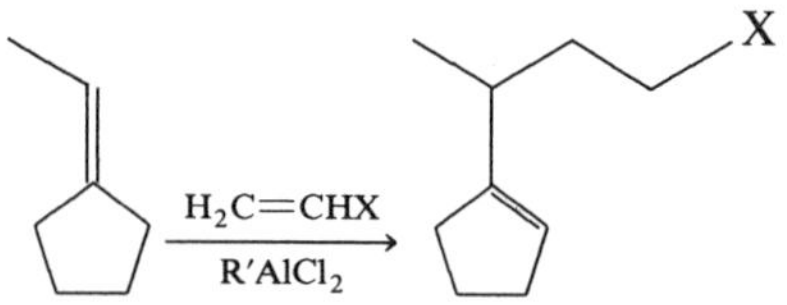

$$\text{(ethylidenecyclopentane)} \xrightarrow{\substack{H_2C{=}CHX \\ R'AlCl_2}} \text{(product with X)}$$

JOC *39* 255 (1974) (X = CHO, COMe, CO$_2$Me, CN); *47* 745 (1982) (X = CHO, COR)
Helv *64* 1682 (1981) (X = CO$_2$Me)
JACS *104* 1930 (1982) (α-substituted acrylate esters)

HC≡CCO₂Me over Et$_n$AlCl$_{3-n}$ ($n = 0, 1, 2$) giving product with CO₂Me

JOC *41* 3061 (1976); *45* 4773 (1980); *47* 3921 (1982)
JACS *101* 5283 (1979); *103* 237, 1293 (1981); *104* 2945 (1982)

See page 158, Section 45.4.

2. ELIMINATION

For epoxide → allylic alcohol see page 116, Section 5.

1. Dehydrogenation

Review:

P. N. Rylander, "Organic Synthesis with Noble Metal Catalysts," Academic Press (1973), Chpt 1

PdCl$_2$	JOC *36* 752 (1971) J Prakt Chem *314* 170 (1972) Syn 240 (1976); 773 (1977)
PdCl$_2$, Pd(OAc)$_2$	JACS *105* 2435 (1983) JOC *50* 3957 (1985)
(PhSeO)$_2$O	CC 130 (1978); 1044 (1981) JCS Perkin I 2209 (1980)

, HCl/O$_3$ or NaIO$_4$ TL *23* 2105 (1982)

See also page 149, Section 30.

Austral J Chem *33* 1537 (1980)

Ann 712 (1983)

X	Base	
Cl	LiCl, DMF, Δ	TL *28* 333 (1987)
	LiBr, Li$_2$CO$_3$, DMF, Δ	JACS *101* 4003 (1979)
		Syn Commun *11* 7 (1981)
Br	LiCl, DMF, Δ	TL *23* 3405 (1982)
	LiBr, Li$_2$CO$_3$, DMF, Δ	CL 3 (1973)
		JOC *44* 71 (1979); *52* 4792
		(1987)
		CC 1319 (1986)
		JACS *106* 3539 (1984)
	CaCO$_3$, CH$_3$CONMe$_2$, Δ	JCS 2532 (1961)
		JOC *35* 186 (1970); *51* 3059
		(1986)
	KOH	JACS *108* 4556 (1986)
	collidine	JACS *72* 362 (1950)
	DBU	JOC *52* 1962 (1987)
		TL *28* 503 (1987)

See also page 149, Section 30.

X	Reagent(s)	
SiMe$_3$	Pd(OAc)$_2$	JACS *107* 2474, 5495 (1985);
		108 3841 (1986); *109* 3017
		(1987)
		JOC *52* 3346, 4647 (1987)
	cat Pd(OAc)$_2$, benzoquinone	JOC *43* 1011 (1978); *51* 4323,
		5232 (1986); *52* 5588 (1987)
		JACS *108* 3443 (1986); *109*
		6199 (1987)
		TL *28* 585 (1987)
	cat Pd(OAc)$_2$,	CL 1133 (1984)
	H$_2$C=CHCH$_2$OCO$_2$Me,	Tetr *42* 2971 (1986)
	CH$_3$CN,	
	(cat Ph$_2$PCH$_2$CH$_2$PPh$_2$)	
	cat Pd(OAc)$_2$-	TL *24* 5635 (1983);
	Ph$_2$PCH$_2$CH$_2$PPh$_2$,	*28* 2397 (1987)
	(H$_2$C=CHCH$_2$O)$_2$CO,	Tetr *42* 2971 (1986)
	CH$_3$CN	
	PhSeCl/m-ClC$_6$H$_4$CO$_3$H	JACS *107* 268 (1985)
	DDQ	TL 3455 (1978); *28* 31, 4943
		(1987)
		Syn 736 (1979)
		JOC *50* 2981 (1985)

$^1O_2$

Ph$_3$CBF$_4$ Angew *16* 413 (1977)
JOC *42* 3961 (1977)

Ac   cat Pd(OAc)$_2$,
    H$_2$C=CHCH$_2$OCO$_2$Me,
    n-Bu$_3$SnOMe, CH$_3$CN,
    (cat Ph$_2$PCH$_2$CH$_2$PPh$_2$)
    TL *24* 5635 (1983)
    CL 1133 (1984)
    Tetr *42* 2971 (1986)

CO$_2$CH$_2$CH=CH$_2$ cat Pd(OAc)$_2$,
    (cat Ph$_2$PCH$_2$CH$_2$PPh$_2$)
    TL *24* 1797 (1983)
    CL 1133 (1984)
    Acct Chem Res *20* 140 (1987)
    (review)

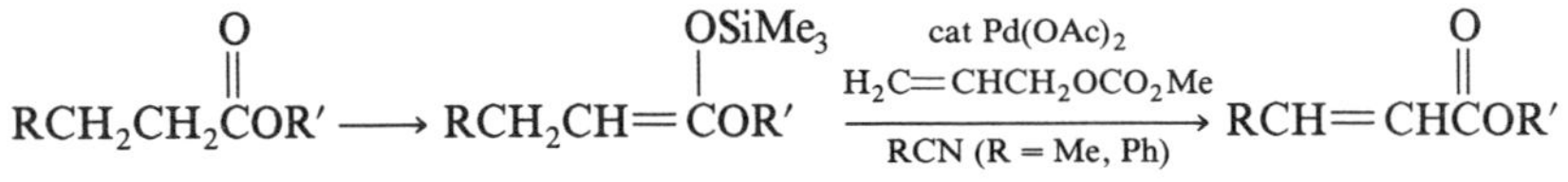

$$\underset{\text{O}}{\overset{\text{O}}{RCH_2CH_2\overset{\|}{C}OR'}} \longrightarrow RCH_2CH=\underset{OSiMe_3}{\overset{|}{C}OR'} \xrightarrow[\underset{RCN\ (R\ =\ Me,\ Ph)}{H_2C=CHCH_2OCO_2Me}]{cat\ Pd(OAc)_2} RCH=\overset{\text{O}}{\overset{\|}{C}}COR'$$

TL *25* 4783 (1984)
Tetr *42* 2971 (1986)

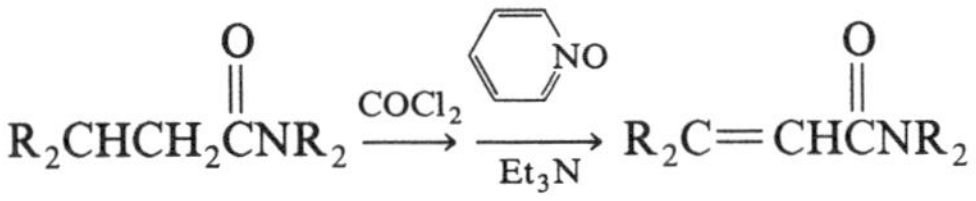

$$R_2CHCH_2\overset{\text{O}}{\overset{\|}{C}}NR_2 \xrightarrow[Et_3N]{\underset{COCl_2}{}} R_2C=CH\overset{\text{O}}{\overset{\|}{C}}NR_2$$

JACS *101* 4381 (1979)

$$\xrightarrow{(PhSeO)_2O}$$

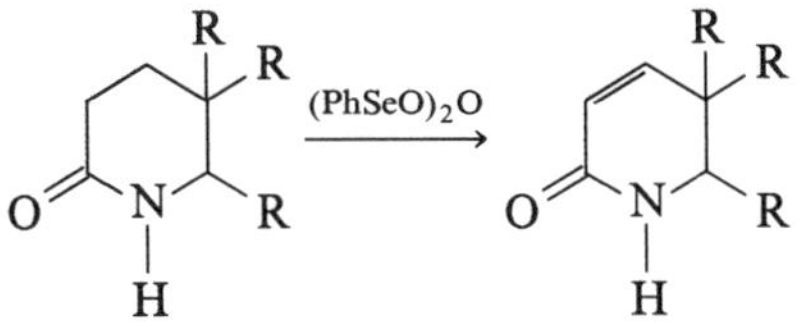

JOC *46* 1442 (1981)

2. Dehydrohalogenation of Alkyl Halides

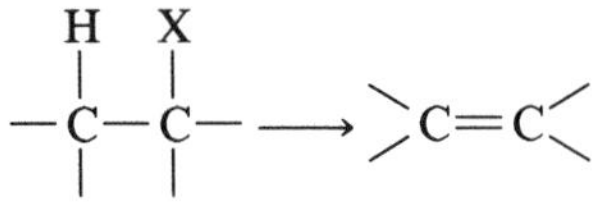

$$-\underset{|}{\overset{\overset{\text{H}}{|}}{C}}-\underset{|}{\overset{\overset{\text{X}}{|}}{C}}- \longrightarrow\ \diagup C=C\diagdown$$

Reviews:

    Chem Rev *45* 347 (1949); *80* 453 (1980)
    Houben-Weyl, "Methoden der Organischen Chemie," Vol V/1b, pp 9–44, 134–180
    Acct Chem Res *12* 198, 430 (1979)

DMF   JACS *86* 2309 (1964)

LiF, Li$_2$CO$_3$, HMPA   JOC *51* 3407 (1986)

KF-alumina   BCSJ *56* 1885 (1983)

Ag$_2$O   Helv *34* 1176 (1951)

NaOAc   Org Syn Coll Vol *3* 125 (1955)

NaOH, (*n*-Bu$_4$N)HSO$_4$ Syn 688 (1979)

NaOH, HO(CH$_2$CH$_2$O)$_n$H   JOC *47* 2493 (1982)

NaO-*i*-Pr, triglyme   Org Syn Coll Vol *5* 285 (1973)

KOH   Helv *60* 3060 (1977)

KOH, HO(CH$_2$CH$_2$O)$_n$H	JOC 47 2493 (1982)
KOR	JACS 75 4112 (1953); 78 2199, 2203 (1956); 88 1425 (1966) Helv 60 3060 (1977) Coll Czech Chem Commun 46 833, 850 (1981)
KO-t-Bu, DMSO	JOC 30 2054 (1964); 32 510 (1967); 52 5218 (1987) JOC USSR 15 853 (1979) JACS 108 468, 1251 (1986)
KO-t-Bu, 18-crown-6	Syn 372 (1979)

JOC 45 900 (1980)

chiral LiNRR'	TL 28 5517 (1987)
LiNEt$_2$	Organomet Chem Syn 1 375 (1972)

JOC 32 510 (1967)

Et$_3$N	Org Syn 45 22 (1965)
PhNH$_2$	J Gen Chem USSR 25 2017 (1955)
PhNMe$_2$	Helv 40 130 (1957)
pyridine	Org Syn Coll Vol 4 980 (1963)
collidine	Ann 585 132 (1954)
quinoline	Rec Trav Chim 69 535 (1950) Org Syn Coll Vol 4 608 (1963) Org Syn 51 115 (1971)
DBN	Ber 99 2012 (1966) Angew 79 53 (1967) TL 2543 (1977) JOC 47 4358 (1982); 52 5624 (1987)
DBU	Angew 79 53 (1967) Syn 591 (1972) TL 2543 (1977) JOC 47 1944, 4358 (1982); 52 5067 (1987) JCS Perkin I 2379 (1982) (polyene)
DBU, NiCl$_2$(PPh$_3$)$_2$-PPh$_3$-n-BuLi	CC 1621 (1986) (1° RX → terminal alkene, X = Br, I)

Ber 91 380 (1958)

Ph$_3$CLi	JCS Perkin I 1820 (1980)
Mg or Li/NiCl$_2$, H$_2$C=CH$_2$	Ber 113 171 (1980)

3. 1,1-Dihalides

$$\underset{\underset{Cl}{|}}{\overset{\overset{Cl}{|}}{R}}CCH_3 \xrightarrow{\text{NaOEt}} \overset{\overset{Cl}{|}}{R}C{=}CH_2$$

Org Mag Res *10* 192 (1977)

$$R_2CX_2 \longrightarrow R_2C{=}CR_2$$

Mg	JACS *108* 1265 (1986)
LiAlH$_4$, TiCl$_3$ or TiCl$_4$	Syn 607 (1976)
LiAlH$_4$, CrCl$_3$	TL 3829 (1977)
Ni	JOC *48* 4904 (1983)
cat Pd(PPh$_3$)$_4$, (ClMe$_2$Si)$_2$	CL 613 (1982)
Zn-Cu	Syn 652 (1970) JACS *106* 8174 (1984)
CuCl	Tetr *25* 3461 (1969)

4. 1,2-Dihalides

$$\underset{\underset{}{}}{\overset{\overset{X\ \ X}{|\ \ \ |}}{Ar}CRCHR} \xrightarrow{\text{silica gel}} \overset{\overset{X}{|}}{Ar}CR{=}CR$$

X = Cl, Br

JOC *52* 1145 (1987)

$$-\overset{\overset{X}{|}}{C}-\overset{\overset{Y}{|}}{C}- \longrightarrow \underset{}{\overset{}{>}}C{=}C\underset{}{\overset{}{<}}$$

Review: Can J Chem *42* 1294 (1964)

Reagent(s)	X	Y	Stereochemistry of Elimination	
electrolysis	Br	Br	anti	Coll Czech Chem Commun *28* 1664 (1963) JOC *39* 2408 (1974) Syn 964 (1979) TL *22* 623 (1981)
Li	Br	Br	anti	TL 4269 (1968)
Na, NH$_3$	Br	Br	nonstereospecific	JACS *74* 4590 (1952)
Na naphthalenide	Cl	Cl	?	JOC *52* 3595 (1987)
Na naphthalenide	Br	Br	anti	JOC *37* 507 (1972)
Mg	Br	Br	anti	JACS *74* 4590 (1952) TL 4269 (1968)

Reagent(s)	X	Y	Stereochemistry of Elimination	(Continued)
LiAlH$_4$	Br	Br	anti	JACS *71* 1675 (1949) Can J Chem *42* 1294 (1964)
LiAlH$_4$, TiCl$_3$ or TiCl$_4$	Cl	Cl	?	Syn 607 (1976)
LiAlH$_4$, TiCl$_3$ or TiCl$_4$	Br	Br	?	Syn 607 (1976)
LiAlH$_4$, CrCl$_3$	Br	Br	nonstereospecific	TL 3829 (1977)
SnCl$_2$	Br	Br	nonstereospecific	JACS *92* 4599 (1970)
SnCl$_2$, *i*-Bu$_2$AlH	Br	Br	?	CL 2069 (1984)
n-Bu$_3$SnH	Br	Br	nonstereospecific or anti	JACS *92* 2849 (1970)
CrCl$_2$	Br	Br	?	JACS *67* 1728 (1945)
CrSO$_4$	Cl	Cl	nonstereospecific	JACS *86* 4603 (1964)
CrSO$_4$	Cl	I	nonstereospecific	JACS *86* 4603 (1964)
CrSO$_4$	Br	Br	anti	JACS *86* 4603 (1964)
Cr(ClO$_4$)$_2$, (CH$_2$NH$_2$)$_2$	Cl	Br	nonstereospecific	JACS *90* 1582 (1968)
Cr(ClO$_4$)$_2$, (CH$_2$NH$_2$)$_2$	Br	Br	anti	JACS *90* 1582 (1968)
Fe graphite	Br	Br	anti	JOC *47* 876 (1982)
(Et$_4$N)HFe$_3$(CO)$_{11}$	Br	Br	?	JOMC *171* 85 (1979)
Co(CN)$_5^{3-}$	Br	Br	?	JACS *87* 5361 (1965)
Pd(PPh$_3$)$_4$	Br	Br	anti	BCSJ *52* 3629 (1979)
cat Pd(PPh$_3$)$_4$, (ClMe$_2$Si)$_2$	Cl	Cl	?	CL 613 (1982)
cat Pd(PPh$_3$)$_4$, (ClMe$_2$Si)$_2$	Br	Br	?	CL 613 (1982)
Cu	Br	Br	nonstereospecific	JOC *27* 4523 (1962)
CuCl	Br	Br	?	Tetr *25* 3461 (1969)
Zn	Br	Br	anti or nonstereospecific	JACS *63* 22 (1941); *74* 4590 (1952); *80* 182 (1958); *87* 838 (1965); *107* 516 (1987) Org Syn Coll Vol *3* 526 (1955) JOC *27* 4523 (1962); *50* 2356 (1985) TL 4269 (1968) BCSJ *52* 1752 (1979)

Zn, H$_2$NCSNH$_2$	Br	Br	?	Syn Commun *11* 901 (1981)
Zn, cat TiCl$_4$	Br	Br	anti	Syn 1025 (1982)
Zn-Cu	Br	Br	anti	JCS 3057 (1931) JACS *59* 403 (1937); *106* 5295 (1984)
NaI	Cl	Cl	nonstereospecific	JOC *27* 4523 (1962)
NaI	Cl	Br	syn	JOC *41* 3284 (1976)
NaI	Cl	Br	nonstereospecific	JOC *27* 4523 (1962)
NaI	Cl	I	nonstereospecific	JOC *27* 4523 (1962)
NaI	Br	Br	?	JACS *72* 362 (1950) JOC *30* 1658 (1965)
NaI	Br	Br	nonstereospecific	JOC *27* 4523 (1962) JACS *92* 4602 (1970)
NaI	Br	Br	syn	JOC *41* 3284 (1976)
NaI	Br	Br	anti	JACS *87* 838 (1965)
KI	Br	Br	anti	TL 4269 (1968)
Na$_2$S·9 H$_2$O	Cl	Cl	anti	Syn 879 (1981)
Na$_2$S·9 H$_2$O	Br	Br	anti	Syn 879 (1981)
NaSPh	Br	Br	?	J Prakt Chem *53* 1 (1896) Ber *30* 1799 (1897)
NaO$_2$SPh	Br	Br	?	J Prakt Chem *53* 1 (1896) Ber *30* 1799 (1897)
Na$_2$Se	Cl	Cl	?	JOC *31* 4292 (1966)
Na$_2$Se	Br	Br	?	JOC *31* 4292 (1966)
Na$_2$Se	I	I	?	JOC *31* 4292 (1966)
RSeNa (R = Me, Ph)	Cl	Br	syn	TL *21* 1877 (1980)
RSeNa (R = Me, Ph)	Cl	I	anti	TL *21* 1877 (1980)
RSeNa (R = Me, Ph)	Br	Br	anti	TL *21* 1877 (1980)
cat [thiophene]—TeNa, NaBH$_4$	Br	Br	anti	TL *23* 3601 (1982)
PPh$_3$	Br	Br	nonstereospecific	JOC *28* 1353, 1521 (1963)

5. β-Halo Ethers

$$\begin{array}{cc} RO & X \\ | & | \\ -C-C- \\ | & | \end{array} \longrightarrow \ \ >C=C<$$

X	Reagent(s)	Stereochemistry of Elimination	
Cl	Li, NH$_3$	?	JOC *43* 786 (1978); *45* 48 (1980)
			JACS *107* 3271, 3279 (1985)
	Li; 4,4'-*t*-BuC$_6$H$_4$C$_6$H$_4$-*t*-Bu	?	JACS *107* 3285 (1985)
	Na	?	JCS 1707 (1950)
			Ann 536, 1478 (1982)
	Na, EtNH$_2$	?	JACS *107* 516 (1985)
	Cr(ClO$_4$)$_2$	?	JACS *109* 4752 (1987)
Cl, Br	Na or K	?	JCS Perkin I 595 (1978)
	Na$^+$ C$_{10}$H$_8^-$	?	JCS Perkin I 595 (1978)
Br	Li	?	JOC *9* 310 (1944)
	Na	nonstereospecific	Ber *43* 2175 (1910)
			JACS *72* 2120 (1950); *80* 182 (1958); *104* 1116 (1982)
			JCS 1707 (1950)
			CC 552 (1977)
	Na(Hg)	?	JOC *50* 5465 (1985)
	Mg	?	JACS *52* 651 (1930), *56* 126 (1934)
	Zn	nonstereospecific	JACS *52* 3396 (1930); *53* 1505, 2427 (1931); *54* 751 (1932); *55* 3293 (1933); *80* 182 (1958)
			JOC *17* 807 (1952); *52* 1803 (1987)
			Org Syn Coll Vol *4* 748 (1963)
			CC 1462 (1987)
			TL *28* 3839 (1987)
	Zn(Ag)	?	BCSJ *53* 3383 (1980)
			JACS *104* 1114 (1982); *105* 1058 (1983)
			JOC *50* 3224 (1985); *52* 598, 603 (1987)
			CC 1642 (1986)
	n-BuLi	?	JACS *107* 3271 (1985)
I	Zn	?	TL *28* 6497 (1987)
	t-BuLi	?	JACS *109* 4390 (1987)
	Me$_2$BBr, *n*-Bu$_4$NI	?	TL *28* 5985 (1987)

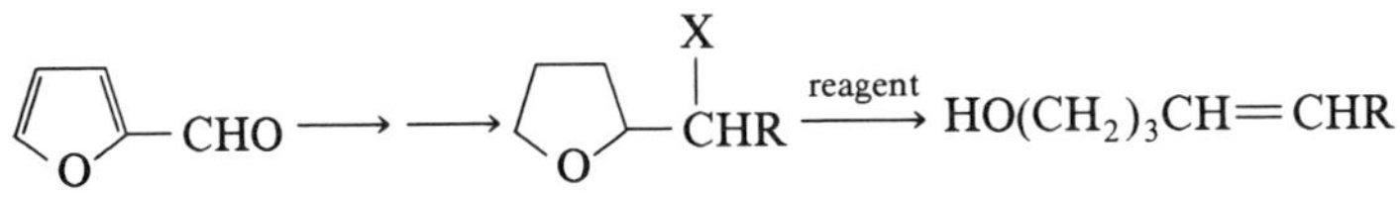

$$X = Cl, Br, I$$

Reagent

Na	BSCF *2* 745 (1935); *10* 484 (1943) Org Syn *25* 84 (1945) JCS 1707 (1950)
Mg	JCS 195 (1936)
Me$_2$BBr, *n*-Bu$_4$NI	TL *28* 5985 (1987)

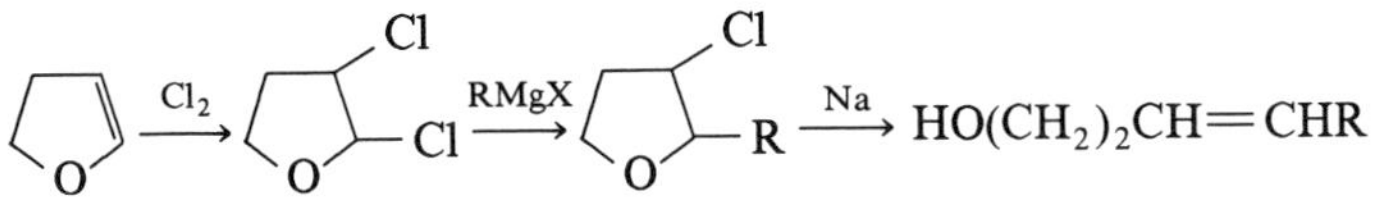

JCS 1152, 1714 (1950)

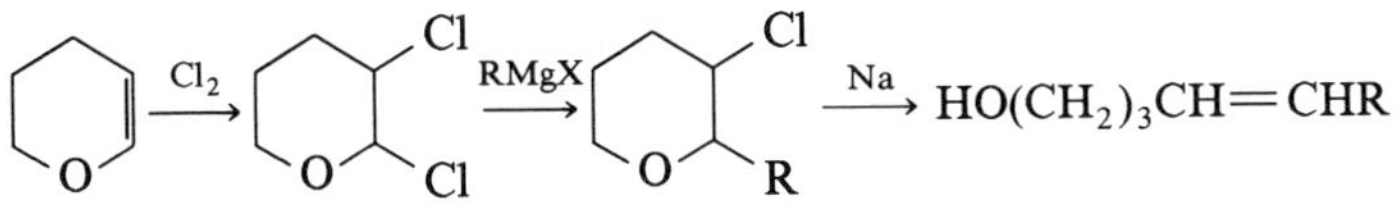

JCS 1707 (1950)
Ann 536, 1478 (1982)

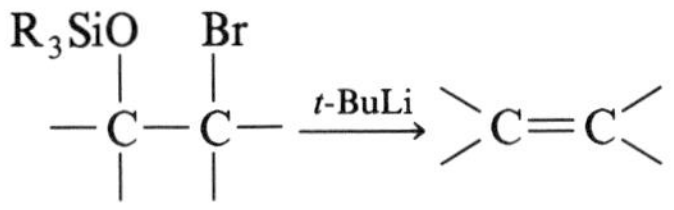

Ber *115* 1990 (1982)

6. Halohydrins

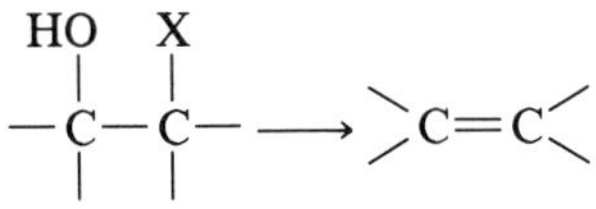

X	Reagent(s)	Stereochemistry of Elimination	
Cl	Na, NH$_3$	?	JACS *105* 7358 (1983)
	n-BuLi/Li$^+$ C$_{10}$H$_8^-$	?	CC 1153 (1982)
	electrolysis	?	Angew Int *16* 57 (1977)
	Cr(ClO$_4$)$_2$, EDA	?	JACS *105* 2435 (1983)

$$-\underset{|}{\overset{HO}{\underset{|}{C}}}-\underset{|}{\overset{X}{\underset{|}{C}}}- \longrightarrow \;\;\overset{}{\underset{}{>}}C{=}C\overset{}{\underset{}{<}} \quad (\textit{Continued})$$

X	Reagent(s)	Stereochemistry of Elimination	
Br	Zn	?	JCS 1370 (1955)
	$Cr(ClO_4)_2$, EDA	nonstereospecific	JACS 90 1582 (1968)
	$TiCl_3$, $LiAlH_4$	nonstereospecific	JOC 43 3249 (1978)
	$(CF_3SO_2)_2O$, py/HMPA	anti	JACS 102 1433 (1980)
	$KSCN/K_2CO_3/MeI$	cis	TL 2709 (1975)
	$KSeCN/K_2CO_3$	cis	TL 2709 (1975)
I	electrolysis	?	JOC 44 1404 (1979)
	$SnCl_2$, $POCl_3$, py	anti	JCS 112, 2539 (1959)

$$R^1R^2\underset{|}{\overset{Cl}{C}}-\overset{O}{\overset{\|}{C}}X \quad X = Cl, OEt$$

1. R^3MgX / 2. Li $\longrightarrow R^1R^2C{=}CR^3_2$

1. $LiAlH_4$ or $LiAlH_4\text{-}AlCl_3$ / 2. Li $\longrightarrow R^1R^2C{=}CH_2$

$$R^1R^2\underset{|}{\overset{Cl}{C}}-\overset{O}{\overset{\|}{C}}R^3 \xrightarrow[\text{2. Li}]{\text{1. } LiAlH_4, AlCl_3} R^1R^2C{=}CHR^3$$

JOC 46 2721 (1981)

$$\text{(cyclohexanone-2-Cl)} \xrightarrow[\text{2. } Li^+ C_{10}H_8^-]{\text{1. RLi}} \text{(1-R-cyclohexene)}$$

CC 1153 (1982)

7. β-Halo Esters

$$-\underset{|}{\overset{X}{C}}-\underset{|}{\overset{O_2CR}{C}}- \longrightarrow \;\;>C{=}C<$$

X	Reagent(s)	Stereochemistry of Elimination	
Cl	electrolysis	anti	Acta Chem Scand 17 2139 (1963)

Cl, Br	NaI	syn	JOC 44 1404 (1979)
	Zn	anti or nonstereospecific	JACS 80 183 (1958) JOC 44 1404 (1979) JCS Perkin I 1523, 1535 (1980)
Cl, Br, I	electrolysis	?	JACS 94 5139 (1972)
Br	electrolysis	?	JOC 45 154 (1980)
	Zn/Ag-graphite	?	CC 1149 (1986)
	$Cr(ClO_4)_2$, $(CH_2NH_2)_2$	nonstereospecific	JACS 90 1582 (1968)
I	Zn	?	TL 27 3297 (1986) JOC 52 1051 (1987)

8. β-Halo Mesylates

$$-\overset{\overset{\displaystyle Cl}{|}}{\underset{|}{C}}-\overset{\overset{\displaystyle OMs}{|}}{\underset{|}{C}}-\ \xrightarrow[NH_3]{Na}\ \diagup\!\!\!\!\diagdown C\!=\!C\diagdown\!\!\!\!\diagup$$

JOC 50 3957 (1985); 52 4772 (1987)

$$R-\overset{\overset{\displaystyle Cl}{|}}{\underset{\underset{\displaystyle Cl}{|}}{C}}-\overset{\overset{\displaystyle OMs}{|}}{\underset{\underset{\displaystyle H}{|}}{C}}-R\ \xrightarrow[NH_3]{Na}\ RCH\!=\!CHR$$

JOC 50 3957 (1985); 52 4772 (1987)

9. Amine Oxides (Cope Elimination)

$$-\overset{\overset{\displaystyle H}{|}}{\underset{|}{C}}-\overset{\overset{\displaystyle \overset{\displaystyle O}{\uparrow}}{\underset{\displaystyle NR_2}{}}}{\underset{|}{C}}-\ \xrightarrow{\Delta}\ \diagup\!\!\!\!\diagdown C\!=\!C\diagdown\!\!\!\!\diagup$$

JACS 81 2799 (1959); 108 1039 (1986)
Org Rxs 11 361 (1960) (review)
Org Syn Coll Vol 4 612 (1963)

10. Quaternary Ammonium Salts

$$\left[-\overset{\overset{\displaystyle H}{|}}{\underset{|}{C}}-\overset{\overset{\displaystyle \overset{+}{N}Me_3}{|}}{\underset{|}{C}}-\right]OH^-\ \xrightarrow{\Delta}\ \diagup\!\!\!\!\diagdown C\!=\!C\diagdown\!\!\!\!\diagup$$

Org Rxs 11 317 (1960) (Hofmann elimination, review)
Org Syn Coll Vol 5 608 (1973)
Org Syn 55 3 (1976)

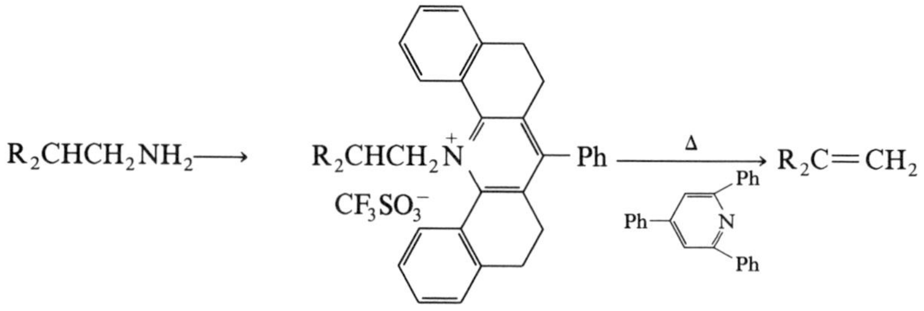

$$R_2CHCH_2NH_2 \longrightarrow$$

$$R_2CHCH_2\overset{+}{N} \quad \text{—Ph} \quad \xrightarrow{\Delta} \quad R_2C{=}CH_2$$

$$CF_3SO_3^-$$

CC 96 (1981)
JOC *47* 3506 (1982)
JCS Perkin I 2347 (1982)

11. *N*-Alkyl-*N*,*N*-Disulfonimides

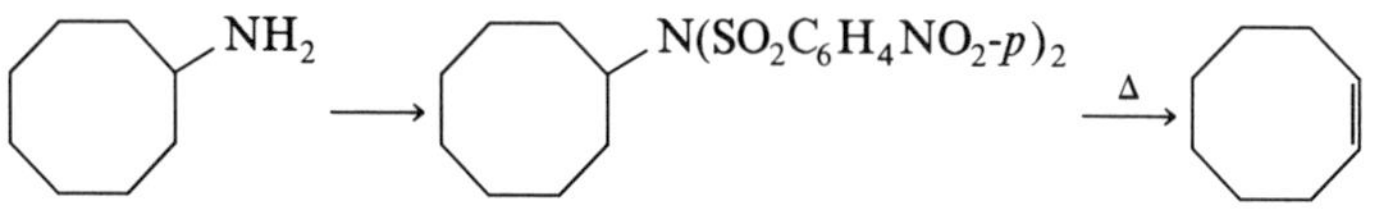

NH$_2$ →   N(SO$_2$C$_6$H$_4$NO$_2$-*p*)$_2$ $\xrightarrow{\Delta}$

TL *22* 199 (1981)

12. Diazo Compounds

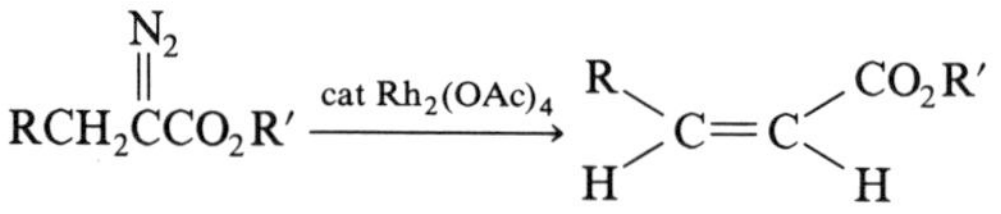

$$\overset{\overset{N_2}{\|}}{RCH_2CCO_2R'} \xrightarrow{\text{cat Rh}_2(\text{OAc})_4} \overset{R}{\underset{H}{}}{>}C{=}C{<}\overset{CO_2R'}{\underset{H}{}}$$

TL *22* 4163 (1981)

$$2\ ArCHN_2 \xrightarrow{\text{Rh catalyst}} ArCH{=}CHAr$$

$$Z > E$$

TL *23* 2277 (1982)

13. Deoxygenation of Epoxides

For ring opening of epoxides to allylic alcohols see page 116, Section 5.

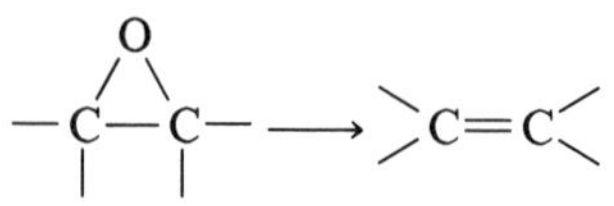

$$-\overset{}{\underset{|}{C}}-\overset{O}{\overset{/\backslash}{\underset{|}{C}}}- \longrightarrow\ >C{=}C<$$

Review: Tetr *36* 575 (1980)

13.1. Inversion

Ph$_2$PLi/HOAc, H$_2$O$_2$/NaH   CC 142 (1974)
                                          JACS *109* 1248 (1987)

Ph$_2$PLi/MeI

JACS *93* 4070 (1971); *95* 822 (1973)
JOC *38* 1178 (1973)

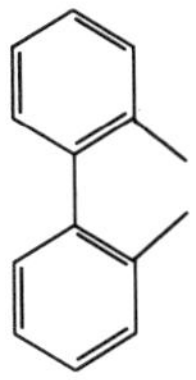

PLi/H$^+$/EtI/NaNH$_2$/Δ

TL *28* 3445 (1987)

Me$_3$SiSiMe$_3$, KOCH$_3$

JACS *98* 1265 (1976)
JOC *41* 3063 (1976)
CC 168 (1980)

PhSiMe$_2$Li

Syn 199 (1976)

CpFe(CO)$_2$Na

JACS *94* 7170 (1972)
TL 4009 (1975)

NaBH$_2$S$_3$/LiAlH$_4$/PhCHO, H$^+$/LiNR$_2$

JCS Perkin I 433 (1974)

13.2. Retention

Li

TL *21* 1173 (1980)

WCl$_6$, *n*-BuLi

Org Syn *60* 29 (1981)
CC 1084, 1642 (1986)
JOC *52*, 598 (1987)

WCl$_6$; Li, LiI or *n*-BuLi

JACS *94* 6538 (1972)

N$_2$C(CO$_2$Me)$_2$, cat Rh$_2$(OAc)$_4$

TL *25* 251 (1984)

N$_2$C(CO$_2$Me)$_2$, cat [(*n*-C$_7$H$_{15}$CO$_2$)$_2$Rh]$_2$

JOC *51* 5503 (1986)

NaI, 18-crown-6, BF$_3$·OEt$_2$

JOC *43* 2076 (1978)

Me$_3$SiI

Nouv J Chim *3* 705 (1979)

Me$_3$SiCl, NaI

TL *22* 3551 (1981)

(CF$_3$CO)$_2$O, NaI

JOC *43* 1841 (1978)

PI$_3$, (py or Et$_3$N)

Nouv J Chim *3* 705 (1979)

P$_2$I$_4$, (py or Et$_3$N)

Syn 905 (1978)
Nouv J Chim *3* 705 (1979)

Ph$_3$P·HI, Ph$_3$PI$_2$

Syn 828 (1980)

[(PhO)$_3$PCH$_3$]I, BF$_3$·OEt$_2$

JOC *43* 2076 (1978)

KS$_2$CO-*n*-Bu

Chem Ind 460 (1964)

Ph$_3$PS, H$^+$/Ph$_3$P

JACS *94* 2880 (1972)

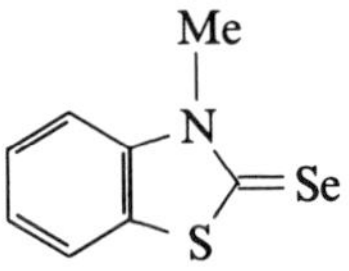

=Se, CF$_3$CO$_2$H

Syn 200 (1976)

Ph_3PSe, CF_3CO_2H	CC 253 (1973) TL 2091 (1974)
KSeCN	JCS Perkin I 1216 (1975) JOC *51* 2712 (1986) TL *27* 4813 (1986)
RSeH (R = Me, Ph)/$(CF_3CO)_2O$ or H^+	TL 1385 (1976)
$(EtO)_2PONa(Li)$, cat Te	CC 658 (1977) JOC *45* 2347 (1980)

13.3. Non-stereospecific or Unknown Stereochemistry

R_3P (R = *n*-Bu, Ph)	Chem Ind 330 (1959) JACS *87* 2683 (1965)
$(EtO)_3P$	JOC *22* 1118 (1957)
KSCN or H_2NCSNH_2, $H^+/(EtO)_3P$	JOC *26* 3467 (1961)
H_2NCSNH_2, $NaHCO_3/PPh_3$	CC 1434 (1987)
NaTeH/TsCl, py	TL *26* 6197 (1985)
Ph_3PBr_2/Zn	JOC *41* 3279 (1976)
Zn, HOAc	CC 1450 (1970)
Zn, NaI, HOAc, NaOAc	JCS 112, 2539 (1959)
Zn-Cu, EtOH	JOC *36* 1187 (1971)
Mg(Hg), $MgBr_2$	CC 144 (1970)
$Fe(CO)_5$, $Me_2NCONMe_2$	TL 4155 (1977)
$FeCl_3$, *n*-BuLi	CL 883 (1974)
$Co_2(CO)_8$	CC 384 (1974)
Cr	JOC *41* 3647 (1976)
$Cr(ClO_4)_2$, $(CH_2NH_2)_2$	Tetr *24* 3503 (1968)
$TiCl_3$, $LiAlH_4$	JOC *40* 2555 (1975); *43* 3249 (1978)
$TiCl_4$, Zn	JOC *43* 3249 (1978)
WCl_6, $LiAlH_4$	JOC *43* 2477 (1978)
$NbCl_5$, $NaAlH_4$	CL 157 (1982)
Cp_2MX_2, Na(Hg) [M = Zr, Ti $\gg$ Mo > W; $X_2 = Cl_2$, O]	CC 99 (1978)
SmI_2	JACS *102* 2693 (1980)
YbI_2	JACS *102* 2693 (1980)
MeMnCl or *n*-Bu_3MnLi	TL *25* 294 (1984)
Me_3SiLi	JACS *108* 2090 (1986) (probably retention)
Et_3SiH, 300°C	JACS *109* 7534 (1987)

14. β-Mesyl or Thio Esters

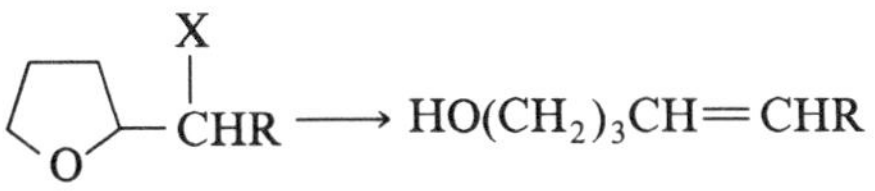

$$\text{(tetrahydrofuranyl)} \overset{X}{\underset{}{\underset{}{CHR}}} \longrightarrow HO(CH_2)_3CH{=}CHR$$

X	Reagent(s)	
MsO	Na or Li, NH$_3$	Syn Commun *12* 915 (1982)
PhS	Li$^+$ C$_{10}$H$_8^-$	Syn Commun *12* 915 (1982)

15. Sulfides

$$RCH_2X \xrightarrow{LiCH_2SCH_3} RCH_2CH_2SCH_3 \xrightarrow[\Delta]{n\text{-BuLi}} RCH{=}CH_2$$

JACS *101* 3283 (1979)

16. α-Halosulfides

$$R_2CHSH + XCHYZ \longrightarrow R_2CHSCHYZ \xrightarrow[2.\ \Delta]{1.\ SO_2Cl_2\ or\ Br_2} R_2C{=}CYZ$$

X = Cl, Br; Y = H, Cl, Br; Z = CO$_2$R, COR, CN

Angew Int *20* 585 (1981)

17. β-Halosulfides

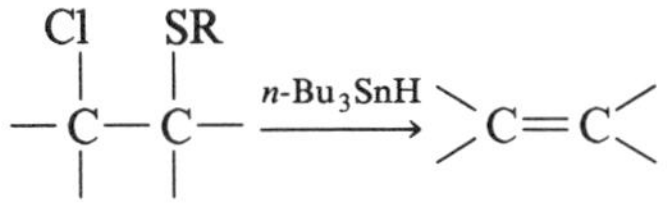

$$\underset{\overset{|}{}}{\overset{Cl}{\underset{|}{C}}}\ \underset{}{\overset{SR}{\underset{|}{C}}} \xrightarrow{n\text{-Bu}_3SnH} {>}C{=}C{<}$$

TL 4223 (1977)

18. β-Hydroxysulfides

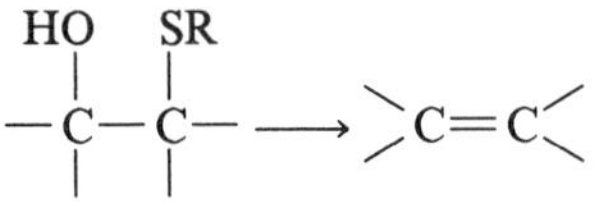

$$\underset{\overset{|}{}}{\overset{HO}{\underset{|}{C}}}\ \underset{}{\overset{SR}{\underset{|}{C}}} \longrightarrow {>}C{=}C{<}$$

n-BuLi/(PhCO)$_2$O/Li-NH$_3$ or   JACS *94* 4758 (1972)
Na$^+$ C$_{10}$H$_8^-$ TL *27* 1343 (1986)

CH$_3$Li/(catechol cyclic phosphate) PCl/Δ   TL 737 (1972)

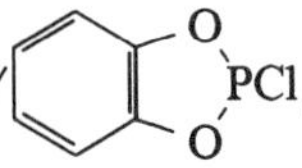

base/CS_2/CH_3I/n-Bu_3SnH TL 4223 (1977)

electrolysis   TL 2807 (1978)

PI_3, P_2I_4 or $SOCl_2$ TL 4111 (1979)

$$\left[\underset{\overset{+}{N}Et}{\bigcirc}\!\!-F \right] BF_4^{\,-},\ Et_3N/LiI$$

     CL 413 (1978)
     JOC *45* 3549 (1980)

$TiCl_4$/Zn   CL 37, 1161, 1523 (1974)

$TiCl_4$, $LiAlH_4$, R_3N CL 871 (1975)

19. Sulfoxides

$$-\underset{|}{\overset{|}{C}}\,\text{H}-\underset{|}{\overset{SOR}{C}}- \xrightarrow{\Delta} \;\text{>}C=C\text{<}$$

JOC *52* 1471 (1987)
TL *28* 221 (3,4-dihydrofuran), 3901 (vinyl fluorides), 4959, 5509 (α,β-unsaturated oxazolines) (1987)

$$-\underset{|}{\overset{H|}{C}}-\underset{|}{\overset{H|}{C}}-\overset{\overset{O}{\|}}{C}- \xrightarrow[\substack{\text{1. base} \\ \text{2. RSCl or RSSR} \\ \text{3. [O]}}]{} -\underset{|}{\overset{H|}{C}}-\underset{|}{\overset{SOR}{C}}-CO- \xrightarrow{\Delta} \;\text{>}C=C-\overset{\overset{O}{\|}}{C}-$$

JACS *95* 6840 (1973); *98* 4887 (1976); *106* 721, 4038, 6414 (1984)
TL 5113 (1973); 1097 (1974); 4197 (1975)
Chem Rev *78* 363 (1978)
Acct Chem Res *11* 453 (1978)
JOC *44* 71 (1979); *50* 2764 (1985); *52* 1218 (1987) (on tosylhydrazone)
Syn 56 (1982)

$$\xrightarrow[\text{2. RX}]{\text{1. NaH}} \qquad \xrightarrow[\text{2. }\Delta]{\text{1. NaIO}_4}$$

JACS *108* 3385 (1986)

$$H-\underset{|}{\overset{H|}{C}}-\overset{\overset{O}{\|}}{C}- \longrightarrow RCH-\underset{|}{\overset{\overset{R'S\ \ H}{|\ \ |}}{C}}-\overset{\overset{O}{\|}}{C}- \xrightarrow[\text{2. }\Delta]{\text{1. [O]}} RCH=\underset{}{\overset{}{C}}-\overset{\overset{O}{\|}}{C}-$$

TL 993, 995, 2179 (1979)
Syn 1003 (1981)
Syn Commun *11* 315 (1981)

$$\begin{array}{c} \text{RSOCHLiR} \xrightarrow{\text{R'CH}_2\text{X}} \\ \\ \text{RSCHLiR} \xrightarrow[\text{2. [O]}]{\text{1. R'CH}_2\text{X}} \end{array} \Bigg\rangle \text{RSOCHRCH}_2\text{R'} \xrightarrow{\Delta} \text{RCH}=\text{CHR'}$$

JACS *96* 7165 (1974)
JOC *40* 2014 (1975); *47* 4801 (1982) (diene)
Syn Commun *9* 317 (1979) (diene)

20. β-Hydroxysulfoxides

$$\underset{\begin{array}{c}|\\ \text{HO} \quad \text{SOR} \\ | \quad\;\; | \\ -\text{C}-\text{C}- \\ | \quad\;\; | \end{array}}{} \longrightarrow \quad \Big\rangle\text{C}=\text{C}\Big\langle$$

(catechol) PCl structure with O, O, PCl   TL 649 (1972)

NBS, NCS or SO_2Cl_2 JACS *95* 3420 (1973)

21. Sulfones

$$\begin{array}{c} \text{RSO}_2 \quad\;\; \text{O} \\ | \qquad\quad \| \\ \text{RCHCH}_2\text{CR} \end{array} \longrightarrow \text{RCH}=\overset{\overset{\displaystyle\text{O}}{\|}}{\text{CHCR}}$$

basic Al_2O_3 TL *27* 3733 (1986)

DBU   CC 1226 (1987)

$$\begin{array}{c}\text{SO}_2\text{Tol}\\ |\\ \text{O}\text{-}\text{C}\text{-}\text{O}\,\text{CH}_2-\text{C}-\\ |\\ \text{H}\end{array} \xrightarrow[\text{2. RX}]{\text{1. LDA}} \begin{array}{c}\text{SO}_2\text{Tol}\\ |\\ \text{O}\text{-}\text{C}\text{-}\text{O}\,\text{CH}_2-\text{C}-\\ |\\ \text{R}\end{array} \xrightarrow[\text{2. Na}_2\text{CO}_3]{\text{1. H}_3\text{O}^+} \begin{array}{c}\text{O} \quad\;\; \text{R}\\ \| \qquad\;\; |\\ -\text{C}-\text{CH}=\text{C}-\end{array}$$

JOC *42* 1349 (1977)
CL 165 (1982)

$$\underset{\text{R}_2\text{CH}}{\overset{\text{SO}_2\text{Ph}}{|}} \xrightarrow[\text{2. BrCH}_2\text{CO}_2\text{Et}]{\text{1. }n\text{-BuLi}} \underset{\text{R}_2\text{CCH}_2\text{CO}_2\text{Et}}{\overset{\text{SO}_2\text{Ph}}{|}} \xrightarrow{\text{Na}_2\text{CO}_3} \text{R}_2\text{C}=\text{CHCO}_2\text{Et}$$

BSCF 525 (1976)

$$\text{RCH}_2\text{SO}_2\text{Ar} \xrightarrow[\text{2. BrCH}_2\text{CR}=\text{CRCO}_2\text{R}]{\text{1. base}} \underset{\text{RCHCH}_2\text{CR}=\text{CRCO}_2\text{R}}{\overset{\text{SO}_2\text{Ar}}{|}} \xrightarrow[\text{or LDA}]{\text{KO-}t\text{-Bu}} \text{RCH}=\text{CHCR}=\text{CRCO}_2\text{R}$$

BSCF 746 (1973)
TL *27* 3337 (1986)

$$\text{[cyclohexane-fused ring with CH}_3\text{, SO}_2\text{, CH}_3] \xrightarrow[\text{2. LiAlH}_4]{\text{1. } n\text{-BuLi}} \text{[bicyclic product with CH}_3\text{, CH}_3]$$

Org Syn *57* 53 (1977)
JACS *109* 3730 (1987)

22. α-Halosulfones (Ramberg-Bäcklund Reaction)

$$RCH_2\overset{\overset{\displaystyle O}{\|}}{\underset{\underset{\displaystyle O}{\|}}{S}}CHXR \xrightarrow{\text{base}} RCH{=}CHR$$

X = halogen

F. G. Bordwell in "Organosulfur Chemistry," Ed. M. J. Janssen, Interscience, New York (1967), Chpt 16 (review)
TL 4645 (1967)
Acct Chem Res *1* 209 (1968) (review)
JACS *91* 3870 (1969); *92* 2581 (1970); *96* 3332 (1972); *109* 2857 (1987)
Tetr *30* 3177 (1974)
Org Rxs *25* 1 (1977) (review)
Ber *114* 909 (1981)
Syn 504 (1982) (phase transfer)
Ann 98 (1983)
JOC *51* 2397 (1986); *52* 1703 (1987)

23. β-Hydroxysulfones

$$\overset{\overset{\displaystyle O}{\|}}{R C R} + R_2\overset{\overset{\displaystyle O}{\|}}{\underset{\underset{\displaystyle O}{\|}}{C}}SR \xrightarrow{H_2O} R{-}\overset{\overset{\displaystyle HO}{|}}{\underset{\underset{\displaystyle R}{|}}{C}}{-}\overset{\overset{\displaystyle SO_2R}{|}}{\underset{\underset{\displaystyle R}{|}}{C}}{-}R \longrightarrow \overset{R}{\underset{R}{>}}C{=}C\overset{R}{\underset{R}{<}}$$

electrolysis	CL 69 (1978)
Na(Hg), MeOH	JOC *42* 2036 (1977)
Na(Hg), MeOH, Na$_2$HPO$_4$ or KH$_2$PO$_4$	JOC *50* 5465 (1985); *52* 3759 (1987) TL *27* 2095 (1986) JACS *108* 2776 (1986)
n-BuLi/PhCSCl/*n*-Bu$_3$SnH	TL 4223 (1977)
MsCl or TsCl/Na(Hg), MeOH	TL 4833 (1973); *23* 1963 (1982)
SOCl$_2$, py/Na(Hg), MeOH	JACS *108* 2776 (1986)

$$\underset{RCHCH=CH_2}{\overset{SO_2Ar}{|}} \xrightarrow[\text{2. }(CH_2O)_n]{\text{1. }n\text{-BuLi}} \underset{\underset{CH_2OH}{|}}{\overset{SO_2Ar}{|}}RCCH=CH_2 \xrightarrow[\text{2. }\Delta]{\text{1. }n\text{-Bu}_3SnH} \overset{R}{\underset{}{H_2C=CCH=CH_2}}$$

TL *22* 2675 (1981)

24. β-Acyloxysulfones

$$\underset{-C-C-}{\overset{RCO_2 \quad SO_2R}{| \quad |}} \longrightarrow \underset{}{>}C=C\underset{}{<}$$

Na, EtOH

Na(Hg), MeOH, (Na$_2$HPO$_4$)

JOC *50* 2948 (1985)

TL 4833 (1973); 4419 (1979); *27* 3903, 6345
   (1986); *28* 5205, 5759, 5763 (1987)
JCS Perkin I 829, 834 (1978); 1045, 1400 (1980)
JOC *49* 3503 (1984); *52* 2838 (1987)
JACS *107* 2996 (1985); *108* 284, 4603 (1986)
CC 479 (1986); 1342 (1987)

25. β-Nitrosulfones

$$\underset{-C-C-}{\overset{NO_2 \quad SO_2Ar}{| \quad |}} \xrightarrow[\text{Na}_2\text{S or NaTeH}]{n\text{-Bu}_3\text{SnH,}} \underset{}{>}C=C\underset{}{<}$$

JOC *52* 5111 (1987)

26. β-Silylsulfones

$$\underset{O}{\overset{O}{\underset{\|}{\overset{\|}{R_2CHSPh}}}} \xrightarrow[\text{2. ICH}_2\text{SiMe}_3]{\text{1. }n\text{-BuLi}} \underset{}{\overset{CH_2SiMe_3}{\underset{|}{R_2CSO_2Ph}}} \xrightarrow{n\text{-Bu}_4\text{NF}} R_2C=CH_2$$

TL *23* 2223 (1982)

27. Disulfones

$$CF_3SO_2CH_2SO_2CH_3 \longrightarrow \underset{\underset{R^2}{|}}{\overset{R^1}{\underset{|}{CF_3SO_2CSO_2CHR^3R^4}}} \xrightarrow{KO\text{-}t\text{-Bu}} R^1R^2C=CR^3R^4$$

TL *25* 4617 (1984)
JOC *50* 2110 (1985)
JACS *108* 2358 (1986)

$$\begin{array}{c} \text{ArSO}_2 \\ | \\ \text{RCHCHR} \xrightarrow[\text{MeOH}]{\text{Na(Hg)}} \text{RCH}{=}\text{CHR} \\ | \\ \text{SO}_2\text{Ar} \; \text{(NaH}_2\text{PO}_4) \end{array}$$

CC 914 (1982)
TL 1653 (1983)
Phosphorus and Sulfur *14* 229 (1983)
JOC *49* 596 (1984); *50* 4340 (1985); *52* 3250, 4732, 4740 (1987)
Can J Chem *62* 2487 (1984)
JACS *107* 4789, 6400 (1985); *108* 3453 (1986)
Tetr *42* 1789 (1986)

28. β-Hydroxyselenides

$$\begin{array}{c} \text{HO} \quad \text{SeR} \\ | \quad\;\; | \\ -\text{C}-\text{C}- \quad\longrightarrow\quad {>}\text{C}{=}\text{C}{<} \\ | \quad\;\; | \end{array}$$

TsOH	TL 1385, 3227, 3743 (1976)
HClO$_4$	TL 1385, 3743 (1976)
(CF$_3$CO)$_2$O, Et$_3$N	TL 1385, 3743 (1976)
MsCl, Et$_3$N	CC 790 (1975)
SOCl$_2$, Et$_3$N	TL 3227, 3743 (1976)
	CC 564 (1982); 1540 (1987)
	JOC *51* 3108 (1986)
POCl$_3$, Et$_3$N	TL 2693 (1978)
PI$_3$, Et$_3$N	CC 565 (1982)
NaH/ [benzodioxaphosphole]PCl	TL 3743 (1976)
Me$_3$SiCl, NaI	JOC *46* 231 (1981)

29. β-Oxoselenides

$$\begin{array}{c} \text{RO} \quad \text{SePh} \qquad \text{RCO}_2 \quad \text{SePh} \\ | \quad\;\; | \qquad\qquad | \quad\;\; | \\ -\text{C}-\text{C}- \;\; \text{or} \;\; -\text{C}-\text{C}- \;\longrightarrow\; {>}\text{C}{=}\text{C}{<} \\ | \quad\;\; | \qquad\qquad | \quad\;\; | \end{array}$$

Na/NH$_3$	CC 83 (1979)
Me$_3$SiCl, NaI	JOC *46* 231 (1981)

30. Selenoxides

See also page 116, Section 5.

Reviews:

Tetr *34* 1049 (1978)
Acct Chem Res *12* 22 (1979)

$$-\overset{H}{\underset{|}{\overset{|}{C}}}-\overset{\overset{O}{\|}}{\underset{|}{\overset{|}{\underset{SeR}{C}}}}- \longrightarrow \rangle C=C\langle$$

JOC *36* 2561 (1971); *43* 1697 (1978)
TL 1979 (1973); 1141 (1978); *22* 1809 (1981); *27* 2949 (1986); *28* 1550, 4917, 5119 (1987)
CC 1578 (1987)

$$RCH_2CH_2X \xrightarrow{\text{Se reagent}} RCH_2CH_2SeR' \xrightarrow{[O]} RCH=CH_2$$

X	Se Reagent	Oxidant	
Br	NaSePh	H_2O_2	JOC *52* 2337 (1987)
OH	o-NO$_2$C$_6$H$_4$SeCN, n-Bu$_3$P	H_2O_2	JOC *41* 1485 (1976)
			JACS *105* 6723 (1983);
			106 4186, 5335 (1984);
			108 468, 1019 (1986)
			TL *27* 4813 (1986); *28* 3061 (1987)
	NSePh, n-Bu$_3$P	?	JOC *46* 1215 (1981)
OMs	NaSePh	KIO$_4$	TL *28* 3671 (1987)
	o-NO$_2$C$_6$H$_4$SeCN, NaBH$_4$	H_2O_2	JACS *98* 1612 (1976)

X = OR, O$_2$CCF$_3$

JOC *39* 428, 429 (1974)

$$-\overset{H}{\underset{|}{\overset{|}{C}}}-\overset{H}{\underset{|}{\overset{|}{C}}}-X \xrightarrow{\text{reagent(s)}} -\overset{H}{\underset{|}{\overset{|}{C}}}-\overset{SeR}{\underset{|}{\overset{|}{C}}}-X \xrightarrow{[O]} \rangle C=\overset{|}{C}-X$$

X	Reagent(s)	
CHO	(PhSe)$_2$, SeO$_2$, H$_2$SO$_4$	TL *23* 4813 (1982)
	PhSeCl (on enamine)	TL *21* 4417 (1980)
COR	PhSeX (X = Cl, Br)	JACS *95* 6137 (1973)
		JOC *50* 2981 (1985)
		TL *28* 5755 (1987)

$$-\overset{\displaystyle H}{\underset{\displaystyle |}{\underset{|}{C}}}-\overset{\displaystyle H}{\underset{\displaystyle |}{\underset{|}{C}}}-X \xrightarrow{\text{reagent(s)}} -\overset{\displaystyle H}{\underset{|}{C}}-\overset{\displaystyle SeR}{\underset{|}{C}}-X \xrightarrow{\text{[O]}} \underset{}{>}C=\overset{}{\underset{|}{C}}-X \quad (\textit{Continued})$$

X	Reagent(s)	
COR (*continued*)	(PhSe)$_2$, SeO$_2$, H$_2$SO$_4$	TL *23* 4813 (1982)
	LiNR$_2$/PhSeX (X = Cl, Br)	JACS *95* 5813 (1973); *97* 5434 (1975); *107* 7745 (1985); *108* 3443 (1986)
		JOC *39* 2133 (1974); *47* 1598 (1982); *51* 2416, 5232 (1986)
		Tetr *37* 3981 (1981)
		TL *28* 6021 (1987)
	PhSeO$_2$CCF$_3$ (on enol acetate)	CC 695 (1973) JACS *97* 5434 (1975)
	MeLi/PhSeBr (on enol acetate)	JOC *39* 2133 (1974)
	PhSeCl, *n*-Bu$_4$NF (on enol silane)	CC 880 (1987)
CO$_2$R	LDA/PhSeX (X = Cl, Br)	JACS *95* 5813, 6137 (1973); *97* 5434 (1975)
		JOC *39* 2133 (1974); *52* 2639 (1987)
lactone	LDA/PhSeX (X = Cl, Br)	JACS *97* 5434 (1975) JOC *51* 4836 (1986)
	LDA/(PhSe)$_2$	JOC *39* 120 (1974); *43* 3693 (1978); *52* 4792 (1987)
CN	LiNR$_2$/PhSeX (X = Br, OAc) or (PhSe)$_2$	TL 2279 (1974)
PhSO$_2$	*n*-BuLi/PhSeCl	TL *28* 5763 (1987)
β-dicarbonyl compounds	PhSeCl, py	JOC *46* 2920 (1981)
	(PhSe)$_2$, SeO$_2$, H$_2$SO$_4$	TL *23* 4813 (1982)
	amide base/PhSeX (X = Cl, Br)	JOC *39* 2133 (1974) JACS *97* 5434 (1975) TL *27* 2691 (1986)
	NaH/Se/MeI	TL *22* 3043 (1981)

$$\overset{\quad O \; Li}{\underset{\quad \| \; \|}{PhSeCR_2}} + XCH_2R' \longrightarrow \overset{O}{\underset{\|}{PhSeCR_2}}CH_2R' \longrightarrow R_2C=CHR'$$

JACS *97* 3250 (1975)

$$\overset{O}{\underset{\|}{C}}-C=\overset{H}{\underset{|}{C}} \xrightarrow[\substack{2.\ PhSeBr\ or \\ (PhSe)_2 \\ 3.\ H_2O_2}]{1.\ R_2CuLi} \overset{O}{\underset{\|}{C}}-C=\overset{R}{\underset{|}{C}}$$

JOC *39* 2133 (1974); *50* 2539 (1985)
JACS *97* 5434 (1975); *108* 6276 (1986)

31. Tellurides

$$RCH_2X \xrightarrow{PhTeCH_2Li} RCH_2CH_2TePh \xrightarrow{chloramine\text{-}T} RCH=CH_2$$

CL 447 (1981)

32. β-Halosilanes

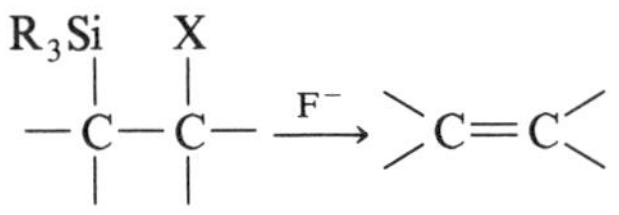

$$\begin{array}{cc} R_3Si & X \\ | & | \\ -C-C- \\ | & | \end{array} \xrightarrow{F^-} \begin{array}{c} \\ \end{array} C=C \begin{array}{c} \\ \end{array}$$

TL *23* 3455 (1982)

33. β-Oxysilanes

See page 178, Section 12.

34. Dehydration of Alcohols

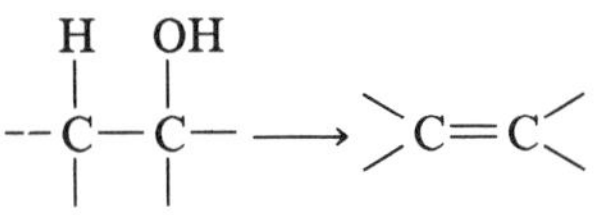

$$\begin{array}{cc} H & OH \\ | & | \\ -C-C- \\ | & | \end{array} \longrightarrow \begin{array}{c} \\ \end{array} C=C \begin{array}{c} \\ \end{array}$$

DMSO	JOC *27* 2377 (1962); *29* 123, 221 (1964) TL *28* 1175 (1987)
HMPA	TL 567 (1971) BCSJ *47* 1693 (1974)
HBr	Org Syn Coll Vol *3* 312 (1955)
H_2SO_4	Org Syn Coll Vol *1* 183, 430 (1941); *2* 606 (1943); *4* 771 (1963)
H_3PO_4	Org Syn Coll Vol *2* 151 (1943) JOC *50* 2179 (1985)
P_2O_5	JACS *69* 2022 (1947) JCS 2154 (1954)
$(CO_2H)_2$	JACS *69* 50 (1947) JCS 2154 (1954)
TsOH	JOC *52* 5574 (1987)
$NaHSO_4$	JACS *69* 2022 (1947)
$KHSO_4$	JACS *69* 2022 (1947); *109* 7122 (1987)
$BF_3 \cdot OEt_2$	JACS *108* 3835 (1986)

CuSO$_4$ Acta Chem Scand B *31* 721 (1977)
                                         JOC *45* 917 (1980); *47* 2590 (1982)

CuSO$_4$-silica gel   TL *28* 4565 (1987)

I$_2$ JCS 588 (1947)

FeCl$_3$, SiO$_2$ JOC *43* 1020 (1978)

KOH   JACS *69* 2022 (1947); *70* 1646 (1948)
                                         Ber *93* 2591 (1960)

Al$_2$O$_3$, Δ   Org Syn Coll Vol *3* 312 (1955)

ThO$_2$, Δ   JACS *85* 2180 (1963)

NBS, py   Chem Rev *63* 21 (1963)

(PhO)$_3$PBr$_2$/K$_2$CO$_3$ JCS Perkin I 1136 (1980)

[(PhO)$_3$PMe]I, HMPA   JOC *37* 4190 (1972); *52* 2644 (1987)

POCl$_3$, py   JACS *83* 5003 (1961); *106* 6690 (1984); *107* 4964
                                             (1985); *108* 3443 (1986)

POCl$_3$, HMPA   JACS *102* 7910 (1980)

SOCl$_2$, py   JACS *77* 1028 (1955)
                                         JCS Perkin I 1136 (1980)
                                         JOC *51* 5463 (1986)

Ph$_2$S[OC(CF$_3$)$_2$Ph]$_2$ JACS *93* 4327 (1971); *94* 5003 (1972); *107* 4964
                                             (1985)
                                         JOC *49* 2682 (1984); *51* 3098 (1986)

MeO$_2$C$\bar{\text{N}}$SO$_2$$\overset{+}{\text{N}}Et_3$ JACS *92* 5224 (1970); *106* 1518 (1984); *107* 1421
                                             (1985); *108* 2343, 3731 (1986)
                                         JOC *35* 2594 (1970); *38* 26 (1973); *39* 2124
                                             (1974); *52* 3614 (1987)
                                         TL *28* 31, 4965, 5643 (1987)

p-MeC$_6$H$_5$OCSCl/Δ   CC 1215 (1972)
                                         JOC *45* 3149 (1980)

$$
\underset{\text{H}}{|}\underset{\text{OH}}{|} \quad \underset{\text{H}}{|}\underset{\text{OM}}{|}
$$

$$-\overset{|}{\underset{|}{C}}-\overset{|}{\underset{|}{C}}- \longrightarrow -\overset{|}{\underset{|}{C}}-\overset{|}{\underset{|}{C}}- \overset{\Delta}{\longrightarrow} \rangle C = C \langle$$

<u>M</u>

Mg, Zn   JOC *44* 1221 (1979)

Al   JOC *44* 1221, 1340 (1979)

$$RCH_2CH_2OH \longrightarrow RCH_2CH_2SeAr \xrightarrow{[O]} RCH = CH_2$$

See page 149, Section 30.

$$\underset{\overset{|}{C}}{\overset{\overset{H}{|}}{}}-C=C-\underset{\overset{|}{C}}{\overset{\overset{OH}{|}}{}} \longrightarrow C=C-C=C$$

See page 242, Section 2.

X	Reagent(s)	
F	HF-py, KHF$_2$, i-Pr$_2$NH, PhCl	TL *28* 663 (1987)
Cl	MgCl$_2$, ether, Δ	JOC *45* 2566 (1980)
	MgCl$_2$, ZnCl$_2$, ether, Δ	JOC *45* 2566 (1980)
	Me$_3$SiCl	TL *27* 1907 (1986)
Br	HBr	Compt Rend *248* 820 (1959)
		BSCF 1072 (1960); 1805 (1970)
		JACS *90* 2882 (1968)
		Syn 37 (1979)
		JOC *50* 2719 (1985)
	HBr, ZnBr$_2$	JACS *90* 2882 (1968)
		TL 1281 (1973)
	MgBr$_2$, ether, Δ	CC 303 (1975)
		JOC *45* 2566 (1980)
	MgBr$_2$, ZnBr$_2$, ether, Δ	JOC *45* 2566 (1980)
	Me$_3$SiCl, LiBr	TL *27* 1907 (1986)
	PBr$_3$, LiBr, collidine/ZnBr$_2$	JACS *90* 2882, 6225 (1968)
I	HI	Syn 37 (1979)
	MgI$_2$, ether, Δ	CC 303 (1975)
		JOC *45* 2566 (1980)
	MgI$_2$, ZnI$_2$, ether, Δ	JOC *45* 2566 (1980)
	Me$_3$SiCl, LiI	TL *27* 1907 (1986)

35. Sulfonate Ester Eliminations

$$-\underset{\overset{|}{}}{\overset{\overset{H}{|}}{C}}-\underset{\overset{|}{}}{\overset{\overset{O_3SR}{|}}{C}}- \longrightarrow {>}C=C{<}$$

NaI, HMPA, Δ	JACS *109* 2212 (1987)
Al$_2$O$_3$	JOC *42* 3173 (1977)
MOMe (M = Li, Na, K)	JOC *26* 4199 (1961)
KO-t-Bu	TL *28* 6489 (1987)
KO-t-Bu, DMSO	JOC *29* 742 (1964); *30* 2054 (1965)
	Org Syn *64* 50 (1985)
DBN	JOC *52* 2644 (1987)
DBU	JOC *52* 4044 (1987)

PhCH(Me)NMe$_2$ (chiral)   TL *28* 6489 (1987)

MNH$_2$ (M = Li, K), NH$_3$ JOC *26* 4199 (1961)

Et$_2$NMgBr   JOC *26* 4199 (1961)

o-NO$_2$C$_6$H$_4$SeNa/H$_2$O$_2$ JOC *52* 4142 (1987)

36. Xanthate Pyrolysis (Chugaev Elimination)

Org Rxs *12* 57 (1962) (review)
TL *28* 2795 (1987)

37. *p*-Tolylthiocarbonate Pyrolysis

CC 1215 (1972)
JOC *45* 3149 (1980)
JACS *105* 7352 (1983); *108* 800 (1986)

38. Carbamate Pyrolysis

JOC *34* 3604 (1969); *46* 2804 (1981)

39. *N*-Methyl-4-Alkoxypyridinium Iodide Pyrolysis

Can J Chem *50* 1181 (1972)

40. 1,2-Diols

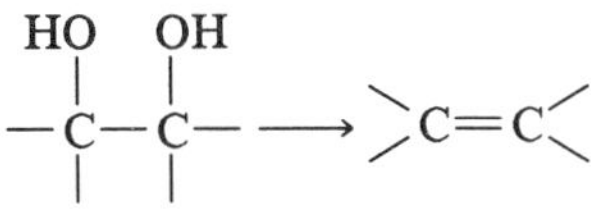

$$\begin{array}{cc} HO & OH \\ | & | \\ -C-C- \\ | & | \end{array} \longrightarrow \enspace >C=C<$$

Review: Org Rxs *30* 457 (1984)

40.1. Stereospecific cis Elimination

thionocarbonate
  P(OR)$_3$, Δ

JACS *85* 2677 (1963); *87* 934 (1965); *93* 4516
  (1971); *94* 8627 (1972)
Carbohydr Res *1* 214 (1965); *1* 444 (1966)
TL 4645 (1967); 3655 (1978); 853 (1972)
JOC *35* 3558 (1970)
Can J Chem *48* 383 (1970)

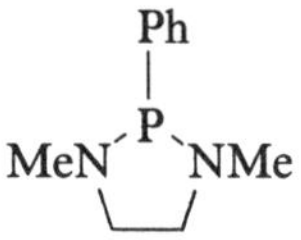

MeN—P(Ph)—NMe (cyclic)

TL *23* 1979 (1982)
JACS *108* 512 (1986)

  Raney Ni

JACS *105* 1988 (1983)

  Ni(COD)$_2$

TL 2667 (1973)

benzylidene acetal, *n*-BuLi

CC 1593 (1968)
Tetr *26* 4339 (1970)
JCS C 886 (1971)
JCS Perkin I 2332 (1973)
JOC *45* 261 (1980)

2-alkoxy-1,3-dioxolane
  Δ

TL *22* 1471 (1981)
Tetr *38* 2395 (1982)
JCS Perkin I 2279 (1986)
JACS *109* 7495 (1987)

  H$^+$, Δ

Austral J Chem *17* 1392 (1964); *21* 2013 (1968)
Carbohydr Res *7* 161 (1968)
JACS *108* 1265 (1986)

2-dialkylamino-1,3-dioxolane
  MeI, Δ

TL 737 (1978)
Rec Trav Chim *104* 266 (1985)

  Ac$_2$O, Δ

TL 5223 (1970); *27* 2575 (1986)
JACS *96* 5254 (1974)

40.2. Non-stereospecific or Unknown Stereochemistry

thionocarbonate, MeI or *i*-PrI, Δ/(Zn)

JOC *39* 3641 (1974)

thionocarbonate, PhCH$_2$N—P(Me)—NCH$_2$Ph, Δ

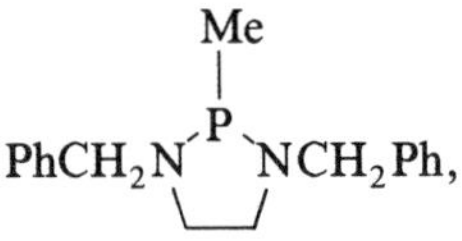

JOC *39* 3641 (1974)

thionocarbonate, $Fe(CO)_5$, Δ	TL 4435 (1972)
carbonate tosylhydrazone pyrolysis	TL 3161 (1973)
2 $MeLi/K_2WCl_6$	CC 370 (1972)
$TiCl_3$, K	JOC *43* 3255 (1978) TL *28* 4965 (1987)
Me_3SiCl, NaI	TL *23* 1365 (1982)
$PBr_3/CuBr$, Zn	BCSJ *52* 1752 (1979)

Ph_3P, I_2, (imidazole) — Syn 469 (1979)

Ph_3P, (diiodoimidazole), (imidazole) — Syn 813 (1979)

NaOH or $NaH/CS_2/MeI/n$-Bu_3SnH	CC 866 (1977) Chem Pharm Bull *26* 1786 (1978) JCS Perkin I 2378 (1979)
Cl_2POX (X = OEt, NMe_2)/Li-NH_3 or Ti-THF	JOC *42* 1311 (1977)

41. 1,2-Dimesylates

$$\underset{\text{MsO}\quad\text{OMs}}{-\overset{|}{\underset{|}{C}}-\overset{|}{\underset{|}{C}}-} \longrightarrow\ \text{>C=C<}$$

Na naphthalenide	TL 3447 (1972) JACS *109* 5524 (1987)
NaI, Zn	TL *28* 2183 (1987) JACS *109* 6403 (1987)

42. 1,3-Diols

$$\underset{\text{HO}\quad\text{HO}\ \text{H}}{C-C-C-C}\ \xrightarrow{(MeO)_2CHNMe_2}\ \underset{\text{NMe}_2}{O\quad O}\ \xrightarrow[\text{2. OH}^-]{\text{1. MeI, }\Delta}\ \underset{\text{HO}}{C-C-C=C}$$

CC 1756 (1987)

43. Acetals

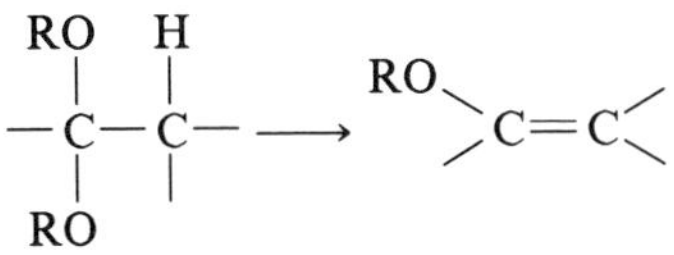

cat TsOH	Syn 38 (1974)
$AlCl_3$ or $MgBr_2$, Et_3N	Helv *62* 1451 (1979)
Me_3SiI, $(Me_3Si)_2NH$	TL *23* 323 (1982)
$Me_3SiMn(CO)_5$	Organomet *1* 1467 (1982)
$TiCl_4$/DBU	Tl *27* 3053 (1986)

44. Dithioacetals

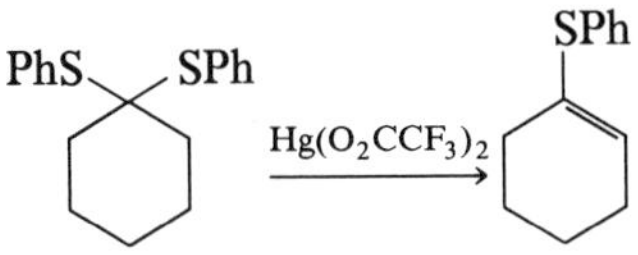

JACS *105* 5075 (1983)

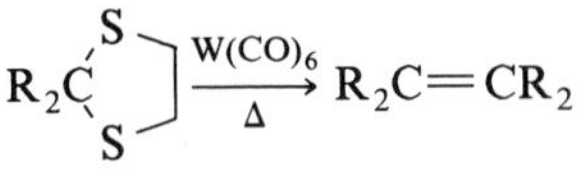

CC 981 (1987)

45. Aldehydes and Ketones

45.1. Direct Elimination

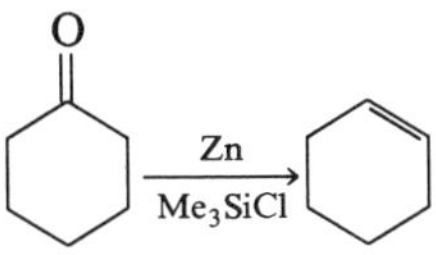

CC 935 (1973)
JCS Perkin I 809 (1975)

$$\mathrm{ArC(=O)CH_2R} \xrightarrow[2.\ \Delta]{1.\ \mathrm{HPO(OEt)_2,\ cat\ NaNH_2}} \mathrm{ArCH{=}CHR}$$

Angew Int *2* 98 (1963)

45.2. Tosylhydrazone Elimination (Bamford-Stevens)

Review: Org Rxs *23* 405 (1976)

JCS 4735 (1952)
Tetr *12* 168 (1961); *31* 1035 (1975)
Ann *691* 41 (1966)
JACS *89* 5734, 5736, 7112 (1967); *90* 4762 (1968); *106* 6006 (1984); *107* 3971, 7352, 7724 (1985)
TL 345 (1968), 2947 (1976)
Syn 595 (1970)
Austral J Chem *23* 857 (1970)
JOC *43* 147, 1404 (1978); *44* 3976 (1979); *50* 5460 (1985); *52* 3346 (1987)
Chem Pharm Bull *28* 984 (1980)

45.3. Shapiro Reaction

$E^+ = H_2O, D_2O, RX, DMF, CH_2O, RCHO, R_2CO, CO_2, Br_2, NCBr, BrCF_2CF_2Br, Me_3SiCl,$
$n\text{-}Bu_3SnCl$

Reviews: Org Rxs *23* 405 (1976)
Acct Chem Res *16* 55 (1983)

JACS *89* 5734 (1967); *90* 4762 (1968); *106* 3539 (1984); *107* 256, 1293, 3971, 4964 (1985); *108* 4586
  (1986); *109* 3174, 5731, 7838 (1987)
TL 1811, 1815 (1975); 2287 (1976); *21* 945, 3849 (1980); *23* 3733 (1982); *27* 2761, 5467 (1986);
  28 1985, 4629, 6159 (1987)
JOC *43* 147, 1404, 1409 (1978); *46* 1315 (1981); *50* 2438 (1985); *52* 569, 2644, 3541 (1987)
Syn 44 (1979)
CC 65, 1121 (1981)
JCS Perkin I 2848 (1981)
Tetr *37* 3935 (1981)

45.4. Enone Conversions

Tetr *20* 957 (1964)

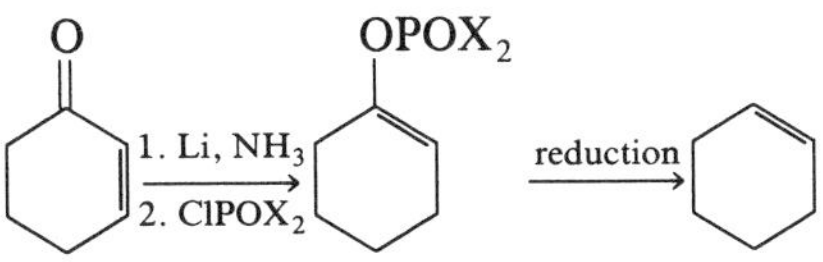

X = OEt, NMe$_2$

TL 2145 (1969)
JACS *94* 5098 (1972)
JOC *43* 2715 (1978)

For methods of reduction see page 160, Section 45.5.

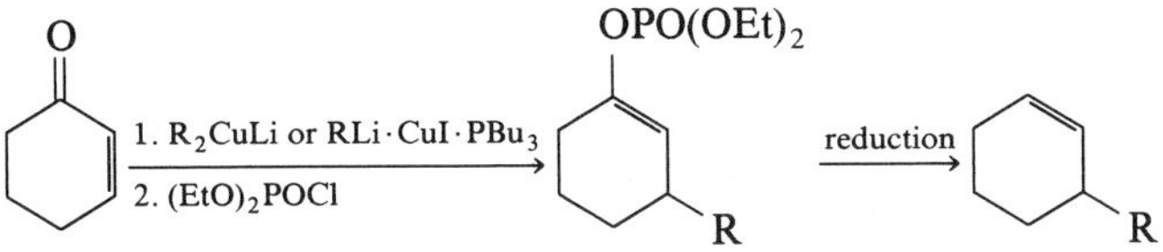

TL 2145 (1969)
Org Syn *52* 109 (1972)
JOC *43* 2715 (1978)
JACS *104* 1907 (1982); *106* 3353 (1984); *108* 5650 (1986)

For methods of reduction see page 160, Section 45.5.

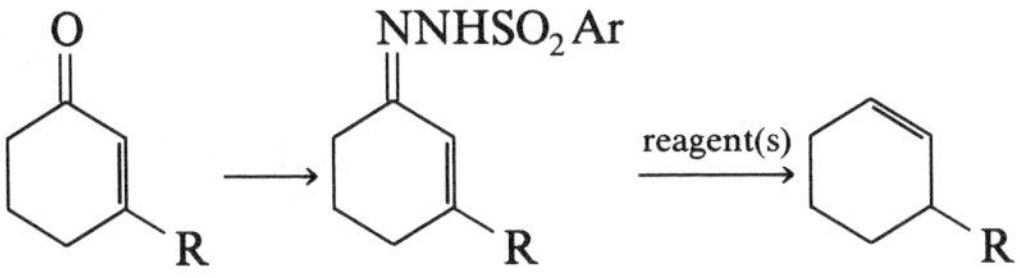

Reagent(s)

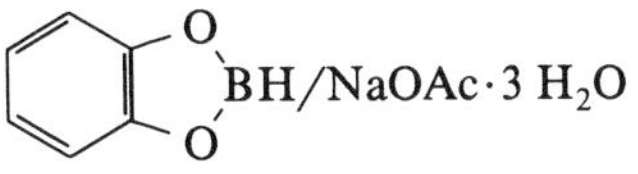

BH/NaOAc·3 H$_2$O

JOC *41* 574 (1976)
Org Syn *59* 42 (1980)
JACS *108* 6276, 7686 (1986)

NaBH$_4$, HOAc

JOC *43* 2299 (1978)

NaBH$_3$CN

JOC *40* 923 (1975); *50* 2798 (1985)
JACS *98* 2275 (1976); *106* 3353 (1984)
Org Prep Proc Int *11* 201 (1979) (review)

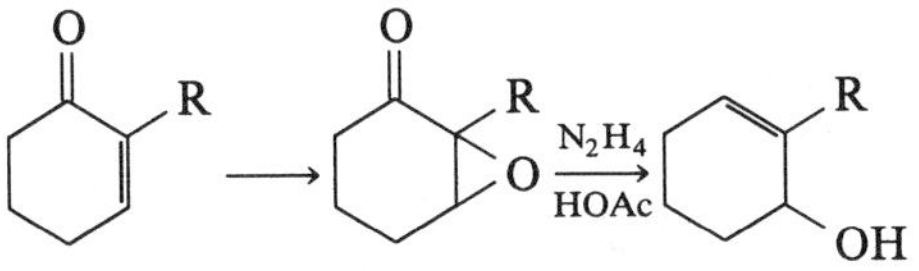

JOC *26* 3615, 4781 (1961); *27* 2205 (1962)
JACS *86* 269 (1964); *102* 862 (1980)
TL *28* 2099 (1987)

45.5. Reduction of Carbonyl Derivatives

enamine
    BH_3 TL 2039 (1964)
    $LiAlH_4$, $AlCl_3$ Proc Chem Soc 19 (1963)
                                              Tetr *24* 4489 (1968)

enol ether
    i-Bu_2AlH JOC *31* 329 (1966)

enol acetate
    BH_3 Gazz Chim Ital *92* 309 (1962)

enol silane
    BH_3/HCl (cyclic only)   TL 4005 (1975)

enol triflate
    cat $Pd(OAc)_2(PPh_3)_2$, HCO_2H, *n*-Bu_3N TL *25* 4821 (1984)
    n-Bu_3SnH or Et_3SiH, cat $Pd(PPh_3)_4$, LiCl   JACS *106* 4630 (1984); *108* 3033 (1986)

enol phosphate
    Li, *t*-BuOH, NH_3 CC 112 (1969)
                                         JACS *106* 3353, 5025 (1984)
    Li, *t*-BuOH, $MeNH_2$ Helv *66* 522 (1983)
                                       JACS *108* 3435 (1986)
    Li, *t*-BuOH, $EtNH_2$ TL 2145 (1969)
                                       Org Syn *52* 109 (1972)
                                       JACS *104* 1907 (1982); *106* 721 (1984); *107* 1308
                                            (1985); *108* 5650 (1986)
                                       JOC *50* 2668 (1985)
    K, $TiCl_3$/EtOH   JOC *43* 2715 (1978)

enol N,N,N',N'-tetramethylphosphorodiamidates
    Li, NH_3 JACS *94* 5098 (1972)
    Li, $EtNH_2$ TL *28* 31 (1987)
    Li, $EtNH_2$, *t*-BuOH   JACS *94* 5098 (1972)

α-halocarbonyl compounds
    $LiAlH_4$/Zn, HOAc   JCS 1370 (1955)
    N_2H_4 JOC *29* 958 (1964)

dithioketal
    Raney Ni   JOC *28* 1443 (1963)
                                       Tetr *25* 2823 (1969)

45.6. Dimerization of Aldehydes and Ketones

$$2\,R_2CO \longrightarrow R_2C{=}CR_2$$

    Reviews:

        Org Prep Proc Int *12* 361 (1980)
        JOC *47* 248 (1982)
        Acct Chem Res *16* 405 (1983) (Ti reagents)

Ti-graphite	JOMC *280* 307 (1985)
TiCl$_3$, Li	JOC *41* 3929 (1976) (unsymmetrical coupling); *43* 3255 (1978) (unsymmetrical coupling); *51* 5446 (1986) Syn Commun *11* 895 (1981) Org Syn *60* 113 (1981) TL *23* 3227 (1982) (unsymmetrical coupling) JACS *106* 723, 6006 (1984) (both intramolecular)
TiCl$_3$, K	JOC *41* 896 (1976) (inter- and intramolecular); *43* 3255 (1978); *51* 2969 (1986); *52* 2905 (1987) (intramolecular) TL *28* 4965 (1987) (intramolecular)
TiCl$_3$, Mg	BSCF 2147 (1973)
TiCl$_3$, Zn-Cu	JOC *42* 2655 (1977); *43* 3255 (1978) (intramolecular); *46* 4293 (1981) TL *23* 1777 (intramolecular), 2723 (intramolecular), 3227 (unsymmetrical coupling) (1982); *28* 3091, 3209 (1987) (both intramolecular) JACS *106* 5018 (1984); *108* 515, 2932, 3513 (intramolecular) (1986); *52* 4885 (1987) (intramolecular)
TiCl$_3$, Zn-Ag	JOC *47* 5229 (1982) JACS *108* 1239 (1986) (intramolecular)
TiCl$_3$, LiAlH$_4$	JACS *96* 4708 (1974) Acct Chem Res *7* 281 (1974) TL 3265 (1976); *22* 3965 (1981) (enones) JOC *43* 3609 (1978); *52* 5636 (1987) Ber *115* 1234 (1982) Rec Trav Chim *101* 112 (1982)
TiCl$_4$, Mg	TL *22* 3965 (1981) (enones)
TiCl$_4$, Zn	CL 1041 (1973) JACS *106* 7514 (1984) (intramolecular); *108* 3460 (1986) TL *26* 1981 (1985) (intramolecular) CC 1072 (1987) (intramolecular)
TiCl$_4$, Zn, py	Syn 553 (1977) Ber *115* 3697 (1982)
TiCl$_4$, LiAlH$_4$	JACS *96* 4708 (1974)
NbCl$_5$, NaAlH$_4$	CL 158 (1982)
Mo(CO)$_6$ or W(CO)$_6$ or WCl$_6$, LiAlH$_4$	JOC *43* 2477 (1978)
WCl$_6$, 2 n-BuLi	JACS *94* 6538 (1972)
Zn, Me$_3$SiCl	CC 1803 (1986)

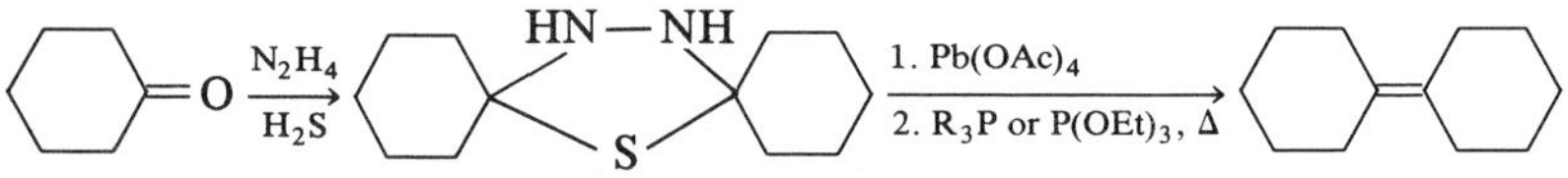

CC 1226 (1970)
JCS Perkin I 305 (1972)
JOC *38* 3061 (1973)
JACS *106* 5018, 8174 (1984); *108* 515, 2932 (1986)

45.7. Miscellaneous Reactions

$$RCH_2CHO \xrightarrow[\text{2. } H_2O_2]{\text{1. ArSeCN, } n\text{-Bu}_3P} RCH{=}CHCN$$

JACS *99* 5210 (1977)

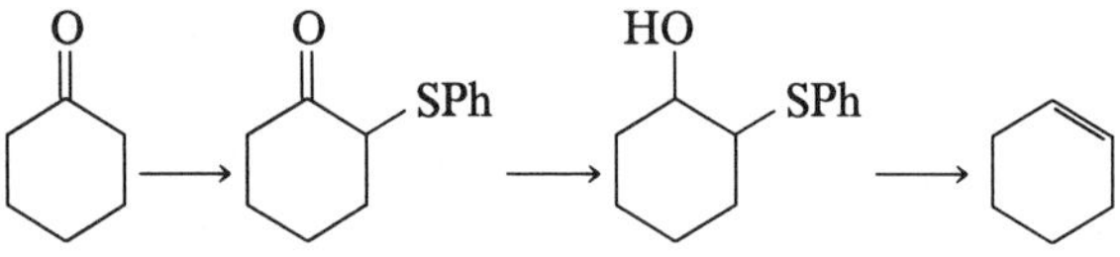

TL 4223 (1977)

46. Carboxylic Acids

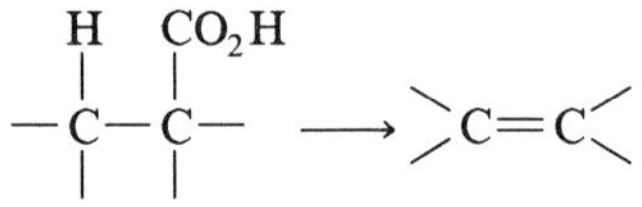

PhI(OAc)$_2$, Cu(OAc)$_2$ JOC *51* 402 (1986)

Pb(OAc)$_4$ JACS *83* 927 (1961)

Pb(OAc)$_4$, cat Cu(OAc)$_2\cdot$H$_2$O, py   JACS *87* 1811, 3609 (1965); *107* 2149 (1985); *108* 4603 (1986)
                                             JOC *32* 2045 (1967); *39* 2217 (1974)
                                             TL 405, 5173 (1968); 399 (1974); *26* 6397 (1985)
                                             Tetr *24* 2215 (1968)
                                             JCS C 1047 (1969)
                                             Org Rxs *19* 279 (1972) (review)
                                             Syn 541 (1973); 889 (1974)

$$RCH{=}CHCO_2H \xrightarrow{Br_2} RCH{-}CHCO_2H \xrightarrow{\text{base}} RCH{=}CHBr$$

See page 319, Section 3.

$$\text{R}_2\text{CO}+\text{R}_2\bar{\text{C}}\text{CO}_2{}^- \longrightarrow \text{R}_2\overset{\overset{\displaystyle \text{OH}}{|}}{\text{C}}\text{CR}_2\text{CO}_2\text{H} \longrightarrow \text{R}_2\text{C}=\text{CR}_2$$

syn elimination
   MsCl, $\text{Na}_2\text{CO}_3/\Delta$ TL 4569 (1968)
   $\text{ArSO}_2\text{Cl, py}/\Delta$ JACS *94* 2000 (1972)

Can J Chem *51* 981 (1973) (vinyl ethers)
JOC *39* 1322, 1650 (1974); *43* 4574 (1978) (vinyl
   ethers); *46* 3359 (1981); *52* 3143 (1987)
CC 52 (1979) (alkenes, vinyl ethers); 1199 (1986)
   (allylic silanes)
Syn 388 (1979) (vinyl ethers)
Helv *62* 2825 (1979)
TL *28* 2753, 3103, 5921 (allylic silane) (1987)

anti elimination
   $\text{EtO}_2\text{CN}{=}\text{NCO}_2\text{Et, PPh}_3$
   $(\text{CH}_3\text{O})_2\text{CHN}(\text{CH}_3)_2, \Delta$

CC 52 (1979) (alkenes and vinyl ethers)
TL 1545 (1975); 2953 (1978) (dienes); 1909 (1979)
   (mechanism); *27* 5417 (1986)
Helv *62* 2825 (1979)
Ber *115* 3453 (1982) (dienes)
CC 1199 (1986) (allylic silanes)

unknown stereochemistry
   Δ TL *23* 5271 (1982)

$$\text{RCH}{=}\text{CRCHO} \xrightarrow[\text{2. AcCl}]{\text{1. R}\bar{\text{C}}\text{HCO}_2{}^-} \text{RCH}{=}\text{CRCH}\overset{\overset{\displaystyle \text{OAc}}{|}}{\text{C}}\text{HRCO}_2\text{H} \xrightarrow[\text{Et}_3\text{N}]{\text{cat Pd(PPh}_3)_4} \text{RCH}{=}\text{CRCH}{=}\text{CHR}$$

JACS *102* 2841 (1980)

$$\text{HO}_2\text{C}-\overset{|}{\underset{|}{\text{C}}}-\overset{|}{\underset{|}{\text{C}}}-\text{CO}_2\text{H} \longrightarrow \text{>C}{=}\text{C<}$$

Review: Tetr *40* 2585 (1984)

electrolysis   JACS *85* 165 (1963); *89* 3922 (1967)
Ber *100* 2427 (1967)
TL 5117, 5123 (1968)

PbO_2 JACS *74* 4370 (1952)

Pb(OAc)_4, py   Helv *41* 1191 (1958)
Angew *70* 343 (1958)
JACS *83* 1705 (1961); *85* 165 (1963); *90* 113
   (1968); *98* 628 (1976)
JOC *30* 1431 (1965); *42* 1654 (1977)
CC 214 (1965); 899 (1974)
Ber *100* 2427 (1967)
Org Rxs *19* 279 (1972) (review)

Pb(OAc)$_2$, DMSO or dioxane   CC 214 (1965); 899 (1974)

Cu$_2$O, quinoline, bipy   TL 4447 (1976)
                                             JACS *109* 4626 (1987)

47. Acid Halides

$$\underset{\text{O}}{\overset{\text{O}}{\underset{\|}{\text{RCH}_2\text{CH}_2\text{CCl(Br)}}}} \xrightarrow{\text{cat Pd}} \text{RCH}=\text{CH}_2$$

JACS *90* 94 (1968)
Syn 157 (1969)
Tetr *30* 11 (1974)
JOC *41* 3452 (1976)
TL *27* 4615 (1986)

48. Acid Anhydrides

$$\text{O}=\underset{\text{R}\quad\text{R}}{\diagdown}\underset{}{\overset{\text{O}}{\diagup}}\text{=O} \longrightarrow \text{RCH}=\text{CHR}$$

Review: Tetr *40* 2585 (1984)

PbO$_2$ JACS *74* 4370 (1952)
                                             Ann *585* 154 (1954)
                                             JOC *23* 141 (1958)
                                             Org Rxs *19* 279 (1972) (review)

Pb(OAc)$_4$, py   Helv *41* 1191 (1958)
                                             JACS *83* 1705 (1961); *85* 3297 (1963); *90* 113
                                                (1968); *93* 6092 (1971)
                                             Org Rxs *19* 279 (1972) (review)

Cu$_2$O, bipy, quinoline, H$_2$O   JACS *109* 4626 (1987)

Ni(CO)$_2$(PPh$_3$)$_2$ TL 2603 (1971)
                                             JOC *41* 887 (1976)

$$\text{O}=\underset{\text{R}\quad\text{R}}{\diagdown}\text{O} \xrightarrow{\text{Na}_2\text{S}} \text{O}=\underset{\text{R}\quad\text{R}}{\diagdown}\overset{\text{S}}{\diagup}\text{=O} \xrightarrow[\substack{\text{Fe}_2(\text{CO})_9 \text{ or} \\ \text{ClRh(PPh}_3)_3}]{(\text{Ph}_3\text{P})_2\text{Ni(CO)}_2 \text{ or}} \text{RCH}=\text{CHR}$$

TL 2603 (1971)

49. Esters

See page 162, Section 46, for elimination of β-hydroxy acids via β-lactones.

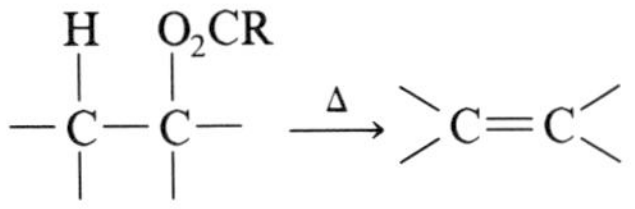

Chem Rev *60* 431 (1960) (review)
Org Syn Coll Vol *4* 746 (1963)
JOC *52* 5034 (1987)
TL *28* 1519 (1987)

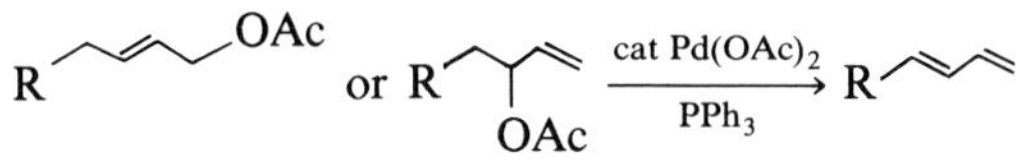

TL 2075 (1978)

$$R_2CHCH_2OAc \text{ or } R_2CHCH_2CH_2CO_2Me \xrightarrow{h\nu} R_2C{=}CH_2$$

TL *22* 1441 (1981)

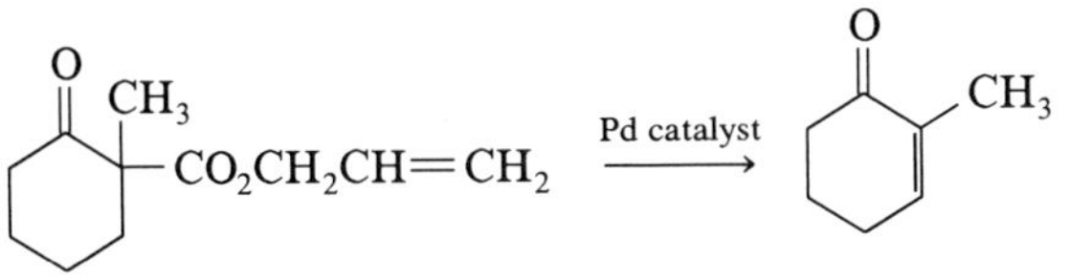

JACS *104* 5844 (1982)
TL *24* 1797 (1983)
CL 1133 (1984)
Syn 1009 (1984)

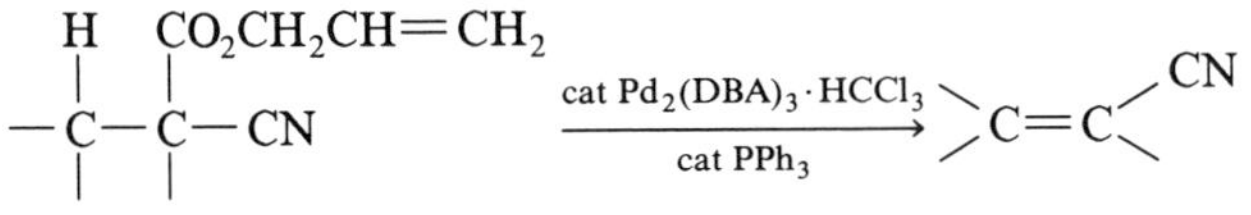

CC 118 (1986)

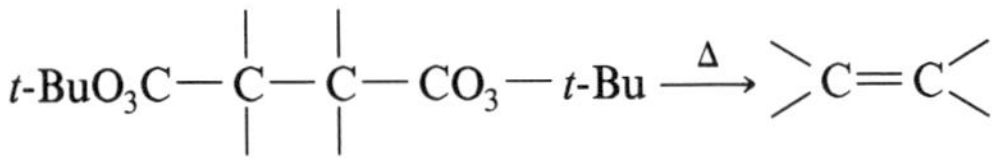

CC 98 (1969)

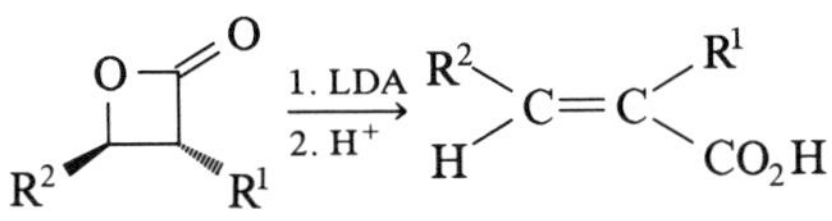

JACS *102* 3620 (1980)

$$R_2CHCOR' \xrightarrow[\substack{\text{3. Li, NH}_3 \text{ or MeNH}_2, \text{ } t\text{-BuOH} \\ \text{or Et}_3\text{Al, cat Pd(PPh}_3)_4}]{\substack{\text{1. LDA} \\ \text{2. ClPO(OEt)}_2}} R_2C=CHOR'$$

JOC *52* 2303 (1987)

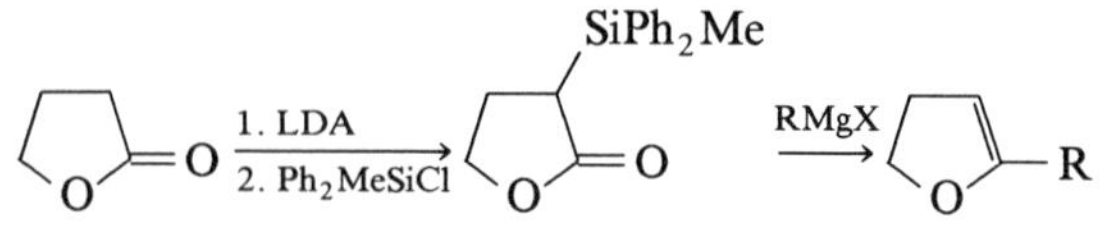

$$\xrightarrow[\text{2. Ph}_2\text{MeSiCl}]{\text{1. LDA}} \xrightarrow[\text{SiPh}_2\text{Me}]{} \xrightarrow{\text{RMgX}} R$$

TL *23* 271 (1982)

$$RCO_2Me \xrightarrow[\substack{\text{2. NaBH}_4 \\ \text{3. electrolysis}}]{\text{1. TsCH}_2\text{MgI}} RCH=CH_2$$

CL 69 (1978)

50. Nitriles

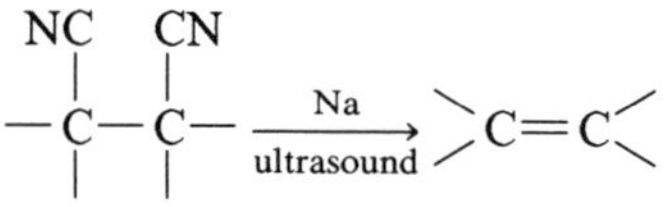

$$\underset{\text{NC \quad CN}}{-\overset{|}{\underset{|}{C}}-\overset{|}{\underset{|}{C}}-} \xrightarrow[\text{ultrasound}]{\text{Na}} \ >C=C<$$

TL *27* 4347 (1986)

51. Miscellaneous Reactions

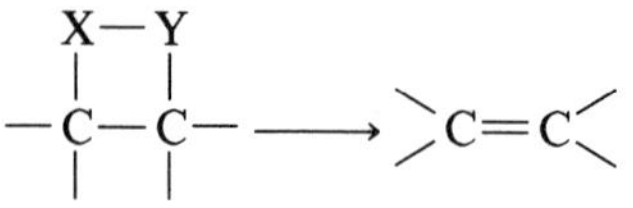

$$\underset{\text{X—Y}}{-\overset{|}{\underset{|}{C}}-\overset{|}{\underset{|}{C}}-} \longrightarrow \ >C=C<$$

Angew Int *21* 225 (1982) (review)
J Chem Ed *59* 313 (1982) (review)

3. ALKYLIDENATION OF CARBONYL AND RELATED COMPOUNDS

$$\underset{\overset{\displaystyle \|}{O}}{-C-CH_2-} \longrightarrow \underset{\overset{\displaystyle \|}{O}}{-C-C=CR_2}$$

RCHO, cat Cp$_2$ZrH$_2$, cat NiCl$_2$	JOC *52* 2239 (1987)
ArCHO, cat Ba(OH)$_2$-C200, sonication	TL *28* 4541 (1987)
LDA/H$_2$CO/TsCl, py/DBU	JOC *52* 4647 (1987)
LDA/H$_2$CO/	Syn Commun *16* 1593 (1986)
	JOC *52* 2378 (1987)
	TL *28* 5081 (1987)

$$\left[\begin{array}{c} \text{O} \underset{\underset{\displaystyle CH_2N=C=NC_6H_{11}}{}}{\overset{\overset{\displaystyle CH_3}{}}{N^+}} \end{array} \right] OTs^-,$$

CuCl$_2$ (on lactone)	
LDA/CH$_3$COCH$_3$/I$_2$	JACS *108* 4556 (1986)
NaOMe, HCO$_2$Me/H$_2$CO, K$_2$CO$_3$	Syn 665 (1983)
base, HCO$_2$R/H$_2$CO; py, Et$_3$N or K$_2$CO$_3$	JOC *32* 3434 (1967)
MeOCO$_2$MgOMe/H$_2$CO, Et$_2$NH, NaOAc, HOAc	JOC *38* 2489 (1973)
LiN(SiMe$_3$)$_2$/HCO$_2$Me/H$_2$CO, K$_2$CO$_3$	JACS *105* 2435 (1983)
	JOC *50* 3957 (1985)
H$_2$CO, (R$_2\overset{+}{N}$H$_2$)Cl$^-$	JACS *95* 4873 (1973)
(CH$_2$O)$_3$, PhNHMe, CF$_3$CO$_2$H, Δ	TL 2111, 2955 (1978)
	Org Syn *60* 88 (1981)
	Syn 952 (1983)
LDA/(Me$_2\overset{+}{N}$=CH$_2$)I$^-$/NaOH	JACS *98* 6715 (1976); *99* 944 (1977)
	TL 1621 (1977)
	JOC *44* 1391 (1979); *45* 524 (1980)

BrCH$_2$SO$_2$Br, hν (on enol silane)/DBN   JOC 49 3664 (1984)

RCHISO$_2$Br, hν (on enol silane)/DBN   JACS 108 4568 (1986)

LDA/PhCH$_2$SCH$_2$Br/NaIO$_4$/Δ JOC 51 2981 (1986)
   (on acids and ketones)

LDA or Et$_3$N, Me$_3$SiCl/PhSCH$_2$Cl,   TL 993, 995 (1979)
   TiCl$_4$/NaIO$_4$/Δ

LDA/(PhSe)$_2$/LDA/RCHO/SOCl$_2$,   JOC 51 3108 (1986)
   Et$_3$N (on amide)

H$_2$C=C(OAc)CH$_3$—H$^+$ or
   Me$_3$SiCl—Et$_3$N/RCHClOMe,   JACS 101 984 (1979)
   Zn(Cu), CH$_2$I$_2$/KHSO$_4$

JOC 52 2239 (1987)

JOC 43 4248 (1978)

Syn 34 (1982)

$$RCH_2\overset{O}{\overset{\|}{C}}CHR'\overset{O}{\overset{\|}{C}}OMe \xrightarrow[\substack{Me_2NH \\ (Me_2NH_2)Cl}]{H_2CO} \xrightarrow{\substack{\Delta \\ MeI}} RCH_2\overset{O}{\overset{\|}{C}}-\overset{CH_2}{\overset{\|}{C}}-R'$$

TL 5037 (1973)

$$RC\overset{O}{\overset{\|}{}}CHRC\overset{O}{\overset{\|}{}}OCH_2CH=CH_2 \xrightarrow{\substack{1.\ H_2CO,\ KHCO_3 \\ 2.\ Ac_2O,\ py \\ 3.\ cat\ Pd_2(DBA)_3,\ PPh_3}} RC\overset{O}{\overset{\|}{}}-\overset{CH_2}{\overset{\|}{C}}-R$$

TL 27 2483 (1986)

$$R^1-\overset{O}{\overset{\|}{C}}-CH_2-R^2 \longrightarrow R^1-\overset{O}{\overset{\|}{C}}-CHR^2-\overset{O}{\overset{\|}{C}}-COEt \xrightarrow[OH^-]{R^3CHO} R^1-\overset{O}{\overset{\|}{C}}-\overset{CHR^3}{\overset{\|}{C}}-R^2$$

JOC 42 1180 (1977)

$$2\ RCH_2CHO \xrightarrow{base} RCH_2\overset{OH}{\overset{|}{C}}HCHRCHO \longrightarrow RCH_2CH=CRCHO$$

See page 647, Section 7.

$$RCHO + CH_3CHO \xrightarrow{NaOH} RCH=CR'CHO$$

Ber *76* 676 (1943)

$$ArCHO + CH_3\overset{O}{\overset{\|}{C}}R \longrightarrow ArCH=CH\overset{O}{\overset{\|}{C}}R$$

NaOH

Org Syn Coll Vol *1* 77 (1941)
Ber *76* 676 (1943)
Syn 647 (1980)

M(OAc)$_2$ (M = Co, Ni, Cu, Zn), bipy   BCSJ *53* 1366 (1980)

$$Ar\overset{O}{\overset{\|}{C}}CH_3 + Ar'CHO \longrightarrow Ar\overset{O}{\overset{\|}{C}}CH=CHAr'$$

NaOH

Syn 647 (1980)
JOC *52* 5560 (1987)

cat Ba(OH)$_2$-C200   TL *28* 4541 (1987)

TiCl$_4$, Et$_3$N   TL *28* 4135 (1987)

cat Co(II)-polymer

CL 1401 (1979)
BCSJ *55* 3208 (1982)

$$RCH_2\overset{O}{\overset{\|}{C}}CH_3 + R'\overset{O}{\overset{\|}{C}}H \xrightarrow{LiI} RCH_2\overset{O}{\overset{\|}{C}}CH=CHR'$$

CC 486 (1980)

$$R\overset{O}{\overset{\|}{C}}CH_2HgI + R'\overset{O}{\overset{\|}{C}}H \xrightarrow{Ni(CO)_4} R\overset{O}{\overset{\|}{C}}CH=CHR'$$

CL 1435 (1979)

$$R_2C=O \xrightarrow[Zn]{BrCH_2C\equiv CR} R_2\overset{OH}{\overset{|}{C}}CH_2C\equiv CR \xrightarrow{H^+} R_2C=CH\overset{O}{\overset{\|}{C}}CH_2R$$

Syn Commun *10* 637 (1980)

$$2\,R\overset{O}{\overset{\|}{C}}CH_3 \xrightarrow{base} R\overset{CH_3}{\overset{|}{C}}=CH\overset{O}{\overset{\|}{C}}R$$

Base

Al(O-t-Bu)$_3$ Org Syn Coll Vol *3* 367 (1955)

Al$_2$O$_3$ Syn 60 (1982)

$$RCHO + R'CH_2CO_2H \xrightarrow{base} RCH=CR'CO_2H$$

Org Syn Coll Vol *4* 777 (1963)

$$RCH_2COH \xrightarrow[\substack{2.\ H_2CO,\ \Delta \\ 3.\ H_3O^+}]{1.\ H_2NCMe_2CH_2OH,\ \Delta} H_2C=C-COH$$

JOC *46* 4147 (1981)

$$RCHO + H_2C(CO_2H)_2 \xrightarrow[base]{\Delta} RCH=CHCO_2H$$

Org Syn Coll Vol *3* 425 (1955)
Org Rxs *15* 204 (1967) (review)
Heterocycles *20* 1541 (1983) (mechanism)

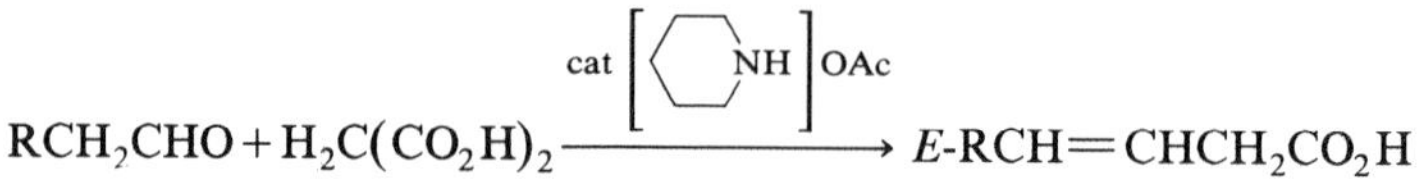

$$RCH_2CHO + H_2C(CO_2H)_2 \xrightarrow{\text{cat}\ [\ \text{NH}\ ]\text{OAc}} E\text{-}RCH=CHCH_2CO_2H$$

JCS 740 (1931); 557 (1933)
TL *28* 93 (1987)

$$RCHO + (CH_3CO)_2O \xrightarrow{base} RCH=CHCO_2H$$

Org Rxs *1* 210 (1942) (review)
Org Syn Coll Vol *3* 426 (1955)

$$RCHO + R'CH_2CO_2R \xrightarrow{base} RCH=CR'CO_2R$$

Org Syn Coll Vol *1* 252 (1941)
Ber *76* 676 (1943)
Org Rxs *6* 1 (1951)

$$RCHO + RO_2CCH_2CH_2CO_2R \xrightarrow{base} RCH=\overset{\overset{\displaystyle CO_2R}{|}}{C}CH_2CO_2R$$

Org Rxs *6* 1 (1951) (review)

$$RCHO + HO_2CCH_2CO_2R \xrightarrow{base} RCH=CHCO_2R$$

JACS *68* 376 (1946)

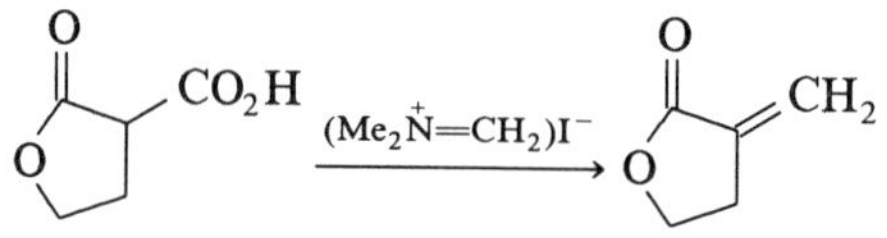

Tetr *38* 2797 (1982)

$$RCHO + R'CH_2CN \xrightarrow{base} RCH=CR'CN$$

Org Syn Coll Vol *3* 715 (1955)

$$(RCHO)\ R_2CO + H_2CXY \xrightarrow{base} R_2C=CXY$$

Reviews: Org Rxs *15* 204 (1967)
          Can J Chem *45* 1001 (1967); *47* 3137 (1969)

$\underline{X}$	$\underline{Y}$	
Ph	CN	BCSJ *56* 1885 (1983)
COR	COR	Syn 667 (1974) Syn Commun *13* 1203 (1983)
COR	CO_2R	Tetr *28* 663 (1972)
CO_2R	CO_2R	JOC *4* 493 (1939); *48* 3603 (1983) Ber *76* 676 (1943) Org Syn Coll Vol *3* 377 (1955) TL 4723 (1970) Tetr *29* 635 (1973) Heterocycles *20* 1541 (1983) Org Syn *64* 63 (1985)
CO_2R	NO_2	Tetr *28* 663 (1972)
CO_2R	$PO(OR)_2$	Tetr *30* 301 (1974)
CO_2R	CN	JACS *95* 4873 (1973)
CN	CN	Chem Rev *69* 591 (1969) (review) Syn 165 (1978) (review) JOC *48* 1366, 3852 (1983) BCSJ *56* 1885 (1983) Tetr *39* 1161, 1167 (1983)
$PO(OR)_2$	$PO(OR)_2$	Tetr *30* 301 (1974)

4. WITTIG AND RELATED REACTIONS

See also page 167, Section 3.

1. Wittig Reaction

1.1. General

Quart Rev *17* 406 (1963) (general)

Angew Int *3* 250 (1964) (unsaturated fatty acids); *16* 423 (1977) (industrial practice); *21* 545 (1982) (synthesis of $R^1R^2C{=}PPh_3$)

Pure Appl Chem *9* 255 (general), 271 (stereochemistry) (1964); *51* 515 (1979) (polyenes)

Org Rxs *14* 270 (1965) (general); *25* 73 (1977) (phosphoryl-stabilized anions)

A. W. Johnson, "Ylid Chemistry," Academic Press, New York (1966)

Topics Stereochem *5* 1 (1970) (stereochemistry)

Syn 765 (1975) (bis-Wittig reaction)

I. Gosney, A. G. Rowley in "Organophosphorus Reagents in Organic Synthesis," Ed. J. I. G. Cadogan, Academic Press, New York (1979), Chpt 2

JCS Perkin I 1 (1979) (polymer-supported reagents)

Ann 1705, 2117 (1981) (pheromones and dienes)

Syn Commun *11* 125 (1981) (phase transfer); *12* 107 (phase transfer), 469 (preparation of ylid via decarboxylation) (1982)

J Chem Res (S) 142 (1981) (phase transfer); 188 (1982) (D labelling in vinylic positions)

JACS *104* 5821 (1982) (stereochemistry); *106* 7514 (1984) (bis-Wittig reaction); *107* 1068 (1985) (stereochemistry)

Chimia *36* 396 (1982) (preparation of ylid)

Russ Chem Rev *51* 1 (1982) (fluorine-containing ylids)

Topics Curr Chem *109* 85 (natural products), 165 (industrial applications) (1983)

Phosphorus and Sulfur *18* 171 (1983) (general)

Ann 2135 (1983) (promotion by high pressure)

JOC *49* 4293 (1984) (preparation of phosphonium salts); *51* 3302 (1986) (concentration effects on stereochemistry)

TL *28* 2191, 4377 (1987) (both promotion by high pressure)

1.2. Intramolecular Wittig

Helv *60* 68, 81 (1977)
Tetr *36* 1717 (1980)
CC 14 (1981) (indole synthesis)
TL *23* 3543 (1982) (bicycloalkenones)
JOC *47* 5372 (1982) (bicycloalkenes)

1.3. Mechanism

Angew Int *7* 650 (1968)
JACS *95* 5778 (1973); *103* 2823 (1981); *104* 5821 (1982); *106* 1873 (1984); *107* 1068 (1985); *108* 7664
    (1986)
TL 2707 (1979)
CC 1072 (1979)
Pure Appl Chem *52* 771 (1980)
JOC *52* 4637 (1987)

1.4. Miscellaneous Reactions

$$R_2CuLi + (H_2C\!\!=\!\!CH\overset{+}{P}Ph_3)Br^- \xrightarrow{R'CHO} RCH_2CH\!\!=\!\!CHR'$$

R = alkyl, aryl, vinylic   cis > trans

TL *26* 1799 (1985)

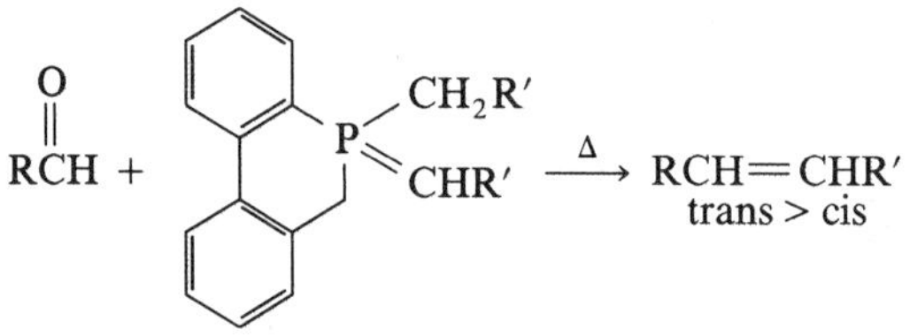

$$RCH\!\!=\!\!CHR'$$
trans > cis

TL *28* 3445 (1987)

$$R\overset{O}{\overset{\|}{C}}H + \left[HO_2C(CH_2)_2\overset{+}{P}Ph_2CH_2R'\right]Br^- \xrightarrow{base} RCH\!\!=\!\!CHR'$$

trans > cis

TL *28* 1165 (1987)

$$R\overset{O}{\overset{\|}{C}}H + (Ph_3\overset{+}{P}CH_2R')X^- \xrightarrow[\text{18-crown-6}]{\text{KO-}t\text{-Bu or K}_2CO_3} RCH\!\!=\!\!CHR'$$

trans > cis

Syn 784 (1975)

$$R\overset{O}{\overset{\|}{C}}H + Ph_3P\!\!=\!\!CH(CH_2)_nX \longrightarrow RCH\!\!=\!\!CH(CH_2)_nX$$

$X = O^-, CO_2^-, NR^-$ trans > cis

TL *22* 4185 (1981); *26* 311 (1985)
JACS *107* 217 (1985)

$$\underset{\text{RCH}}{\overset{\text{O}}{\|}} + Ph_3P=CHC\overset{-}{C}HCOEt \longrightarrow RCH=CHCCH_2COEt$$

cis > trans

TL *27* 739 (1986)

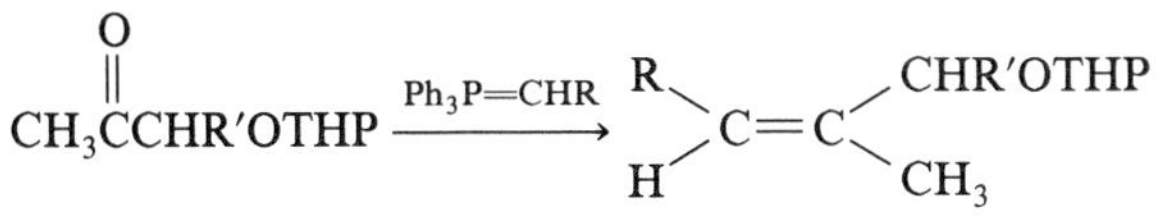

$$CH_3\overset{O}{\overset{\|}{C}}CHR'OTHP \xrightarrow{Ph_3P=CHR} \underset{H}{\overset{R}{>}}C=C\overset{CHR'OTHP}{\underset{CH_3}{<}}$$

JOC *45* 4260 (1980)
CL 1711 (1981)

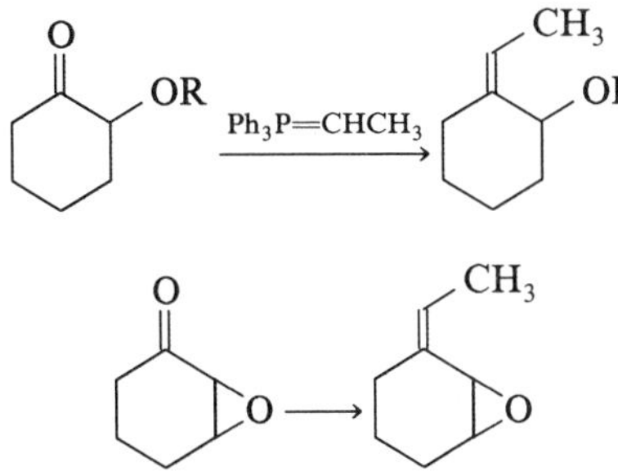

JOC *50* 5910 (1985)

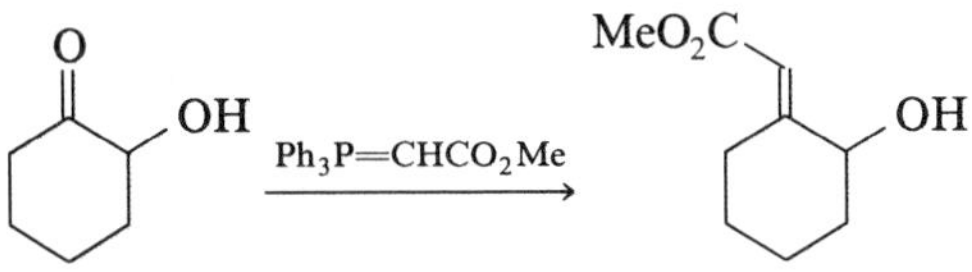

JOC *52* 2629 (1987)

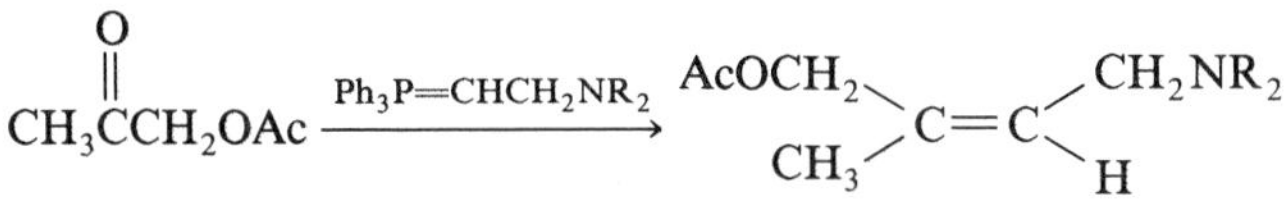

$$CH_3\overset{O}{\overset{\|}{C}}CH_2OAc \xrightarrow{Ph_3P=CHCH_2NR_2} \underset{CH_3}{\overset{AcOCH_2}{>}}C=C\overset{CH_2NR_2}{\underset{H}{<}}$$

> 99% Z

TL *23* 2219 (1982)

$$\underset{\text{HCOR}}{\overset{\text{O}}{\|}} \xrightarrow{Ph_3P=CR'_2} R'_2C=CHOR$$

CL 967 (1981)
TL *23* 427 (1982)
JOC *49* 3595 (1984)

$$R^1\overset{O}{\overset{\|}{C}}OR^2 \xrightarrow{2\ Ph_3P=CHR^3} R^3CH_2CR^1=CHR^3$$

TL 1439 (1975)
JOC *43* 3306 (1978); *44* 3157 (1979)

$$(RCO)_2O \xrightarrow[\text{2. RC}\equiv\text{CLi}]{\text{1. Ph}_3\text{P}=\text{CR}_2} R_2C=CRC\equiv CR$$

CC 703 (1987)

2. β-Oxido Ylids

Angew Int *5* 126 (1966)
Ann *708* 1 (1967)
Syn 38 (1969)
Ber *103* 2814 (1970)
TL 447 (1970); 3231 (1977)
JACS *92* 226, 6635, 6636, 6637 (1970); *93* 4835 (1971)
JOC *45* 3350 (1980); *50* 3111 (1985)

3. Ph₃P=CHLi

JACS *104* 4724 (1982)
TL *26* 555 (1985)

4. Phosphonates

4.1. General

Ber *92* 2499 (1959); *95* 581 (1962)
JACS *83* 1733 (1964); *89* 5292 (1967)
TL 1821 (1971); *21* 2161 (1980) (KF · 2 H₂O as base); *28* 2951 (1987) [cat Ba(OH)₂, EtO₂CCH₂PO(OEt)₂, ultrasound]
Org Syn Coll Vol *5* 509 (1973)
Syn 869 (1974) (phase transfer); 884 (1979) (phase transfer); 117 (1981) (crown ether use); 300 (1983) (heterogeneous); 1097 (1985) [Ba(OH)₂, phase transfer]
Chem Rev *74* 87 (1974) (review)
Org Rxs *25* 73 (1977) (review)
JCS Perkin I 2516 (1980) (polymer-supported phosphonates)
Syn Commun *14* 701 (1980) (phase transfer)
J Chem Res (S) 143 (1981) (phase transfer)
Z Chem *22* 117 (1982) (heteroatom-substituted olefins)
Phosphorus and Sulfur *14* 385 (1983) (reaction conditions)
Tetr *40* 5153 (1984) (stereochemistry); *41* 1259 (1985) (alumina)
JOC *50* 2624 (1985) (α,β-unsaturated esters); *52* 3875 (1987) (ultrasound, phase transfer)
CC 1509 (1987) (enone synthesis)

4.2. Intramolecular

JACS *90* 5926 (1968) (enones); *100* 7069 (1978) (lactone); *103* 1222 (1981); *104* 2027, 2030 (1982); *106* 260, 1148 (1984); *107* 7967 (1985) (enone); *108* 1035, 3110, 3112, 6389 (1986); *109* 2208 (1987) (enone)

Angew Int *7* 300 (1968) (lactones); *20* 286 (1981)

CC 445 (1970) (lactams); 413 (1986) (enone)

JOC *44* 4010, 4011 (1979) (lactones); *52* 1375 (1987) (enone)

Helv *62* 2661 (1979) (enones)

TL *27* 2157, 4873 (1986); *28* 2727 (1987) (lactone)

$$\text{RCHO} + (\text{EtO})_2\overset{\overset{\text{O}}{\|}}{\text{P}}\overset{\overset{\text{Li}}{|}}{-}\text{CHCH}=\text{N}(t\text{-Bu}) \xrightarrow{\text{H}_3\text{O}^+} \text{RCH}=\text{CHCH}\overset{\overset{\text{O}}{\|}}{}$$

JOC *51* 5111 (1986)

5. Phosphonic Acid bis Amides [(R$_2$N)$_2$POCHLiR]

JACS *88* 5652, 5653 (1966); *90* 6816 (1968)

6. Phosphinothioic Amides [PhPS(NR$_2$)CHLiR]

JACS *104* 7041 (1982)

TL *23* 5005 (1982)

JOC *52* 34 (1987)

7. Phosphine Oxides

$$\text{R}_2\overset{\overset{\text{O}}{\|}}{\bar{\text{P}}}\text{CHR} + \text{R}'\text{CHO} \ (\text{R}'_2\text{CO}) \longrightarrow \text{RCH}=\text{CHR}'$$

Ber *92* 2499 (1959)

JCS Perkin I 639 (1976) (dienes); 550 (1977) (dienes); 3099 (1979) (vinyl ethers); 2893 (1983)

B. J. Walker, "Organophosphorus Reagents in Organic Synthesis," Ed. J. I. G. Cadogan, Academic Press, New York (1979)

TL 2433 (1979) (enamines); 2671 (1980) (enamines); *23* 4505 (1982) (allylic amines); *24* 111 (1983); *28* 5559 (1987) (alkylidene cyclopropanes)

CC 100 (1981); 1196 (1987)

Tetr *37* 3911 (1981)

JOC *46* 459 (1981); *51* 1264, 1269 (1986)

Ann 99 (1981)

CL 1143 (1982)

JACS *108* 2662 (1986)

$$\text{R}_2\overset{\overset{\text{O}}{\|}}{\bar{\text{P}}}\text{CHR} + \text{ROCR}'\overset{\overset{\text{O}}{\|}}{} \longrightarrow \text{R}_2\overset{\overset{\text{O}}{\|}}{\text{P}}\text{CHRCR}'\overset{\overset{\text{O}}{\|}}{} \xrightarrow[\text{NaBH}_4/\text{NaH}]{\text{LiBH}_4 \text{ or}} \text{RCH}=\text{CHR}'$$

CC 100 (1981)

TL *24* 111, 5293 (1983)

JOC *52* 4303 (1987)

8. Arsenic Ylids

Org Prep Proc Int *14* 373 (1982) (Ph$_3$As=CHCOCH$_3$)
TL *26* 6447 (1985) (Ph$_3$As=CHCHO); *28* 2155 (Ph$_3$As=CHCH=CHCOR, R = H, Me), 2159
  (Ph$_3$As=CHCONR$_2$, Ph$_3$As=CHCHO, Ph$_3$As=CHCH=CHCHO) (1987)
JOC *52* 3558 (1987) (Ph$_3$As=CHCHO)

9. Sulfinamides

JACS *88* 5656 (1966); *90* 5548, 5553 (1968)

10. Tellurium Compounds

$$ArCHO + BrCH_2X \xrightarrow[\Delta]{n\text{-}Bu_2Te} ArCH=CHX$$

$$X = CO_2R, COR, CN$$

TL *28* 801 (1987)

$$RCHO\,(R_2CO) + R_2Te=CHCO_2Et \longrightarrow RCH=CHCO_2Et$$

TL *24* 2599 (1983)

11. Boron Anions

$$Mes_2BCHLiR + R'CHO\,(R'_2CO) \xrightarrow[\text{or }(CF_3CO)_2O]{Me_3SiCl/HF} RCH=CHR'$$

$$E > Z$$

TL *24* 635 (1983)
CC 297 (1987)

12. Peterson Reaction (Oxysilane Elimination)

$$RCHO + R'_3Si\overline{C}HX \longrightarrow RCH=CHX$$

JOC *33* 780 (1968) (alkenes, vinylic sulfides and phosphines); *37* 939 (vinylic sulfides and phosphonates), 1926 (ketene thioacetals) (1972); *39* 3264 (1974); *43* 1947 (1978); *45* 2013, 2713, 3451 (1980);
  51 2863, 5111 (1986); *52* 2314 (1987) (vinyltriazoles)
TL 1137 (1970); 1403, 4005 (X = CN) (1974); 7 (1976); *22* 1575, 1595, 2751, 2923, 4705 (1981);
  23 1279 (1982); *25* 5177 (1984); *27* 3729, 4189, 5829 (vinylic phosphonates) (1986); *28* 259, 803
  (1987)
CC 526 (1972) (ketene thioacetals); 537 (1975); 877 (1981); 98 (1986)
Angew Int *11* 443 (1972) (ketene thioacetals); *15* 161 (1976)
Ber *106* 2277 (1973) (ketene thioacetals)
JACS *96* 1620 (1974); *102* 3964 (1980); *103* 474 (1981); *104* 5708 (1982); *106* 3245, 3252 (1984);
  107 2474 (1985) (X = OMe)

CL 853 (1975) (X = SR); 1093 (1982) (X = CN)
Acct Chem Res *10* 442 (1977) (review)
Chem Soc Rev *7* 15 (1978)
JCS Perkin I 26 (1979)
E. Colvin, "Silicon in Organic Synthesis," Butterworths, London (1981), Chpt 12 (review)
W. P. Weber, "Silicon Reagents for Organic Synthesis," Springer Verlag, New York (1983), Chpt 6 (review)
Syn Commun *13* 833 (1983)
Syn 384 (1984) (review); 734 (1986)

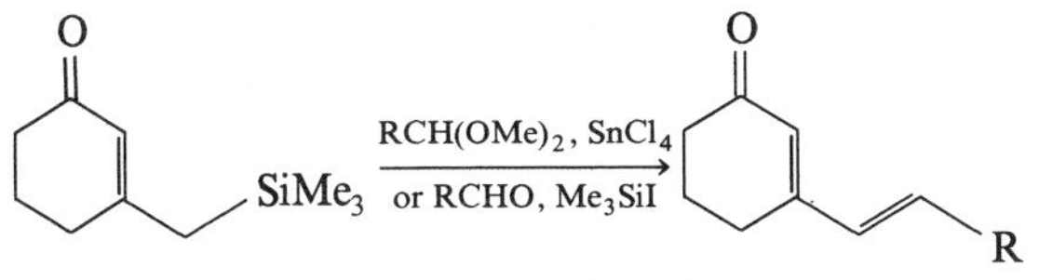

JOC *51* 1932 (1986)

$$RCH + Me_3SiCH = COMe \xrightarrow[AlCl_3/80°C]{TiCl_4/-78°C} RCH = CHCO_2Me$$

95% *Z* olefin
96% *E* olefin

TL *22* 1805 (1981)

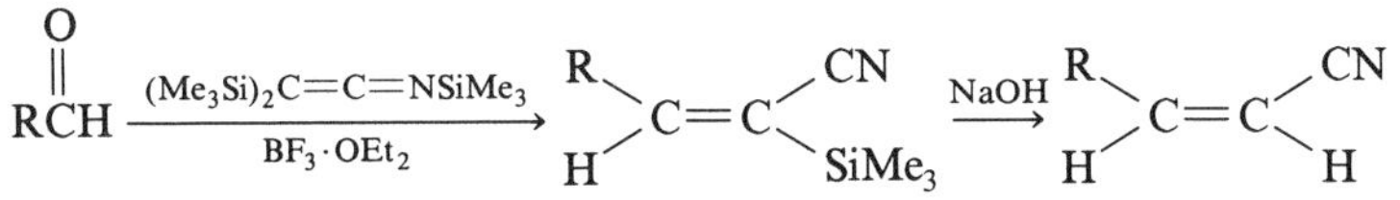

CC 56 (1982)

$$R_2CO + R_3SiCRCH = NR' \longrightarrow R_2C = CRCH = NR$$

TL 7 (1976); *24* 2481 (1983); *26* 2391 (1985); *27* 6177 (1986); *28* 259 (1987)
JOC *45* 2013 (1980); *50* 2798 (1985)

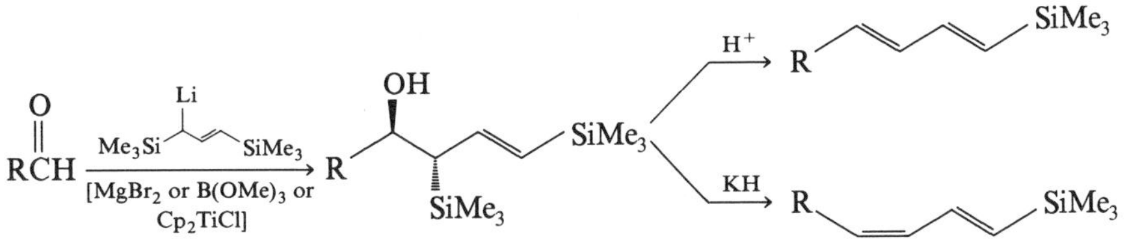

CC 969 (1982); 921 (1983)
JOMC *264* 207 (1984)
TL *28* 211 (1987)

$$RCH + R'_3SiCH = C = CMSiR'_3 \longrightarrow E \text{ or } Z\text{-}RCH = CHC \equiv CSiR'_3$$

M = Li, MgBr

JACS *103* 5568 (1981)
TL *23* 719 (1982)

$$Me_3SiCH_2CO_2Et \xrightarrow[\substack{2.\ H_2SO_4\ or \\ BF_3 \cdot OEt_2}]{1.\ 2\ RMgX} H_2C{=}CR_2$$

TL *23* 1035 (1982)

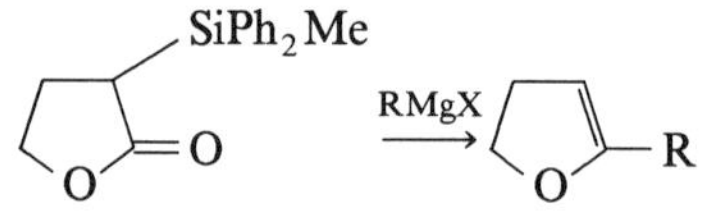

SiPh$_2$Me

$\xrightarrow{RMgX}$ R

TL *23* 271 (1982)

$$\overset{O}{\overset{\|}{RCNR_2}} + XY\overline{C}SiMe_3 \longrightarrow \overset{NR_2}{\overset{|}{RC}}{=}CXY$$

X	Y	
H	CONMe$_2$	TL 709 (1978)
H	RS, RSO$_2$	BCSJ *55* 1205 (1982)
RS	RS	Angew Int *11* 443 (1972)

13. Isonitriles

Angew Int *7* 805 (1968)

14. Miscellaneous Reactions

$$R_2C{=}O \longrightarrow R_2C{=}CH_2$$

ClCH$_2$I, CH$_3$Li/Li	CC 1665 (1986)
H$_2$CI$_2$, Mg(Hg)	TL 5153 (1967) Tetr *26* 1281 (1970) Ber *107* 3486 (1974) JACS *107* 5739 (1985)
H$_2$CX$_2$ (X = Br, I) or ClCH$_2$I, Zn-Cu	JOMC *10* 518 (1967); *12* 263 (1968)
H$_2$CI$_2$, Zn, Me$_3$Al	TL 2417 (1978); *26* 5581 (1985) BCSJ *53* 1698 (1980) JACS *107* 4964 (1985)
H$_2$CBr$_2$, Zn, TiCl$_4$	TL 2417 (1978); *23* 4293 (1982); *27* 5467 (1986); *28* 1893 (1987) JOC *45* 2005 (1980); *48* 2298 (1983); *50* 5898 (1985); *52* 34, 5583 (1987) BCSJ *53* 1698 (1980) Helv *65* 293 (1982); *69* 865 (1986) JACS *108* 3513, 7791 (1986); *109* 3147, 6937 (1987) Org Syn *65* 81 (1987) CC 1008 (1987)

H_2CI_2, Zn, $TiCl_4$	TL *26* 5579 (1985)
H_2CI_2, Zn, Ti(O-i-Pr)$_4$	TL *26* 5581 (1985)
$Cp_2TiCH_2 \cdot ZnX_2$	TL *24* 2043 (1983)

JACS *100* 3611 (1978); *102* 3270 (1980)
TL *23* 3143 (1982); *25* 5733 (1984); *27* 6189 (1986); *28* 3209 (1987)
Pure Appl Chem *55* 1733 (1983) (review)
JOC *50* 1212, 2386 (1985); *52* 1780 (1987)

TL *25* 5733 (1984)

$Cl_2Mo(Me){=}CH_2$	Angew Int *23* 532 (1984) TL *27* 5355 (1986)
$ClMo(Me)_2{=}CH_2$	Angew Int *23* 532 (1984)
$ClMo(O){=}CH_2$	TL *27* 5355 (1986)
$H_2C{=}PPh_3$	Org Rxs *14* 270 (1965) (review) JOC *46* 1105 (1981) Org Syn *64* 164 (1985) Syn Commun *15* 855 (1985)

TL *27* 1909 (1986)

Ph_2POCH_2Na	Ber *91* 61 (1958)
$(Me_2N)_2POCH_2Li$	JACS *88* 5653 (1966) Org Rxs *25* 73 (1977)
$Me_3SiCH_2PO(OMe)_2/\Delta$	CL 1385 (1978)
$Me_3SiCH_2MgCl/H_2O/NaH$	JOC *51* 5311 (1986)
$Me_3SiCH_2MgCl/H_2O/(HO_2C)_2$	JOC *52* 3745 (1987)
$Me_3SiCH_2Li(MgX)/HCl$	JOC *33* 780 (1968) TL 4193 (1973); *21* 3451 (1980); *27* 4873 (1986) Acct Chem Res *10* 442 (1977)
$Me_3SiCH_2Li(MgX)/SOCl_2$ or CH_3COCl	JOC *39* 3264 (1974)
Me_3SiCH_2Li, $CeCl_3/HF$ or KH	JOC *52* 281 (1987)
$Me_3SiCH_2TiCl_3$ (RCHO only)	TL *22* 5031 (1981)
$Me_3SiCH_2CrCl_2/H_3O^+$ (RCHO only)	TL *22* 5031 (1981)
$Me_3GeCH_2TiCl_3$ (RCHO only)	TL *22* 5031 (1981)
$[(Me_3Si)_2N]_2\overline{UCH_2SiMe_2N}SiMe_3$	JOC *52* 688 (1987)
XCH_2Li (X = Ph_3Pb, Ph_3Sn, Ph_2Sb)/Δ or H^+	TL 4399 (1978)

PhSCH$_2$Li/TiCl$_4$-LiAlH$_4$-3°amine   CL 871 (1975)

PhSCH$_2$Li/ [benzene-fused 1,3,2-dioxaphosphole] PCl, Δ   TL 737 (1972)

PhSCH$_2$Li/(PhCO)$_2$O/Li, NH$_3$ JACS *94* 4728 (1972)

PhSCH$_2$Li/(PhCO)$_2$O/TiCl$_4$, Zn   CL 1523 (1974)

PhSOCH$_2$Li/ [benzene-fused 1,3,2-dioxaphosphole] PCl, Δ   TL 649 (1972)

p-MeC$_6$H$_4$SO$_2$CH$_2$MgI/NaBH$_4$/electrolysis   CL 69 (1978)

$$\overset{\overset{O}{\|}}{PhSCH_2Li}/Al(Hg), HOAc \\ \underset{NMe}{\|}$$

JACS *95* 6462 (1973); *101* 3602 (1979);
103 7667 (1981); *106* 4547 (1984)

$$\overset{\overset{S}{\|}}{PhPCH_2Li}/MeI, py \\ \underset{NMe_2}{\|}$$

JACS *104* 7041 (1982); *106* 8217 (1984)

m-CF$_3$C$_6$H$_4$SeCHLi(OMe)/MsCl, Et$_3$N   JACS *101* 6638 (1979)

$$\overset{\overset{O}{\|}}{RCH} + \overset{\overset{Li}{|}}{ArSO_2CHR} \longrightarrow \overset{HO\ SO_2Ar}{\underset{|\ \ \ |}{RCHCHR}} \longrightarrow RCH{=}CHR$$

See page 146, Section 23.

$$\overset{\overset{O}{\|}}{RCH} \xrightarrow[\text{2. Zn, HOAc}]{\text{1. LiCHBr}_2} RCH{=}CHBr$$

TL *22* 3745 (1981)

$$\overset{\overset{O}{\|}}{RCH} \xrightarrow[\text{CrCl}_2]{\text{HCY}_2X\ (Y = \text{halogen})} \underset{H}{\overset{R}{\diagdown}}C{=}C\underset{X}{\overset{H}{\diagup}}$$

$\underline{X}$

alkyl   JACS *109* 951 (1987)

Cl, Br, I   JACS *108* 7408 (1986)

PhS, Me$_3$Si   TL *28* 1443 (1987)

$$\overset{\overset{O}{\|}}{RCH}\ (\overset{\overset{O}{\|}}{RCR}) \xrightarrow[\text{Zn, Et}_2\text{AlCl}]{\text{YCX}_2\text{CO}_2\text{R}'} RCH{=}CYCO_2R'$$

X = Cl, Br; Y = H, Cl

BCSJ *53* 1698 (1980)

$$RCHO \xrightarrow[\substack{\text{2. } ClCO_2Et \\ \text{3. } P(OEt)_3}]{\text{1. } LiSCHLiCO_2Et} E\text{-}RCH{=}CHCO_2Et$$

Syn 127 (1983)

$$\underset{RCR}{\overset{O}{\|}} + \underset{BrCH_2COR}{\overset{O}{\|}} \xrightarrow{n\text{-}Bu_3Sb} R_2C{=}\underset{\|}{\overset{O}{C}}HCOR$$

TL 27 2903 (1986)

$$\underset{RCR}{\overset{O}{\|}} \xrightarrow[\substack{\text{3. } KOH \\ \text{4. } H^+}]{\substack{\text{1. } LiC{\equiv}COEt \\ \text{2. } H_2SO_4}} R_2C{=}\underset{\|}{\overset{O}{C}}HCOH$$

JACS 108 2690 (1986)

$$\underset{RCH}{\overset{O}{\|}} \longrightarrow \longrightarrow \underset{RCH}{\overset{NNLiTs}{\|}} \xrightarrow{R'CHLiX} RCH{=}CHR'$$

X = CN, SPh, SO_2R

JACS 101 249 (1979)

$$\underset{RCR}{\overset{O}{\|}} \xrightarrow{Ph_2C(SH)CO_2H} \underset{R}{\overset{R}{\underset{\|}{}}} \overset{S}{\underset{O}{\times}} \overset{Ph}{\underset{Ph}{\underset{\|}{\overset{}{C}}}} {=}O \xrightarrow[\Delta]{(Et_2N)_3P} R_2C{=}CPh_2$$

CC 1225 (1970)

$$R^1R^2C(OEt)_2 + H_2C{=}CHOEt \xrightarrow[\text{2. } H_3O^+]{\text{1. cat acidic clay}} R^1R^2C{=}\underset{\|}{\overset{O}{C}}HCH$$

Syn 137 (1981)

$$\underset{R^1CR^2}{\overset{O}{\|}} \xrightarrow[X = \text{halogen or } OCH_2\text{-}t\text{-Bu}]{R_2C{=}W(OCH_2\text{-}t\text{-Bu})_2X_2} R^1R^2C{=}CR_2$$

R^1, R^2 = H, Ph; Me, Me; H, OMe; R, OR; Me, NMe_2

CC 531 (1986)

$$\underset{RCOR}{\overset{O}{\|}} + \underset{\underset{SO_2Ph}{|}}{\overset{Li}{\underset{|}{PhSO_2CHR'}}} \longrightarrow \underset{\underset{SO_2Ph}{|}}{\overset{O}{\underset{|}{RCCHR'}}} \xrightarrow{NaBH_4} \underset{\underset{SO_2Ph}{|}}{\overset{HO}{\underset{|}{RCHCHR'}}} \xrightarrow[\text{2. } Na(Hg)]{\text{1. } Ac_2O} RCH{=}CHR'$$

JACS 108 284 (1986)

See page 146, Section 23, for other reagents to effect elimination.

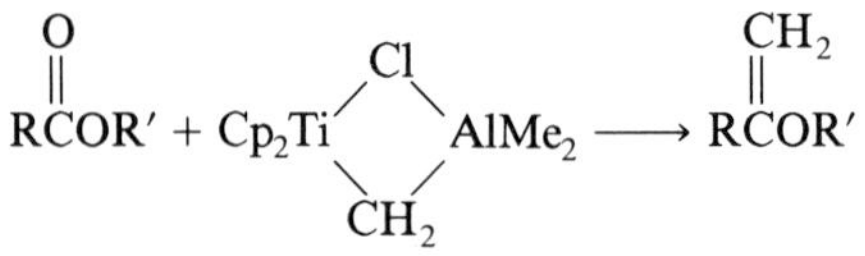

$$\underset{\substack{\| \\ O}}{RCOR'} + Cp_2Ti\underset{CH_2}{\overset{Cl}{\diamond}}AlMe_2 \longrightarrow \underset{\substack{\| \\ CH_2}}{RCOR'}$$

TL 1439 (1975); *23* 3143 (1982); *25* 395 (1984); *28* 4773 (1987)
JACS *102* 3270 (1980); *106* 6868 (1984); *107* 7352 (1985)
Pure Appl Chem *55* 1733 (1983)
JOC *48* 1829 (1983); *50* 1212, 2386 (1985); *51* 5458 (1986)

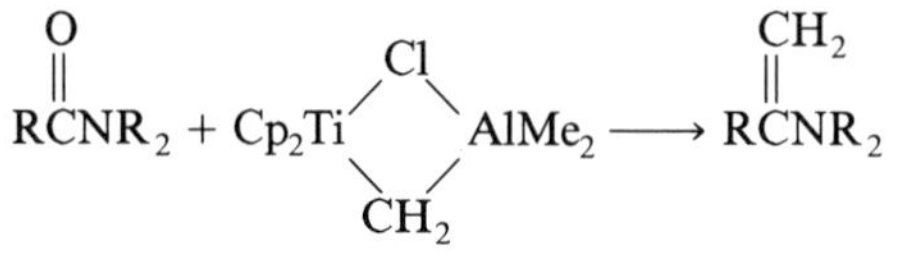

$$\underset{\substack{\| \\ O}}{RCNR_2} + Cp_2Ti\underset{CH_2}{\overset{Cl}{\diamond}}AlMe_2 \longrightarrow \underset{\substack{\| \\ CH_2}}{RCNR_2}$$

JOC *50* 1212 (1985)

$$RCHO\ (R_2CO) \xrightarrow{Ta[CH_2C(CH_3)_3]_3[CHC(CH_3)_3]} RCH{=}CHC(CH_3)_3$$

$$RCO_2R \xrightarrow{\hspace{3cm}} \underset{OR}{RC}{=}CHC(CH_3)_3$$

$$HCONR_2 \xrightarrow{\hspace{3cm}} R_2NCH{=}CHC(CH_3)_3$$

JACS *98* 5399 (1976)

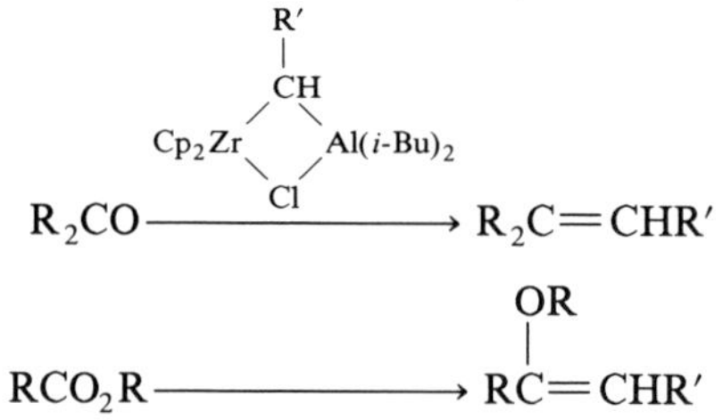

$$R_2CO \xrightarrow{\quad Cp_2Zr\overset{CH\,R'}{\underset{Cl}{\diamond}}Al(i\text{-}Bu)_2 \quad} R_2C{=}CHR'$$

$$RCO_2R \xrightarrow{\hspace{3cm}} \underset{OR}{RC}{=}CHR'$$

JACS *105* 640 (1983)

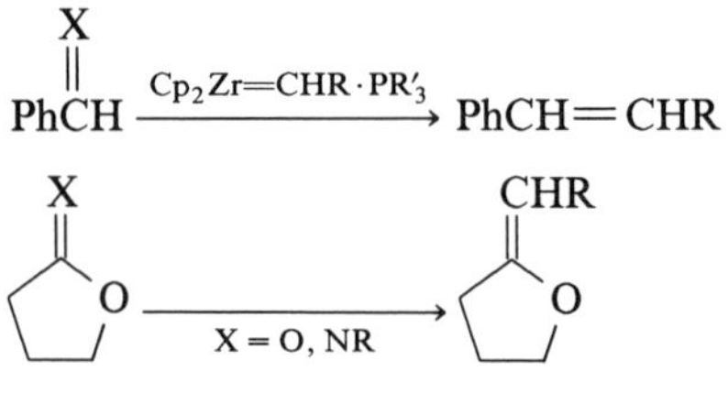

$$\underset{\substack{\| \\ X}}{PhCH} \xrightarrow{Cp_2Zr{=}CHR\cdot PR_3'} PhCH{=}CHR$$

X = O, NR

JACS *106* 8300 (1984)

$$\underset{\substack{\| \\ O}}{RCOR} + R'CHBr_2 \xrightarrow[Me_2NCH_2CH_2NMe_2]{Zn,\ TiCl_4} \underset{RO}{RC}{=}CHR'$$

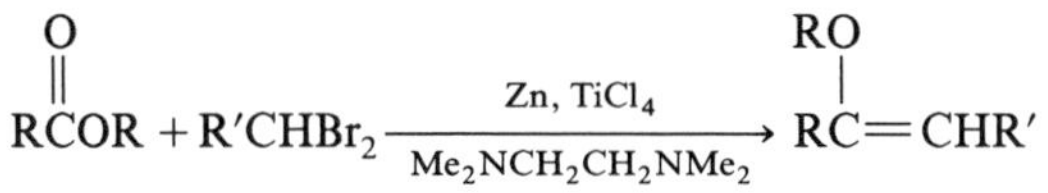

JOC *52* 4410 (1987)

5. METAL-PROMOTED COUPLING REACTIONS

See also page 109, Section 6; and page 211, Section 6.

1. Lithium Reagents

For transition metal-catalyzed cross-coupling of organolithium compounds see the appropriate transition metal.

For other ways of generating vinylic lithium reagents see page 193, Section 6; and page 197, Section 7.

$$\underset{R}{\overset{R}{\diagup}}C=C\underset{X}{\overset{R}{\diagdown}} \xrightarrow[\text{or } 2 \text{ } t\text{-BuLi}]{\text{Li or } sec\text{-BuLi} \quad E^+} \underset{R}{\overset{R}{\diagup}}C=C\underset{E}{\overset{R}{\diagdown}}$$

$$E^+ = D_2O, RX, I_2, Me_3SiCl, RSSR, R_2CO, RCHO, ROOR, ClCO_2R, R\overset{O}{\overset{\diagup\diagdown}{CH-CH_2}}$$

TL 3809 (1974); 4839 (1976); 4661 (1978); *22* 3745 (1981); *28* 5145 (1987)
Syn 434 (1975)
Ber *111* 2785 (1978)
Syn Commun *9* 831 (1979)
Organomet *1* 667 (1982)
JOC *52* 3860, 4495 (1987)

$$\text{(structure)} - Br \xrightarrow{\text{MeLi}} \text{(structure)} - CH_3$$

JOC *52* 2674 (1987)

$$Br(CH_2)_n \diagup\diagdown I \xrightarrow{n\text{- or } t\text{-BuLi}} \widehat{(CH_2)_n}$$

JACS *105* 6344 (1983); *106* 6105 (1984)
TL *26* 5671 (1985) (cyclopropenes); *28* 5793 (1987)

$$\underset{H}{\overset{R}{>}}C=C\underset{H}{\overset{OR}{<}} \quad \xrightarrow[\text{2. E}^+]{\text{1. RLi}} \quad \underset{H}{\overset{R}{>}}C=C\underset{E}{\overset{OR}{<}}$$

$$E^+ = RX, RCHO, R_2CO, RCO_2R, RCN, CO_2$$

Ann *763* 208 (1972)
Syn 888 (1974); 748 (1982)
JACS *96* 7125 (1974); *97* 3822 (1975) (cuprate); *103* 5259 (1981)
TL 4187 (1977); *24* 3905 (1983) (cuprate); *27* 5975, 6201 (1986)
Compt Rend C *284* 281 (1977)
Tetr *37* 3997 (1981)
Syn Commun *12* 579 (1982)
JOC *51* 4492 (1986)

$$RCH=CHCO_2R \quad \xrightarrow[\substack{\text{2. R'X} \\ \text{3. silica gel, }\Delta}]{\text{1. LiNR}_2} \quad RCH=\overset{\overset{\displaystyle R'}{|}}{C}CO_2R$$

CC 1410 (1987)

$$PhCH=CHCN \quad \xrightarrow[\text{2. 2 E}^+]{\text{1. 2 LDA}} \quad PhCH=\overset{\overset{\displaystyle E}{|}}{C}CN \text{ or } Ph\overset{\overset{\displaystyle E}{|}}{C}=\overset{\overset{\displaystyle E}{|}}{C}CN$$

$$E^+ = MeOD, RX, RCHO, (MeS)_2$$

JOC *52* 3825 (1987)

$$YCR=CHX \quad \xrightarrow[\text{2. E}^+]{\text{1. base}} \quad RCY=\overset{\overset{\displaystyle E}{|}}{C}X$$

X	Y	R	E⁺	
Cl	CH₃—(dioxane)	H	RX, RCHO, R₂CO, lactone	JOC *52* 2335 (1987)
SR	RS	H	D⁺, CH₃OSO₂F, RCHO, (RS)₂	TL 3583 (1977)
SPh	N O (morpholine)	H, Me	D₂O, 1° RI, RCHO	J Chem Res (S) 48 (1982)
SOR	RS	H	MeOD, MeI, (MeS)₂	TL 4277 (1979)
SO₂Ph	MeO	H	CH₃OD, MeI, R₂CO	JOC *52* 4760 (1987)
SO₂Ph	RO	H, Me	D₂O, RX, RCHO, R₂CO	TL *28* 4127 (1987)
CO₂R	Ph	Ph	D₂O, MeI, RCHO, R₂CO, CO₂, Michael acceptor	JCS Perkin I 1329 (1981)

CO_2R	OMe	R	D^+, RCHO, R_2CO, RCOX, $(RCO)_2O$, RCO_2R, $(MeS)_2$	TL *23* 1793 (1982) Tetr *39* 2043 (1983)
CO_2R	OMe	CO_2R	D^+, RCHO, Ac_2O, $(MeS)_2$	Tetr *39* 2043 (1983)
CO_2R	RNCOR	H	D_2O	TL *22* 4259 (1981)
CN	EtO	H	D_2O	TL *22* 4259 (1981)
CN	R_2N	H	MeOD, RI, RCHO	Angew Int *16* 853 (1977)
CN	RS	H	MeOD, MeI, RCHO, $(RS)_2$	TL 4277 (1979)
CN	Ph	H	D^+ D^+, MeI, RCHO, R_2CO, enone (1,4-addition)	TL 4273 (1979) Syn 797 (1979) Tetr *37* 2143 (1981)
CN	Ar	H	D^+, MeI, $(MeS)_2$	Tetr *39* 2043 (1983)
CN	Ph	Ph	RI, R_2CO, CO_2, Michael acceptor	JCS Perkin I 1228, 1232 (1978)

$$\text{HCY=CRX} \xrightarrow[\text{2. } E^+]{\text{1. base}} \text{ECY=CRX}$$

Review: Bull Soc Chim Belg *92* 825 (1983)

X	Y	R	E^+	
COR	R_2N	H	D^+	TL 4273 (1979)
CO_2H	H	OMe	D^+, RCHO	Syn 160 (1985)
CO_2H	Br	*n*-Bu	Ac_2O, RCHO	JOC *50* 2195 (1985)
CO_2R	Ph	H	R_2CO	JCS Perkin I 1329 (1981)
CO_2R	RO	H	D^+, RCHO, R_2CO, RCO_2R, $(RS)_2$	TL 4273 (1979); *22* 4259 (1981); *23* 1793 (1982) Tetr *39* 2043 (1983)
CO_2R	RO	OMe	PhNCO, PhNCS	Ber *115* 2674 (1982)
CO_2R	RO	OR	RCHO	Syn 748 (1982)
CO_2R	PhS	Me	Michael acceptors	CL 815 (1982) CC 496 (1982)
CO_2R	R_2N	H	D^+	TL 4273 (1979)
CO_2R	R_2N	H	RCHO, R_2CO, Michael acceptor	Angew Int *15* 171 (1976)
CO_2R	R_2N	H	CO_2, RNCO, RNCS	Ber *115* 2674 (1982)
CO_2R	R_2N	OMe	RCHO, PhNCO, PhNCS	Ber *115* 2674 (1982) Syn 748 (1982)

$$HCY=CRX \xrightarrow[\text{2. E}^+]{\text{1. base}} ECY=CRX \quad (\textit{Continued})$$

X	Y	R	E⁺	
CO_2R	R_2N	SMe	D⁺, RCHO	TL *22* 4259 (1981) Syn 748 (1982)
CO_2R	CO_2R	OMe	D⁺, RCHO, Ac_2O, $(MeS)_2$	Tetr *39* 2043 (1983)
CONHR	H	OMe	D⁺, RX, RCHO, Me_3SiCl, RSSR	Syn 160 (1985)
CONHR	PhS	OCH_2Ph	RCHO	TL *27* 5591 (1986)
$CONR_2$	Ar	H	D⁺, MeI, RCHO	Tetr *39* 2043 (1983)
$CONR_2$	RO	H	D⁺	TL *22* 4259 (1981)
$CONR_2$	R_2N	H	D⁺	TL 4273 (1979); *22* 4259 (1981)
$CONR_2$	R_2N	H	D⁺, MeI, RCO_2R	Angew Int *17* 204 (1978)
$CONR_2$	R_2N	H	$\overset{\overset{\text{X}}{\|\|}}{ArCR}$ (X = O, NPh)	Syn 869 (1977)
$CONR_2$	R_2N	H	CO_2, RNCO, RNCS	Ber *115* 2674 (1982)
CN	RO	Me	D⁺	TL *22* 4259 (1981)
CN	R_2N	H	D⁺	TL 4273 (1979)
CN	R_2N	H	D⁺, RI, RCHO	Angew Int *16* 853 (1977)
CN	R_2N	SMe	D⁺	TL *22* 4259 (1981)

$$\underset{RC=CHCR(H)}{\overset{R_2N \quad\quad O}{\overset{\|}{\underset{\|}{}}}} \xrightarrow[\text{2. H}_2\text{O}]{\text{1. R'Li}} \underset{RC=CHCR(H)}{\overset{R' \quad\quad O}{\overset{\|}{\underset{\|}{}}}}$$

JOC *43* 4248 (1978)
CC 75 (1986)

See also page 191, Section 2, for analogous reactions.

$$\xrightarrow{\text{RLi}} \underset{H}{\overset{R}{>}}C=C\underset{(CH_2)_3OH}{\overset{H}{<}}$$

Can J Chem *41* 2600 (1963)

$$RLi + (H_2C=CR'CH_2\overset{+}{N}Me_3)I^- \longrightarrow RCH_2CR'=CH_2$$

JOC *52* 3683 (1987)

$$RCH{=}CRCH_2X \xrightarrow[\text{2. E}^+]{\text{1. base}} RCH{=}CR\overset{\overset{\displaystyle E}{|}}{C}HX \xrightarrow[\substack{\text{of sulfur}\\\text{compounds}}]{\text{reduction}} RCH{=}CRCH_2E$$

Review: Org Rxs *27* 1 (1982)

See also page 253, Section 3.6.

$\underline{X}$	$\underline{E^{\,+}}$	
H	H_2O, Me_3SiCl	TL 669 (1973)
	MeI	Helv *57* 1567 (1974)
	3° RBr	JACS *109* 3391 (1987)
	R'X, RCHO, R_2CO	TL 4115 (1973)
	D_2O, RX, R_2CO,	JACS *105* 6350 (1983)
	$\quad Me_3SiCl$	
	D_2O, R'X, RCHO,	TL 3047 (1975)
	$\quad R_2CO$, CO_2,	JACS *101* 3340 (1979)
	$\quad Me_3SiX$, Me_3SnCl	
	R'X, RCHO, R_2CO, O_2,	Chem Pharm Bull *32* 4632
	$\quad (RS)_2$	$\quad$ (1984)
	$H_2\overset{\displaystyle\overset{O}{\diagup\,\backslash}}{C}{-}CH_2$	Helv *57* 1567, 2261 (1974)
		JACS *108* 3385 (1986)
Cl	R'X	JOC *46* 1504 (1981)
OR	RCHO, R_2CO	JACS *96* 5560 (1974)
		$\quad$ (via Zn compounds)
		TL 833 (1979)
$OSiR_3$	RCHO, R_2CO	JOC *41* 3620 (1976)
	R_3SiCl	JOC *43* 2551 (1978)
(benzimidazolone structure with N–R and C=O)	RCHO	CL 1279 (1979)
SH	RCHO, R_2CO	Ber *112* 1420 (1979) (via Mg compounds)
SAr	R'X	TL 5629 (1968); 3707 (1969); *24* 5531 (1983); *27* 2157 (1986)
		Proc Natl Acad Sci USA *68* 1294 (1971)
		Tetr *27* 5861 (1971)
		CC 1311 (1972)
		JACS *95* 4444 (1973); *103* 4615 (1981)
		Helv *57* 2261 (1974)
		Syn 129 (1974)
		JOC *45* 4097 (1980)

$$\underset{X}{RCH=CRCH_2X} \xrightarrow[\text{2. E}^+]{\text{1. base}} RCH=CRCHX \xrightarrow[\substack{\text{of sulfur}\\\text{compounds}}]{\text{reduction}} RCH=CRCH_2E \quad (\textit{Continued})$$

(with E on the CHX carbon)

$\underline{X}$	$\underline{E^+}$	
	epoxide	TL *28* 5665 (1987)
	$H_2C=CHCONR_2$,	JOC *52* 218 (1987)
	$PhN=NPh$	

S—(pyridin-2-yl)	R'X	BCSJ *44* 2285 (1971)
SO_2Ar	R'X	BSCF 746 (1973)
		JOC *39* 2135 (1974); *51* 5100 (1986)
		JCS Perkin I 761 (1981)
		CL 25 (1981); 725 (1983)
		JACS *107* 396 (1985)
		TL *27* 2683 (1986); *28* 187, 3193, 6045 (1987)
		CC 1761 (1986); 1036 (1987)

$$\underset{}{H_2C\overset{O}{\overset{\diagup\diagdown}{-}}CHR}$$
enone (1,4-addition)

BSCF 513 (1976)
TL *28* 6045 (1987)

$\underset{\|}{\overset{S}{SCNR_2}}$	RX	JACS *105* 2909 (1983)
		Syn 100 (1975)
		CL 249 (1975)
	RX, RCHO, R_2CO, epoxide	TL 4027 (1975)
SePh	R'X	JOC *52* 3759 (1987)
$SiMe_2N(i\text{-}Pr)_2$	RCHO, $ZnCl_2$	JOC *52* 957 (1987)

$$\underset{\underset{CH_3}{|}}{H_2C=CC\equiv CH} \xrightarrow[\substack{\text{2. E}^+\\\text{3. H}_2O}]{\text{1. }n\text{-BuLi, KO-}t\text{-Bu}} \underset{\underset{CH_2E}{|}}{H_2C=CC\equiv CH}$$

$$E^+ = RX,\ RCH\overset{O}{\overset{\diagup\diagdown}{-}}CH_2,\ RCHO,\ R_2CO,\ RCONR_2,\ CO_2,\ (MeS)_2,\ RSCN$$

JOC *52* 5261 (1987)

$$\text{(isoprenol)} \xrightarrow[\substack{\text{2. E}^+\\\text{3. H}_2O}]{\text{1. KO-}t\text{-Bu, }n\text{-BuLi}} \text{(E-substituted product)}$$

$$E^+ = RCHO,\ R_2CO,\ H_2C\overset{O}{\overset{\diagup\diagdown}{-}}CHR$$

TL 111 (1978)
Syn Commun *13* 237 (1983)

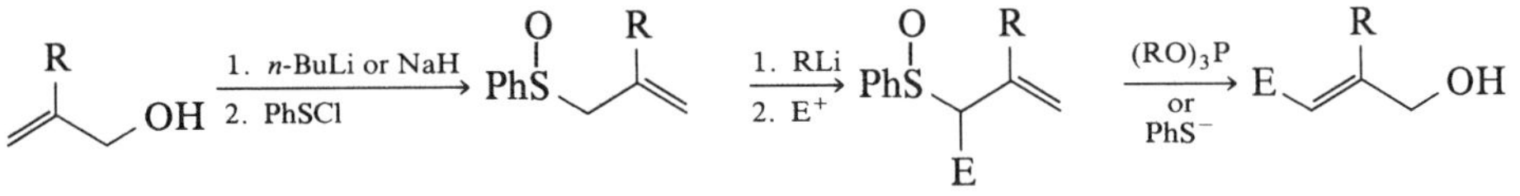

Electrophile

RX

JACS *93* 4956 (1971); *94* 3672 (1972)
CC 702 (1972)
TL 1385, 1389 (1973)
JOC *38* 2245 (1973)
Acct Chem Res 7 147 (1974)

epoxide

Syn Commun *11* 723 (1981)

2. Grignard Reagents

For transition metal-catalyzed cross-coupling reactions of Grignard reagents see the appropriate transition metal.

$$RMgY + H_2C=CHCH_2X \longrightarrow RCH_2CH=CH_2$$

X	R	
Cl	1° alkyl	Org Mag Res *10* 192 (1977)
	1° alkyl, Ph	Ind Eng Chem *33* 115 (1941)
Br	*t*-Bu	JACS *55* 4555 (1933)
	c-C$_5$H$_{11}$	JACS *68* 1101 (1946)
	n-Bu, Ph	Helv *17* 351 (1934)
	Me	JOC *14* 505 (1949)
OR	1° alkyl, aryl	JACS *75* 5408 (1953)
OPO(OEt)$_2$	1° alkyl, allyl, benzyl, aryl, alkynyl	CC 285 (1982)
$\overset{+}{N}Me_3$ I$^-$	Me, Ph	JOC *52* 3683 (1987)

$$RMgX \xrightarrow[\substack{\text{2. KH (cis) or} \\ BF_3 \cdot OEt_2 \text{ (trans)}}]{\substack{t\text{-BuSiMe}_2 \\ | \\ 1. \quad R'CHCHO}} \underset{\substack{\text{cis or trans}}}{R'CH=CHR}$$

JACS *103* 6251 (1981)

$$ArMgX + RCH=\overset{OEt}{\underset{|}{C}}C_6H_5 \longrightarrow RCH=\overset{Ar}{\underset{|}{C}}C_6H_5$$

JACS *73* 1663 (1951)

$$1° \ RMgX + \text{(tetrahydropyran)} \longrightarrow RCH=CH(CH_2)_3OH$$

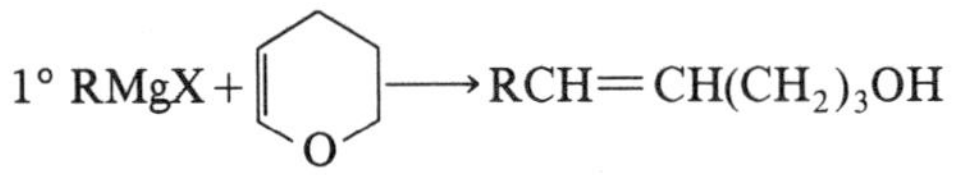

JACS *76* 4538 (1954)

$$\underset{RC=CHCR(H)}{\overset{R_2N \quad O}{\underset{| \qquad \|}{}}} \xrightarrow[\text{2. H}_2\text{O}]{\text{1. R'MgX}} \underset{RC=CHCR(H)}{\overset{R' \quad O}{\underset{| \qquad \|}{}}}$$

Ber *64* 2543 (1931)
BSCF 515 (1960); 1294 (1964)

See also page 185, Section 1, for analogous reactions.

$$\text{Me}_2\text{NCH} \xrightarrow{\text{RMgX}} \text{RCH}$$

Syn Commun *10* 661 (1980)

$$\underset{\text{CH}_3}{\overset{\text{H}}{>}}\text{C}=\text{C}\underset{\text{MgBr}}{\overset{\text{CH}_3}{<}} \xrightarrow{\text{1}° \text{RI}} \underset{\text{CH}_3}{\overset{\text{H}}{>}}\text{C}=\text{C}\underset{\text{R}}{\overset{\text{CH}_3}{<}}$$

TL 3225 (1976)

$$\text{H}_2\text{C}=\text{CHCH}_2\text{MgBr} + \text{ROTs} \longrightarrow \text{H}_2\text{C}=\text{CHCH}_2\text{R}$$

TL 1181 (1977)
JACS *109* 7488 (1987)

3. Boron Reagents

See also page 206, Section 14.

$$\text{R}_3\text{B} \xrightarrow{\text{PhOCH}=\text{CHCH}_2\text{MgCl}} \text{RCH}=\text{CHCH}_2\text{BR}_2 \xrightarrow[\text{H}_2\text{O}_2]{\text{NaOH}} \text{RCH}_2\text{CH}=\text{CH}_2$$

Syn Commun *12* 813 (1982)

$$\underset{\text{R}}{\overset{\text{R}}{>}}\text{C}=\text{C}\underset{\text{R}}{\overset{\text{X}}{<}} \longrightarrow \underset{\text{R}}{\overset{\text{R}}{>}}\text{C}=\text{C}\underset{\text{R}}{\overset{\text{R}'}{<}}$$

n-BuLi/R$_3'$B/I$_2$ JOC *43* 1279 (1978)
                                         Syn 945 (1980)

R$_3'$B or R'B⊂⊃ (R' =1° alkyl), cat   TL *27* 6379 (1986)
   PdCl$_2$[bis(diphenylphosphino)ferrocene],
   NaOH or NaOMe

$$(\text{R}_3\text{BMe})\text{Li} + \text{BrCH}=\text{CHCO}_2\text{Et} \xrightarrow{\text{CuI}} \text{RCH}=\text{CHCO}_2\text{Et}$$
$$\text{retention}$$

TL 3369 (1977)

$$RC\equiv CB\hspace{-0.5em}\text{(} + \underset{\text{MeO}}{\overset{\text{O}}{RC=CHCR}} \longrightarrow RC\equiv CC=CHCR$$

(with MeO on one position and O on the carbonyl; product with R and O substituents)

JOC *42* 3106 (1977)

$$R_3B \xrightarrow{\text{LiC}\equiv CCH_2Cl} RC\equiv CCH_2BR_2 \xrightarrow[\text{LiOMe, CuI}]{H_2C=CR'CH_2Br} RC\equiv CCH_2CH_2CR'=CH_2$$

CL 1289 (1982)

4. Aluminum Reagents

See also page 197, Section 8; page 198, Section 12; and page 201, Section 13.

$$\underset{SO_2Ph}{(CH_3)_2C=CHCHR} \xrightarrow{R_2AlR'} \underset{R'}{(CH_3)_2C=CHCHR}$$

R′ = vinyl, alkynyl

JACS *108* 1098 (1986)

5. Thallium Reagents

$$ArTl(O_2CCF_3)_2 + H_2C=CHX \xrightarrow{\text{cat Li}_2PdCl_4} ArCH=CHX$$

X = COR, CO_2R, CN

JOMC *99* C8 (1975)
BCSJ *53* 553 (1980)
JOC *51* 3708 (1986)

6. Silicon, Germanium and Tin Reagents

For other reactions of allylic silanes and stannanes look under the functional group introduced; for silicon and tin reagents, see also page 201, Section 13.

$$ArSnMe_3 + BrCH_2CH=CR_2 \xrightarrow{\text{cat ZnCl}_2} ArCH_2CH=CR_2$$

TL *24* 1905 (1983)

$$\underset{R}{\overset{R}{>}}C=C\underset{SnR_3}{\overset{R}{<}} \xrightarrow[\text{2. E}^+]{\text{1. RLi}} \underset{R}{\overset{R}{>}}C=C\underset{E}{\overset{R}{<}}$$

E^+ = RX, RCHO, Me_3SiCl

CC 926 (1986)

$$RX + R''_3SnCH{=}CHR' \xrightarrow{\text{Pd catalyst}} RCH{=}CHR'$$

RX

$R_f I$	TL *28* 5857 (1987)
ArX	Proc Acad Sci USSR, Chem Sec *272* 333 (1983)
	Macromolecules *18* 321 (1985)
	JOC *52* 422 (1987)
	JACS *109* 7223 (1987)
$ArCl \cdot Cr(CO)_3$	CC 1755 (1987)
ArOTf	JACS *109* 5478 (1987)

$$(ArN_2)BF_4 + n\text{-}Bu_3SnCH{=}CH_2 \xrightarrow{\text{Pd catalyst}} ArCH{=}CH_2$$

JOC *48* 1333 (1983)

$$ArH + H_2C{=}CHCH_2MR_3 \xrightarrow{\text{Tl(O}_2\text{CCF}_3)_3} ArCH_2CH{=}CH_2$$

$$R_3M = Me_3Si,\ Me_3Ge,\ n\text{-}Bu_3Sn$$

TL *22* 4491 (1981)

$$RCH{=}CHCH_2SiR_3 \xrightarrow{E^+} R\overset{\displaystyle E}{\overset{|}{C}}HCH{=}CH_2$$

Review: Pure Appl Chem *55* 1707 (1983) (stereochemistry)

E^+

$R_3CCl,\ TiCl_4$	TL 4925 (1978)
	Syn 446 (1979)
	JOC *45* 3559 (1980)
	JACS *104* 4962 (1982)
R_3CX (X = OTf, OSiMe$_3$, OAc), Lewis acid	JOC *47* 3219 (1982)
RX (R = allylic, benzylic; X = halogen, OR), TiCl$_4$	TL *23* 2953 (1982)

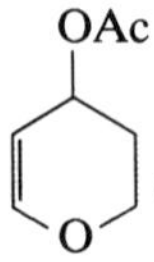

, Lewis acid

JOC *47* 3803 (1982)
TL *28* 4951 (1987)
JACS *109* 2082, 8117 (1987)

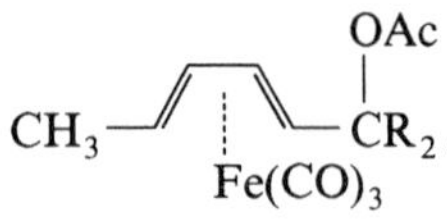

TL *28* 641 (1987)

RNO$_2$ (R = allylic, benzylic), SnCl$_4$ CC 1285 (1986)

ArCR$_2$OH, BF$_3\cdot$OEt$_2$ — JOC *47* 2125 (1982)

epoxides, Lewis acid — See page 512, Section 4.1.

, TiCl$_4$ — JOC *50* 2782 (1985)

ROCHClR, cat Me$_3$SiX (X = I, OTf) — CL 409 (1983)

ROCHRCHBrR, AgBF$_4$ — JOC *48* 1557 (1983)

, Me$_3$SiOTf — JOC *52* 892 (1987)

, BF$_3\cdot$OEt$_2$ — JOC *52* 2335 (1987)

, Lewis acid — TL *24* 1563 (1983)
JOC *50* 3017 (1985)

, BF$_3\cdot$OEt$_2$ — CC 1245 (1987)

, Lewis acid — JACS *104* 7371 (1982); *105* 2088 (1983)
TL *25* 3951 (1984)

, ZnBr$_2$ — JOC *49* 2513 (1984)

, TiCl$_4$ — Angew Int *25* 178 (1986)

PhCH(OMe)$_2$, BF$_3\cdot$OEt$_2$ — JOC *48* 3351 (1983)

RCH(SAr)$_2$, SnCl$_4$ — JACS *109* 7199 (1987)

PhSCH$_2$Cl, TiCl$_4$ — TL *23* 723 (1982)
CL 961 (1982)

PhSCHClR, Lewis acid — TL *24* 1711 (1983)

PhSCH(NO$_2$)R, SnCl$_4$ — CC 947 (1987)

PhSCHRCHRNO$_2$, TiCl$_4$ (E = PhSCHRCHR) — JOC *52* 4133 (1987)

PhX—C—C—Cl (X = S, Se), ZnBr$_2$ — TL *24* 5911 (1983)

$$RCH=CHCH_2SiR_3 \xrightarrow{E^+} \overset{\overset{\textstyle E}{\textstyle |}}{R}CHCH=CH_2 \quad (\textit{Continued})$$

$$E^+ \atop \left[\left\langle\overset{\textstyle S}{\underset{\textstyle S}{+)}}\right\rangle\right]BF_4^- \qquad\qquad TL\ 23\ 4835\ (1982)$$

JOC *51* 5148 (1986)

$$RCH=CHCH_2SnR_3 \longrightarrow \overset{\overset{\textstyle R'}{\textstyle |}}{R}CHCH=CH_2$$

R'X [R' = 1°, 2°, 3° alkyl;   JOMC *26* C4 (1971); *56* C11 (1973);
  X = Cl, Br, I, OC(= S)OPh, SPh,   *61* C33 (1973); *96* 225 (1975)
  SePh], free radical initiation or hν   JACS *104* 5829 (1982)
                                JOC *47* 3590 (1982); *49* 1462 (1984);
                                52 3659 (1987)
                                TL *24* 1357 (1983); *25* 1867 (1984);
                                26 3311 (1985) (intramolecular);
                                27 4857 (1986)
                                BCSJ *56* 2480 (1983)
                                CC 1339 (1986)

R_fI, cat $Pd(PPh_3)_4$ TL *28* 5857 (1987)

$RCOCH_2X$, AIBN (R' = $RCOCH_2$)   CL 795 (1978)

, hν   JOC *51* 5148 (1986)

ArX, cat $Pd(PPh_3)_4$ (R' = Ar)   CL 301 (1977)
                                TL *28* 3935 (1987)

$R_2C(SMe)_2$, $(Me_2SSMe)BF_4$ JACS *107* 719 (1985)
  (R' = R_2CSMe)

immonium salts (R' = CH_2NR_2)   JOC *52* 1378 (1987)

$$R_2C=O\ (RCHO) \xrightarrow[\text{2. } BF_3\cdot HOAc]{\text{1. } Me_3SiCH_2CH_2M\ (M=Li,\ MgX)} R_2CHCH=CH_2$$

JOC *47* 1983 (1982)

M	E^+	
Si	BF_3-HOAc; $TiCl_4$ + RX, RCOCl, enone, or epoxide	Syn 446 (1979)
Sn	H^+	CC 630 (1975)

7. Selenium Reagents

$$RCH=C(SeMe)_2 \xrightarrow[\text{2. } E^+]{\text{1. } n\text{-BuLi}} RCH=\overset{\displaystyle SeMe}{\underset{\displaystyle |}{C}}E \longrightarrow RCH=\overset{\displaystyle X}{\underset{\displaystyle |}{C}}E$$

$E^+ = H_2O, D_2O, MeI, RCHO, R_2CO, Me_2NCHO, ClCO_2Me, CO_2$
$X = H$ (n-Bu_3SnH), Br (Br_2)

TL *23* 3411 (1982)

8. Titanium Reagents

$$H_2C=CHCH_2CH_2OH \xrightarrow[\text{Me}_3\text{Al-TiCl}_4]{\text{Me}_2\text{TiCl}_2 \text{ or}} \underset{H}{\overset{CH_3}{>}}C=C\underset{CH_2CH_2OH}{\overset{H}{<}}$$

CL 1819 (1982)
Angew Int *21* 309 (1982)

9. Zirconium Reagents

See also page 198, Section 12.

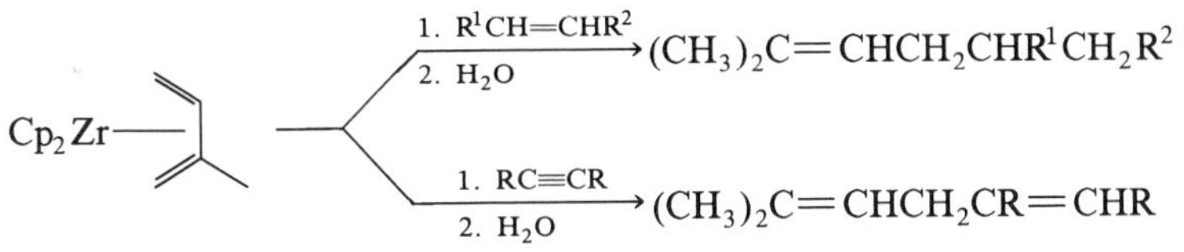

$$Cp_2Zr\!-\!\!\!<\begin{array}{l} \xrightarrow[\text{2. } H_2O]{\text{1. } R^1CH=CHR^2} (CH_3)_2C=CHCH_2CHR^1CH_2R^2 \\[2em] \xrightarrow[\text{2. } H_2O]{\text{1. } RC\equiv CR} (CH_3)_2C=CHCH_2CR=CHR \end{array}$$

CL 719 (1981)

10. Iron Reagents

$$RMgX + XCH=CHR \xrightarrow{\text{cat FeY}_3} RCH=CHR$$

JACS *93* 1487 (1971) (Y = Cl)
Syn 303 (1971) (Y = Cl)
JOC *40* 599 (1975); *41* 502 (1976) (both Y = PhCOCHCOPh)

$$R_2C=CRSO_2R' \xrightarrow[\text{cat Fe(acac)}_3]{R''MgX} R_2C=CRR''$$

$R' = t\text{-Bu, Ph}$ $\qquad\qquad$ $R'' = 1° \text{ alkyl, Ph}$

TL *23* 2469 (1982)

$$\left[\begin{array}{c} \overset{\displaystyle CpFe(CO)_2}{\underset{\displaystyle |}{ROCH=CHOR}} \end{array}\right]^+ BF_4^- \xrightarrow[\substack{\text{3. Nuc}_2 \\ \text{4. NaI}}]{\substack{\text{1. Nuc}_1 \\ \text{2. HBF}_4}} Nuc_1-CH=CH-Nuc_2$$

Nuc = R_2CuLi, $R_2Cu(CN)Li_2$, PhMgBr, ketone enolate

JACS *106* 7264 (1984)

$$H_2C=CHCH_2X + NaCH(CO_2Et)_2 \xrightarrow{\text{Fe catalyst}} H_2C=CHCH_2CH(CO_2Et)_2$$

X	Catalyst	
OAc	$Fe_2(CO)_9$	JOMC *285* C13 (1985)
OCO_2Et	$n\text{-Bu}_4N[Fe(CO)_3NO]$	JOC *52* 974 (1987)

11. Cobalt Reagents

$$RCo(salophen)py + H_2C=CHX \xrightarrow{h\nu} RCH=CHX$$

$R = 1°, 2° \text{ alkyl}$ $\qquad$ $X = \text{Ph, COR, CO}_2R, \text{CN}$

CC 871 (1987)

12. Nickel Reagents

$$H_2C=CHCH_2Br \xrightarrow{Ni(CO)_4} \left[\begin{array}{c} Br \\ Ni\overbrace{}Ni \\ Br \end{array}\right] \xrightarrow[\text{DMF}]{RX} RCH_2CH=CH_2$$

R = alkyl, aryl, vinyl

Review: Org Rxs *19* 115 (1972)

JACS *89* 2755 (1967); *97* 459 (1975); *100* 5800 (1978); *107* 5574, 5663 (mechanism) (1985)
CC 235, 2289 (1973)
BCSJ *47* 3098 (1974); *49* 3351 (1976)
JCS Perkin I 2411 (1981)
Organomet *1* 259 (1982)

$$R^1 \underset{O}{\overline{}}{)}_n \xrightarrow[\text{cat NiCl}_2\text{L}_2]{R^2\text{MgX}} \underset{R^1}{\overset{R^2}{>}}C=C\underset{H}{\overset{(CH_2)_{n+1}OH}{<}}$$

$$n = 1, 2$$

JACS *101* 2246 (1979)
JOC *49* 4894 (1984); *50* 719 (1985)
CC 241, 429 (1987)

$$RCH\!=\!CHX + R'M \xrightarrow{\text{Ni catalyst}} RCH\!=\!CHR'$$

$\underline{X}$	$\underline{R'M}$	
halogen	RO$_2$CCH$_2$Li	JACS *99* 4833 (1977)
	RMgX	Pure Appl Chem *52* 669 (1980) (review)
	RMgX (R =1°, 2°, 3° alkyl)	JACS *94* 4374 (1972); *104* 180 (1982)
		JOMC *55* C91 (1973); *168* 227 (1979)
		Helv *56* 460 (1973)
		TL 3 (1974); *22* 137, 315 (1981); *24* 3209 (1983); *28* 6351 (1987)
		BCSJ *49* 1958 (1976)
		CL 767 (1980)
		CC 647 (1982)
		JOC *48* 2195 (1983); *50* 3261 (1985); *52* 678 (1987)
	RMgX (R = benzylic)	Helv *56* 460 (1973)
		TL 3 (1974); *21* 79, 4623 (1980); *27* 2049 (1986)
		JACS *98* 3718 (1976); *104* 180 (1982)
		BCSJ *49* 1958 (1976)
		JOMC *209* Cl (1981)
		CC 647 (1982); 1746 (1987)
		JOC *48* 2195 (1983); *50* 3261 (1985); *51* 5169 (1986)
	RMgX (R = vinylic)	BCSJ *49* 1958 (1976)
	ArMgX	JACS *94* 4374 (1972)
		CC 144 (1972); 647 (1982); 883 (1986)
		JOMC *55* C91 (1973); *168* 227 (1979)
		BCSJ *49* 1958 (1976)
		JOC *52* 678 (1987)
		TL *28* 6351 (1987)
	Me$_3$SiCH$_2$MgCl	TL *23* 27 (1982); *28* 6351 (1987)
	RC≡CMgX	JOC *49* 4733 (1984)
		JACS *108* 4685 (1986)
	RZnX (R =1° alkyl)	Bull Acad Sci USSR, Div Chem Sci 620 (1986)
	EtO$_2$CCH$_2$ZnBr	JOMC *209* 109 (1981)

$$RCH{=}CHX + R'M \xrightarrow{\text{Ni catalyst}} RCH{=}CHR' \quad (\textit{Continued})$$

X	R'M	
OSiMe$_3$	RMgX (R = 1° alkyl, aryl, benzylic)	TL 21 3915 (1980)
OPO(OEt)$_2$	Me$_3$SiCHRMgCl	Syn 1001 (1981)
OR	RMgX (R = Ph, 1° alkyl)	JACS 101 2246 (1979)
		JOC 49 4894 (1984); 50 719 (1985)
		CC 241, 429 (1987)
SR	RMgX (R = 1° alkyl, aryl)	TL 43 (1979); 25 5177 (1984); 27 6301 (1986)
		CC 637 (1979); 647, 840 (1982)
		CL 1209 (1980)
O$_2$SR	RMgX (R = Me, Ph)	TL 23 2469 (1982)
SePh	RMgX (R = n-Bu, Ph)	TL 21 87 (1980)

$$\underset{RC{=}CHCO_2R}{\overset{OPO(OR'')_2}{|}} \xrightarrow{R'M} \underset{RCH{=}CHCO_2R}{\overset{R'}{|}}$$

R'M	
Me$_3$SiCH$_2$MgCl, cat Ni(acac)$_2$	Can J Chem 60 673 (1982)
	TL 28 2753 (1987)
Me$_2$Zn, cat Ni(acac)$_2$	JACS 109 1564 (1987) (lactone)

See also page 206, Section 14, for related reactions.

$$RCH{=}CHM + R'X \xrightarrow{\text{Ni catalyst}} RCH{=}CHR'$$

M	R'X	
MgX	ArX	BCSJ 49 1958 (1976)
		JOC 50 5370 (1985)
Al(i-Bu)$_2$	ArX	CC 596 (1976)
		JACS 109 2393 (1987)
	ArOPO(OEt)$_2$	TL 22 4449 (1981)
Cp$_2$ZrCl	ArX	JACS 103 4466 (1981); 109 2393 (1987)

$$RM + H_2C{=}CHCH_2X \xrightarrow{\text{Ni catalyst}} RCH_2CH{=}CH_2$$

X	RM	
Cl	(i-PrO$_2$CCH$_2$CH$_2$)$_2$Zn	JACS 109 8056 (1987)
OH	RMgX (R = 1° alkyl, benzylic, aryl)	JOMC 127 371 (1977); 134 265 (1977)
		Helv 63 987 (1980)
		CC 313, 681 (chiral) (1981)
		TL 28 4547 (1987)

OPh	RMgBr (R = Et, Ph)	CC 112 (1983)
Cl, OR, OSiR$_3$	PhMgBr	CC 313 (1981)
Br, OH, OPh, S-t-Bu	RMgX (R = Me, Et)	Tetr 42 2043 (1986)

CC 1515 (1987)

13. Palladium Reagents

See also page 193, Sections 5 and 6; and page 210, Section 17.

$$R_f I + RCH{=}CHBr \xrightarrow[\substack{\text{Pd catalyst} \\ \text{ultrasound}}]{\text{Zn}} RCH{=}CHR_f$$

$$R_f I + RCH{=}CHCH_2 Br \longrightarrow \overset{\overset{\textstyle R_f}{\textstyle |}}{R}CHCH{=}CH_2$$

CL 137 (1982)
JACS 107 5186 (1985)

$$RCH{=}CHX + R'M \xrightarrow{\text{Pd catalyst}} RCH{=}CHR'$$

X	R'M	
halogen	RLi (R =1° alkyl, aryl)	JOMC 91 C39 (1975) JOC 44 2408 (1979)
	RMgX (R =1°, 2° alkyl)	JOMC 91 C39 (1975) TL 191 (1978); 1073, 1871 (1979); 27 2529 (1986) JOC 44 2408 (1979); 48 2195 (1983); 51 3772 (1986) CC 647 (1982) JACS 104 180 (1982); 106 158 (1984)
	RMgX (R = benzylic)	CC 647 (1982); 1746 (1987) JOC 51 3772 (1986)

$$RCH{=}CHX + R'M \xrightarrow{\text{Pd catalyst}} RCH{=}CHR' \quad (\textit{Continued})$$

$\underline{X}$	$\underline{R'M}$	
	RMgX (R = vinylic)	JOMC *91* C39 (1975)
		TL 191 (1978)
		JOC *44* 2408 (1979)
	RC≡CMgX	TL 191 (1978)
		Tetr *37* 2617 (1981)
	ArMgX	TL 191 (1978)
		CC 647 (1982)
		JACS *109* 1257 (1987)
	Me$_3$SiCH$_2$MgCl	TL *23* 27 (1982)
	PhSnMe$_3$	Proc Acad Sci USSR, Chem Sec
		272 333 (1983)
	RC≡CSnR$_3$	JACS *109* 2138 (1987)
	RZnX (R = 1° alkyl)	JACS *102* 3298 (1980);
		109 1257 (1987)
		JOC *46* 4093 (1981); *51* 4080
		(1986); *52* 4885 (1987)
		JOMC *285* 109 (1985)
		Bull Acad Sci USSR, Div Chem
	R.	Sci 620 (1986)
	▷—ZnCl	TL *28* 5075 (1987)
	EtO$_2$CCH$_2$ZnBr	JOMC *209* 109 (1981)
	EtO$_2$C(CH$_2$)$_n$ZnI	TL *27* 955 (1986)
	(n = 2, 3)	
	(RO$_2$CCHRCH$_2$)$_2$Zn	JACS *109* 8056 (1987)
	ArCH$_2$ZnX	TL *22* 2715 (1981)
	ArZnCl	JOC *50* 2121 (1985)
		J Med Chem *29* 2053 (1986)
	RC≡CZnX	CC 683 (1977)
		TL *27* 4351, 5533 (1986)
		JACS *108* 4685 (1986)
OPO(OPh)$_2$	Me$_3$SiCH$_2$MgCl	Syn 1001 (1981)
	R$_3$Al (R = 1° alkyl,	TL *21* 2531 (1980);
	vinylic, alkynyl)	*22* 1609 (1981)
O$_3$SCF$_3$	R'SnR$_3$ (R' = alkyl,	JACS *106* 4630 (1984);
	allyl, alkynyl,	*108* 3033 (1986)
	vinylic), LiCl	CC 809 (1985) (intramolecular)
		TL *27* 1523 (1986)
	EtO$_2$C(CH$_2$)$_n$ZnI	TL *27* 955 (1986)
	(n = 2, 3)	
	(RO$_2$CCH$_2$CH$_2$)$_2$Zn	JACS *109* 8056 (1987)
	ArZnX	TL *28* 701 (1987)

$$R^1CH=CCl_2 \xrightarrow[\text{Pd catalyst}]{R^2MgBr(ZnCl)} \underset{R^1}{\overset{H}{\diagdown}}C=C\underset{Cl}{\overset{R^2}{\diagup}} \xrightarrow[\text{Pd catalyst}]{R^3MgBr(ZnCl)} \underset{R^1}{\overset{H}{\diagdown}}C=C\underset{R^3}{\overset{R^2}{\diagup}}$$

R^1, R^2 or R^3 = aryl, 1° alkyl

JACS *109* 1257 (1987)

$$\underset{C=C-C-}{\overset{X \quad\quad O}{| \quad\quad\quad ||}} \xrightarrow[\text{cat Pd(PPh}_3)_4]{RCH=CH(CH_2)_2ZnX} RCH=CH(CH_2)_2C=C-\overset{O}{\overset{||}{C}}-$$

X = Cl, Br

JOC *45* 5223 (1980)

See also page 206, Section 14, for related reactions.

$$RCH=CHX + HC\equiv CR' \xrightarrow{\text{Pd catalyst}} RCH=CHC\equiv CR'$$

<u>X</u>

O_3SCF_3

JOC *50* 2302 (1985)
TL *27* 1523 (1986)
Heterocycles *26* 355 (1987)

halogen

TL 4467 (1975); *22* 421 (1981); *27* 2033, 3589,
   5857 (1986); *28* 1127, 1649, 3857, 3859, 4875,
   4879, 5751, 5849 (1987)
JOMC *93* 253, 259 (1975)
Syn Commun *11* 917 (1981)
J Chem Res (S) 93 (1982)
Tetr *38* 631 (1982)
JACS *106* 3548, 5734 (1984); *107* 7515 (1985);
   108 5589 (1986); *109* 1879 (1987)
CC 1816 (1986)

$$H_2C=CHCH_2X + RM \xrightarrow{\text{Pd catalyst}} RCH_2CH=CH_2$$

<u>X</u>	<u>RM</u>	
OH	RMgX	JOMC *186* C1 (1980)
		CC 313 (1981)
Cl	*n*-Bu$_3$SnR	TL *28* 1191 (1987)
	(R = heterocyclic)	
Br	[furan]—Li	Syn 51 (1987)
Cl, Br	RSnR$'_3$ (R = aryl,	JACS *105* 7173 (1983);
	vinylic)	*106* 4833 (1984)
Cl, OR, OSiR$_3$	RMgX	CC 313 (1981)
RCO$_2$ (lactone)	PhZnCl	CC 160 (1982)

$$RX + H_2C{=}CHY \xrightarrow{\text{Pd catalyst}} RCH{=}CHY$$

R = aryl, heterocyclic, benzylic, vinylic

X = halogen

Y = H, CO_2R, CO_2H, $CONH_2$, CN, Ar, alkyl, alkenyl, OR

Reviews:

Pure Appl Chem *50* 691 (1978)
Acct Chem Res *12* 146 (1979)
J. Tsuji, "Organic Synthesis with Palladium Compounds," Springer, New York (1980)
Org Rxs *27* 345 (1982)
R. F. Heck, "Palladium Reagents in Organic Syntheses," Academic Press, New York (1985)

JOC *37* 2320 (1972); *40* 1083 (1975); *43* 2454, 2941, 2947, 2949, 5018 (1978); *44* 4078 (1979); *46* 1067, 2767, 5414 (1981); *52* 3319 (1987)
JACS *96* 1133 (1974); *101* 4743 (1979); *109* 4335 (1987)
Syn 365 (1981)
CC 541 (1981); 1287 (1984); 1755 (1987)
CL 1993 (1982) ($H_2C{=}CHSiMe_3 \longrightarrow ArCH{=}CH_2$)
JOMC *258* 101 (1983)
TL *26* 2667 (1985); *28* 3039 (1987)

$$ArI + H_2C{=}CR^1CHR^2OSiMe_3 \xrightarrow[\text{LiCl}]{\text{Pd(OAc)}_2} ArCH{=}CR^1\overset{\overset{\displaystyle O}{\|}}{C}R^2$$

CL 403 (1981)

$$ArO_3SR_f + H_2C{=}CHX \xrightarrow[\text{Et}_3N]{\text{cat PdCl}_2(PPh_3)_2 \;\Delta} ArCH{=}CHX$$

X = CN, CO_2Et

TL *27* 1171 (1986)

$$ArH + H_2C{=}CHX \xrightarrow[\text{[Cu(OAc)}_2]]{\text{Pd(OAc)}_2} ArCH{=}CHX$$

X = R, Ar, CN, CO_2R

Tetr *25* 4809, 4815, 4819 (1969)
Syn 524 (1973) (review)
CL 1230 (1979); 239 (1986) (ArH = uracil)
CC 859 (1981) (olefin = quinone)
JOC *46* 851 (1981) (ArH = heterocycles)

$$o\text{-}X{-}Ar{-}H \longrightarrow o\text{-}X{-}Ar{-}PdOAc(Cl) \xrightarrow{H_2C{=}CHR} o\text{-}X{-}Ar{-}CH{=}CHR$$

X
—
NHCOR   JOC *46* 4416 (1981)

CHRNR'$_2$ Acct Chem Res *2* 144 (1969)
                                     JOMC *102* 239 (1975); *179* 301 (1979);
                                        182 537 (1979)
                                     TL 355 (1977); *21* 2757 (1980); *27* 2169 (1986)
                                     BCSJ *51* 663 (1978); *52* 142, 957 (1979)
                                     Tetr *37* 173 (1981)

CH=N—t-Bu

TL *23* 1957 (1982)

CH$_2$CH$_2$NMe$_2$

TL *27* 1971 (1986)

$$RM + H_2C=CHX \xrightarrow{\text{Pd(II) salts}} RCH=CHX$$

RM	X	
ArMgX	Ar	CC 918 (1978)
ArTl(O$_2$CCF$_3$)$_2$	—	See page 193, Section 5.
K$_2$(PhSiF$_5$)	CHO, CO$_2$R	Organomet *1* 542 (1982)
Ph$_2$CuM (M = Li, MgX)	Ph, CO$_2$R	TL 4657 (1979)
RHgX (R = Me, CO$_2$R′, Ar)	—	See page 210, Section 17.

$$ArN_2^+ + H_2C=CHR \xrightarrow{\text{Pd catalyst}} ArCH=CHR$$

CL 159 (1977)
Tetr *37* 31 (1981)

$$ArNH_2 + H_2C=CHX \xrightarrow[\text{Pd(O) or Pd(II)}]{t\text{-BuONO}} ArCH=CHX$$

X = R, Ar, CO$_2$R

CL 551 (1980)
JOC *45* 2359 (1980); *46* 4885 (1981)

$$ArCOCl + H_2C=CHX \xrightarrow[\text{R}_3\text{N}]{\text{cat Pd(OAc)}_2} ArCH=CHX$$

X = Ar, COR, CO$_2$R, CONR$_2$, CN

JOMC *233* 267 (1982); *240* 209 (1982)

$$\overset{O}{\overset{\|}{Ar}}\overset{}{C}Cl + H_2C=CHOR \xrightarrow[\text{Et}_3\text{N}]{\text{cat Pd(OAc)}_2 \, \Delta} \overset{O}{\overset{\|}{Ar}}\overset{}{C}CH=CHOR$$

TL *28* 4215 (1987)

$$RCH=CH_2 + {}^-CRX_2 \xrightarrow[2 \text{ Et}_3\text{N}]{\text{PdCl}_2 \quad 25°C} \overset{CRX_2}{\overset{|}{R}}C=CH_2$$

X = CO$_2$R, COR

JACS *99* 7093 (1977)

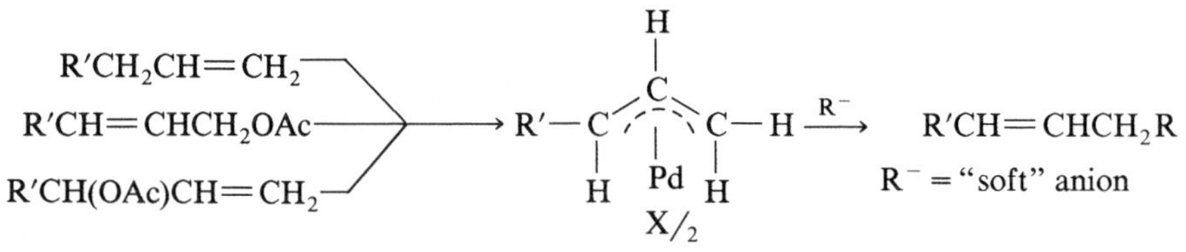

$$R'CH_2CH=CH_2 \atop R'CH=CHCH_2OAc \atop R'CH(OAc)CH=CH_2} \longrightarrow \quad$$

R′CH₂CH=CH₂
R′CH=CHCH₂OAc ⟶
R′CH(OAc)CH=CH₂

$$R'-C \cdots \overset{\overset{\displaystyle H}{|}}{\underset{\underset{\displaystyle Pd}{|}}{C}} \cdots C-H \xrightarrow{R^-} \quad R'CH=CHCH_2R$$

H   H   X/₂

R⁻ = "soft" anion

Reviews:

Tetr *33* 2615 (1977)
Acct Chem Res *13* 385 (1980)
J. Tsuji, "Organic Synthesis with Palladium Compounds," Springer Verlag, New York (1980)
R. Heck, "Palladium Reagents in Organic Syntheses," Academic Press, New York (1985)

14. Copper Reagents

See also page 197, Section 10; and page 201, Section 13.

$$R'X \xrightarrow[\text{reagent}]{\text{organocopper}} R'CH=CHR$$

RCH=CHMgX, cat CuI	TL 1073 (1979)
	BCSJ *54* 2831 (1981)
	JOC *51* 4726 (1986)
RCH=CHLi, CuI	Organomet *1* 667 (1982)
	CC 925 (1986)
R$_f$CF=CFCu	JACS *108* 4229 (1986)
(H₂C=CH)₂CuLi	J Ind Chem Soc *45* 1026 (1968)
	Helv *54* 1939 (1971)
	JACS *93* 7016 (1971)
[(H₂C=CH)₂CuCN]Li₂	JACS *109* 1186 (1987)
[**RCH=CHCuC≡CC(CH₃)₃**]Li	TL *23* 739 (1982)

$$R'CH=CHX \xrightarrow[\text{reagent}]{\text{organocopper}} R'CH=CHR$$

X = halogen

RLi or RMgX, cat CuBr	Compt Rend C *278* 967 (1974)
RMgX, cat Li₂CuCl₄	JOMC *128* 1 (1977)
	JACS *109* 1856 (1987)
RCu (R = Ph, C≡CR, CH₂C≡CSiMe₃, CH₂CMe₃)	JOMC *93* 415 (1975)

R_2CuLi

JACS *89* 3911 (1967); *90* 5615 (1968); *94* 2520 (1972)
JOC *35* 1715 (1970)
Org Rxs *19* 1 (1972) (review)

$$\underset{}{-\overset{\overset{\displaystyle Y}{|}}{C}=\overset{}{\underset{}{C}}-X} \quad \xrightarrow{\text{organocopper reagent}} \quad \underset{}{-\overset{\overset{\displaystyle R}{|}}{C}=\overset{}{\underset{}{C}}-X}$$

X	Y	Reagent	
H or R	O_3SCF_3	$R_2CuLi(MgX)$	TL *21* 4313 (1980)
		$(R_2CuCN)Li_2$	Heterocycles *26* 355 (1987)
	$OPO(OPh)_2$	R_2CuLi	TL 4405 (1976)
COR	halogen	RMgX, CuX	JCS Perkin I 593, 599 (1981)
			CC 671 (1986)
		R_2CuLi	JOC *41* 636, 3629 (1976);
			46 2089 (1981)
			Can J Chem *60* 1256 (1982)
		(RCuSPh)Li	JOC *40* 2694 (1975)
			TL 3233, 3237 (1976)
			CC 1033 (1978) (R = Me_3Sn)
			JCS Perkin I 593 (1981)
			Can J Chem *60* 1256,
			2965 (1982)
	OAc	R_2CuLi	TL 2071 (1973)
			JOC *41* 3629 (1976)
	OCO_2R	R_2CuLi	Ind J Chem *12* 325 (1974)
	$OPO(OEt)_2$	R_2CuLi	Can J Chem *57* 1431 (1979)
	OR	R_2CuLi	Ind J Chem *12* 325 (1974)
	SR	RMgX, CuBr	Syn 320 (1976)
		RMgX, CuCN	Syn Commun *6* 209 (1976)
			JOC *51* 4807 (1986)
		R_2CuLi	CC 907 (1973)
			TL 3817 (1973)
			Acta Chem Scand B *33* 460 (1979)
			CL 815 (1982)
			JOC *48* 2786 (1983);
			51 4687 (1986);
			52 110 (1987)
			JACS *107* 4679 (1985)
		(RCuSPh)Li	TL *23* 3751 (1982)
			JACS *107* 4679 (1985)
		$(RCuC{\equiv}CCMe_2OMe)Li$	JOC *48* 2786 (1983)
			JACS *107* 4679 (1985)
		$(R_2CuSCN)Li_2$	JOC *51* 4687 (1986)
		$(R_2CuCN)Li_2$	JOC *51* 4687 (1986)
CO_2R	Cl	R_2CuLi	JOC *43* 3974 (1978);
			46 2089 (1981)
	Br	$(R_3BMe)Li$, CuI	TL 3369 (1977)

$$\underset{\underset{\displaystyle C=C-X}{|\quad|}}{\overset{\displaystyle Y}{|}} \xrightarrow{\text{organocopper reagent}} \underset{\underset{\displaystyle C=C-X}{|\quad|}}{\overset{\displaystyle R}{|}} \quad (\textit{Continued})$$

X	Y	Reagent	
O_2CR	R_2CuLi		TL 2071 (1973);
			27 959 (1986)
			Syn Commun 3 321 (1973)
			JOC 43 3974 (1978)
			Ann 1173 (1982)
	$R_2CuLi \cdot n\text{-}Bu_3P$		TL 925 (1974)
OCO_2R	R_2CuLi		JOC 43 3974 (1978)
$OPO(OEt)_2$	MeMgCl, MeCu		TL 25 1333, 1643 (1984);
			28 731 (1987)
	R_2CuLi		Can J Chem 57 1431 (1979);
			59 2239 (1981)
			Ann 1173 (1982)
			CC 421 (1982); 24 (1987)
			JOC 49 1707 (1984);
			52 1106 (1987)
			JACS 107 2712 (1985)
			TL 27 5555 (1986)
O_3SCF_3	R_2CuLi		JOC 51 3247 (1986)
SR	RMgX, CuX		CL 1097 (1973); 705 (1974)
			Syn 320 (1976)
	R_2CuLi		CC 907 (1973)
			JOC 50 2730 (1985)
	$(RCuSPh)Li$		JACS 107 4679 (1985)

For related reactions see page 198, Section 12; and page 201, Section 13.

$$\underset{O}{\langle\!\!\!\square\rangle} \xrightarrow[\text{1 CuI}]{\text{5 1° RLi}} \underset{H}{\overset{R}{\diagdown}}C=C\underset{CH_2CH_2OH}{\overset{H}{\diagup}}$$

CL 1641 (1982)

$$R'CH=CHCH_2X \xrightarrow{RM} R'CH=CHCH_2R$$

For allylic rearrangements see page 119, Section 8.

X	RM	
Cl	RMgBr, cat CuX (X = Cl, I)	Helv 65 684 (1982)
	RMgBr, cat Cu(acac)$_2$	Tetr 39 3283 (1983)
	RMgX (X = Cl, Br),	TL 23 3115 (1982)
	cat Li$_2$CuCl$_4$	Helv 65 684 (1982)
		CC 590 (1986)
	R_2CuLi (R =1° alkyl)	TL 4439 (1976)
	$R_2CuLi \cdot SMe_2$ (R =1°	TL 2705 (1978)
	alkyl, vinyl)	

Br	RMgBr, cat CuX	CL 177 (1982)
	(X = Br, I)	BCSJ *56* 1446 (1983)
	ArMgBr, CuI	JOC *50* 2427 (1985)
	RCu (R = *sec*-Bu, *t*-Bu)	JOC *49* 1840 (1984)
	Ar$_2$CuLi	JCS Perkin I 2909 (1982)
		JOC *49* 1840 (1984); *52* 3394 (1987)
	[**Ar**CuC≡CCMe$_2$OMe]Li	TL *27* 1607 (1986)
OTs	RMgBr, cat Li$_2$CuCl$_4$	Helv *65* 684 (1982)
OAc	RMgBr, cat CuBr	Syn 804 (1983)
	RMgBr, cat CuI	Helv *65* 684 (1982)
	RMgBr, cat Cu(acac)$_2$	Tetr *39* 3283 (1983)
	RMgX, cat Li$_2$CuCl$_4$	Angew Int *13* 82 (1979)
	(R = 1° alkyl, aryl)	Helv *65* 684 (1982)
		TL *24* 5103 (1983); *28* 1175 (1987)
		Syn 804 (1983)
		CC 827 (1987)
	R$_2$CuLi (R = Me, *n*-Bu)	JOMC *136* 103 (1977)
		JOC *46* 5304 (1981); *48* 721, 3986 (1983)
		TL *23* 3093 (1981)
	R$_2$CuLi (R = Ph)	JACS *101* 4413 (1979)
	Li$_2$Cu$_3$Me$_5$	TL *23* 3093 (1982)
OAc (R' = CO$_2$R)	RCu·(AlCl$_3$)$_n$	Syn Commun *10* 119 (1980)
	(R = *n*-Bu)	
O$_2$CCMe$_3$	R$_2$CuLi	Can J Chem *61* 632 (1983)
		JOC *52* 897 (1987)
OCO$_2$Et	RMgBr, cat CuBr	Syn 804 (1983)
OCONMe$_2$	RMgX, cat CuI	JOC *51* 5456 (1986)
OPO(OEt)$_2$	RMgBr, cat CuX	CL 177 (1982)
	(X = Br, I)	BCSJ *56* 1446 (1983)

benzoxazole-O	1° RMgX, CuBr	JOMC *231* 179 (1982)
		JCS Perkin I 2953 (1983)
SO$_2$Ph	RMgX (R = 1° alkyl),	TL 2393 (1979)
	cat Cu(acac)$_2$	Tetr *39* 3283, 3289 (1983)
	R$_2$CuLi	Tetr *39* 3283 (1983)
$\overset{+}{S}$Me$_2$ Br$^-$	RMgBr, cat CuBr	Syn 804 (1983)
$\overset{+}{N}$R'$_3$ X$^-$	RMgBr, cat CuBr	Syn 804 (1983)
	RMgX, cat Li$_2$CuCl$_4$	JOC *52* 2947 (1987)
	R$_2$CuLi	TL *21* 67 (1980)
		JOC *52* 3683 (1987)

$$RCH=CHCH_2OH \xrightarrow[\text{2. R'MgX}]{\text{1. Me}_2C=CCIN} RCH=CHCH_2R'$$

$$R' = Ph, CH=CH_2, Et (R = Ph)$$

TL *24* 5745 (1983)

$$RCH=CH_2 \xrightarrow[\text{Cp}_2\text{TiCl}_2]{\text{LiAlH}_4} \xrightarrow[\text{Cu(OAc)}_2]{\text{H}_2\text{C}=\text{CHCH}_2\text{X}} R(CH_2)_3CH=CH_2$$

CL 1155 (1978)

$$\xrightarrow[\text{cat CuI}]{\text{RX}}$$

MgX   R

TL 1181 (1977)
Syn Commun *11* 859 (1981)

15. Silver Reagents

$$\begin{array}{c} R \\ R \end{array}C=C\begin{array}{c} Ar \\ Br \end{array} + Ar'H \xrightarrow[(X = BF_4, OTf)]{\text{AgX}} \begin{array}{c} R \\ R \end{array}C=C\begin{array}{c} Ar \\ Ar' \end{array}$$

JOC *47* 5003 (1982)

16. Zinc Reagents

For transition metal-catalyzed cross-coupling of organozinc compounds see the appropriate transition
   metal.

$$R_2C=CRCH_2ZnX + H_2C=CHR' \xrightarrow{\text{H}_2\text{O}} H_2C=CRCR_2\overset{\text{CH}_3}{\underset{|}{C}}HR'$$

JOMC *221* 123, 131 (1981)

17. Mercury Reagents

See also page 201, Section 13.

$$RCH=CHX + ArHgX \xrightarrow{\text{cat ClRh(PPh}_3)_3} RCH=CHAr$$

X = halogen

JOC *48* 4377 (1983)

$$RCH=CHI + R'HgX \xrightarrow{h\nu} RCH=CHR'$$

JACS *106* 4622 (1984)
TL *26* 4975 (1985); *28* 6113 (1987)

$$RHgCl + H_2C=CHX \xrightarrow{\text{PdX}_2} RCH=CHX$$

R = Me, CO_2R', Ar

X = CHO, COR, CO_2R, CN, Ar

R. C. Larock, "Organomercury Compounds in Organic Synthesis," Springer Verlag, New York (1985),
   Chpt 7 (review)

6. ALKYNE AND DIENE ADDITION REACTIONS

See also page 241, Section 7.

1. Hydrogen Addition

diene $\longrightarrow$ monoene

H_2, cat $K_3[Co(CN)_5H]$	TL 115 (1979)
	JOC 45 3860 (1980)
H_2, cat $ClRh(PPh_3)_3$	Discuss Faraday Soc 46 60 (1968)
H_2, cat $Ni(acac)_2$-$Al_2Et_3Cl_3$-PPh_3	BCSJ 55 343 (1982)
H_2, cat Pd(O)-polymer	Israel J Chem 17 269 (1978)

$$RCH{=}CHCH{=}CHX \longrightarrow RCH_2CH{=}CHCH_2X$$

X	Reagent(s)	
H or R'	H_2, cat $C_6H_5CO_2CH_3 \cdot Cr(CO)_3$	TL 1919 (1968)
		JOC 34 3930, 3936 (1969);
		49 4096 (1984)
	H_2, cat naphthalene $\cdot Cr(CO)_3$	TL 28 1893 (1987)
COR	H_2, cat $C_6H_5CO_2CH_3 \cdot Cr(CO)_3$	JACS 106 3875 (1984) (cis product)
COR, CO_2R	i-Bu_2AlH, cat MeCu, HMPA/ H_2O	JOC 51 537 (1986)
CO_2R	H_2, cat $C_6H_5CO_2CH_3 \cdot Cr(CO)_3$	TL 1919 (1968); 28 5841 (1987)
		JACS 90 2446 (1968)
	Et_3SiH, cat $ClRh(PPh_3)_3$	Syn Commun 15 965 (1985)
	$Na_2S_2O_4$	CL 715 (1982) (trans product)

$$RC \equiv CR' \longrightarrow \underset{H}{\overset{R}{\diagdown}} C = C \underset{H}{\overset{R'}{\diagup}}$$

Review: Syn 457 (1973) (hydrogenation)

Li/H$_2$O	JACS *92* 2268 (1970) (R, R' = Ph) JOC *35* 1702 (1970) (R, R' = Ph) Angew Int *25* 167 (1986) [R, R' = (CH$_2$)$_6$]
Zn(Cu), MeOH	TL *21* 1069 (1980)
Zn, KCN, H$_2$O, ROH	Helv *58* 1016 (1975) JACS *109* 8051 (1987)
Zn, HOAc, MeOH	TL *28* 5395 (1987)
H$_2$, cat [Rh(NBD)(PR$_3$)$_n$]PF$_6$	JACS *98* 2143 (1976)
H$_2$, cat nickel boride P-2	JACS *85* 1005 (1963) CC 553 (1973) JOC *38* 2226 (1973); *46* 1263 (1981)
H$_2$, cat Ni graphite, EDA	JOC *46* 5340, 5344 (1981)
H$_2$, cat Ni$_4$[CNC(CH$_3$)$_3$]$_7$	Pure Appl Chem *50* 941 (1978)
H$_2$, cat NaH-Ni(OAc)$_2$-*t*-AmOH	TL 3965 (1977) JOC *45* 1937, 1946 (1980)
H$_2$, cat Pd polymer	J Catalysis *57* 315 (1979)
H$_2$, cat Pd graphite, EDA	CC 540 (1981)
H$_2$, cat Pd on tungsten film	JOC *52* 3132 (1987)
H$_2$, cat Pd-CaCO$_3$, Pb(OAc)$_2$, quinoline (Lindlar)	Helv *35* 446 (1952) JACS *78* 2518 (1956); *107* 1028 (1985) Org Syn Coll Vol *5* 880 (1973) JOC *41* 3497 (1976); *43* 3435 (1978); *52* 3126 (1987)
H$_2$, cat Pd-CaCO$_3$, cat MnCl$_2$, Pb(OAc)$_2$, quinoline	Tetr *39* 2315 (1983)
H$_2$, cat Pd-BaSO$_4$, py	JOC *50* 2309 (1985)
H$_2$, cat Pd-BaSO$_4$, quinoline	JACS *78* 2518 (1956) Org Syn *64* 108 (1985) JOC *51* 4158 (1986)
H$_2$, cat PdCl$_2$, DMF	Ber *109* 531 (1976)
H$_2$, cat NaH-Pd(OAc)$_2$-*t*-AmOH, quinoline	JOC *49* 4058 (1984)
(Et$_3$NH)O$_2$CH, cat Pd-C (R = aryl, vinyl)	JOC *45* 4926 (1980)
MgH$_2$, CuI or CuO-*t*-Bu	JOC *43* 757 (1978)
2 *n*-BuMgX, CuI/H$_2$O	JOC *41* 4089 (1976)

i-BuMgX (X = Cl, Br), Cp$_2$TiCl$_2$/H$_3$O$^+$	TL *22* 85 (1981) CC 1126 (1982) JACS *109* 1469 (1987)
BH$_3$ or R$_2$BH/HOAc	JACS *81* 1512 (1959); *83* 3834 (1961); *93* 3395 (1971); *109* 2138 (1987) Tetra *37* 2617 (1981) JOC *51* 4512, 4514 (1986)
ClBH$_2$/HOAc	JOC *38* 1617 (1973)

⬡ BH /HOAc (catecholborane)	JACS *94* 4370 (1972)
⬡ BH /H$_2$O/ HOCH$_2$CH$_2$OH/HOAc	JACS *107* 3626 (1985)
⬡BH/MeOH	JOC *51* 4512 (1986)
R$_2$BH/n-BuLi/NaOH	JOC *41* 3484 (1976)
R$_2$BH/cat Pd(OAc)$_2$	CC 702 (1978) BCSJ *53* 1670 (1980)
NaBH$_4$, cat PdCl$_2$	CC 515 (1983)
i-Bu$_2$AlH/H$_2$O	Ann *629* 222 (1960) JOC *43* 2739 (1978)
LiAlH$_4$, TiCl$_4$	TL 15 (1976)
LiAlH$_4$, NiCl$_2$	TL 4481 (1977)
NaAlH$_4$, NbCl$_5$	CL 157 (1982)
Cl$_3$SiH/KF/CuF$_2\cdot$2 H$_2$O	TL 1141 (1979)
LaNi$_5$H$_6$	JOC *52* 5695 (1987)

$$RC\equiv CR' \longrightarrow \underset{\underset{H}{|}}{\overset{\overset{R}{|}}{C}}=\underset{\underset{R'}{|}}{\overset{\overset{H}{|}}{C}}$$

Li/H$_2$O	Angew Int *25* 167 (1986)
Li, CH$_3$NH$_2$	CC 634 (1968)
Na, NH$_3$	JACS *63* 216, 2683 (1941); *65* 2020 (1943); *74* 3643 (1952) JCS 3558 (1955) Syn 567 (1972); 616 (1973); 114 (1979)
Yb, NH$_3$	JOC *43* 4555 (1978)
i-Bu$_2$AlH/H$_2$O	TL 3145 (1979) (3° propargylic amines)
LiHAlMe(i-Bu)$_2$/H$_2$O	JACS *89* 5085 (1967)
CrSO$_4$, H$_2$O	JACS *86* 4358 (1964) JOC *51* 253 (1986)

$$RC{\equiv}CX \longrightarrow \underset{H}{\overset{R}{\diagdown}} C{=}C \underset{H}{\overset{X}{\diagup}}$$

X	Reagent(s)	
I	$\left(\diagup \diagdown \right)_2$ BH/HOAc	JACS *107* 713 (1985)
	$KO_2CN{=}NCO_2K$, HOAc	JOC *40* 1083 (1975)
CR_2OH	*i*-BuMgX, cat Cp_2TiCl_2/H_2O	CC 718 (1981)
CO_2R	$NaH_2Al(OCH_2CH_2OCH_3)_2$, CuBr	JOC *40* 3619 (1975)

$$RC{\equiv}CX \longrightarrow \underset{H}{\overset{R}{\diagdown}} C{=}C \underset{X}{\overset{H}{\diagup}}$$

X	Reagent(s)	
Cl	$LiAlH_4$	JACS *101* 5101 (1979)
OR	$LiAlH_4$	JOC *52* 2919 (1987)
CH_2NH_2, CH_2NHR	$LiAlH_4$	JOC *52* 5044 (1987)
CH_2OH	Na, NH_3	JOC *41* 3497 (1976)
	$LiAlH_4$	JACS *71* 4140 (1949); *89* 4245 (1967); *92* 4898 (1970); *106* 7614 (1984); *107* 1028 (1985)
		JCS 1094 (1952); 1584 (1953); 1854 (1954); 2754 (1957)
		JOC *31* 528 (1966); *39* 968 (1974); *47* 4595 (1982); *51* 4158 (1986); *52* 3798 (1987)
		CC 1017 (1968)
		TL *27* 5857 (1986)
	$NaAlH_2(OCH_2CH_2OMe)_2$	JOC *41* 3497 (1976); *43* 3435 (1978); *47* 4595 (1982); *51* 1155 (1986)
		Org Syn *64* 182 (1985)
		JACS *109* 1186, 1469, 2205 (1987)
		TL *28* 527, 803, 2041 (1987)
	$NaAlH_2(OCH_2CH_2OEt)_2$ (?)	JOC *51* 863 (1986)
$(CH_2)_nOH$	$LiAlH_4$	Syn 561 (1977) ($n = 2, 6, 7$) Ber *114* 292 (1981) ($n = 7$)
COR, CO_2Me	*i*-Bu_2AlH, HMPA, (cat MeCu)	JOC *52* 1624 (1987)
CN	$LiAlH_4$	Syn 430 (1979)

2. Alkylation and Functionalization

See also page 333, Section 4, for the synthesis of vinylic halides; and page 273, Section 9.

2.1. Thermal

TL *28* 1501 (1987)

Helv *62* 852 (1979); *65* 13, 2413, 2517 (1982)

2.2. Free Radical

$$RC\equiv C-C_n-X \longrightarrow RCH=C \quad C_n \quad \text{or}$$

X	Reagent(s)	Ring Size	
halogen	n-Bu$_3$SnH	5	JOC *47* 5382 (1982) (3-methylene tetrahydrofurans); *52* 4943 (1987) JACS *105* 3720 (1983) (3-alkylidene tetrahydrofurans); *107* 1448 (1985) (serial cyclization) CL 1437 (1984) (3-methylene tetrahydrofurans) TL *26* 6001 (1985) (methylene dihydrobenzofuran); *27* 3715 (1986) (3-methylene tetrahydrofuran); *28* 2009 (3-methylene tetrahydrofurans), 2887 (carbapenam) (1987) CC 587 (1987) (3-methylene tetrahydrofurans)
		6	JOC *44* 546 (1979)
		6, 7	TL *23* 2505 (1982) JOC *48* 1841 (1983) (both bicyclic β-lactams)
	n-Bu$_3$SnCl, NaBH$_3$CN	5	TL *28* 5203, 6393 (1987) (both 3-methylene tetrahydrofurans)

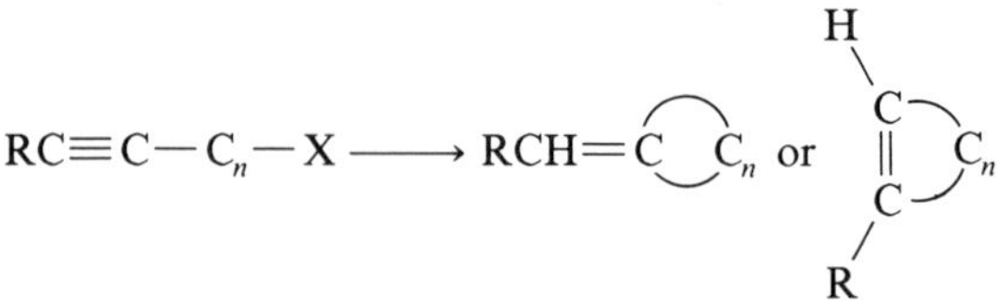

$$RC\equiv C-C_n-X \longrightarrow RCH=C \underset{C_n}{\overset{}{\diagup}} \text{ or } \begin{array}{c} H \\ \| \\ C \end{array} \diagup C_n$$

(*Continued*)

X	Reagent(s)	Ring Size	
	Co(I)	5	JOC *47* 1775, 5382 (1982); *50* 5875 (1985) (all 3-methylene tetrahydrofurans)
	vitamin B$_{12}$, reductant	5	Chimia *39* 203 (1985)
NO$_2$	*n*-Bu$_3$SnH	5	CL 635 (1985) (3-methylene tetrahydrofurans) Tetr *41* 4013 (1985) (3-methylene tetrahydrofurans)
O$_2$CC$_6$H$_4$CF$_3$	hν	5	JOC *52* 5583 (1987)
OCS$_2$CH$_3$	*n*-Bu$_3$SnH	5	JOC *52* 5583 (1987)
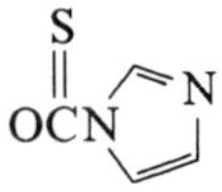 OCN (imidazole, C=S)	*n*-Bu$_3$SnH	5, 6	JOC *49* 1313 (1984)
SPh	*n*-Bu$_3$SnH	5, 6	TL *23* 4765 (1982) Tetr *41* 3959 (1985) (pyrrolizidinones and indolizidinones)
SePh	Ph$_3$SnH	5	CC 941, 1205 (1985); 878 (1986) JOC *52* 4943 (1987)

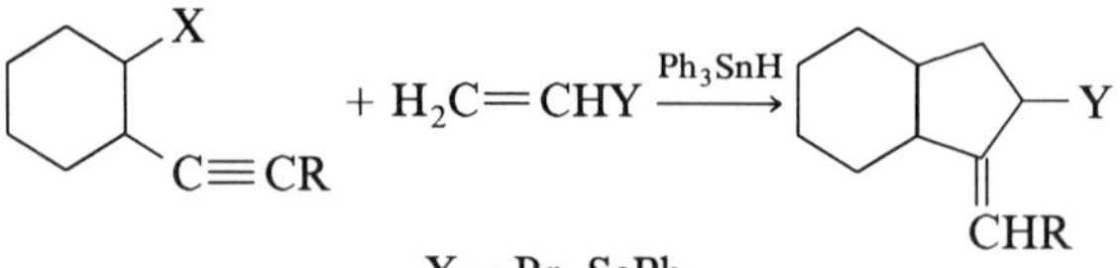

X = Br, SePh

Y = CO$_2$Me, CN, SO$_2$Ph

CC 980 (1985)

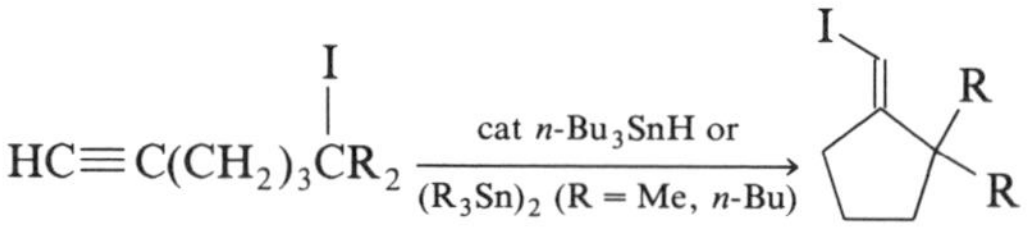

$$HC\equiv C(CH_2)_3CR_2 \xrightarrow[\text{(R}_3\text{Sn)}_2 \text{ (R = Me, }n\text{-Bu)}]{\text{cat }n\text{-Bu}_3\text{SnH or}}$$

JACS *108* 2489 (1986)
TL *28* 2477 (1987)

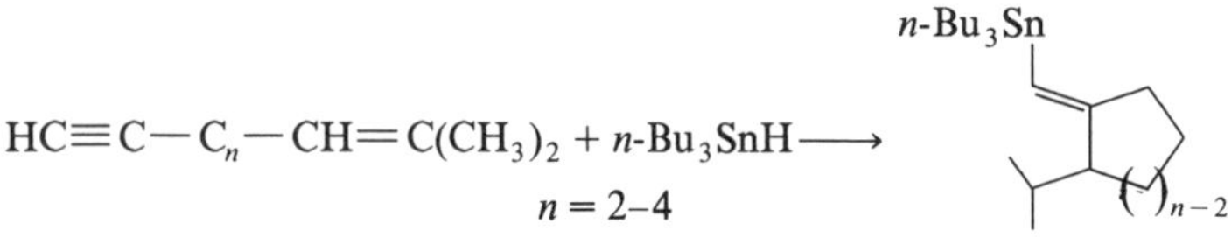

$$HC\equiv C-C_n-CH=C(CH_3)_2 + n\text{-Bu}_3\text{SnH} \longrightarrow$$

$$n = 2-4$$

JACS *109* 2829 (1987)
TL *28* 1503 (1987)

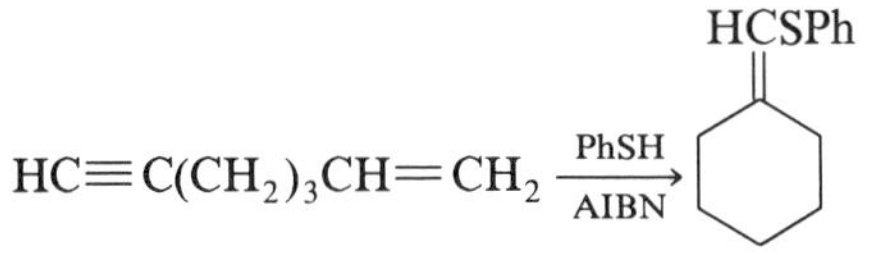

$$HC{\equiv}C(CH_2)_3CH{=}CH_2 \xrightarrow[\text{AIBN}]{\text{PhSH}}$$

HCSPh

TL *28* 1503 (1987)

2.3. Michael Additions

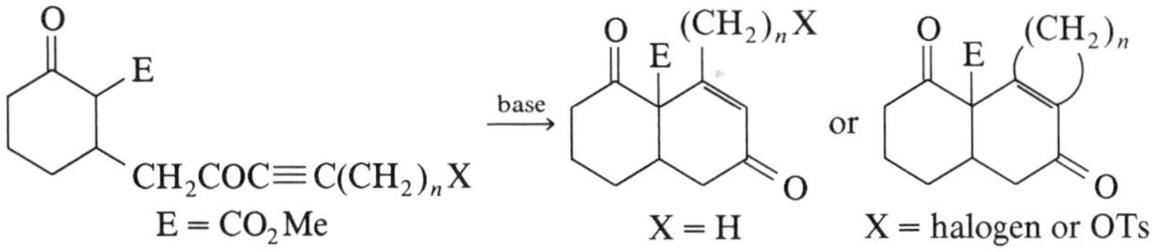

$$E = CO_2Me$$

$$CH_2COC{\equiv}C(CH_2)_nX$$

$$\text{base} \longrightarrow$$

X = H   or   X = halogen or OTs

TL *27* 5455 (1986); *28* 3457 (1987)

2.4. Organolithium Compounds

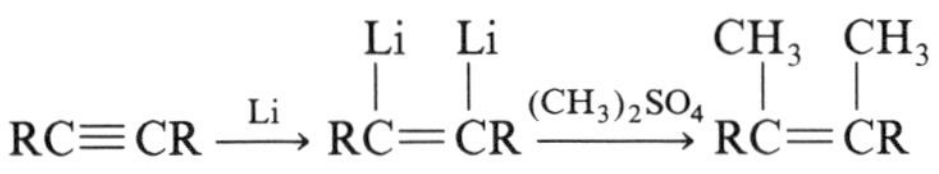

$$RC{\equiv}CR \xrightarrow{Li} \overset{Li\;\;\;Li}{RC{=}CR} \xrightarrow{(CH_3)_2SO_4} \overset{CH_3\;\;\;CH_3}{RC{=}CR}$$

Angew Int *25* 167 (1986)

$$XC{\equiv}CCNHMe \xrightarrow[\text{2. H}_2\text{O}]{\text{1. RLi}} \overset{X\;\;\;\;R}{\underset{H\;\;\;\;CONHMe}{C{=}C}}$$

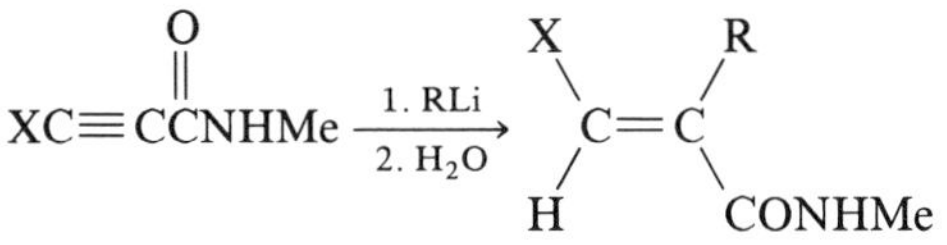

X = Ph, SiMe₃

JACS *107* 6740 (1985)

2.5. Organomagnesium Compounds

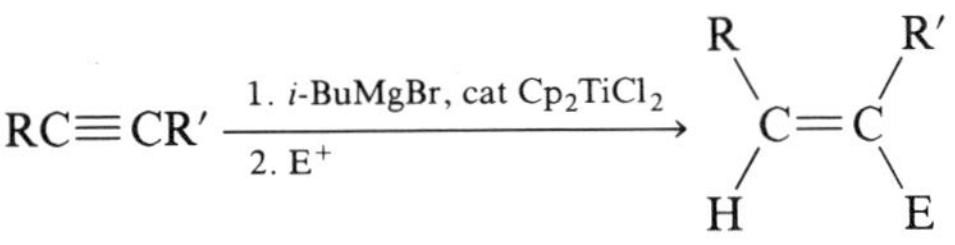

$$RC{\equiv}CR' \xrightarrow[\text{2. E}^+]{\text{1. }i\text{-BuMgBr, cat Cp}_2\text{TiCl}_2} \overset{R\;\;\;\;R'}{\underset{H\;\;\;\;E}{C{=}C}}$$

E⁺ = H₂O, D₂O, I₂, MeI, RCHO

R, R′ = Me, Ph; *n*-Bu, Me₃Si; Me₃Si, Ph

TL *22* 85 (1981)

$$RC \equiv CCH_2OH \xrightarrow[\text{cat } Cp_2TiCl_2]{i\text{-BuMgCl}} \xrightarrow{E^+} \overset{R}{\underset{E}{>}}C=C\overset{CH_2OH}{\underset{H}{<}}$$

$\underline{E^+}$

MeI   CC 718 (1981)

CO_2/H^+ JOC *52* 3860 (1987)

$$Me_3SiC \equiv CCH_2OH \xrightarrow[\text{2. MeI or } n\text{-BuI-CuI}]{\text{1. } i\text{-BuMgCl, cat } Cp_2TiCl_2} \overset{Me_3Si}{\underset{R}{>}}C=C\overset{H}{\underset{CH_2OH}{<}}$$

CC 1126 (1982)

$$RC \equiv CSiMe_3 \xrightarrow[\text{cat } Ni(acac)_2\text{-Me}_3Al]{MeMgBr} \left[\overset{R}{\underset{Me}{>}}C=C\overset{SiMe_3}{\underset{MgBr}{<}} \right] \xrightarrow{E^+} \overset{R}{\underset{Me}{>}}C=C\overset{SiMe_3}{\underset{E}{<}}$$

$E^+ = D_2O, CO_2, RCHO, I_2, H_2C=CHBr, CH_3I, H_2C=CHCH_2Br$

JACS *100* 4624 (1978)
TL 1682 (1979)
Israel J Chem *24* 108 (1984)

$$RC \equiv CCH_2OH \xrightarrow[\text{cat CuI}]{R'MgX} \xrightarrow[\text{2. } H_2O]{\text{1. } E^+} \overset{R}{\underset{E}{>}}C=C\overset{R'}{\underset{CH_2OH}{<}}$$

R' = 1°, 2°, 3° alkyl; aryl; allylic; benzylic

$E^+ = H_2O, I_2, H_2C=CHCH_2Br, RCHO, R_2CO, CO_2$

JOMC *168* 1, 227, 233 (1979)
JOC *51* 4080 (1986)

2.6. Organoboron Compounds

$$HC \equiv CCH_2Cl \xrightarrow[\text{3. HOAc}]{\substack{\text{1. } R_2BH \\ \text{2. 2 MeLi}}} RCH_2CH=CH_2$$

Syn 672 (1973)

$$RC \equiv CH \longrightarrow \overset{R}{\underset{H}{>}}C=C\overset{H}{\underset{B}{<}}\underset{O}{\overset{O}{\diagdown}}C_6H_4 \xrightarrow[\substack{\text{cat Pd black} \\ Et_3N}]{R'X} \overset{R}{\underset{R'}{>}}C=CH_2$$

R' = aryl, vinylic

JOMC *213* C53 (1981)

$$RC\equiv CH \xrightarrow{R_2'BH} \underset{H}{\overset{R}{>}}C=C\underset{BR_2'}{\overset{H}{<}}$$

$$\xrightarrow[\text{NaOH or NaOMe}]{I_2} \underset{H}{\overset{R}{>}}C=C\underset{H}{\overset{R'}{<}}$$

JACS *89* 3652 (1967);
93 6309 (1971)
JOC *41* 3947 (1976);
47 1792 (1982)
JOMC *225* C1 (1982)
Tetr *38* 2355 (1982)

$$\underset{H}{\overset{R}{>}}C=C\underset{R'}{\overset{H}{<}}$$

NCBr

Pd(OAc)$_2$, Et$_3$N

JACS *94* 6560 (1972)
CC 852 (1977)
BCSJ *53* 1670 (1980)

$$RC\equiv CH \xrightarrow[\substack{2.\ NaOMe \\ 3.\ I_2}]{1.\ R'BHBr\cdot SMe_2} \underset{H}{\overset{R}{>}}C=C\underset{H}{\overset{R'}{<}}$$

JOC *47* 171, 3806, 5407 (1982)

$$RC\equiv CH \xrightarrow[2.\ X_2]{1.\ RLi} RC\equiv CX \xrightarrow[\substack{\text{organoborane}\ 1.\ OH^-\ \text{or}\ OMe^- \\ 2.\ HOAc}]{} \underset{H}{\overset{R}{>}}C=C\underset{R'}{\overset{H}{<}}$$

X = Br, I

Organoborane

R$_2'$BH

JACS *89* 5086 (1967); *94* 4013 (1972)
Syn 555 (1972); 195 (1982)
JOC *47* 754 (1982); *51* 5270 (1986)
TL *23* 2785 (1982)

R'BHBr·SMe$_2$

JOC *47* 3808 (1982); *51* 5270 (1986)

(CH$_3$)$_2$CHC(CH$_3$)$_2$BHR'

JOC *51* 5270 (1986)

$$RC\equiv CBr \xrightarrow[\substack{2.\ NaOMe \\ 3.\ NaOMe,\ I_2}]{1.\ R_2'BH} \underset{H}{\overset{R}{>}}C=C\underset{R'}{\overset{R'}{<}}$$

JOC *47* 754 (1982)

$$RC\equiv CBr \longrightarrow \longrightarrow \underset{H}{\overset{R}{>}}C=C\underset{H}{\overset{BR_2'}{<}} \xrightarrow[\substack{\text{cat Pd(PPh}_3)_4 \\ NaOEt}]{R''X} \underset{H}{\overset{R}{>}}C=C\underset{H}{\overset{R''}{<}}$$

R'' = aryl, vinylic

TL *27* 3745 (1986)

$$R^1C\equiv CBr \xrightarrow[\substack{3.\ KHB(O\text{-}i\text{-}Pr)_3 \\ 4.\ R^2Li \\ 5.\ KOH\ or\ NaOR}]{\substack{1.\ Br_2BH\cdot SMe_2 \\ 2.\ i\text{-}PrOH}} \underset{HR^2}{\overset{R^1B(O\text{-}i\text{-}Pr)_2}{C=C}} \xrightarrow[\substack{cat\ Pd(PPh_3)_4 \\ KOH}]{R^3X} \underset{HR^2}{\overset{R^1R^3}{C=C}}$$

R^3 = allylic, aryl, vinylic

CL 1329 (1986)

$$R^1C\equiv CR^2 \xrightarrow{R^3_2BH} \underset{HBR^3_2}{\overset{R^1R^2}{C=C}} \xrightarrow{reagents} \underset{HR}{\overset{R^1R^2}{C=C}}$$

R	Reagents	
$CH_2CH=CH_2$	$NaOMe/CuBr\cdot SMe_2/$ $H_2C=CHCH_2Br$	JOC *45* 550 (1980)
	$MeLi/CuI/H_2C=CHCH_2Br$	BCSJ *53* 1471 (1980)
	$MeCu/H_2C=CHCH_2Br$	JOC *45* 1640 (1980)
	$Pd(OAc)_2, H_2C=CHCH_2Br$	BCSJ *53* 1670 (1980)
	cat $Pd(PPh_3)_4,$ $NaOH, H_2C=CHCH_2Br$	TL *21* 2865 (1980)
$CH_2C\equiv CH$	$MeLi/CuI/HC\equiv CCH_2Br$	BCSJ *53* 1471 (1980)
CH_2Ar	cat $Pd(PPh_3)_4,$ $NaOH, ArCH_2X$ $(X = Cl, Br)$	Tl *21* 2865 (1980) JOC *47* 2117 (1982)
Ar	cat $Pd(PPh_3)_4,$ $NaOH, ArI$	JOC *47* 2117 (1982)
	cat $Pd(PPh_3)_4,$ $NaOEt, ArX$ $(X = Br, I)$	CC 866 (1979)
$CH=CHR'$	—	See page 248, Section 3.2.
$C\equiv CR$	$LiC\equiv CR/I_2, OH^-$	CC 874 (1973) TL *27* 539 (1986)
	$NaOMe/CuCN/BrC\equiv CR$	JACS *106* 462 (1984)
	cat $Pd(PPh_3)_4, BrC\equiv CR$, base	TL 3437 (1979); *22* 127 (1981) Tetr *37* 2617 (1981) JACS *107* 972 (1985) Pure Appl Chem *57* 1749 (1985)

$$R^1C\equiv CH \xrightarrow[2.\ R^3_3B]{1.\ RLi} (R^1C\equiv C\bar{B}R^2_3)Li^+ \xrightarrow{reagent(s)} alkene$$

Alkene	Reagent(s)	
$Z\text{-}R^1CH=CHR^2$	RCO_2H	TL 2961 (1974); *28* 1007 (1987)
$R^1CH=CR^2_2$	$HCl/I_2, OH^-$	Syn 376 (1975)

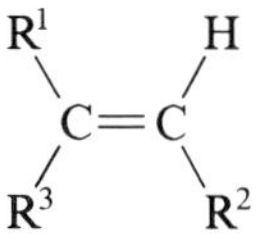

$$\underset{R^3}{\overset{R^1}{\diagup}}C=C\underset{R^2}{\overset{H}{\diagdown}}$$

R³X/HOAc

TL 1633 (1975);
28 1003, 1007 (1987)

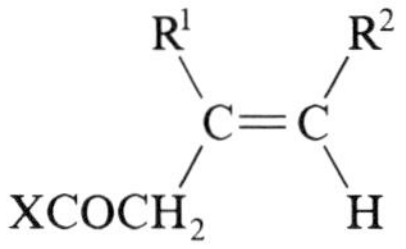

$$\underset{XCOCH_2}{\overset{R^1}{\diagup}}C=C\underset{H}{\overset{R^2}{\diagdown}}$$

BrCH₂COX (X = R, OR)/
HOAc

TL 4491 (1973)

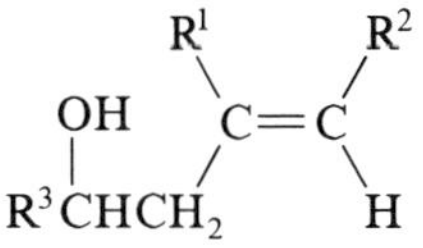

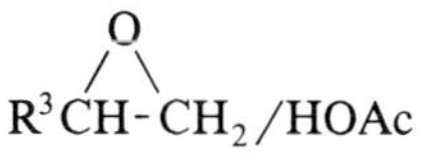

$$\underset{R^3CHCH_2}{\overset{OH}{|}}\underset{}{\overset{R^1}{\diagup}}C=C\underset{H}{\overset{R^2}{\diagdown}}$$

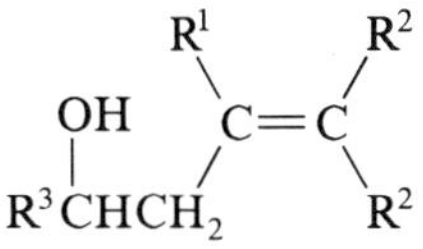

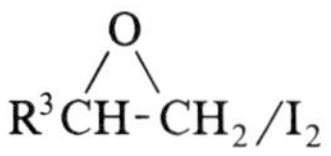

R³CH-CH₂/HOAc

TL 2741 (1973)

$$\underset{R^3CHCH_2}{\overset{OH}{|}}\underset{}{\overset{R^1}{\diagup}}C=C\underset{R^2}{\overset{R^2}{\diagdown}}$$

R³CH-CH₂/I₂

TL 2741 (1973)

$$\underset{H}{\overset{R^1}{\diagup}}C=C\underset{CH}{\overset{R^2}{\diagdown}}\underset{S}{\overset{S}{}}$$

$$\left[\overset{S}{\underset{S}{\diagup}}{}_{+}\right]BF_4^-/RCO_2H$$

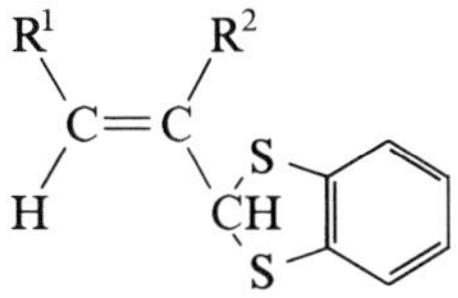

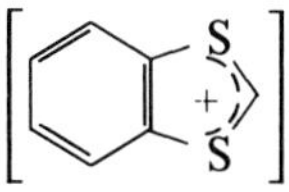

CC 164 (1981)

R²CH=CR′CHR³CHXCO₂Et

R³CH=CXCO₂Et
(X = COCH₃, CO₂Et)

TL *22* 797 (1981)

$$R'_3B + RC{\equiv}C\overset{O}{\overset{||}{C}}CH_3 \xrightarrow[H_2O]{O_2} RR'C{=}CH\overset{O}{\overset{||}{C}}CH_3$$

E + Z

JACS *92* 3503 (1970)

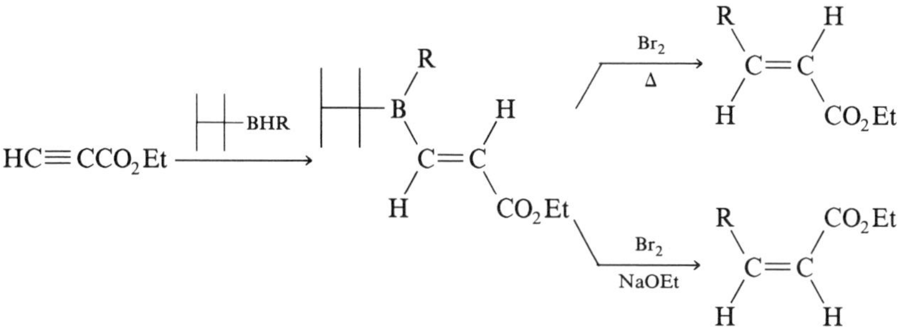

HC≡CCO₂Et

JOC *39* 2321 (1974)

$$RC{\equiv}CH \longrightarrow (R'_3\bar{B}C{\equiv}CR)Li^+ \xrightarrow[H_2SO_4]{CH_3COCl \quad CrO_3} R'_2C{=}CR\overset{O}{\overset{||}{C}}CH_3$$

TL 795 (1973)

$$RC\equiv CH \longrightarrow RCH=CHX$$

Stereochemistry	X	Reagents	
cis	Br	$(Sia)_2BH/Br_2/H_2O$	JACS *89* 4531 (1967)
	Br	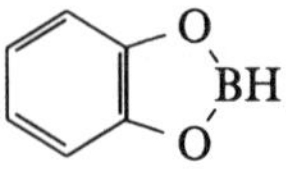 [catecholborane] BH/Br_2, NaOMe	JACS *95* 6456 (1973)
	Br, I	$RLi/X_2/$[cyclohexyl]$_2 BH/HOAc$ 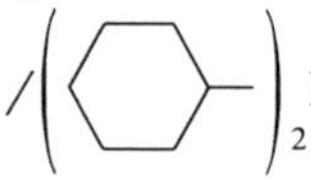	JACS *89* 5086 (1967)
trans	Br	$(Sia)_2BH/\Delta$	JACS *89* 4531 (1967)
	I	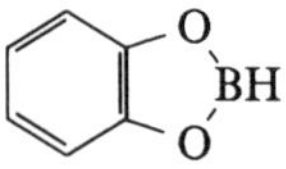 [catecholborane] BH/I_2, NaOH	JACS *95* 5785 (1973)
	I	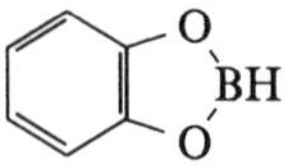 [catecholborane] BH/ICl, NaOAc	Syn Commun *11* 247 (1981)

$$RC\equiv CH \longrightarrow RC\equiv CX \xrightarrow[\text{2. Pb(OAc)}_4 \text{ or PhI(OAc)}_2]{\text{1. } R_2'BH} RCH=CXR'$$

$$X = Cl\,(Z)$$
$$X = Br\,(E \text{ or } Z)$$

CL 665 (1978)
BCSJ *53* 1652 (1980)

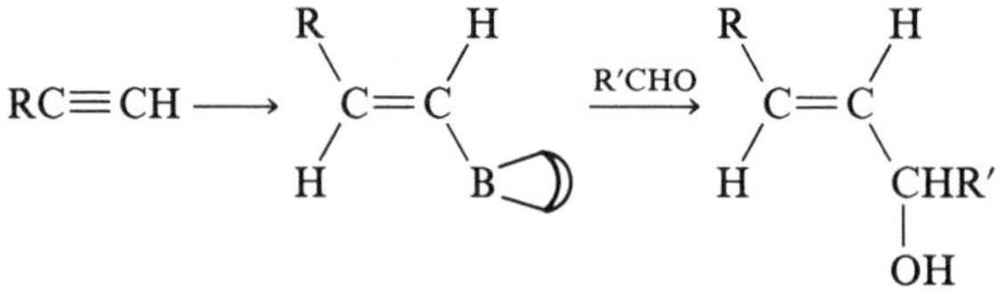

JOC *42* 579 (1977)

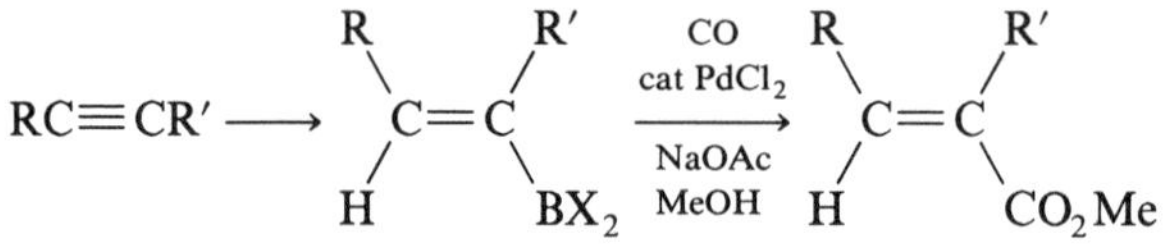

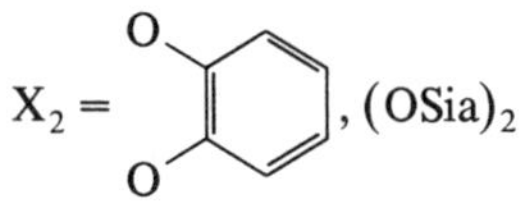

$$X_2 = \text{[o-phenylene dioxy]}, (OSia)_2$$

CL 879 (1981); 1329 (1986)

2.7. Organoaluminum Compounds

$$RC\equiv CH \xrightarrow{i\text{-Bu}_2\text{AlH}} \begin{array}{c} R \\ \backslash \\ C=C \\ / \quad \backslash \\ H \quad Al(i\text{-Bu})_2 \end{array} \xrightarrow[\text{2. R}'_2\text{CO}]{\text{1. Cp}_2\text{TiCl}_2} RCH_2CH=CR'_2$$

CL 429 (1982)

$$RC\equiv CH(R) \xrightarrow{i\text{-Bu}_2\text{AlH}} \begin{array}{c} R \quad H(R) \\ \backslash \quad / \\ C=C \\ / \quad \backslash \\ H \quad Al(i\text{-Bu})_2 \end{array} \longrightarrow \begin{array}{c} R \quad H(R) \\ \backslash \quad / \\ C=C \\ / \quad \backslash \\ H \quad X \end{array}$$

$\underline{X}$	Reagent(s)	
Br, I	Br_2 or I_2	JACS *89* 2753 (1967)
I	I_2	JACS *108* 3403, 7791 (1986)
CH_2OMe	$ClCH_2OMe$	Syn 816 (1976) TL *23* 2087 (1982)
$CH_2N\langle\rangle$	$i\text{-BuOCH}_2N\langle\rangle$	TL *21* 3763 (1980)
$CH=CHR'$	$R'CH=CHX$, cat Ni(O) $R'CH=CHX$, cat Pd(O)	JACS *109* 2393 (1987) JACS *98* 6729 (1976); *109* 2393 (1987) TL *28* 1649, 4875 (1987)

$$RC\equiv CH(R) \xrightarrow[\text{2. MeLi}]{\text{1. }i\text{-Bu}_2\text{AlH}} \begin{array}{c} R \quad H(R) \\ \backslash \quad / \\ C=C \\ / \quad \backslash \\ H \quad \overset{-}{Al}Me(i\text{-Bu})_2 \end{array} \xrightarrow{\text{reagent(s)}} \begin{array}{c} R \quad H(R) \\ \backslash \quad / \\ C=C \\ / \quad \backslash \\ H \quad X \end{array}$$

$\underline{X}$	Reagent(s)	
R'	R'X	TL 1927 (1976)
$CH_2N\langle\rangle$	$i\text{-BuOCH}_2N\langle\rangle$	TL *21* 3763 (1980)
CHOHR'	$R'CHO/H^+$	JACS *89* 2754 (1967) TL *23* 2087 (1982); *28* 5129 (1987)
CH_2CHOHR'	$R'\overset{O}{\overset{\frown}{CH\text{-}CH_2}}/H^+$	Syn 632 (1975) CC 17 (1976)
$\langle\rangle{=}O$	$\langle\rangle{=}O/H^+$	TL 4083 (1972)
CO_2H	CO_2/H^+	JACS *89* 2754 (1967)
CN	$(CN)_2$	JACS *90* 7139 (1968)

$$RC\equiv CSiMe_3 \xrightarrow[\text{2. reagent(s)}]{\text{1. }i\text{-Bu}_2\text{AlH}} \begin{array}{c} R \\ \diagdown \\ C \\ \diagup \quad \diagdown \\ H \quad\quad X \end{array} C = C \begin{array}{c} SiMe_3 \end{array}$$

$\underline{X}$	Reagent(s)	
H	H_2O	JOC *36* 3520 (1971)
		Syn 803 (1980)
Cl, Br, I	NCS, Br_2 or I_2	JOC *43* 2739 (1978)
R'	MeLi/R'X (R' = Me, allyl)	JOC *41* 2214, 2215 (1976)
CH_2OH	H_2CO/H_2O	TL *28* 3547 (1987)
CH_2CHOHR	MeLi/$\overset{O}{\overset{\diagup\diagdown}{RCH-CH_2}}$/$H_2O$	JACS *106* 4192 (1984)

$$HC\equiv CCO_2Me \xrightarrow[\text{2. }\bigcirc\text{-Br}]{\text{1. }i\text{-Bu}_2\text{AlH, HMPA}} H_2C=CCO_2Me$$

JOC *52* 1624 (1987)

$$RC\equiv CR \xrightarrow{i\text{-Bu}_2\text{Me}\overline{\text{Al}}\text{H}} \begin{array}{c} H \quad\quad R \\ \diagdown \quad\quad \diagup \\ C=C \\ \diagup \quad\quad \diagdown \\ R \quad \overline{\text{Al}}\text{Me}(i\text{-Bu})_2 \end{array} \xrightarrow{E^+} \begin{array}{c} H \quad\quad R \\ \diagdown \quad\quad \diagup \\ C=C \\ \diagup \quad\quad \diagdown \\ R \quad\quad E \end{array}$$

$\underline{E^+}$	
CO_2, H_2CO, I_2	JACS *89* 5085 (1967)
$(CN)_2$	JACS *90* 7139 (1968)

$$RC\equiv CCH_2OH \xrightarrow[\text{2. }I_2]{\text{1. aluminum hydride}} \begin{array}{c} R \quad\quad I \\ \diagdown \quad\quad \diagup \\ C=C \\ \diagup \quad\quad \diagdown \\ H \quad\quad CH_2OH \end{array}$$

Aluminum hydride	
$LiAlH_4$, cat $AlCl_3$	JACS *89* 4245 (1967); *91* 4318 (1969)
n-BuLi/i-Bu$_2$AlH	JACS *92* 6314 (1970)

$$RC\equiv CCH_2OH \xrightarrow[\text{2. }I_2]{\text{1. aluminum hydride}} \begin{array}{c} R \quad\quad H \\ \diagdown \quad\quad \diagup \\ C=C \\ \diagup \quad\quad \diagdown \\ I \quad\quad CH_2OH \end{array}$$

Aluminum hydride	
$LiAlH_4$, $NaOCH_3$	JACS *89* 4245 (1967); *90* 5618 (1968)
	TL 1821 (1971); *28* 5793 (1987)
$NaAlH_2(OCH_2CH_2OCH_3)_2$	JOC *51* 4316 (1986); *52* 1236, 3860, 3883 (1987)
	TL *28* 723 (1987)

$$RC{\equiv}CR \xrightarrow{R'_3Al} \underset{R'}{\overset{R}{>}}C{=}C\underset{AlR'_2}{\overset{R}{<}} \xrightarrow{H_2O} \underset{R'}{\overset{R}{>}}C{=}C\underset{H}{\overset{R}{<}}$$

Ann *629* 222 (1960)

$$RC{\equiv}CH(R) \xrightarrow[Cp_2ZrCl_2]{R^1_3Al \text{ or } R^1_2AlCl} \underset{R^1}{\overset{R}{>}}C{=}C\underset{AlR^1_2}{\overset{H(R)}{<}} \xrightarrow{reagent(s)} \underset{R^1}{\overset{R}{>}}C{=}C\underset{X}{\overset{H(R)}{<}}$$

Review: Pure Appl Chem *53* 2333 (1981)

X	Reagent(s)	
H or D	H_2O or D_2O	JACS *100* 2252 (1978); *103* 4985 (1981); *107* 6639 (1985) JOC *46* 4093 (1981) TL *27* 3311 (1986)
I	I_2	JACS *100* 2252 (1978); *102* 3298 (1980); *104* 4708 (1984); *107* 6639 (1985); *108* 7791 (1986) Syn 501 (1979) TL *23* 27 (1982) JOC *46* 4094 (1981); *51* 2230 (1986) CL 293 (1982) CC 1237 (1986)
R^2	R^2X, cat Pd(O) or Ni(O), $ZnCl_2$ (R^2 = aryl, vinylic, alkynyl, allylic, benzylic)	JACS *100* 2254 (1978); *103* 2882 (1981) TL *22* 2715 (1981); *28* 2221 (1987) Acct Chem Res *15* 340 (1982) (review)
CH_2OH	n-BuLi/CH_2O	TL 2357 (1978)
CH_2OMe	n-BuLi/$ClCH_2OMe$	TL 2357 (1978)
CO_2H	n-BuLi/CO_2/H^+	TL 2357 (1978) JOC *46* 4094 (1981)
CO_2Et	$ClCO_2Et$	TL 2357 (1978); *23* 27 (1982) JOC *46* 4094 (1981)
CH_2CHOHR^2	n-BuLi/$H_2C\overset{O}{\overset{\triangle}{-}}CHR^2$	Syn 1034 (1980) JOC *45* 5223 (1980)
$CH_2CH_2COCH_3$	$H_2C{=}CHCOCH_3$	Pure Appl Chem *53* 2333 (1981)

$$Br(CH_2)_nC\equiv CM \xrightarrow[\text{(Cp}_2\text{ZrCl}_2)]{\text{RAlR}'_2} \underset{(CH_2)_n}{\overset{R \quad M}{C=C}} \xrightarrow[\substack{(M = AlMe_2, \\ n = 2)}]{I_2} \overset{R \quad I}{\square}$$

$$n = 2, 4$$

$$M = AlMe_2, SiMe_3$$

$$R = H, alkyl$$

JACS *105* 6344 (1983); *106* 6105 (1984)

$$RC\equiv CR' \xrightarrow[\text{2. NBS or } I_2]{\text{1. Et}_2\text{AlCl, Cp}_2\text{TiCl}_2} \underset{Et}{\overset{R}{C}}=\underset{X}{\overset{R'}{C}}$$

R' = Ph, SiMe₃; X = Br, I

JOC *50* 2121 (1985)

$$CH_3CH_2C\equiv C(CH_2)_2OH \xrightarrow[\text{2. MeOH}]{\text{1. Me}_3\text{Al, TiCl}_4} \underset{CH_3}{\overset{CH_3CH_2 \quad (CH_2)_2OH}{C=C}}\underset{H}{}$$

JOC *46* 807 (1981); *50* 2124 (1985)

$$RCH=CHCH=CHC\overset{O}{\overset{\|}{X}} \xrightarrow[\text{2. MeLi/H}_2\text{C}=\text{CHCH}_2\text{Br}]{\text{1. }i\text{-Bu}_2\text{AlH, cat MeCu, HMPA}} RCH_2CH=CHCH\overset{H_2C=CHCH_2}{\overset{|}{C}OX}$$

X = Me, OMe

JOC *52* 439 (1987)

2.8. Organosilicon Compounds

Review: Syn 761 (1979) (electrophilic substitution of organosilicon compounds)

$$RC\equiv CH \longrightarrow \underset{H}{\overset{R \quad H}{C=C}}\underset{SiF_5^{2-}}{} \xrightarrow{\text{reagent(s)}} \underset{H}{\overset{R \quad H}{C=C}}\underset{X}{}$$

X	Reagent(s)	
Cl, Br	CuCl₂, CuBr₂	Organomet *1* 369 (1982)
Cl, Br, I	Cl₂, Br₂, NBS, I₂	Organomet *1* 355 (1982)
Ph	PhI, Pd catalyst	Organomet *1* 542 (1982)
CO₂R	CO, ROH, PdCl₂, NaOAc	Organomet *1* 542 (1982)
E-CH=CHR	PdCl₂	Organomet *1* 542 (1982)

E-CH=CHX (X = CHO, CO_2R, CN)	H_2C=CHX, $Pd(OAc)_2$	Organomet 1 542 (1982)
CHR^1CR^2=CHR^3	R^1CH=CR^2CHR^3X, $Pd(OAc)_2$ (X = Cl, Br, OTs)	Organomet 1 542 (1982)
OR′	R′OH, O_2, cat $Cu(OAc)_2$	TL 21 4105 (1980)

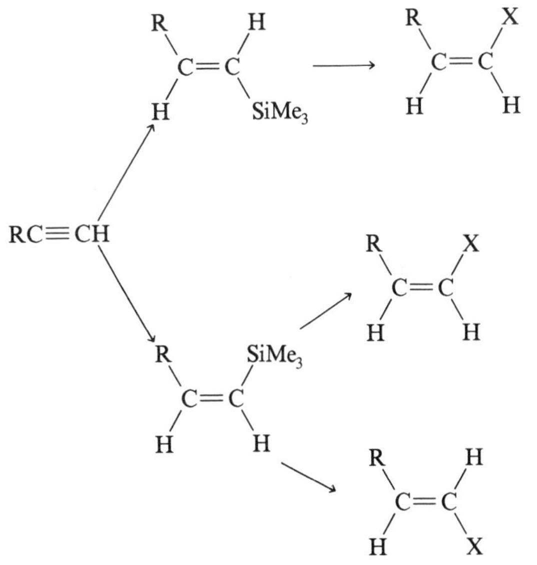

X = Cl, Br
  X_2/NaOMe or
    KF or alumina

Syn Commun 7 475
  (1977); 8 291 (1978)
JOC 43 4424 (1978)

X = Br
  m-ClC$_6$H$_4$CO$_3$H/HBr,   JACS 99 1993 (1977)
  BF$_3$

X = I
  ICl/KF

JOMC 285 109 (1985)

X = I
  I_2 or I_2-
    AgO$_2$CCF$_3$/KF

TL 543 (1976); 3317
  (1977)

X = Cl, Br
  X_2/NaOMe or KF or   Syn Commun 7 475
  alumina

(1977); 8 291 (1978)
JOC 43 4424 (1978)

X = Br
  m-ClC$_6$H$_4$CO$_3$H/HBr,   TL 1453 (1976)
  BF$_3$

JACS 99 1993 (1977)

Reviews:

Syn 761 (1979) (halogenation of organosilanes)
TL 27 883 (1986) (I_2, $AlCl_3$ or $SnCl_4$- variable stereochemistry)
JOC 52 1100 (1987) (stereochemistry of halogenation)

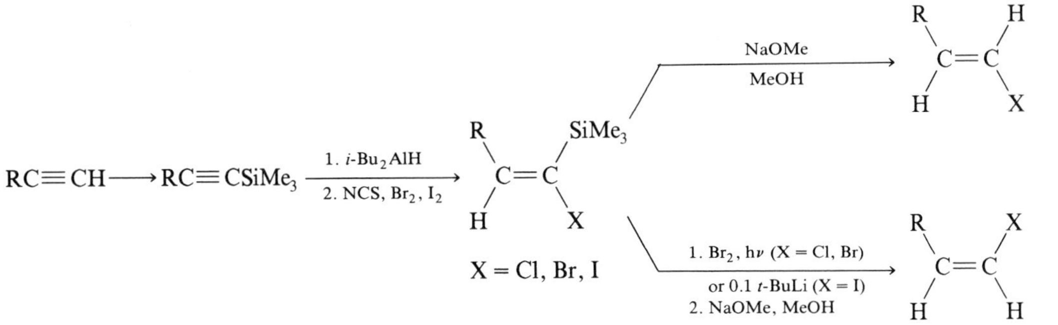

JOC 43 2739 (1978)
Syn 999 (1981)

$$R^1CH=CBrSiMe_3 \xrightarrow[\substack{sec\text{-BuLi}/R^2X \\ (\text{retention})}]{(R^2CuSPh)Li \text{ or}} \xrightarrow[\substack{2. \text{ NaOMe} \\ (\text{inversion})}]{1. Br_2} R^1CH=CR^2Br \xrightarrow[\substack{\text{or } R_2^3CuLi \\ (\text{retention})}]{RLi/R^3X} R^1CH=CR^2R^3$$

E or Z

JOC *44* 4623 (1979)

$$R'CC\equiv CR \xrightarrow[\substack{2. H_2O}]{1. Me_3SiX} R'CCH=CR$$

R = H or alkyl
X = Br, I

TL *27* 4759, 4763 (1986)

$$\substack{R^2 \\ \diagdown \\ \\ / \\ H} C=C \substack{R^1 \\ / \\ \\ \diagdown \\ SiMe_3} \xrightarrow{E^+} \substack{R^2 \\ \diagdown \\ \\ / \\ H} C=C \substack{R^1 \\ / \\ \\ \diagdown \\ E}$$

E^+ = HI; BrCN, $AlCl_3$ (E = Br); I_2; Cl_2CHOMe, $AlCl_3$ (E = CHO);

CH_3COCl, $AlCl_3$ (E = CH_3CO); Br_2 (inversion of stereochemistry)

TL 3317 (1977)

2.9. Organotin Compounds

$$RC\equiv CH \longrightarrow \substack{R \\ \diagdown \\ \\ / \\ H} C=C \substack{H \\ / \\ \\ \diagdown \\ X}$$

X	Reagents	
Br	n-Bu$_3$SnH/Br$_2$	JOC *43* 3450 (1978)
		JACS *106* 5734 (1984);
		107 7515 (1985)
I	n-Bu$_3$SnH/I$_2$	JOC *43* 3450 (1978)
		TL *23* 3851 (1982)
	n-Bu$_3$SnMgMe/CuCN/	JACS *109* 2138 (1987)
	H$_2$O/I$_2$	
R'	n-Bu$_3$SnH/n-BuLi/R'X	JOC *40* 2265 (1975)
		TL 3847 (1977)

n-Bu$_3$SnH/n-BuLi/   JOC *40* 2265 (1975);
43 3450 (1978)
TL 4705 (1976)

n-C$_3$H$_7$C$\equiv$CCu/ /
H$_2$O

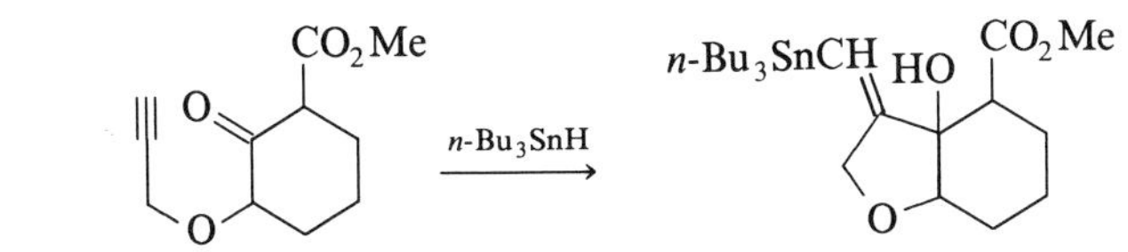

TL *28* 2001 (1987)

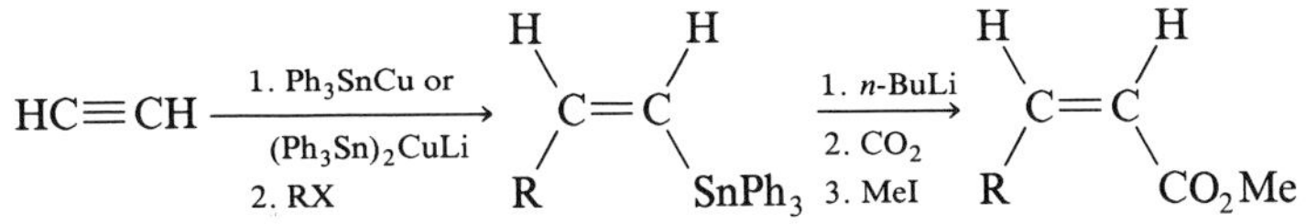

TL *23* 2797 (1982)

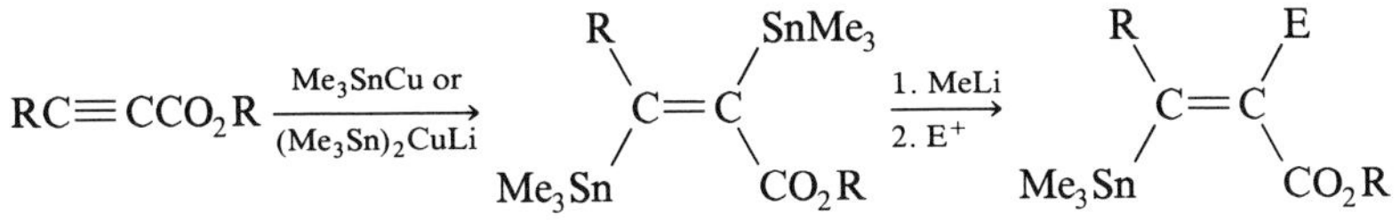

$E^+ = RX$ (1° alkyl, allylic, benzylic), R_2CO

JOC *47* 1602 (1982)

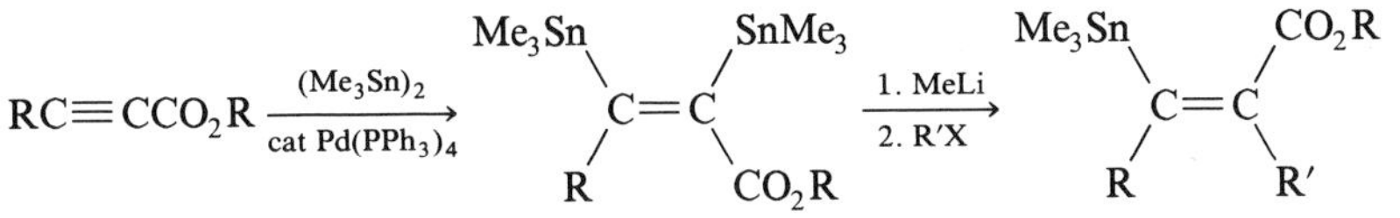

JOC *52* 4421 (1987)

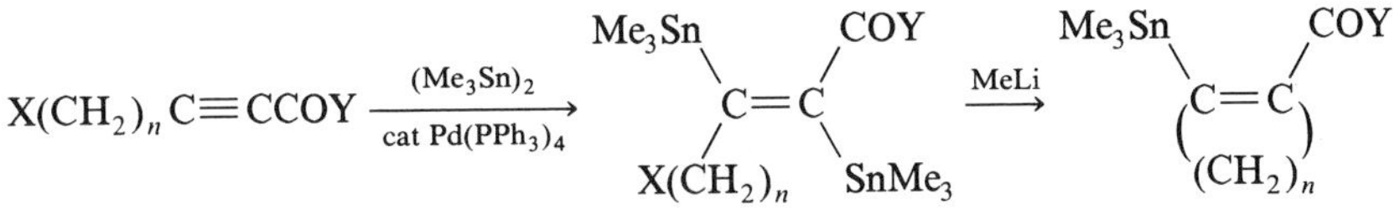

X = halogen
Y = OR, NR_2
n = 3–5

CC 626 (1986)

2.10. Organosulfur, -selenium and -tellurium Compounds

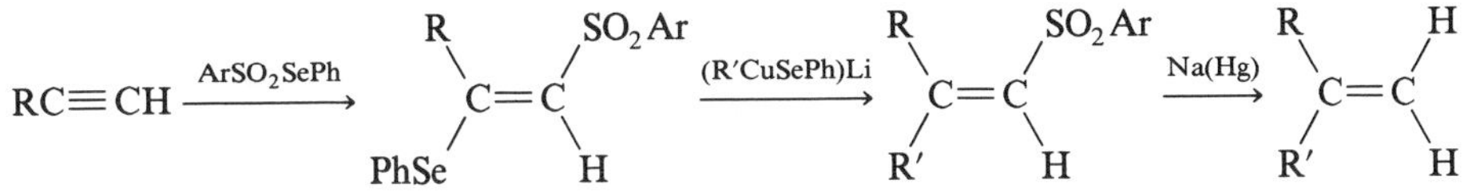

TL *27* 2187 (1986)
JOC *52* 4258 (1987)

$$RC\equiv CH \xrightarrow[NaBH_4]{(PhTe)_2} \underset{H}{\overset{R}{>}}C=C\underset{H}{\overset{TePh}{<}}$$

PhMgBr
Ni or Co catalyst

$$\underset{H}{\overset{R}{>}}C=C\underset{H}{\overset{Ph}{<}}$$

TL *23* 1181 (1982)

CO
cat PdCl$_2$
CuCl$_2$
R'OH

$$\underset{H}{\overset{R}{>}}C=C\underset{H}{\overset{CO_2R'}{<}}$$

JOC *52* 4859 (1987)

2.11. Organotitanium Compounds

See also page 217, Section 2.5; and page 223, Section 2.7.

$$R_f X + H_2C=CHC(CH_3)=CH_2 \xrightarrow[ultrasound]{Zn,\ Cp_2TiCl_2} R_f CH_2CH=C(CH_3)CH_3$$

X = Br, I

JACS *107* 5186 (1985)

2.12. Organozirconium Compounds

See also page 223, Section 2.7.

$$RC\equiv CH(R) \xrightarrow{Cp_2ZrHCl} \underset{H}{\overset{R}{>}}C=C\underset{ZrCl(Cp)_2}{\overset{H(R)}{<}} \xrightarrow{reagent(s)} \underset{H}{\overset{R}{>}}C=C\underset{X}{\overset{H(R)}{<}}$$

X	Reagent(s)	
Ar	ArI, cat Ni(PPh$_3$)$_4$	JACS *99* 3168 (1977) JOC *52* 3319 (1987)
CH=CHR^1	R^{1}CH=CHX, cat Pd(O)	TL 1027 (1978); *23* 62 (1982); *28* 4875, 4879 (1987) JACS *109* 2393 (1987)
CH(R)=CHR	CuCl$_2$	TL 1303 (1977)
CH$_2$CH=CH$_2$	(π-C$_3$H$_5$PdCl)$_2$	JACS *102* 7381 (1980) TL *22* 4655 (1981)
	H$_2$C=CHCH$_2$Cl, cat (π-C$_3$H$_5$PdCl)$_2$	TL *22* 2629 (1981)
CH$_2$CH$_2$COR^1	H$_2$C=CHCOR^1, cat Ni(acac)$_2$	JACS *99* 8045 (1977)
	H$_2$C=CHCOR^1, CuOTf, LiI	TL 1303 (1977)

CN	RN=C (R = t-Bu, SiMe$_3$)/I$_2$	TL *28* 295 (1987)
Cl	NCS	JACS *97* 679 (1975)
Br	NBS	JACS *97* 679 (1975); *108* 5559 (1986)
I	I$_2$	JACS *97* 679 (1975); *109* 2138 (1987) Agric Biol Chem *46* 717 (1982)

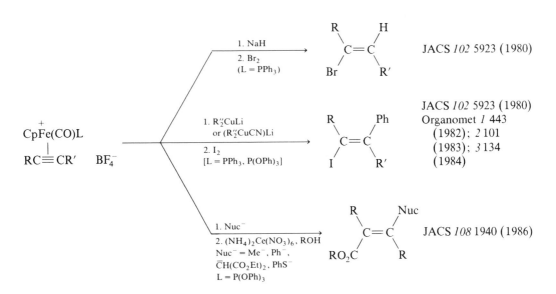

SePh   (phthalimide-NSePh structure)   JOC *52* 2334 (1987)

JACS *107* 2568 (1985)
TL *27* 2829 (1986); *28* 917 (1987)
JOC *51* 4080 (1986)

2.13. Organoiron Compounds

CpFe(CO)L
|
RC≡CR'   BF$_4^-$

1. NaH
2. Br$_2$
(L = PPh$_3$)
→ (vinyl product)   JACS *102* 5923 (1980)

1. R''$_2$CuLi or (R''$_2$CuCN)Li
2. I$_2$
[L = PPh$_3$, P(OPh)$_3$]
→ (vinyl product)   JACS *102* 5923 (1980) Organomet *1* 443 (1982); *2* 101 (1983); *3* 134 (1984)

1. Nuc$^-$
2. (NH$_4$)$_2$Ce(NO$_3$)$_6$, ROH
Nuc$^-$ = Me$^-$, Ph$^-$,
C̄H(CO$_2$Et)$_2$, PhS$^-$
L = P(OPh)$_3$
→ (vinyl product)   JACS *108* 1940 (1986)

$$Fe(CO)_3$$

$$\xrightarrow[\text{3. E}^+]{\begin{array}{l}\text{1. CO}\\ \text{2. RLi}\end{array}}$$

$$R = \text{stabilized anion}; \; E^+ = CF_3CO_2H, \; MeI, \; EtOSO_2F \; (E' = OEt)$$

JACS *105* 2497 (1983)

2.14. Organocobalt Compounds

$$RX \xrightarrow[\begin{array}{l}\text{1. NaCo(CO)}_4\\ \text{2. CO}\\ \text{3. H}_2\text{C}=\text{CHCH}=\text{CH}_2\end{array}]{} RCCH_2-C \cdots C-H \xrightarrow{\text{NaCHXY}} RCCH_2CH=CHCH_2CHXY$$

X	Y
CO$_2$R	COR
CO$_2$R	CO$_2$R
CO$_2$R	CN

JACS *104* 4917 (1982)

2.15. Organonickel Compounds

See also page 217, Section 2.5; and page 223, Section 2.7.

$$RC\equiv CR' \xrightarrow[\text{2. reagent(s)}]{\text{1. MeNi(acac)PPh}_3} RC=CR'$$

X	Reagent(s)
H	TsOH or LiAlH$_4$
Me	Me$_3$Al or MeLi
I	I$_2$
CO$_2$Me	CO, MeOH

JACS *103* 3002 (1981)

$$RC\equiv CR' \xrightarrow[\text{Ni catalyst}]{\text{HCN}} \underset{H}{\overset{R}{\diagdown}}C=C\underset{CN}{\overset{R'}{\diagup}}$$

CC 1231 (1982)

2.16. Organopalladium Compounds

See also page 218, Section 2.6; page 223, Section 2.7; page 226, Section 2.8; page 228, Section 2.9; and page 239, Section 2.19.

$$RC\equiv CR' + ArI \xrightarrow[\text{R}_3\text{N, HCO}_2\text{H}]{\text{cat Pd(OAc)}_2(\text{PPh}_3)_2} \underset{R}{\overset{Ar}{\diagdown}}C=C\underset{R'}{\overset{H}{\diagup}}$$

TL *25* 3137 (1984) (R = R' = Ph); *27* 6397 (1986) (R' = SiMe$_3$)
Tetr *41* 5121 (1985) (R = Ar, R' = CR$_2$OH)

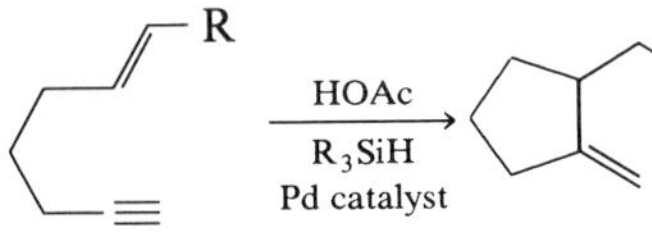

$$\text{JACS } 109 \text{ 3061 (1987)}$$

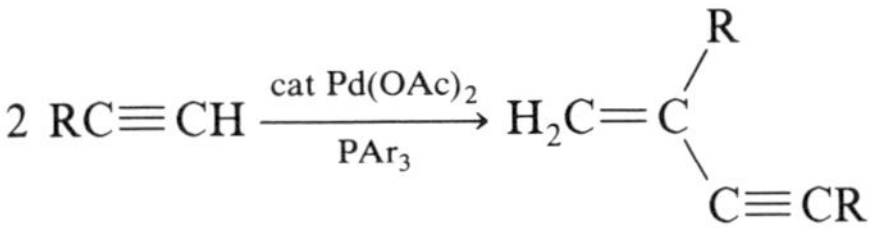

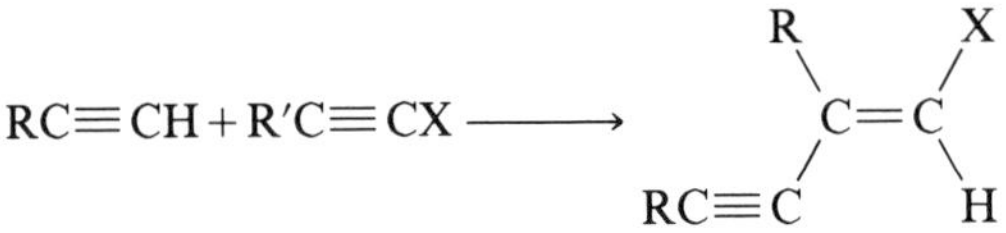

$$2\ RC\equiv CH \xrightarrow[\text{PAr}_3]{\text{cat Pd(OAc)}_2} H_2C=C\begin{smallmatrix}R\\\\C\equiv CR\end{smallmatrix}$$

$$RC\equiv CH + R'C\equiv CX \longrightarrow$$

$$X = CO_2Me,\ SO_2Ph$$

$$\text{JACS } 109 \text{ 3486 (1987)}$$

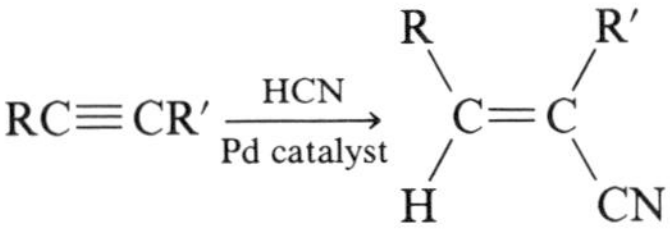

$$RC\equiv CR' \xrightarrow[\text{Pd catalyst}]{\text{HCN}}$$

$$\text{CC 1231 (1982)}$$

2.17. Organocopper Compounds

See also page 217, Section 2.5; page 218, Section 2.6; page 228, Section 2.9; and page 231, Section 2.13.

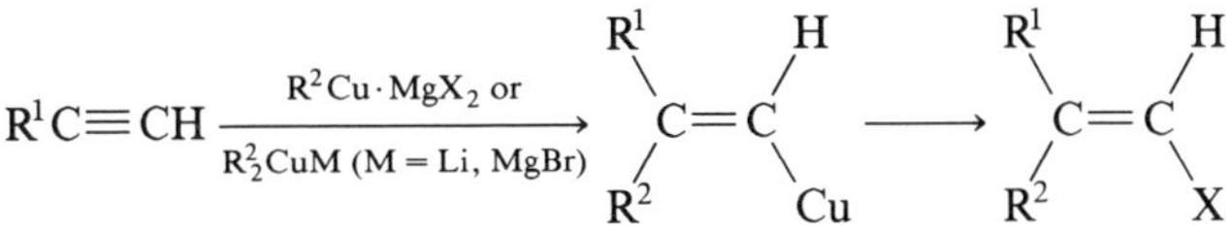

$$R^1C\equiv CH \xrightarrow[R_2^2CuM\ (M = Li,\ MgBr)]{R^2Cu\cdot MgX_2\ \text{or}} \longrightarrow$$

Reviews:

Pure Appl Chem *50* 709 (1978)
Syn 841 (1981)
Tetr *40* 641 (1984)
"Current Trends in Organic Synthesis. Proceedings of the 4th International Conference on Organic Synthesis," Ed. H. Nozaki, Pergamon, New York (1982), pp 291–302

Formation of vinylcopper reagent:

TL 2583 (1971); 3461 (1976); 2023 (1977); 1363 (1978)
BSCF 1656 (1974)
Rec Trav Chim *95* 299, 304 (1976); *100* 98, 249 (1981)
JOC *44* 3888 (1979)
JOMC *177* 293 (1979); *215* C1 (1981)

Use of functionally substituted alkynes:

TL 2583 (1971); 2313 (1976); *24* 5077 (1983); *28* 2363 (1987)
JOMC *57* C99 (1973); *144* 13 (1983)
J Mol Catal *1* 43 (1975/76)
BSCF 693 (1977)
Rec Trav Chim *100* 337 (1981)

<u>X</u>	Reagent(s)	
H(D)	$H^+ (D^+)$	TL 2583 (1971); 2023 (1977); 1363 (1978); *23* 5155 (1978); *24* 5077 (1983) JOMC *77* 269 (1974) BSCF 1656 (1974) Rec Trav Chim *95* 299, 304 (1976); *100* 98, 337 (1981) JOC *44* 3888 (1979); *52* 5419 (1987) Syn Commun *11* 157 (1981) J Chem Res (S) 354 (1981)
Cl	NCS, HMPA	TL 2023, 3545 (1977) Rec Trav Chim *96* 168 (1977)
Br	$HgBr_2/Br_2$, py NBS, HMPA BrCN	Syn 803 (1974) TL 2023, 3545 (1977) Rec Trav Chim *96* 168 (1977) Rec Trav Chim *96* 168 (1977)
I	I_2 ICN	TL 2583 (1971); 3461 (1976); *23* 5155 (1982) JOMC *77* 269 (1974); *177* 293 (1979) Rec Trav Chim *96* 168 (1977) JOC *43* 1279 (1978); *52* 1381 (1987) J Chem Res (S) 354 (1981) JACS *106* 6105 (1984) Org Syn *62* 1 (1984) Rec Trav Chim *96* 168 (1977) JOC *52* 4885 (1987)
R^3	R^3X, HMPA, $P(OR^4)_3$ $R^3 = 1°$ alkyl, allylic, benzylic, R^5OCHR^6X; X = Cl, Br, I; R^4 = Me, Et	TL 2583 (1971); 1363, 3125 (1978); 1433 (1980); *23* 3587, 5155 (1982) JOMC *40* C49 (1972); *77* 269 (1974) Syn 826 (1979) JOC *44* 1345, 3888 (1979) Rec Trav Chim *100* 249 (1981) J Chem Res (S) 354 (1981) Org Syn *64* 1 (1986)
Ar	ArI, $ZnBr_2$, cat $Pd(PPh_3)_4$	TL *22* 3851 (1981) BSCF II 321 (1983)
$CH(OMe)R^3$	$R^3CH(OMe)_2$, $BF_3 \cdot OEt_2$	TL *25* 3075 (1984)
CH_2OH	H_2CO/H^+	Tetr *36* 1961 (1980)

$CHOHR^3$	$R_3CHXO(CH_2)_2Cl$ $(X = Cl, Br)/$ $n\text{-BuLi}/H^+$	TL 2407 (1973) JOMC 77 269 (1974)
CHR^3SR^4	$ClCHR^3SR^4$	JOMC 77 269 (1974) Syn 43 (1984)
CH_2NMe_2	$(Me_2\overset{+}{N}{=}CH_2)I^-$	BSCF II 377 (1984)
$CHR^4NR_2^3$	$R^5SCHR^4NR_2^3$	TL 21 3763 (1980); 23 5155 (1982) BSCF II 377 (1984)
	$R^5OCHR^4NR_2^3$	BSCF II 377 (1984)
$CH_2N(CH_3)CHO$	$ClCH_2N(CH_3)CHO$	Syn 40 (1984)
CH_2N (succinimide)	$ClCH_2N$ (succinimide)	Syn 40 (1984)

COR^3	R^3COCl, HMPA or
	   cat $Pd(PPh_3)_4$
		TL 1363 (1978); 24 5081
		   (1983)
		 JOC 44 3888 (1979)
		 Pure Appl Chem 56 91 (1984)
		 Tetr 42 1369 (1986)

CO_2H	CO_2, HMPA,
	   cat $P(OEt)_3/H^+$
		JOMC 54 C53 (1973); 77 281
		   (1974); 144 13 (1978)
		 TL 3461 (1976); 23 5155
		   (1982)
		 JOC 44 1345 (1979)
		 Tetr 36 1961 (1980)
		 Rec Trav Chim 100 249
		   (1981)

CS_2CH_3	CS_2/CH_3I	Syn 432 (1979)
CONHPh	$PhNCO/H^+$	JOMC 54 C53 (1973); 77 281 (1974)
CN	XCN (X = Cl, $PhSO_2$, Ts)	Syn 784 (1977) JOMC 206 257 (1981)
CH_2CHR^3OH	$(LiC{\equiv}CR^4)/$ $H_2C{-}CHR^3$ (epoxide) $/H^+$	TL 3461 (1976); 1363, 2465 (1978) JOC 44 1345, 3888 (1979) Tetr 36 1961 (1980) JACS 104 1774 (1982)
$CH_2CH_2COR^3$	$H_2C{=}CHCOR^3/H^+$	JACS 99 253 (1977) TL 1363 (1978) JOC 44 3888 (1979) Tetr 36 1961 (1980) Pure Appl Chem 56 91 (1984)

$CH_2CH_2CH(CO_2Et)_2$ (cyclopropane) $\begin{smallmatrix}CO_2Et\\CO_2Et\end{smallmatrix}$/$H^+$ Tetr *36* 1961 (1980)

$\begin{smallmatrix}H\\\end{smallmatrix}C=C\begin{smallmatrix}R^2\\R^1\end{smallmatrix}$ O_2 or Δ TL 2583 (1971)
JOMC *77* 269 (1974)
BSCF 1656 (1974)
Rec Trav Chim *95* 299, 304 (1976)

$CH=CHR^3$ $XCH=CHR^3$ (X = Br, I), $ZnBr_2$, cat $Pd(PPh_3)_4$ TL *22* 959 (1981); *23* 1589, 5155 (1982)
BSCF II 321, 332 (1983)
Tetr *40* 2741 (1984)

Z-$CH=CHE$ $HC\equiv CH$/E^+ (E = 1° R^3, CO_2H, CH_2NEt_2)   BSCF 1656 (1974)
TL *23* 5151 (1982)

E-$CH=CHCOR$ $XCH=CHCOR$, cat $Pd(PPh_3)_4$ Tetr *42* 1369 (1986)

E-$CH=CHCO_2Et$ $HC\equiv CCO_2Et$/H^+ (other alkynes also)   TL 3461 (1976); *23* 5151, 5155 (1982)
Tetr *36* 1961 (1980)

Z,Z-$CH=CHCH=CHE$ 2 $HC\equiv CH$/E^+ (E^+ = RX, I_2, RCHO, CO_2, enone, $HC\equiv CCO_2Et$)   JCS Perkin I 1809 (1986)

Z-$CH_2CH=CHR'$ $(H_2C=CHPPh_3)Br$/R'CHO   TL *26* 1799 (1985)
JOC *52* 1801 (1987)

E-$CH_2CH=CHCH_2OH$ $H_2C=CHCH\overset{O}{\overbrace{\quad}}CH_2$/$H^+$ Syn 528 (1978)
TL 2027 (1978)

$C\equiv CR^3$ $XC\equiv CR^3$ (X = Br, I), TMEDA   TL 1465 (1975)
Syn 826 (1979)
Tetr *36* 1215 (1980)

    $TsO(Ph)IC\equiv CR^3$ JACS *109* 7561 (1987)

$$RC\equiv CH + R_f X \xrightarrow[\text{ultrasound}]{Zn,\ CuI} RCH=CHR_f$$

$$X = Br,\ I$$

CL 1453 (1982)
JACS *107* 5186 (1985)

$$RC\equiv CH \xrightarrow[\text{2. } E^+]{\text{1. } (Me_2PhSi)_2CuLi\cdot LiCN} \begin{smallmatrix}R\\E\end{smallmatrix}C=C\begin{smallmatrix}H\\SiMe_2Ph\end{smallmatrix}$$

$$E^+ = H^+,\ I_2,\ RX,\ RCOCl,\ CO_2,\ \text{enone, epoxide}$$

JCS Perkin I 2527 (1981)

$$RC\equiv CX \xrightarrow[\text{2. H}_2\text{O}]{\text{1. organocopper reagent}} \underset{R'}{\overset{R}{\diagdown}}C=C\underset{H}{\overset{X}{\diagup}}$$

X	Reagent(s)	
COCF$_3$	(R'CuCN)Li, (R$_2'$CuCN)Li$_2$	TL *28* 5271 (1987)
COR	MeLi, CuI	JOC *51* 5320 (1986) (on HC≡CCOR)
	R'Cu	JACS *97* 1197 (1975)
	R$_2'$CuLi	JOC *41* 3629 (1976)
		Tetr *37* 4027 (1981)
		TL *28* 5081 (1987)
CO$_2$H	R'Cu	JACS *91* 6186 (1969)
		JCS C 1380 (1970)
	R$_2'$CuLi	JACS *91* 6186 (1969)
CO$_2$R	R'Cu	JCS C 1380 (1970)
		JACS *94* 4395, 5374 (1972); *97* 1197 (1975)
		TL 1811 (1979); *27* 5471 (1986); *28* 2964 (1987)
	R$_2'$CuLi	JACS *91* 1851, 1853 (1969); *94* 4395 (1972); *97* 1197 (1975); *103* 7007 (1981); *109* 6199 (1987)
		Helv *54* 1939 (1971)
		Acta Chem Scand *25* 1471 (1971)
		TL 1277, 1281 (1973); 1811 (1979); *21* 2057 (1980); *28* 2963, 5473 (1987)
		JOC *41* 3629 (1976); *51* 1077 (1986); *52* 398 (1987)
		CC 1220 (1987)
	[**R'**CuC≡C(CH$_2$)$_3$CH$_3$]Li	JACS *94* 7210 (1972)
		TL *28* 2963 (1987)
	OSiMe$_2$(t-Bu) | **(RC=CHCH$_2$CuC≡CSiMe$_3$)Li**	JACS *107* 5495 (1985)
	OSiMe$_2$(t-Bu) | **(RC=C=CHCuC≡CSiMe$_3$)Li**	TL *28* 1299 (1987)
	(RCH=CRCuC≡CCMe$_2$OMe)Li	JACS *108* 5559 (1986)
	(Z-RCH=CHCuSPh)Li	TL 3461 (1976)
CN	R'Cu·MgX$_2$, R$_2'$CuLi	Syn 454 (1978)
	R$_2'$CuLi	JACS *106* 462 (1984)
CO$_2$H, CO$_2$R, CHO, COR	R'Cu·BR$_3$	JOC *44* 1744 (1979)

$$MeO_2CC \equiv CCO_2Me \longrightarrow \begin{array}{c} MeO_2C \\ R \end{array} C=C \begin{array}{c} CO_2Me \\ H \end{array}$$

MeCu/H$_2$O   JACS *95* 6149 (1973)
RCu·MgX$_2$/H$_2$O   CL 905 (1981)
RCu·BR'$_3$/H$_2$O   JOC *44* 1744 (1979)

$$(MeCuR)Li + HC \equiv CCO_2Et \xrightarrow{R'COCl} \begin{array}{c} H \\ Me \end{array} C=C \begin{array}{c} CO_2Et \\ COR \end{array}$$

JOC *46* 3696 (1981)

$$(Me_3Si)_2NCH_2C \equiv CCO_2Me \xrightarrow[2.\ RCOCl]{1.\ (CH_3CuX)Li\ (X = ?)} \begin{array}{c} CH_3 \\ (Me_3Si)_2NCH_2 \end{array} C=C \begin{array}{c} CO_2Me \\ COR \end{array}$$

TL *28* 2963 (1987)

$$RC \equiv CCO_2Et \xrightarrow[EtOH]{PhSNa} \begin{array}{c} PhS \\ R \end{array} C=C \begin{array}{c} CO_2Et \\ H \end{array} \xrightarrow[CuI]{R'MgI} \begin{array}{c} R' \\ R \end{array} C=C \begin{array}{c} CO_2Et \\ H \end{array}$$

CL 705 (1974)
TL *28* 5473 (1987)

$$\begin{array}{c} R \\ H \end{array} C=C \begin{array}{c} \\ H \end{array})_2CuLi + H_2C=CHCH=CHCO_2R \longrightarrow \begin{array}{c} R \\ H \end{array} C=C \begin{array}{c} CH_2CH=CHCH_2CO_2R \\ H \end{array}$$

$E + Z$

Helv *55* 82 (1972)

$$\xrightarrow[2.\ H_2O]{1.\ RCu} R$$

TL 1611 (1973)

$$\xrightarrow[2.\ H_2O]{1.\ RCu} R$$

TL 1611 (1973)

$$\xrightarrow[2.\ H_2O]{1.\ R_2CuLi} R$$

JOC *38* 2100 (1973)

2.18. Organosilver Compounds

$$RC \equiv CCN \xrightarrow[2.\ H_2O]{1.\ R'_2AgMgX \cdot 2\ LiBr} \begin{array}{c} R \\ R' \end{array} C=C \begin{array}{c} CN \\ H \end{array}$$

TL 3327 (1979)
JOMC *206* 257 (1981)

$$\text{HC}{\equiv}\text{CC}{\equiv}\text{CR} \xrightarrow[\text{2. E}^+]{\text{1. R}'_2\text{AgMgX or R}'\text{AgMgX}_2} \text{R}'\text{CH}{=}\overset{\overset{\displaystyle E}{\displaystyle |}}{\text{C}}\text{C}{\equiv}\text{CR}$$

$$\text{E}^+ = \text{NBS, I}_2\text{, H}_2\text{C}{=}\text{CHCH}_2\text{Br}\text{, CO}_2$$

Rec Trav Chim *100* 337 (1981)

2.19. Organozinc Compounds

See also page 230, Section 2.11; and page 233, Section 2.17.

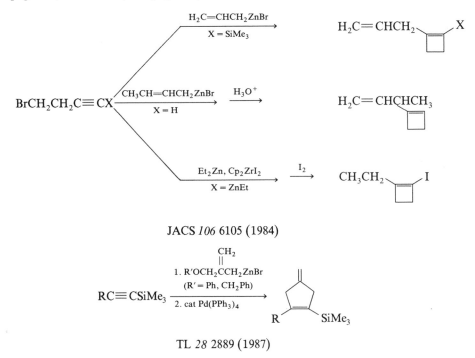

JACS *106* 6105 (1984)

$$\text{RC}{\equiv}\text{CSiMe}_3 \xrightarrow[\text{2. cat Pd(PPh}_3)_4]{\overset{\substack{\text{CH}_2 \\ \| \\ \text{1. R}'\text{OCH}_2\text{CCH}_2\text{ZnBr} \\ (\text{R}' = \text{Ph, CH}_2\text{Ph})}}{}}$$

TL *28* 2889 (1987)

2.20. Organomercury Compounds

See also page 260, Section 3.16.

$$\text{RC}{\equiv}\text{CH(R)} \xrightarrow[\substack{\text{2. Hg(OAc)}_2 \\ \text{3. NaCl}}]{\text{1. R}'_2\text{BH}} \overset{\text{R}}{\underset{\text{H}}{>}}\text{C}{=}\text{C}\overset{\text{H(R)}}{\underset{\text{HgCl}}{<}} \longrightarrow \overset{\text{R}}{\underset{\text{H}}{>}}\text{C}{=}\text{C}\overset{\text{H(R)}}{\underset{\text{X}}{<}}$$

X	Reagent(s)	
CH$_3$	CH$_3$RhI$_2$(PPh$_3$)$_2$	TL *22* 2443 (1981)
		JOMC *225* 31 (1982)
R^l	R$^l_2$CuLi	Organomet *1* 74 (1982)
	(R^l = alkyl, vinyl)	
COR^l	R^{l}COCl, AlCl$_3$	JOC *43* 710 (1978)

CO_2R^1	CO, R^1OH, $CuCl_2$, cat $PdCl_2$ ($R^1 = H$, alkyl)	JOC *40* 3237 (1975)
COCH(R)=CHR	CO, cat $[ClRh(CO)_2]_2$	JOC *45* 3840 (1980)
OAc	$Hg(OAc)_2$, cat $Pd(OAc)_2$	JACS *102* 1966 (1980)

$$RHgCl + XC{\equiv}CCO_2R \xrightarrow{NaBH_4} RCX{=}CHCO_2R$$

$$X = H, CO_2R$$

Angew Int *21* 768 (1982); *26* 479 (1987)

2.21. Miscellaneous Reactions

$$RC{\equiv}CR + N_2CHCO_2R \xrightarrow{cat\ Rh_2(O_2CR)_4}$$

TL 1239 (1978)

See page 574, Section 2.2.

$$RC{\equiv}CR \longrightarrow RC{=}CR$$
(with I and X substituents)

X	Reagent(s)	
H, F, Cl, Br, I, OAc, SCN, pyridinium salt, Ar	$I(py)_2BF_4$; Et_3SiH, F^-, Cl^-, Br^-, I^-, HOAc, SCN^-, py or Ar-H	TL *27* 3303 (1986)
Br, SCN	I_2, CuO-HBF$_4$, LiBr or KSCN	CC 1491 (1987)
SCN	ISCN	BSCF 2569 (1964) Compt Rend *258* 3878 (1964)

$$RC{\equiv}CH + HOCR' \longrightarrow RCH{=}CHOCR'$$
(with O double bonds)

See page 830, Section 5.

7. DIENES

Reviews:

A. S. Onishchenko, "Diene Synthesis," D. Davey Co., New York (1964)
Houben-Weyl, Vol V/1c, G. Thieme, Stuttgart (1970)
Tetr *33* 1845 (1977) (insect pheromones)
Syn 817 (1977) (insect pheromones)

1. Rearrangement

$$\square\!\!\!\!\square \xrightarrow{h\nu \text{ or } \Delta} H_2C{=}CHCH{=}CH_2$$

Angew *66* 640 (1954)
Ann *615* 14 (1958); *627* 1 (1959)
JACS *84* 4141 (1962); *87* 3996 (1965); *88* 1073 (1966); *90* 5310, 6896 (1968); *91* 5404, 7557 (1969); *93* 4616 (1971); *94* 4262 (1972); *101* 3340 (1979); *102* 3548, 6353 (1980); *103* 1256 (1981); *107* 2099, 3921 (1985); *109* 6086 (1987)
Proc Chem Soc 334 (1962)
Trans Faraday Soc *58* 957 (1962)
Ber *96* 2362 (1963); *97* 1811, 2934 (1964)
JOC *29* 257 (1964); *43* 2726, 4559 (1978); *52* 3708 (1987)
TL 1207 (1965); 3387 (1974); 3047 (1975); *21* 1997 (1980); *28* 1501 (1987)
Adv Phys Org Chem *4* 183 (1966)
Angew Int *7* 559 (1968) (review)
Acct Chem Res *7* 65 (1974) (mechanism)
Syn Commun *12* 167 (1982)

(Cope reaction)

Org Rxs *22* 1 (1975) (review)
Helv *60* 978 (1977) (aza-Cope)
TL *22* 1583 (1981) (acid catalyzed 2-acyl-1,5-dienes); *23* 2733, 2737, 2741 (1982) (all aza-Cope)
JACS *104* 7225 (1982) (PdCl$_2$ catalyzed)

Ber *115* 2309 (1982) (heteroatom substituted)
Angew Int *21* 198 (1982) (aza-Cope)
R. K. Hill in "Asymmetric Synthesis," Ed. J. D. Morrison, Academic Press, New York (1984), Vol 3,
    Chpt 8 (review)
Chem Rev *84* 205 (1984) (catalysis, review)

OR ⟶ cat ClRh(PPh$_3$)$_3$ ⟶ OR

TL 3797 (1968)

R⌒⌒⌒CO$_2$R' 1. LDA / 2. H$_2$O ⟶ R⌒⌒⌒CO$_2$R'

R⌒⌒⌒CO$_2$R' ⟶ R⌒⌒⌒CO$_2$R'

CL 1541 (1984)

⟶ hν or Δ ⟶

CC 173 (1970)
E. N. Marvel, "Thermal Electrocyclic Reactions," Academic Press, New York (1980) (review)
JACS *107* 1034 (1985)
JOC *51* 4131 (1986)
JCS Perkin I 225 (1987)
TL *28* 1469 (1987)

2. Elimination Reactions

$$\underset{\overset{|}{C}}{\overset{H}{|}}-C=C-\underset{\overset{|}{C}}{\overset{H}{|}} \longrightarrow \underset{\overset{|}{C}}{\overset{H}{|}}-\underset{\overset{|}{C}}{\overset{Br}{|}}-\underset{\overset{|}{C}}{\overset{Br}{|}}-\underset{\overset{|}{C}}{\overset{H}{|}} \xrightarrow{base} C=C-C=C$$

LiCl, Li$_2$CO$_3$, HMPA, Δ   JACS *99* 7899 (1977)
                                     Syn 449 (1977)

KO-*t*-Bu   TL *28* 4965 (1987)
                                     JACS *109* 2857 (1987)

DBN   TL *28* 5833 (1987)

$$\underset{\overset{|}{C}}{\overset{H}{|}}-C=C-\underset{\overset{|}{C}}{\overset{H}{|}} \xrightarrow{NBS} \underset{\overset{|}{C}}{\overset{H}{|}}-C=C-\underset{\overset{|}{C}}{\overset{Br}{|}} \xrightarrow{DBU} C=C-C=C$$

JACS *109* 2212 (1987)

$$\underset{\overset{|}{C}-C=C-\overset{\overset{Br}{|}}{C}}{\overset{Br}{|}} \longrightarrow C=C-C=C$$

Zn	CC 508 (1966)
Zn(Hg)	Syn 742 (1982)
Zn-Cu, KI, I_2	JACS *99* 8 (1977); *108* 512 (1986)
cat NaBH$_4$—$\left(\underset{S}{\langle \ \rangle}-Te \right)_2$	JOC *50* 3170 (1985)

$$\underset{RCHCH=CHCH_2OR'}{\overset{OCH_2OCH_3}{|}} \xrightarrow{LDA} RCH=CHCH=CHOR'$$

JOC *52* 4817 (1987)

$$C=C-\overset{\overset{HO}{|}}{C}-\overset{\overset{H}{|}}{C} \longrightarrow C=C-C=C$$

Al$_2$O$_3$, Δ	JOC *52* 3541 (1987)
Tf$_2$O, lutidine	JOC *52* 3956 (1987)

$$C=C-C-\overset{\overset{OTs}{|}}{C} \xrightarrow{KO\text{-}t\text{-}Bu} C=C-C=C$$

JOC *52* 4732 (1987)

$$\overset{\overset{H}{|}}{C}-C=C-\overset{\overset{OH}{|}}{C} \longrightarrow C=C-C=C$$

(C$_5$H$_5$NH)OTs	TL *23* 4747 (1982) JOC *52* 3541 (1987)
2,4-(NO$_2$)$_2$C$_6$H$_3$SCl, Et$_3$N	JACS *100* 5981 (1978); *104* 7051 (1982) JOC *51* 5232 (1986)
t-BuO$_2$H, VO(acac)$_2$/Me$_3$SiCl/	JACS *97* 3252 (1975) BCSJ *52* 1752 (1979)

Et$_2$AlN $\rangle$ /PBr$_3$/Zn

$$RCH_2CH=CHCH_2OR' \text{ or } RCH_2CHOR'CH=CH_2 \xrightarrow[Ph_3P]{cat\ Pd(OAc)_2} RCH=CHCH=CH_2$$

R' = Ac, Ph

TL 2075 (1978)

$$\overset{\overset{H}{|}}{C}-C=C-\overset{\overset{OAc}{|}}{C} \xrightarrow[HC\equiv CCH_2ZnBr]{cat\ Pd(PPh_3)_4} C=C-C=C$$

JOC *47* 4161 (1982)

$$\underset{\overset{\displaystyle |}{\text{C}}}{\overset{\displaystyle \text{HO}}{}}-\text{C}=\text{C}-\underset{\overset{\displaystyle |}{\text{C}}}{\overset{\displaystyle \text{OH}}{}}\longrightarrow \text{C}=\text{C}-\text{C}=\text{C}$$

LiAlH$_4$ JCS 4006 (1954)

TiCl$_3$, LiAlH$_4$ JACS *104* 5807 (1982)

MsCl, py/NaI   JCS C 15 (1967)
                                               Syn 956 (1982)

$$\underset{\overset{\displaystyle |}{\text{C}}}{\overset{\displaystyle \text{HO}}{}}-\text{C}=\text{C}-\underset{\overset{\displaystyle |}{\text{C}}}{\overset{\displaystyle \text{OMe}}{}} \xrightarrow{\text{LiAlH}_4,\ \text{TiCl}_3} \text{C}=\text{C}-\text{C}=\text{C}$$

JOC *52* 3560 (1987)

$$\underset{\overset{\displaystyle |}{\text{C}}}{\overset{\displaystyle \text{HO}}{}}-\text{C}\equiv\text{C}-\underset{\overset{\displaystyle |}{\text{C}}}{\overset{\displaystyle \text{OTHP}}{}} \xrightarrow{\text{LiAlH}_4} \text{C}=\text{C}-\text{C}=\text{C}$$

Acta Chem Scand *26* 2540 (1972)
JACS *108* 1338 (1986)

Helv *58* 1450 (1975)

$$\underset{\overset{\displaystyle |}{\text{C}}}{\overset{\displaystyle \text{H}}{}}-\text{C}=\text{C}-\underset{\overset{\displaystyle |}{\text{C}}}{\overset{\displaystyle \text{OSPh}}{}} \xrightarrow{\Delta} \text{C}=\text{C}-\text{C}=\text{C}$$

TL *22* 4137 (1981)
JOC *51* 4594 (1986)

JOC *47* 4801 (1982)

$$\begin{array}{c}\text{R}_2\text{C}=\text{CRCH}_2\text{SAr}\\ \text{or}\\ \text{R}_2\text{C}=\text{CRCH}_2\text{SO}_2\text{Ar}\end{array} \xrightarrow[\text{2. } n\text{-Bu}_3\text{SnCH}_2\text{I}]{\text{1. } n\text{-BuLi}} \text{R}_2\text{C}=\text{CRCH}=\text{CH}_2$$

TL *23* 2205 (1982)

$$\underset{\underset{RCHSO_2Ph}{|}}{\overset{\overset{Li}{|}}{}} + \underset{}{\overset{\overset{O}{\|}}{HCCH_2CH_2R'}} \longrightarrow \underset{\underset{OR}{|}}{\overset{\overset{PhSO_2}{|}}{RCHCHCH_2CH_2R'}} \xrightarrow{KO\text{-}t\text{-Bu}} RCH=CHCH=CHR'$$

R = alkyl, Ac

JACS *106* 3670 (1984)
JOC *51* 3830 (1986)

$$\underset{\underset{RCHSO_2Ph}{|}}{\overset{\overset{Li}{|}}{}} + \overset{\overset{O}{\|}\quad\overset{O}{\|}}{HCCH_2CH_2CNR_2} \xrightarrow[\text{2. KO-}t\text{-Bu}]{\text{1. Ac}_2\text{O}} RCH=CHCH=CHCNR_2\overset{O}{\|}$$

TL *27* 603 (1986)
JOC *51* 3896 (1986)

$$\underset{\underset{RCHSO_2Ph}{|}}{\overset{\overset{Li}{|}}{}} + \underset{\underset{R}{|}}{\overset{\overset{O}{\|}}{HCC}}=CHCH_2CH_2R \longrightarrow \underset{\underset{X}{|}}{\overset{\overset{PhSO_2\ \ R}{|\quad|}}{RCHCHC}}=CHCH_2CH_2R$$

$$\xrightarrow{KO\text{-}t\text{-Bu}} RCH=CHC\overset{\overset{R}{|}}{=}CHCH=CHR$$

X = Cl, Br, OR

JACS *106* 3670 (1984)
CL 1883 (1985)
JOC *51* 3834 (1986)

$$R_2C=CHCHO \xrightarrow[\text{2. Ac}_2\text{O}]{\text{1. RCHLiCO}_2\text{Li}} R_2C=CH\overset{\overset{OAc}{|}}{CH}CHRCO_2H \xrightarrow[Et_3N]{\text{cat Pd(PPh}_3)_4} R_2C=CHCH=CHR$$

JACS *102* 2841 (1980)

$$RCH_2CH=CH_2 \xrightarrow[(CF_3CO)_2O]{MeSOCH_2CO_2Et} RCH=CHCH_2\overset{\overset{SMe}{|}}{CH}CO_2Et \xrightarrow[\Delta]{NaIO_4} RCH=CHCH=CHCO_2Et$$

TL *22* 1343 (1981)

TL *23* 3277 (1980)

$$\text{(cyclic } SO_2\text{)} \longrightarrow R\text{-(cyclic } SO_2\text{)}\text{-}R \longrightarrow RCH{=}CHCH{=}CHR$$

alkylation	CL 1003 (1983)
	CC 1323 (1984); 236 (1985); 934 (1987)
	JCS Perkin I 515 (1985); 1039 (1986)
	Heterocycles *23* 2913 (1985)
	JOC *51* 1000, 4718, 4934 (1986); *52* 244, 3394, 4468, 5082 (1987)

elimination

| hν | JOC *39* 2366 (1974) |

Δ	JACS *88* 2857, 2858 (1966); *97* 3666 (1975)
	JOC *39* 2366 (1974); *52* 244, 3394 (1987)
	JCS Perkin II 1470 (1976)
	CL 1003 (1983)
	CC 1323 (1984); 236 (1985)
	Heterocycles *23* 2913 (1985)

| base, Δ | CL 1003 (1983) |
| | JOC *51* 4934 (1986) |

LiAlH$_4$	TL 947 (1977)
	CL 1003 (1983)
	JOC *51* 4934 (1986); *52* 4468, 5082 (1987)

| K, ultrasound | JOC *52* 2224 (1987) |

$$RCH_2CH{=}CH_2 \xrightarrow[\text{2. } R_3N]{\text{1. } BrCH_2SO_2Br,\ h\nu} RCH_2CH{=}CHSO_2CH_2Br \xrightarrow{\text{KO-}t\text{-Bu}} RCH{=}CHCH{=}CH_2$$

JACS *105* 6164, 6165 (1983); *108* 4568 (1986)
TL *25* 5469 (1984)
Org Syn *65* 90 (1987)

3. Organometallic Approaches

For symmetrical coupling of allylic and vinylic halides or alcohols see also page 45, Section 1.

3.1. Organolithium and -magnesium Compounds

See also page 257, Section 3.14.

$$\underset{R^1}{\overset{R^2}{>}}C{=}C\underset{MgX}{\overset{R^3}{<}} + \underset{H}{\overset{I}{>}}C{=}C\underset{R^5}{\overset{R^4}{<}} \xrightarrow{\text{cat Pd(PPh}_3)_4} \underset{R^1}{\overset{R^2}{>}}C{=}C\underset{\underset{H}{>}C{=}C\underset{R^5}{\overset{R^4}{<}}}{\overset{R^3}{<}}$$

TL 191 (1978)

$$H_2C=C(CH_3)MgBr + H_2C=C(CH_3)Br \xrightarrow{\text{cat NiCl}_2L_2} H_2C=C\overset{\displaystyle CH_3}{\underset{}{|}}-\overset{\displaystyle CH_3}{\underset{}{|}}C=CH_2$$

BCSJ *49* 1958 (1976)

$$H_2C=CH\overset{\displaystyle MgCl}{\underset{}{|}}C=CH_2 \xrightarrow[\substack{\text{cat CuI or} \\ \text{Pd(PPh}_3)_4}]{RX} H_2C=CH\overset{\displaystyle R}{\underset{}{|}}C=CH_2$$

R = 1° alkyl, aryl, 1° benzylic

BCSJ *54* 2831 (1981)

$$\overset{R}{\underset{H}{>}}C=C\overset{H}{\underset{X}{<}} \longrightarrow \overset{R}{\underset{H}{>}}C=C\overset{H}{\underset{M}{<}} \xrightarrow{\text{reagent(s)}} \overset{R}{\underset{H}{>}}C=C\overset{H}{\underset{\overset{\displaystyle H}{\underset{H}{>}C=C\underset{R}{<}}}{}}$$

M	Reagent(s)	
Li	CoCl$_2$	JOMC *24* 537 (1970)
	(n-Bu$_3$P·CuI)$_4$/O$_2$	JACS *89* 5302 (1967)
	CuI or (n-Bu$_3$P·CuI)$_4$	JACS *93* 1379 (1971)
	or (n-Bu$_3$P·AgI)$_4$	
	AgI	JOC *36* 1694 (1971)
MgBr	SOCl$_2$	JOC *37* 3749 (1972)
MgX (X = Cl, Br)	CuCl	Angew Int *6* 85 (1967)

$$RCH=CHC\overset{\displaystyle CH_3OCH_2O}{\underset{}{|}}=CH_2 \xrightarrow[\text{2. R'CHO}]{\text{1. }sec\text{-BuLi}} RCH=CHC\overset{\displaystyle CH_3OCH_2O}{\underset{}{|}}=CH\overset{\displaystyle OH}{\underset{}{|}}CHR'$$

JOC *51* 4492 (1986)

$$RCH=CHCH=CHOR \xrightarrow[\text{2. Me}_3\text{SiCl}]{\text{1. }sec\text{-BuLi}} RCH=CHCH=C\overset{\displaystyle OR}{\underset{}{|}}SiMe_3$$

JOC *52* 4818 (1987)

$$R_2\overset{\displaystyle THPO}{\underset{}{|}}CCH=CHCH=CH_2 \xrightarrow{R'Li} R_2C=CHCH=CHCH_2R'$$

R' = 1°, 2° alkyl; NEt$_2$; SiMe$_2$Ph; SnMe$_3$

JOC *52* 4416 (1987)

$$R_2C=CRCH=CH\overset{\displaystyle O}{\overset{\displaystyle \|}{C}}R \xrightarrow[\substack{NH_3 \\ EtOH}]{R'Li \quad Li} R_2C=CRCH_2CH=CHRR'$$

JOC *44* 1159 (1979)

3.2. Organoboron Compounds

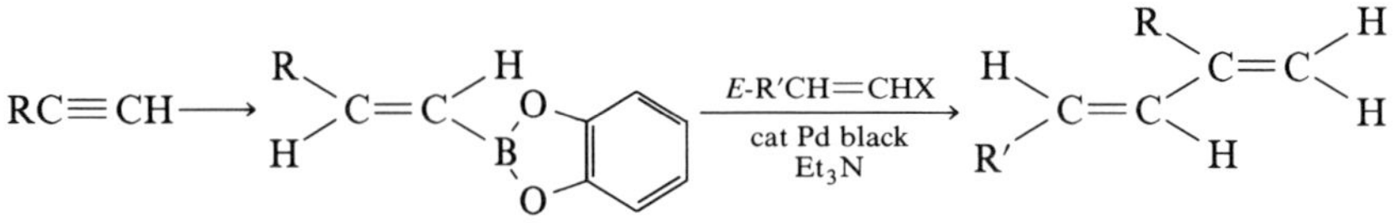

$$RC\equiv CH \longrightarrow \underset{H}{\overset{R}{>}}C=C\underset{B}{\overset{H}{<}} \quad \xrightarrow[\substack{cat\ Pd\ black \\ Et_3N}]{E\text{-}R'CH=CHX} \quad$$

(diene product shown) cis-cis arrangement

JOMC *213* C53 (1981)

$$RC\equiv CH \xrightarrow[O_2,\ NH_3]{Cu(I)} RC\equiv CC\equiv CR \xrightarrow[\substack{1.\ 2R'_2BH \\ 2.\ HOAc}]{}$$

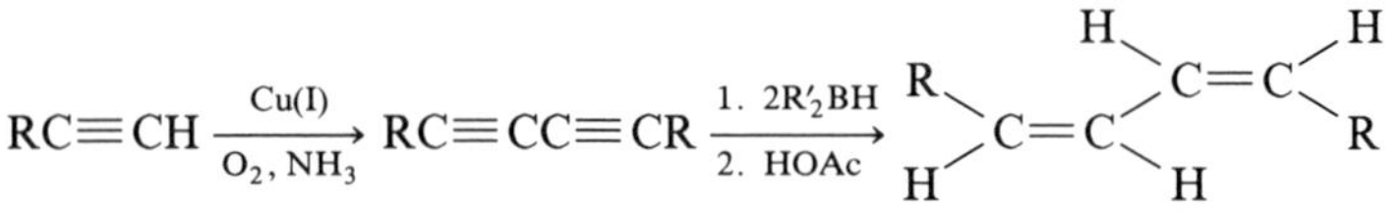

cis-cis

JACS *92* 4068 (1970)

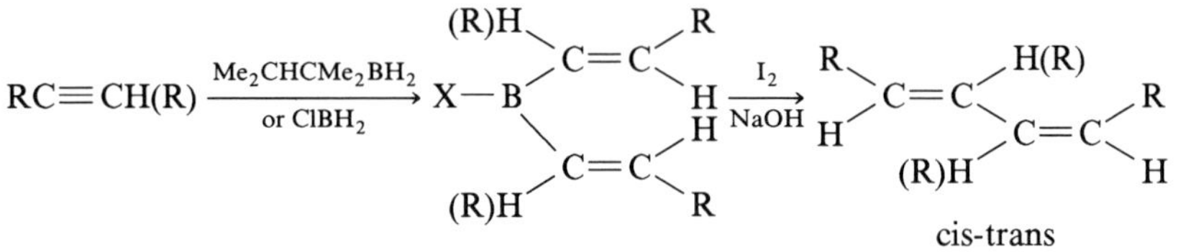

$$RC\equiv CH(R) \xrightarrow[\text{or ClBH}_2]{Me_2CHCMe_2BH_2} X-B \xrightarrow[NaOH]{I_2}$$

cis-trans

JACS *90* 6243 (1968)
JOC *38* 1617 (1973)

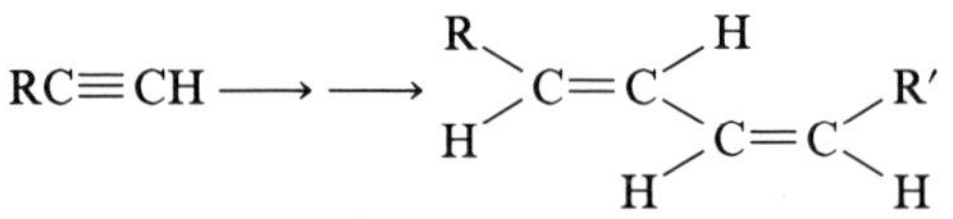

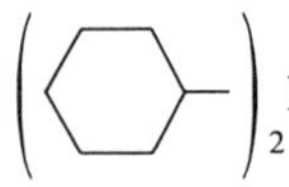

$$RC\equiv CH \longrightarrow \longrightarrow$$

cis-trans

(Sia)$_2$BH/R′C≡CLi/I$_2$, NaOH/   CC 874 (1973)
  (Sia)$_2$BH/HOAc

$$\left(\bigcirc-\right)_2 BH/R'C\equiv CLi/BF_3\cdot OEt_2 \qquad \text{JOMC } 156 \text{ 159 (1978)}$$
or *n*-Bu$_3$SnCl/HOAc

$$2\ RC\equiv CH \longrightarrow \longrightarrow$$

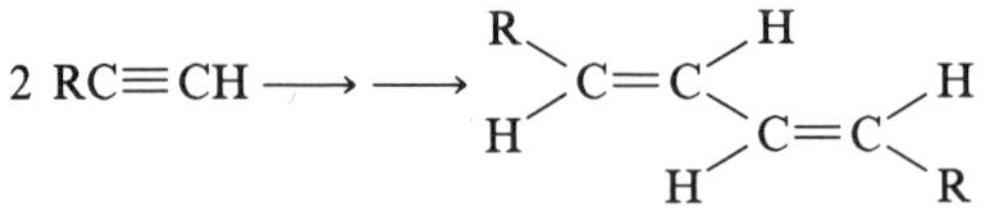

trans-trans

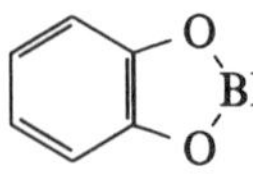

BH/H$_2$O/cat PdCl$_2$, LiCl, Et$_3$N   JOMC *179* C7 (1979)

R′$_2$BH/CuBr·SMe$_2$, NaOCH$_3$ JOC *45* 549 (1980)

R′$_2$BH/NaOCH$_3$/ZnCl$_2$ Organomet *5* 2161 (1986)

ClBH$_2$/CH$_3$Cu   JACS *97* 5606 (1975);
       99 5652 (1977)
       BCSJ *50* 3427 (1977)

$$RC{\equiv}CH \xrightarrow[\text{or } R_2BH]{\substack{O \\ BH \\ O}} \xrightarrow[\substack{\text{Pd catalyst} \\ OH^- \text{ or } OR^-}]{R'CH=CHBr} \underset{H}{\overset{R}{>}}C{=}C\underset{CH=CHR'}{\overset{H}{<}} \quad \text{retention}$$

TL 3437 (1979); *22* 127 (1981); *28* 3959, 5849 (1987)
Tetr *37* 2617 (1981); *39* 3271 (1983)
BCSJ *55* 2221 (1982)
Pure Appl Chem *57* 1749 (1985)
JACS *107* 972 (1985); *109* 4756 (1987)
JOC *52* 292 (1987)
CL 27 (1987)

$$RC{\equiv}CR' \xrightarrow[\substack{2.\ BBr_3 \\ 3.\ HC{\equiv}CR''}]{1.\ HBBr_2 \cdot SMe_2} \xrightarrow[\substack{2.\ t\text{-BuLi} \\ 3.\ MeOH}]{1.\ I_2,\ KOAc} \underset{R}{\overset{H}{>}}C{=}C\underset{H}{\overset{R'}{<}}\!\!C{=}C\underset{R''}{\overset{H}{<}}$$

TL *27* 977 (1986)

$$RC{\equiv}CCl \xrightarrow[\substack{2.\ HC{\equiv}CR'}]{1.\ (CH_3)_2CHC(CH_3)_2BH} \xrightarrow[\substack{2.\ (CH_3)_2CHCO_2H \\ \Delta}]{1.\ NaOCH_3} \underset{H}{\overset{R}{>}}C{=}C\underset{H}{\overset{H}{<}}\!\!C{=}C\underset{R'}{\overset{H}{<}}$$

trans-trans

CC 606 (1973)

$$RC{\equiv}CBr \longrightarrow \underset{H}{\overset{R}{>}}C{=}C\underset{H(R'')}{\overset{BR'_2}{<}} \xrightarrow[\substack{\text{cat Pd(PPh}_3)_4 \\ KOH \text{ or } NaOEt}]{XCH=CHR'} \underset{H}{\overset{R}{>}}C{=}C\underset{H(R'')}{\overset{CH=CHR'}{<}} \quad \text{retention}$$

TL *27* 3745 (1986)
CL 1329 (1986)

$$RCHO \xrightarrow{Me_3SiCH=CHCH_2B\overset{O}{\underset{O}{<}}} \begin{cases} \xrightarrow{KH} \underset{R}{\overset{H}{>}}C{=}C\underset{CH=CH_2}{\overset{H}{<}} \\ \xrightarrow{H_2SO_4} \underset{H}{\overset{R}{>}}C{=}C\underset{CH=CH_2}{\overset{H}{<}} \end{cases}$$

TL *22* 2751 (1981)

$$ClCH_2C{\equiv}CCH_2Cl \xrightarrow[\substack{2.\ RLi}]{1.\ R_2BH} H_2C{=}\overset{R}{\underset{|}{C}}CH=CH_2$$

CC 531, 1629 (1987)

$$R^1C{\equiv}CR^2 \longrightarrow \underset{H}{\overset{R^1}{>}}C{=}C\underset{BR_2}{\overset{R^2}{<}} \longrightarrow \underset{H}{\overset{R^1}{>}}C{=}C\underset{CH_2CH=CH_2}{\overset{R^2}{<}}$$

See page 218, Section 2.6.

$$RCH=CH_2 \xrightarrow[\text{2. } PhOCH=CHCH_2MgCl]{\text{1. } BH_3} RCH=CHCH_2BR'_2 \xrightarrow[H_2C=CHCH_2Br]{CuI} RCH=CHCH_2CH_2CH=CH$$

Syn Commun *12* 813 (1982)

$$R^1CH=CHCH_2\bar{B}R_3 + R^2CH=CHCH_2X \longrightarrow H_2C=CHCHR^1CH_2CH=CHR^2$$

X = Cl, Br

JACS *100* 6282 (1978); *103* 1969 (1981)

3.3. Organoaluminum Compounds

$$RC\equiv CH(R) \xrightarrow{i\text{-}Bu_2AlH} \underset{H}{\overset{R}{\diagup}}C=C\underset{Al(i\text{-}Bu)_2}{\overset{H(R)}{\diagdown}} \xrightarrow{CuCl} \cdots$$

trans-trans

JACS *92* 6678 (1970)

$$2\,RC\equiv CR \xrightarrow[\Delta]{i\text{-}Bu_2AlH} \cdots \xrightarrow{H_2O} \cdots$$

cis-cis

Ann *629* 222 (1960)

$$RC\equiv CH \xrightarrow[Mn(acac)_3]{R'_3Al} \cdots$$

JOC *44* 1496 (1979)

$$RC\equiv CH \xrightarrow{i\text{-}Bu_2AlH} \underset{H}{\overset{R}{\diagup}}C=C\underset{Al(i\text{-}Bu)_2}{\overset{H}{\diagdown}} \xrightarrow[\text{metal catalyst}]{R'CH=CHX \;(X = Cl, Br, I)} \underset{H}{\overset{R}{\diagup}}C=C\underset{CH=CHR'}{\overset{H}{\diagdown}}$$

<u>metal</u>

Ni   JACS *109* 2393 (1987)

Pd   JACS *98* 6729 (1976); *109* 2393 (1987)
                         TL *28* 1649, 4875 (1987)

$$\xrightarrow[\substack{cat\,Pd(PPh_3)_4\\ZnCl_2}]{R^2CH=CHX\;(X = Br, I)} \underset{R^1}{\overset{R}{\diagup}}C=C\underset{CH=CHR^2}{\overset{H}{\diagdown}}$$

retention

JACS *100* 2254 (1978)

TL *28* 2221 (1987)

$$RC\equiv CH \xrightarrow[Cp_2ZrCl_2]{R'_3Al} \underset{R^1}{\overset{R}{\diagup}}C=C\underset{AlR'_2}{\overset{H}{\diagdown}}$$

$$\xrightarrow[cat\,Pd(PPh_3)_4]{XCH_2CH=CR_2\;(X = Cl, OAc)} \underset{R^1}{\overset{R}{\diagup}}C=C\underset{CH_2CH=CR_2}{\overset{H}{\diagdown}}$$

JACS *103* 2882 (1981)

CC 160 (1982)

$$(CH_3)_2C=CHCHR \xrightarrow[\text{ }]{i\text{-}Bu_2AlCH=CHR'} (CH_3)_2C=CHCHRCH=CHR'$$

with SO₂Ph on the CHR carbon

JACS *108* 1098 (1986)

$$HC\equiv CCO_2Me \xrightarrow[\text{2.}]{\text{1. } i\text{-}Bu_2AlH, \text{ HMPA}} H_2C=CCO_2Me$$

JOC *52* 1624 (1987)

3.4. Organosilicon Compounds

$$RC\equiv CH \xrightarrow[\text{2. KF}]{\text{1. HSiCl}_3, \text{ cat } H_2PtCl_6} \begin{array}{c} R \\ H \end{array}C=C\begin{array}{c} H \\ SiF_5^{2-} \end{array}$$

With H₂C=CHCH₂Cl, Pd(OAc)₂:

$$\begin{array}{c} R \\ H \end{array}C=C\begin{array}{c} H \\ CH_2CH=CH_2 \end{array}$$

TL 2161 (1978)
Organomet *1* 542 (1982)

With H₂C=CHX, Pd(OAc)₂:

$$\begin{array}{c} R \\ H \end{array}C=C\begin{array}{c} H \\ \end{array}C=C\begin{array}{c} H \\ X \end{array}$$

X = CHO, CO₂R, CN
Organomet *1* 542 (1982)

$$\begin{array}{c} R \\ H \end{array}C=C\begin{array}{c} H \\ \end{array}C=C\begin{array}{c} H \\ R \end{array}$$

AgF or AgNO₃ TL 1137 (1979)
CuCl TL 1141 (1979)
PdCl₂ Organomet *1* 542 (1982)

$$\overset{O}{\overset{\|}{R}CCH_2SiMe_3} \xrightarrow{H_2C=CHMgBr} \overset{OH}{\underset{|}{H_2C=CHCRCH_2SiMe_3}} \xrightarrow[\text{HOAc}]{\text{NaOAc}} \overset{R}{\underset{|}{H_2C=CHC=CH_2}}$$

TL *28* 693, 697 (1987)

$$RCHO + Me_3Si\overset{Li}{\underset{}{\diagup}}SiMe_3 \xrightarrow[\text{Cp}_2\text{TiCl}]{\text{MgBr}_2 \text{ or } B(OMe)_3 \text{ or}} R\overset{OH}{\underset{SiMe_3}{\diagup}}SiMe_3$$

$$\xrightarrow{H^+} R\diagdown\diagup SiMe_3$$

$$\xrightarrow{KH} R\diagdown\diagup SiMe_3$$

CC 969 (1982); 921 (1983)
JOMC *264* 207 (1984)
TL *28* 211 (1987)

$$Me_3SiCH_2CH{=}CHS(\textit{t-Bu}) \xrightarrow[\text{2. RCHO}]{\text{1. }\textit{t-BuLi}} \underset{R}{\overset{H}{>}}C{=}C\underset{H}{\overset{H}{<}}\underset{H}{\overset{H}{>}}C{=}C\underset{S(\textit{t-Bu})}{\overset{H}{<}} \xrightarrow[\text{cat NiCl}_2L_2]{R'MgI} \underset{R}{\overset{H}{>}}C{=}C\underset{H}{\overset{H}{<}}\underset{H}{\overset{H}{>}}C{=}C\underset{R}{\overset{H}{<}}$$

TL *25* 5177 (1984)

$$C{=}C{-}C{-}X + C{=}C{-}C{-}SiMe_3 \xrightarrow{\text{TiCl}_4} C{=}C{-}C{-}C{-}C{=}C$$

X = Br, OR, OAc

TL *23* 2953 (1982)
JOC *47* 3803 (1982)

3.5. Organotin Compounds

$$RCH{=}CHSn(\textit{n-Bu})_3 \xrightarrow{\text{Cu(NO}_3)_2 \cdot 3H_2O} \underset{\text{retention}}{RCH{=}CHCH{=}CHR}$$

JOC *52* 4296 (1987)

$$RCH{=}CHX + R'CH{=}CHSnR_3 \xrightarrow{\text{Pd catalyst}} RCH{=}CHCH{=}CHR'$$

$\underline{X}$	
OTf	JACS *108* 3033 (1986); *109* 3785 (1987)
	(intramolecular, macrocyclic lactones)
	TL *27* 5595 (1986)
Br, I	JACS *108* 1359 (1986); *109* 813 (1987)

$$H_2C{=}CHCH_2X + \textit{n-Bu}_3SnC\overset{CH_3}{=}CHCH_3 \xrightarrow{\text{Pd catalyst}} H_2C{=}CHCH_2C\overset{CH_3}{=}CHCH_3$$

X = Cl, Br

JACS *105* 7173 (1983); *106* 4833 (1984)

$$RCH{=}CHCH_2X \xrightarrow[\text{electrolysis}]{R_3SnCl} (RCH{=}CHCH_2)_2$$

X = halogen, OAc

TL *27* 4469 (1986)

$$RCH{=}CHCH_2X + R'CH{=}CHCH_2SnR_3 \xrightarrow{\text{Pd catalyst}} RCH{=}CHCH_2\overset{R'}{C}HCH{=}CH_2$$

$\underline{X}$	
OAc	TL *21* 2595 (1980)
Br, OAc	TL *21* 2599 (1980)

$$\underset{H_2C{=}\overset{H_2C{=}CH}{C}CH_2SnMe_3}{} + BrCH_2CH{=}CR_2 \xrightarrow{\text{cat ZnCl}_2} \underset{H_2C{=}\overset{H_2C{=}CH}{C}CH_2CH_2CH{=}CR_2}{}$$

TL *24* 1905 (1983)

3.6. Sulfur, Tellurium and Phosphorus Reagents

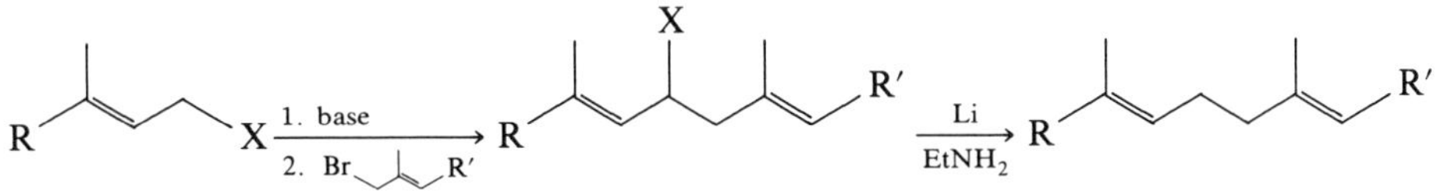

X	
SPh	TL 3707 (1969); *27* 2157 (1986) Tetr *27* 5861 (1971) Proc Natl Acad Sci USA *68* 1294 (1971) JACS *95* 4444 (1973) Syn 129 (1974) JOC *45* 4097 (1980)
SO$_2$Ph	BSCF 746 (1973) JOC *39* 2135 (1974) JCS Perkin I 761 (1981) CL 25 (1981) (intramolecular); 725 (1983) CC 1761 (1986); 1036 (1987)

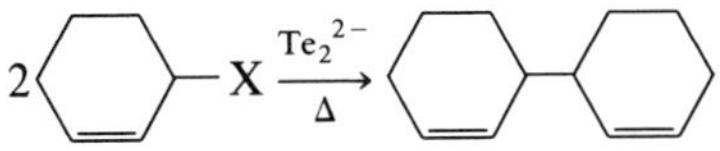

JOC *47* 1641 (1982)

$$RCH=CHCH=PPh_3 + R'CH=CHCH_2X \xrightarrow[\text{EtNH}_2]{\text{Li}} RCH=CHCH_2CH_2CH=CHR'$$

JACS *92* 2139 (1970)

$$RCH=CHCH=CHCH_2Br \xrightarrow[\substack{2.\ n\text{-BuLi} \\ 3.\ R'X}]{1.\ P(OEt)_3} RCH=CHCH=CHCHR' \xrightarrow{\text{LiAlH}_4} RCH=CHCH_2CH=CHR'$$

with OP(OEt)$_2$ substituent

TL 245 (1979)
Ann 536 (1982)

3.7. Organotitanium Compounds

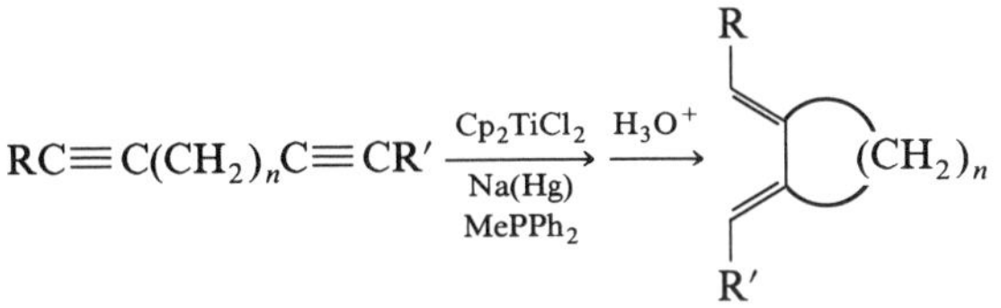

$$RC\equiv C(CH_2)_nC\equiv CR' \xrightarrow[\substack{\text{Na(Hg)} \\ \text{MePPh}_2}]{\text{Cp}_2\text{TiCl}_2\ \text{H}_3\text{O}^+} $$

$n = 3-5$

JACS *106* 6422 (1984); *109* 2788 (1987)

3.8. Organozirconium Compounds

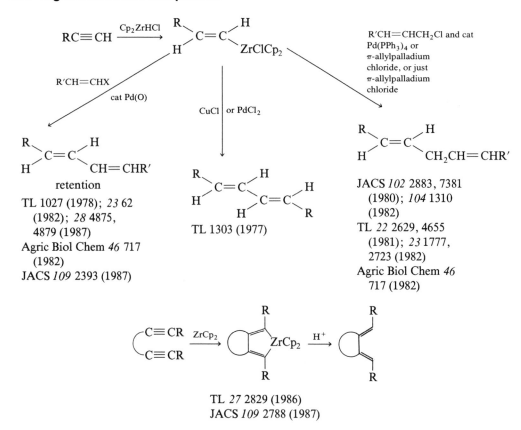

$$RC\equiv CH \xrightarrow{Cp_2ZrHCl} \underset{H}{\overset{R}{>}}C=C\underset{ZrClCp_2}{\overset{H}{<}}$$

R'CH=CHCH$_2$Cl and cat Pd(PPh$_3$)$_4$ or π-allylpalladium chloride, or just π-allylpalladium chloride

R'CH=CHX

cat Pd(0)

CuCl or PdCl$_2$

$$\underset{H}{\overset{R}{>}}C=C\underset{CH=CHR'}{\overset{H}{<}}$$

retention

TL 1027 (1978); *23* 62
(1982); *28* 4875,
4879 (1987)
Agric Biol Chem *46* 717
(1982)
JACS *109* 2393 (1987)

$$\underset{H}{\overset{R}{>}}C=C\underset{H}{\overset{H}{<}}C=C\underset{R}{\overset{H}{<}}$$

TL 1303 (1977)

$$\underset{H}{\overset{R}{>}}C=C\underset{CH_2CH=CHR'}{\overset{H}{<}}$$

JACS *102* 2883, 7381
(1980); *104* 1310
(1982)
TL *22* 2629, 4655
(1981); *23* 1777,
2723 (1982)
Agric Biol Chem *46*
717 (1982)

$$\underset{C\equiv CR}{\overset{C\equiv CR}{(}}\xrightarrow{ZrCp_2}\xrightarrow{H^+}$$

TL *27* 2829 (1986)
JACS *109* 2788 (1987)

3.9. Organoiron Compounds

$$R\diagup\diagdown\xrightarrow[\substack{2.\ R'COCl, \\ AlCl_3}]{1.\ Fe(CO)_5} R\diagup\diagdown\diagup\underset{Fe(CO)_3}{\overset{\overset{O}{\parallel}{CR'}}}\xrightarrow[\substack{2.\ Me_3NO}]{1.\ LiAlH_4,\ AlCl_3} R\diagup\diagup\diagdown CH_2R'$$

CC 373 (1981)

$$2\ C=C-C-X \xrightarrow{Fe} C=C-C-C-C=C$$

Can J Chem *47* 1238 (1969)
JOC *52* 5560 (1987)

3.10. Organocobalt Compounds

$$2\ RC\equiv CH \xrightarrow{CoCl_2-NaBH_4-PPh_3} \underset{H}{\overset{R}{>}}C=C\underset{H}{\overset{H}{<}}C=C\underset{R}{\overset{H}{<}}$$

TL *27* 6253 (1986)

$$2 \ RC \equiv CR + RCH = CHR \xrightarrow[\text{2. } FeCl_3]{\text{1. } CpCo(CO)_2}$$

JACS *102* 4839 (1980) (intramolecular); *107* 1664 (1985) (intramolecular)
CC 53 (1981); 388 (1986) (intramolecular)
Angew Int *20* 802 (1981) (intramolecular); *23* 539 (1984) (review)
JOC *47* 3447 (1982) (intramolecular)

3.11. Organorhodium Compounds

See also page 260, Section 3.16.

$$RCH = CHI + H_2C = CHR' \xrightarrow{\text{cat } ClRh(PPh_3)_3} RCH = CHCH = CHR'$$

JOMC *219* C16 (1981)

3.12. Organonickel Compounds

See also page 246, Section 3.1.

$$RCH = CHX \longrightarrow RCH = CHCH = CHR$$

$$X = halogen$$

Ni(COD)$_2$, DMF	JACS *94* 9234 (1973); *103* 6460 (1981)
NaH, *t*-AmONa, Ni(OAc)$_2$, bipy	JOMC *264* 263 (1984) TL *27* 3517 (1986)
Zn, cat NiCl$_2$(PEt$_3$)$_2$, KI	BCSJ *57* 1887 (1984)

$$RCH = CHI + H_2C = CHR' \xrightarrow{\text{cat } Ni(PPh_3)_4} RCH = CHCH = CHR'$$

JOMC *219* C16 (1981)

$$\overset{\overset{\displaystyle SO_2Ph}{|}}{R_2C = CHCHMgX} \xrightarrow{\text{cat } Ni(acac)_2} R_2C = CHCH = CHCH = CR_2$$

TL *23* 2457 (1982)

$$RCH = CHBr + \left\langle\!\!\!\left\langle\text{—}NiBr \middle/ _2 \right.\right. \xrightarrow[DMF]{} RCH = CHCH_2CH = CH_2$$

JACS *89* 2755 (1967)
Org Rxs *19* 115 (1972) (review)

$$2 \ RCH = CHCH_2Br \xrightarrow[DMF]{Ni(CO)_4} (RCH = CHCH_2)_2$$

Org Rxs *19* 115 (1972) (review)
TL *28* 2795 (1987)

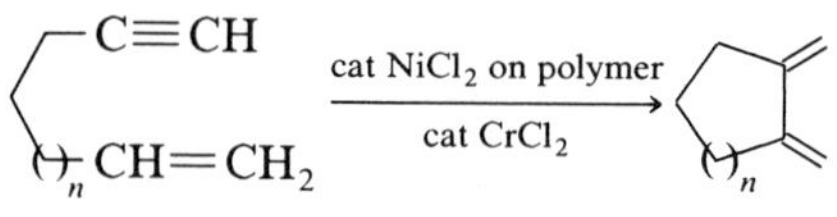

JACS *109* 5268 (1987)

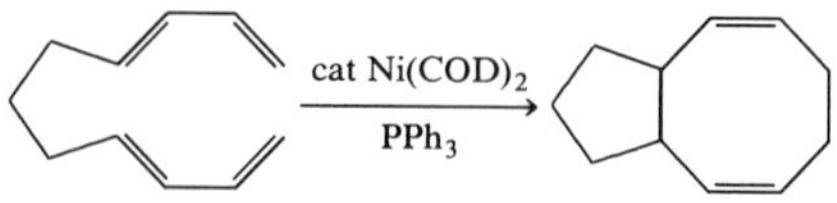

JACS *108* 4678 (1986)
TL *28* 2221, 2451 (1987)

3.13. Organopalladium Compounds

See also page 246, Section 3.1; page 248, Section 3.2; page 250, Section 3.3; page 251, Section 3.4; page 252, Section 3.5; page 254, Section 3.8; and page 260, Section 3.16.

$$RCH{=}CHX + H_2C{=}CHR' \xrightarrow{\text{Pd catalyst}} RCH{=}CHCH{=}CHR'$$

$\underline{X}$

halogen

JOC *37* 2320 (1972); *40* 1083 (1975);
  46 1061, 1067 (1981)
JACS *96* 1133 (1974)
Pure Appl Chem *53* 2323 (1981)
Tetr *37* 4035 (1981)
Syn 365 (1981)
TL *26* 2667 (1985)

O_3SCF_3

JOC *50* 2302 (1985)
Heterocycles *26* 355 (1987)

$$ArX + H_2C{=}CHCH{=}CHR \xrightarrow[\text{amine}]{\text{cat Pd(OAc)}_2(\text{PAr}_3)_2} ArCH{=}CHCH{=}CHR$$

JOC *43* 5018 (1978)
Pure Appl Chem *53* 2323 (1981)

$$RCHO + H_2C{=}CHCHR' \text{ or } HOCH_2CH{=}CHR' \xrightarrow[\text{PPh}_3]{\overset{\Delta}{\text{cat Pd(acac)}_2}} RCH{=}CHCH{=}CHR'$$

(with OH on the CHR')

TL *22* 3109 (1981)
Syn Commun *16* 1003 (1986)

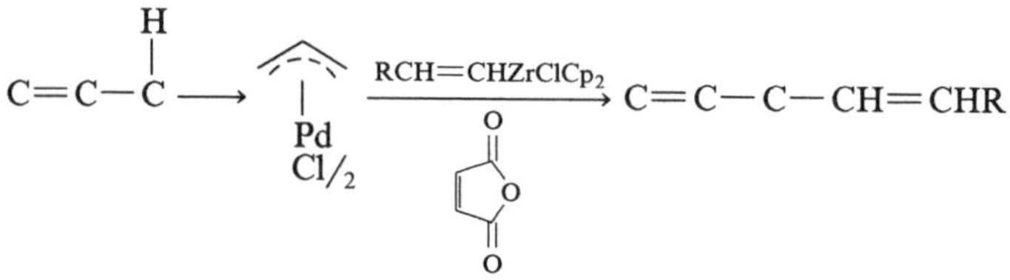

JACS *104* 1310 (1982)

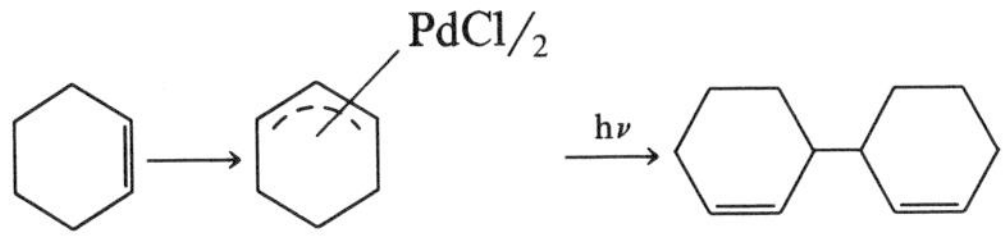

CC 258 (1980)

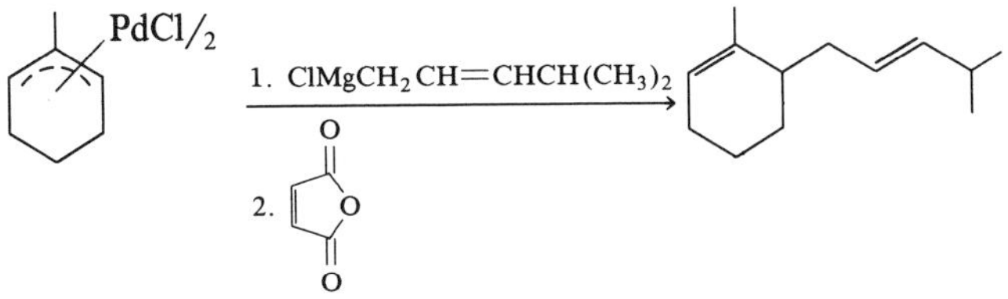

JACS *106* 5029 (1984)

X
|
PhCHCH=CH₂ or PhCH=CHCH₂X $\xrightarrow[\text{cat PPh}_3]{\overset{n\text{-Bu}_3\text{SnCl}}{\underset{\text{electrolysis}}{\text{cat PdCl}_2}}}$ H₂C=CHCHCH₂CH=CHPh

$\overset{\text{Ph}}{\underset{|}{}}$

X = Cl, OAc

TL *27* 4469 (1986)

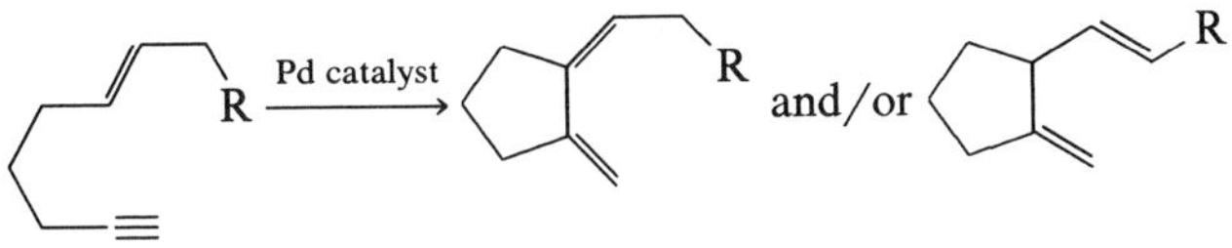

JACS *107* 1781, 4586 (1985); *108* 6053 (1986)
TL *26* 4887 (1985); *28* 1611 (1987)
See page 255, Section 3.12, for a similar reaction

3.14. Organocopper Compounds

See also page 233, Section 2.17; page 246, Section 3.1; page 248, Section 3.2; page 250, Section 3.3; and page 251, Section 3.4.

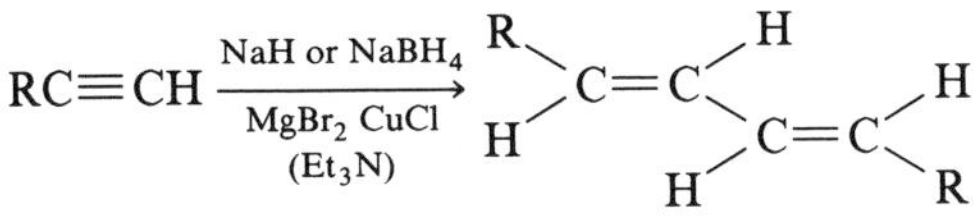

CC 495 (1987)

RCH=CHMgX $\xrightarrow{\text{CuCl}}$ RCH=CHCH=CHR

Angew Int *6* 85 (1967)
BCSJ *44* 3063 (1971)

$$\underset{Li}{\overset{SiMe}{\diagdown}}C=CH_2 \xrightarrow[\text{2. CuCl}_2]{\text{1. CuI}} \text{SiMe}_3 \diagup C=CH_2 \diagdown \text{SiMe}_3$$

TL *27* 2761 (1986)

$$\underset{R^2}{\overset{R^1}{\diagdown}}C=C\underset{Cu}{\overset{H}{\diagup}} \quad or \quad \underset{R^2}{\overset{R^1}{\diagdown}}C=C\underset{)_2CuLi}{\overset{H}{\diagup}} \xrightarrow{O_2 \text{ or } \Delta} \underset{R^2}{\overset{R^1}{\diagdown}}C=C\underset{}{\overset{H}{\diagup}} \underset{H}{\overset{}{\diagdown}}C=C\underset{R^1}{\overset{R^2}{\diagup}}$$

JACS *88* 4541 (1966); *89* 5302 (1967); *93* 1379 (1971)
TL 2583 (1971)
Angew Int *13* 291 (1974)
JOMC *77* 269 (1974)
BSCF 1656 (1974)
Rec Trav Chim *95* 299, 304 (1976)
Organomet *1* 667 (1982)

$$RC{\equiv}CCO_2R$$

1. (RCH=CH)$_2$CuLi
2. H$_2$O

$$\underset{RCH=CH}{\overset{R}{\diagdown}}C=C\underset{H}{\overset{CO_2R}{\diagup}}$$

Helv *54* 1939 (1971)
JACS *94* 4395 (1972); *99* 7365 (1977)
TL 1611 (1973); 3461 (1976)

1. (H$_2$C=CHCH$_2$)$_2$CuLi
   or H$_2$C=CHCH$_2$Cu
2. H$_2$O

$$\underset{H_2C=CHCH_2}{\overset{R}{\diagdown}}C=C\underset{H}{\overset{CO_2R}{\diagup}}$$

JACS *94* 4395 (1972)
TL 1811 (1979)

$$\diagup\diagdown\diagup CO_2CH_3 \xrightarrow[\text{2. H}_2O]{\text{1. H}_2C=CHCu} \diagup\diagdown\diagup\diagdown CO_2CH_3$$

Helv *55* 82 (1972)
JACS *94* 4395 (1972)
TL 1611 (1973)

$$RCH=CHX \xrightarrow{\text{vinylic copper reagent}} RCH=CHCH=CHR'$$

X	Reagent	
halogen	(R'CH=CH)$_2$CuLi	JACS *91* 6470 (1969)
		TL 191 (1978)
OTf	[H$_2$C=C(CH$_3$)]$_2$CuMgBr	CC 1452 (1985)
	(H$_2$C=CH)$_2$CuCN(MgBr)$_2$	TL *28* 3201 (1987)

$$(RCH=CH)_2CuLi \cdot SMe_2 + R'CH=CHCH_2Cl \longrightarrow RCH=CHCH_2CH=CHR'$$

TL 2705 (1978)

$$R_2CuLi \xrightarrow[\text{2. (MgCl}_2)]{\text{1. HC}\equiv\text{CH}} \xrightarrow[\substack{\text{5\% Pd(PPh}_3)_4 \\ \text{ZnBr}_2}]{\text{XCH}=\text{CHR}' \text{ (X = Br, I)}} \underset{H}{\overset{R}{>}}C=C\underset{H}{\overset{CH=CHR'}{<}} \text{ retention}$$

TL *22* 959 (1981); *23* 1589, 5155 (1982)
BSCF II 321, 332 (1983)
Tetr *40* 2741 (1984)

$$R_2CuLi \xrightarrow[\text{2. E}^+]{\text{1. 2 HC}\equiv\text{CH}} R\diagup\!\!\!=\!\!\diagdown\!\!\!\diagup^E$$

$E^+ = R'X$, enone, $HC\equiv CCO_2Me$, $RCHO$, CO_2, I_2

JCS Perkin I 1809 (1986)

$$RCH=CHC\equiv CH \xrightarrow[\text{2. H}_2\text{O}]{\text{1. R}'_2\text{CuMgX}} RCH=CHC\overset{\overset{\textstyle R'}{|}}{=}CH_2$$

JOMC *144* 13 (1978)

$$R_2CuLi \xrightarrow[\substack{\text{2. (H}_2\text{C}=\text{CHPPh}_3\text{)Br} \\ \text{3. R'CHO}}]{\text{1. HC}\equiv\text{CH}} \underset{R}{\overset{H}{>}}C=C\underset{CH_2}{\overset{H}{<}}\underset{}{\overset{H}{>}}C=C\underset{R'}{\overset{H}{<}}$$

TL *26* 1799 (1985)
JOC *52* 1801 (1987)

$$CH_3CH=CHCH=CHCH_2OAc \xrightarrow[\text{Li}_2\text{CuCl}_4]{\text{RMgX}} CH_3CH=CHCH=CHCH_2R$$

TL *28* 1175 (1987)

$$RCH=CHCH_2X \xrightarrow[\text{CuI}]{\overset{\displaystyle \begin{array}{c}\text{NLi}\end{array}}{}} (RCH=CHCH_2)_2$$

TL 783 (1979)

3.15. Organozinc Compounds

See also page 256, Section 3.13.

$$RC\equiv CSiMe_3 \xrightarrow[\text{2. cat Pd(PPh}_3)_4]{\overset{\overset{\textstyle CH_2}{\|}}{\text{1. R'OCH}_2\text{CCH}_2\text{ZnBr (R' = Ph, CH}_2\text{Ph)}}} \underset{R}{\diagup}\overset{}{\diagdown}SiMe_3$$

TL *28* 2889 (1987)

$$CF_3CF=CFZnX + ICF=CF_2 \xrightarrow{\text{cat Pd(PPh}_3)_4} CF_3CF=CFCF=CF_2$$

TL *27* 4387 (1986)

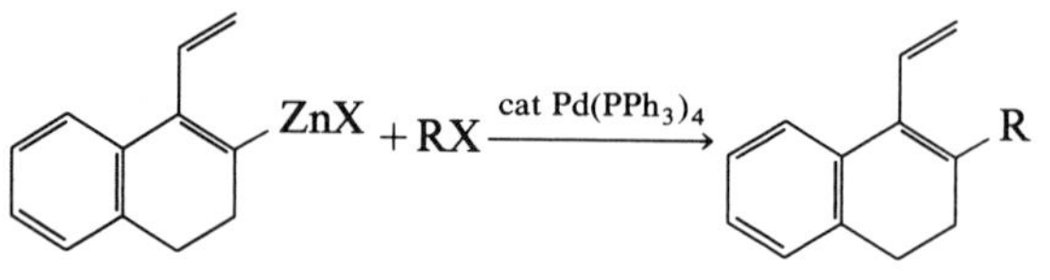

R = aryl, vinylic

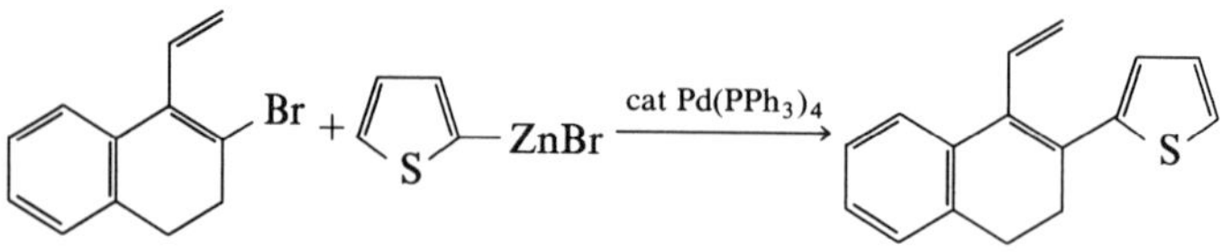

TL *28* 1469 (1987)

$$H_2C{=}CHCH_2ZnBr + BrCH_2CH{=}C(CH_3)_2 \longrightarrow H_2C{=}CHCH_2CH_2CH{=}C(CH_3)_2$$

TL *24* 1905 (1983)

3.16. Organomercury Compounds

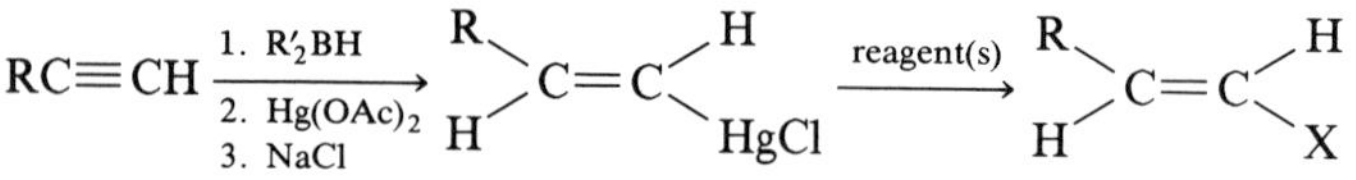

X	Reagent(s)	
$CH{=}CH_2$	$(H_2C{=}CH)_2CuLi$	TL *22* 3435 (1981)
		Organomet *1* 74 (1982)
$CR{=}CH_2$	$PdCl_2$, C_6H_6	JOC *43* 1468 (1978)
$CH{=}CHR$ (*E*)	cat $Pd(PPh_3)_4$	TL 3207 (1974)
	Li_2PdCl_4, HMPA	JOC *41* 2241 (1976)
	cat $[ClRh(CO)_2]_2$	JOC *42* 1680 (1977)
$CH{=}CHR'$	$R'CH{=}CH_2$, cat $LiPdCl_3$, $CuCl_2$	Bull Korean Chem Soc 7 142, 235, 472 (1986)
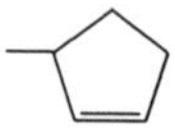	 , Li_2PdCl_4	TL *22* 5231 (1981)
$CH_2CH{=}CH_2$	$H_2C{=}CHCH_2Cl$, Li_2PdCl_4	JOMC *156* 45 (1978)

4. Birch Reduction

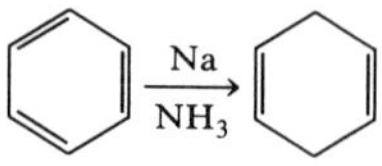

Quart Rev *4* 69 (1950); *12* 17 (1958)
Chem Rev *46* 317 (1950)
"Steroid Reactions," Holden-Day, Inc., San Francisco (1963), pp 267–88, 299–325
"Organic Reactions in Liquid Ammonia, Chemistry in Non-Aqueous Ionizing Solvents," Wiley, New York (1963), Vol 1, part 2
"Reduction," Marcel Dekker, New York (1968), pp 95–170, 186–194

Syn 161 (1970)
Adv Org Chem *8* 1 (1972)
H. O. House, "Modern Synthetic Reactions," Benjamin, Inc., Menlo Park, California (1972), pp
   190–205
Natl Prod Repts *3* 35 (1986)
JOC *51* 4983 (1986)

TL *28* 1173 (1987)

JOC *46* 788 (1981)
TL *27* 1469 (1986)

X	E$^+$	
COR	RX	JOC *38* 3887 (1973)
		JCS Perkin I 3214 (1981); 2399 (1983); 383, 735 (1985)
		CC 1292 (1986)
		TL *27* 5253, 5303 (1986)
CO$_2$H(R)	RCH=CHCO$_2$R (1,4-addition)	CC 315 (1980)
	RX	JOC *34* 126 (1969); *41* 2649 (1976); *42* 1794 (1977); *45* 1722 (1980); *49* 4429 (1984); *50* 915, 3086 (1985); *51* 2844 (1986); *52* 5482 (1987)
		Rec Trav Chim *90* 137 (1971)
		JCS Perkin II 851 (1976)
		TL 2079 (1976); *21* 3309 (1980); *22* 3683, 4115 (1981); *23* 1095 (1982); *24* 1369 (1983); *27* 1481, 3923 (1986)
		Austral J Chem *30* 1045 (1977); *31* 1157 (1978); *34* 675 (1981)
		Ind J Chem B *16* 1027 (1978)
		Syn 374 (1979)
		JACS *102* 5085, 6628 (1980); *107* 5574 (1985); *108* 5893 (1986); *109* 3991 (1987)
		Ber *114* 214 (1981)
		Tetr *38* 2831 (1982); *39* 4221 (1983) (naphthalene)
		Natl Prod Repts *3* 35 (1986) (review)

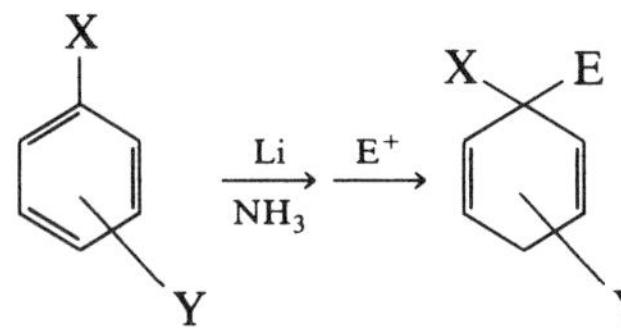

(*Continued*)

X	E+	
CONR$_2$	RX	JOC *51* 4983 (1986)
CN	RX	JOC *51* 4983 (1986)
		JACS *109* 3991 (1987)

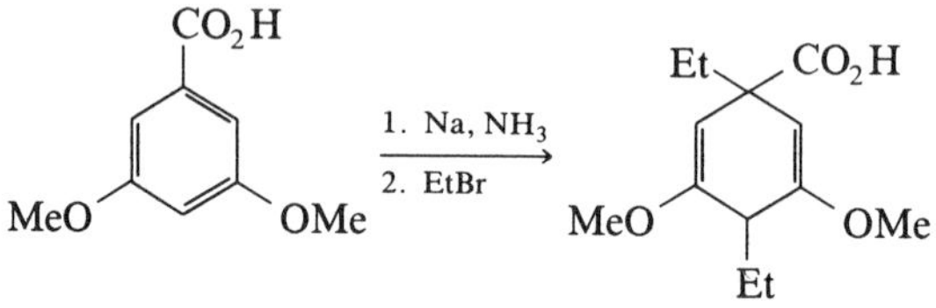

Syn Commun *11* 223 (1981)

5. Cyclization

See also page 253, Section 3.7; page 254, Section 3.10; page 255, Section 3.12; and page 256, Section 3.13.

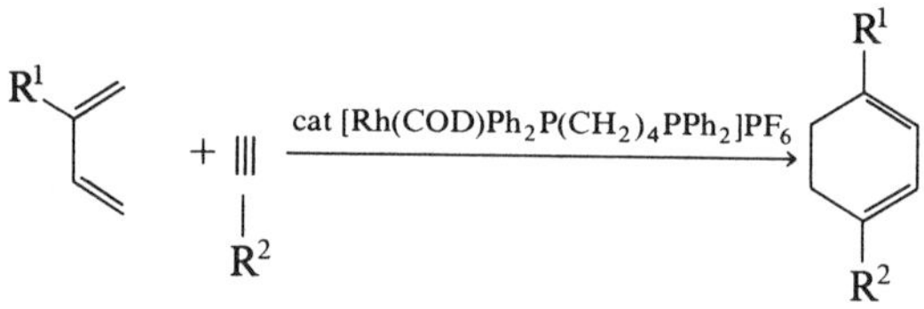

TL *28* 3361 (1987)

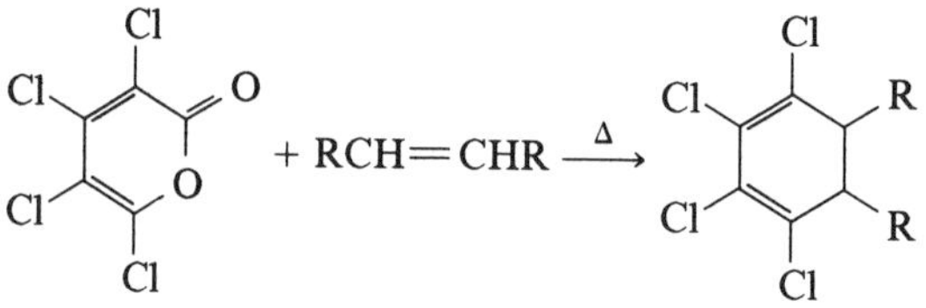

JACS *109* 7074 (1987)

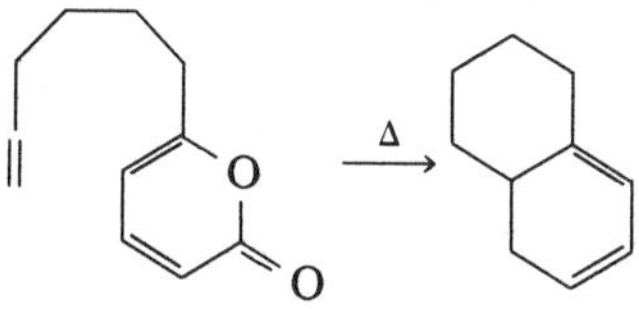

JACS *109* 6124 (1987)

6. Miscellaneous Reactions

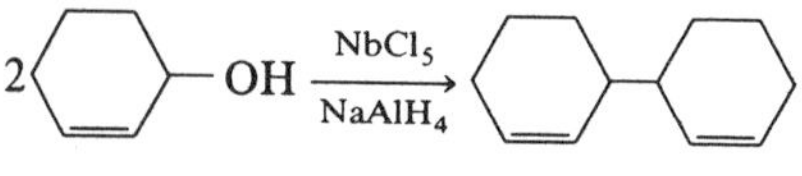

CL 157 (1982)

8. DIELS-ALDER REACTION

For Diels-Alder reactions affording arenes see page 101, Section 6.

1. General Reviews

Org Rxs *4* 1, 60 (1948); *5* 136 (1949)
Chem Rev *61* 537 (1961)
A. S. Onishchenko, "Diene Synthesis," D. Davey Co., New York (1964)
R. Huisgen, R. Grashey and J. Sauer, "The Chemistry of Alkenes," Ed. S. Patai, Interscience,
   New York (1964), Chpt 11
A. Wasserman, "Diels-Alder Reactions," Elsevier, New York (1965)
Angew Int *5* 211 (1966); *6* 16 (1967)
Houben-Weyl, "Methoden der Organischen Chemie," Vol 5/1c, G. Thieme, Stuttgart (1970),
   p 977

2. Regioselectivity

Facial selectivity	JACS *109* 663 (1987)
	JOC *52* 3050, 4726, 4732 (1987)
Frontier orbitals	JACS *95* 4092 (1973); *108* 7381 (1986)
	Acct Chem Res *8* 361 (1975) (review)
	JOC *43* 1864 (1978); *46* 2338 (1981)
	(quinones)
	Pure Appl Chem *55* 237 (1983)
Diene polarity	JACS *100* 7098 (1978)
	TL *22* 2043, 2047 (1981)
Dienophile substituents	JOC *47* 5009 (1982)
Lewis acids	JACS *86* 3899 (1964); *95* 4094 (1973)
	TL 5127 (1970)
	Can J Chem *50* 2377 (1972)
	JOC *45* 5012 (1980)

Cation-radical	JACS *103* 718 (1981); *104* 2665 (1982); *105* 2378 (theory), 3584 (1983) TL *28* 927 (1987)
Micelles	JOC *47* 3186 (1982); *48* 3137 (1983)

3. Endo Addition

Angew *50* 510 (1937) (review)
Chem Rev *61* 537 (1961) (review)
TL 731 (1966)
JACS *92* 6548 (1970); *93* 4606 (1971); *94* 3633 (1972)
JOC *48* 276 (1983)

4. Asymmetric Diels-Alder Reaction

Angew Int *4* 989 (1951); *23* 876 (1984) (review); *24* 112, 784 (intramolecular, heterodienophile) (1985)
JOC *26* 4778 (1961); *31* 2418 (1966); *48* 1137, 4441 (1983); *49* 1527 (1984); *51* 1457 (1986)
Tetr *19* 2333 (1963); *42* 4035, 5045 (heterodienophile), 5157 (review), 6477 (heterodienophile) (1986)
TL 6359 (1966); *23* 4875 (1982); *24* 3451 (1983) (heterodienophile); *26* 1631, 3095, 5437 (intramolecular) (1985); *27* 667, 4507, 4895 (chiral catalyst), 5509 (1986); *28* 107, 183, 777 (chiral catalysts), 2681, 4045, 5687 (chiral catalysts), 5481 (1987)
JACS *97* 6908 (1975); *102* 7595 (1980); *104* 2269 (1982); *106* 4261 (1984); *108* 3510 (chiral catalyst), 7373 (1986)
CC 437 (1979) (chiral catalysts); 540 (1983) (heterodienophile), 1786 (1985); 609 (1986); 423, 676 (chiral catalysts) (1987)
Helv *67* 1397 (1984)
L. A. Paquette in "Asymmetric Synthesis," Ed. J. D. Morrison, Academic Press, New York (1984), Chpt 7 (review)
CL 1109, 1967 (1986) (both chiral catalysts)

5. Mechanism

Angew Int *6* 16 (1967); *19* 779 (1980) (both reviews)
JOC *37* 2181 (1972) (diradical)
JACS *95* 4092, 4094 (1973) (concerted); *106* 203 (1984) (unsymmetrical transition state)

6. Effect of High Pressure

JACS *96* 3664 (1974); *98* 1992 (1976); *102* 6893 (1980); *103* 7007 (1981); *108* 3040 (1986)
JOC *42* 282 (1977); *44* 3347 (1979); *50* 2576, 3239 (1985)
Chem Rev *78* 407 (1978) (review)
Heterocycles *16* 1287, 1367 (review) (1981)
Helv *65* 1021 (1982)
TL *23* 2611, 4875 (1982); *26* 2229 (1985); *27* 1015 (1986); *28* 4045, 5267 (1987)
Syn Commun *13* 537 (1983)
Tetr *42* 5045, 6477 (1986)

7. Use of Microwave Oven

TL *27* 4945 (1986)

8. Use of Ultracentrifuge

TL *28* 707 (1987)

9. Retro Diels-Alder Reaction

Angew Int *5* 211 (1966) (review); *25* 414 (1986) (review)
Chem Rev *68* 415 (1968) (review)
Tetr *34* 19 (1978)
TL *22* 4553, 4557 (1981); *23* 5463 (1982); *25* 3459, 3461 (1984); *26* 2103 (1985); *27* 3045 (1986);
   28 183, 357, 1329, 3027, 3147, 5819, 6503, 6507 (1987)
JOC *47* 598 (1982); *49* 1033, 4091 (1984); *51* 2851 (1986); *52* 4603, 5746 (aza) (1987)
JACS *106* 4862 (1984); *108* 1019 (1986); *109* 5859 (1987) (aza)
Can J Chem *62* 2019 (1984)
Syn 121 (1985) (review); 207 (1987) (review)
Austral J Chem *38* 1339 (1985)
Syn Commun *15* 959 (1985)

10. Intramolecular Diels-Alder Reaction

Angew Int *16* 10 (1977) (review); *21* 620 (1982) (allene + arene); *24* 784 (1985) (asymmetric,
   heterodienophile)
Syn 793 (1978) (*o*-quinodimethane review)
JACS *100* 6289, 8034 (1978); *102* 6353 (1980); *103* 2090, 4948, 5200, 6261, 6696 (1981); *104* 1033,
   1140, 2269, 3216, 4725 (isobenzofuran as diene), 5708 (bridgehead alkene synthesis), 5715 (bridgehead
   enol lactone synthesis), 5719, 5728, 5853 (1982); *105* 3732 (1983); *106* 1133, 6085 (acid catalyzed),
   6422, 6735, 7641, 8327 (1984); *107* 2149, 2573, 4072, 4791 (1985); *108* 4953 (1986); *109* 447, 1186,
   4390, 6124 (1987)
JOC *45* 4264, 4267 (1980); *46* 1506, 1509, 2273, 3763 (1981); *47* 180, 337, 610 [via bis(orthoquinone)],
   1789, 2682 (vinyl allene as diene), 4337 (pyrone as diene), 4611, 4786, 4825 (1982); *50* 725
   (bridgehead dienes), 1770, 2626, 2719, 2807, 3086 (1985); *51* 1150, 1155, 1633, 2487, 3075 (ionic),
   3553 (iminium ions), 3740, 4023 (1986); *52* 564, 1236, 1638, 2040, 4369, 4661, 5548, 5700 (1987)
Chem Rev *80* 63 (1980) (review)
Chem Soc Rev *9* 41 (1980) (review)
Helv *64* 478, 1387 (1981); *65* 2212 (1982)
TL *22* 97, 101, 361, 3929, 3933, 4877 (benzyne), 5141 (1981); *23* 303, 1523, 2361 (1982); *24* 3295, 3701
   (1983); *25* 19 (1984); *26* 591, 1249, 2229, 2689, 3307, 5109, 5437 (asymmetric) (1985); *27* 1837, 3045,
   3227, 4877, 6291 (PhNMe$_2$ as solvent) (1986); *28* 735, 927 (cation radical), 931, 1059, 1175, 2087,
   2447, 2937, 2969, 3423, 3819 (photochemical), 5249, 5253, 5255, 5339, 5895 (allene), 6035, 6253
   (1987)
CL 29 (1981)
Ann 973 (1982)
Can J Chem *62* 183 (1984) (review)

Org Rxs *32* 1 (1984) (review)

D. F. Taber, "Intramolecular Diels-Alder and Alder Ene Reactions," Springer, New York (1984) (review)

CC 1423 (1984); 421 (1985); 671, 724, 1447, 1449, 1797 (solvent effect) (1986); 418, 898, 1449 (allenic ester), 1540, 1786 (1987)

Acct Chem Res *18* 16 (1985) (imino review)

11. Inverse Electron Demand Diels-Alder Reaction

See also page 101, Section 6.

JOC *30* 3679 (1965); *51* 5328 (1986); *52* 4610 (1987)

TL 4651 (1972); *27* 667 (1986); *28* 2681 (1987)

JACS *94* 2891 (1972); *103* 6261 (1981); *107* 5745 (1985); *108* 7373 (1986) (asymmetric)

Ann 306 (1981)

Helv *64* 478 (1983)

12. Use of Heterodienophiles

JOC *44* 3347 (1979); *47* 1981, 3183, 3649 (1982); *51* 3553 (1986) (iminium ions)

Heterocycles *12* 949 (1979) (review)

JCS Perkin I 1795 (1979)

Tetr *37* 1825 (1981) (singlet oxygen review); *38* 1299, 3087 (review) (1982); *42* 5045, 6477 (1986) (both asymmetric aldehyde, high pressure)

TL *22* 4807 (1981); *23* 3739 (1982); *24* 3451 (1983) (aldehyde); *27* 1975, 2981 (imine), 6181 (1986); *28* 1059 (azo), 2103 (iminium ions), 4951 (aldehyde), 5103 (azetinone) (1987)

JACS *103* 6387 (1981); *106* 2453, 2455, 2456 (all aldehydes), 3240 (intramolecular acylimino), 4294 (aldehyde), 8217 (aldehyde, intramolecular) (1984); *107* 1246, 1256, 1269, 1274, 1280, 1285 (all aldehydes), 1768 (iminium ions), 6647, 7761, 7762 (all aldehydes) (1985); *108* 2486, 4145 (both aldehydes) (1986); *109* 1572, 2082, 4390 (1987) (all aldehydes)

CC 540 (1983) (asymmetric aldehyde)

Acct Chem Res *18* 16 (1985) (imino review)

D. L. Boger, S. M. Weinreb, "Hetero Diels-Alder Methodology in Organic Synthesis," Academic Press, New York (1987) (review)

Adv Heterocyclic Chem *42* 245 (1987) (review)

13. Use of Heteroatom-Containing Dienes

JACS *72* 3079 (1950); *73* 913, 5267 (1951); *75* 1312 (1953) (all 1-oxa)

Chem Rev *62* 405 (1962) (general review); *75* 651 (1975) (general review)

TL 129 (1965) (1-oxa); 3417 (1977) (1-thia); *21* 1133 (1980) (1-oxa); *25* 721 (1984) (1-oxa); *26* 5273 (1985) (1-oxa); *27* 6181 (1986) (1-oxa); *28* 4045 (1987) (1-oxa)

JOC *42* 282 (1977) (1-oxa); *50* 2719 (1-aza), 5678 (2-aza) (1985); *52* 1638 (1987) (1-oxa)

Tetr *39* 2869 (1983) (aza review)

Angew Int *24* 784 (1985) (1-oxa, intramolecular, asymmetric)

Ann 2261 (1985) (1-oxa)

D. L. Boger, S. M. Weinreb, "Hetero Diels-Alder Methodology in Organic Synthesis," Academic Press,
  New York (1987) (general review)
Adv Heterocyclic Chem *42* 245 (1987) (review)
CC 884 (1987) (1-oxa)

14. Use of 1,3-Dienes with Attached Heteroatom Substituents

Reviews:

  Syn 753 (1981)
  Acct Chem Res *14* 400 (1981) (R_3SiO)

Helv *58* 590, 593 (1975) (1-MeO$_2$CNR); *65* 2563 (1982) (1,1-MeO, 3-Me$_3$SiO); *66* 2769 (1983)
  (1-ArNH, 2-CO$_2$Me); *68* 1133 (1985) (1-EtO, 4,4-Me)
Chem Listy *70* 1266 (1976) (1-R$_2$N)
JACS *98* 3027 (1-MeO, 3-Me$_3$SiO), 5017 (2-MeO, 3-PhS) (1976); *101* 6996, 7001, 7008 (1979) (1-MeO,
  3-Me$_3$SiO and derivatives); *102* 3554 (2-RO or RCO$_2$, 3-RS) (1980); *103* 2816 (1981) (1-RCONH);
  104 2059 (1-PhS, 2-MeO, 3-Me), 2308 (1-AcO, 4-Me$_3$Si), 2923 (1-PhS, 2-Me$_3$SiO, 3-H or Me), 3511
  (1-Me, 3-Me$_3$SiO), 4299 (2,3-Me$_3$SiCH$_2$) (1982); *105* 6335 (1983) (1-RO$_2$CNH, 4-PhS or PhSO or
  PhSO$_2$) (1983); *109* 6396 (1987) (2-PhSO$_2$, 4-H or Me)
CC 681 (1976) (1-Me$_3$Si); 178 (1978) (1-Me$_3$Si; 1-Me$_3$Si, 3-Me$_3$SiO)
JOC *41* 2625 (1976) (1,4-AcO, 2-Me); *42* 282 (1-AcO); 1819 (1-PhSe, 2-Me$_3$SiO, 4-MeO) (1977);
  43 379 (1,1-MeO, 3-Me$_3$SiO), 2726 (2,3-Me$_3$SiO), 4559 (1-RCO$_2$, 4-Et) (1978); *46* 1810, 4161 (1980)
  (both various MeO, Me$_3$SiO); *47* 2051 (2-Me$_3$SiO; 2-Me$_3$SiCH$_2$; 1-MeO, 3-Me$_3$SiO), 3649 (various
  Me$_3$SiO), 4005 (1-PhS, 2-MeO), 4774 (1,3-MeO), 5009 (1-MeO, 3-Me), 5083 (1,4-MeO) (1982); *48*
  3096 (1,4-*t*-BuO), 4986 (1-MeO, 3-Me$_3$SiO, 2-H or Me), 5051 (1-PhS, 2-MeO; 2-PhS, 3-MeO; 2-ArS,
  3-AcO or MeO; 1-PhS, 4-AcO) (1983); *49* 1898 (1,4-*t*-BuMe$_2$SiO), 2954 (2-Cl or Br, 3-PhS or PhSe;
  1,2-Cl, 3-PhS or PhSe; 1-Cl, 2-Br, 3-PhS or PhSe), 3595 (1-RO, 3-Me; 1-RO, 3-RSCH$_2$), 3628
  (1-*t*-BuO, 3-Me$_3$SiO) (1984); *50* 141 (2-PhS, 3-Me), 1955 (1,3-AcO; 1-EtO$_2$CO, 3-AcO) (1985); *51*
  2210 (2-Ph$_2$PO) (1986); *52* 2334 (1987) (1-PhSe, 4-MeO)
TL 2113 (1976) (1,4-OCH$_2$O), 3323 (1978) (2-Et$_3$Si); 4537 (1979) (1-RCONR); *22* 3381 (1981) (1-PhS,
  2-RO, 3-Me); *23* 551 (2-Me$_3$SiCH$_2$), 775 (2-MeO), 2155 (1,3-MeO) (1982); *24* 1171 (1983) (1-RO,
  3-RSCH$_2$); *25* 5715 (1984) (1-MeO, 4-CH$_2$CO$_2$Na); *26* 381 (1-EtO, 1-*t*-BuMe$_2$SiO), 2555, 2559
  (both 1-Ar or 1,2-Ar), 5175 (1-RO, 3-Me$_3$SiCH$_2$; 1-MeO, 3-Me$_3$Si; 2-Me$_3$SiCH$_2$), 6519 (1-MeO,
  1-Me$_3$SiO, 3-MeO or Me$_3$SiO) (1985); *27* 2761 (2,3-Me$_3$Si), 2913 (2-*n*-Bu$_3$Sn), 5075 (1-MeO, 3-Cl)
  (1986); *28* 4169 (1987) [1-B(OR)$_2$]
Org Syn *58* 163 (1978) (2-Me$_3$SiO); *61* 147 (1983) (1-MeO, 3-Me$_3$SiO)
JCS Perkin I 1582 (1,1-RO); 2415 (1-Me$_3$Si; 1-Me$_3$Si, 3-Me or 3-Me$_3$SiO or 4-Me or 4-Me$_3$Si) (1981)
Tetr *37* 4081 (1981) (1-MeO, 1-AcO, 3-Me$_3$SiO); *40* 5039 (1984) (1-Me$_3$SiO, 1,2-MeO, 4-Me)
Syn Commun *11* 481 (1981) (1,2-MeS); *14* 483 (2-COMe), 797 (1-PhSeCH$_2$CH$_2$O, 3-Me) (1984)
Syn 273 (1,4-EtO$_2$CNH), 380 (1,1,2,3-Me$_3$SiO); 958 (1-EtO$_2$CNH, 4-PhCO$_2$) (1982); 899 (1985)
  (3-*n*-Bu$_3$Sn; 3-PhMe$_2$Si)
CL 1057 (1982) [1-RO (R = *t*-Bu, Me$_3$Si, Ac), 3-Me$_3$SiO]
Can J Chem *62* 2676 (1984) [2-OPO(OEt)$_2$, 4-Me)]

JACS *109* 6124 (1987)

15. o-Quinodimethanes

Syn 793 (1978) (review)
Helv *62* 2017 (1979)
CC 1119 (1979); 158, 430, 699 (1982)
Chem Soc Rev *9* 41 (1980) (review)
Acct Chem Res *13* 270 (1980) (review); *17* 35 (1984) (indole-2,3-quinodimethane review)
JACS *102* 863, 6885 (1980); *103* 476, 1992 (1981); *104* 1106, 1140 (indole-2,3-quinodimethane), 3511
    (1982); *105* 1586 (1983)
Heterocycles *14* 1615 (1980) (review)
Tetr *37* 3, 2547, 2555, 3813 (1981)
TL *22* 1357 (1981); *23* 2973 (1982)
JCS Perkin I 1383, 1386 (1981); 2239, 2249 (1982)
Pure Appl Chem *53* 1181 (1981) (review)
JOC *47* 2331 (1982); *50* 2764 (1985) (intramolecular); *51* 3490, 4160, 4749 (intramolecular) (1986);
    52 28, 4833 (1987)
Ann 1999 (1982)
Can J Chem *62* 183 (1984) (review)

16. Arynes

R. W. Hoffman, "Dehydrobenzene and Cycloalkynes," Academic Press, New York (1967)
BCSJ *50* 3338 (1977)
JACS *102* 6649 (1980) (bisaryne); *108* 4932 (1986)
JOC *46* 4874 (1981) (bisaryne); *51* 979 (1986) (bisaryne); *52* 792, 3835 (bisaryne) (1987)
TL *22* 4877 (1981) (intramolecular); *25* 2073 (1984) (bisaryne); *27* 5319 (1986)
Tetr *38* 427 (1982) (hetarynes review)

17. Some Other Dienophiles

$R^1CH=CR^2CR^3(OR')_2$ JACS *109* 2182 (1987)
  (R^1, R^2 or R^3 = H or Me)
  (as allyl cation)

vinylic sulfoxides   TL *27* 6041 (1986)

$PhSOCH=CHCO_2R$ Austral J Chem *37* 1677 (1984)

vinylic sulfones   TL *21* 3339 (1980); *23* 5127 (1982)
                                            JACS *102* 853 (1980)
                                            JOC *48* 4976, 4986 (1983)

E- or *Z*- $PhSO_2CH=CHSO_2Ph$ See page 270, Section 18, $HC\equiv CH$.

$H_2C=C(SO_2Ph)_2$ Syn 757 (1984)

$H_2C=CHNO_2$ J Chem Res (S) 78 (1983)

$Me_3SiCH=CHNO_2$ JOC *49* 3235 (1984)

Z- $Me_3SiCH=CHCO_2H$ Angew Int *23* 233 (1984)

$H_2C\!=\!C(COR)CO_2R'$

JOC *47* 4152 (1982)

$H_2C\!=\!C(O_2CR)COCH_3$

Helv *64* 188 (1981)
JOC *47* 409 (1982)
TL *28* 865 (1987)

$(MeO)_2CHCH_2CH\!=\!CHCOCH_3$

Can J Chem *60* 2760 (1982)

cyclic enones

JOC *47* 5056 (1982); *48* 2802 (1983);
   50 4686, 4691, 4696 (1985);
   51 2642, 2649, 5177 (1986)
Tetr *41* 349 (1985)

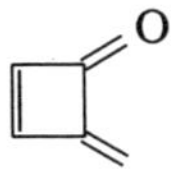

TL *23* 841 (1982)

2,5-cyclohexadienones

Can J Chem *57* 377 (1979)
JOC *48* 1810 (1983)

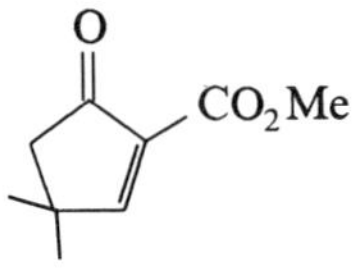

TL *24* 1353 (1983)

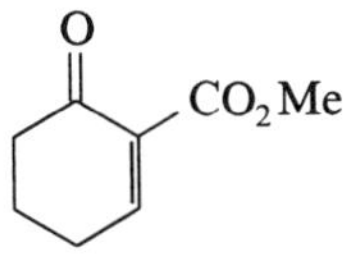

Syn Commun *12* 715 (1982)

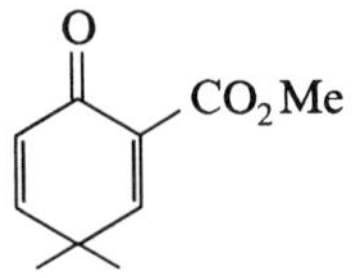

TL 2919 (1977)
Can J Chem *59* 601 (1981)

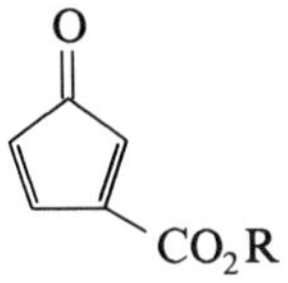

JOC *52* 5298 (1987)

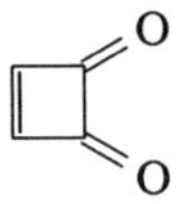

TL *23* 841 (1982)

quinones

CC 837 (1983)

butenolides

TL 5317 (1972); *25* 19 (1984)
   (intramolecular); *27* 1081 (1986);
   28 3405 (1987)
CC 737 (1985)
JCS Perkin I 2279 (1986)

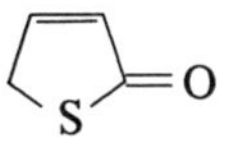

JOC *47* 4777 (1982)

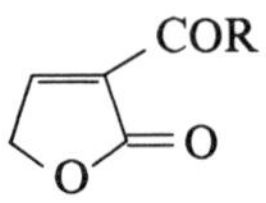

JOC *47* 4152 (1982)

Can J Chem *60* 921 (1982)

JOC *47* 1451 (1982)

JOC *47* 3647 (1982)

$H_2C=C=CHF$ JACS *107* 7183 (1985)

$H_2C=C=CHSO_2Ph$ JOC *50* 512 (1985)

allenic ketones   J Chem Res (S) 300 (1982)

allenic esters   Helv *65* 2563 (1982)

$n\text{-Bu}_3SnC\equiv CCO_2R(CN)$ CC 1452 (1984)

$ClCOC\equiv CCOCl$ Ber *115* 804 (1982)

18. Dienophile Equivalents

Equivalent	Reagent(s)	
$H_2C=CH_2$	$H_2C=CHSO_2Ph$ /Na(Hg)	JACS *102* 853 (1980)
		JOC *48* 4976 (1983)
		CC 451 (1986)
	$BrCH=CHBr$ /Zn	Ber *93* 1888 (1960)
$RCH=CHR$	$RCH=CRNO_2$ /$n\text{-Bu}_3SnH$	CC 33 (1982)
$H_2C=C=CH_2$	$H_2C=CBrCHO$ /$NaBH_4$/Zn	JOC *38* 3961 (1973)
	$H_2C=CBrSO_2CH_3$ /NaOMe	JACS *94* 1012 (1972)
	$H_2C=C(SiMe_3)CH_2Cl$ /MeMgBr	Organomet Chem Syn *1* 253 (1971)
$H_2C=C=CHR$	$(H_2C=CHPPh_3)Br$/LDA/RCHO	TL 2095 (1974)
		JOC *42* 4095 (1977)
$H_2C=C=O$	review	Syn 289 (1977)
	$H_2C=CClCOCl$ /NaN_3/Δ/H_3O^+	JACS *93* 4326 (1971)
	$H_2C=C(OAc)CN$ /base	JACS *78* 2473 (1956); *82* 627 (1960)
		JOC *31* 3787 (1966); *46* 4152 (1981)
		TL 121 (1972)
	$H_2C=CClCN$ /(S^{2-}), KOH	CA *64* 4965g (1966); *71* 112496p (1969)
		JACS *91* 5675 (1969); *93* 1489 (1971)
		Ber *105* 1840 (1972)
		TL 121 (1972)
		CC 642 (1974); 39 (1975)
		JOC *40* 2565 (1975); *46* 4152 (1981)

	$H_2C=C(CN)N\underset{\diagup}{\overset{\diagdown}{}}O$ /hydrolysis	TL *23* 953 (1982)
	$H_2C=C(CN)SMe$ /hydrolysis	TL *23* 953 (1982)
	$H_2C=C(SMe)CO_2Me$ /hydrolysis	TL *23* 953 (1982)
	$H_2C=CHCN$ /PCl$_5$/OH$^-$	JOC *33* 2111 (1968)
		CA *71* 112496p (1969)
	$H_2C=CHB(OBu)_2$ /[O]	TL 121 (1972)
	$H_2C=CHNO_2$ /NaOH/HCl	JACS *96* 5261 (1974)
		Tetr *32* 961 (1976)
	$H_2C=CHCO_2H$ /LDA/O$_2$	TL 4611 (1975)
	$H_2C=CHCO_2H$ /LDA/ (MeS)$_2$/NCS	JACS *97* 3528 (1975)
	$H_2C=CHSO_2Ar$ /LDA/ MoO$_5\cdot$py$\cdot$HMPA	TL *21* 3339 (1980)
MeO$_2$CCH=C=O	MeOC$\equiv$CCO$_2$Me /H$_3$O$^+$	CC 1227 (1982)
	PhSC$\equiv$CCO$_2$Me /H$_2$O, HgCl$_2$	CC 1227 (1982)
	BrC$\equiv$CCO$_2$Me /NaOMe/H$_3$O$^+$	TL 383 (1979)
	MeS$\diagdown$ structure /NaHCO$_3$,	JACS *99* 7079 (1977)
	MeOH/NCS/H$_3$O$^+$	
O=C=O	(H$_2$C=CHPPh$_3$)Br/LDA/O$_2$	TL 2095 (1974)
		JOC *42* 4095 (1977)
	OC(CO$_2$Et)$_2$/OH$^-$/H$^+$/Pb(OAc)$_4$ or Curtius	JACS *97* 6892 (1975)
		JOC *42* 4095 (1977)
(H$_2$C=CH)$_2$CO	H$_2$C=CHCO(CH$_2$)$_2$Cl /base	JACS *91* 2806 (1969)
	H$_2$C=CHCOCH(SePh)CH$_3$ /[O]	Syn Commun *8* 211 (1978)
HC$\equiv$CH	review	Tetr *40* 2585 (1984)
	TsC$\equiv$CH /Na(Hg)	CC 639 (1980)
		JACS *107* 686 (1985)
	PhSOCH=CH$_2$ /Δ	JACS *100* 1597 (1978)
	ClCH=CHCl /Zn or Na	JACS *74* 2193 (1952)
		Tetr *28* 3031 (1972)
	structure—Ph /n-BuLi	CC 1593 (1968)
		JCS C 886 (1971)
		JACS *95* 7161 (1973)
		JCS Perkin I 2332 (1973)
	structure=S /P(OMe)$_3$, Fe(CO)$_5$ or Ni(COD)$_2$	JACS *85* 2677 (1963); *95* 7161 (1973)
		TL 2667 (1973)
		JOMC *69* 423 (1974)
	Z-PhSO$_2$CH=CHSO$_2$Ph /Na(Hg)	CC 914 (1982)
		JOC *49* 596 (1984); *52* 3250, 4732 (1987)
		Can J Chem *62* 2487 (1984)
		JACS *107* 4789, 6400 (1985); *108* 3453 (1986)

Equivalent	Reagent(s)	
	E-PhSO$_2$CH=CHSO$_2$Ph /Na(Hg)	TL *24* 1653 (1983) Can J Chem *62* 2487 (1984) JACS *107* 6400 (1985) JOC *50* 4340 (1985); *52* 4740 (1987) Tetr *42* 1789 (1986)

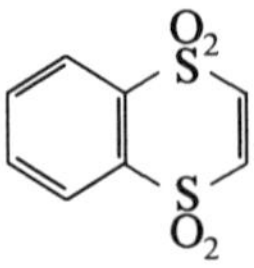

/Na(Hg)

Heterocycles *23* 1119 (1985)

	E-PhSO$_2$CH=CHSiMe$_3$ / n-Bu$_4$NF	TL *22* 4643 (1981) JOC *48* 4976 (1983)
	Z- Me$_3$SiCH=CHCO$_2$H / electrolysis	Angew Int *23* 233 (1984)
	ClCOCH=CHCOCl /Pb(OAc)$_4$	Syn 117 (1979)
	MeO$_2$CC≡CCO$_2$Me /NaOH/Cu	Tetr *28* 3031 (1972)
	MeO$_2$CC≡CCO$_2$Me /NaOH/CuO	TL 4447 (1976)
	MeO$_2$CC≡CCO$_2$Me /NaOH/ electrolysis	JACS *96* 4671 (1974)

RC≡CH

$$\underset{\text{or Na}_2\text{S}}{\overset{\overset{\displaystyle NO_2}{|}}{RC}=CHSO_2Ph} / n\text{-Bu}_3\text{SnH}$$

TL *27* 1595 (1986)

N≡N

RO$_2$CN=NCO$_2$R /base/[O]

TL 2433 (1974)
JACS *97* 918 (1975)

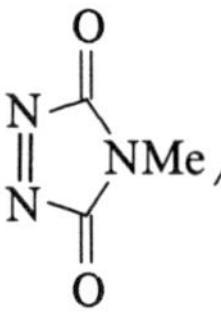

NMe /KOH/CuCl$_2$

JACS *94* 3658 (1972)

9. CYCLIZATION AND RELATED REACTIONS

For cyclizations that annulate an alkene-containing moiety onto the carbon-carbon double bond of an enone see also page 653, Section 6.

1. Three-Membered Rings

$$RC{\equiv}CR + N_2CHCO_2R \xrightarrow{\text{cat } Rh_2(O_2CR)_4} \underset{R \qquad R}{\overset{CO_2R}{\triangle}}$$

TL 1239 (1978)

2. Four-Membered Rings

Thermal reactions

JCS 3880 (1955)
JACS *84* 4600 (1962); *86* 1676 (1964); *87* 3996 (1965); *88* 1073, 4800 (1966); *89* 112 (1967); *90* 3582, 5310 (1968); *92* 399 (1970)
JOC *30* 3524 (1965)
TL 3719, 3723 (1969)
Tetr *25* 4375 (1969)
CC 1583 (1970)

Photochemical reactions

JOC *27* 1910 (1962); *52* 2644 (1987)
JACS *84* 1220 (1962); *88* 2742 (1966); *94* 4794 (1972); *97* 2272 (1975)

Proc Chem Soc 334 (1962)
Chem Weekb *19* 381 (1964); *21* 1001 (1965)
Rev Pure Appl Chem *16* 117 (1966)
Fortschr Chem Forsch *7* 445 (1967)
Adv Alicyclic Chem *2* 185 (1968)
Intra-Science Chem Reports, Vol 2 (1968)

$$\| + \||| \quad \xrightarrow{h\nu} \quad \square$$

Ber *97* 2942 (1964); *102* 3974 (1969)
TL 3409 (1966); *27* 2703 (1986) (intramolecular); *28* 267 (1987)
JOC *49* 832 (1984) (intramolecular); *51* 5232 (1986) (intramolecular)

$$\text{(cyclohexene)} + HC\equiv CCO_2Et \xrightarrow{AlCl_3} \text{(bicyclic)} CO_2Et$$

JOC *41* 3061 (1976)

$$\text{(cyclohexene)} \xrightarrow[MeC\equiv CCO_2Me]{cat\ CpFe(CO)_2BF_4(H_2C=CMe_2)} \text{(bicyclic)} CO_2Me$$

Organomet *1* 397 (1982)

$$\text{(cyclohexene)} + \text{(cyclobutylidene)} \overset{Br}{\underset{Br}{}} \xrightarrow{PhLi} \text{(tricyclic)}$$

TL *23* 1661 (1982)

3. Five-Membered Rings

$$RCH=CHR + \underset{\triangle}{\overset{R'\quad R'}{C}} \xrightarrow{Ni\ or\ Pd\ catalyst} \overset{R'\quad R'}{\underset{R\quad R}{\text{(cyclopentane)}}}$$

R′ = H, Me, Ph

Angew Int *16* 249 (1977); *21* 622 (1982); *25* 1 (1986) (review)
Ber *113* 3334 (1980); *114* 3313 (1981); *116* 2920 (1983)
JOMC *221* C33 (1981)
TL *26* 1045 (1985)

$$RCH=CR'X \longrightarrow \underset{R\quad X}{\text{(methylenecyclopentane)}} R'$$

X = electron-withdrawing group

$$\underset{AcOCH_2\overset{\|}{C}CH_2SiMe_3,\ Pd\ catalyst}{\overset{CH_2}{}}$$

JACS *101* 6429, 6432 (1979); *102* 6359 (1980);
103 5972, 5974 (1981); *104* 3733, 6668 (1982);
105 2315, 2326 (1983); *107* 721, 1075, 1293
(1985); *108* 284, 6051 (1986)

Organomet *1* 1543 (1982)
TL *27* 1445, 4137 (1986); *28* 4547 (1987)
Angew Int *25* 1 (1986) (review)

$MsOCH_2C(=CH_2)CH_2SiMe_3$, cat $Ni[P(OEt)_3]_4$ CC 1201 (1986)

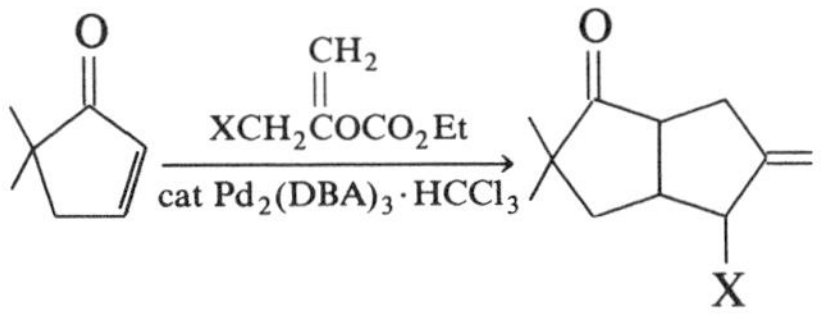
O
CH2
XCH2COCO2Et
cat Pd2(DBA)3·HCCl3
O
X

X	
$ArSO_2$	TL *25* 5183 (1984); *28* 6053 (1987)
CN	TL *25* 5183 (1984)

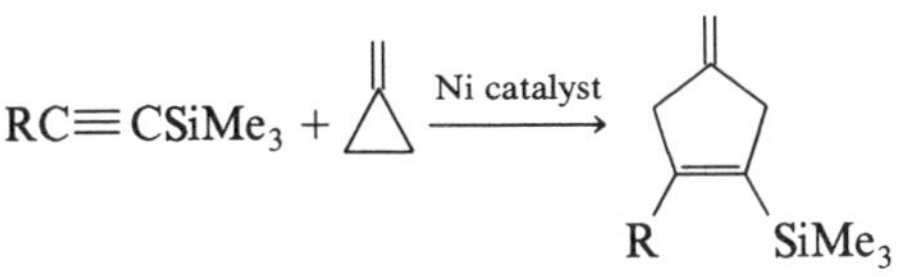
RC≡CSiMe3 +
Ni catalyst
R
SiMe3

Angew Int *24* 316 (1985)

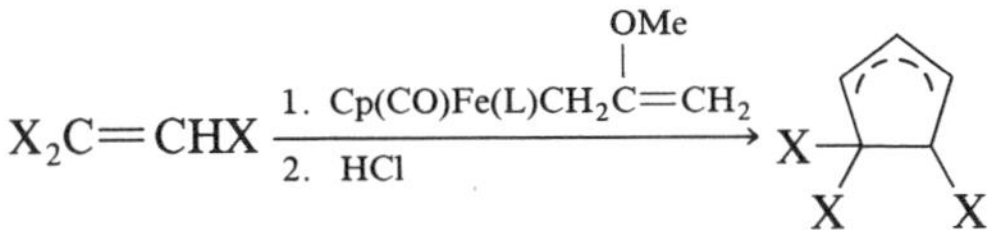
OMe
X2C=CHX
1. Cp(CO)Fe(L)CH2C=CH2
2. HCl
X
X
X

X = Ph, CO_2R, CN

JOMC *243* 451 (1983)

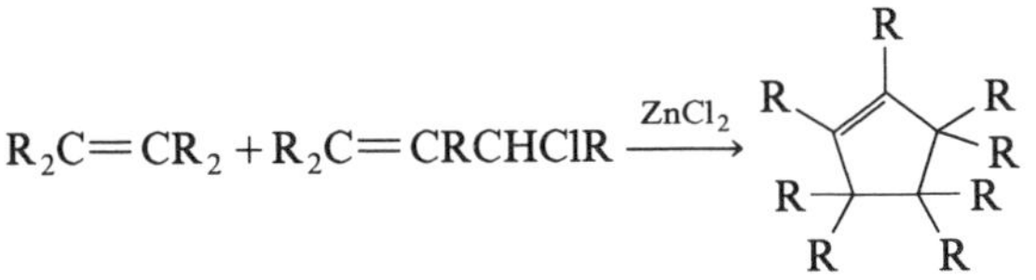
R2C=CR2 + R2C=CRCHClR
ZnCl2
R

Angew Int *20* 1027 (1981)

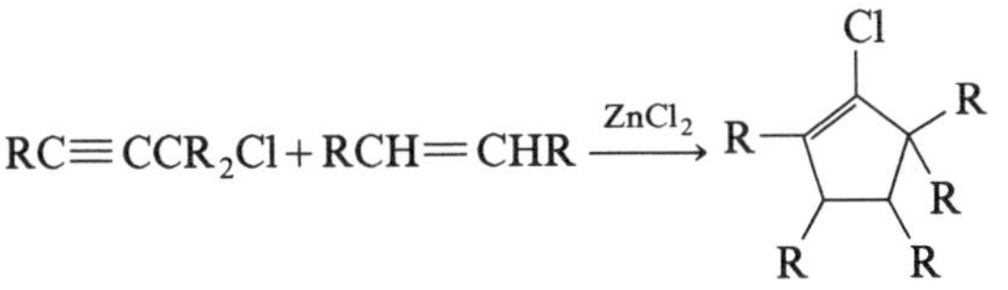
RC≡CCR2Cl + RCH=CHR
ZnCl2
Cl
R

JOC *46* 1041 (1981)

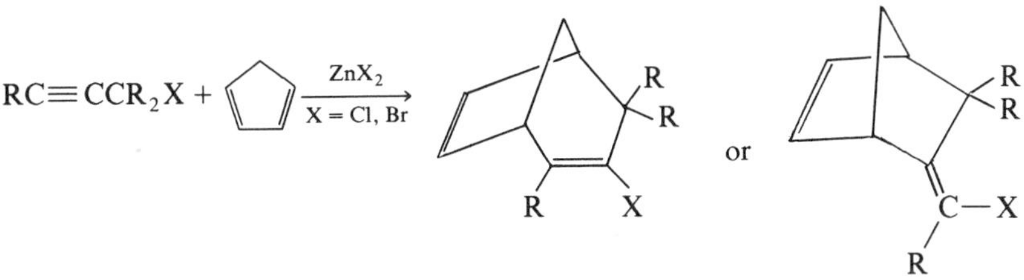
RC≡CCR2X +
ZnX2
X = Cl, Br
R
X
or
C—X

Angew Int *19* 814 (1980)
Ber *115* 3479 (1982)

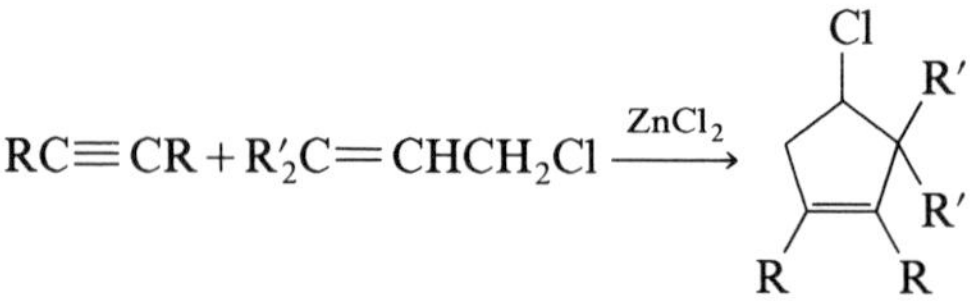
RC≡CR + R′2C=CHCH2Cl
ZnCl2
Cl
R′
R′
R
R

TL *21* 577 (1980); *28* 235, 689 (1987)

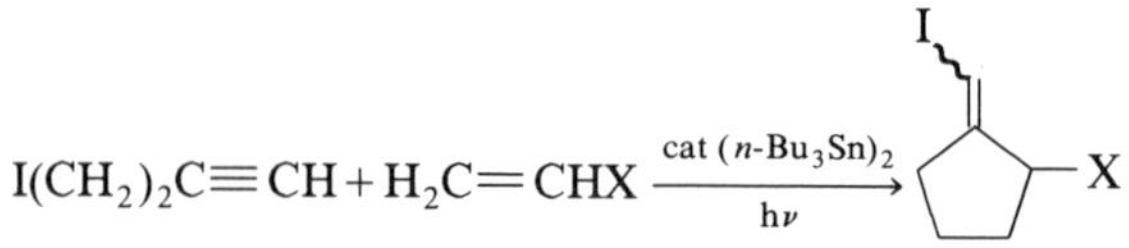
I(CH2)2C≡CH + H2C=CHX
cat (n-Bu3Sn)2
hν
I
X

X = CN, CO_2R, SO_2Ph, COR, CHO

JACS *109* 6558 (1987)

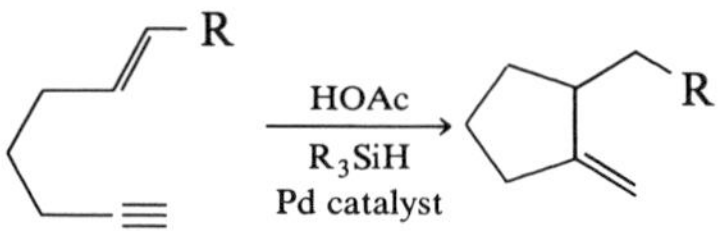
R
HOAc
R3SiH
Pd catalyst
R

JACS *109* 3061 (1987)

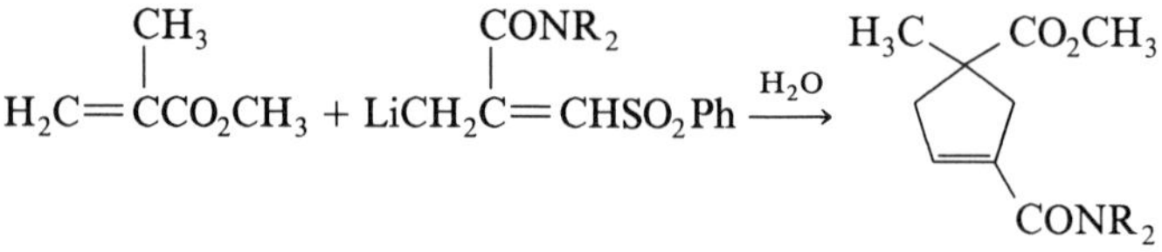
CH3
H2C=CCO2CH3 + LiCH2C=CHSO2Ph
CONR2
H2O
H3C
CO2CH3
CONR2

TL *27* 5911 (1986)

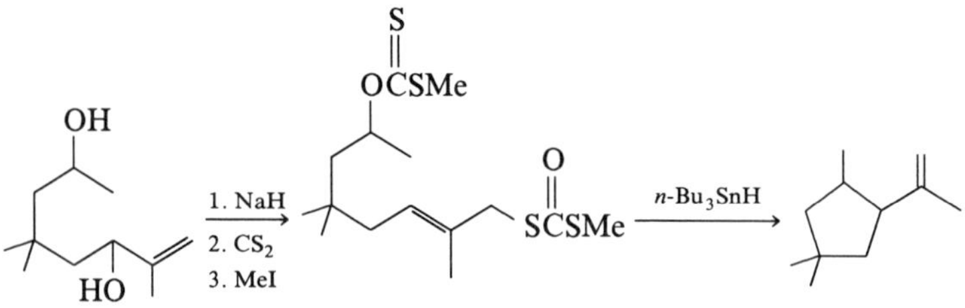
OH
HO
1. NaH
2. CS2
3. MeI
S
OCSMe
O
SCSMe
n-Bu3SnH

TL *28* 5973 (1987)

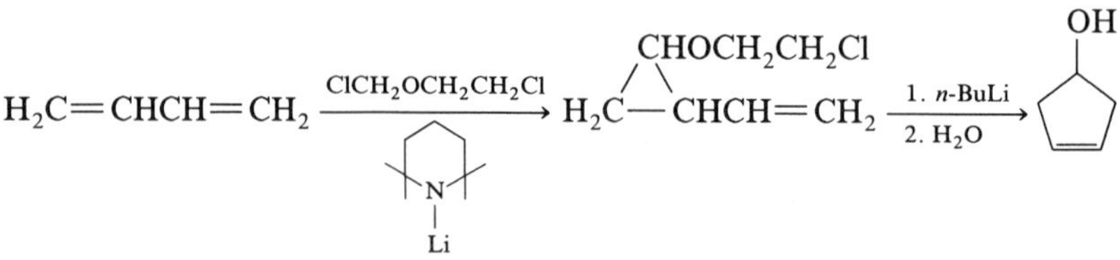
H2C=CHCH=CH2
ClCH2OCH2CH2Cl
N
Li
CHOCH2CH2Cl
H2C—CHCH=CH2
1. n-BuLi
2. H2O
OH

JOC *45* 1340 (1980)
JACS *103* 2443 (1981)

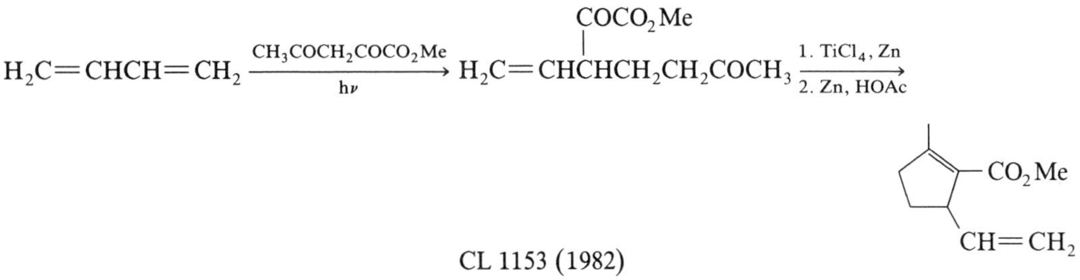

$$H_2C=CHCH=CH_2 \xrightarrow[h\nu]{CH_3COCH_2COCO_2Me} \overset{\overset{\textstyle COCO_2Me}{|}}{H_2C=CHCHCH_2CH_2COCH_3} \xrightarrow[\text{2. Zn, HOAc}]{\text{1. TiCl}_4\text{, Zn}}$$

CL 1153 (1982)

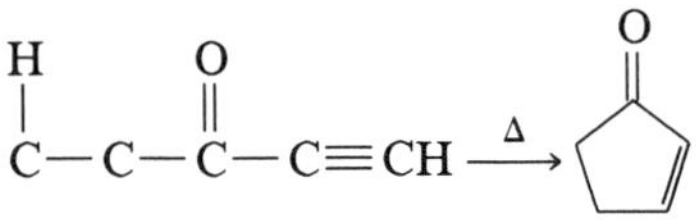

$$\overset{\overset{\textstyle H}{|}}{C}-C-\overset{\overset{\textstyle O}{\|}}{C}-C\equiv CH \xrightarrow{\Delta}$$

Helv 62 852 (1979); 65 13, 2413, 2517 (1982)

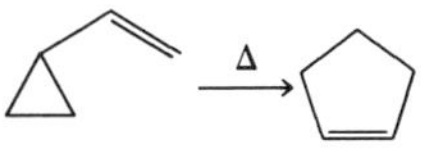

$$\xrightarrow{\Delta}$$

Org Rxs 33 247 (1985) (review)
JOC 52 4641 (1987)

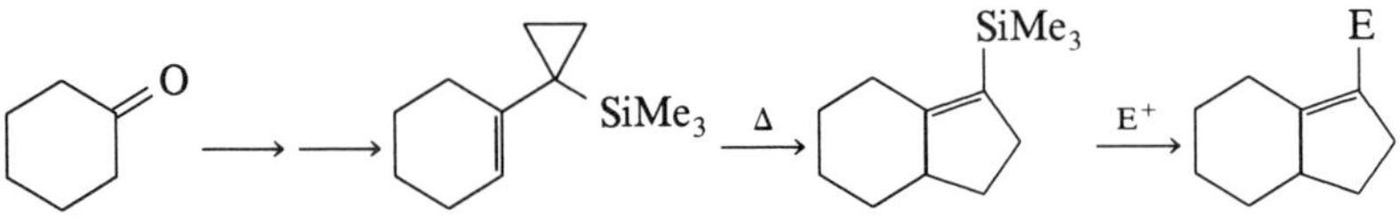

$$\xrightarrow{} \xrightarrow{} \quad SiMe_3 \xrightarrow{\Delta} \quad SiMe_3 \xrightarrow{E^+} \quad E$$

$$E^+ = Br_2 \text{ or RCOCl-AlCl}_3$$

TL 23 263 (1982)

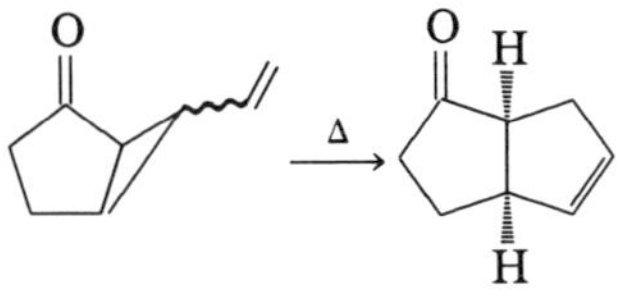

$$\xrightarrow{\Delta}$$

JOC 40 2265 (1975); 45 5020 (1980)
TL 27 2885 (1986)

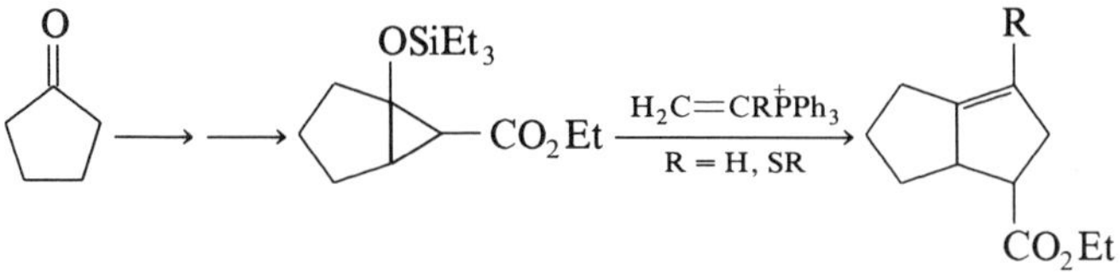

$$\xrightarrow{} \xrightarrow{} \quad \overset{OSiEt_3}{\underset{CO_2Et}{}} \xrightarrow[R = H, SR]{H_2C=CR\overset{+}{P}Ph_3} \quad \overset{R}{\underset{CO_2Et}{}}$$

JACS 107 734 (1985)
JOC 52 1, 4139 (1987)

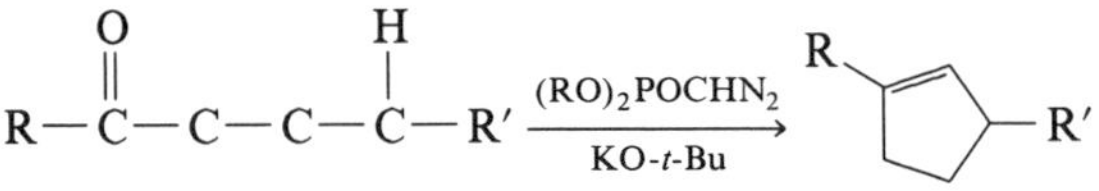

$$R-\overset{\overset{\textstyle O}{\|}}{C}-C-\overset{\overset{\textstyle H}{|}}{C}-R' \xrightarrow[\text{KO-}t\text{-Bu}]{(RO)_2POCHN_2} \quad \overset{R}{\underset{}{}}-R'$$

JOC 48 5251 (1983); 50 2557 (1985)

$$\underset{\text{RC—CR}}{\overset{\text{O O}}{\overset{\| \ \|}{}}} \xrightarrow[\text{SnF}_2]{\text{ICH}_2\text{CCH}_2\text{SiMe}_3}\ \text{HO}\underset{\text{R R}}{\boxed{}}\text{OH}$$

JACS *108* 4683 (1986)

$$\text{Cl—C—}\overset{\overset{\text{C}}{\|}}{\text{C}}\text{—C—X—C}_n\text{—C}{=}\text{C}\xrightarrow[\substack{\text{2. ZnX}_2 \\ \text{3. E}^+}]{\text{1. Mg}}\ \ \boxed{}\ \xrightarrow[\text{X = O, E = ZnX}]{\text{cat Pd(PPh}_3)_4}\xrightarrow{\text{H}^+}\ \boxed{}\text{C}_n\text{—OH}$$

X = O, NMe; *n* = 1, 2; E⁺ = NH₄Cl, Me₃SnCl

TL *28* 5929 (1987)

4. Five-, Six- and Seven-Membered Rings

$$\text{RC}{\equiv}\text{C—C}_n\text{—X}\longrightarrow\text{RCH}{=}\text{C}\boxed{}\text{C}_n\ \ \text{or}\ \ \overset{\text{H}}{\underset{\text{R}}{}}\boxed{}\text{C}_n$$

See page 215, Section 2.2.

$$\underset{\text{C}=\text{C—C}_n\text{—C}{=}\text{C}}{\overset{\overset{\text{X}}{|}}{}}\xrightarrow{n\text{-Bu}_3\text{SnH}}\ \boxed{}_n\ \ \text{or}\ \ \boxed{}_n\text{C}$$

X = Br, I   Ring size = 5–7

JACS *104* 2321 (1982)
TL *24* 1871 (1983); *26* 957 (1985) (pyrrolidines); *26* 5927 (1985); *27* 1355 (cyclic alcohols from enol silanes), 4525 (kinetics, mechanism), 4529 (1986)
CC 1445 (1983); 78 (1986) (lactams)
CL 1437 (1984) (3-methylene tetrahydrofurans)

5. Six-Membered Rings

$$\underset{\text{MeO}\quad\text{OMe}}{\overset{\text{SiMe}_3}{\boxed{}}}\xrightarrow{\text{SnCl}_4}\underset{\text{OMe}}{\boxed{}}$$

JCS Perkin I 251 (1981)

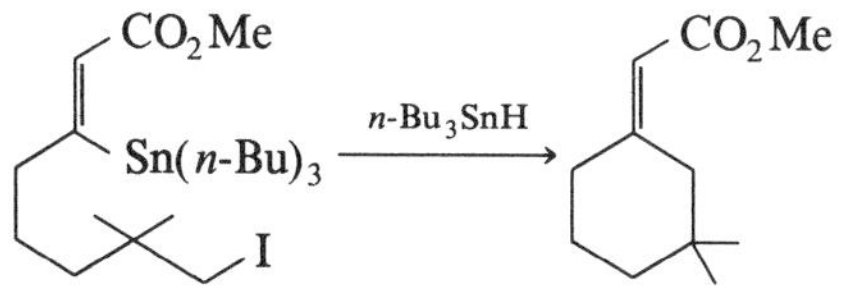

TL *28* 2941 (1987)

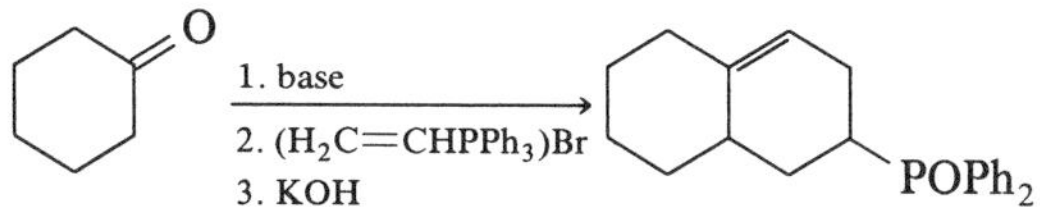

JACS *107* 1424 (1985)

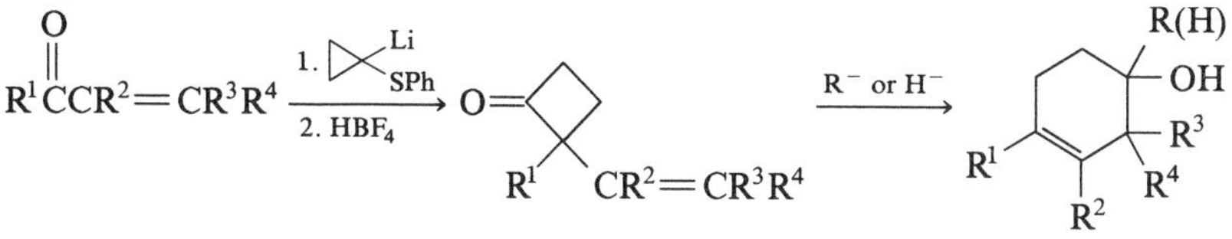

Tetr *37* 3943 (1981)

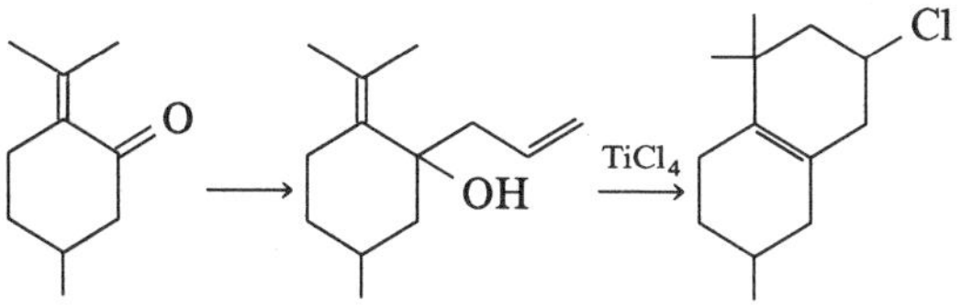

TL *28* 2583 (1987)

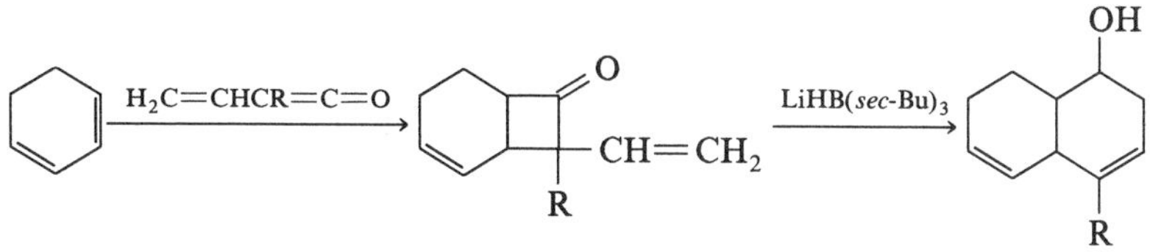

Tetr *37* 3943 (1981)

6. Seven-Membered Rings

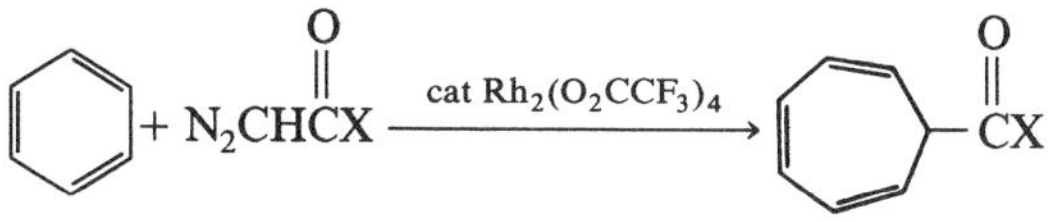

X	
R	CC 491 (1985)
OR	CC 765 (1980)
	JOC *46* 873 (1981)

X

$$CH_2$$
$$+ AcOCH_2\overset{\|}{C}CH_2SiMe_3 \xrightarrow{\text{Pd catalyst}}$$

X = CO_2R, SO_2Ph

JACS *109* 3483 (1987)

7. Seven- and Eight-Membered Rings

$$CH_3\overset{O}{\underset{\|}{C}}(CH_2)_n\overset{O}{\underset{\|}{C}}CH_3 \xrightarrow[\text{SnF}_2]{Me_3SiCH_2\overset{CH_2}{\underset{\|}{C}}CH_2I} \quad CH_3-\underset{(CH_2)_n}{\overset{O}{\bigcirc}}-CH_3$$

n = 2, 3

JACS *109* 6877 (1987)

8. Eight-Membered Rings

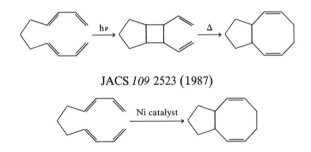

JACS *109* 2523 (1987)

See page 232, Section 2.15.

10. MISCELLANEOUS REACTIONS

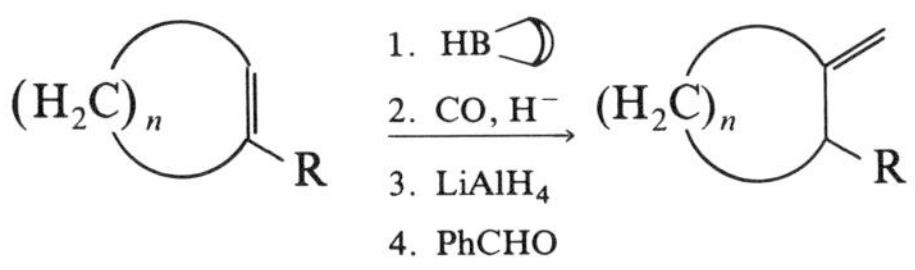

$$(H_2C)_n \underset{R}{\diagdown} \quad \begin{array}{l} \text{1. HB} \\ \xrightarrow{\text{2. CO, H}^-} \\ \text{3. LiAlH}_4 \\ \text{4. PhCHO} \end{array} \quad (H_2C)_n \underset{R}{\diagdown}$$

JOC *46* 647 (1981)

$$RCH{=}CH_2 + CH_3OR' \xrightarrow{\text{cat IrH}_5(i\text{-Pr}_3P)_2} RCH_2CH{=}CHOR'$$

TL *28* 3249 (1987)

$$\underset{RCCHR_2}{\overset{O}{\|}} \xrightarrow[\text{2. KO-}t\text{-Bu}]{\text{1. LiCH}_2\text{SOPh}} \underset{RC-CR_2}{\overset{H_2C \quad OH}{\| \quad |}}$$

Syn 640 (1980)
Ber *113* 819, 831, 845, 856 (1980)

$$R_2CHSO_2Ar \xrightarrow[\text{3. }n\text{-Bu}_3\text{SnCH}_2\text{I}]{\substack{\text{1. }n\text{-BuLi} \\ \text{2. 0.2 }i\text{-Pr}_2\text{NH}}} \underset{R_2CCH_2SnBu_3}{\overset{SO_2Ar}{|}} \xrightarrow{n\text{-Bu}_4\text{NF}\cdot 3\text{H}_2\text{O}} R_2C{=}CH_2$$

JOC *50* 3622 (1985)

$$R_2CHCN \xrightarrow[\text{2. }n\text{-Bu}_3\text{SnCH}_2\text{I}]{\text{1. LiNR}_2} \underset{R_2CCH_2SnBu_3}{\overset{CN}{|}} \xrightarrow{\text{MeLi}} R_2C{=}CH_2$$

JOC *50* 3625 (1985)

$$H_2C{=}CHX \xrightarrow[\text{DABCO}]{E^+} \underset{H_2C{=}CX}{\overset{E}{|}}$$

X	E⁺	
CHO	RCHO	TL *27* 2031, 4307 (1986)
COR	H₂C=CHCOR (1,4-addition)	TL *28* 4591 (1987)

281

$$H_2C{=}CHX \xrightarrow[\text{DABCO}]{E^+} H_2C{=}\overset{\overset{\displaystyle E}{|}}{C}X \quad (\textit{Continued})$$

$\underline{X}$	$\underline{E}^+$	
CO_2R	RCHO	JOC *50* 3849 (1985)
CO_2R, CN	$RCOCO_2R$	TL *28* 4351 (1987)
CN	RCHO	Syn Commun *17* 587 (1987)
	$H_2C{=}CHCN$ (1,4-addition)	TL *28* 4591 (1987)

$$RCH{=}CH\overset{\overset{\displaystyle O}{||}}{C}R \xrightarrow[\substack{2.\ R'CHO \\ 3.\ H_2O_2}]{1.\ R_2BSePh} RCH{=}\overset{\overset{\displaystyle O}{||}}{\underset{\underset{\displaystyle R'CHOH}{|}}{C}}CR$$

JOC *50* 730 (1985)

ALKYNES

GENERAL REFERENCES

A. W. Johnson, "The Chemistry of Acetylenic Compounds," E. Arnold and Co., London (1946)

Org Rxs 5 1 (1949)

R. A. Raphael, "Acetylenic Compounds in Organic Synthesis," Butterworths Scientific Publications, London (1955)

Angew 72 391 (1960)

T. F. Rutledge, "Acetylenic Compounds," Reinhold Book Corp., New York (1968)

"Chemistry of Acetylenes," Ed. H. G. Viehe, Marcel Dekker, New York (1969)

L. Brandsma, "Preparative Acetylenic Chemistry," Elsevier, Amsterdam (1971)

Houben-Weyl, "Methoden der Organischen Chemie," 4th ed, Vol V/2a, G. Thieme, Stuttgart (1977)

"The Chemistry of the Carbon-Carbon Triple bond," Ed. S. Patai, J. Wiley, New York (1978)

J Chem Res (S) 106 (1978); 190 (1979); 270 (1981)

L. Brandsma and H. D. Verkruijsse, "Synthesis of Acetylenes, Allenes and Cumulenes," Elsevier, New York (1981)

1. ISOMERIZATION

$$RC\equiv C(CH_2)_nCH_3 \longrightarrow R(CH_2)_{n+1}C\equiv CH$$

Review: Org Rxs 5 1 (1949)

NaNH$_2$	BSCF [4] *35* 481 (1924)
	Ann Chim [10] *3* 191, 325 (1925)
	JOC *34* 222 (1969)
Na(K)NH$_2$, H$_2$N(CH$_2$)$_3$NH$_2$	Rec Trav Chim *96* 160 (1977)
LiNHCH$_2$CH$_2$NH$_2$, KO-*t*-Bu	Can J Chem *62* 1333 (1984)
NaNHCH$_2$CH$_2$NH$_2$	JOC *45* 734 (1980)
	Can J Chem *60* 1238 (1982)
LiNH(CH$_2$)$_3$NH$_2$, KO-*t*-Bu	Can J Chem *62* 1333 (1984)
	JOC *52* 1835 (1987)
NaNH(CH$_2$)$_3$NH$_2$	Can J Chem *58* 2567 (1980)
KNH(CH$_2$)$_3$NH$_2$	JACS *97* 891 (1975); *109* 2138 (1987)
	TL 2565 (1976); 411 (1977); *22* 4171 (1981); *28* 3857 (1987)
	CC 959 (1976); 318 (1977)
KNH(CH$_2$)$_3$NH$_2$, NH$_3$	TL *28* 2409 (1987)

$$HOCH_2C\equiv CCH_2R \xrightarrow{NaNH(CH_2)_3NH_2} HOCH_2CH_2C\equiv CR$$

Can J Chem *58* 2567 (1980)

$$RCH_2C\equiv CH \longrightarrow RC\equiv CCH_3$$

KOH, EtOH	Org Rxs *5* 1 (1949) (review)
	JACS *73* 1273 (1951)
	Quart Rev *24* 585 (1970)
KOEt, EtOH	BSCF 805 (1962)
NaNH$_2$, DMSO	Tetr *26* 2127 (1970)

$$RCH=C=CH(CH_2)_nCH_3 \xrightarrow{LiNH(CH_2)_3NH_2} R(CH_2)_{n+2}C\equiv CH$$

JOC *52* 1835 (1987)

2. ELIMINATION REACTIONS

Review: Angew Int *4* 49 (1965)

$$RCH = C(CH_3)_2 \xrightarrow[\text{aq HOAc}]{\text{NaNO}_2} RC \equiv CCH_3$$

TL *27* 267 (1986)
JOC *51* 2687 (1986)

$$RCH = CH_2 + 4 Li \xrightarrow{\text{THF}} RC \equiv CLi + 3LiH \xrightarrow{\text{H}_2\text{O}} RC \equiv CH$$

JOC *32* 105 (1967)

$$RCH = CHR' \xrightarrow{\text{Br}_2} RCH \overset{\overset{\displaystyle Br}{|}}{-} CHR' \overset{\overset{\displaystyle Br}{|}}{} \xrightarrow{\text{base}} RC \equiv CR'$$

Base

KF-alumina (R = R' = Ph)	BCSJ *56* 1885 (1983)
NaOH, (n-Bu$_4$N)HSO$_4$	TL 4723 (1976)
NaOH or KOH, HO(CH$_2$CH$_2$O)$_n$H	JOC *47* 2493 (1982)
KOH, ROH	Org Syn *27* 76 (1947)
	Org Syn Coll Vol *3* 350 (1955)
	Ber *89* 1786 (1956)
	JACS *85* 3492 (1963)
	J Chem Res (S) 270 (1981)
KOH, phase transfer	Tetr *37* 1653 (1981)
KO-*t*-Bu	Helv *54* 2060 (1971)
	JOC *51* 1088 (1986)
	JACS *109* 1868 (1987)
	TL *28* 6359 (1987)
KO-*t*-Bu, 18-crown-6	Ann 1 (1980)
Na, NH$_3$	JACS *56* 2064, 2120 (1934)
	Org Rxs *5* 1 (1949)
	Rec Trav Chim *85* 123 (1960)
	Org Syn Coll Vol *4* 763 (1963)
	TL 41 (1970)
	J Chem Res (S) 106 (1978)
NaNH$_2$, DMSO	Tetr *26* 2127 (1970)

$$ArCH_2Br \xrightarrow[\text{(PhCH}_2\text{NEt}_3)\text{Cl}]{\text{NaOH, HCBr}_3} ArC\equiv CBr$$

CC 563 (1979)

$$p\text{-}XC_6H_4CHBr_2 \xrightarrow[\text{KO-}t\text{-Bu}]{\text{KOH or}} p\text{-}XC_6H_4C\equiv CC_6H_4X\text{-}p$$

JOC USSR *15* 579 (1979)

$$2°, 3° \; RCl + H_2C=CHCl \xrightarrow{\text{AlCl}_3} RCH_2CHCl_2 \xrightarrow{\text{NaNH}_2} RC\equiv CH$$

Rec Trav Chim *84* 31 (1965)
J Chem Res (S) 106 (1978)

$$RCH_2X \xrightarrow{\text{LiCHCl}_2} RCH_2CHCl_2 \xrightarrow[\text{2. H}^+ \text{ or R'X}]{\text{1. 3 }n\text{-BuLi}} RC\equiv CH(R')$$

Syn 502 (1979)

$$\underset{\text{O}}{\overset{\text{O}}{\underset{\|}{R}}}CCH_2R' \xrightarrow{\text{PCl}_5} \underset{\text{Cl}}{\overset{\text{Cl}}{R}}CCH_2R' \; (\text{or } RCCl=CHR') \xrightarrow{\text{base}} RC\equiv CR'$$

Review: Angew *72* 391 (1960)

Base

Base	Reference
KOH	BSCF *35* 357 (1924)
	Bull Soc Chim Belg *34* 427 (1925)
	Ann Chim (10) *16* 421 (1931)
	JACS *64* 543 (1942)
	J Chem Res (S) 270 (1981)
KO-*t*-Bu, 18-crown-6	Ann 1 (1980)
NaNH$_2$, mineral oil	JACS *56* 1207 (1934)
NaNH$_2$, NH$_3$	Org Rxs *5* 1 (1949)
	Rec Trav Chim *84* 31 (1965)
	J Chem Res (S) 106 (1978)
	JACS *106* 5312 (1984)
NaNH$_2$, DMSO	Tetr *25* 4249 (1969)
n-BuLi	Syn 502 (1979)

$$RMgX\,(Li) \xrightarrow[\text{2. E}^+]{\text{1. H}_2\text{C}=\text{CF}_2} RC\equiv CE$$

R = 1°, 2°, 3° alkyl; allyl; aryl; vinyl

E$^+$ = H$_2$O, *n*-BuBr, CH$_3$CHO, Me$_3$SiCl

TL *23* 4325 (1982)

$$RMgX \text{ (Li)} \xrightarrow{Cl_2C=CF_2} RCF=CCl_2 \xrightarrow[\text{2. } E^+]{\text{1. 2 } n\text{-BuLi}} RC\equiv CE$$

$$E^+ = H_2O, CO_2$$

JOC *41* 1487 (1976)

$$\overset{\overset{\displaystyle X}{\displaystyle |}}{RC}=CHR' \xrightarrow{\text{base}} RC\equiv CR'$$

Review: Angew *72* 391 (1960)

X	Base	
Cl	NaOH, (n-Bu$_4$N)HSO$_4$	TL 4723 (1976)
	KOH	JACS *63* 1175 (1941)
	NaOMe, MeOH	JOC *26* 2619 (1961)
	NaOEt	Ann *308* 264 (1899)
	KO-t-Bu	JCS Perkin II 795 (1982)
	NaNH$_2$, toluene	JOC *34* 222 (1969)
Cl, Br	KF or Et$_4$NF	JCS Perkin I 340 (1974)
	(PhCH$_2$NMe$_3$)OH	BSCF II 125 (1976)
	KO-t-Bu, 18-crown-6	Ann 1 (1980)
	NaNH$_2$, mineral oil	JACS *56* 1207 (1934)
	NaNH$_2$, NH$_3$	Ann Chim [10] *3* 325 (1925);
		7 785 (1962)
		JACS *56* 2064, 2120 (1934)
		Org Syn Coll Vol *1* 185 (1941)
		Org Rxs *5* 1 (1949)
		Org Syn *65* 68 (1987)
Br	K$_2$CO$_3$, EtCOMe	JOC *47* 2484 (1982)
	(R = CH$_2$OR)	
	NaOH, EtOH	JACS *76* 4558 (1954)
		J Chem Res (S) 270 (1981)
	KOH	Org Syn Coll Vol *1* 438 (1941)
	KO-t-Bu	Angew Int *4* 953 (1965)
	KO-t-Bu, t-BuOH	Ann *707* 66 (1967)
	DBU	JOC *47* 2484 (1982)
	NaNH$_2$, DMSO	Tetr *26* 2127 (1970)
	NaH or NaNH$_2$, HMPA	BSCF 1293 (1966)
	LDA	JOC *52* 5622 (1987)
I	KOH, EtOH	JOC *34* 3502 (1969)
	KO-t-Bu, t-BuOH	JOC *34* 3502 (1969)
O$_3$SCF$_3$	2,6-(t-Bu)$_2$C$_6$H$_3$OK or LDA	CC 459 (1981)
O$_3$SR, O$\overset{+}{P}$Ar$_3$, (RO)$_2$PO$_2$	—	Angew Int *8* 429 (1969)

$$\overset{\overset{\displaystyle Cl}{\displaystyle |}}{RCH}=CCH_3 \xrightarrow{KNH(CH_2)_3NH_2} RCH_2C\equiv CH$$

TL *28* 4547 (1987)

$$RC = CH_2 \xrightarrow[\text{2. } R_2CO]{\text{1. 2 LDA}} RC \equiv CCR_2$$
(Cl on left structure, OH on right structure)

TL *28* 5793 (1987)

$$Ar_2C = CHBr \xrightarrow{\text{base}} ArC \equiv CAr$$

Ann *279* 319, 324, 337 (1894)
Angew Int *4* 49 (1965) (review)

$$RC = CHCl \xrightarrow{n\text{-BuLi}} RC \equiv CH$$
(Cl on left structure)

Org Syn *64* 73 (1985)

$$H_2C = CHSeAr \xrightarrow[\text{2. } E^+]{\text{1. LDA}} H_2C = CSeAr \xrightarrow[\text{2. DABCO, } \Delta]{\text{1. } m\text{-ClC}_6H_4CO_3H} HC \equiv CE$$
(E on middle structure)

$$E^+ = RX, R_2CO$$

JACS *102* 5967 (1980)

$$RCH_2CHO \xrightarrow[\text{2. } CBr_4, PPh_3]{\text{1. } n\text{-Bu}_3SnLi} RCH_2CHSnBu_3 \xrightarrow[\text{2. } Pb(OAc)_4]{\text{1. DBU}} RC \equiv CH$$
(Br on middle structure)

TL *23* 4607 (1982)

$$RCCH_3 \longrightarrow RCCH_2Br \xrightarrow{P(OEt)_3} RC = CH_2 \xrightarrow[NH_3]{NaNH_2} RC \equiv CH$$
(O double bonds on first two structures, OPO(OEt)$_2$ on third structure)

JCS 3712 (1963)

$$RCCH_3 \xrightarrow[\text{2. ClPO(OEt)}_2]{\text{1. LiNR}_2} RC = CH_2 \xrightarrow[\text{2. } E^+]{\text{1. 2.5LiNR}_2} RC \equiv CE$$
(O double bond on first structure, OPO(OEt)$_2$ on second structure)

$$E^+ = H_2O, CH_2O, ClCO_2Me$$

JOC *45* 2526 (1980); *52* 398, 4885 (1987)
Org Syn *64* 44 (1985)
JACS *107* 2568 (1985); *108* 3835 (1986)
TL *28* 5793 (1987)

$$R_3CCCH_3 \xrightarrow[\Delta]{MeMgX} R_3CC \equiv CH$$
(O double bond on first structure)

JCS Perkin I 1079 (1982)

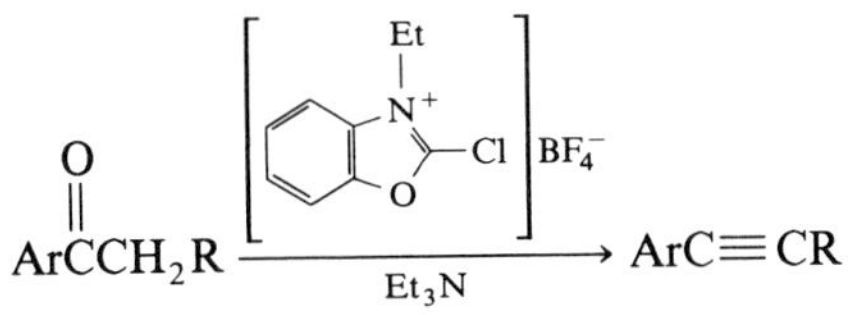

$$\underset{\text{Et}_3\text{N}}{\overset{\left[\begin{array}{c}\text{Et}\\ \text{N}^+\\ \text{C}-\text{Cl}\\ \text{O}\end{array}\right]\text{BF}_4^-}{\longrightarrow}}$$

ArCCH$_2$R $\xrightarrow[\text{Et}_3\text{N}]{}$ ArC≡CR

CL 481 (1979)

RCCH$_2$CR′ ⟶ RC≡CCR′

Et$_2$NCF$_2$CHFX (X = Cl, CF$_3$)
  (R′ = R or OR)

CL 1327 (1980)

H$_2$NNH$_2$/Br$_2$/NaOH/HCl (R′ = OR)

Syn 72 (1981)

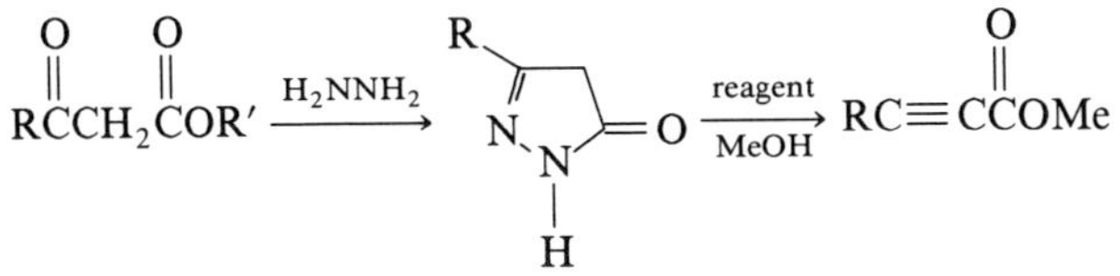

RCCH$_2$COR′ $\xrightarrow{\text{H}_2\text{NNH}_2}$ (pyrazolone) $\xrightarrow[\text{MeOH}]{\text{reagent}}$ RC≡CCOMe

Reagent

Tl(NO$_3$)$_3$ — Angew Int *11* 48 (1972)

Pb(OAc)$_4$ — Syn 1100 (1982)

R—C—C—R ⟶ R—C—C—R $\xrightarrow{[0]}$ RC≡CR

HgO — JACS *74* 3636, 3643 (1952)
Helv *35* 1598 (1952)
Org Syn Coll Vol *4* 377 (1963)

Pb(OAc)$_4$ — TL 4511 (1968)

AgO$_2$CCF$_3$, Et$_3$N — JOC *23* 665 (1958)

(EtO)$_3$P, Δ — JOC *29* 2243 (1964)

O$_2$, CuCl, py — TL 4573 (1973)

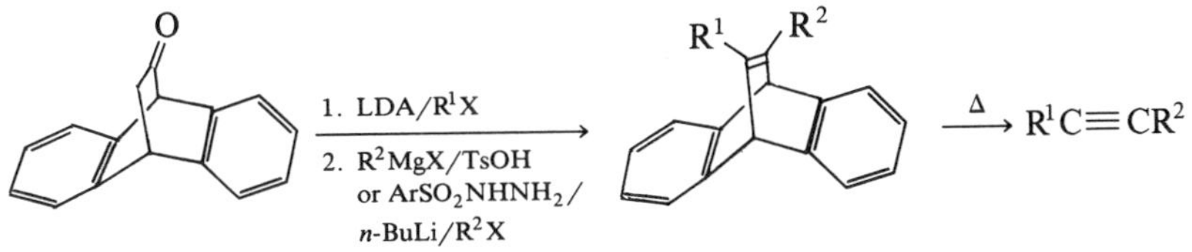

$\xrightarrow[\substack{\text{2. R}^2\text{MgX/TsOH}\\ \text{or ArSO}_2\text{NHNH}_2/\\ n\text{-BuLi/R}^2\text{X}}]{\text{1. LDA/R}^1\text{X}}$ $\xrightarrow{\Delta}$ R^{1}C≡CR^2

CL 1241 (1981)

R$_3$CCO$_2$Me $\xrightarrow[\Delta]{\text{excess MeMgX}}$ R$_3$CC≡CH

JCS Perkin I 1079 (1982)

$$\underset{\text{RCHSO}_2\text{Ph}}{\overset{\text{Li}}{|}} + \underset{\text{HCR}'}{\overset{\text{O}}{\|}} \xrightarrow{\text{Ac}_2\text{O}} \underset{\underset{\text{OAc}}{|}}{\overset{\overset{\text{PhSO}_2}{|}}{\text{RCHCHR}'}} \xrightarrow{\text{KO-}t\text{-Bu}} \text{RC}\equiv\text{CR}'$$

JACS *106* 3670 (1984)
JOC *51* 3830 (1986)

$$\underset{\text{RCOR}''}{\overset{\text{O}}{\|}} \xrightarrow{\text{R}'\bar{\text{C}}\text{HSO}_2\text{Ph}}$$

$$\underset{\text{RCH}}{\overset{\text{O}}{\|}} \xrightarrow[\text{2. [0]}]{\text{1. R}'\bar{\text{C}}\text{HSO}_2\text{Ph}}$$

$$\longrightarrow \underset{\text{RCCHR}'\text{SO}_2\text{Ph}}{\overset{\text{O}}{\|}}$$

$$\downarrow \begin{array}{l}\text{1. base} \\ \text{2. ClPOX}_2\end{array}$$

$$\underset{\text{RC}=\text{CR}'\text{SO}_2\text{Ph}}{\overset{\text{OPOX}_2}{|}} \xrightarrow[\text{or Na(Hg)}]{\text{Na, NH}_3} \text{RC}\equiv\text{CR}'$$

X = OEt, OPh, NMe$_2$

JACS *100* 4852 (1978)
TL 2625 (1978)
JCS Perkin I 2429 (1979)

$$\underset{\text{RCOLi}}{\overset{\text{O}}{\|}} \xrightarrow[\text{R}'=\text{H, Ph}]{\text{LiCHR}'\text{SMe}} \underset{\text{RC}-\text{CHR}'}{\overset{\text{O}\quad\text{SMe}}{\|\quad|}} \xrightarrow[\text{2. MeLi or } n\text{-BuLi}]{\text{1. TsNHNH}_2} \text{RC}\equiv\text{CR}'$$

JOC *43* 4366 (1978)

$$\text{RC}\equiv\text{CCO}_2\text{H} \xrightarrow{\Delta} \text{RC}\equiv\text{CH}$$

Helv *54* 2060 (1971)

3. WITTIG AND RELATED APPROACHES

$$RCHO \xrightarrow[\text{2 KO-}t\text{-Bu}]{(Ph_3\overset{+}{P}CH_2X)X^-} \xrightarrow{H_2O} RC\equiv CH$$

R = aryl > alkyl

X = Cl   CC 446 (1978)

X = Br   TL *21* 4021 (1980);

                   27 5853 (1986)

$$RCHO \xrightarrow{Ph_3P=CHCl} RCH=CHCl \xrightarrow[\text{2. E}^+]{\text{1. MeLi or } n\text{-BuLi}} RC\equiv CE$$

$\underline{E^+}$

H_2O JACS *91* 4318 (1969)

                       TL 1495 (1973)

$ClCO_2Me$ TL *28* 5473 (1987)

$$RCHO \xrightarrow{\text{ylid}} RCH=CX_2 \xrightarrow[\text{2. E}^+]{\text{1. base}} RC\equiv CE$$

$$E^+ = H_2O, CO_2, ClCO_2Me$$

$(Me_2N)_3P=CCl_2/n\text{-BuLi}/H_2O$ Gazz Chim Ital *110* 195 (1980)

Ph_3P, CBr_4, (Zn)/n-BuLi or Li(Hg)/   TL 3769 (1972); *27* 3059 (1986);
    H_2O or CO_2 or $ClCO_2Me$ *28* 1143, 2099 (1987)

                                 Gazz Chim Ital *110* 195 (1980)

                                 Tetr *37* 3873 (1981)

                                 Ber *115* 828 (1982)

                                 BCSJ *55* 2221 (1982)

                                 JACS *106* 1501 (1984); *108* 2776 (1986)

                                 JOC *52* 5243 (1987)

$$RCHO \xrightarrow[\text{PPh}_3]{CBr_4} RCH=CBr_2 \xrightarrow[\text{PEG-400}]{NaOH} RC\equiv CBr$$

JOC *51* 4354 (1986)

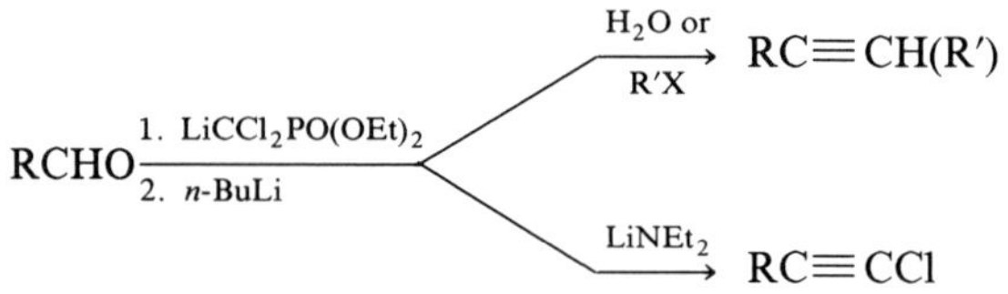

$$RCHO \xrightarrow[\text{2. } n\text{-BuLi}]{\text{1. } LiCCl_2PO(OEt)_2}$$

$$\xrightarrow[R'X]{H_2O \text{ or}} RC{\equiv}CH(R')$$

$$\xrightarrow{LiNEt_2} RC{\equiv}CCl$$

Syn 458 (1975)
JOC *42* 28 (1977)
Gazz Chim Ital *110* 195 (1980)
TL *27* 87 (1986)

$$RCHO \xrightarrow{Ph_3P{=}CF_2} RCH{=}CF_2 \xrightarrow{R'Li} RC{\equiv}CR'$$

CL 935 (1980)

$$RCHO \xrightarrow[KO\text{-}t\text{-Bu}]{N_2CHPO(OMe)_2} RC{\equiv}CH$$

$$ArCOMe \xrightarrow{\hspace{2cm}} ArC{\equiv}CCH_3$$

JOC *44* 4997 (1979); *47* 1837 (1982)

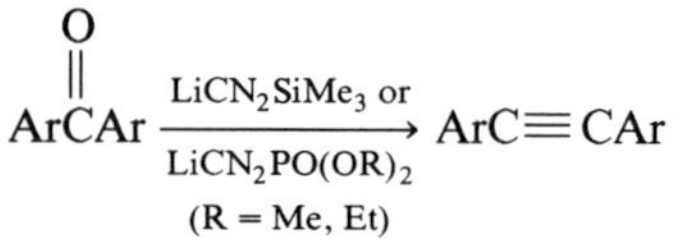

$$\underset{ArCAr}{\overset{\overset{\displaystyle O}{\|}}{}} \xrightarrow[LiCN_2PO(OR)_2]{LiCN_2SiMe_3 \text{ or}} ArC{\equiv}CAr$$
(R = Me, Et)

JCS Perkin I 869 (1977)

$$\underset{RCCl}{\overset{\overset{\displaystyle O}{\|}}{}} + 2Ph_3P{=}CHX \longrightarrow \underset{RCCX}{\overset{\overset{\displaystyle O}{\|}}{}}{=}PPh_3 \xrightarrow{\Delta} RC{\equiv}CX$$

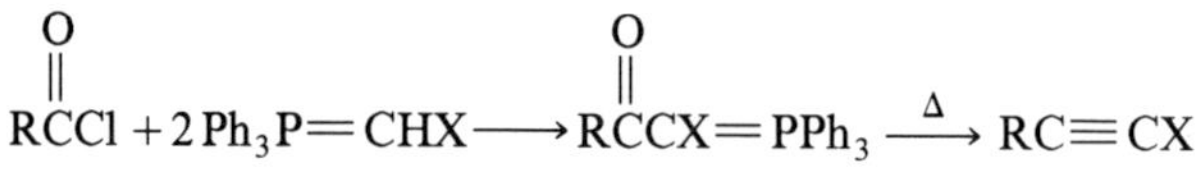

(best if X = Ar, CO_2R, CN)

Proc Chem Soc 302 (1961)
Ber *94* 3005 (1961)
Angew Int *1* 160 (1962)
JCS 2333 (1962); 543 (1964)
TL *23* 343 (1982)

4. ALKYLATION, ALKENYLATION AND ARYLATION OF ALKYNES AND ALLENES

1. Organolithium, -sodium and -magnesium Compounds

See also page 302, Sections 7 and 8.

$$RC{\equiv}CH \longrightarrow RC{\equiv}CM \xrightarrow{R'X} RC{\equiv}CR'$$

$$M = Li, Na, MgX$$

JACS 58 796 (1936); 60 1882 (1938) (R'X = EtOTs)
JOC 21 (1937); 24 840 (1959); 51 4158 (1986)
Org Rxs 5 1 (1949) (review)
JCS 893 (1951)
Org Syn Coll Vol 4 117 (1963)
BSCF 2000 (1964); 1525 (1965)
Syn 567 (1972); 441 (1974); 423 (1975)
Syn Commun 3 245 (1973)
J Chem Res (S) 106 (1978); 190 (1979); 270 (1981) (reviews)
CL 669 (1980)
TL 27 5445 (1986)

$$HO(CH_2)_n C{\equiv}CH \xrightarrow[NH_3]{2\,LiNH_2} \xrightarrow{RX} HO(CH_2)_n C{\equiv}CR$$

Helv 61 2275 (1978)

$$HO(CH_2)_n C{\equiv}CH + X(CH_2)_m CO_2H \xrightarrow[\text{or } LiNH_2,\ NH_3]{2n\text{-BuLi}} HO(CH_2)_n C{\doteq}C(CH_2)_m CO_2H$$

$$n = 1, 2;\ m = 4, 5, 10, 11;\ X = Br, I$$

TL 27 573, 2279, 2369 (1986)

$$HC{\equiv}CLi{\cdot}EDA + I(CH_2)_n CO_2H \xrightarrow{HMPA} HC{\equiv}C(CH_2)_n CO_2H$$

Syn Commun 10 653 (1980)

$$RC{\equiv}CMgBr + TsOCH_2C{\equiv}CH \longrightarrow RC{\equiv}CCH_2C{\equiv}CH$$

Syn 292 (1979)

$$RC \equiv CM + R_2CO \xrightarrow{H_2O} RC \equiv CCR_2\overset{OH}{\underset{|}{}}$$

$$M = Li, MgX$$

A. W. Johnson, "The Chemistry of Acetylenic Compounds," Vol 1, "Acetylenic Alcohols," Edward Arnold and Co., London (1946) (review)

JOC *22* 1611 (1957)

CL 447 (1979); 255 (1980) (asymmetric)

CC 363 (1981) (stereochemistry)

$$HC \equiv CCR_2\overset{OH}{\underset{|}{}} \xrightarrow[CuBr_2]{HBr} BrCH = C = CR_2 \xrightarrow{LiAlH_4} HC \equiv CCHR_2$$

JOC *33* 3655 (1968)

J Chem Res (S) 106 (1978)

$$RC \equiv CSO_2Ar + R'M \longrightarrow RC \equiv CR'$$

$$R' = alkyl, aryl; M = Li, MgX$$

JOC *44* 3444 (1979)

$$RX + LiCH_2C \equiv CSiR_3' \longrightarrow RCH_2C \equiv CSiR_3' \longrightarrow RCH_2C \equiv CH$$

$$R' = Me, i\text{-}Pr$$

TL 5041 (1968); 2247 (1970); *23* 719 (1982); *27* 2187 (1986)

JACS *90* 5618 (1968); *92* 6314 (1970)

$$RX \xrightarrow{LiC \equiv CCHLiSPh} \xrightarrow{Li, NH_3} \xrightarrow{H^+} RCH_2C \equiv CH$$

JOC *46* 5041 (1981)

$$RX \xrightarrow[2. \ BrCH_2CBr=CH_2]{1. \ Mg} RCH_2\overset{Br}{\underset{|}{C}}=CH_2 \xrightarrow{NaNH_2} RCH_2C \equiv CH$$

BSCF [4] *29* 528 (1921)

Ann Chim [10] *3* 325 (1925)

Org Syn Coll Vol *1* 180, 185 (1941)

$$RCH = CHCH_2X \xrightarrow{"BrMgCH_2C \equiv CH"} RCH = CHCH_2CH_2C \equiv CH$$

JCS 4244 (1955)

TL 2249 (1970)

$$RCH_2C \equiv CH \xrightarrow{2 \ R'Li} \xrightarrow{E_1^+} \xrightarrow{E_2^+} RCHC \equiv CE_2\overset{E_1}{\underset{|}{}}$$

$$E_1^+ \ or \ E_2^+ = RX, \ H_2\overset{O}{\overset{/\backslash}{C}}-CH_2, \ RCHO, \ R_2CO, \ CO_2, \ Me_3SiCl, \ I_2$$

JOMC *3* 165 (1965)

JACS *91* 3094 (1969); *98* 8413 (1976); *103* 3112 (1981) (R = SePh)

Tetr *26* 2345 (1970); *28* 5385 (1972)

Israel J Chem *10* 827 (1972)

CC 1030 (1974); 817 (1975); 215 (1979)

Syn 321 (1976)

JCS Perkin I 1218 (1979); 2338 (1980)

Rec Trav Chim *99* 113 (1980)

JOC *46* 5041 (1981) (R = SPh)

$$RCH_2C{\equiv}CCH_3 \xrightarrow[\text{TMEDA}]{t\text{-BuLi}} \xrightarrow{BrCH_2CH{=}CH_2} RCH_2C{\equiv}CCH_2CH_2CH{=}CH_2$$

JOC *47* 3364 (1982)

$$RCH_2C{\equiv}CSiMe_3 \xrightarrow[\text{2. R'X}]{1.\ t\text{-BuLi}} \overset{\overset{\displaystyle R'}{|}}{R}CHC{\equiv}CSiMe_3$$

JACS *108* 1359 (1986)

$$BrCH_2C{\equiv}CSiMe_3 \xrightarrow{ArMgX} ArCH_2C{\equiv}CSiMe_3$$

JOC *52* 1889 (1987)

$$\overset{\overset{\displaystyle Cl}{|}}{R_2}CC{\equiv}CR \xrightarrow{R'MgX} \overset{\overset{\displaystyle R'}{|}}{R_2}CC{\equiv}CR$$

JACS *62* 1798 (1940); *68* 1202 (1946); *86* 5244 (1964); *89* 6177 (1967)
J Chem Res (S) 106 (1978); 190 (1979)

2. Organoboron Compounds

$$RC{\equiv}CH \xrightarrow[\text{2. R'}_3\text{B}]{1.\ \text{base}} [\,RC{\equiv}\bar{C}BR'_3\,]Li^+ \xrightarrow{\text{reagent}} RC{\equiv}CR'$$

Reagent

I_2

JACS *95* 3080 (1973)
JOC *39* 731 (1974); *46* 2311 (1981); *51* 162,
    4507, 4514 (1986); *52* 2919 (1987)
TL 1961 (1975)
Syn 679 (1977)
Ber *115* 828 (1982)

CH_3SOCl TL 1847 (1973)

$$RCH{=}CH_2 \xrightarrow[\substack{\text{3. LiC}{\equiv}\text{CR'} \\ \text{4. I}_2}]{\substack{\text{1. (CH}_3\text{)}_2\text{CHC(CH}_3\text{)}_2\text{BHCl} \\ \text{2. MeOH}}} RCH_2CH_2C{\equiv}CR'$$

JOC *51* 4518, 4521 (1986)

$$RC{\equiv}CH + R'_3B \longrightarrow RC{\equiv}CR'$$

electrolysis   CL 999 (1977); 461 (1980)
$Pb(OAc)_4$ CL 413 (1980)

$$RC{\equiv}CH \xrightarrow[\text{2. }E\text{-(Sia)}_2\text{BCH}{=}\text{CHR'}]{1.\ R''\text{Li}} [\,E\text{-}RC{\equiv}CB(Sia)_2CH{=}CHR'\,]Li$$

$$\xrightarrow[\text{2. H}_2\text{O}_2,\,\text{OH}^-]{1.\ I_2,\,\text{OH}^-} E\text{-}RC{\equiv}CCH{=}CHR'$$

CC 874 (1973)
TL 411 (1977); *27* 539 (1986)

$$RC\equiv CH \longrightarrow RC\equiv CX \xrightarrow{\text{reagents}} E\text{-}RC\equiv CCH=CHR'$$

Reagents

$E\text{-}RCH=CHB(\text{Sia})_2$, NaOMe, cat $Pd(PPh_3)_4$ TL 3437 (1979)

$E\text{-}RCH=CHB$⟨⟩, NaOMe, CuI   JOC *46* 645 (1981)

$$R_3B \xrightarrow{\text{LiC}\equiv CCH_2Cl} RC\equiv CCH_2BR_2 \xrightarrow[\text{LiOMe, CuI}]{H_2C=CR'CH_2Br} RC\equiv CCH_2CH_2CR'=CH_2$$

CL 1289 (1982)

$$\overset{\overset{\displaystyle OAc}{|}}{HC\equiv CCR_2} \xrightarrow[\substack{\text{2. } R_3'B \\ \text{3. } H_2O}]{\text{1. } n\text{-BuLi}} R'C\equiv CCHR_2$$

JOC *42* 2650 (1977)

3. Organoaluminum Compounds

$$RC\equiv CH \longrightarrow (RC\equiv C)_3Al \xrightarrow{R'X} RC\equiv CR'$$

JACS *97* 7385 (1975)

$$RC\equiv CAlEt_2 + R'\overset{\overset{\displaystyle SO_2Ph}{|}}{CH}CH=C(CH_3)_2 \longrightarrow RC\equiv CCHR'CH=C(CH_3)_2$$

JACS *108* 1098 (1986)

$$RC\equiv CBr + R_3'Al \xrightarrow{\text{cat Ni(mesal)}_2} RC\equiv CR'$$

TL 2831 (1978)

4. Organosilicon Compounds

$$Me_3SiC\equiv CSiMe_3 \xrightarrow[\text{10\% AlCl}_3]{3°RX\ (X = Cl,\ Br)} RC\equiv CR$$

CC 959 (1982)

$$RC\equiv CSiMe_3 \xrightarrow[\text{AlCl}_3]{3°R'Cl} RC\equiv CR'$$

CC 959 (1982)
JACS *107* 6546 (1985)

$$RC\equiv CSiMe_3 \xrightarrow[\text{cat PdCl}_2(\text{PhCN})_2]{ClCH_2CH=CH_2} \overset{R}{\underset{Cl}{>}}C=C\overset{SiMe_3}{\underset{CH_2CH=CH_2}{<}} \xrightarrow{n\text{-Bu}_4NF} RC\equiv CCH_2CH=CH_2$$

CL 1485 (1982)

5. Organotin Compounds

$$n\text{-Bu}_3\text{SnCH}=\text{C}=\text{CH}_2 + \text{ArCH}=\overset{\overset{\displaystyle\text{OAc}}{|}}{\text{CHCHAr}} \xrightarrow{\text{cat Pd(PPh}_3)_4} \text{ArCH}=\overset{\overset{\displaystyle\text{CH}_2\text{C}\equiv\text{CH}}{|}}{\text{CHCHAr}}$$

JOC *48* 5302 (1983)

$$\text{R}_3\text{SnC}\equiv\text{CR}' + \text{RX} \xrightarrow{\text{Pd catalyst}} \text{RC}\equiv\text{CR}'$$

RX	
R_fI	TL *28* 5857 (1987)
ArX	Izv Akad Nauk SSSR, Ser Khim 479 (1980) [CA *93* 26019h (1980)] JOC USSR *17* 18 (1981) JOMC *250* 551 (1983) Proc Acad Sci USSR, Chem Sec *272* 333 (1983); *274* 39 (1984)
heterocyclic halides	JACS *109* 2138 (1987) (thiophene)
ArOTf	JACS *109* 5478 (1987)
$\text{RCH}=\text{CHI}$	JACS *109* 2138 (1987)
$\text{RCH}=\text{CHOTf}$	JACS *106* 4630 (1984); *108* 3033 (1986) TL *27* 1523 (1986)

For RX = RCOCl see also page 686, Section 2.4.

$$\text{R}_3\text{SnC}\equiv\text{CR}' + \text{RHgX} \xrightarrow{h\nu} \text{RC}\equiv\text{CR}'$$

TL *27* 3479 (1986)

6. Organocobalt Compounds

$$\text{RC}\equiv\overset{\overset{\displaystyle\text{OH}}{|}}{\text{CCR}_2} \xrightarrow[\text{2. HBF}_4]{\text{1. Co}_2(\text{CO})_8} \text{RC}\equiv\overset{\overset{\displaystyle\text{Co}_2(\text{CO})_6}{|}}{\overset{+}{\text{CCR}}_2} \xrightarrow[\text{2. decomplexation}]{\text{1. Nuc}} \text{RC}\equiv\overset{\overset{\displaystyle\text{Nuc}}{|}}{\text{CCR}_2}$$

Nuc	
NaBH_4	JACS *107* 4999 (1985)
anisole	TL 4163 (1977)
allylic silanes	TL *21* 1595 (1980)
ketones, enol silanes, enol acetates	JACS *102* 2508 (1980)
enol silane	CC 1353 (1987)
isopropenyl acetate	Syn Commun *10* 503 (1980)
β-diketones, β-keto esters	TL 4349 (1978)

See also TL *23* 2555 (1982); *26* 1269 (1985)

$$RC\equiv CCR_2 \overset{OR'}{|} \xrightarrow[\substack{2.\ Nuc \\ 3.\ decomplexation}]{1.\ Co_2(CO)_8} RC\equiv CCR_2 \overset{Nuc}{|}$$

R'	Nuc	
Me	allylic silane, $BF_3 \cdot OEt_2$; enol silane, Lewis acid	JACS *108* 3128 (1986)

$$CH_3CH=CN \overset{n\text{-}Bu_2BO}{\underset{Me}{|}} \overset{O}{\underset{}{\parallel}} C \overset{O}{\underset{}{\diagdown}} O \quad \text{Ph}$$

JACS *109* 5749 (1987)

Ac	$AlMe_3$	JOMC *212* 115 (1981)
	$Al(C\equiv CR)_3$	TL *24* 2239 (1983)

7. Organonickel Compounds

$$RCH=CHX + Me_3SiC\equiv CMgBr \xrightarrow{Ni\ catalyst} RCH=CHC\equiv CSiMe_3$$

JOC *49* 4733 (1984)

8. Organopalladium Compounds

See also page 301, Section 5.

$$RC\equiv CH + R'X \xrightarrow[\substack{base\ (CuI)}]{Pd\ catalyst} RC\equiv CR'$$

R'X	
ArX	TL 4467 (1975); *27* 1653 (1986); *28* 2887, 3857, 4879, 5395, 5981 (1987) JOMC *93* 253, 259 (1975) Syn 627 (1980) JOC *46* 2280 (1981) JOC USSR *18* 308 (1982) JACS *107* 5670 (1985); *108* 2481, 3150 (1986)
heterocyclic halides	TL 4467 (1975) (pyridines) JOMC *93* 259 (1975) (thiophene) Chem Pharm Bull *26* 3843 (1978) (pyrimidines); *27* 270 (1979) (quinolines, isoquinolines and acridines); *28* 3488 (1980) (pyridazines); *30* 1865 (1982) (pyrimidines); *34* 1447 (1986) (pyrazines) Syn 627 (1980) (pyridine); 364 (1981) (various heterocycles); 312 (1983) (various heterocycles) Heterocycles *19* 329 (1982) (pyrazines) JOC *50* 2462 (1985) (pyridines); *52* 2469 (pyridines), 3997 (pyrazines and pterin), 5243 (thiophenes) (1987) Tetr *41* 621 (1985) (thiophenes)

ArO_3SR_f

TL *27* 1171 (1986)

RCH=CHX

TL 4467 (1975); *22* 421 (1981); *27* 2033, 3589,
5857 (1986); *28* 1127, 1649, 3857, 3959, 4875,
4879, 5751, 5849 (1987)
JOMC *93* 253, 259 (1975)
Syn Commun *11* 917 (1981)
J Chem Res (S) 93 (1982)
Agric Biol Chem *46* 717 (1982)
Tetr *38* 631 (1982)
JACS *106* 3548, 5734 (1984); *107* 7515 (1985);
108 5589 (1986); *109* 1879 (1987)
CC 1816 (1986)

$RCH=CHO_3SCF_3$

JOC *50* 2302 (1985)
TL *27* 1523 (1986)

Heterocycles *26* 355 (1987)

R″X, CO (R′ = R″CO; R″ = aryl,
  heterocyclic, vinylic)

CC 333 (1981)

R″COCl (R′ = R″CO; R″ = alkyl, aryl,
  alkenyl, NMe_2)

Syn 777 (1977)

$$RC\equiv CM + R'X \xrightarrow{\text{Pd catalyst}} RC\equiv CR'$$

M	R′X	
ZnCl	RCH=CHX	CC 683 (1977) TL *27* 4351, 5533 (1986)
	ArX	JOC *43* 358 (1978) JACS *108* 3403 (1986)
	ArO_3SR_f	TL *28* 2387 (1987)
MgX	ArX RCH=CHX	JOMC *118* 349 (1976) Tetr *37* 2617 (1981) JACS *107* 1028 (1985)
SnR_3	—	See page 301, Section 5.
Cu	ArX	JACS *107* 1028 (1985) See also page 304, Section 9 for related reactions

$$RC\equiv CH + CO + R'X \xrightarrow[\substack{Et_3N \\ \Delta}]{\text{Pd catalyst}} RC\equiv C\overset{\overset{\displaystyle O}{\|}}{C}R'$$

R′ = aryl, heterocyclic, vinylic

CC 333 (1981)

$$RC\equiv CX + R'MgX \xrightarrow{\text{Pd catalyst}} RC\equiv CR'$$

JOC *51* 3772 (1986)

9. Organocopper Compounds

Review: Tetr *40* 1433 (1984)

$$RC{\equiv}CH + ArX \xrightarrow[Cu]{K_2CO_3} RC{\equiv}CAr$$

Bull Acad Sci USSR, Div Chem Sci 2539 (1968); 1079 (1970); 1209, 1488 (1971)

$$RC{\equiv}CH + XCH_2C{\equiv}CR' \longrightarrow RC{\equiv}CCH_2C{\equiv}CR'$$

CuI, DBU or DBN   Ann 658 (1978)

CuCl, NH$_3$ BSCF 913 (1974)

$$2\ RC{\equiv}CH \longrightarrow RC{\equiv}CC{\equiv}CR$$

Reviews: Russ Chem Rev *32* 229 (1963)
Adv Org Chem *4* 225 (1963)

cat CuCl, NH$_4$Cl, H$_2$O	Adv Org Chem *4* 225 (1963) (review) JOC *29* 2051 (1964)
cat CuCl, cat TMEDA, acetone, O$_2$	JOC *27* 3320 (1962)
cat Cu(OAc)$_2$, py, O$_2$	JOC *25* 1275 (1960) Adv Org Chem *4* 225 (1963) (review)
CuCl, NH$_4$Cl, NH$_3$, H$_2$O, O$_2$	JCS 1998 (1952)
Cu(OAc)$_2$, py, MeOH, Et$_2$O, O$_2$	JCS 889 (1959)

$$RC{\equiv}CX + HC{\equiv}CR' \xrightarrow[\substack{NH_2OH \cdot HCl \\ EtNH_2}]{cat\ CuCl} RC{\equiv}CC{\equiv}CR'$$

Adv Org Chem *4* 225 (1963) (review)
J Chem Res (S) 199 (1982)
Agric Biol Chem *46* 717 (1982)

$$RC{\equiv}CH \longrightarrow RC{\equiv}CMgX \longrightarrow RC{\equiv}CR'$$

R$_2$C=CHCH$_2$OMs, cat Li$_2$CuCl$_4$	JACS *108* 806 (1986)
RC≡CCMe$_2$Cl, cat CuCl	JACS *107* 6546 (1985)

$$RC{\equiv}CH \longrightarrow RC{\equiv}CCu \xrightarrow{R'X} RC{\equiv}CR'$$

<u>R'X</u>

ArX   Russ Chem Rev *37* 748 (1968) (review)
Bull Acad Sci USSR, Div Chem Sci 2043 (1963)
JOC *28* 2163, 3313 (1963); *30* 3857 (1965); *31*
   4071 (1966); *42* 2626 (1977); *52* 1339 (1987)
JACS *86* 4358 (1964); *88* 3027 (1966); *91* 6464
   (1969)
Chem Ind 2101 (1964)
JCS C 578 (1967); 2173 (1969)
CC 718 (1967)
J Label Compds *6* 197 (1970)

RCH=CHX

CC 1259 (1967)
JOC USSR *4* 21 (1968)
JOC *47* 2109 (1982)

H_2C=CHCH$_2$X

JACS *91* 6464 (1969); *109* 3684 (1987) (NaCN
    added)
Compt Rend *270* 354 (1970)
BSCF 913 (1974)
Ann 658 (1978)

Me$_3$SiC≡CCH$_2$I

TL *28* 3547 (1987)

$$MCH_2C≡CSiR_3 \xrightarrow[\text{2. E}^+]{\text{1. CuI}} ECH_2C≡CSiR_3 \longrightarrow ECH_2C≡CH$$

$\underline{M}$	$\underline{E^+}$	
Li	RCHO	JACS *109* 5437 (1987)
MgBr	R_2C=CHCH$_2$Cl	JOC *52* 3860, 3883 (1987) TL *28* 527, 723 (1987)
	R_2C=CRCH$_2$OPO(OEt)$_2$	JACS *106* 6006 (1984) TL *28* 527 (1987)

$$R'C≡CX + RM \longrightarrow R'C≡CR$$

$$X = Br, I$$

$\underline{RM}$	
ArCu	TL 5209 (1972); 1441 (1978)
RCu·MgBr$_2$ (R = alkyl, vinyl)	TL 1466 (1975) Tetr *36* 1215 (1980)
R$_2$CuLi	JOMC *251* 133 (1983)
(RCuCN)Li	JOMC *251* 133 (1983)

$$\underset{RC≡CCR_2}{\overset{OAc}{|}} \xrightarrow{R_2'CuLi} \underset{RC≡CCR_2}{\overset{R'}{|}} (\text{or } RR'C=C=CR_2)$$

CC 876 (1978)
JOC *45* 4740 (1980)

$$\underset{C=C=C-Y}{\overset{X}{\overset{|}{}}} \xrightarrow{RM} R-\overset{|}{C}-C≡C-Y$$

$\underline{X}$	$\underline{Y}$	$\underline{RM}$	
Br	H	PhCu·MgBr$_2$·LiBr	JOC *52* 3920 (1987)

Br	H	(PhCuBr)MgBr·LiBr,
		    (RCuBr)MgX·LiBr,
		    (RCuCN)Li
		    (R = 1° alkyl)
			TL *28* 6073 (1987)

$$C=C=\overset{\overset{\displaystyle X}{|}}{C}-Y \xrightarrow{RM} R-C-C\equiv C-Y \quad (\textit{Continued})$$

X	Y	RM	
Br	alkyl	Me$_3$Cu$_2$Li, MeCu·n-Bu$_3$P, (RCuCN)Li (R =1°, 2° alkyl; aryl)	TL *25* 3059 (1984)
Br	alkyl	RCH=CHLi, CuCN	JOC *51* 2230 (1986)
Br	alkyl	(RC≡CCuCN)Li	JACS *103* 4618 (1981)
I	alkyl	RCH=CHCu	TL *23* 1651 (1982)
I	alkyl	(ArCuCN)Li	TL *24* 3291 (1983)
I	OMe	RMgX, CuBr	JOC *45* 1158 (1980)
OMe	H	RMgX, CuX (X = Cl, Br, I)	Rec Trav Chim *93* 183 (1974)

10. Organozinc Compounds

See page 302, Section 8.

11. Organomercury Compounds

$$PhC\equiv CX + RHgCl \xrightarrow{h\nu} PhC\equiv CR$$

$$X = I, PhS, PhSO_2, HgC\equiv CPh, n\text{-Bu}_3Sn$$

$$TL\ 27\ 3479\ (1986)$$

12. Miscellaneous Reactions

$$RC\equiv CH + H_2C=CHCH=CH_2 \xrightarrow{cat\ H_2Ru(PR_3)_4} RC\equiv CCH=CHCH_2CH_3$$

$$CC\ 496\ (1981)$$

$$R_fI + HC\equiv CCR_2OH \xrightarrow{electrolysis} R_fCH=CICR_2OH \xrightarrow[\Delta]{OH^-} R_fC\equiv CH$$

$$CC\ 433\ (1982)$$

HALIDES

GENERAL REFERENCES

Houben-Weyl, "Methoden der Organischen Chemie," 4th ed, Vol V/4, G. Thieme Verlag, Stuttgart (1960) (bromine and iodine compounds)

Houben-Weyl, "Methoden der Organischen Chemie," 4th ed, Vol V/3, G. Thieme Verlag, Stuttgart (1962) (fluorine and chlorine compounds)

Quart Rev *16* 44 (1962) (The Fluorination of Organic Compounds)

W. A. Sheppard and C. M. Sharts, "Organic Fluorine Chemistry," W. A. Benjamin, New York (1969)

R. D. Chambers, "Fluorine in Organic Chemistry," Wiley-Interscience, New York (1973)

"The Chemistry of the Carbon-Halogen Bond," Ed. S. Patai, J. Wiley, New York (1973)

Org Rxs *21* 125 (1974) (Modern Methods to Prepare Monofluoroaliphatic Compounds)

Tetr *34* 3 (1978) (Introduction of Fluorine into Organic Molecules: Why and How)

Angew Int *20* 647 (1981) (Methods of Fluorination in Organic Chemistry)

"The Chemistry of Halides, Pseudo-Halides and Azides," Eds. S. Patai and Z. Rappoport, J. Wiley, New York (1983)

1. HALOGENATION OF HYDROCARBONS

Reviews:

C. Walling, "Free Radicals in Solution," J. Wiley, New York (1957), pp 347–396
Quart Rev *14* 336 (1960)
JACS *87* 2161, 2172 (1965)
W. A. Pryor, "Free Radicals," McGraw-Hill, New York (1966), pp 179–213
M. L. Poutsma, "Methods in Free Radical Chemistry," Ed. E. S. Huyser, Vol 1, Marcel Dekker, New York (1969), pp 79–193
W. A. Thaler, "Methods in Free Radical Chemistry," Ed. E. S. Huyser, Marcel Dekker, New York (1969), Vol 2, pp 121–227
Syn 7 (1970)

1. Aliphatic Halogenation

Fluorination

F_2	JACS *98* 3034, 3036 (1976)
	TL *21* 5067 (1980)
	Nouv J Chim *4* 239 (1980)
	J Fluorine Chem *20* 689 (1982)
	JOC *51* 3522 (1986); *52* 2769, 4928 (1987)
ClF_3	JOC *52* 798 (1987)
CF_3OF	JACS *92* 7494 (1970); *98* 3034, 3036 (1976)
	Nouv J Chim *4* 239 (1980)

Chlorination

Cl_2	JCS 144 (1960)
SO_2Cl_2	JACS *61* 2142 (1939); *62* 927 (1940)
	JCS 1851 (1951)
Cl_3CSO_2Cl	JACS *82* 5246 (1960)
NCS	JOC *18* 649 (1953)

$C_6H_5ICl_2$	JOC _29_ 3692 (1964); _32_ 1517 (1967)
	CL 961 (1979)
_t_-C_4H_9OCl	JACS _89_ 4891, 4895 (1967)
PCl_5	Angew _77_ 506 (1955)
R_2NCl, CF_3CO_2H, $FeSO_4 \cdot 7\,H_2O$	JOC _44_ 3728 (1979)

Bromination

Br_2	JCS 144 (1960)
	Rec Trav Chim _83_ 67 (1964)
Cl_3CSO_2Br	JOC _30_ 38 (1965)
NBS	JCS 2240 (1952)
	JOC _18_ 649 (1953)
$(C_6H_5)_2C{=}NBr$	JOC _32_ 223 (1967)
Cl_3CBr, DBU (activated CH)	CL 73 (1978)
$(NC)_2CBr_2$ (activated CII)	BCSJ _37_ 547 (1964)

Iodination

_t_-C_4H_9OI	JACS _90_ 808 (1968)

2. Allylic Halogenation

Reviews:

Chem Rev _43_ 271 (1948) (NBS); _63_ 21 (1963) (NBS)
Angew _71_ 349 (1959) (NBS)

Chlorination

$C_6H_5SO_2NClC_6H_{13}$	Ann _703_ 34 (1967)
_t_-C_4H_9OCl	JACS _83_ 3877 (1961)
	TL _21_ 781 (1980)
NCS, TsNSO	JOC _44_ 4204 (1979)
NCS, cat $(ArSe)_2$	JOC _44_ 4204, 4208 (1979)
$PhSeCl_3/H_2O$, $NaHCO_3$	JOC _52_ 4086 (1987)
HOCl	TL _21_ 441 (1980)
NaCl, H_2O, H_2SO_4, CH_2Cl_2, electrolysis	TL _22_ 3193 (1981)

Bromination

NBS	JOC _14_ 375 (1945); _50_ 2007 (1985)
	Chem Rev _43_ 271 (1948) (review)
	JACS _73_ 5153 (1951); _106_ 3297 (1984);
	_108_ 1251 (1986)
	Org Syn Coll Vol _4_ 108 (1963)
$(C_6H_5)_2C{=}NBr$	JOC _32_ 223 (1967)

3. Benzylic Halogenation

Reviews:

Chem Rev *43* 271 (1948) (NBS); *63* 21 (1963) (NBS)
Angew *71* 349 (1959) (NBS)

Fluorination

CF_3OF JACS *92* 7494 (1970)

Chlorination

SO_2Cl_2 JACS *61* 2142 (1939)
JCS 1851 (1951)
Tetr *25* 4363 (1969); *26* 2041 (1970)

SO_2Cl_2, cat $Pd(PPh_3)_4$ CL 223 (1978)

Cl_2, PCl_5 Org Syn Coll Vol *2* 133 (1943)

PCl_5 JOC *34* 3655 (1969)

$C_6H_5ICl_2$ JOC *29* 3692 (1964)

t-C_4H_9OCl JACS *82* 6108, 6113 (1960);
89 4885, 4891, 4895 (1967)

Cl_2O $(ArCH_3 \longrightarrow ArCCl_3)$ JACS *104* 4680 (1982)

Bromination

Br_2 Org Syn Coll Vol *2* 443 (1943); *3* 788 (1955);
4 984 (1963)

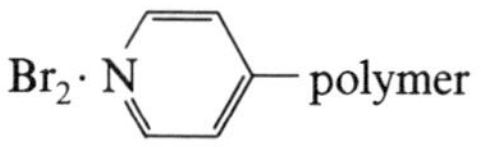

$Br_2 \cdot N$⟨⟩—polymer JOC *51* 929 (1986)

$CuBr_2$, t-BuO_2H, Ac_2O Syn Commun *11* 669 (1981)

NBS Chem Rev *43* 271 (1948) (review)
Org Syn Coll Vol *4* 921 (1963)
JOC *50* 2128, 2557, 2939 (1985); *51* 3407 (1986)

$BrCCl_3$ JACS *82* 391 (1960)

CBr_4 JACS *54* 2025 (1932)

$(C_6H_5)_2C{=}NBr$ JOC *32* 223 (1967)

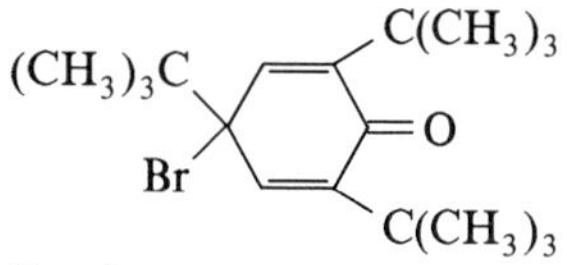

(structure with $(CH_3)_3C$, $C(CH_3)_3$, Br, $=O$, $C(CH_3)_3$) Tetr *25* 4357 (1969)

Iodination

t-C_4H_9OI JACS *90* 808 (1968)

2. AROMATIC HALOGENATION

See page 365, Section 9, for substitution of OH by halogen.

$$Ar-H \longrightarrow Ar-X$$

Reviews:

E. T. McBee and H. B. Hass, Ind Eng Chem *33* 137 (1941)

P. H. Groggins, "Unit Processes in Organic Synthesis," McGraw-Hill, New York (1958), p 204

P. B. D. de la Mare and J. H. Ridd, "Aromatic Substitution, Nitration and Halogenation," Butterworths Scientific Publications, London (1959)

H. P. Braendlin and E. T. McBee, "Friedel-Crafts and Related Reactions," Ed. G. Olah, Interscience Publishers, New York (1964), Vol III, Pt. 2, p 1517

R. O. C. Norman and R. Taylor, "Electrophilic Substitution in Benzenoid Compounds," Elsevier, New York (1965), pp 119–155

Fluorination

review	Chem Eng News, July 9, 72 (1962)
F_2	JOC *35* 723, 4020 (1970)
	TL *28* 255 (1987) (pyridines)
XeF_2	Israel J Chem *17* 71 (1978)
XeF_2, HF	JACS *91* 1563 (1969)
CF_3OF	CC 806 (1968)
	JACS *92* 7494 (1970)
	JCS Perkin I 2889 (1972)
	Anal de Quim *70* 871 (1974)
	JOC *39* 2120 (1974); *41* 3413 (1976)
	Israel J Chem *17* 60 (1978)
CF_3COF	JOC *46* 4629 (1981); *49* 806 (1984)
$Tl(O_2CCF_3)_3/KF/BF_3$	JOC *42* 362 (1977)
$CsSO_4F$	JACS *103* 1964 (1981)
$CsSO_4F$, cat BF_3	CC 148 (1981)
	Tetr *40* 189 (1984)
	JOC *50* 3609 (1985); *51* 3242 (1986)

$(CF_3SO_2)_2NF$ JACS *109* 7194 (1987)

$$\left[\begin{array}{c} \overset{CO_2Me}{\underset{\underset{CO_2Me}{N^+F}}{\bigcirc}} \end{array} \right] OTf^-$$ TL *27* 4465 (1986)

(and other substituted pyridinium salts)

Chlorination

Cl_2 JOC *50* 2145 (1985) (phenols)

Cl_2, H_2O TL *28* 4805 (1987) (aryl ethers)

Cl_2, silica   TL 3395 (1974)
                                                       CL 1423 (1980)

Cl_2, $FeCl_3$ CL 1423 (1980)

SO_2Cl_2 Org Syn Coll Vol *3* 267 (1955) (phenol)
                                                       Austral J Chem *28* 1113 (1975) (phenol)
                                                       JOC *46* 4486 (1981); *50* 2145 (1985); *52* 4485
                                                          (1987) (all phenols only)
                                                       TL *23* 4569 (1982) (phenol)

SO_2Cl_2, cat $(PhS)_2$, cat $AlCl_3$ or $FeCl_3$ JOC *50* 2145 (1985) (phenols)

S_2Cl_2, SO_2Cl_2, $AlCl_3$ JCS *127* 2677 (1925)
                                                       JACS *82* 4254 (1960)

$CuCl_2$ Org Syn Coll Vol *5* 206 (1973)

$TiCl_4$, CF_3CO_3H TL 2611 (1970)

Cl_2O, H_2SO_4 or CF_3SO_3H JACS *104* 4680 (1982)

Cl_2O, $POCl_3$, $(CF_3SO_2)_2O$ Ber *112* 1677 (1979)

NaOCl, HOAc   JCS 1056 (1909)
                                                       JACS *108* 1000 (1986)

Bromination

Br_2 vapor   Org Syn Coll Vol *4* 256 (1963)

Br_2, CCl_4 Org Syn Coll Vol *1* 121 (1941); *2* 95 (1943); *3*
                                                          134 (1955); *5* 147 (1973)

Br_2, CS_2 Org Syn Coll Vol *4* 128 (1941)

Br_2, H_2O JOC *52* 4485 (1987)

Br_2, H_2O, H^+ Org Syn Coll Vol *2* 97, 592 (1943); *4* 947 (1963)

Br_2, HOAc   Org Syn Coll Vol *2* 100, 173 (1943)
                                                       JOC *52* 4485 (1987)

Br_2, HOAc, NaOAc   JOC *52* 4485 (1987)

Br_2, KBr, H_2O JOC *52* 4485 (1987)

Br_2, H_2O_2, n-Bu_4NBr CC 1421 (1987)

Br_2, t-BuNH$_2$	JOC *32* 2358 (1967) (phenols) JACS *108* 806 (1986)
Br_2, silica	TL 3395 (1974)
Br_2, Fe	Org Syn Coll Vol *1* 123 (1941); *3* 138 (1955); *4* 114 (1963)
Br_2, AlCl$_3$	Org Syn Coll Vol *5* 117 (1973)
Br_2, Tl(OAc)$_3$	TL 1623 (1969) JOC *37* 88 (1972)
Br_2, AgO$_3$SR	Syn 693 (1978)
Br_2, AgNO$_3$, HOAc, HNO$_3$	JCS 573 (1950)
Br_2, AgSO$_4$, H$_2$SO$_4$	JCS 573 (1950)
Br_2, Hg(NO$_3$)$_2$, HOAc	JCS 573 (1950)
Br_2, KBrO$_3$, HOAc, H$_2$SO$_4$	JCS 573 (1950)
KBrO$_3$, H$_2$SO$_4$	JOC *46* 2169 (1981)
HOBr, HClO$_4$ or H$_2$SO$_4$	JCS 997 (1962)
HOBr, HOAc	JCS 2317 (1954)
HBr, H$_2$O$_2$, n-Bu$_4$NBr	CC 1421 (1987)
Hg(OAc)$_2$/Br$_2$	JCS 637 (1926)
NBS	JACS *80* 4327 (1958) JOC *30* 304 (1965); *44* 4733 (1979)
CBr$_4$	JACS *54* 2025 (1932)
CuBr$_2$	JACS *86* 427 (1964)

Iodination

I_2	Org Syn Coll Vol *2* 347 (1943) TL *27* 5963 (1986)
I_2, alumina	TL *27* 2207 (1986)
I_2, HNO$_3$	Org Syn Coll Vol *1* 323 (1932)
I_2, AgO$_2$CCF$_3$	Org Syn Coll Vol *4* 547 (1963)
I_2, SO$_3$, H$_2$SO$_4$	Org Syn Coll Vol *3* 796 (1955) JCS C 1480 (1970)
I_2, NaNO$_2$, H$_2$SO$_4$	JACS *65* 1273 (1943)
I_2, KIO$_3$, H$_2$SO$_4$	JACS *65* 1273 (1943)
I_2, Ag$_2$SO$_4$	JCS 150 (1952)
I_2, AlCl$_3$, CuCl$_2$	CL 1481 (1982)
I_2, CuCl$_2$	JOC *35* 3436 (1970)
I_2, HgO	Org Syn Coll Vol *2* 357 (1943)

I$_2$, HIO$_4 \cdot 2$ H$_2$O   BCSJ *39* 128 (1966)
                                              CC 1476 (1987)

I$_2$, PhI(O$_2$CCF$_3$)$_2$ Syn 486 (1980)

I$_2$, CF$_3$CO$_2$H, electrolysis   Acta Chem Scand B *34* 47 (1980)

ICl   Org Syn Coll Vol *2* 196, 343, 349 (1943)

MI (M = *n*-Bu$_4$N, Li, Na, K, I),   BCSJ *54* 2847 (1981)
  (NH$_4$)$_2$[Ce(NO$_3$)$_6$], CH$_3$CN

FeI$_2 \cdot 4$ H$_2$O, CuCl$_2$ JOC *35* 3436 (1970)

Tl(O$_2$CCF$_3$)$_3$/KI   TL 2427 (1969)
                                              JACS *93* 4841 (1971)

$$\text{ArH} \xrightarrow{\text{RLi}} \text{ArLi} \xrightarrow{\text{RSO}_2\text{NFR}'} \text{ArF}$$

JACS *106* 452 (1984)

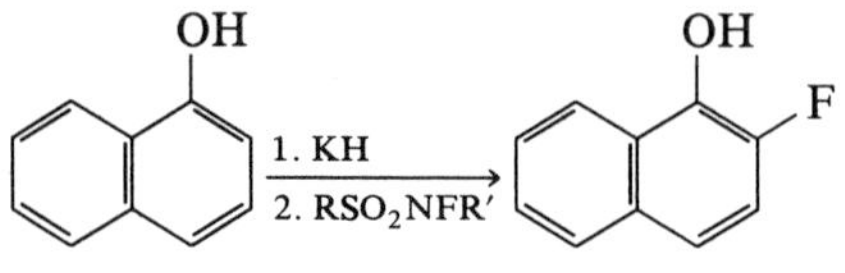

JACS *106* 452 (1984)

$$\text{ArH} + \text{RCHO} + \text{HCl} \longrightarrow \text{Ar}\overset{\overset{\textstyle \text{Cl}}{|}}{\text{CH}}\text{R}$$

Org Syn Coll Vol *3* 195 (1955); *4* 980 (1963)

3. HALOGENATION OF ALKENES

See also page 340, Section 4; and page 215, Section 2 (halogenation of vinyl metallics).

1. Halogen Addition

$$\text{>C=C<} \longrightarrow \overset{X}{\underset{|}{-}}\overset{X}{\underset{|}{C}}-\overset{|}{\underset{|}{C}}-$$

Review: H. O. House, "Modern Synthetic Reactions," 2nd ed, W.A. Benjamin, Inc. (1972), pp 422–429

X	Reagent(s)	
F	F_2	JOC *31* 1859, 3871 (1966); *51* 3607 (1986)
		JACS *88* 1822 (1966); *89* 609 (1967)
		JCS Perkin I 1105 (1982)
	XeF_2	JACS *86* 5021 (1964)
		JOC *42* 1559 (1977)
		JCS Perkin I 2169 (1977)
		Israel J Chem *17* 71 (1979)
		J Fluorine Chem *20* 13 (1982)
	XeF_2, H^+	TL 1015 (1974)
		JOC *41* 4002 (1976); *43* 696 (1978); *44* 1255 (1979)
		Tetr *33* 1017 (1977)
	XeF_2, $BF_3 \cdot OEt_2$	TL 363 (1977)
	XeF_4	JACS *86* 5021 (1964)
	$CsSO_4F$	JOC *52* 919 (1987)
	CF_3OF	JCS Perkin I 1105 (1982)

$$\text{C=C} \longrightarrow -\underset{|}{\overset{X}{C}}-\underset{|}{\overset{X}{C}}- \quad (\textit{Continued})$$

X	Reagent(s)	
Cl	Cl$_2$	JACS *61* 940 (1939); *63* 2541 (1941); *73* 3329 (1951)
	SO$_2$Cl$_2$	JACS *61* 940, 3432 (1939)
	PCl$_5$	JACS *61* 940 (1939)
	NCS, HCl	JACS *81* 2191 (1959)
	HCl, CaCl$_2$, H$_2$O$_2$, (PhCH$_2$NEt$_3$)Cl	Syn 676 (1977)
	[polymer-C$_6$H$_4$CH$_2\overset{+}{N}$Me$_3$]ICl$_2^-$	CC 1278 (1980)
Br	Br$_2$	Org Syn Coll Vol *1* 521 (1941); *2* 171, 177, 270 (1943); *4* 195 (1963)
	Br$_2 \cdot$ (dioxane)	Zh Obshch Khim *24* 610 (1954)
	(pyridinium) Br$_3^-$	Ber *56* 1262 (1923) J Chem Ed *31* 291 (1954) Syn 966 (1979) (selective diene addition) TL *22* 623 (1981)
	[polymer-C$_6$H$_4$CH$_2\overset{+}{N}$Me$_3$]Br$_3^-$	Syn 143 (1980)
	HBr, CaBr$_2$, H$_2$O$_2$, (PhCH$_2$NEt$_3$)Cl	Syn 676 (1977)
	NaBr, H$_2$O, H$_2$SO$_4$, CH$_3$CN, electrolysis	JOC *46* 3312 (1981)
I	I$_2$, CuO–HBF$_4$, NaI	CC 1491 (1987)

$$\text{RCH=CHCO}_2\text{H} \xrightarrow{\text{Br}_2} \underset{|}{\overset{\text{Br}}{\text{RCH}}}-\underset{|}{\overset{\text{Br}}{\text{CHCO}_2\text{H}}} \xrightarrow{\text{base}} \text{RCH=CHBr}$$

JCS 2012 (1950)
JACS *75* 2645 (1953)
JOC *27* 2339 (1962)
BSCF 1736 (1971)
J Chem Res (S) 270 (1981)

$$\text{C=C} \longrightarrow -\underset{|}{\overset{X}{C}}-\underset{|}{\overset{Y}{C}}-$$

X	Y	Reagent(s)	
F	Cl	(1,3,5-triazine with NCl$_2$ groups), HF	Bull Acad Sci USSR, Div Chem Sci 1016 (1966) JOC USSR *5* 1879 (1969); *7* 1382, 1876 (1971); *8* 1139 (1972)

F	Cl, Br, I	review	Russ Chem Rev *41* 740 (1972)
F	Br	XeF$_2$, Br$_2$	Israel J Chem *10* 271 (1977)
		AcNHBr, HF	JACS *81* 2191, 4107 (1959); *82* 4001 (1960)
			Chem Ind 452 (1963)
			JOC *29* 1202 (1964)
			Can J Chem *43* 1689 (1965)
			Syn 217 (1978)
		NBS, HF	TL *28* 4003 (1987)
		1,3-dibromo-5,5-dimethylhydantoin, HF	Can J Chem *43* 1689 (1965)
			J Fluorine Chem *31* 99 (1986)
			JOC *52* 658 (1987)
			TL *28* 4003 (1987)
		1-bromo-3,5,5-trimethylhydantoin, HF	Can J Chem *43* 1689 (1965)
		NBS, R$_3$N, HF	TL *27* 4449 (1986)
F	I	I(py)$_2$BF$_4$, BF$_4^-$	Angew *24* 319 (1985)
		I(collidine)$_2$BF$_4$	Syn 551 (1987)
		NIS, HF	JACS *82* 4001, 4007 (1960)
Cl	Br	AcNHBr, LiCl, HCl	JACS *81* 2191 (1959)
Cl	I	ICl	JCS Perkin I 226 (1977)
			CC 1577 (1987)
		I(py)$_2$BF$_4$, Cl$^-$	Angew Int *24* 319 (1985)
Br	I	I(py)$_2$BF$_4$, Br$^-$	Angew Int *24* 319 (1985)

$$C{=}C{-}C{=}C \longrightarrow \overset{\overset{X}{|}}{C}{-}\overset{\overset{X}{|}}{C}{-}C{=}C + \overset{\overset{X}{|}}{C}{-}C{=}C{-}\overset{\overset{X}{|}}{C}$$

X		
Cl		JACS *52* 4043 (1930); *73* 244 (1951)
		JCS 829 (1934); 3204 (1962)
		J Gen Chem USSR *27* 253 (1957)
		JOC *31* 4167 (1966); *33* 2946 (1968); *37* 2228 (1972); *39* 736 (1974); *41* 334 (1976)
Br		Helv *5* 756 (1922)
		JCS 729 (1928)
		JACS *75* 2512 (1953); *81* 5943 (1959)
		J Gen Chem USSR *24* 453 (1954)
		JOC *33* 2342 (1968); *34* 2779 (1969); *35* 2967 (1970); *37* 2228 (1972); *38* 4109 (1973)

2. Hydrohalogenation

2.1. Markovnikov Addition

$$RCH{=}CH_2 \longrightarrow R\overset{\displaystyle X}{\underset{\displaystyle |}{C}}HCH_3$$

Reviews:

Houben-Weyl, "Methoden der Organischen Chemie," 4th ed, Vol V/3, pp 99–108 (X = F) and pp 812–822 (X = Cl); Vol V/4, pp 102–124 (X = Br) and pp 535–539 (X = I); G. Thieme, Stuttgart (1962 and 1960 respectively)

P. B. D. de la Mare and R. Bolton, "Electrophilic Additions to Unsaturated Systems," Elsevier, New York (1966), Chpt 5

C. A. Buehler and D. E. Pearson, "Survey of Organic Chemistry," Wiley Interscience, New York (1970), Vol 1, p 356

H. O. House, "Modern Synthetic Reactions," W. A. Benjamin, Inc., New York (1972), pp 446–452

Examples:

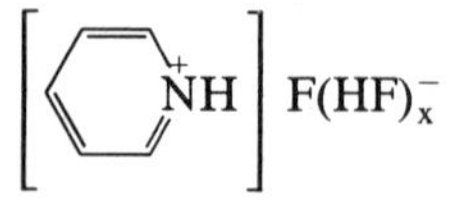 $\left[\underset{\displaystyle \overset{+}{N}H}{\bigcirc}\right] F(HF)_x^-$	Syn 779 (1973) JOC *44* 3872 (1979)
aq HX (X = Cl, Br, I) (phase transfer)	JOC *45* 3527 (1980)
HCl	Org Syn Coll Vol *2* 336 (1943) JOC *51* 5191 (1986)
HBr	JACS *68* 1805 (1946) Org Syn Coll Vol *3* 576 (1955) JOC *21* 1362 (1956)
KI, H_3PO_4	Org Syn Coll Vol *4* 543 (1963)
I_2, Al_2O_3	TL *28* 4497 (1987)

$$R\overset{\displaystyle Br}{\underset{\displaystyle |}{C}}{=}CH_2 + HBr \longrightarrow R\overset{\displaystyle Br}{\underset{\displaystyle |}{\underset{\displaystyle Br}{\overset{\displaystyle |}{C}}}}CH_3$$

JACS *58* 1806 (1936)

2.2. Anti-Markovnikov Addition

$$RCH{=}CH_2 \longrightarrow RCH_2CH_2X$$

X	Reagents	
Cl	BH_3 or 9-BBN/Cl_2NTs	J Chem Res (S) 376 (1981)
Br	HBr, peroxides	JOC *11* 281 (1946); *47* 5372 (1982) JACS *68* 1101 (1946); *109* 6937 (1987)

		Chem Rev *62* 599 (1962) (review)
		Org Rxs *13* 150 (1963) (review)
	HBr, hν	JOC *21* 1362 (1956); *42* 1709 (1977)
		Org Rxs *13* 150 (1963) (review)
		Tetr *25* 5149 (1969)
	$BH_3/Hg(OAc)_2/Br_2$	JACS *92* 3221 (1970)
	BH_3/Br_2, $NaOCH_3$	JACS *92* 6660 (1970)
		JOC *51* 5291 (1986)
	BH_3/Br_2	JACS *92* 7212 (1970)
		JOC *46* 3113 (1981)
	$BH_3/BrCl$	JOC *46* 3113 (1981)
	$BH_3/NaBr$, chloramine T	JOC *46* 3113 (1981)
	9-BBN/Br_2	JOMC *26* C51 (1971)
	i-Bu$_3$Al, cat Cp$_2$ZrCl$_2$/NBS	TL *28* 5793 (1987)
	LiAlH$_4$, Cp$_2$TiCl$_2$/Br$_2$	CL 1117 (1977)
Cl, Br	LiAlH$_4$, TiCl$_4$/CuX$_2$	CL 833 (1978)
	LiAlH$_4$, TiCl$_4$ or ZrCl$_4$/X$_2$	JOMC *142* 71 (1977)
	HSiCl$_3$, cat H$_2$PtCl$_6$/KF/	TL 1809 (1978)
	CuX$_2$	Organomet *1* 369 (1982)
Br, I	Cl$_2$AlH, cat Et$_3$B/X$_2$, py	JACS *108* 6036 (1986)
Cl, Br, I	HSiCl$_3$, cat H$_2$PtCl$_6$/KF/	JACS *100* 290 (1978)
	X$_2$ or NBS	Organomet *1* 355 (1982)
I	(Sia)$_2$BH/I$_2$, NaOH	JACS *90* 5038 (1968)
	BH$_3$/I$_2$, NaOCH$_3$	JACS *98* 1290 (1976)
		Syn 114 (1976)

$$\left(\langle \bigcirc \rangle - \right)_2 BH/I_2, \ NaOCH_3 \qquad \text{JACS } 107 \ 3915 \ (1985)$$

BH$_3$/NaI, chloramine T   JOC *46* 2582 (1981); *52* 28 (1987)

BH$_3$/ICl, NaOAc   JOC *45* 3578 (1980)

$$\left(\langle \bigcirc \rangle - \right)_2 BH/ICl, \ NaOAc \qquad \text{Syn Commun } 11 \ 521 \ (1981)$$

i-Bu$_3$Al, cat Cp$_2$ZrCl$_2$/I$_2$ TL *21* 1501 (1980)

$$\underset{\substack{|\\ RCH=CHBr}}{Br} \text{ or } \underset{\substack{|\\ RC=CH_2}}{Br} \xrightarrow[\text{peroxides}]{HBr} \underset{\substack{|\\ RCHCH_2Br}}{Br}$$

JACS *58* 1806 (1936)

2.3. Unknown Regiochemistry

$$\bigcirc\!\!= \longrightarrow \bigcirc\!\!-I$$

I$_2$, CuO-HBF$_4$, Et$_3$SiH   CC 1491 (1987)

I(py)$_2$BF$_4$, Et$_3$SiH   Angew *24* 319 (1985)

2.4. Isomerization—Addition

$$RCH{=}CH(CH_2)_nCH_3 \xrightarrow[\text{2. E}^+]{\text{1. HZrClCp}_2} R(CH_2)_{n+2}CH_2X$$

$$E^+ = NCS, PhICl_2, NBS, Br_2, I_2 \qquad X = Cl, Br, I$$

JACS *96* 8115 (1974); *98* 262 (1976)
Angew Int *15* 333 (1976)
JOC *46* 1821 (1981)

3. Hydrogen Substitution

$$R_2C{=}CHR \xrightarrow[\text{HF}]{\text{AcNHBr}} \xrightarrow{\text{KOH}} R_2C{=}CRF$$

Syn 217 (1978)

$$RCH{=}CHR \xrightarrow{\text{PhSeCl}_3} \underset{\underset{\text{PhSeCl}_2}{|}}{\overset{\overset{\text{Cl}}{|}}{RCH-CHR}} \xrightarrow[\text{NaHCO}_3]{\overset{\Delta}{H_2O}} \overset{\overset{\text{Cl}}{|}}{RC}{=}CHR$$

JOC *52* 4086 (1987)

$$RCH{=}CH_2 \xrightarrow[\text{2. SO}_2\text{Cl}_2]{\text{1. PhSeCl}} \underset{}{\overset{\overset{\text{Cl}}{|}}{RCHCH_2SeCl_2Ph}} \xrightarrow[\text{NaHCO}_3]{\overset{\Delta}{H_2O}} \overset{\overset{\text{Cl}}{|}}{RC}{=}CH_2$$

TL *28* 1463 (1987)

$$RCH{=}CH_2 \xrightarrow[\substack{\text{2. O}_3 \\ \text{3. py, }\Delta}]{\text{1. PhSeCl}} \overset{\overset{\text{Cl}}{|}}{RC}{=}CH_2$$

JACS *109* 7543 (1987)

$$RR'_2CCH{=}CH_2$$

$$\xrightarrow[\text{CH}_3\text{CN}]{\text{PhSeBr}} \xrightarrow{\text{O}_3} \overset{\overset{\text{Br}}{|}}{RR'_2CC}{=}CH_2$$

$$\xrightarrow[\text{CCl}_4]{\text{PhSeBr}} \xrightarrow{\text{H}_2\text{O}_2}$$

$$\xrightarrow{R'=H} \underset{H}{\overset{R}{>}}C{=}C\underset{CH_2Br}{\overset{H}{<}}$$

$$\xrightarrow{R'=\text{alkyl}} \underset{H}{\overset{RR'_2C}{>}}C{=}C\underset{Br}{\overset{H}{<}}$$

TL 3909 (1977)

$$\underset{C=C-C}{\overset{H\quad O}{\underset{|\quad\;||}{}}} \longrightarrow \underset{C=C-C}{\overset{X\quad O}{\underset{|\quad\;||}{}}}$$

X	Reagent(s)	
Cl, Br	PhSeX, py	TL *22* 3301 (1981)
Br	Br_2/NaHCO$_3$ Br_2/Et$_3$N	Syn 389 (1982) TL *28* 6485 (1987)

$$H_2C=CRCO_2R \xrightarrow[\text{2. NaO-}i\text{-Pr}]{\text{1. Br}_2} BrCH=CRCO_2\text{-}i\text{-Pr}$$

JOC *50* 2195 (1985)

4. Halofunctionalization

For the synthesis of vinylic halides via halofunctionalization of alkynes see page 240, Section 2.21.

$$\underset{/}{\overset{\backslash}{}}C=C\underset{\backslash}{\overset{/}{}} \longrightarrow -\underset{|}{\overset{|}{C}}-\underset{|}{\overset{|}{C}}- \quad \overset{X\quad OH}{\underset{}{}}$$

Review: Russ Chem Rev *41* 740 (1972)

X	Reagents	
Cl	Cl_2, H_2O	JCS *119* 1774 (1921); *121* 2595 (1922); 1817 (1949) Ber *61* 510, 518 (1928) JACS *72* 4608 (1950) Org Syn *33* 15 (1953) Ann *596* 140 (1955)
	Cl_2, H_2O, HgCl$_2$ Cl_2, H_2O, HgCl$_2$, NaOH/HNO$_3$	Ann *596* 138 (1955) Org Syn Coll Vol *1* 158 (1941)
	Cl_2, H_2O, NaHCO$_3$, CO$_2$ HOCl, H_2O	JCS 1817 (1949) Helv *35* 1263 (1952); *39* 423 (1956) Ber *89* 2424 (1956) JACS *79* 2341 (1957)
	HOCl, H_2O, CuCl$_2$ EtOCl/H_2O *t*-BuOCl, H_2O, HOAc *t*-BuOCl, HOAc *t*-BuOCl, HOAc, H_2SO_4 NaOCl, H_2O, CO$_2$ KOCl, H_2O, CO$_2$ Ca(OCl)$_2$, HOAc, H_2O	Rec Trav Chim *50* 261 (1931) Ber *58* 572 (1925) JCS 114 (1946); 239 (1949) Helv *35* 1263 (1952) JACS *63* 2541 (1941) JCS 1817 (1949) JACS *44* 148 (1922) JACS *58* 2396 (1936)

$$>C=C< \longrightarrow -\overset{X}{\underset{|}{C}}-\overset{OH}{\underset{|}{C}}- \quad (Continued)$$

X	Reagents	
	Ca(OCl)$_2$, CO$_2$, H$_2$O	JACS 67 516 (1945); 71 2666 (1949)
		BSCF 1240 (1964)
	Me$_3$SiCl, H$_2$O$_2$	Syn Commun 9 37 (1979)
	TiCl$_4$, t-BuO$_2$H or (t-BuO)$_2$	JOC 50 915 (1985)
	ClNHCONH$_2$, HOAc, H$_2$O	Org Syn Coll Vol 4 157 (1963)
	TsNClNa, H$_2$O	Syn 362 (1981)
Br	Br$_2$, H$_2$O	Ann 348 285 (1906)
		JCS 111 240 (1917);
		117 359 (1920);
		119 1774 (1921)
	Br$_2$, H$_2$O, KBr	JCS 101 758 (1912); 1487 (1928)
		JOC 18 1586 (1953)
	Br$_2$, H$_2$O, Na$_2$CO$_3$	Ber 54 1945 (1921)
	Br$_2$, H$_2$O, NaHCO$_3$, CO$_2$	JCS 1817 (1949)
	Br$_2$, H$_2$O, HgO	Helv 7 108 (1924)
		JCS 285 (1936); 1817 (1949)
	HOBr, H$_2$O	Helv 35 1263 (1952)
		JACS 79 2341 (1957)
	Ca(OBr)$_2$, H$_3$BO$_3$	Ber 32 3490 (1899)
	CH$_3$CONHBr, H$_2$O	Ber 59 1279 (1926)
		JACS 64 2780 (1942);
		74 1160 (1952);
		78 1740 (1956)
		Helv 26 562, 586, 705, 721, 1799 (1943)
		Gazz Chim Ital 108 643 (1978)
		JOC 52 4384 (1987)
	CH$_3$CONHBr, H$_2$O, H$^+$	JACS 61 1576 (1939);
		79 1130 (1957)
		JOC 18 1586 (1953);
		51 3407 (1986)
	CH$_3$CONHBr, H$_2$O, HOAc, NaOAc	Helv 27 821 (1944);
		28 1420 (1945);
		30 1616 (1947)
	NBS, H$_2$O	JACS 74 1160 (1952); 76 4373 (1954); 77 2549 (1955)
		Can J Chem 43 2398 (1965)
		JOC 35 2670 (1970)
	NBS, H$_2$O, DMSO	JACS 90 5498 (1968)
	NBS, H$_2$O, H$^+$	JCS 401 (1952)
		JACS 76 5017 (1954)
I	I$_2$, H$_2$O	Ber 58 794, 1064, 1071 (1925);
		59 113, 375 (1926);
		60 991 (1927)
		JCS C 928 (1970)

I_2, H_2O, sulpholane, $HCCl_3$	JCS Perkin I 226 (1977)
I_2, H_2O/H_2O_2	Can J Chem *42* 2710 (1964)
I_2, H_2O, HgO	Compt Rend *130* 1766 (1900); *131* 528 (1900); *135* 1055 (1902)
	JCS 1817 (1949)
I_2, H_2O, CuO-HBF_4	CC 1491 (1987)
I_2, H_2O, $NaNO_2$, H^+	JCS C 846 (1970)
I_2, H_2O, KIO_3, H^+	JCS C 846 (1970)
I_2, PDC, molecular sieves	Tetr *39* 1765 (1983)
I_2, AgO_2CCF_3/MeOH	TL *27* 4245 (1986)
$I(py)_2BF_4$, H_2O	Angew Int *24* 319 (1985)

$$\underset{}{\overset{}{>}}C=C\underset{}{\overset{}{<}} \longrightarrow -\overset{\overset{X}{|}}{\underset{|}{C}}-\overset{\overset{OR}{|}}{\underset{|}{C}}-$$

For intramolecular reactions see page 445, Section 2.

Review: Russ Chem Rev *41* 740 (1972)

X	Reagents	
F	$CsSO_4F$, ROH	JOC *52* 919 (1987)
	XeF_2, CH_3OH	JOC *50* 2751 (1985)
	CF_3OF (R = CF_3)	CC 227 (1969)
		JCS Perkin I 739 (1974)
	CH_3OH, $Hg(OAc)_2$ /NaCl/CH_3CO_2F	JOC *52* 2588 (1987)
Cl	Cl_2, ROH	JACS *48* 2166 (1926)
	t-BuOCl, ROH	JACS *63* 858, 1624 (1941); *72* 4608 (1950); *73* 2302 (1951)
		Chem Rev *54* 925 (1954) (review)
	$PhSO_2NCl_2$, ROH	J Gen Chem USSR *8* 370 (1938) [CA *32* 519 (1938)]
		JACS *76* 693 (1954)

, CH_3OH JOC USSR *5* 1879 (1969); *7* 1876 (1971)

Br	Br_2, ROH	JACS *46* 1727 (1924); *48* 2166 (1926); *71* 1096 (1949); *74* 5518 (1952)
		Helv *35* 762 (1952)
	Br_2, ROH, $AgNO_3$	Ann *516* 231 (1935)
	$CH_3CONHBr$, ROH	Ber *59* 1279 (1926)
		JACS *65* 2196 (1943); *74* 1160 (1952)
		JOC *17* 233 (1952)
	NBS, ROH	JACS *76* 4368, 4373 (1954)
		TL *28* 2009 (1987)

$$\text{>C=C<} \longrightarrow \underset{\underset{}{\overset{X}{\mid}}}{-\text{C}-}\underset{\underset{}{\overset{OR}{\mid}}}{\text{C}-} \quad (\textit{Continued})$$

$\underline{X}$	Reagent(s)	
	PhSO$_2$NBr$_2$, ROH	JACS *69* 2001 (1947); *76* 693 (1954)
	BrC(NO$_2$)$_3$, ROH	Ber *55* 2099 (1922); *56* 1239 (1923); *57* 2039 (1924); *59* 1279, 1876 (1926)

	(EtO$_2$C)$_2$CBrNO$_2$ or (NO$_2$)$_2$CBr$_2$; ROH	Ber *59* 1876 (1926)
I	ICl, ROH	JACS *50* 2249 (1928)
	I$_2$, MeOH	JCS Perkin I 226 (1977)
	I$_2$, MeOH, HgO	Compt Rend *135* 1055 (1902)
	I$_2$, ROH, Cu(OAc)$_2$·H$_2$O	Syn 402 (1978)
	I$_2$, ROH/Cl$_2$	JACS *50* 2249 (1928)
	NIS, ROH	JACS *109* 8119 (1987)
	I(py)$_2$BF$_4$, ROH	Angew Int *24* 319 (1985)

$$\text{C=C}-\text{C=C} \longrightarrow \text{X}-\text{C}-\underset{\underset{}{\overset{OMe}{\mid}}}{\text{C}}-\text{C=C}$$

$\underline{X}$	Reagents	
Br	Br$_2$, MeOH	JOC *35* 539 (1970)
I	I$_2$, CuO·HBF$_4$, MeOH	CC 1491 (1987)
	I(py)$_2$BF$_4$, MeOH	TL *27* 1715 (1986)

$$\text{>C=C<} \longrightarrow \underset{\underset{}{\overset{X}{\mid}}}{-\text{C}-}\underset{\underset{}{\overset{O_2CR}{\mid}}}{\text{C}-}$$

For iodolactonization see page 941, Section 8.

$\underline{X}$	Reagent(s)	
F	CH$_3$CO$_2$F	CC 443 (1981)
	CF$_3$CO$_2$F	JOC *45* 672 (1980)
Cl	Cl$_2$, RCO$_2$H	JACS *61* 1457 (1939)
		J Gen Chem USSR *25* 709 (1955)
	Cl$_2$, Ac$_2$O, NaOAc	JACS *61* 1457 (1939)
	Cl$_2$, AgO$_2$CR	Ber *68* 824 (1935)

	t-BuOCl, RCO_2H	JACS 63 858 (1941)
	CrO_2Cl_2, AcCl	TL 3523 (1977)
	AcOCl	Ann 519 165 (1935)
	$TsNClNa \cdot 3\ H_2O$, RCO_2H	TL 21 1709 (1980)
Br	$Br_2 / NaOAc$	JACS 76 693 (1954)
	Br_2, AgO_2CR	Ber 68 824 (1935)
		JCS 1515, 2934 (1952); 762 (1954)
	Br_2, $Hg(OAc)_2$, HOAc	J Gen Chem USSR 25 709 (1955)
	RCO_2Br	Ann 519 165 (1935)
	$CH_3CONHBr$, HCO_2H	Ber 59 1279 (1926)
	$CH_3CONHBr$, HOAc, LiOAc	TL 27 4253 (1986)

$$\text{Br}\!-\!\underset{\text{Br}}{\overset{\text{Br}}{\bigcirc}}\!-\!NBrCOCH_3,$$

JCS 762 (1954)

	RCO_2H, py	
	$BrC(NO_2)_3$, HCO_2H	Ber 56 1239 (1923)
I	ICl, TlO_2CR	JCS Perkin I 226 (1977)
	I_2, NaO_2CR	JCS Perkin I 226 (1977)
	I_2, KOAc, Al_2O_3	CC 1301 (1987)
	I_2, AgO_2CR	Ber 65 1339 (1932); 67 1729 (1934)
	$I(py)_2BF_4$, RCO_2H	Angew Int 24 319 (1985)

$$C{=}C{-}C{=}C \xrightarrow[X = Cl,\ Br]{XO_2CR} \overset{X}{\underset{}{C}}{-}\overset{O_2CR}{\underset{}{C}}{-}C{=}C$$

JOC 37 2228 (1972)

$$C{=}C{-}C{=}C \longrightarrow \overset{X}{\underset{}{C}}{-}C{=}C{-}\overset{O_2CR}{\underset{}{C}}$$

$\underline{X}$	Reagents	
Cl	cat $Pd(OAc)_2$, LiCl, LiOAc$\cdot$2 H_2O, benzoquinone	TL 23 1617 (1982); 28 4199 (1987) Tetr 41 5761 (1985) JACS 107 3676 (1985)
I	I_2, HOAc, $CuO\text{-}HBF_4$	CC 1491 (1987)

$$C{=}C{-}C{=}C \xrightarrow[\text{Nuc or Nuc}^-]{I(py)_2BF_4} I{-}C{-}\overset{Nuc}{\underset{}{C}}{-}C{=}C \text{ or } I{-}C{-}C{=}C{-}C{-}Nuc$$

Nuc = H, Cl, OMe, O_2CH, NO_2, $NHCOCH_3$

TL 27 1715 (1986)

$$\text{>C=C<} \longrightarrow \underset{\overset{|}{C}}{\overset{\overset{X}{|}}{}} - \underset{\overset{|}{C}}{\overset{\overset{Y}{|}}{}} -$$

X	Y	Reagent(s)	
Cl	N$_3$	Cl$_2$, NaN$_3$	TL 3309 (1969)
			Acct Chem Res *4* 9 (1971) (review)
Br	N$_3$	Br$_2$, NaN$_3$	JACS *90* 216 (1968)
			TL 3309 (1969)
			Acct Chem Res *4* 9 (1971) (review)
I	N$_3$	ICl, NaN$_3$	JACS *87* 4203 (1965); *89* 2077 (1967)
			TL 3309 (1969)
			Acct Chem Res *4* 9 (1971) (review)
		I$_2$, NaN$_3$, Al$_2$O$_3$	CC 1301 (1987)
		I$_2$, NaN$_3$, Adogen 464, H$_2$O, HCCl$_3$	Syn 462 (1977)
		I$_2$, NaN$_3$, sulpholane, HCCl$_3$	JCS Perkin I 226 (1977)
I	NO$_2$	I$_2$, CuO, HBF$_4$, NaNO$_2$	CC 1491 (1987)
		I(py)$_2$BF$_4$, NO$_2^-$	Angew Int *24* 319 (1985)
I	NCO	I(py)$_2$BF$_4$, NCO$^-$	Angew Int *24* 319 (1985)
I	NHCOCH$_3$	I(py)$_2$BF$_4$, CH$_3$CN	Angew Int *24* 319 (1985)
I	O$_2$SAr	I$_2$, CuO-HBF$_4$, NaO$_2$SAr	CC 1491 (1987)
I	SCN	ICl, KSCN, HCCl$_3$, sulpholane	TL 1531 (1976)
		I$_2$, (SCN)$_2$	TL 3567 (1972)
		I$_2$, KSCN, HCCl$_3$, sulpholane	TL 1531 (1976)
		I$_2$, KSCN, 18-crown-6, HCCl$_3$	Syn 462 (1977)
		I$_2$, KSCN, Adogen 464, H$_2$O, HCCl$_3$	Syn 462 (1977)
		I$_2$, KSCN, Al$_2$O$_3$	CC 1301 (1987)
		I$_2$, TlSCN, HCCl$_3$, sulpholane	TL 1531 (1976)
I	C$_6$H$_4$OMe	I$_2$, CuO-HBF$_4$, C$_6$H$_5$OMe	CC 1491 (1987)
I	Ar	I(py)$_2$BF$_4$, Ar-H	Angew Int *24* 319 (1985)

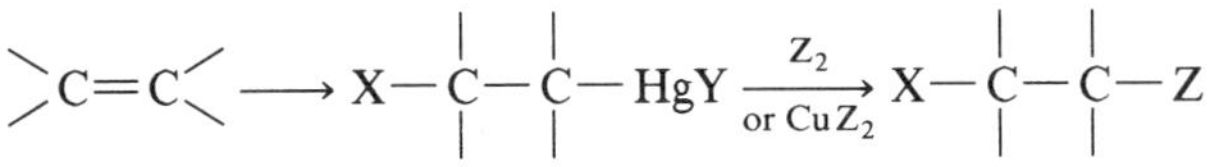

$$\diagdown C = C \diagup \longrightarrow X - \overset{|}{\underset{|}{C}} - \overset{|}{\underset{|}{C}} - HgY \xrightarrow[\text{or } CuZ_2]{Z_2} X - \overset{|}{\underset{|}{C}} - \overset{|}{\underset{|}{C}} - Z$$

X = HO, RO, RO$_2$, RCO$_2$, R$_2$N, RCONH, N$_3$, F

Y = OAc, O$_2$CCF$_3$, NO$_3$, F, Cl

Z = Cl, Br, I

R. C. Larock, "Solvomercuration/Demercuration Reactions in Organic Synthesis," Springer, New York (1986) (review)

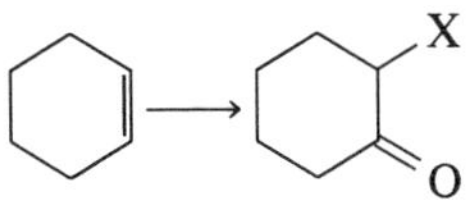

X	Reagent(s)	
Cl	O$_2$, FeCl$_3$, py, hν	JOC *46* 509 (1981)
	CrO$_2$Cl$_2$	Syn Commun *11* 7 (1981)
I	I$_2$, PCC	TL *21* 4521 (1980)

5. Miscellaneous Reactions

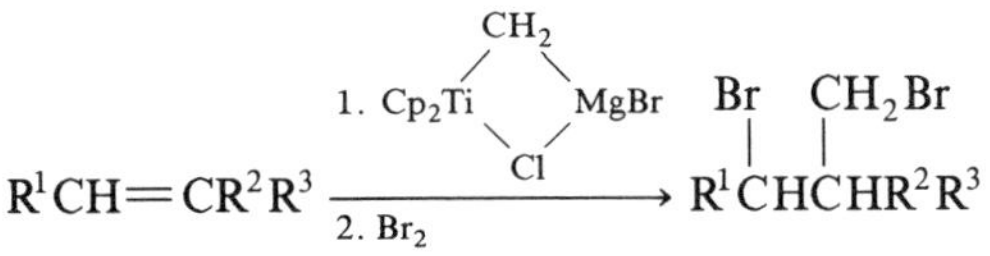

$$R^1CH = CR^2R^3 \xrightarrow[\text{2. Br}_2]{\text{1. } Cp_2Ti \diagdown \overset{CH_2}{\underset{Cl}{\diagup}} MgBr} R^1\overset{Br}{\underset{|}{C}}HCHR^2R^3$$

TL *24* 3935 (1983)

4. HALOGENATION OF ALKYNES

See also page 215, Section 2; and page 337, Section 5, for additional preparations of vinylic halides.

$$RC\equiv CH \longrightarrow RC\equiv CX$$

X	Reagent(s)	
Cl	NaOH, CCl$_4$ (phase transfer)	Rocz *49* 1779 (1975)
	n-BuLi/NCS	Syn 296 (1979)
		JACS *101* 5101 (1979)
	PhSO$_2$Cl (on RC≡CNa)	Ann Chim [10] *16* 309 (1931)
		JACS *56* 1106 (1934)
		BSCF 1447 (1957)
	n-BuLi/Cl$_2$	Tetr *22* 965 (1966)
	HOCl	Ber *63* 1868, 1886 (1930)
		Can J Chem *32* 500 (1954)
		JACS *85* 1648 (1963)
Br	DBU, Cl$_3$CBr	CL 73 (1978)
	NaOH, Br$_2$	JCS 2295 (1963)
	HOBr	Ber *63* 1868 (1930)
		BSCF 1447 (1957)
		Arch Pharm *290* 118 (1959)
		JACS *85* 1648 (1963)
	n-BuLi/NBS	JACS *94* 4013 (1972)
	n-BuLi/Br$_2$	Tetr *22* 965 (1966)
		JACS *89* 5086 (1967)
	BrCN (on RC≡CMgBr)	Ann Chim [10] *5* 5 (1926)
I	I$_2$, NH$_3$	JACS *54* 787 (1932); *55* 2150 (1933); *56* 1207 (1934)
		JCS 741 (1933)
	I$_2$, O⟨⟩NH	JOC *27* 3305 (1962)
	HOI	Ber *37* 4412 (1904); *63* 1886 (1930)

$$RC \equiv CH \longrightarrow RC \equiv CX \quad (\textit{Continued})$$

$\underline{X}$	Reagent(s)	
I	I_2, $Hg(OAc)_2$	Bull Acad Sci USSR, Div Chem Sci 869 (1968)
	$NaOCH_3/I(py)_2BF_4$	Syn 661 (1987)
	$Na, NH_3/I_2$	JACS 55 2150 (1933); 56 1106 (1934)
	RLi/I_2	Tetr 22 965 (1966)
		JACS 89 5086 (1967); 107 1028 (1985)
		JOC 51 5320 (1986)
	Na/I_2	Ann 308 264 (1899)
	ICN (on $RC \equiv CMgBr$)	Ann Chim [10] 5 5 (1926)

$$RC \equiv CR' \longrightarrow \overset{\overset{\displaystyle X \quad X}{\displaystyle | \quad |}}{RC = CR'}$$

$\underline{X}$	Reagent(s)	
Cl	SO_2Cl_2	BCSJ 54 2843 (1981)
Br	Br_2	TL 1629 (1970)
	$[polymer\text{-}C_6H_5CH_2\overset{+}{N}Me_3]Br_3^-$	Syn 143 (1980)
I	I_2, alumina	TL 29 35 (1988)

$$RC \equiv CH + HBr \xrightarrow[\text{peroxides}]{} RCH = CHBr + \overset{\overset{\displaystyle Br}{\displaystyle |}}{R}CHCH_2Br$$

JACS 58 1806 (1936)

$$RC \equiv CH \longrightarrow \overset{\overset{\displaystyle X}{\displaystyle |}}{RC} = CH_2$$

$\underline{X}$	Reagent(s)	
Cl	HCl	JOC 62 1367 (1940); 39 1124 (1974)
		Chem Ber 106 2001 (1973)
	$(Et_4N)HCl_2$	JCS Perkin I 1797 (1977)
Br	HBr	JACS 58 1806 (1936)
		Helv 54 2060 (1971)
	$(Et_4N)HBr_2$	Syn 805 (1980)
	$RLi/Me_3SiCl/HBr$	JOC 39 3307 (1974)
	$B\text{-}Br\text{-}9\text{-}BBN/HOAc$	TL 24 731 (1983)
I	I_2, Al_2O_3	TL 28 4497 (1987)
	$B\text{-}I\text{-}9\text{-}BBN/HOAc$	TL 24 731 (1983)

$$\text{RC} \equiv \text{CR} \longrightarrow \overset{\overset{\displaystyle H}{|}}{R}C = \overset{\overset{\displaystyle Cl}{|}}{C}R$$

HCl, Lewis acid

CC 857 (1972)
JCS Perkin I 2491 (1973)
JACS *98* 3295 (1976)

$(\text{Et}_4\text{N})\text{HCl}_2$

JCS Perkin I 1797 (1977)

$$\overset{\overset{\displaystyle O}{\|}}{R'C}\text{C} \equiv \text{CR} \xrightarrow[\text{HOAc or CF}_3\text{CO}_2\text{H}]{\text{LiBr or NaI}} \overset{\overset{\displaystyle O}{\|}}{R'C}\text{CH} = \overset{\overset{\displaystyle X}{|}}{C}R$$

TL *27* 4763 (1986)

$$\text{RC} \equiv \text{CR}' \longrightarrow \text{RC} \equiv \text{CR}' \ \text{BF}_4^- \overset{\underset{\displaystyle \text{CpF\overset{+}{e}(CO)PPh}_3}{|}}{}$$

1. NaH
2. Br$_2$

$$\overset{R}{\underset{Br}{}} C = C \overset{H}{\underset{R'}{}}$$

1. LiCuPh$_2$
2. I$_2$

$$\overset{R}{\underset{I}{}} C = C \overset{Ph}{\underset{R'}{}}$$

JACS *102* 5923 (1980)

$$\text{RC} \equiv \text{CH} \longrightarrow \text{RC} \equiv \text{CX} \xrightarrow[\substack{\text{2. Pb(OAc)}_4 \text{ or} \\ \text{PhI(OAc)}_2}]{\text{1. HBR}_2'} \text{RCH} = \text{CXR}'$$

X = Cl, Br

BCSJ *53* 1652 (1980)

$$\overset{\overset{\displaystyle O}{\|}}{R}\text{CC} \equiv \text{CR} \xrightarrow[\substack{\text{2. R'CHO} \\ \text{3. H}_2\text{O}}]{\substack{\text{1. Me}_3\text{SiI, Et}_2\text{AlI or} \\ n\text{-Bu}_4\text{NI-TiCl}_4}} \overset{\overset{\displaystyle O \ \ OH}{\| \ \ |}}{R}\text{CC}\underset{\underset{\displaystyle RCI}{\|}}{}\text{CHR}'$$

TL *27* 4767 (1986)

$$\text{RC} \equiv \text{CR} \longrightarrow R - \overset{\overset{\displaystyle I}{|}}{C} = \overset{\overset{\displaystyle Nuc}{|}}{C} - R$$

See page 215, Section 2.

$$\overset{\overset{\displaystyle OH}{|}}{R_2}\text{CC} \equiv \text{CH} \xrightarrow[\text{2. PDC}]{\text{1. I}_2} R_2C = \overset{\overset{\displaystyle I}{|}}{C}\text{CHO}$$

TL *22* 1041 (1981)

5. INTERCONVERSION OF HALIDES

1. Aliphatic Halides

$$RY \longrightarrow RX$$

X	Y	Reagent(s)	
F	—	review	Org Rxs *2* 49 (1944)
	Cl	KF (phase transfer)	Syn 428 (1974)
			JOC *50* 879 (1985)
		KF, HOCH$_2$CH$_2$OH	JACS *79* 2311 (1957)
		KF, (HOCH$_2$CH$_2$)$_2$O	Coll Czech Chem Commun
			39 2616 (1974)
		KF, CH$_3$CN	CL 761 (1981)
		KF, CaF$_2$, CH$_3$CN	CC 791, 793 (1986)
		KF, CaF$_2$, sulpholane	CC 791 (1986)
		KF, *n*-Bu$_4$NBr, HCONH$_2$	TL *27* 1499 (1986)
		CsF, CaF$_2$, sulpholane	CC 791 (1986)
		n-Bu$_4$NF (no solvent)	JOC *49* 3216 (1984)
		n-Bu$_4$NF, $\left[\begin{array}{c} \overset{+}{N}H \end{array} \right] OTs^-$	TL *28* 4419 (1987)
		(*n*-Bu$_4$N)HF$_2$	TL *28* 4733 (1987)
		[resin-$\overset{+}{N}$Me$_3$]F$^-$	Syn 472 (1976)
		n-Bu$_3$MePF	TL 353 (1979)
			Tetr *36* 1931 (1980)
		AgBF$_4$	TL *28* 5347 (1987) (gem
			dihalides and trihalides)
		Mg/TsNFR	JACS *106* 452 (1984)
	Br	KF (phase transfer)	Syn 428 (1974)
		KF, cat 18-crown-6	JACS *96* 2250 (1974)
			JOC *40* 782 (1975)
		KF, HOCH$_2$CH$_2$OH	Org Syn Coll Vol *4* 525
			(1963)

RY $\longrightarrow$ RX　(*Continued*)

X	Y	Reagent(s)	
		KF, $(HOCH_2CH_2)_2O$	Coll Czech Chem Commun *39* 2616 (1974)
		KF, $H(OCH_2CH_2)_nOH$	CL 283 (1978)
		KF, HMPA	BSCF 334 (1969)
		KF, CH_3CN	CL 761 (1981)
		KF, CaF_2, CH_3CN	CC 793 (1986)
		KF, n-Bu_4NBr, $HCONH_2$	TL *27* 1499 (1986)
		n-Bu_4NF (no solvent)	JOC *49* 3216 (1984)
		n-Bu_4NF, $\left[\underset{+}{\langle\!\!\!\!\bigcirc\!\!\!\!\rangle}NH\right]OTs^-$	TL *28* 4419 (1987)
		$(n$-$Bu_4N)HF_2$	TL *28* 4733 (1987)
		[resin-$\overset{+}{N}Me_3$]F^-	Syn 472 (1976) JCS Perkin I 2248 (1979)
		n-Bu_3MePF	TL 353 (1979) Tetr *36* 1931 (1980)
		$(Ph_4P)HF_2$	CC 672 (1985)
		KF, $(Ph_4P)Br$	TL *28* 111 (1987)
		Mg/TsNFR	JACS *106* 452 (1984)
		$Li/FClO_3$	Ber *102* 1944 (1969)
		AgF	JOC *40* 782 (1975) TL *28* 4003 (1987)
		HgF_2	JACS *70* 2310 (1948) JOC *40* 782 (1975)
	I	[resin-$\overset{+}{N}Me_3$]F^-	JCS Perkin I 2248 (1979)
		$(n$-$Bu_4N)HF_2$	TL *28* 4733 (1987)
		AgF	JOC *40* 782 (1975)
		$AgBF_4$	TL *28* 5347 (1987) (gem dihalides)
		HgF_2	JOC *40* 782 (1975)
Cl	Br	LiCl, CH_3COCH_3	JCS 3173 (1955)
		NaCl, H_2O, phase transfer	CC 1250 (1986)
		cat NaCl, i-PrCl	BCSJ *49* 1989 (1976)
		[resin-$\overset{+}{N}Me_3$]Cl^-	Syn 472 (1976)
		graphite-$SbCl_5$	TL 763 (1974)
		ClF_2CCO_2Ag	TL 3447 (1970)
	I	cat NaCl, n-PrCl	BCSJ *49* 1989 (1976)
		Cl_2	TL *27* 6055 (1986)
		$PhICl_2$	TL *27* 6055 (1986)
		graphite-$SbCl_5$	TL 763 (1974)
Br	—	review	Houben-Weyl, Vol V/4, p 354
	Cl	HBr, cat $FeBr_3$	CC 1013 (1987)
		LiBr, CH_3COCH_3	JACS *77* 4903 (1955)
		LiBr, phase transfer	CC 1250 (1986)

		NaBr, CH$_3$OH	JACS *39* 1730 (1917); *77* 165 (1955)
		cat NaBr, EtBr	BCSJ *49* 1989 (1976)
		CaBr$_2$	BSCF *1* 860 (1934)
		CaBr$_2$, R$_4$NBr	Syn 34 (1984)
		AlBr$_3$	JACS *78* 491 (1956)
		[resin-$\overset{+}{N}$Me$_3$]Br$^-$	Syn 472 (1976)
	I	cat NaBr, EtBr	BCSJ *49* 1989 (1976)
		Mg/Br$_2$	JACS *41* 287 (1919)
		Br$_2$	TL *27* 6055 (1986)
I	—	review	Houben-Weyl, Vol V/4, p 595
	F	Me$_3$SiI	JOC *46* 3727 (1981)
	Cl	HI, cat FeI$_3$	CC 1013 (1987)
		NaI, CH$_3$COCH$_3$	JACS *109* 5491 (1987)
		KI, *n*-Bu$_4$PI, silica	Syn 952 (1979)
		NaI, ZnCl$_2$	JCS Perkin I 416 (1976)
		NaI, FeCl$_3$	JCS Perkin I 416 (1976)
		NaI, FeCl$_3$ or HgCl$_2$	TL 2691 (1974)
		cat NaI, MeI	BCSJ *49* 1989 (1976)
		Me$_3$SiI	JOC *46* 3727 (1981)
	Br	NaI, phase transfer catalyst	JOC *50* 5828 (1985)
		NaI, CH$_3$COCH$_3$	JCS 1605 (1936)
		NaI, HMPA	JACS *90* 6225 (1968)
		NaI, benzo-15-crown-5	CC 879 (1974)
		cat NaI, MeI	BCSJ *49* 1989 (1976)
		KI, DMF	JOC *34* 3519 (1969)
		KI, crown ether	CC 879 (1974)
		KI, *n*-Bu$_4$PI, silica	Syn 952 (1979)
		[resin-$\overset{+}{N}$Me$_3$]I$^-$	Syn 472 (1976)
		Mg/I$_2$	JACS *41* 287 (1919)
		Mg/HgBr$_2$/I$_2$	JACS *61* 1585 (1939); *91* 5774 (1969)

2. Vinylic Halides

X	Y	Reagent(s)	
F	I	*t*-BuLi/PhSO$_2$NF-*t*-Bu	JACS *108* 2445 (1986)
Cl, Br	I	CuX	JOMC *93*, 415 (1975)
I	Br	KI, Zn, NiBr$_2$	CL 1435 (1978)
		KI, NiBr$_2$, electrolysis	TL *27* 3497 (1986)

3. Aromatic Halides

$$ArX \xrightarrow[\text{pyridine}]{\text{CuY}} ArY$$

$$X = Br, I \qquad Y = Cl, Br, I$$

Proc Chem Soc 113 (1962)
JCS 1097, 1108 (1964)
Tetr *40* 1433 (1984) (review)

$$ArY \longrightarrow ArX$$

X	Y	Reagent(s)	
F	Cl	KF, DMF or DMSO	JOC *28* 1666 (1963)
		KF, sulfolane	JCS 6264 (1965)
		KF, (Ph$_4$P)Br	TL *28* 111 (1987)
	Br	Li/FClO$_3$	Ber *102* 1944 (1969)
		Mg/TsNF-t-Bu	JACS *106* 452 (1984)
		KF, sulfolane	JCS 6264 (1965)
Cl	Br	Mg/Ph$_3$PCl$_2$	JOC *32* 3710 (1967)
		CuCl	BSCF 720 (1973)
	I	CCl$_4$, hν	JOC *35* 528 (1970)
Br	I	Mg/Br$_2$	JACS *41* 287 (1919)
I	Cl	NaI, DMF	JOC *23* 305 (1958)
		CuI·alumina or charcoal	CC 1409 (1987)
	Br	Mg/I$_2$	JACS *41* 287 (1919)
			JOC *3* 55 (1938)
		Mg/ICN	Ann Chim *4* 28 (1915)
		NaI, CH$_3$COCH$_3$	JOC *28* 218 (1963)
		KI, I$_2$, Ni, DMF	JOC *52* 691 (1987)
		KI, NiBr$_2$, Zn	CL 191 (1978)
		KI, NiBr$_2$, electrolysis	TL *27* 3497 (1986)
		KI, CuI, HMPA	JACS *108* 3150 (1986)
		KI, cat Pd(OAc)$_2$,	BCSJ *48* 3298 (1975)
		NaOEt, HMPA	
		CuI·alumina or charcoal	CC 1409 (1987)

4. Homologation of Halides

$$1° \ RX + LiCH_2SPh \longrightarrow RCH_2SPh \xrightarrow[\substack{\text{NaI} \\ \text{DMF}}]{\text{CH}_3\text{I}} RCH_2I$$

$$X = Br, I$$

TL 5787 (1968)

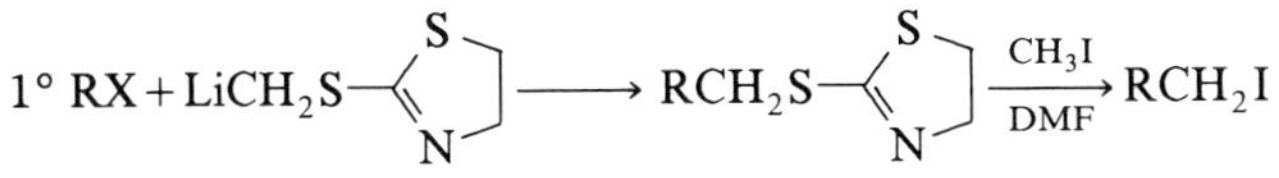

$$1° \ RX + LiCH_2S-\underset{N}{\overset{S}{\langle}} \longrightarrow RCH_2S-\underset{N}{\overset{S}{\langle}} \xrightarrow[DMF]{CH_3I} RCH_2I$$

X = Cl, Br

TL 2743 (1972)

$$RX \xrightarrow[\text{2. ROCH}_2\text{NR}_2]{\text{1. Mg}} RCH_2NR_2 \xrightarrow{ClCO_2R} RCH_2Cl$$

TL 28 427 (1987)

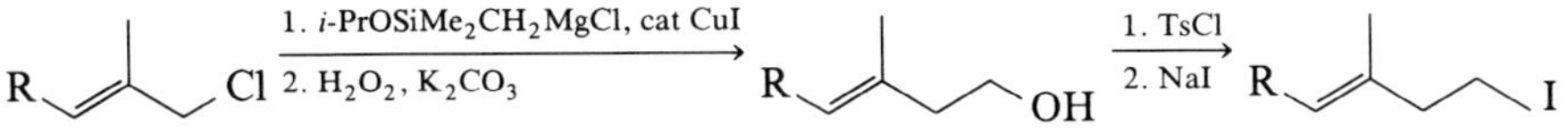

$$R \diagdown = \diagup Cl \xrightarrow[\text{2. H}_2\text{O}_2, \text{K}_2\text{CO}_3]{\text{1. }i\text{-PrOSiMe}_2\text{CH}_2\text{MgCl, cat CuI}} R \diagdown = \diagup OH \xrightarrow[\text{2. NaI}]{\text{1. TsCl}} R \diagdown = \diagup I$$

JOC 52 4885 (1987)

$$1° \ RX + LiCH_2TePh \longrightarrow RCH_2TePh \xrightarrow{reagent} RCH_2X$$

X = Cl, Br, I

Reagent: $SOCl_2$, Br_2, I_2, MeI-NaI

$$1° \ RX + LiCH(TePh)_2 \longrightarrow RCH(TePh)_2 \xrightarrow{Br_2} RCHBr_2$$

CL 1081 (1982)

$$1° \ RX \xrightarrow{Ph_2AsOCH_2Li} RCH_2\overset{O}{\underset{||}{As}}Ph_2 \xrightarrow[\substack{\text{2. SO}_2\text{Cl}_2, \\ \text{Br}_2, \text{ or I}_2}]{\text{1. LiAlH}_4} RCH_2X$$

X = Cl, Br, I

Ber 115 645 (1982)

$$RCH_2X \xrightarrow[\text{2. }\Delta]{\text{1. PhSOCHYLi}} RCH{=}CHY$$

R = aryl, X = Br

R = alkyl, X = I

Y	
F	TL 24 725 (1983)
Cl	TL 617 (1979)

$$RCH_2X \xrightarrow[\text{2. }\Delta]{\text{1. PhSOCCl}_2\text{Li}} RCH{=}CCl_2$$

TL 24 527 (1983)

$$RX + H_2C = CHR' \xrightarrow[\text{acid}]{\text{Lewis}} RCH_2\overset{\overset{\displaystyle X}{|}}{C}HR'$$

$$X = Cl, Br$$

Review: Angew Int *20* 184 (1981)

RX

RX
JACS *67* 1152 (1945); *75* 6217 (1953)
JOC *48* 1159 (1983)

ArCHRX
JOC *29* 2685 (1964); *44* 3022 (1979); *48* 1159 (1983); *50* 2995 (1985)

Me$_2$C=CHCH$_2$Cl
Angew Int *20* 1027 (1981); *21* 82 (1982)

RCHClOR
Ann *525* 151 (1936)
Angew Int *6* 335 (1967) (review)
JOC *48* 1159 (1983)

TL *28* 4517 (1987)

$$R_fI + H_2C = CR_2 \xrightarrow{\text{cat Raney Ni}} R_f CH_2\overset{\overset{\displaystyle I}{|}}{C}R_2$$

CC 498 (1986)

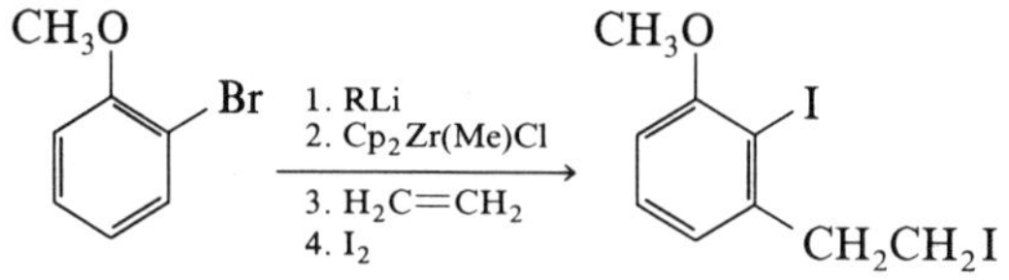

JACS *109* 7137 (1987)

$$RX + H_2C = CHCH = CH_2 \xrightarrow[\text{acid}]{\text{Lewis}} RCH_2\overset{\overset{\displaystyle X}{|}}{C}HCH = CH_2 + RCH_2CH = CHCH_2X$$

Angew Int *20* 184 (1981)
Ber *115* 3528 (1982) (R = propargylic)
JOC *48* 1159 (1983)

$$RX + R'C \equiv CR' \xrightarrow[\text{acid}]{\text{Lewis}} RR'C = CXR'$$

CC 857 (1972)
JCS Perkin I 2491 (1973); 353 (1974)
JCS Perkin II 1517 (1976)
JACS *98* 3295 (1976)
JOC *44* 3022 (1979)
Angew Int *20* 184 (1981)

5. Rearrangement of Halides

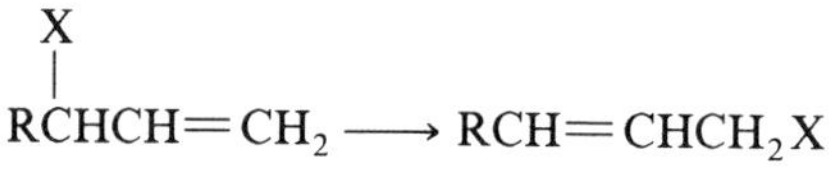

$$\overset{\overset{\text{X}}{|}}{\text{RCHCH}}=\text{CH}_2 \longrightarrow \text{RCH}=\text{CHCH}_2\text{X}$$

JCS 2720 (1959)

$$\overset{\overset{\text{Br}}{|}}{\text{R}^1\text{CH}}-\overset{\overset{\text{O}}{\|}}{\text{C}}-\text{CH}_2\text{R}^2 \xrightarrow{\text{HBr}} \text{R}^1\text{CH}_2-\overset{\overset{\text{O}}{\|}}{\text{C}}-\overset{\overset{\text{Br}}{|}}{\text{CHR}^2}$$

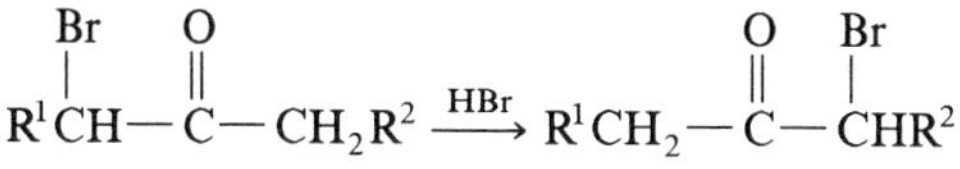

JOC *19* 538 (1954)
JCS 1342 (1955)

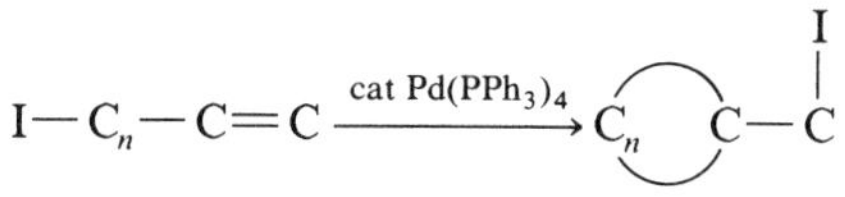

$$\text{I}-\text{C}_n-\text{C}=\text{C} \xrightarrow{\text{cat Pd(PPh}_3)_4} \text{C}_n \overset{\overset{\text{I}}{|}}{\underset{\text{C}}{\text{C}}}-\text{C}$$

TL *23* 5315 (1982); *26* 1519 (1985)
Tetr *41* 5465 (1985)
CC 1375 (1986)

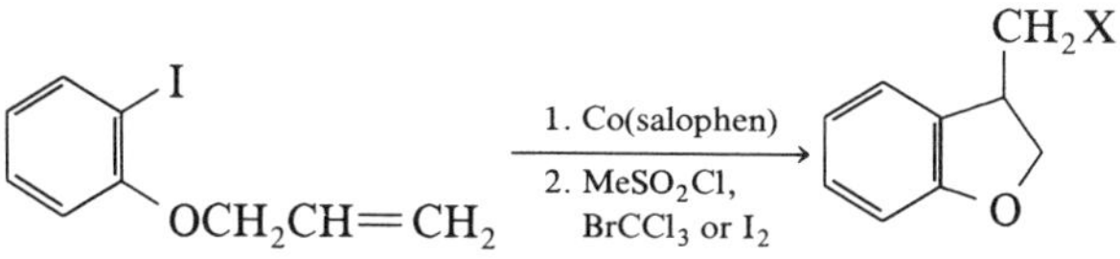

$$\xrightarrow[\substack{2.\ \text{MeSO}_2\text{Cl,} \\ \text{BrCCl}_3\ \text{or}\ \text{I}_2}]{1.\ \text{Co(salophen)}}$$

X = Cl, Br, I

TL *28* 1451 (1987)

$$\xrightarrow[\text{h}\nu\ \text{or}\ \Delta]{\text{cat (}n\text{-Bu}_3\text{Sn})_2}$$

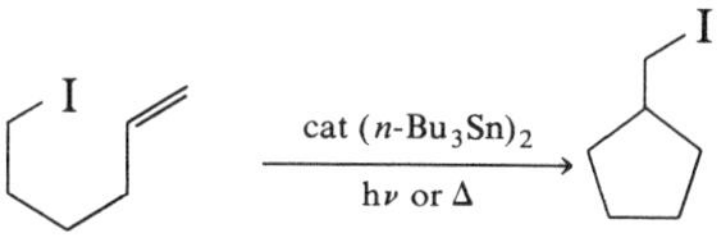

TL *27* 5821 (1986); *28* 2477 (1987)

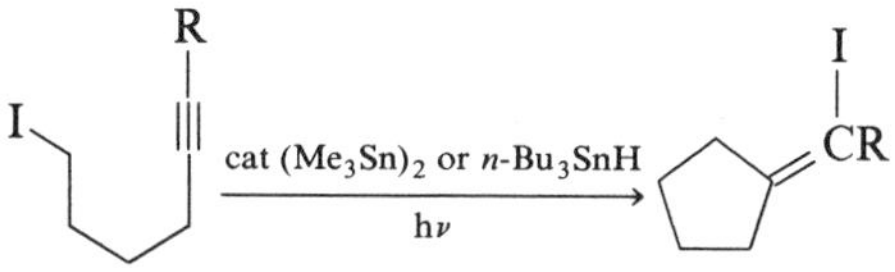

$$\xrightarrow[\text{h}\nu]{\text{cat (Me}_3\text{Sn})_2\ \text{or}\ n\text{-Bu}_3\text{SnH}}$$

JACS *108* 2489 (1986)
TL *28* 2477 (1987)

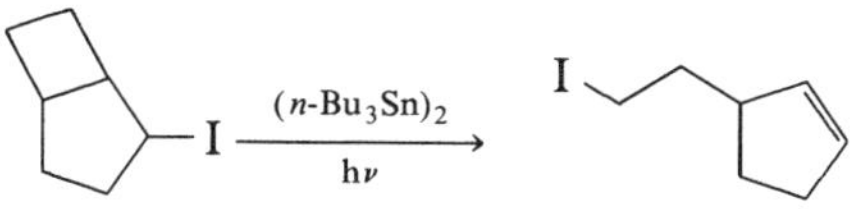

$$\xrightarrow[\text{h}\nu]{(n\text{-Bu}_3\text{Sn})_2}$$

TL *28* 5063 (1987)

6. HALOGENATION OF NITROGEN COMPOUNDS

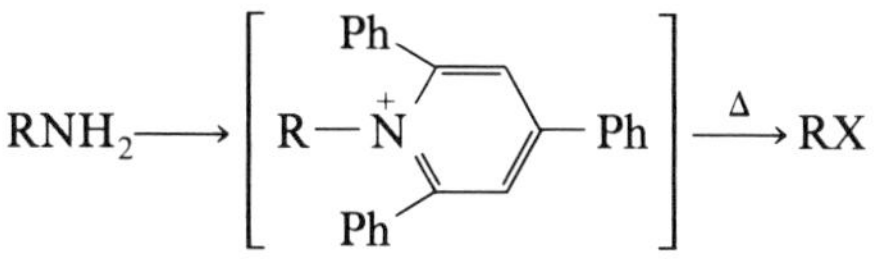

$$RNH_2 \longrightarrow \left[R-\overset{+}{N} \underset{Ph}{\overset{Ph}{\diagdown}} Ph \right] \overset{\Delta}{\longrightarrow} RX$$

X	R	
F	1° alkyl, benzylic	JCS Perkin I 2901 (1980)
Br	1° alkyl, benzylic	JCS Perkin I 1890 (1980)
I	1°, 2° alkyl; benzylic; aryl; heterocyclic	Syn 634 (1977) JCS Perkin I 433 (1979)

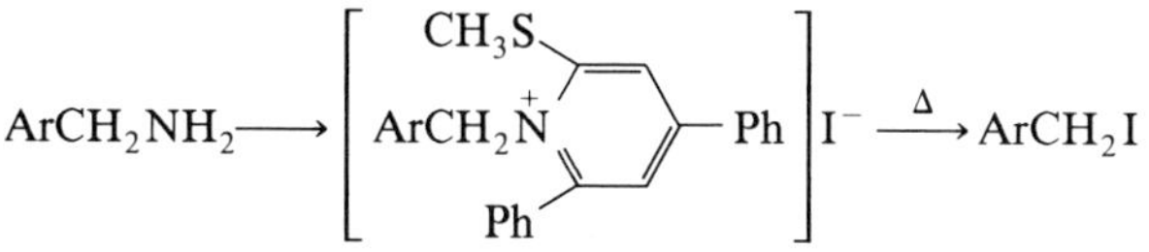

$$ArCH_2NH_2 \longrightarrow \left[ArCH_2\overset{+}{N} \underset{Ph}{\overset{CH_3S}{\diagdown}} Ph \right] I^- \overset{\Delta}{\longrightarrow} ArCH_2I$$

Syn 853 (1980)

$$ArNH_2 \longrightarrow ArX$$

Reviews:

Chem Rev *40* 251 (1947)
Quart Rev *6* 358 (1952)
"The Chemistry of Diazonium and Diazo Groups," Ed. S. Patai, J. Wiley, New York (1978), Part 1, pp 288–290

X	Reagents	
F	$HNO_2/BF_4^-/\Delta$	Ber *60* 1186 (1927) Org Rxs *5* 193 (1949)
	$[C_5H_5\overset{+}{N}H]F(HF)_x^-/NaNO_2$	JOC *44* 3872 (1979)
	$HCl, NaNO_2/HPF_6, \Delta$	Org Syn Coll Vol *5* 133 (1973)
Cl	$HCl, NaNO_2/Cu$	Ber *23* 1220 (1890)
	$HCl, NaNO_2/CuCl$	Org Syn Coll Vol *1* 170 (1941); *2* 130 (1943)

345

$$ArNH_2 \longrightarrow ArX \quad (Continued)$$

X	Reagents	
	H_2SO_4, $NaNO_2$/CuCl	Org Syn Coll Vol *4* 160 (1963)
		JACS *108* 1000 (1986)
Br	HBr, $NaNO_2$/Cu	Org Syn Coll Vol *1* 135 (1941)
	HBr, $NaNO_2$/CuBr	Org Syn Coll Vol *3* 185 (1955)
		JACS *109* 3378 (1987)
	H_2SO_4, $NaNO_2$/CuBr	JOC *51* 1339 (1987)
	AmONO, $HCBr_3$	JCS C 1249 (1966)

X		
I	HNO_2/⬠NH/KI, H^+	Syn 572 (1980)
	HCl, $NaNO_2$/KI	JOC *51* 1339 (1987)
	H_2SO_4, $NaNO_2$/KI, Cu	Org Syn Coll Vol *2* 355 (1943)
	AmONO, I_2	JOC *33* 1636 (1968)
	I_2, DMSO on $(ArN_2)BF_4$	Syn Commun *11* 639 (1981)
Cl, Br	t-BuONO, CuX_2	JOC *42* 2426 (1977)
	t-BuSNO, CuX_2	BCSJ *53* 1065 (1980)
	HgX_2, Δ	Ber *65* 1605 (1932)
Cl, Br, I	t-BuSNO$_2$; CCl_4, $HCBr_3$ or I_2	BCSJ *53* 2023 (1980)
Br, I	KX, H_2SO_4/$NaNO_2$/Cu	Ber *23* 1220 (1890)
	HNO_2/Et_2NH/Me_3SiCl,	
	LiBr or NaI	JOC *46* 5239 (1981)

$$ArNH_2 + H_2C{=}CHR \xrightarrow[CuX_2]{t\text{-BuSNO}} ArCH_2\overset{\overset{X}{|}}{C}HR$$

$$X = Cl, Br$$

BCSJ *53* 1065 (1980)

$$\underset{RCCHN_2}{\overset{O}{\overset{\|}{}}} \longrightarrow \underset{RCCH_2X}{\overset{O}{\overset{\|}{}}}$$

X	Reagent(s)	
F	$[C_5H_5\overset{+}{N}H]F(HF)_x^-$	JOC *44* 3872 (1979)
	HF	Bull Acad Sci USSR, Div Chem
		Sci 363 (1956)
		Ber *89* 864 (1956)
		JACS *79* 1959 (1957)
		Chem Ind 394 (1957)
Cl	HCl	Helv *27* 1108 (1904)
		JCS 1310 (1928)
		JACS *68* 2220 (1946); *72* 3477
		(1950); *74* 2082 (1952); *76*
		1185 (1954)
		Org Syn Coll Vol *3* 119 (1955)
Br	HBr	JOC *12* 767, 776 (1947); *18* 868
		(1953)
		JCS 278 (1948)
		JACS *74* 2082, 2550 (1952)

$$R_2CN_2 \xrightarrow{X_2} R_2CX_2$$

$\underline{X}$	
F	JOC *46* 3917 (1981)
Cl	J Prakt Chem [2] *38* 433 (1888) Ber *86* 1467 (1953); *87* 971 (1954)
Br	J Prakt Chem [2] *38* 433 (1888) Ber *66* 1541 (1933); *86* 1467 (1953); *87* 971 (1954) JACS *72* 3655 (1950)
I	Ber *18* 1283 (1885); *19* 2460 (1886); *27* 1888 (1894); *28* 2374 (1895); *35* 897 (1902); *60* 1364 (1927); *66* 1541 (1933) J Prakt Chem [2] *38* 422, 433 (1888) Ann *325* 143 (1902); *394* 36 (1912) JACS *65* 1458, 1516 (1943)

$$R_2C{=}NNH_2 \xrightarrow{IF} R_2CF_2$$

JACS *109* 896 (1987)

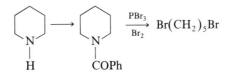

$$\xrightarrow[Br_2]{PBr_3} Br(CH_2)_5Br$$

JACS *68* 2402 (1946)
JOC *51* 2206 (1986); *52* 5466 (1987)

7. HALOGENATION OF ETHERS

$$R-O-R' \longrightarrow RX$$

X	Reagent(s)	
Cl	BCl_3	Chem Ind 609 (1963) (R' = Me)
	PhCOCl, cat n-Bu_3SnCl,	JOC 47 1215 (1982)
	cat $PhCH_2PdCl(PPh_3)_2$	(R = $PhCH_2$)
Br	HBr, H_2SO_4	JOC 52 5480 (1987)
	HBr, HOAc	Syn 771 (1978)
	HBr, cat $[n\text{-}C_{16}H_{33}P(n\text{-}Bu)_3]Br$	Syn 771 (1978)
	BBr_3	Chem Ind 609 (1963) (R' = Me)
	B-Br-9-BBN	JOMC 156 221 (1978)
		(R = 3° > 2° > 1° alkyl)
	Ph_3PBr_2	Chem Ind 200 (1969)
		(R' = t-Bu)
I	Me_3SiI	JOC 42 3761 (1977)
	H_2SiI_2	JOC 52 4846 (1987)

See also page 501, Section 6, for cleavage of ethers to alcohols plus organic halides.

$$\underset{\text{OMe}}{R_2CC}\equiv CR \longrightarrow \underset{\text{Cl}}{R_2CC}\equiv CR$$

HCl, $ZnCl_2$ JACS 107 6546 (1985)

CH_3COCl, cat $SnCl_4$ JACS 107 6546 (1985)

$$O\!\!-\!\!(CH_2)_n \longrightarrow X(CH_2)_nOH$$

See also page 508, Section 3.

$$O\!\!-\!\!(CH_2)_n \longrightarrow RCO_2(CH_2)_nX$$

X	n	Reagent(s)	
Cl	2	PhCOCl,	JOC 47 1215
		cat n-Bu_3SnCl,	(1982)
		cat $PhCH_2PdCl(PPh_3)_2$	

$$O\!\!-\!\!(CH_2)_n \longrightarrow RCO_2(CH_2)_n X \quad (\textit{Continued})$$

X	n	Reagent(s)	
	2	PhCOCl, cat n-Bu$_2$SnCl$_2$, cat Ph$_3$P	TL *27* 3021 (1986)
	3–5	RCOCl, cat K[Pt(C$_2$H$_4$)Cl$_3$]	JOC *48* 751 (1983)
	4	CH$_3$COCl, cat Rh$_2$(C$_2$H$_4$)$_4$Cl$_2$	JOC *48* 751 (1983)
	4	RCOCl, cat ZnCl$_2$	JACS *61* 2667 (1939)
	4	RCOCl, cat Mo(CO)$_6$	JOC *38* 64 (1973)
Br	2, 4, 5	(CH$_3$CO)$_2$O, MgBr$_2$	JOC *40* 3571 (1975)
	4	CH$_3$COBr	TL 3813 (1965)
	4	CH$_3$COBr, cat ZnCl$_2$	JCS 1536 (1954) JOC *51* 3372 (1986)
	4	CH$_3$COBr, cat K[Pt(C$_2$H$_4$)Cl$_3$]	JOC *48* 751 (1983)
I	4,5	RCOCl, NaI	TL *23* 681 (1982)

$$ROSiR'_3 \longrightarrow RX$$

X	Reagent(s)	R	
F	PhPF$_4$	1°, 2°, 3° alkyl	TL 847 (1972) Tetr *29* 1877 (1973) BSCF 2861 (1974) JOC *44* 3406 (1979)
	n-Bu$_4$NF or (PhCH$_2$NMe$_3$)F, RSO$_2$F (R = Me, p-Tol), molecular sieves	1° alkyl, 1° benzylic	TL *26* 4207 (1985)
Br	Ph$_3$P·Br$_2$	1°, 2° alkyl	JOC *51* 4941 (1986)
	CBr$_4$, PPh$_3$, CH$_3$COCH$_3$	1° alkyl, allylic, benzylic; 2° alkyl	TL *28* 1697 (1987)
I	Me$_3$SiCl, NaI	1°, 2°, 3° alkyl; 1° benzylic	Syn 379 (1979)

$$\text{ROTHP} \longrightarrow \text{RX}$$

X	Reagents	R	
Br	Br_2, Ph_3P	1° alkyl	JOC *40* 2410 (1975); *44* 4603 (1979) Syn Commun *6* 21 (1976) JACS *101* 4773 (1979); *107* 686 (1985)
	Br_2, $Ph_2PCH_2CH_2PPh_2$	2° alkyl, 1° allylic, 1° propargylic	TL *28* 767 (1987)
I	I_2, $Ph_2PCH_2CH_2PPh_2$	1°, 2° alkyl; 1° allylic	TL *28* 767 (1987)

8. HALOGENATION OF ALCOHOLS

Review: Tetr *36* 1901 (1980) (allylic alcohols)

1. Direct Methods

$$ROH \longrightarrow RX$$

$\underline{X}$	Reagent(s)	
F	HF, KF	JACS *85* 1609 (1963) (3° benzylic)
	[C$_5$H$_5$NH]F(HF)$_x^-$	JOC *44* 3872 (1979) (2°, 3° alkyl)
	Et$_2$NCF$_2$CHClF	J Gen Chem USSR *29* 2125 (1959) (1° alkyl)
		TL 1065 (2° alkyl), 1249 (1°, 2° alkyl) (1962)
		JOC *29* 2187 (1964) (1°, 2° alkyl); *44* 3406 (1979) (2° alkyl)
		Coll Czech Chem Commun *39* 2616 (1974) (1° alkyl)
	Et$_2$NSF$_3$	JOC *40* 574 (1975) (1°, 2°, 3° alkyl; allylic; benzylic); *43* 1090 (1978) (2° alkyl); *44* 3406 (1979) (2° alkyl); *52* 4804 (1987) (2° alkyl)
		TL 573 (1977) (1° alkyl); 1823 (1979) (2° alkyl); *28* 3891 (1987) (2° alkyl)
		JACS *109* 3046 (1987) (2° allylic)
	(Me$_2$N)$_2$SF$_2$	JOC *40* 574 (1975) (2° alkyl, allylic)
	n-Bu$_4$NF, RSO$_2$F (R = Me, *p*-Tol), molecular sieves	TL *26* 4207 (1985) (1° alkyl)
	PhPF$_4$	Chem Pharm Bull *21* 867 (1973) (1°, 2° alkyl); *23* 196 (1975) (2° alkyl)
		BSCF 2861 (1974) (2° alkyl)

$$\text{ROH} \longrightarrow \text{RX} \quad (\textit{Continued})$$

X	Reagent(s)	
F	Ph$_2$PF$_3$	Chem Pharm Bull *16* 1784 (1968) (1°, 2° alkyl; benzylic)
	Ph$_3$PF$_2$	Chem Pharm Bull *16* 1009, 1784 (1968) (both 1°, 2° alkyl)
Cl	HCl	Org Syn Coll Vol *1* 286, 288, 519 (1932) (all 1° alkyl)
		Org Syn Coll Vol *1* 144 (1941) (3° alkyl)
		JACS *78* 653 (1956) (benzylic); *82* 6163 (1960) (allylic); *92* 4461 (1970) (allylic)
		BSCF 632 (1974) (allylic)
		JOC *47* 2590 (1982) (2° benzylic)
	aq HCl (phase transfer)	Syn 37 (1974) (1° alkyl)
	HCl, ZnCl$_2$	Helv *20* 1462 (1937) (1° alkyl)
		Org Syn Coll Vol *1* 142 (1941) (1° alkyl)
		JCS 636 (1943) (1°, 2°, 3° alkyl)
	HCl, CaCl$_2$, CuCl, Cu	JOC *26* 725 (1961) (3° propargylic)
	TiCl$_4$, PhNHMe	BCSJ *54* 1456 (1981) (1° allylic)
	[C$_5$H$_5$$\overset{+}{\text{N}}$H]F(HF)$_x^-$, NaCl	JOC *44* 3872 (1979) (1°, 2°, 3° alkyl; benzylic)
	(COCl)$_2$, DMF	JACS *107* 3285 (1985) (1° alkyl)
	AcCl	JACS *108* 3762 (1986) (3° benzylic)
	H$_2$C=CHOCOCl	TL *28* 2933 (1987) (1° benzylic)
	(CF$_3$CO)$_2$O/LiCl	Syn 511 (1987) (1° alkyl)
	Ph$_3$P·Cl$_2$	JACS *86* 964 (1964) (1°, 2°, 3° alkyl)
	Ph$_3$P, NCS	TL 3937 (1973) (2° alkyl)
		JACS *106* 3286 (1984) (1° benzylic)
	Ph$_3$P, Cl$_3$CCOCCl$_3$	JOC *44* 359 (1979) (allylic); *46* 824 (1981) (1°, 2° alkyl; allylic)
	Ph$_3$P, CCl$_4$	Angew Int *14* 801 (1975) (review)
		Can J Chem *44* 1331 (1966) (1°, 2° alkyl)
		Tetr *23* 359 (1967) (1°, 2° alkyl)
		CC 1358 (1968) (1°, 2° alkyl)
		JOC *37* 1466 (1972) (allylic); *43* 1150 (2° alkyl), 2821 (mechanism) (1978); *51* 5291 (1986) (1° alkyl)
		JACS *96* 3684 (1974) (propargylic); *98* 5016 (1976) (allylic); *108* 1019 (1986) (1° allylic)
		BSCF 607 (1975) (allylic)

Ph$_3$P, CCl$_4$, py

Carbohydr Res *61* 511 (1978)
  (1° alkyl)

Ph$_3$P, Cl$_3$CCCl$_3$
polymer-Ar$_3$P, CCl$_4$

Syn 139 (1983) (2° alkyl)
CC 622 (1975) (1° alkyl)
JOC *40* 1669 (1975) (1°, 2° alkyl; benzylic)

[pyridine ring]–N–CH$_2$CH$_2$PPh$_2$,
CCl$_4$, HCCl$_3$

JOC *52* 4999 (1987)
  (1°, 2° alkyl)

(PhO)$_3$P, HCl

JCS 2224 (1953) (1°, 2°, alkyl; allylic)

(PhO)$_3$P, Cl$_2$

JCS 2281 (1954) (1°, 2°, 3° alkyl)

(PhO)$_3$P, NCS

TL 3937 (1973) (2° alkyl, 1° benzylic)

(PhO)$_3$P, CCl$_4$

JCS 2224 (1953) (1° alkyl)
Chem Ind 900 (1966) (1°, 2° alkyl; benzylic)

(*n*-C$_8$H$_{17}$)$_3$P, CCl$_4$

Can J Chem *46* 86 (1968) (1°, 2° alkyl; benzylic)

(PhO)$_3$P, C$_6$H$_5$CH$_2$Cl

JCS 2224 (1953) (1°, 2° alkyl; allylic)
JOC *30* 2635 (1965) (2° alkyl)

(Me$_2$N)$_3$P, CCl$_4$

CC 1350 (1968) (1° alkyl)
BSCF 607 (1975) (allylic)

PCl$_3$, DMF

Syn 398 (1976) (1°, 2° alkyl; allylic)

POCl$_3$

JOC *27* 1449 (1962) (1° alkyl)

POCl$_3$, DMF and other amides

Syn 746 (1980) (1°, 2°, 3° alkyl; benzylic)

Me$_2$NPOCl$_2$

CL 923 (1978) (1°, 2° alkyl)

[benzene ring fused to O–PCl$_3$–O (catechol) structure]

Z Chem *22* 126 (1982)
  (1°, 2° alkyl; 1° benzylic)

PCl$_5$

JCS 1138 (1946); 1709 (1953) (both 2° alkyl)
JCS C 1124 (1970) (2° alkyl)

SOCl$_2$

JACS *51* 3475 (1929) (1° benzylic); *62* 2295 (1940) (1° benzylic); *73* 2239 (1951) (1° alkyl); *75* 2053 (1953) (1° alkyl, benzylic); *107* 3950 (1985) (1° alkyl)
Ann *479* 211 (1930) (1°, 2° allylic)
JCS 684 (1942) (1° benzylic); 636 (1943) (1°, 2° alkyl); 1138 (1946) (2° alkyl); 3650 (1950) (1° alkyl); 2005 (1952) (1° propargylic); 2720 (1959) (2° allylic)
Org Syn Coll Vol *4* 333 (1963) (1° alkyl)
JCS C 1124 (1970) (2° alkyl)
TL 2931 (1970) (2° alkyl); 87 (1971) (1° alkyl)

ROH ⟶ RX   (*Continued*)

X	Reagent(s)	

Cl

		BSCF 632 (1974) (allylic)
		JOC *52* 547 (1987) (1° alkyl)
	NCS, Me$_2$S	TL 4339 (1972); 3857 (1976)
		Tetr *32* 1675 (1976)
		JOC *51* 3098 (1986)
		(all references allylic or benzylic)

MsCl, LiCl, 2,6-lutidine
   or *s*-collidine

JOC *36* 3044 (1971)
   (1°, 2° allylic);
     51 858 (1986) (2° allylic); *52* 3860,
   3883 (1987) (both 1° allylic)
TL *28* 723 (1987) (1° allylic)

TsCl, py, Δ

Compt Rend C *276* 683 (1973)
   (allylic)

TsCl, DMAP

TL 393 (1984) (1° allylic)
CC 590 (1986) (1° allylic)

Me$_3$SiCl

TL *27* 1907 (1986) (cyclopropylcarbinyl)

ZnCl$_2$, PPh$_3$, EtO$_2$CN=NCO$_2$Et

JOC *49* 3027 (1984)
   (1°, 2° alkyl; 1° allylic)
Syn Commun *16* 611 (1986) (2° alkyl)

PdCl$_2$(PhCN)$_2$

TL 4575 (1978) (2° alkyl)

BF$_4^-$, Et$_4$NCl

CL 383 (1977) (1°, 2° alkyl)

Br

HBr

Helv *20* 1462 (1937) (1° alkyl)
JCS 636 (1943) (1°, 2° alkyl)
Org Syn Coll Vol *2* 246 (1943); *3* 227
   (1955) (both 1° alkyl)
JOC *50* 2759 (1985) (3° alkyl); *52*
   5560 (1987) (1°, 2° benzylic;
   2° allylic)
JACS *109* 3098 (1987) (1° benzylic)

HBr (phase transfer)

TL *28* 1223 (1987) (1°, 2° alkyl)

HBr, H$_2$SO$_4$

Org Syn Coll Vol *1* 25–35 (1941)
   (1° alkyl; allylic)
JCS 636 (1943) (1° alkyl); 132 (1953)
   (1°, 2° alkyl)

HBr, LiBr or CaBr$_2$

BCSJ *53* 1181 (1980) (3° alkyl)

[C$_5$H$_5$NH]F(HF)$_x^-$, NH$_4$Br

JOC *44* 3872 (1979) (1°, 2°, 3° alkyl;
   benzylic)

TiBr$_4$, PhNHMe

BCSJ *54* 1456 (1981) (1° allylic)

$ZnBr_2$, Ph_3P, $EtO_2CN{=}NCO_2Et$	JOC *49* 3027 (1984) (1° alkyl, 1° allylic); *52* 2960 (1° alkyl), 4191 (1° alkyl) (1987)
P, Br_2	Org Syn Coll Vol *1* 36 (1941) (1°, 2°, 3° alkyl)
PBr_3	JACS *48* 1080 (1926) (1°, 2° alkyl); *51* 280 (1929) (1°, 2°, 3° alkyl); *62* 2295 (1940) (2° alkyl); *68* 2513 (1946) (2°, 3° alkyl); *82* 410 (1960) (2° alkyl); *93* 7016 (1971) (allylic); *98* 4925 (1976) (allylic); *107* 2712 (1985) (1° allylic)
	JCS 636 (1943) (1°, 2° alkyl); 2005 (1952) (1° propargylic)
	Org Syn Coll Vol *2* 358 (1943) (1° alkyl); *3* 370 (1955) (1° alkyl)
	Helv *29* 1133 (1946) (3° allylic, rearrangement)
	JOC *26* 3645 (1961) (2° alkyl); *39* 2135 (1974) (allylic); *52* 4047 (1° allylic), 5594 (1° alkyl) (1987)
PBr_3, py	Ann *479* 211 (1930) (1°, 2° allylic)
	JACS *70* 3314 (1948) (1° alkyl); *71* 1292 (1949) (1° propargylic); *76* 803 (1954) (1° alkyl)
	JCS 3650 (1950) (1° alkyl)
	Org Syn Coll Vol *3* 793 (1955) (1° alkyl)
	JOC *41* 1262 (1976) (allylic)
PBr_3, CaH_2	TL 3857 (1976) (allylic)
	JACS *99* 3513 (1977) (allylic)
PBr_5	JCS 1709 (1953) (2° alkyl)
	JOC *26* 3645 (1971) (2° alkyl)
Ph_3P, Br_2	Ann 626 26 (1959) (1°, 2° alkyl)
	JACS *86* 964 (1964) (1°, 2° alkyl); *99* 1612 (1977) (1° alkyl); *107* 5238 (1985) (1°, 2° alkyl)
	JOC *30* 2635, 2639, 3469 (1965) (all 2° alkyl)
	TL 341 (1965) (2° alkyl)
	Syn 901 (1981) (1° allylic)
$Ph_2PCH_2CH_2PPh_2$, Br_2	TL *28* 767 (1987) (1° allylic, 1° propargylic)
n-Bu_3P, Br_2	JACS *86* 964 (1964) (1°, 2° alkyl)
$(PhO)_3P$, Br_2	JCS 2281 (1954) (1°, 2°, 3° alkyl)
	TL 483 (1963) (1°, 2° alkyl)
	JCS C 2260 (1967) (1°, 2° propargylic and allenic)
$(PhO)_3P$, NBS	TL 3937 (1973) (2° alkyl)
$(PhO)_3P$, $C_6H_5CH_2Br$	JCS 2224 (1953) (1° alkyl, allylic)

$$\text{ROH} \longrightarrow \text{RX} \quad (\textit{Continued})$$

$\underline{X}$	Reagent(s)	
Br	Ph$_3$P, NBS	JCS 2337 (1962) (1° alkyl)
		JOC *34* 212 (1969) (1° alkyl); *51* 2637 (1986) (1° alkyl)
		TL 3937 (1973) (2° alkyl)
		JACS *99* 3167 (1977) (1° alkyl); *107* 7515 (1985) (1° allylic); *108* 1035 (1986) (1° alkyl)
	Ph$_3$P, NBA	TL 3937 (1973) (2° alkyl)
	Ph$_3$P, CBr$_4$	JACS *92* 2139 (1970) (allylic); *95* 8749 (1973) (1° alkyl); *109* 2738 (1987) (1° benzylic)
		JOC *42* 353 (1977) (1° alkyl); *51* 5291 (1986) (1° alkyl)
		TL *28* 675 (1° allylic), 3225 (2° allylic), 6425 (1° alkyl) (1987)
	Ph$_3$P, CBr$_4$, py	Carbohydr Res *61* 511 (1978) (1° alkyl)
	Ph$_2$PCH$_2$CH$_2$PPh$_2$, CBr$_4$	JOC *51* 789 (1986) (1° allylic)
	(n-C$_8$H$_{17}$)$_3$P, CBr$_4$	Can J Chem *46* 86 (1968) (1° alkyl, benzylic)
		CL 85 (1986) (1° allylic)
		TL *27* 1607 (1986) (1° allylic)
	Ph$_3$P, BrCl$_2$CCCl$_2$Br	Syn 139 (1983) (1°, 2° alkyl)
		TL *28* 5751 (1987) (1° allylic)
	Ph$_3$P, Br$_3$CCBr$_3$	Syn 139 (1983) (2° alkyl)
	Ph$_3$P; 2,4,5-tribromoimidazole	JCS Perkin I 681 (1982) (1°, 2° alkyl)
	SOBr$_2$	JOC *26* 3645 (1961) (2° alkyl)
		TL 87 (1971) (1° alkyl)
	Me$_2$S, NBS	TL 4339 (1972) (allylic, benzylic)
		JACS *108* 1019 (1986) (1° allylic)
	BH$_3$, Br$_2$	Chem Ind 223 (1965) (?)
	NaBH$_4$, Br$_2$	Chem Ind 223 (1965) (?)
	Me$_3$SiBr	TL 4483 (1978) (1°, 2°, 3° alkyl; benzylic)
	Me$_3$SiCl, LiBr	JOC *45* 1638 (1980) (1°, 2°, 3° alkyl; allylic; benzylic)
	(Me$_3$Si)$_2$, (C$_5$H$_5$NH)Br$_3$	JOC *45* 1638 (1980) (1°, 2° alkyl; allylic; benzylic)
	(CF$_3$CO)$_2$O/LiBr	Syn 511 (1987) (1° alkyl; 1° allylic)
		TL *28* 4745 (1987) (1° allylic)
	PhSeCN, n-Bu$_3$P/Br$_2$, Et$_3$N	CC 656 (1980) (2° alkyl)
I	HI	Helv *20* 1462 (1937) (1° alkyl)
		JCS 636 (1943) (1°, 2° alkyl)
	KI, H$_3$PO$_4$	JOC *15* 491 (1950) (1°, 2°, 3° alkyl)
		Org Syn Coll Vol *4* 323 (1963) (1° alkyl)
	HI, LiI or CaI$_2$	BCSJ *53* 1181 (1980) (3° alkyl)

$[C_5H_5\overset{+}{N}H]F(HF)_x^-$, KI	JOC *44* 3872 (1979) (1°, 2°, 3° alkyl; benzylic)
NaI, $BF_3 \cdot OEt_2$	TL *28* 4969 (1987) (1° benzylic)
P, I_2	Org Syn Coll Vol *2* 322, 399 (1943) (both 1° alkyl)
P_2I_4	TL 1801 (1979) (1°, 2°, 3° alkyl and benzylic)
Ph_3P, I_2	JACS *86* 964 (?), 2093 (1° alkyl) (1964)
$R_3P \cdot I_2$ (R = *n*-Bu, Ph), HMPA	Austral J Chem *35* 517 (1982) (1°, 2° alkyl; 1° propargylic and allylic)
Ph_3P, I_2, imidazole	CC 978 (1979) (1°, 2° alkyl)
	JCS Perkin I 2866 (1980) (1°, 2° alkyl); 681 (1982) (1°, 2° alkyl)
	TL *24* 4883 (1983) (1° propargylic); *28* 3091 (1° alkyl), 5391 (1° alkyl), 5457 (1° alkyl) (1987)
	JOC *51* 858 (1° alkyl), 4726 (1° alkyl) (1986)
	JACS *109* 1186 (1° alkyl), 4717 (1° alkyl), 6187 (1° alkyl) (1987)
	CC 877 (1987) (1° alkyl)
Ph_3P; 2,4,5-triiodoimidazole	CC 978 (1979) (1°, 2° alkyl)
	JCS Perkin I 2866 (1980) (1°, 2° alkyl); 681 (1982) (1° alkyl)
Ph_3P, NIS	TL 3937 (1973) (2° alkyl)
Ph_3P, ICH_2CH_2I	Syn 139 (1983) (1°, 2° alkyl)
Ph_3P, CI_4, py	Carbohydr Res *61* 511 (1978) (1° alkyl)
Ph_3P, ZnI_2, $EtO_2CN{=}NCO_2Et$	JOC *49* 3027 (1984) (1° alkyl)
	Syn Commun *16* 611 (1986) (2° alkyl)
	TL *28* 2977 (1987) (1° alkyl)
$(PhO)_3P$, I_2	JCS 2281 (1954) (1°, 2° alkyl)
$(PhO)_3P$, CH_3I	JCS 2224 (1°, 2°, 3° alkyl; allylic; benzylic), 3002 (1° alkyl) (1953)
	JACS *86* 2093 (1964); *88* 5684 (1966) (both 1° alkyl)
	Austral J Chem *21* 513 (1968) (2° alkyl)
	JOC *35* 2319 (1° alkyl), 2868 (2° alkyl) (1970)
	Org Syn *51* 44 (1971) (1° alkyl)
	TL *28* 3123 (1987) (1° alkyl)
PCl, I_2	JOC *32* 4160 (1967) (1°, 2° alkyl)
$(Me_2N)_3P \cdot I_2$, HMPA	Austral J Chem *35* 517 (1982) (1° allylic)
BH_3, I_2	Chem Ind 1582 (1964) (1°, 2° alkyl; benzyl)
$NaBH_4, I_2$	Chem Ind 223 (1965) (?)
$MeSiCl_3$, NaI	JOC *48* 3667 (1983) (2°, 3° alkyl; benzylic)

$$\text{ROH} \longrightarrow \text{RX} \quad (\textit{Continued})$$

<u>X</u>   <u>Reagent(s)</u>

Me$_3$SiI   TL 2659 (1977) (1°, 2°, 3° alkyl;
                                                              benzylic)

Me$_3$SiSiMe$_3$, I$_2$ Angew Int 18 612 (1979) (1°, 2° alkyl;
                                                              benzylic)

Me$_3$SiCl, NaI   JOC 44 1247 (1979) (1°, 2°, 3° alkyl;
                                                              allylic; benzylic)
                                                            TL 28 5063 (1987) (2° alkyl)

MgI$_2$ TL 28 6441 (1987) (3° alkyl)
(CF$_3$CO)$_2$O/LiCl   Syn 511 (1987) (1° alkyl)

$$\underset{\underset{\displaystyle C_6H_6}{\xrightarrow{\hspace{0.6cm}MgI_2\hspace{0.6cm}}}}{R-\overset{\displaystyle OH}{\overset{|}{C}}-C{=}C} \quad R-C{=}C-C-I$$

Syn 608 (1984)

2. Via Sulfonates

$$\text{ROH} \longrightarrow \text{RO}\overset{\displaystyle O}{\underset{\displaystyle O}{\overset{\|}{\underset{\|}{S}}}}\text{R}' \longrightarrow \text{RX}$$

$$R' = CH_3, C_6H_5, p\text{-}C_6H_4CH_3, CF_3$$

<u>X</u>   <u>Reagent(s)</u>

F   KF (phase transfer)   Syn 428 (1974) (1°, 2° alkyl); 430
                                            (1975) (2° alkyl)

         KF, CH$_3$OH   JACS 80 5559 (1958) (1° alkyl)
         KF, HOCH$_2$CH$_2$OH   JCS 872 (1958) (1° alkyl)
                                          Chem Ind 157 (1958) (1°, 2° alkyl)
         KF, O(CH$_2$CH$_2$OH)$_2$ JACS 77 4899 (1955) (1°, 2° alkyl)
                                          Can J Chem 34 757 (1956) (1° alkyl)
         KF, dicyclohexyl-18-crown-6   CC 879 (1974) (1° alkyl)
         n-Bu$_4$NF (no solvent)   JOC 49 3216 (1984) (1°, 2° alkyl)
         n-Bu$_4$NF, solvent?   TL 28 4003 (1987) (2° alkyl)
         n-Bu$_4$NF, Et$_2$O   JACS 109 3046 (1987) (1° alkyl)
         n-Bu$_4$NF, THF   JOC 44 2194 (1979) (2° alkyl)
         n-Bu$_4$NF, CH$_3$COCH$_3$ JCS 954 (1962) (2° alkyl)
            or CH$_3$COCH$_2$CH$_3$
         n-Bu$_4$NF, CH$_3$CN   Carbohydr Res 5 292 (1967)
                                            (1°, 2° alkyl)

         n-Bu$_4$NF$\cdot$3 H$_2$O; o-Cl$_2$C$_6$H$_4$,   JOC 52 658 (1987) (1° alkyl)
            CH$_3$CN or THF

	[resin-$\overset{+}{N}Me_3$]F$^-$	Syn 472 (1976) (1°, 2° alkyl)
		JCS Perkin I 2248 (1979) (1°, 2° alkyl)
	(n-Bu$_4$N)HF$_2$	TL *28* 4733 (1987) (1°, 2° alkyl)
	n-Bu$_3$MePF	TL 353 (1979) (2° alkyl)
		JOC *44* 3406 (1979) (2° alkyl)
		Tetr *36* 1931 (1980) (1° alkyl)

Cl   LiCl, ROH (CH$_3$COCH$_2$CH$_3$)

JCS 315 (2° alkyl), 326 (1° alkyl)
  (1949); 3650 (1950) (1° alkyl); 2005
  (1952) (1° propargylic); 954 (1962)
  (2° alkyl)

LiCl, HMPA

TL 1393 (1969) (allylic)
JOC *44* 2369 (1970) (1° alkyl)
JACS *99* 1612 (1977) (allylic)

LiCl, CH$_3$CONEt$_2$

or (pyrrolidinone NEt)

JOC *44* 2369 (1970) (1° alkyl)

LiCl, DMF
LiCl, DMF, collidine

JOC *51* 5291 (1986) (1° alkyl)
Tetr *27* 5979 (1971)
JOC *36* 3044 (1971)
BSCF 607 (1975)
TL 3681 (1976)
(all allylic)

NaCl or KCl,
  dicyclohexyl-18-crown-6

CC 879 (1974) (1° alkyl)

KCl (phase transfer)
n-Bu$_4$NCl, CH$_3$COCH$_2$CH$_3$
[resin-$\overset{+}{N}Me_3$]Cl$^-$
(C$_5$H$_5$NH)Cl

Syn 430 (1975) (2° alkyl)
JCS 954 (1962) (2° alkyl)
JCS Perkin I 2248 (1979) (2° alkyl)
Monatsh *83* 1398 (1952) (1° alkyl)
JACS *80* 2726 (1958) (1° alkyl)
JOC *26* 2883 (1961) (2° alkyl)

TsCl, py, Δ (direct)

Compt Rend C *276* 683 (1973)
  (1° allylic)

Br   HBr
LiBr, THF
LiBr, CH$_3$COCH$_3$

JOC *27* 4349 (1962) (1° alkyl)
JACS *107* 2712 (1985) (1° allylic)
JCS 954 (1962) (2° alkyl)
JOC *35* 2803 (1970) (1° alkyl);
  52 2337 (1987) (1° alkyl)
JACS *107* 2730 (1985) (1° allylic);
  109 6937 (1987) (1° alkyl)

LiBr, CH$_3$CONEt$_2$
NaBr

JOC *44* 2369 (1979) (1° alkyl)
JOC *26* 3645 (1961) (2° alkyl)
Org Syn Coll Vol *4* 753 (1963)
  (1° alkyl)

NaBr, dicyclohexyl-18-crown-6
KBr (phase transfer)
KBr, dicyclohexyl-18-crown-6
CaBr$_2$

CC 879 (1974) (1° alkyl)
Syn 430 (1975) (2° alkyl)
CC 879 (1974) (1°, 2° alkyl)
JCS 3650 (1950) (1° alkyl); 2005 (1952)
  (1° propargylic)
JOC *27* 624 (1962) (2° alkyl)

$$ROH \longrightarrow RO\overset{\underset{\parallel}{O}}{\underset{\parallel}{S}}R' \longrightarrow RX \quad (\textit{Continued})$$

$$R' = CH_3, C_6H_5, p\text{-}C_6H_4CH_3, CF_3$$

X	Reagent(s)	
Br	MgBr$_2$	J Lipid Res 7 568 (1966) (1° alkyl)
	n-Bu$_4$NBr, C$_6$H$_6$	JOC 45 4387 (1980) (1°, 2° alkyl)
	n-Bu$_4$NBr, CH$_3$COCH$_2$CH$_3$	JCS 954 (1962) (2° alkyl)
	n-Bu$_4$NBr, HMPA	JACS 107 3271 (1985) (1° alkyl)
	(C$_5$H$_5$NH)Br	Monatsh 83 1398 (1952) (1° alkyl)
I	NaI	JCS 315 (2° alkyl), 326 (1°, 2° alkyl), 2542 (1° alkyl) (1949); 3650 (1950) (1° alkyl); 2005 (1952) (1° propargylic); 955 (1962) (2° alkyl)
		JACS 72 561 (1950); 83 1460 (1961); 91 4771 (1969) (all 1° alkyl); 107 2124 (1° propargylic), 7967 (1° alkyl) (1985); 108 468 (1986) (1° alkyl)
		JOC 21 1487 (1956) (1° alkyl); 52 4517 (1° alkyl); 4810 (1° propargylic) (1987)
		Can J Chem 34 757 (1956) (1° alkyl)
	NaI, copper bronze	JACS 107 196 (1985) (1° alkyl)
	NaI or KI, dicyclohexyl-18-crown-6	CC 879 (1974) (1° alkyl)
	KI	Monatsh 82 970 (1951) (1°, 2° alkyl)
	KI (phase transfer)	Syn 430 (1975) (2° alkyl)
	(C$_5$H$_5$NH)I	Monatsh 83 1398 (1952) (1° alkyl)
	n-Bu$_4$NI, C$_6$H$_6$	JOC 45 4387 (1980) (1°, 2° alkyl)
		TL 28 2941 (1987) (1° alkyl)
	ZnI$_2$	JACS 109 4717 (1987) (1° allylic)

3. Via Haloformates

$$ROH \longrightarrow RO\overset{\underset{}{O}}{\overset{\parallel}{C}}X \overset{\text{reagent}}{\longrightarrow} RY$$

X	Y	Reagent	
F	F	BF$_3$·OEt$_2$	JACS 77 5033 (1955)
Cl	F	TlF	JACS 77 3099 (1955)
Cl	Cl	BF$_3$·OEt$_2$	JACS 77 5033 (1955)
Cl	Cl	NaCl	JOC 32 2633 (1967)
Cl	Cl	PPh$_3$	Ann 698 106 (1966)
Cl	Br	NaBr	JOC 32 2633 (1967)
Cl	I	NaI	JOC 32 2633 (1967)

4. **Other Approaches**

For approaches via silyl ethers see page 349, Section 7.

$$ROH \longrightarrow RX$$

$\underline{X}$ $\underline{Reagent(s)}$

F Ph_2NCN, KO-t-Bu/HF or   Can J Chem *43* 3173 (1965)
               HBF_4/Δ

Cl $NO_2 \underset{}{\overset{NO_2}{\bigcirc}} F/LiCl$ BSCF 405 (1975)

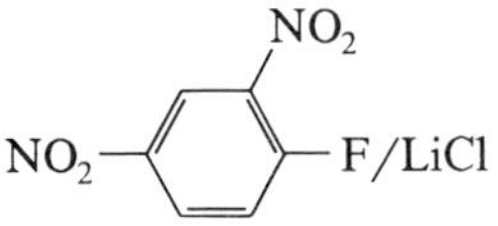

OR

$$ROH \xrightarrow[CuCl]{i\text{-}PrN=C=N\text{-}i\text{-}Pr} i\text{-}PrN=\overset{|}{C}NH\text{-}i\text{-}Pr \xrightarrow[CF_3SO_3H]{n\text{-}Bu_4NX} RX$$

X = Br, I

TL *28* 4445 (1987)

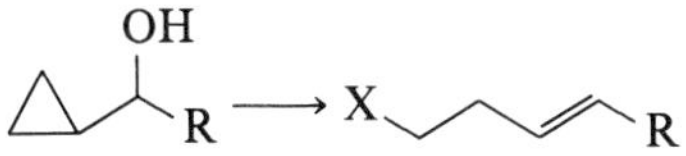

OH

$\triangle\!\!\diagup\!\!\overset{OH}{\underset{}{}}R \longrightarrow X\diagdown\diagup\diagdown\diagup R$

See page 151, Section 34.

9. HALOGENATION OF PHENOLS

For ring halogenation see page 315, Section 2.

$$ArOH \longrightarrow ArX$$

X	Reagent(s)	
Cl	POCl$_3$	JCS 678 (1944)
		JACS 76 6073 (1954)
		Org Syn Coll Vol 3 272 (1955)
		JOC 27 1462 (1962)
	POCl$_3$, PCl$_5$	JACS 76 1109 (1954)
		JCS 1666 (1964)
	PCl$_5$	JCS 1666 (1964)
	PCl$_5$, C$_6$H$_5$COCl	JCS 1666 (1964)
	COCl$_2$, Ph$_3$P	Ann 698 106 (1966)
Br	Ph$_3$P, Br$_2$	JACS 86 964 (1964)

10. HALOGENATION OF SULFUR AND SELENIUM COMPOUNDS

$$RSO_3H \text{ or } RSH \xrightarrow[PPh_3]{I_2} RI$$

Syn 371 (1981)

$$RCH(SPh)_2 \xrightarrow{HgF_2} RCHSPh$$

with F substituent shown above SPh carbon

TL *28* 3901 (1987)

$$RSeR' \xrightarrow{Br_2} RBr$$

TL 2647 (1976)

11. HALOGENATION OF CARBONYL AND RELATED COMPOUNDS

1. Aldehydes and Ketones

$$
\begin{array}{cc}
\text{H} & \text{O} \\
| & \| \\
-\text{C}-\text{C}- \\
|
\end{array}
\longrightarrow
\begin{array}{cc}
\text{X} & \text{O} \\
| & \| \\
-\text{C}-\text{C}- \\
|
\end{array}
$$

Review:

H. O. House, "Modern Synthetic Reactions," W. A. Benjamin, Inc., Menlo Park, California (1972), pp 459–478

X	Reagent(s)	
F	F$_2$ (on enol silane)	TL *27* 2715 (1986)
		JOC *52* 4307 (1987)
		(2-fluoro-1,3-diketones)
	F$_2$/NaO$_2$CCF$_3$ (on enol acetate)	CC 479 (1979)
		JACS *101* 2782 (1979)
	FClO$_3$, py (on enol ether)	JACS *81* 5259 (1959)
	FClO$_3$ (on enamine)	JOC *23* 1406 (1958)
	XeF$_2$, cat HF (on enol ester)	JOC *47* 573 (1982)
	CF$_3$OF (on enol silane)	JACS *102* 4845 (1980)
	CF$_3$OF (on enol ester)	CC 804 (1968); 1497 (1969);
		122 (1972)
		Israel J Chem *17* 60 (1978)
	CF$_3$OF (on enol ether and enamine)	CC 804 (1968)
		Israel J Chem *17* 60 (1978)
	R$_f$OF (on enol ester)	CC 122 (1972)
	CF$_2$(OF)$_2$ (on enol ester)	CC 122 (1972)
	SF$_5$OF (on enol ester)	CC 122 (1972)
	(C$_5$H$_5$NF)OTf (on enol silane)	TL *27* 4465 (1986)
	MeLi/RSO$_2$NFR' (on enol acetate)	JACS *106* 452 (1984)
	KH/RSO$_2$NFR'	JACS *106* 452 (1984)

$$\underset{|}{\overset{\text{H}\ \ \ \text{O}}{\underset{|}{-\text{C}-\text{C}-}}} \longrightarrow \underset{|}{\overset{\text{X}\ \ \ \text{O}}{\underset{|}{-\text{C}-\text{C}-}}} \quad (\textit{Continued})$$

X   Reagent(s)

Cl   Syn 1018 (1982)

(aryl ketones only)
SO_2Cl_2 (ketones only)   JOC *37* 4090 (1972); *46* 4486 (1981)
                                          Syn Commun *11* 7 (1981)
$CuCl_2 \cdot 2 H_2O$, LiCl, H_2O JACS *77* 5274 (1955)
$CuCl_2$ or $FeCl_3$ JOC *45* 2022 (1980)
  (on enol silane)
$TlCl_3$ CC 1336 (1986)
[polymer-$C_6H_4CH_2\overset{+}{N}Me_3$]$ICl_2^-$ CC 1278 (1980)
NCS (on enol silane)   TL *27* 2563 (1986)
t-BuOCl, HOAc   JOC *51* 3380 (1986)
LDA/CF_3SO_2Cl JACS *107* 7771 (1985)

Br Br_2 (ketones)   BSCF 4169 (1969)
                                          JACS *108* 1251 (1986)

          Br_2, O   O (aldehydes)   BSCF 1465 (1973)

          Br_2, hν,   JCS Perkin I 501 (1977)

          Br_2, HOAc   JACS *72* 362 (1950)
                                          JOC *35* 186 (1970); *51* 3380 (1986)
          Br_2 (on enol ether)   BSCF 4169 (1969)
          Br_2 (on enol silane)   JACS *109* 2435 (1987)
          Br_2 (on enolate)   JOC *38* 2576 (1973)
          Br_2, $AlCl_3$ JOC *48* 2520 (1983)
          $(C_5H_5\overset{+}{N}H)Br_3^-$ JACS *70* 417 (1948)
                                          TL *23* 3405 (1982)
          (pyrrolidone)$_3 \cdot HBr_3$ Can J Chem *47* 706 (1969)
                                          JOC *52* 5624 (1987)

                                          Syn 488 (1980)

          $(Ph\overset{+}{N}Me_3)Br_3^-$ TL (9) 24 (1959)
                                          BSCF 1822 (1961); 90 (1962)
                                          JACS *106* 3539 (1984)
                                          JOC *52* 4792 (1987)
          [polymer-$C_6H_4CH_2\overset{+}{N}Me_3$]$Br_3^-$ Syn 64 (1979); 143 (1980)
          [$Ph_3\overset{+}{P}(CH_2)_2CO_2H]Br_3^-$ TL 373 (1975)
          NBS   JACS *109* 7230 (1987)
          NBS (on enol acetate)   JACS *75* 3513 (1953)

NBS (on enol silane)

JOC *39* 1785 (1974);
52 3346 (1987)
Syn 194 (1976)

(enones)

Tetr *29* 1625 (1973)

(aldehydes, ketones, enones)

NCCHBrCON(CH$_3$)$_2$
or BrCH(CN)$_2$

Tetr *31* 231 (1975)

NCCBr$_2$CON(CH$_3$)$_2$

Ind J Chem B *17* 305 (1980)

PhSeBr

JOC *51* 3380 (1986)

CuBr$_2$

JOC *27* 4397 (1962); *29* 3459
(1964); *40* 1990 (1975)

I	I$_2$, CaO or KOH	JACS *80* 250 (1958); *81* 439 (1959) JOC *25* 1966 (1960)

I$_2$ (on enolate)

TL 2817 (1979)

I$_2$, PCC, molecular sieves
(on enol silane)

Syn Commun *12* 1127 (1982)

I$_2$, Cu(OAc)$_2$, HOAc, Δ
(ketones and enol acetates)

Syn 312 (1981)

I$_2$, AgOAc/(Et$_3$NH)F
(on enol silane)

JOC *44* 1731 (1979)

I$_2$, TlOAc (on enol acetate)

JCS Perkin I 126 (1978)

NIS (on enol acetate)

JACS *75* 3493 (1953); *76* 1722
(1954); *77* 3826 (1955)

Me$_3$SiI, (Me$_3$Si)$_2$NH/NaI,
m-ClC$_6$H$_4$CO$_3$H

JOC *52* 3919 (1987)

NaI, *m*-ClC$_6$H$_4$CO$_3$H/H$_3$O$^+$
(on enol silane or enamine)

JOC *52* 3919 (1987)

Cl, Br	X$_2$ (on enol silane)	JOC *39* 1785 (1974) Syn 194 (1976)

X$_2$ (on enamine)

Acta Chem Scand B *32* 646 (1978)

NBS or NCS (on enol borinate)

Can J Chem *50* 2387 (1972)

Cl, Br, I	NH$_4$X, electrolysis (on enol acetate or silane)	JOC *45* 2731 (1980)

Pb(OAc)$_4$, X$^-$ (on enol ether,
acetate or silane)

Syn 1021 (1982)

JOC *47* 573 (1982)

JOC *52* 4307 (1987)

TL *27* 4465 (1986)

$$CH_3\overset{O}{\underset{||}{C}}CH\overset{O}{\underset{||}{C}}CH_3 \quad \underset{R}{|} \quad \xrightarrow[\text{NaOCH}_3,\ \text{CH}_3\text{OH}]{\text{NCS}} \quad R\overset{Cl}{\underset{|}{CH}}COCH_3$$

TL *28* 5505 (1987)

$$R^1\overset{O}{\underset{||}{C}}CH_2R^2 \xrightarrow[\text{HOCH}_2\text{CH}_2\text{OH, THF}]{(\text{PhNMe}_3)\text{Br}_3} R^1\overset{}{C}\overset{}{\underset{|}{CHR^2}}$$
$$\underset{Br}{}$$

Syn 309 (1982)

$$\underset{}{\overset{H\ \ O}{C=C-C-}} \longrightarrow \underset{}{\overset{X\ \ O}{C=C-C-}}$$

X	Reagents	
Cl, Br	PhSeX, py	TL *22* 3301 (1981)
Br	Br$_2$/Et$_3$N	JOC *33* 1454 (1968); *47* 5088 (1982)

$$(CH_3)_2C=CH\overset{O}{\underset{||}{C}}CH_3 \xrightarrow{\text{HOCl}} H_2C=\overset{Cl\ \ O}{\underset{\underset{CH_3}{|}}{C}CH\overset{||}{C}CH_3}$$

TL *22* 5019 (1981)

X = Me$_3$Si, Ac

TL *27* 4465 (1986)

$$R_2CO \xrightarrow{R'SeH} R_2C(SeR')_2 \xrightarrow[\text{2. R''X}]{\text{1. RLi}} R_2\overset{SeR'}{\underset{|}{C}R''} \xrightarrow[\text{Et}_3\text{N}]{Br_2} R_2\overset{Br}{\underset{|}{C}R''}$$

TL 2647 (1976)

$$R_2C=O \longrightarrow R_2CX_2$$

X	Reagent(s)	
F	SF$_4$, (HF)	JACS *81* 3165 (1959); *82* 543 (1960); *89* 88 (1967); *91* 1386 (1969); *94* 2020 (1972)

		JOC *26* 2436 (1961); *27* 3164, 4086 (1962); *31* 991 (1966); *36* 818 (1971)
		J Med Chem *6* 174 (1963)
		Tetr *20* 1625 (1964)
		Org Rxs *21* 1 (1974) (review)
		JOC USSR *17* 1260 (1981)
	PhSF$_3$	JACS *84* 3058 (1962); *91* 1386 (1969)
	MoF$_6$, cat BF$_3$	Tetr *27* 3965 (1971)
	Et$_2$NSF$_3$	Syn 787 (1973)
		JOC *40* 574 (1975)
		JCS Perkin I 1354 (1979)
		JACS *109* 3046 (1987)
	HSCH$_2$CH$_2$SH, BF$_3$·2 HOAc/	JOC *51* 3508 (1986)

Cl	PCl$_5$	Ann *115* 29 (1860); *196* 259 (1879); *314* 369 (1901); *401* 121 (1913); *612* 1 (1958); 1 (1980)
		Compt Rend *136* 94 (1903)
		Ber *40* 2160 (1907); *43* 2940 (1910)
		JACS *50* 172 (1928); *59* 2434 (1937); *71* 3439 (1949); *72* 3952 (1950); *73* 1007, 5382 (1951); *108* 3762 (1986)
		Org Syn Coll Vol *2* 573 (1943)
		JCS 1425 (1952); 3116 (1954)
	PCl$_5$, PCl$_3$	Ann *435* 190 (1924); *501* 219 (1933)
Br	PCl$_3$Br$_2$	Ber *7* 759 (1874); *8* 406 (1875)
		JCS *45* 522 (1884)
		Ann *235* 299 (1886)
		Rec Trav Chim *50* 316, 837 (1931)
	HBr (on PhCHO)	Ann *341* 15 (1905)
	BBr$_3$ (on ArCHO)	Syn Commun *9* 341 (1979)
I	H$_2$NNH$_2$/I$_2$, Et$_3$N	JCS 470 (1962)
		Austral J Chem *23* 989 (1970)
	Me$_2$NNH$_2$/H$_2$NNH$_2$/I$_2$, Et$_3$N	JOC *33* 4317 (1968)

CL 465 (1978)

$$\underset{\text{RCCH}_2\text{R}}{\overset{\overset{\displaystyle O}{\|}}{\text{RCCH}_2\text{R}}} \longrightarrow \underset{\text{RC}=\text{CHR}}{\overset{\overset{\displaystyle X}{|}}{\text{RC}=\text{CHR}}}$$

X	Reagent(s)	
F	⬡NSF$_3$	JOC *50* 2007 (1985)
Cl	PCl$_5$	Ann *308* 264 (1899); 1 (1980)
		JACS *56* 1207 (1934); *59* 2434 (1937); *63* 1175 (1941); *65* 2208 (1943); *77* 1691 (1955); *90* 395 (1968); *109* 7838 (1987)
		Org Mag Res *10* 192 (1977)
		JOC *43* 2839 (1978); *47* 1084 (1982)
	PCl$_5$, PCl$_3$, POCl$_3$	JACS *65* 2208 (1943)
	(benzodioxaphosphole)PCl$_3$	Z Chem *22* 126 (1982)
I	N$_2$H$_4$/I$_2$, Et$_3$N	JCS 470 (1962)
		JACS *91* 2115 (1969); *103* 6526 (1981); *104* 6646 (1982)
		CC 174 (1969)
		JOC *34* 3502 (1969)
		Austral J Chem *23* 989 (1970); *24* 1425 (1971)
		BCSJ *53* 3252 (1980)
		JCS Perkin I 3017 (1982)
	N$_2$H$_4$/I$_2$, ⬡NH	JACS *103* 7122 (1981)
	N$_2$H$_4$/I$_2$, quanidine base	TL *24* 1605 (1983); *27* 1811 (1986)
		JOC *50* 2438 (1985)
	N$_2$H$_4$, I$_2$, DABCO	JOC *48* 4058 (1983)

See also page 158, Section 45.3.

$$\underset{\text{(cyclohexanone)}}{\overset{\displaystyle O}{}} \longrightarrow \underset{\text{(OTf enol)}}{\overset{\displaystyle OTf}{}} \xrightarrow{\text{reagent(s)}} \underset{\text{(X)}}{\overset{\displaystyle X}{}}$$

X	Reagent(s)	
Br	Me$_3$SnSnMe$_3$, cat Pd(PPh$_3$)$_4$/Br$_2$	JOC *51* 277 (1986)
I	MgI$_2$, Et$_3$N	Syn 222 (1986)

$$\underset{\text{O}}{\overset{\text{O}}{\parallel}}\text{C}-\underset{\text{H}}{\overset{\text{H}}{\underset{\mid}{\text{C}}}}-\underset{\text{O}}{\overset{\text{O}}{\parallel}}\text{C}- \longrightarrow -\underset{\text{O}}{\overset{\text{O}}{\parallel}}\text{C}-\text{C}=\underset{\text{X}}{\overset{\text{X}}{\text{C}}}-$$

X	Reagent(s)	
Cl	PCl$_3$	JCS *83* 110, 494 (1903); 1996 (1955); 2431 (1956)
		JACS *72* 1645 (1950)
		Can J Chem *37* 1870 (1959)
		J Gen Chem USSR *31* 3719 (1961)
		JOC *30* 1129 (1965)
		JCS Perkin II 142 (1972)
	PCl$_5$	JACS *77* 1136 (1955)
		(β-keto ester)
	POCl$_3$	JCS *83* 110 (1903)
		Ann *612* 158 (1958)
	COCl$_2$	Coll Czech Chem Commun *24* 2378 (1959)
		J Gen Chem USSR *30* 563 (1960)
	CH$_3$COCl	Ber *94* 96 (1961)
	ClCOCOCl	Syn 47 (1974)
		JOC *41* 636 (1975)
	PPh$_3$, CCl$_4$	Syn 708 (1975)
	Ph$_3$P·Cl$_2$, Et$_3$N	Syn Commun *5* 193 (1975)
		Can J Chem *60* 210 (1982)
Br	PBr$_3$	JCS *83* 110, 494 (1903)
		JOC *30* 1129 (1965)
		JCS Perkin II 142 (1972)
	Ph$_3$P, CBr$_4$	Syn 708 (1975)
	Ph$_3$P·Br$_2$, Et$_3$N	Syn Commun *5* 193 (1975)
		Can J Chem *60* 210 (1982)
I	Ph$_3$P·I$_2$, Et$_3$N	Syn Commun *5* 193 (1975)
		Can J Chem *60* 210 (1982)

X = Cl, Br, I

JOC *46* 197 (1981)

$$R_2C=O \xrightarrow[\text{2. H}_2\text{O}]{\text{1. LiCH}_2\text{AsOPh}_2} R_2\underset{\overset{\mid}{\text{OH}}}{\text{C}}\text{CH}_2\text{AsPh}_2 \xrightarrow[\text{2. Br}_2]{\text{1. LiAlH}_4} R_2\underset{\overset{\mid}{\text{OH}}}{\text{C}}\text{CH}_2\text{Br}$$

Ber *115* 645 (1982)

$$R'_2CO \longrightarrow R'_2C=CRX$$

X	Reagent(s)	
F	Ph$_3$P=CHF	Syn 75 (1969) Ber *104* 2885 (1971) JOC *40* 2796 (1975)
	Ph$_3$P=CHR /PhLi/FClO$_3$ [(*n*-Bu$_3$P)$_2$CF]Cl/NaOH (R = H)	Syn 38 (1969) JACS *107* 2811 (1985); *109* 3046 (1987)
Cl	Ph$_3$P=CHCl	JACS *82* 1510 (1960); *83* 1617 (1961) Ber *94* 1373 (1961); *99* 689 (1966) CC 446 (1978)
	Ph$_3$P=CHR /*n*-BuLi or PhLi/Cl$_2$IC$_6$H$_5$ Ph$_3$P=CHR /*n*-BuLi/NCS	Syn 38 (1969) TL 447 (1970) TL 447 (1970)
Br	Ph$_3$P=CHBr	Angew Int *1* 51 (1962) Ber *99* 689 (1966) TL *21* 4021 (1980)
	Ph$_3$P=CBrR	JOC *43* 2833 (1978) TL *27* 1995 (1986)
	Ph$_3$P=CHR /PhLi/Br$_2$ LiCHBr$_2$/Zn, HOAc	Syn 38 (1969) TL *22* 3745 (1981)
I	Ph$_3$P=CHR /*n*-BuLi/ Hg(OAc)$_2$/I$_2$	TL 447 (1970)

$$R_2CO \longrightarrow R_2C=CXY$$

X	Y	Reagent(s)	
F	F	NaO$_2$CCF$_2$Cl, PPh$_3$	TL 1461 (1964)
F	Cl	NaO$_2$CCFCl$_2$, PPh$_3$ HCFCl$_2$, KO-*t*-Bu, PPh$_3$	TL 71 (1968) JACS *84* 854 (1962) TL 71 (1968)
Cl	Cl	CCl$_4$, PPh$_3$	JACS *84* 1312 (1962) Compt Rend C *276* 903 (1973)
		CCl$_4$, P(NMe$_2$)$_3$	BSCF 2047 (1971) TL 1035 (1971)
		BrCCl$_3$, P(NMe$_2$)$_3$	TL 1239 (1977); *28* 5473 (1987)
		HCCl$_3$, KO-*t*-Bu, PPh$_3$	JACS *84* 854 (1962)
		LiCCl$_2$PO(OEt)$_2$	JOMC *59* 237 (1973) Syn 458 (1975)
		LiCCl$_2$SiMe$_3$	TL *24* 4727 (1983)
Br	Br	CBr$_4$, PPh$_3$	JACS *84* 1745 (1962) BSCF 3145 (1972) Compt Rend C *274* 1357 (1972)

TL 3769 (1972); 1373
  (1975); *28* 5145 (1987)

CBr$_4$, PPh$_3$, Zn	TL 3769 (1972)
CBr$_4$, P(NMe$_2$)$_3$	TL 1035 (1971)
HCBr$_3$, KO-*t*-Bu, PPh$_3$	JACS *84* 854 (1962)

2. Acetals

$$\underset{\underset{\text{OR}}{|}}{\overset{\overset{\text{H   OR}}{|\ \ \ |}}{-\text{C}-\text{C}-}} \longrightarrow \underset{\underset{\text{OR}}{|}}{\overset{\overset{\text{X   OR}}{|\ \ \ |}}{-\text{C}-\text{C}-}}$$

Cl$_2$ or Br$_2$	BSCF 2735 (1973)
Br$_2$	BSCF 4169 (1969)
Br$_2$, hν	Ber *95* 803 (1962)
Br$_2$, 2-methoxynaphthalene	JOC *52* 3018 (1987) (chiral)
Br$_2$, py	JACS *64* 1963 (1942)
Br$_2$, CaCO$_3$	JACS *49* 2517 (1927); *64* 1966 (1942)
Br$_2$, cat HBr	Angew *25* 259 (1986)
Br$_2$, (HBr), electrolysis	Syn Commun *10* 821 (1980)
(PhN̈Me$_3$)Br$_3^-$	BSCF 1822 (1961) Tetr *19* 861 (1963)
(C$_5$H$_5$N̈H)Br$_3^-$	JACS *84* 2344 (1962)
NBS, hν	JACS *73* 973 (1951)
NBS, (PhCO$_2$)$_2$	Ber *95* 803 (1962)

$$R^1CCH_2R^2 \xrightarrow[\text{HOCH}_2\text{CH}_2\text{OH, THF}]{(\text{PhNMe}_3)\text{Br}_3} R^1\underset{\underset{\text{Br}}{|}}{\overset{\overset{\text{O   O}}{\diagup\diagdown}}{\text{C}}}CHR^2$$

Syn 309 (1982)

X	Y	Reagent(s)	
OR	Cl	BCl$_3$	TL *28* 5595 (1987)
	Br	Me$_2$BBr	JOC *49* 3912 (1984)
		Me$_3$SiBr	Ber *113* 3058 (1980)
	I	Me$_3$SiI	Ber *113* 3075 (1980) Carbohydr Res *115* 95 (1983)
		Me$_3$SiCl, NaI	Ann 1052 (1984)

(*Continued*)

X	Y	Reagent(s)	
O$_2$CR	Br	Me$_3$SiBr	Ber *113* 3075 (1980)
	I	Me$_3$SiI	Ber *113* 3075 (1980)
		Me$_3$SiCl, NaI	Ann 1052 (1984)

$$R_2C\underset{O}{\overset{O}{\diagdown}}\quad\xrightarrow{PBr_5}\quad R_2CBr_2$$

JACS *108* 1265 (1986)

3. Carboxylic Acids

$$\underset{|}{\overset{H}{\underset{|}{C}}}-\overset{O}{\overset{\|}{C}}OH \longrightarrow \underset{|}{\overset{X}{\underset{|}{C}}}-\overset{O}{\overset{\|}{C}}OH$$

Review: Chem Rev *52* 237 (1953)

X	Reagent(s)	
F	n-BuLi/ArSO$_2$NFR′	JACS *106* 452 (1984)
	CF$_3$OF (on silyl ketene acetal)	JACS *102* 4845 (1980)
Cl	Cl$_2$, O$_2$, ClSO$_3$H, (chloranil)	JOC *40* 2960 (1975)
		Chem Ind 538 (1977)
		BCSJ *52* 255 (1979)
		Org Syn *59* 20 (1980)
	Cl$_2$; ClSO$_3$H; 7,7,8,8-tetra- cyanoquinodimethane	JOC *48* 1364 (1983)
	SO$_2$Cl$_2$	JACS *69* 86 (1947)
	SOCl$_2$, hν	JOC *38* 3919 (1973)
	LDA/CCl$_4$	JOC *52* 307 (1987)
Br	Br$_2$, P, (H$_2$O)/H$_2$O	Ber *14* 891 (1881); *20* 2026 (1887); *22* 1745 (1889); *24* 938, 2388 (1891)
		Ann *242* 141 (1887)
		Org Syn Coll Vol *2* 74 (1943)
		JACS *71* 2581 (1949)
I	I$_2$, Cu(OAc)$_2$, HOAc	CL 1509 (1984)
	I$_2$, CuCl, CuCl$_2$	CL 1509 (1984)

$$RCO_2H \xrightarrow{SF_4} RCF_3$$

JACS *82* 543 (1960)

4. Acid Halides

$$\underset{|}{\overset{H}{\underset{|}{\text{C}}}}-\overset{O}{\overset{||}{\text{CCl}}} \longrightarrow \underset{|}{\overset{X}{\underset{|}{\text{C}}}}-\overset{O}{\overset{||}{\text{CCl}}}$$

X	Reagent(s)	
Cl	NCS, SOCl$_2$	JOC *40* 3420 (1975)
Br	NBS, SOCl$_2$	JOC *40* 3420 (1975)
I	I$_2$, SOCl$_2$	JOC *40* 3420 (1975)

5. Esters

$$\underset{|}{\overset{H}{\underset{|}{\text{C}}}}-\overset{O}{\overset{||}{\text{COR}}} \longrightarrow \underset{|}{\overset{X}{\underset{|}{\text{C}}}}-\overset{O}{\overset{||}{\text{COR}}}$$

X	Reagent(s)	
F	CF$_3$OF (on silyl ketene acetal)	JACS *102* 4845 (1980)
	$\left(\text{C}_5\text{H}_5\text{NF}\right)$OTf	TL *27* 4465 (1986)
	(on silyl ketene acetal)	
Cl	Cl$_2$, PCl$_5$	JACS *66* 2074 (1944)
Cl, Br	LDA/CX$_4$	JOC *43* 3687 (1978)
Br	P, Br$_2$/ROH	Syn 39 (1969)
Br, I	LiNR$_2$/X$_2$	TL 3995 (1971); *28* 2477 (1987)
		JOC *43* 3687 (1978)
I	ICl (on silyl ketene acetal)	JOC *52* 4414 (1987)
	LDA/ZnCl$_2$/I$_2$	JOC *52* 4414 (1987)

$$\underset{R}{\overset{O\quad O}{\underset{|}{\overset{||\quad ||}{\text{CH}_3\text{CCHCOR}}}}} \xrightarrow[\text{NaOCH}_3,\ \text{CH}_3\text{OH}]{\text{NCS}} \overset{Cl}{\overset{|}{\text{RCHCO}_2\text{R}}}$$

TL *28* 5505 (1987)

$$\text{NO}_2\text{CHRCO}_2\text{Et} \xrightarrow[\text{2. FClO}_3]{\text{1. spray dried KF}} \overset{F}{\overset{|}{\text{NO}_2\text{CRCO}_2\text{Et}}}$$

JOC *52* 5061 (1987)

$$RCH(CO_2Et)_2 \xrightarrow[\text{2. reagent}]{\text{1. base}} R\overset{\overset{\displaystyle F}{|}}{C}(CO_2Et)_2$$

Reagent

ArSO$_2$NFR   JACS *106* 452 (1984)

$$\left[\begin{array}{c} CH_3-\underset{\underset{\displaystyle CH_3}{}}{\overset{\overset{\displaystyle CH_3}{}}{\underset{NF}{\bigcirc}}} \end{array} \right] OTf$$ TL *27* 4465 (1986)

$$C_n \overset{\displaystyle O}{\underset{\displaystyle O}{\diagdown}} C=O \xrightarrow[\text{2. ROH}]{\text{1. Me}_3\text{SiI}} I-C_n-\overset{\overset{\displaystyle O}{||}}{C}OR$$

$$n = 2-5$$

Syn Commun *11* 763 (1981)

$$\underset{O}{\overset{\overset{\displaystyle O}{||}}{\diagdown}}=O \xrightarrow[(\text{PhCH}_2\text{NEt}_3)\text{Cl}]{\text{HBr}} Br(CH_2)_3\overset{\overset{\displaystyle O}{||}}{C}CH_3$$

Syn Commun *10* 897 (1980)

6. Amides

$$-\overset{\overset{\displaystyle H}{|}}{\underset{\displaystyle |}{C}}-\overset{\overset{\displaystyle O}{||}}{C}NR_2 \longrightarrow -\overset{\overset{\displaystyle F}{|}}{\underset{\displaystyle |}{C}}-\overset{\overset{\displaystyle O}{||}}{C}NR_2$$

CF$_3$OF (on enol silane)   JACS *102* 4845 (1980)

R$_f$OF (on enol silane)   JOC *51* 1482 (1986)

base/ArSO$_2$NFR   JACS *106* 452 (1984)

$$RNH\overset{\overset{\displaystyle O}{||}}{C}R' \xrightarrow[\text{PCl}_5]{\text{SOCl}_2 \text{ or}} RCl$$

JACS *84* 769 (1962)

$$\underset{\underset{\displaystyle COPh}{N}}{\bigcirc} \xrightarrow[\text{Br}_2]{\text{PBr}_3} Br(CH_2)_5Br$$

Org Syn Coll Vol *1* 428 (1941)
JACS *68* 2402 (1946); *71* 2808 (1949)
JOC *51* 2206 (1986); *52* 5466 (1987)

12. HALODECARBOXYLATION OF ACIDS AND DECARBONYLATION OF ACID HALIDES

$$RCO_2H \longrightarrow RX$$

Review: Chem Rev 56 219 (1956) (Hunsdieker reaction)

X	Reagent(s)	
F	XeF$_2$, HF	JOC 48 4158 (1983)
Cl	NCS/Pb(OAc)$_4$	Syn 493 (1973) JACS 107 516 (1985)
	Pb(OAc)$_4$, LiCl	JACS 87 2500 (1965); 107 199, 7978 (1985) JOC 30 3265 (1965) Org Rxs 19 279 (1972) (review)
	Pb(O$_2$CR)$_4$, CCl$_4$, hν	Ann 735 47 (1970)
Br	AgO or KOH- AgNO$_3$/Br$_2$	Org Syn Coll Vol 3 578 (1955) Org Rxs 9 332 (1957)
	Tl$_2$CO$_3$/Br$_2$ TlOEt/Br$_2$	JCS Perkin I 2608 (1981) JOC 34 1172 (1969) JCS Perkin I 2608 (1981)
	HgO, Br$_2$, (hν)	JOC 26 280 (1961); 30 415 (1965); 34 3216 (1969); 37 664 (1972); 44 3405 (1979); 52 460 (1987) Org Syn 43 9 (1963) JACS 106 2194 (1984); 109 7230 (1987) Syn Commun 14 983 (1984) (gem-dibromocycloalkanes) TL 27 5371 (1986); 28 5263 (1987)
I	I$_2$ (on AgO$_2$CR) t-C$_4$H$_9$OI, (hν)	JCS 368 (1941) JCS 2438 (1965) JOC 45 4226 (1980)

$$RCO_2H \longrightarrow RX \quad (\textit{Continued})$$

X	Reagent(s)	
Pb(OAc)$_4$, I$_2$		JCS 2438 (1965)
		JOC *31* 1857 (1966)
		Org Rxs *19* 279 (1972) (review)
PhI(OAc)$_2$, I$_2$, (hν)		JOC *51* 402 (1986)
		CC 675 (1987)
		TL *28* 6381 (1987)
HgO, I$_2$		JOC *37* 664 (1972)

$$\overset{\displaystyle O}{\underset{\displaystyle \parallel}{}}$$
$$RCX \longrightarrow RX$$

Review: Syn 157 (1969)

X	R	Reagent(s)	
Cl	alkyl, aryl	cat ClRh(PPh$_3$)$_3$	JACS *88* 3452 (1966); *90* 99 (1968)
	aryl	cat Pd/C, PdCl$_2$ or Pd(PPh$_3$)$_4$	TL *23* 371 (1982)
Br	aryl, vinylic	/BrCCl$_3$, AIBN	TL *26* 5939 (1985)
	aryl	cat ClRh(CO)(PPh$_3$)$_2$	JACS *90* 99 (1968)
I	aryl	cat ClRh(PPh$_3$)$_3$	JOC *33* 1928 (1968)
Cl, Br	aryl	cat ClRh(PPh$_3$)$_3$	JACS *89* 2338 (1967)

13. HALOGENATION OF ORGANOMETALLICS

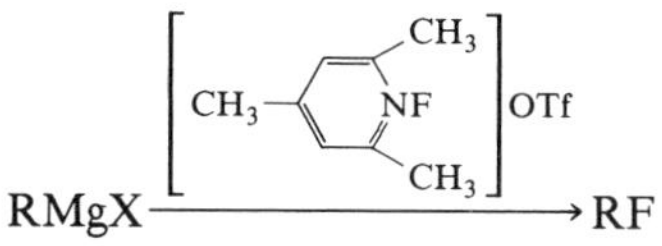

$$RMgX \xrightarrow{\left[\begin{array}{c} CH_3 \\ CH_3--NF \\ CH_3 \end{array}\right] OTf} RF$$

TL *27* 4465 (1986)

$$ArB(OH)_2 \xrightarrow[\text{chloramine-T}]{NaX} ArX$$

X = Br, I

Org Prep Proc Int *14* 359 (1982)

$$ArSnR_3 \xrightarrow{CsSO_4F} ArF$$

CC 1623 (1986)

$$RCH=CHSn(n\text{-}Bu)_3 \xrightarrow[X = Br, I]{X_2} RCH=CHX$$

TL *28* 2033 (1987)
JOC *52* 3687 (1987)

See also page 215, Section 2, for additional preparations of vinylic halides via hydrometallation-halogenation.

$$R_3CSeR' \xrightarrow[Et_3N]{Br_2} R_3CBr$$

TL 2647 (1976)

AMINES

GENERAL REFERENCES

Houben-Weyl, "Methoden der Organischen Chemie," 4th ed, Vol XI/1, G. Thieme, Stuttgart (1957)

P. A. S. Smith, "The Chemistry of Open-Chain Organic Nitrogen Compounds," W. A. Benjamin, New York (1965), Vol 1, pp 60–78 and 115–122

L. Spialter and J. A. Pappalardo, "The Acyclic Aliphatic Tertiary Amines," Macmillan, New York (1965)

"Chemistry of the Amino Group," Ed. S. Patai, Interscience, New York (1968)

"The Chemistry of Amino, Nitroso and Nitro Compounds and their Derivatives," Ed. S. Patai, J. Wiley, New York (1982)

1. FROM ALKANES

$$R_3CH + AlCl_3 + NCl_3 \longrightarrow R_3CNH_2$$

Tetr *23* 3563 (1967)

2. FROM ALKENES

Review: Tetr *39* 703 (1983)

For analogous syntheses of amides and sulfonamides see also page 824, Section 3.

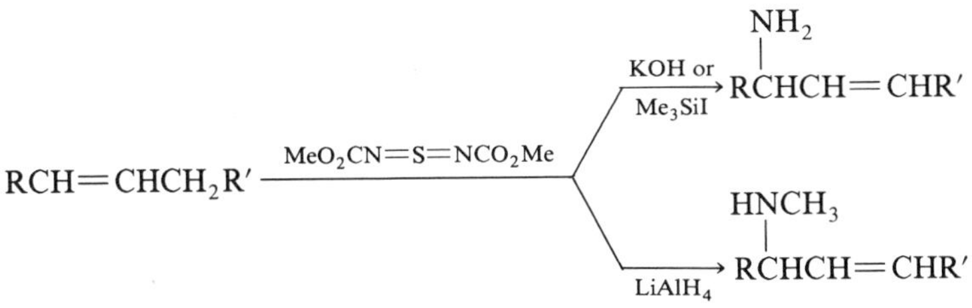

$$RCH{=}CHCH_2R' \xrightarrow{MeO_2CN{=}S{=}NCO_2Me}$$

$$\xrightarrow[Me_3SiI]{KOH \text{ or}} \overset{NH_2}{\underset{|}{R}CHCH{=}CHR'}$$

$$\xrightarrow[LiAlH_4]{} \overset{HNCH_3}{\underset{|}{R}CHCH{=}CHR'}$$

JOC *48* 3561 (1983)

$$RNH_2\ (R_2NH) + H_2C{=}CH_2 \xrightarrow[\Delta]{cat\ NaNHR} RNHCH_2CH_3$$

JOC *22* 646 (1957)
Org Syn Coll Vol *5* 575 (1973)

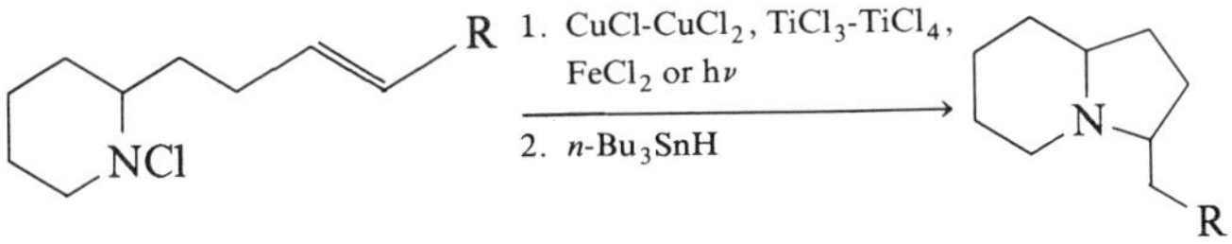

$$\xrightarrow[2.\ n\text{-}Bu_3SnH]{1.\ CuCl\text{-}CuCl_2,\ TiCl_3\text{-}TiCl_4,\ FeCl_2\ or\ h\nu}$$

BSCF 111, 115 (1970)
TL 3107 (1970); 903 (1971); 2191 (1974); *22* 61 (1981)
Acct Chem Res *8* 165 (1975)
JOC *51* 5043 (1986)

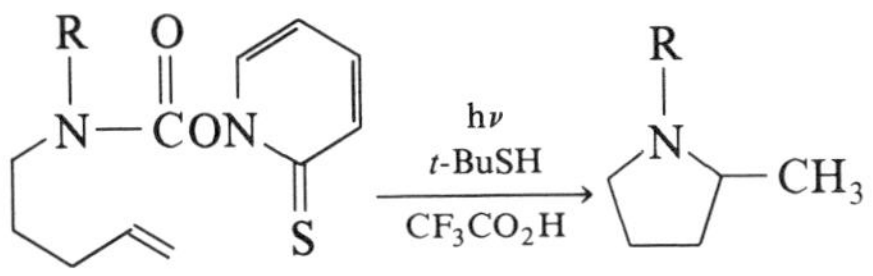

$$\xrightarrow[CF_3CO_2H]{h\nu \\ t\text{-}BuSH}$$

JACS *109* 3163 (1987)

$$\underset{}{C}{=}C{-}\underset{}{C_n}{-}\overset{\displaystyle\overset{\textstyle HNR}{|}}{C} \longrightarrow \overset{\displaystyle\overset{\textstyle X}{|}}{C}{-}\overset{\displaystyle\overset{\textstyle RN}{|}}{C}{-}\overset{\displaystyle\overset{\textstyle C}{|}}{C_n}$$

Review:

P. A. Bartlett, "Asymmetric Synthesis," Ed. J. D. Morrison, Academic Press, New York (1984), Vol 3, Part B, Chpt 6

<u>X</u>   Reagent

PhSe   PhSeBr   JOC *44* 287 (1979)

JACS *107* 3891 (1985)

HgX   HgX$_2$ R. C. Larock, "Solvomercura-
                                                                           tion/Demercuration Reac-
                                                                           tions in Organic Synthesis,"
                                                                           Springer, New York (1986),
                                                                           Chpt 6 (review)

$$RCH{=}CH_2 \xrightarrow[HgX_2]{R_2NH} \xrightarrow[LiAlH_4]{NaBH_4 \text{ or}} \overset{\displaystyle\overset{\textstyle NR_2}{|}}{R}CHCH_3$$

$$X = NO_3, OAc, Cl$$

Compt Rend C *262* 1591 (1966); *272* 1141 (1971)
TL 5165 (1967); 2289 (1969); 4399 (1971)
BSCF 583 (1970)
Syn 806 (1974); 116, 467 (1975)
R. C. Larock, "Solvomercuration/Demercuration Reactions in Organic Synthesis," Springer, New York (1986), Chpt 6 (review)

$$RCH{=}CH_2 \xrightarrow[\text{2. NaBH}_4, \text{NaOH}]{\text{1. H}_2\text{NPO(OEt)}_2, \text{Hg(NO}_3)_2} \overset{\displaystyle\overset{\textstyle HNPO(OEt)_2}{|}}{R}CHCH_3 \xrightarrow{HCl} \left[\overset{\displaystyle\overset{\textstyle \overset{+}{N}H_3}{|}}{R}CHCH_3 \right]Cl^-$$

Syn 918 (1982)

$$RCH{=}CH_2 \longrightarrow \overset{\displaystyle\overset{\textstyle RNH}{|}}{R}CHCH_2HgCl \longrightarrow \overset{\displaystyle\overset{\textstyle RNM}{|}}{R}CHCH_2M \xrightarrow[\text{2. H}_2\text{O}]{\text{1. E}^+} \overset{\displaystyle\overset{\textstyle RNH}{|}}{R}CHCH_2E$$

M = alkali metal

$E^+ = D_2O, O_2, CO_2, RCHO, R_2CO, Me_3SiCl, RX, (MeS)_2, R_2C{=}NR$

TL 2015 (1978)
JOC *44* 4798 (1979); *46* 1281 (1981); *47* 1560 (1982)

$$RCH{=}CH_2 \xrightarrow{BH_3} \xrightarrow{\text{reagent(s)}} RCH_2CH_2NH_2$$

Reagent(s)

NH_2OSO_3H | JACS *88* 2870 (1966)
JOC *32* 3199 (1967)
Syn 196 (1974)

NH_4OH, NaOCl | CC 62 (1982)

chiral   chiral

JACS *108* 6761 (1986)

$$RCH{=}CH_2 \xrightarrow{HBX_2} RCH_2CH_2BX_2 \xrightarrow{R'N_3} \xrightarrow{H_2O} RCH_2CH_2NHR'$$

X = Cl or alkyl

JACS *93* 4329 (1971); *94* 2114 (1972); *95* 2394 (1973)

$$XCH{=}CH_2 + HNR_2 \longrightarrow XCH_2CH_2NR_2$$

X = COR, CO_2H, CO_2R, $CONR_2$, CN

Org Syn Coll Vol *1* 196 (1941); *3* 93, 258 (1955)
JCS 399 (1945); 4166 (1957)
JACS *68* 1259 (1946); *69* 971 (1947); *71* 1901, 2532 (1949); *108* 8112 (1986)
Org Rxs *5* 79 (1949) (review)
JOC *23* 94 (1958); *24* 1096 (1959); *42* 1650 (1977); *52* 5742 (1987)
Org Syn Coll Vol *4* 146 (1963) [$Cu(OAc)_2 \cdot H_2O$ catalyzed]
BSCF 1717 (1971)
Chem Pharm Bull *25* 1319 (1977)
TL *28* 1757 (intramolecular), 1761 (intramolecular), 3103 (1987)

$$R_2C{=}CH_2 + HNR'_2 \xrightarrow[\text{Rh catalyst}]{CO, H_2O} R_2CHCH_2CH_2NR'_2$$

JOC *47* 445 (1982)

$$RCH{=}CH_2 \xrightarrow[\text{2. HCl, EtOH}]{\text{1. NBS, } H_2NCN} RCHCH_2Br \xrightarrow{\text{base}} RCHCH_2NH_2$$

JACS *107* 2931 (1985)

$$XCH{=}CHY \xrightarrow[Me_3SiOTf]{Me_3SiCH_2N{=}CHPh}$$

X, Y = H, CO_2R, CONMeCO

TL *23* 2589 (1982)

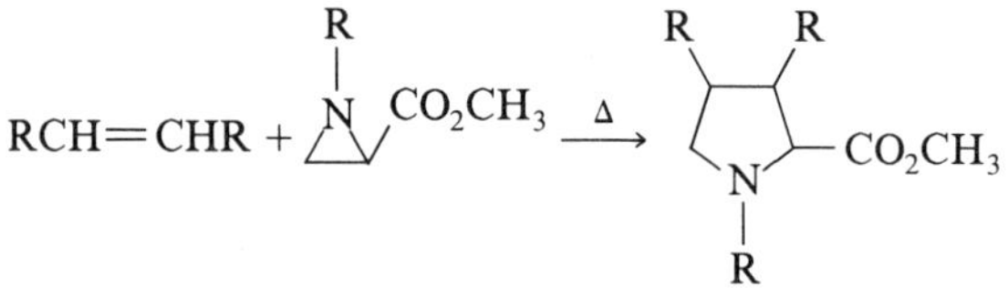

$$RCH{=}CHR + \overset{R}{\underset{}{\overset{|}{N}}}\diagdown CO_2CH_3 \xrightarrow{\Delta} \text{(pyrrolidine)} CO_2CH_3$$

JACS *89* 1753 (1967)
Angew Int *8* 602 (1969)
CC 1187, 1188 (1971)
BSCF 709 (1974)
JOC *50* 2309 (1985)

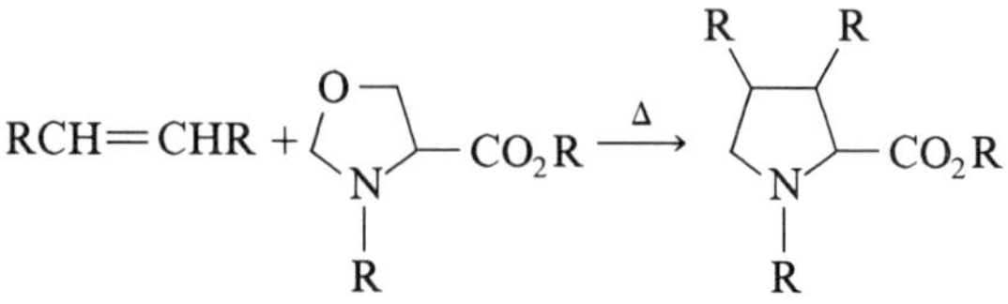

$$RCH{=}CHR + \text{(oxazolidine)}{-}CO_2R \xrightarrow{\Delta} \text{(pyrrolidine)}{-}CO_2R$$

TL *28* 2973 (intramolecular), 2975 (1987)

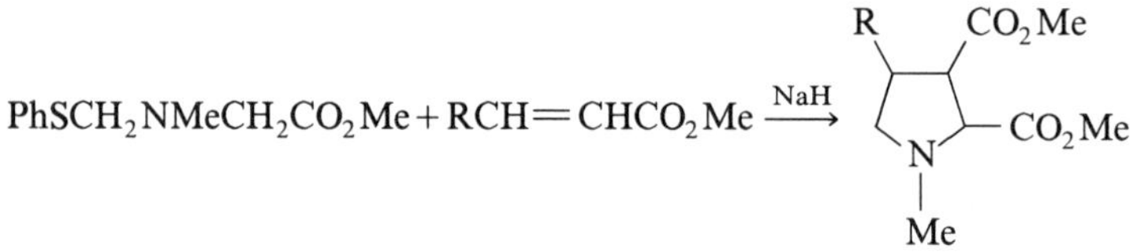

$$PhSCH_2NMeCH_2CO_2Me + RCH{=}CHCO_2Me \xrightarrow{NaH} \text{(pyrrolidine)}$$

TL *25* 1579 (1984)

$$R^1CHO + R^2NHCHR^3CO_2R + R^4CH{=}CHR^5 \xrightarrow{\Delta} \text{(pyrrolidine)}$$

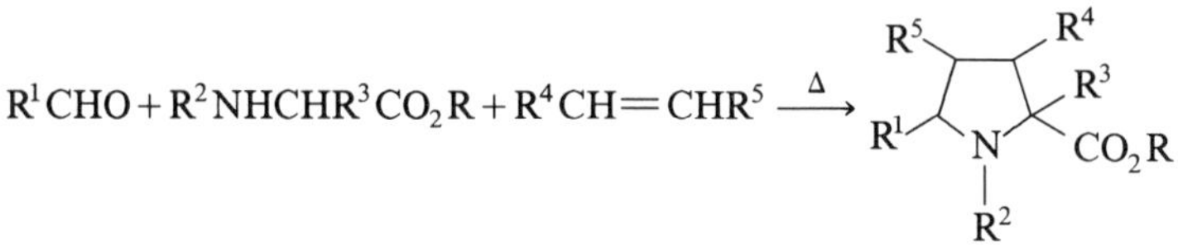

CC 109 (1978); 180, 182 (1984); 1566 (1985)
TL 2823, 2885 (1978); *25* 4613 (1984); *26* 2775 (1985)
JOC *48* 2994 (1983) (intramolecular)
Bull Soc Chim Belg *93* 593 (1984) (review)
CL 973 (1986)

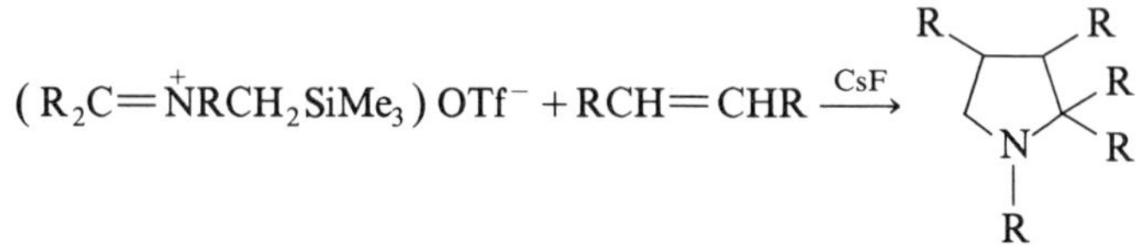

$$(R_2C{=}\overset{+}{N}RCH_2SiMe_3)\,OTf^- + RCH{=}CHR \xrightarrow{CsF} \text{(pyrrolidine)}$$

JACS *101* 6452 (1979); *102* 7993 (1980)
TL *24* 4303 (1983)
JOC *48* 1554 (1983)
Chem Rev *86* 941 (1986) (review)

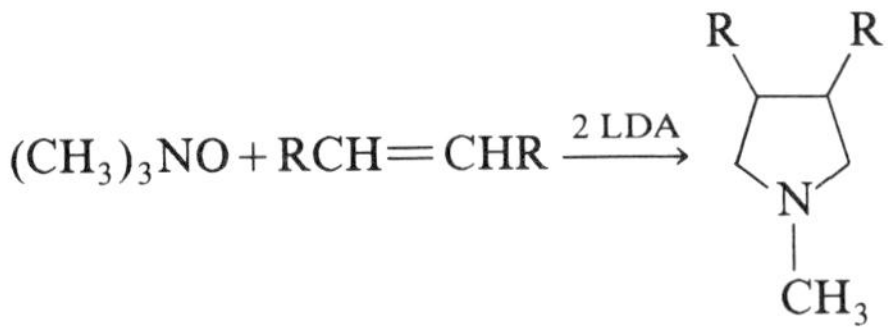

$$(CH_3)_3NO + RCH\!=\!CHR \xrightarrow{2\,LDA}$$

CC 31 (1983)
Heterocycles *23* 653 (1985)
Can J Chem *63* 725 (1985)

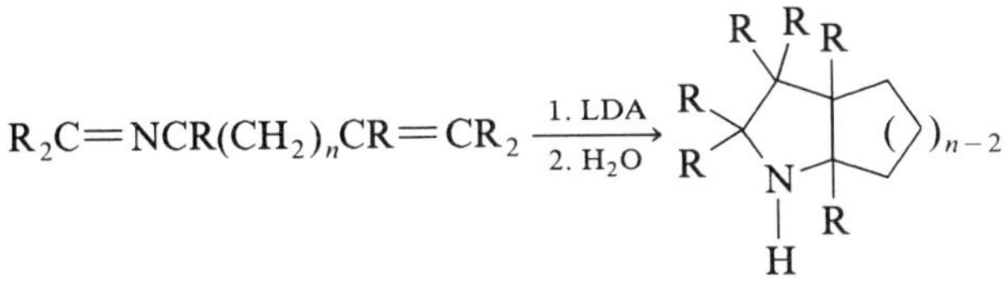

$$R_2C\!=\!NCR(CH_2)_nCR\!=\!CR_2 \xrightarrow[2.\ H_2O]{1.\ LDA}$$

$n = 1, 2$

JACS *108* 2769 (1986)

$$C\!=\!C\!-\!C\!=\!C\!-\!C_n\!-\!N_3 \xrightarrow{\Delta}$$

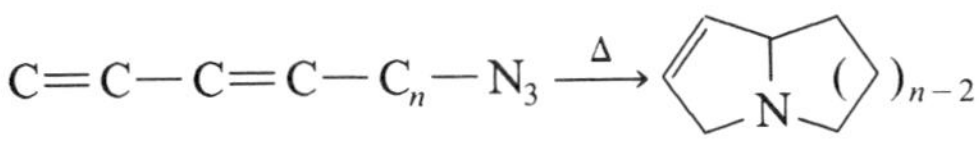

$n = 1, 2$

3. FROM ALKYL AND ARYL HALIDES

See also page 401, Section 4.

$$RX + R'NH_2 \longrightarrow R'NHR$$

$\underline{X}$

halide

> JCS 992 (1930)
> JACS *54* 1499, 3441, 4457 (1932); *82* 6163 (1960)
> Org Syn Coll Vol *1* 102 (1941); *2* 290 (1943); *3*
> 256 (1955); *4* 466 (1963); *5* 88 (1973)

tosylate

> JACS *55* 345 (1933)
> JCS 694 (1955)

sulfate

> Org Syn Coll Vol *5* 1018 (1973)

phosphate

> Org Syn Coll Vol *5* 1085 (1973)

$$RX + LiNR'_2 \longrightarrow RNR'_2$$

Ind Eng Chem *29* 1361 (1937); *33* 218 (1941)

$$RC{\equiv}CX + LiNR_2 \longrightarrow RC{\equiv}CNR_2$$

Angew *3* 506 (1964); *6* 767 (1967)

$$ArX + HNMe_2 \xrightarrow{\Delta} ArNMe_2$$

JACS *71* 740 (1949)

$$ArX + HNR_2 \xrightarrow{\text{Cu catalyst}} ArNR_2$$

Org Syn Coll Vol *2* 15 (1943); *3* 307 (1955)
JACS *72* 888 (1950) (intramolecular); *109* 1496 (1987)
Tetr *40* 1433 (1984) (review)

$$ArX + (K)NaNR_2 \longrightarrow ArNR_2$$

JOC *48* 4397 (1983); *50* 1334 (1985); *51* 5157 (1986); *52* 2619 (1987)

$$ArX + n\text{-}Bu_3SnNR_2 \xrightarrow{\text{cat } PdCl_2(PAr_3)_2} ArNR_2$$

CL 927 (1983)

$$R_3N + R'X \longrightarrow \left[R_3\overset{+}{N}R'\right]X^-$$

Org Syn Coll Vol *4* 85, 98, 582, 585 (1963); *5* 315, 608, 825, 883, 989 (1973)

$$RNH_2 \xrightarrow{\text{PhCHO}} PhCH{=}NR \xrightarrow[\text{2. NaOH}]{\text{1. R'X}} RNHR'$$

$$R'X = RI, R_2SO_4$$

Org Syn Coll Vol *5* 736, 758 (1973)

$$RX + H_2NN(CH_3)_2 \longrightarrow \left[\begin{array}{c} CH_3 \\ | \\ RN\overset{+}{N}H_2 \\ | \\ CH_3 \end{array}\right]X^- \xrightarrow{HNO_2} RN(CH_3)_2$$

R = 1°, 2° alkyl; benzylic; propargylic

Syn Commun *12* 801 (1982)

$$RX \longrightarrow RN_3 \longrightarrow RNH_2$$

See page 409, Section 5.

$$ArNH_2 \xrightarrow[\text{Et}_3N]{\text{Ph}_3PBr_2} ArN{=}PPh_3 \xrightarrow[\text{2. OH}^-]{\text{1. 1° RX}} ArNHR$$

Syn 295 (1980)

$$R'NH_2 \longrightarrow R'NHY \xrightarrow[\text{2. RX}]{\text{1. base}} R'NY \overset{\overset{R}{|}}{} \longrightarrow RNHR'$$

Y	
COCF$_3$	JOC *39* 1315 (1974)
	TL 4987 (1978)
SO$_2$CF$_3$	Tetr *31* 2517 (1975)
SO$_2$CH$_2$COPh	Tetr *31* 2517 (1975)
PO(OEt)$_2$	Angew Int *16* 107 (1977)

$$RX + ArNH\overset{\overset{O}{\|}}{C}R' \xrightarrow[\text{crown ether}]{KO\text{-}t\text{-}Bu} \xrightarrow{2\ H_2O} RNHAr$$

$$RX = MeI, H_2C{=}CHCH_2Br$$

Syn Commun *9* 757 (1979)

$$RX + KN \underset{O}{\overset{O}{\bigcirc}} \longrightarrow RN \underset{O}{\overset{O}{\bigcirc}} \xrightarrow{N_2H_4} RNH_2$$

Org Syn Coll Vol *2* 83 (1943)
JACS *72* 2786 (1950)
Angew Int *7* 919 (1968)
Syn 389 (1976) (phase transfer)
BCSJ *55* 1671 (1982) (phase transfer)

$$ArX + CuN \underset{O}{\overset{O}{\bigcirc}} \longrightarrow ArN \underset{O}{\overset{O}{\bigcirc}} \xrightarrow{N_2H_4} ArNH_2$$

X = Br, I

CC 578 (1969)

$$RX + (EtO)_2 \overset{O}{\underset{\|}{P}} - \overset{Na}{\underset{|}{N}} - \overset{O}{\underset{\|}{C}}O\text{-}t\text{-}Bu \longrightarrow (EtO)_2 \overset{O}{\underset{\|}{P}} - \overset{R}{\underset{|}{N}} - \overset{O}{\underset{\|}{C}}O\text{-}t\text{-}Bu \xrightarrow{HCl} (R\overset{+}{N}H_3)Cl^-$$

Syn 922 (1982)

$$RX + NaN(SiMe_3)_2 \xrightarrow{HCl} RNH_2$$

Ber *94* 2311 (1961)

$$ArI + CuN(SiMe_3)_2 \xrightarrow{CH_3OH} ArNH_2$$

CC 256 (1974)

$$RX \xrightarrow{LiN(SAr)_2} RN(SAr)_2 \xrightarrow{HCl} RNH_2$$

TL 3411 (1970)
BCSJ *44* 2797 (1971)

$$RX \xrightarrow[]{(CH_2)_6N_4 \quad H^+} RNH_2$$

Org Rxs *8* 204 (1954)
Syn 161 (1979)

$$RX \xrightarrow[Mg]{Li \text{ or}} RM \xrightarrow{reagent(s)} RNH_2$$

M = Li, MgX

Reagent(s)

NH$_2$Cl	JACS *58* 27 (1936); *63* 1692 (1941)
NH$_2$OCH$_3$	JCS 781 (1946)
NH$_2$OCH$_3$, CH$_3$Li (M = Li; R =1°, 2°, 3° alkyl; aryl; benzylic)	JOC *47* 2822 (1982) JACS *108* 6016 (1986)

$$RX \xrightarrow[\text{Mg}]{\text{Li or}} RM \xrightarrow{\text{reagent(s)}} RNH_2 \quad (\textit{Continued})$$

$$M = Li, MgX$$

Reagent(s)

$NH_2O_2PPh_2$ TL *23* 5399 (1982)

$N_3CH_2SiMe_3$ (R = aryl only)   CC 1322 (1983)
                                                         TL *27* 6193 (1986)

N_3CH_2SPh/KOH or HCO_2H (R = aryl only)   JACS *103* 2484 (1981)

$R^1C(N_3){=}CHR^2$ (M = Li; R = aryl, benzylic)   TL *23* 699 (1982)

$$RX \longrightarrow RLi \xrightarrow[\text{2. H}_2\text{O}]{\text{1. LiNMeOMe}} RNHMe$$

$$R = 1°, 2°, 3° \text{ alkyl; aryl}$$

JACS *108* 6016 (1986)

$$RX \xrightarrow[\text{Mg}]{\text{Li or}} \xrightarrow{(Me_2\overset{+}{N}{=}CH_2)X^-} RCH_2NMe_2$$

$$R = 1° \text{ alkyl, aryl}$$

<u>X</u>

O_2CCF_3 Syn Commun *6* 539 (1976)

I   TL 1299 (1977); *28* 3241 (1987)

For related reactions see page 401, Section 4.

$$2 RX \xrightarrow[\text{2. (MeO)}_2\text{CHNR}'_2]{\text{1. 2 Mg}} \xrightarrow{\text{H}_2\text{O}} R_2CHNR'_2$$

Syn 757 (1978)

$$RX \xrightarrow[\text{2. Ph}_2\text{PON}{=}\text{CHOEt}]{\text{1. Mg}} \overset{\displaystyle O}{\overset{\displaystyle \|}{Ph_2PNHCHR_2}} \xrightarrow[\text{2. NaOH}]{\text{1. HCl}} R_2CHNH_2$$

$$R = 1° \text{ alkyl, aryl}$$

Syn 691 (1979)

$$H_2C{=}CR^1CH_2Br \xrightarrow[\substack{\text{2. Zn(Cu)} \\ \text{3. Al(Hg)}}]{\text{1. R}^2\text{C}{\equiv}\text{NO}} \overset{\displaystyle NH_2}{\overset{\displaystyle |}{R^2CHCH_2CR^1}}{=}CH_2$$

Angew Int *18* 78 (1979)

4. FROM OTHER AMINES

For imines → amines see also page 421, Section 9.

1. Alkylation on Nitrogen

See also page 397, Section 3; page 419, Section 8; and page 421, Section 9.

$$R_2NH \longrightarrow R_2NR'$$

$[(R'O)_3\overset{+}{P}CH_3]BF_4^-$	JOC *49* 4877 (1984)
R'_2CuLi, O_2	JOC *45* 2739 (1980)
$(Ar_2I)X, cat\ Cu(OAc)_2\ (R' = Ar)$	JOC *45* 2127 (1980)
$Ph_3Bi, cat\ Cu(OAc)_2\ (R' = Ph)$	TL *28* 887 (1987)
$Ph_3Bi(OAc)_2, cat\ Cu\ (R' = Ph)$	TL *27* 3615 (1986)
$Ph_3Bi(OAc)_2, cat\ Cu(OAc)_2\ (R' = Ph)$	J Gen Chem USSR *55* 413, 2232 (1985)
$PhPb(OAc)_3, cat\ Cu(OAc)_2$ or $Cu(O_2CCF_3)_2\ (R' = Ph)$	TL *28* 3111 (1987)

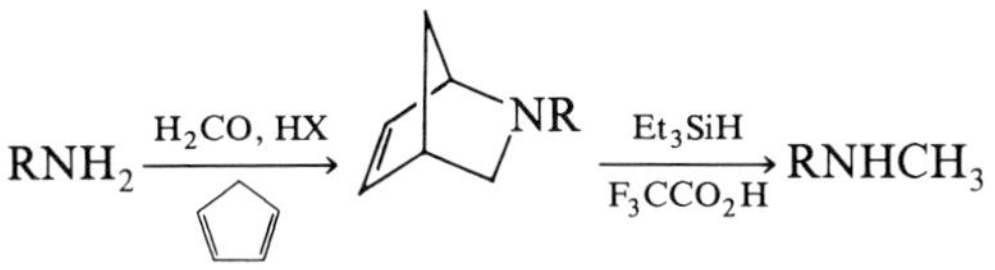

$$RNH_2 \xrightarrow[]{H_2CO, HX} \quad \xrightarrow[F_3CCO_2H]{Et_3SiH} RNHCH_3$$

JOC *52* 5746 (1987)

$$RNH_2 \cdot HX + H_2CO + n\text{-}Bu_3SnCH_2CH = CH_2 \longrightarrow RNH(CH_2)_2CH = CH_2$$

JOC *52* 1378 (1987)

$$RNH_2 + H_2C=CHCH_2O_2CR' \xrightarrow{\text{cat Pd(PPh}_3)_4} RNHCH_2CH=CH_2$$

Tetr *33* 2615 (1977) (review)

J. Tsuji, "Organic Synthesis with Palladium Compounds," Springer Verlag, Berlin (1980) (review)

R. F. Heck, "Palladium Reagents in Organic Synthesis," Academic Press, London (1985) (review)

$$RNH_2 + RCH=CHX \longrightarrow RNHCHRCH_2X$$

$$X = COR, CO_2H, CO_2R, CONR_2, CN$$

See page 391, Section 2.

$$RNH_2 \xrightarrow[\text{H}_2\text{O}_2]{\text{Na}_2\text{WO}_4 \cdot 2\,\text{H}_2\text{O}} RNO \xrightarrow[\text{2. Na, NH}_3 \text{ or } \text{C}_{10}\text{H}_8]{\text{1. R'NHNH}_2, \text{ PbO}_2} RNHR'$$

JOC *50* 5391 (1985)

Org Syn *65* 166 (1987)

$$R_2CHNH_2 \xrightarrow[\Delta]{\text{Raney Ni}} (R_2CH)_2NH$$

Syn 70 (1979)

$$R_2CHNH_2 \xrightarrow[\text{PPh}_3]{\overset{\Delta}{\text{cat RuCl}_3}} (R_2CH)_2NH \text{ or } (R_2CH)_3N$$

JOC *46* 1759 (1981)

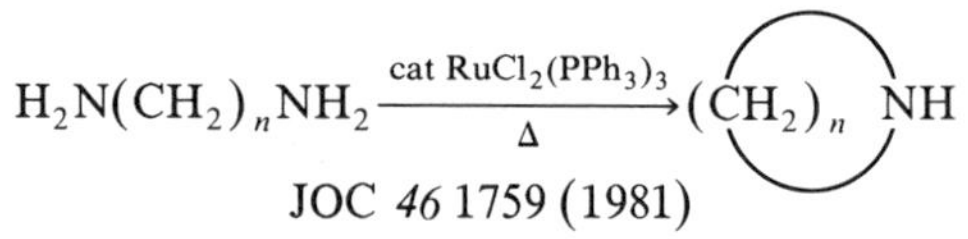

$$H_2N(CH_2)_nNH_2 \xrightarrow[\Delta]{\text{cat RuCl}_2(\text{PPh}_3)_3} (CH_2)_n\ NH$$

JOC *46* 1759 (1981)

2. Alkylation on Carbon

For a review of metalation and electrophilic substitution of amine derivatives adjacent to nitrogen see Chem Rev *84* 471 (1984).

See page 900, Section 4, for alkylation on carbon next to an amide nitrogen.

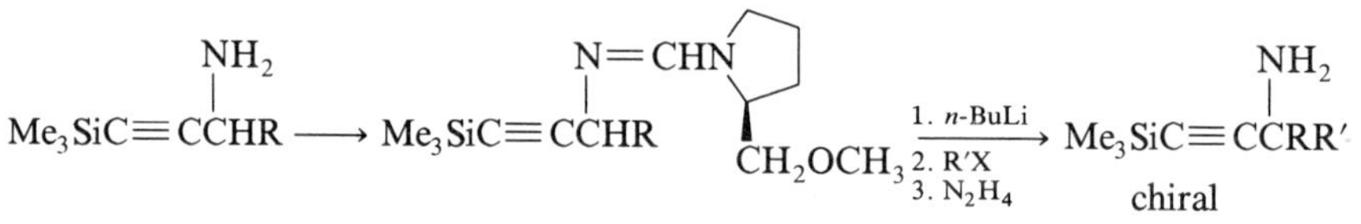

Ann 1668 (1983)

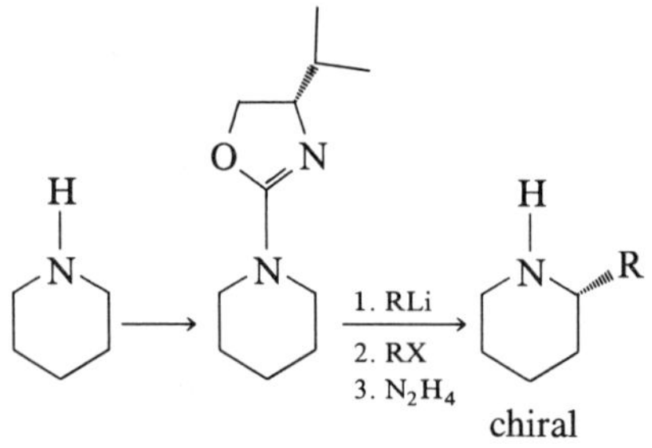

JOC *51* 3076 (1986)

JACS *109* 1265 (1987) (mechanism)

$$RNHCHR_2 \longrightarrow \overset{\overset{\displaystyle HC=NR}{|}}{RNCHR_2} \xrightarrow[\substack{2.\ E^+ \\ 3.\ H_3O^+}]{1.\ base} \overset{\overset{\displaystyle E}{|}}{RNHCR_2}$$

$$E^+ = RX, RCHO, R_2CO$$

JACS *102* 7125 (1980); *105* 117 (1983); *106* 3270 (1984); *107* 7974 (1985) (chiral); *109* 1262 (1987)
  (chiral, mechanism)
TL *22* 5115, 5119 (1981); *27* 1465 (1986) (chiral)
JOC *51* 3108 (1986) (chiral)

$$RNHCHR_2 \longrightarrow \overset{\overset{\displaystyle NO}{|}}{RNCHR_2} \xrightarrow[2.\ E^+]{1.\ base} \overset{\overset{\displaystyle ON}{|}\ \ \overset{\displaystyle E}{|}}{RN-CR_2} \xrightarrow[\substack{H_2\text{-Ni or} \\ NaBH_4,\ NiCl_2\ \text{or}\ TiCl_4}]{LiAlH_4\ \text{or}} \overset{\overset{\displaystyle E}{|}}{RNHCR_2}$$

$$E^+ = RX, R_2CO$$

Angew Int *14* 15 (1975)
Ber *108* 1293 (1975); *110* 1852 (1977); *111* 2630 (1978)
Can J Chem *53* 2473 (1975)
Syn 548 (1976); 423 (1979); 741 (1980) (reduction)
Org Syn *58* 113 (1978)
Chem Rev *78* 275 (1978) (review)
JCS Perkin I 579 (1979)

$$RCH_2NH_2 \longrightarrow \underset{\underset{\displaystyle OMe}{|}}{\overset{\overset{\displaystyle NO}{|}}{RCH_2NCHCH_3}} \xrightarrow[2.\ E^+]{1.\ LDA} \underset{\underset{\displaystyle OMe}{|}}{\overset{\overset{\displaystyle E\ \ NO}{|\ \ \ |}}{RCHNCHCH_3}}$$

$$\xrightarrow{Al\text{-}Ni} \overset{\overset{\displaystyle E}{|}}{RCHNHEt}$$

$$\xrightarrow[MeOH]{HCl} \overset{\overset{\displaystyle E}{|}}{RCHNH_2}$$

$$E^+ = RX, RCHO, R_2CO$$

JOC *48* 2388 (1983)

$$R_2CHN\equiv C \xrightarrow[2.\ E^+]{1.\ base} \overset{\overset{\displaystyle NC}{|}}{R_2CE} \xrightarrow{H_3O^+} \overset{\overset{\displaystyle NH_2}{|}}{R_2CE}$$

$$E^+ = RX, R_2CO$$

Angew Int *10* 491 (1971)
Ann 183 (1976); 40 (1977)

$$\text{NPO(NMe}_2)_2 \xrightarrow[2.\ E^+]{1.\ n\text{-BuLi}} \underset{\underset{\displaystyle E}{|}}{\text{NPO(NMe}_2)_2}$$

$$E^+ = RX, RCHO, R_2CO, \text{epoxide}, ArH\cdot CrO_3, I_2, Me_3SiCl, n\text{-Bu}_3SnCl$$

Helv *64* 643 (1981)
Tetr *39* 1963 (1983)

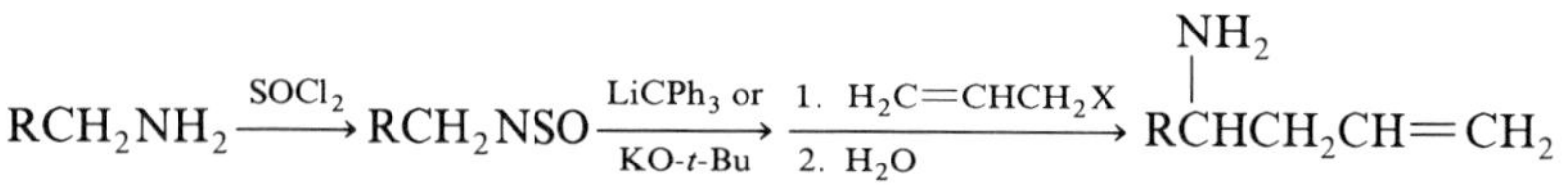

$$RCH_2NH_2 \xrightarrow{SOCl_2} RCH_2NSO \xrightarrow[\text{KO-}t\text{-Bu}]{\text{LiCPh}_3 \text{ or}} \xrightarrow[\text{2. H}_2\text{O}]{\text{1. H}_2\text{C=CHCH}_2\text{X}} RCHCH_2CH=CH_2 \quad (NH_2)$$

JACS *100* 2894 (1978)

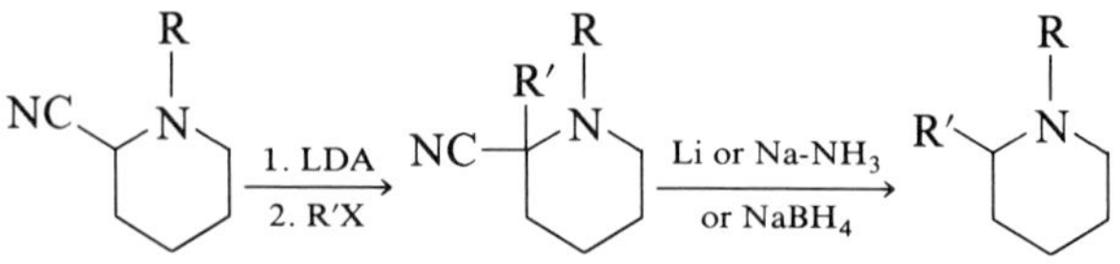

TL *23* 3369 (1982); *28* 547 (1987)

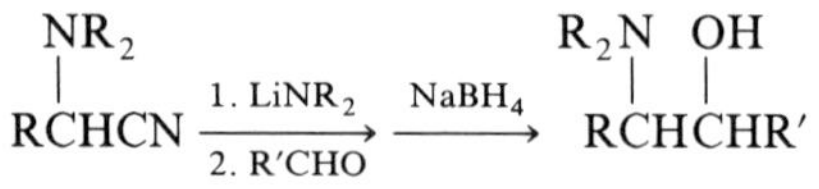

$$RCHCN \xrightarrow[\text{2. R'CHO}]{\text{1. LiNR}_2} \xrightarrow{\text{NaBH}_4} RCHCHR' \quad (NR_2) \quad (R_2N \ OH)$$

TL 771 (1979)

$$R_2CHNH_2 \xrightarrow[\text{PhNO}]{\text{t-BuOCl}} R_2CHN=NPh \xrightarrow[\text{2. Zn, HOAc}]{\text{1. R'Li}} R_2CR' \quad (O^-) \quad (NH_2)$$

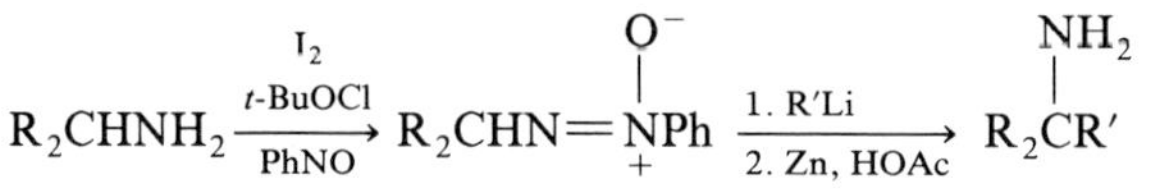

JCS Perkin I 2030 (1979)

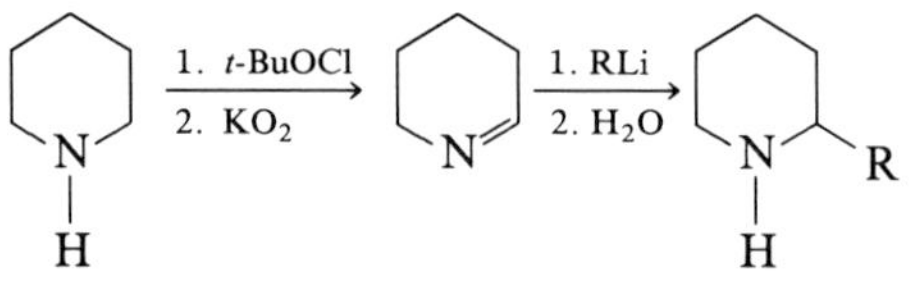

JOC *45* 1515 (1980)

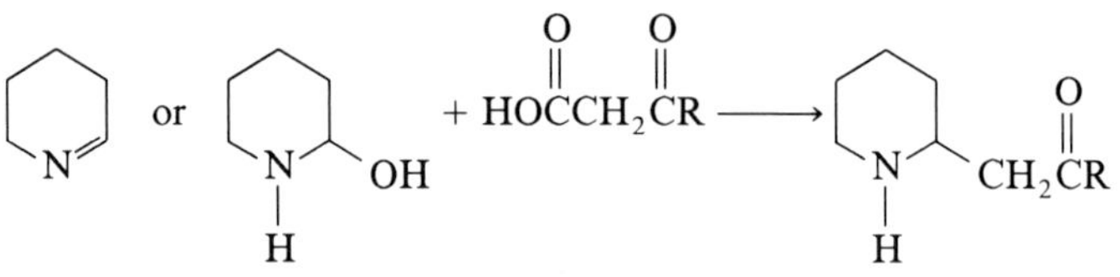

Ann *626* 123 (1959)

$$RM + YCH_2NR'_2 \longrightarrow RCH_2NR'_2$$

RM	Y	R'	
CH₃CH₂CH=CHCH₂M (M = Li, MgX, ZnX)	n-BuO, n-BuS	alkyl	BSCF 2544 (1968)
n-BuMgBr	OR	alkyl	TL *28* 547 (1987)
RMgX (R = 1° alkyl)	n-BuO	alkyl	BSCF II 148 (1983)
RMgBr (R = Me, PhCH₂, allyl, propargyl, alkynyl, aryl)	MeO, c-C₆H₁₃O	SiMe₃	Angew Int *23* 53 (1984)
RMgX (1° alkyl, allyl, benzyl, aryl)	n-BuO	alkyl	JCS *123* 532 (1923)

RMgX (1°, 2°, 3° alkyl; allyl; benzyl)	EtO	alkyl	JACS *62* 1450 (1940)
RMgX (1°, 2° alkyl; benzylic; aryl)	n-BuO	alkyl	JACS *77* 1098 (1955)
RMgX (R = 1°, 2° alkyl; allylic; aryl)	PhS	alkyl	JOC *32* 272 (1967)
PhMgBr	CN	alkyl	BSCF 3803 (1968) TL *28* 741 (1987)
RMgX (R = 1° alkyl, CH_2CO_2-t-Bu, vinylic), EtO_2CCH_2ZnBr, Et_2NCOCH_2ZnBr, $(EtO_2C)_2CRZnBr$ (R = H, Me)	n-BuO	alkyl	BSCF II 395 (1982)
$RC{\equiv}CCH_2M$ (M = MgX, AlX_2, ZnX); $CH_3CH{=}CHCH_2M$ (M = Li, MgX, AlX_2, ZnX, Cu)	n-BuO	alkyl	JOMC *198* 1 (1980)
$RC{\equiv}CMgBr$	n-BuO	alkyl	BSCF II 21 (1983)
$RC{\equiv}CMgX$, Et_2NCOCH_2ZnBr, $(CH_3CH{=}CHCH_2)_nAlBr_{3-n}$ or $(H_2C{=}C{=}CMe)_nAlBr_{3-n}$	OR	i-Pr, SiMe₃	TL *28* 1659 (1987)
E-RCH=CHAl(i-Bu)₂, R₂CuLi (R = 1° alkyl, vinylic)	n-BuO, i-BuO, PhS	alkyl	TL *21* 3763 (1980)

$$R_2Zn + MeOCHCO_2CH_3 \longrightarrow RCHCO_2CH_3$$

with NEt_2 substituents

JOMC *256* 193 (1983)

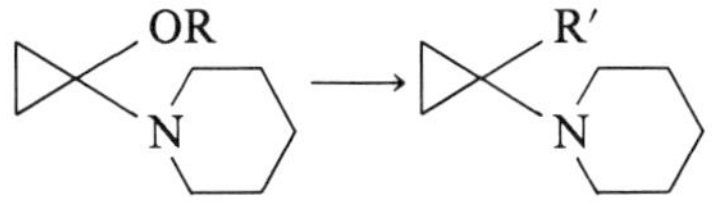

R	Reagent(s)	
H	TiCl₄; ArH, enol silane or $(RO_2C)_2CH_2$ [R' = Ar, CH_2COR or $CH(CO_2R)_2$]	TL *23* 785 (1982)
SiMe₃	R'MgX (R = 1° alkyl, aryl, vinylic)	TL *23* 785 (1982)
H or SiMe₃	KCN, HOAc (R' = CN)	TL *23* 1413 (1982)

$$2\ RX \xrightarrow[\text{2. (MeO)}_2\text{CHNR}'_2]{\text{1. 2 Mg}} \xrightarrow{\text{H}_2\text{O}} R_2\text{CHNR}'_2$$

Syn 757 (1978)

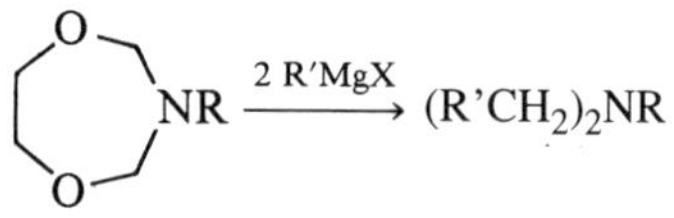

$$\text{NR} \xrightarrow{\text{2 R'MgX}} (R'\text{CH}_2)_2\text{NR}$$

TL *24* 1597 (1983)

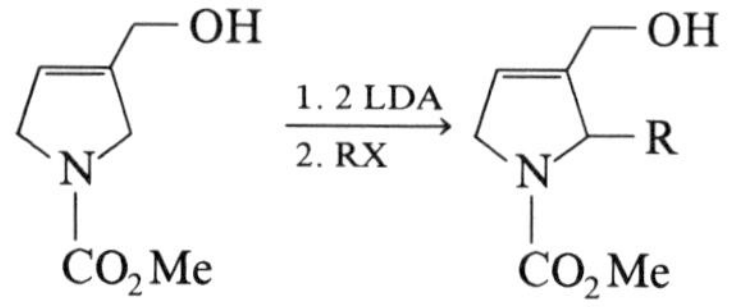

$$\xrightarrow[\text{2. RX}]{\text{1. 2 LDA}}$$

JOC *48* 1129 (1983)

$$\xrightarrow[\substack{\text{2. BrCH}_2\text{CH}_2\text{C(=CH}_2)\text{CH}_2\text{Br} \\ \text{3. NaI}}]{\text{1. } n\text{-BuLi}}$$

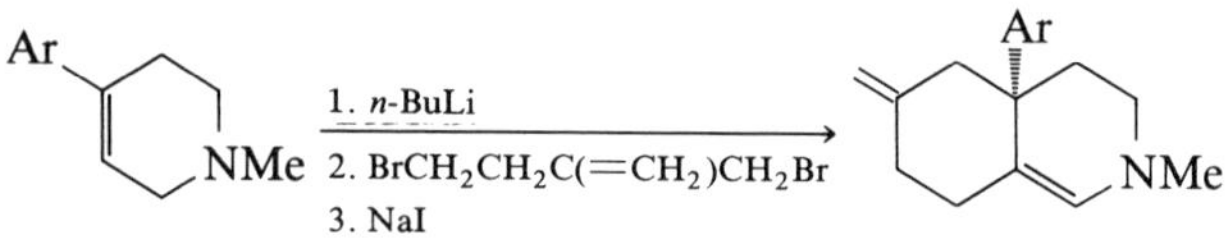

TL *23* 285 (1982)

3. Dealkylation

Review: Houben-Weyl, Vol XI/1, p 961

$$R_2\text{NCH}_2\text{Ar} \xrightarrow[\text{catalyst}]{\text{H}_2} R_2\text{NH}$$

Catalyst

review	Org Rxs 7 263 (1953)
Pd-C	Helv *35* 1162 (1952) J Chem Res (S) 164 (1981)
PdCl$_2$-C	JACS *65* 1984 (1943); *72* 3410 (1950); *75* 5598 (1953)

$$R_2\text{NCH}_2\text{CH}_2\text{CN} \xrightarrow{250-275°C} R_2\text{NH} + \text{H}_2\text{C=CHCN}$$

JOC *17* 1043 (1952)

$$\langle N-R \xrightarrow[\text{2. HCl, HBr or Br}_2]{\text{1. H}_2\text{C=CHOCOCl}} \langle NH$$

TL 1567 (1977)

$$\left(R_3\overset{+}{N}R'\right)X^- \longrightarrow R_3N$$

Reagent(s)	R'	X	
Δ	Me, benzylic, allylic	halide	Org Rxs 7 142 (1953) (review)
	Me	OAc	Tetr 24 5493 (1968)
	alkyl	OH	Org Rxs 11 317 (1960) (review)
LiAlH$_4$	Me	I	JACS 82 4651 (1960)
	Me, Et	I	JCS 1729 (1965)
LiI	Me	OTs	JOC 28 2407 (1963)
KO$_2$CH, HCO$_2$H	Me	Br	Coll Czech Chem Commun 26 471 (1961)
KOAc	Me	Br	Coll Czech Chem Commun 26 471 (1961)
NaSPh	Me	Cl	TL 1375 (1966)
	Me, Et	I	JCS 1729 (1965)
H$_2$NCH$_2$CH$_2$OH	various alkyl	halide, ClO$_4$	Ber 90 395, 403 (1957)

4. Rearrangement

$$\text{RNCH}_2\text{CH=CH}_2 \quad \xrightarrow[\text{Lewis acid}]{\Delta, \text{H}^+ \text{ or}} \quad \text{RNH} \quad \text{CH}_2\text{CH=CH}_2$$

JOC 22 1418 (1957)
TL 4661 (1971); 25 3159 (1984)
Helv 56 105 (1973); 60 978 (1977)
Yakugaku Zasshi 97 553 (1977)
JOC USSR 15 2350 (1979); 19 920 (1983)
Bull Acad Sci USSR, Div Chem Sci 1910 (1982)
Chem Pharm Bull 31 2220 (1983)

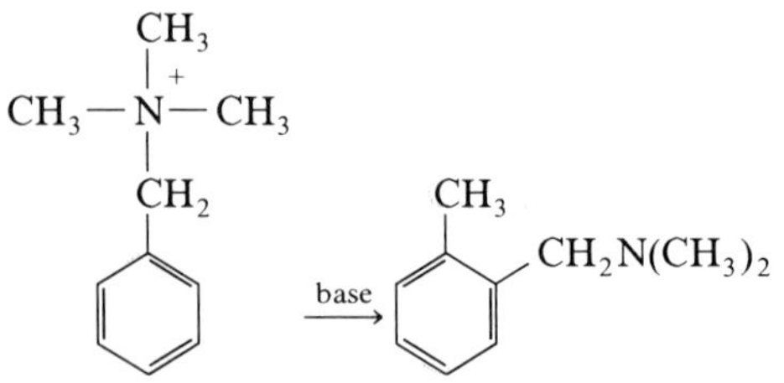

$$CH_3-\overset{\overset{\displaystyle CH_3}{|}}{\underset{\underset{\displaystyle C_6H_5}{|}}{\overset{+}{N}}}-CH_3$$

CH₃

CH₃—N⁺—CH₃

CH₂

(phenyl ring)

—base→

CH₃

CH₂N(CH₃)₂

(phenyl ring)

Sommelet–Hauser rearrangement

JACS *73* 4122 (1951)

Org Syn *34* 61 (1954)

Org Rxs *18* 403 (1970) (review)

A. R. Lepley, A. G. Giumanini in "Mechanisms of Molecular Migrations," Ed. B. S. Thyagarajan, Wiley Interscience, New York (1971), Vol 3, p 297 (review)

JOC *36* 984 (1971); *44* 2348 (1979)

I. Zugravescu, M. Petrovanu, "N-Ylid Chemistry," McGraw-Hill, New York (1976), Chpt 2 (review)

CH₃

CH₃—N⁺—CH₃

CH₂

(phenyl ring)

—base→

CH₂CH₂N(CH₃)₂

(phenyl ring)

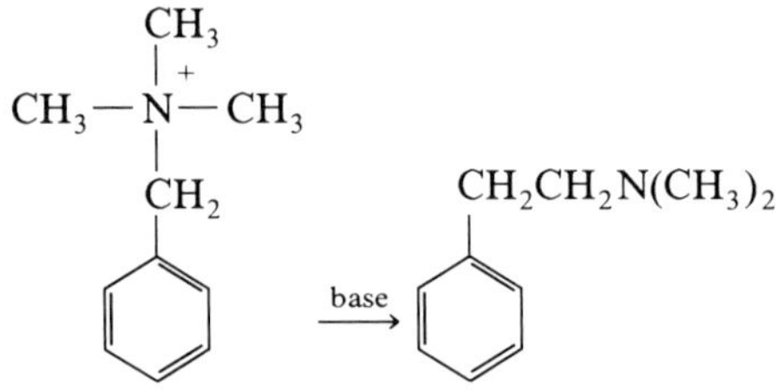

Stevens rearrangement

Org Rxs *18* 403 (1970) (review)

A. R. Lepley, A. G. Giumanini in "Mechanisms of Molecular Migrations," Ed. B. S. Thyagarajan, Wiley Interscience, New York (1971), Vol 3, p 297 (review)

JOC *36* 984 (1971); *44* 2348 (1979)

I. Zugravescu, M. Petrovanu, "N-Ylid Chemistry," McGraw-Hill, New York (1976), Chpt 2 (review)

5. FROM AZIDES

See also page 391, Section 2.

$$RN_3 \longrightarrow RNH_2$$

H_2, cat Pd or Pt	Chem Rev 54 1 (1954)
H_2, cat Pd-C	J Med Chem 12 658 (1969)
	CC 915 (1970); 1738 (1987)
	JOC 36 250 (1971); 37 335 (1972);
	50 3095 (1985); 51 1069, 5373 (1986)
	Ber 105 1524 (1972)
	TL 28 4601 (1987)
H_2, cat Pd-CaCO$_3$	Syn 590 (1975)
H_2, cat PdO	JACS 76 2887 (1954)
H_2, cat PtO$_2$	JOC 27 3045 (1962); 40 1659 (1975)
	TL 28 4499 (1987)
$(NH_4)O_2CH$, cat Pd-C	TL 24 1609 (1983); 28 6133 (1987)
BH_3	JACS 87 4203 (1965)
$NaBH_4$	JOC 23 127 (1958)
$NaBH_4$, MeOH, THF	Syn 48 (1987)
$NaBH_4$, i-PrOH, Δ	JACS 84 485 (1962); 86 1427 (1964)
	Chem Pharm Bull 18 2368 (1970)
$NaBH_4$ (phase transfer)	JOC 47 4327 (1982)
$LiAlH_4$	JACS 73 5865 (1951); 91 2961 (1969);
	92 6302 (1970); 108 2034 (1986)
	Chem Rev 54 1 (1954)
	JOC 27 2925 (1962); 31 684 (1966);
	51 5373 (1986)
	Carbohydr Res 3 318 (1967)
	CL 635 (1977)
n-Bu$_3$SnH	JACS 107 519 (1985)
	TL 28 6381 (1987)

n-Bu$_2$SnH$_2$	TL *28* 5941 (1987)
Zn, HCl	CC 64 (1970) JCS C 414 (1971)
Zn, HOAc	JCS C 414 (1971)
SnCl$_2$, MeOH	TL *27* 1423 (1986); *28* 4597 (1987)
TiCl$_3$, H$_2$O	Syn 65 (1978)
VCl$_2$, H$_2$O (R = aryl only)	Syn 815 (1976)
CrCl$_2$, H$_2$O	CC 64 (1970) JCS C 414 (1971) Tetr *29* 1801 (1973)
MoCl$_5$, Zn, H$_2$O, THF	Syn 830 (1980)
P(OEt)$_3$/TsOH, EtOH	TL *28* 6513 (1987)
P(OEt)$_3$/HCl	Syn 202 (1985)
PPh$_3$, H$_2$O	TL *24* 763 (1983); *28* 379, 1757, 1761 (1987) BSCF 815 (1985) JOC *52* 5044 (1987)
PPh$_3$/HBr	Ann *591* 117 (1955)
PPh$_3$/NH$_4$OH or NaOH	JOC *40* 1659 (1975) TL *27* 227 (1986)
H$_2$S, solvent?	Chimia *22* 141 (1968)
H$_2$S, H$_2$O	JACS *73* 2327 (1951)
H$_2$S, H$_2$O, py	Syn 45 (1977)
H$_2$S, EtOH	JOC *33* 2910 (1968) Tetr *23* 387 (1967); *25* 3313 (1969) Monatsh *101* 724 (1970)
HS(CH$_2$)$_3$SH, Et$_3$N	TL 3633 (1978)
PhSH, Sn(SPh)$_2$, Et$_3$N	TL *28* 5941 (1987)
Na$_2$S·9 H$_2$O, Et$_3$N, MeOH	JOC *44* 4712 (1979)
NaTeH	CL 1733 (1984)
CH$_3$COCH$_2$COCH$_3$, Et$_3$N	J Heterocyclic Chem *10* 565 (1973) JOC *40* 1066 (1975) Syn 491 (1977) TL *28* 5941 (1987)

$$\underset{\displaystyle \text{RCH=CR}'}{\overset{\displaystyle \overset{\textstyle N_3}{|}}{}} \xrightarrow{\text{LiAlH}_4} \underset{\displaystyle \text{RCH}_2\text{CHR}'}{\overset{\displaystyle \overset{\textstyle NH_2}{|}}{}}$$

JOC *42* 2935 (1977)

6. FROM NITRO COMPOUNDS

Reviews:

R. L. Augustine, "Catalytic Hydrogenation," Marcel Dekker, New York (1965), Chpt 5
M. Freifelder, "Practical Catalytic Hydrogenation," Wiley Interscience, New York (1971), Chpt 10
M. Freifelder, "Catalytic Hydrogenation in Organic Synthesis: Procedures and Commentary," J. Wiley and Sons, New York (1978), Chpt 5
P. N. Rylander, "Hydrogenation Methods," Academic Press, New York (1985), Chpt 8

$$RNO_2 \longrightarrow RNH_2$$

$$R = alkyl$$

H_2, Raney Ni	Ber *86* 939 (1953)
H_2, Raney Ni, HOAc	Org Syn Coll Vol *4* 221 (1963)
H_2, cat Pd-C	Ber *86* 939 (1953)
	JACS *81* 505 (1959)
	CL 797 (1981)
$(NH_4)O_2CH$, cat Pd-C	TL *25* 3415 (1984)
	JOC *51* 4856 (1986)
cat Pd,	JCS 3586 (1954)
H_2, cat PtO_2	JOC *37* 335 (1972)
H_2, cat $RhCl_3(py)_3$-$NaBH_4$	JCS Perkin I 2509 (1973)
H_2, KOH, cat $RuCl_2(PPh_3)_3$	JOC *40* 519 (1975)
CO, $Ru_3(CO)_{12}$, NaOH (phase transfer)	TL *21* 2603 (1980)
$N_2H_4 \cdot 2\ H_2O$, Ni_2B	JOC *51* 4294 (1986)
$LiBH_4$, MeOH, diglyme	JOC *51* 4000 (1986)
$NaBH_4$, cat Pd-C	Syn 713 (1987)

NaBH$_4$, cat NiCl$_2$·6 H$_2$O	TL *26* 6413 (1985)
	JOC *51* 4856 (1986)
NaBH$_4$, CoCl$_2$·6 H$_2$O	TL 4555 (1969)
	CC 344 (1986)
LiAlH$_4$	JACS *70* 3738 (1948); *73* 1293 (1951)
	TL *28* 6281 (1987)
Mg(Hg)-TiCl$_4$, *t*-BuOH, THF	Syn Commun *13* 495 (1983)
Al(Hg), H$_2$O	JACS *90* 3245 (1968)
	TL *28* 6281 (1987)
Sn, HCl	Org Syn Coll Vol *2* 617 (1943)
Zn, HCl	TL *28* 577 (1987)
Fe, HCl	JACS *73* 1293 (1951)

$$\text{ArNO}_2 \longrightarrow \text{ArNH}_2$$

Review: Org Rxs *20* 455 (1973) (sulfides and polysulfides)

H$_2$, cat Ni	JACS *61* 3564 (1939)
H$_2$, cat Raney Ni	JOC *18* 1506 (1953); *51* 3903 (1986)
	JACS *76* 5149 (1954)
	Acta Chem Scand *9* 1079 (1955)
	Org Syn Coll Vol *3* 59, 63 (1955); *4* 357 (1963);
	5 1130 (1973)
H$_2$, cat KBH$_4$-NiCl$_2$·6 H$_2$O-C	Carbohydr Res *88* 323 (1981)
H$_2$, cat Pd-C	JOC *37* 335 (1972); *51* 3308 (1986);
	52 1844, 5717 (1987)
	Org Syn Coll Vol *5* 829 (1973)
	Syn 940 (1982)
	JACS *107* 3328 (1985)
H$_2$, cat Pd(acac)$_2$, py	Chem Ind 1057 (1975)
H$_2$, cat PtO$_2$	Org Syn Coll Vol *1* 240 (1941)
H$_2$, cat PtO$_2$, K$_2$CO$_3$, CH$_3$OH	JOC *44* 409 (1979)
H$_2$, cat Pt salt-NaBH$_4$	JACS *84* 2828 (1962)
H$_2$, cat CoS$_x$, *i*-PrOH	JOC *44* 3671 (1979)
H$_2$, cat RhCl$_3$(py)$_3$-NaBH$_4$	JCS Perkin I 2509 (1973)
H$_2$, cat Ru	JOC *44* 1233 (1979)
H$_2$, cat RuCl$_2$(PPh$_3$)$_3$	TL 2163 (1975)
cat Pd-C, ⬡	JCS Perkin I 1300 (1975)
cat Pd, ⬡	JCS 3586 (1954)
PhNHNH$_2$	JCS 330 (1929)

N_2H_4, (catalysts)	Chem Rev 65 51 (1965)
N_2H_4, graphite	TL 26 6233 (1985)
N_2H_4, Raney Ni	JACS 75 4334 (1953); 76 5149 (1954) J Gen Chem USSR 27 261 (1957) Can J Chem 38 2363 (1960) TL 23 147 (1982) BCSJ 56 3159 (1983)
N_2H_4, cat Pd-C	Org Syn Coll Vol 5 30 (1973) JCS Perkin I 444 (1977)
N_2H_4, $FeCl_3 \cdot 6 H_2O$, C, CH_3OH	CL 259 (1975) JOC 50 5092 (1985); 52 1339 (1987)
N_2H_4, Fe(III) oxide	Syn 834 (1978)
$(NH_4)O_2CH$, cat Pd-C	TL 25 3415 (1984)
$(Et_3NH)O_2CH$, cat $RuCl_2(PPh_3)_3$, cat Pd-C	JOMC 309 C63 (1986)
$(Et_3NH)O_2CH$, cat $Pd(OAc)_2(PAr_3)_2$	JOC 42 3491 (1977)
$(Et_3NH)O_2CH$, cat Pd-C	JOC 42 3491 (1977); 45 4992 (1980) JACS 108 1000 (1986)
HCO_2H, cat Pd-C	JCS Perkin I 443 (1977)
phosphinic acid or sodium phosphinate or phosphorous acid or sodium phosphite, cat Pd-C	JCS Perkin I 443 (1977)
NaO_2CH, KH_2PO_4,	Syn Commun 11 925 (1981)

CO, H_2O, Se, Et_3N	Angew Int 19 1008 (1980)
$Fe(CO)_5$	Can J Chem 48 1543 (1970)
$Fe(CO)_5$, CO, H_2O, Et_3N	JACS 100 3969 (1978)
$Fe_2(CO)_9$, aq NaOH, C_6H_6	JACS 99 98 (1977)
$Fe_3(CO)_{12}$, CH_3OH, Δ	JOC 37 930 (1972)
$Fe_3(CO)_{12}$, Al_2O_3	CC 821 (1980)
$Fe_3(CO)_{12}$, aq NaOH, C_6H_6, cat $(PhCH_2NEt_3)Cl$	JACS 99 98 (1977)
$Fe_3(CO)_{12}$, KOH, 18-crown-6	Angew Int 16 41 (1977)
$(Et_4N)HFe_3(CO)_{11}$	JOMC 171 85 (1979)
CO, cat $Pd(OAc)_2$, PR_3, HOAc, H_2O	Bull Acad Sci USSR, Div Chem Sci 1223 (1986)
CO, H_2O, Rh catalyst, Δ	TL 3385 (1971)

CO, $Ru_3(CO)_{12}$, NaOH (phase transfer)	TL *21* 2603 (1980)
$LiBH_4$, MeOH, diglyme	JOC *51* 4000 (1986)
$NaBH_2S_3$	Can J Chem *49* 2990 (1971)
$NaBH_4$, cat Pd-C	Syn 713 (1987)
$NaBH_4$, cat Pd-C, H_2O	JCS 371 (1962)
$NaBH_4$, $SnCl_2 \cdot 2\ H_2O$, EtOH	Chem Pharm Bull *29* 1443 (1981)
$NaBH_4$, $TiCl_4$	Syn 695 (1980)
$NaBH_4$, $FeCl_2$	Chem Ind 480 (1983)
$NaBH_4$, $CoCl_2 \cdot 6\ H_2O$	TL 4555 (1969)
$NaBH_4$, $NiCl_2 \cdot 6\ H_2O$	Chem Pharm Bull *29* 1159 (1981)
$NaBH_4$, $NiX_2(PPh_3)_2$ (X = Cl, Br, I)	JCS Japan *92* 1225 (1971)
$NaBH_4$, CuCl	Chem Ind 75 (1984)
$NaBH_4$, $Cu(OAc)_2$	TL *27* 1205 (1986)
$NaBH_4$, cat $Cu(acac)_2$	JCS Perkin I 2409 (1979)
KBH_4, cat $PdCl_2$	BSCF 1996 (1959)
$LaNi_5H_6$	JOC *52* 5695 (1987)
$LaNi_{4.5}Al_{0.5}H_5$	JOC *52* 5695 (1987)
Al_2Te_3, H_2O	Angew Int *19* 1009 (1980)
Mg(Hg)-$TiCl_4$, *t*-BuOH, THF	Syn Commun *13* 495 (1983)
$TiCl_3$, H_2O	Chem Pharm Bull *28* 2515 (1980)
$CrCl_2$, CH_3OH	Syn 792 (1977)
Li, NH_3, CH_3OH	Syn Commun *12* 293 (1982)
Na, NH_3, CH_3OH	JACS *69* 1657 (1947) Syn Commun *12* 293 (1982) JOC *51* 3904 (1986)
Zn, H_2O	Org Syn Coll Vol *2* 447 (1943)
Zn, HCl	Proc Ind Acad Sci A *44* 331 (1956)
Zn, NaOH	Org Syn Coll Vol *2* 501 (1943)
Zn, NH_3	Tetr *5* 340 (1959)
Fe	JACS *66* 1781 (1944) JOMC *65* 289 (1974)
Fe, HOAc	Org Syn Coll Vol *2* 471 (1943)
Fe, HCl	Org Syn Coll Vol *2* 160 (1943); *5* 346 (1973)
$FeSO_4$, HCl	Org Syn Coll Vol *3* 56 (1955)

Sn, HCl	JACS *61* 1001 (1939); *91* 3544 (1969); *109* 3098 (1987) Org Syn Coll Vol *1* 455 (1941); *2* 175 (1943) JCS 1133 (1949)
$SnCl_2 \cdot 2\ H_2O$, EtOH	TL *25* 839 (1984)
$SnCl_2$, HCl	Ber *62* 3035 (1929) Org Syn Coll Vol *2* 130, 254 (1943); *3* 239, 453 (1955) Chem Ind 888 (1972)
$SnBr_2$, HBr	Org Syn Coll Vol *2* 132 (1943)
Na_2S	Org Syn Coll Vol *3* 86 (1955)
Na_2S, S, NaOH	Org Syn Coll Vol *4* 31 (1963)
Na_2S, S, NH_4Cl	JOC *51* 3903 (1986)
Na_2S_2	Org Syn Coll Vol *5* 1067 (1973)
$Na_2S_2O_4$	Org Syn Coll Vol *3* 69 (1955)
$(NH_4)_2S$	Org Syn Coll Vol *1* 52 (1941); *3* 82, 242 (1955)
NaTeH (sterically hindered Ar only)	CL 1373 (1983)
paraffin, 360–390°C	Syn 23 (1978)

$$NO_2\!-\!Ar\!-\!NO_2 \longrightarrow NO_2\!-\!Ar\!-\!NH_2$$

H_2, cat Pd	JOC *8* 331 (1943)
Raney Ni, *i*-PrOH, HOAc	Chem Ind 477 (1983)
Ni, HOAc, maleic acid	Chem Ind 826 (1983)
HCO_2H, Et_3N, cat Pd-C	JOC *45* 4992 (1980)
cat Pd, cyclohexene	JCS 3586 (1954)
cat Pd-C, cyclohexene	JCS Perkin I 1300 (1975)
Fe, HOAc	Syn 924 (1978)
$SnCl_2$, HCl	Ber *19* 2161 (1886) Rec Trav Chim *65* 207, 331 (1946)
sulfides and polysulfides	Org Rxs *20* 455 (1973) (review)

$$ArNO_2 \longrightarrow ArNMe_2$$

H_2, CH_2O, CH_3OH, Raney Ni	Ind J Chem B *14* 904 (1976)
H_2, CH_2O, EtOH, cat Pd-C	Org Syn Coll Vol *5* 552 (1973)

7. HETEROATOM AND METAL DISPLACEMENT

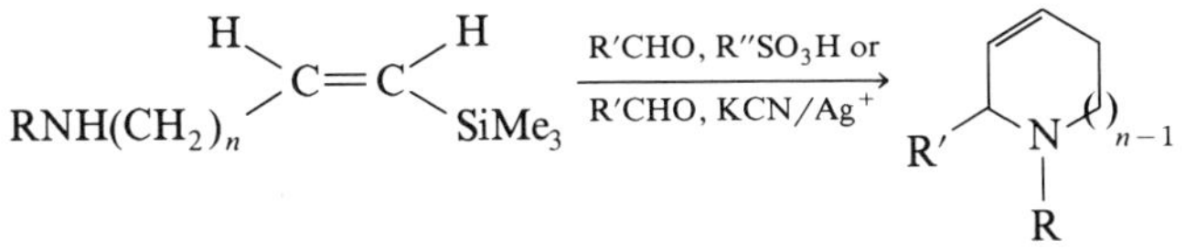

$$RNH(CH_2)_n \overset{H}{\underset{}{C}}=C\overset{H}{\underset{SiMe_3}{}} \quad \xrightarrow[R'CHO, KCN/Ag^+]{R'CHO, R''SO_3H \text{ or}}$$

$$n = 2, 3$$

JACS *109* 6097, 6107 (1987)

$$\xrightarrow{CF_3CO_2H}$$

$$n = 1, 2$$

JACS *109* 6097, 6115 (1987)

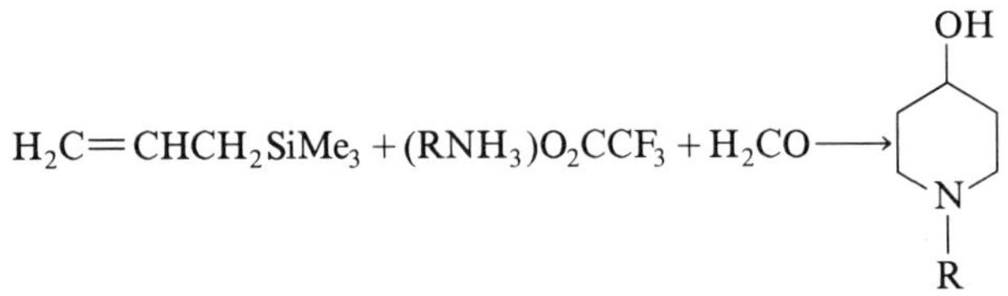

$$H_2C{=}CHCH_2SiMe_3 + (RNH_3)O_2CCF_3 + H_2CO \longrightarrow$$

JACS *108* 3513 (1986)

$$PhCH_2NH(CH_2)_3CH{=}CHCH_2SiMe_3 \xrightarrow[CF_3CO_2H]{H_2CO}$$

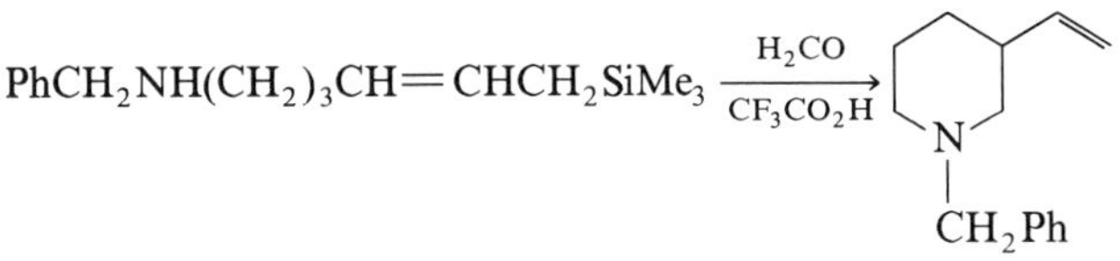

TL *27* 5067 (1986)

$$ArSnR'_3 + (R_2\overset{+}{N}{=}CH_2)\,Cl^- \longrightarrow ArCH_2NR_2$$

TL *27* 5011 (1986)

$$R_2C{=}CRCH_2SePh \xrightarrow[2.\ R'NH_2]{1.\ NCS} R_2\overset{HNR'}{\underset{}{C}}CR{=}CH_2$$

JOC *51* 5243 (1986)

417

8. FROM ALCOHOLS

See page 397, Section 3, for amination via tosylates.

Review: Russ Chem Rev *49* 14 (1980)

$$ROH + HNMe_2 \xrightarrow[\text{cat CuO, Cr}_2\text{O}_3, \text{Na}_2\text{O, SiO}_2]{\text{H}_2, \Delta} RNMe_2$$

TL 1937 (1977)
Syn Commun *8* 27 (1978)

$$1° \; ROH + ArNH_2 \longrightarrow ArNHR$$

Raney Ni

JACS *77* 4052 (1955)
Org Syn Coll Vol *4* 283 (1963)

KOH, Δ

Org Syn Coll Vol *4* 91 (1963)

$$1° \; ROH + PhNH_2 \xrightarrow[\text{cat RuCl}_2(\text{PPh}_3)_3]{\Delta} PhNR_2$$

TL *22* 2667 (1981)

$$ROH + HNR^1R^2 \xrightarrow[\text{Raney Ni}]{\text{Al(O-}t\text{-Bu)}_3} RNR^1R^2$$

Syn 722 (1977)

$$1° \; ROH \xrightarrow{\text{NaH}} \xrightarrow[\text{R}^1\text{R}^2\text{NH}]{(\text{Ph}_3\overset{+}{\text{P}}\text{NMePh})\text{I}^-} RNR^1R^2$$

TL 471 (1975)

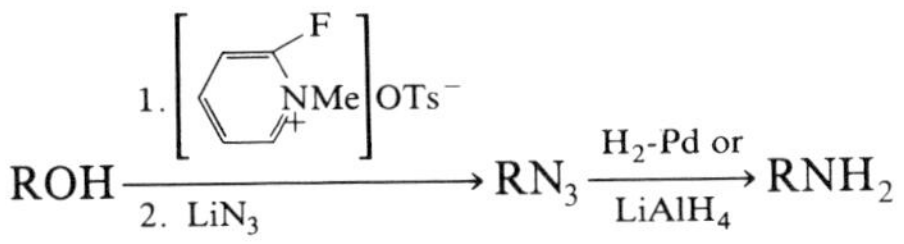

$$ROH \xrightarrow[\text{2. LiN}_3]{1. \left[\underset{\overset{+}{\text{NMe}}}{\overset{\text{F}}{\bigcirc}}\right]\text{OTs}^-} RN_3 \xrightarrow[\text{LiAlH}_4]{\text{H}_2\text{-Pd or}} RNH_2$$

R = 1°, 2° alkyl; allylic; benzylic

CL 635 (1977)

$$\underset{\text{RCHCO}_2\text{R}}{\overset{\text{OH}}{|}} + \underset{\text{RCNHOR}}{\overset{\text{O}}{\|}} \xrightarrow[\text{Ph}_3\text{P}]{\text{RO}_2\text{CN}=\text{NCO}_2\text{R}} \underset{\text{RCHCO}_2\text{R}}{\overset{\text{RONCOR}}{|}}$$

JOC *52* 4978 (1987)

$$\text{ROH} + \text{HN}(\text{COR}')_2 \xrightarrow[\text{Ph}_3\text{P}]{\text{EtO}_2\text{CN}=\text{NCO}_2\text{Et}} \text{RN}(\text{COR}')_2 \xrightarrow{\text{N}_2\text{H}_4} \text{RNH}_2$$

JACS *94* 679 (1972)
JOC *52* 5127 (1987)

For further alkylations of imides see page 397, Section 3.

$$\text{ROH} \xrightarrow[\text{2. ClSO}_2\text{NMe}_2]{\text{1. NaH}} \xrightarrow{\text{H}_3\text{O}^+} \text{RNMe}_2$$

R = allylic, benzylic

JACS *87* 5261 (1965)

$$\text{ArOH} \xrightarrow[\Delta]{\text{NH}_4\text{OH, SO}_2} \text{ArNH}_2$$

JOC *25* 214 (1960)

9. FROM ALDEHYDES AND KETONES

1. Transamination

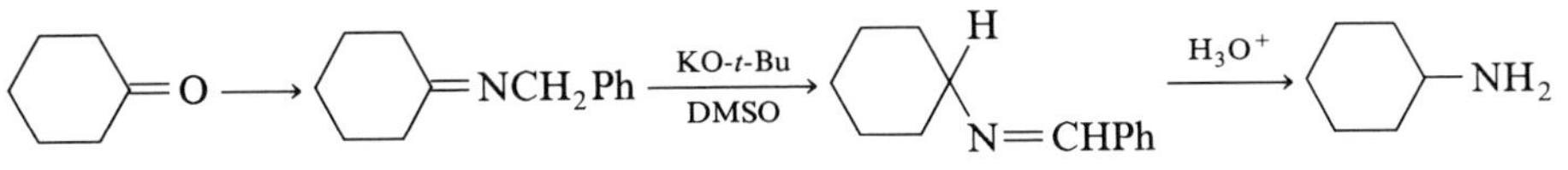

Can J Chem *48* 570 (1970)

2. Reductive Amination

$$R_2C=O + R'NH_2 \rightleftharpoons R_2C=NR' \xrightarrow[\text{agent}]{\text{reducing}} R_2CHNHR'$$

Reviews:

Org Rxs *4* 174 (1948)

M. Freifelder, "Practical Catalytic Hydrogenation," Wiley Interscience, New York (1971), Chpt 16

M. Freifelder, "Catalytic Hydrogenation in Organic Synthesis: Procedures and Commentary," J. Wiley and Sons, New York (1978), Chpt 10

Russ Chem Rev *49* 14 (1980)

JOC *48* 3412 (1983) (stereoselectivity of various metal hydride reagents)

Reducing agent

H_2, cat Pd-C	JOC *51* 3635 (1986)
H_2, cat Pt-C	JOC *31* 3875 (1966)
Raney nickel	JOC *19* 1054 (1954)
	J Chem Res (S) 164 (1981)
i-PrOH, Na_2CO_3, cat ClRh(PPh$_3$)$_3$	Syn 442 (1981)
HCO_2H	Org Rxs *5* 301 (1949)
Zn, acid	JACS *62* 2159 (1940); *63* 972, 2843 (1941)

Reducing agent   (*continued*)

BH_3	JOC *30* 2877 (1965)
	BSCF 4439 (1970)
$BH_3 \cdot py$, HOAc	JCS Perkin I 717 (1984)
$BH_3 \cdot Me_2NH$	JOC *26* 1437 (1961)
$BH_3 \cdot Me_2NH$, HOAc	JOC *48* 3412 (1983)
$BH_3 \cdot t\text{-BuNH}_2$	TL *22* 3447 (1981)
$BH_3 \cdot t\text{-BuNH}_2$, HOAc	JOC *48* 3412 (1983)
	TL *25* 695 (1984)
$(CF_3CO_2)_2BH$	JOC *46* 355 (1981)
$NaBH_4$	JOC *22* 1068 (1957); *28* 3259 (1963); *48* 3412
	(1983); *51* 486 (1986)
	Chem Pharm Bull *17* 98 (1969)
	JACS *107* 7524 (1985)
	TL *28* 749 (1987)
$NaBH_4$, HOAc	JACS *96* 7812 (1974)
	Org Prep Proc Int *17* 317 (1985) (review)
$NaBH_3CN$	JACS *91* 3996 (1969); *93* 2897 (1971); *108* 1039
	(1986); *109* 1814 (1987)
	Syn 135 (1975) (review)
	CC 1088 (1979)
	Org Prep Proc Int *11* 201 (1979) (review)
	TL *21* 789 (1980); *22* 3447 (1981)
	JOC *48* 3412 (1983); *52* 5044 (1987)
$NaBH_3CN$, HOAc	Org Prep Proc Int *17* 317 (1985) (review)
$NaBH_3CN$, HCl	TL *23* 1929 (1982)
$NaBH_3CN$, $ZnCl_2$	JOC *50* 1927 (1985)
$(n\text{-Bu}_4N)BH_3CN$	JOC *46* 3571 (1981)
$NaHBR_2CN$	TL *22* 3447 (1981)
$LiHBEt_3$	TL *22* 3447 (1981); *25* 695 (1984)
$LiHB(sec\text{-Bu})_3$	TL *22* 3447 (1981); *25* 695 (1984)
	JOC *48* 3412 (1983)
$NaHB(O_2CCHR^1NR^2COR^3)_3$ (chiral)	TL *22* 3869 (1981)
$i\text{-Bu}_2AlH$	TL *23* 1929 (1982)
$i\text{-Bu}_2AlH$, $n\text{-BuLi}$	TL *23* 1929 (1982)
$LiAlH_4$	Org Rxs *6* 469 (1951)
	JOC *23* 535 (1958)
	BSCF 4439 (1970)
	TL *23* 1929 (1982)
$LiAlH_4$; $NaOCH_3$, $Ti(O\text{-}i\text{-Pr})_4$, LiCl, $NiCl_2$, $TiCl_3$, $TiCl_4$, $BF_3 \cdot OEt_2$, or R_3Al (R = Me, $i\text{-Bu}$)	TL *23* 1929 (1982)

$Mg(AlH_4)_2$ TL *23* 1929 (1982)

$NaH_2Al(OCH_2CH_2OCH_3)_2$ JOC *48* 3412 (1983)

Et_3SiH, CF_3CO_2H Bull Acad Sci USSR, Div Chem Sci 1345 (1968)

$MFe(CO)_4H$ (M = Na, K) Syn 733 (1974)
BCSJ *49* 1378 (1976)
Tetr *42* 259 (1986)

$LaNi_5H_6$ JOC *52* 5695 (1987)

$LaNi_{4.5}Al_{0.5}H_5$ JOC *52* 5695 (1987)

$Na_2S_2O_4, H_2O, \Delta$ Austral J Chem *32* 201 (1979)

PhSeH TL *21* 3385 (1980)

C_8K Syn 30 (1979)

$$R_2NH \xrightarrow[\text{NaBH}_4\text{-CF}_3\text{CO}_2\text{H or NaBH}_3\text{CN-HOAc}]{\text{paraformaldehyde}} R_2NCH_3$$

Syn 709 (1987)

$$RNH_2 \xrightarrow[\text{reducing agent}]{\text{H}_2\text{CO}} RNMe_2$$

Reducing agent

$MFe(CO)_4H$ (M = Na, K) Syn 733 (1974)

$NaBH_4$, MeOH TL 261 (1973)

$NaBH_3CN, ZnCl_2$, MeOH JOC *50* 1927 (1985)

$NaBH_3CN, CH_3CN$ JOC *37* 1673 (1972)

$NaBH_4, H_2SO_4$ Anal Chem *48* 484 (1976)
Syn 743 (1980)

$$R_2C{=}O \xrightarrow[\text{NaBH}_3\text{CN or LiBH}_3\text{CN}]{\text{NH}_4\text{OAc}} R_2CHNH_2$$

JACS *93* 2897 (1971); *107* 8066 (1985)
Syn 135 (1975) (review)
JOC *52* 2615, 4274 (1987)
TL *28* 773 (1987)

$$R_2C{=}O \xrightarrow[\text{HCO}_2\text{H}]{(\text{NH}_4)\text{O}_2\text{CH}} \xrightarrow{\text{H}_2\text{O}} R_2CHNH_2$$

$$RNH_2 \xrightarrow[\text{HCO}_2\text{H}]{\text{H}_2\text{CO}} RN(CH_3)_2$$

Reviews:

Org Rxs *5* 301 (1949)
"The Acyclic Aliphatic Tertiary Amines," p 44

$$\underset{\text{RCR}'}{\overset{\text{O}}{\overset{\|}{}}} \longrightarrow \underset{\text{RCR}'}{\overset{\text{NOX}}{\overset{\|}{}}} \xrightarrow[\text{agent}]{\text{reducing}} \underset{\text{RCHR}'}{\overset{\text{NH}_2}{\overset{|}{}}}$$

$\underline{X}$	Reducing Agent	
H	H_2, cat PtO_2	JOC *31* 1342, 1346 (1966); *37* 335 (1972)
	H_2, cat Pd-C	Org Syn Coll Vol *5* 376 (1973)
	H_2, cat Pd-C, HCl	JACS *50* 3370 (1928)
	Raney nickel	JCS C 531 (1966)
	Na, EtOH	Org Syn Coll Vol *2* 318 (1943)
	Na, *n*-PrOH	JOC *31* 1342, 1346 (1966)
	Na(Hg), HOAc	JACS *71* 2257 (1949)
	Al(Hg), H_2O	Angew Int *18* 78 (1979)
	BH_3	JOC *34* 1817 (1969); *51* 105 (1986) JCS Perkin I 643 (1979)
	$NaBH_2S_3$	Can J Chem *48* 735 (1970)
	$NaBH_4$, $TiCl_4$	Syn 695 (1980)
	$NaBH_4$, $NiCl_2 \cdot 6\,H_2O$	Ber *117* 856 (1984)
	$NaBH_4$, MoO_3	Chem Ber *117* 856 (1984) TL *28* 5497 (1987)
	$LiAlH_4$	JOC *17* 294 (1952); *50* 3948 (1985); *52* 4717 (1987) J Biol Chem *211* 725 (1954)
CH_3	BH_3	JOC *34* 1817 (1969) JCS Perkin I 643 (1979)
	$NaBH_3(O_2CCF_3)$	Chem Pharm Bull *26* 2897 (1978)
$PhCH_2$	$LiAlH_4$	JOC *52* 3211 (1987)
R, $SiMe_3$	BH_3, $R-\overset{\overset{\text{H}}{|}}{\underset{\underset{\text{NH}_2}{|}}{C}}-CPh_2OH$	JCS Perkin I 2039, 2615 (polymer-supported reagent) (1985)
$POPh_2$	$NaBH_4/H^+$	TL *28* 5619 (1987)
	$LiHB(sec\text{-Bu})_3/H^+$	TL *28* 5619 (1987)
Ac, Ts	BH_3	CC 590 (1967)

$$RR'C{=}NOH \longrightarrow RR'C{=}NPOPh_2 \xrightarrow[\text{agent}]{\text{reducing}} \underset{\text{chiral}}{RR'CHNHPOPh_2}$$

Reducing agent: $LiAlH_4$-(R) or (S)-bi-2-naphthol; $LiAlH_4$-1,2(S)-diphenyl-3(R)-methyl-4-(dimethylamino)-2-butanol; or potassium 9-0-(1,2:5,6-di-0-isopropylidene-α-D-glucofuranosyl)-9-boratabicyclo[3.3.1]nonane

JOC *52* 702 (1987)

$$R_2C{=}O \xrightarrow{Me_3SiN_3} \overset{\overset{\displaystyle OSiMe_3}{|}}{R_2CN_3} \xrightarrow{LiAlH_4} R_2CHNH_2$$

TL 2737 (1977)

$$R_2C{=}O \longrightarrow \overset{\overset{\displaystyle CN}{|}}{R_2CNR'_2} \xrightarrow[\text{agent}]{\text{reducing}} R_2CHNR'_2$$

Reducing agent

Na, NH$_3$ .... TL 61 (1976)
Chem Pharm Bull 25 2689 (1977)

NaBH$_4$ .... TL 3105 (1969)

$$RCHO \longrightarrow \overset{\overset{\displaystyle NSiMe_3}{||}}{RCH} \xrightarrow[\text{2. KOH}]{\text{1. NbCl}_4\text{(THF)}_2} \overset{\overset{\displaystyle H_2N\ \ NH_2}{|\ \ \ \ |}}{RCHCHR}$$

JACS 109 3152 (1987)

3. Alkylative Amination

See also page 401, Section 4.

$$R^1CHO \longrightarrow R^1CH{=}NR^2 \xrightarrow[\text{2. H}_2\text{O}]{\text{1. R}^3\text{M}} R^1R^3CHNHR^2$$

Reviews:

Chem Rev 63 489 (1963)
K. Harada, "The Chemistry of the Carbon-Nitrogen Double Bond," Ed. S. Patai, Interscience, New York (1970)

R³M	
RLi	K. Harada, "The Chemistry of the Carbon-Nitrogen Double Bond," Ed. S. Patai, Interscience, New York (1970), pp 271–272 (review)
CpFe(CO)(PPh$_3$)COCH$_2$M (M = Li, AlEt$_2$)	JACS 106 441 (1984)
RCHLiCO$_2$R	See page 873, Section 2.
LiCH(NCS)CO$_2$R, BF$_3$·OEt$_2$	TL 24 4503 (1983) JACS 105 5946 (1983)
RLi [R = Me, RC≡C, RCOCH$_2$, CH$_2$CO$_2$R, RCHNO$_2$, CH(NCO)CO$_2$Et]; BF$_3$·OEt$_2$	TL 24 4503 (1983)
CH$_3$CH=CHCH$_2$M [M = Li, MgCl, B⟨⟩, Sn(n-Bu)$_3$ (BF$_3$·OEt$_2$)]	JOC 50 3115 (1985) (diastereoselectivity)

$$R^1CHO \longrightarrow R^1CH = NR^2 \xrightarrow[2.\ H_2O]{1.\ R^3M} R^1R^3CHNHR^2 \quad (\textit{Continued})$$

$\underline{R^3M}$

$RC \equiv CLi$, $BF_3 \cdot OEt_2$	TL *25* 1083 (1984)
RMgX	Org Syn Coll Vol *4* 605 (1963)
	K. Harada, "The Chemistry of the Carbon-Nitrogen Double Bond," Ed. S. Patai, Interscience, New York (1970), pp 266–271
	Syn 223 (1987)
RMgBr (R = Et, Ph); $BF_3 \cdot OEt_2$	TL *24* 4503 (1983)
$H_2C = CHCH_2M$ [M = MgCl, B(OMe)$_2$, B⟩, (AlEt$_3$)MgCl, Ti(O-i-Pr)$_3$, ZnBr]	CC 814 (1985)
$RCH = CHCH_2B$⟩	JACS *106* 5031 (1984); *108* 7778 (1986) (both stereoselective)
	CC 1131 (1985)
$H_2C = CHCH_2B(OR)_2$	Ann 2000 (1983)
$Me_3SiC \equiv CCHRB$⟩ or $Me_3SiC = C = CHR$	JOC *50* 2193 (1985)
n-Bu$_3$SnCH$_2$CH$=$CHR (R = H, Me); $BF_3 \cdot OEt_2$ or TiCl$_4$	JOC *50* 146 (1985)
Cp$_2$TiCHRCH$=$CH$_2$	JOMC *224* 327 (1982)
RCu, $BF_3 \cdot OEt_2$	TL *25* 1079 (1984)
R$_2$CuLi, $BF_3 \cdot OEt_2$	TL *25* 1079 (1984)
RCH(ZnBr)CO$_2$ZnBr	JOMC *231* 185 (1982)

$$R^1CH = NR^2 \xrightarrow[2.\ R_2CO]{1.\ NbCl_3(DME)} R^1CHCR_2$$

with R^2NH above and OH below the product.

JACS *109* 6551 (1987)

$$ArCH = NAr + H_2C = CR \text{ (OSiMe}_3) \xrightarrow{cat\ Me_3SiOTf} ArCHCH_2CR \text{ (HNAr, } O\text{)}$$

CC 1053 (1987)

$$RCH = NR \longrightarrow R\text{-pyrrolidine}$$

Me$_3$SiCH$_2$C($=$CH$_2$)CH$_2$OMs, cat Ni[P(OEt)$_3$]$_4$ or Pd$_2$(DBA)$_3 \cdot$HCCl$_3$-8 P(OEt)$_3$	CC 1201 (1986)
n-Bu$_3$SnCH$_2$C($=$CH$_2$)CH$_2$OAc, $BF_3 \cdot OEt_2$/ Pd catalyst	JACS *107* 1778 (1985)

$$R_2C=NX \xrightarrow[\text{2. } H_3O^+]{\text{1. } R'M} R_2\overset{\overset{\displaystyle R'}{\displaystyle |}}{C}NH_2$$

$\underline{X}$	$\underline{R'M}$	
OCH_2Ph	$R'Li$	TL *28* 4973 (1987)
SAr	$R'Li$	JOC *42* 398 (1977)
	$H_2C=CHCH_2B(OR)_2$	TL *27* 2079 (1986)
$SiMe_3$	$R'Li$	Syn 461 (1982)
		JOC *48* 289 (1983)
	$R'MgX$	JOC *48* 289 (1983)
COR	$R'Li$	Angew *21* 203 (1982)

$$ArCHO \xrightarrow[H^+]{H_2NSO_2NH_2} (ArCH=N)_2SO_2 \xrightarrow[\substack{\text{2. } H_2O, \text{ py} \\ \text{3. NaOH}}]{\text{1. } RLi(MgX)} Ar\overset{\overset{\displaystyle R}{\displaystyle |}}{C}HNH_2$$

TL *27* 3957 (1986)

$$RCHO \longrightarrow RCH=NOR'' \xrightarrow[\text{2. } R'MgX]{\text{2 } R'Li \text{ or}} \xrightarrow{BH_3} RR'CHNH_2$$

R = aryl > alkyl

TL *27* 3033 (1986)

$$H_2C=NOCH_2Ph \xrightarrow[\substack{\text{2. } R^2Li \\ \text{3. } R^3X}]{\text{1. } R^1Li} R^1CH_2NR^2R^3$$

CC 305 (1987)

$$R'CH=\overset{+}{N}\!\bigcirc Cl^- \xrightarrow{RM} RR'CHN\!\bigcirc$$

$\underline{RM}$	
$RMgX$ (R = 1° alkyl)	BSCF II 148 (1983)
$RC\equiv CMgX$	BSCF II 21 (1983)
$RMgX$ (R = 1° alkyl, vinylic, CH_2CO_2R'), $RZnX$ [R = CH_2COX (X = OEt, NEt_2), $CR(CO_2Et)_2$]	BSCF II 395 (1982)

$$RCHO \longrightarrow RCH=NNMe_2 \xrightarrow[\text{2. } H_2, \text{ cat PtO}_2]{\text{1. } R'Li} RCHR'NH_2$$

JACS *108* 8265 (1986)

$$\underset{ArCH}{\overset{O}{\overset{\|}{}}} \longrightarrow \underset{\substack{ArCH \\ \text{chiral}}}{\overset{NNR'_2}{\overset{\|}{}}} \xrightarrow[\text{2. } H_2, \text{ cat Pd-C}]{\text{1. } RMgX} \underset{\substack{ArCHR \\ \text{chiral}}}{\overset{NH_2}{\overset{|}{}}}$$

CC 668 (1979)

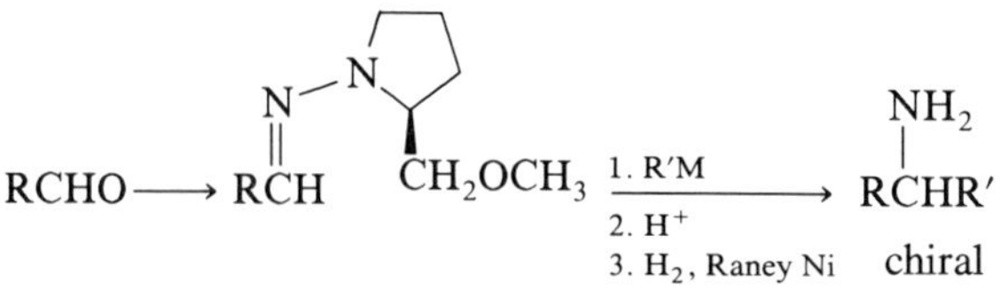

$$RCHO \longrightarrow RCH \overset{N-N}{=} CH_2OCH_3 \xrightarrow[\substack{2.\ H^+ \\ 3.\ H_2,\ Raney\ Ni}]{1.\ R'M} \underset{chiral}{RCHR'} \overset{NH_2}{|}$$

R'M

RLi   Angew Int *25* 1109 (1986)

RCeCl$_2$ JACS *109* 2224 (1987)

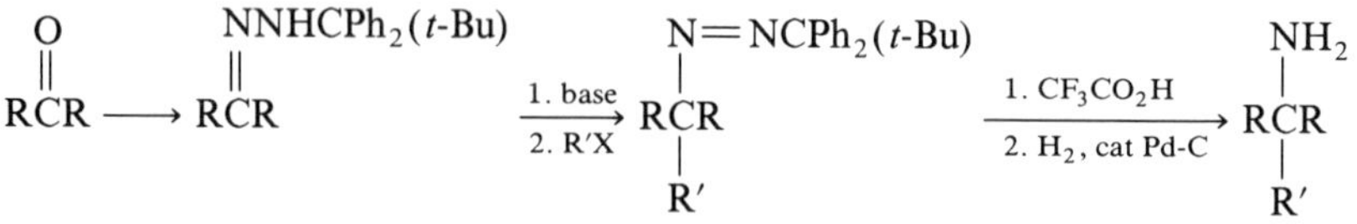

$$\underset{RCR}{\overset{O}{\overset{\|}{}}} \longrightarrow \underset{RCR}{\overset{NNHCPh_2(t\text{-}Bu)}{\overset{\|}{}}} \xrightarrow[\substack{2.\ R'X}]{1.\ base} \underset{\substack{| \\ R'}}{\overset{N=NCPh_2(t\text{-}Bu)}{RCR}} \xrightarrow[\substack{2.\ H_2,\ cat\ Pd\text{-}C}]{1.\ CF_3CO_2H} \underset{\substack{| \\ R'}}{\overset{NH_2}{RCR}}$$

CC 176 (1986)

$$ArCHO \xrightarrow[\substack{2.\ RLi\ or\ RMgX \\ 3.\ H_2O}]{1.\ LiN(SiMe_3)_2} \underset{ArCHR}{\overset{NH_2}{|}}$$

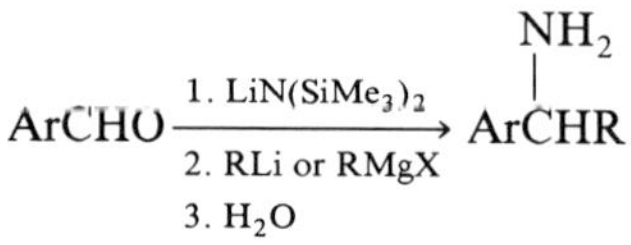

JOC *48* 289 (1983) ; *52* 4665 (1987)

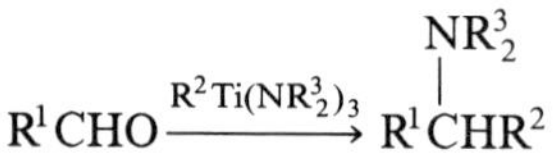

$$R^1CHO \xrightarrow{R^2Ti(NR_2^3)_3} \underset{R^1CHR^2}{\overset{NR_2^3}{|}}$$

R^1 = *t*-Bu, aryl, vinylic

Helv *65* 2598 (1982)

$$R_2C{=}O \xrightarrow[\substack{2.\ NaBH_4 \\ 3.\ H_3O^+}]{1.\ Me_3SiCHLiNMeCH{=}N\text{-}t\text{-}Bu} R_2CHCH_2NHMe$$

JACS *104* 877 (1982)

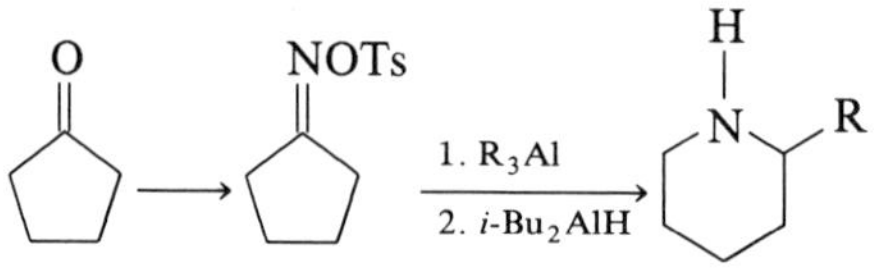

JACS *103* 7368 (1981)
JOC *51* 5043 (1986)

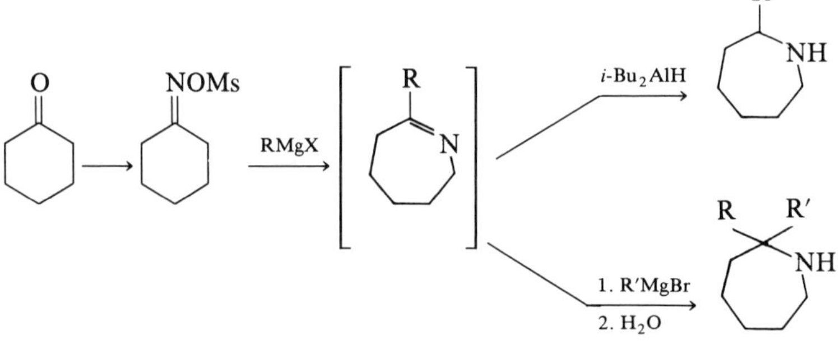

R' = allyl, propargyl
TL *23* 3395 (1982)

4. Mannich and Related Reactions

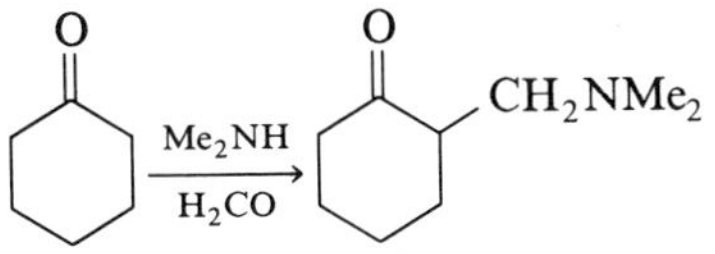

$$\text{cyclohexanone} \xrightarrow[\text{H}_2\text{CO}]{\text{Me}_2\text{NH}} \text{2-(CH}_2\text{NMe}_2\text{)cyclohexanone}$$

Reviews:

Org Rxs *1* 303 (1942); *7* 99 (1953)
Org Syn Coll Vol *3* 305 (1955)
Angew *68* 265 (1956)
B. Riechert, "Die Mannich Reaktion," Springer, Berlin (1959)
H. Hellman and G. Opitz, "α-Aminoalkylierung," Verlag Chemie, Weinheim, Germany (1960)
Syn 703 (1973)

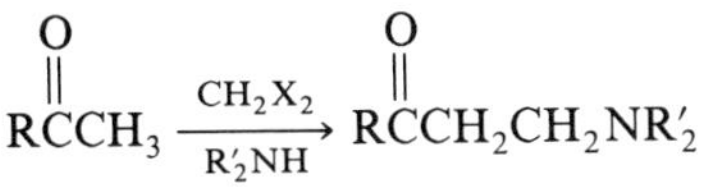

$$\underset{\text{RCCH}_3}{\overset{\text{O}}{\parallel}} \xrightarrow[\text{R}_2'\text{NH}]{\text{CH}_2\text{X}_2} \underset{\text{RCCH}_2\text{CH}_2\text{NR}_2'}{\overset{\text{O}}{\parallel}}$$

Angew Int *21* 922 (1982)
BCSJ *55* 1331 (1982)

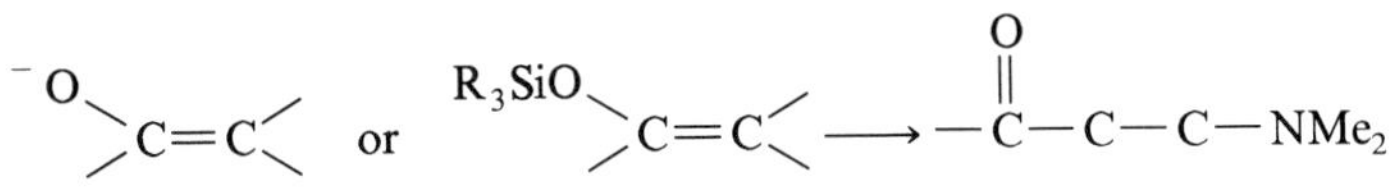

$$\underset{}{\overset{^-\text{O}}{\diagdown}}\text{C}=\text{C}\diagup \quad \text{or} \quad \overset{\text{R}_3\text{SiO}}{\diagdown}\text{C}=\text{C}\diagup \longrightarrow \; -\overset{\overset{\text{O}}{\parallel}}{\text{C}}-\text{C}-\text{C}-\text{NMe}_2$$

$(\text{Me}_2\overset{+}{\text{N}}{=}\text{CH}_2)\text{O}_2\text{CCF}_3^-$

JACS *99* 944 (1977)
Tetr *35* 613 (1979)
Org Rxs *59* 153 (1980)

$(\text{Me}_2\overset{+}{\text{N}}{=}\text{CH}_2)\text{Cl}^-$

TL *23* 703, 1419 (1982)
JACS *107* 2474 (1985) (on enol carbonate)

$(\text{Me}_2\overset{+}{\text{N}}{=}\text{CH}_2)\text{I}^-$

JACS *95* 602 (1973) (on enol borinates); *98* 6715
  (1976); *99* 6066 (1977)
TL 1621 (1977); *21* 805 (1980)

$\text{Me}_2\text{NCH}_2\text{NMe}_2, \text{ClCH}_2\text{I}$

CL 1213 (1980)
BCSJ *55* 534 (1982)

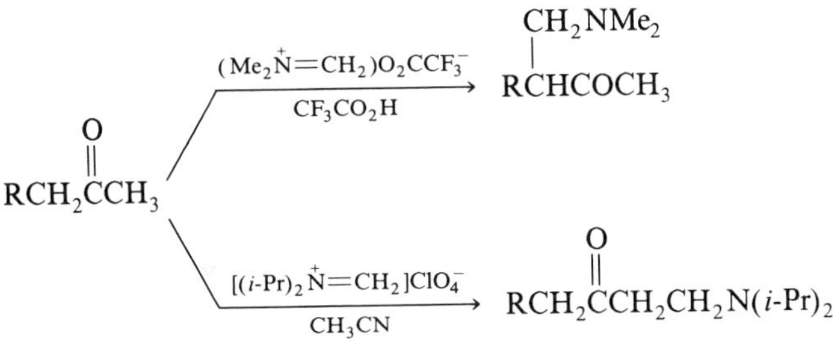

$$\underset{\text{RCH}_2\text{CCH}_3}{\overset{\text{O}}{\parallel}} \diagup \xrightarrow[\text{CF}_3\text{CO}_2\text{H}]{(\text{Me}_2\overset{+}{\text{N}}{=}\text{CH}_2)\text{O}_2\text{CCF}_3^-} \underset{\text{RCHCOCH}_3}{\overset{\text{CH}_2\text{NMe}_2}{\mid}}$$

$$\diagdown \xrightarrow[\text{CH}_3\text{CN}]{[(i\text{-Pr})_2\overset{+}{\text{N}}{=}\text{CH}_2]\text{ClO}_4^-} \underset{\text{RCH}_2\text{CCH}_2\text{CH}_2\text{N}(i\text{-Pr})_2}{\overset{\text{O}}{\parallel}}$$

Tetr *33* 295 (1977)

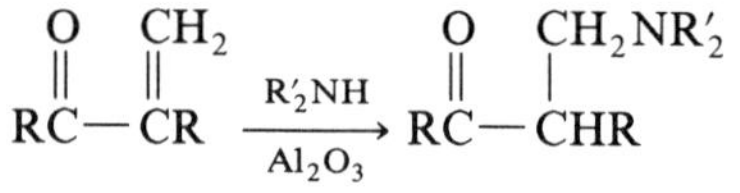

$$\underset{\substack{\| \quad \| \\ RC-CR}}{\overset{\substack{O \quad CH_2 \\}}{}} \xrightarrow[Al_2O_3]{R'_2NH} \underset{\substack{\| \quad | \\ RC-CHR}}{\overset{\substack{O \quad CH_2NR'_2 \\}}{}}$$

TL *21* 809 (1980)

$$ArH + H_2CO + HNR_2 \longrightarrow ArCH_2NR_2$$

JACS *70* 4232 (1948)

10. FROM CARBOXYLIC ACIDS AND ACID DERIVATIVES

1. Molecular Rearrangements

Hoffman

$$\underset{\text{RCNH}_2}{\overset{\overset{\displaystyle O}{\|}}{}} \longrightarrow \text{RNH}_2$$

NaOX (X = Br, I)	Org Rxs *3* 267 (1946)
	JACS *71* 3929 (1949)
PhIO, HCO$_2$H, H$_2$O	Syn 538 (1983)
PhI(O$_2$CCF$_3$)$_2$	Biochem Biophys Res Commun *80* 1 (1978)
	JOC *44* 1746 (1979); *49* 4272, 4277 (1984)
PhI(OH)OTs/OH$^-$	JOC *51* 2669 (1986)
Pb(OAc)$_4$/R'XH (X = O, NH)/H$^+$ or OH$^-$	CC 161 (1965)
	TL 4039 (1965)
	JACS *87* 1141 (1965)
	JOC *40* 3554 (1975)

Curtius

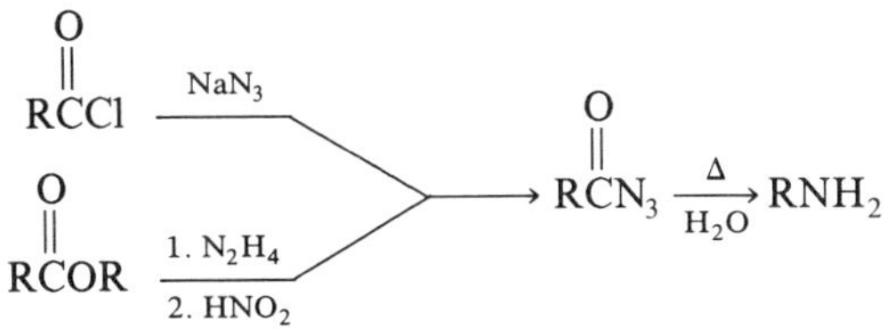

$$\underset{\text{RCCl}}{\overset{\overset{\displaystyle O}{\|}}{}} \xrightarrow{\text{NaN}_3}$$

$$\underset{\text{RCOR}}{\overset{\overset{\displaystyle O}{\|}}{}} \xrightarrow[\text{2. HNO}_2]{\text{1. N}_2\text{H}_4}$$

$$\longrightarrow \underset{\text{RCN}_3}{\overset{\overset{\displaystyle O}{\|}}{}} \xrightarrow[\text{H}_2\text{O}]{\Delta} \text{RNH}_2$$

Org Rxs *3* 337 (1947)
Org Syn Coll Vol *4* 819 (1963)

$$\underset{\text{RCOH}}{\overset{\overset{\displaystyle O}{\|}}{}} \xrightarrow[\text{H}_2\text{SO}_4]{\text{NaN}_3} \xrightarrow{\text{NaOH}} \text{RNH}_2$$

Org Syn Coll Vol *5* 273 (173)

$$\underset{\substack{\| \\ RCOH}}{O} \longrightarrow \longrightarrow \underset{\substack{\| \\ RCN_3}}{O} \xrightarrow[R'OH]{\Delta} \underset{\substack{\| \\ RNHCOR'}}{O}$$

Org Rxs *3* 337 (1947) (review)
TL *25* 3515 (1984)
JOC *51* 3007, 5123 (1986); *52* 4875 (1987)

$$\underset{\substack{\| \\ RCOH}}{O} \xrightarrow[R'OH]{N_3PO(OPh)_2} RNHCO_2R' \xrightarrow{NaOH} RNH_2$$

JACS *94* 6203 (1972)
Tetr *30* 2151 (1974)

$$\underset{\substack{\| \\ RCOH}}{O} \xrightarrow[\substack{2.\ NaN_3}]{1.\ ClCO_2Et} \underset{\substack{\| \\ RCN_3}}{O} \xrightarrow{R'XH} \underset{\substack{\| \\ RNHCXR'}}{O}$$

X = O, NH
JOC *26* 3511 (1961); *43* 2164 (1978)

Lossen

$$\underset{\substack{\| \\ RCOH}}{O} \xrightarrow{H_2NOH} \underset{\substack{\| \\ RCNHOH}}{O} \xrightarrow[reagent]{\Delta} RNH_2$$

Reagent: $SOCl_2$, Ac_2O, P_2O_5, or PPA

Chem Rev *33* 242 (1943)

Schmidt

$$\underset{\substack{\| \\ RCOH}}{O} \xrightarrow[\Delta]{HN_3} RNH_2$$

Org Rxs *3* 307 (1946)

2. Reduction of Amides and Imides

$$\underset{\substack{\| \\ RCNR'_2}}{O} \longrightarrow RCH_2NR'_2$$

BH₃

JACS *86* 3566 (1964); *107* 1421 (1985)
JOC *31* 3867 (1966); *33* 3637 (1968);
  38 912 (1973); *41* 149 (1976); *42*
  4148 (1977); *51* 4856, 5373 (1986); *52* 5742
  (1987)
J Heterocyclic Chem *5* 875 (1968)
Syn 752 (1978)

BH₃·SMe₂

Syn 441 (1981)
JOC *47* 3153 (1982)
TL 3315 (1982); *28* 4601 (1987)

$BH_3 \cdot SMe_2$, $BF_3 \cdot OEt_2$	Syn 996 (1981)
	JOC *42* 512 (1977)
$NaBH_4$, $TiCl_4$	Syn 695 (1980)
$NaBH_4$, $CoCl_2 \cdot 6\ H_2O$	TL 4555 (1969)
$NaBH_4$, $AlCl_3$	JACS *78* 2582 (1956)
$NaBH_4$, HOAc	TL 763 (1976)
$NaBH_4$, $MeSO_3H$	JOC *46* 2579 (1981)
$NaBH_4$, py	Chem Pharm Bull *17* 98 (1969)
$NaBH_4$, CF_3CH_2OH	JOC *50* 3948 (1985)
$(Et_3O)BF_4 / NaBH_4$	TL 61 (1968)
$(Et_3O)BF_4 / NaBH_4$, $SnCl_4$	Syn 652 (1977)
$POCl_3 / NaBH_4$	TL 219 (1976); *27* 2103 (1986) JOC *42* 2082 (1977)
$POCl_3 / Zn$, EtOH	Experientia *33* 101 (1977)
$HSCH_2CH_2SH / NaBH_4$	Chem Ind 322 (1976)
$LiBH_4$, MeOH, diglyme	JOC *51* 4000 (1986)
$LiBH_3CN$	JACS *91* 3996 (1969)
AlH_3	JACS *88* 1464 (1966); *90* 2927 (1968); *109* 6124 (1987)
$i\text{-}Bu_2AlH$	Bull Acad Sci USSR, Div Chem Sci 2046 (1959) Ann *623* 9 (1959) Syn 617 (1975) JOC *50* 2443 (1985); *52* 5745 (1987)
$LiAlH_4$	Org Rxs *6* 469 (1941) Helv *31* 1397 (1948); *38* 2036 (1955) Org Syn Coll Vol *4* 339, 354, 564 (1963) JACS *107* 5717, 5739 (1985); *108* 2034 (1986) JOC *51* 3140, 5373 (1986); *52* 1844, 2018, 5320 (1987)
$NaH_2Al(OCH_2CH_2OCH_3)_2$	TL 3303 (1968)
$LiHAl(OCH_3)_3$	JACS *87* 5614 (1965)
$P_4S_{10} / $ Raney nickel	JOC *16* 131 (1951)
$P_2S_5 / MeI / NaBH_4$ or $NaBH_3CN$	JOC *46* 3730 (1981)
$(p\text{-}MeOC_6H_4)_2P_2S_4 /(Et_3O)BF_4 / NaBH_4$	TL *21* 4061 (1980)

$$R_2NCOCH_3 \longrightarrow R_2NCH_3$$
(with O double bond above the C)

BH$_3$ JOC *46* 2431 (1981)

LiAlH$_4$ J Label Compds *6* 261 (1970)
                                               JOC *46* 2431 (1981); *51* 3295 (1986)

$$RN(CR')_2 \xrightarrow[BF_3 \cdot OEt_2]{NaBH_4} RN(CH_2R')_2$$
(with O double bond above the C)

Ann 461 (1979)

3. Reductive Alkylation

$$RNH_2 + HO_2CR' \longrightarrow RNHCH_2R'$$

NaBH$_4$ JACS *96* 7812 (1974)
                                               JOC *40* 3453 (1975)
                                               Syn 766 (1978)

BH$_3 \cdot$NMe$_3$ Syn 1013 (1983)

$$R_2NH \xrightarrow[2.\ R'CO_2R]{1.\ LiAlH_4} R_2NCH_2R$$

JOC *25* 1033 (1960); *27* 1042 (1962)
Syn 608 (1975)

$$RNH_2 \xrightarrow[2.\ BH_3 \cdot SMe_2]{1.\ Ac_2O,\ HCO_2H} RNHCH_3$$

TL *23* 3315 (1982)

$$R^1NHCOR^2 \xrightarrow[2.\ R^3CO_2R^4]{1.\ LiAlH_4} R^1N(CH_2R^2)CH_2R^3$$

$$R^1NHCO_2R^2 \xrightarrow[2.\ R^3CO_2R^4]{1.\ LiAlH_4} R^1NMeCH_2R^3$$

TL 3395 (1979)

$$RCNR_2 \longrightarrow RCHNR_2$$
(with O double bond above the C; R' above the CH)

n-PrMgBr/NaBH$_3$CN (R′ = n-Pr; lactam)   JACS *107* 5534 (1985)

ArLi/BH$_3 \cdot$SMe$_2$ JOC *50* 2719 (1985)

ArLi/LiAlH$_4$ or BH$_3 \cdot$SMe$_2$ or   JOC *50* 3885 (1985)
    NaBH$_4$-CF$_3$CO$_2$H (R′ = Ar)

ArLi/NaBH$_3$CN (R′ = Ar; lactams)   JOC *45* 3664 (1980)

RC≡CBF$_2$/LiAlH$_4$ (R′ = RC≡C; lactams)   TL *24* 1719 (1983)

$$\underset{\text{RCNR}_2}{\overset{\text{S}}{\|}} \xrightarrow[\text{2. LiAlH}_4]{\text{1. R'Li}} \underset{\text{RCHNR}_2}{\overset{\text{R'}}{|}}$$

TL *28* 1529 (1987)

$$\text{CH}_2\text{C}\overset{\text{O}}{\diagdown}\text{NR} \xrightarrow[\text{2. H}_3\text{O}^+]{\text{1. R'MgX}} \text{CH}=\text{C}\overset{\text{R'}}{\diagup}\text{NR}$$

Coll Czech Chem Commun *7* 482 (1935); *8* 533 (1936)
JACS *55* 295, 2543 (1933); *64* 2588 (1942); *79* 5279 (1957)

$$\overset{\text{O}}{\overset{\|}{\text{C}}}\text{NCH}=\text{CH}_2 \xrightarrow[\text{2. H}_3\text{O}^+]{\text{1. ArLi}} \overset{\text{Ar}}{\overset{|}{\text{C}}}\text{N}$$

Syn 242 (1977)
JOC *47* 3652 (1982)

$$\overset{\text{MeO}}{\diagdown}\text{N} \xrightarrow[\text{2. H}_2\text{O}]{\text{1. 2 RLi}} \overset{\text{R} \quad \text{R}}{\diagdown\diagup}\text{NH}$$

JACS *109* 4940 (1987)

4. α-Amination

$$\text{RCH(CN)CO}_2\text{R} \xrightarrow[\text{2. Ph}_2\text{P(O)ONH}_2]{\text{1. base}} \overset{\text{NH}_2}{\overset{|}{\text{RC(CN)CO}_2\text{R}}}$$

TL *23* 5399 (1982); *28* 4385 (1987)

5. Ring Opening of *N*-Tosyl Aziridines

$$\text{Nuc}^- + \overset{\overset{\text{Ts}}{|}}{\underset{\text{N}}{\text{H}_2\text{C}-\text{CHR'}}} \xrightarrow{\text{H}_2\text{O}} \text{Nuc}-\text{CH}_2\overset{\overset{\text{HNTs}}{|}}{\text{CHR'}}$$

Nuc⁻ = R₂CuLi; R₂CuLi, BF₃·OEt₂; (R₂CuCN)Li₂; H₂C=CHCH₂MgCl; RC≡CLi;
NaN₃, BF₃·OEt₂; LiBr, BF₃·OEt₂; Li₂NiBr₄; Li₂CuCl₄; NaSPh; NaSH

TL *27* 4157 (1986); *28* 3341 (1987)

11. FROM NITRILES

$$RCN \longrightarrow RCH_2NH_2$$

Reviews:

M. Freifelder, "Practical Catalytic Hydrogenation," Wiley Interscience, New York (1971), Chpt 12

M. Freifelder, "Catalytic Hydrogenation in Organic Synthesis: Procedures and Commentary," J. Wiley and Sons, New York (1978), Chpt 6

P. N. Rylander, "Hydrogenation Methods," Academic Press, New York (1985), Chpt 7

H_2, cat $HRh(PPr_3^i)_3$	CC 870 (1979)
H_2, cat Ni	Org Syn Coll Vol *3* 229, 358, 720 (1955) JOC *25* 1658 (1960) JCS C 531 (1966)
H_2, cat Raney Ni	TL *28* 6015 (1987)
H_2, cat Pd	JACS *50* 3370 (1928) JCS 426 (1942)
H_2, cat PtO_2	JOC *37* 335 (1972)
BH_3	JACS *82* 681 (1960) JOC *51* 4856 (1986)
$BH_3 \cdot SMe_2$	Syn 605 (1981) JOC *47* 3153 (1982)
$LiBH_4$, MeOH, diglyme	JOC *51* 4000 (1986)
$NaBH_4$, $AlCl_3$	JACS *78* 2582 (1956)
$NaBH_4$, $CoCl_2 \cdot 6 H_2O$	TL 4555 (1969) JACS *104* 6801 (1982); *108* 67 (1986)
KBH_4, cat $PdCl_2$	BSCF 1996 (1959)
$NaBH_3O_2CCF_3$	TL 2875 (1976)

$$RCN \longrightarrow RCH_2NH_2 \quad (\textit{Continued})$$

NaBH$_3$OH	JOC 42 3963 (1977)
AlH$_3$	JACS 90 2927 (1968)
i-Bu$_3$Al	Ann 623 9 (1959)
LiAlH$_4$	JACS 70 3738 (1948); 77 2544 (1955)
	Org Rxs 6 469 (1951)
	JOC 52 2301 (1987)
LiAlH$_4$, AlCl$_3$	JACS 77 2544 (1955)
	Chem Pharm Bull 32 873 (1984)
	Syn 40 (1987)
LaNi$_{4.5}$Al$_{0.5}$H$_5$	JOC 52 5695 (1987)

$$RCN \xrightarrow[\text{2. NaBH}_4]{\text{1. (Et}_3\text{O)BF}_4} RCH_2NHEt$$

JOC 34 627 (1969)

$$RCN \xrightarrow{\text{R'MgX}} \left[\begin{array}{c} NH \\ \parallel \\ RCR' \end{array} \right] \xrightarrow[\text{agent}]{\text{reducing}} \begin{array}{c} NH_2 \\ | \\ RCHR' \end{array}$$

Reducing agent

Li, NH$_3$	JOC 51 5338 (1986); 52 3901 (1987)
	(CuBr catalyzed Grignard addition)
H$_2$, Pt catalyst	JACS 72 876 (1950); 74 4607 (1952)
NaBH$_4$	JACS 109 3378 (1987)
LiAlH$_4$	JACS 75 5898 (1953)

$$RCN \xrightarrow[\substack{\text{2. NbCl}_4\text{(THF)}_2 \\ \text{3. KOH}}]{\text{1. } n\text{-Bu}_3\text{SnH}} \begin{array}{c} H_2N \quad NH_2 \\ | \qquad | \\ RCHCHR \end{array}$$

JACS 109 3152 (1987)

$$\underset{RCCN}{\overset{O}{\overset{\parallel}{}}} \xrightarrow[\text{2. NaBH}_4\text{, CoCl}_2]{\text{1.}} \begin{array}{c} OH \\ | \\ RCHCH_2NH_2 \\ \text{chiral} \end{array}$$

JOC 50 3237 (1985)

ETHERS

GENERAL REFERENCES

A. Resowsky, "The Chemistry of Heterocyclic Compounds," Interscience, New York (1964), Vol 19, pt 1 (epoxides)

Houben-Weyl, "Methoden der Organischen Chemie," 4th ed, Vol VI/3, G. Thieme, Stuttgart (1965): ethers, p 1; epoxides, p 367; oxetanes, p 489; tetrahydrofurans, p 517

"The Chemistry of the Ether Linkage," Ed. S. Patai, Interscience, New York (1967)

"The Chemistry of Ethers, Crown Ethers, Hydroxyl Groups and their Sulphur Analogues," Ed. S. Patai, J. Wiley, New York (1980)

Tetr *39* 2323 (1983) (Recent Advances in the Preparation and Synthetic Applications of Oxiranes)

Syn 629 (1984) (Synthetically Useful Reactions of Epoxides)

1. FROM OTHER ETHERS

See also page 455, Sections 1 and 2; and page 467, Section 3.

$$ArOSiMe_2(t\text{-}Bu) + RX \xrightarrow[\text{DMF}]{\text{KF}} ArOR$$

$$RX = PhCH_2Br,\ MeI$$

TL *28* 4139 (1987)

$$ArCH_2OMe \xrightarrow[\text{2. E}^+]{\text{1. base}} ArCHOMe \overset{\displaystyle E}{|}$$

$$E^+ = RX,\ R_2\overset{\circ}{C}O,\ RCO_2Et$$

TL 4155 (1979)
JCS Perkin I 1652 (1981)

$$RM + XCH_2OR \longrightarrow RCH_2OR$$

$$X = \text{halogen}$$

<u>RM</u>

RMgX

Ber *97* 636 (1964)
Angew Int *6* 335 (1967) (review)
TL *28* 2225 (1987)

R$_2$CuLi

TL *28* 2225 (1987)

$$RCHOR \overset{\displaystyle Cl}{|} + H_2C=CHCH_2SiMe_3 \xrightarrow[\text{(X = I, OTf)}]{\text{cat Me}_3\text{SiX}} RCHOR \overset{\displaystyle CH_2CH=CH_2}{|}$$

CL 409 (1983)

$$Br-\overset{|}{\underset{|}{C}}-\overset{|}{\underset{|}{C}}-OR + H_2C=CHCH_2SiMe_3 \xrightarrow{\text{AgBF}_4} H_2C=CHCH_2-\overset{|}{\underset{|}{C}}-\overset{|}{\underset{|}{C}}-OR$$

JOC *48* 1557 (1983)

2. ALKYLATION OF ALCOHOLS AND PHENOLS

See also page 462, Section 5.

$$\underset{\text{H}}{\overset{\text{H}}{\underset{|}{\text{C}}}}-\text{C}_n-\text{OH} \longrightarrow \overset{\text{O}}{\overset{\triangle}{\text{C}-\text{C}_n}}$$

Review: Syn 501 (1971) (hypoiodite reaction)

Reagent(s)	n	
Cl_2, Ag_2CO_3	3	Tetr *29* 3675 (1973)
Cl_2, HgO	3	Tetr *29* 3675 (1973)
Br_2, AgOAc	3	TL 1049 (1973)
		Tetr *29* 3675 (1973)
		JOC *47* 3559 (1982)
Br_2, Ag_2CO_3	3	Tetr *29* 3675 (1973)
Br_2, Ag_2O	3	JACS *86* 3905, 5503 (1964)
		JOC *30* 3216 (1965)
		CC 981 (1969); 976 (1970);
		451 (1973)
		Tetr *29* 3675 (1973)
Br_2, HgO	3	Tetr *29* 3675 (1973)
I_2, HgO	3, 4	JACS *87* 1807 (1965);
		109 8117 (1987)
		CC 981 (1969)
		TL *24* 5915 (1983); *25* 2035
		(1984); *28* 4951 (1987)
NIS, hν	3	JOC *50* 3015 (1985)
DDQ	3, 4	TL *28* 5175 (1987)

$$\text{MeO}-\!\!\!\left\langle\bigcirc\right\rangle\!\!\!-\text{C}=\text{C}-\underset{\underset{\text{H}}{|}}{\overset{\text{H}}{\overset{|}{\text{C}}}}-\text{C}_n-\text{OH} \xrightarrow{\text{DDQ}} \text{MeO}-\!\!\!\left\langle\bigcirc\right\rangle\!\!\!-\text{C}=\text{C}-\!\!\overset{()_{n-2}}{\underset{\text{O}}{\boxed{}}}$$

n = 3, 4

Heterocycles *23* 553 (1985)

$$ROH + R'X \xrightarrow{\text{reagent(s)}} ROR'$$

R'X	Reagent(s)	
MeI, EtX (X = Br, I), PhCH$_2$X (X = Cl, Br)	Ag$_2$O	JCS *83* 1021, 1037 (1903) JACS *73* 4043 (1951); *108* 4603 (1986); *109* 6124 (1987) JOC *26* 4553 (1961); *27* 290 (1962); *52* 3889, 4647 (1987) JCS C 2372 (1969) TL *28* 4019, 5353 (1987)
MeCl, Me$_2$SO$_4$	NaOH, C$_6$H$_6$	Syn 123 (1979)
Me$_2$SO$_4$	NaOH, *n*-Bu$_4$NI	Angew Int *12* 846 (1973)
1° alkyl, benzylic chloride, BrCH$_2$CO$_2$R	NaOH, cat (*n*-Bu$_4$N)HSO$_4$	TL 3251 (1975); *28* 4143 (1987)
1° alkyl bromide and iodide	KOH, DMSO	Tetr *35* 2169 (1979)
R$_2$SO$_4$ (R = Me, Et)	KOH, DMSO	Syn 428 (1979)
MeI	NaH	TL 21 (1973) Syn 434 (1974) JACS *109* 3353 (1987)
PhCH$_2$Br	NaH, DMF	JOC *52* 4665 (1987)
1° alkyl bromide and iodide, PhCH$_2$Cl	NaH, HMPA	BSCF 1866 (1965)
1° alkyl bromide, benzylic chloride	cat Ni(acac)$_2$, Δ	Syn 803 (1977)
2,4-Me$_2$C$_6$H$_3$CH$_2$Cl	K$_2$CO$_3$, KI, CH$_3$CN	Syn 987 (1982)
1° alkyl iodide, PhCH$_2$Br	KF, alumina	BCSJ *55* 2504 (1982)
⟨⟩—NHC=N—⟨⟩ with OR'	—	Ber *99* 1479 (1966)
[(R'O)$_3$P⁺CH$_3$]BF$_4^-$	—	JOC *49* 4877 (1984)

$$HO(CH_2)_nX \xrightarrow{\text{base}} (H_2C)_n O$$

"Heterocyclic Compounds," Ed. R. C. Elderfield, Wiley, New York (1950), Vol 1, p 8
JACS *72* 1593 (1950); *75* 4778 (1953)
JOC *52* 3860, 3883 (1987)
TL *28* 723, 1781, 2709 (1987)

$$R'X + MOR \longrightarrow R'OR$$
$$M = Li, Na, K, MgX$$

"Chemistry of the Ether Linkage," pp 445–498 (review)
JACS *69* 2451 (1947)
JCS 616 (1948)
JOC *43* 4682 (1978); *52* 4495 (1987)

$$ArX + KOR \xrightarrow{\Delta} ArOR$$

Org Syn Coll Vol 5 926 (1973)

$$ArI + ArOH \xrightarrow[\text{DMSO}]{2 \text{ KO-}t\text{-Bu}} ArOAr$$

JOC 47 4374 (1982)

$$ArX + HOR \xrightarrow{\text{Cu catalyst}} ArOR$$

Tetr 40 1433 (1984) (review)

$$ArX + NaOAr \xrightarrow[\Delta]{Cu} ArOAr$$

Org Syn Coll Vol 2 445 (1943)
Chem Rev 38 405 (1946) (review)
JOC 29 3624 (1964); 32 2501 (1967)
Tetr 40 1433 (1984) (review)

$$R'X + CuOR \longrightarrow R'OR$$

R' = 1° alkyl, aryl, vinylic

JACS 96 2829 (1974)

$$RX + (ArO)_3PO \xrightarrow[\text{DMF}]{\text{KOH}} ArOR$$

$RX = n\text{-}C_8H_{17}Br, PhCH_2Br, ArCl$

Syn 828 (1982)

$$ROH \longrightarrow ROR'$$

CH_2N_2, cat $BF_3 \cdot OEt_2$	JACS 80 2584 (1958)
	Angew 70 105 (1958)
CH_2N_2 or CH_3CHN_2, cat $BF_3 \cdot OEt_2$	Z Naturforsch B 14 209 (1959)
$RCHN_2$, cat $AlCl_3$ or $BF_3 \cdot OEt_2$	Ann 677 55 (1964)
CH_2N_2, cat HBF_4	JACS 80 2584 (1958)
	Tetr 6 36 (1959)
CH_2N_2, silica gel	TL 4405 (1979)
$RCOCHN_2$, cat $BF_3 \cdot OEt_2$	JACS 72 5161 (1950)

$$ROH + (R'_3O)BF_4 \longrightarrow ROR'$$

R' = Me, Et

JOC 42 1801 (1977)

$$ROH + Ph_3Bi(OAc)_2 \xrightarrow[X = Cl, OAc]{\text{cat } CuX_2} ROPh$$

J Gen Chem USSR 55 2232 (1985)

$$ArOH \longrightarrow ArOR$$

MeI, KOH, DMSO	Tetr 35 2169 (1979)
MeI, NaOH, HMPA	JOC 39 1968 (1974)
Me_2SO_4, NaOH	Org Syn Coll Vol 1 58 (1941)

1° RBr, K_2CO_3	Org Syn Coll Vol *3* 140 (1955)
RX (1° RCl, RBr, RI; 2° RBr), KF-Al_2O_3	CL 755 (1959) BCSJ *55* 2504 (1982)
RX (1° RI, allyl or benzyl Br), KF-Celite	CL 45 (1979)
RX (R =1°, 2° alkyl; benzylic), Et_4NF, DMF	Can J Chem *57* 1887 (1979)
RX (R =1°, 2° alkyl; allylic; benzylic), NaOH, $(PhCH_2NEt_3)Br$, H_2O, CH_2Cl_2	Tetr *30* 1379 (1974)
RX (R =1°, 2° alkyl; allylic; benzylic), anion exchange resin	Syn 113 (1977)
RX (3° RCl), cat $Ni(acac)_2$, $NaHCO_3$	Syn 186 (1982)
$(Me_3O)BF_4$	J Prakt Chem 867 (1964)
$Me_2\overset{+}{S}O\overset{-}{C}H_2$	TL 867 (1964)
$(Me_3S)I$, Ag_2O	Ann *611* 117 (1958)
Me_3SOH	JOC *44* 638 (1979)
Me_3SeOH	TL 1787 (1979)
ROH (R =1°, 2° alkyl), $EtO_2CN\!=\!NCO_2Et$, PPh_3	JCS Perkin I 461 (1975) Chem Ind 281 (1975) TL 2243 (1978) Heterocycles *20* 1975 (1983) JOC *50* 3095 (1985)
ROH (R = 2° allylic), $EtO_2CN\!=\!NCO_2Et$, *n*-Bu_3P	JACS *107* 3891 (1985)
CH_2N_2, cat HBF_4	JACS *80* 2584 (1958) Tetr *6* 36 (1959) Org Syn Coll Vol *5* 245 (1973)
$Ph_4BiO_2CCF_3$ (R = Ph)	CC 503 (1981)
$Ph_3Bi(OAc)_2$, cat Cu or $Cu(OAc)_2$ (R = Ph)	J Gen Chem USSR *55* 2232 (1985) TL *27* 3619 (1986)

$$2\ ROH \longrightarrow ROR$$

Reagent	R	
H_3PO_4	benzhydryl	JACS *73* 2630 (1951)
$ZnCl_2$	2° allylic; 1° or 2° benzylic	JOC *52* 3917 (1987)

CuSO$_4$, 180°C	1° alkyl	Acta Chem Scand B *31* 721 (1977)
DMSO, 175°C	1°, 2° benzylic	JOC *42* 2012 (1977)

$$ROH + R'OH \longrightarrow ROR'$$

H$^+$	JACS *54* 2088 (1932); *70* 2400 (1948) Org Syn Coll Vol *4* 72 (1963)
DMSO, 175°C (R = benzylic)	JOC *42* 2012 (1977)
ZnCl$_2$ (R = allylic or benzylic)	JOC *52* 3917 (1987)

$$HO(CH_2)_n OH \longrightarrow (H_2C)_n\ O$$

DMSO, Δ (*n* = 4, 5)	JOC *28* 1388 (1963); *29* 123 (1964)
various Brönsted and Lewis acids (*n* = 4)	Tetr *37* 2149 (1981)
ZnCl$_2$ (*n* = 4, 5, 11)	JOC *52* 3917 (1987)
Nafion-H (*n* = 4–7)	Syn 474 (1981)
Amberlyst 15, Δ (*n* = 5)	Syn 208 (1973)
(CH$_3$)$_2$C(OAc)COBr/Amberlite IRA-400, CH$_3$OH (*n* = 2)	TL *24* 367 (1984); *28* 4959 (1987)
n-BuLi/TsCl/*n*-BuLi (*n* = 3)	Syn 550 (1981)
TsOH, Δ (*n* = 4)	Org Syn Coll Vol *4* 534 (1963) JOC *37* 1947 (1972)
Me$_2$SO$_4$ (*n* = 4)	JACS *103* 7398 (1981)
TsCl, py (*n* = 4, 5)	JACS *72* 1593 (1950) TL 2731 (1975) JOC *45* 1828 (1980)
(Me$_2\overset{+}{N}$=CCl$_2$)Cl$^-$/NaOMe or MeLi (*n* = 2)	TL *27* 4697 (1986) (from cis diols)
[(CH$_3$)$_3$CN=C=NC(CH$_3$)$_3$]BF$_4^-$ /Et$_3$N, Δ (*n* = 4–6)	JACS *97* 464 (1975)
EtO$_2$CN=NCO$_2$Et, PPh$_3$ (*n* = 2–6)	BCSJ *49* 510 (1976) TL 5153 (1978) JOC *48* 5396 (1983)
CCl$_4$, PPh$_3$, (K$_2$CO$_3$) (*n* = 2, 4, 5)	JOC *46* 3361 (1981); *47* 3980 (1982); *48* 5396 (1983) TL *24* 661 (1983)
CCl$_4$, P(NMe$_2$)$_3$/(NaOR) (*n* = 2, 3)	TL 4459 (1973); 3459 (1975) Tetr *32* 1283 (1976)
P(OEt)$_5$ (*n* = 4)	JACS *93* 4004 (1971)

$Ph_3P(OEt)_2$ ($n = 2, 4, 5$)

Phosphorus *1* 151 (1971)
JACS *93* 4004 (1971); *107* 5210 (1985)
JOC *48* 5396 (1983); *52* 525 (1987)
  (1,4-oxathianes)
Phosphorus and Sulfur *26* 15 (1986)

$Ph_3P(OCH_2CMe_3)_2$ ($n = 2, 4, 5$)

JOC *51* 5490 (1986)

polymer-$PPh_2(OEt)_2$ ($n = 2, 4, 5$)

JOC *50* 5007 (1985)

$Ph_2S[OC(CF_3)_2Ph]_2$ ($n = 2, 4, 5$)

JACS *96* 4604 (1974)
Helv *61* 822 (1978)

Al_2O_3 ($n = 4\text{--}6$)

BCSJ *53* 3031 (1980)

$$\underset{-\overset{|}{\underset{|}{C}}-\overset{|}{\underset{|}{C}}-}{\overset{\text{HO}\quad\text{O}_3\text{SR}}{}} \xrightarrow{\text{base}} \underset{-\overset{|}{C}-\overset{|}{C}-}{\overset{\text{O}}{\overbrace{}}}$$

R = Me, *p*-Tol

Helv *30* 1929 (1947)
Angew Int *18* 958 (1979)
JACS *102* 7984 (1980); *109* 6205 (1987)
TL *24* 4539 (1983); *28* 2619, 2627, 2863, 6191 (1987)
JOC *50* 5687 (1985); *52* 2378 (1987)

$$\underset{-\overset{|}{\underset{|}{C}}-\overset{|}{\underset{|}{C}}-}{\overset{\text{HO}\quad\text{SR}'}{}} \xrightarrow[\text{2. base}]{\text{1. }(R_3O)BF_4\ (R = Me, Et)} \underset{-\overset{|}{C}-\overset{|}{C}-}{\overset{\text{O}}{\overbrace{}}}$$

JOC *50* 5687 (1985)
TL *27* 6329 (1986); *28* 797, 5677 (1987)

$$ROH + R'CH\!=\!CH_2 \xrightarrow{H^+} R'\overset{\overset{\text{OR}}{|}}{C}HCH_3$$

R = alkyl, aryl

Ind Eng Chem *43* 1596 (1951)
JOC *20* 1232 (1955); *21* 247 (1956)

$$ROH + R'CH\!=\!CH_2 \xrightarrow[\text{2. NaBH}_4, \text{NaOH}]{\text{1. Hg(O}_2\text{CCF}_3)_2} R'\overset{\overset{\text{OR}}{|}}{C}HCH_3$$

TL 5165 (1967)
JACS *91* 5646 (1969); *105* 7407 (1983)
R. C. Larock, "Solvomercuration/Demercuration Reactions in Organic Synthesis," Springer, New York (1986), Chpt 3 (review)

$$R\overset{\overset{\text{O}}{\|}}{C}CH\!=\!CH_2 + HOR' \xrightarrow{H_2SO_4} R\overset{\overset{\text{O}}{\|}}{C}CH_2CH_2OR'$$

JOC *17* 962 (1952)

$$NCCH{=}CH_2 + ROH \xrightarrow{\text{NaOR}} NCCH_2CH_2OR$$

JCS 535 (1945)

$$RCH{=}CH_2 \xrightarrow{\text{BH}_3} \xrightarrow[\text{NaClO}_4,\,\text{NaOCH}_3,\,\text{CH}_3\text{OH}]{\text{electrolysis}} RCH_2CH_2OCH_3$$

CL 1021 (1974)

$$C{=}C{-}C_n{-}OH \longrightarrow X{-}C{-}\overset{\displaystyle O}{\overset{\triangle}{C}}{-}C_n$$

Review:

P. A. Bartlett, "Asymmetric Synthesis," Ed. J. D. Morrison, Academic Press, New York (1984), Vol 3, Part B, Chpt 6

For intermolecular reactions see page 325, Section 4.

X	Reagent(s)	
Br	Br$_2$ NBS	JOC *52* 4191 (1987) CC 264 (1969) JACS *93* 5813 (1971); *101* 260 (1979); *106* 2668 (1984) JOC *52* 1686, 4191 (1987) TL *28* 3065 (1987)
		JACS *106* 2668 (1984)
I	I$_2$ I$_2$, NaHCO$_3$	JOC *51* 2230 (1986) JACS *103* 3963 (1981); *107* 3271 (1985) TL *26* 2885 (1985) JOC *52* 4062, 4191, 4449 (1987)
	I$_2$, KI	Proc Acad Sci USSR, Chem Sec *146* 787 (1962)
	I$_2$, KI, NaHCO$_3$ I$_2$, HgO, hν NIS iodonium dicollidine perchlorate	BCSJ *53* 3383 (1980) TL *28* 4011 (1987) JOC *51* 2230 (1986) JOC *52* 4062 (1987)
OH	*m*-ClC$_6$H$_4$CO$_3$H *t*-BuO$_2$H, cat VO(acac)$_2$/HOAc	TL *28* 5501 (1987) TL 2741 (1978); *28* 5501, 5665 (1987)
MeS	MeSCl, *i*-Pr$_2$NEt (MeSSMe$_2$)BF$_4$, *i*-Pr$_2$NEt (MeSSMe$_2$)SbCl$_6$	TL *28* 523 (1987) TL *26* 6159 (1985) JCS Perkin I 3106 (1981)

$$C{=}C{-}C_n{-}OH \longrightarrow X{-}C{-}\overset{\displaystyle O}{\overset{\triangle}{C{-}C}}C_n \quad (\textit{Continued})$$

X	Reagent(s)	
ArS	n-BuLi/PhSCl	TL *28* 523 (1987)
	PhSCl, i-Pr$_2$NEt	TL *28* 523 (1987)
	PhSN⌒O, CF$_3$SO$_3$H	CC 1280 (1987)
	ArSSCN, LiClO$_4$	JOC USSR *14* 2265 (1978)
	(Ar = Ph, p-Tol)	
PhSe	(PhSe)$_2$, electrolysis	TL *28* 4343 (1987)
	PhSeCl	CC 725 (1977)
		TL 1257 (1977); *21* 129 (1980)
		(allenic alcohols)
		Can J Chem *55* 3894 (1977)
		JACS *102* 3784 (1980); *106* 3353
		(1984); *109* 2082 (1987)
		Tetr *37* 4097 (1981) (review)
		JOC *51* 495 (1986);
		52 4191 (1987)
	PhSeBr	Helv *61* 3075 (1978)
	PhSeOTf	TL *28* 4297, 4415 (1987)

PhSeN (succinimide)   Tetr *37* 4097 (1981) (review)

PhSeN (phthalimide)   Tetr *37* 4097 (1981) (review)
JACS *107* 7792 (1985)
JOC *52* 4191 (1987)

ArTe	(ArTeO)$_2$O/NH$_2$NH$_2$	TL *28* 1281 (1987)
ArTeX$_2$	ArTeX$_3$ (X = Cl, Br)	TL *28* 5611 (1987)
HgX	HgX$_2$	R. C. Larock,
		"Solvomercuration/Demercuration
		Reactions in Organic Synthesis,"
		Springer, New York
		(1986), Chpt 3 (review)
		JOC *52* 4191 (1987)

$$(CH_3)_2C{=}CH(CH_2)_2\overset{\displaystyle OH}{\overset{|}{C}}HR \longrightarrow YC(CH_3)_2\text{—}\underset{O}{\diagup}\text{—}R$$

Y   Reagents

OH   (tetrabromocyclohexadienone) /AgBF$_4$, MeOH   JACS *106* 2668 (1984)

OAc, OH, OCH_3,
ONO$_2$ or NHAc

TlX_3 $(X = O_2CCF_3, NO_3)$;
HOAc, H_2O, CH_3OH,
THF or CH_3CN

JOC *50* 2416 (1985)

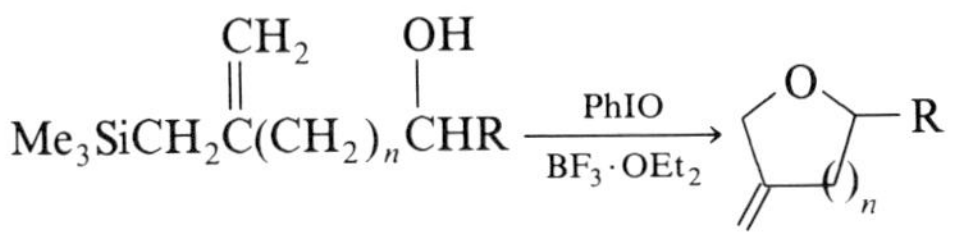

$$Me_3SiCH_2\overset{\overset{\textstyle CH_2}{\|}}{C}(CH_2)_n\overset{\overset{\textstyle OH}{|}}{C}HR \xrightarrow[BF_3 \cdot OEt_2]{PhIO}$$

$n = 1, 2$

CC 1108 (1982)

$$\xrightarrow{(C_5H_5NH)OTs}$$

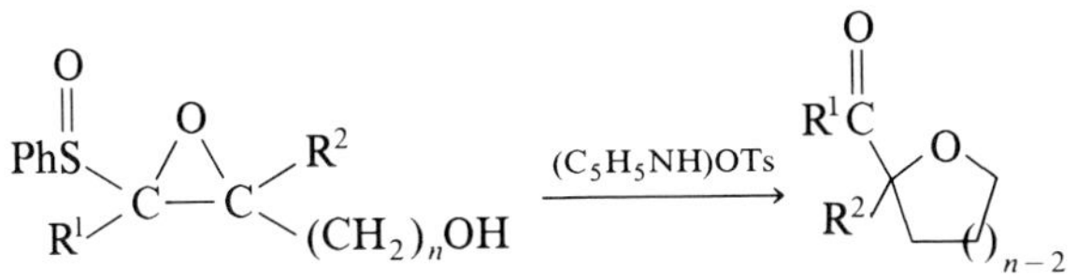

$n = 3, 4, 11$

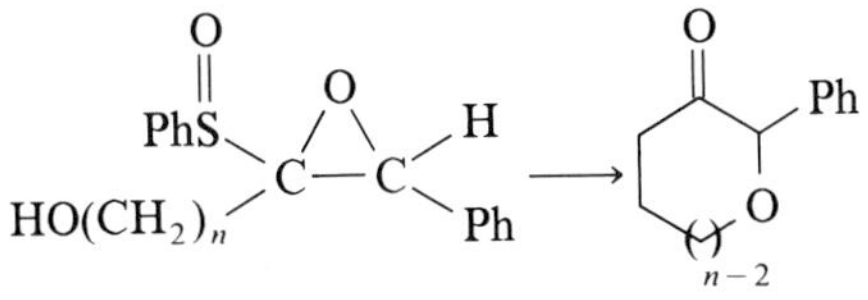

$$\longrightarrow$$

$n = 3, 4$

TL *28* 2603 (1987)

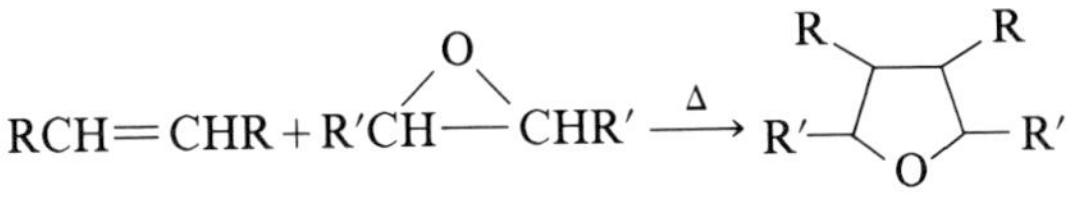

$$RCH{=}CHR + R'CH{-\!\!\!-}CHR' \xrightarrow{\Delta}$$

JACS *87* 3657, 3665 (1965)
CC 1190, 1192 (1971); 134 (1984)
JOC *41* 2654 (1976); *43* 4256 (1978)
Angew Int *16* 572 (1977) (review)
TL *21* 4909 (1980) (intramolecular); *23* 4665 (1982) (intramolecular); *25* 1137 (1984) (intramolecular);
    28 3155 (1987)
Ber *117* 2157 (1984)

3. ALKENE AND ALKYNE ADDITIONS AND SUBSTITUTIONS

1. Alkene Substitution

$$R_2C=CHCH_2R \xrightarrow[\substack{MgSO_4,\ MeOH \\ \text{electrolysis}}]{(PhSe)_2} R_2\overset{\overset{\displaystyle OMe}{|}}{C}CH=CHR$$

JACS *103* 4606 (1981)

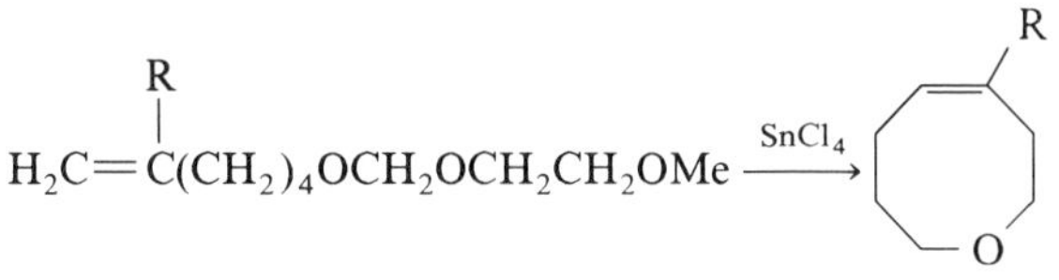

$$H_2C=\overset{\overset{\displaystyle R}{|}}{C}(CH_2)_4OCH_2OCH_2CH_2OMe \xrightarrow{SnCl_4}$$

R = H, SiMe₃

JACS *108* 3516 (1986)

2. Alkene Addition

See also page 445, Section 2.

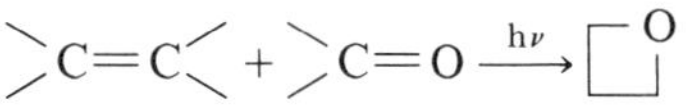

$$\text{\Large$>$}C=C\text{\Large$<$} + \text{\Large$>$}C=O \xrightarrow{h\nu} \square\!=\!O$$

Adv Photochem *6* 301 (1968) (review)
Org Photochem *5* 1 (1981) (review)
TL *24* 3217 (1983); *28* 5933, 6151 (1987) (both intramolecular)
H. A. J. Carless, "Synthesis Organic Photochemistry," Ed. W. M. Horspool, Plenum Press, New York (1984), p 425 (review)
JACS *106* 4186, 7200 (1984)
CC 667 (1984)
Ber *118* 1485 (1985); *120* 307 (1987)
Science *227* 857 (1985)
Angew Int *24* 877 (1985)

$$\underset{\overset{|}{\text{RCH}=\text{CR(CH}_2)_2\text{CHR}}}{\overset{\text{OCH}_2\text{Ar}}{}} \quad \xrightarrow[\text{CH}_3\text{CN}]{\text{I}_2} \quad$$

JACS *103* 3963 (1981)
TL *27* 2195 (1986)
JOC *52* 320, 5067 (1987)

$$\xrightarrow[\text{NaHCO}_3]{\text{NBS}}$$

CC 1462 (1987)

3. Epoxidation of Alkenes

$$>\text{C}=\text{C}< \longrightarrow >\overset{\overset{\text{O}}{\diagup\diagdown}}{\text{C}-\text{C}}<$$

enzymes	JOC *46* 3129 (1981)
RCO$_3$H	Org Rxs *7* 378 (1953) (review)
m-ClC$_6$H$_4$CO$_3$H	TL 849 (1965); 4347, 4733 (1979)
	JOC *32* 1363 (1967); *35* 251 (1970); *36* 3832 (1970); *38* 1380 (allylic alcohols), 1385 (1973); *50* 2179 (1985); *51* 2505 (1986); *52* 4495, 4511 (1987)
	JCS C 731 (1970)
	JACS *92* 6914 (1970); *96* 5254 (1974); *106* 7854 (1984); *109* 5437, 7838 (1987)
	CC 421 (1976)
	JCS Perkin I 2885 (1982)
m-ClC$_6$H$_4$CO$_3$H, NaHCO$_3$	JOC *38* 2267 (1973)
	TL 427 (1977)
m-ClC$_6$H$_4$CO$_3$H, Na$_2$CO$_3$	JOC *52* 4647 (1987)
m-ClC$_6$H$_4$CO$_3$H, K$_2$CO$_3$ or K$_2$HPO$_4$	JOC *43* 610 (1978)
m-ClC$_6$H$_4$CO$_3$H, phosphate	JOC *43* 610 (1978); *44* 1351 (1979)
m-ClC$_6$H$_4$CO$_3$H, KF	TL *22* 3895 (1981)
3,5-(NO$_2$)$_2$C$_6$H$_3$CO$_3$H	JOC *43* 3163 (1978)
	JACS *106* 7854 (1984)
	TL *28* 2771 (1987)
3,5-(NO$_2$)$_2$C$_6$H$_3$CO$_3$H, Na$_2$HPO$_4$	JACS *109* 5167, 5280 (1987)
o-HO$_2$CC$_6$H$_4$CO$_3$H	JACS *82* 6373 (1960)
CH$_3$CO$_3$H, NaOAc	JCS Perkin I 2909 (1982)

CF_3CO_3H | JACS *106* 7854 (1984)

$PhCH_2OCO_3H$ | JOC *39* 3054 (1974)

CH_3 $OSiMe_2(t\text{-}Bu)$

, cat $Cu(O_2CCF_3)_2$ | TL *28* 1909 (1987)

OCO_2Ph

H_2O_2, $ArSeO_2H$ | JOC *42* 2035 (1977)
Syn 299 (1978)

H_2O_2, cat $[C_5H_5N(CH_2)_{15}CH_3]_3PMo_{12}O_{40}$ | Syn Commun *14* 865 (1984)
(allylic alcohols)

H_2O_2, $[C_5H_5N(CH_2)_{15}CH_3]_3PMo_{12}O_{40}$ | JOC *52* 1868 (1987)

H_2O_2, cat tungstic acid, Et_3N | JACS *87* 734 (1965)
(allylic alcohols)

H_2O_2, cat tungstic acid, NaOAc or Me_3NO | TL *27* 707, 711 (1986)

H_2O_2, cat $Na_2WO_4 \cdot 2 H_2O$, H_3PO_4, H_2SO_4, | JOC *48* 3831 (1983)
phase transfer catalyst

H_2O_2; $[(n\text{-}C_8H_{17})_3NCH_3]HWO_4$ or | TL *28* 2237 (1987)
$[(n\text{-}C_8H_{17})_3NCH_3]_2WO_4$; $(n\text{-}C_8H_{17})_3PO$,
$PhPO_3H_2$ or $C_{12}H_{25}PO_3H_2$

H_2O_2, cat $(PhCH_2PPh_3)_2W_2O_{11}$ | TL *27* 2617 (1986)

H_2O_2, cat $(n\text{-}Bu_4N)_2W_2O_{11}$ | TL *27* 2617 (1986)

H_2O_2, cat chloromanganese(tetra-2,6- | CC 888 (1985)
dichlorophenylporphyrin), imidazole

H_2O_2, $Fe(acac)_3$ | TL 948 (1978)

H_2O_2, $Cl_2CHCOCHCl_2$, Na_2HPO_4 | TL *22* 2089 (1981)

H_2O_2, CF_3COCF_3 | JACS *101* 2484 (1979)

H_2O_2, RCN (R = Me, Ph) | JOC *26* 659 (1961); *32* 1363 (1967); *36* 3832
(1971); *50* 2179 (1985)
Tetr *18* 763 (1962)
JCS C 731 (1970)
Org Syn *60* 63 (1981)
JACS *106* 3539 (1984)

H_2O_2, $(EtO)_2POCN$, [pyridine with OH] | Chem Pharm Bull *29* 1774 (1981)

90% H_2O_2, $CH_3C(OEt)_3$ | TL 1001 (1979)

H_2O_2, [imidazole–CO–imidazole structure] | CC 711 (1974)
JOC *43* 180 (1978); *44* 1485 (1979)

H_2O_2, EtO_2CCl, Na_2HPO_4 or $Na_3PO_4 \cdot 12 H_2O$ | JOC *44* 2569 (1979)

H_2O_2, ArNCO (Ar = Ph, p-ClC$_6$H$_4$)	TL 2029 (1970)
H_2O_2, PhCONCO	CC 711 (1974)
	JOC 43 180 (1978); 44 1485 (1979)
t-BuO$_2$H, Al(O-t-Bu)$_3$ (allylic alcohols)	TL 27 3387 (1986)
t-BuO$_2$H, n-Bu$_2$SnO (allylic alcohols)	TL 27 3387 (1986)
t-BuO$_2$H, cat Mo(CO)$_6$ (allylic and homoallylic alcohols)	JACS 95 6136 (1973); 96 5254 (1974)
	TL 2741 (1978); 4733 (1979)
	JOC 51 2505 (1986)
t-BuO$_2$H, cat VO(acac)$_2$ (allylic and homoallylic alcohols)	JACS 95 6136 (1973); 96 5254 (1974);
	103 7690 (1981); 107 256 (1985)
	CC 421 (1976)
	TL 2741 (1978); 4733 (1979); 27 3353, 3387, 6035, 6071 (1986); 28 1439, 2099, 6191 (1987)
	JOC 51 4728 (1986); 52 34, 4495, 4511, 4898 (1987)
PhCMc$_2$O$_2$H, cat VO(acac)$_2$, 2,6-lutidine (allylic alcohol)	JACS 109 6187 (1987)
t-BuO$_2$H, cat VOSO$_4$ or MoO$_2$(acac)$_2$, (+)-3-trifluoroacetylcamphor	Z Chem 18 218 (1978)
t-BuO$_2$H, Ti(O-i-Pr)$_4$ (allylic alcohols)	TL 27 3387 (1986)
t-BuO$_2$H, Ti(O-i-Pr)$_4$, dialkyl tartrate (allylic alcohols, Sharpless, enantioselective)	JACS 102 5974 (1980); 103 464, 6237 (1981); 107 1691, 7515, 7967 (1985); 108 2776 (1986); 109 1525, 2205, 4718 (1987)
	Tetr 37 3873 (1981)
	Pure Appl Chem 55 589 (1983)
	Org Syn 63 66 (1984)
	JOC 49 1707 (1984); 50 3752 (1985); 51 934, 1077, 4726, 4728 (1986); 52 685, 940, 1106 (1987)
	"Asymmetric Synthesis," Ed. J. D. Morrison, Academic Press, New York (1985), Vol 5, Chpts 7, 8 (reviews)
	CC 1759 (1985); 1237, 1732 (1986)
	TL 27 3535, 4913 (1986); 28 375, 1139, 2033, 3075, 4019, 4985, 6351 (1987)
	Syn 89 (1986) (review)
	Angew Int 25 87 (1986)
	CL 1523 (1987)
t-BuO$_2$H, Ti(O-i-Pr)$_4$, diethyl tartrate, cat CaH$_2$, cat silica gel (enantioselective)	TL 26 6221 (1985)
t-BuO$_2$H, cat Ti(O-i-Pr)$_4$, cat dialkyl tartrate, molecular sieves (allylic alcohols, enantioselective)	TL 27 5791 (1986); 28 131, 5129, 5205 (1987)
	JOC 51 1922 (1986); 52 940, 2596 (1987)
	JACS 109 5765, 8120 (1987)
PhCMe$_2$O$_2$H, cat Ti(O-i-Pr)$_4$, cat diisopropyl tartrate, molecular sieves (allylic alcohols, enantioselective)	JOC 51 3710, 5413 (1986); 52 4973 (1987)
	JACS 109 5765, 8120 (1987)

t-BuO$_2$H, Zr(O-i-Pr)$_4$, dicyclohexyltartramide (homoallylic alcohols, enantioselective)	CL 85 (1987)
Ph$_3$SiO$_2$H	TL 4337 (1979)

Ph, Ph, Br, N=N, O$_2$H — JOC *47* 1141 (1982)

$$\overset{O}{\overset{/\ \backslash}{RSO_2N---CHAr}}\ \text{(enantioselective)}$$
JACS *105* 3123 (1983)
TL *27* 5079 (1986)

O$_2$, CH$_3$CHO	Syn 711 (1977)
O$_2$, PhCOCHOHPh, hν	JACS *103* 2049 (1981)
O$_2$, cat [Fe$_3$O(O$_2$CR)$_6$L$_3$]$^+$	JACS *104* 6450 (1982)
O$_2$, cat Mn porphyrin	JACS *106* 6871 (1984)
MO$_5 \cdot$HMPT (M = Mo, W)	JOC *51* 2374 (1986) TL *28* 6191 (1987) (homoallylic i-Pr$_3$Si ethers)
PhIO, Co catalyst	JOC *52* 4545 (1987)
NBS, H$_2$O/base	TL 121 (1962) JACS *85* 3295 (1963); *90* 5618 (1969) JOC *38* 1385 (1973); *51* 5447 (1986); *52* 4505 (1987) JCS Perkin I 2909 (1982)
NIS, H$_2$O/DBU	JOC *51* 2505 (1986)
CH$_3$CONHBr, H$_2$O/NaOH	JOC *51* 3407 (1986)
CH$_3$CONHBr, H$_2$O/K$_2$CO$_3$	JOC *52* 2860 (1987)
1,3-dibromo-5,5-dimethylhydantoin, H$_2$O/NaOH	TL *27* 4403 (1986)
I$_2$, H$_2$O, HI/NaOH	JCS C 928 (1970)
I$_2$, Ag$_2$O, H$_2$O	TL 207 (1976) JOC *51* 3023 (1986)
NaOCl, cat Mn porphyrin acetate, R$_4$NCl, py	JACS *106* 6668 (1984)
KHSO$_5$, acetone, 18-crown-6	JOC *45* 4758 (1980)
KHSO$_5$, CH$_3$COCH$_3$, Na$_2$(EDTA), 18-crown-6, CH$_2$Cl$_2$, H$_2$O	JOC *47* 2670 (1982)

H$_2$O, NaBr, electrolysis

TL 4661 (1979)
JOC *46* 3312 (1981); *50* 3160 (1985)

H$_2$O, [polymer-$\overset{+}{N}R_3$]Br$^-$, electrolysis

JOC *47* 3575 (1982)

$$C{=}C{-}C_n{-}\underset{\underset{}{|}}{\overset{\overset{OH}{|}}{C}} \xrightarrow[\substack{\text{2. CO}_2\\ \text{3. I}_2}]{\text{1. }n\text{-BuLi}} \underset{I}{\overset{}{C}}{-}\underset{\underset{O}{|}}{\overset{\overset{O}{|}}{C}}{-}C_n \longrightarrow C{-}C{-}C_n{-}\overset{OH}{\underset{|}{C}}$$

$$n = 0, 1$$

CC 465 (1981)
JOC *47* 4626 (1982); *52* 3560 (1987)

$$C{=}C{-}C{-}\underset{\underset{R}{|}}{\overset{\overset{O}{\|}}{C}}{-}O{-}\overset{\overset{O}{\|}}{C}{-}O{-}t\text{-Bu} \xrightarrow{I_2} I \underset{R}{} \xrightarrow[\text{MeOH}]{K_2CO_3} C{-}C{-}C{-}\overset{X}{\underset{|}{C}}{-}R$$

$$X = OH \text{ or } OCO_2Me$$

JOC *47* 4013 (1982)

$$C{=}C{-}C{-}\overset{\overset{OPO(OEt)_2}{|}}{\underset{|}{C}} \xrightarrow[\text{2. NaOEt}]{\text{1. I}_2} C{-}C{-}C{-}\overset{OPO(OEt)_2}{\underset{|}{C}}$$

JACS *99* 4829 (1977)

$$\underset{\underset{|}{}}{>}C{=}C{-}\overset{\overset{O}{\|}}{C}{-} \longrightarrow {>}C{-}C{-}\overset{\overset{O}{\|}}{C}{-}$$

H$_2$O$_2$, NaHCO$_3$

JACS *103* 3460 (1981)
JOC *52* 4647 (1987)

H$_2$O$_2$, K$_2$CO$_3$

JOC *52* 3560 (1987)

H$_2$O$_2$, LiOH

TL *28* 2099 (1987)

H$_2$O$_2$, NaOH

Ber *54* 2327 (1921)
Helv *25* 836 (1942)
JACS *79* 1488 (1957); *105* 2435 (1983);
  109 4690 (1987)
Org Syn Coll Vol *4* 552 (1963)
Tetr *19* 1091 (1963)
JOC *50* 2981, 3957 (1985); *51* 3098 (1986)

H$_2$O$_2$, NaOH, poly-L or D-alanine
  (enantioselective)

TL *28* 4857 (1987)

H$_2$O$_2$, cat Na$_2$WO$_4\cdot$2 H$_2$O, H$_2$O

JOC *24* 54 (1959); *50* 1979 (1985)

H_2O_2 or t-BuO_2H, NaOH, R$_4$NCl (enantioselective)	TL 1831 (1976) Chimia *30* 445 (1976) CC 427 (1978) JOC *45* 2498 (1980)
H_2O_2 or t-BuO_2H, n-Bu_4F	CL 285 (1987)
t-BuO_2H, cat bovine serum albumin, buffer (enantioselective)	TL *28* 1577 (1987)
t-BuO_2H, NaOH	JACS *106* 4558 (1984)
t-BuO_2H, Triton-B	JACS *80* 5845 (1958); *99* 5773 (1977)
t-BuO_2H, DBU	JACS *107* 1777 (1987)
t-BuO_2H, RLi	CC 1378 (1986)
t-BuO_2H, KH	JACS *101* 2493 (1979); *106* 4038 (1984) TL *27* 1343, 6189 (1986) JOC *51* 3393 (1986)
NaOCl, py	JOC *28* 250 (1963)
NaOCl, (n-Bu_4N)HSO$_4$	Gazz Chim Ital *110* 267 (1980)
KHSO$_5$, CH_3COCH_3, NaHCO$_3$	JOC *51* 1925 (1986) (α,β-unsaturated acids)
KHSO$_5$, CH_3COCH_3, (18-crown-6)	JOC *45* 4758 (1980) (α,β-unsaturated acids)

4. Diene Addition

X	Reagents	
PhSe	PhSeCl, H_2O	JOC *44* 1742 (1979) Tetr *41* 5301 (1985)
HgX	HgX$_2$, H_2O	R. C. Larock, "Solvomercuration/ Demercuration Reactions in Organic Synthesis", Springer, New York (1986), Chpt 2 (review)

Tetr *21* 2353 (1965)
JACS *101* 4396 (1979)
TL *21* 3531 (1980); *28* 731 (1987)

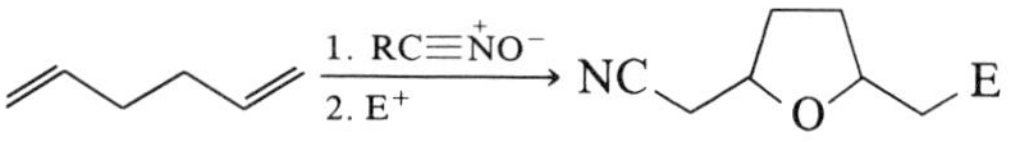

$$\text{(pentadienyl)} \xrightarrow[\text{2. E}^+]{\text{1. RC} \equiv \overset{+}{\text{N}}\text{O}^-} \text{NC} \diagdown\!\!\diagup\!\!\bigcirc\!\!\diagdown \text{E}$$

$$R = Ph_3C, Me_3C, Me_3Si, R'CH(OSiMe_3) ; E^+ = Br_2, I_2, PhSeBr$$

JACS *109* 7577 (1987)

5. Alkyne Addition

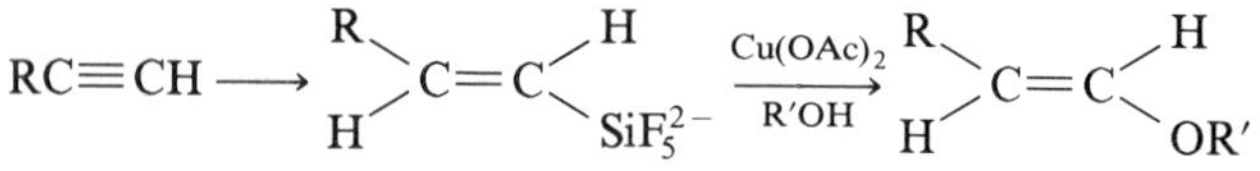

$$RC \equiv CH \longrightarrow \underset{H}{\overset{R}{\diagdown}} C = C \underset{SiF_5^{2-}}{\overset{H}{\diagup}} \xrightarrow[\text{R'OH}]{\text{Cu(OAc)}_2} \underset{H}{\overset{R}{\diagdown}} C = C \underset{OR'}{\overset{H}{\diagup}}$$

Organomet *1* 369 (1982)

$$HC \equiv C - C_3 - OH \longrightarrow \bigcirc\!\!\!O$$

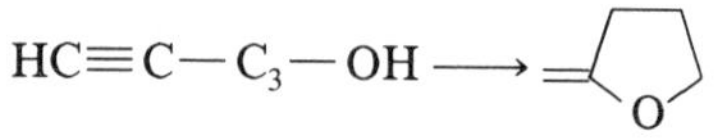

Ag$_2$CO$_3$

TL *28* 6447 (1987)

HgCl$_2$, Et$_3$N

JACS *104* 5842 (1982)

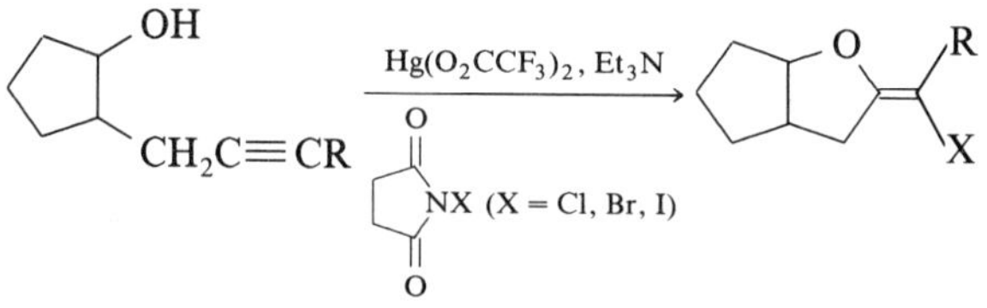

$$\xrightarrow[\text{O}]{\text{Hg(O}_2\text{CCF}_3)_2, \text{Et}_3\text{N}}$$

NX (X = Cl, Br, I)

JACS *104* 5842 (1982)

4. ACETAL, CARBONYL AND ORTHO ESTER CONVERSIONS

1. Acetals

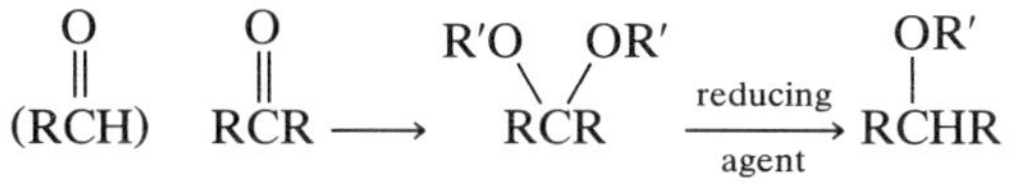

$$\underset{\text{(RCH)}}{\overset{O}{\overset{\|}{\text{}}}} \quad \underset{\text{RCR}}{\overset{O}{\overset{\|}{\text{}}}} \longrightarrow \underset{\text{RCR}}{\overset{R'O \quad OR'}{\overset{\diagdown \diagup}{\text{}}}} \xrightarrow[\text{agent}]{\text{reducing}} \underset{\text{RCHR}}{\overset{OR'}{\overset{|}{\text{}}}}$$

Reducing agent

H$_2$, cat Rh-alumina, H$^+$	JOC *26* 1026 (1961)
H$_2$, cat Ni	JACS *54* 1651 (1932)
H$_2$, cat PtO$_2$, H$^+$	JCS 5598 (1963)
H$_2$, CO, cat Co$_2$(CO)$_8$ (R = aryl)	Can J Chem *54* 685 (1976)
BH$_3$	Curr Sci 404 (1963) Can J Chem *52* 888 (1974)
ClBH$_2 \cdot$OEt$_2$	CL 9 (1984)
Cl$_2$BH	JCS Perkin I 1807 (1981)

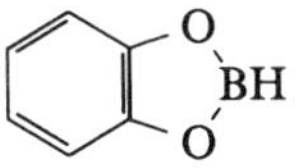

BH

JOC *42* 512 (1977)

NaBH$_4$, CF$_3$CO$_2$H (R = aryl)	Org Prep Proc Int *17* 11 (1985)
NaBH$_3$CN, HCl, CH$_3$OH	TL 1357 (1978)
Zn(BH$_4$)$_2$, Me$_3$SiCl	JOC *52* 2594 (1987)
i-Bu$_2$AlH	Izv Akad Nauk SSSR, Ser Khim 2255 (1959) [CA *54* 10837h (1960)] Ber *113* 3697 (1980) TL *24* 4581 (1983) (chiral); *27* 983 (1986) (chiral) JOMC *285* 83 (1985) (chiral)

$$
\overset{O}{\underset{RCR}{\|}} \longrightarrow \overset{R'O\quad OR'}{\underset{RCR}{\diagdown\diagup}} \xrightarrow[\text{agent}]{\text{reducing}} \overset{OR'}{\underset{RCHR}{|}} \quad (\textit{Continued})
$$

Reducing agent

i-Bu$_3$Al	TL *28* 4181 (1987) (chiral)
Cl$_2$AlH	CC 334 (1987) (chiral)
Br$_2$AlH	Tl *24* 4581 (1983); *27* 983, 987 (1986) (all chiral)
	JOMC *285* 83 (1985) (chiral)
	CC 334 (1987) (chiral)
LiAlH$_4$, AlCl$_3$	JOC *23* 1088 (1958); *26* 4553 (1961);
	27 67 (1962)
	JACS *81* 6087 (1959); *84* 2371 (1962);
	107 3891 (1985)
	Org Syn Coll Vol 5 303 (1973)
LiAlH$_4$, TiCl$_4$ (R = alkyl)	BCSJ *51* 2059 (1978)
LiAlH$_4$/BF$_3\cdot$OEt$_2$	JOC *27* 67 (1962)
Et$_3$SiH, CF$_3$CO$_2$H	JOC USSR *8* 902 (1972)
Et$_3$SiH, Nalfion-H	JOC *51* 2826 (1986)
Et$_3$SiH, BF$_3\cdot$OEt$_2$	Carbohydr Res *128* C9 (1984)
Et$_3$SiH, CF$_3$CO$_2$H, BF$_3\cdot$OEt$_2$	JACS *104* 3539 (1982); *109* 8117 (1987)
R$_3$SiH (R = Me, Et),	TL 4679 (1979)
cat Me$_3$SiOTf	JOC *52* 892 (1987)
R$_3$SiH (R = Me, Et, *n*-Pr), cat	Compt Rend *254* 1814 (1962)
ZnCl$_2$	
Et$_3$SiH, TiCl$_4$	TL *27* 987 (1986) (chiral)
	JACS *109* 527 (1987) (chiral)

$$
\overset{O}{\underset{(RCR)}{\|}}\ \overset{O}{\underset{RCH}{\|}} \longrightarrow \overset{R''O\quad OR''}{\underset{RCH}{\diagdown\diagup}} \xrightarrow[]{\text{reagent(s)}} \overset{OR''}{\underset{RCHR'}{|}}
$$

Reagent(s)

R'Li or R'$_2$CuLi(MgX), TiCl$_4$	TL *25* 3947 (1984) (chiral)
R'MgX	JACS *73* 4893 (1951)
	Rec Trav Chim *81* 238 (1962)
	Organomet *1* 1670 (1982)
R'MgX, TiCl$_4$	BCSJ *54* 776 (1981)
	TL *25* 3947 (1984) (chiral)
H$_2$C=CHCH$_2$M (M = B, SnR$_3$),	JACS *108* 7116 (1986) (chiral)
Lewis acid?	

$H_2C\!=\!CHCH_2B$⟩), $TiCl_4$	CC 1218 (1987) (chiral)
$RCH\!=\!CRCH_2Al_{2/3}Br$	JOMC *170* 1 (1979)
$RCH\!=\!CRCHRSiMe_3$,	
$\quad BF_3\!\cdot\!OEt_2$	CL 941 (1976) TL 2589 (1978); *21* 951 (1980) JOC *48* 335 (1983)
$\quad AlCl_3$	CL 575 (1978)
$\quad Me_3SiI$	TL *22* 745 (1981)
$\quad Me_3SiOTf$	TL *21* 71, 2527 (1980); *23* 723 (1982) JOC *52* 892 (1987)
$\quad SnCl_4$	JCS Perkin I 251 (1981) (intramolecular) JACS *104* 7371 (1982) (chiral)
$\quad TiCl_4$	CL 941 (1976); 575 (1978) TL 2589 (1978); 429 (1979); *21* 951, 2049 (1980); $\quad$ *23* 725 (1982); *28* 6343 (1987) JACS *105* 2088 (1983) (chiral); *108* 7116 (1986) $\quad$ (chiral) CC 1218 (1987) (chiral)
$\quad TiCl_4$, $Ti(O\text{-}i\text{-}Pr)_4$	TL *25* 3951 (1984) (chiral) JOC *50* 2598 (1985) (chiral)
$R'CH\!=\!CHCH\!=\!CHCH_2SiMe_3$; $\quad TiCl_4$, $BF_3\!\cdot\!OEt_2$ or cat Me_3SiI	TL *21* 2049, 3783 (1980); *22* 745 (1981) Organomet *1* 1651 (1982)
$\qquad\quad CH_2$ $\qquad\quad \|\|$ $H_2C\!=\!CHCCH_2SiMe_3$, $TiCl_4$ or Me_3SiI	Tetr *39* 883 (1983)
$RC\!\equiv\!CSiMe_3$, $TiCl_4$	JACS *105* 2904 (1983) (chiral)
$RC\!\equiv\!CM$ ($M = Me_3Si$, $n\text{-}Bu_3Sn$), $\quad$ Lewis acid?	JACS *108* 7116 (1986) (chiral)
$n\text{-}Bu_3SnC\!\equiv\!CR$, $TiCl_4$	TL *28* 4589 (1987)
Me_3SiCN, $SnCl_2$ or $\quad BF_3\!\cdot\!OEt_2$ ($R' = CN$)	TL *22* 4279 (1981) Tetr *39* 967 (1983)
Me_3SiCN, $TiCl_4$ ($R' = CN$)	JOC *48* 2294 (1983) (chiral) TL *25* 591 (1984) (chiral)
$H_2C\!=\!CHCH_2SnMe_3$, $(Et_2Al)_2SO_4$	CL 977 (1979)
$H_2C\!=\!CHCH_2Sn(n\text{-}Bu)_3$, $TiCl_4$	CC 1218 (1987) (chiral)
$RTiCl_3$	JOMC *285* 83 (1985)
RCu or R_2CuLi, $BF_3\!\cdot\!OEt_2$	TL *25* 3083 (1984) (chiral)

For reactions with enol silanes and Lewis acids see page 750, Section 7.2.

$$\text{RCH}\!\!=\!\!\text{CHCH(OEt)}_2 + \text{Zn(CH}_2\text{CH}_2\text{CO}_2\text{-}i\text{-Pr})_2 \xrightarrow[\substack{\text{Me}_3\text{SiCl or} \\ \text{BF}_3\cdot\text{OEt}_2}]{\text{cat CuBr}\cdot\text{SMe}_2} \overset{\displaystyle \overset{\text{OEt}}{|}}{\text{RCH}\!\!=\!\!\text{CHCHCH}_2\text{CH}_2\text{CO}_2\text{-}i\text{-Pr}}$$

JACS *109* 8056 (1987)

$$\text{PhCH(OMe)}_2 + \underset{\text{OEt}}{\overset{\text{OSiMe}_3}{\triangleright\!\!<}} \xrightarrow{\text{cat ZnI}_2} \overset{\displaystyle\overset{\text{OMe}}{|}}{\text{PhCHCH}_2\text{CH}_2\text{CO}_2\text{Et}}$$

JACS *109* 8056 (1987)

$$\overset{\displaystyle\overset{\text{OR}'}{|}}{\text{RCHOCH}_2\text{CH}_2\text{OMe}} + \text{H}_2\text{C}\!\!=\!\!\text{CHCH}_2\text{SiMe}_3 \xrightarrow{\text{TiCl}_4} \overset{\displaystyle\overset{\text{OR}'}{|}}{\text{RCHCH}_2\text{CH}\!\!=\!\!\text{CH}_2}$$

JOC *47* 2496 (1982)

$$\overset{\displaystyle\overset{\text{OR}}{|}}{\text{RCHSR}} + \text{R}_2\text{C}\!\!=\!\!\text{CHCH}_2\text{MMe}_3 \xrightarrow[\text{acid}]{\text{Lewis}} \overset{\displaystyle\overset{\text{X}}{|}}{\text{RCHCR}_2\text{CH}\!\!=\!\!\text{CH}_2}$$

X = OR or SR

M	
Si	CC 459 (1982)
Sn	TL *28* 6299 (1987)

$$\overset{\displaystyle\overset{\text{OCH}_3}{|}}{\text{RCHSPh}} + n\text{-Bu}_3\text{SnCH}\!\!=\!\!\text{C}\!\!=\!\!\text{CH}_2 \xrightarrow{} \overset{\displaystyle\overset{\text{X}}{|}}{\text{RCHCH}_2\text{C}\!\!\equiv\!\!\text{CH}}$$

BF$_3\cdot$OEt$_2$ X = OCH$_3$

TiCl$_4$ X = SPh

TL *28* 6299 (1987)

$$\text{CH}_3\text{OCH}_2\text{CH}_2\text{OCH}_2\text{O(CH}_2)_n\overset{\displaystyle\overset{\text{SiMe}_3}{|}}{\text{C}}\!\!=\!\!\text{CR}_2 \xrightarrow{\text{SnCl}_4} \underset{\text{CR}_2}{\overset{\text{O}}{\Big(\,\Big)_n}}$$

n = 2–4

JACS *108* 1303 (1986)

$$\text{CH}_3\text{OCH}_2\text{CH}_2\text{OCH}_2\text{O(CH}_2)_{n+1}\overset{\displaystyle\overset{\text{R}^1}{|}}{\text{C}}\!\!=\!\!\text{CHR}^2 \xrightarrow{\text{SnCl}_4} \text{(pyran)}_n$$

R^1 = H, SiMe$_3$; R^2 = H, Me, *n*-Bu; *n* = 3, 4

JACS *108* 3516 (1986)

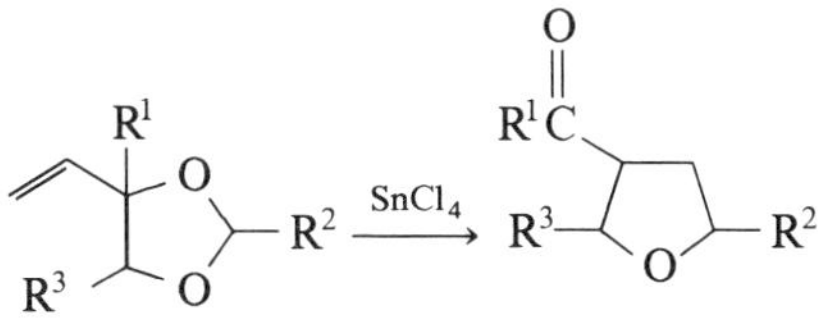

JACS *109* 4748 (1987)

$$R^1CH=CR^2OCR_2^3OR^4 \xrightarrow{\ i\text{-Bu}_3\text{Al}\ } R^2\overset{\underset{\textstyle OH}{|}}{C}HCHR^1CR_2^3OR^4$$

JOC *52* 5700 (1987)

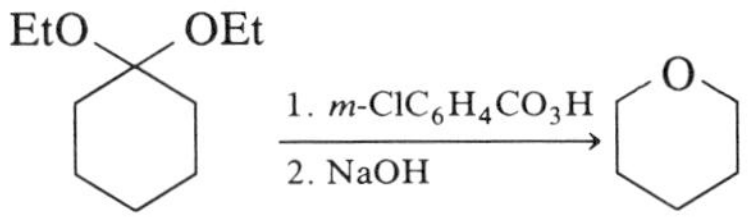

JOC *50* 3009 (1985)

2. Aldehydes and Ketones

$$\underset{\text{O}}{ArCCH_2R} \longrightarrow \underset{\text{OSiMe}_3}{ArC=CHR} \xrightarrow[\substack{BF_3\cdot OEt_2 \\ MeOH}]{(PhIO)_n} \underset{\text{O \quad OMe}}{ArC-CHR}$$

JOC *52* 150 (1987)

$$2\ RCHO \longrightarrow (RCH_2)_2O$$

$BH_3 \cdot C_5H_5N$, CF_3CO_2H	CL 415 (1979)
	Chem Pharm Bull *27* 2405 (1979)
Et_3SiH, CF_3CO_2H	Proc Acad Sci USSR, Chem Sec *179* 328 (1968)
	JOC *39* 2740 (1974)
Et_3SiH, $BF_3 \cdot OEt_2$	JOMC *117* 129 (1976)
Et_3SiH, cat Ph_3CClO_4	CL 743 (1985)
Et_3SiH, cat Me_3SiOTf	JOC *52* 4314 (1987)

$$2\ R_2CO \longrightarrow (R_2CH)_2O$$

n-$BuSiH_3$, CF_3CO_2H	JOC *39* 2740 (1974)
Et_3SiH, cat Me_3SiOTf	JOC *52* 4314 (1987)

$$RCHO + R'OH \longrightarrow RCH_2OR'$$

$BH_3 \cdot C_5H_5N$, CF_3CO_2H	Chem Pharm Bull *27* 2405 (1979)

Et$_3$SiH, CF$_3$CO$_2$H   Proc Acad Sci USSR, Chem Sec *179* 328 (1968)
                                                JACS *94* 3659 (1972)
                                                JOC USSR *8* 902 (1972)

Et$_3$SiH, H$_2$SO$_4$ JACS *94* 3659 (1972)

$$R_2CO + R'OH \xrightarrow[CF_3CO_2H]{Et_3SiH} R_2CHOR'$$

JACS *94* 3659 (1972)

$$RCHO\ (R_2CO) \longrightarrow RCH_2OR'\ (R_2CHOR')$$

HC(OR')$_3$, Nalfion-H/Et$_3$SiH,   JOC *51* 2826 (1986)
  Nalfion-H

Me$_3$SiOR', cat Ph$_3$CClO$_4$/Et$_3$SiH   CL 743 (1985)

Me$_3$SiOR', cat Me$_3$SiI/Me$_3$SiH   JOC *52* 4314 (1987)

$$RCHO \xrightarrow[2.\ R'X]{1.\ R_3SnLi} \overset{\overset{\displaystyle OR'}{|}}{R}CHSnR_3 \xrightarrow[2.\ E^+]{1.\ n\text{-BuLi}} \overset{\overset{\displaystyle OR'}{|}}{R}CHE$$

$$E^+ = RX,\ Me_2SO_4,\ RCHO,\ R_2CO$$

JACS *100* 1481 (1978); *106* 3376 (1984)

$$RCHO \xrightarrow[2.\ H_2C=CHCH_2SiMe_3]{1.\ Cl_2Ti(OR')_2} \overset{\overset{\displaystyle OR'}{|}}{R}CHCH_2CH=CH_2$$

Angew Int *24* 765 (1985)

$$\overset{\overset{\displaystyle O}{\|}}{R}CR \xrightarrow[2.\ MeI]{1.\ R'C\equiv CLi} \overset{\overset{\displaystyle OMe}{|}}{R}_2CC\equiv CR'$$

Syn 459 (1981)

$$(R\overset{\overset{\displaystyle O}{\|}}{C}H)\quad R\overset{\overset{\displaystyle O}{\|}}{C}R \longrightarrow R_2C\!\!\overset{O}{\overset{\diagdown\diagup}{-}}\!\!CR'R''$$

Review: Tetr *36* 2531 (1980)

(CH$_3$)$_2\overset{+}{S}\overset{-}{C}H_2$ JACS *84* 3782 (1962); *87* 1353 (1965); *106* 723,
                                                  4038 (1984); *109* 3353 (1987)
                                                TL 661 (1962); 169 (1963); *23* 5283 (1982);
                                                  28 1877 (1987)
                                                Ber *96* 1881 (1963)
                                                JCS C 731 (1970)
                                                Angew Int *12* 845 (1973)
                                                Cancer Lett *1* 339 (1976)
                                                Heterocycles *8* 397 (1977)
                                                JOC *43* 3425 (1979); *46* 2731 (1981); *51* 3393,
                                                  5311 (1986); *52* 603 (1987)
                                                Helv *63* 1665 (1980)
                                                Syn Commun *12* 613 (1982)
                                                CC 1642 (1986)

$Me_2\overset{+}{S}\overset{-}{C}HAr$	JOC 46 2731 (1981)
$Me_2\overset{+}{S}\overset{-}{C}HCH=CHR$	JOC 47 1698 (1982)
$Me_2\overset{+}{S}\overset{-}{C}HCO_2^-$	JOC 35 1600 (1970) TL 28 2095 (1987)
$PhEt\overset{+}{S}\overset{-}{C}HR'$ (intramolecular)	JOC 47 5372 (1982)
$[Ar\overset{+}{S}(CH_2R')_2]O_3SF^-$, KO-$t$-Bu (Ar = polymer; R' = H, Me)	TL 203 (1979)
$R\overset{+}{S}MeCH_2^-$ (R = Me, n-$C_{12}H_{25}$)	Helv 63 1665 (1980)
$Ph_2\overset{+}{S}\overset{-}{C}R'_2$	TL 2325 (1967)
$(CH_3)_2\overset{+}{S}O\overset{-}{C}H_2$	JACS 84 867 (1962); 87 1353 (1965); 109 1269 (1987) JCS C 731 (1970) Org Syn Coll Vol 5 755 (1973) JOC 52 4044 (1987)
PhSOCHClR', base/n-BuLi/H_2O (R'' = H)	TL 27 2379 (1986)
$\overset{\displaystyle O}{\underset{\displaystyle NTs}{\overset{\displaystyle \|}{\underset{\displaystyle \|}{PhSCH_2^-}}}}$	JACS 92 5753 (1970) Acct Chem Res 6 341 (1973)
$\overset{\displaystyle O}{\underset{\displaystyle NMe_2}{\overset{\displaystyle \|}{\overset{+}{R}S\overset{-}{C}R'_2}}}$	JACS 92 6594 (1970); 95 7692 (1973) JOC 38 1793 (1973) Acct Chem Res 6 341 (1973)
$\underset{\displaystyle PhSCHLiR'}{\overset{\displaystyle NTs}{\overset{\displaystyle \|}{}}}$ (R' = H, Ph)	JOC 44 2065 (1979)
$PhSCH_2Li/(Me_3O)BF_4/OH^-$	JACS 95 3429 (1973)
$MeSCH_2Li/MeI/KO$-t-Bu	JOC 50 3988, 5887 (1985)
$EtSCH=CHCH_2Ti(O$-i-$Pr)_3/(Me_3O)BF_4/$ NaOH (R' = $CH=CH_2$, R'' = H)	JACS 104 7663 (1982)
$EtSCH=CHCH_2Ti(O$-i-$Pr)_3/MeO_3SCF_3$, 2,6-(t-Bu)$_2C_5H_3N/Cs_2CO_3$ (R' = $CH=CH_2$, R'' = H)	TL 27 5691 (1986)
$LiSCH=CHCH_2Li$, $MgBr_2/MeI/NaOH$ (R' = $CH=CH_2$, R'' = H)	Angew Int 15 437 (1976)
$Ph_2S=\triangleleft^{R}$ (R = H, Me)	JACS 95 2038, 5311 (1973); 108 4965 (1986)
$Ph\overset{+}{S}eMeCH_2^-$	Angew Int 20 671 (1981)
$MeSeCH_2Li/MeOSO_2F/KOH$	TL 23 4389 (1982)
$Me_2SeC(Li)Me_2/MeOSO_2F/KOH$	CC 564 (1982)
$MeSeCR'R''Li/MeI$ or Me_2SO_4/KO-t-Bu	Angew Int 14 700 (1982)

$$RCR \xrightarrow{\quad\quad} R_2C - CR'R'' \quad (\textit{Continued})$$

with carbonyl (O) on left and epoxide (O) on product

Ph$_3$As=CHR'	JACS *103* 1283 (1981)
CH$_3$CH=CBrCO$_2$Et, LDA (R' = CH=CH$_2$, R'' = CO$_2$Et)	JOC *51* 4746 (1986)
H$_2$CBr$_2$, Li	CC 1047 (1969) Tetr *27* 6109 (1971)
H$_2$CBr$_2$, Li(Hg)	Tetr *27* 6109 (1971)
H$_2$CBr$_2$, n-BuLi	Tetr *27* 6109 (1971)
H$_2$CBr$_2$, sec-BuLi, LiBr	TL *25* 835 (1984)
H$_2$CCII, MeLi	TL *27* 795 (1986)

$$RCR \xrightarrow{\quad\quad} \begin{array}{c} O-CH_2 \\ | \quad\; | \\ R_2C-CH_2 \end{array}$$

CH$_3$SCH$_2$Na (with O and NTs on sulfur)	JACS *101* 6135 (1979)
H$_2$C=CH$_2$, hν	See page 455, Section 2.

$$(RCR) \quad RCH \xrightarrow{\quad\quad} \begin{array}{c} R \\ \text{(tetrahydrofuran with exocyclic methylene)} \\ (R)H \end{array}$$

R'$_3$SnCH$_2$C(=CH$_2$)CH$_2$OAc, cat Pd(OAc)$_2$, cat PPh$_3$	JACS *107* 1778, 8277 (1985)
Me$_3$SiCH$_2$C(=CH$_2$)CH$_2$OAc, cat n-Bu$_3$SnOAc, cat Pd(OAc)$_2$, cat PPh$_3$	TL *27* 5971 (1986)

cyclopentanone derivative
1. m-ClC$_6$H$_4$CO$_3$H
2. i-Bu$_2$AlH
3. HgO, I$_2$, hν
4. NaBH$_4$ or MeLi
→ tetrahydrofuran product

TL *25* 3995 (1984)
JOC *50* 2489 (1985)

3. Lactols

$$HO - \underset{O}{\bigcirc}{)}_n \xrightarrow{\quad\quad} \underset{O}{\bigcirc}{)}_n$$

Reagents	n	
Et$_3$SiH, CF$_3$CO$_2$H	2	CC 1568 (1986)
Et$_3$SiH, BF$_3$·OEt$_2$	1,2	JOC *46* 2417 (1981)
	2	JACS *104* 4976 (1982)
	2	Carbohydr Res *128* C9 (1984)
	1	CC 1512 (1985)

$$HO\text{-}\underset{O}{\overset{}{\bigcirc}}{)_n} \xrightarrow[BF_3 \cdot OEt_2]{RM} R\text{-}\underset{O}{\overset{}{\bigcirc}}{)_n}$$

$$n = 1, 2$$

RM

R_2Zn (R = Me, Et, Ph),
R_3Al (R = Me, Et),
$(H_2C\text{=}CHCH_2)_2SnBr_2$
or $n\text{-}Bu_2Sn(CH\text{=}CH_2)_2$ TL *28* 6339 (1987)

$H_2C\text{=}CHCH_2SiMe_3$ JACS *104* 4976 (1982)
                                                             Angew Int *25* 556 (1986)

Me_3SiCN (R = CN)   Angew Int *25* 556 (1986)

4. Carboxylic Acids

$$\underset{Cl}{\overset{|}{R}CHCO_2H} \xrightarrow[2.\ KOH]{1.\ LiAlH_4} R\overset{O}{\overset{/\ \backslash}{CH}\text{---}CH_2} \ (chiral)$$

Syn 316 (1982)

5. Esters

$$\overset{O}{\overset{\|}{R}COR'} \longrightarrow RCH_2OR'$$

H_2, cat Pt oxide (δ-lactones)   Chem Ind 975 (1964)

BH_3 JOC *26* 1685, 4553 (1961); *27* 2127 (1962);
                                                               36 3485 (1971)
                                                             Can J Chem *44* 1097 (1966)

$i\text{-}Bu_2AlH/Et_3SiH$, $BF_3 \cdot OEt_2$ JOC *46* 2417 (1981)
  (γ- and δ-lactones)

$LiAlH_4$, $BF_3 \cdot OEt_2$ JOC *25* 875 (1960)
                                                             Tetr *18* 953 (1962)

$LiAlH_4$, $AlCl_3$ (lactones only)   Chem Ind 230 (1977)

$HSiCl_3$, hν   JACS *91* 4587 (1969)
                                                             JOC *37* 76, 4349 (1972); *38* 795 (1973); *39* 2470
                                                               (1974); *40* 3885 (1975)
                                                             BCSJ *47* 932 (1974)

$(ArPS_2)_2S$, Raney Ni   JOC *46* 831 (1981)

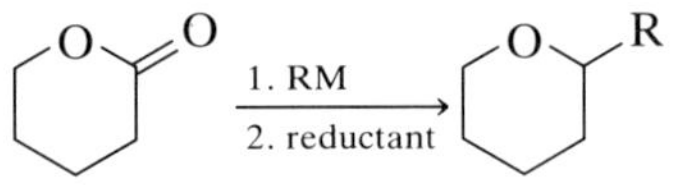

RM	Reductant	
$LiCH_2CO_2Et$, $H_2C{=}CHCH_2MgBr$	Et_3SiH, $BF_3 \cdot OEt_2$	JACS *104* 4976 (1982)
MeLi, PhMgCl	Et_3SiH, CF_3CO_2H	CC 1568 (1986)

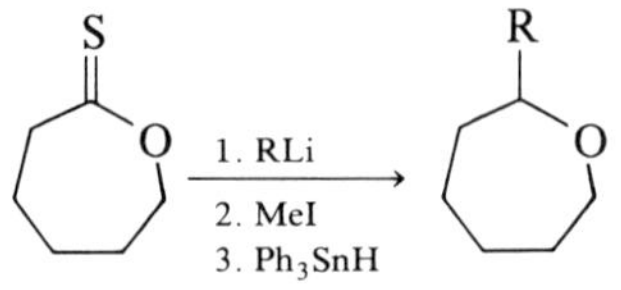

JACS *109* 2504 (1987)

$$\underset{RCOR'}{\overset{O}{\overset{\|}{}}} \longrightarrow \underset{RCOR'}{\overset{CH_2}{\overset{\|}{}}}$$

CH_2N_2 TL *28* 3011 (1987)

$Cp_2Ti\overset{CH_2}{\underset{Cl}{<}}AlMe_2$ JACS *102* 3270 (1980)

$$PhOAc + RX \xrightarrow[\text{18-crown-6}]{K_2CO_3} PhOR$$

$R = 1°$ alkyl, allylic, benzylic

CC 815 (1982)

6. Peresters

$$\underset{RCOOR'}{\overset{O}{\overset{\|}{}}} \xrightarrow{RMgX} ROR'$$

JACS *81* 4230 (1959)
Org Syn *43* 55 (1963)
Org Syn Coll Vol *5* 924 (1973)

7. Ortho Esters

$$RM + (PhO)_2CHOEt \longrightarrow R_2CHOEt$$

RM	
PhMgBr	Ber *103* 643 (1970)
$RCH{=}CR'CH_2Al_{2/3}Br$	JOMC *222* 1 (1981)

5. MISCELLANEOUS REACTIONS

$$\text{(cyclopentanol with methyl substituent)} \xrightarrow[\text{2. NaBH}_4 \text{ or MeLi}]{\text{1. HgO, I}_2, \text{h}\nu} \text{(methyl tetrahydrofuran)}$$

CL 55 (1983)
JOC *49* 3753 (1984)

ALCOHOLS AND PHENOLS

GENERAL REFERENCES

"The Chemistry of the Hydroxyl Group," Ed. S. Patai, Interscience, New York (1971)

Houben-Weyl, "Methoden der Organischen Chemie," 4th ed, Vol VI/1c, Parts I and II, G. Thieme, Stuttgart (1976) (phenols)

Houben-Weyl, "Methoden der Organischen Chemie," 4th ed, Vol VI/1a, Parts I and II, G. Thieme, Stuttgart (1979 and 1980 respectively) (alcohols)

Syn 501 (1981) (preparation of acetals from alcohols or oxiranes and carbonyl compounds)

Houben-Weyl, "Methoden der Organischen Chemie," 4th ed, Vol VI/1b, Part III, G. Thieme, Stuttgart (1984) (alcohols)

1. ALCOHOL TRANSPOSITION

Resolution of alcohols by lipase-catalyzed transesterification:

JACS *107* 7072 (1985)
JOC *52* 256 (1987)
TL *28* 1647 (1987)

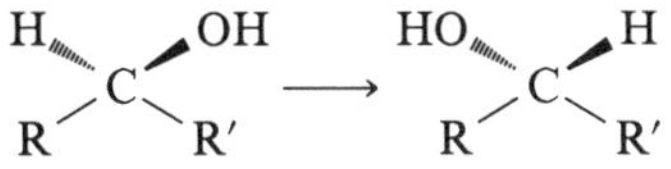

$$H_{\cdots}\underset{R}{\overset{OH}{\underset{|}{C}}}R' \longrightarrow HO_{\cdots}\underset{R}{\overset{H}{\underset{|}{C}}}R'$$

TsCl/(n-Bu$_4$N)OAc/LiAlH$_4$	JCS C 1605 (1969)
TsCl/(Et$_4$N)OAc[O$_2$CH]/saponify	TL 3265, 3269 (1972)
TsCl/NaO$_2$CPh/K$_2$CO$_3$	JACS *96* 5876 (1974)
p-ClC$_6$H$_4$SO$_2$Cl/KO$_2$CH(OAc)/saponify	Ann 901 (1974)
TsCl/DMF/saponify	JACS *80* 2906 (1958)
p-XC$_6$H$_4$SO$_2$Cl (X = Cl, Me)/KNO$_2$	Ann 901 (1974) Syn 292 (1980)
TsCl or MsCl/(n-Bu$_4$N)NO$_3$/reduction	TL *26* 3369 (1985)
MsCl/KOH	JACS *107* 2138 (1985)
MsCl/KO$_2$	TL 3183 (1975)
MsCl, Et$_3$N/CsO$_2$CEt, DMF/hydrolysis	JOC *46* 4321 (1981)
mesylation/CsOAc, 18-crown-6/ hydrolysis	CL 1555 (1984)
(Ms)$_2$O, Et$_3$N/CsOAc, DMF/K$_2$CO$_3$, MeOH	TL *28* 503 (1987)
PhCOCl/Al$_2$O$_3$	Helv *42* 2177 (1959)
RCO$_2$H, EtO$_2$CN=NCO$_2$Et, R$_3$P/base	BCSJ *40* 2380 (1967); *44* 3427 (1971); *49* 510 (1976) TL 1619 (1973); *27* 3535 (1986); *28* 723, 1143, 3151 (1987) Helv *59* 2100 (1976); *60* 417 (1977)

Angew Int *18* 958 (1979)
Syn 1 (1981) (review)
JOC *48* 5083 (1983); *50* 2981 (1985);
   52 3468, 3883, 4235 (1987)
JACS *107* 4339 (1985); *109* 4690 (1987)

HOAc, (Me₃CCH₂O)₂CHNMe₂/saponify   Ann 821 (1974)

TL *26* 607 (1985)

t-BuO₂H, VO(acac)₂/MeSO₂Cl/Na, NH₃   TL 2621 (1976)
                                               BCSJ *52* 1757 (1979)

p-NO₂C₆H₄SeCN, *n*-Bu₃P/H₂O₂   CC 770 (1978)

NaH, CS₂, MeI/*n*-Bu₃SnH/   Syn 1011 (1980)
   m-ClC₆H₄CO₃H/HCl

ClCONMe₂/Hg(O₂CCF₃)₂/PPh₃/cleavage   JACS *100* 4822 (1978)

H₂O, cat HCl or HOAc   JCS 396 (1946)

2. SUBSTITUTION

1. Hydrogen Substitution

$$RCH_2OK \xrightarrow[\substack{n\text{-BuLi}}]{(CH_3)_2C=CHCl} \xrightarrow{H_2O} R\overset{\overset{\displaystyle OH}{|}}{C}HCH=C(CH_3)_2$$

JACS *107* 2189 (1985)

2. Halide and Sulfonate Substitution

$$RX \longrightarrow ROH$$

H_2O	JOC *52* 4592 (1987) (2° benzylic bromide)
Al_2O_3, H_2O	CC 1136 (1987) (2° alkyl chloride)
$CaCO_3$, H_2O	JOC *51* 3762 (1986) (1° benzylic bromide)
polymer-supported carbonate	Syn 793 (1981) (1° alkyl chloride, bromide, iodide; 1° allylic bromide; 1° benzylic chloride, bromide)
$(n\text{-Bu}_4N)OAc/LiAlH_4$	JCS C 1605 (1969) (2° alkyl tosylates)
KO_2, 18-crown-6/H_2O	TL 3183 (1975) (1° alkyl, allylic, benzylic bromides; 2° alkyl tosylate; 2° allylic mesylate)
Li or Mg/O_2 or $t\text{-BuO}_2Li$	JOC *41* 1459 (1976) (2° cyclopropyl bromide)
$Li/MoO_5 \cdot py \cdot HMPA$	JOMC *59* 293 (1973) (1° alkyl lithium)
$Mg/MoO_5 \cdot py \cdot HMPA$	JOC *42* 1479 (1977) (aryl bromides) Austral J Chem *31* 2091 (1978) (1°, 2°, 3° alkyl bromides; aryl bromides)

NaOH, DMF   JOC *47* 4024 (1982)

N_2H_4, H_2O/H_2SO_4, MeOH   JOC *32* 3723 (1967)

1. *n*-BuLi
2. R_2BX
3. H_2O_2, NaOH

X = H, R

JOC *50* 2401 (1985)

$$RX \longrightarrow RCH_2OH$$

n-Bu$_3$SnCH$_2$OH, 2 *n*-BuLi/H$_2$O   Angew Int *15* 438 (1978)
                                                  Ber *113* 1290 (1980)

(*i*-PrO)$_2$MeSiCH$_2$MgCl; Cu,   JOC *48* 2120 (1983)
    Ni or Pd catalyst/
    H$_2$O$_2$ or CH$_3$CO$_3$H

(Mes)$_2$BCH$_2$Li/[O]   Chem Soc Rev *11* 191 (1982)

◯BCH$_2$Li/H$_2$O$_2$, NaOH   JACS *94* 6854 (1972)

2,4,6-(*i*-Pr)$_3$C$_6$H$_2$CO$_2$CH$_2$Li/LiAlH$_4$ JACS *99* 5213 (1977)

$$RX \xrightarrow[\text{2. LiAlH}_4]{\text{1. 2,4,6-(}i\text{-Pr)}_3\text{C}_6\text{H}_2\text{CO}_2\text{CHLiCH}_3} \overset{\text{OH}}{\underset{|}{R}}CHCH_3$$

JOC *43* 4255 (1978)

For analogous conversions see page 553, Section 8.

$$Cl-\overset{|}{\underset{|}{C}}-\overset{|}{\underset{|}{C}}-OH \xrightarrow[\substack{3.\ E^+ \\ 4.\ H_2O}]{\substack{1.\ n\text{-BuLi} \\ 2.\ Li^+C_{10}H_8^-}} E-\overset{|}{\underset{|}{C}}-\overset{|}{\underset{|}{C}}-OH$$

E$^+$ = D$_2$O, RX, RCHO, R$_2$CO, CO$_2$, O$_2$, Me$_2$S$_2$

CC 1153 (1982)
JCS Perkin I 3019 (1983)

$$\overset{\text{Br}}{\underset{|}{R}}CHCH_2OH + R'MgCl \xrightarrow{\text{(cat CuBr)}} \overset{\text{R'}}{\underset{|}{R}}CHCH_2OH$$

R' = allyl, vinyl, aryl

BSCF II 289 (1980)

3. Amine Substitution

$$1°, 2° \ RNH_2 \xrightarrow[\text{2. Zn, HOAc}]{\text{1. } N_2O_4} ROH$$

JCS Perkin I 1114 (1977)

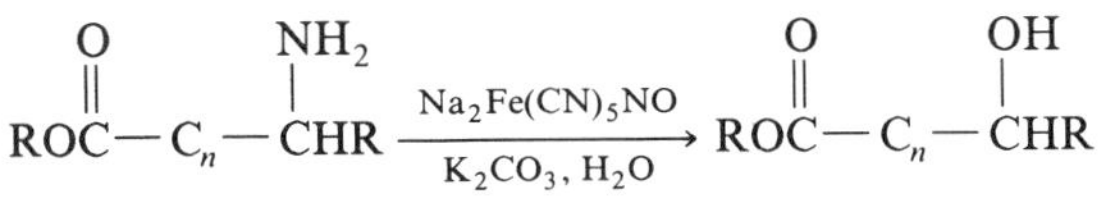

$$\underset{\text{ROC}}{\overset{O}{\parallel}}-C_n-\underset{\overset{|}{\text{NH}_2}}{\text{CHR}} \xrightarrow[\text{K}_2\text{CO}_3, \text{H}_2\text{O}]{\text{Na}_2\text{Fe(CN)}_5\text{NO}} \underset{\text{ROC}}{\overset{O}{\parallel}}-C_n-\underset{\overset{|}{\text{OH}}}{\text{CHR}}$$

$$n = 1, 3$$

JOC 51 3913 (1986)

RCH$_2$NH$_2$ ⟶ [structure] BF$_4^-$ $\xrightarrow[(n\text{-Bu}_4\text{N})\text{BF}_4]{}$ RCH$_2$OH

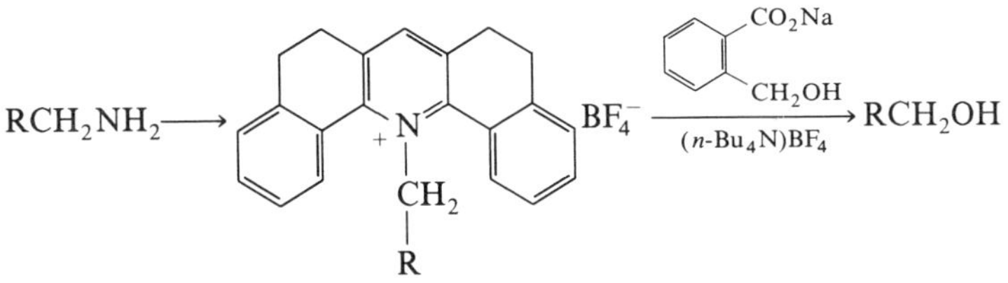

JCS Perkin I 1492 (1981)

$$ArNH_2 \longrightarrow ArOH$$

NaHSO$_3$/NaOH/HCl	JACS 83 5015 (1961)
NaNO$_2$, H$_2$SO$_4$/H$_2$O, H$_2$SO$_4$, Na$_2$SO$_4$, Δ	Org Syn Coll Vol 3 130 (1955)
DMSO [on (ArN$_2^+$)BF$_4^-$]	Helv 59 1427 (1976)

4. Acetate Substitution

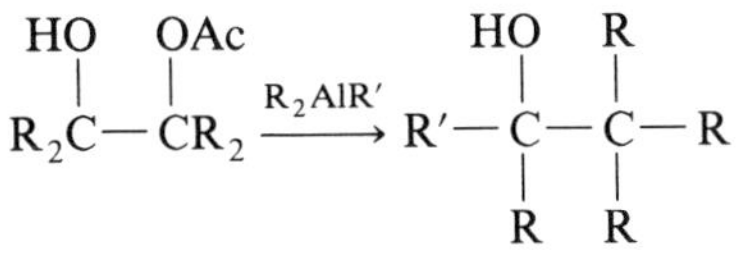

$$\underset{\text{R}_2\text{C}}{\overset{\text{HO}}{\underset{|}{|}}}-\underset{\text{CR}_2}{\overset{\text{OAc}}{\underset{|}{|}}} \xrightarrow{\text{R}_2\text{AlR}'} \text{R}'-\underset{\overset{|}{\text{R}}}{\overset{\text{HO}}{\overset{|}{\text{C}}}}-\underset{\overset{|}{\text{R}}}{\overset{\text{R}}{\overset{|}{\text{C}}}}-\text{R}$$

BCSJ 55 3941 (1982)

3. OXIDATION

1. Alkanes

Review: Angew Int *17* 909 (1978)

$$R_3CH \longrightarrow R_3COH$$

O_3, silica gel (3° H)	JOC *40* 2141 (1975); *50* 2759 (1985)
$p\text{-}NO_2C_6H_4CO_3H$, Δ (3° H)	Angew Int *18* 407 (1979)
CF_3CO_3H (2° H)	CC 1051 (1976)
$(PhNEt_3)MnO_4$ (3° H)	Angew Int *18* 68 (1979)
RuO_4 (3° H)	Acta Chem Scand B *40* 430 (1986)

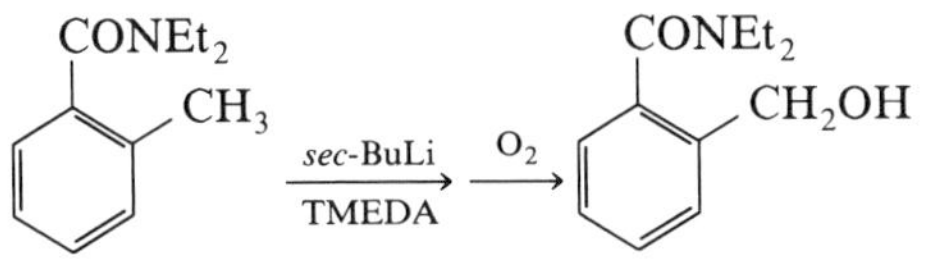

JOC *52* 674 (1987)

2. Arenes

$$ArH \longrightarrow ArOH$$

Reviews:

Adv Org Chem *1* 103 (1960)
R. O. C. Norman, R. Taylor, "Electrophilic Substitution in Benzenoid Compounds," Elsevier, Amsterdam (1965)
"Applications of Biochemical Systems in Organic Chemistry," Wiley, New York (1976), Part I, pp 69–106
"Comprehensive Organic Chemistry," Pergamon, Oxford (1979), Vol 1

H_2O_2, HF	JOC *35* 4028 (1970)
H_2O_2, HF, py	Syn 536 (1979)

$$\text{ArH} \longrightarrow \text{ArOH} \quad (\textit{Continued})$$

H_2O_2, HF, BF_3	JOC *46* 4305 (1981)
H_2O_2, HF, SbF_5	Nouv J Chim *6* 477 (1982) TL *27* 4565 (1986)
H_2O_2, HSO_3F, SO_2ClF, (SbF_5)	JOC *43* 865 (1977)
H_2O_2, $AlCl_3$	JOC *36* 3184 (1971)
H_2O_2, $FeSO_4$ (on ArOH)	J Prakt Chem *152* 46 (1939)
t-BuO_2H, $AlCl_3$	BCSJ *43* 293 (1970)
CF_3CO_3H	JCS 1804 (1959); 5404 (1964) JOC *27* 627 (1962)
CF_3CO_3H, BF_3	JACS *85* 2177 (1963) JOC *29* 2397 (1964); *30* 331 (1965)

i-PrOCOOCO-i-Pr (with two C=O), $AlCl_3$ — JACS *87* 1566, 4811 (1965)

$NaOH/K_2S_2O_8$ (on ArOH)	JCS 2303 (1948)
(cyclopentadienyl ring), $H_3PO_4/PdCl_2(CH_3CN)_2/H_2O_2$, HCl (on ArOH)	TL *22* 2327 (1981)
O_2, $HO_2CC(OH){=}C(OH)CO_2H$, peroxidase catalyst (on ArOH)	JACS *103* 6263 (1981)

$$o\text{-}X\!-\!Ar\!-\!H \longrightarrow o\text{-}X\!-\!Ar\!-\!OH$$

Reagents	X	
n-BuLi/$MoO_5 \cdot$py$\cdot$HMPA	MeO	TL *28* 2643 (1987)
sec-BuLi, TMEDA/O_2	MeO, $CONEt_2$	JOC *52* 674 (1987)

3. Alkenes

See page 493, Section 1, for addition to the double bond; and page 116, Section 5, for oxidation with allylic transposition.

$$\text{RCH}_2\text{CH}{=}\text{CH}_2 \longrightarrow \underset{\underset{\text{RCHCH}{=}\text{CH}_2}{|}}{\overset{\text{OH}}{}}$$

SeO_2

Org Rxs *5* 331 (1949); *24* 261 (1976)
JOC *35* 1646, 1653 (1970); *44* 4683 (1979);
   51 256 (1986); *52* 3468 (1987)
B. S. Thyagarajan, "Selective Organic
   Transformations," Vol 1 (1970), p 307
JACS *93* 4835 (1971)
Syn 215 (1978)
CL 85 (1986)

SeO$_2$, t-BuO$_2$H

JACS *99* 5526 (1977)
JOC *48* 1404 (1983); *50* 1602 (1985);
   52 3860 (1987)
TL *27* 2279 (1986); *28* 1561, 5945 (1987)

SeO$_2$, t-BuO$_2$H, silica gel

CL 1703 (1981)

SeO$_2$,

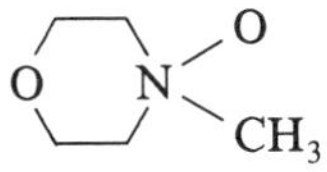

JOC *51* 1635 (1986)

RLi, TMEDA/O$_2$

JACS *94* 4298 (1972)
Chem Pharm Bull *32* 4632 (1984)
JOC *51* 4315 (1986)

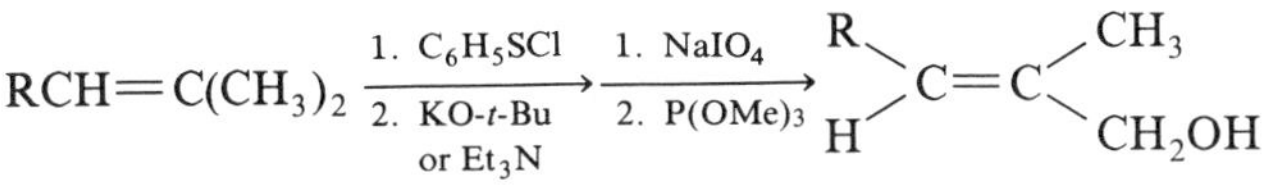

$$RCH{=}C(CH_3)_2 \xrightarrow[\substack{\text{2. KO-}t\text{-Bu} \\ \text{or Et}_3N}]{\text{1. C}_6\text{H}_5\text{SCl}} \xrightarrow[\text{2. P(OMe)}_3]{\text{1. NaIO}_4}$$

TL 4539, 5123 (1978)

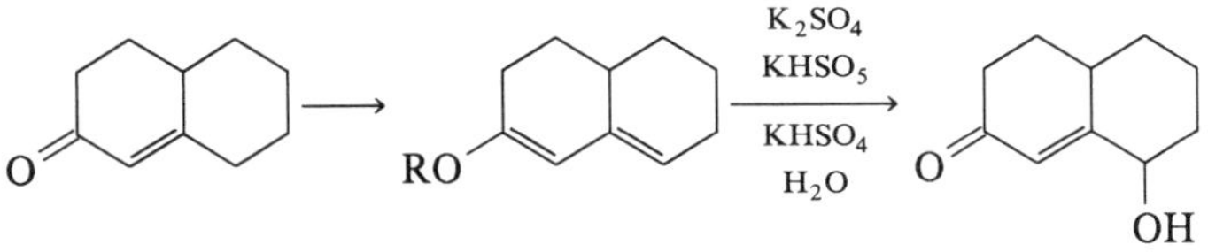

R = Me, Ac, SiMe$_3$

TL *22* 4201 (1981)

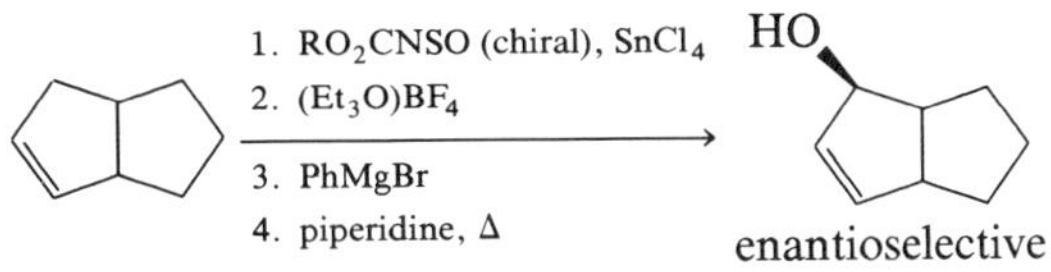

enantioselective

JACS *109* 2839 (1987)

4. Alcohols

$$ArCR_2OH \longrightarrow ArOH$$

H$_2$O$_2$, HOAc, HClO$_4$

JOC *15* 775 (1950)

H$_2$O$_2$, TsOH

JOC *51* 5436 (1986); *52* 5283 (1987)

H$_2$O$_2$, BF$_3$·OEt$_2$

JOC *51* 5436 (1986)
JACS *109* 2717 (1987)
TL *28* 1027 (1987)

5. Carbonyl Compounds

$$\begin{array}{c} \text{H} \\ | \\ -\text{C}-\text{X} \\ | \end{array} \longrightarrow \begin{array}{c} \text{OH} \\ | \\ -\text{C}-\text{X} \\ | \end{array}$$

X	Reagent(s)	
CHO	m-ClC$_6$H$_4$CO$_3$H (on enol silane)/H$_3$O$^+$	JOC *40* 3427 (1975)
COR	PhIO, NaOH, MeOH	TL *22* 1283 (1981)
		CC 641 (1981)
	PhI(OAc)$_2$, NaOH, MeOH	TL *22* 1283 (1981); *25* 4745 (1984); *28* 5709 (1987)
		JACS *103* 686 (1981)
	o-HO$_2$CC$_6$H$_4$IO, KOH, MeOH/H$_3$O$^+$	TL *25* 691, 4745 (1984)
		Org Syn *64* 138 (1985)
	NaH, P(OEt)$_3$/O$_2$	JOC *51* 1478 (1986)
	NaO-t-Bu, O$_2$, P(OEt)$_3$	JOC *33* 3294 (1968)
	KO-t-Bu, O$_2$/Zn, HOAc	JCS 1578 (1962)
		JACS *90* 2448 (1968)
	LiN(SiMe$_3$)$_2$/O$_2$, P(OEt)$_3$	JOC *51* 3393 (1986)
	O$_2$, hν (on enol silane)/ PPh$_3$/MeOH	TL 2375 (1977)
	m-ClC$_6$H$_4$CO$_3$H (on enol silane)/H$_3$O$^+$ or (Et$_3$NH)F	TL 4319 (1974); 2935 (1976); 4603 (1978); *28* 435, 581, 3723, 5017 (1987)
		JOMC *77* C19 (1974)
		JOC *40* 3427 (1975); *42* 1581 (1977); *43* 1599 (1978); *52* 3745 (1987)
		CC 27 (1977)
		Org Syn *64* 118 (1985)
		JACS *109* 7575 (1987)
	OsO$_4$, O$\underset{}{\diagup}$N$\diagdown$·H$_2$O (on enol silane)/H$_3$O$^+$	TL *22* 607 (1981)
		JACS *108* 5549 (1986)
	CrO$_2$Cl$_2$ (on enol silane)	TL *23* 2917 (1982)
	(PhSeO)$_2$O	CL 763 (1979)
		Tetr *37* 473 (1981)
	LDA/MoO$_5$·py·HMPA	JACS *96* 5944 (1974); *106* 3539, 4547 (1984)
		JOC *43* 188 (1978)
		Org Syn *64* 127 (1985)
	NaN(SiMe$_3$)$_2$/ PhSO$_2$N$-$CHPh (oxaziridine)	JOC *52* 4592 (1987)
	KN(SiMe$_3$)$_2$/ PhSO$_2$N$-$CHPh (oxaziridine)	JOC *49* 3241 (1984)

	NaN(SiMe$_3$)$_2$/chiral	JOC *51* 4083 (1986)
	RSO$_2$N$\overset{\overset{\displaystyle O}{\diagup\diagdown}}{\quad}CR_2$	
	chiral RSO$_2$N$\overset{\overset{\displaystyle O}{\diagup\diagdown}}{\quad}$CHAr (on enol silane)/H$_3O^+$	JOC *52* 954 (1987)
CO$_2$H	*m*-ClC$_6$H$_4$CO$_3$H (on R$_2$C=C(OSiMe$_3$)$_2$)/ H$_3$O$^+$	JOC *40* 3783 (1975)
	LDA/O$_2$	JOC *40* 3253 (1975) Syn Commun *9* 63 (1979) TL *27* 2199 (1986)
CO$_2$R	LDA/O$_2$	TL 1731 (1975) JOC *52* 3323 (1987)
	LiNR$_2$/MoO$_5$·py·HMPA	JACS *96* 5944 (1974) JOC *43* 188 (1978); *46* 4825 (1981); *52* 1170, 3346 (1987) TL *26* 203 (chiral), 5631 (1985); *27* 1833 (1986); *28* 1147, 6381 (lactone) (1987)
	KN(SiMe$_3$)$_2$, KO-*sec*-Bu/ MoO$_5$·py·HMPA (Li)KN(SiMe$_3$)$_2$/	TL *27* 3999 (1986) Helv *69* 615 (1986) JOC *49* 3241 (1984)
	PhSO$_2$N$\overset{\overset{\displaystyle O}{\diagup\diagdown}}{\quad}$CHPh	
	chiral RSO$_2$N$\overset{\overset{\displaystyle O}{\diagup\diagdown}}{\quad}$CHAr (on silyl ketene acetal)/ H$_3$O$^+$	JOC *52* 954 (1987)
CO$_2$R, CONR$_2$	LDA/chiral RSO$_2$N$\overset{\overset{\displaystyle O}{\diagup\diagdown}}{\quad}CR_2$	JOC *51* 2402 (1986); *52* 5288 (1987)
CONR$_2$	LDA/O$_2$ LDA/P(OEt)$_3$/O$_2$ LDA/MoO$_5$·py·HMPA	TL 1731 (1975) JOC *52* 4352 (1987) JOC *50* 2170 (1985)
CON$\overset{\displaystyle O}{\diagdown}$O Me$\diagdown$Ph	NaN(SiMe$_3$)$_2$/ PhCH$\overset{\overset{\displaystyle O}{\diagdown}}{\quad}NSO_2$Ph	JACS *107* 4346 (1985) (chiral)
CN	LDA/O$_2$/SnCl$_2$ LDA/MoO$_5$·py·HMPA	JOC *40* 269 (1975) JOC *41* 740 (1976)

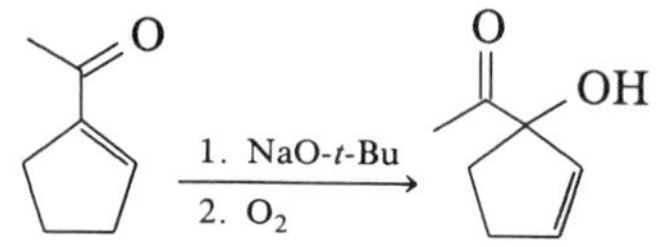

JOC *33* 3695 (1968)

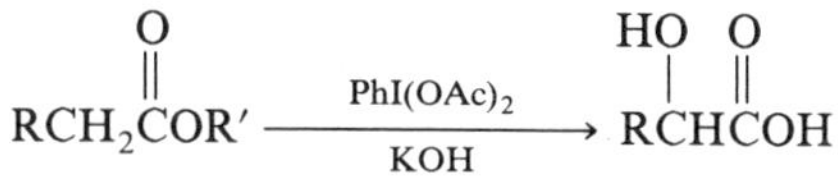

TL *22* 2747 (1981)

$$ArCHO \xrightarrow[\text{NaOH}]{H_2O_2} ArOH$$

TL *28* 455 (1987)

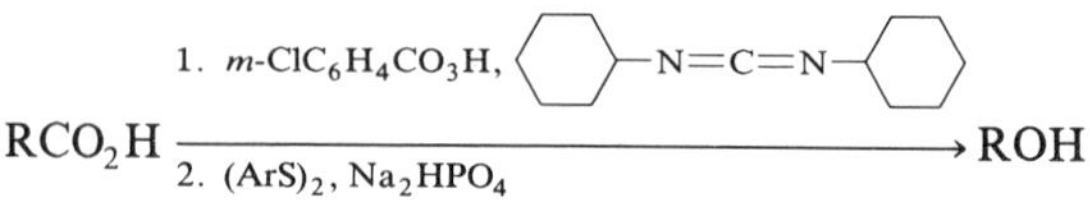

TL *28* 4489 (1987)

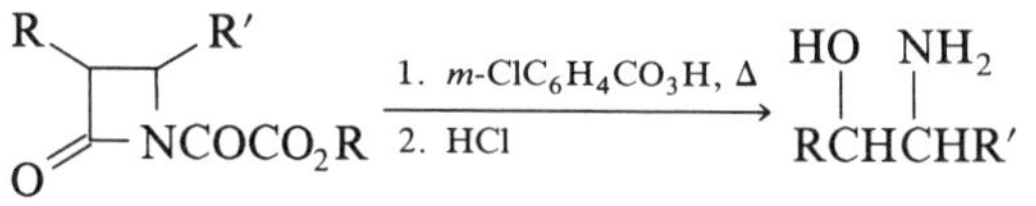

J Antibiotics *33* 796 (1980)

6. Organometallics

$$RM \longrightarrow ROH$$

M	Reagent(s)	
Li	—	See also page 485, Section 1.
	$MoO_5 \cdot py \cdot HMPA$	JOC *50* 5660 (1985)
		(R = CH_2-oxazoline)
	$(Me_3SiO)_2/HCl$	Syn 633 (1986)
		(R =1°, 2° alkyl; aryl)
	$Me_3CN\!-\!CHPh$	JACS *101* 1044 (1979)
		(R = aryl)
	$PhSO_2N\!-\!CHPh$	JACS *101* 1044 (1979)
		(R =1° alkyl, aryl)
		JOC *50* 5660 (1985)
		(R = CH_2-oxazoline)
		TL *28* 5115 (1987) (R = aryl)

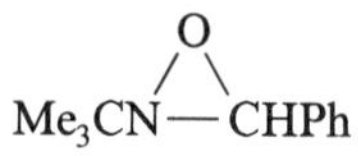

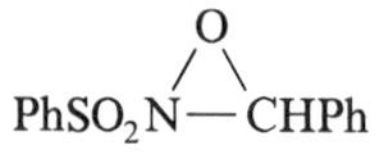

MgX	MoO$_5$·py·HMPA	JOC 42 1479 (1977) (R = aryl)
	PhSO$_2$N—CHAr (epoxide: O)	JACS 101 1044 (1979) (R = 1° alkyl, aryl) TL 28 5115 (1987) (R = 1°, 2° alkyl; aryl)
	chiral RSO$_2$N—CHR (epoxide: O)	TL 28 5115 (1987) (R = aryl)
B	—	See page 497, Section 5.
SiMe$_2$Ph	KBr, CH$_3$CO$_3$H, NaOAc, HOAc	TL 28 4229 (1987)
	Hg(OAc)$_2$, CH$_3$CO$_3$H, HOAc	TL 28 4229 (1987)
	cat Hg(OAc)$_2$, cat Pd(OAc)$_2$, CH$_3$CO$_3$H, HOAc	TL 28 4229 (1987)
	m-ClC$_6$H$_4$CO$_3$H, KF	TL 26 397 (1985)
	HBF$_4$/m-ClC$_6$H$_4$CO$_3$H, Et$_3$N	CC 29 (1984); 318 (1985)
	BF$_3$·2 HOAc/m-ClC$_6$H$_4$CO$_3$H, Et$_3$N	CC 305 (1986)
	BF$_3$·2 HOAc/CH$_3$CO$_3$H, Et$_3$N	CC 1198 (1986)
	BF$_3$·2 HOAc/m-ClC$_6$H$_4$CO$_3$H, KF	CC 29 (1984)
SiMe(OEt)$_2$	H$_2$O$_2$, HCO$_3^-$ or CO$_3^{2-}$	Organomet 2 1694 (1983)
	H$_2$O$_2$, KF or KHF$_2$	Organomet 2 1694 (1983)
	t-BuO$_2$H, KHF$_2$	Organomet 2 1694 (1983)
	m-ClC$_6$H$_4$CO$_3$H	Organomet 2 1694 (1983)
SiMe$_2$OR	H$_2$O$_2$, NaHCO$_3$	TL 27 3377 (1986)
SiR$_2$OEt	m-ClC$_6$H$_5$CO$_3$H, KF	Tetr 39 983 (1983)
SiR$_n$F$_{3-n}$ ($n = 1, 2$)	m-ClC$_6$H$_4$CO$_3$H, KF	Tetr 39 983 (1983)
SiF$_5^{2-}$	m-ClC$_6$H$_4$CO$_3$H	JACS 100 2268 (1978) Tetr 39 983 (1983)

4. ALKENE AND ALKYNE ADDITIONS

1. Alkenes

$$RCH{=}CH_2 \xrightarrow[H_2O]{cat\ H^+} R\overset{\displaystyle OH}{\underset{\displaystyle |}{C}}HCH_3$$

JACS *40* 822, 1950 (1918); *46* 1512 (1924); *49* 873 (1927); *56* 460, 2138 (1934); *61* 940 (1939); *73* 3792 (1951); *74* 5372 (1952); *77* 1584 (1955); *79* 3724 (1957); *93* 4907 (1971)
JCS 4203 (1960)

$$RCH{=}CH_2 \xrightarrow[H_2O]{Hg(OAc)_2} R\overset{OH}{\underset{|}{C}}HCH_2HgOAc \xrightarrow[NaOH]{NaBH_4} R\overset{OH}{\underset{|}{C}}HCH_3$$

Review:

    R. C. Larock, "Solvomercuration/Demercuration Reactions in Organic Synthesis," Springer Verlag, New York (1986), Chpt 2

JACS *89* 1522, 1524, 1525 (1967); *102* 7798 (1980) (diene regioselectivity)
JOC *35* 1844 (1970); *39* 1474 (1974); *44* 1910 (1979); *46* 531, 930, 3810 (1981)

$$RCH{=}CH_2 \longrightarrow R\overset{OH}{\underset{|}{C}}HCH_2HgCl \xrightarrow{Li_2CuMe_3} R\overset{OH}{\underset{|}{C}}HCH_2CH_3$$

Organomet *1* 74 (1982)

$$RCH{=}CH_2 \longrightarrow R\overset{OH}{\underset{|}{C}}HCH_2HgCl \longrightarrow R\overset{OM}{\underset{|}{C}}HCH_2M \xrightarrow[2.\ H_2O]{1.\ E^+} R\overset{OH}{\underset{|}{C}}HCH_2E$$

    M = alkali metal

    $E^+ = D_2O, O_2, CO_2, RCHO, R_2CO, Me_3SiCl, (MeS)_2, R_2C{=}NR$

            TL 2015 (1978)
            JOC *44* 4798 (1979); *47* 1560 (1982)

$$RCH=CHCH_3 \xrightarrow[\text{2. } O_2 \text{ or } t\text{-BuO}_2H]{\text{1. (Cp)}_2\text{ZrHCl}} RCH_2CH_2CH_2OH$$

TL 3041 (1975); *23* 157 (1982)

$$RCH=CH_2 \longrightarrow RCH_2CH_2OH$$

NaBH$_4$, SnCl$_4$	CC 796 (1979)
NaBH$_4$, TiCl$_4$	CC 414 (1980)
NaBH$_4$, cat Rh(III) porphyrin, O$_2$	JOC *52* 2555 (1987)
LiBH$_4$, Cp$_2$TiCl$_2$/H$_2$O$_2$, NaOMe	CL 1069 (1979)
Cl$_2$AlH, cat Et$_3$B/O$_2$	JACS *108* 6036 (1986)
LiAlH$_4$, TiCl$_4$/O$_2$	JOMC *142* 71 (1977)
LiAlH$_4$, Cp$_2$TiCl$_2$/O$_2$ or *m*-ClC$_6$H$_4$CO$_3$H	CL 1117 (1977)
Cl$_3$SiH, cat H$_2$PtCl$_6$/KF/*m*-ClC$_6$H$_4$CO$_3$H	JACS *100* 2268 (1978) Tetr *39* 983 (1983)
Cl$_3$SiH, cat PdCl$_2$L$_2$ (chiral)/CuF$_2$/ *m*-ClC$_6$H$_4$CO$_3$H	JOC *51* 3772 (1986)
(EtO)$_2$SiMeH, cat H$_2$PtCl$_6$/H$_2$O$_2$	Organomet *2* 1694 (1983)
(EtO)$_2$SiMeH, cat ClRh(PPh$_3$)$_3$/ Me$_3$NO, KHF$_2$	TL *27* 75 (1986)

See also page 497, Section 5.

$$R-\overset{\overset{\displaystyle OH}{|}}{C}=C-C_n-C \xrightarrow[\substack{\text{2. cat } H_2PtCl_6 \cdot 6 H_2O \\ \text{3. } H_2O_2}]{\text{1. (HMe}_2\text{Si)}_2\text{NH}} R-\overset{\overset{\displaystyle OH}{|}}{C}-\overset{\overset{\displaystyle OH}{|}}{C}-C_n-C$$

JACS *108* 6090 (1986)
TL *27* 3377 (1986)

$$H_2C=CHCH_2\overset{\overset{\displaystyle OH}{|}}{C}HR \longrightarrow H_2C=CHCH_2\overset{\overset{\displaystyle O_2CCMe_3}{|}}{C}HR \xrightarrow[\substack{\text{2. } n\text{-Bu}_3\text{SnH} \\ \text{3. hydrolysis}}]{\text{1. } I_2} CH_3\overset{\overset{\displaystyle OH}{|}}{C}HCH_2\overset{\overset{\displaystyle OH}{|}}{C}HR$$

JACS *106* 5304 (1984)

$$\overset{>}{}C=C\overset{<}{} \longrightarrow -\overset{\overset{\displaystyle HO}{|}}{C}-\overset{\overset{\displaystyle OH}{|}}{C}-$$

Reviews:

Adv Org Chem *1* 103 (1960)
Chem Rev *80* 187 (1980) (OsO$_4$)

cis hydroxylation

KMnO$_4$
                              Org Syn Coll Vol *2* 307 (1943)
                              JACS *78* 5342 (1956)

KMnO$_4$ (phase transfer)	TL *21* 177 (1980)
KMnO$_4$, NaOH (phase transfer)	TL 4907 (1972)
(n-C$_{16}$H$_{33}$NMe$_3$)MnO$_4$	Syn 431 (1984)
(Ph$_3$PMe)MnO$_4$	Tetr *35* 1109 (1979)
OsO$_4$	JOC *25* 257 (1960)
OsO$_4$, py	Tetr *40* 2247 (1984) (on allylic alcohols) JOC *52* 2301 (1987) TL *28* 4955 (1987)
OsO$_4$, dihydroquinidine acetate (chiral)	JACS *102* 4263 (1980) TL *28* 3139 (1987)
OsO$_4$, dihydroquinine acetate (chiral)	JACS *102* 4263 (1980) TL *28* 3139 (1987)
OsO$_4$, quinuclidine (chiral)	TL *28* 3139 (1987)
OsO$_4$, (-)-(R,R)-N,N,N',N'-tetramethyl-cyclohexane-1,2-*trans*-diamine (chiral)	TL *27* 3951 (1986)
OsO$_4$, chiral diamines derived from tartaric acid	CL 131 (1986)

OsO$_4$,

JACS *109* 6213 (1987)

cat OsO$_4$, H$_2$O$_2$	JACS *70* 1484 (1948)
cat OsO$_4$, t-BuO$_2$H, (Et$_4$N)OH	JACS *98* 1986 (1976)
cat OsO$_4$, t-BuO$_2$H, (Et$_4$N)OAc	JOC *43* 2063 (1978)
cat OsO$_4$, Me$_3$NO·2 H$_2$O	JACS *109* 3402 (1987)
cat OsO$_4$, Me$_3$NO·2 H$_2$O, py	JACS *109* 3353 (1987)

cat OsO$_4$,

TL 1973 (1976); *28* 1603, 5473 (diene), 5755 (1987)
Tetr *40* 2247 (1984) (on allylic alcohols)
JACS *107* 2712, 3891 (1985); *108* 1094 (1986); *109* 5437 (1987)
JOC *51* 2637 (1986); *52* 3745, 3784, 4505, 5624, 5700 (1987)

cat OsO$_4$, PhSeOR (R = Me, Ph)	TL *22* 2051 (1981)
NBS, H$_2$O or m-ClC$_6$H$_4$CO$_3$H-HBr/ NCCH$_2$CO$_2$H, TsCl, py/NaH/H$_3$O$^+$/ K$_2$CO$_3$, MeOH	TL *23* 4217 (1982)
I$_2$, AgOAc, H$_2$O, HOAc/KOH	JACS *80* 209 (1958) JCS 770 (1963)

$$>C=C< \longrightarrow -\overset{\overset{\text{HO}}{|}}{C}-\overset{\overset{\text{OH}}{|}}{C}- \quad (Continued)$$

cis hydroxylation (*continued*)

I_2, AgOAc, H_2O, HOAc/LiAlH$_4$	JCS C 1327 (1966)
I_2, KIO$_3$, HOAc/KOH	TL 4485 (1973)
I_2, TlOAc, HOAc/H_2O, Δ/NaOH	Org Syn *59* 169 (1980)

trans hydroxylation

I_2, TlOAc, HOAc/NaOH	Org Syn *59* 169 (1980)
I_2, AgO$_2$CPh/KOH	Org Rxs *9* 350 (1957) (review)
RCO$_3$H/H_3O^+	Org Rxs *7* 378 (1953) (review)
HCO$_2$H, H_2O_2, H_2O	JCS 3634 (1950) Org Syn Coll Vol *3* 217 (1955) Syn 449 (1979)
CF$_3$CO$_3$H, (Et$_3$NH)O$_2$CCF$_3$/HCl	JACS *76* 3472 (1954)
HO$_2$CCH$_2$CH$_2$CO$_3$H, H_2O	BSCF 2800 (1963)
KHSO$_5$, KHSO$_4$, K$_2$SO$_4$, HOAc, H_2SO_4, H_2O	JOC *25* 1901 (1960)

$$>C=C< \longrightarrow -\overset{\overset{\text{HO}}{|}}{C}-\underset{\underset{\text{X}}{|}}{\overset{|}{C}}-$$

X	Reagent(s)	
halogen	—	See page 325, Section 4.
OAc	I_2, AgOAc, H_2O, HOAc	JACS *109* 6403 (1987)
NHPh	PhNH$_2$, HgO·2 HBF$_4$	Syn 376 (1981)

2. Alkynes

$$RC\equiv CH \xrightarrow[\text{cat Rh(III) porphyrin}]{\text{NaBH}_4, O_2} RCH_2CH_2OH$$

$$RC\equiv CR \xrightarrow{\hspace{3cm}} R\overset{\overset{\text{OH}}{|}}{C}HCH_2R$$

$$\text{JOC } 52 \text{ } 2555 \text{ } (1987)$$

5. ORGANOBORANES

General References

H. C. Brown, "Hydroboration," W. A. Benjamin, New York (1962)
Organomet Chem Syn *1* 305 (1972) (alkylation of halo compounds)
Angew Int *11* 692 (1972) (free radical reactions)
H. C. Brown, "Boranes in Organic Chemistry," Cornell Univ. Press, Ithaca, New York (1972)
G. M. L. Cragg, "Organoboranes in Organic Synthesis," Marcel Dekker, New York (1973)
See Intra-Science Chemistry Reports (1973)

1. Hydroboration–Oxidation

$$RCH{=}CH_2 \xrightarrow{\text{H}-\text{B}\langle} RCH_2CH_2B{\langle} \xrightarrow[\text{NaOH}]{\text{H}_2\text{O}_2} RCH_2CH_2OH$$

Reviews:

H. C. Brown, "Hydroboration," W. A. Benjamin, New York (1962)
Org Rxs *13* 1 (1963)
JOC *51* 445 (1986) (cyclic dienes)

$BH_3 \cdot THF$	JACS *82* 4708 (1960); *88* 5851 (1966)
$BH_3 \cdot SMe_2$	JOC *39* 1437 (1976); *42* 1392 (1977)
	Org Prep Proc Int *13* 225 (1981) (review)
$MeBH_2$	JOC *51* 4925 (1986)
Me_2BH	JOC *51* 4925 (1986)
9-BBN	JACS *96* 7765 (1974)
	JOC *44* 2328 (1979); *46* 4599 (1981);
	50 5583 (1985)
$(Sia)_2BH$	JACS *83* 1241 (1961); *84* 190 (1962) (dienes)
	Org Syn *64* 164 (1985)
$\langle\hspace{-0.3em}\bigcirc\hspace{-0.3em}\rangle$ BH	JACS *106* 3768 (1984)

Me$_2$CHCMe$_2$BHCl·SMe$_2$

JOC 45 4540 (1980); 47 863, 872 (1982)

H$_2$BCl·OEt$_2$

JOC 38 182 (1973)
JACS 98 1785 (1976)

H$_2$BCl·SMe$_2$

JOC 42 2533 (1977); 44 2417 (1979); 51 895
   (1986)

H$_2$BBr·SMe$_2$

Syn 695 (1977)
JOC 44 2417 (1979); 51 895 (1986)

H$_2$BI·SMe$_2$

JOC 44 2417 (1979); 51 895 (1986)

HBCl$_2$·OEt$_2$

JACS 98 1798 (1976)

HBCl$_2$·SMe$_2$

JOC 42 2533 (1977)

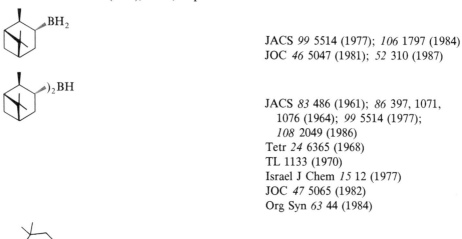

JACS 93 1816 (1971); 97 5249 (1975)
Tetr 32 981 (1976)

JOC 42 3243 (1977)

Syn 214 (1981)

1.1. Asymmetric Hydroboration

Reviews:

Tetr 37 3547 (1981)
H. C. Brown, P. K. Jadhav in "Asymmetric Synthesis," Ed. J. D. Morrison, Academic Press,
   New York (1983), Vol 2, Chpt 1

JACS 99 5514 (1977); 106 1797 (1984)
JOC 46 5047 (1981); 52 310 (1987)

JACS 83 486 (1961); 86 397, 1071,
   1076 (1964); 99 5514 (1977);
   108 2049 (1986)
Tetr 24 6365 (1968)
TL 1133 (1970)
Israel J Chem 15 12 (1977)
JOC 47 5065 (1982)
Org Syn 63 44 (1984)

JOC 46 2988 (1981)

Heterocycles *18* 169 (1982)

JACS *107* 4549 (1985)

$$RC\equiv CH \xrightarrow[\substack{\text{2. OH}^- \\ \text{3. [O]}}]{\text{1. BH}_3} RCH_2CH_2OH$$

JACS *89* 291 (1967)

2. Carbonylation and Related Reactions

$$\diagdown B-R \longrightarrow RCH_2OH$$

Organoborane	Reagents	
R_3B	$\overline{C}H_2\overset{+}{S}Me_2/[O]$	CC 505 (1967)
	$CO, NaBH_4/OH^-$	JACS *89* 2740 (1967)
	$CO, MHBR_3$ (M = Li, Na, K)$/OH^-$	JOC *44* 467 (1979)
⟨BR	$CO, LiHAl(OMe)_3/OH^-$	JACS *91* 2144 (1969)
	$CO, LiHAl(OMe)_3/LiAlH_4/[O]$	JOC *45* 4067 (1980)
	$CO, KHB(O\text{-}i\text{-}Pr)_3/OH^-$	Syn 701 (1979)

$$R_3B \longrightarrow R_2CHOH$$

$CO, H_2O, \Delta/KOH$	JACS *89* 2738 (1967)
$CO, LiHAl(OMe)_3/HCl/H_2O_2, NaOH$	Syn 676 (1978)

$$R_3B \longrightarrow R_3COH$$

$CO, HOCH_2CH_2OH, \Delta/H_2O_2, NaOH$	JACS *89* 2737, 5478 (1967); *90* 5283 (1968); *91* 1224, 1226 (1969); *92* 2460, 6648 (1970); *93* 1818 (1971) JOMC *26* C67 (1971) Syn 197 (1972) CC 607 (1979) Tetr *37* 2441 (1981) JOC *51* 4925 (1986) TL *28* 3771 (1987)
$NaCN/(CF_3CO)_2O/[O]$	CC 1048 (1971); 186 (1973) JCS Perkin I 653 (1981)
$HCClF_2, LiOCEt_3/[O]$	JACS *93* 2070 (1971)

$HCCl_2OCH_3$, $LiOCEt_3/[O]$

JOC *38* 2422, 3968 (1973)
Angew Int *20* 1038 (1981)
Organomet *1* 212 (1982)

$LiC(SPh)_3/HgCl_2$ or $MeOSO_2F/[O]$

CC 1149 (1981)

3. Other Reactions

$$(RCH_2)_3B \xrightarrow[h\nu]{Br_2} R\overset{Br}{\overset{|}{C}}HB(CH_2R)_2 \xrightarrow[\text{2. [O]}]{\text{1. } H_2O} R\overset{OH}{\overset{|}{C}}HCH_2R$$

$$(RCH_2)_3B \xrightarrow[\text{[O]}]{Br_2 \quad H_2O} R\overset{OH}{\overset{|}{C}}(CH_2R)_2$$

$$Me_2CHCMe_2BH_2 \xrightarrow[\text{2. } H_2C=CHR]{\text{1. }} \xrightarrow[\text{2. [O]}]{\text{1. } Br_2} \text{(cyclopentane with HO and } CH_2CH_2R\text{)}$$

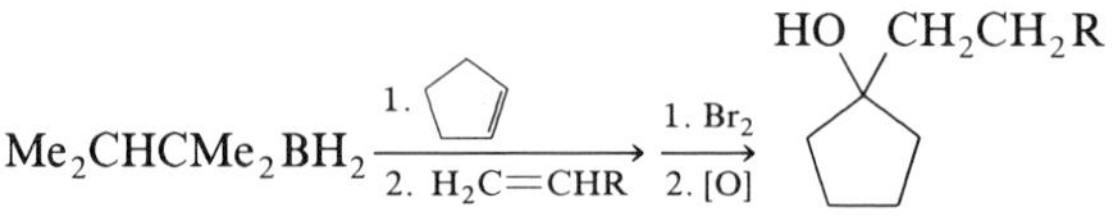

JACS *93* 1025, 2796 (1971)
Syn 303, 304 (1972)
JOC *44* 2417 (1979)

$$R_3B \longrightarrow R_2\overset{OH}{\overset{|}{C}}R'$$

$LiCR'(SPh)_2/HgCl_2$, H_2O/H_2O_2, NaOH   CC 863 (1974)

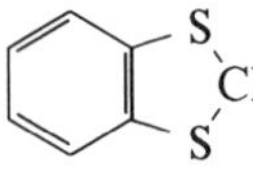

(benzo-1,3-dithiole)CHR'Li/$HgCl_2$, H_2O/H_2O_2, NaOH   JCS Perkin I 1172 (1977)
TL 1895 (1979)

$$R_3B \xrightarrow{LiC(OMe)=CH_2} \xrightarrow[NaOH]{E^+ \quad H_2O_2} R_2\overset{OH}{\overset{|}{C}}CH_2E$$

$$E^+ = HCl, R'X$$

JOMC *156* 123 (1978)

6. ETHER CLEAVAGE

Reviews:

Chem Rev *54* 615 (1954); *59* 737 (1959) (epoxides)
Houben-Weyl, "Methoden der Organischen Chemie," 4th ed, Vol 6/3, G. Thieme, Stuttgart
   (1965), p 143
"The Chemistry of the Ether Linkage," Ed. S. Patai, Interscience, New York (1967), pp 21–80
Syn 249 (1983); 629 (1984) (epoxides)
Pure Appl Chem *55* 589 (1983) (chiral 2,3-epoxy alcohols)

1. Cleavage of Acyclic Ethers

$$ROR' \longrightarrow ROH$$

See also page 349, Section 7.

Review: Chem Rev *54* 615 (1954)

Reagent(s)	R	R′	
H_2, cat Raney Ni	1°, 2°, 3° alkyl; aryl	benzylic	Org Rxs 7 263 (1953)
H_2, cat Pd-C	aryl 1°, 2° alkyl	$2,4\text{-Me}_2\text{C}_6\text{H}_3\text{CH}_2$ PhCH_2	Syn 987 (1982) CC 1234 (1980) JOC *47* 1373 (1982)
cat Pd(OH)$_2$, C,  ⬡, Δ	1°, 2° alkyl	PhCH_2	Syn 396 (1981) TL *28* 4131 (1987)
HCO_2H, cat Pd-C	1°, 2° alkyl	PhCH_2	Carbohydr Res *83* 175 (1980)
$(NH_4)O_2CH$, cat Pd-C, MeOH or CH_3COCH_3	1°, 2° alkyl	PhCH_2, Ph_3C	Syn 76 (1985)

ROR′ ⟶ ROH (*Continued*)

Reagent(s)	R	R′	
(NH₄)O₂CH, cat Pd-C, MeOH, HOAc	serine or threonine	PhCH₂	Syn 929 (1980)
N₂H₄, H₂O, cat Pd-C	1°, 2° alkyl	PhCH₂	Syn 317 (1986) TL 28 6381 (1987)
Na-K	aryl	PhCH₂	JOC 30 1610 (1965)
Na, NH₃	aryl	alkyl	JCS 102 (1947)
i-Bu₂AlH or i-Bu₃Al	aryl	Me	Syn 617 (1975)
LiAlH₄, C₆H₆	aryl	Me	CC 507 (1980)
LiAlH₄, CCl₄, THF	3° alkyl	Me	JOC 51 1620 (1986)
HCl	aryl	MeOCH₂	JACS 104 5551 (1982)
HBr, HOAc	aryl	Me	JACS 109 3098 (1987)
CF₃CO₂H	aryl	MeOCH₂CH₂OCH₂	JACS 100 8031 (1978)
BF₃·OEt₂, n-Bu₄NI	1°, 2° alkyl; aryl	Me, PhCH₂, CH₂CH=CH₂	Syn 274 (1985)
BCl₃	aryl	Me	JCS 2929 (1960) TL 4153 (1966)
BBr₃	1°, 2° alkyl; aryl	1°, 2° alkyl; PhCH₂	JACS 64 1128 (1942) JCS 2929 (1960) Chem Ind 1658 (1963) Tetr 24 2289 (1968) JOC 41 170 (1976); 44 4444 (1979); 51 1620 (1986) J Med Chem 20 164 (1977); 24 773 (1981) Syn Commun 9 341, 407 (1979)
BBr₃, NaI, 15-crown-5	1°, 2° alkyl	Me	TL 22 4239 (1981)
BI₃	aryl	Me	Syn Commun 9 341 (1979)
BX₃·SMe₂ (X = Cl, Br)	aryl	Me, CH₂	TL 21 3731 (1980)
BF₃·OEt₂, Me₂S	1°, 2° alkyl; aryl	PhCH₂	Chem Pharm Bull 28 3662 (1980) CC 1802 (1987)
BF₃·OEt₂, RSH	1°, 2° alkyl; aryl	Me, PhCH₂	JCS Perkin I 2237 (1976); 611 (1977) CL 97 (1979) JOC 44 1661 (1979) Chem Pharm Bull 28 3662 (1980)

BF$_3$·OEt$_2$, PhSH	2° alkyl	MeOCH$_2$	JACS *100* 1938 (1978)
BF$_3$·OEt$_2$, HSCH$_2$CH$_2$SH, HCl	2° alkyl	Me	JACS *106* 3539 (1984)
Me$_2$BBr	1°, 2° alkyl; aryl 1°, 2°, 3° alkyl	Me, PhCH$_2$ MeOCH$_2$, MeOCH$_2$CH$_2$OCH$_2$, MeSCH$_2$	TL *24* 2969 (1983) TL *24* 3969 (1983) JOC *49* 3912 (1984)
Ph$_2$BBr	1° alkyl	EtOCH$_2$, MeOCH$_2$CH$_2$OCH$_2$	TL *24* 3969 (1983) JOC *49* 3912 (1984)
AlX$_3$ (X = Cl, Br), EtSH	1°, 2° alkyl; aryl	Me, CH$_2$	CL 97 (1979) JOC *45* 4275 (1980)
R$_2$AlCl (R = Et, *i*-Bu)	1° alkyl	Ph$_3$C	TL *23* 2641 (1982)
Me$_3$SiSR (R = Me, Ph), ZnI$_2$, *n*-Bu$_4$NI	1°, 2° alkyl; aryl	Me, PhCH$_2$	TL *21* 2305 (1980)
H$_2$SiI$_2$	1° alkyl, aryl	Me	JOC *52* 4846 (1987)
SiCl$_4$, NaI	1°, 2° alkyl; aryl	1°, 2° alkyl; allyl; benzyl	Syn 1048 (1982)
MeSiCl$_3$, NaI	1°, 2° alkyl	Me, Et, PhCH$_2$, Ph$_3$C, THP	Angew Int *20* 690 (1981) JOC *48* 3667 (1983)
Me$_3$SiCl, NaI	1°, 2° alkyl; aryl	Me, Et, PhCH$_2$, Ph$_3$C	CC 874 (1978) JOC *44* 1247 (1979) Ann 1416 (1983) Chem Ind 34 (1984)
Me$_3$SiI	1°, 2° alkyl; aryl	Me, *t*-Bu	Angew Int *15* 774 (1976) JOC *42* 3761 (1977); *44* 4444 (1979) Org Syn *59* 35 (1980) CC 876 (1987)
(Me$_3$Si)$_2$, I$_2$	2° alkyl, aryl	Me, PhCH$_2$, Ph$_3$C	Angew Int *18* 612 (1979) Ann 1416 (1983)
PhSiMe$_3$, I$_2$	aryl	1° alkyl, PhCH$_2$	Syn 417 (1977)
P$_2$I$_4$	aryl	MeOCH$_2$, MeOCH$_2$CH$_2$, Me$_3$SiCH$_2$CH$_2$OCH$_2$	TL *27* 1607 (1986)
FeCl$_3$	1°, 2° alkyl	ArCH$_2$	TL *28* 3823 (1987)
ZnBr$_2$ or TiCl$_4$	2°, 3° alkyl; allylic; aryl	MeOCH$_2$CH$_2$OCH$_2$	TL 809 (1976) JACS *100* 8034 (1978)
HgCl$_2$ or AgNO$_3$, H$_2$O	1° alkyl, 1° allylic, 1° benzylic	MeSCH$_2$	TL 3269 (1975)

ROR′ ⟶ ROH   (*Continued*)

Reagent(s)	R	R′	
n-Bu$_4$NF	1°, 2° alkyl; allylic; aryl	Me$_3$SiCH$_2$CH$_2$OCH$_2$	TL *21* 3343 (1980)
n-Bu$_4$NI, BF$_3$·OEt$_2$	1°, 2° alkyl; aryl	Me, allyl, benzyl	Syn 274 (1985)
LiI; 2,4,6-collidine	aryl	Me	CC 616 (1969)
NaNH$_2$, ⬡NH	aryl	1° alkyl, PhCH$_2$	Chem Ind 80 (1957)
PhNMeNa, HMPA	aryl	Me	Syn 638 (1980)
NaCN, DMSO	aryl	Me	TL 5183 (1978)
Na$_2$S, *N*-methyl-pyrrolidone	aryl	Me	JACS *98* 3237 (1976)
MeSLi, DMF	aryl	Me	TL 3859 (1977) JACS *101* 1353 (1979) JOC *45* 4407 (1980)
MeSNa, DMF or HMPA	aryl	9-anthrylmethyl	JACS *96* 590 (1974)
EtSNa, DMF	aryl	Me	TL 1327 (1970) Austral J Chem *25* 1719, 1731 (1972) JACS *98* 3237 (1976); *108* 1311, 2662, 3040 (1986) JOC *43* 2553 (1978); *52* 1072 (1987)
PrSNa, DMF	aryl	Me	TL *28* 4015 (1987)
n-PrSLi, HMPA	aryl	Me	TL 4459 (1970)
n-BuSLi, HMPA	aryl	Me	TL 505 (1977)
t-BuSLi, HMPA	aryl	Me	JACS *107* 2712 (1985)
PhSLi, HMPA	aryl	Me	TL *27* 2687 (1986)
p-MeC$_6$H$_4$SNa, HMPA	aryl	Me, PhCH$_2$	Syn 191 (1976)
PhCH$_2$SeNa, DMF	aryl	Me	JOC *42* 1228 (1977)
Ph$_2$PLi	aryl	Me, PhCH$_2$, allyl	Chem Ind 1386 (1964) JCS 4120 (1965) TL 1071 (1976) JACS *107* 7967 (1985)
DDQ, H$_2$O	1° alkyl or allylic, 2° alkyl	4-MeOC$_6$H$_4$CH$_2$ or 3,4-(MeO)$_2$C$_6$H$_3$CH$_2$	TL *23* 885 (1982); *28* 5125 (1987)
(NH$_4$)$_2$Ce(NO$_3$)$_6$, H$_2$O, CH$_3$CN	1°, 2° alkyl	*p*-MeOC$_6$H$_4$	CC 201 (1984)

2. Reduction of Epoxides

$$RCH\overset{O}{\underset{\diagdown}{}}CH_2 \longrightarrow RCH\overset{OH}{\underset{|}{}}CH_3$$

H_2, Raney nickel, NaOH	JACS *71* 3362 (1949)
Li, RNH_2	Quart Rev *12* 22–23 (1958)
Li, NH_3	JACS *107* 1015 (1985)
	JOC *51* 5450 (1986)
Li, $H_2NCH_2CH_2NH_2$	JOC *35* 3243 (1970); *51* 3391 (1986)
	JACS *92* 6914 (1970); *107* 1015 (1985)
Ca, $H_2NCH_2CH_2NH_2$	JOC *51* 3391 (1986)
Li, biphenyl/H_2O	Angew Int *25* 653 (1986)
Li or Mg, naphthalene/H_2O	Angew Int *25* 653 (1986)

	JOC *42* 512 (1977)
BH_3, $NaBH_4$	JOC *45* 3836 (1980)
$LiBH_4$	JCS C 928 (1970)
$LiBH_4$, MeOH	JOC *51* 4000 (1986)
$NaH_2B[(OCH_2CH_2)_nOH]_2$ ($n = 8, 9$)	JOC *52* 671 (1987)
$LiHBEt_3$	JACS *95* 8486 (1973); *106* 3252 (1984);
	109 8117 (1987)
	Angew Int *18* 958 (1979)
	JOC *45* 1 (1980)
	TL *28* 4959 (1987)
$KHBPh_3$	JOC *52* 5564 (1987)
$Li\left[n\text{-}Bu_2B \right]$	CC 672 (1976)
$i\text{-}Bu_2AlH$	JOC *50* 2443 (1985)
$LiAlH_4$	JACS *70* 3738 (1948); *71* 1675 (1949);
	74 923, 5917 (1952)
	Org Rxs *6* 469 (1951)
$NaH_2Al(OCH_2CH_2OCH_3)_2$	JOC *38* 1385 (1973)
$LiHAl(i\text{-}Bu)_2(n\text{-}Bu)$	JOC *49* 1717 (1984)
NaTeH/Ni boride	TL *26* 6197 (1985)

$$RCH\overset{O}{\underset{\diagdown}{}}CH_2 \longrightarrow RCH_2CH_2OH$$

H_2, Raney nickel	JACS *71* 3362 (1949)
Li, naphthalene/H_2O (R = Ph)	Angew Int *25* 653 (1986)

$$\text{RCH}\overset{O}{\overset{\diagdown}{\underset{}{-}}}\text{CH}_2 \longrightarrow \text{RCH}_2\text{CH}_2\text{OH} \quad (\textit{Continued})$$

K, biphenyl/H_2O (R = CO_2Et)	Angew Int *25* 653 (1986)
BH_3, $BF_3\cdot OEt_2$	CC 1549 (1968)
BH_3, $NaBH_4$	JACS *90* 2686 (1968)
$NaBH_3CN$, $BF_3\cdot OEt_2$	JOC *46* 5214 (1981) TL *27* 6071 (1986)
$KHBPh_3$, cat Ph_3B	JOC *52* 5564 (1987)
$LiAlH_4$, $AlCl_3$	JACS *78* 3226 (1956); *84* 2356 (1962); *90* 6495 (1968) Rec Chem Prog *22* 129 (1961) TL 981 (1978)

$$\text{RCH}\overset{O}{\overset{\diagdown}{\underset{}{-}}}\text{CHR}' \longrightarrow \overset{OH}{\underset{|}{\text{RCHCH}_2\text{R}'}}$$

Li, NH_3	BCSJ *52* 2640 (1979); *53* 2642 (1980)
Li, K or Mg; naphthalene or biphenyl/H_2O	Angew Int *25* 653 (1986)
i-Bu$_2$AlH	JACS *107* 1015 (1985) TL *28* 2021 (1987)

$$\text{C}{=}\text{C}-\overset{O}{\overset{\diagup\diagdown}{\text{C}-\text{C}}}-\text{C} \xrightarrow{\text{LiAlH}_4} \text{C}{=}\text{C}-\overset{OH}{\underset{|}{\text{C}}}-\underset{\underset{H}{|}}{\text{C}}-\text{C}$$

JACS *74* 5917 (1952); *81* 1643 (1959); *86* 2657 (1964); *107* 7978 (1985)
JOC *33* 423 (1968); *50* 5910 (1985); *52* 4898 (1987)

cat Pd-BaSO$_4$, (cyclohexene)	Helv *54* 2775, 2785 (1971)
Al(Hg), H_2O, HCl	Steroids *27* 603 (1976)

$$\underset{\text{RCHCH}}{\overset{HO}{\underset{|}{}}}\overset{O}{\overset{\diagup\diagdown}{-\text{CH}_2}} \xrightarrow{\text{LiAlH}_4} \underset{\text{RCHCHCH}_3}{\overset{HO\ \ OH}{\underset{|\ \ \ |}{}}}$$

JOC *50* 5687 (1985)

$$\text{RCH}\overset{O}{\overset{\diagup\diagdown}{\underset{}{-}}}\text{CHCH}_2\text{OH} \longrightarrow \overset{OH}{\underset{|}{\text{RCHCH}_2\text{CH}_2\text{OH}}}$$

$NaBH_3CN$, $BF_3\cdot OEt_2$	TL *28* 4569 (1987)
$LiAlH_4$	JOC *47* 1378 (1982) TL *28* 2629 (1987)

$NaH_2Al(OCH_2CH_2OCH_3)_2$

TL *23* 2719, 4541 (1982); *27* 3535 (1986);
  28 703 (1987)
JOC *47* 1378 (1982)
JACS *104* 1109, 5521 (1982); *109* 2205 (1987)
CC 1292 (1982); 1408 (1986); 311 (1987)
CL 175 (1983)

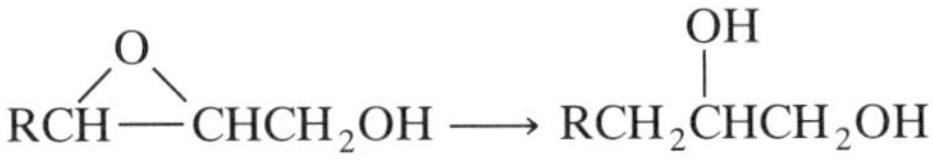

$LiBH_4$, $Ti(O\text{-}i\text{-}Pr)_4$

TL *27* 4343 (1986)

$i\text{-}Bu_2AlH$

TL *23* 2719, 3597 (1982)

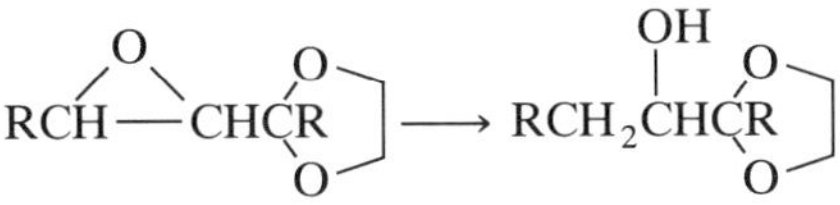

$LiAlH_4$

JOC *50* 5696 (1985)

Zn, Me_3SiCl

TL *28* 551 (1987)

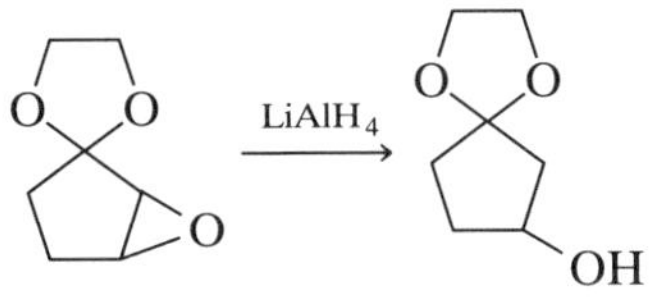

TL *28* 551 (1987)

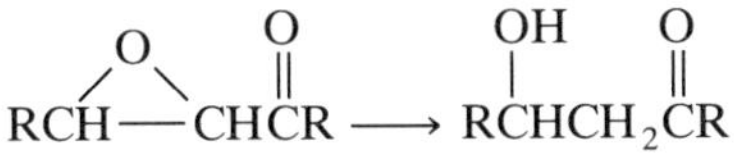

PhSeNa, HOAc, EtOH

TL *28* 4293 (1987)

NaTeH

CL 271 (1984)

NaI, NaOAc, HOAc

TL 4377 (1974)

Al(Hg)

CC 254 (1973)
JOC *38* 3187 (1973); *43* 3942 (1978);
  52 3560 (1987)
Helv *57* 781 (1974)
JACS *100* 4618 (1978)
Steroids *34* 683 (1979)

$CrCl_2$

JOC *19* 131 (1954)

$Cr(OAc)_2$

JACS 77 4601 (1955)
Helv *41* 1667 (1958)
Coll Czech Chem Commun *26* 1207 (1961)
JOC *37* 565 (1972); *43* 3946 (1978)

Zn, CH_3OH, HOAc

Chem Pharm Bull *24* 2810 (1976)

SmI_2, CH_3OH

JOC *51* 2596 (1986)

electrolysis

JOC *46* 5017 (1981)

$$\underset{RCH-CHCO_2R}{\overset{O}{\triangle}} \longrightarrow \underset{RCHCH_2CO_2R}{\overset{OH}{|}}$$

K, biphenyl/H_2O Angew Int *25* 653 (1986)

MgI_2/*n*-Bu_3SnH TL *28* 4435 (1987)

SmI_2, $Me_2NCH_2CH_2OH$, THF, HMPA   TL *28* 4437 (1987)

3. Nucleophilic Opening of Cyclic Ethers

$$\underset{O}{\overset{}{\frown}}(CH_2)_n \longrightarrow HO(CH_2)_n X$$

X	n	Reagent(s)	
Cl	2	HCl	Ber *49* 477, 2778, 2782 (1916)
			JACS *63* 2541 (1941); *76* 1235, 1259 (1954)
			JCS 915 (1963); 644 (1965)
		($\overset{+}{NH}$) Cl^-	Syn Commun *11* 287 (1981)
		NaCl	JCS 915 (1963); 644 (1965)
		$AlCl_3$	Tetr *25* 1807 (1969) BCSJ *49* 1063 (1976)
		$SnCl_4$	JACS *76* 1235 (1954)
		$FeCl_3$	JOC *42* 343 (1977)
		Li_2CuCl_4	TL *27* 3697 (1986)
Br	2	HBr, $MgBr_2$, NH_4Br or NaBr	JACS *76* 1259 (1954)
		$(Me_2N)_2BBr$	TL *27* 827 (1986)
		Me_3SiBr, Et_3N (forms silyl ether)	Nouv J Chim *3* 705 (1979)
		Li_2NiBr_4	TL *25* 2061 (1984)
	2–4	Me_2BBr	TL *24* 2969 (1983) JOC *52* 1680 (1987)
		Me_3SiBr (forms silyl ether)	Ber *95* 174 (1962) TL 4483 (1978) Syn 383 (1981)
	2, 4–6	BBr_3	Heterocycles *18* 163 (1982)
	3	$MgBr_2$	JACS *73* 124 (1951)
	4	R_4NBr (R = Et, *n*-Bu)	JOC *51* 3372 (1986)
I	2	HI	JACS *76* 1259 (1954)
		Me_3SiI, Et_3N (forms silyl ether)	Nouv J Chim *3* 705 (1979)
		MgI_2	TL *28* 4435 (1987)
	2, 4	*n*-Bu_4NI, $BF_3 \cdot OEt_2$	Syn 274 (1985)
	4	NaI, $AlCl_3$	TL *25* 219 (1984)
		Me_3SiI (forms silyl ether)	Ber *95* 174 (1962) TL 4483 (1978)

Cl, Br, I	2	R_3SiX, cat PPh_3 or $n\text{-}Bu_4NCl$ (forms silyl ether)	TL 22 3803 (1981)
		X_2, PPh_3	TL 24 1307 (1983); 26 2011 (1985)

$$\underset{}{\text{>C}\overset{\displaystyle O}{-}\text{C<}} \xrightarrow[\text{2. (hydrolysis)}]{\text{1. reagent(s)}} -\underset{}{\overset{\displaystyle HO}{C}}-\underset{X}{C}-$$

Review: Chem Rev 59 737 (1959)

X	\	Reagent(s)	
RO		ROH, cat H_2SO_4	JACS 108 3040 (1986)
		$NaOCH_3$, CH_3OH	JOC 52 4812 (1987)
		$n\text{-}Bu_3SnOR$ (intramolecular)	TL 2249 (1976)
OH, OR		H_2O or ROH, Nalfion-H	Syn 280 (1981)
OR, SR, O_2CR		ROH, RSH or RCO_2H; Al_2O_3	TL 3597 (1975) Angew Int 17 487 (1978) (review)
SR		Me_3SiSR	TL 21 1485 (1980)
		RSH; cat $SmCl_3$, $CeCl_3$ or $Eu(fod)_3$	TL 28 6065 (1987)
		Li, K or Mg; naphthalene or biphenyl/RSSR	Angew Int 25 653 (1986)
O_3SCF_3		$Me_3SiO_3SCF_3$	JACS 101 2738 (1979)

SePh		(K)NaSePh	JACS 95 2697 (1973); 99 8208
			  (1977); 106 7854 (1984);
			  107 1691, 1777 (1985)
			TL 5087 (1978)
			Tetr 34 1049 (1978)
			JOC 51 2148 (1986)

		$Me_3SiSePh$, cat ZnI_2 or $n\text{-}BuLi$	CL 909 (1979)
		$i\text{-}Bu_2AlSePh$	JOC 50 5897 (1985)
NHR		RNH_2	TL 28 1603 (1987)
		Me_3SiNHR	TL 27 2451 (1986) (aryl epoxides only)
NR_2		Et_2AlNR_2	TL 22 195 (1981)
		R_2NH, Al_2O_3	Angew Int 17 487 (1978) (review)
		Me_3SiNR_2	JOC 32 2210 (1967)
		$n\text{-}Bu_3SnNR_2$, cat $LiNR_2$	JOMC 8 255 (1967)
N_3		Me_3SiN_3	Ann 266 (1975)
		Me_3SiN_3, cat $Ti(O\text{-}i\text{-}Pr)_4$	TL 27 4423 (1986)
NRCOR		$Me_3SiNRCOR$, cat $NaOSiMe_3$	Ber 102 14 (1969)

$$\text{>C}\overset{\displaystyle O}{\overset{\diagup\diagdown}{-}}\text{C<} \quad\xrightarrow[\text{2. (hydrolysis)}]{\text{1. reagent(s)}}\quad \overset{\displaystyle HO}{\underset{\displaystyle X}{-C-\overset{|}{C}-}} \quad (\textit{Continued})$$

X	Reagent(s)	
$-N{\equiv}C$	Me$_3$SiCN; cat Pd(CN)$_2$, SnCl$_2$ or Me$_3$Ga	JOC *52* 1013 (1987)
	Me$_3$SiCN, cat ZnCl$_2$	TL *24* 655 (1983)
	Me$_3$SiCN, cat ZnI$_2$	JACS *104* 5849 (1982) TL *25* 3259 (1984); 27 6307 (1986)
$-C{\equiv}N$	Me$_3$SiCN; Al(O-*i*-Pr)$_3$, *i*-Bu$_2$AlO-*i*-Pr or Et$_2$AlCl	JOC *52* 1013 (1987)
	Me$_3$SiCN, cat AlCl$_3$	TL 1449 (1973)
	Me$_3$SiCN, cat Et$_2$AlCl	JOC *47* 2873 (1982)

$$\overset{\displaystyle HO}{\underset{\displaystyle RCHCH}{|}}\overset{\displaystyle O}{\overset{\diagup\diagdown}{-}}CH_2 + Nuc \longrightarrow \overset{\displaystyle HO\ \ OH}{RCHCHCH_2Nuc}$$

Nuc = NaN$_3$; *t*-BuSH, NaOH; KCN

JOC *50* 5687 (1985)

$$RCH\overset{\displaystyle O}{\overset{\diagup\diagdown}{-}}CHCH_2OH + Nuc \longrightarrow \underset{\displaystyle Nuc}{RCH\overset{\displaystyle OH}{C}HCH_2OH}\quad or \quad \underset{\displaystyle Nuc}{R\overset{\displaystyle HO}{C}HCHCH_2OH}$$

Nuc = NaN$_3$, NaSPh, NaSePh

JOC *50* 5696 (1985)

$$RCH\overset{\displaystyle O}{\overset{\diagup\diagdown}{-}}CHCH_2OH \longrightarrow \underset{\displaystyle X}{RCH\overset{\displaystyle OH}{C}HCH_2OH}$$

X	Reagents	
Cl	(*i*-PrO)$_2$TiCl$_2$, diethyl tartrate	TL *27* 6031 (1986)
N$_3$	NaN$_3$, zeolite	CL 1327 (1986)

$$RCH\overset{\displaystyle O}{\overset{\diagup\diagdown}{-}}CHCH_2OH \xrightarrow[\text{NH}_4\text{Cl}]{\text{NaN}_3} \underset{\displaystyle N_3}{RCH\overset{\displaystyle HO}{C}HCH_2OH}$$

JOC *52* 3337 (1987)

$$\text{RCH}\overset{\displaystyle O}{\overbrace{}}\text{CHCH}_2\text{OH} + \text{Nuc} \xrightarrow{\text{Ti(O-}i\text{-Pr)}_4} \underset{\underset{\displaystyle \text{Nuc}}{|}}{\text{RCHCHCH}_2\text{OH}} \ \overset{\text{OH}}{\overset{|}{}}$$

Nuc = R_2NH, ROH, NaOAr, PhSH, PhSeH, PhSNa, N_3^-, CN^-,

NH$_4$X (X = Cl, Br, SCN, O$_2$CR), RCO$_2$H

JOC *50* 1557 (1985); *51* 3710, 5413 (1986); *52* 667 (on *p*-nitrobenzoate ester), 4973 (1987)
TL *27* 4987, 4991 (1986)

$$\text{RCH}\overset{\displaystyle O}{\overbrace{}}\text{CHCH}_2\text{OH} + \text{Nuc} \longrightarrow \underset{\underset{\displaystyle \text{Nuc}}{}}{\text{RCHCHCH}_2\text{Nuc}} \ \overset{\text{HO OH}}{\overset{| \ \ |}{}}$$

Nuc	
RSH, NaOH; R$_2$NH	JOC *50* 5687 (1985)
RC≡CLi, BF$_3$·OEt$_2$	CC 202 (1984)

$$\text{RCH}\overset{\displaystyle O}{\overbrace{}}\text{CHCH}_2\text{O}_3\text{SR}' \xrightarrow[\text{cat HClO}_4]{\text{R}''\text{OH}} \text{RCHCHCH}_2\text{O}_3\text{SR}' \ \overset{\text{R}''\text{O OH}}{\overset{| \ \ |}{}}$$

R′ = Me, *p*-C$_6$H$_4$Me; R″ = H, Me

JOC *50* 5687 (1985)

$$\text{RCH}\overset{\displaystyle O}{\overbrace{}}\text{CHCH} + \text{Nuc} \longrightarrow \text{RCHCHCH}$$

Nuc = NaN$_3$, NaSPh

JOC *50* 5696 (1985)

$$\text{RCH}\overset{\displaystyle O}{\overbrace{}}\text{CHCX} \xrightarrow{\text{Nuc}} \underset{\underset{\displaystyle \text{Nuc}}{}}{\text{RCHCHCX}} \ \overset{\text{HO O}}{\overset{| \ \ \|}{}}$$

X	Nuc	
OH	R$_2$NH, Ti(O-*i*-Pr)$_4$	JOC *50* 1560 (1985)
	PhSH or PhSNa, Ti(O-*i*-Pr)$_4$	JOC *50* 1560 (1985)
OR	NH$_3$	Arch Pharm *307* 871 (1974); *308* 135 (1975) JCS Perkin I 1618 (1980)
NR$_2$	NaN$_3$, MgSO$_4$	JOC *50* 5696 (1985)
	R$_2$NH	BCSJ *54* 2136 (1981) Arch Pharm *314* 464 (1981)
	PhSH or PhSNa, Ti(O-*i*-Pr)$_4$	JOC *50* 1560 (1985)
OH, OR, NR$_2$	ArNH$_2$	Arch Pharm *312* 26 (1979)
	R$_2$NH	Arch Pharm *312* 138 (1979)

$$\underset{\text{RCH}-\text{CHCX}}{\overset{\text{O} \quad \text{O}}{\diagdown \quad \parallel}} \xrightarrow{\text{Nuc}} \underset{\underset{\text{Nuc}}{|}}{\overset{\text{HO} \quad \text{O}}{\underset{|}{\text{RCHCHCX}}{\overset{|}{}}}}$$

X	Nuc	
OH	PhCH$_2$NH$_2$ or NH$_3$	JCS 1116 (1962)
	R$_2$NH	JOC *50* 1560 (1985)
	NaSPh	JOC *50* 1560 (1985)
OR	NH$_3$	JACS *80* 187 (1958)
NR$_2$	NaSPh	JOC *50* 1560, 5696 (1985)

$$\underset{}{\overset{\text{O}}{\square^{\parallel}}} \longrightarrow \text{X(CH}_2)_3\text{OSiMe}_3$$

X	Reagents	
—CN	Me$_3$SiCN, cat Et$_2$AlCl	JOC *47* 2873 (1982)
—N≡C	Me$_3$SiCN, cat ZnI$_2$	Syn Commun *15* 775 (1985)

4. Alkylation of Cyclic Ethers

4.1. Epoxides

$$\underset{\text{RCH}-\text{CHR}'}{\overset{\text{O}}{\diagup \diagdown}} \xrightarrow[\text{naphthalene or biphenyl}]{\text{Li, K or Mg}} \underset{\text{RCHCHR}'}{\overset{\text{M} \quad \text{OM}}{\underset{|}{}\underset{|}{}}} \xrightarrow[\text{2. H}_2\text{O}]{\text{1. E}^+} \underset{\text{RCHCHR}'}{\overset{\text{E} \quad \text{OH}}{\underset{|}{}\underset{|}{}}}$$

$$\text{E}^+ = \text{H}_2\text{O, D}_2\text{O, RX, CO}_2, \text{RSSR}$$

Angew Int *25* 653 (1986)

$$\text{RM} + \underset{\text{H}_2\text{C}-\text{CHR}'}{\overset{\text{O}}{\diagup \diagdown}} \longrightarrow \underset{\text{RCH}_2\text{CHR}'}{\overset{\text{OH}}{\underset{|}{}}}$$

RM	
Li, naphthalene/n-Bu$_3$P·CuI/RX (R = 1° alkyl, aryl)	JOC *52* 5057 (1987)
CH$_3$Li	JACS *92* 4979 (1970)
CH$_3$Li, CH$_3$MgCl	TL *27* 5335 (1986)
	JOC *52* 4505 (1987)

RLi, BF$_3$·OEt$_2$ (R =1° alkyl, aryl, vinylic)	JACS _106_ 3693 (1984)

R—▷—Li (intramolecular) — JOC _47_ 3211 (1982)

▷<Li / SPh — TL _23_ 2379 (1982)

S / >—Li / S — JACS _106_ 2949 (1984)

PhSCH(Li)SiMe$_3$	TL _27_ 4403 (1986)
RCHLiSO$_2$Ph, BF$_3$·OEt$_2$	TL _27_ 6345 (1986) CC 1226 (1987)
RC(Li)$_2$SO$_2$Ph	JACS _109_ 6205 (1987)
LiCH$_2$CO$_2$Li or LiCH$_2$CO$_2$R	JOC _42_ 1688 (1977)
LiCH$_2$CN	JOC _51_ 2230 (1986) TL _28_ 1611 (1987)
H$_2$C=CHCH$_2$Li	JACS _108_ 8015 (1987) JOC _52_ 862 (1987)
Me$_3$SiC≡CCH$_2$Li	JACS _97_ 3258 (1975)
(_i_-Pr)$_3$SiC≡CCH$_2$Li	TL _23_ 719 (1982)
ArLi	JOC _52_ 5668 (1987) TL _28_ 861 (1987)
PhLi, CuCN	TL _28_ 1783 (1987)
RCH=CHLi, MgBr$_2$	JOC _52_ 4495 (1987)
RCH=CHLi, cat CuI	JACS _109_ 5437 (1987)
HC≡CLi(EDA), DMSO	Compt Rend C _265_ 839 (1967) JCS D 674 (1969) CC 223 (1973) JACS _101_ 5364 (1979); _103_ 7520 (1981) JOC _51_ 4840 (1986)
RC≡CLi	Syn 26 (1978); 490 (1981) Helv _65_ 385 (1982) TL _23_ 1331 (1982) JACS _106_ 3548 (1984); _109_ 6176 (1987)
LiO$_2$CC≡CLi	TL 3173 (1976)
RC≡CLi, BF$_3$·OEt$_2$	TL _24_ 391 (1983); _27_ 4991 (1986); _28_ 391, 5457 (1987) JOC _52_ 622, 2860 (1987) JACS _109_ 7495 (1987)

$$RM + H_2C\overset{O}{\overset{\diagup\diagdown}{-}}CHR' \longrightarrow RCH_2\overset{OH}{\underset{|}{C}}HR' \quad (\textit{Continued})$$

RM

RMgX	Org Syn Coll Vol *1* 306 (1941)
	Chem Rev *49* 413 (1951); *59* 737 (1959)
	JOC *16* 673 (1951); *46* 4608 (1981)
	Rec Trav Chim *87* 1249 (1968)
	JACS *92* 4979 (1970)
	J Chem Res (S) 162 (1982)
	TL *27* 6071 (1986)
$H_2C{=}CHCH_2MgCl$	TL *27* 5335, 6071 (1986)
Me_2Mg	TL *23* 1763 (1982)
RMgX, cat CuCN	TL *27* 2679 (1986)
$MeMgCl$, cat $CuCl{\cdot}SMe_2$	TL *28* 3835 (1987)
RMgX, cat CuI	TL 1503 (1979); *23* 1267 (1982)
	Can J Chem *61* 1166 (1983)
	CC 620 (1986)
RMgX, CuI	CC 836 (1986); 429 (1987)
	JOC *51* 2230 (1986)
	TL *27* 6071 (1986); *28* 2627 (1987)
$H_2C{=}CHCH_2MgX$, CuI	TL *27* 6071 (1986)
$H_2C{=}CHMgBr$, CuI	TL *27* 4485, 5791 (1986)
	CC 877 (1987)
R_2CuMgX	TL *28* 1781, 1993 (1987)
$(RCH{=}CH)_2CuMgX$	TL *28* 1993 (1987)
R_2CuLi	JACS *92* 3813, 4979 (1970); *109* 8105 (1987)
	JOC *38* 4263 (1973); *40* 2263 (1975)
	(α-silyl epoxides); *48* 546 (1983);
	52 4495 (1987)
	Org Rxs *22* 253 (1975)
	JCS Perkin I 2885 (1982)
	TL *27* 6071 (1986); *28* 1781, 1993, 6191 (1987)
$R_2CuLi{\cdot}2\,RLi$	TL *21* 4365 (1980)
$(R_2C{=}CR)_2CuLi$	TL *26* 5837 (1985); *27* 2519, 5277 (1986);
	28 1781, 1993 (1987)
$[(RC{\equiv}C)Cu(\mathbf{CH{=}CHR})]\,Li$	JCS Perkin I 2954 (1979), 852 (1980)
$(Me_3SiC{\equiv}CCu\mathbf{CH_2CH{=}CROSiMe_3})\,Li$	JACS *107* 5495 (1985)
$(RCuCN)Li$	TL 3407 (1977); 2399 (1978)
$(R_2CuCN)Li_2$	JACS *104* 2305 (1982); *109* 6176 (1987)
	JOC *49* 3928 (1984)

$[(H_2C{=}CH)_2CuCN]Li_2$	JACS *106* 5304 (1984) JOC *52* 2838 (1987)
$[(H_2C{=}CH)_2CuCN]Li_2$, $BF_3 \cdot OEt_2$	TL *27* 5791 (1986)
$\left[\begin{smallmatrix} \square \\ S \end{smallmatrix} {-}CuR(CN) \right]Li_2$, $BF_3 \cdot OEt_2$	TL *25* 5959 (1984)
$\left[\begin{smallmatrix} \square \\ S \end{smallmatrix} {-}CuR(CN) \right]Li_2$	TL *25* 5959 (1984); *27* 4825 (1986) (R = vinylic); *28* 945 (1987) JOMC *285* 437 (1985)
$(R_2CuSCN)_2Li_2$	JOC *48* 546 (1983)
$\left[RCuN({-}\bigcirc{-})_2 \right]Li$	JACS *104* 5824 (1982)
$(RCuPPh_2)Li$	JACS *104* 5824 (1982)
$Et_2AlCH_2CO_2\text{-}t\text{-}Bu$	JOC *41* 1669 (1976) Syn 284 (1983)
$E\text{-}i\text{-}Bu_2AlCH{=}CHR$	JOC *42* 2712 (1977)
$(R_3AlCH{=}CRR')Li$	Syn 632 (1975); 1034 (1980) CC 17 (1976)
$RC{\equiv}CAlR'_2$ (R' = Me, Et)	CC 634 (1968); 287 (1975); 907 (1976) TL 1379 (1969); 2695 (1970); 3899 (1973); 1769 (1976) JACS *94* 4342, 4343 (1972); *95* 7171 (1973); *100* 3950 (1978); *103* 7520 (1981); *108* 5559 (1986); *109* 2205 (1987) JOC *41* 1669 (1976); *42* 394 (1977); *43* 353 (1978); *50* 3923 (1985) JCS Perkin I 2954 (1979); 852 (1980)
$(t\text{-}Bu)Me_2SiCH_2C{\equiv}CAlEt_2$	JOC *52* 4495 (1987)
$MeOAl(Me)C{\equiv}CR$	TL 3899 (1973)
$[Me_2Al(C{\equiv}CR)_2]Li$	JOC *50* 3923 (1985)
$RSiMe_3$ (R = allylic), $TiCl_4$	Syn 446 (1978) TL *24* 5661 (1983) CC 585 (1984) (intramolecular)
$RSiMe_3$ (R = allylic), $BF_3 \cdot OEt_2$	Can J Chem *61* 214 (1983) (intramolecular) TL *28* 2753 (1987) (intramolecular)
R_3MnLi	TL *27* 5351 (1986)

JACS *106* 6093 (1984)

$$\text{ArH} + \text{H}_2\text{C} \overset{O}{\underset{\triangle}{-}} \text{CHR} \xrightarrow{\text{Lewis acid}} \text{ArCHRCH}_2\text{OH}$$

JCS 5404 (1964) (intramolecular)
Tetr *25* 1807 (1969)
JOC *48* 2449, 4572 (1983); *52* 425 (1987) (all intramolecular)

$$\text{RCH} \overset{O}{\underset{\triangle}{-}} \text{CHSiMe}_2(\text{O-}i\text{-Pr}) \xrightarrow[\text{cat CuCN}]{\text{R'MgX}} \overset{OH}{\underset{R'}{\text{RCHCHSiMe}_2(\text{O-}i\text{-Pr})}} \xrightarrow[\text{KF}]{\text{H}_2\text{O}_2} \overset{\text{HO}\ \ \text{OH}}{\text{RCHCHR'}}$$

JOC *52* 4412 (1987)

$$\overset{\text{HO}}{\text{RCHCH}} \overset{O}{\underset{\triangle}{-}} \text{CH}_2 \xrightarrow{\text{R'M}} \overset{\text{HO}\ \ \text{OH}}{\text{RCHCHCH}_2\text{R'}}$$

R'M	
Me$_2$CuLi, LiC≡CCH$_2$OTHP	JOC *50* 5687 (1985)
PhSeCHSiMe$_3$, LiC≡CCH$_2$OTHP	TL *24* 4539 (1983)

$$\text{RCH} \overset{O}{\underset{\triangle}{-}} \text{CHCH}_2\text{OH} \xrightarrow{\text{R'M}} \overset{OH}{\underset{R'}{\text{RCHCHCH}_2\text{OH}}}$$

R'M	
LiC≡CH·EDA	CC 874 (1986) (mixture)
(1,3-dithian-2-yl)Li (S—⟨⟩—Li with S)	TL *27* 4825 (1986)
RMgX, cat CuI	JOC *48* 3131 (1983) JACS *108* 1035 (1986)

R'$_2$CuLi	JOC *38* 4346 (1973) (cyclohexyl
	  system); *52* 2596 (1987)
	TL 4343 (1979); *23* 707 (1982); *24* 1377 (1983);
	  28 5009 (1987)
	Tetr *37* 3873 (1981)

Me$_2$CuLi, BF$_3$·OEt$_2$	TL *28* 1139 (1987) (on trityl ether)
Me$_3$CuLi$_2$	JOC *52* 685 (1987) (mixture)
(CH$_3$CuCN)Li	JOC *51* 956 (1986) (mixture)

$(Me_2CuCN)Li_2$

JACS *108* 2776 (1986)
TL *28* 5009 (1987)
JOC *52* 3784 (1987)

$$R^1\!-\!\overset{\displaystyle O}{\overset{\diagup\;\backslash}{C}}\!-\!C\!-\!C_n\!-\!OR^2 \xrightarrow{\ RM\ } R^1\!-\!\overset{\displaystyle OH}{\underset{\underset{\displaystyle R}{|}}{\overset{|}{C}}}\!-\!\overset{|}{C}\!-\!C_n\!-\!OR^2$$

n	R^2		RM	
1	H		R_3Al	TL *23* 3597 (1982); *24* 1377 (1983); *27* 3369, 4601 (1986); *28* 1139 (1987) JOC *52* 2596 (1987)
			$RC\!\equiv\!CAlEt_2$	TL *23* 3597 (1982)
			Me_3CuLi_2	JOC *52* 685 (1987) (mixture)
			$(MeCuCN)Li$	JOC *51* 956 (1986) (mixture)
			$(Me_2CuCN)Li_2,$ $BF_3\!\cdot\!OEt_2$	TL *28* 5009 (1987)
	H, Me		Me_2CuLi	JOC *38* 4346 (1973) (cyclohexyl system)
	$CH_2OCH_3,$ CH_2Ph		$Me_2AlC\!\equiv\!CR$	JOC *48* 409 (1983)
	SiR_3		RLi, CuCN	CC 1237 (1986)
1,2	CH_2Ph		Me_3Al, cat *n*-BuLi or LiOMe	Angew Int *21* 71 (1982)

$$CH_3\overset{\displaystyle O}{\overset{\diagup\;\backslash}{CH}}\!-\!\overset{\displaystyle OMs}{\overset{|}{CH}}CHCH_3 \xrightarrow[\substack{\text{2. }n\text{-BuLi}\\ \text{3. }(R^2_2CuCN)Li_2,\,BF_3\cdot OEt_2}]{\text{1. }(R^1_2CuCN)Li_2,\,BF_3\cdot OEt_2} CH_3\overset{\displaystyle OH}{\overset{|}{CH}}\underset{\underset{\displaystyle R^1}{|}}{CH}\underset{\underset{\displaystyle R^2}{|}}{CH}CH_3$$

TL *28* 5631 (1987)

$$CH_3\overset{\displaystyle O}{\overset{\diagup\;\backslash}{CH}}\!-\!\overset{\displaystyle OX}{\overset{|}{CH}}\underset{\underset{\displaystyle CH_3}{|}}{CH}CHR' \xrightarrow{Li_2Cu(CN)R_2} CH_3\overset{\displaystyle HO}{\overset{|}{CH}}\underset{\underset{\displaystyle R}{|}}{CH}\overset{\displaystyle OX}{\underset{\underset{\displaystyle CH_3}{|}}{CH}}CHR'$$

$X = H, SiR_3$

TL *28* 6195 (1987)

$$RCH\overset{\displaystyle O}{\overset{\diagup\;\backslash}{}}\!CHCH\!\!\left.\begin{array}{c} O \\ | \\ O \end{array}\right] \xrightarrow[\text{2. }H_2O]{\text{1. }Me_2CuLi} R\overset{\displaystyle HO}{\overset{|}{CH}}\underset{\underset{\displaystyle CH_3}{|}}{CH}CH\!\!\left.\begin{array}{c} O \\ | \\ O \end{array}\right.$$

JOC *50* 5696 (1985)

$$R^1-\overset{\overset{\displaystyle O}{\diagdown\!\diagup}}{C}-C-\overset{\overset{\displaystyle O}{\|}}{COR^2} \xrightarrow{RM} R^1-\overset{\overset{\displaystyle HO}{|}}{\underset{\underset{\displaystyle R}{|}}{C}}-C-\overset{\overset{\displaystyle O}{\|}}{COR^2}$$

R^2	RM	
H	$n\text{-Bu}_2\text{CuLi}$ or $(R_2\text{CuCN})\text{Li}_2$ ($R = n\text{-Bu, Ph}$)	TL *26* 4683 (1985)
Et	Me_2CuLi	JACS *106* 2949 (1984) ($R^1 = CO_2Et$)
alkyl	$\text{Et}_2\text{AlC}\equiv\textbf{CR}$	TL *23* 2331 (1982); *27* 5397 (1986) JOC *47* 3941 (1982) JACS *108* 5559 (1986)

$$R^1-\overset{\overset{\displaystyle O}{\diagdown\!\diagup}}{C}-C-\overset{\overset{\displaystyle O}{\|}}{COR^2} \xrightarrow{RM} R^1-\overset{\overset{\displaystyle OH}{|}}{\underset{\underset{\displaystyle R}{|}}{C}}-C-\overset{\overset{\displaystyle O}{\|}}{COR^2}$$

R^2	RM	
H	$n\text{-Bu}_2\text{CuLi}$, $(R_2\text{CuCN})\text{Li}_2$ ($R = n\text{-Bu, Ph}$)	TL *26* 4683 (1985)
alkyl	$R_2\text{CuLi}$	JACS *92* 3813 (1970); *106* 2949 (1984) ($R^1 = CO_2Et$) JOC *38* 4263, 4346 (1973)

$$\overset{\overset{\displaystyle O}{\diagdown\!\diagup}}{C}-C-C_n-X \xrightarrow{\text{reagent(s)}} \overset{\overset{\displaystyle HO}{|}}{C}-C \underset{}{\overset{}{\bigcirc}} C_n \quad \text{or} \quad \overset{\overset{\displaystyle HO}{|}}{C}-\overset{\overset{}{}}{\underset{\underset{\displaystyle C_n}{\diagdown\!\diagup}}{C}}$$

$$X = \text{Br, I}$$

Reagent(s)

n-BuLi	JOC *43* 3800 (1978); *45* 922 (1980); *47* 3211 (1982)
sec-BuLi	TL *25* 4323 (1984)
t-BuLi	TL *26* 3643 (1985)
Cu, *n*-Bu$_3$P	JOC *52* 5057 (1987)

$$\overset{\overset{\displaystyle O}{\diagdown\!\diagup}}{C}-C-C_n-\overset{\overset{\displaystyle H}{|}}{C}-X \xrightarrow{\text{base}} \overset{\overset{\displaystyle HO-C}{|}}{\underset{\underset{\displaystyle C_n}{\diagdown\!\diagup}}{C}}-\overset{\overset{\displaystyle X}{|}}{C} \quad \text{or} \quad \overset{\overset{\displaystyle HO}{}}{\boxed{}}\begin{smallmatrix}C_n\\ \\X\end{smallmatrix}$$

Reviews:

Russ Chem Rev *41* 403 (1972)
Tetr *39* 2323 (1983)
Syn 629 (1984)

X	Ring size	
Ph$_2$	3	JOC *41* 885 (1976)
SO$_2$Ph	3–6	JOC *41* 3648 (1976); *44* 4603 (1979); *47* 2564 (1982); *50* 2943 (1985); *51* 5311 (1986); *52* 4614 (1987) TL 503 (1976); *22* 4339 (1981); *26* 6301 (1985) Can J Chem *56* 505 (1978); *59* 1415 (1981)
(SO$_2$Ph)$_2$	3, 5–7	CC 406 (1987)
COR	3–6	JCS 2988 (1951) JACS *79* 3519 (1957); *86* 3162 (1964); *94* 7132 (1972); *107* 1438 (1985) (β-lactam) TL 3731 (1970); 3683 (1972) Tetr *28* 5525, 5533 (1972) JOC *37* 2911 (1972) Can J Chem *55* 1629 (1977)
CO$_2$R	3	JOC *41* 885 (1976) TL 2441 (1976)
(CO$_2$R)$_2$	5	JOC *34* 4060 (1969)
CONR$_2$	3	TL *23* 1343 (1982) JOC *49* 2682 (1984)
	6	JOC *52* 4044 (1987)
CN	3–6	JACS *96* 5268, 5270 (1974) TL 585 (1975) JOC *41* 3648 (1976)

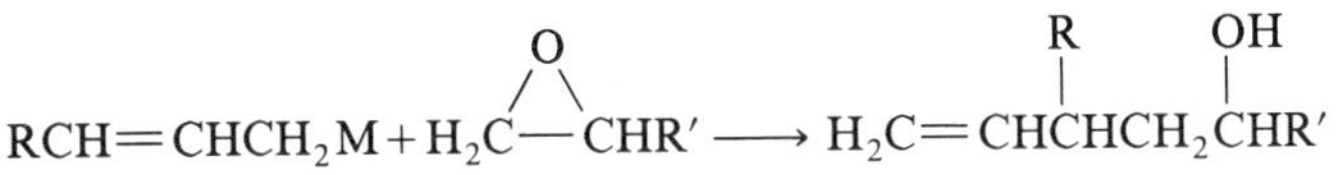

$$RCH{=}CHCH_2M + H_2C\!-\!CHR' \longrightarrow H_2C{=}CHCHCH_2CHR'$$

$$M = Li, Na, MgX, ZnX$$

JOMC *69* 1 (1974)

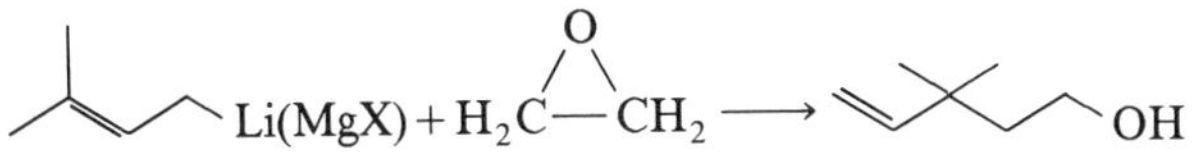

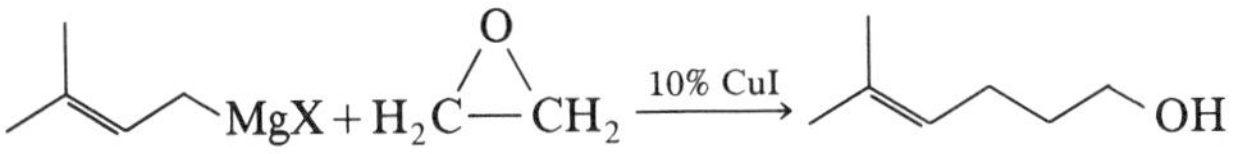

TL 4069 (1978)

$$\text{(cyclohexylidene)}=CHCH_2SiMe_3 + H_2C\overset{O}{-}CH_2 \xrightarrow{\text{Lewis acid}} \text{(cyclohexane)}\overset{CH_2CH_2OH}{\underset{CH=CH_2}{}}$$

Lewis acid

$BF_3 \cdot OEt_2$ 　　　　　　　　Can J Chem *61* 214 (1983) (intramolecular)
　　　　　　　　　　　　　　TL *28* 2753 (1987) (intramolecular)

$TiCl_4$ 　　　　　　　　　　Syn 446 (1978)
　　　　　　　　　　　　　　TL *24* 5661 (1983)
　　　　　　　　　　　　　　CC 585 (1984) (intramolecular)

$$RM + H_2C=CHCH\overset{O}{-}CH_2 \longrightarrow RCH_2CH=CHCH_2OH$$

See page 119, Section 8.

$$C=C-\overset{O}{C}-C \xrightarrow{RM} C=C-\overset{OH}{\underset{R}{C}}-C$$

RM

RMgX 　　　　　　　　　　CC 248 (1969)
　　　　　　　　　　　　　　BSCF 2556 (1970)
　　　　　　　　　　　　　　TL 2005 (1971)
　　　　　　　　　　　　　　JOC *51* 1687 (1986)

$Et_2AlCH_2CO_2\text{-}t\text{-Bu}$ ($R = CH_2CO_2\text{-}t\text{-Bu}$) 　　　Syn 284 (1983)

$LiAlR_4$ 　　　　　　　　　　TL *23* 4697 (1982)

R_2CuLi 　　　　　　　　　　TL *23* 4697 (1982)

$(R_2CuCN)Li_2$ 　　　　　　　　TL *23* 4697 (1982)

4.2. Oxetanes

$$\text{(oxetane)} \longrightarrow R(CH_2)_3OH$$

RLi 　　　　　　　　　　　　JACS *73* 124 (1951)

RMgX 　　　　　　　　　　JACS *73* 124 (1951); *109* 7280 (1987)

RMgX, cat CuI ($R = 1°$ alkyl, 　　　TL 1503 (1979)
　allylic, aryl)

RLi, $BF_3 \cdot OEt_2$ ($R = 1°$ alkyl, 　　　JACS *106* 3693 (1984)
　aryl, vinylic)

$$\text{(oxetane)} + Me_2C=CHCH_2SiMe_3 \xrightarrow{TiCl_4} H_2C=CHCMe_2(CH_2)_3OH$$

JOC *50* 2782 (1985)

5. Rearrangement

$$R_2CHOR \xrightarrow[\text{2. } H_2O]{\text{1. } R'Li} R_3COH$$

Angew Int *9* 763 (1970)

$$ArOH \longrightarrow ArOCH_2SiMe_3 \xrightarrow[\text{2. KOH}]{\text{1. RLi}} ArCH_2OH$$

JOC *47* 5051 (1982)

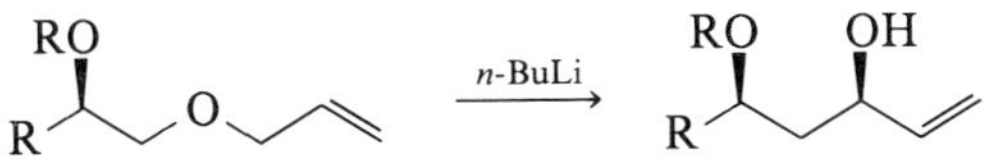

TL *28* 1043 (1987)

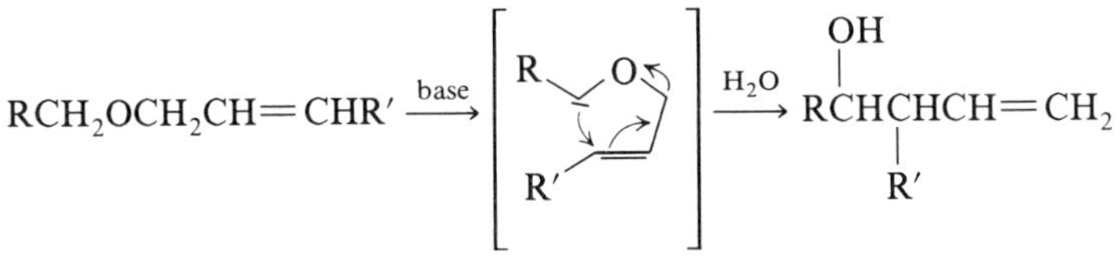

$$RCH_2OCH_2CH=CHR' \xrightarrow{\text{base}} \left[\begin{array}{c} R \\ R' \end{array} \right] \xrightarrow{H_2O} \underset{\underset{R'}{|}}{R}CHCHCH=CH_2 \;\; \overset{\overset{OH}{|}}{}$$

Reviews:

Angew Int *18* 563 (1979) (stereochemistry of 2,3-sigmatropic rearrangements)
Chem Rev *86* 885 (1986)

R	
—	JACS *100* 1927 (1978); *108* 3841 (1986); *109* 3017, 6199 (1987) TL *26* 5013, 5017 (1985); *28* 2099, 4993 (1987) JOC *52* 2960 (1987)
aryl	CC 4 (1970) JACS *93* 3556 (1971) JOC *48* 279 (1983); *49* 1842 (1984)
vinylic	CC 4 (1970) CL 557 (1977) JACS *103* 6492 (1981); *107* 3915 (1985) TL *23* 3931 (1982) JOC *48* 279 (1983); *49* 1707, 1842 (1984); *51* 4315 (1986)
alkynyl	JOC *48* 279 (1983); *49* 1842 (1984); *51* 4316 (1986); *52* 3860, 3883 (1987) CL 1379 (1983) TL *25* 565 (1984); *28* 723, 3323 (1987) Tetr *40* 2303 (1984)
heterocyclic	TL *24* 513 (1983); *25* 6011 (1984)
COR	Helv *57* 2084 (1974)

R
—

CO$_2^-$ TL 22 69 (1981)
                                                JOC 51 4090 (1986)

CO$_2$R   TL 27 4511 (Me$_3$SiOTf added), 4581 (Cp$_2$ZrCl$_2$
                                                added) (1986); 28 803 (1987) (Cp$_2$ZrCl$_2$ added)

CONR$_2$ CL 1729 (1985)
                                                TL 27 4577 (1986) (Cp$_2$ZrCl$_2$ added)

CN   TL 24 2077 (1974)

Reagent	R^3
i-Bu$_3$Al or i-Bu$_2$AlH	H
R$_3^3$Al	R^3 (R^3 = Me, Et)
Et$_2$AlC≡CPh	C≡CPh
E-(i-Bu)$_2$AlCH=CHC$_6$H$_{13}$	CH=CHC$_6$H$_{13}$

TL 22 3985 (1981)

Org Rxs 2 1 (1944) (review)
Org Syn Coll Vol 3 418 (1955)
Helv 52 337 (1967)
Adv Heterocyclic Chem 8 143 (1967) (review, nitrogen heterocycles); 42 203 (1987) (review, heteroaro-
    matics)
Chimia 24 89 (1970) (review)
Austral J Chem 34 819, 1079 (1981)
CL 1131 (1982)
CC 1120 (1982); 182, 750 (1986)
TL 23 4407 (1982), 27 4945 (1986) (in microwave ovens); 28 3075 (1987)
Tetr 38 3079 (1982)
Chem Rev 84 205 (1984) (review, catalysis)
JACS 107 3891 (1985)

Org Rxs 2 1 (1944) (review)
JACS 70 1747 (1948)

6. Acetal Cleavage

$$\overset{\overset{\displaystyle O \diagup \diagdown O}{|}}{RCR} + R'Li \xrightarrow{H_2O} \overset{\overset{\displaystyle OH}{|}}{\underset{\underset{\displaystyle R'}{|}}{RCR}}$$

Syn Commun *13* 769 (1983)

$$\overset{R \diagdown \diagup \diagdown R}{\underset{\underset{R^1-\overset{\displaystyle O \quad O}{C}-R^2}{}}{}} \xrightarrow{reagent(s)} R^1-\overset{\overset{\displaystyle OCHRCH_2CHROH}{|}}{\underset{\underset{\displaystyle R^2}{|}}{C}}-R^3 \longrightarrow R^1-\overset{\overset{\displaystyle OH}{|}}{\underset{\underset{\displaystyle R^2}{|}}{C}}-R^3 \text{ (chiral)}$$

R = Me, Ph

R^3	Reagent(s)	
H	Cl_2AlH	CC 334 (1987)
	Br_2AlH	TL *24* 4581 (1983);
		27 983, 987 (1986)
		JOMC *285* 83 (1985)
		CC 334 (1987)
	$i\text{-}Bu_2AlH$	TL *24* 4581 (1983);
		27 983 (1986)
		JOMC *285* 83 (1986)
	$TiCl_4$, Et_3SiH	TL *27* 987 (1986)
1° alkyl, Ph	$R_2CuLi \cdot BF_3$	TL *25* 3083 (1984)
1° R, $H_2C=CHCH_2$	RMgX or RLi or	TL *25* 3947 (1984)
	$R_2CuLi(MgX)$, $TiCl_4$	
$H_2C=CHCH_2$	$H_2C=CHCH_2SiMe_3$, $SnCl_4$	JACS *104* 7371 (1982)
	$H_2C=CHCH_2SiMe_3$, $TiCl_4$	JACS *108* 7116 (1986)
	$H_2C=CHCH_2M$ (M = B$\diagdown$)),	JACS *108* 7116 (1986)
	$n\text{-}Bu_3Sn$, Ph_3Sn),	
	Lewis acid?	
$H_2C=CRCH_2$	$H_2C=CRCH_2SiMe_3$ (R = H,	JACS *105* 2088 (1983)
	Me), $TiCl_4$, $Ti(O\text{-}i\text{-}Pr)_4$	TL *25* 3951 (1984)
		JOC *50* 2598 (1985)
$RC\equiv C$	$RC\equiv CSiMe_3$, $TiCl_4$	JACS *105* 2904 (1983)
	$RC\equiv CM$ (M = Me_3Si,	JACS *108* 7116 (1986)
	$n\text{-}Bu_3Sn$), Lewis acid?	
$CH_3CH_2COCH_2$	$CH_3CH_2COCH_2SiMe_3$, $SnCl_4$	JACS *104* 7371, 7372 (1982)
	$\overset{\overset{\displaystyle OSiMe_3}{|}}{}$	
$RCOCH_2$	$RC=CH_2$, $TiCl_4$	JACS *104* 7371, 7372 (1982);
		106 7588 (1984)
		JOC *52* 180 (1987)

$$
\underset{\substack{| \quad | \\ O \quad O \\ R^1 - C - R^2}}{\overset{R}{\diagdown}} \xrightarrow{\text{reagent(s)}} \underset{\substack{| \\ R^2}}{R^1 - \overset{\text{OCHRCH}_2\text{CHROH}}{\underset{|}{C}} - R^3} \longrightarrow \underset{\substack{| \\ R^2}}{R^1 - \overset{\text{OH}}{\underset{|}{C}} - R^3}\,(\textit{chiral}) \quad (\textit{Continued})
$$

R = Me, Ph

$\underline{R^3}$	$\underline{\text{Reagent(s)}}$	
CH_2CO_2H	$\underset{\substack{| \\ CF_3CO_2H}}{H_2C = \overset{OSiMe_2\text{-}t\text{-Bu}}{\underset{|}{C}}\text{-}t\text{-Bu}}, \quad TiCl_4 /$	TL *26* 2535 (1985)
CN	$NCSiMe_3$, $TiCl_4$	JOC *48* 2294 (1983)
		TL *25* 591 (1984)

$$
RCHO \longrightarrow R \overset{\overset{\displaystyle O \diagdown}{\diagdown}}{\underset{O \diagup}{\diagup}} \underset{CH_3}{\overset{\displaystyle \| }{\diagup}} \xrightarrow[\text{base}]{R'M} \underset{\text{chiral}}{R\overset{OH}{\underset{|}{C}}HR'}
$$

$\underline{R'M}$

R'Li (R' =1° alkyl, aryl, vinyl),   TL *27* 2945 (1986)
  CuBr·SMe₂

$H_2C = CHCH_2Li$, CuBr, $BF_3 \cdot OEt_2$ TL *27* 2945 (1986)

Me₃SiR' (R' = allyl, alkynyl, CN),   Angew Int *25* 178 (1986)
  Cl₃TiX (X = Cl, O-i-Pr)

$$
R \overset{\overset{\displaystyle O}{\diagdown}}{\underset{O}{\diagup}} \underset{Ph}{\overset{\displaystyle \| }{\diagup}} \xrightarrow[\text{or } H_2C = CHCH_2SiMe_3,\, ZnBr_2]{H_2C = CROSiMe_3,\, BF_3 \cdot OEt_2} \xrightarrow{Pb(OAc)_4} \underset{\text{chiral}}{R\overset{OH}{\underset{|}{C}}HR'}
$$

R' = CH_2COR or $CH_2CH = CH_2$

JOC *49* 2513 (1984)

$$
R^1CH = CR^2OCR^3_2OR^4 \xrightarrow{i\text{-Bu}_3Al} R^2\overset{OH}{\underset{|}{C}}HCHR^1CR^3_2OR^4
$$

JOC *52* 5700 (1987)

$$
\underset{\substack{| \\ R^2 \quad O}}{\overset{\substack{R^1 \quad O \\ |}}{\diagdown C \diagup}} \!\!\!\! C_n \longrightarrow R^1R^2CHO - C_n - OH
$$

$\underline{R^1}$	$\underline{R^2}$	$\underline{\text{Reagent(s)}}$	$\underline{n}$	
Ph	H	i-Bu₂AlH	2,3	CL 1593 (1983)
				Chem Pharm Bull *32* 791
				(1984)

Ph	H	LiAlH$_4$, AlCl$_3$	2–4	JACS *84* 2371 (1962)
				JOC *27* 67 (1962)
				Can J Chem *47* 1195 (1969)
				Carbohydr Res *44* 1 (1975);
				51 C19 (1976)
				TL 3551 (1976); *22* 3919
				(1981); *23* 3507 (1982)
p-MeOC$_6$H$_4$	H	NaBH$_3$CN, CF$_3$CO$_2$H or Me$_3$SiCl	3	CC 201 (1984)
H or alkyl	H or alkyl	LiAlH$_4$, AlCl$_3$	2,3	JACS *84* 2371 (1962)
				JOC *27* 67 (1962)
				Can J Chem *42* 990 (1964);
				43 1030 (1965); *47* 1195 (1969)
				Org Syn Coll Vol *5* 303 (1973)

$$\begin{array}{c} R^1 \\ \diagdown \\ R^2 \end{array} C \begin{array}{c} O \\ \diagdown \\ O \end{array} C_n \longrightarrow HO-C_n-OH$$

Review:

T. W. Greene, "Protective Groups in Organic Synthesis," Wiley-Interscience, New York (1981), pp 72–86

R^1	R^2	Reagent(s)	n	
Ph	H	N$_2$H$_4$, H$_2$O, cat Pd-C	2	Syn 317 (1986)
Ph	H	HCO$_2$H, cat Pd-C	3	Carbohydr Res *83* 175 (1980)
Ph	H	⬡ , cat Pd(OH)$_2$—C	3	Syn 396 (1981)
p-CH$_3$OC$_6$H$_4$	CH$_3$	SnCl$_4$/*n*-Bu$_4$NOH	2,3	JOC *46* 2419 (1981)

$$RO\!-\!\!\left(\!\!\begin{array}{c}\\O\end{array}\!\!\right)_{\!n} \longrightarrow RO(CH_2)_{n+3}OH$$

n	Reagents	
1,2	LiAlH$_4$, AlCl$_3$	JOC *30* 2441 (1965)
		Can J Chem *45* 2547 (1967)
2	LiAlH$_4$, HCl	JACS *73* 5917 (1951)

TL *23* 889 (1982)

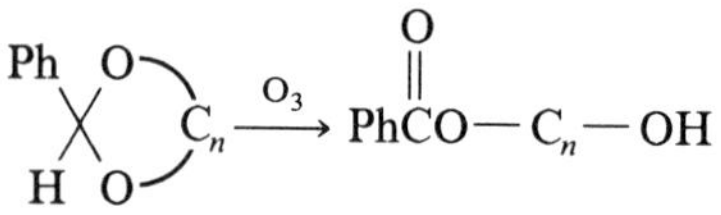

$$\underset{\underset{H}{Ph}}{\overset{\overset{O}{\diagdown}}{\diagup}}\underset{O}{\overset{O}{\bigtriangleup}}C_n \xrightarrow{\ O_3\ } PhCO-\overset{\overset{O}{\|}}{C}_n-OH$$

$$n = 2,3$$

Can J Chem *53* 1204 (1975)

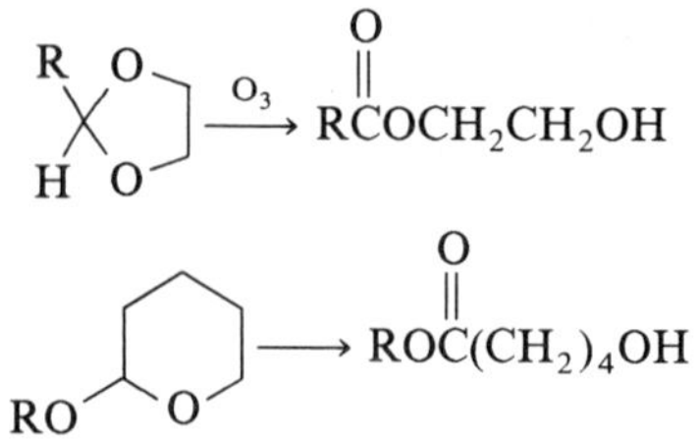

$$\underset{\underset{H}{R}}{\overset{\overset{O}{\diagdown}}{\diagup}}\underset{O}{\overset{O}{\bigtriangleup}} \xrightarrow{\ O_3\ } R\overset{\overset{O}{\|}}{C}OCH_2CH_2OH$$

$$\longrightarrow RO\overset{\overset{O}{\|}}{C}(CH_2)_4OH$$

Can J Chem *49* 2465 (1971)

7. REDUCTION OF CARBONYL COMPOUNDS

1. Aldehydes and Ketones

Review: Syn 605 (1982) (stereoselectivity of reduction of chiral aminocarbonyl compounds)

1.1. Hydrogenation

Reviews:

C. A. Buehler and D. E. Pearson, "Survey of Organic Syntheses," Wiley Interscience, New York (1970), p 201

M. Freifelder, "Practical Catalytic Hydrogenation," Wiley Interscience, New York (1971), Chpt 14

H. O. House, "Modern Synthetic Reactions," W. A. Benjamin, New York (1972), p 1

M. Freifelder, "Catalytic Hydrogenation in Organic Synthesis: Procedures and Commentary," J. Wiley and Sons, New York (1978), Chpt 9

P. N. Rylander, "Hydrogenation Methods," Academic Press, New York (1985), Chpt 4

H_2, $FeCl_2$ (δ-hydroxy-β-keto ester)	Helv 69 803 (1986)
H_2, cat [Rh(1,5-hexadiene)Cl]$_2$ $+ R_3P$ (chiral) (asymmetric)	JOMC 94 C47 (1975)
H_2, cat [Rh(1,5-hexadiene)L$_2$]X (L$_2$ = 2,2'-bipyridine, phenanthrolines; X = PF$_6$, BPh$_4$) (C=O > C=C)	JOMC 140 63 (1977)
H_2, cat [Rh(bipy)S$_2$]$^+$ or [Rh(bipy)$_2$]$^+$ (S = solvent) (C=O > C=C)	JOMC 157 345 (1978)
H_2, cat RuCl$_2$(CO)$_2$(PPh$_3$)$_2$, Δ (aldehydes only)	JOMC 145 189 (1978)
H_2, cat NaH-NaO-t-Am-Ni(OAc)$_2$	JOC 45 1946 (1980)
H_2, cat nickel boride or Raney nickel	JOC 46 1263 (1981)

1.2. Reduction by Metals

Review: Tetr 42 6351 (1986) (alkali metals in protic solvents)

Li, NH$_3$	JACS *90* 6486 (1968); *108* 800 (1986)
	JCS C 968 (1969)
	Helv *64* 2109 (1981)
	Acct Chem Res *16* 399 (1983)
Li, Na, or K; NH$_3$	BSCF 4399 (1970)
	JCS Perkin I 999 (1972)
Li, Na or K; NH$_3$; NH$_4$Cl	BSCF 4404 (1970)
	JCS Perkin I 999 (1972)
	Helv *64* 2109 (1981)
	CC 1558 (1986)
Li, NH$_3$, ROH	BSCF 4404 (1970)
	Pure Appl Chem *49* 1049 (1977)
	JACS *108* 3443 (1986)
Li(Na), ROH	JCS C 968 (1969)
Na, H$_2$O, EtOH	JACS *106* 4547 (1984)
Na, *i*-PrOH	JOC *51* 4779 (1986)
	Org Syn *65* 203 (1987)
Al(Hg), H$_2$O	Arch Pharm *280* 361 (1942)
Fe, HOAc	Org Syn Coll Vol *1* 304 (1932)
	JACS *61* 2134 (1939)

1.3. Metal Hydride Reduction

Reviews:

N. G. Gaylord, "Reduction with Complex Metal Hydrides," Interscience, New York (1956)
H. C. Brown, "Boranes in Organic Chemistry," Cornell Univ. Press, Ithaca, NY (1972), pp 209–250
H. O. House, "Modern Synthetic Reactions," Benjamin, New York (1972), p 49
Chem Soc Rev *5* 23 (1976)
Tetr *35* 449 (1979) (stereochemistry and mechanism)
Topics Stereochem *11* 53 (1979) (stereochemistry)
JACS *103* 4540 (1981) (stereochemistry of cyclohexanone reductions)

LiH	JOC *52* 4299 (1987) (ynal)
MgH$_2$	JOC *43* 1557 (1978)
ROMgH	JOC *43* 1560 (1978)
ArOMgH	JOC *43* 1557, 1560 (1978)
ROMg$_2$H$_3$ (R = Me, Ar)	TL 3133 (1977)
R$_2$NMgH	JOC *43* 1564 (1978)
BH$_3$	JACS *92* 1637 (1970)
BH$_3$·py	JOC *23* 1561 (1958)
BH$_3$·pyridine polymer	JOC *45* 2724 (1980)
BH$_3$·py, CF$_3$CO$_2$H	Chem Pharm Bull *27* 2405 (1979)

$BH_3 \cdot NH_3$	TL *21* 693 (1980)
$BH_3 \cdot t\text{-}BuNH_2$	TL *21* 693 (1980)
$BH_3 \cdot NMe_3$	CL 61 (1986) (ketone on silica gel)
NBH_2	Syn 214 (1981)
$BH_3 \cdot SMe_2$	Syn 733 (1982)
$(CH_3)_2CHC(CH_3)_2BH_2$	JOC *37* 2942 (1972); *39* 1631 (1974)
$(CH_3)_2CHC(CH_3)_2BHCl \cdot SMe_2$	JOC *51* 5264 (1986)
$(Sia)_2BH$	JACS *92* 7161 (1970) JOC *39* 1631 (1974)
BH $\Big)_2$	JOC *39* 1631 (1974)
	JACS *88* 2871 (1966) JOC *39* 1631 (1974)
9-BBN	JOC *40* 1864 (1975); *41* 1778 (1976); *42* 1197 (1977)
9-BBN · py	JOC *42* 4169 (1977)
	JOC *42* 512 (1977)
$(CF_3CO_2)_2BH$	JOC *46* 355 (1981)
SiaB	JOC *43* 1470 (1978)
$LiBH_4$	JACS *107* 6046 (1985) (ynone) Ber *118* 722 (1985) (β-diketone) JOC *52* 5067 (1987)
$NaBH_4$	JACS *71* 122 (1949); *75* 1286 (1953); *84* 373 (1962); *88* 2811 (1966) Tetr *26* 2411 (1970); *40* 2233 (1984) (on B chelate of β-hydroxy ketones) Ber *118* 722 (1985) (β-diketone) TL *26* 2951 (1985) (on β-chelate of δ-keto-β-hydroxy ester); *28* 155 (1987) (on B chelate of β-hydroxy ketone) Helv *69* 803 (1986) (on B chelate of δ-hydroxy-β-keto ester) JOC *52* 4062 (1987) (β-hydroxy ketone)
$NaBH_4$, alumina	Syn 891 (1978)
$NaBH_4$, Fontainebleau sand	CC 1066 (1981)

NaBH$_4$, HOAc	CC 535 (1975) (RCHO only) TL 28 4725 (1987)
NaBH$_4$, tartaric acid	Syn Commun 14 955 (1984)
NaBH$_4$, n-Bu$_3$B	CL 1415 (1980) (β-hydroxy ketone diastereoselectivity)
NaBH$_4$, Ti(O-i-Pr)$_4$	TL 28 703 (1987) (β-hydroxy ketone)
NaBH$_4$, NiCl$_2$/Me$_3$SiCl	TL 28 5741 (1987) (aldehydes only)
NaBH$_4$, PdCl$_2$	CL 1029 (1981) (hindered ketones)
NaBH$_4$; LnCl$_3$, SmCl$_3$ or CeCl$_3$	JACS 100 2226 (1978)
NaBH$_4$, CeCl$_3$	Syn Commun 10 623 (1980) JOC 51 5320 (1986) (ynones)
KBH$_4$	JOC 50 2668 (1985)
(Et$_4$N)BH$_4$	TL 21 3963 (1980) (aldehyde only)
(n-Bu$_4$N)BH$_4$	JOC 41 690 (1976); 45 216 (1980) TL 28 5661 (1987)
Zn(BH$_4$)$_2$	JACS 82 6074 (1960); 106 1154 (1984) (β-keto imide); 107 1421 (1985); 108 4603 (1986) (β-keto ester) Proc Natl Acad Sci USA 72 3355 (1975) TL 21 1641 (1980) (β-keto ester); 22 4723 (1981) (α,β-epoxy ketone); 24 2653 (α-hydroxy ketone), 4805 (β-keto imide), 5385 (β-keto amide) (1983); 25 6015 (1984) (chiral, β-keto amides); 26 6465 (1985) (β-keto imide); 27 6341 (1986) (α-hydroxy ketone); 28 5129 (1987) Syn Commun 13 901 (1983) (α,β-epoxy ketones) Acct Chem Res 17 338 (1984) (review) Chem Pharm Bull 32 1411 (1984) (β-hydroxy or alkoxy ketone) Ber 118 722 (1985) (β-diketone) Helv 69 803 (1986) (δ-hydroxy-β-keto ester)
Zn(BH$_4$)$_2$, ZnCl$_2$	JACS 109 7488 (1987) (β-keto ester)
Cp$_2$ZrClBH$_4$	TL 4985 (1978)
(Ph$_3$P)$_2$CuBH$_4$; HCl, AlCl$_3$, BF$_3$·OEt$_2$ or ZnCl$_2$	TL 22 675 (1981)
NaBH$_4$, PhCHOHCO$_2$H	Ind J Chem B 21 212 (1982) (ketones only?)
NaBH$_3$OH	JOC 42 3963 (1977)
NaBH$_2$S$_3$	Syn 526 (1972)
NaBH$_3$CN	JOC 40 2530 (1975) Syn 135 (1975)
NaBH$_3$CN, CF$_3$CO$_2$H	JOC 45 216 (1980)
NaBH$_3$CN, ZnCl$_2$	JOC 50 1927 (1985)
LiBH$_3$CN	JACS 91 3996 (1969)

$[(Ph_3P)_2CuBH_3CN]_2$	TL *21* 813 (1980)
LiH_3BMe	Syn Commun *12* 723 (1982)
$LiH_3B(n\text{-}Bu)$	JOC *47* 3311 (1982)
$LiH_2B(\text{—}\langle\text{aryl}\rangle\text{—})_2$	JACS *96* 274 (1974)
$K[RB(H)(O{-}O)]$	JOC *51* 337 (1986)
$NaHB(OR)_3$ (R = Me, *i*-Pr, *t*-Bu)	JCS 3426 (1955) JACS *78* 3616 (1956) Tetr *26* 2411 (1970)
$KHB(O\text{-}i\text{-}Pr)_3$	JOC *49* 885 (1984) Ber *118* 722 (1985) (β-diketone)
$NaHB(OAc)_3$	CC 535 (1975) (aldehydes only) TL *24* 273 (1983) (β-hydroxy ketones); *25* 5449 (1984) (β-hydroxy ketones) JACS *108* 2476 (1986) CC 880 (1987)
$KHB(OAc)_3$	TL 4851 (1979) (aldehydes only)
$(Me_4N)HB(OAc)_3$	TL *27* 5939 (1986) (β-hydroxy ketone) JOC *52* 3211 (1987) (β-hydroxy ketone)
$(n\text{-}Bu_4N)HB(OAc)_3$	TL *24* 4287 (1983) (aldehydes only)
$LiHBEt_3$	JOC *45* 1 (1980); *52* 5700 (1987) JACS *108* 3385 (1986) TL *27* 6233 (1986)
$KHBEt_3$	TL *26* 4643 (1985) (β-keto amides)
$Li^+[\,\,\overset{-}{B}\,\text{H}\,]$	JACS *92* 709 (1970) JOC *52* 4898 (1987)
$LiHB(sec\text{-}Bu)_3$	JACS *94* 7159 (1972); *95* 4100 (1973) TL 141 (1973); 4487 (1978); *28* 35 (1987) JOC *50* 3957 (1985)
$KHB(sec\text{-}Bu)_3$	JACS *95* 4100 (1973) TL 4487 (1978); *28* 2999 (1987) CC 1239 (1982)
$LiHB(Sia)_3$	JACS *98* 3383 (1976) TL 4487 (1978)
$KHB(Sia)_3$	JOC *51* 238 (1986)
$KHBPh_3$	JOC *51* 226 (1986); *52* 5564 (1987)

Li(n-Bu)$_2$B⬭	JACS *98* 1965 (1976)
(n-Bu$_4$N)B$_3$H$_8$	TL *23* 3337 (1982)

i-Bu$_2$Al—O—(aryl, 2,6-di-t-Bu, 4-CH$_3$)

JOC *44* 1363 (1979)
JACS *108* 284 (1986)

i-Bu$_2$Al—O—(aryl, 2,6-di-t-Bu, 4-CH$_3$), i-Bu$_2$AlH

JACS *109* 7488 (1987) (β-keto ester)

MeAl[—O—(aryl, 2,6-di-t-Bu, 4-CH$_3$)]$_2$, t-BuMgCl

TL *26* 3853 (1985) (gives equatorial alcohols)

(HAlN-i-Pr)$_6$	TL 2369 (1977) Z Chem *17* 18 (1977)
AlH$_3$	JACS *88* 1464 (1966); *90* 2927 (1968) JCS B 581 (1967)

i-Bu$_2$AlH	Ann *623* 9 (1959)
	JOC *24* 627 (1969); *47* 2590 (1982);
	  50 2443 (1985)
	CC 213 (1970)
	Syn 617 (1975) (review); 732 (1982)
	JACS *107* 2730, 7524 (1985)
	Ber *118* 722 (1985) (β-diketone)
	TL *26* 435 (1985) (α-sulfoxide); *27* 3009
	  (β-hydroxy ketone), 6233 (1986);
	  28 4861, 4865 (1987) (both α-sulfoxide)

i-Bu$_2$AlH, ZnCl$_2$	TL *26* 435 (1985) (α-sulfoxide); *28* 797
	  (α-sulfoxide), 4861 (α-sulfoxide),
	  6481 (β-hydroxy ketone) (1987)
	JOC *52* 304 (1987)

i-Bu$_3$Al	Ann *623* 9 (1959)
i-Bu$_3$Al, amine	BCSJ *51* 2664 (1978)

LiAlH$_4$	JACS *71* 1675 (1949); *78* 2579 (1956); *88* 1458
	  (1966); *98* 8114 (1976)
	Org Rxs *6* 469 (1951)
	Tetr *26* 2411 (1970); *35* 449, 567 (1979)
	JOC *43* 2173 (1978); *44* 2760 (1979); *52* 1425
	  (β-diketones), 4062 (β-hydroxy ketones) (1987)
	TL 4487 (1978); *26* 435 (1985) (α-sulfoxide)
	Ber *118* 704 (1,2-; 1,3-; 1,4-; and 1,5- diketones),
	  722 (1,2- and 1,3-diketones) (1985)

LiAlH$_4$, silica gel	TL *23* 4585 (1982) (α,β,γ and δ-keto ester to hydroxy ester)

LiAlH$_4$, py	JACS *84* 1756 (1962)
LiAlH$_4$, AlCl$_3$	JOC *30* 3809 (1965)
LiAlH$_4$, TiCl$_4$	JOC *52* 1425 (1987) (β-diketones)
LiHAl(OCH$_3$)$_3$	JACS *87* 5614, 5620 (1965); *106* 6414 (1984) Tetr *26* 2411 (1970) Ber *118* 722 (1985) (β-diketone)
LiHAl(O-*t*-Bu)$_3$	JACS *80* 5372 (1958); *87* 5620 (1965); *107* 686, 2730 (1985) Coll Czech Chem Commun *24* 2284 (1959) Tetr *24* 2039 (1968); *26* 2411 (1970) JOC *52* 1425 (β-diketones), 1429, 4495 (1987) TL *28* 1439, 3839 (1987)
LiHAl(O-*t*-Bu)$_3$, TiCl$_4$	JOC *52* 1425 (1987) (β-diketones)
LiHAl(OCEt$_3$)$_3$	JOC *46* 4628 (1981)
LiHAl(*i*-Bu)$_2$(*t*-Bu)	Syn 171 (1977) JOC *47* 4581 (1982)
NaH$_2$Al(OCH$_2$CH$_2$OCH$_3$)$_2$	TL *24* 2653 (1983) (α-silyloxy ketones)
PhMe$_2$SiH, cat *n*-Bu$_4$NF	JACS *106* 4629 (1984) JOC *51* 2267 (1986) (mechanism); *52* 3218 (1987) (α-fluoro-β-keto ester)
PhMe$_2$SiH, CF$_3$CO$_2$H or [(Et$_2$N)$_3$S]F$_2$SiMe$_3$	JACS *107* 8294 (1985) (β-keto amides)
Ph$_2$SiH$_2$; KF, CsF, KO$_2$CH or (PhCH$_2$NMe$_3$)F	JOMC *172* 143 (1979)
Ph$_3$SiH, AlCl$_3$	JOC *52* 3218 (1987) (α-fluoro-β-keto ester)
Et$_3$SiH, HCl or H$_2$SO$_4$	JOC *39* 2740 (1974)
Et$_3$SiH, BF$_3$	JOC *43* 374 (1978)
Et$_3$SiH, BF$_3$·OEt$_2$	JOMC *117* 129 (1976)
Et$_3$SiH, TiCl$_4$	TL *28* 6331 (1987)
R$_2$SiH$_2$ or R$_3$SiH (R = Et, Ph), cat RuCl$_2$(PPh$_3$)$_3$	JOC *47* 2469 (1982)
LiHSi$\left[\begin{array}{c} O \\ O \end{array}\right]_2$	JOC *52* 948 (1987)
LiHSi$\left[\begin{array}{c} O \\ O \end{array}\right]_2$	JOC *52* 948 (1987)

(MeO)$_3$SiH, LiOMe   CC 1411 (1986)

(MeO)$_3$SiH, LiOCMe$_2$CMe$_2$OLi   CC 1411 (1986)

(EtO)$_3$SiH, KF or CsF   CC 121 (1981)
                                             Tetr *37* 2165 (1981); *39* 999 (1983)

(EtO)$_2$SiMeH or   Syn 981 (1982)

$$Me_3SiO(-\overset{\overset{\displaystyle H}{|}}{\underset{\underset{\displaystyle Me}{|}}{Si}}-O)_n-SiMe_3,$$

    KF or KO$_2$CH

(MeHSiO)$_n$, *n*-Bu$_2$Sn(OAc)$_2$ JOC *38* 162 (1973)

R$_2$SnH$_2$ or R$_3$SnH (R = *n*-Bu, Ph)   JACS *80* 3798 (1958); *83* 1246 (1961)

n-Bu$_3$SnH, pressure   JOC *51* 1672 (1986)

n-Bu$_3$SnH, silica   JOC *43* 3977 (1978)

n-Bu$_3$SnH, ZnCl$_2$ TL *27* 3009 (1986) (β-hydroxy ketone)

Ph$_3$SnH   JOC *50* 2149 (1985)

cat Cp$_2$ZrH$_2$, *i*-PrOH   JOC *51* 240 (1986) (ketones only)

LiH, VCl$_3$ JOC *43* 4804 (1978)

NaH, FeCl$_2$ or FeCl$_3$ JOC *41* 1667 (1976)

NaH, NaO-*t*-Am, Ni(OAc)$_2$ TL 1069 (1977)

LaNi$_5$H$_6$ CC 163 (1984)
                                             JOC *52* 5695 (1987)

Li$_n$CuH$_{n+1}$ (*n* = 1–5)   JOC *43* 183 (1978)

[(Ph$_3$P)$_2$N]HM(CO)$_4$L   JACS *107* 2428 (1985)
    [M = Cr, W; L = CO,
    P(OMe)$_3$]/CH$_3$CO$_2$H or C$_6$H$_5$OH

1.4. Miscellaneous Reagents

RMgX, cat Cp$_2$TiCl$_2$ TL *21* 2171 (1980)

Na$_2$S$_2$O$_4$, H$_2$O, DMF   TL *22* 179 (1981)

SmI$_2$, H$_2$O, THF   JACS *109* 6187 (1987)

SmI$_2$, MeOH   JACS *102* 2693 (1980)

LiO-*i*-Pr   JCS C 804 (1969)

cat LiO-*i*-Pr, cat NiCl$_2$, *i*-PrOH   JOC *50* 3082 (1985)

Al(O-*i*-Pr)$_3$ Org Rxs *2* 178 (1944)
                                             JACS *78* 2579 (1956)

H$_2$Se, hν Angew Int *19* 1008 (1980)

$NaH_2PO_2 \cdot H_2O$, Na_2CO_3, cat Pd-C, H_2O	JOC *50* 3408 (1985) (aryl aldehydes only)
NaO_2CH, KH_2PO_4 1-methyl-2-pyrrolidinone	JOC *46* 3367 (1981)
HCO_2H, cat $RuCl_2(PPh_3)_3$	BCSJ *55* 2441 (1982)

1.5. Reduction of Aldehydes in the Presence of Ketones

H_2, cat $RuCl_2(PPh_3)_3$	CL 1085 (1977)
H_2, cat $H_3Ir(PPh_3)_3$, HOAc	JOMC *129* C43 (1977)
$BH_3 \cdot t\text{-}BuNH_2$	TL *21* 697 (1980) JOC *51* 4047 (1986)
9-BBN·py	JOC *42* 4169 (1977)
	JOC *42* 512 (1977)
SiaB	JOC *43* 1470 (1978)
BH_4 exchange resin	TL *24* 5367 (1983)
$NaBH_4$	Austral J Chem *28* 1383 (1975)
$NaBH_4$, $n\text{-}Bu_4NBr$, H_2O, C_6H_6	Ind J Chem B *25* 626 (1986)
$NaBH_4$, $t\text{-}BuSH$	TL 263 (1977)
$NaBH_4$, $NiCl_2/Me_3SiCl$	TL *28* 5741 (1987)
$NaBH_4$, $ErCl_3$, aq EtOH	TL *22* 4077 (1981)
$LiBH_4$	Austral J Chem *28* 1383 (1975)
$LiBH_4$, molecular sieves	JOC *44* 3969 (1979)
$(Et_4N)BH_4$	TL *21* 3963 (1980)
$(Ph_3P)_2CuBH_4$, HCl	TL *22* 675 (1981)
$(n\text{-}Bu_4N)BH_3CN$	JACS *95* 6131 (1973)
$NaHB(OAc)_3$	CC 535 (1975)
$KHB(OAc)_3$	TL 4851 (1979)
$(n\text{-}Bu_4N)HB(OAc)_3$	TL *24* 4287 (1983)
$NaHB$ (R = Me, t-Bu)	CL 461 (1981)
$KHBPh_3$	JOC *51* 226 (1986)
$Li\left[n\text{-}Bu_2B\right.$$\left.\right]$	JACS *98* 1965 (1976)
$LiHAl(O\text{-}t\text{-}Bu)_3$	Austral J Chem *28* 1383 (1975) JACS *108* 4586 (1986)

LiHAl(OCEt$_3$)$_3$ JOC *46* 4628 (1981)

(MeO)$_3$SiH, LiOMe   CC 1411 (1986)

(EtO)$_3$SiH, KF or CsF   CC 121 (1981)
                                                       Tetr *37* 2165 (1981); *39* 999 (1983)

(EtO)$_2$SiMeH or   Syn 981 (1982)

$$Me_3SiO(-\underset{\underset{Me}{|}}{\overset{\overset{H}{|}}{Si}}-O)_n-SiMe_3 , KO_2CH$$

n-Bu$_3$SnH, silica   JOC *43* 3977 (1978)

EtCH(OMgBr)$_2$ TL *22* 621 (1981)

SmI$_2$, MeOH   Nouv J Chim *1* 5 (1977)
                                                       JACS *102* 2693 (1980)

i-PrOH or (*i*-Pr)$_2$CHOH, alumina   TL 3601 (1975)
                                                       JOC *42* 1202 (1977)

1.6. Reduction of Ketones in the Presence of Aldehydes

NaBH$_4$, MCl$_3$·6 H$_2$O   JACS *101* 5848 (1979)
    (M = Ce, Er, Cr, La), EtOH   JOC *44* 4187 (1979)
                                                       TL *22* 4077 (1981)

t-BuNH$_2$/LiHAl(O-*t*-Bu)$_3$/H$_3$O$^+$ or   TL *21* 5085 (1980)
    alumina   Tetr *38* 1827 (1982)

Ti(NEt$_2$)$_4$/LiAlH$_4$/H$_2$O   CC 406 (1983)

1.7. Selective Reduction between Aldehydes

BH$_4$ exchange resin   TL *24* 5367 (1983)

1.8. Reduction of Conjugated Aldehydes in the Presence of Non-conjugated Aldehydes

NaBH$_4$, ErCl$_3$ TL *22* 4077 (1981)

1.9. Selective Reduction between Ketones

BH$_3$·*t*-BuNH$_2$ TL *21* 697 (1980)

KHBPh$_3$ JOC *51* 226 (1986)

Li$\left[n\text{-Bu}_2\,B\overset{\frown}{\underset{\smile}{\,}} \right]$ JACS *98* 1965 (1976)

NaBH$_4$ JACS *74* 2814 (1952)

NaBH$_4$, ErCl$_3$·6 H$_2$O   JOC *44* 4187 (1979)

BH_4 exchange resin   TL *24* 5367 (1983)

$MeAl\left(-O-\underset{t\text{-Bu}}{\overset{t\text{-Bu}}{\bigcirc}}-Me\right)_2$, t-BuMgCl   TL *26* 3853 (1983) (cyclic > acyclic)

1.10. Reduction of Enones in the Presence of Aldehydes

$NaBH_4$, $CeCl_3$, EtOH   JOC *47* 381 (1982)

1.11. Reduction of Enones in the Presence of Ketones

(HAlN-i-Pr)$_6$ TL 2369 (1977)

$NaBH_4$, $CeCl_3$ JOC *44* 4187 (1979)

1.12. Reduction of Ketones in the Presence of Enones

$BH_3 \cdot SMe_2$ TL *22* 4929 (1981)

$BH_3 \cdot NH_3$ TL *21* 693 (1980)

$NaBH_4$ JCS 3426 (1955); 2680 (1960)
                                                          TL *21* 693 (1980); *27* 4461 (1986)

$LiHAl(O$-t-Bu)$_3$ Tetr *24* 2039 (1968)

1.13. Reduction of Enones to Allylic Alcohols

For enone to saturated ketone reductions see page 8, Section 3.6.

$$>C=\overset{|}{C}-\overset{O}{\overset{||}{C}}- \longrightarrow >C=\overset{|}{C}-\overset{OH}{\underset{H}{\overset{|}{C}}}-$$

H_2, cat Pt-Ge   CC 1729 (1986)

H_2, cat [H_2Ir(Et$_2$PPh)$_4$]$^+$ CC 746 (1986)

$BH_3 \cdot SMe_2$ JOC *43* 1829 (1978)
                                                          Org Prep Proc Int *13* 225 (1981) (review)

$BH_3 \cdot t$-BuNH$_2$ TL *21* 693 (1980)

$\overset{}{\underset{}{\bigcirc}}NBH_2$ Syn 214 (1981)

9-BBN   JOC *40* 1864 (1975); *41* 1778 (1976);
                                                          42 1197 (1977)
                                                          TL *23* 3405 (1982)
                                                          JACS *108* 4561 (1986)

$\underset{\text{(chiral)}}{CH_3\diagdown\underset{}{\overset{H}{\diagup}}B\supset}$ JOC *50* 1384 (1985)

LiBH$_4$, N,N'-dibenzoylcystine,   CC 413 (1984)
   t-BuOH (chiral)

LiBH$_4$, TbCl$_3$ TL *27* 4759 (1986)

NaBH$_4$	JOC *35* 1041 (1970); *52* 5560 (1987)
	TL 2441 (1976); *28* 5945 (1987)
NaBH$_4$, CeCl$_3$	CC 601 (1978)
	JACS *100* 2226 (1978); *101* 5848 (1979); *103* 5454 (1981); *104* 1750 (1984); *107* 268, 1763, 5219 (1985); *108* 2090, 3110, 3731, 3739 (1986); *109* 2082, 3017, 3025, 3987, 4690, 6199, 8119 (1987)
	JOC *44* 689, 4187 (1979); *47* 381, 1855 (1982); *49* 2152 (1984); *50* 2981 (1985); *51* 491, 789, 1622 (1986); *52* 943, 3250, 3541, 4135, 5233, 5457, 5624 (1987)
	Syn Commun *12* 167 (1982)
	TL *27* 6341 (1986); *28* 333, 4951, 5655, 5977, 6253, 6485 (1987)
Zn(BH$_4$)$_2$	JACS *109* 3981 (1987)
NaBH$_3$CN, H$^+$, MeOH	JOC *40* 2530 (1975)
NaBH$_3$CN, CF$_3$CO$_2$H	JOC *45* 216 (1980)
(*n*-Bu$_4$N)BH$_3$CN, H$^+$, MeOH	JOC *40* 2530 (1975)
LiH$_3$B(*n*-Bu)	JOC *47* 3311 (1982)
LiHBEt$_3$	JOC *52* 4665 (1987)
LiHB(*sec*-Bu)$_3$	TL 4487 (1978)
	CC 1226 (1987)
LiHB(Sia)$_3$	TL 4487 (1978)
KHB(*sec*-Bu)$_3$	TL 4487 (1978)
KHBPh$_3$	JOC *52* 5564 (1987)

JACS *94* 8616 (1972)

JOC *52* 4020 (1987)

NaHB(OAc)$_3$	TL *28* 5755 (1987)
LiAlH$_4$	JCS 5280 (1965)
	JOC *34* 2206 (1969); *35* 1041 (1970)
LiAlH$_4$, LiBr or MgBr$_2$	TL *28* 5681 (1987)
LiAlH$_4$; 2S, 3R-Darvon alcohol	TL *27* 4759 (1986)

LiAlH$_4$, ArNHCH$_2$CH(NHCH$_3$)CH$_2$CH$_2$OH   TL 24 4123 (1983)
  (Ar = Ph; 2,6-Me$_2$C$_6$H$_3$) (chiral)

LiAlH$_4$, PhCHOHCHMeNMe$_2$,   CL 239 (1984)
  (chiral)

LiAlH$_4$, PhCHOHCHMeNMe$_2$, EtNHPh   CC 1026 (1980)
  (chiral)

LiAlH$_4$, PhCH$_2$COHPhCHMeCH$_2$NMe$_2$ TL 27 4759 (1986)
  (chiral)

LiHAl(i-Bu)$_2$(n-Bu)   JACS 102 7910 (1980)
                                       JOC 49 1717 (1984)

LiHAl(OMe)$_3$ JOC 34 2206 (1969)

JACS 106 6717 (1984)
CC 1226 (1987)

AlH$_3$ JOC 34 2206 (1969); 38 1380 (1973)
                                       TL 28 503 (1987)

i-Bu$_2$AlH   CC 213 (1970)
                                       JACS 102 7910 (1980); 109 3017, 3025, 3981
                                         (1987)
                                       JOC 50 2443 (1985); 51 5232 (1986);
                                         52 1907, 3841 (1987)

i-Bu$_2$AlH, ZnCl$_2$ TL 28 61 (1987)

i-Bu$_3$Al   JOC 47 4640 (1982)

JOC 44 1363 (1979)
BCSJ 54 3033 (1981)

(HAlN-i-Pr)$_6$ TL 2369 (1977)

PhMe$_2$SiH or Ph$_2$MeSiH, cat n-Bu$_4$NF   JACS 106 4629 (1984)

JOC 52 948 (1987)

JOC 52 948 (1987)

R_2SiH_2 (R = Et, Ph), cat ClRh(PPh$_3$)$_3$, MeOH	Organomet 1 1390 (1982)

RSiH$_3$, cat ClRh(PPh$_3$)$_3$ $\left(R = Ph, -\bigcirc \right)$ Bull Acad Sci USSR, Div Chem Sci
                                                                                                                                  26 995 (1977)

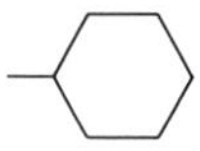

LiH, VCl$_3$	JOC 45 1041 (1980)
NaH, NaO-t-Am, ZnCl$_2$	JOC 44 2203 (1979)
NaH, ZnCl$_2$, MgBr$_2$	JOC 44 2203 (1979)
Li$_4$CuH$_5$	JOC 43 183 (1978)
i-PrOH, alumina (enals only)	TL 3601 (1975)

1.14. Reduction of Enones to Saturated Alcohols

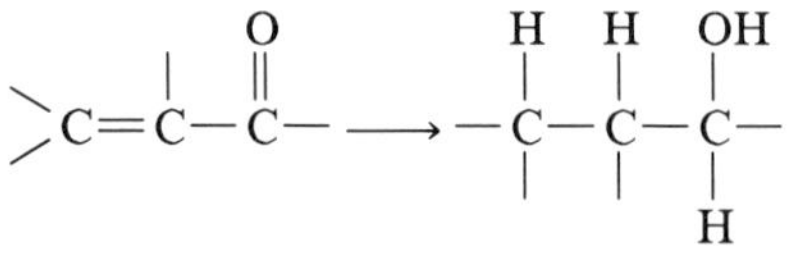

$$\underset{\diagup}{\overset{\diagdown}{>}}C=\overset{|}{C}-\overset{\overset{O}{\|}}{C}- \longrightarrow -\overset{\overset{H}{|}}{C}-\overset{\overset{H}{|}}{\underset{|}{C}}-\overset{\overset{OH}{|}}{\underset{|}{C}}-$$

H$_2$, cat Rh-alumina	TL 28 5615 (1987)
Li, NH$_3$/Li, MeOH/NH$_4$Cl	JACS 108 800 (1986)
Li, NH$_3$, EtOH	JOC 50 2981 (1985)
Li, NH$_3$, NH$_4$Cl	CC 1044 (1987)
NaBH$_4$, MeOH	JACS 108 3443 (1986)
KHB(sec-Bu)$_3$	Austral J Chem 34 745 (1981)
KHBPh$_3$	JOC 52 5564 (1987)
Et$_3$SiH, CF$_3$CO$_2$H	JOC USSR 7 2145 (1971)
HFe(CO)$_4^-$	BCSJ 55 1329 (1982)
LaNi$_5$H$_6$	JOC 52 5695 (1987)
LaNi$_{4.5}$Al$_{0.5}$H$_5$	JOC 52 5695 (1987)

1.15. Asymmetric Reduction of Aldehydes and Ketones

Reviews:

Syn 329 (1978) (asymmetric hydrogenation and hydrosilation)
Tetr 37 3547 (1981) (organoboron reagents); 42 5157 (1986) (chiral metal hydride complexes)
JACS 105 3725 (1983) (Cram and anti-Cram reduction of α-chiral ketones)
"Asymmetric Synthesis," Ed. J. D. Morrison, Academic Press, New York (1983), Chpt 2 (chiral boron reagents), Chpt 3 (LiAlH$_4$ derived reagents)
Topics Stereochem 14 231 (1983) (complex aluminum hydrides and tricoordinate aluminum reagents)
JOC 52 5406 (1987) (reagent comparison)

H_2, chiral Rh catalyst	CC 428 (1977) (α-keto esters) Org Syn *63* 18 (1984) (α-keto lactones) CL 1603 (1984) (α-keto esters, amides and lactones) TL *27* 4477 (1986) (α-keto lactone); *28* 3675 (1987) (α-keto lactone)
H_2, cat RuX_2(BINAP) (X = Cl, Br)	JACS *109* 5856 (1987) (β-keto esters)
H_2, Raney Ni, tartaric acid	BCSJ *36* 155 (1963) (β-keto ester); *53* 3367 (1980) (β-diketone to diol); *55* 2186 (1982) (β-keto ester); *56* 1414 (1983) (β-keto ester)

BH_3, cat

JACS *109* 5551 (1987)

BH_3, cat

R = H   JACS *109* 5551 (1987)
R = Me   JACS *109* 7925 (1987)

BH_3, various chiral β-amino alcohols

CC 315 (1981) (aryl ketones)
JCS Perkin I 1673 (1983) (aryl ketones)

BH_3, R—C—CPh_2OH

CC 469 (1983) (aryl ketones)
JOC *49* 555 (1984) (aliphatic ketones); *52* 5406
(1987) (summary)
JCS Perkin I 2039 (aliphatic and aryl ketones,
α-halo, α-OH, α-OSiMe$_3$, α-OAc ketones,
α-keto ester); 2615 (polymer-supported reagent,
aryl and α-halo ketones) (1985)

Chem Pharm Bull *27* 1479 (1979)

BH_3, chiral Li amide

TL *27* 635 (1986)

JACS *108* 7402, 7404 (1986) (alkyl ketones)

JOC *49* 2558 (1984) (low ee's)

JOC *49* 3646 (1984) (low ee's)

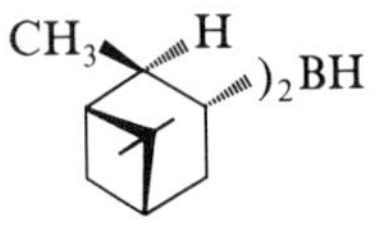

JACS *83* 3166 (1961)
Tetr *24* 6365 (1968)
JOC *42* 2996 (1977)

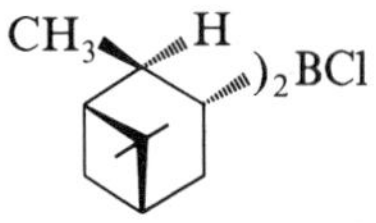

JOC *50* 5446 (1985) (aryl ketones); *51* 3394 (1986)
(3° alkyl ketones); *52* 5406 (1987) (summary)

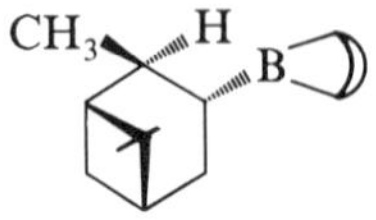

JACS *99* 5211 (1977) (1-deuteroaldehydes);
101 2352 (1979) (1-deuteroaldehydes); *102* 867
(1980) (ynones); *106* 1531 (α-keto esters), 3548
(ynone), 4192 (ynone), 7217 (1-deuteroaldehyde)
(1984); *107* 3915 (1985) (ynone)
JOMC *156* 203 (1978) (mechanism)
JOC *46* 4107 (1981) (ynones);
47 1606 (alkyl ketones) (1982);
48 1784 (1983) (α-halo ketones);
49 1316 (1984) (alkyl ketones);
50 1384 (1985) (alkyl ketones, enones, ynones,
α-halo ketones); *52* 1372 (ynone), 2860 (ynone),
5406 (summary) (1987)
Tetr *40* 1371 (1984) (ynones)
Org Syn *63* 57 (1985) (ynone)

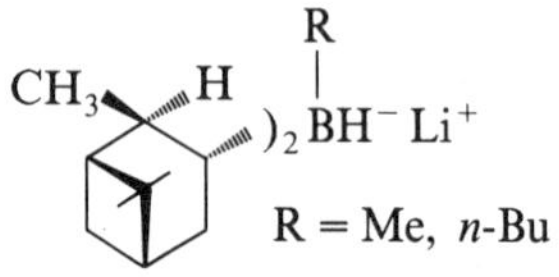

R = Me, *n*-Bu

JCS C 2557 (1971)
JACS *93* 1491 (1971)

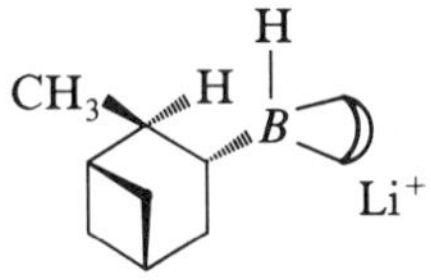

JOC *42* 2534 (1977)

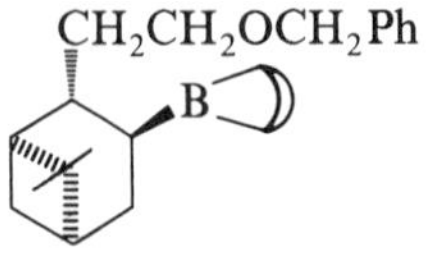

JOC *47* 2814 (1982) (ynones)
Tetr *40* 1371 (1984) (ynones)

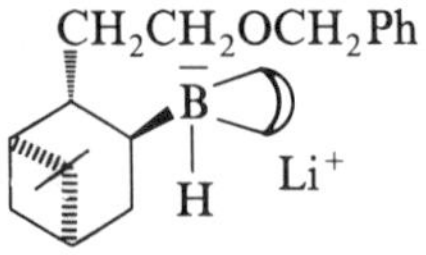

JOC *47* 2495 (1982); *52* 5406 (1987) (summary)

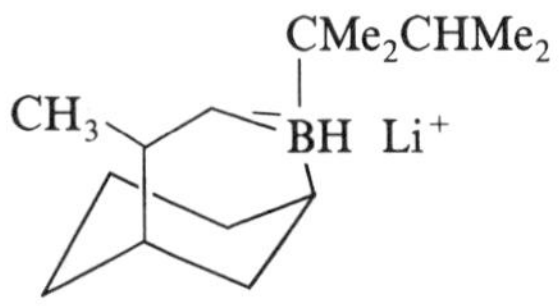

JACS *93* 1491 (1971)

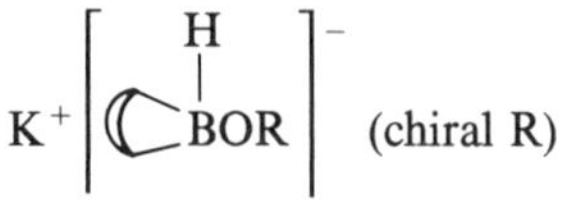

JOC *51* 3278 (1986)

JOC *51* 1934 (RCOAr, hindered RCOR),
  3278 (aliphatic and aryl ketones),
  3396 (α-keto esters) (1986);
  52 5406 (1987) (summary)

JOC *52* 4020 (1987) (ketones, enones)

LiBH$_4$; *N,N'*-dibenzoylcystine; ROH

• CC 413 (1984) (aryl ketones and enones);
  138 (1985) (3-aryl-3-oxo esters); 801 (1987)
  (acetylpyridines, α- and β-amino ketones)
TL *28* 2837 (1987) (β-keto ester)

NaBH$_4$, bovine serum albumin

CC 926 (1978)
JOC *51* 5423 (1986) (δ-keto acids)

NaBH$_4$, hydroxymonosaccharide
  derivatives

JOC *44* 1720 (1979)

NaBH$_4$, R'CO$_2$H,

JOC *45* 4229 [R$_2$ = (CH$_2$)$_5$],
  4231 (R = Me) (1980)
JCS Perkin I 900 (1981) [R = Me; R$_2$ = (CH$_2$)$_5$]

NaBH$_4$, ZnCl$_2$,

CC 807 (1979)
BCSJ *54* 1424 (1981) (aryl ketones)

i-Bu$_2$AlH, SnCl$_2$,

CL 2071 (1984); 813 (1985)
  (α-, β- and γ-keto esters)

LiAlH$_4$, quinine

Coll Czech Chem Commun *30* 2487 (1965);
  32 3897 (1967); *39* 1869 (1974)

LiAlH$_4$, (Me$_2$NCH$_2$CHOH)$_2$

Ber *107* 1748 (1974); *113* 1691 (1980)

LiAlH$_4$, PhCHOHCHMeNMe$_2$, EtNHPh

CL 981 (1980)
CC 1026 (1980) (enones only)

LiAlH$_4$, PhCHOHCHMeNMe$_2$,

CL 239 (1984) (cyclic enones and aryl ketones)

LiAlH$_4$, PhCHOHCHMeNMe$_2$,
  2,5-Me$_2$C$_6$H$_3$OH

Tetr *32* 939 (1976)
TL 2683 (1979); *21* 1735, 1739 (1980);
  24 4477 (1983) (all ynones)
JOC *51* 1264 (1986)

LiAlH$_4$, PhCHOHCHMeNMeCH$_2$-polymer,
  2,5-Me$_2$C$_6$H$_3$OH

JOC *51* 3462 (1986)

LiAlH$_4$, PhCH$_2$COHPhCHMeCH$_2$NMe$_2$	JACS *94* 9254 (1972); *99* 8339, 8341 (1977); *107* 1034 (1985) (4,6-alkadien-1-yn-3-one) JOC *38* 1870 (1973); *45* 582 (1980) TL *27* 4759 (1986) (enone)
LiAlH$_4$, ArNHCH$_2$CH(NHMe)CH$_2$CH$_2$OH	TL *23* 4111 (1982) (Ar = Ph); *24* 4123 (Ar = Ph; 2,6-Me$_2$C$_6$H$_3$) (enones)
LiAlH$_4$, PhCHMeNHMe·HCl	TL 3195 (1973)
LiAlH$_4$, O$_2$S(NHCHMePh)$_2$, MeNHPh	JOC *49* 3861 (1984)
LiAlH$_4$, *o*-Me$_2$NC$_6$H$_4$CH$_2$NHCHMePh	JOC *42* 1578 (1977)

LiAlH$_4$, CH$_2$NHPh   CL 783 (1977)
                                                  BCSJ *51* 1869 (1978)

LiAlH$_4$,   Heterocycles *12* 499 (1979)

LiAlH$_4$, terpenic glycols	JOC *42* 2073 (1977)
LiAlH$_4$, *cis*-2,3-pinanediol	Ber *106* 1312 (1973) (low ee's)
LiAlH$_4$; 1,4:3,6-dianhydro-D-mannitol or 1,3:4,6-di-O-benzylidene-D-mannitol	JCS Perkin I 1123 (1977) (very low ee's)

LiAlH$_4$, (X = NH$_2$, NHR)   JOC *50* 3013 (1985) (aryl ketones)

                               JCS C 1822, 2280 (1966); 197 (1967)

X = H, OEt

                               TL 1337 (1974)

R = Me, Et, *i*-Pr, PhCH$_2$

                               JACS *101* 3129, 5843 (1979); *106* 6709, 6717 (1984) (enones, ynones and aryl ketones); *108* 6384 (1986)
Pure Appl Chem *53* 2315 (1981) (review)
TL *22* 247 (1981) (ynones)
JOC *52* 5406 (1987) (summary)
CC 1226 (1987)

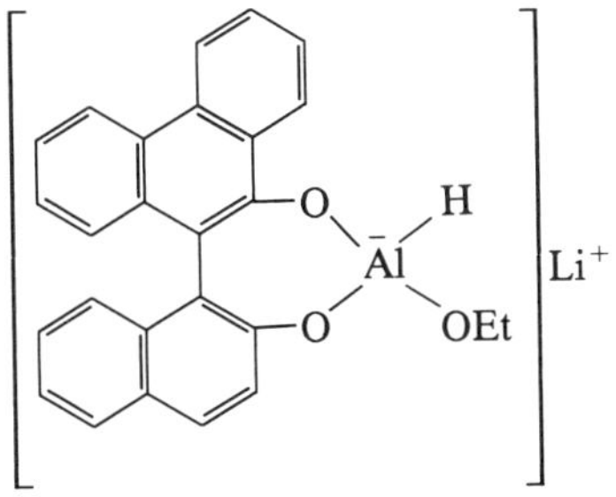

CC 1490 (1984)

R_2SiH_2 or R_3SiH, cat Rh(I), opt act phosphines

J Chem Res (S) 320 (1980)
CC 1238 (1982)

Ph_2SiH_2, cat Rh-diop

CC 1238 (1982)

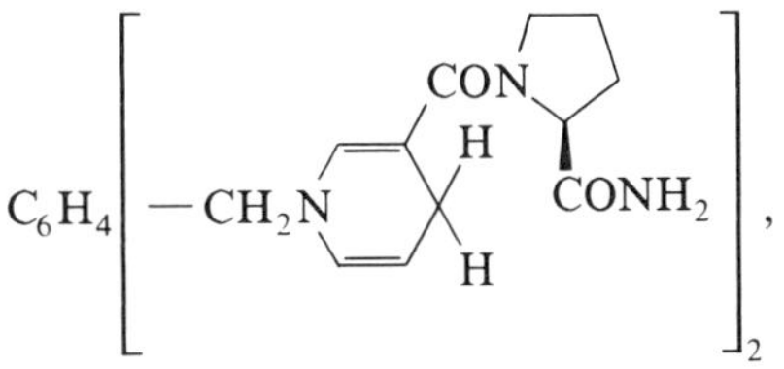

$Mg(ClO_4)_2$ *o*, *m*, or *p*

JACS *103* 4613 (1981)

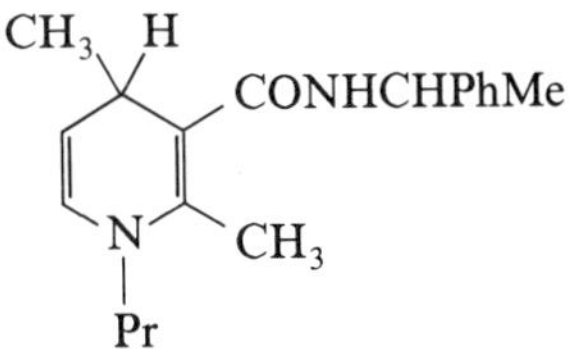

, $Mg(ClO_4)_2$

JACS *101* 7036 (1979) (α-keto esters, aryl ketones)

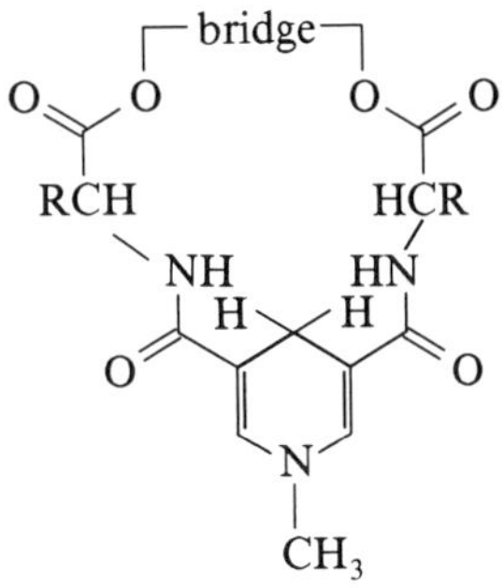

, $Mg(ClO_4)_2$

JACS *101* 2759 (1979); *103* 2091 (1981); *107* 3981 (1985)

bakers yeast

Adv Carbohydr Chem *4* 75 (1949) (review)
Org Syn Coll Vol *2* 545 (1949) (α-hydroxy ketone)
Can J Chem *29* 678 (1951) (β-keto carboxylates)
Biochem *3* 838 (1964) (ketones)
Biochem J *95* 633 (1965) (5-oxodecanoic acid)
Coll Czech Chem Commun *31* 2615 (1966) (methyl ketones)
JACS *88* 3595 (1966) (1-deuteroaldehydes); *102* 870 (1980) (β-keto ester); *104* 4251 (β-keto ester), 5473 (ketone) (1982); *105* 5925 (1983) (β-keto ester); *108* 4912 (1986) (ketone); *109* 8102 (1987) (β-keto ester)

bakers yeast (*continued*)

CC 400 (1975) (α-hydroxy, halo and formyloxy
   ketones; α-keto acid; α- and β-keto esters);
   315 (1977) (α-PhSO, Cl, OAc ketones); 456
   (1978) (trifluoromethyl ketones); 908 (1979)
   (bicyclic ketone); 599 (1983) (β-keto
   carboxylates); 138 (1986)
   (2,2-dithioalkan-1-ones); 1368 (1987)
   (β-keto ester)
Austral J Chem *29* 2459 (1976) (α- and β-keto
   esters and amides); *31* 1965 (1978)
   (α-RS, α-RSO, α-RSO$_2$ ketones)
Helv *60* 1175 (1977); *62* 2829 (1979); *63* 1383
   (1980); *65* 495 (1982); *66* 485 (1983); *67* 1843
   (1984) (all β-keto esters)
BSCF II 215 (1978) (α-phosphate)
Tetr *37* 1341 (1981); *41* 919 (1985)
   (both β-keto esters)
JOC *47* 2820 (1982) (β-diketones to hydroxy
   ketones); *50* 127 (β-keto carboxylate), 3411
   (β-diketones to hydroxy ketones) (1985);
   51 1253 (2,2-dithioalkan-1-ones), 2795 (α-fluoro
   ketones) (1986); *52* 192 (β-keto esters), 256,
   1141, 1359 (β-keto esters) 2036 (β-diketones to
   hydroxy ketones), 2086 (α-acetoxy ketone),
   2244 (β-keto ester), 3223 (cyclic β-diketones to
   hydroxy ketones), 4363 (γ- and δ-keto acids)
   (1987)
Angew Int *22* 1012 (1983) (α-formyl ester)
Agric Biol Chem *47* 1431 (1983) (γ- and δ-keto
   esters)
TL *24* 2009 (1983) (β-keto ester); *25* 1241
   (β-diketones to hydroxy ketones), 4623 (3-keto
   glutarate and adipate esters), 5083 (α-sulfenyl-
   β-keto esters) (1984); *26* 101 (β-keto esters),
   771 (α-PhS ketone), 4213 (β-keto esters) (1985);
   27 565 (β-diketones to hydroxy ketones), 1915
   (α-keto esters); 2091 (β-keto esters), 2657
   (β-keto esters), 3547 (2-acyl-1,3-dithianes;
   2-acylthiazoles), 4737 (α-chloro ketones), 4817
   (α-PhSO$_2$ ketone), 5275 (β-keto esters and
   amides), 5281 (β-keto esters), 5397 (α-chloro
   ketone, β-keto ester), 5405 (β-keto thioester)
   (1986); *28* 2709 (β-chloro-α-keto ester), 3189
   (β-keto ester) (1987)
Syn 897 (1983) (CH$_3$ and CF$_3$ ketones)
Chem Pharm Bull *31* 4384 (1983)
   (MeO$_2$CCOCHMeCO$_2$Me)
Org Syn *63* 1 (1984) (β-keto esters)
CL 1475 (1985) (δ-keto acids)

Corynebacterium equi IFO 3730

JOC *52* 2735 (1987) (α-PhS, α-PhSO and α-PhSO$_2$
   ketone)

Sporobolomycetes pararoseus

JOC *43* 2357 (1978) (aryl ketones)

Cryptococcus macerans

JOC *45* 3352 (1980) (α-halo aryl ketones)

Mucor rammanianus	JACS *97* 865 (1975)
Dipodascus uninucleatus	JACS *97* 865 (1975)
Geotrichum candidum	TL *28* 5037 (1987) (β-diketone to hydroxy ketone)
Candida albicans	Chem Pharm Bull *31* 4384 (1983) (MeO$_2$CCOCHMeCO$_2$Me)
Kloeckera corticis ATCC 20109	TL *23* 5489 (1982) (2,2-dithia-1-alkanone)
Thermoanaerobium brockii alcohol dehydrogenase	Angew Int *23* 151 (1984) (β-keto esters) JACS *108* 162, 3474 (4,5 and 6 chloro ketones) (1986) JOC *52* 256 (1987)
horse liver alcohol dehydrogenase	JACS *98* 8476 (1976) Can J Chem *59* 1574 (1981) (3-thiacycloalkanones) TL *28* 3059 (1987)
lactate dehydrogenase	JACS *104* 4458 (1982) (α-keto acid) JOC *52* 2608 (1987) (α-keto acid)
glycerol dehydrogenase	JOC *51* 25 (1986); *52* 2608 (1987) (both α-hydroxy ketones)
various enzymes, yeasts, molds, fungi or bacteria	Adv Carbohydr Chem *4* 75 (1949) (review) Appl Microbiol *11* 389 (1963) (γ- and δ-keto acids) BSCF 4217 (1972) Angew Int *23* 570 (1984); *24* 539 (1985) (both reviews) JACS *107* 2992, 4028 (1985); *108* 284 (1986) TL *27* 2631 (β-keto esters), 2657 (β-keto esters), 4453 (α-diketones to diol) (1986); *28* 1487 (aryl ketone), 3939 (β-keto esters), 5033 (4-*O*-benzyl-2-methyl-3-oxo-butyrate esters) (1987) JOC *52* 256 (1987)

1.16. Pinacol Reaction

$$2\,R_2CO \longrightarrow \underset{\displaystyle R_2\overset{\textstyle |}{C}-\overset{\textstyle |}{C}R_2}{\overset{\displaystyle HO\quad OH}{}}$$

electrolysis	JACS *72* 3797 (1950); *74* 4260 (1952) JOC *15* 435 (1950); *31* 3755 (1966); *33* 294, 2145 (1968); *34* 2807 (1969) Acta Chem Scand *11* 283 (1957) JCS 863 (1958) Chem Rev *62* 19 (1962) (review) JCS C 653 (1966); 2388 (1968)
photolysis	Chem Rev *40* 181 (1947) (review)
Li, THF	TL *28* 1813 (1987)
Li, NH$_3$	TL *28* 1813 (1987)
Na	TL 3613 (1970) JACS *108* 1265 (1986)
Na(Hg)	JCS 2423 (1928) JACS *55* 1179, 2827 (1933); *73* 2586 (1951); *108* 4561 (1986) TL 1879 (1964)

Mg-graphite	CC 1802 (1986)
Mg, MgI$_2$	JACS *49* 236 (1927); *51* 306 (1929) Chem Rev *57* 417 (1957) (review)
Mg, Me$_3$SiCl, HMPA	TL 75 (1972)
Mg, HgCl$_2$	Org Syn Coll Vol *1* 459 (1941)
Me$_8$Sn$_3$	TL 2847 (1978)
Cp$_2$TiCl$_2$, *sec*-BuMgCl	TL *28* 5717 (1987)
TiCl$_3$, NaOH	TL *23* 3517 (1982)
TiCl$_3$, K	TL *23* 5485 (1982)
TiCl$_4$, *n*-BuLi	Chimia *40* 12 (1986) (ArCHO)
TiCl$_4$, Mg(Hg)	JOC *41* 260 (1976); *47* 1657 (1982); *51* 2969 (1986) TL *28* 4965 (1987)
TiCl$_4$, Zn	CL 1041 (1973) TL *26* 1983 (1985) (intramolecular); *28* 1799 (1987) (intramolecular)
Ce, I$_2$	TL *23* 1353 (1982)
SmI$_2$	JOMC *250* 227 (1983) TL *24* 765 (1983)
Zn, HCl	JACS *107* 686 (1985)
Zn, Me$_3$SiCl, lutidine	TL *24* 2821 (1983) (intramolecular)

2. Carboxylic Acids and Derivatives

$$RCO_2H \longrightarrow RCH_2OH$$

BH$_3$	JACS *92* 1637 (1970); *106* 2160 (1984); *109* 3098 (1987) J Med Chem *13* 203 (1970) JOC *38* 2786 (1973) Org Syn *64* 104 (1985)
BH$_3$·SMe$_2$	JACS *101* 6710 (1979) Org Prep Proc Int *13* 225 (1981) (review)
BH$_3$·SMe$_2$, BF$_3$·OEt$_2$	TL 3527 (1977) (amino acid) Org Syn *63* 136 (1984) (amino acids)
BH$_3$·SMe$_2$, B(OMe)$_3$	JOC *39* 3052 (1974)
9-BBN	JOC *41* 1778 (1976)
(catechol borane structure) BH	JOC *42* 512 (1977)
LiBH$_4$, MeOH, diglyme, Δ	JOC *51* 4000 (1986)

$NaBH_4$, $AlCl_3$	JACS *78* 2582 (1956)
$NaBH_4$, $TiCl_4$	Syn 695 (1980)
$NaBH_4$, $MeSO_3H$, DMSO	JOC *46* 2579 (1981) (R = alkyl only)
i-Bu_2AlH	Syn 617 (1975)
$LiAlH_4$	Org Rxs *6* 469 (1951) JACS *81* 610 (1959); *108* 4138 (1986); *109* 7816 (1987) JOC *51* 5019 (1986); *52* 2337 (1987)
$LiAlH_4$, $AlCl_3$	JACS *81* 610 (1959)
$NaH_2Al(OCH_2CH_2OCH_3)_2$	JOC *47* 5201 (1982)

$NaBH_4$	JCS Perkin I 2470 (1980)

$$RCOCl \longrightarrow RCH_2OH$$

9-BBN	JOC *41* 1778 (1976)

	JOC *42* 512 (1977)
$NaBH_4$	JACS *71* 122 (1949)
$NaBH_4$, alumina	Syn 912 (1979)
$NaBH_4$, $AlCl_3$	JACS *78* 2582 (1956)
$NaBH_4$, $TiCl_4$	Syn 695 (1980)
$Zn(BH_4)_2$, TMEDA	TL *27* 4213 (1986)
$LiH_3B(n\text{-}Bu)$	JOC *47* 3311 (1982)
$LiHBEt_3$	JOC *45* 1 (1980)
$NaBH_3CN$, $ZnCl_2$	JOC *50* 1927 (1985)
$(n\text{-}Bu_4N)B_3H_8$	TL *23* 3337 (1982)
i-Bu_2AlH	JOC *50* 2443 (1985)
$LiAlH_4$	Org Rxs *6* 469 (1951) JACS *81* 610 (1959) Syn 901 (1981)
$LiAlH_4$, $AlCl_3$	JACS *81* 610 (1959)
$LiHAl(i\text{-}Bu)_2(n\text{-}Bu)$	JOC *49* 1717 (1984)

/$NaBH_4$	JCS Perkin I 2470 (1980)

$$RCO_2R' \longrightarrow RCH_2OH$$

electrolysis, NH_3	TL *28* 1173 (1987)

$$RCO_2R' \longrightarrow RCH_2OH \quad (\textit{Continued})$$

Li or Na, NH_3	JOC *44* 2810 (1979)
Na, NH_3, EtOH	Ann 1532 (1982)
Na, EtOH	Org Syn Coll Vol *2* 372 (1943)
Na, *t*-BuOH, HMPA	CC 567 (1978)
$BH_3 \cdot SMe_2$	Syn 439 (1981)
	JOC *47* 3153 (1982)
	TL *28* 1147 (lactone), 3091 (1987)
$BH_3 \cdot SMe_2$, cat $NaBH_4$	CL 1389 (1984) (α-hydroxy ester)
	CC 992 (1987) (α-hydroxy ester)
9-BBN	JOC *41* 1778 (1976)

	JOC *42* 512 (1977)
$LiBH_4$	Carbohydr Res *4* 504 (1967)
	JOC *47* 4702 (1982)
	TL *23* 4991 (1982); *28* 535, 4681 (lactone),
	5458 (lactone) (1987)
	JACS *109* 1186 (1987)
$LiBH_4$, MeOH	JOC *51* 4000 (1986)
$LiBH_4$; cat $Li\left[H_2B \right]$, $LiHBEt_3$,	JOC *47* 1604 (1982); *49* 3891 (1984)
MeOB or $B(OMe)_3$	
$NaBH_4$, H_2O, THF	TL *28* 1147 (1987) (lactone)
$NaBH_4$, polyethylene glycols, Δ	JOC *46* 4584 (1981)
$NaBH_4$, MeOH	JOC *28* 3261 (1963); *52* 3777 (1987)
	TL *28* 6069 (1987)
$NaBH_4$, EtOH	Chem Pharm Bull *13* 995 (1965) (amino acid
	ester hydrochloride salts)
	TL *28* 2709 (1987) (α-hydroxy ester)
$NaBH_4$, MeOH, *t*-BuOH, Δ	Syn Commun *12* 463 (1982)
$NaBH_4$, LiCl, ROH	JOC *52* 1252 (1987)
	TL *28* 3671 (1987)
$NaBH_4$, $AlCl_3$	JACS *78* 2582 (1956)
$NaBH_4$, $MeSO_3H$, DMSO (R = alkyl)	JOC *46* 2579 (1981)
$Ca(BH_4)_2$	JOC *47* 4702 (1982)
	TL *28* 5161 (1987)
$NaBH_3OH$	JOC *42* 3963 (1977)
$LiH_3B(n\text{-}Bu)$	JOC *47* 3311 (1982)
$LiHBEt_3$	JOC *45* 1 (1980); *52* 4352 (1987)
	CC 1786, 1797 (1987)
AlH_3	BCSJ *55* 3555 (1982)

i-Bu$_2$AlH

Ann *623* 9 (1959)
Syn 617 (1975) (review)
JOC *49* 1707 (1984); *50* 2443 (1985);
  52 1201 (1987)
TL *27* 5799 (1986) (lactone)

LiHAl(*i*-Bu)$_2$(*n*-Bu)

JOC *49* 1717 (1984)
JACS *109* 5280 (1987)

LiAlH$_4$

JACS *69* 1197 (1947); *107* 2730 (1985); *108* 468,
  1019 (1986)
Helv *31* 1617 (1948); *32* 1156 (1949)
  (both α-amino esters)
Org Rxs *6* 469 (1951)
Org Syn *63* 140 (1984)
JOC *50* 2026 (1985); *52* 5419, 5480 (1987)

LiAlH$_4$, AlCl$_3$

JACS *81* 610 (1959); *105* 3252 (1983);
  109 6719 (1987)
JOC *51* 2863 (1986)

HSi(OEt)$_3$, CsF/H$_3$O$^+$

Syn 558 (1981)

(EtO)$_2$SiMeH or

Syn 981 (1982)

$$Me_3SiO(-\overset{\overset{\displaystyle H}{|}}{\underset{\underset{\displaystyle Me}{|}}{Si}}-O)_n-SiMe_3,$$

KF or KO$_2$CH

i-BuMgBr, cat Cp$_2$TiCl$_2$

TL *21* 2175 (1980)

LiH, VCl$_3$

JOC *45* 1041 (1980)

$$RCH{=}CHCO_2R \xrightarrow[\text{EtOH}]{\text{Li, NH}_3} RCH_2CH_2CH_2OH$$

JOC *37* 2871 (1972)

$$\overset{\overset{\displaystyle O}{\|}}{RCSR'} \longrightarrow RCH_2OH$$

NaBH$_4$

CC 330 (1978)
Syn Commun *11* 599 (1981)
TL *23* 3151 (1982)

LiAlH$_4$

JACS *108* 4603 (1986)

$$(RCO)_2O \longrightarrow RCH_2OH$$

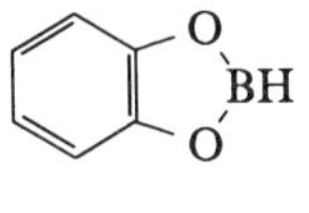

JOC *42* 512 (1977)

LiH$_3$B(n-Bu)   JOC *47* 3311 (1982)

LiHAl(i-Bu)$_2$(n-Bu)   JOC *49* 1717 (1984)

$$RCO_2H \longrightarrow RCO_2CO_2Et \longrightarrow RCH_2OH$$

NaBH$_4$ J Med Chem *7* 483 (1964)
                                    Chem Pharm Bull *16* 492 (1968); *27* 816 (1979)
                                    TL *27* 6349 (1986)

NaBH$_4$, SmI$_3$ TL *28* 5977 (1987)

LiBH$_4$ Z Naturforsch B *10* 252 (1955)

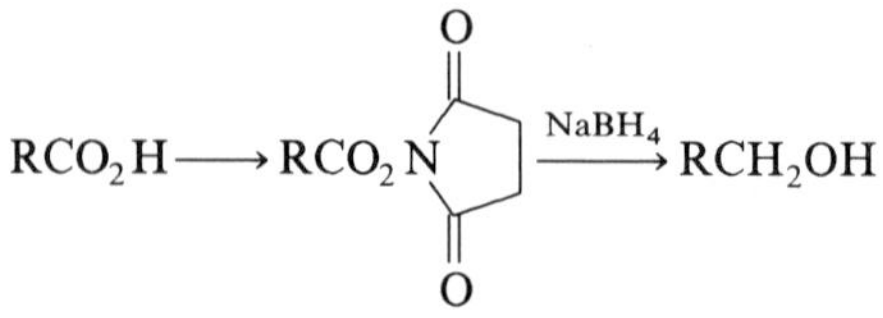

CL 981 (1979)

$$RCONR'_2 \longrightarrow RCH_2OH$$

9-BBN   JOC *41* 1778 (1976)

LiHBEt$_3$ Syn 635 (1977)
                                   JOC *45* 1 (1980)

$$RCO_2Na \longrightarrow RCH_2OH$$

BH$_3$·THF   TL *23* 2475 (1982)

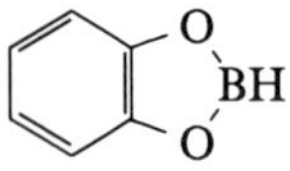

JOC *42* 512 (1977)

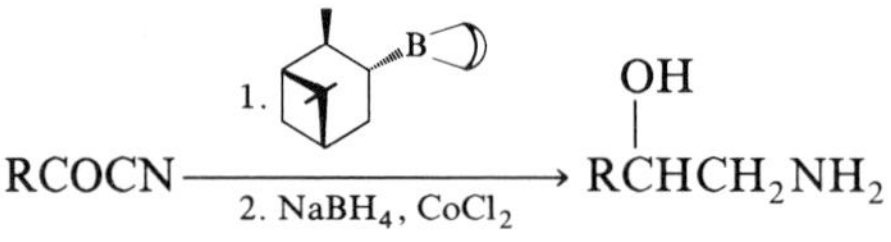

chiral

JOC *50* 3237 (1985)

8. ALKYLATION OF CARBONYL COMPOUNDS

1. Addition to Carbon Monoxide

$$2\ ArLi + RBr + CO \xrightarrow{H_2O} Ar_2\overset{\displaystyle OH}{\underset{\displaystyle |}{C}}R$$

JOC *46* 4625 (1981)

2. Addition to Aldehydes and Ketones

2.1. Organometallic Additions

$$RX + M + R'_2CO \xrightarrow{H_2O} R\overset{\displaystyle OH}{\underset{\displaystyle |}{C}}R'_2$$

Review: Syn 18 (1977) (M = Mg, Li, Ca, Zn)

M	RX	
Li	1°, 2°, 3° alkyl; allylic; benzylic; vinylic; aryl halides	JACS *102* 7926 (1980) (ultrasound)
	1°, 2°, 3° alkyl; allylic; aryl halides	JCS Perkin I 1655 (1972)
	1° alkyl, allylic, aryl halides	CC 1160 (1970)
	2° alkyl chloride	CC 1319 (1986) (ultrasound)
	2° alkyl chloride, bromide, iodide	TL *28* 2013 (1987) (ultrasound)
	3° benzylic chloride	CC 225 (1987) (4,4'-*t*-BuC$_6$H$_4$-*t*-Bu added)
	H$_2$C=CHCH$_2$Br	JACS *102* 7926 (1980) (ultrasound)
	bromobenzene	JCS Perkin II 378 (1976)

553

M	RX	
Mg	1° alkyl, allylic, benzylic, propargylic, aryl halides	Syn 18 (1977) (review)
	$Me_2N(CH_2)_3Cl$	Syn Commun *11* 241 (1981)
Ca	1° alkyl iodides	JOMC *66* 219 (1974)
Al, Sn	$RCH{=}CHCH_2X$ (X = Cl, Br)	Organomet *2* 191 (1983) TL *26* 6121 (1985)
	$RCH{=}\overset{\underset{\mid}{CO_2Et}}{C}CH_2Br$	CL 541 (1986)
Al, $SnCl_2$, H_2O	$RCH{=}CHCH_2Cl$	JOC *50* 5396 (1985) TL *27* 2395 (1986)
Al, $BiCl_3$, H_2O	$RCH{=}CHCH_2Br$	CC 708 (1987)
Sn	XCR_2CO_2R (X = Cl, Br) $RCOCR_2Br$ allylic bromides and iodides	CL 161, 929 (1982) CL 467 (1982); 1727 (1983) CL 1527 (1981); 541 (1986)
Sn, ultrasound	$H_2C{=}CHCH_2Br$	TL *26* 1449 (1985)
Sn, electrolysis	$H_2C{=}CHCH_2Br$	TL *25* 6017 (1984)
Sn, H_2O	$H_2C{=}CXCH_2Br$ (X = H, Br, OAc)	Organomet *2* 191 (1983) JOC *49* 172 (1984)
	$RCH{=}\overset{\underset{\mid}{CO_2Et}}{C}CH_2Br$	CL 541 (1986)
SnF_2	$H_2C{=}CHCH_2I$ CBr_4	CL 1507 (1980); 1109 (1981); 929 (1982) CL 1505 (1981)
$SnCl_2$	$H_2C{=}CHCH_2I$	Tetr *37* 3873 (1981)
$SnCl_2$, $\quad\overset{\underset{\mid}{NaO}\ \ \overset{\mid}{ONa}}{}$ $EtO_2CHCHCO_2Et$ (chiral)	$H_2C{=}CHCH_2Br$	CC 685 (1986) JOC *52* 5447 (1987)
Sb	$H_2C{=}CHCH_2I$	TL *28* 3707 (1987)
Sb, LiI	$H_2C{=}CHCH_2X$ $[X = Br, OPO(OPh)_2]$	TL *28* 3707 (1987)
Bi	$H_2C{=}CHCH_2X$ (X = Br, I)	TL *26* 4211 (1985)
$BiCl_3$, Zn or Fe	$R_2C{=}CHCH_2X$ (X = Cl, Br, I)	TL *27* 4771 (1986)
$CrCl_2$	$RSCH_2Cl$ ArI, $RCH{=}CRX$ (X = Br, I) $RC{\equiv}CX$ (X = Br, I)	JOC *51* 5045 (1986) TL *24* 5281 (1983) TL *26* 5585 (1985)

CrCl$_2$, cat NiCl$_2$	RCH$=$CRO$_3$SCF$_3$ RCH$=$CHI RC$\equiv$CI	JACS *108* 6048 (1986) JOC *52* 4823 (1987) TL *28* 3463 (1987)
CrCl$_2$, cat NiCl$_2$ or Pd(OAc)$_2$	RCH$=$CHI	JACS *108* 5644 (1986)
CrCl$_2$ or CrCl$_3$-LiAlH$_4$	allylic bromides, iodides and tosylates	JACS *99* 3179 (1977); *107* 5219 (1985) TL 1685 (1978); *22* 1037 (1981); *23* 2343 (1982); *27* 4957, 5091 (1986); *28* 5615 (1987) Tetr *37* 3873 (1981) BCSJ *55* 561 (1982) CL 85 (1986) (intramolecular) JOC *52* 316 (1987)
Mn	RCH$=$CHCH$_2$Br	Organomet *1* 1249 (1982)
Mn, I$_2$	H$_2$C$=$CHCH$_2$Br	CL 1237 (1983)
cat NiBr$_2$(bipy), electrolysis	ClCH$_2$CO$_2$Me, H$_2$C$=$C(Me)CH$_2$Cl	TL *28* 55 (1987)
Zn	review RCHBrCOR RCHXCO$_2$R	"Methods of Elemento-Organic Chemistry," North-Holland, Amsterdam (1967), Vol 3 JACS *89* 5727 (1967) Org Rxs *1* 1 (1942); *22* 423 (1975) (both reviews) Org Syn Coll Vol *4* 444 (1963) JOC *34* 3689 (1969); *35* 3966 (1970) [B(OMe)$_3$ added]; *39* 269 (1974); *46* 4323 (1981); *47* 5030 (1982) (ultrasound); *48* 4108 (1983); *52* 4796 (Me$_3$SiCl added), 5745 (intramolecular) (1987) Organomet Chem Rev A *8* 183 (1972) (review) Syn 452 (1975) BSCF II 145 (1980) Acta Chem Scand B *35* 273 (1981) (chiral RO$_2$CCH$_2$Br) JACS *107* 3891 (1985) [B(OMe)$_3$ added]; *108* 1617 (1986) TL *28* 6145 (1987)
	BrCF$_2$CO$_2$Et RCHBrCONR$_2$ BrCH$_2$C(Me)$=$CHCO$_2$R CF$_3$Br CF$_3$I	TL *25* 2301 (1984) JOC *52* 5745 (1987) (intramolecular) TL *27* 5193 (1986) CC 642 (1987) CL 1679 (1981) (ultrasound)

$\underline{M}$	$\underline{RX}$	
	allylic halides	JOC *41* 551 (1976); *47* 3148 (1982); *48* 4108 (1983); *50* 2011 (1985) TL *28* 3151, 4551 (1987)
	$\overset{\displaystyle SO_2Ph}{\underset{\displaystyle \vert}{}}$ $RCH{=}CCH_2Br$	TL *27* 5091 (1986)
	$H_2C{=}CHCH{=}CHCH_2Br$	TL *27* 5211 (1986)
	propargylic halides	Syn Commun *10* 637 (1980)
Zn, ultrasound	$H_2C{=}CHCH_2Br$	TL *26* 1449 (1985)
Zn, ultrasound, H_2O	$R_2C{=}CHCH_2X$ (X = Cl, Br)	JOC *50* 910 (1985)
Zn, ultrasound, (Cp_2TiCl_2)	perfluoroalkyl bromides and iodides	JACS *107* 5186 (1985)
Zn, cat $MCl_2(PPh_3)_2$ (M = Ni, Pd)	perfluoroalkyl iodides	CL 517 (1984)
Zn, cat $Pd(PPh_3)_4$	$H_2C{=}CHCH_2OAc$	JOC *52* 3702 (1987) (RCHO only)
Zn-Cu	$BrCH_2CO_2Et$	Syn 698 (1977)
	$BrCH_2CH{=}CHCO_2R$	JOC *52* 4397 (1987)
Zn-CuCl or Zn-AgOAc-Et_2AlCl	$ClCF_2COR$	TL *28* 6481 (1987)
Zn-Ag, ultrasound	(ring structure with Br and SO_2)	CC 1552 (1987)
Zn-Ag graphite	$RCHXCO_2R$ (X = Cl, Br)	CC 775 (1986)
Zn-Ag, Et_2AlCl	$BrCH_2CO_2R$	BCSJ *53* 3301 (1980)
Zn, cat $HgCl_2$	$RC{\equiv}CCF_2Br$	TL *28* 659 (1987)
Ce(Hg)	benzylic and propargylic iodides	JOC *49* 3904 (1984)
	allylic bromides and iodides	TL *22* 4987 (1981) JOC *49* 3904 (1984)
Sm	CH_2I_2	TL *27* 3243 (1986)
SmI$_2$	1° alkyl iodides, bromides, tosylates; allylic; propargyl; benzylic halides; $RCHBrCO_2R$	JACS *102* 2693 (1980)
	benzylic bromides, allylic halides	JOMC *250* 227 (1983)
	allylic bromides and iodides, benzylic bromides	TL *23* 3497 (1982)
	CH_2X_2 (X = Br, I)	TL *27* 3891 (1986)

	PhCH$_2$OCH$_2$Cl	TL 25 3225 (1984)
		JACS 109 4424 (1987)
	BrCHRCO$_2$R	TL 27 3889 (1986)
		(intramolecular)
		JACS 109 6559 (1987)
		(intramolecular)
	RCOCl	TL 25 2869 (1984)
SmI$_2$, cat	allylic acetates	TL 27 1195 (1986)
Pd(PPh$_3$)$_4$		

$$RCCHR_2 \xrightarrow[\text{2. H}_2\text{Cl}_2,\ \text{SmI}_2]{\text{1. LDA}} RC\underset{\underset{CH_2}{\diagdown\diagup}}{-}CR_2$$

(with O double bond on RCCHR$_2$ and OH on product RC—CR$_2$)

TL 28 1307 (1987)

$$X-C_n-\overset{\overset{O}{\|}}{C}-R \longrightarrow C_n\underset{R}{\overset{OH}{\diagup}}C$$

n	X	Reagent(s)	
3	Br	Mg	CC 890 (1968)
			BSCF 359 (1968)
3, 4	Br	Mg(Hg)	Compt Rend C 265 1472 (1967)
4	Br	Mg or Ca, NH$_3$	Can J Chem 58 2524 (1980)
	I	Mg	Ber 35 2684 (1902)
			JOC 47 5368 (1982)
		Mg, HgCl$_2$	TL 28 3963 (1987)
		n-BuLi	TL 28 3963 (1987)
		n-Bu$_2$CuLi	TL 28 5071 (1987)
		SmI$_2$	TL 28 3963 (1987)
			Syn Commun 17 901 (1987)
4, 5	I	SmI$_2$,	TL 25 3281 (1984)
		[cat Fe(PhCOCHCOPh)$_3$]	JOC 51 1778 (1986)
			JACS 109 453 (1987)
5	Br	Mg	CC 1287 (1968)
			Angew Int 20 576 (1981)
		SmI$_2$	JACS 109 6559 (1987)
	I	n- or t-BuLi	TL 26 4987 (1985)
			JOC 51 1778 (1986)
		SmI$_2$	JOC 50 2759 (1985)
5, 6	I	nickel tetraphenyl-porphine, Li naphthalenide	JACS 92 395 (1970)

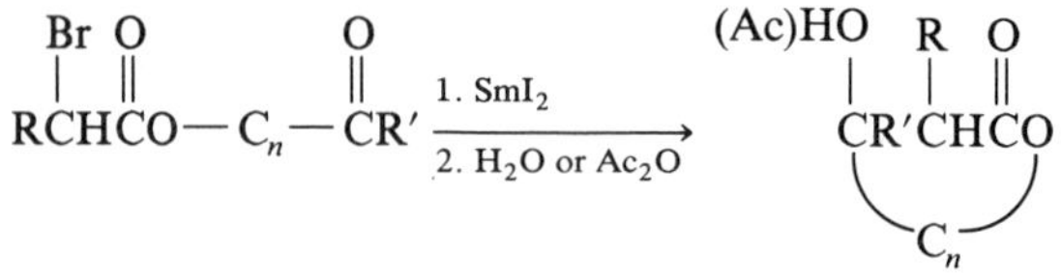

$$\underset{\substack{| \quad || \\ RCHCO}}{\overset{\substack{Br \quad O}}{}} - C_n - \overset{O}{\underset{||}{C}}R' \xrightarrow[\text{2. H}_2\text{O or Ac}_2\text{O}]{\text{1. SmI}_2} \quad \overset{(Ac)HO \quad R \quad O}{\underset{\substack{| \quad | \quad ||}}{CR'CHCO}}$$

<u>n</u>

2   JACS *109* 6556 (1987)

8–14   TL *27* 3889 (1986)

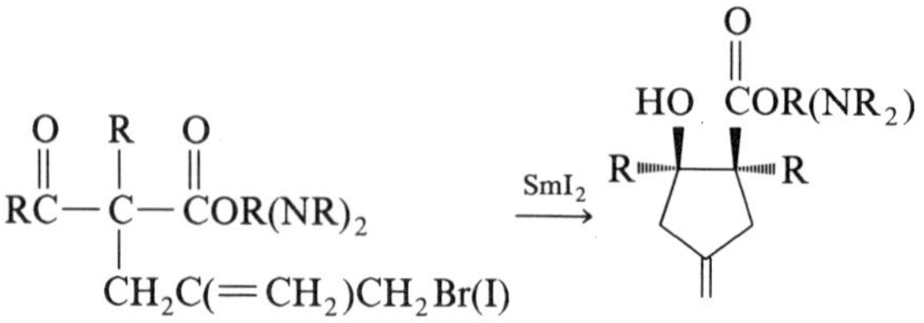

$$\underset{\substack{| \\ CH_2C(=CH_2)CH_2Br(I)}}{\overset{\substack{O \quad R \quad O \\ || \quad | \quad ||}}{RC - C - COR(NR)_2}} \xrightarrow{\text{SmI}_2}$$

JACS *109* 453 (1987)

$$I - C_n - \overset{O}{\underset{||}{C}}H \xrightarrow{n\text{-Bu}_3\text{SnH}}$$

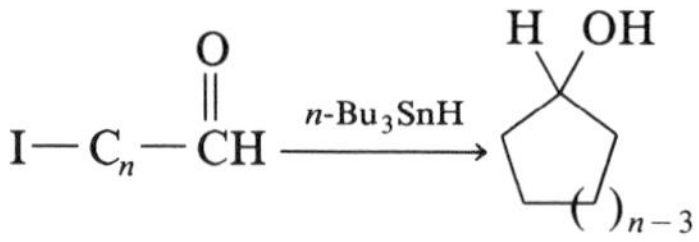

$$n = 4, 5$$

JACS *108* 2116, 8102 (1986); *109* 3484 (1987)

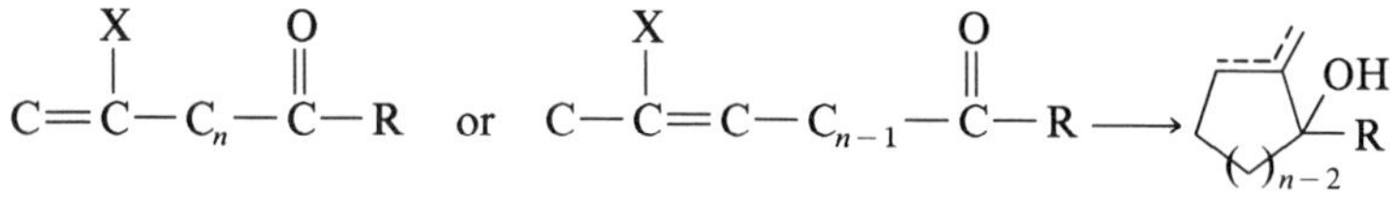

$$\overset{\substack{X \\ |}}{C=C} - C_n - \overset{O}{\underset{||}{C}} - R \quad \text{or} \quad \overset{\substack{X \\ |}}{C - C} = C - C_{n-1} - \overset{O}{\underset{||}{C}} - R \longrightarrow$$

JOC *51* 3405 (1986)

X	Ring Size	Reagent	
Br	6	Li	JACS *104* 6879 (1982)
I	5	n-BuLi	TL *23* 4987 (1985)
		R$_2$CuLi	JACS *92* 395 (1970)
	6	Mg	JOC *51* 3405 (1986)

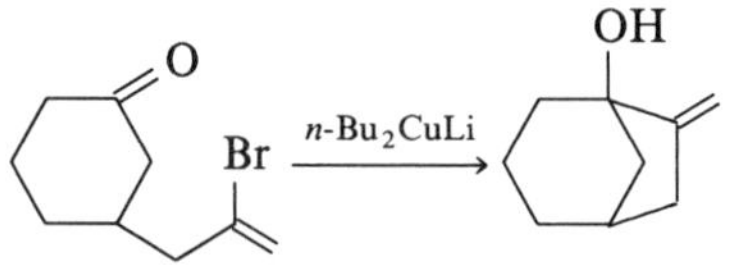

$$\text{Br} \xrightarrow{n\text{-Bu}_2\text{CuLi}}$$

JACS *92* 396 (1970)
TL *23* 3291 (1982)

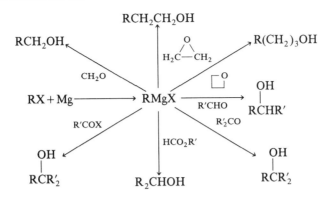

Review:

> M. S. Kharasch and O. Reinmuth, "Grignard Reactions of Non-Metallic Substances," Prentice Hall, Englewood Cliffs, New Jersey (1954)

$$R^1CR^2 + R^3M \xrightarrow{H_2O} R^1 - \underset{\underset{R^3}{|}}{\overset{\overset{OH}{|}}{C}} - R^2$$

Reviews:

> M. S. Kharasch and O. Reinmuth, "Grignard Reactions of Non-Metallic Substances," Prentice Hall, Englewood Cliffs, New Jersey (1954)
> JOC *39* 3258 (1974) (organometallic "ate" complexes of Li, B, Al, Mg, Zn)
> Chem Rev *75* 521 (1975) (stereochemistry)
> "Organometallic Chemistry Reviews," Vol 3 (1977)
> J. C. Stowell, "Carbanions in Organic Synthesis," J. Wiley, New York (1979)
> R. L. Augustine, "Carbon-Carbon Bond Formation," Vol 1, Marcel Dekker, New York (1979)
> E. Negishi, "Organometallics in Organic Synthesis," J. Wiley, New York (1980)
> Pure Appl Chem *52* 545 (1980) (mechanism of Grignard addition to ketones)
> JACS *103* 4540 (1981) (stereochemistry of cyclohexanone additions)
> Syn 605 (1982) (stereoselectivity of addition to chiral aminocarbonyl compounds)
> CC 452 (1982) (stereoselectivity of various R^3M on α-chiral ketone)
> E. L. Eliel in "Asymmetric Synthesis," Ed. J. D. Morrison, Academic Press, New York, Vol II, Part A (1983), p 125
> Angew Int *23* 556 (1984) (chelation and non-chelation control in addition reactions of chiral α- and β-alkoxy carbonyl compounds)

R^3M

RLi	JACS *81* 2748 (1959) (stereochemistry); *97* 5280 (1975) (stereochemistry); *101* 1455 (1979) (asymmetric); *103* 4540 (stereochemistry); 4585 (chiral catalysis) (1981); *107* 6411 (1985) (stereochemistry)
	CC 1160 (1970); 1600 (1986) (α-alkoxy or -siloxy ketone diastereoselectivity)
	JCS Perkin I 1655 (1972)
	JOC *38* 904 (1973); *40* 1469 (1975) (stereochemistry); *43* 1064, 2173 (stereochemistry) (1978); *46* 2798 (1981) (chiral catalysis)

R³M

JCS Perkin II 378 (1976)
TL 2659 (1976) (stereochemistry); 1709 (1977)
    (stereochemistry); 28 2629 (1987)
CL 219 (1978) (asymmetric)
Helv 62 1701, 1710, 2695 (1979)
Tetr 38 2725 (1982) (stereoselectivity using
    chiral ligand)

RLi, LiClO$_4$

JOC 44 4371 (1979)

RLi, n-Bu$_4$NBr

CC 923 (1987) (Cram selective)

RLi, crown ether

JACS 107 6411 (1985) (stereochemistry)

RLi, chiral

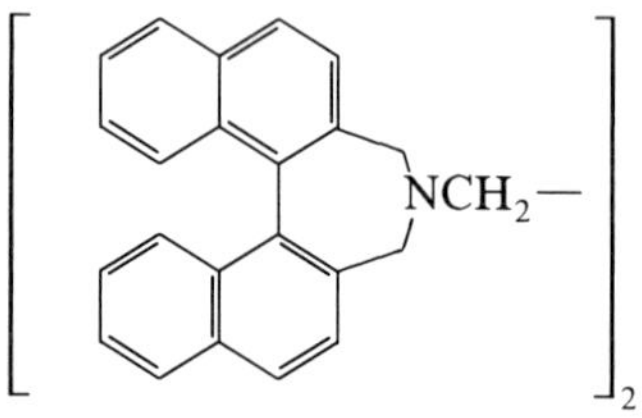

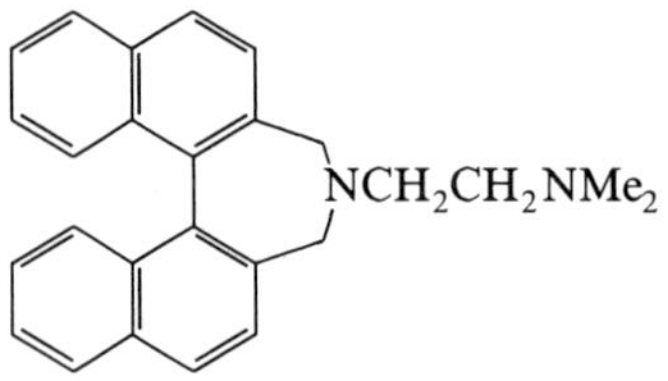

JACS 103 4585 (1981) (enantioselective)

or

RLi, chiral

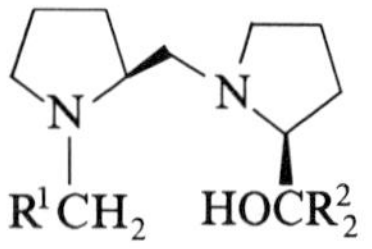

JACS 101 1455 (1979) (enantioselective)

n-BuLi, chiral lithium amides

TL 25 5187 (1984) (enantioselective)

MeLi

JACS 107 4573 (1985) (stereochemistry)
JOC 50 422 (1985)

MeLi, MeAl

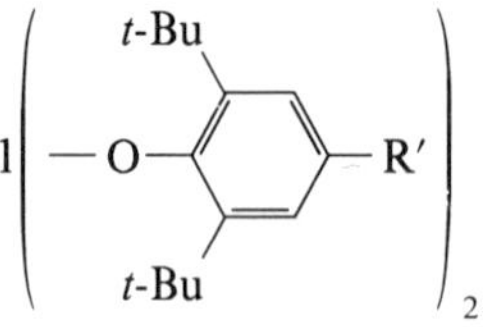

(R′ = Me, t-Bu)

JACS 107 4573 (1985)
    (aldehydes and cyclic ketones only, axial
    addition)
TL 28 1439 (1987)

R$_2$C(OR′)Li

TL 21 1285 (1980); 28 4131 (1987)
JACS 106 3376 (1984); 107 3285 (1985)

LiCH$_2$OLi	Angew Int *15* 738 (1978) Ber *113* 1290, 2055 (1980)
Li(CH$_2$)$_3$OLi	Ber *113* 1290 (1978)
(EtO)$_2$CHCH$_2$CH$_2$Li	CC 425 (1987)
LiCH$_2$CN	JACS *109* 613 (1987) (diastereoselective)
RCH=CRLi	TL 4839 (1976); *22* 3745 (1981) Ber *111* 2785 (1978)
LiCH=CHCH$_2$OLi	Ber *113* 1290 (1978)
H$_2$C=C(Li)CHO	JOC *52* 2625 (1987)
HC≡CLi	CC 363 (1981) (stereochemistry) JACS *107* 1034 (1985)
R$_3$SiC≡CLi	CL 929 (1982) JOC *52* 4191 (1987)

RC≡CLi	JACS *87* 5632 (1965); *107* 3915 (1985)
	JOC *38* 3588 (1973); *50* 2390 (1985); *52* 3241
	   (enediones), 3662, 3798, 4124 (1987)
	CL 447 (1979) (asymmetric); 255 (1980)
	   (asymmetric)
	TL *28* 1019 (1987)

EtOC≡CLi	JOC *52* 2332 (1987)
LiO$_2$CC≡CLi	Syn Commun *12* 977 (1982)
RO$_2$CC≡CLi	JOC *45* 28 (1980); *52* 3342 (1987) Syn Commun *12* 977 (1982)
R$_3$SiC≡CCH$_2$Li	TL *23* 719 (1982)
RCHLiSOAr	JOC *46* 4101, 5452 (1981) JCS Perkin I 1278 (1981)
LiCH$_2$Cl	TL *25* 835 (1984); *27* 795 (1986)
LiCHX$_2$ (X = Cl, Br, I)	JACS *96* 3010 (1974); *104* 4708 (1982); *106* 3030 (1984); *108* 3102 (1986) TL *22* 3745 (1981) Topics Curr Chem *106* 55 (1982) (review)
LiCX$_3$ (X = Cl, Br)	JACS *96* 3010 (1974) Topics Curr Chem *106* 55 (1982) (review)
LiCF$_2$CF$_3$	TL *26* 5243 (1985) JOC *52* 2481 (1987)

Ber *115* 1990 (1982)

Me$_3$SiCH$_2$CH$_2$M (M = Li, MgBr)   JOC *47* 1983 (1982)

$\underline{R^3M}$

$\triangleright\!\!<\!\!\begin{smallmatrix}Li\\SiMe_3\end{smallmatrix}$	TL *23* 259 (1982)
$H_2C{=}C(Li)SiMe_3$	TL *23* 259 (1982)
$HC{\equiv}CNa$	Org Syn Coll Vol *3* 416 (1955) Syn 211 (1982)
$RC{\equiv}CNa$	JOC *38* 3588 (1973)
$RC{\equiv}CH$, NaOH (phase transfer)	Ann 1750 (1982)
NCNa	Org Syn Coll Vol *2* 7 (1943)
O_2NCH_2Na	Org Syn Coll Vol *4* 221 (1963)
O_2NCH_2K	Org Syn Coll Vol *5* 833 (1973)
RMgX	Org Syn Coll Vol *1* 188 (1941); *3* 200 (1955) JACS *73* 4462 (1951); *74* 4954 (1952); *81* 2748 (1959) (stereochemistry); *103* 4540 (1981) (stereochemistry); *107* 4573, 6411 (1985) (both stereochemistry) M. S. Kharasch and O. Reinmuth, "Grignard Reactions of Non-Metallic Substances," Prentice Hall, Englewood Cliffs, New Jersey (1954) (review) JOC *27* 3860 (1962) (stereochemistry); *37* 1918 (1972) (stereochemistry); *50* 2179 (1985) (bicyclic ketones); *51* 5353 (1986) (2-alkoxy aldehyde stereoselectivity) BSCF 219 (1971) (stereochemistry) TL 4383 (1979); *21* 1031 (1980) (asymmetric); *23* 5211 (1982) (δ-keto ester stereoselectivity); *28* 2629 (1987) JCS Perkin II 597 (1981) (stereochemistry) JCS Perkin I 197 (stereochemistry using chiral alkoxide ligand), 979 (stereochemistry) (1982) CC 1600 (1986) (α-alkoxy or -siloxy ketone diastereoselectivity)
RMgX, crown ether	JACS *107* 6411 (1985) (stereochemistry)
RMgX, n-Bu$_4$NBr	CC 923 (1987) (Cram selective)
$R_2C{=}CRMgX$	JOC *52* 3798 (1987)
RMgX, MeAl$\left(-O-\!\!\!\!\begin{smallmatrix}t\text{-Bu}\\ \\t\text{-Bu}\end{smallmatrix}\!\!\!\!-R'\right)_2$ (R' = Me, t-Bu)	JACS *107* 4573 (1985) (aldehydes and cyclic ketones only, axial addition, anti-Cram)
RMgOR' or RMgNR'$_2$	JOC *43* 4094 (1978) (stereoselective)

R_2Mg	JOC *37* 1918 (1972) (stereochemistry) CL 601 (1978) (asymmetric) JACS *101* 1455 (1979) (asymmetric); *107* 6411 (1985) (stereochemistry)
Me_2Mg, $(LiClO_4)$	JOC *44* 4371 (1979)
R_2Mg, chiral	JACS *101* 1455 (1979) (enantioselective)

$$R^1CH_2 \quad HOCR_2^2$$

Me_3SiCH_2MgCl	TL *23* 3249 (1982) Ber *115* 1132 (1982)
$XMgCX_2CO_2R$	JCS Perkin I 2063 (1982)
$H_2C=CHCH=CHMgBr$	JOC *52* 4416 (1987)
$RC\equiv CMgBr$	TL *28* 1019 (1987)
$BrMgC\equiv CMgBr$	CC 363 (1981)
$RCaX$	JOMC *66* 219 (1974)
R_nAlX_{3-n} (X = halogen, OR, NR_2)	JOC *44* 4792 (1979)
R_3Al	JOC *37* 1918 (1972); *40* 1469 (1975) BCSJ *47* 1971 (1974)
$RC\equiv CAlR_2$	Compt Rend *261* 1992 (1965) BSCF 205 (1968)
$(RC\equiv C)_3Al$	Compt Rend C *262* 289 (1966)
$LiAlR_4$	JOC *39* 3258 (1974) TL *26* 4181 (1985) (on menthyl phenylglyoxalate)

$RSiR'_3$, cat $[(Me_2N)_3S]Me_3SiF_2$
  (R = $CHCl_2$, CCl_3, CCl_2Me, $CF_2CH=CH_2$,
  $CF=CF_2$, $CF=CF$-*n*-Bu, $CF=CFSEt$,
  $CCl=CF_2$, $CF=CFH$)
	JACS *107* 4085 (1985)

$HC\equiv CCH_2SiMe_3$, *n*-Bu_4NF (allenic product))	TL *22* 455 (1981)
$ArCH_2SiMe_3$, *n*-Bu_4NF	TL *24* 4217 (1983) BSCF II 90 (1985) JOC *52* 4381 (1987)
$RSiMe_3$ (R = benzylic, 4-picolyl, phenylallyl), KF-18-crown-6 or *n*-Bu_4NF-silica	TL *23* 577 (1982)
$RSiMe_3$ (R = benzyl, allyl, heterocyclic); CsF or *n*-Bu_4NF-silica	TL 5079 (1982)
$ArSiMe_3$, cat KO-*t*-Bu	Angew Int *20* 265 (1981)
$ArSiMe_3$, CsF	TL *28* 2933 (1987) (intramolecular)

$$R^3M$$

Cr(CO)$_3$·ArSiMe$_3$, cat CsF	Angew Int *20* 266 (1981)
SiMe$_3$ | RCH=CCO$_2$R, *n*-Bu$_4$NF (R^3 = RCH=CCO$_2$R)	Syn 734 (1986)
Me$_3$SiC≡CSiMe$_3$, AlCl$_3$	TL 2449 (1976)
Me$_3$SiC≡CSiMe$_3$, KF, 18-crown-6	CC 840 (1979)
RC≡CSiMe$_3$, *n*-Bu$_4$NF	Angew Int *15* 498 (1976) Tetr *39* 975 (1983)
PhCOSiMe$_3$, *n*-Bu$_4$NF/H$_2$O (R^3 = PhCO)	TL *22* 1881 (1981)
R$_f$SnX$_2$I (X = F, Cl)	CL 1337 (1981)
n-Bu$_3$SnCH$_2$CH$_2$COX (X = OMe, NHPh), TiCl$_4$	JOC *50* 5907 (1985)
RC≡CSn(*n*-Bu)$_3$, TiCl$_4$	CC 102 (1986)
RC≡CSnEt$_3$, ZnCl$_2$	TL 495 (1967)
R$_4$Pb, TiCl$_4$	JACS *109* 4395 (1987) (aldehydes only)
RTi review	Topics Curr Chem *106* 1 (1982)
RTiCl$_3$	Angew Int *19* 1011 (1980); *22* 989 (1983) Syn Commun *11* 647 (1981) JACS *105* 4833 (1983) CC 1600 (1986) (α-alkoxy or -siloxy ketone diastereoselectivity) Tetr *42* 2931 (1986)
RO$_2$CCH$_2$CH$_2$TiCl$_3$	JACS *99* 7360 (1977); *105* 651 (1983); *108* 3745 (1986)
RO$_2$CCH$_2$CH$_2$TiCl$_3$, Ti(O-*i*-Pr)$_4$	JACS *107* 2138 (1985); *108* 3745 (1986)
R$_2$TiCl$_2$	Angew Int *19* 1011 (1980) Ber *118* 1441 (1985)
	TL *27* 5711 (1986) (enantioselective)

RTi(OR')$_3$	Angew Int *19* 1011 (1980); *21* 135 (1982);
	  22 31 (1983)
	Helv *63* 2451 (1980); *64* 357, 2485 (1981);
	  65 249 (1982)
	JACS *105* 4833 (1983); *107* 4577 (1985)
	Ber *118* 1421, 1441 (1985)
	JOC *50* 422 (1985)

	CC 1600 (1986) (α-alkoxy or -siloxy ketone diastereoselectivity) Tetr *42* 2931 (1986) TL *27* 5351 (1986) (β-Me$_2$N ketone); *28* 2959 (1987)
R$_2$Ti(OR')$_2$	Angew *21* 135 (1982) Ber *118* 1421 (1985)
Me$_4$Ti	Angew *21* 135 (1982) Ber *118* 1421 (1985)
RZr(OR')$_3$	Helv *64* 1552 (1981) Angew Int *21* 135 (1982); *22* 31 (1983)
Me$_4$Zr	Angew Int *21* 135 (1982)
MeHfX$_3$ (X = Cl, OEt)	TL *27* 5351 (1986) (α-hydroxy ketone)
Me$_2$VOCl	TL *27* 5351 (1986) (β-hydroxy ketone)
(CH$_3$)$_2$Nb(OEt)$_3$	TL *23* 2301 (1982) (RCHO > R$_2$CO)
CH$_3$NbCl$_4$	TL *23* 2301 (1982) (RCHO > R$_2$CO)
(CH$_3$)$_2$Ta(OEt)$_3$	TL *23* 2301 (1982) (RCHO > R$_2$CO)
CH$_3$Ta(OEt)$_4$	TL *23* 2301 (1982) (RCHO > R$_2$CO)
CH$_3$TaCl$_4$	TL *23* 2301 (1982) (RCHO > R$_2$CO)
MeCrCl	TL *27* 5351 (1986) (β-hydroxy ketone)
MeCrCl$_2$	TL *27* 5355 (1986) (RCHO and β-Me$_2$N ketone)
MeMnCl	TL *27* 5351 (1986) (α- and β-Me$_2$N ketone)
RMnX	TL 3383 (1977); *27* 4445 (1986) (RCHO > R$_2$CO)
(Cl$_2$CH)$_3$MnLi	TL *27* 5351 (1986) (α-Me$_2$N ketone)
RCu·BF$_3$	JOC *47* 119 (1982)
RCu·MgBr$_2$	JOC *50* 422 (1885)
R$_2$CuLi	JOC *31* 3128 (1966); *40* 1460 (1975); *50* 422 (1985) CC 320 (1973) TL 3128 (1974); *21* 1035 (1980) (asymmetric); *27* 3059, 5351 (β-Me$_2$N ketone) (1986) JACS *107* 6411 (1985) (stereoselectivity)
$\left(\text{H}_2\text{C}{=}\overset{\overset{\text{CH}_3}{\|}}{\text{C}}{-} \right)_2 \text{CuLi}$	TL *28* 4147 (1987)
R$_2$CuLi, crown ether	JACS *107* 6411 (1985) (anti-Cram)
R$_2$CuLi, *n*-Bu$_4$NBr	CC 923 (1987) (anti-Cram)

$$\underline{R^3M}$$

$Me_3SiI/R_2CuLi/H_2O$	Syn Commun *13* 213 (1983)

$$\left[\underset{\underset{\displaystyle CO_2Me}{|}}{H_2C=CCuC\equiv C(CH_2)_3CH_3} \right] Li \qquad \text{TL 3897 (1975)}$$

R_3CuLi_2	JACS *97* 5280 (1975) (stereochemistry)
	TL 2659 (1976) (stereochemistry); 1709 (1977)
	(stereochemistry); *28* 5547 (1987)
$(RC\equiv C)_3CuLi_2$	CC 892 (1975) (on enones)
$R_5Cu_3Li_2$, crown ether	JACS *107* 6411 (1985) (anti-Cram)
$Me_2Cu(CN)Li_2$	JACS *106* 6414 (1984) (on RCHO)
$R_2Cu(CN)Li_2$, crown ether	JACS *107* 6411 (1985) (anti-Cram)
RZnX	"Methods of Elemento-organic Chemistry," Vol 3, North-Holland, Amsterdam (1967) (review)
	Adv Organometal Chem *12* 83 (1974) (review)
CF_3CX_2ZnX (X = Cl, Br)	TL *27* 2135, 2139 (1986)
$RO_2CCHRZnX$	See the earlier equation in this section under Zn
$(RO_2CCHRCH_2)_2Zn$, Me_3SiCl	JACS *109* 8056 (1987)
$BrZnCH_2CR=CHCO_2SiMe_3$	JOMC *219* C1 (1981) (α or γ)
R_2Zn, $TiCl_4$	Syn Commun *11* 261 (1981)
	Angew Int *22* 989 (1983)
	JACS *105* 4833 (1983)
	JOC *52* 320 (1987)
R_2Zn, cat bis[(−)-camphorquinone-α-dioximato) cobalt(II) or palladium(II)]	CL 841 (1983) (ArCHO only, enantioselective)
R_2Zn, cat cinchona alkaloids	JOC *52* 135 (1987) (ArCHO only, enantioselective)
	TL *28* 6163 (1987)

R_2Zn,

JACS *108* 6071 (1986)
    (RCHO only, enantioselective)

R_2Zn, cat
 (X = H, Li)

TL *28* 4841 (1987)
    (ArCHO only, enantioselective)

R_2Zn, cat $PhCHOHCHMeNR_2$	CC 1690 (1987) (enantioselective)
	TL *28* 3013 (1987) (ArCHO only, enantioselective)
R_2Zn, chiral polymeric amino alcohol	JOC *52* 4140 (1987) (ArCHO only, enantioselective)

R$_2$Zn, several chiral amino alcohols	TL *25* 2823 (1984) (PhCHO only, enantioselective); *28* 6163 (1987) (ArCHO only, enantioselective)
R$_2$Zn, chiral pyrrolidinylmethanols	CC 467 (1987) (enantioselective) JACS *109* 7111 (1987) (enantioselective)

R$_2$Zn, cat

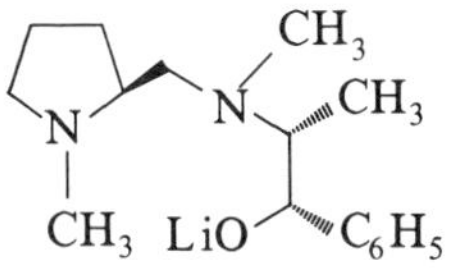

TL *28* 5233 (1987) (RCHO only, enantioselective)

R$_2$Zn, cat PhCHCH(CH$_3$)NCH$_2$CH$_2$N(CH$_3$)$_2$ (with OLi and CH$_3$ substituents)

TL *28* 5233 (1987) (ArCHO only, enantioselective)

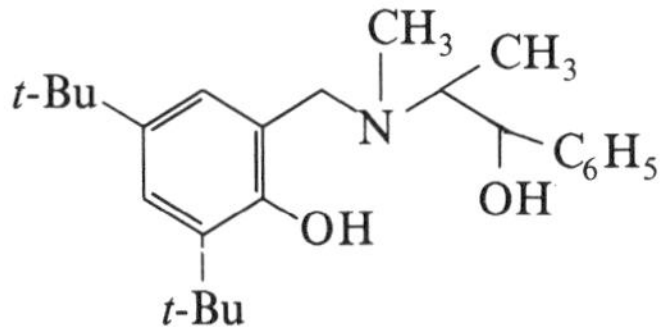

R$_2$Zn, cat

TL *28* 5237 (1987) (ArCHO only, enantioselective)

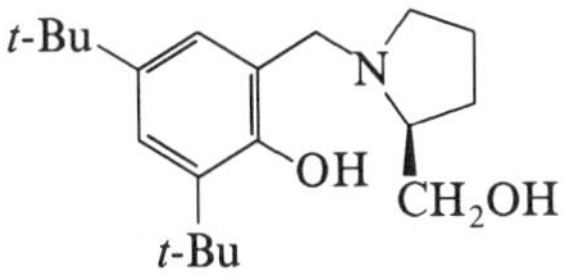

R$_2$Zn, cat

TL *28* 5237 (1987) (ArCHO only, enantioselective)

RC≡CZnX	TL *28* 1019 (1987)
(ArCH$_2$)$_2$Cd, MgX$_2$	TL 1581 (1979)
RCeX$_2$ (X = Cl, I)	CC 1042 (1982) TL *25* 4233 (1984); *27* 3329 (1986); *28* 31 (1987) JOC *49* 3904 (1984); *52* 281, 4142 (1987) JACS *108* 7873 (1986); *109* 7908 (1987)
RC≡CCeCl$_2$	CL 1543 (1984) TL *26* 1549 (1985); *27* 195 (1986); *28* 3971, 4583 (1987) CC 1290, 1474 (1987) JOC *52* 4135 (1987)
(RC≡C)$_3$Ce	CL 1543 (1984)
RYbI	JOC *52* 3524 (1987)

$$\underset{\text{R}^1\text{CH (R}_2\text{CO)}}{\overset{\text{O}}{\overset{\|}{}}} \longrightarrow \underset{\text{R}^1\text{CHCHR}^2\text{CH}=\text{CHR}^3}{\overset{\text{OH}}{\overset{|}{}}}$$

See also the direct reaction of aldehydes and ketones with allylic halides plus metals, as well as page 119, Section 8, for ene-type reactions

Reviews:

JOMC *69* 1 (1974) (Li, Na, Mg, Zn, Cd, Al)
Heterocycles *18* 357 (1982)
Angew Int *21* 555 (1982) (diastereoselectivity)
Topics Curr Chem *106* 55 (1982) (Ti)
JOMC *285* 31 (1985)

allylic lithium compounds

Tetr *42* 2803 (1986)

ClCH=CHCH$_2$Li (no rearrangement)

JOMC *236* 139 (1982)

Me$_3$SiCH=CHCH$_2$Li (no rearrangement)

CL 713 (1979)

R$_2$C=CHCH(Li)SO$_2$Ph (no rearrangement)

CL 1331 (1982)

R^{2}CH=CHCHR^{3}MgX

TL 3829 (1975); *21* 365 (1980); *26* 823 (1985); *28* 869 (1987)
Austral J Chem *37* 65 (1984)
JOC *51* 3290 (1986)

H$_2$C=C(Me)CH$_2$MgCl

JOC *52* 1603 (1987)

(H$_2$C=CHCH$_2$)$_2$Mg

BSCF 4038 (1969)

H$_2$C=CHCH$_2$MgCl, ZnCl$_2$

TL *27* 4961 (1986)

R^{2}CH=CHCHR^{3}B(OR)$_2$

Angew Int *17* 768 (1978); *18* 306 (1979); *19* 218 (1980); *21* 309 (1982)
TL 4653 (1979); *21* 4883 (1980); *22* 2751 (1981); *26* 4327 (1985) (chiral); *27* 3349 (1986) (chiral); *28* 869, 931 (chiral), 4169, 5303 (chiral) (1987)
JOMC *195* 137 (1980)
Ber *114* 359, 375, 2786, 2802 (1981); *115* 2357 (1982) (all chiral)
JOC *46* 1309 (1981); *47* 2498 (1982); *49* 3429 (1984); *52* 316 (1987) (chiral)
Helv *65* 1258 (1982)
JACS *107* 8186 (1985) (chiral); *108* 294 (chiral), 3422 (chiral aldehydes) (1986); *109* 953 (chiral), 8117 (chiral) (1987)

RCH=CHCHClB(OR)$_2$

Angew Int *23* 437 (1984); *25* 189 (1986)

H$_2$C=CHCHXB(OR)$_2$
(X = Cl, Br, OMe, SR)

TL *24* 3209 (1983)

H$_2$C=CHCH(SiMe$_3$)B(OR)$_2$

Organomet *2* 236 (1983)

R'OCH=CHCH$_2$B(OR)$_2$

TL *23* 845 (1982); *24* 2227 (1983); *27* 3353 (1986)
JOC *47* 2498 (1982)
Syn Commun *12* 779 (1982)
JACS *109* 7575 (1987)

R^{2}CH=CHCHR^{1}BR^{1}OH

JACS *104* 2330 (1982)

RCH=CHCH$_2$BR$_2$

JOC *42* 2292 (1977); *49* 4089 (1984) (chiral); *51* 432 (chiral), 886 (1986); *52* 319 (chiral), 3701 (chiral), 4831 (chiral) (1987)
CL 993 (1980)
JACS *105* 2092 (1983) (chiral); *107* 2564 (1985); *108* 293, 5919 (chiral) (1986); *109* 7553 (chiral), 8120 (chiral) (1987)
TL *25* 1215, 5111 (1984) (both chiral); *27* 5135 (1986)
CC 877 (1987) (chiral)

$CH_3CH=CHCHXB$⟩ (X = H, Me_3Si, Me_3Sn)	CC 191 (1983)
$\overset{\displaystyle SiMe_3}{\underset{\displaystyle |}{Me_3SiCH_2C}}=CHCH_2B$⟩	JOC 51 4733 (1986)
$Me_3SiCH=CHCH_2Li$, R_2BCl	CC 1326 (1982)
$CH_3CH=CHCH(MMe_3)B$⟩ (M = Si, Sn)	JACS 103 3229 (1981)
$R^2CH=CHCH_2\bar{B}R_3$	CC 1072 (1980) JACS 103 1969 (1981)
$RCH=CHCH_2AlX_2$	BSCF 1475 (1963)
$Me_3SiCH=CHCH_2Li$, $EtAlCl_2$	CC 1326 (1982)
$R^2CH=CHCH_2AlEt_2$	JACS 102 2118 (1980)
$ROCH=CHCH_2AlEt_2$	CC 845 (1982)
$H_2C=CHCH_2Al(i\text{-}Bu)_2$, $Sn(OTf)_2$, chiral diamine	CL 97 (1986) (chiral)
$CH_3CH=CHCH(OCON\text{-}i\text{-}Pr_2)Al(i\text{-}Bu)_2$	Angew Int 21 372 (1982)
$(CH_3CH=CHCH_2AlMe_3)Li$	JACS 108 4943 (1986)
$(XCH=CHCH_2AlEt_3)Li$ (X = Me_3Si, PhSe)	TL 23 4597 (1982)
$(ROCH=CHCH_2AlEt_3)Li$	JOC 45 195 (1980) TL 23 4959 (1982)
$(RSCH=CHCH_2AlEt_3)Li$	JOC 45 195 (1980) TL 23 4959 (1982); 27 5947 (1986)
$Me_2C=CHCH_2SiF_3$, CsF	TL 28 4081 (1987)
$R^2CH=CHCH_2SiMe_3$, $BF_3 \cdot OEt_2$	TL 3513 (1978) (intramolecular); 26 823 (1985); 28 4959 (1987) Helv 66 1655 (1983) (intramolecular) JOC 49 4214 (1984) CC 396 (1986) JACS 109 2512 (1987)
$R^2CH=CHCHR^3SiMe_3$, $AlCl_3$	JOMC 84 199 (1975) TL 2449 (1976); 429 (1979) Helv 66 1655 (1983) (intramolecular)
$H_2C=CRCH_2SiMe_3$, $EtAlCl_2$	JACS 104 6879 (1982) (intramolecular)
$R^2CH=CHCH_2SiMe_3$, SnX_4 (X = Cl, Br)	TL 3513 (1978) (intramolecular); 24 4765 (1983) Helv 66 1655 (1983) (intramolecular) JOC 49 4214 (1984); 51 3290 (1986); 52 316 (1987) JACS 109 2512 (1987)
$R^2CH=CHCHR^3SiMe_3$, $TiCl_4$	CC 927 (1976); 102 (1986)

$R^2CH{=}CHCHR^3SiMe_3$, $TiCl_4$ (*continued*)	TL 1295 (1976); 1385 (1977); 2589 (1978); 429 (1979); *23* 723 (1982); *24* 2865, 5661 (1983); *26* 823 (1985); *27* 4485 (1986); *28* 655 (intramolecular), 969 (chiral) (1987) JACS *101* 3340 (1979); *104* 4962, 4963 (1982); *105* 4833 (1983) CL 961 (1982) JOC *48* 281 (1983) (chiral); *49* 4214 (1984); *51* 3290 (1986); *52* 5492 (1987) Angew Int *22* 989 (1983)
$H_2C{=}CBrCH_2SiMe_3$, $TiCl_4$	JACS *104* 6879 (1982)
$R^2CH{=}CHCHR^3SiMe_3$, R_4NF	TL 3043, 3513 (intramolecular) (1978); *23* 577 (1982) Helv *66* 1655 (1983) (intramolecular)
$R_2C{=}CH{-}\overset{\displaystyle Me_3Si}{\triangleleft}$, $n\text{-}Bu_4NF$	TL *23* 3227 (1982)
$Me_3SiCH{=}CHCH_2SiMe_3$, $n\text{-}Bu_4NF$	JOMC *264* 207 (1984)
$H_2C{=}CH\overset{\displaystyle CH_2}{\overset{\|}{C}}CH_2SiMe_3$, $n\text{-}Bu_4NF$	Tetr *39* 883 (1983)
$R^2CH{=}CHCH{=}CHCHR^3SiMe_3$, $TiCl_4$	JOC *45* 1721 (1980) Organomet *1* 1651 (1982) JACS *106* 3240 (1984) TL *28* 5921 (1987)
$\left[R_2C{=}CHCH_2Si\left(\!\!\underset{O}{\overset{O}{\diagdown}}\!\!\diagup\!\!\bigcirc\right)_{\!2}\right]^{-} Et_3\overset{+}{N}H$	CC 1517 (1987) (RCHO only)
$RCH{=}CHCH_2SnCl_n(n\text{-}Bu)_{3-n}$	JOMC *162* 37 (1978); *197* 45 (1980); *226* 149 (1982)
$H_2C{=}CHCHMeSnCl(n\text{-}Bu)_2$	JOMC *231* 307 (1982)
$(R^2_2C{=}CHCH_2)_2SnBr_2$ (R^2 = H, Me)	JOMC *280* 307 (1985)
$R^2CH{=}CHCH_2SnR_3$, Δ (R = Et, $n\text{-}Bu$)	TL 495 (1967); *28* 5343 (1987) JOMC *35* C20 (1972) JACS *102* 4548 (1980); *106* 7970 (1984) (intramolecular) Pure Appl Chem *53* 2401 (1981) (review)
$R^2CH{=}CHCH(OR)Sn(n\text{-}Bu)_3$, Δ	CC 1115 (1982); 800 (1984) TL *28* 527 (1987)
$R^2CH{=}CHCH(OR)Sn(n\text{-}Bu)_3$, $BF_3{\cdot}OEt_2$	TL *28* 527 (1987) (intramolecular)
$Me_3SiCH{=}CHCH_2Li$, $n\text{-}Bu_3SnCl$, BF_3	CC 1326 (1982)
$R^2CH{=}CHCH_2SnR_3$, MgX_2 (X = Br, I)	TL *25* 265, 1879, 3927 (1984); *28* 4381 (1987) JOC *51* 5478 (1986); *52* 316 (1987) JACS *109* 7553 (1987)

$R^2CH{=}CR^4CHR^3SnR_3$, $BF_3 \cdot OEt_2$	CL 919, 977 (1979); 1297, 1299 (1982)
	JACS *102* 4548, 7107 (1980); *106* 6835, 7970
	(intramolecular) (1984); *107* 1778 (1985);
	108 4645 (1986)
	CC 683 (1980)
	TL *23* 4959 (1982); *25* 265, 1879, 1883, 3927
	(1984); *26* 6235 (1985); *27* 5423 (1986);
	28 139, 2629, 3939, 5343 (1987)
	Tetr *40* 2239 (1984)
	Organomet *4* 1213 (1985) (chiral)
	JOC *51* 863, 886, 1856, 2621, 5478 (1986); *52* 316
	(1987)
$R^2CH{=}CR^4CH_2SnR_3$, $SnCl_4$	TL *25* 3927 (1984); *26* 6235 (1985); *28* 2629
	(1987)
	JACS *106* 7970 (1984) (intramolecular)
	JOC *51* 3290, 5478 (1986); *52* 316 (1987)
$R^2CH{=}CR^4CH_2SnR_3$, $TiCl_4$	TL *25* 265, 1879, 1883, 3927 (1984); *28* 2629
	(1987)
	JACS *106* 7970 (1984) (intramolecular)
	CC 102 (1986)
	JOC *51* 863, 1856, 3290, 5478 (1986)
$R^2CH{=}CHCH_2SnR_3$, ZnX_2 (X = Cl, Br, I)	TL 495 (1967); *25* 265, 1879, 1883, 3927 (1984)
$R^2CH{=}CHCH_2SnR_3$, other Lewis acids	CL 977 (1979)
	TL *25* 265, 1883 (1984)
	JACS *106* 7970 (1984) (intramolecular)
$ROCH{=}CHCH_2SnR_3$, $MgBr_2$	TL *28* 139 (1987)
$ROCH{=}CHCH_2SnR_3$, $BF_3 \cdot OEt_2$	TL *28* 143 (1987)
$(H_2C{=}CHCH_2)_2SnR_2$, $BF_3 \cdot OEt_2$	CL 1529 (1983) (chiral R)
	Organomet *4* 1213 (1985)

$$\underset{\displaystyle H_2C{=}CHCCH_2SnMe_3,\ BF_3 \cdot OEt_2}{\overset{\displaystyle \overset{CH_2}{\|}}{}}$$

	CL 977 (1979)
$R^2CH{=}CHCH_2TiCp_2$	TL *23* 4589 (1982)
	JOMC *224* 327 (1982)
	CC 921 (1983)
$R^2CH{=}CHCH_2TiCp_2X$ (X = Br, I)	CC 342, 1140 (1981); 98 (1986)
	TL *22* 243 (1981)
$R^2CH{=}CHCH_2Ti(OR)_3$	Angew Int *19* 1011 (1980); *21* 135 (1982)
	Helv *65* 1085, 1972 (1982)
	JACS *104* 7663 (1982); *107* 1691 (1985)
	Ber *118* 1441 (1985)
	Tetr *42* 2803 (1986)
	TL *28* 869 (1987)
$R^2CH{=}CHCH_2Ti(NEt_2)_3$	Ber *118* 1441 (1985)
$CH_3CH{=}CHCH(OCON\text{-}i\text{-}Pr_2)Ti(NEt_2)_3$	Angew Int *21* 372 (1982)

[Me$_3$SiCH=CHCH$_2$Ti(O-i-Pr)$_4$]Li   Ber 118 1441 (1985)

[H$_2$C=CHCH$_2$Ti(O-i-Pr)$_4$]MgCl   TL 23 5259 (1982) (RCHO > R$_2$CO)
                                                     Ber 118 1441 (1985)

[H$_2$C=CHCH$_2$Ti(NMe$_2$)$_4$]Li   TL 23 5259 (1982) (R$_2$CO > RCHO)

R^{2}CH=CHCH$_2$ZrCp$_2$Cl   TL 22 2895 (1981)

(H$_2$C=CHCH$_2$)$_4$Zr   JACS 105 4833 (1983)

H$_2$C=CHCH$_2$CrCl$_2$ TL 27 5351 (1986) (β-hydroxy ketone)

H$_2$C=CHCH$_2$Cu·SMe$_2$ /MgBr$_2$ TL 28 869 (1987)

R^{2}CH=CHCHR^{3}ZnX   BSCF 1475 (1963)
                                                     TL 3829 (1975) (with and without rearrangement)

(H$_2$C=CHCH$_2$)$_2$Zn   BSCF 974 (1962); 4038 (1969)
                                                     Pure Appl Chem 53 1163 (1981)
                                                     JOC 51 3742 (1986); 52 1141 (1987)
                                                     TL 28 869 (1987)

H$_2$C=CHC(=CH$_2$)CH$_2$ZnBr   Syn 742 (1979)

(H$_2$C=CHCH$_2$)$_2$Cd   BSCF 4038 (1969)

JACS 109 4710 (1987)

JACS 109 576 (1987)

CC 570 (1987)

TL 21 365 (1980)
CC 342 (1981)

$$H_2C = \overset{\overset{\displaystyle CH_3}{\mid}}{C}CH = CH_2 \longrightarrow Cp_2Zr \xrightarrow[\text{2. } H_2O]{\text{1. RCHO } (R_2CO)} H_2C = \overset{\overset{\displaystyle CH_3}{\mid}}{C}CH\overset{\overset{\displaystyle OH}{\mid}}{C}H_2\overset{\mid}{C}HR$$

CL 671 (1981)

$$MCHR^1C \equiv CR^2 \quad \text{or} \quad R^1CH = C = CR^2M + R\overset{\overset{\displaystyle O}{\parallel}}{C}H \ (R_2CO) \xrightarrow{H_2O} R\overset{\overset{\displaystyle OH}{\mid}}{C}HCHR^1C \equiv CR^2$$

$\underline{M}$	
Li	JOC *47* 2225 (1982)
MgX	JACS *90* 6141 (1968); *109* 8051 (1987)
	Compt Rend C *269* 342 (1969)
	TL 37 (1972); *27* 3311 (1986)
	BSCF 3371 (1973)
	JOC *47* 2225 (1982)
B(OR)$_2$	JOMC *92* 17 (1975); *195* 137 (1980)
	JACS *104* 7667 (1982); *108* 483 (1986) (both
	chiral)
	TL *27* 1175 (1986)
	JOC *51* 886 (1986)
BEt$_2$	JACS *106* 3875 (1984)
B⟩	JOC *50* 1577 (1985) (RCHO > R$_2$CO)
Al$_{2/3}$Br	Ann *682* 62 (1965)
	TL 37 (1972)
	BSCF 3371 (1973)
SiMe$_3$, (TiCl$_4$)	JOC *45* 3925 (1980); *51* 3870 (1986)
Ti(OR)$_3$	JOC *47* 2225 (1982); *51* 886 (1986)
Cu	JACS *109* 5437 (1987)
ZnBr	BSCF 3371 (1973); 1248 (1975)
	TL *22* 1579 (1981)
	JOC *47* 2225 (1982)

$$R\overset{\overset{\displaystyle O}{\parallel}}{C} - C_n - CHR\overset{\overset{\displaystyle O}{\parallel}}{C}H \xrightarrow{Ti(NEt_2)_4} \xrightarrow[M = Li, MgX]{R'M} \xrightarrow{H_2O} R\overset{\overset{\displaystyle OH}{\mid}}{\underset{\underset{\displaystyle R'}{\mid}}{C}} - C_n - CHR\overset{\overset{\displaystyle O}{\parallel}}{C}H$$

CC 406 (1983)

$$R'CH = CH\overset{\overset{\displaystyle O}{\parallel}}{C}H \longrightarrow R'CH = CHCH \overset{O \diagup CONMe_2}{\underset{O \diagdown CONMe_2}{\diagup}} \xrightarrow[\text{2. KO-}t\text{-Bu}]{\text{1. } R_3Al} R'CH = CH\overset{\overset{\displaystyle OH}{\mid}}{C}HR$$
chiral

JACS *106* 5004 (1984)

$$R^1CHO \xrightarrow[\text{cat }[(Et_2N)_3S](Me_3SiF_2)]{R_3SiCCl_2R^2} \xrightarrow[\text{MeOH}]{HCl} R^1\overset{\overset{\displaystyle OH}{|}}{C}HCCl_2R^2$$

$$R^2 = H, Cl, Me$$

$$R^1CHO \xrightarrow[\text{cat }[(Et_2N)_3S](Me_3SiF_2)]{R_3SiCF=CFR^2} R^1\overset{\overset{\displaystyle OSiR_3}{|}}{C}HCF=CFR^2$$

$$R^2 = H, F, R, SR$$

JACS *107* 4085 (1985)

$$R^1\overset{\overset{\displaystyle O}{||}}{C}R^2 \xrightarrow[\text{2. Al(Hg)}]{\overset{\displaystyle MgBr}{\overset{|}{\text{1. chiral }p\text{-TolSOCHCO}_2R}}} R^1R^2\overset{\overset{\displaystyle OH}{|}}{C}CH_2CO_2R$$

chiral

CC 162 (1977)

$$RLi(MgCl) + H_2C=CF_2 \xrightarrow[\text{2. H}_2\text{O}]{\text{1. R'CHO}} RC\equiv C\overset{\overset{\displaystyle OH}{|}}{C}HR'$$

$$R = 1°, 2°, 3° \text{ alkyl; allyl; vinyl}$$

TL *23* 4325 (1982)

2.2. Miscellaneous Additions

JCS Perkin II 7, 14 (1976)
JACS *108* 3841 (1986); *109* 3017 (1987)
JOC *52* 5521 (1987)

$$RC\overset{\overset{\displaystyle O}{||}}{C}H_2C\overset{\overset{\displaystyle O}{||}}{N}(CH_2R')_2 \xrightarrow{h\nu} \text{(structure)}$$

CC 743 (1974)
JOC *43* 1005 (1978)

$$R_2CO + \text{(structure with OSiMe}_3\text{, OR)} \xrightarrow[\text{X = Cl, I}]{\text{cat ZnX}_2} R_2\overset{\overset{\displaystyle OSiMe_3}{|}}{C}CH_2CH_2CO_2R$$

JOC *50* 2802 (1985)
JACS *109* 8056 (1987)
TL *28* 337 (1987)

$$R'CHO \longrightarrow \overset{\overset{\displaystyle OH}{|}}{R}CHR'$$

R	Reagents	
CHX$_2$ (X = Cl, Br, I)	H$_2$CX$_2$, LiNR$_2$	JACS *96* 3010 (1974)
		BCSJ *50* 1588 (1977)
CCl$_3$	HCCl$_3$, KOH	JACS *70* 1189 (1948); *72* 5012 (1950)
		JOC *50* 2527 (1985); *52* 944 (1987)
	HCCl$_3$, NaOH, (PhCH$_2$NEt$_3$)Cl	Ber *110* 96 (1977)
	HCCl$_3$, KO-*t*-Bu	JOC *50* 2527 (1985)
	HCCl$_3$, NaH	JOC *50* 2527 (1985)
CX$_3$ (X = Cl, Br)	HCX$_3$, KO-*t*-Bu	Ber *91* 2664 (1958)
	HCX$_3$, LiNR$_2$	JACS *96* 3010 (1974)
		BCSJ *50* 1588 (1977)
	HCX$_3$, NaNH$_2$	Ber *96* 420 (1963)

(furan-2(3H)-one ring), PhI, SmI$_2$ — TL *28* 5877 (1987)

H$_2$C(1,3-dioxolane), PhI, SmI$_2$ — TL *28* 5877 (1987)

$$R'CHO + R'X \xrightarrow{\text{electrolysis}} \overset{\overset{\displaystyle OH}{|}}{R}CHR'$$

R'	R'X	
CCl$_3$	CCl$_4$	TL 1521 (1978)
	CCl$_4$, HCCl$_3$	TL *22* 871 (1981)
		JACS *106* 259 (1984)
		JOC *50* 2527 (1985)
CCl$_2$CO$_2$Me	Cl$_3$CCO$_2$Me, HCCl$_2$CO$_2$Me	JACS *106* 259 (1984)
		JOC *50* 2527 (1985)
CHPhCO$_2$Me	BrCHPhCO$_2$Me, PhCH$_2$CO$_2$Me	JOC *50* 2527 (1985)
CH(CO$_2$Me)$_2$	BrCH(CO$_2$Me)$_2$, H$_2$C(CO$_2$Me)$_2$	JOC *50* 2527 (1985)
allylic	allylic chloride	TL *22* 1895 (1981)
allylic, benzylic, α-chloro ester, CF$_3$, CCl$_3$, aryl	R'X	TL *27* 3129 (1986)

$$RCR + H_2C{=}CHR' \longrightarrow R_2\overset{\overset{\displaystyle OH}{|}}{C}CH_2CH_2R'$$

(with $\overset{O}{\overset{||}{}}$ on RCR)

electrolysis — JACS *93* 5284 (1971) (intramolecular); *100* 545 (1978) (intramolecular)

JOC *50* 2202 (1985) (intramolecular, R' = CO$_2$Et); *51* 1041 (1986) (intramolecular)

hν, Et$_3$N	TL *26* 4591 (1985) (intramolecular); *28* 4545 (1987) (intramolecular)
hν, HMPA	TL *26* 4591 (1985) (intramolecular) JOC *51* 4196 (1986) (intramolecular)
Li, NH$_3$	JCS 3154 (1965) (intramolecular)
Na, moist ether	CC 206 (1965) (intramolecular)
Na, *t*-BuOH	JOC *44* 2369 (1979)
Mg, Me$_3$SiCl	JOC *50* 5193 (1985) (intramolecular)
Zn, Me$_3$SiCl, lutidine	TL *24* 2821 (1983) (intramolecular)
SmI$_2$	TL *28* 4367 (1987) (intramolecular)

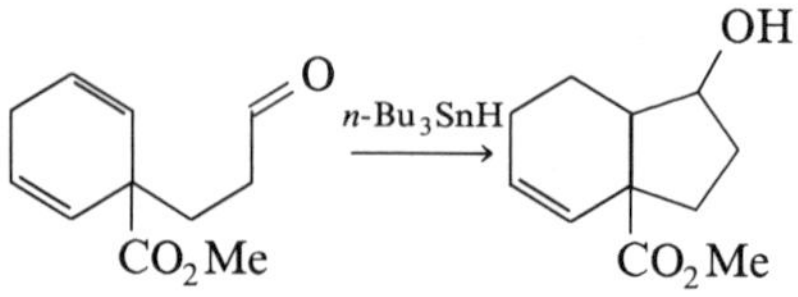

JACS *108* 5893 (1986)

$$\overset{O}{\overset{\|}{RCH}} + H_2C=CHCH_2R' \xrightarrow{\text{(acid)}} \overset{OH}{\overset{|}{RCHCH_2CH}}=CHR'$$

See page 119, Section 8.

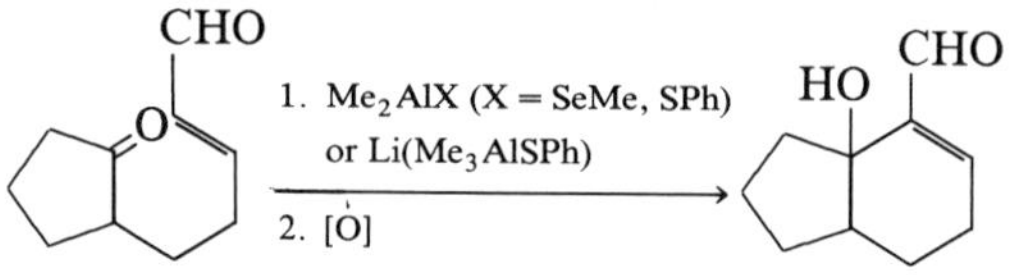

TL *21* 361 (1980)
BCSJ *54* 274 (1981)
JACS *109* 8017 (1987)

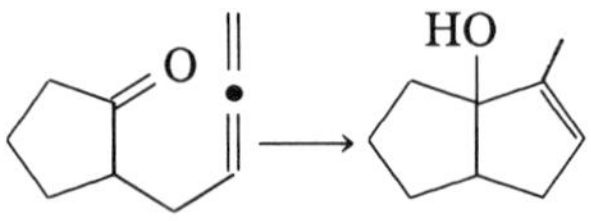

electrolysis	Tetr *41* 4001 (1985)
Na, naphthalene	Tetr *41* 4001 (1985) TL *27* 2243 (1986)

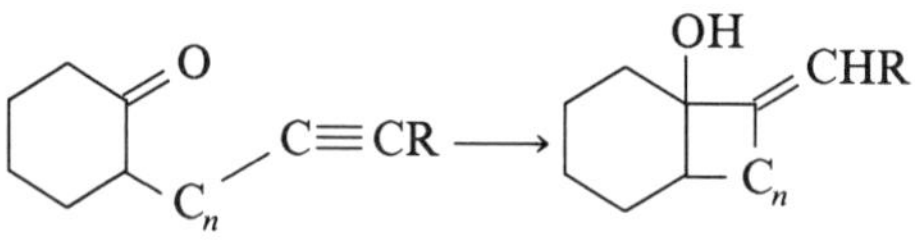

electrolysis	CL 1233 (1976)
hν, HMPA	JOC *51* 4196 (1986)
hν, Et$_3$N	TL *28* 4545, 4547 (1987)

Li, NH$_3$, (NH$_4$)$_2$SO$_4$ JACS *87* 1148 (1965)

Na   JOC *46* 2622 (1981)

Na, naphthalene   JOC *41* 1943 (1976); *46* 2622 (1981)
TL *23* 5471 (1982); *27* 399 (1986); *28* 5945 (1987)

K, NH$_3$, (NH$_4$)$_2$SO$_4$ JACS *101* 7107 (1979)

Zn, Me$_3$SiCl   TL *24* 2821 (1983); *28* 2001 (1987)

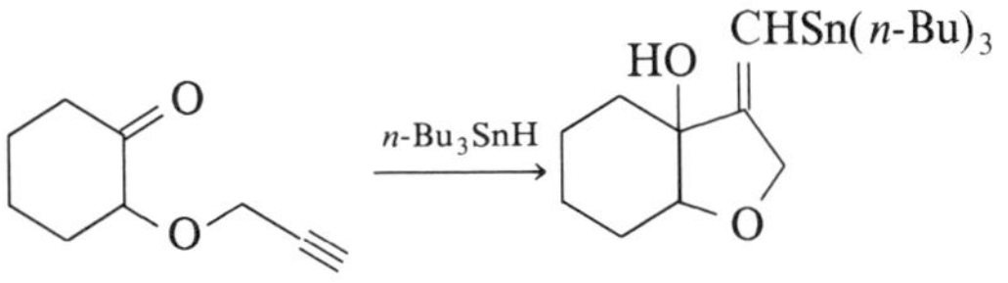

TL *28* 2001 (1987)

$$R^1CH{=}NR^2 \xrightarrow[\text{2. } R_2CO]{\text{1. NbCl}_3\cdot\text{DME}} R_2\overset{\overset{\displaystyle HO}{|}}{C}-\overset{\overset{\displaystyle NHR^2}{|}}{C}HR^1$$

JACS *109* 6551 (1987)

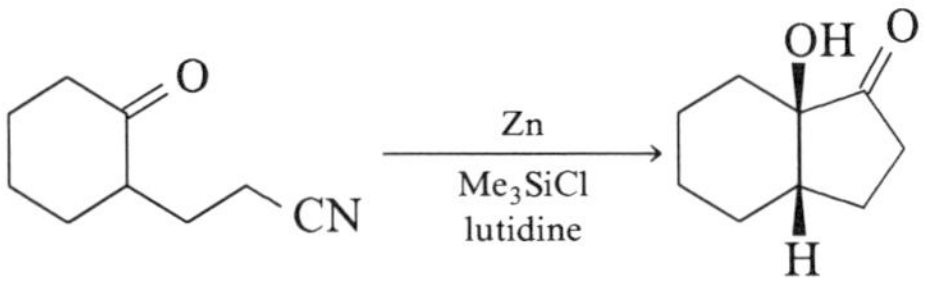

TL *24* 2821 (1983)

$$\underset{R^1\overset{\overset{\displaystyle O}{\|}}{C}H}{} + Ar_3P{=}CR^2CH_2MR_3^3 \longrightarrow R^1\overset{\overset{\displaystyle R_3^3SiO}{|}}{C}H\overset{\overset{\displaystyle R^2}{|}}{C}{=}CH_2$$

M
—
Si   TL *26* 4471 (1985); *27* 6373 (1986); *28* 2379, 4561 (1987)
CC 880 (1986)

Sn   TL *28* 4561 (1987)

$$R\overset{\overset{\displaystyle O}{\|}}{C}H + H_2C{=}CHX \xrightarrow[\text{(high pressure)}]{\text{DABCO}} R\overset{\overset{\displaystyle OH}{|}}{C}H\underset{\underset{\displaystyle CH_2}{\|}}{C}X$$

X = CHO, COMe, CO$_2$Me, CN, SO$_2$Ph

Angew Int *22* 795 (1983)
JCS Perkin I 2293 (1983)
Helv *67* 413 (1984)
JOC *50* 3849 (1985)
TL *27* 2031, 4307, 5007, 5095 (1986)
Syn Commun *17* 587 (1987)

3. Alkylation of α-Haloketones

$$\underset{\text{O}}{\overset{\text{O}}{R^1\overset{\|}{C}CH_2Cl}} \xrightarrow{R^2Li} R^1R^2\overset{\overset{\text{OLi}}{|}}{C}CH_2Cl \xrightarrow[\begin{array}{l}\text{2. } E^+\\\text{3. } H_2O\end{array}]{\text{1. } Li^+C_{10}H_8^-} R^1R^2\overset{\overset{\text{OH}}{|}}{C}CH_2E$$

$$E^+ = D_2O, RCHO, R_2CO, CO_2, O_2, (MeS)_2$$

CC 1153 (1982)

$$\xrightarrow[\begin{array}{l}\text{1. } RC\equiv CLi(MgBr)\\\text{2. } LiAlH_4\\\text{3. } H_2O\end{array}]{}$$

JACS *105* 3348 (1983)
Org Syn *64* 10 (1985)

4. Addition to Acetals

$$HO\!\!-\!\!\underset{O}{\overset{}{\diamond}}\!\!)_n \xrightarrow{RM} R\overset{\overset{\text{OH}}{|}}{C}HCH_2CH_2(CH_2)_nOH$$

$$n = 1, 2$$

$$RM = MeLi, RMgX, Me_3Al, MeTiCl_3, MeTi(O\text{-}i\text{-}Pr)_3$$

JOC *52* 4603 (1987)
TL *28* 423, 6335 (1987)

5. Addition to Carboxylic Acid Derivatives

$$R'COCl \xrightarrow{RMnI} R'\overset{\overset{\text{OH}}{|}}{C}R_2$$

Can J Chem *58* 287 (1980)

$$R^1COCl \xrightarrow[\begin{array}{l}\text{1. } R^2MnI\\\text{2. } R^3Li(MgX)\end{array}]{} R^1\!\!-\!\!\overset{\overset{\text{OH}}{|}}{\underset{\overset{|}{R^2}}{C}}\!\!-\!\!R^3$$

$$R^1COCl \xrightarrow[\begin{array}{l}\text{1. } R^2MnI\\\text{2. } NaBH_4 \text{ or}\\\quad\ LiAlH_4\end{array}]{} R^1\overset{\overset{\text{OH}}{|}}{C}HR^2$$

TL *27* 4441 (1986)

$$RCO_2Et \longrightarrow \overset{\overset{\displaystyle OH}{\displaystyle |}}{RCHR'}$$

$i\text{-}Bu_2AlH/R'MgX$	JOC *48* 2775 (1983) TL *28* 3905 (1987)
$R'MgX$, $LiBH_4$	TL *25* 1321 (1984); *28* 3905, 4143 (1987)
2 $R'MgX$, Cp_2TiCl_2	TL *21* 2175 (1980)

$$ArLi + HCO_2Et \xrightarrow{H_2O} Ar_2CHOH$$

JOC *47* 4347 (1982)

$$2\ RM + R'CO_2R'' \xrightarrow{H_2O} \overset{\overset{\displaystyle OH}{\displaystyle |}}{R_2CR'}$$

RM
———

CF_3CF_2Li	JOC *52* 2481 (1987)
ArLi	JACS *108* 3762 (1986)
RMgX	M. S. Kharasch and O. Reinmuth, "Grignard Reactions of Non-metallic Substances," Prentice Hall, New York (1954), pp 549–708 (review) Org Syn Coll Vol *3* 839 (1955) JACS *108* 4119 (1986) JOC *52* 4511 (1987) (lactone)
Me_3SiCH_2MgCl, $CeCl_3$	TL *28* 6261 (1987)

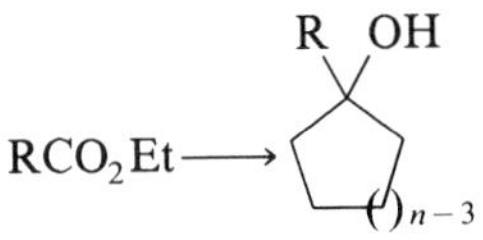

$$RCO_2Et \longrightarrow$$

Reagent(s)	n	
$Br(CH_2)_5Br$, Li, ultrasound	5	TL *28* 2013 (1987)
BrMg-C_n-MgBr	4, 5	Compt Rend *144* 1358 (1907) JACS *72* 3483 (1950) TL *22* 4995 (1981) JOC *51* 2147 (1986); *52* 4025 (1987)

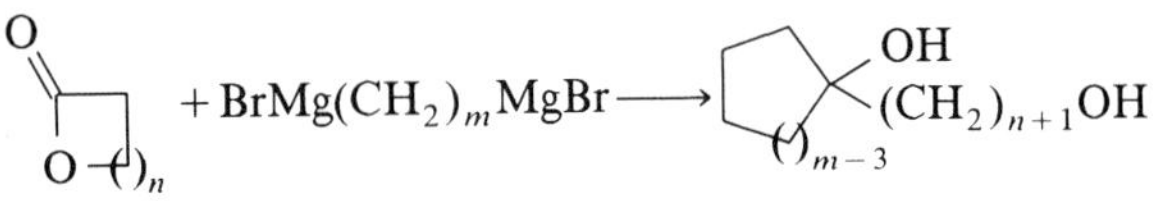

$$+ BrMg(CH_2)_mMgBr \longrightarrow$$

$n = 1–4 \qquad m = 4, 5$

JOC *45* 1828 (1980); *52* 569 (1987)
Tetr *40* 865 (1984)
TL *28* 4997 (1987)

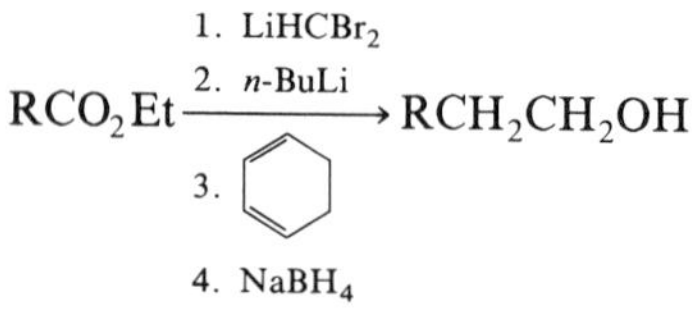

$$RCO_2Et \xrightarrow[\substack{3. \\ 4.\ NaBH_4}]{\substack{1.\ LiHCBr_2 \\ 2.\ n\text{-BuLi}}} RCH_2CH_2OH$$

JACS *108* 1324 (1986)

$$EtO\overset{\overset{\textstyle O}{\|}}{C}OEt \xrightarrow{RMgX} R_3COH$$

Org Syn Coll Vol *2* 602 (1943)

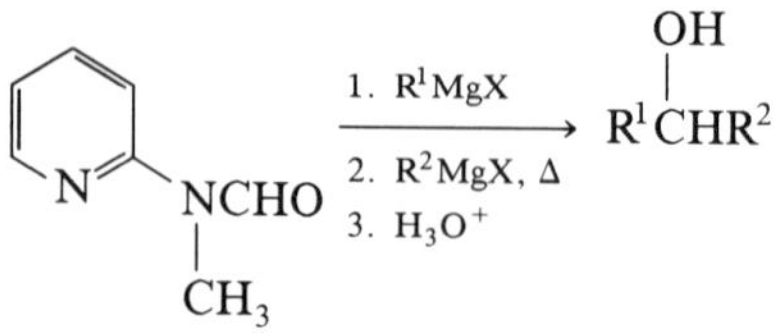

$$\xrightarrow[\substack{2.\ R^2MgX,\ \Delta \\ 3.\ H_3O^+}]{1.\ R^1MgX} \overset{\overset{\textstyle OH}{|}}{R^1CHR^2}$$

TL *22* 1085 (1981)

9. MISCELLANEOUS REACTIONS

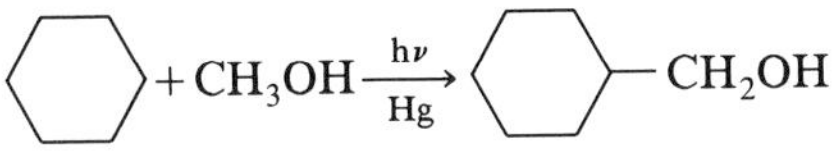

$$\text{cyclohexane} + CH_3OH \xrightarrow[\text{Hg}]{h\nu} \text{cyclohexyl}-CH_2OH$$

CC 970 (1987)
TL *28* 5599 (1987)

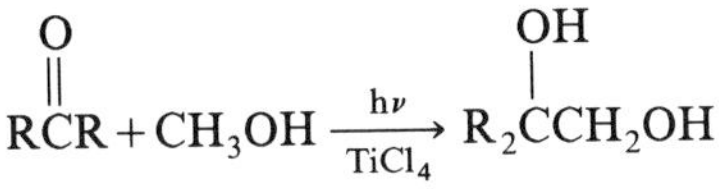

$$\underset{\text{RCR}}{\overset{O}{\|}} + CH_3OH \xrightarrow[\text{TiCl}_4]{h\nu} R_2\overset{OH}{\underset{|}{C}}CH_2OH$$

JOC *45* 3778 (1980)

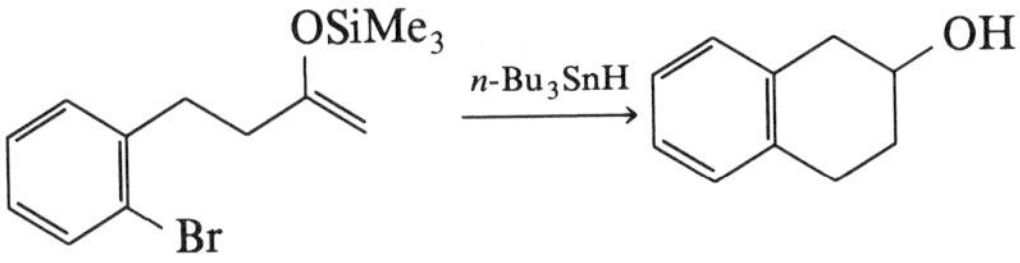

OSiMe$_3$ ... $\xrightarrow{n\text{-Bu}_3\text{SnH}}$ ... OH

TL *27* 1355 (1986)

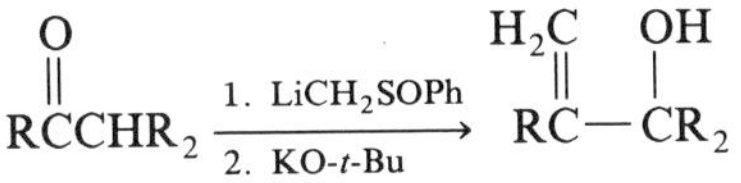

$$\underset{\text{RCCHR}_2}{\overset{O}{\|}} \xrightarrow[\text{2. KO-}t\text{-Bu}]{\text{1. LiCH}_2\text{SOPh}} \underset{\text{RC}-\text{CR}_2}{\overset{\text{H}_2\text{C}\quad\text{OH}}{\| \qquad |}}$$

Syn 640 (1980)

$$R-C=C-C-\text{SOPh} \xrightarrow[\text{or PhS}^-]{\text{P(OR)}_3} R-\overset{OH}{\underset{|}{C}}-C=C$$

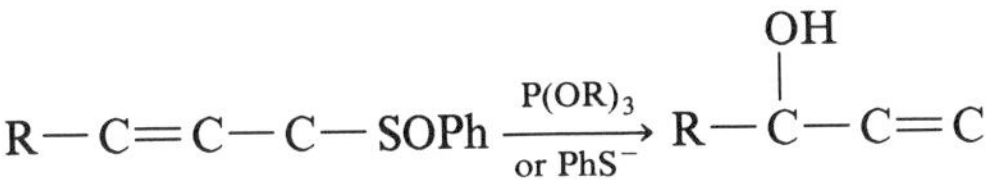

JOC *52* 4634 (1987)
TL *28* 1925, 5865 (1987)

See also page 185, Section 1.

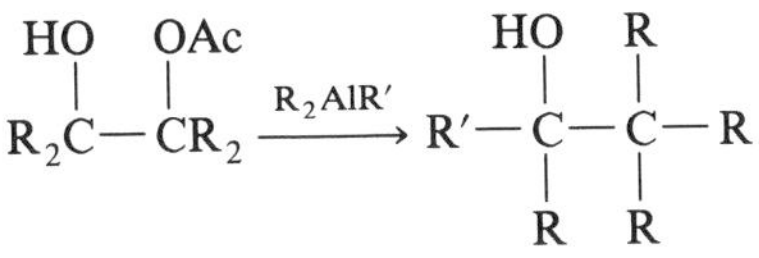

$$\underset{R_2C-CR_2}{\overset{\text{HO}\quad\text{OAc}}{| \qquad |}} \xrightarrow{R_2\text{AlR}'} \underset{R'-\overset{|}{\underset{\underset{R}{|}}{C}}-\overset{|}{\underset{\underset{R}{|}}{C}}-R}{\overset{\text{HO}\quad R}{| \quad\;|}}$$

BCSJ *55* 3941 (1982)

581

ALDEHYDES AND KETONES

GENERAL REFERENCES

Chem Rev *38* 227 (1946)

Quart Rev *20* 169 (1966)

"The Chemistry of the Carbonyl Group," Ed. S. Patai, Interscience, New York (1966–70)

C. D. Gutsche, "The Chemistry of Carbonyl Compounds," Prentice Hall, Englewood Cliffs, New Jersey (1967)

Houben-Weyl, "Methoden der Organischen Chemie," 4th ed, Vol VII/2a, G. Thieme, Stuttgart (1973) (ketones)

Houben-Weyl, "Methoden der Organischen Chemie," 4th ed, Vol VII/2b, G. Thieme, Stuttgart (1976) (ketones)

Houben-Weyl, "Methoden der Organischen Chemie," 4th ed, Vol VII/2c, G. Thieme, Stuttgart (1977) (ketones)

Syn 633 (1979) (Synthesis of aldehydes, ketones and carboxylic acids from lower carbonyl compounds by C—C coupling reactions)

Tetr *36* 2531 (1980) (Carbonyl homologation and masked homoenolate equivalents)

Houben-Weyl, "Methoden der Organischen Chemie," 4th ed, Vol E3, G. Thieme, Stuttgart, New York (1983) (aldehydes)

1. CARBONYL TRANSPOSITION

$$\underset{\text{O}}{\overset{\text{O}}{\|}}\text{RCCH}_2\text{R}' \longrightarrow \underset{\text{O}}{\overset{\text{O}}{\|}}\text{RCH}_2\text{CR}'$$

PhCHO, OH$^-$/LiAlH$_4$, AlCl$_3$/O$_3$	CC 898 (1967)
	JCS C 244 (1970)
PhCHO, OH$^-$/NaBH$_4$/Ac$_2$O/O$_3$/Li, NH$_3$	JACS *104* 1907 (1982)
t-BuONO$_2$, KO-*t*-Bu/NaBH$_4$/Zn, HOAc	JOC *33* 1733 (1968)
AmONO, KO-*t*-Bu/N$_2$H$_4$, KOH/NaHSO$_3$	CC 1350 (1968)
LiN(*i*-Pr)(*c*-C$_6$H$_{11}$)/(PhS)$_2$/NaBH$_4$/MsCl-py	JACS *97* 438 (1975)
or TsOH/HgCl$_2$-H$_2$O or TiCl$_4$-HOAc	
TsNHNH$_2$/2 *n*-BuLi/(MeS)$_2$/*n*-BuLi/	TL 531 (1979)
NH$_4$Cl/H$_2$O, HgCl$_2$	CL 931, 1099 (1980)
PhSO$_2$NHNH$_2$/2 *n*-BuLi/Me$_3$SiCl/	JOC *43* 1620 (1978); *45* 3028 (1980)
m-ClC$_6$H$_4$CO$_3$H/LiAlH$_4$/	
Na$_2$Cr$_2$O$_7$, H$_2$SO$_4$	

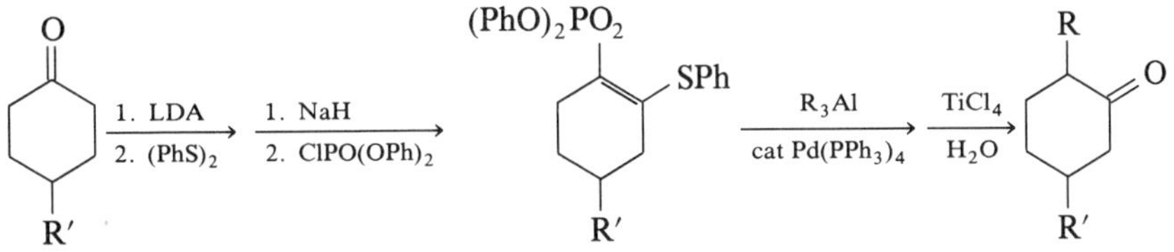

TL *22* 1609 (1981)

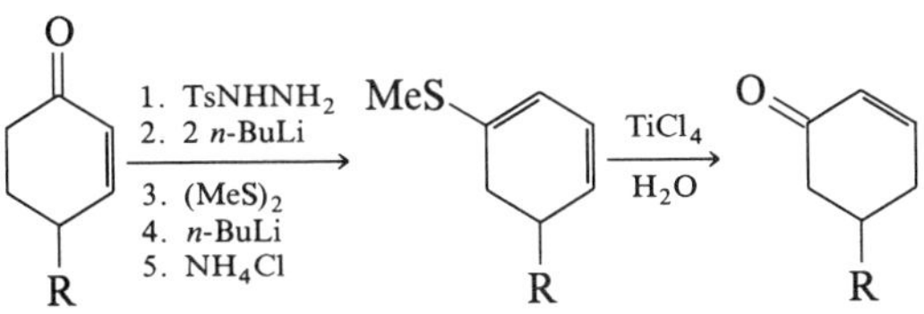

CL 1099 (1980)

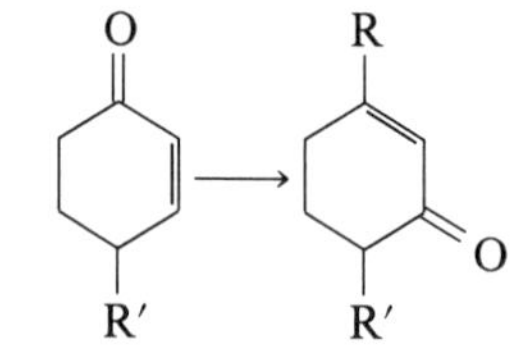

RMgX/PCC	JACS *108* 3443 (1986)
RLi/PCC	Syn Commun *6* 649 (1976) JOC *42* 682, 813 (1977); *50* 2557 (1985); *51* 4497 (1986) JACS *105* 7352 (1983) CC 358 (1987)
RC≡CLi/PDC	TL *28* 1069 (1987)
RLi/CrO₃, H₂SO₄	JOC *36* 2021 (1971); *39* 2317 (1974) Syn Commun *11* 7 (1981)
RMgX/CrO₃, H₂SO₄	JACS *108* 3385 (1986)
RLi/PhSeCl/O₃/Hg(OAc)₂ or HCl or HCO₂H	JOC *45* 2551 (1980), *47* 1258 (1982)
RLi/PhSCl/(PhS)₂/H₂O, HgCl₂	JACS *97* 4018 (1975)

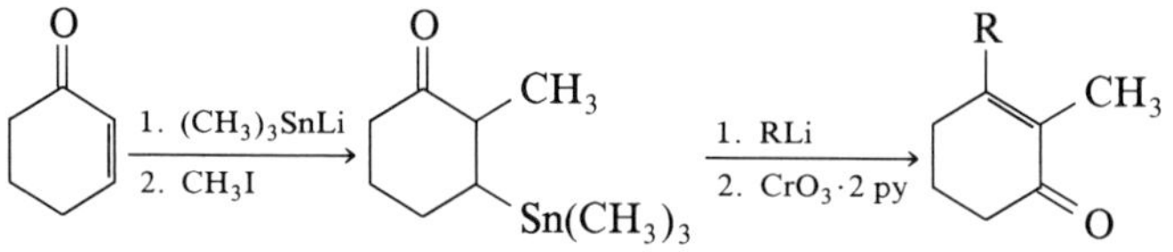

JACS *99* 4836 (1977)

X	R'M	
OEt	R'MgBr	Angew Int *19* 816 (1980)
OEt	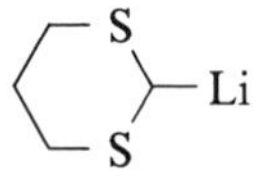 (dithiane)–Li	Ber *115* 3898 (1982)
NMe₂	RLi	JOC *51* 879, 1631 (1986)

JOC *51* 4743 (1986)

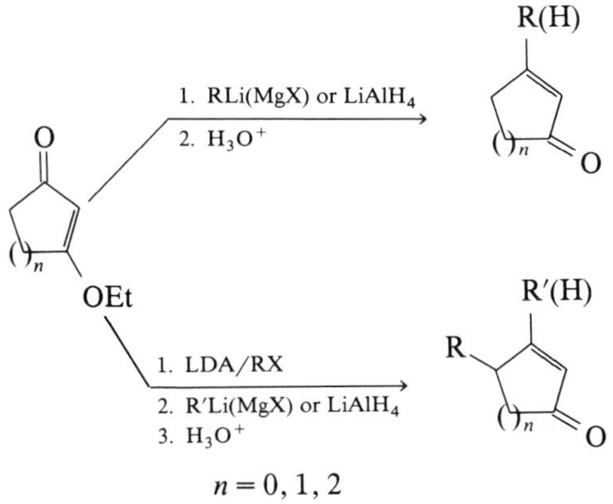

$n = 0, 1, 2$

JACS 72 1645 (1950); 103 82 (1981); 105 1292 (1983); 106 2115, 3353 (1984); 107 268, 7732, 7745 (1985); 108 806, 1106, 3110, 6276 (1986); 109 4690, 5491 (1987)
JCS 1779 (1953)
JOC 38 1451, 1775 (1973); 43 3968 (1978); 46 2400 (1981); 47 381, 3297, 4820, 5096 (1982); 48 2318 (1983); 49 2152 (1984); 50 3155, 5550, 5727 (1985); 51 879, 1490, 4323 (1986); 52 120, 3491 (1987)
TL 22 15, 97 (1981); 23 3283 (1982); 27 2087, 5232 (1986); 28 1329, 3065, 3107 (1987)
Tetr 38 3527 (1982)
Syn Commun 12 167, 521 (1982)
Can J Chem 60 2965 (1982)
Org Syn 64 68, 73 (1985)
CC 75 (1986)

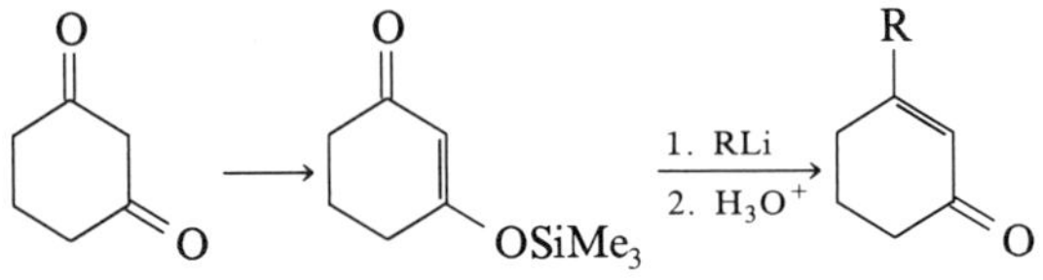

Syn Commun 12 795 (1982)

Heterocycles 19 1211 (1982)

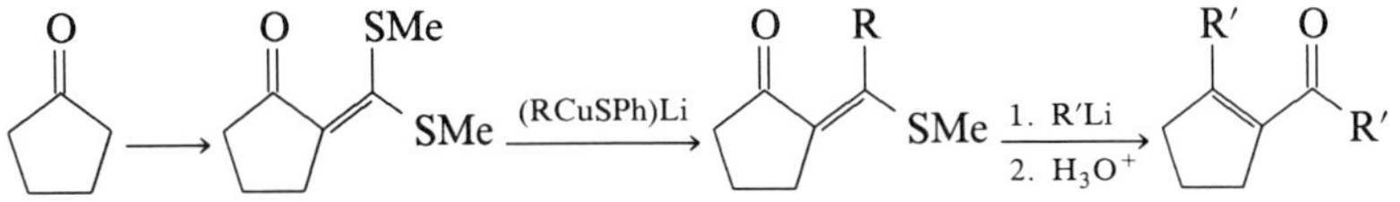

TL 23 3751 (1982)

TL *23* 561 (1982)

Organomet *1* 1240, 1243 (1982)
JOC *52* 110 (1987)

2. OXIDATION

1. Alkanes

$$(R = H \text{ or alkyl})$$

CrO_3, HOAc	JCS 727 (1940)
	JACS *69* 576 (1947)
	Tetr *20* 409 (1964)
	Ind J Chem *2* 229 (1964)
	JCS C 2603 (1968); 1240 (1969)
	Acta Chem Scand *24* 2252 (1970)
	Steroids *18* 593 (1971); *22* 327 (1973)
	Org Prep Proc Int *4* 67 (1972)
	Syn Commun *3* 89 (1973)
	JOC *39* 1416 (1974); *46* 1752 (1981);
	50 2435 (1985)
CrO_3, H_2SO_4, H_2O, CH_3COCH_3	JCS C 2603 (1968); 1240 (1969)
	JOC *50* 2435 (1985)
CrO_3, 3,5-dimethylpyrazole	JACS *108* 3040 (1986)
CrO_3, HCl, bipy	JOC *50* 2435 (1985)
cat CrO_3, t-BuO_2H	TL *28* 2131 (1987)
cat [image: CrO_2], t-BuO_2H	Bull Soc Chim Belg *94* 651 (1985)
	TL *27* 3139 (1986)
PCC, Celite	Syn Commun *16* 1493 (1986)
PCC, t-BuO_2H, Celite	JOC *52* 5048 (1987)

591

cat $Cr(CO)_6$, t-BuO_2H	JOC *50* 2791 (1985)
$Ce(O_3SCH_3)_3 \cdot 2\ H_2O$, electrolysis	TL *28* 1067 (1987) (ArCHO > ArCOR)
$K_2S_2O_8$, $CuSO_4$, H_2O	TL *22* 2605 (1981) (electron-rich arenes)
$(PhSeO)_2O$	TL 3331 (1979)
O_2, cat $ClRh(PPh_3)_3$	TL 3665 (1967); 2917 (1968)
DDQ	Syn 144 (1979) (requires *p*-MeO)

2. Alkenes

See also page 325, Section 4.

$$RCH{=}CHCH_2R \longrightarrow RCH{=}CHCR\overset{\displaystyle O}{\overset{\|}{}}$$

SeO_2	Tetr *6* 217 (1959) CC 1277 (1968) JACS *92* 3429 (1970); *93* 4835, 5311 (1971) Org Rxs *24* 261 (1976) (review)
t-BuO_2H, cat $Cr(CO)_6$	TL *25* 1235 (1984) JCS Perkin I 267 (1985)
t-BuO_2H, PCC, Celite	JOC *52* 5048 (1987)
t-BuO_2H, cat CrO_3	TL *28* 4665 (1987)
Na_2CrO_4, HOAc, Ac_2O	JOC *35* 192, 4068 (1970) JACS *106* 6690 (1984)
t-$C_4H_9OCrO_3H$	JOC *34* 3587 (1969)
CrO_3, t-BuOH	JACS *79* 6308 (1957) Helv *35* 284 (1962)
CrO_3, py, Celite	JOC *52* 3346 (1987)
$CrO_3 \cdot 2$ py	JOC *34* 3587 (1969); *52* 3573 (1987) Syn Commun *6* 217 (1976) JACS *107* 7724 (1985); *108* 6276 (1986)
$CrO_3 \cdot 1$ py	J Chem Res (S) 42 (1979)
$CrO_3 \cdot 3,5$-dimethylpyrazole	JOC *43* 2057 (1978); *52* 1686, 2960 (1987) JACS *108* 2090, 3513 (1986) TL *27* 3411, 5245 (1986)
CrO_3, HOAc	JACS *63* 758 (1941) BCSJ *52* 184 (1979)

CrO$_3$, HOAc, Ac$_2$O   BCSJ *52* 184 (1979)
                                              (4-oxo-2-alkenoates)
                                              JOC *47* 5093 (1982)
                                              (4-oxo-2-alkenoates)

$$\xrightarrow[\text{H}_2\text{O, EtOH}]{\text{Na}_2\text{O}_2}$$

TL *22* 5127 (1981)

(C$_5$H$_5$NH)$_2$Cr$_2$O$_7$ TL *27* 1481 (1986)
                                              JACS *109* 3991 (1987)

Na$_2$CrO$_4$, HOAc, Ac$_2$O   JACS *109* 3991 (1987)

CrO$_3$, HOAc, Ac$_2$O   JACS *109* 3991 (1987)
                                              JOC *52* 5482 (1987)

$$\xrightarrow[\text{KOH, EtOH}]{\text{tetrazolium salt}}$$

TL *28* 4323 (1987)

$$\text{MeO}_2\text{CCH}_2\text{CH}=\text{CHCO}_2\text{Me} \xrightarrow[\text{carbon}]{\text{activated}} \text{MeO}_2\text{CCCH}=\text{CHCO}_2\text{Me}$$

CC 882 (1987)

$$\xrightarrow[\text{cat PdCl}_2, \text{CuCl}]{\text{O}_2}$$

TL *27* 5955 (1986)

$$\text{RCH}=\text{CH}_2 \longrightarrow \text{RCCH}_3$$

cat PdCl$_2$, CuCl or CuCl$_2$, O$_2$ Angew Int *1* 80 (1962) (review)
                                              JOC *29* 241 (1964); *39* 3276 (1974); *52* 4592
                                                (1987)
                                              TL 2975 (1976); *21* 955 (1980); *27* 2529, 4431
                                                (1986)
                                              JACS *101* 5070, 5072 (1979); *105* 7352 (1983);
                                                108 3474 (internal alkene), 8015 (1987)
                                              Syn Commun *10* 273 (1980)

	Topics Curr Chem *91* 29 (1980) (review) Syn 369 (1984) (review) Israel J Chem *24* 153 (1984) CC 1578 (1987)
cat $PdCl_2$, $CuCl_2$, O_2, PEG-400	TL *26* 2263 (1985)
cat $PdCl_2$, $CuCl_2$, H_2O, C_6H_6, $(n\text{-}C_{16}H_{33}NMe_3)Br$	TL *24* 5159 (1983)
cat $Pd(OAc)_2$, H_2O_2	JOC *45* 5387 (1980)
cat $Pd(OAc)_2$, cat benzoquinone, electrolysis	TL *28* 3683 (1987)
$Hg(OAc)_2$, CH_3OH/cat Li_2PdCl_4, $CuCl_2$	CC 818 (1971)
cat $Hg(O_2CEt)_2$, CrO_3 or $Na_2Cr_2O_7$, H_3O^+	JOC *40* 3577 (1975) Syn Commun *11* 7 (1981)

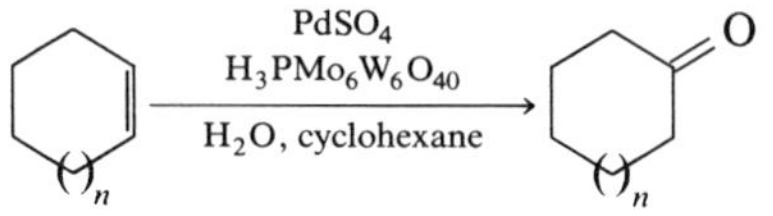

$n = 0, 1$

CC 1274 (1981)

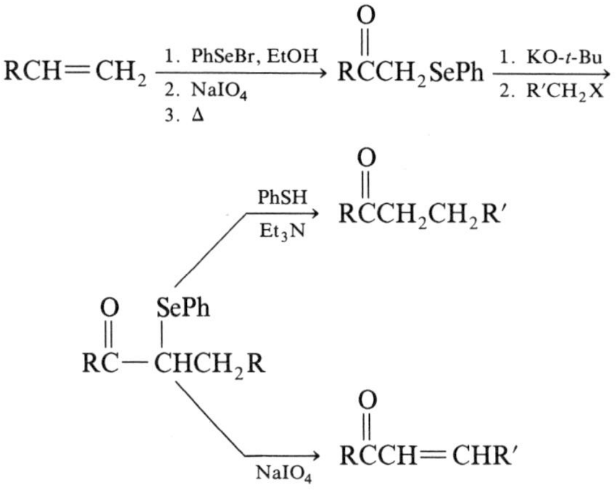

TL 799 (1978)

$$RCH{=}CH_2 \xrightarrow[O_2,\ py]{R_fI(Ph)O_3SCF_3} RC(O)CH_2R_f$$

TL *23* 4101 (1982)

$$RCH{=}CH_2 \longrightarrow RCH_2CHO$$

BH_3 or R_2BH/PCC	JOMC *172* C20 (1979) Syn 151 (1980)
PhSH, AIBN/NCS/CuO, $CuCl_2$, H_2O	JOC *41* 3261 (1976)

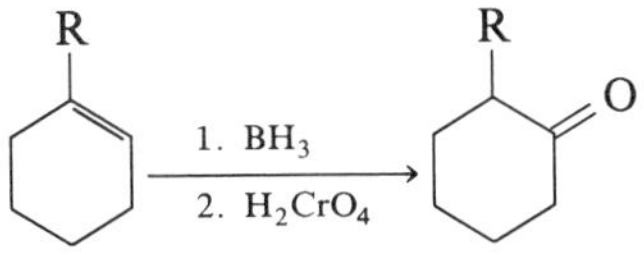

R ... R ... O

1. BH$_3$
2. H$_2$CrO$_4$

JACS *83* 2951 (1961)
JOC *50* 2179 (1985)

$$RCH{=}CHX \longrightarrow RC{\overset{\overset{\displaystyle O}{\|}}{C}}H_2X$$

X	Oxidant	
COR, CO$_2$R	H$_2$O$_2$ or t-BuO$_2$H, cat Na$_2$PdCl$_4$	CL 257 (1980)
CH$_2$OR, CH$_2$OAc	benzoquinone or O$_2$-CuCl, cat PdCl$_2$	TL *23* 2679 (1982)
CH$_2$COCH$_3$, CH$_2$CO$_2$R	O$_2$, CuCl, cat PdCl$_2$	CL 859 (1982)

$$RCH{=}CR_2 \longrightarrow RC{\overset{\overset{\displaystyle O}{\|}}{H}} + RC{\overset{\overset{\displaystyle O}{\|}}{R}}$$

Reviews:

Acct Chem Res *1* 313 (1968)
W. Carruthers, "Some Modern Methods of Organic Synthesis," Cambridge Univ. Press (1971), Chpt 6
Russ Chem Rev *50* 636 (1981) (ozonolysis)

O$_3$, reducing agent	Chem Rev *58* 925 (1958) (review)
O$_3$/H$_2$, cat Pd-C	Tetr *2* 203 (1958); *3* 230 (1958)
O$_3$/Zn, HOAc	JACS *77* 1212 (1955); *108* 4603 (1986)
O$_3$/I$^-$, HOAc	Org Syn *41* 41 (1961)
O$_3$/Me$_2$S	TL 4273 (1966); 1391 (1969); 2417 (1973) JACS *108* 1039, 1239 (1986) JOC *52* 1603 (1987)
O$_3$/SC(NH$_2$)$_2$	Tetr *38* 3013 (1982)
O$_3$/Ph$_3$P	JOC *30* 1976 (1965)
O$_3$/(MeO)$_3$P	JOC *25* 2703 (1963) JACS *108* 1039 (1986)
O$_3$/py	JACS *80* 915 (1958) Compt Rend *250* 1078 (1960)
KMnO$_4$	JOC *51* 3213 (1986)
NaIO$_4$, cat KMnO$_4$	Can J Chem *33* 1710, 1714 (1955)
NaIO$_4$, cat RuCl$_3$·(H$_2$O)$_n$	JOC *51* 3247 (1986)
NaIO$_4$, cat RuO$_2$	JOC *52* 2875 (1987)

$NaIO_4$, cat OsO_4

Soc Chem Ind *62* 90 (1943)
JOC *21* 478 (1956)
JACS *108* 4149 (1986)

$NaIO_4$, cat OsO_4, [morpholine N-oxide structure: O—N—Me]

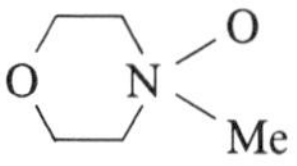

TL *28* 3225 (1987)

(bipy)H_2CrOCl_5

Org Prep Proc Int *14* 362 (1982) (aryl alkenes)

$$R_2C{=}CHR \longrightarrow R_2C{=}O + RCO_2H$$

$KMnO_4$, $NaIO_4$

Can J Chem *33* 1714 (1955)
Syn Commun *12* 1063 (1982)

$KMnO_4$, silica gel

JOC *52* 3698 (1987)

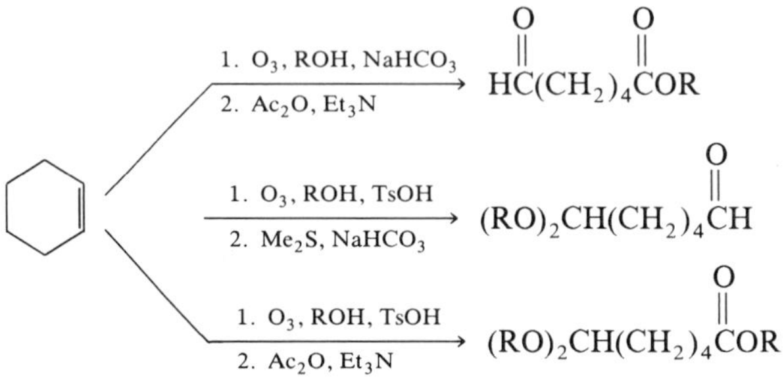

1. O_3, ROH, $NaHCO_3$
2. Ac_2O, Et_3N

$$HC(CH_2)_4COR$$ (each C=O)

1. O_3, ROH, TsOH
2. Me_2S, $NaHCO_3$

$$(RO)_2CH(CH_2)_4CH$$ (terminal C=O)

1. O_3, ROH, TsOH
2. Ac_2O, Et_3N

$$(RO)_2CH(CH_2)_4COR$$ (C=O)

TL *23* 3867 (1982)
Org Syn *64* 150 (1985)

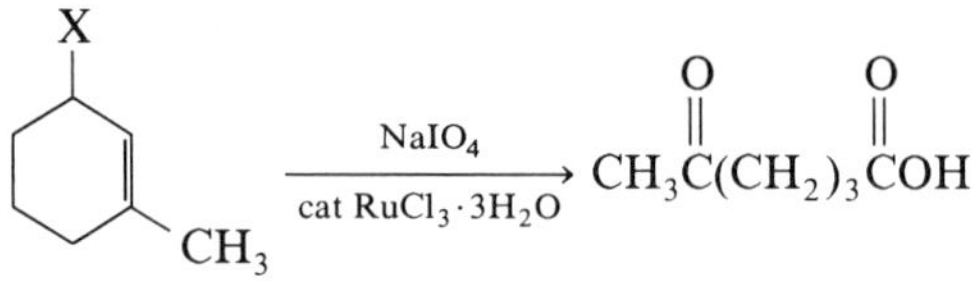

$NaIO_4$
cat $RuCl_3 \cdot 3H_2O$

$$CH_3C(CH_2)_3COH$$ (each C=O)

X = OH, =O

$$HOC(CH_2)_4COH$$ (each C=O)

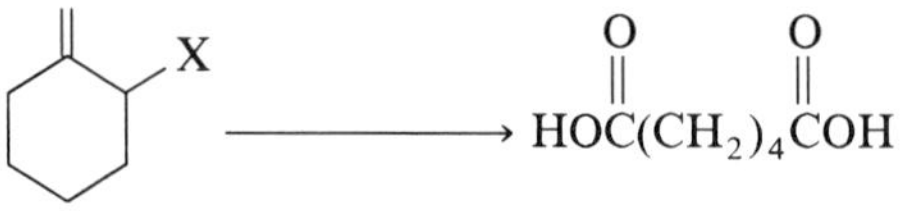

X = OH, =O

JOC *52* 689 (1987)

3. Alkynes

$$RC{\equiv}CR' \longrightarrow RCH_2CR'$$ (C=O)

BH_3 or R_2BH/H_2O_2, NaOH

JACS *83* 3834 (1961); *89* 5086 (1967); *91* 4771 (1969)
Org Rxs *13* 1 (1963)
TL 41 (1970)
Organomet Chem Syn *1* 249 (1971)

$ClBH_2/H_2O_2$, NaOH or NaOAc

JOC *38* 1617 (1973)

BH /H$_2$O$_2$, NaOH · · · · · JACS *94* 4370 (1972)

HSiMe(OEt)$_2$, cat ClRh(PPh$_3$)$_3$/Me$_3$NO, KHF$_2$ (R' = H) · · · · · TL *27* 75 (1986)

cat PdCl$_2$(CH$_3$CN)$_2$, H$_2$O, ultrasound · · · · · TL *28* 3127 (1987) (1,4- and 1,5-diketones)

cat NaAuCl$_4$·2H$_2$O, H$_2$O, ultrasound · · · · · TL *28* 3127 (1987) (1,5-diketone)

H$_2$O, H$_2$SO$_4$, Hg(II) · · · · · JCS 3257 (1954)
JACS *76* 524 (1954); *86* 935, 936 (1964) (both diketones)
Can J Chem *50* 1105 (1972)
JOC *40* 2250 (1975); *47* 3331, 3707 (1982)
Syn 671 (1978)
Z Chem *22* 185 (1982)
R. C. Larock, "Solvomercuration/Demercuration Reactions in Organic Synthesis," Springer, New York (1986) (review)
TL *28* 5709 (1987)

PhHgOH/H$_2$O · · · · · JOC *47* 3331 (1982) (RC≡CH → RCOCH$_3$ only)

Syn 473 (1981)

TL *23* 1051 (1982)

TL 2981 (1976); *23* 3193 (1982)

$$RC≡CR' \longrightarrow RC-CR'$$

KMnO$_4$, NaHCO$_3$, MgSO$_4$, H$_2$O, CH$_3$COCH$_3$ · · · · · JOC *44* 1574 (1979)

KMnO$_4$, HOAc, CH$_2$Cl$_2$, phase transfer agent · · · · · Syn 462 (1978)
JOC *44* 2726 (1979)

KMnO$_4$, H$_2$O, HOAc, CH$_2$Cl$_2$, phase transfer agent · · · · · Syn 462 (1978)
JOC *44* 2726 (1979)

Zn(MnO$_4$)$_2$, silica gel · · · · · JACS *105* 7755 (1983)

MoO$_5$(HMPA), cat Hg(OAc)$_2$ · · · · · TL *27* 5139 (1986)

NaOCl, cat RuO$_2$ · · · · · TL 2941 (1971)

NaIO$_4$, cat RuO$_2$ TL 2941 (1971); *27* 6133 (1986)
                                                     JACS *109* 6176 (1987)

PhIO, cat RuCl$_2$(PPh$_3$)$_3$ Helv *64* 2531 (1981)

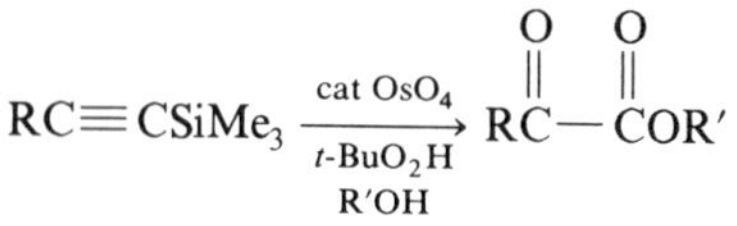

$$RC\equiv CSiMe_3 \xrightarrow[\substack{t\text{-BuO}_2\text{H} \\ \text{R'OH}}]{\text{cat OsO}_4} RC\overset{\overset{O}{\|}}{—}\overset{\overset{O}{\|}}{C}OR'$$

R' = H or alkyl

TL *27* 1947 (1986)

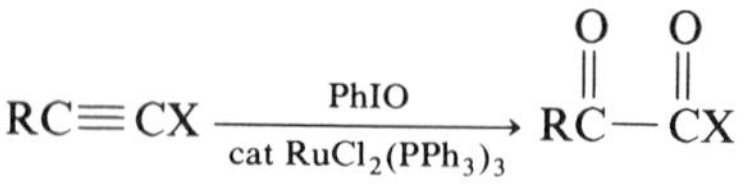

$$RC\equiv CX \xrightarrow[\text{cat RuCl}_2(\text{PPh}_3)_3]{\text{PhIO}} RC\overset{\overset{O}{\|}}{—}\overset{\overset{O}{\|}}{C}X$$

X = OR, NR$_2$

TL *23* 3661 (1982)

$$ArC\equiv CAr \xrightarrow{\text{(bipy)H}_2\text{CrOCl}_5} 2\ Ar\overset{\overset{O}{\|}}{C}H$$

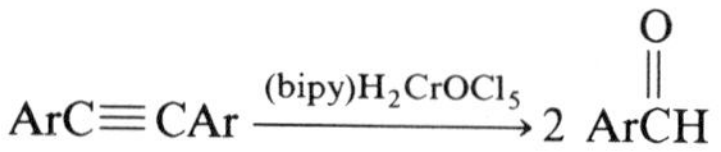

Org Prep Proc Int *14* 362 (1982)

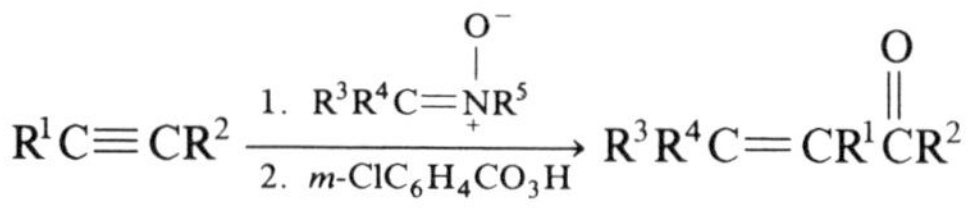

$$R^1C\equiv CR^2 \xrightarrow[\substack{\\ 2.\ m\text{-ClC}_6\text{H}_4\text{CO}_3\text{H}}]{1.\ R^3R^4C=\overset{+}{N}R^5, \overset{O^-}{|}} R^3R^4C=CR^1\overset{\overset{O}{\|}}{C}R^2$$

TL *28* 913 (1987)

$$R^1C\equiv CH \xrightarrow[2.\ R^2_3B]{1.\ RLi} [\,R^2_3\bar{B}C\equiv CR^1\,]Li^+ \xrightarrow{R^3X} R^2_2BC=CR^1R^3 \xrightarrow{[O]} R^2\overset{\overset{O}{\|}}{C}CHR^1R^3$$

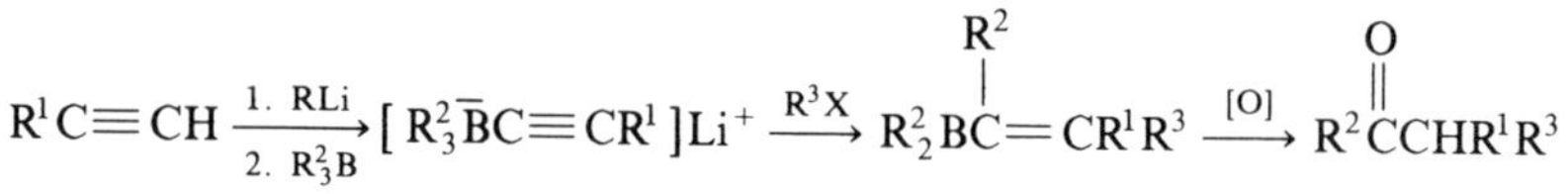

CC 544 (1973)
TL 2741 (1973); *28* 1003 (1987)

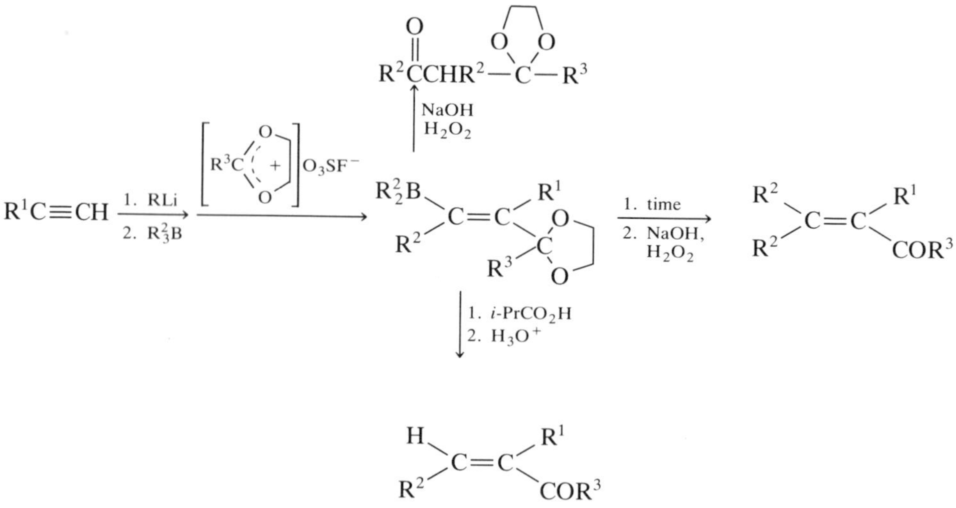

TL *27* 1935 (1986)

$$R^1C\equiv CH \xrightarrow[\text{2. } R_3^2B]{\text{1. RLi}} [R_3^2\bar{B}C\equiv CR^1]Li^+ \xrightarrow[\text{2. [O]}]{\text{1. } R^3CH=CXY} R^2\overset{\overset{\displaystyle O}{\parallel}}{C}CHR^1CHR^3CHXY$$

$\underline{X}$	$\underline{Y}$	
H, Me	NO_2	CC 913 (1977)
COMe, CO_2Et	CO_2Et	TL *22* 797 (1981)

$$RC\equiv CX \xrightarrow{\text{organoborane}} \xrightarrow[\text{NaOH}]{\text{NaOMe} \quad H_2O_2} RCH_2\overset{\overset{\displaystyle O}{\parallel}}{C}R'$$

$$X = Cl, Br, I$$

Organoborane

$R'BHBr\cdot SMe_2$	JOC *47* 3808 (1982); *51* 5270 (1986)
R'_2BH	JOC *47* 754 (1982); *51* 5270 (1986)
$(CH_3)_2CHC(CH_3)_2BHR'$	Syn 193 (1982)
	JOC *51* 5270 (1986)

$$RC\equiv CBr \xrightarrow[\text{2. ROH}]{\text{1. } HBBr_2\cdot SMe_2} \xrightarrow{R'Li(MgX)} \xrightarrow[\text{NaOH}]{H_2O_2} RCH_2\overset{\overset{\displaystyle O}{\parallel}}{C}R'$$

JOC *51* 5277 (1986)

4. Halides, Nitrates and Sulfonates

$$RCH_2X \text{ or } R_2CHX \longrightarrow R\overset{\overset{\displaystyle O}{\parallel}}{C}H \text{ or } R\overset{\overset{\displaystyle O}{\parallel}}{C}R$$

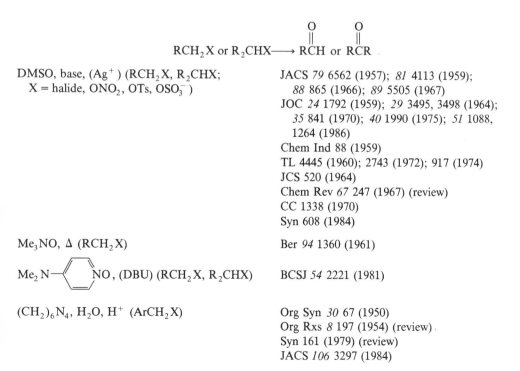

DMSO, base, (Ag^+) $(RCH_2X, R_2CHX;$ X = halide, ONO_2, OTs, $OSO_3^-)$	JACS *79* 6562 (1957); *81* 4113 (1959); *88* 865 (1966); *89* 5505 (1967) JOC *24* 1792 (1959); *29* 3495, 3498 (1964); *35* 841 (1970); *40* 1990 (1975); *51* 1088, 1264 (1986) Chem Ind 88 (1959) TL 4445 (1960); 2743 (1972); 917 (1974) JCS 520 (1964) Chem Rev *67* 247 (1967) (review) CC 1338 (1970) Syn 608 (1984)
Me_3NO, Δ (RCH_2X)	Ber *94* 1360 (1961)
Me_2N—⟨ ⟩—NO, (DBU) (RCH_2X, R_2CHX)	BCSJ *54* 2221 (1981)
$(CH_2)_6N_4$, H_2O, H^+ $(ArCH_2X)$	Org Syn *30* 67 (1950) Org Rxs *8* 197 (1954) (review) Syn 161 (1979) (review) JACS *106* 3297 (1984)

NaSPh/NCS/HgCl$_2$, CdCO$_3$ (RCH$_2$X)   Syn Commun 6 575 (1976)

KO$_2$SePh, 18-crown-6 (ArCH$_2$Cl)   TL 28 2933 (1987)

KO$_2$SePh, K$_2$HPO$_4$ (ArCH$_2$X; X = Cl, Br)   Syn 747 (1984)

Me$_2$SeO, K$_2$HPO$_4$ [ArCH$_2$X (X = Cl, Br)]   Syn 747 (1984)

NaN(SOPh)—⬡—/Δ/H$_2$O   JOC 46 4617 (1981)

  (1° allylic or benzylic)

p-AcOC$_6$H$_4$N(Tf)Na/NaOEt/HCl (RCH$_2$X)   JOC 44 1835 (1979)

PhNHTf, K$_2$CO$_3$/HCl (RCHXCOR′)   JOC 44 1835 (1979)

IF$_5$/H$_2$O (RCH$_2$I, R$_2$CHI)   Syn 419 (1977)

(n-Bu$_4$N)$_2$Cr$_2$O$_7$ (1° allylic RBr, 1°   Chem Ind 213 (1979)
  benzylic RCl and RBr, 2° alkyl RBr)   JOC 51 3347 (1986)

Ph—[ring with O, NONa]—Ph /Δ or hν (RCH$_2$X, R$_2$CHX)   JCS Perkin I 2493 (1979)

$$PhCH_2Br \xrightarrow[\substack{2.\ n\text{-BuLi} \\ 3.\ Me_3SiCl}]{1.\ PhSNa} PhCHSiMe_3 \xrightarrow[2.\ RX]{1.\ n\text{-BuLi}} \xrightarrow[2.\ H_3O^+]{1.\ m\text{-ClC}_6H_4CO_3H} PhCR$$

with SPh on the silane carbon and O on PhCR.

TL 21 4759 (1980)

$$>C=C<^X \longrightarrow -\overset{H}{\underset{|}{C}}-\overset{O}{\overset{\|}{C}}-$$

X	Reagent(s)	
F	Hg(OAc)$_2$, CF$_3$CO$_2$H/NaHCO$_3$	CL 651, 935 (1980)
Cl	H$_2$SO$_4$	Coll Czech Chem Commun 12 93, 101, 129 (1947)
	Hg(OAc)$_2$, HCO$_2$H	BSCF 1941 (1976) TL 3489 (1979)
	Hg(OAc)$_2$, HOAc, BF$_3$·OEt$_2$	TL 1943 (1978)
	Hg(OAc)$_2$, CF$_3$CO$_2$H	TL 1943 (1978)
	Hg(O$_2$CCF$_3$)$_2$, CH$_3$NO$_2$	TL 3489 (1979); 22 425 (1981) Helv 63 1383 (1980)

5. Epoxides

$$(CH_3)_2C\overset{O}{\overset{\diagup\diagdown}{-}}CHR \xrightarrow{H_5IO_6} R\overset{O}{\overset{\|}{C}}H$$

JOC 51 5447 (1986)

$$RCH\overset{O}{\overset{\diagup\diagdown}{-}}CH_2 \xrightarrow[H_2O,\ THF]{NaIO_4} R\overset{O}{\overset{\|}{C}}H$$

CC 1434 (1987)

6. Sulfides

$$RCH_2SPh \xrightarrow[\text{2. CuO, CuCl}_2\text{, H}_2\text{O}]{\text{1. NCS}} RCH\overset{O}{\underset{\|}{}}$$

JOC *41* 3261 (1976)

7. Sulfones

$$(RCH_2SO_2Ar)\ R_2CHSO_2Ar \longrightarrow RCR\overset{O}{\underset{\|}{}}\quad (RCH\overset{O}{\underset{\|}{}})$$

LDA/MoO$_5$·py·HMPA	TL *21* 3339 (1980); *27* 5951 (1986)
	JACS *104* 6115 (1982); *107* 1034 (1985) (enal)
	JOC *51* 5311 (1986)
n-BuLi/(Me$_3$SiO)$_2$	JOC *48* 4432 (1983)
n-BuLi/ClB(OMe)$_2$/m-ClC$_6$H$_4$CO$_3$H	TL *26* 2333 (1985)
n-BuLi/CCl$_4$/AgClO$_4$, CF$_3$CO$_2$H	BCSJ *53* 3027 (1980) (aryl ketones only)

8. Amines

$$R^1 - \underset{\underset{NH_2}{|}}{C}H - R^2 \longrightarrow R^1 - \overset{O}{\underset{\|}{C}} - R^2$$

(aldehydes or ketones)

KMnO$_4$, aq t-BuOH (R$_2$CO > RCHO)	JOC *32* 3129 (1967)
Ag(O$_2$CC$_5$H$_4$N)$_2$ (RCHO, R$_2$CO)	JCS 4962 (1965)
AgNO$_3$, Na$_2$S$_2$O$_8$, NaOH (RCHO, R$_2$CO)	JCS C 1384, 1388 (1966)
IF$_5$/H$_2$O (RCHO, R$_2$CO)	Syn 419 (1977)
t-BuOCl, NaHCO$_3$/NaOEt/aq H$_2$SO$_4$ (RCHO, R$_2$CO)	JACS *76* 5554 (1954)
NCS/NaOMe/aq H$_2$SO$_4$ (R$_2$CO)	Coll Czech Chem Commun *24* 2975 (1959)
NBS/base/H$_3$O$^+$ (R$_2$CO)	JACS *90* 3245 (1968)

$$\left[Ph - \underset{\underset{Ph}{}}{\overset{\overset{Ph}{}}{\bigcirc}}O^+ \right] BF_4^- / \qquad \text{JCS Perkin I 2500 (1979)}$$

$$Ph - \underset{\underset{Ph}{}}{\bigcirc}\overset{O}{\diagup}NONa/\Delta \text{ (RCHO)}$$

$$\left[\begin{array}{c} \text{Ph} \\ | \\ \text{Br—C} \overset{\oplus}{\underset{}{}} \text{N—SMe} \\ \text{N—N} \\ | \\ \text{Ph} \end{array} \right] \text{Br}^-, \text{Et}_3\text{N}/$$
　　　　　　　　　　　　　　　　　　　　　　　TL 2131 (1978)

$\text{Et}_3\text{N}/\text{EtO}_2\,\text{CN}{=}\text{NCO}_2\text{Et}/\text{H}_3\text{O}^+$ (RCHO)

(cyclohexylidene)=O /ArSOCl/Δ/H₂O (RCHO only)　　　JOC *46* 4617 (1981)

t-Bu / quinone =O, =O /H₃O⁺ (R₂CO)　　　　　JACS *91* 1429 (1969)
t-Bu

$\begin{array}{c} \text{Y}\quad\text{CH}_3 \\ \text{CH}_3 \text{—C}_6\text{—COCHO} \ (\text{X, Y = H, NO}_2)/ \\ \text{X}\quad\text{CH}_3 \end{array}$　　　JACS *91* 1429 (1969)

H_3O^+ (RCHO, R₂CO)

benzothiazole—C—CHO (X = H, NO₂)/　　　JCS Perkin I 1652 (1972)
X

DABCO or NaOMe/aq (HO₂C)₂
(RCHO, R₂CO)

pyridyl—CHO, DBU or NaOH/H₃O⁺　　　　Syn 756 (1982)

(RCHO, R₂CO)

pyridyl-CHO /m-ClC₆H₄CO₃H/KOH　　　　JACS *97* 6900 (1975)
(R₂CO, ArCHO)

pyridyl-CHO /LDA/H⁺ (RCHO)　　　　JOC *46* 1937 (1981)

$$\left[\begin{array}{c} \text{CHO} \\ | \\ \text{pyridinium} \\ \overset{+}{\text{N}} \\ | \\ \text{CH}_3 \end{array} \right] \text{PhSO}_3^-/\text{DBU or Et}_3\text{N}/\text{H}^+$$
　　　　　　　　　　　　　　　　　　　JACS *104* 4446 (1982)

(RCHO, R₂CO)

$$\text{R}_2\text{C}{=}\text{CRCH}_2\text{NR}_2 \xrightarrow[\text{2. Ac}_2\text{O}]{\text{1. H}_2\text{O}_2} \text{R}_2\text{C}{=}\text{CR}\overset{\displaystyle O}{\overset{\|}{\text{CH}}}$$

CL 1987 (1982)

9. Nitro Compounds

$$\underset{RCHR'}{\overset{NO_2}{|}} \longrightarrow \underset{RCR'}{\overset{O}{\|}} \text{ (aldehydes and/or ketones)}$$

aldehydes only

NaO-t-Bu/KMnO$_4$	JOC 46 1037 (1981)

ketones only

KMnO$_4$, silica gel	CC 635 (1982)
NaNO$_2$, n-C$_3$H$_7$ONO	JOC 38 1418 (1973)
VCl$_2$, HCl, H$_2$O	TL 2533 (1976)
(NH$_4$)$_2$Ce(NO$_3$)$_6$, Et$_3$N, H$_2$O	TL 23 3521 (1982)
R$_3$SiCl, DBU/m-ClC$_6$H$_4$CO$_3$H	TL 28 5361 (1987)
KO-t-Bu/t-BuO$_2$H, VO(acac)$_2$ or Mo(CO)$_6$	TL 331 (1977)
LDA/MoO$_5$·py·HMPA	TL 22 5235 (1981)
Triton B/O$_3$	TL 28 2883 (1987)

aldehydes and ketones

TiCl$_3$, NH$_4$OAc	JACS 93 5309 (1971) JOC 38 4367 (1973)
CrCl$_2$, CH$_3$OH, HCl	Syn 792 (1977)
OH$^-$/H$^+$ (Nef reaction)	Chem Rev 55 137 (1955) (review) TL 1331 (1972); 3215 (1974) JACS 99 3862 (1977); 108 1039 (1986)
KOH or NaH-t-BuOH/KMnO$_4$	Ber 69 1789 (1936) JOC 27 3699 (1962); 47 4534 (1982)
KOH/KMnO$_4$, H$_2$O, MgSO$_4$	JOC 50 4971 (1985)
NaOCH$_3$/O$_3$	JOC 39 259 (1974)
NaOCH$_3$/^{1}O$_2$	JOC 43 1271 (1978)
NaOCH$_3$/TiCl$_3$	JOC 51 4368 (1986)
NaOCH$_3$/silica gel	JACS 99 3861 (1977)
Et$_3$N/(NH$_4$)$_2$Ce(NO$_3$)$_6$	Syn 44 (1980)
Et$_3$N/(n-C$_{16}$H$_{33}$NMe$_3$)MnO$_4$	Syn Commun 17 195 (1987)
H$_2$O$_2$, K$_2$CO$_3$	Syn 662 (1980)

$$\underset{RCH=CR}{\overset{NO_2}{|}} \longrightarrow \underset{RCH_2CR}{\overset{O}{\|}}$$

n-Bu$_3$SnH/O$_3$ or m-ClC$_6$H$_4$CO$_3$H	TL 28 5365 (1987)
LiHB(sec-Bu)$_3$/H$_2$SO$_4$, H$_2$O	Syn 654 (1985)

CrCl$_2$, HCl, H$_2$O TL *26* 3777 (1985)

Raney Ni, NaH$_2$PO$_2$·H$_2$O TL *24* 417 (1983)

electrolysis CL 607 (1983)

10. Ethers

$$ArCH_2OR \longrightarrow Ar\overset{\overset{\displaystyle O}{\|}}{C}H$$

$$R_2CHOCH_3 \longrightarrow R_2C{=}O$$

IF$_5$/H$_2$O Syn 419 (1977)

UF$_6$/H$_2$O JACS *98* 6717 (1976); *100* 5396 (1978)

(NO)BF$_4$ (silyl or stannyl ethers) Syn 609 (1976)

(NO$_2$)BF$_4$ JOC *42* 3097 (1977)

(NH$_4$)$_2$Ce(NO$_3$)$_6$, NaBrO$_3$, H$_2$O Syn 897 (1980)
   (Me, Et, PhCH$_2$ and silyl ethers)

$$R_2CHOCH_2Ph \xrightarrow[H_2SO_4]{CrO_3} R\overset{\overset{\displaystyle O}{\|}}{C}R$$

JOC *46* 1492 (1981)

11. Alcohols

$$\overset{\overset{\displaystyle OH}{|}}{R}CHR' \longrightarrow R\overset{\overset{\displaystyle O}{\|}}{C}R' \text{ (aldehydes and/or ketones)}$$

Reviews: "activated" DMSO reagents Chem Rev *67* 247 (1967)
Tetr *34* 1651 (1978)
Syn 165 (1981)

Aldehydes only

$\left(Me_2N\!-\!\!\left\langle \bigcirc \right\rangle\!\!-\!NH \right) CrO_3Cl$ (allylic, benzylic) JOC *47* 1787 (1982)

CrO$_3$-graphite Can J Chem *50* 3058 (1972)

cat RuCl$_3$, NaIO$_4$, H$_2$O, CH$_3$CN JOC *52* 4592 (1987)

RuCl$_2$(PPh$_3$)$_3$ TL *22* 1605 (1981)

cat RuCl$_2$(PPh$_3$)$_3$, (Me$_3$SiO)$_2$ TL *24* 2185 (1983)

cat RuCl$_2$(PPh$_3$)$_3$, PhI(OAc)$_2$ TL *22* 2361 (1981)

K$_2$FeO$_4$ (1° benzylic)   CL 1397 (1978)

Pb(OAc)$_4$, Mn(OAc)$_2$ TL *27* 2287 (1986)

t-BuO$_2$H, cat (ArSe)$_2$ TL 2801 (1979)

DMSO, air, Δ (allylic, benzylic)   JACS *86* 298 (1964)

DMSO, py·SO$_3$, *i*-Pr$_2$NEt   TL *28* 1603 (1987)

O$_2$, cat   JACS *106* 3374 (1984)

cat CuCl (allylic, benzylic)

O$_2$, cat   JACS *106* 3374 (1984)

CuCl$_2$ (alkyl, allylic, benzylic)

[pyridine structure] OTs⁻,   CL 369 (1978)

Et$_3$N/hexamine (benzylic)

[NCN=NCN structure], *t*-BuOMgBr   JACS *106* 7970 (1984)
                                                           TL *28* 527 (1987)

n-PrMgBr/*t*-BuOMgBr, *m*-ClC$_6$H$_4$CO$_3$H   BCSJ *50* 2773 (1977)

n-PrMgBr/*t*-BuOMgBr, NCS   BCSJ *50* 2773 (1977)

Ketones only

O$_3$ JOC *41* 889 (1976)

O$_2$, PdCl$_2$, NaOAc   CC 157 (1977)

cat PdCl$_2$, K$_2$CO$_3$, CCl$_4$, Δ   CL 1171 (1981)

cat Pd(PPh$_3$)$_4$, K$_2$CO$_3$, PhBr, DMF   TL 1401 (1979)

t-BuO$_2$H, (*t*-BuO)$_3$Al or Me$_3$Al   TL *21* 1657 (1980)

t-BuO$_2$H, cat SeO$_2$ (allylic)   TL *28* 3831 (1987)

t-BuO$_2$H, cat VO(acac)$_2$ TL *24* 5009 (1983)

t-BuO$_2$H, cat Mo(CO)$_6$, cat   Syn 59 (1986)
    [C$_5$H$_5$N(CH$_2$)$_{15}$CH$_3$]Cl, MgSO$_4$

t-BuO$_2$H, cat (PhCH$_2$NMe$_3$)OMoBr$_4$ TL *25* 4417 (1984)

H_2O_2, cat $(NH_4)_6Mo_7O_{24} \cdot 4\,H_2O$, n-Bu$_4$NCl, K_2CO_3	TL *25* 173 (1984)
t-BuO$_2$H, cat $[C_5H_5N(CH_2)_{15}CH_3]_3PMo_{12}O_{40}$	Syn Commun *16* 537 (1986)
m-ClC$_6$H$_4$CO$_3$H, 10% HCl	TL 4115 (1975)
PhICl$_2$, py	TL 3635 (1973)
NBS (on ROSiMe$_3$)	JOC *43* 371 (1978)
NaOCl, HOAc	JOC *45* 2030 (1980); *52* 2602 (1987) TL *23* 4647 (1982)
Ca(OCl)$_2$, HOAc, CH$_3$CN	TL *23* 35 (1982)
NaOCl, cat RuCl$_3$	CC 1420 (1970) JOC *48* 1366 (1983)
cat RuCl$_2$(PPh$_3$)$_3$, PhIO	TL *22* 2361 (1981)
NaIO$_4$, cat RuO$_2$	JOC *50* 2759 (1985)
NaIO$_4$, cat RuO$_2$, cat (PhCH$_2$NEt$_3$)Cl	JOC *52* 1149 (1987)
cat RuO$_2 \cdot 2\,H_2O$, electrolysis	JOC *51* 151 (1986)
cat K$_2$RuO$_4$, K$_2$SO$_5$	CC 58 (1979)
cat R$_4$N(RuO$_4$) (R = n-Pr, n-Bu),	CC 1625 (1987)

cat IrH$_5$(i-Pr$_3$P)$_2$	TL *28* 3115 (1987)
CrO$_3$, HOAc	JACS *75* 422 (1953)
cat CrO$_3$, t-BuO$_2$H, H$_2$O (benzylic)	TL *28* 2133 (1987)
Na$_2$Cr$_2$O$_7 \cdot 2\,H_2O$-H$_2$SO$_4$-silica gel	Tetr *35* 1789 (1979)
K$_2$Cr$_2$O$_7$, H$_2$SO$_4$, ether (Brown) (2° alkyl, allylic)	JACS *83* 2952 (1961) JCS C 1972 (1966) JOC *36* 387 (1971)
(C$_5$H$_5$NH)$_2$Cr$_2$O$_7$, HOAc, EtOAc (2° allylic only)	JOC *51* 5472 (1986)
(C$_5$H$_5$NH)$_2$Cr$_2$O$_7$, (C$_5$H$_5$NH)O$_2$CCF$_3$	JACS *109* 4690 (1987)

Cu(MnO$_4$)$_2 \cdot 8\,H_2O$	JOC *47* 2790 (1982)
KMnO$_4$, CuSO$_4 \cdot 5\,H_2O$	JOC *44* 3446 (1979)
NaMnO$_4 \cdot H_2O$ (solid)	TL *22* 1655 (1981)
(NH$_4$)$_2$Ce(NO$_3$)$_6$ or Ce(SO$_4$)$_2 \cdot 2\,H_2SO_4$, NaBrO$_3$	TL *23* 539 (1982) JOC *50* 2759 (1985)

DMSO, Ac$_2$O	JOC *30* 1107 (1965)
	JACS *89* 2416 (1967); *109* 7477 (1987) (acyloin)
	Chem Rev *67* 247 (1967) (review)
DMSO, PhSO$_2$Cl/Et$_3$N	JOC *39* 1977 (1974)
DMSO, TsCl/Et$_3$N	JOC *39* 1977 (1974)
DMSO, (TsO)$_2$/Et$_3$N	JOC *39* 1977 (1974)
DMSO, P$_2$O$_5$	JACS *87* 4651 (1965)

DMSO, Cl/Et$_3$N   JOC *39* 1977 (1974)

DMSO, *i*-Pr N=C=N-*i*-Pr, cat Cl$_2$CHCO$_2$H	JACS *107* 3285 (1985)
DDQ (2° allylic)	JOC *50* 5897 (1985)
DDQ, HIO$_4$ (2° allylic)	Syn 848 (1978)
(Ph$_3$C)BF$_4$	TL 2771 (1978)
Raney Ni, (1-octene)	JOC *51* 5482 (1986)
tetrazolium salt, KOH, EtOH	TL *28* 4323 (1987) (on α-hydroxy ketone)
electrolysis, KI, H$_2$O	TL 165 (1979)
electrolysis, poly-4-vinylpyridine·HBr	JOC *45* 5269 (1980)
electrolysis, *n*-C$_8$H$_{17}$SCH$_3$, Et$_4$NBr	TL *21* 1867 (1980)
electrolysis, NiOOH	Tetr *38* 3299 (1982)

Aldehydes and ketones

BF$_4$, AgBF$_4$ JOC *51* 5454 (1986)

Cl$_2$, py	TL 3059 (1974)
Cl$_2$ or Br$_2$, HMPA (1° benzylic)	Syn 811 (1976)
(*n*-Bu$_3$Sn)$_2$O/Br$_2$ (1° benzylic)	TL 4597 (1976)
Et$_3$SnOMe/Et$_3$SnOMe, Br$_2$ (2° alkyl, 1° allylic, 1° or 2° benzylic)	CL 145 (1975)
NBS (on ROSnR′$_3$) (1° allylic or benzylic)	JACS *98* 1629 (1976)
NBA, NBS, or NCS (1° benzylic)	Chem Rev *63* 21 (1963)

NIS, n-Bu$_4$NI

Syn 394 (1981)

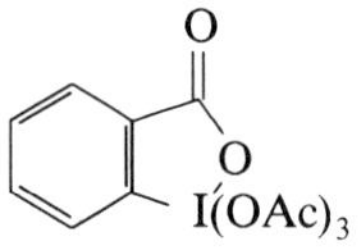

JOC *48* 4155 (1983)
TL *28* 2921, 4259 (trifluoromethyl ketones)
  (1987)

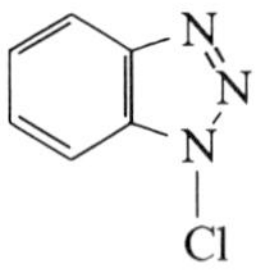

(1° benzylic)

JCS C 1474 (1969)

NaOCl (phase transfer) (1° benzylic)

TL 1641 (1976)

NaOCl, cat , KBr, NaHCO$_3$

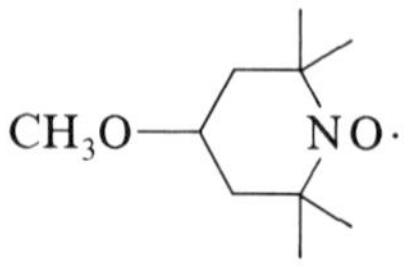

JOC *52* 2559 (1987)

Ca(OCl)$_2$ (1° benzylic)

TL 35 (1982)

CrO$_3$, HMPA (2° benzylic)

Syn 394 (1976)

CrO$_3$, Celite, Et$_2$O, CH$_2$Cl$_2$

Syn 815 (1979)

CrO$_3$-polymer

JACS *98* 6737 (1976)

CrO$_3$, cat n-Bu$_4$NX

TL *21* 4653 (1980)

CrO$_3$-H$_2$O-SiO$_2$

Syn 534 (1978)

CrO$_3$·2 py (Collins)

JOC *26* 4814 (1961); *35* 4000 (1970)
TL 3363 (1968); 399 (1979)
Org Syn *52* 5 (1972); *55* 84 (1975)
Syn 567 (1981) (β-hydroxy ketones)

CrO$_3$·2 py, Celite

JOC *52* 5419 (1987)

CrO$_3$·3,5-dimethylpyrazole

TL 4499 (1973)

CrO$_3$-py-HCl (PCC)

TL 2647 (1975)
Carbohydr Res *67* 491 (1978); *81* 187 (1980)
Syn 245 (review), 881 (1982)
JOC *50* 2095, 2607, 2764 (1985); *52* 5452 (1987)
JACS *108* 4603 (1986)

PCC-polymer

JOC *43* 2618 (1978)

PCC, molecular sieves (1° benzylic,
  2° alkyl or benzylic)

CC 561 (1980)
JCS Perkin I 1967 (1982)
TL *26* 3731 (1985); *28* 675 (1987)

PCC, molecular sieves, py

TL *28* 703 (1987)

PCC, py

Chem Phys Lipids *27* 281 (1980)

PCC, alumina

Syn 223 (1980)
JOC *47* 564 (1982)
TL *28* 703 (1987)

PCC, Celite	JOC *50* 2626 (1985); *52* 3662 (1987) TL *28* 1791 (1987)
PCC, Celite, molecular sieves	TL *28* 171 (1987)
PCC, NaOAc	TL 2647 (1975) JCS Perkin I 599 (1981) Syn 245 (1982) (review) JACS *108* 4603 (1986); *109* 5167 (1987) JOC *50* 2707 (1985); *52* 5452 (1987)
PCC, NaOAc, py	Chem Phys Lipids *25* 381 (1979)
PCC, NaOAc, alumina	TL *28* 869, 6437 (1987)
PCC, NaOAc, molecular sieves	TL *28* 1073 (1987) JOC *52* 4885 (1987)
CrO_3-DMAP-HCl (1° allylic or benzylic, 2° allylic)	JOC *47* 1787 (1982)
CrO_3-bipy-HCl	Syn 691 (1980)
CrO_3-py-HF	Syn 588 (1982)
CrO_3, H_2SO_4, CH_3COCH_3 (Jones) (1° allylic or benzylic)	JCS 39 (1946); 2548 (1953) JOC *21* 1547 (1956); *25* 1434 (1960); *40* 1664 (1975) Org Syn *45* 28 (1965)
$(C_5H_5NH)HCrO_4$-silica gel	Tetr *35* 1789 (1979)
$(n\text{-Bu}_4N)HCrO_4$, $HCCl_3$ (1° allylic or benzylic)	Syn 356 (1979)
$Na_2Cr_2O_7$, H_2O (1° allylic, 1° or 2° benzylic)	JOC *35* 3589 (1970)
$Na_2Cr_2O_7$, H_2SO_4, DMSO	JOC *39* 3304 (1974)
$K_2Cr_2O_7$, H_2SO_4 (phase transfer)	TL 1601 (1978) JCS Perkin II 788 (1979) Syn 134 (1979)
$K_2Cr_2O_7$, DMSO or polyethylene glycol (1° allylic, 1° or 2° benzylic)	Syn 646 (1980)
$K_2Cr_2O_7$, benzene (phase transfer) (1° allylic or benzylic, 2° benzylic)	TL 4167 (1977)
$(n\text{-Bu}_4N)_2Cr_2O_7$, CH_2Cl_2 (1° allylic, 1° or 2° benzylic)	Syn Commun *10* 75 (1980)
$(PhCH_2NEt_3)_2Cr_2O_7$, HMPA (1° allylic or benzylic; 2° benzylic)	Syn 1091 (1982)
$(C_5H_5NH)_2Cr_2O_7$ (PDC), CH_2Cl_2	TL 2647 (1975); 399 (1979) JOC *41* 380 (1976) Ind J Chem B *22* 69 (1983)
PDC, cat $(C_5H_5NH)O_2CCF_3$	JOC *50* 2981 (1987)

PDC, HOAc, molecular sieves	TL *26* 1699 (1985); *28* 5129 (1987)
	Syn Commun *16* 11 (1986)
cat PDC, (Me$_3$SiO)$_2$	TL *24* 2185 (1983)
PDC-polymer	JOC *46* 1728 (1981)
PDC, molecular sieves (1° or 2° benzylic, 1° allylic, 2° alkyl)	CC 561 (1980)
	JCS Perkin I 1967 (1982)
	JACS *109* 6389 (1987)
	TL *28* 2959 (1987)
(Ph$_3$PCH$_2$PPh$_3$)Cr$_2$O$_7$ (1° allylic, 1° or 2° benzylic)	TL *27* 1775 (1986)
CrO$_2$Cl$_2$, SiO$_2$-Al$_2$O$_3$	JOC *42* 2182 (1977)
(Phen)CrOCl$_3$	TL *21* 1583 (1980)
(Phen)H$_2$CrOCl$_5$	TL *21* 1583 (1980)
(py)CrO$_5$	TL 3749 (1977)
MnO$_2$ (particularly allylic, benzylic and propargylic)	JACS *77* 4399 (1955); *107* 1028, 1034 (1985); *108* 2691 (1986); *109* 4690, 6719 (1987)
	JCS 1430 (1952); 4685 (1956); 4983 (1963)
	Quart Rev *13* 61 (1959)
	JOC *24* 1051 (1959); *35* 3971 (1970); *51* 2863, 4158 (1986); *52* 398, 3662, 5067, 5700 (1987)
	Proc Chem Soc 110 (1964)
	Syn 65, 133 (1976) (reviews)
	CC 1237 (1986)
	TL *28* 5403 (1987)
KMnO$_4$ (on solid support) (1° allylic or benzylic)	JACS *99* 3837 (1977)
	TL *22* 4889 (1981)
BaMnO$_4$ (1° or 2° alkyl, 1° or 2° benzylic, 1° or 2° allylic, 2° propargylic)	TL 839 (1978); *23* 3283 (1982); *28* 5655, 5865 (1987)
	BCSJ *56* 914 (1983)
	JACS *106* 7861 (1984); *108* 4953 (1986)
	CC 1319 (1986)
	JOC *51* 4169 (1986)
[*n*-C$_{16}$H$_{33}$N(CH$_3$)$_3$]MnO$_4$ (benzylic)	CL 2131 (1984)
AgMnO$_4$ (1° alkyl, 1° or 2° benzylic)	TL *23* 1847 (1982)
Pb(OAc)$_4$, py	TL 3071 (1964); 4427 (1967)
Ag$_2$CO$_3$, Celite	Compt Rend C *267* 900 (1968)
	CC 1102, 1118 (1969)
	Tetr *29* 1011 (1973)
	Syn 401 (1979)
	TL *22* 3721, 3725 (1981)
Ni(O$_2$CPh)$_2$, Br$_2$	Syn Commun *10* 881 (1980)
NiBr$_2$, (PhCO$_2$)$_2$	JOC *44* 2955 (1979)

NiO$_2$ (1° or 2° allylic, benzylic, allenic)

JOC *27* 1597 (1964)
Chem Rev *75* 491 (1975)
TL 1595 (1979)
JACS *104* 2642 (1982)

cat Pd(OAc)$_2$, PhI, *n*-Bu$_4$NCl, NaHCO$_3$, DMF (1° alkyl or benzylic, 2° alkyl)

TL *26* 6257 (1985)

RuO$_4$ (1° benzylic, 2° alkyl)

JACS *80* 6682 (1958)
Rev Pure Appl Chem *22* 47 (1972)
(review)

KIO$_4$, cat RuO$_2$, K$_2$CO$_3$ (2° alkyl)

Carbohydr Res *24* 192 (1972)
Methods Carbohydr Chem *7* 3 (1976)

NaOCl, cat RuO$_2$

Methods Carbohydr Res *6* 337 (1972)

cat H$_2$Ru(PPh$_3$)$_4$, H$_2$C=CHCH$_2$OCO$_2$Me (1° allylic, 2° alkyl or allylic)

TL *27* 1805 (1986)

O$_2$, CuCl, phenanthroline, K$_2$CO$_3$ (1° allylic or benzylic, 2° benzylic)

TL 1215 (1977)

O$_2$, cat *n*-Bu$_4$N$\left[\begin{array}{c} \end{array}\right]$

JOC *52* 5467 (1987)

H$_2$O$_2$, cat WO$_4^{2-}$ or MoO$_4^{2-}$, phase transfer catalyst (1° benzylic, 2° alkyl)

JOC *51* 2661 (1986)

H$_2$O$_2$, [C$_5$H$_5$N(CH$_2$)$_{15}$CH$_3$]$_3$PMo$_{12}$O$_{40}$

JOC *52* 1868 (1987)

Fe(NO$_3$)$_3$ on clay

Syn 849 (1980)

FeCl$_3$, hν

JOC *42* 171 (1977)

K$_2$FeO$_4$, Al$_2$O$_3$, CuSO$_4$·5 H$_2$O (1° or 2° benzylic and allylic, 2° alkyl)

TL *27* 2875 (1986)

Ag$_2$FeO$_4$ (1° or 2° benzylic and allylic)

Syn Commun *16* 211 (1986)

(NH$_4$)$_2$Ce(NO$_3$)$_6$, NaBrO$_3$ (1° benzylic)

Syn 936 (1978)
TL 539 (1982)

[Ce(NO$_3$)$_3$]$_2$CrO$_4$ (1° or 2° benzylic)

Syn Commun *14* 631 (1984)
JOC *50* 5902 (1985)

(Ph$_3$BiCl)$_2$O

CC 1099 (1978)

Me$_2$SeO, molecular sieves

Syn 747 (1984)

(PhSeO)$_2$O (1° allylic or benzylic)

CC 952 (1978)

(ArSe)$_2$, *t*-BuO$_2$H

JOC *47* 837 (1982)

Al(O-*t*-Bu)$_3$, CH$_3$COCH$_3$ (Oppenauer)

Org Rxs *6* 207 (1951)

EtMgBr/PhCHO (1° or 2° allylic, 2° alkyl)

TL *28* 769 (1987)

cat $Cp_2Zr(O\text{-}i\text{-}Pr)_2$, PhCHO or PhCOPh

JOC *52* 4855 (1987)

cat Cp_2ZrH_2, RCOR
(1° or 2° alkyl, allylic, benzylic)

JOC *51* 240 (1986); *52* 4855 (1987)

DMSO, ⟨ ⟩—N=C=N—⟨ ⟩, H^+

(Pfitzner-Moffatt)

JACS *85* 3027 (1963); *87* 5661, 5670 (1965);
   88 1762 (1966)
Org Syn *47* 25 (1967)
Chem Rev *67* 247 (1967)

DMSO, Ac_2O

JACS *87* 4214 (1965)

DMSO, P_2O_5

JACS *87* 4214, 4651 (1965)

DMSO, $SOCl_2/Et_3N$

Tetr *34* 1651 (1978)

DMSO, $(COCl)_2/Et_3N$ (Swern)

JOC *43* 2480 (1978); *44* 4148 (1979); *50* 2198
   (1985); *51* 5282 (1986); *52* 5700 (1987)
Tetr *34* 1651 (1978)
Syn 165 (review), 567 (β-hydroxy ketones)
   (1981)
Org Syn *64* 164 (1985)
JACS *108* 1035 (1,2-diol), 4603 (1986);
   109 3353 (1987)

DMSO, $(MeSO_2)_2O/Et_3N$

JOC *39* 1977 (1974)

DMSO, $ClSO_2NCO/Et_3N$ (1° allylic or benzylic)

Syn 141 (1980)

DMSO, $(CF_3CO)_2O/Et_3N$

JOC *41* 957, 3329 (1976); *52* 4851 (1987)
   (α-dicarbonyl compounds)
Syn 297 (1978)

DMSO, $py \cdot SO_3$, Et_3N

Chem Pharm Bull *30* 1921 (1982)
JOC *52* 1252 (1987)

DMSO, $py \cdot SO_3/Et_3N$

JACS *89* 5505 (1967)

DMSO/$COCl_2/Et_3N$

JCS 1855 (1964)

DMSO, Cl_2/Et_3N

TL 919 (1973)

DMSO, [pyridinium structure with CH_3, F, N, Me] OTs^-, Et_3N

CL 369 (1978)

$PhSCH_3$ or $(CH_3)_2S$, Cl_2 or
NCS/Et_3N or $i\text{-}Pr_2NEt$

JACS *94* 7586 (1972)
JOC *38* 1233 (1973); *51* 5282 (1986);
   52 4505 (1987)

$CH_3COCOCl$, py/$h\nu$

JOC *41* 3030 (1976)

$m\text{-}ClC_6H_4CO_3H$, ⟨ ⟩NH·HCl (1° benzylic)

JOC *40* 1860 (1975)

DDQ (1° or 2° benzylic)

JOC *45* 1596 (1980)

5-deazaflavin, KOH (1° benzylic)

CC 825 (1977)

$EtO_2CN=NCO_2Et$, $EtO_2CCH_2NO_2$, PPh_3 TL 2295 (1981)

NPh (1° benzylic)   CC 744 (1966)

n-PrMgBr/t-BuOMgBr,   BCSJ *50* 2773 (1977)
JACS *106* 7970 (1984)

n-PrMgBr/t-BuOMgBr, PhI(OAc)$_2$ BCSJ *50* 2773 (1977)

(NO)BF$_4$ (1° or 2° benzylic)   Syn 609 (1976)

hν, platinised TiO$_2$ TL *25* 3363 (1984)
(1° alkyl or allylic, 2° benzylic)

Selective oxidation, 1° > 2° ROH

  RuCl$_2$(PPh$_3$)$_3$ TL *22* 1605 (1981)

  cat RuCl$_2$(PPh$_3$)$_3$, (Me$_3$SiO)$_2$ TL *24* 2185 (1983)

  DMSO, ClCOCOCl/Et$_3$N   JACS *108* 1035 (1986)

  DMSO, py·SO$_3$, i-Pr$_2$NEt   TL *28* 1603 (1987)

Selective oxidation, 2° > 1° ROH

  (Ph$_3$C)BF$_4$ JACS *98* 7882 (1976)
TL 2771 (1978)

  RCHO (R = Ph, CCl$_3$), Al$_2$O$_3$ TL 3499 (1976)

  NBA   JACS *74* 483 (1952); *76* 3682 (1954);
85 1409 (1963)

  NBS, H$_2$O   JACS *109* 3017 (1987)

  Br$_2$, (n-Bu$_3$Sn)$_2$O   TL 4597 (1976)

  Cl$_2$, py   TL 3059 (1974)

  X$_2$, HMPA (X = Cl, Br)   Syn 811 (1976)

  HOCl, HOAc   TL 4647 (1982)

  NaOCl, HOAc   JOC *45* 2030 (1980)
TL *23* 4647 (1982)

  t-BuO$_2$H, cat VO(acac)$_2$ TL *24* 5009 (1983)

  t-BuO$_2$H, cat (PhCH$_2$NMe$_3$)OMoBr$_4$ TL *25* 4417 (1984)

  t-BuO$_2$H, cat [C$_5$H$_5$N(CH$_2$)$_{15}$CH$_3$]PMo$_{12}$O$_{40}$ Syn Commun *16* 537 (1986)

t-BuO$_2$H, cat Mo(CO)$_6$, cat [C$_5$H$_5$N(CH$_2$)$_{15}$CH$_3$]Cl, MgSO$_4$	Syn 59 (1986)
H$_2$O$_2$, cat (NH$_4$)$_6$Mo$_7$O$_{24}$·4 H$_2$O, n-Bu$_4$NCl, K$_2$CO$_3$	TL 25 173 (1984)
H$_2$O$_2$, [C$_5$H$_5$N(CH$_2$)$_{15}$CH$_3$]$_3$PMo$_{12}$O$_{40}$	JOC 52 1868 (1987)
(NH$_4$)$_2$Ce(NO$_3$)$_6$ or Ce(SO$_4$)$_2$·2 H$_2$SO$_4$, NaBrO$_3$	TL 23 539 (1982) JOC 50 2759 (1987)
Raney Ni	JOC 51 5482 (1986)
electrolysis, poly-4-vinylpyridine·HBr	JOC 45 5269 (1980)

Selective oxidation, allylic or benzylic > alkyl ROH

See earlier sections

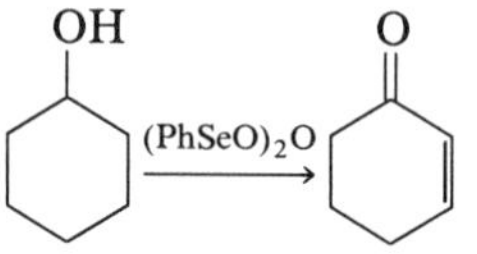

CC 952 (1978)

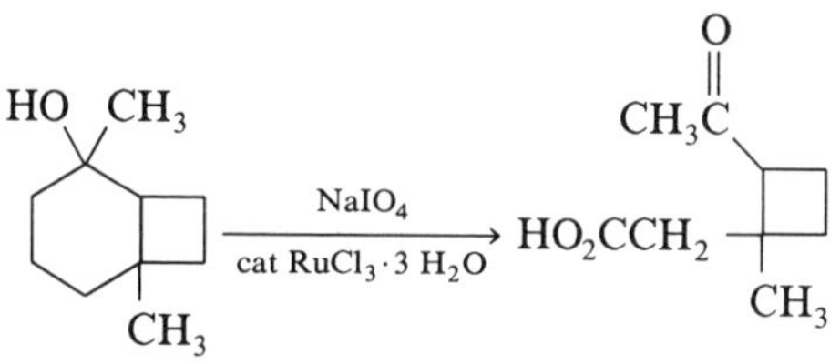

JOC 52 689 (1987)

12. Allylic Alcohols

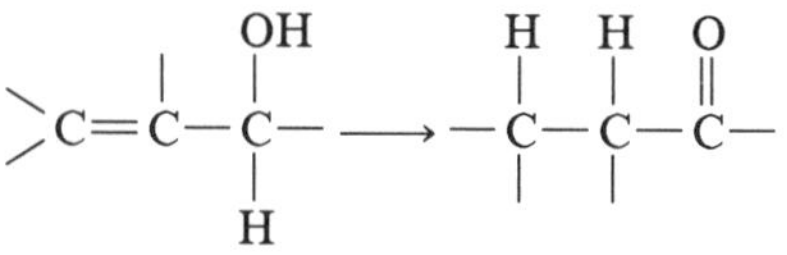

See page 627, Section 1.

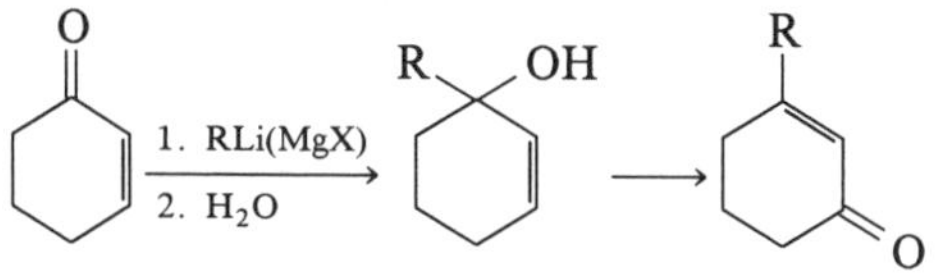

See page 587, Section 1.

$$\text{(cyclohexenol with CH}_3) \xrightarrow[\text{cat RuCl}_3 \cdot 3\,\text{H}_2\text{O}]{\text{NaIO}_4} CH_3\overset{\overset{\displaystyle O}{\|}}{C}(CH_2)_3CO_2H$$

$$\text{(structure)} \longrightarrow CH_3\overset{\overset{\displaystyle O}{\|}}{C}CH_2\overset{\overset{\displaystyle CH_3}{|}}{C}HCH_2CH_2CO_2H$$

JOC *52* 689 (1987)

$$H_2C\!\!=\!\!CH\overset{\overset{\displaystyle OH}{|}}{C}HR \longrightarrow R'CH_2CH_2\overset{\overset{\displaystyle O}{\|}}{C}R$$

See page 817, Section 16.

13. Diols

$$\overset{\overset{\displaystyle HO\ \ OH}{|\ \ \ |}}{RCHCHR} \longrightarrow 2\ R\overset{\overset{\displaystyle O}{\|}}{C}H$$

Ca(OCl)$_2$, H$_2$O, CH$_3$CN, HOAc	TL *23* 3135 (1982)
(NH$_4$)$_4$[Ce(SO$_4$)$_4$]·2 H$_2$O, H$_2$SO$_4$	JACS *81* 1494 (1959)
NaBiO$_3$, H$_3$PO$_4$, H$_2$O	JCS 1907 (1950)
I(OAc)$_3$	JCS Perkin I 1483 (1978)
HIO$_4$	Org Rxs *2* 341 (1944)
NaIO$_4$, H$_2$O	TL *24* 1377 (1983); *28* 3035 (1987) JACS *109* 3353 (1987)
NaIO$_4$, NaOH	JOC *50* 2095 (1985)
Amberlite 904-NaIO$_4$	JCS Perkin I 509 (1982)
Pb(OAc)$_4$	Ann *599* 81 (1956) Newer Methods Prep Org Chem *2* 367 (1963) TL *27* 5853 (1986); *28* 2999 (1987)
HgO, I$_2$	JOC *51* 5446 (1986)
MnO$_2$	Angew *85* 401 (1973)
H$_2$CrO$_4$	JACS *82* 1401 (1960); *84* 1252, 2241 (1962)
PCC	Syn Commun *12* 833 (1982)

$$\underset{\substack{| \ \ | \\ \text{RCHCHR}}}{\text{HO} \ \ \text{OH}} \longrightarrow 2 \ \overset{\overset{\displaystyle O}{\|}}{\text{RCH}} \quad (\textit{Continued})$$

NIS   JOC *46* 1927 (1981)

electrolysis   JACS *97* 2546 (1975)

$$\underset{\substack{| \ \ | \\ \text{RCHCHR}}}{\text{HO} \ \ \text{OH}} \xrightarrow[\text{MeOH, H}_2\text{SO}_4]{\text{electrolysis}} 2 \ \text{RCH(OMe)}_2$$

JOC *46* 3312 (1981)

$$\text{HO}-\text{C}_n-\text{OH} \longrightarrow \text{HO}-\underset{\text{O}}{\overset{}{\diagdown}}\!\!\!\diagup\!\!)_{n-3}$$

Reagent	n	
$RuCl_2(PPh_3)_3$	4	JACS *106* 1148 (1981)
	5	TL *28* 3123 (1987)
$BaMnO_4$	4	JCS Perkin I 1579 (1983)

14. Halohydrins

$$\underset{\substack{| \ \ | \\ \text{RCHCHR}'}}{\text{X} \ \ \text{OH}} \xrightarrow[\substack{\text{PAr}_3 \\ \text{K}_2\text{CO}_3}]{\text{cat Pd(OAc)}_2} \text{RCH}_2\overset{\overset{\displaystyle O}{\|}}{\text{C}}\text{R}'$$

TL *23* 3085 (1982)

15. α-Hydroxyacids

$$\underset{\substack{| \ \ \| \\ \text{RCHCOH}}}{\text{HO} \ \ \text{O}} \longrightarrow \overset{\overset{\displaystyle O}{\|}}{\text{RCH}}$$

NIS   JOC *47* 3006 (1982)

$NaIO_4$, H_2O, HOAc, CH_3COCH_3 TL 1725 (1968)

16. α-Thioacids

See page 617, Section 18.

17. Ketones

$$\overset{\overset{\displaystyle O}{\|}}{\text{RC}}\text{CH}_2\text{R}' \longrightarrow \text{R}-\overset{\overset{\displaystyle O}{\|}}{\text{C}}-\overset{\overset{\displaystyle O}{\|}}{\text{C}}-\text{R}'$$

SeO_2 Org Rxs *5* 331 (1949); *24* 261 (1976) (reviews)
                               JOC *52* 2602 (1987)

CuBr$_2$/KI, DMSO, Na$_2$CO$_3$ JOC *40* 1990 (1975)

t-BuOCH(NMe$_2$)$_2$/O$_2$, hν JOC *50* 3573 (1985)

18. Carboxylic Acids

$$\underset{\text{RR'CHCOH}}{\overset{\overset{\textstyle O}{\|}}{}} \longrightarrow \underset{\text{RCR'}}{\overset{\overset{\textstyle O}{\|}}{}}$$

2 LDA/(MeS)$_2$/NCS, ROH   JACS *99* 3101 (1977); *107* 8066 (1985); *109* 4626 (1987)

2 LDA/(PhS)$_2$/electrolysis   TL 1045 (1979)

2 LDA/O$_2$/Me$_2$NCH(OMe)$_2$ TL 4611 (1975)

(n-Bu$_4$N)IO$_4$ (R = Ar)   TL *21* 2655 (1980)

(n-Bu$_4$N)IO$_4$ Syn 563 (1980)
   (R = OH in acid, product is R'CHO)

19. Esters

$$\underset{\text{RCH}_2\text{COR}}{\overset{\overset{\textstyle O}{\|}}{}} \xrightarrow[\text{or LDA/DMF-Me}_2\text{SO}_4]{(\text{Me}_2\text{N})_3\text{CH or MeOCH(NMe}_2)_2} \underset{\text{RCCO}_2\text{R}}{\overset{\overset{\textstyle \text{Me}_2\text{NCH}}{\|}}{}} \xrightarrow[h\nu]{O_2} \underset{\text{RC}-\text{COR}}{\overset{\overset{\textstyle O\quad O}{\|\quad\|}}{}}$$

JOC *50* 3573 (1985)

$$\underset{\text{R}_2\text{CHOCOCH}_2\text{CH}=\text{CH}_2}{\overset{\overset{\textstyle O}{\|}}{}} \xrightarrow[\text{CH}_3\text{CN}]{\text{cat Pd(OAc)}_2} \underset{\text{RCR}}{\overset{\overset{\textstyle O}{\|}}{}}$$

R$_2$CH = 1° allylic or benzylic, 2° alkyl

TL *25* 2791 (1984)

20. Amides

$$\underset{\text{ArCH}_2\text{NHCR}}{\overset{\overset{\textstyle O}{\|}}{}} \xrightarrow[(\text{PhCO}_2)_2]{\text{NBS}} \xrightarrow{\text{H}_3\text{O}^+} \underset{\text{ArCH}}{\overset{\overset{\textstyle O}{\|}}{}}$$

TL 3875 (1979)

$$\underset{\text{RCH}_2\text{CNR}_2}{\overset{\overset{\textstyle O}{\|}}{}} \xrightarrow[\text{or LDA/DMF-Me}_2\text{SO}_4]{(\text{Me}_2\text{N})_3\text{CH or MeOCH(NMe}_2)_2} \underset{\text{RCCONR}_2}{\overset{\overset{\textstyle \text{Me}_2\text{NCH}}{\|}}{}} \xrightarrow[h\nu]{O_2} \underset{\text{RC}-\text{CNR}_2}{\overset{\overset{\textstyle O\quad O}{\|\quad\|}}{}}$$

JOC *50* 3573 (1985)

21. Nitriles

$$\text{RR'CHCN} \longrightarrow \text{R}\overset{\overset{\displaystyle O}{\|}}{\text{C}}\text{R'}$$

NaOH, O$_2$, (Et$_3$NCH$_2$Ph)Cl	Syn 1009 (1980)
LDA/t-BuMe$_2$SiCl/I$_2$/Ag$_2$O (R = Ar)	JOC *39* 2799 (1974)
LDA/O$_2$/SnCl$_2$/NaOH	JOC *40* 267 (1975)
LDA/O$_2$/Na$_2$SO$_3$, buffer	TL *28* 2087 (1987)
base, O$_2$ (R = Ar)	Tetr *3* 97 (1958) TL 657 (1964)
K$_2$CO$_3$, air, DMSO (R = R' = Ar)	Org Prep Proc Int *2* 137 (1970) JOC *48* 4097 (1983)

LiN(i-Pr), air, DMSO   JOC *48* 4097 (1983)

LDA/PhSCl/NBS, H$_2$O	TL 3029 (1974)
PCl$_5$/NaOH	TL 437 (1967); 2301 (1973) JOC *33* 2211 (1968) JACS *109* 7270 (1987)

3. REDUCTION OF CARBOXYLIC ACIDS, ACID DERIVATIVES AND NITRILES TO ALDEHYDES

1. Carboxylic Acids

$$\underset{RCOH}{\overset{O}{\parallel}} \longrightarrow \underset{RCH}{\overset{O}{\parallel}}$$

Review: Org Rxs *8* 218 (1954)

Li, RNH$_2$ (R = Me, Et)/NH$_4$Cl/H$_3$O$^+$	JOC *28* 2918 (1963)
	JACS *92* 5774 (1970)
BH$_3$·SMe$_2$/PCC	Syn 704 (1979)
(CH$_3$)$_2$CHC(CH$_3$)$_2$BH$_2$	JOC *37* 2942 (1972)
(CH$_3$)$_2$CHC(CH$_3$)$_2$BHCl·SMe$_2$	JACS *106* 8001 (1984)
	JOC *51* 5264 (1986); *52* 5400 (1987)
(CH$_3$)$_2$CHC(CH$_3$)$_2$BHBr·SMe$_2$	JOC *52* 5030 (1987)
	TL *28* 2389 (1987)
⊂BH/LiH$_2$B⊃	TL *28* 4575 (1987)
⊂BH/*t*-BuLi/⊂BH	TL *28* 6231 (1987)
i-Bu$_2$AlH	J Gen Chem USSR *34* 1021 (1964)

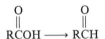

	CL 1447 (1974)
	JOC *49* 2279 (1984)

PhSiH$_2$ CH$_2$NMe$_2$ PhSiH$_2$ NMe$_2$

TL *28* 3941 (1987)

or *o*-PhSiH$_2$C$_6$H$_4$CH$_2$NMe$_2$; Δ

i-BuMgBr, Cp$_2$TiCl$_2$	Syn 871 (1981)
(Me$_2$N=CHCl)Cl/LiHAl(O-*t*-Bu)$_3$, cat CuI	TL *24* 1543 (1983)

o-HSC$_6$H$_4$OH, POCl$_3$,   J Heterocyclic Chem *11* 943 (1974)
  HClO$_4$/LiAlH$_4$/H$_2$O, HgCl$_2$

o-(NH$_2$)C$_6$H$_4$, PPA/NaOEt/MeI/NaBH$_4$ Syn 303 (1981)
  or LiAlH$_4$/H$_3$O$^+$

$$\underset{\text{RR'CHCOH}}{\overset{\text{O}}{\overset{\|}{}}} \xrightarrow[\text{2. HCO}_2\text{Et}]{\text{1. 2 LiNR}_2} \underset{\text{RR'CCHO}}{\overset{\text{CO}_2^-}{\overset{|}{}}} \xrightarrow{\text{H}_2\text{O}} \underset{\text{RR'CHCH}}{\overset{\text{O}}{\overset{\|}{}}}$$

TL 699 (1970)

$$\underset{\text{RCH}_2\text{C}-\text{COH}}{\overset{\text{O   O}}{\overset{\| \;\;\; \|}{}}} \xrightarrow[\text{2. H}_3\text{O}^+]{\text{1. O\hspace{-0.5em}\bigcirc\hspace{-0.5em}NH , cat TsOH}} \underset{\text{RCH}_2\text{CH}}{\overset{\text{O}}{\overset{\|}{}}}$$

TL *23* 459 (1982)

2. Acid Halides

$$\underset{\text{RCX}}{\overset{\text{O}}{\overset{\|}{}}} \longrightarrow \underset{\text{RCH}}{\overset{\text{O}}{\overset{\|}{}}}$$

H$_2$, cat Pd, BaSO$_4$ (Rosenmund)   Ber *51* 585 (1918); *54* 425 (1921)
                                            Org Rxs *4* 362 (1948)
                                            Rec Trav Chim *90* 1323 (1971); *100* 21 (1981)
                                            JACS *108* 2608 (1986)

H$_2$, cat Pd-C, 2,6-lutidine   Syn 767 (1976)
                                            JACS *108* 2608 (1986)

H$_2$, cat PdCl$_2$(PPh$_3$)$_2$, R'$_3$N (R = aryl)   JACS *96* 7761 (1974)

NaBH$_4$, DMF, −70°C   TL *22* 11 (1981)

NaBH$_4$, py, DMF, 0°C   Syn Commun *12* 839 (1982)

NaBH$_4$, CdCl$_2$ CC 354 (1978)
                                            JCS Perkin I 27 (1980)

(Ph$_3$P)$_2$CuBH$_4$ TL 1437, 2473 (1978); 975 (1979)
                                            JOC *45* 3449 (1980)

[(Ph$_3$P)$_2$CuBH$_3$CN]$_2$ TL *21* 813 (1980)

LiHAl(O-*t*-Bu)$_3$ JACS *79* 252 (1956); *80* 5377 (1958)
                                            J Lipid Res *8* 380 (1967)
                                            Syn 217 (1972)

Et$_3$SiH, cat Pd-C   JOC *34* 1977 (1969)

Et$_3$SiH, cat PtCl$_2$(PPh$_3$)$_2$ or   CC 1703 (1970)
  ClRh(CO)(EtPPh$_2$)$_2$ JCS Dalton 2460, 2646 (1975)

polymethylhydrosiloxane, cat RhCl$_3$,   J Mol Catal *37* 359 (1986)
  cat [(*n*-C$_8$H$_{17}$)$_3$NCH$_3$]Cl

n-Bu₃SnH

JOC *25* 284 (1960)
JACS *88* 571 (1966)

n-Bu₃SnH, cat Pd(PPh₃)₄

CC 432 (1980)
JOC *46* 4439 (1981)

Na₂Fe(CO)₄/HOAc

BCSJ *44* 2569 (1971)

(Me₄N)HFe(CO)₄

TL 781 (1977)

o-HOC₆H₄SH, HBF₄/NaBH₄/H₂O, HgCl₂

JCS Perkin I 323 (1976)

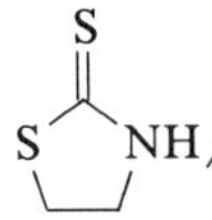

JCS Perkin I 2470 (1980)

JOC *51* 5400 (1986)

$$2 \text{ RCCl} \xrightarrow{\text{SmI}_2} \text{RC-CR}$$

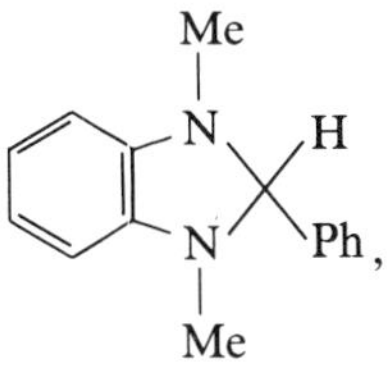

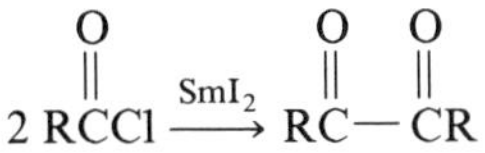

TL *22* 3959 (1981); *25* 2869 (1984)

3. Anhydrides

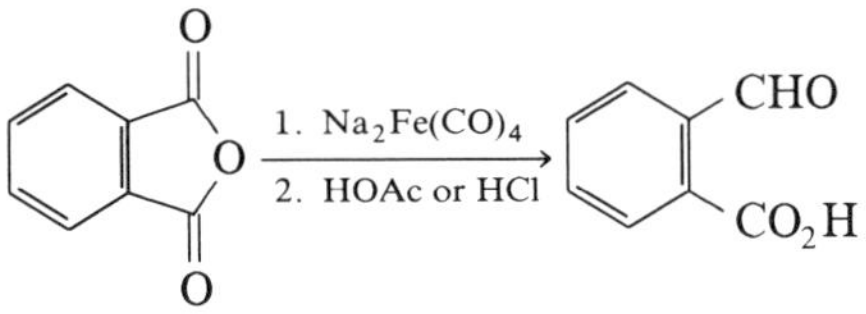

1. Na₂Fe(CO)₄
2. HOAc or HCl

TL 3535 (1973)

4. Esters

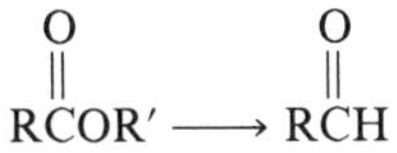

$$\text{RCOR}' \longrightarrow \text{RCH}$$

HAl$\left(-N \underset{2}{\bigcirc} X \right)$ (X = O, NMe)

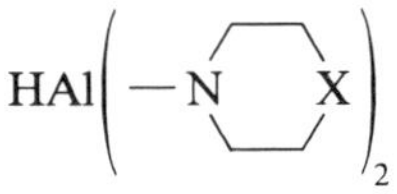

CL 215 (1975)

i-Bu₂AlH

TL 619 (1962); 1779 (1969); *27* 2103 (1986)
Bull Acad Sci USSR, Div Chem Sci 288 (1963)
JOC *40* 3495 (1975); *48* 2775 (1983);
  52 2361 (1987)
Syn 617 (1975) (review)
JACS *107* 199 (1985)

$$\underset{RCOR'}{\overset{O}{\overset{\|}{}}} \longrightarrow \underset{RCH}{\overset{O}{\overset{\|}{}}} \quad (\textit{Continued})$$

LiAlH$_4$, Et$_2$NH	JOC *52* 5486 (1987)
NaAlH$_4$	TL 2087 (1963)
NaH$_2$Al(i-Bu)$_2$	TL 619 (1962)
NaH$_2$Al(OCH$_2$CH$_2$OCH$_3$)$_2$	JACS *95* 7501 (1973)
NaH$_2$Al(OCH$_2$CH$_2$OCH$_3$)$_2$, HN⟨ ⟩NCH$_3$	Syn 526 (1976)
LiHAl(O-t-Bu)$_3$ (R' = C$_6$H$_5$)	JOC *31* 283 (1966)

$$\underset{RCOEt}{\overset{O}{\overset{\|}{}}} \xrightarrow[\substack{\text{3. 1,3-cyclohexadiene} \\ \text{4. Me}_3\text{SiCl}}]{\substack{\text{1. LiCHBr}_2 \\ \text{2. } n\text{-BuLi}}} RCH{=}CHOSiMe_3 \xrightarrow{H_3O^+} \underset{RCH_2CH}{\overset{O}{\overset{\|}{}}}$$

JACS *108* 1325 (1986)

5. Thioesters

$$\underset{RCSR'}{\overset{O}{\overset{\|}{}}} \longrightarrow \underset{RCH}{\overset{O}{\overset{\|}{}}}$$

Raney Ni	Org Rxs *8* 229 (1954)
i-Bu$_2$AlH $\left(R' = -\underset{N}{\overset{S}{\langle\ \ \rangle}} \right)$	CC 330 (1978)

6. Lactones

$$\underset{O}{\overset{O}{\underset{\|}{C}}}{=}O \longrightarrow \underset{O}{\overset{OH}{\underset{\|}{CH}}}$$

(Me$_2$CHCHMe)$_2$BH	JOC *51* 5032 (1986) TL *28* 1073 (1987)
i-Bu$_2$AlH	Helv *46* 2799 (1963) JOC *30* 3564 (1965); *50* 2489 (1985); *52* 4603, 4647, 4898, 5700 (1987) JACS *91* 5675 (1969); *97* 2287 (1975) Syn 617 (1975) (review) TL *27* 6341, 6345 (1986); *28* 221 (1987)
NaH$_2$Al(OCH$_2$CH$_2$OCH$_3$)$_2$, EtOH	Syn 526 (1976)

7. Amides

$$\underset{\overset{\parallel}{O}}{RCNR_2} \longrightarrow \underset{\overset{\parallel}{O}}{RCH}$$

Review: Org Rxs *8* 252 (1954)

Amide	Reagent	
general	LiH$_2$Al(OEt)$_2$	JACS *86* 1089 (1964)
	LiHAl(OEt)$_3$	JACS *86* 1089 (1964)
	HAl$\left(-N\underset{}{\frown}NCH_3\right)_2$	CL 875 (1975)
—N◁	LiAlH$_4$	JACS *83* 2016, 4549 (1961)
—N⟨imidazole⟩	LiHAl(O-*t*-Bu)$_3$	Ann *654* 119 (1962)
		JOC *35* 458 (1970)
	LiAlH$_4$	Org Rxs *8* 252 (1954)
		Angew *70* 165 (1958)
		Ann *654* 119 (1962)
		Angew Int *1* 351 (1962)
—N⟨dimethylpyrazole⟩	LiAlH$_4$	Angew *70* 165 (1958)
		Ann *622* 37 (1959)
		J Med Chem *20* 510 (1977)
		TL *23* 525 (1982); *28* 3123 (1987)
—N⟨dibenzazepine⟩	LiAlH$_4$	Ann *577* 11 (1952)
		JOC *18* 1190 (1953)
—N(CH$_3$)$_2$	(Me$_2$CHCHMe)$_2$BH	JACS *92* 7161 (1970)
	NaAlH$_4$	Tetr *25* 5555 (1969)
	LiAlH$_4$	Ber *84* 625 (1951)
	LiH$_2$Al(OEt)$_2$	JACS *81* 502 (1959); *86* 1089 (1964)
	LiH$_2$Al(OCH$_2$CH$_2$OCH$_3$)$_2$	JOC USSR *13* 1081 (1977)
	LiH$_2$Al$\left(OCH_2\underset{\overset{\parallel}{O}}{C}\underset{\overset{\parallel}{O}}{H}CH_2\right)_2$	JOC USSR *13* 1081 (1977)
	LiHAl(OEt)$_3$	JACS *86* 1089 (1964)

Amide	Reagent	
—N⟨ ⟩NMe	$NaH_2Al(OCH_2CH_2OCH_3)_2$,	JOC *51* 3566 (1986)
	EtOH or HN⟨ ⟩NCH_3?	
—$N(Me)CH_2CH_2NMe_2$	$NaH_2Al(OCH_2CH_2OCH_3)_2$,	JOC *51* 3566 (1986)
	EtOH or HN⟨ ⟩NCH_3?	
—NR_2 [R = Me, Et, $(CH_2)_5$]	$LiHAl(i\text{-}Bu)_2(n\text{-}Bu)$	JOC *49* 1717 (1984)
—NRPh	$LiAlH_4$	Angew *64* 458 (1952); *65* 525 (1953); *66* 174 (1954) Ber *88* 301 (1955)
	$i\text{-}Bu_2AlH$	Bull Acad Sci USSR, Div Chem Sci 2046 (1959)
—N(Me)OMe	$i\text{-}Bu_2AlH$ $LiAlH_4$	TL *22* 3815 (1981) TL *22* 3815 (1981); *28* 1857 (1987) JACS *107* 7790 (1985)
—N⟨S,S⟩ (thione)	$i\text{-}Bu_2AlH$	CL 1443 (1977) BCSJ *52* 555 (1979)
	$LiHAl(O\text{-}t\text{-}Bu)_3$	BCSJ *52* 555 (1979)

$$RCNHNHTs \xrightarrow{\text{base}} RCH \quad (\text{McFadyen-Stevens})$$

(with O double bonds on C of RCNHNHTs and RCH)

JCS 584 (1936)
Org Rxs *8* 232 (1954) (review)
JOC *26* 3664 (1961)
CC 793 (1978)

8. Nitriles

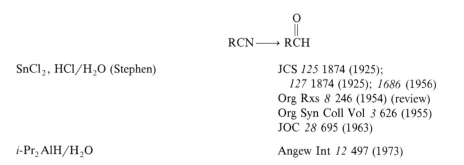

$$RCN \longrightarrow RCH \quad (\overset{O}{\overset{\|}{})}$$

$SnCl_2$, HCl/H_2O (Stephen)

    JCS *125* 1874 (1925); *127* 1874 (1925); *1686* (1956)
    Org Rxs *8* 246 (1954) (review)
    Org Syn Coll Vol *3* 626 (1955)
    JOC *28* 695 (1963)

$i\text{-}Pr_2AlH/H_2O$

    Angew Int *12* 497 (1973)

$i\text{-Bu}_2\text{AlH}/\text{H}_2\text{O}$

Proc Acad Sci USSR, Chem Sec *116* 879 (1958)
JOC *24* 627 (1959); *29* 3046 (1964); *35* 186, 858
    (1970); *37* 2138 (1972); *41* 3512 (1976);
    46 4600 (1981); *50* 2026, 2443 (1985);
    52 28, 4191 (1987)
Rec Trav Chim *85* 343 (1966)
Syn 617 (1975) (review)
JACS *107* 7524 (1985); *108* 4614 (1986)

$\text{LiHAl(OEt)}_3/\text{H}_2\text{O}$

TL (3) 9 (1959)
JACS *86* 1085 (1964)

$\text{LiH}_2\text{Al(OEt)}_2/\text{H}_2\text{O}$

JACS *86* 1085 (1964)
JOC *44* 4603 (1979)

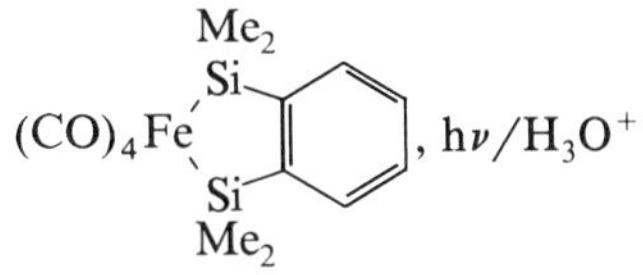

$(\text{CO})_4\text{Fe}$ 〔benzene ring with $\underset{\text{Si}}{\overset{\text{Me}_2}{\text{Si}}}$ and $\underset{\text{Me}_2}{\overset{\text{Si}}{\text{Si}}}$〕, $h\nu/\text{H}_3\text{O}^+$

JOC *46* 3372 (1981)

$(\text{Et}_3\text{O})\text{BF}_4/\text{Et}_3\text{SiH}/\text{H}_2\text{O}$

CC 45 (1974)
JOC *46* 602 (1981)

$i\text{-PrCl}, \text{FeCl}_3/\text{Et}_3\text{SiH}/\text{H}_2\text{O}$

JOC *46* 602 (1981)

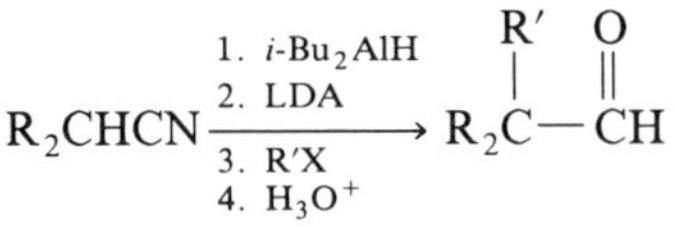

$$\text{R}_2\text{CHCN} \xrightarrow[\substack{3.\ \text{R'X}\\ 4.\ \text{H}_3\text{O}^+}]{\substack{1.\ i\text{-Bu}_2\text{AlH}\\ 2.\ \text{LDA}}} \underset{\text{R}_2\text{C}-\text{CH}}{\overset{\text{R'}\quad\text{O}}{|\qquad||}}$$

JOC *46* 5250 (1981)

4. REARRANGEMENTS

1. Isomerization

$$C=C-\underset{\underset{H}{|}}{\overset{\overset{OH}{|}}{C}} \longrightarrow \underset{\underset{H}{|}}{C}-\underset{\underset{H}{|}}{C}-\overset{\overset{O}{||}}{C}$$

cat Mo(N$_2$)$_2$(dppe)$_2$ JOMC *252* 105 (1983)

cat H$_2$Rh(PPh$_3$)$_4$ JOMC *297* C37 (1985)

cat RuHCl(PPh$_3$)$_3$ TL 4133 (1974)

cat [ClRh(CO)$_2$]$_2$, NaOH (phase transfer)   JOC *45* 2269 (1980)

cat Rh(ClO$_4$)(BINAP)(CH$_3$OH)$_2$ TL *28* 4719 (1987)
   (enantioselective)

JACS *106* 5208 (1984)

2. Epoxides

$$\underset{>}{C}-\underset{<}{C} \xrightarrow{\hspace{1cm}} -\underset{|}{\overset{O}{\underset{||}{C}}}-\underset{|}{C}-$$

Review: Chem Rev 59 737 (1959)

electrogenerated H$^+$	TL 24 2857 (1983)
	JOC 50 3160 (1985)
LiBr·HMPA, benzene	JACS 93 1693 (1971)
LiBr or LiClO$_4$, n-Bu$_3$PO	JACS 90 4193 (1968)
LiI	CC 1200 (1982)
NaI, CH$_3$I, DMF	Acta Chem Scand 18 1551 (1964)
NaI, n-PrI, DMSO	CC 227 (1968)
MgBr$_2$	Helv 31 1077 (1948)
	JACS 76 4402 (1954); 79 6283 (1957)
	TL 23 4389 (1982)
BeCl$_2$/AgBF$_4$	TL 23 4385 (1982)
BF$_3$·OEt$_2$	JACS 76 1235 (1954); 84 867 (1962);
	105 7352 (1983);
	108 3443, 4586 (1986)
	Org Syn Coll Vol 4 375 (1963)
	TL 23 4389 (1982)
AlCl$_3$	JACS 80 1744 (1958)
cat Co$_2$(CO)$_8$	JOC 27 2706 (1962)
cat ClRh(PPh$_3$)$_3$	JOC 42 2299 (1977)
NiBr$_2$(PPh$_3$)$_2$	CL 1323 (1986)
ZnBr$_2$	JOC 30 4271 (1965)

$$-\underset{|}{C}-\underset{|}{C}-\overset{O}{\overset{||}{C}}- \xrightarrow[\text{(Ph}_2\text{PCH}_2)_2]{\text{cat Pd(PPh}_3)_4} -\overset{O}{\overset{||}{C}}-\underset{|}{C}-\overset{O}{\overset{||}{C}}-$$

JACS 102 2095 (1980)

$$RCH_2CH \overset{O}{\underset{}{\diagdown\diagup}} CR'SiMe_3 \xrightarrow{Pd(OAc)_2} RCH=CHCR'\overset{O}{\overset{||}{}}$$

CL 1997 (1982)

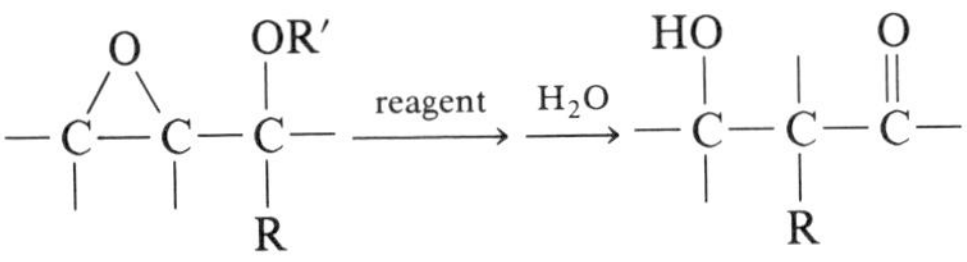

$$-\overset{|}{\underset{|}{C}}\overset{O}{\diagup\diagdown}\overset{OR'}{\underset{|}{C}}-\overset{OR'}{\underset{R}{C}}-\xrightarrow[\text{reagent}]{}\xrightarrow[]{H_2O}-\overset{|}{\underset{|}{C}}-\overset{HO}{\underset{R}{C}}-\overset{O}{\overset{\|}{C}}-$$

R'	Reagent	
H	$BF_3 \cdot OEt_2$	JACS *108* 3827 (1986)
		TL *27* 6233, 6237 (1986);
		28 5891 (1987)
		CL 113 (1987)
$SiMe_3$	$TiCl_4$	JACS *108* 3827 (1986)
	$(i\text{-PrO})_2TiCl_2$	TL *27* 6233, 6237 (1986)
	cat Me_3SiX (X = I, OTf)	TL *28* 3515 (1987)

$$RCH=CHCCH_3 \xrightarrow{Me_2S=CH_2} RCH=CH\overset{O}{\overset{\diagup\diagdown}{C}}-CH_2 \xrightarrow{MgBr_2} RCH=CHCH\overset{O}{\overset{\|}{C}}H \xrightarrow{KOH} RCH_2CH=C\overset{O}{\overset{\|}{C}}H$$
$$\underset{CH_3}{} \qquad \underset{CH_3}{} \qquad \underset{CH_3}{}$$

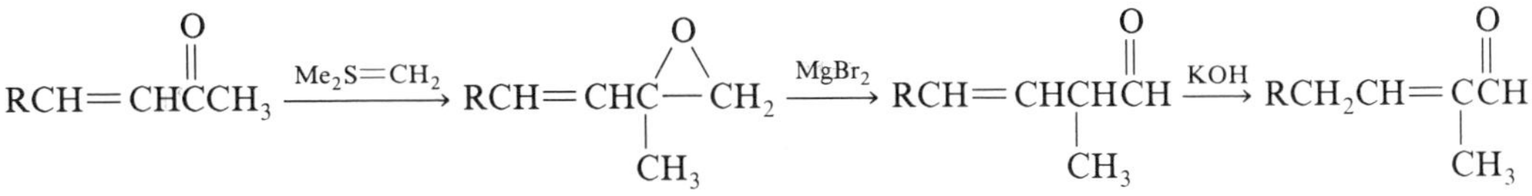

Helv *63* 1665 (1980)

3. Pinacol and Related Rearrangements

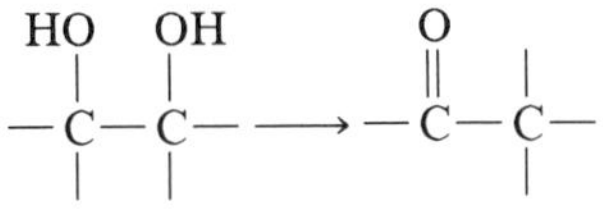

$$-\overset{HO}{\underset{|}{C}}-\overset{OH}{\underset{|}{C}}- \longrightarrow -\overset{O}{\overset{\|}{C}}-\overset{|}{\underset{|}{C}}-$$

Reviews:

Org Syn Coll Vol *1* 462 (1941)
Quart Rev *14* 357 (1960)
"Molecular Rearrangements," Interscience, New York (1963), Vol 1, Chpt 1

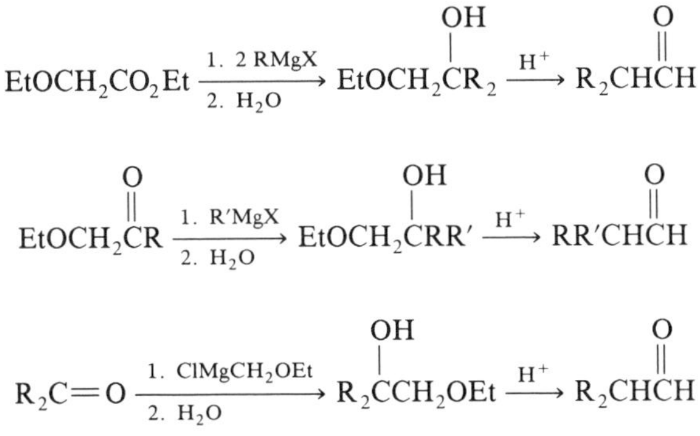

$$EtOCH_2CO_2Et \xrightarrow[2.\ H_2O]{1.\ 2\ RMgX} EtOCH_2\overset{OH}{\underset{|}{C}}R_2 \xrightarrow{H^+} R_2CH\overset{O}{\overset{\|}{C}}H$$

$$EtOCH_2\overset{O}{\overset{\|}{C}}R \xrightarrow[2.\ H_2O]{1.\ R'MgX} EtOCH_2\overset{OH}{\underset{|}{C}}RR' \xrightarrow{H^+} RR'CH\overset{O}{\overset{\|}{C}}H$$

$$R_2C=O \xrightarrow[2.\ H_2O]{1.\ ClMgCH_2OEt} R_2\overset{OH}{\underset{|}{C}}CH_2OEt \xrightarrow{H^+} R_2CH\overset{O}{\overset{\|}{C}}H$$

JACS *61* 2134 (1939); *68* 2339 (1946)
BSCF 459 (1959)

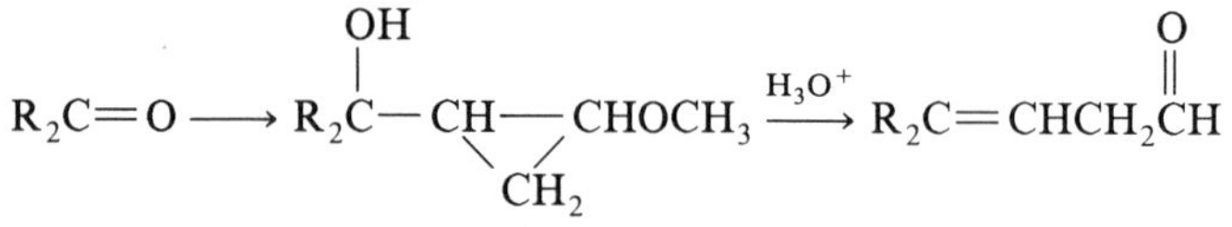

$$R_2C{=}O \longrightarrow R_2\overset{\overset{\displaystyle OH}{|}}{C}{-}CH{-}CHOCH_3 \xrightarrow{H_3O^+} R_2C{=}CHCH_2\overset{\overset{\displaystyle O}{||}}{CH}$$

with CH₂ bridging the CH—CHOCH₃

TL 3685 (1975)

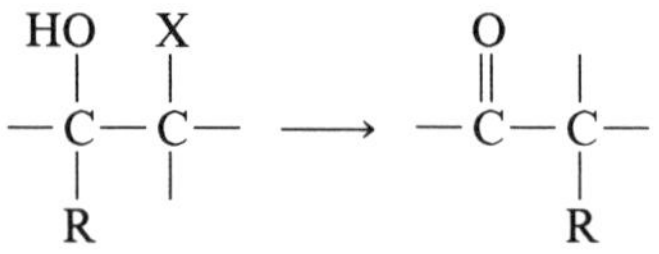

$$\underset{R}{\overset{HO \quad X}{-\overset{|}{C}-\overset{|}{C}-}} \longrightarrow \underset{R}{\overset{O}{-\overset{||}{C}-\overset{|}{C}-}}$$

X = Cl, OTs

JACS *83* 1251 (1961)
CC 565 (1968)
TL 5553 (1968)
JOC *33* 453 (1968); *37* 4090 (1972)
JCS Perkin I 225 (1987)
See also page 787, Section 4.

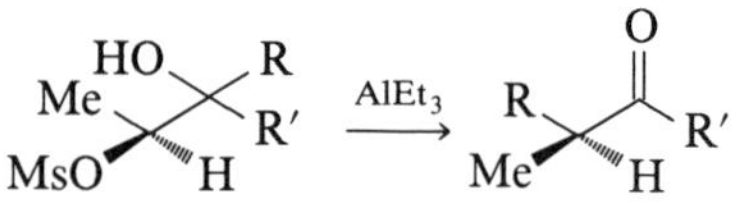

$$\underset{MsO}{\overset{Me}{}}\overset{HO \quad R}{\underset{ \quad H}{\bignotation}}R' \xrightarrow{AlEt_3} \underset{Me}{\overset{R}{}}\overset{O}{\underset{H}{\bignotation}}R'$$

R = aryl, vinylic

TL *24* 4997 (1983); *25* 1817 (1984)

4. Ring Opening

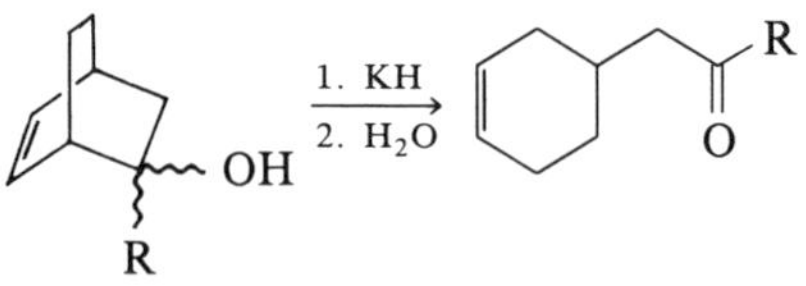

$$\xrightarrow[\text{2. } H_2O]{\text{1. } KH}$$

Helv *64* 2193 (1981)

5. Ring Contraction

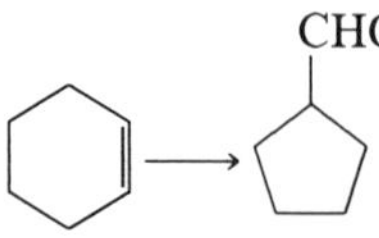

Hg(ClO₄)₂, H₂O   JACS *95* 2591 (1973)

Tl(ClO₄)₃, H₂O   JACS *95* 2591 (1973)

Tl(NO₃)₃, CH₃OH/H₃O⁺   TL 5275 (1970); 4753 (1971)

6. Ring Expansion

See page 733, Section 12 for related reactions.

Review: "Carbocyclic Ring Expansion Reactions," Academic Press, New York (1968)

TL 5327 (1967)
JOC *33* 453 (1968)

Org Rxs *11* 157 (1960) (review)
Org Syn Coll Vol *4* 221 (1963)
JCS 2513 (1965)
"Carbocyclic Ring Expansion Reactions," p 74 (review)
JOC *33* 2069 (1968); *39* 914 (1974); *43* 1050 (1978); *45* 185 (1980); *47* 2685 (1982); *52* 2602 (1987)
Angew Int *10* 491 (1971)
Tetr *29* 1941 (1973)
TL 4929 (1973)
Can J Chem *57* 1557 (1979)

$$\underset{\text{RCH}}{\overset{O}{\|}} \xrightarrow{CH_2N_2} \underset{\text{RCH}_2\text{CH}}{\overset{O}{\|}} + \underset{\text{RCCH}_3}{\overset{O}{\|}} + \text{RCH}\underset{\overset{\diagup O\diagdown}{}}{}\text{CH}_2$$

Newer Methods Prep Org Chem *1* 513 (1948) (review)
Org Rxs *8* 364 (1954) (review)
JACS *82* 4099 (1960); *84* 989 (1962) (enones)
Org Syn Coll Vol *4* 225 (1963)
Can J Chem *46* 1913 (1968); *57* 1557 (1979)
"Carbocyclic Ring Expansion Reactions," Chpt 4 (review)
JOC *33* 4090 (1968)
TL 393 (1971)
Tetr *29* 1941 (1973)
JCS Perkin I 477 (1982)

JACS *101* 4003 (1979); *105* 2435 (1983)
JOC *45* 2036 (1980); *48* 4763 (1983); *50* 3957 (1985)

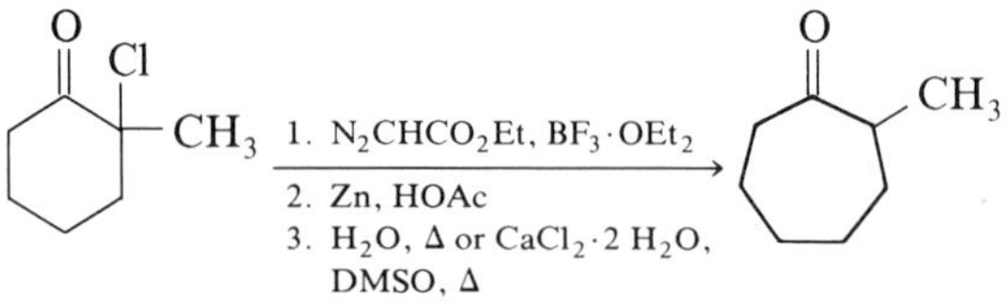

JOC *48* 2590 (1983)

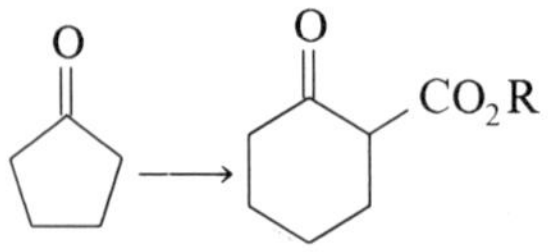

N$_2$CHCO$_2$Et, (Et$_3$O)BF$_4$

JACS *92* 5767 (1970)
JOC *42* 459, 466 (1977)
Can J Chem *57* 1557 (1979)

N$_2$CHCO$_2$R, BF$_3$·OEt$_2$

Can J Chem *42* 1333 (1964)
TL 4937 (1973)
Syn Commun *5* 125 (1975); *8* 413 (1978)
JOC *52* 3455 (1987)

N$_2$CHCO$_2$R, SbCl$_5$

JOC *42* 459 (1977); *50* 3957 (1985)

N$_2$CHCO$_2$R, base/Δ, hν, H$^+$ or
    transition metal catalyst

See page 642, Section 7.16.

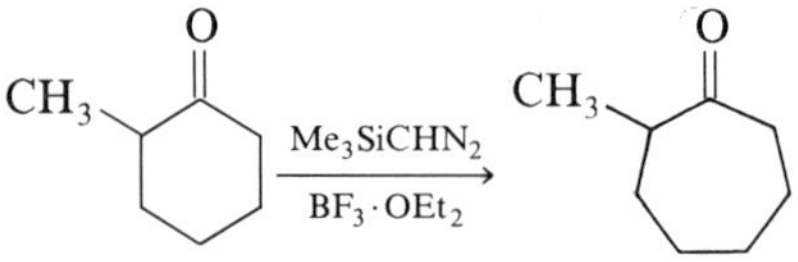

TL *21* 4619 (1980)

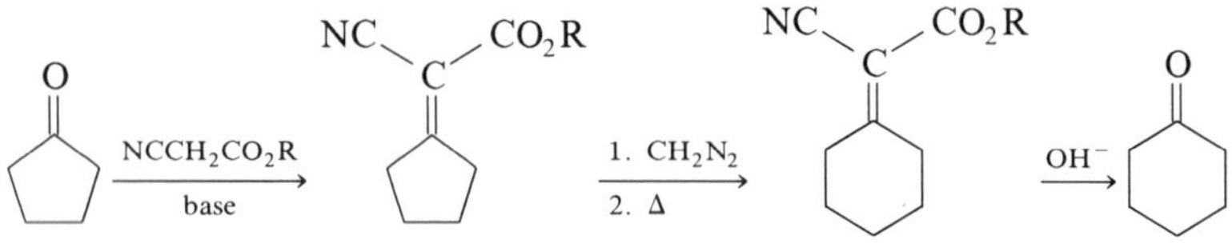

J Chem Res (S) 436 (1978)

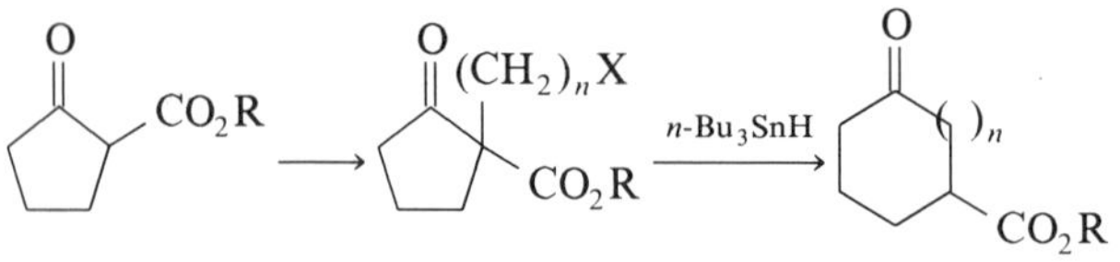

n = 1, 3, 4

JCS Perkin I 1139 (1986)
JACS *109* 3493, 6548 (1987)
CC 666 (1987)

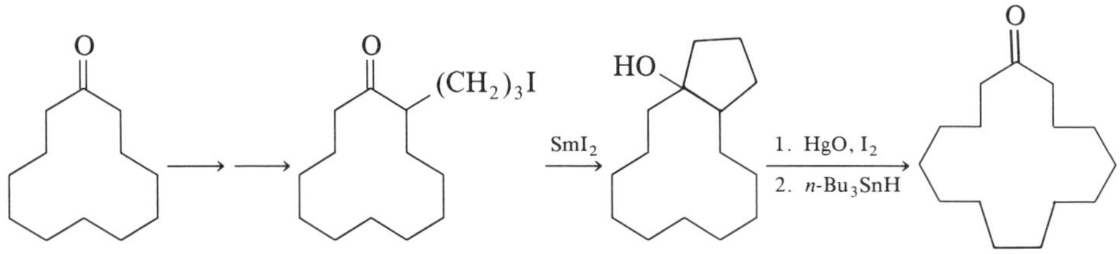

TL *28* 3963 (1987)

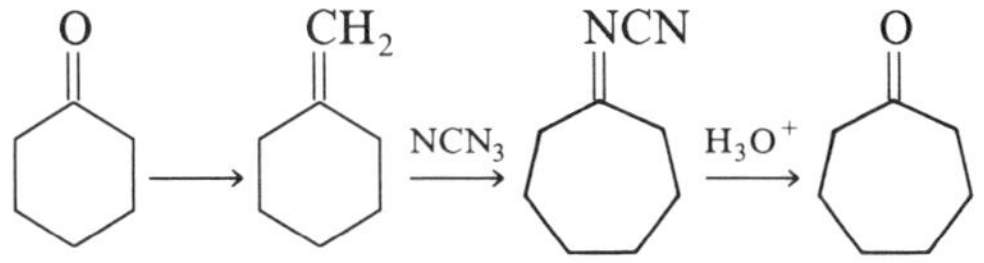

JACS *86* 4506 (1964); *91* 3676 (1969) (diene)
JOC *38* 2821 (1973)
TL *28* 3209 (1987)

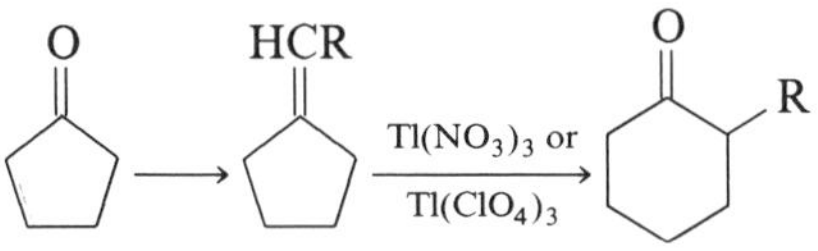

JACS *95* 2591 (1973)
JOC *38* 3455 (1973)
TL 1827 (1977)

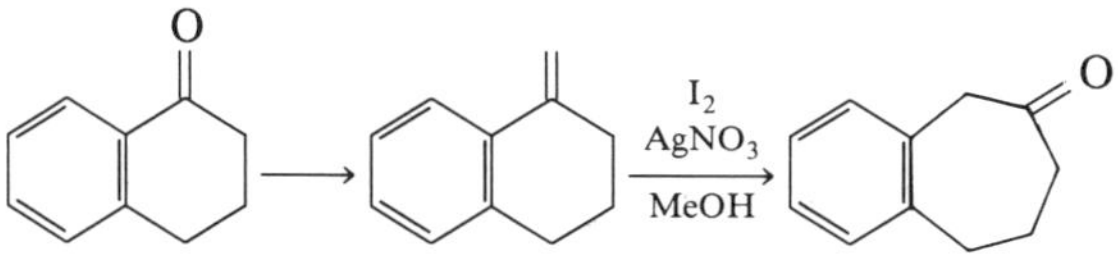

TL *27* 3783 (1986)

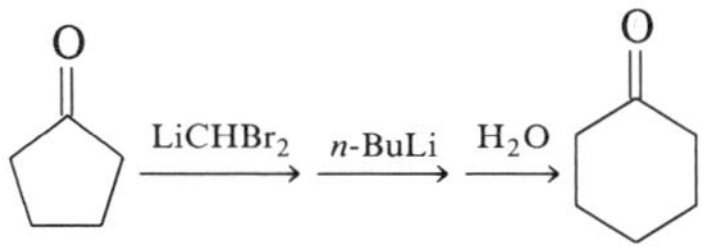

JOMC *40* Cl (1972); *97* 325 (1975)
Compt Rend *276* 433 (1973)
JACS *96* 6510 (1974); *106* 3030 (1984)
TL 2617 (1976); *28* 585 (1987)
BCSJ *50* 1592 (1977)

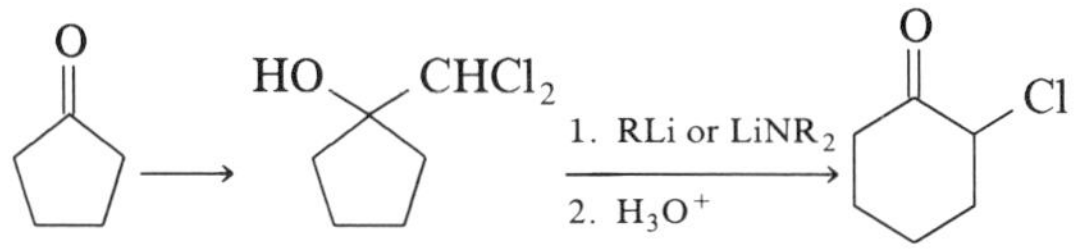

Ber *106* 2626 (1973)
BCSJ *50* 1592 (1977)

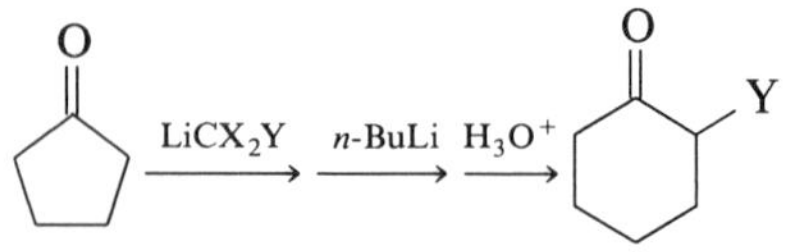

X = Cl, Br
Y = Cl, alkyl, aryl

JOMC *40* Cl (1972); *97* 325 (1975)
Compt Rend C *276* 433 (1973)
Syn 968 (1979)

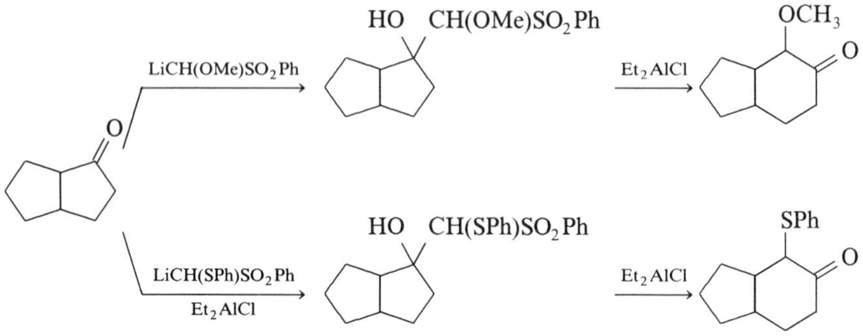

JACS *109* 4124 (1987)

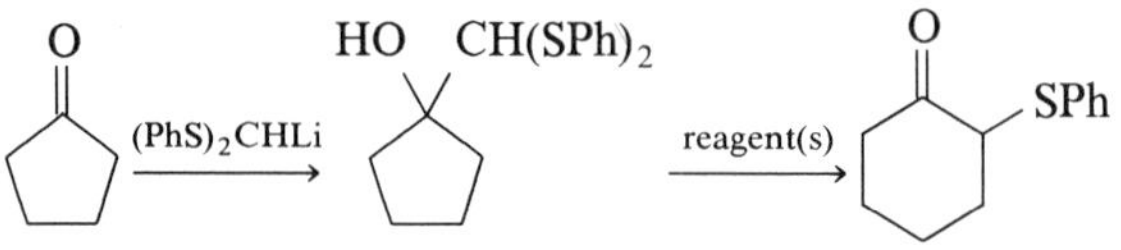

Reagent(s)

CuO$_3$SCF$_3$ JACS *97* 4749 (1975); *107* 4339 (1985)

2 RLi/H$_2$O   TL *28* 2203 (1987)

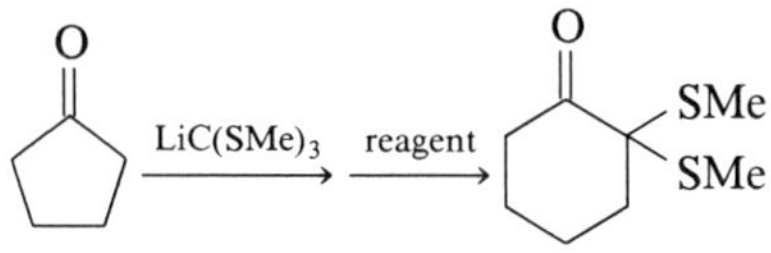

Reagent

HgCl$_2$ TL *21* 4301 (1980)

CuBF$_4$·4 CH$_3$CN   JOC *49* 608 (1984)

CuClO$_4$·4 CH$_3$CN   JOC *49* 608 (1984); *50* 1573 (1985)

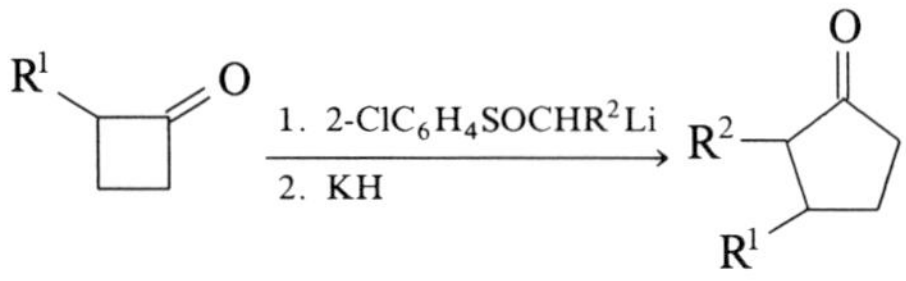

R^1 = aryl, vinylic
JOC *52* 774 (1987)

$$\underset{\substack{\text{R}^1\text{S} \quad \text{O} \\ \text{CR}^2}}{} \xrightarrow[\text{or FeCl}_3]{\text{AlCl}_3, \text{AlBr}_3} \underset{\substack{\text{SR}^1 \\ \text{R}^2}}{\text{O}}$$

TL *24* 79 (1983)

$$\underset{\text{R}^1}{} \xrightarrow[\text{2. }\Delta]{\substack{\text{O} \quad \text{Li} \\ \text{1. PhSe}-\text{CR}^2_2 \quad \text{R}^2}} \underset{\substack{\text{R}^2 \\ \text{R}^1}}{\text{O}}$$

R^1 = alkyl

JOC *48* 2098 (1983); *52* 774 (1987)

$$\underset{\text{R}^1\text{CR}^2}{\overset{\text{O}}{\parallel}} \nearrow \underset{\text{R}^2}{\overset{\text{R}^1}{}} \xrightarrow[\substack{\text{2. MeI, AgBF}_4 \\ \text{3. KOH}}]{\text{1. PhSeCH}_2\text{Li}} \underset{\text{R}^2}{\overset{\text{R}^1}{}} \xrightarrow[\Delta]{\text{LiI}} \underset{\substack{\text{R}^1 \\ \text{R}^2}}{\text{O}}$$

$$\searrow \underset{\substack{\text{MeSe} \quad \text{OH} \\ \text{CR}^1\text{R}^2}}{} \xrightarrow[\text{2. KOH}]{\text{1. MeI}} \underset{\substack{\text{R}^1 \\ \text{R}^2}}{\text{O}} \xrightarrow{\text{LiI}} \underset{\substack{\text{R}^1 \\ \text{R}^2}}{\text{O}}$$

CC 1200 (1982)

$$\underset{}{\overset{\text{O}}{\bigcirc}} \xrightarrow[\text{2. H}_2\text{O}]{\text{1. R}_2\bar{\text{C}}\text{SeR}'} \underset{\substack{\text{HO} \quad \text{CR}_2\text{SeR}' \\ }}{\bigcirc} \xrightarrow{\text{reagent(s)}} \underset{\substack{\text{R} \\ \text{R}}}{\overset{\text{O}}{}}$$

Reagent(s)

MeOSO$_3$F	TL *23* 4385 (1982) CC 564 (1982)
AgBF$_4$	TL *23* 983, 4385 (1982); *28* 1545 (1987)
TlOEt, HCCl$_3$	TL *23* 4385 (1982); *25* 2713, 5043 (1984); *28* 1545, 1549 (1987) JOC *50* 5200 (1985) (enones) CC 702 (1986)
KOH, (PhCH$_2$NEt$_3$)Cl, CH$_2$Cl$_2$	TL *28* 1545, 1549 (1987)

$$\underset{\text{O}}{} \longrightarrow \underset{\text{OH}}{} \xrightarrow{\text{KO-}t\text{-Bu}} \underset{\text{O}}{} + \underset{\text{O}}{}-\text{CH}_3$$

Can J Chem *58* 2730 (1980)

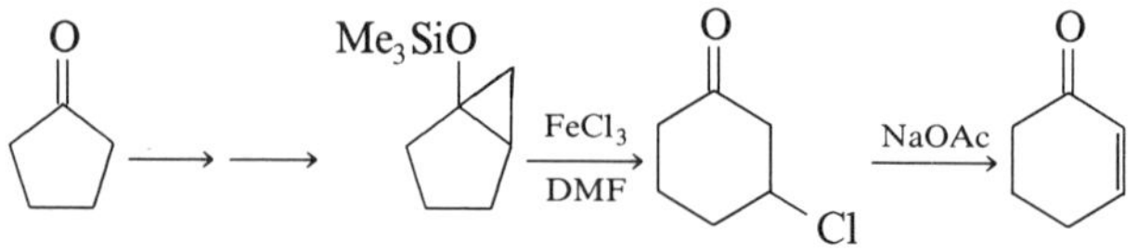

JOC *41* 2073 (1976); *44* 4481 (1979); *51* 3405 (1986)
Org Syn *59* 113 (1980)

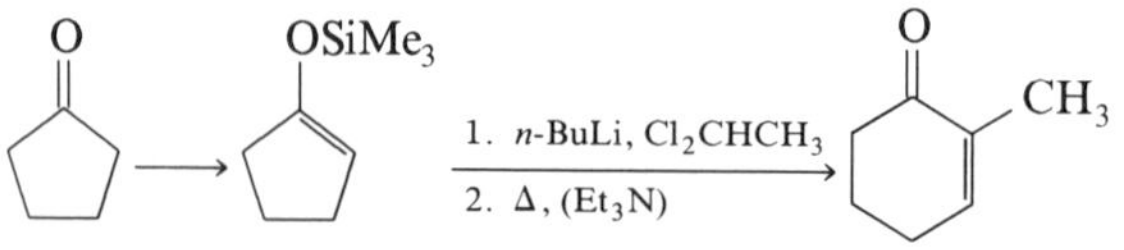

Syn 289, 291 (1981)
JACS *109* 3025 (1987)

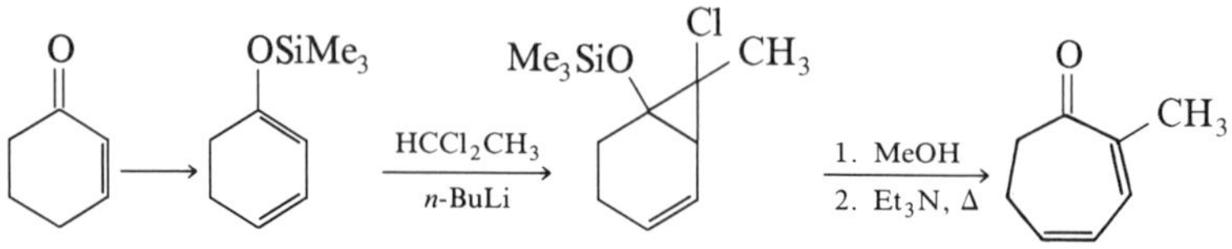

TL *22* 645 (1981)

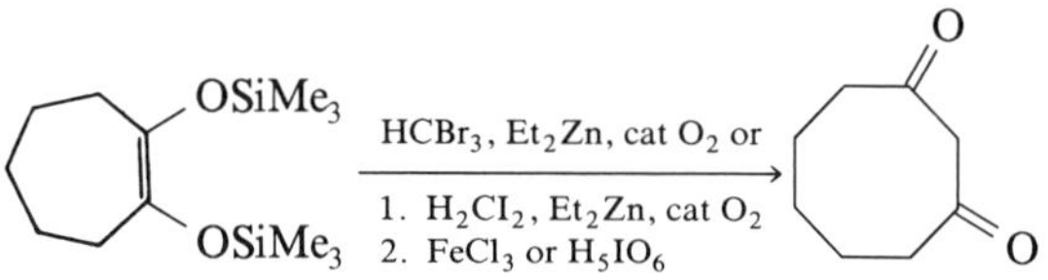

JOC *52* 3603 (1987)

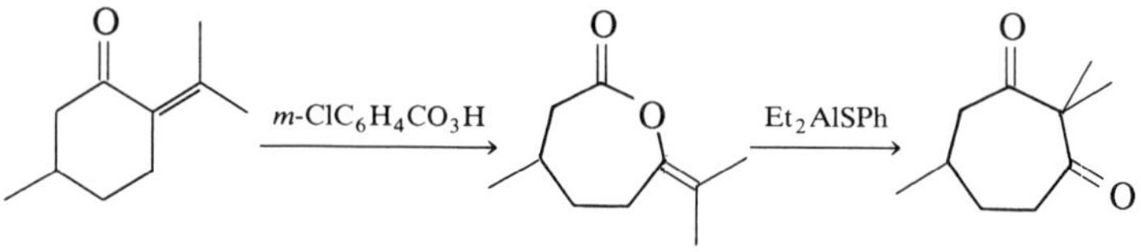

TL *23* 99 (1982)

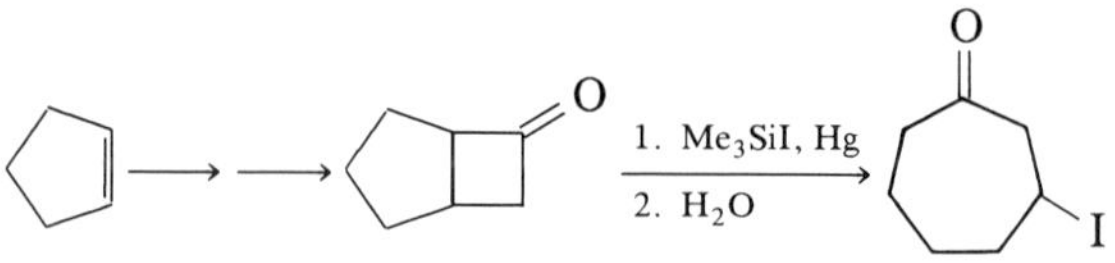

TL *21* 2639 (1980)

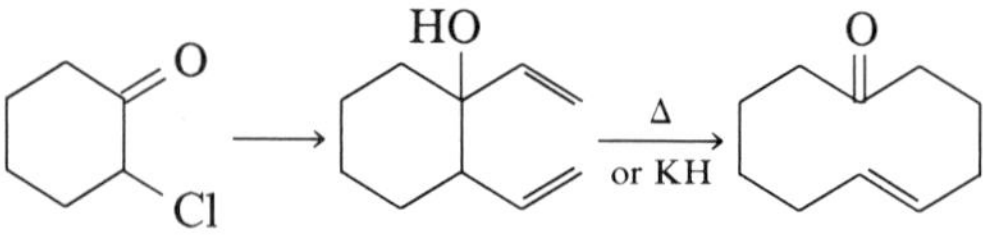

TL 509 (1970)
CL 667 (1973)
JACS *101* 2493 (1979); *105* 625 (1983)
BCSJ *53* 2958 (1980); *58* 146 (1985)
JOC *51* 1124 (1986)

See page 639, Sections 7.7 and 7.8.

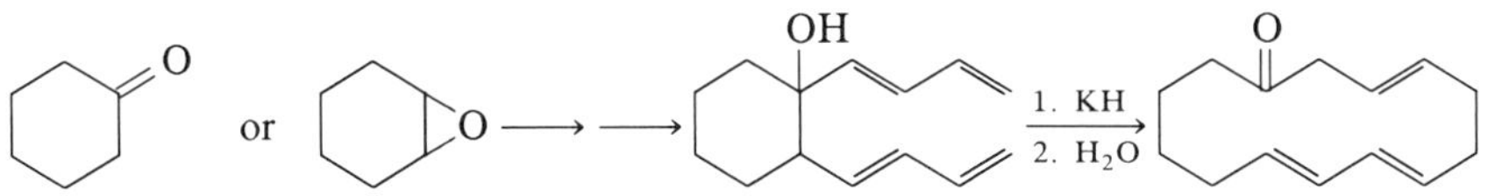

TL *22* 2471 (1981)
Tetr *37* 3967 (1981)
JACS *107* 7771 (1985)

7. Electrocyclic Rearrangements

7.1. Claisen

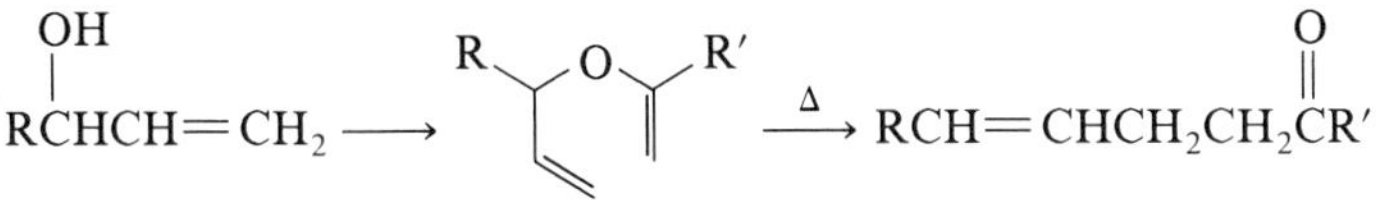

Reviews:

Chem Rev *27* 495 (1940); *84* 205 (1984) (catalysis)
Org Rxs *2* 1 (1944); *22* 1 (1975)
Quart Rev *22* 391 (1968)
Chem in Britain *5* 111 (1969)
Acct Chem Res *10* 227 (1977)
Syn 589 (1977)

Key articles and recent references:

JACS *83* 198 (1961); *89* 4559 (1967); *92* 3126 (1970); *93* 5805, 5813 (1971); *103* 6983, 6984 (1981); *104* 4972 (1982) (carbanion acceleration); *106* 6868 (1984); *107* 2730, 5572 (anionic oxy-Claisen), 7352 (1985); *109* 1160 (alkoxy substitution), 1170 (substituent effects), 3987 (1987)
TL 493 (1961); 2145, 3243, 3253 (1969); *22* 3985 (1981) (organoaluminum promotion); *23* 3143 (1982); *24* 3, 4397 (1983); *25* 1543, 4579 (1984); *26* 3655 (1985); *27* 6267 ($PdCl_2$ catalyzed), 6311 (1986); *28* 3065, 5879 (catalyst diastereoselectivity) (1987)
Helv *50* 2091, 2095 (1967); *61* 3075 (1978)
CC 264 (1969)
Org Syn Coll Vol *5* 25 (1973)
JOC *48* 1829, 3876 (1983); *49* 2347 (1984); *52* 5190, 5742 (carbanion acceleration) (1987)

Heteroatom systems:

Fortschr Chem Forsch *16* 75 (1970)

7.2. Amino-Claisen

$$R_2CO + MeNHCH_2C(R'){=}CH_2 + (EtO)_2POCHN_2 \xrightarrow{\text{KO-}t\text{-Bu}}$$

$$R_2C{=}CHNMeCH_2C(R'){=}CH_2 \xrightarrow[\text{2. }\Delta \text{; 3. } H_2O]{\text{1. } Me_2SO_4} H_2C{=}C(R')CH_2CR_2CH{=}O$$

TL *25* 2303 (1984)

7.3. Thio-Claisen

Review: Phosphorus and Sulfur *7* 69 (1979)

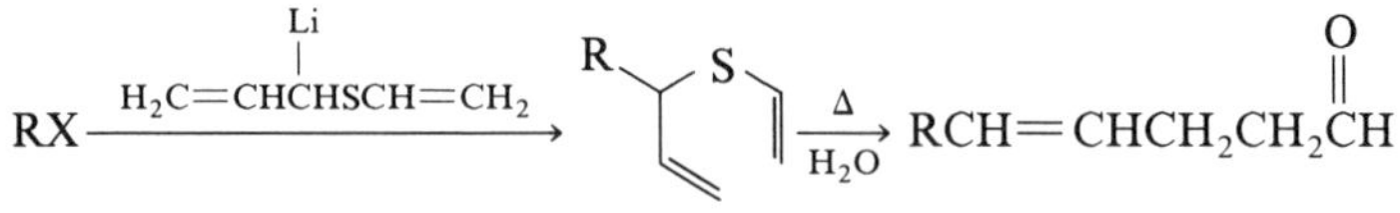

$$RX \xrightarrow{\underset{H_2C=CHCHSCH=CH_2}{Li}} \quad \xrightarrow[H_2O]{\Delta} RCH=CHCH_2CH_2\overset{O}{\overset{\|}{C}}H$$

JACS *95* 2693 (1973)

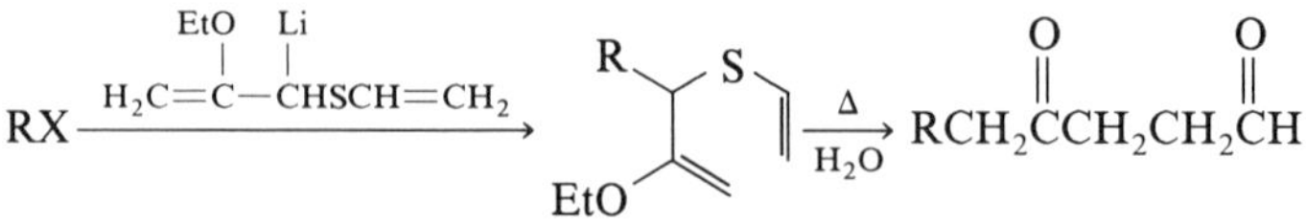

$$RX \xrightarrow{\underset{H_2C=C-CHSCH=CH_2}{EtO \ Li}} \quad \xrightarrow[H_2O]{\Delta} RCH_2\overset{O}{\overset{\|}{C}}CH_2CH_2\overset{O}{\overset{\|}{C}}H$$

JACS *95* 4446 (1973)

$$\text{(cyclohexanone)} \xrightarrow{\underset{H_2C=CHCH_2SCHP(OEt)_2}{Li \ O}} \quad \xrightarrow[HgO]{\Delta} \text{(ring)} \begin{matrix} CHO \\ CH_2CH=CH_2 \end{matrix}$$

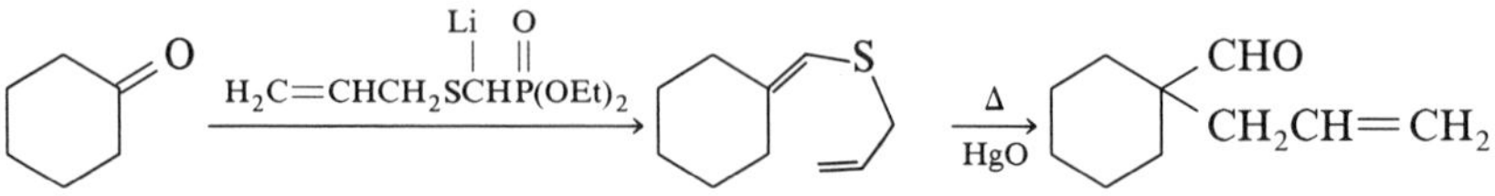

JACS *92* 5522 (1970)

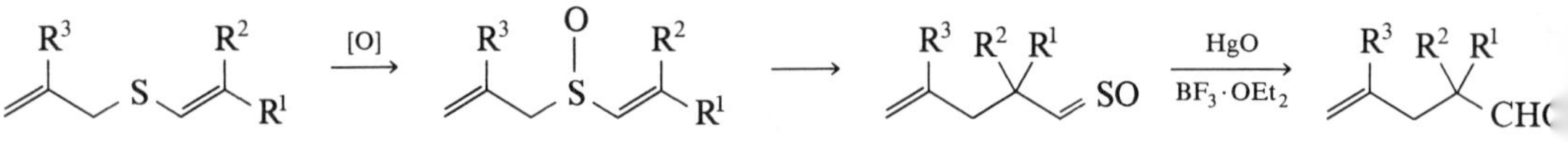

JACS *107* 6731 (1985)

7.4. Claisen–Cope

CC 947 (1967); 1657 (1968); 248 (1972)
JACS *91* 3281 (1969)
Helv *53* 1145 (1970)
JCS C 220 (1970)
JCS Perkin I 2634, 2738, 2741 (1973)
Syn 532 (1978)
TL *23* 3839 (1982)

7.5. Cope–Claisen

JACS *101* 1611 (1979); *102* 880, 6576 (1980); *103* 1853 (1981); *104* 7181 (1982)
JOC *47* 5229 (1982); *49* 3278 (1984)

7.6. 2,3-Wittig-oxy-Cope

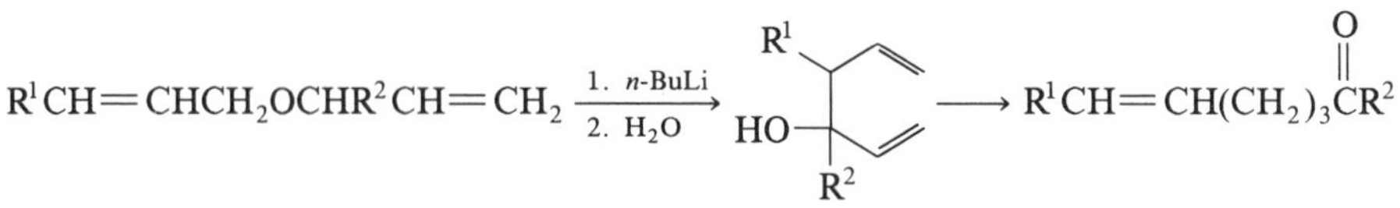

$$R^1CH=CHCH_2OCHR^2CH=CH_2 \xrightarrow[2. \ H_2O]{1. \ n\text{-}BuLi} \quad \xrightarrow{} R^1CH=CH(CH_2)_3\overset{O}{\overset{\|}{C}}R^2$$

CL 1349, 1643 (1982)
TL *23* 3931 (1982)

7.7. 1,3-oxy-Cope

$$C=C-\underset{\underset{R}{|}}{\overset{\overset{OX}{|}}{C}}- \longrightarrow \ C-C=\underset{\underset{R}{|}}{\overset{\overset{OX}{|}}{C}}- \ \xrightarrow{H_2O} \ R-\underset{\underset{H}{|}}{C}-C-\overset{\overset{O}{\|}}{C}-$$

R	X	
benzylic	K	CC 846 (1976) JOC *43* 1050 (1978); *45* 185 (1980); *47* 798 (1982) JACS *103* 7661 (1981); *104* 4411 (1982)
	SiMe₃	CC 846 (1976) JOC *47* 798 (1982)
allylic	H	JACS *86* 5017, 5019 (1964); *94* 7074 (1972); *104* 4411 (1982)
	K	CC 846 (1976) JOC *43* 1050 (1978); *52* 3798 (1987) JACS *104* 891 (1982)
	SiMe₃	JACS *94* 7074 (1972); *96* 200 (1974) CC 846 (1976) JOC *41* 1233 (1976); *42* 280 (1977); *52* 3798 (1987)

S–S ring (with R), K — JOC *43* 4903 (1978)

7.8. 3,3-oxy-Cope

$$\text{XO}\diagdown\overset{R}{\diagup} \ \longrightarrow \ \left[\text{XO}\diagdown\overset{R}{\diagup} \right] \ \longrightarrow \ RCH=CH(CH_2)_3\overset{\overset{O}{\|}}{CH}$$

X	

H	JACS *86* 5017, 5019 (1964); *87* 1150 (1965); *89* 3462
	    (1967); *90* 4729, 4730, 4732, 6141 (enynol) (1968);
	    92 2404 (1970) (enynol); *94* 4779, 7074 (1972);
	    95 5281 (1973); *102* 3972 (1980); *106* 3869 (1984)
	    (enynol); *108* 6343 (1986)
	TL 6115 (1966); 1337, 1341, 2681 (1969); 509 (1970);
	    23 4263 (1982) [Hg(O₂CCF₃)₂ promoted]
	CC 143 (1968) (enynol); 491 (1974)
	BSCF 1491 (1968)
	JOC *35* 856 (1970); *46* 5447 (1981); *47* 4815 (1982)
	CL 667 (1973)
	BCSJ *53* 2958 (1980)

alkali metal	JACS *97* 4765 (1975); *99* 4186 (1977); *100* 2242,
	    4309 (1978); *101* 2493 (1979); *102* 774, 2463, 3972
	    (1980); *103* 6235 (1981); *105* 625 (1983); *106*
	    4038, 7614 (1984); *108* 7873 (1986)

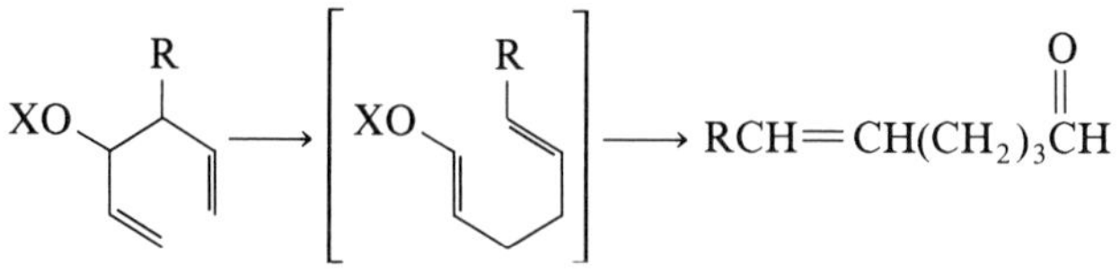

$$\longrightarrow RCH{=}CH(CH_2)_3\overset{\displaystyle O}{\overset{\|}{C}}H \quad (\textit{Continued})$$

X

Angew Int *15* 437 (1976)
Ber *112* 1420 (1979)
JOC *46* 9, 2199, 4272, 5447 (1981); *47* 2268, 3190
   (1982); *50* 5747 (1985); *51* 1124 (1986); *52* 3798
   (1987)
TL *22* 3167, 4651 (1981); *24* 2931 (1983); *25* 5103
   (1984); *28* 31, 5351 (1987)
Tetr *38* 2195 (1982)
J Ind Chem Soc *61* 99 (1984) (review)
BCSJ *58* 146 (1985)

SiR$_3$

JACS *94* 7074 (1972); *95* 5281 (1973); *96* 200 (1974)
JOC *41* 1233 (1976); *46* 5447 (1981); *52* 622, 3798
   (1987)

R

JOC *46* 5447 (1981)

7.9. oxy-Cope–Claisen

JOC *46* 5448 (1981)

7.10. oxy-Cope–Cope

JOC *46* 5448 (1981)

7.11. Carroll

See page 775.

7.12. Nazarov

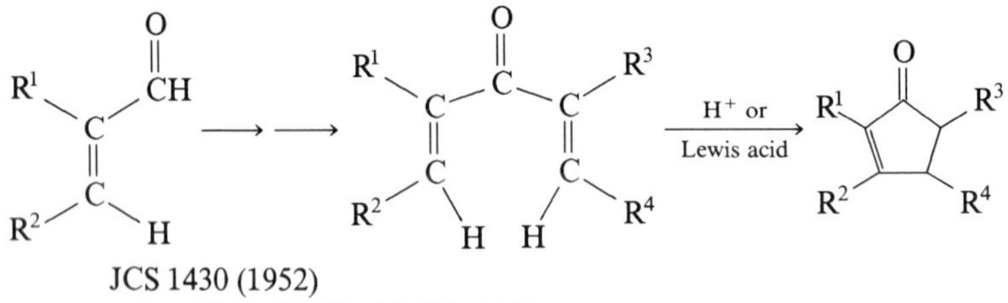

JCS 1430 (1952)
J Gen Chem USSR *27* 693 (1957); *30* 765 (1960)
Helv *54* 2913 (1971); *66* 2377, 2397 (1983) (both Si-directed)
JCS Perkin I 2271 (1972); 1026 (1973)
BCSJ *53* 169 (1980)
JOC *46* 3696 (1981)
JACS *104* 2642 (1982) (Si-directed)
Syn 429 (1983) (review)
TL *27* 2801, 5947 (Sn-directed) (1986)
Tetr *42* 2821 (1986) (Si-directed)

See page 658, Section 4; and page 661, Section 5 for related reactions.

7.13. Electrocyclization

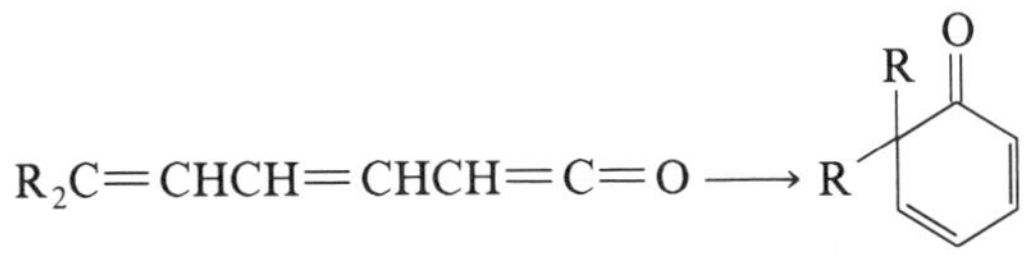

$$R_2C=CHCH=CHCH=C=O \longrightarrow R\text{-cyclohexenone with R and O}$$

CC 1728 (1987)

7.14. 2,3-Sigmatropic Rearrangements

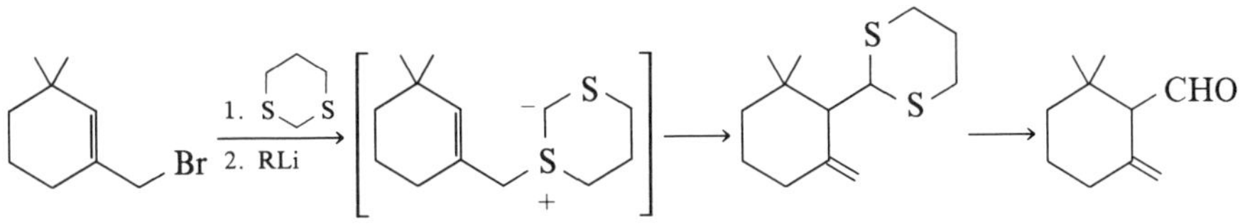

CC 757 (1972)

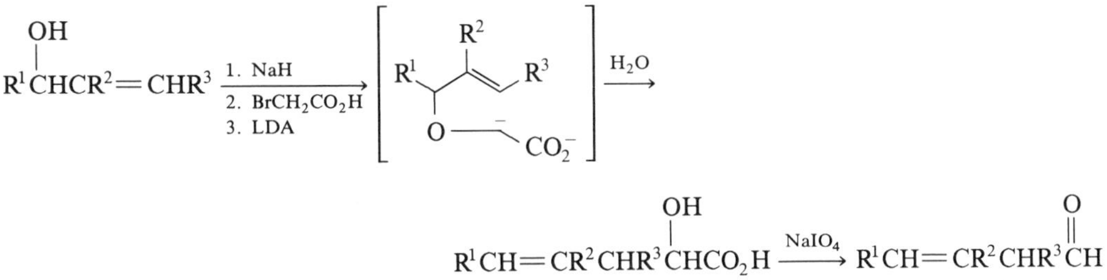

$$R^1CHCR^2=CHR^3 \xrightarrow[\text{3. LDA}]{\substack{\text{1. NaH} \\ \text{2. BrCH}_2\text{CO}_2\text{H}}} \left[\text{allyl intermediate} \right] \xrightarrow{H_2O}$$

$$R^1CH=CR^2CHR^3\overset{OH}{\underset{}{C}}HCO_2H \xrightarrow{NaIO_4} R^1CH=CR^2CHR^3CH(=O)$$

TL *22* 69 (1981)

$$R_2C=CHCH_2O-\bar{C}-CN \longrightarrow H_2C=CHC-C-CN \longrightarrow H_2C=CHC-C-R'$$

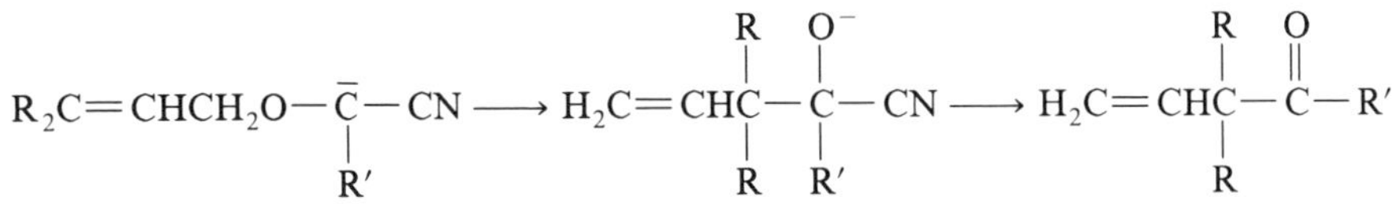

TL *24* 2077 (1974)

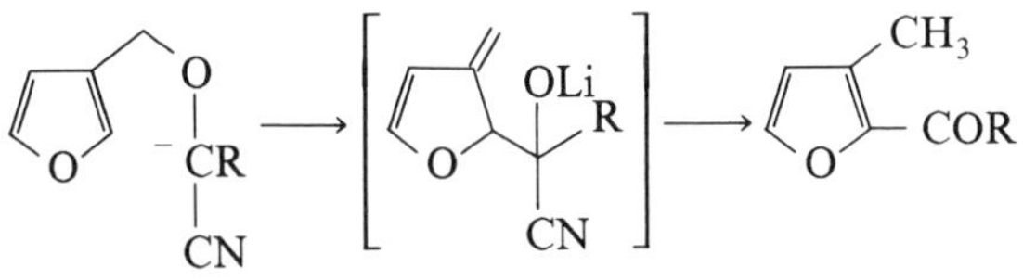

Syn Commun *7* 113 (1977)

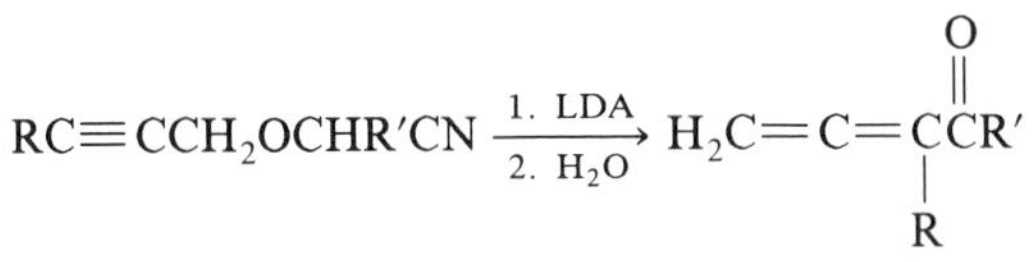

$$RC\equiv CCH_2OCHR'CN \xrightarrow[\text{2. H}_2O]{\text{1. LDA}} H_2C=C=CCR'$$

Syn Commun *7* 273 (1977)

7.15. Fries Rearrangement

7.15.1. Lewis acids

Org Rxs *1* 342 (1942) (review)
Org Syn Coll Vol *3* 280 (1955)
"Friedel-Crafts and Related Reactions," Ed. G. A. Olah, Interscience, New York (1964), Vol 3,
　Chpt 33
TL *23* 3299 (1982) (meta-Fries)
Ber *115* 1089 (1982) (equilibrium)

7.15.2. Photochemical

Proc Chem Soc 217 (1960)
JOC *27* 2293 (1962); *46* 374 (1981); *51* 4432 (1986); *52* 3815 (1987)
Tetr *21* 1015 (1965)
V. I. Sternberg, "Organic Photochemistry," Ed. O. L. Chapman, Arnold, New York (1967), Vol 1,
　Chpt 3 (review)
Chem Rev *67* 599 (1967) (review)
Helv *51* 1980 (1968)
JACS *90* 7249 (1968); *92* 2187 (1970); *94* 2219 (1972); *96* 449 (1974)
TL 3429, 5423 (1968); 2935 (1973)
BCSJ *42* 1831 (1969)
Adv Photochem *8* 109 (1971) (review)
CC 289 (1974)
BSCF 785 (1977)
Syn Commun *9* 877 (1979)
Ber *113* 261 (1980)
Syn 882 (1981)

7.16. Miscellaneous Rearrangements

$$X = R, \, OR$$

Reagent

Δ	JACS *94* 8084 (1972) JOC *37* 2405 (1972)
hν	JOC *37* 2405 (1972)
H_2SO_4	JACS *94* 8084 (1972)
HCl	Ann 1767 (1974) J Chem Res (S) 76, 142 (1979) JCS Perkin I 1822 (1977)
cat $Rh_2(OAc)_4$	CC 959 (1979) JCS Perkin I 2566 (1981) Syn 197 (1983) JACS *109* 6187 (1987)
cat $ClRh(PPh_3)_3$, $PdCl_2$ or $CoCl_2$	Syn 197 (1983)

P_2O_5, $MeSO_3H$	JOC *38* 4071 (1973)
PPA	J Ind Chem Soc *34* 178 (1957)

5. CONDENSATION REACTIONS

See also page 167, Section 3 for condensation reactions which generate a carbon-carbon double bond.

1. Acetoacetic Ester

See also page 768, Section 20.

Org Rxs *1* 266 (1942) (review)
JOC *51* 268 (1986)

2. Acyloin

Reviews:

Org Rxs *4* 256 (1948); *23* 259 (1976)
Chem Rev *64* 573 (1964)
Syn 236 (1971) (Na, Me$_3$SiCl)
JOC *40* 393 (1975) (mechanism)

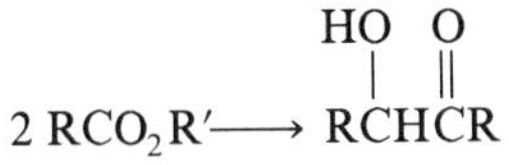

$$2\ RCO_2R' \longrightarrow \overset{\overset{\displaystyle HO}{|}}{R}\overset{\overset{\displaystyle O}{\|}}{CH}CR$$

Na, ether	JACS *52* 3988 (1930); *53* 750 (1931)
Na, NH$_3$	JOC *5* 362 (1940)
Na, zylene	JACS *57* 2303 (1935) Org Syn Coll Vol *2* 114 (1943) JCS C 2617 (1968)

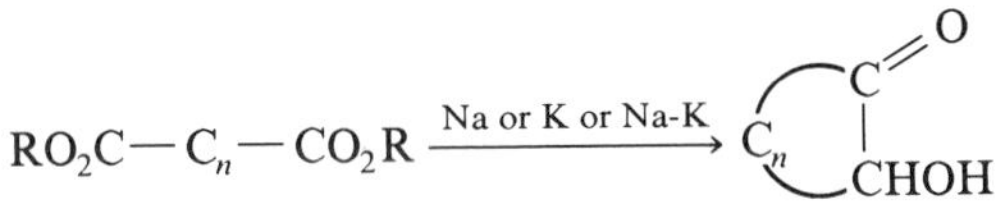

$$RO_2C-C_n-CO_2R \xrightarrow{\text{Na or K or Na-K}} C_n \begin{array}{c} C{=}O \\ | \\ CHOH \end{array}$$

$\underline{n}$

2	JACS *72* 983 (1950) JOC *31* 2017 (1966) TL 1529 (1969)
3	JACS *72* 3376 (1950), *75* 3997, 6231 (1953)
4	JACS *72* 3376 (1950); *79* 6050 (1957)
7, 8, 10, 12, 14, 18	Helv *30* 1741 (1947)
7–12	JACS *88* 4267 (1966)
7–16	Helv *30* 1822 (1947)
8	JACS *74* 3626 (1952) Org Syn Coll Vol *4* 840 (1963)
8, 14, 16	Helv *30* 1815 (1947)

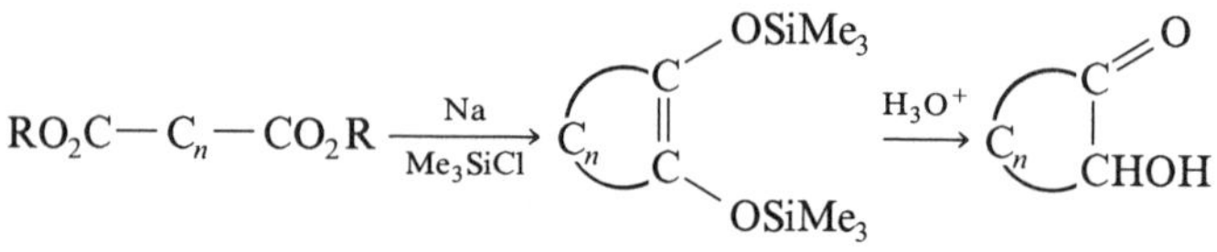

$$RO_2C - C_n - CO_2R \xrightarrow[Me_3SiCl]{Na} \underset{OSiMe_3}{\overset{OSiMe_3}{C_n}} \xrightarrow{H_3O^+} \underset{CHOH}{\overset{O}{C_n}}$$

Review: Syn 236 (1971)

$\underline{n}$

2	Ber *100* 3820 (1967) TL 587, 3319 (1968); *28* 4669 (1987) Org Syn *57* 1 (1977)
3	Ber *97* 1383 (1964) JOC *35* 1272 (1970)
4	Ber *97* 1383 (1964) TL 591 (1968) JACS *107* 686 (1985)
5	Ber *100* 3820 (1967) TL 591 (1968) JOC *52* 3603 (1987)
8	Ber *97* 1383 (1964) JACS *106* 6006 (1984)
10	JACS *106* 6006 (1984)
22	JACS *109* 7477 (1987)

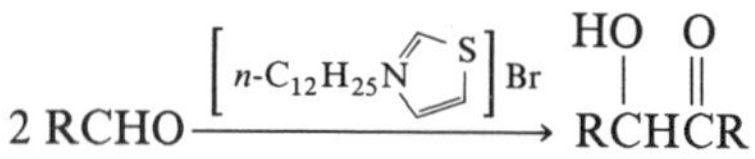

$$2\,RCHO \xrightarrow{\left[n\text{-}C_{12}H_{25}N\overset{S}{\diagup}\right]Br} \underset{RCHCR}{\overset{HO\;\;O}{\mid\;\;\parallel}}$$

CC 891 (1973)

3. Dieckmann

Org Syn Coll Vol *2* 116 (1943)
Org Rxs *15* 1 (1967) (review)

4. Mannich

See page 429, Section 4.

5. Michael

See page 792, Section 2.

6. Thorpe–Ziegler

JACS *81* 4074 (1959); *106* 6006 (1984)
Org Rxs *15* 1 (1967) (review)

7. Aldol and Related Reactions

For aldol condensation followed by dehydration see page 167, Section 3.

General references:

Org Rxs *16* 1 (1968) (aldol); *28* 203 (1982) (directed aldol)
BCSJ *53* 1417 (1980) (inter- and intramolecular aldol promoted by R_2AlOPh, pyridine)
Can J Chem *59* 3303 (1981) (thermal crossed aldol)
TL *22* 429 (1981) (intramolecular regioselectivity); *23* 4891 (1982) (transition state stereoselectivity)
Tetr *38* 2279 (retroaldol-aldol), 2939 (1982) (intramolecular)
JACS *104* 872 (1982) (intramolecular)
JOC *47* 1349 (1982) (intramolecular)
BCSJ *55* 3931 (1982) (retroaldol-aldol)
Syn 294 (1983) [$Si(OMe)_4$, KF as base]
C. H. Heathcock in "Asymmetric Synthesis," Ed. J. D. Morrison, Academic Press, New York (1984), Vol 3, Chpt 2

7.1. Metal Enolates

Li

JOC *45* 1066, 3846 (1980); *46* 191, 1296, 2290 (1981); *48* 4330 (1983); *50* 2095 (1985); *52* 4062, 4681 (1987)
Angew Int *19* 557 (1980)
JACS *103* 3099 (1981); *104* 5526, 5528 (1982); *105* 1667 (1983); *107* 5396 (1985); *108* 4603 (1986); *109* 3353 (1987)

TL *24* 5233 (1983); *27* 3511 (1986); *28* 985, 3835, 3839 (1987)

JCS Perkin I 1809 (1983)

Helv *68* 264 (1985)

CC 1620 (1987) (enantioselective, chiral amide base)

B

CL 153 (1977); 153, 1193 (1981)

JACS *101* 6120 (1979); *103* 1566, 1568, 3099 (1981); *104* 5521, 5523, 5528 (1982); *107* 5292 (1985)

TL 1665, 2225, 2229, 3937 (1979); *21* 4291, 4675 (1980); *22* 3555 (1981); *23* 2387 (1982); *27* 4787 (1986) (enantioselective); *28* 1229 (1987) (enantioselective)

BCSJ *53* 174 (1980)

JCS Perkin I 1809 (1983)

Pure Appl Chem *55* 1749 (1983)

CC 147 (1983); 1102 (1987)

JOC *52* 1347 (1987)

Al

CL 379 (1979)

TL 2257 (1979); *23* 2387 (1982)

Helv *68* 264 (1985)

Sn

CC 162 (1981)

JCS Perkin I 1809 (1983)

Pure Appl Chem *55* 1749 (1983)

TL *24* 3347 (1983)

Ti

TL *22* 4691 (1981)

Topics Curr Chem *106* 55 (1982) (review)

Zr

TL *21* 3975, 4607 (1980)

JACS *104* 5528 (1982)

JCS Perkin I 1809 (1983)

Zn

JOC *52* 3489 (1987)

Ce

TL *24* 5233 (1983); *28* 3817 (1987)

$(Et_2N)_3\overset{+}{S}$

JACS *103* 2106 (1981)

$$\underset{\text{RCCH}_3}{\overset{O}{\overset{\|}{}}} \xrightarrow[i\text{-Pr}_2\text{NEt}]{\text{organoborane}} \xrightarrow{\text{R'CHO}} \underset{\text{chiral}}{\overset{O}{\overset{\|}{\text{RCCH}_2}}}\underset{}{\overset{OH}{\overset{|}{\text{CHR}'}}}$$

Organoborane

TL *27* 4721 (1986)

TL *27* 4787 (1986)

$$\underset{\text{RC}=\text{CH}_2}{\overset{\text{OSiMe}_3}{|}} + \underset{\text{HCR}'}{\overset{\text{O}}{\|}} \longrightarrow \underset{\text{RCCH}_2\text{CHR}'}{\overset{\text{O}}{\|}}\overset{\text{OH}}{\underset{|}{}}$$

high pressure	JACS *105* 6963 (1983)
H$_2$O	JOC *51* 2142 (1986)
BF$_3\cdot$OEt$_2$	JACS *105* 1667 (1983) TL *25* 5973 (1984)
SnCl$_4$	TL *24* 3347 (1983); *25* 5973 (1984)
TiCl$_4$	CL 1223 (1974) JACS *96* 7503 (1974); *107* 5292 (1985) Angew Int *16* 817 (1977); *22* 989 (1983) TL *24* 3341, 3343 (1983); *25* 5973 (1984); *27* 3369 (1986); *28* 985 (1987) JOC *50* 2375 (1985); *52* 120 (1987) Org Syn *65* 6 (1987)
cat *n*-Bu$_4$NF	JACS *99* 1265 (1977) JOC *48* 932 (1983)
[(Et$_2$N)$_3$S]SiMe$_3$F$_2$	JACS *103* 2106 (1981)
n-Bu$_3$SnF, cat PdCl$_2$(PAr$_3$)$_2$	TL *24* 5001 (1983)
cat [(COD)Rh(Ph$_2$P(CH$_2$)$_4$PPh$_2$)]X (X = PF$_6$, ClO$_4$) or Rh$_4$(CO)$_{12}$	TL *27* 5517 (1986)
cat [Rh(NBD)(dppe)]ClO$_4$	TL *28* 793 (1987)

TL *24* 3347 (1983)

CL 353, 1441 (enantioselective with chiral diamine), 1459, 1601 (1982); 595 (1983)

TL *23* 627 (1982)
CL 851 (1983)

$$
\underset{\substack{\text{O}\;\;\text{O}\\ \|\;\;\|\\ CH_3C-CH}}{}\;\xrightarrow{\;Sn\;}\;\underset{\substack{CH_3\;\;\;H}}{\overset{Sn}{\underset{O\;\;\;O}{\bigcirc}}}\;\xrightarrow{\;RCHO\;}\;\underset{\substack{\text{OH}\;\;\text{O}\\ |\;\;\|\\ R\;\;\;\;\;CH_3\\ |\\ OH}}{}
$$

CL 1825 (1983)

$$
\underset{\substack{\text{O}\;\;\text{HgX}\\ \|\;\;|\\ RC-CHR}}{} + \underset{\substack{\text{O}\\ \|\\ R'CH}}{}\;\xrightarrow{\;BF_3\cdot OEt_2\;}\;\underset{\substack{\text{O}\;\;\text{OH}\\ \|\;\;|\\ RCCHCHR'\\ |\\ R}}{}
$$

JACS *104* 2323 (1982)

$$
\underset{\substack{\text{O}\\ \|\\ ArCCH_2R}}{} + Ar'CHO\;\xrightarrow[\;Et_3N\;]{\;TiCl_4\;}\;\underset{\substack{\text{O}\;\;\text{OH}\\ \|\;\;|\\ ArCCHCHAr'\\ |\\ R}}{}
$$

TL *28* 4135 (1987)

7.2. Related Reactions

$$
\underset{\substack{RN}}{\bigcirc}\;\xrightarrow[\;Et_3N\;]{\;BCl_3\;}\;\underset{\substack{RNBCl_2}}{\bigcirc}\;\xrightarrow[\;2.\;H_3O^+\;]{\;1.\;RCHO\;}\;\underset{\substack{\text{O}\;\;\;\;\;\text{OH}\\ \;\;\;\;\;\;|\\ \;\;\;\;CHR}}{\bigcirc}
$$

TL 1423 (1979) (enantioselective)
Syn Commun *9* 515 (1979)

$$
\underset{\substack{\text{O}\\ \|\\ -C-C-X\\ |}}{}\;\xrightarrow[\substack{2.\;R_2CO\\3.\;H_2O}]{\;1.\;reagent(s)\;}\;\underset{\substack{\text{O}\;\;\;\text{OH}\\ \|\;\;\;|\\ -C-C-CR_2\\ |}}{}
$$

X = halogen

Reagent(s)

Sn	CL 467 (1982), 1727 (1983)
CuCl, Zn	TL *28* 6481 (1987)
CuBr, Zn, Et$_2$AlCl	BCSJ *53* 3301 (1980)
Zn	JACS *89* 5727 (1967)
Zn, Et$_2$AlCl, AgOAc	TL *28* 6481 (1987)
electrolysis, LaBr$_3$	JOC *52* 2496 (1987)

$$
\underset{\substack{\text{LiO}\;\;\text{Li}\\ |\;\;\;|\\ RC=CR}}{} + R'_2CO\;(RCHO)\;\xrightarrow{\;H_2O\;}\;\underset{\substack{\text{O}\;\;\text{OH}\\ \|\;\;|\\ RCCHRCR'_2}}{}\;\;or\;\;\underset{\substack{\text{O}\\ \|\\ RCCR=CR'_2}}{}
$$

JACS *104* 1777 (1982)

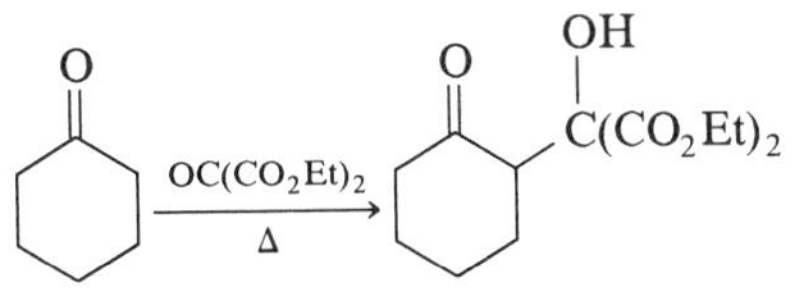

TL *22* 4323 (1981)

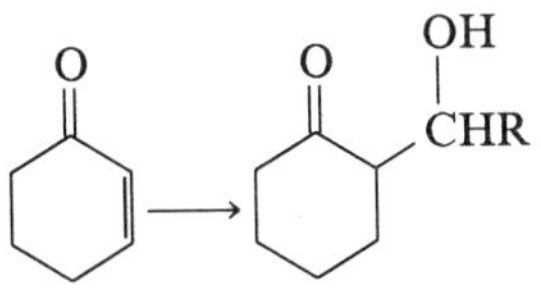

See page 776, Section 21.

$$\underset{\overset{\parallel}{O}}{RCC\equiv CR} \xrightarrow[\substack{2.\ R'CHO \\ 3.\ H_2O}]{1.\ Me_3SiI,\ Et_2AlI\ or\ n\text{-}Bu_4NI\text{-}TiCl_4} \underset{\overset{\parallel}{O}\ \overset{\mid}{OH}}{\underset{\overset{\parallel}{RCI}}{RCCCHR}}$$

TL *27* 4767 (1986)

8. Other Condensation Reactions

$$2\ \underset{\overset{\parallel}{O}}{RCCH_3} \xrightarrow[Zn]{Me_2SiI_2} \underset{\overset{\parallel}{O}}{RCCH_2CHRCH_3}$$

CL 1255 (1980)

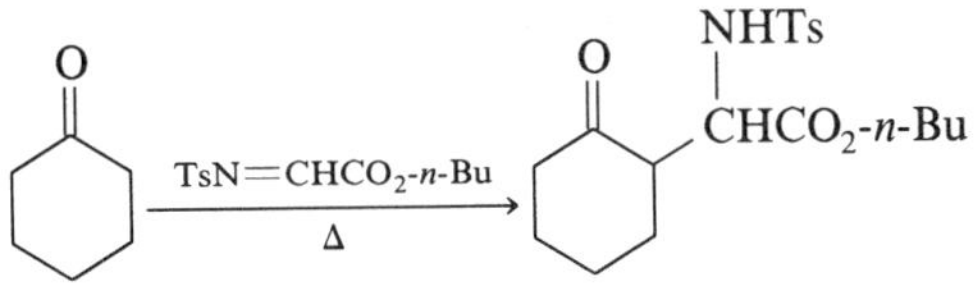

TL *22* 4323 (1981)

$$\underset{\overset{\parallel}{O}}{CH_3CCH_3} + \underset{\overset{\parallel}{O}}{EtOCH} \xrightarrow{NaOMe} \underset{\overset{\parallel}{O}\quad\overset{\parallel}{O}}{CH_3CCH_2CH}$$

Org Syn Coll Vol *4* 210 (1963)

$$\underset{\overset{\parallel}{O}}{RCCH_3} + \underset{\overset{\parallel}{O}}{XCR'} \xrightarrow{base} \underset{\overset{\parallel}{O}\quad\overset{\parallel}{O}}{RCCH_2CR'}$$

X = Cl, OR

Org Rxs *8* 59 (1954) (review)

$$\underset{\overset{\parallel}{O}}{CH_3CCH_3} + \underset{\overset{\parallel}{O}}{(CH_3C)_2O} \xrightarrow{BF_3} \underset{\overset{\parallel}{O}\quad\overset{\parallel}{O}}{CH_3CCH_2CCH_3}$$

Org Rxs *8* 59 (1954) (review)
Org Syn Coll Vol *3* 16 (1955)

$$\underset{RC=CHR}{\overset{\overset{\displaystyle OBR_2}{|}}{}} \xrightarrow[\text{2. HCl}]{\text{1. R'CN}} \underset{\underset{R}{|}}{\overset{\overset{\displaystyle O \quad O}{\parallel \quad \parallel}}{RCCHCR'}}$$

Syn Commun *12* 189 (1982)

$$\overset{\overset{\displaystyle O}{\parallel}}{CH_3CCH_3} + \overset{\overset{\displaystyle O \quad O}{\parallel \quad \parallel}}{EtOC-COEt} \xrightarrow{\text{NaOEt}} \overset{\overset{\displaystyle O \quad O \quad O}{\parallel \quad \parallel \quad \parallel}}{CH_3CCH_2C-COEt}$$

Org Syn Coll Vol *1* 238 (1941)

6. CYCLIZATION AND ANNULATION REACTIONS

Reviews:

Tetr *32* 3 (1976)
Syn 177 (1976)

1. Cyclization

See also page 702, Section 3.

CC 876 (1974); 746 (1981)
Org Rxs *26* 361 (1979) (review)
JCS Perkin I 1924 (1980); 1203 (1981)
TL *21* 691 (1980)

Org Rxs *26* 361 (1979) (review)
Tetr *37* 2407 (1981) (review)
J Chem Res (S) 5 (1981)

Org Rxs *26* 361 (1979) (review)
JACS *103* 1996, 2009, 2017 (1981)
Tetr *37* 2407 (1981) (review)

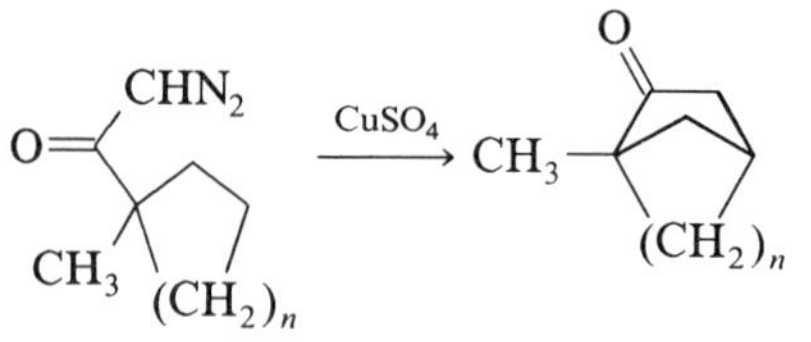

JOC *47* 3242 (1982)

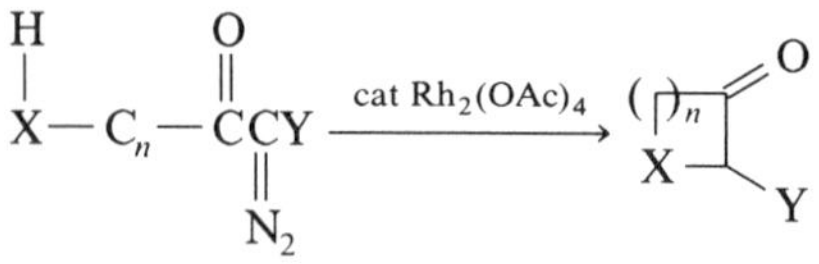

Y	X	n	
H	C	2	JACS *107* 5732 (1985); *109* 5432 (1987)
			TL *28* 165 (2-indanones); 4773 (1987)
		3	CC 129 (1984) (2-tetralones)
			TL *26* 6035 (1985)
CO$_2$R	C	1	JACS *106* 5295 (1984)
			TL *28* 637 (1987)
		2	JOC *47* 4808 (1982); *52* 28 (1987)
			JACS *105* 5935 (1983); *106* 5295 (1984);
			107 196, 5289 (1985); *108* 7686 (1986)
			TL *26* 3059 (1985); *28* 637, 5351 (1987)
		3	TL *24* 5453 (1983) (naphthols)
	N	1–3	TL *21* 31, 2783 (1980); *23* 3105 (1982);
			28 781 (1987)
			JOC *50* 5223 (1985)
			JACS *108* 6054 (1986)
	O	4, 5	TL *27* 1403 (1986)
		5	TL *28* 5351 (1987)
	O or S	2	JOC *50* 5223 (1985)
		3, 4	TL *28* 5351 (1987)
SO$_2$Ph	C	2	TL *28* 3459 (1987)
			JOC *52* 5742 (1987)

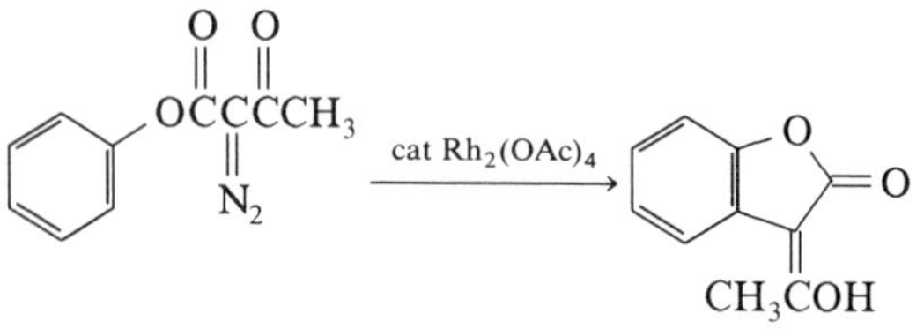

CC 1150 (1987)

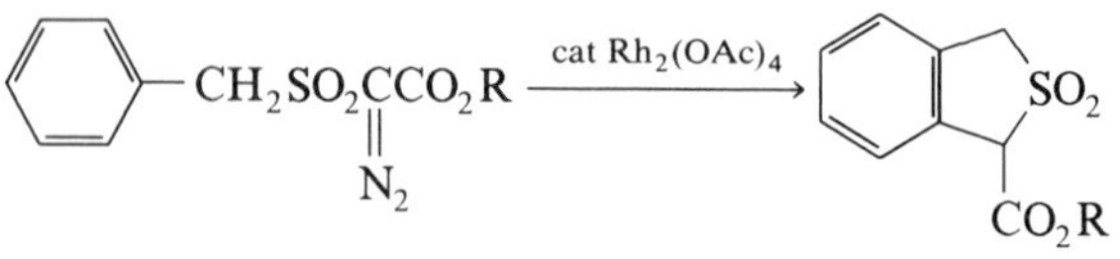

TL *27* 5679 (1986)

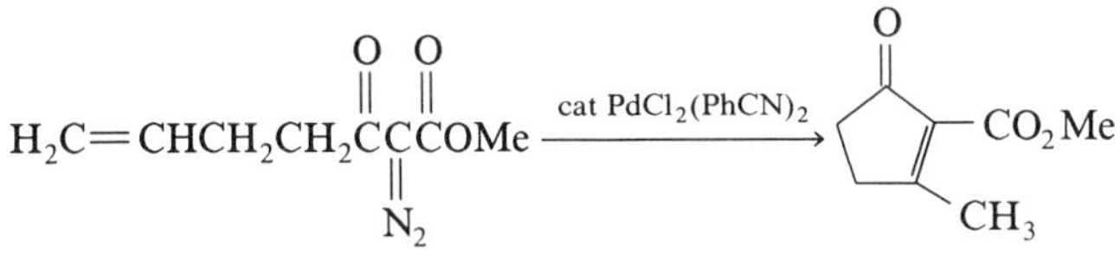

$$H_2C=CHCH_2CH_2\overset{O}{\underset{\underset{N_2}{\|}}{C}}\overset{O}{\underset{}{\overset{\|}{C}}}COMe \xrightarrow{\text{cat PdCl}_2(\text{PhCN})_2}$$

JOC *51* 3382 (1986)

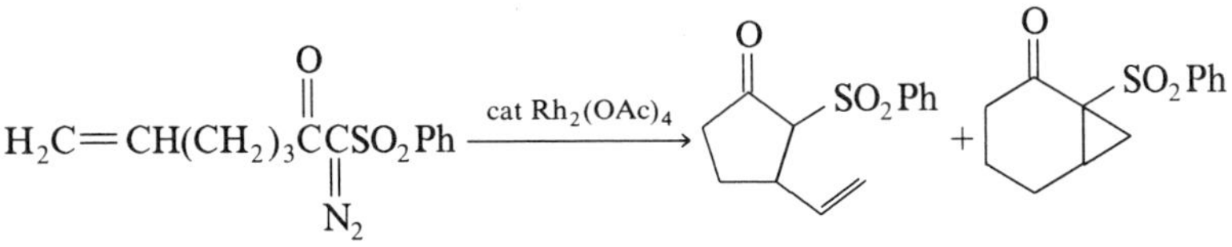

$$H_2C=CH(CH_2)_3\overset{O}{\underset{\underset{N_2}{\|}}{C}}CSO_2Ph \xrightarrow{\text{cat Rh}_2(\text{OAc})_4}$$

TL *28* 3459 (1987)

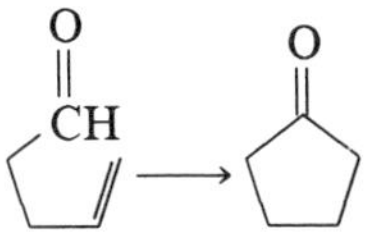

"ClRh(PAr$_3$)$_2$"

SnCl$_4$

TL 1287 (1972)
JACS *98* 1281 (1976); *100* 640 (1978);
101 489 (1979); *102* 190 (1980)

TL 1287 (1972)
CC 146 (1979)

$$(CH_3)_2CHNO_2 \xrightarrow[\text{base}]{H_2C=CHCO_2R} (CH_3)_2\overset{NO_2}{\underset{}{C}}CH_2CH_2\overset{O}{\overset{\|}{C}}OR \xrightarrow[(R=H)]{PPA}$$

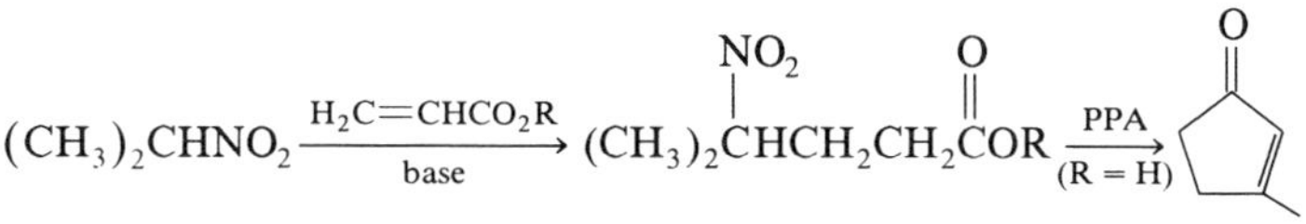

CC 1149 (1980)

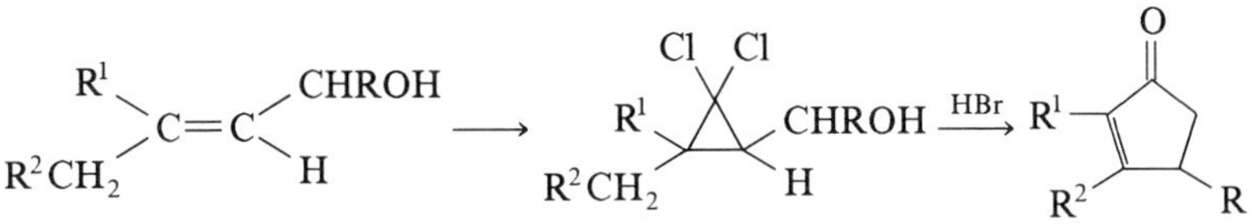

JACS *96* 3713 (1974)
BCSJ *53* 1010 (1980)

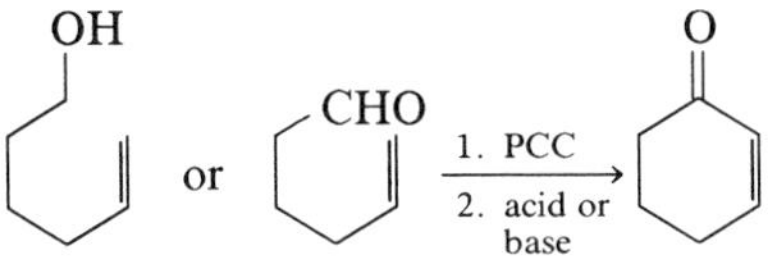

$$\xrightarrow[\text{2. acid or base}]{\text{1. PCC}}$$

TL 2461 (1978)

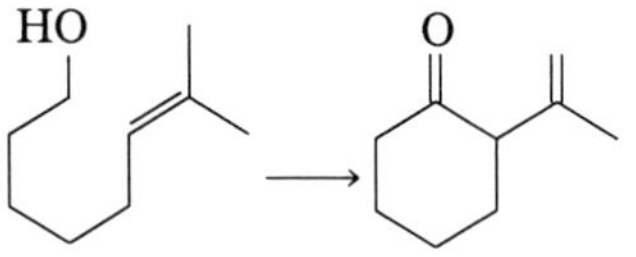

PCC

JOC *41* 380 (1976); *50* 2668 (1985)
TL 2461 (1978)

CrO$_3$, py, CF$_3$CO$_2$H

Can J Chem *55* 1039 (1977)

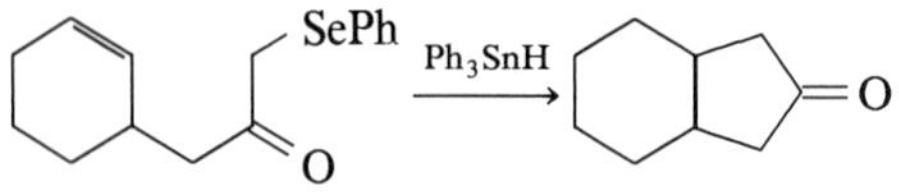

CC 1520 (1987)

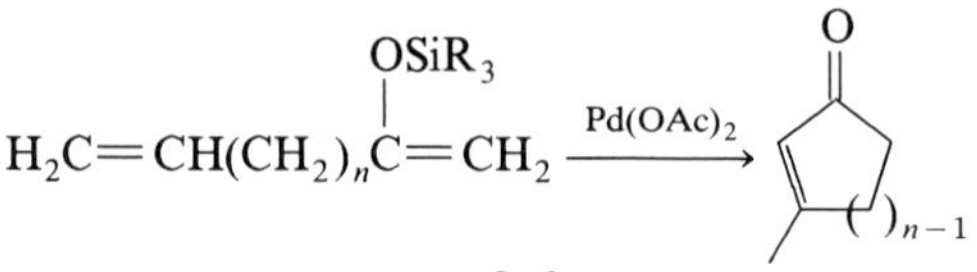

$$H_2C=CH(CH_2)_n\overset{\overset{\displaystyle OSiR_3}{|}}{C}=CH_2 \xrightarrow{Pd(OAc)_2}$$

n = 2, 3

JACS *101* 494 (1979); *102* 4519 (1980); *104* 1784, 5808 (1982); *108* 2090 (1986)
TL *27* 2223 (1986)

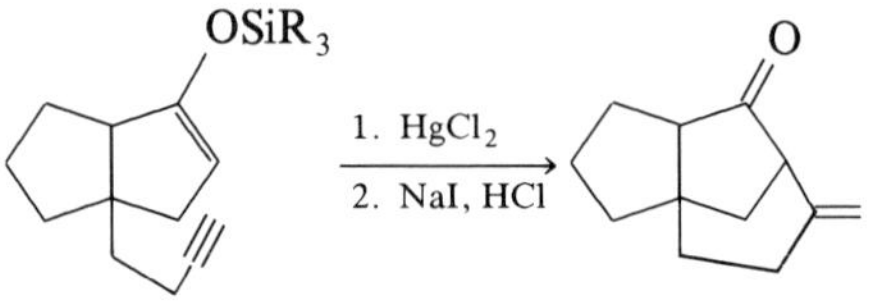

OSiR$_3$

1. HgCl$_2$
2. NaI, HCl

JACS *107* 1726 (1985)
TL *28* 3923 (1987)

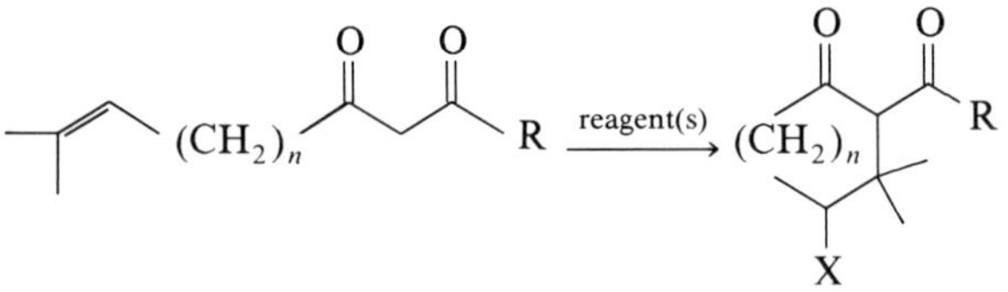

(CH$_2$)$_n$ R $\xrightarrow{\text{reagent(s)}}$ (CH$_2$)$_n$ R

n	R	Reagent(s)	X	
2	Me, OMe	SnCl$_4$	H	Angew Int *20* 687 (1981)
	OMe	PhSCl, silica gel	PhS	Can J Chem *59* 2239 (1981)
	OMe	PhSeCl, AlCl$_3$	PhSe	Can J Chem *59* 2239 (1981)
3	OMe	SnCl$_4$	H	Angew Int *20* 687 (1981)

1. LiAlH$_4$, TiCl$_3$ or TiCl$_4$, Et$_3$N
2. H$_3$O$^+$

CO$_2$Et

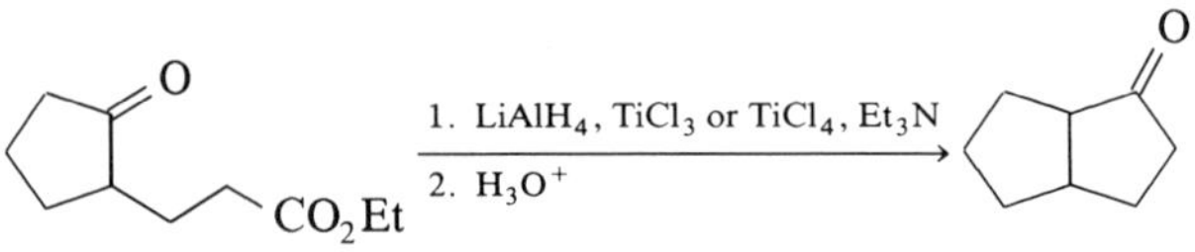

JACS *105* 1660 (1983)
TL *24* 1885 (1983)
CC 1607 (1987)

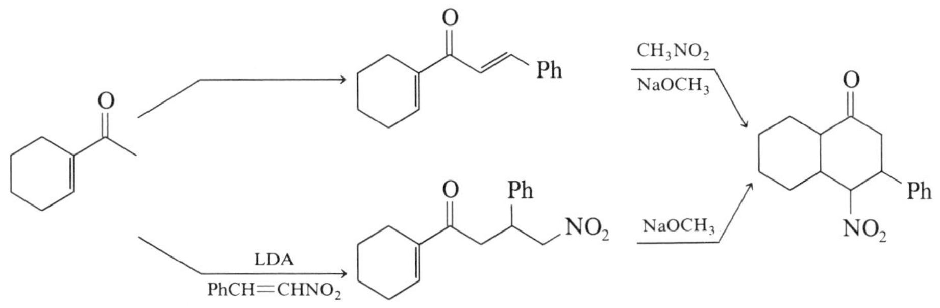

TL *28* 2945 (1987)

2. Spiroannulation

Review:

Syn 383 (1974) (Synthesis of Carbocyclic Spiro Compounds via Intramolecular Alkylation Routes)

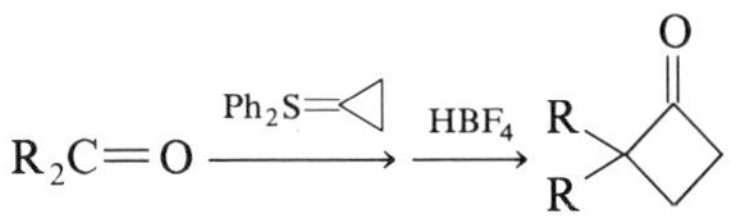

$$R_2C=O \xrightarrow{Ph_2S\triangleleft} \xrightarrow{HBF_4}$$

JACS *95* 5321 (1973)
Acct Chem Res 7 85 (1974)

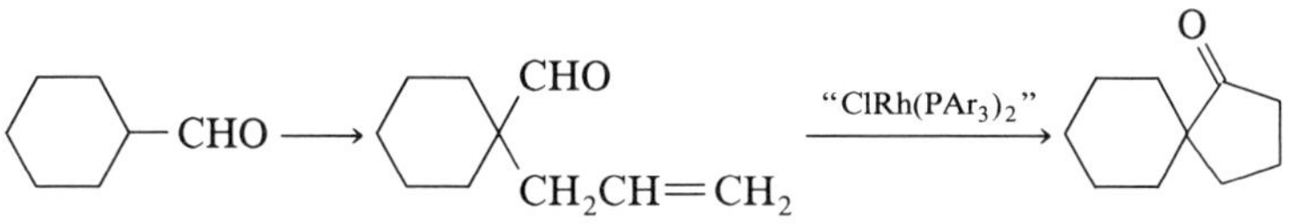

JACS *102* 190 (1980)

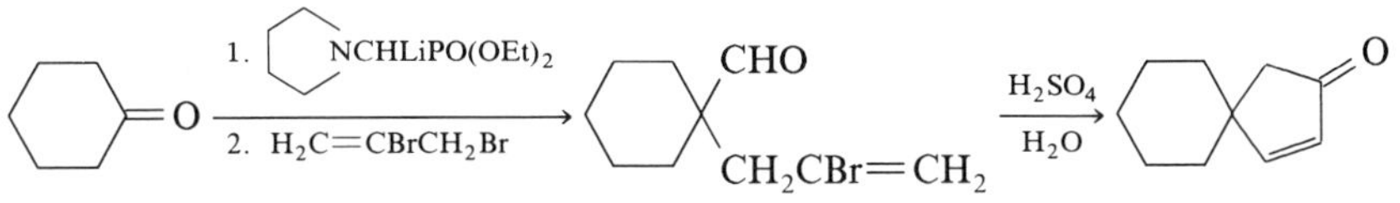

JOC *42* 2520 (1977); *43* 1027 (1978)

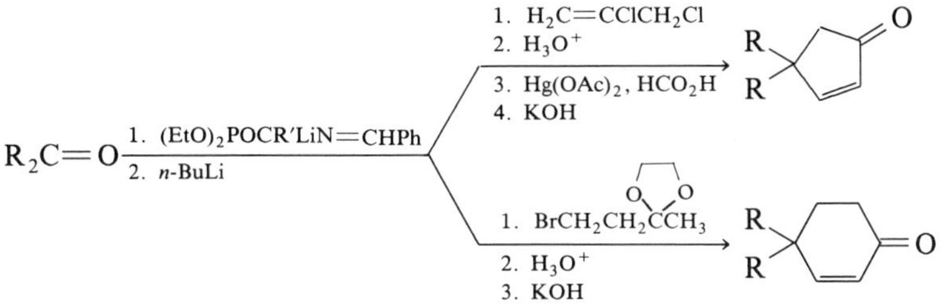

$$R_2C=O \xrightarrow[\text{2. } n\text{-BuLi}]{\text{1. } (EtO)_2POCR'LiN=CHPh}}$$

1. $H_2C=CClCH_2Cl$
2. H_3O^+
3. $Hg(OAc)_2, HCO_2H$
4. KOH

1. $BrCH_2CH_2CCH_3$
2. H_3O^+
3. KOH

JACS *102* 5866 (1980)

3. One-Carbon Annulation

See page 673, Section 7.

4. Two-Carbon Annulation

See also page 640, Section 7.12.

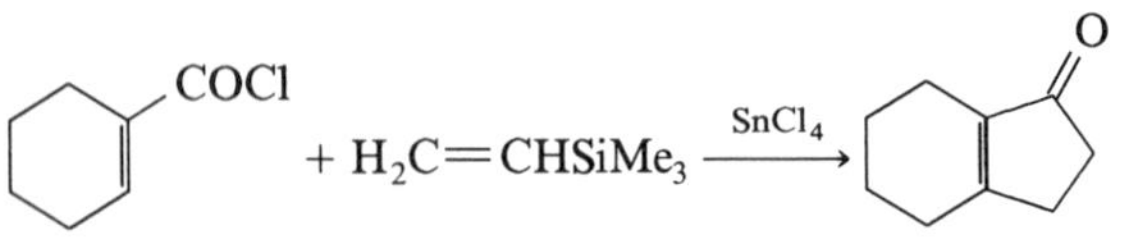

TL 1995, 1999 (1979)
JOC *45* 1046, 3017 (1980)

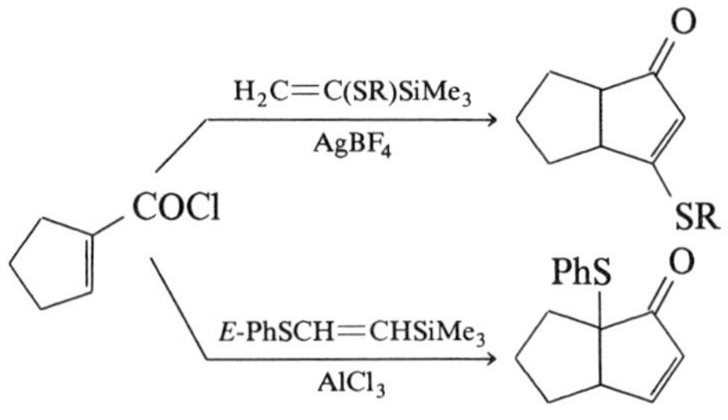

Organomet *1* 1240, 1243 (1982)

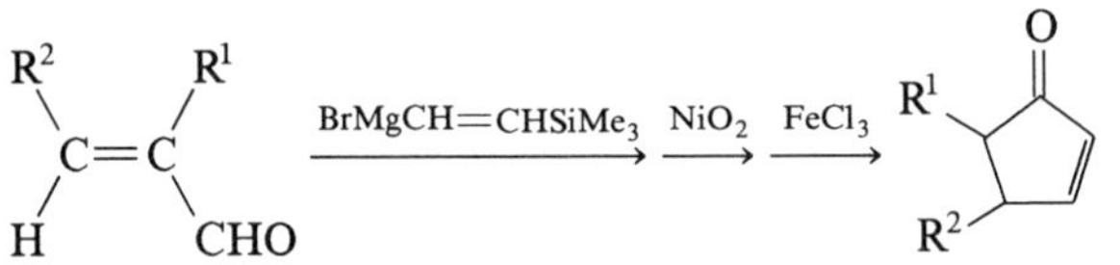

JACS *104* 2642 (1982)

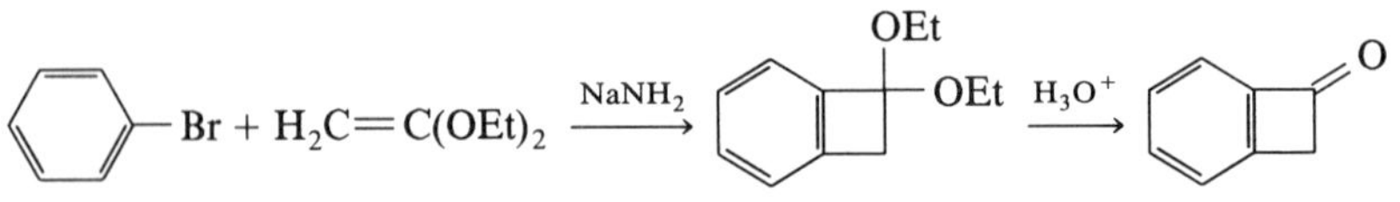

JOC *47* 2393, 2396 (1982)

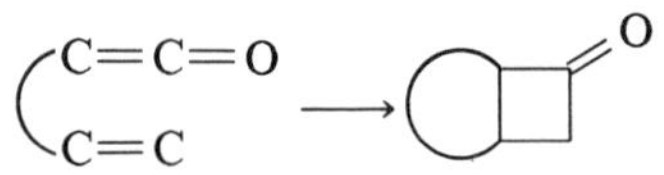

JACS *107* 2192 (alkyl and vinyl ketene), 2194 (alkoxy ketene), 4339 (vinyl ketene) (1985)
TL *26* 3535 (vinyl ketene), 5619 (vinyl ketene) (1985); *27* 1913 (imino ketene), 5471 (vinyl ketene) (1986)
JOC *50* 2809 (vinyl ketenes), 5167 (alkoxy ketene), 5177 (phenoxy ketene) (1985); *52* 307 (chloro ketene), 1568 (vinyl ketene), 5413 (vinyl ketene) (1987)

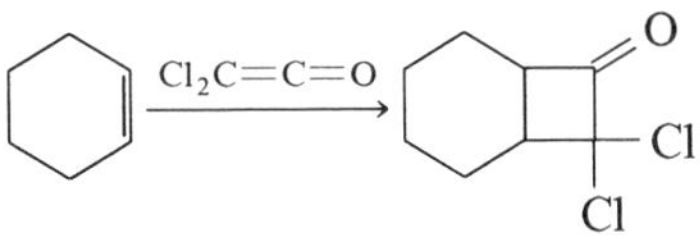

Tetr *27* 615 (1971); *37* 2949 (1981) (review)
Syn 415 (1971) (review)
Helv *64* 64 (1981)
JOC *47* 3871, 3881 (1982); *48* 4763 (1983); *50* 3957 (1985); *52* 3693, 4885 (1987)
JACS *105* 2435 (1983); *106* 5295 (1984); *108* 8015 (1986); *109* 4752 (1987) (asymmetric)

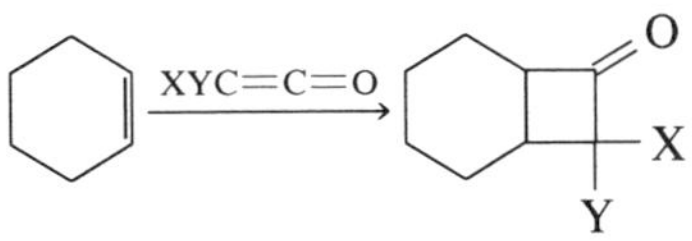

Reviews:

Org Rxs *3* 108 (1946) (preparation of ketenes); *12* 26 (1962)
J Chem Ed *53* 81 (1976)
Chem Soc Rev *10* 289 (1981) (cyano ketenes)
Chem Rev *86* 821 (1986) (conjugated ketenes)

X	Y	
CF_3	CF_3	JACS *87* 4019 (1965)
$H_2C=CH$	H	JOC *41* 3303 (1976)
$H_2C=C(CH_3)$	H	JOC *31* 718 (1966)
vinylic	R	Helv *66* 2331 (1983)
Cl	R	JOC *38* 4106 (1973)
Cl	CN	JOC *49* 2190 (1984)
PhS	H, R	TL *21* 2577 (1980)
$-SCH_2CH_2S-$		TL *21* 2577 (1980)

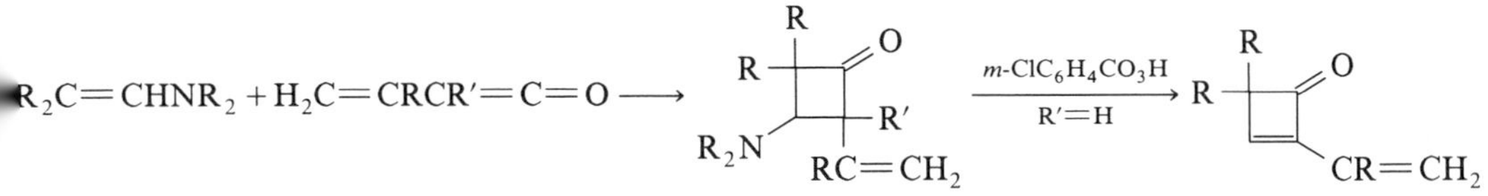

Helv *64* 2231 (1982)

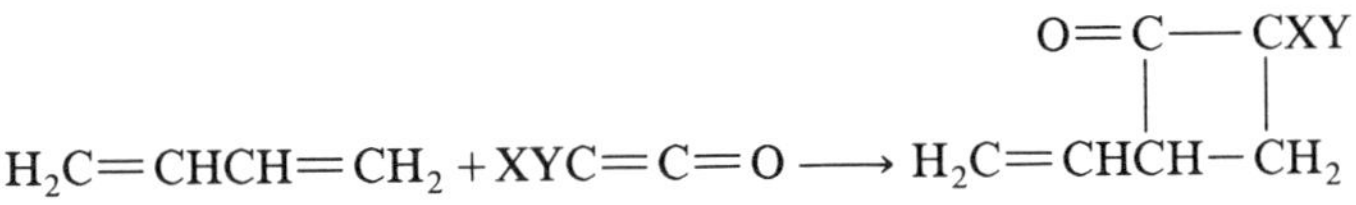

Helv *53* 417 (1953); *65* 451 (vinyl ketenes), 703, 1563 (vinyl ketenes) (1982)
Syn 415 (1971) (review)
Tetr *27* 615 (1971); *37* 2949 (1981) (review)
CC 245 (1978)
JACS *103* 4616 (1981)
JCS Perkin I 684 (1981)
JOC *52* 1284 (1987)

$$R^1CH{=}CHSO_2Ph + R^2C{\equiv}CNR^3_2 \xrightarrow{H_3O^+}$$

JOC *47* 1608 (1982)

$$\text{(cyclohexene)} \xrightarrow[2.\ H_2O]{1.\ R_2C{=}C{=}\overset{+}{N}Me_2}$$

JACS *94* 2870 (1972); *104* 2920 (1982); *107* 2192, 2194 (1985) (both intramolecular)
Angew Int *20* 782, 879 (1981)
JOC *50* 5169 (1985) (intramolecular); *52* 2216 (1987) (intramolecular)
TL *27* 5211 (1986) (intramolecular)

$$H_2C{=}CHCH{=}CH_2 \xrightarrow[2.\ H_2O]{1.\ R_2C{=}C{=}\overset{+}{N}Me_2} H_2C{=}CHCH{-}CH_2 \begin{smallmatrix} R_2C{-}CO \\ | \qquad | \end{smallmatrix}$$

JACS *94* 2870 (1972)
Syn 706 (1981)

$$RC{\equiv}CR \xrightarrow[2.\ H_2O]{1.\ R'_2C{=}C{=}\overset{+}{N}Me_2} \begin{smallmatrix} R'_2C{-}CO \\ | \qquad | \\ C{=}C \\ R \qquad R \end{smallmatrix}$$

Angew Int *14* 569 (1975); *20* 879 (1981)
TL *25* 5043 (1984)

$$RC{\equiv}CR \xrightarrow{XYC{=}C{=}O}$$

X	Y	
H	H	TL 1031 (1962)
		JOC *38* 1451 (1973)
Me	Me	TL 1031 (1962)
		JOC *38* 1451 (1973)
$-(CH_2)_4-$		JOC *38* 1451 (1973)
CF$_3$	CF$_3$	JOC *35* 3308 (1970)
Ph	Ph	TL 1031 (1962)
		JCS 2708 (1962)
		Helv *45* 600 (1962)
Ar	Ar	Tetr *36* 2291 (1980)
Cl	Cl	Ann *722* 232 (1969); 2074 (1974)
		JACS *96* 2267 (1974)
		TL 2715 (1976); *24* 23 (1983)
		Syn 689 (1979)
		JOC *48* 3382 (1983)
Cl	CN	JOC *50* 3226 (1985)

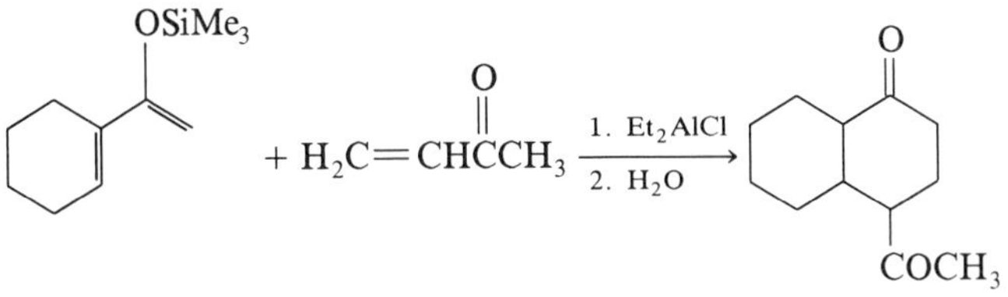

CC 1333 (1987)

5. Three-Carbon Annulation

Review: Syn 529 (1984)

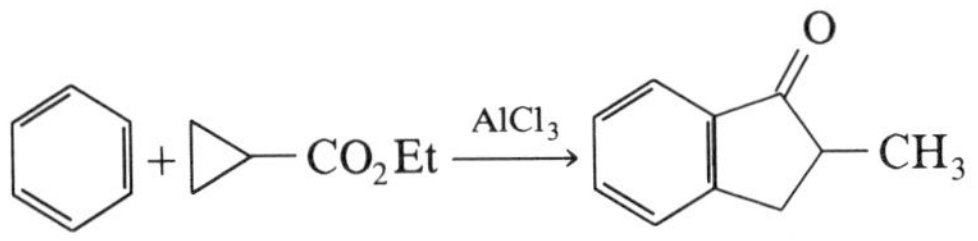

JOC *46* 3758 (1981)

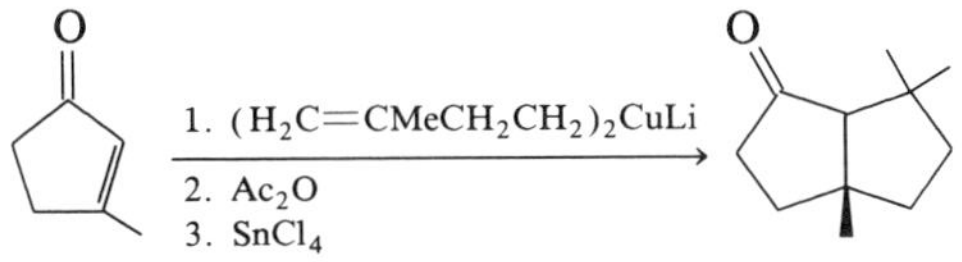

TL *23* 5471 (1982)

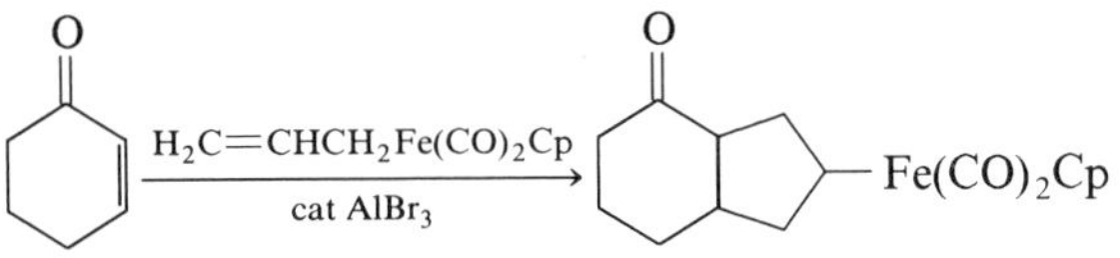

Organomet *1* 1679 (1982)

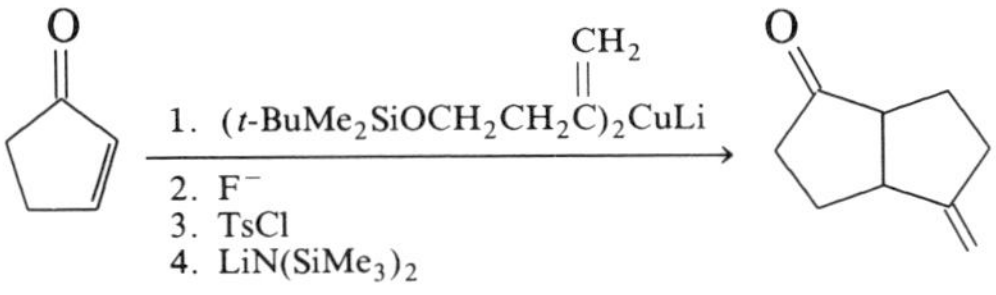

Organomet *1* 1243 (1982)

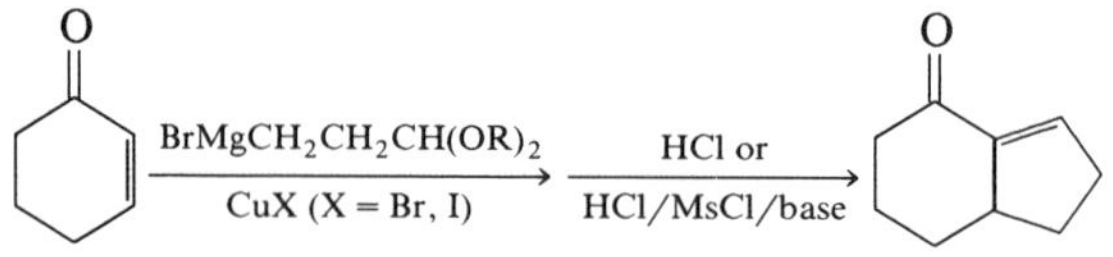

TL 4217 (1978)
JACS *104* 6081 (1982); *105* 7352 (1983); *106* 1442 (1984)
JOC *47* 5045 (1982)

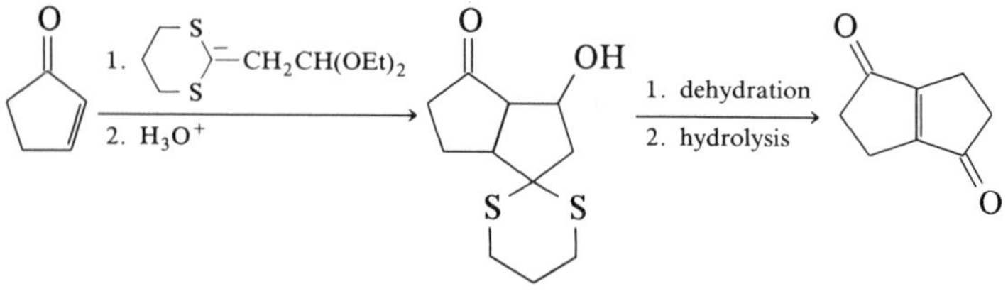

JOC *51* 3861 (1986)

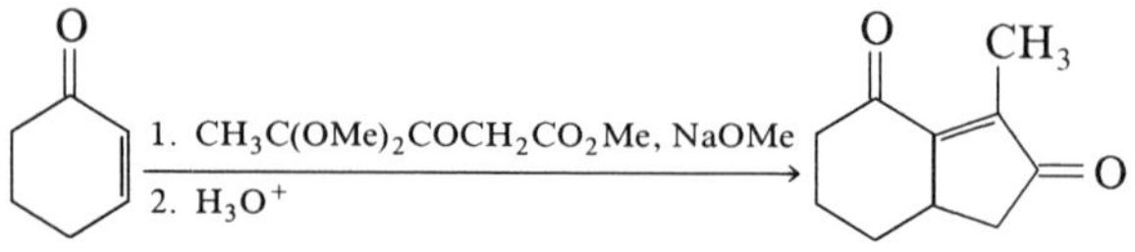

JOC *47* 4791 (1982)

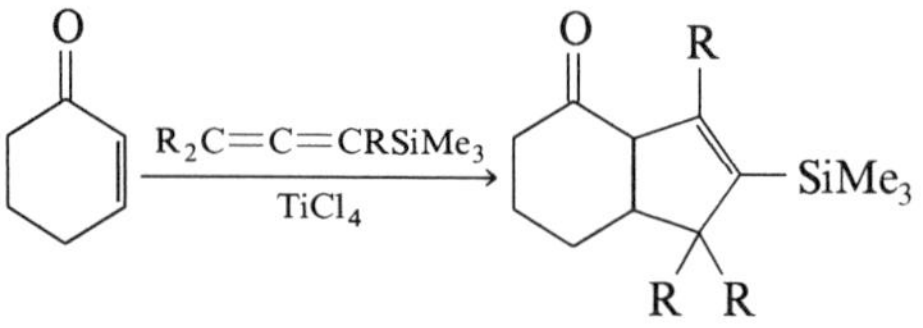

JACS *103* 1604 (1981)
Tetr *39* 935 (1983)

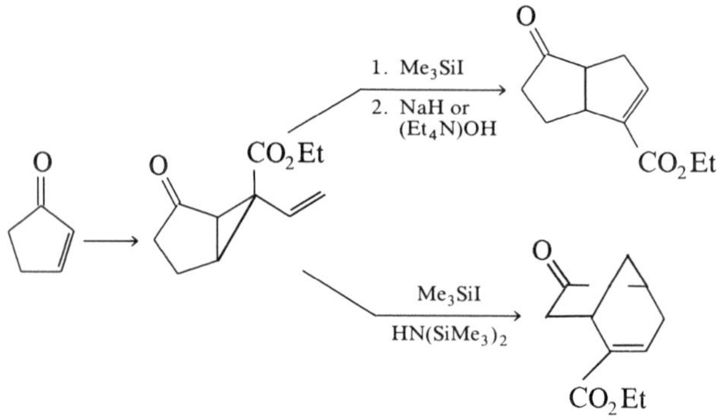

TL *28* 167 (1987)

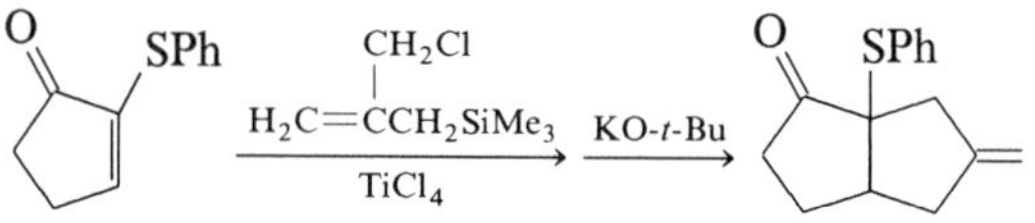

TL *21* 4557 (1980)

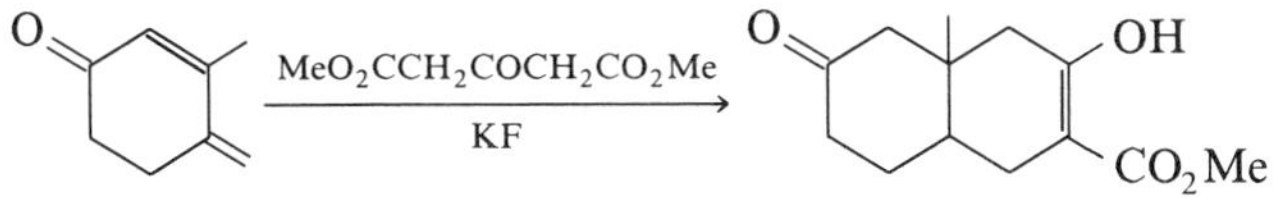

JCS Perkin I 25 (1982)

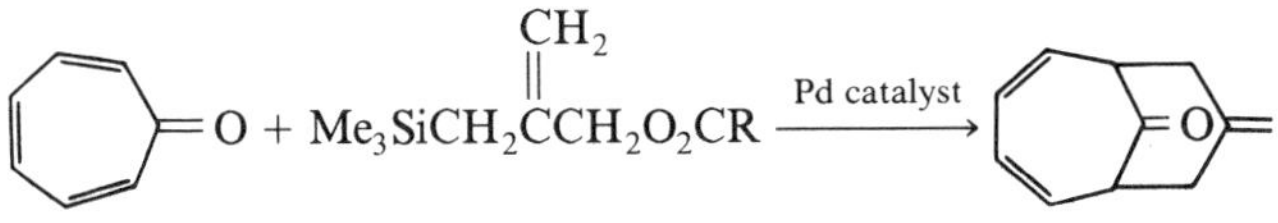

JACS *109* 615 (1987)

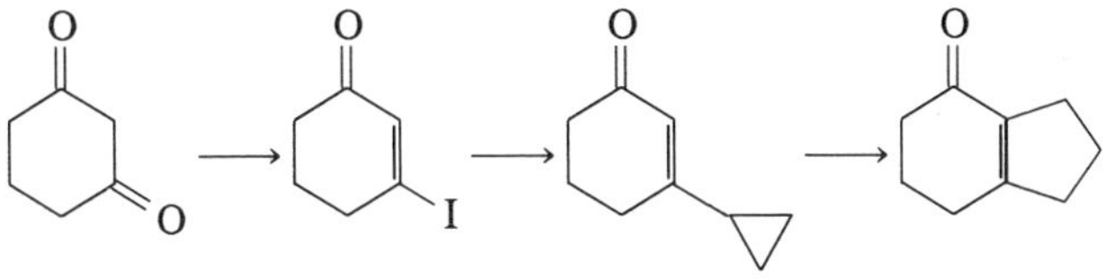

Can J Chem *60* 2965 (1982)

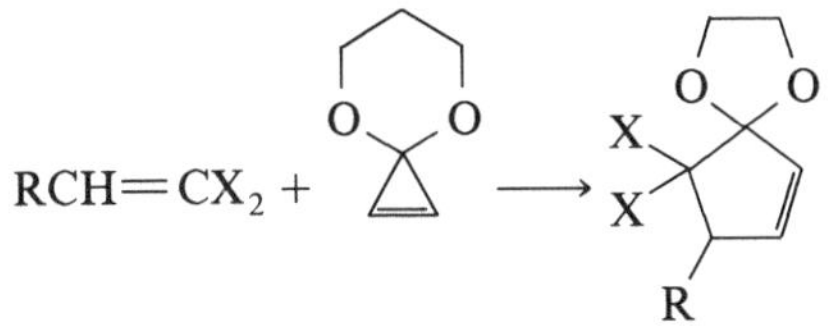

RCH=CX$_2$ +

X = CN, CO$_2$R, COR

JACS *106* 805 (1984)
Org Syn *65* 22 (1987)

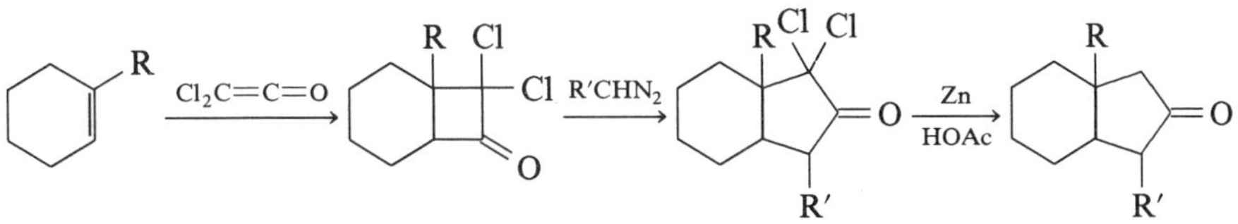

JACS *101* 4003 (1979); *105* 2435 (1983); *106* 5295 (1984); *108* 8015 (1986); *109* 4752 (1987) (asymmetric)
JOC *45* 2036 (1980); *48* 4763 (1983); *50* 3957 (1985)

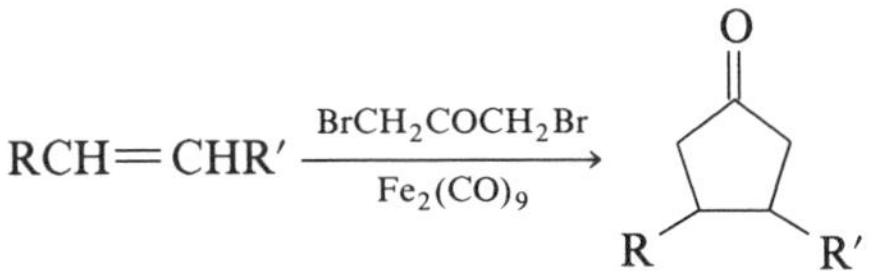

RCH=CHR'

JACS *94* 7202 (1972); *95* 2722 (1973); *99* 5196 (1977); *101* 220 (1979) (intramolecular); *105* 7358 (1983)
TL 4347 (1976); 993 (1978)
Acct Chem Res *12* 61 (1979) (review)
Org Rxs *29* 163 (1983) (review)

$$R'CH\!=\!CHNR_2 \xrightarrow[\text{Fe}_2(CO)_9]{RCHBrCOCHBrR}$$

JACS *94* 1772 (1972); *99* 5196 (1977); *100* 1799 (1978)
Acct Chem Res *12* 61 (1979) (review)
Org Rxs *29* 163 (1983) (review)

$$H_2C\!=\!CHCH\!=\!CH_2 \xrightarrow[\text{reagent}]{BrCH_2COCH_2Br}$$

Review: Org Rxs *29* 163 (1983)

Reagent(s)

Fe$_2$(CO)$_9$ JACS *93* 1272 (1971); *99* 5196 (1977);
                                                    100 1765 (1978); *101* 220 (1979)
                                                    (intramolecular)
                                                    TL 1741 (1973); 1049 (1974); *27* 2881 (1986)
                                                    JOC *40* 806 (1975)
                                                    Acct Chem Res *12* 61 (1979) (review)

Zn, B(OEt)$_3$ TL 4487 (1975)

Zn-Cu   JACS *94* 3940 (1972); *96* 5466 (1974)
                                                    Syn 902 (1978)

Zn-Cu, ultrasound   TL *27* 687 (1986)

Zn-Cu, Me$_3$SiCl   Syn 902 (1978)
                                                    Ber *117* 3325 (1984)

$$H_2C\!=\!CHCH\!=\!CH_2 \xrightarrow[\text{SnCl}_4]{\overset{\text{OSiMe}_3}{\overset{|}{H_2C\!=\!CCHO}}}$$

TL *23* 1693 (1982)

$$ArCR\!=\!CH_2 \xrightarrow[\text{ZnCl}_2 \text{ or AgClO}_4]{\overset{\text{OSiMe}_3}{\overset{|}{H_2C\!=\!CCR'_2X}}}$$

$$H_2C\!=\!CHCH\!=\!CH_2 \longrightarrow$$

Angew Int *18* 163 (1979)
CL 103 (1979)
JACS *104* 1330 (1982)
TL *27* 2881 (1986)

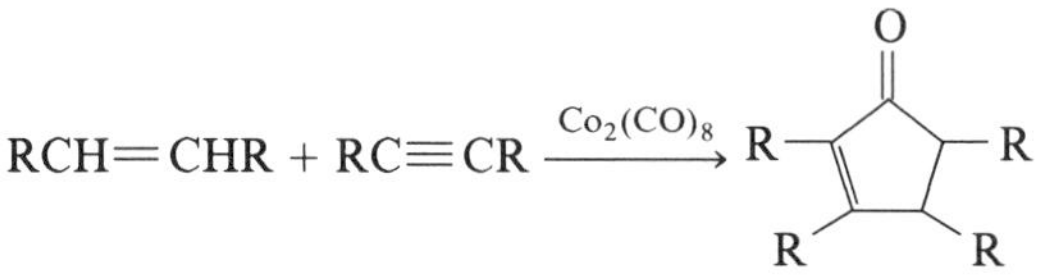

$$RCH=CHR + RC\equiv CR \xrightarrow{Co_2(CO)_8}$$

See page 677, Section 8.

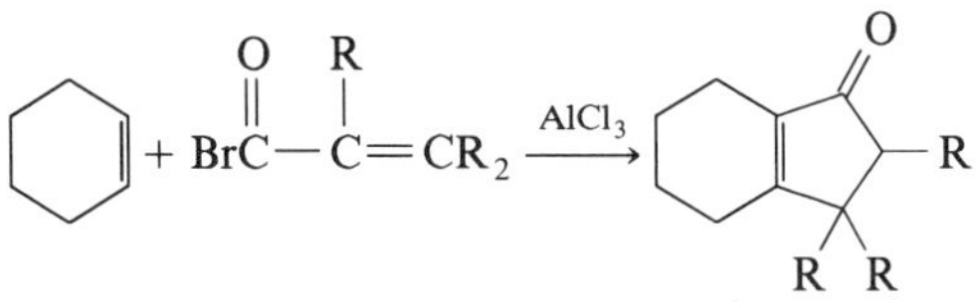

TL 4553 (1979)

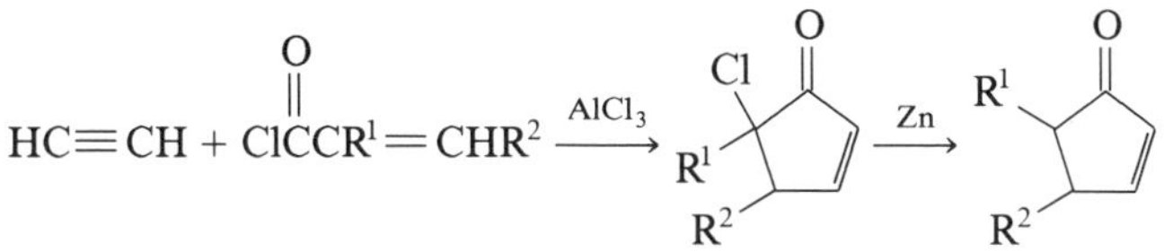

$$HC\equiv CH + ClCCR^1=CHR^2 \xrightarrow{AlCl_3} \xrightarrow{Zn}$$

Compt Rend C *266* 478 (1968)
BSCF 3098 (1970)
TL 3131 (1970); 4413 (1976)
Tetr *28* 4027 (1972)
JOC *45* 2224 (1980); *52* 5280 (1987)

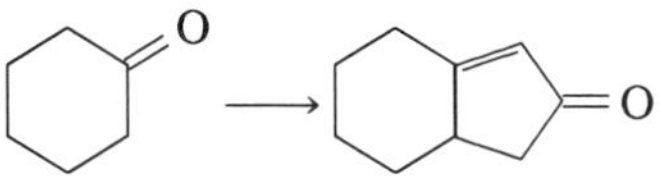

LDA/ BrCH$_2$C(OEt)$=$CHPO(OMe)$_2$/ H$_3$O$^+$/NaH	TL 3279 (1979) Can J Chem *60* 1114 (1982)
LDA/ BrCH$_2$COCH$=$PPh$_3$	Angew Int *18* 940 (1979)
LDA/ClCH$_2$C($=$CH$_2$)O$_2$P(OEt)$_2$, 5% Pd(PPh$_3$)$_4$/NaOH, Δ	TL *27* 1115 (1986) JOC *52* 1440 (1987)
Me$_3$SiCl, Et$_3$N/ H$_2$C$=$C(NO$_2$)CH$_3$, SnCl$_4$/H$_2$O/OH$^-$	JACS *98* 4679 (1976); *106* 2149 (1984)

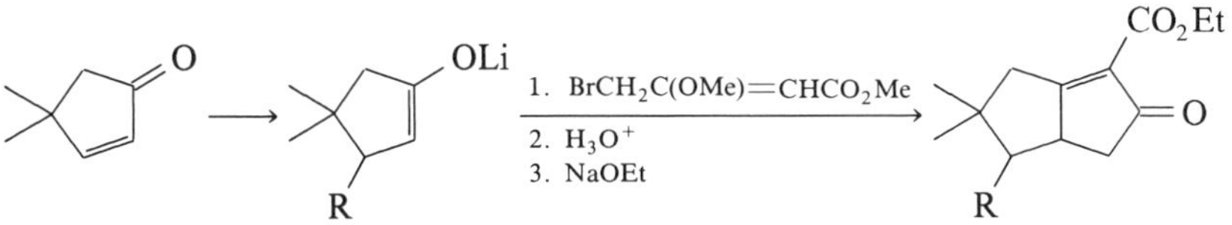

1. BrCH$_2$C(OMe)$=$CHCO$_2$Me
2. H$_3$O$^+$
3. NaOEt

R = H or CH$_2$CH$=$CH$_2$

JACS *104* 6646 (1982)
TL *27* 5245 (1986)

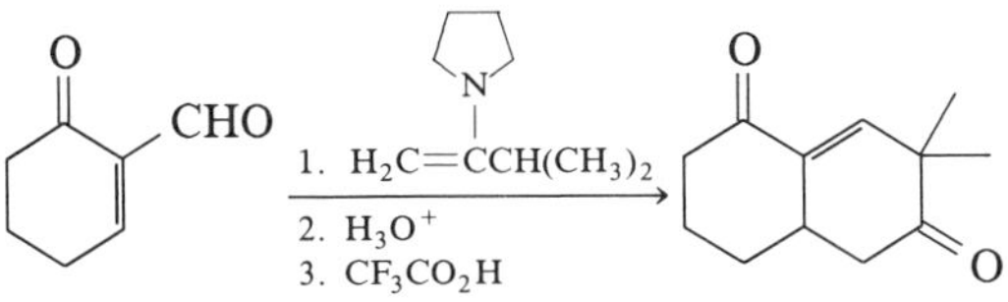

1. H$_2$C$=$CCH(CH$_3$)$_2$
2. H$_3$O$^+$
3. CF$_3$CO$_2$H

TL *27* 1449 (1986)

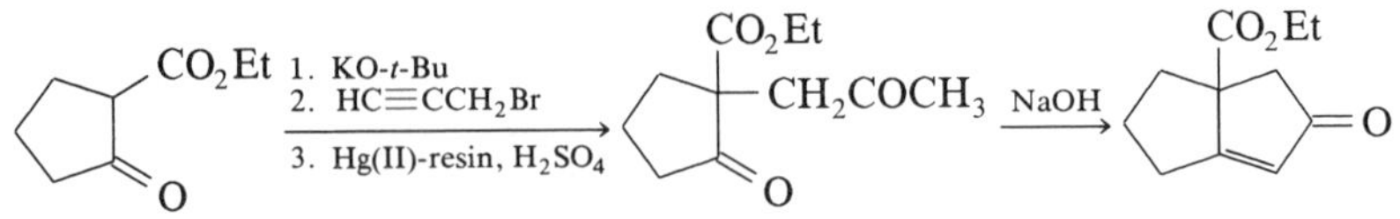

JOC *46* 2815 (1981)

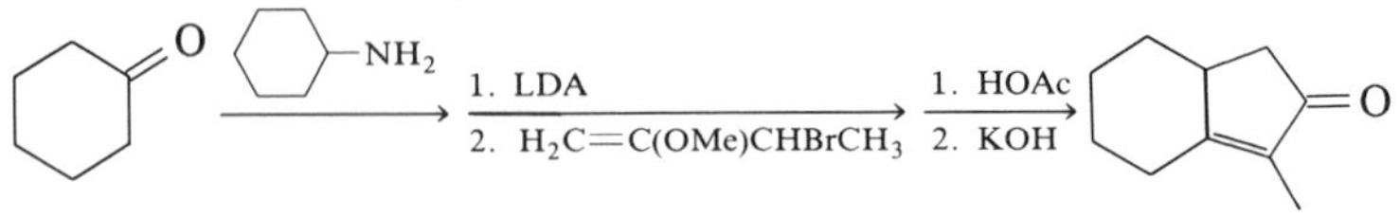

JOC *43* 4650 (1978)

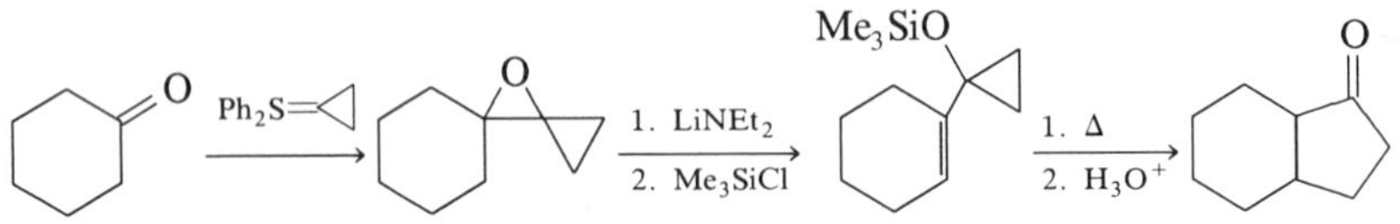

JACS *95* 5311 (1973)
Acct Chem Res *7* 85 (1974)

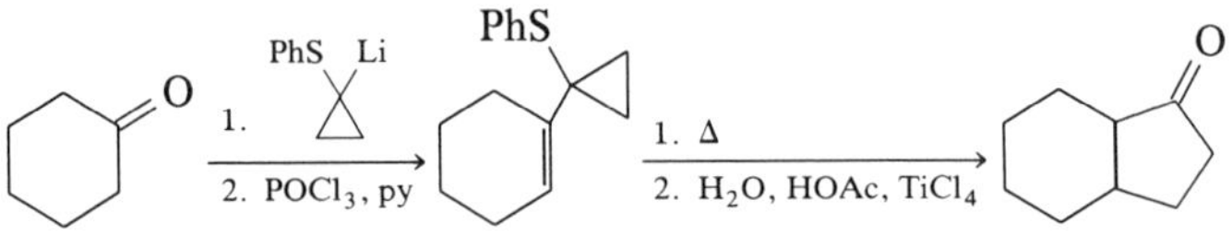

JACS *98* 248 (1976)

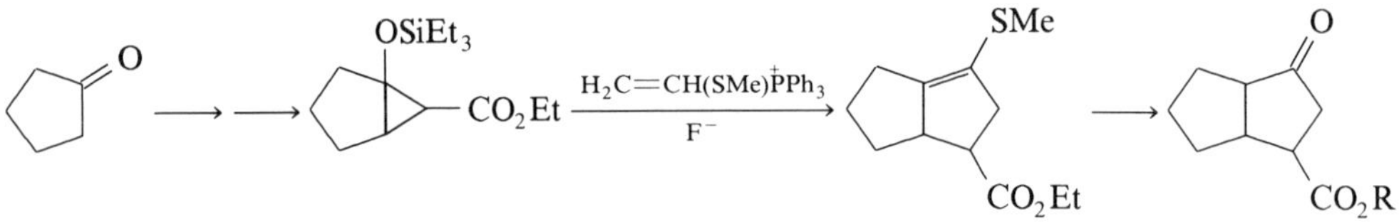

JACS *107* 734 (1985)
JOC *52* 1, 4139 (1987)

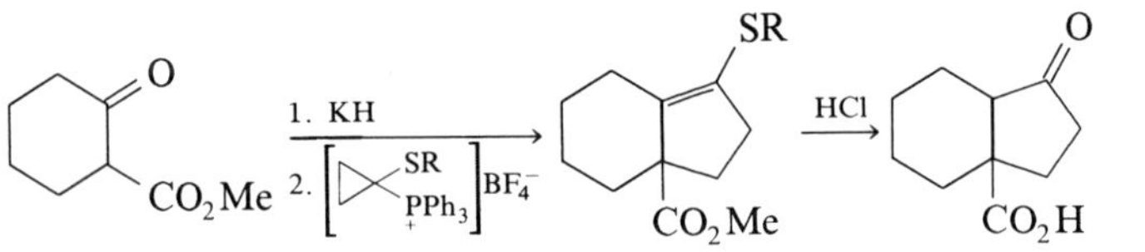

TL 4531 (1975)
JOC *46* 1828 (1981)

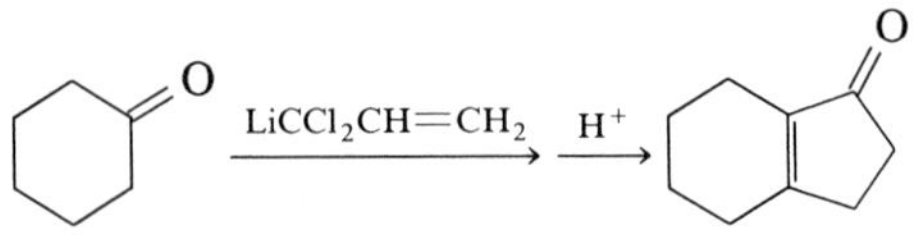

TL 771 (1978)
BCSJ *53* 1010 (1980)

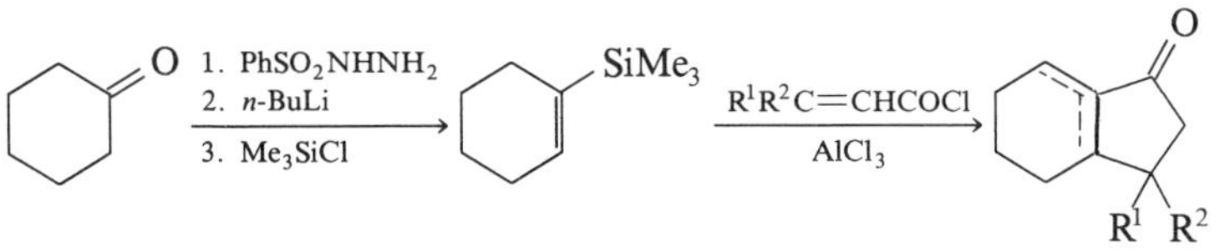

JOC *45* 3017 (1980)

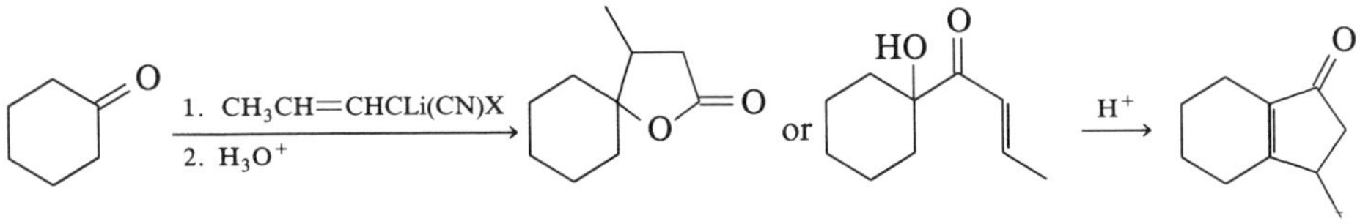

X	
NMe$_2$	TL *21* 1205 (1980)
OCH(OEt)CH$_3$	JOC *45* 395 (1980)
OSiMe$_3$	JOC *44* 462 (1979); *45* 395 (1980)

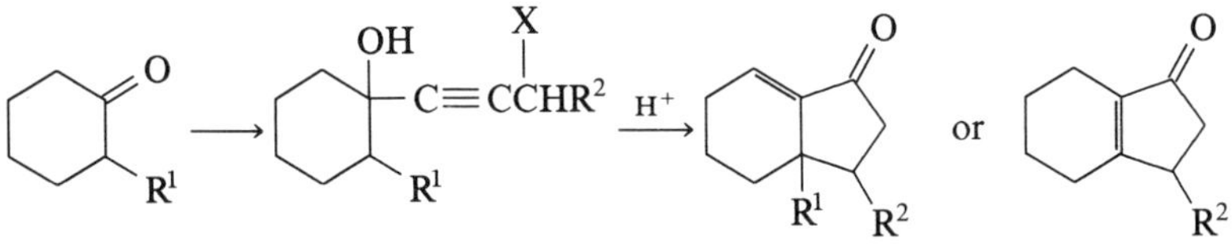

$(R^1 = H)$

X	
OH	Helv *59* 1226 (1976)
	JCS Perkin I 410 (1976)
	TL 3585 (1976); 3529 (1979); 3897 (1980)
	JACS *101* 1599 (1979); *107* 1769 (1985)
	BCSJ *54* 2747 (1981)
NEt$_2$	JCS 2247 (1953)

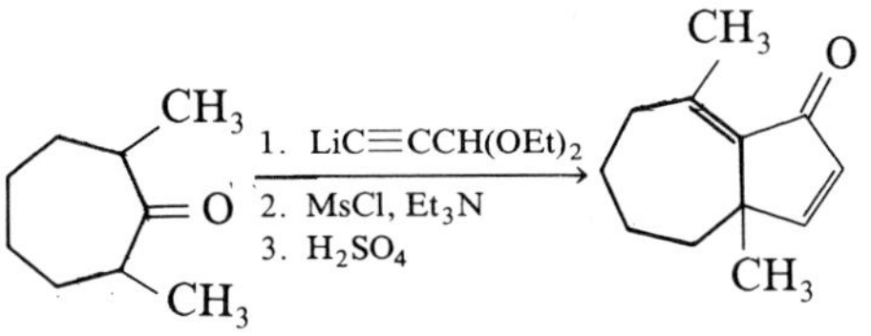

TL *22* 3565 (1981)

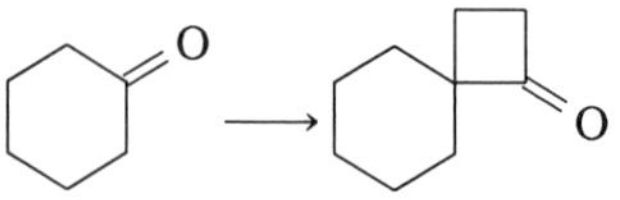

Br
⋈ /H⁺
Li

TL 3295 (1974)
Ber *108* 2368 (1975)

OR
⋈ /HBF₄ or SnCl₄
Li

TL *22* 2455 (1981)
Helv *64* 2598 (1981)
JOC *50* 3255 (1985)

SR
⋈ /acid
Li

Ber *108* 2368 (1975)
JACS *99* 3080 (1977)
TL *22* 2455 (1981); *23* 4385 (1982)

SeR
⋈ (R = Ph, Me)/TsOH
Li

TL *23* 4385 (1982)

6. Four-Carbon Annulation

6.1. Robinson and Related Annulations

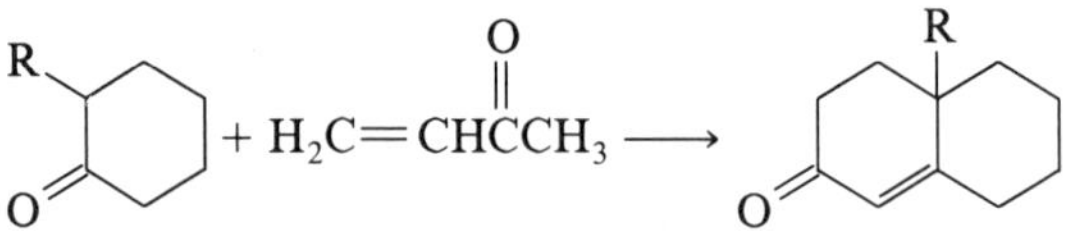

Reviews:

Tetr *32* 3 (1976)
Syn 177 (1976)

Stereochemistry:

JOC *29* 2501 (1964); *31* 2543 (1966); *33* 3913 (1968); *36* 178, 594 (1971)
TL 4807 (1965)
CC 753 (1967); 47, 1249 (1968)
Can J Chem *47* 4307 (1969)
JACS *93* 1539 (1971)

Acid promotion:

TL 4995 (1971)
JOC *45* 5399 (1980)
JACS *108* 4561 (1986)

Amino acid-catalyzed chiral annulation:

Angew Int *10* 496 (1971)
JOC *39* 1612, 1615 (1974); *40* 675 (1975)
JCS Perkin I 2349 (1983)
Org Syn *63* 26, 37 (1984)
TL *27* 1501 (1986) (mechanism)

Aprotic conditions:

JOC *48* 3349 (1983)

Use of enamines:

JACS *92* 5657 (1970)
JOC *40* 862 (1975); *47* 1438 (1982); *50* 2607 (1985)
Tetr *32* 1415 (1976)

Chiral imine annulation:

JACS *107* 273 (1985)

Modified Michael acceptors and their equivalents:

$CH_3CH(OCH_2Ph)CH_2CH_2Br$	JACS *78* 501 (1956)
$(CH_3COCH_2CH_2NMeEt_2)I$	JCS 1855 (1949)
$CH_3COCH_2CH_2Cl$	JCS 1029 (1964)
$ClCH_2COCH_2CH_2Cl$	CC 367 (1971)
$CH_3OCH_2COCH_2CH_2OCH_3$	JACS *100* 1263, 2140, 2150 (1978)
$RCH_2COC(SiMe_3)=CH_2$	JACS *95* 6152, 6867 (1973); *96* 6179 (1974); *107* 7184 (1985)
	Org Syn *58* 158 (1978)
	CC 174 (1982)
	Can J Chem *60* 509 (1982)
	Tetr *39* 925 (1983)
$CH_3O_2CCH_2COCH=CH_2$	JACS *86* 2034 (1964)
	TL 2755 (1972)
$CH_3CH(SePh)COCH=CH_2$	Syn Commun *8* 211 (1978)
$H_2C=CH(CH_2)_3COCH=CH_2$ (bis-annulation)	JACS *101* 5070 (1979)
$CH_3C(SiMe_3)=CHCH_2I$	JACS *96* 3682, 3684 (1974)
$CH_3CCl=CHCH_2Cl$	Coll Czech Chem Commun *13* 300 (1948)
	Helv *32* 1284, 2360 (1949)
	BSCF 780 (1954)
	JACS *82* 4245 (1960)
	JOC *30* 3642 (1965); *57* 4807 (1986)
	Tetr *25* 2159 (1969)
	Acct Chem Res *5* 311 (1972)
	BCSJ *52* 2978 (1979)
$Ph_3\overset{+}{P}CH=C(OEt)CH=CHR$ (R = H, Me)	JOC *43* 4673 (1978)

JACS *89* 5459, 5463, 5464 (1967)
JOC *37* 1652, 1659, 1664 (1972)

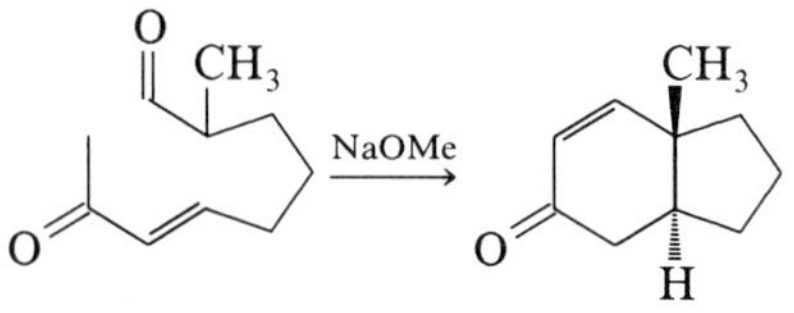

JACS *104* 310, 3767 (1982)

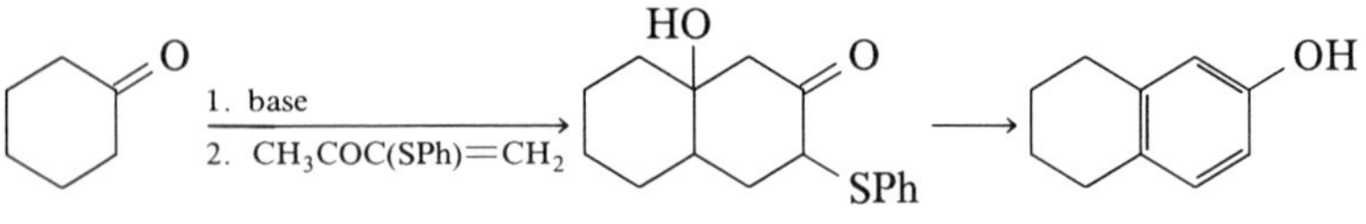

JOC *47* 1200 (1982)

6.2. Reversed Robinson Annulation

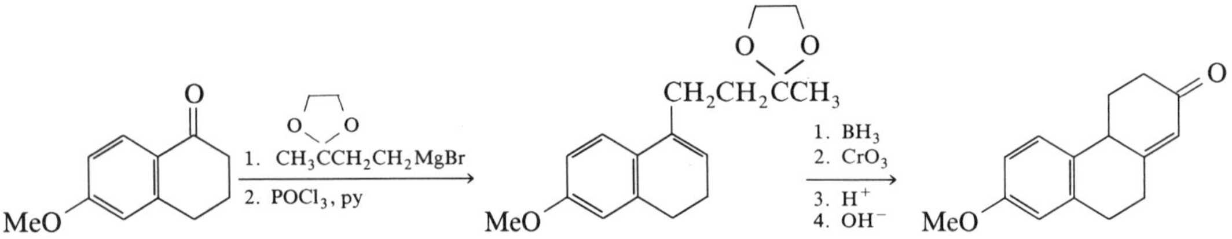

TL 3105 (1976)

6.3. Other Four-Carbon Annulations

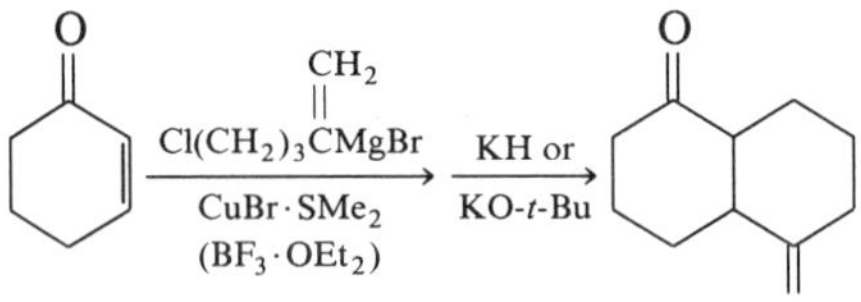

JOC *49* 4567 (1984)
CC 1342 (1987)

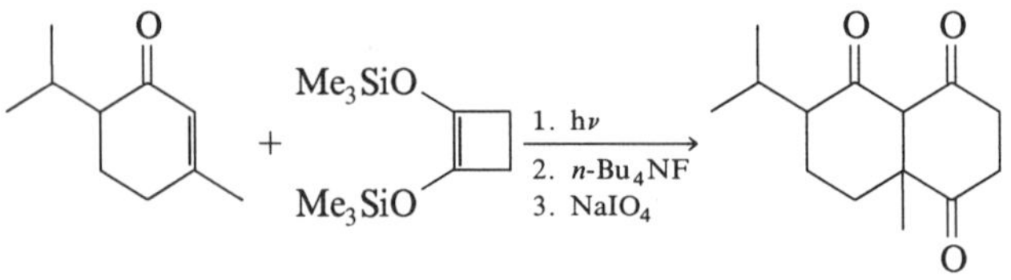

Syn 1024 (1980)

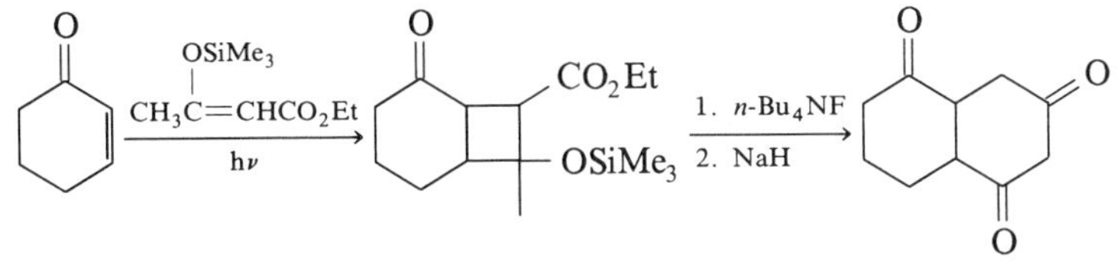

TL *23* 295 (1982)

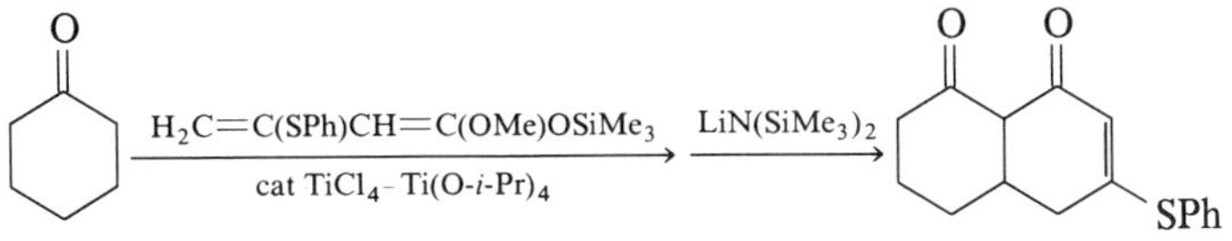

JOC *52* 110 (1987)

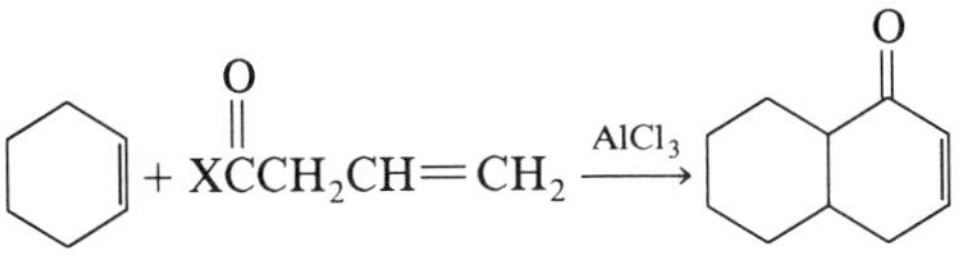

X = Cl, Br

CC 263 (1979)
JOC *44* 4042 (1979)

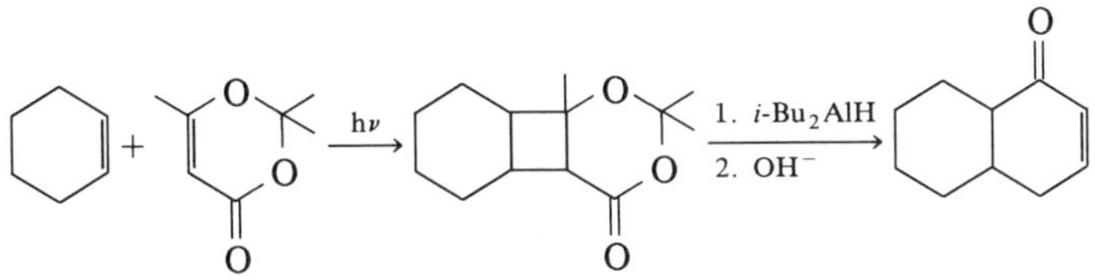

JACS *102* 3634 (1980)

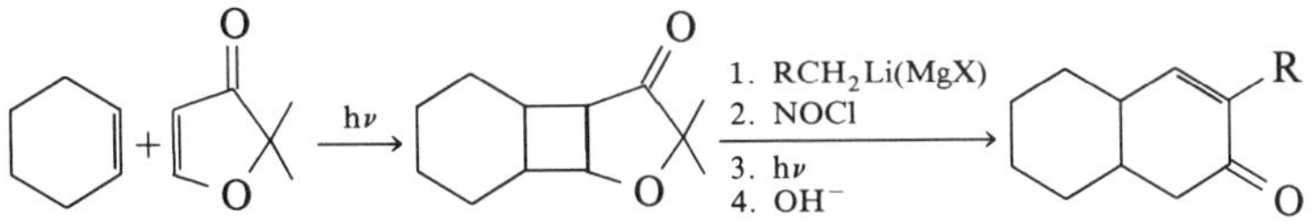

JACS *104* 4990 (1982)

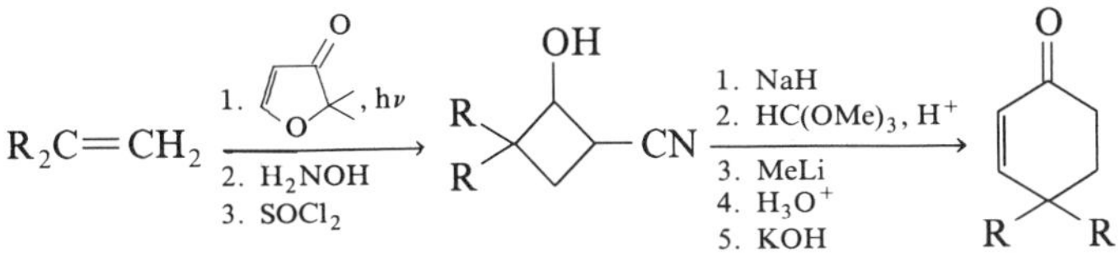

TL 2657 (1979); *23* 1235 (1982)

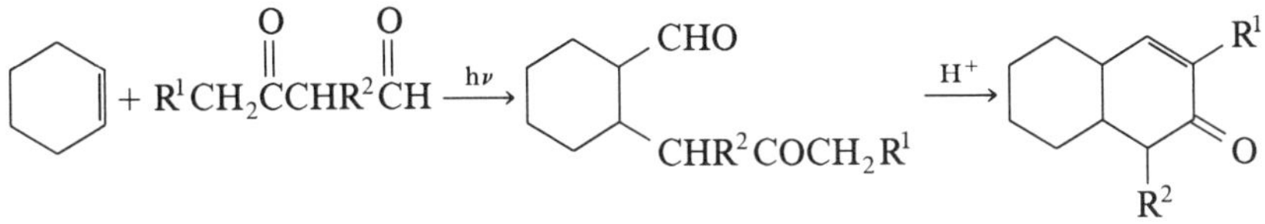

TL 3969 (1975)
JOC *40* 1865 (1975)

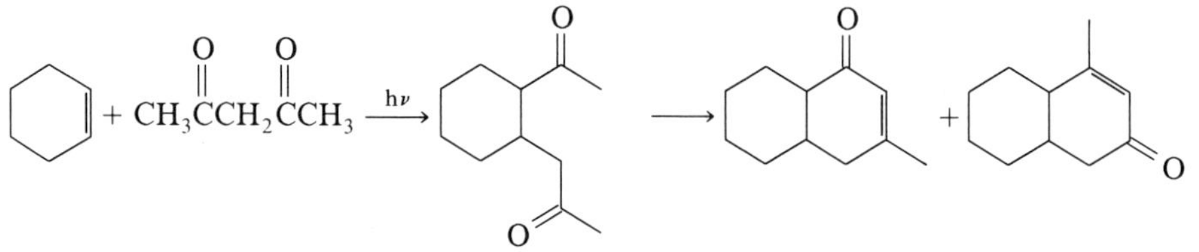

Can J Chem *41* 440 (1963)

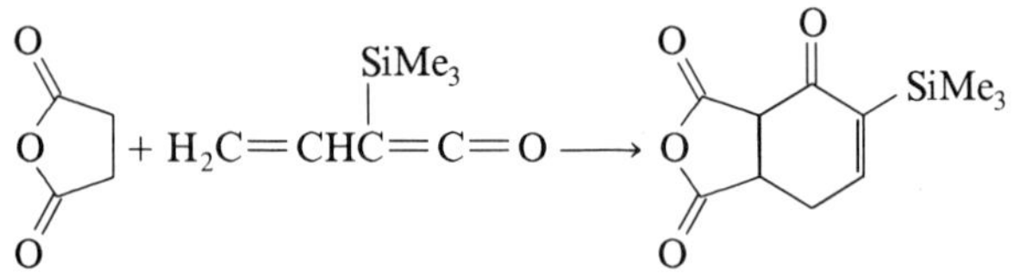

JOC *45* 4810 (1980)

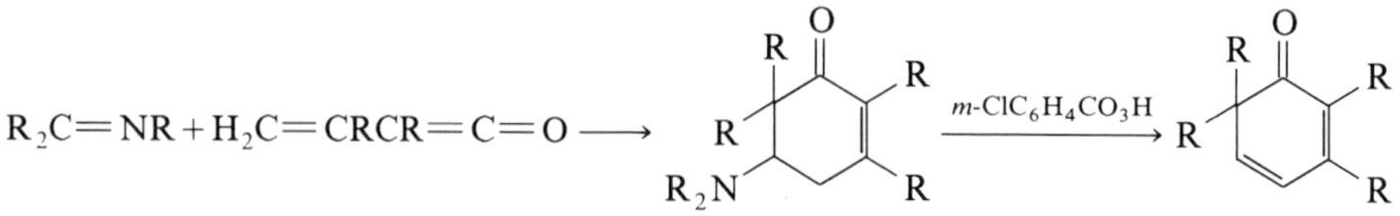

$R_2C{=}NR + H_2C{=}CRCR{=}C{=}O \longrightarrow$

Helv *65* 2230 (1982)

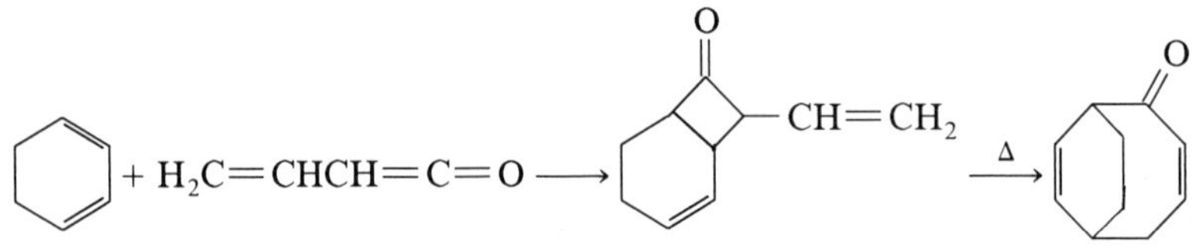

JACS *104* 7670 (1982)

7. Five-Carbon Annulation

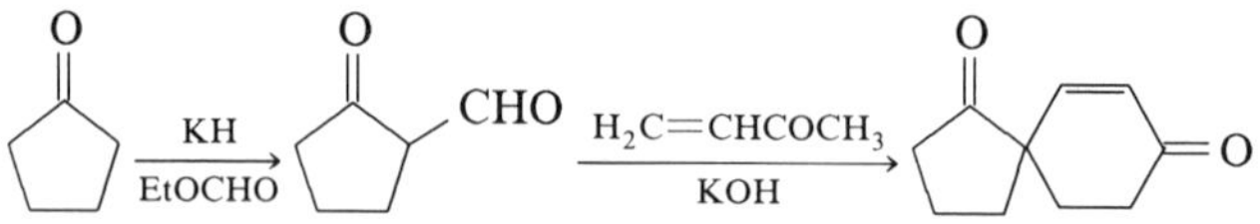

Syn 796 (1983)

8. Six-Carbon Annulation

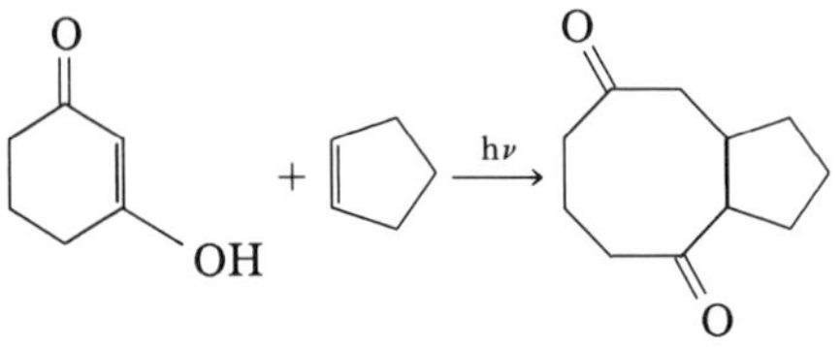

TL *23* 5485 (1982)
JOC *52* 2905 (1987)

7. CARBONYLATION AND RELATED REACTIONS

See also page 681, Section 8.

1. Alkanes

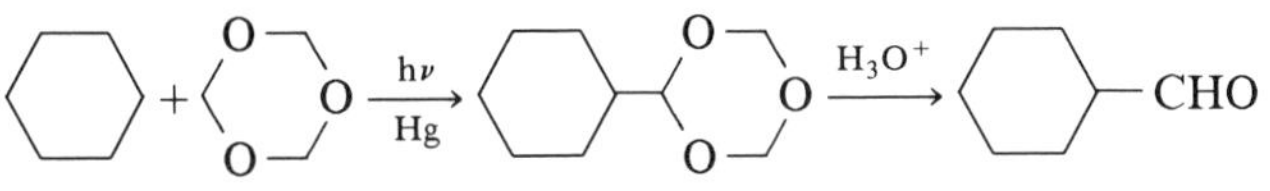

CC 970 (1987)

2. Organolithium Compounds

$$PhLi \xrightarrow[\text{2. H}_2\text{O}]{\text{1. CO}} Ph\overset{\overset{\displaystyle O}{\|}}{C}CHPh_2$$

Org Prep Proc Int *13* 144 (1981)

$$ArLi + CO \longrightarrow Ar\overset{\overset{\displaystyle O}{\|}}{C}-\overset{\overset{\displaystyle O}{\|}}{C}Ar$$

JOC *47* 4347 (1982)

3. Organoboron Compounds

Review:

Acct Chem Res *2* 65 (1969)

$$RCH=CH_2 \xrightarrow[\text{hydride}]{\text{9-BBN}} \xrightarrow{CO} \xrightarrow{[O]} RCH_2CH_2\overset{\overset{\displaystyle O}{\|}}{C}H$$

Hydride

LiHAl(O-*t*-Bu)$_3$ JACS *90* 499 (1968); *91* 2144, 4606 (1969)

KHB(O-*i*-Pr)$_3$ CC 607 (1979); 1273 (1982)
Syn 701 (1979)

JACS *107* 4980 (1985); *109* 7553 (1987)

$$R_3B \xrightarrow{\text{reagent(s)}} \xrightarrow{[O]} R\overset{\overset{\displaystyle O}{\|}}{C}R$$

Reagent(s)

CO, H$_2$O, Δ JACS *89* 2738, 4530 (1967)

LiC(SPh)$_3$ CC 1149 (1981)

$$(CH_3)_2CHC(CH_3)_2BH_2 \xrightarrow{\text{alkene A}} \xrightarrow{\text{alkene B}} \xrightarrow[\Delta]{CO} \xrightarrow{[O]} R_A\overset{\overset{\displaystyle O}{\|}}{C}R_B$$

JACS *89* 5285 (1967)
Syn 196 (1972)

$$H_2C=CH(CH_2)_nCH=CH_2 \xrightarrow{(CH_3)_2CHC(CH_3)_2BH_2} \xrightarrow{\text{reagents}} \xrightarrow{[O]} O=C\!\!\!\left(CH_2\right)_{n+4}$$

Reagents

CO, Δ JACS *89* 5477 (1967)
CC 594 (1968); 607 (1979)

KCN, (CF$_3$CO)$_2$O   TL *21* 2381 (1980); *22* 2075 (1981)
Syn Commun *10* 93 (1980)

$$R_3B \xrightarrow{NaCN} \xrightarrow{(CF_3CO)_2O} \xrightarrow{[O]} R\overset{\overset{\displaystyle O}{\|}}{C}R$$

$$(CH_3)_2CHC(CH_3)_2BH_2 \xrightarrow{\text{alkene A}} \xrightarrow{\text{alkene B}} \xrightarrow[\text{2. (CF}_3\text{CO)}_2\text{O}]{\text{1. NaCN}} \xrightarrow{[O]} R_A\overset{\overset{\displaystyle O}{\|}}{C}R_B$$

CC 1529 (1970); 1048 (1971); 186 (1973)
JCS Perkin I 129 (1975)
Syn Commun *10* 675 (1980)

$$(CH_3)_2CHC(CH_3)_2BHCl \xrightarrow[\substack{2. \ R'Li \ (MgX) \\ . \ [or \ KHB(O\text{-}i\text{-}Pr)_3/ \\ R'CH=CH_2]}]{1. \ RCH=CH_2} \xrightarrow[\substack{2. \ PhCOCl \ or \\ (CF_3CO)_2O \\ 3. \ H_2O_2, \ NaOH}]{1. \ NaCN} RCH_2CH_2\overset{\displaystyle O}{\overset{\|}{C}}R'(CH_2CH_2R')$$

JACS *102* 5919 (1980)

JOC *45* 4542 (1980)

$$R_2BX \xrightarrow[(Li)KOCR'_3 \ (R' = Me, \ Et)]{HCCl_2OCH_3} \xrightarrow{[O]} R_2C=O$$

X = Cl, Br, OR

JACS *95* 6876 (1973); *104* 6844 (1982); *109* 5420 (1987) (chiral)
JOC *44* 2417 (1979); *47* 863 (1982)
Organomet *1* 212 (1982)
Tetr *40* 1325 (1984) (chiral)
TL *28* 3771 (1987)

$$R_3B \ or \ RB\!\!\overset{\frown}{\underset{\smile}{\quad}} + R'X \xrightarrow[\substack{Zn(acac)_2 \\ cat \ PdCl_2(PPh_3)_2 \\ cat \ PPh_3}]{CO} \overset{\displaystyle O}{\overset{\|}{R}}CR'$$

R' = aryl, benzylic

JOMC *301* C17 (1986)

$$R_3B \xrightarrow{reagent} \xrightarrow{[O]} \overset{\displaystyle O}{\overset{\|}{R}}CR'$$

Reagent

LiCH(SPh)$_2$R'   CL 961 (1973)

(o-C$_6$H$_4$(S)$_2$)CR'Li   TL 1893 (1979)

$$R_3B \xrightarrow[\substack{2. \ R'OSO_2F \ (R' = Me, \ Et) \\ 3. \ H_3O^+}]{1. \ LiC(OMe)=CHOMe} \overset{\displaystyle O}{\overset{\|}{R}}CCH_2R'$$

CL 1059 (1981)

4. Organosilicon Compounds

$$RCH=CHSiMe_3 \xrightarrow[TiCl_4]{Cl_2CHOCH_3} RCH=CH\overset{\displaystyle O}{\overset{\|}{C}}H$$

CL 859 (1978)

5. Organozirconium Compounds

$$RCH{=}CHCH_3 \text{ or } RCH_2CH{=}CH_2 \xrightarrow[\substack{2.\ CO \\ 3.\ H^+}]{1.\ Cp_2ZrHCl} RCH_2CH_2CH_2\overset{\displaystyle O}{\overset{\|}{C}}H$$

JACS *97* 228 (1975)

$$RCH{=}CHCH{=}CH_2 \xrightarrow[\substack{2.\ CO \\ 3.\ H^+}]{1.\ Cp_2ZrHCl} RCH{=}CHCH_2CH_2\overset{\displaystyle O}{\overset{\|}{C}}H$$

JACS *98* 262 (1976)

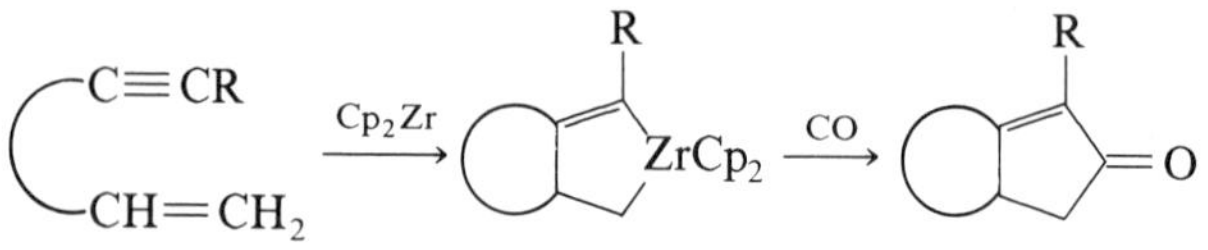

JACS *107* 2568 (1985)
TL *27* 2829 (1986); *28* 917 (1987)

6. Organomanganese Compounds

$$RMn(CO)_5 + CO + H_2C{=}CHX \xrightarrow[\substack{2.\ h\nu,\ O_2}]{1.\ 6\ kbar} R\overset{\displaystyle O}{\overset{\|}{C}}CH_2CH_2X$$

$$X = CO_2Me,\ SO_2Ph$$

TL *28* 2229 (1987)

$$RMn(CO)_5 + CO + HC{\equiv}CR' \xrightarrow{HCl} R\overset{\displaystyle O}{\overset{\|}{C}}CH{=}CHR'$$

TL *28* 2233 (1987)

7. Organoiron Compounds

See page 712, Section 3.

8. Organocobalt Compounds

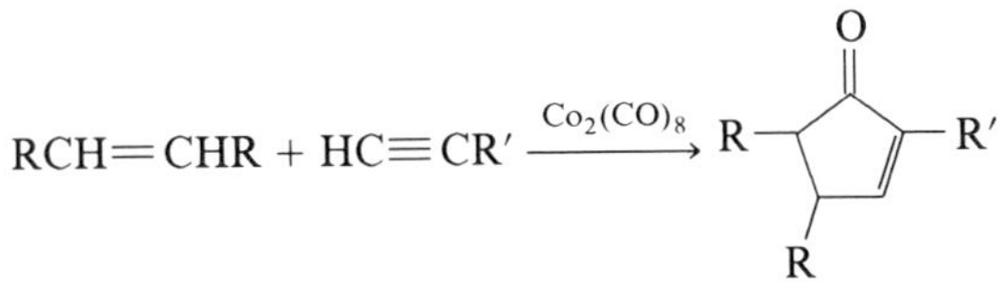

$$RCH = CHR + HC \equiv CR' \xrightarrow{Co_2(CO)_8}$$

Reviews:

Tetr *41* 5855 (1985)
Ann NY Acad Sci *295* 2 (1977)

Intermolecular:

JCS Perkin I 977 (1973); 30 (1976)
J Chem Res (S) 8 (1977); 244 (1983); 344, 3131 (1984)
Organomet *1* 1560 (1982)
TL *24* 2905 (1983)
Tetr *41* 5995 (1985)
JOC *50* 5215 (1985); *52* 3595 (1987)

Intramolecular:

JOC *46* 5357, 5436 (1981); *49* 5025 (1984); *52* 569, 5296 (1987)
JACS *105* 2477 (1983); *108* 3128, 3835 (1986); *109* 7495 (1987)
TL *25* 4041 (1984); *26* 2475, 4851 (1985); *27* 1241, 1245 (1986); *28* 5465 (1987)
Bull Acad Sci USSR, Div Chem Sci 2434 (1984); 2455 (1985); 211, 213 (1987)
Tetr *41* 5861 (1985)

9. Organorhodium Compounds

$$\underset{XCCH=CH_2}{\overset{O}{\|}} \xrightarrow[\text{cat } Rh_4(CO)_{12}]{CO} \underset{(XCCH_2CH_2)_2C=O}{\overset{O}{\|}}$$

$$X = OR, NR_2$$

CL 361 (1982)

$$RCH = CHHgCl \xrightarrow[\text{cat } [ClRh(CO)_2]_2]{CO} (RCH = CH)_2 CO$$

$$ArHgCl \longrightarrow Ar_2CO$$

JOC *45* 3840 (1980)

10. Organonickel Compounds

$$ArX + CO + Sn(CH_3)_4 \xrightarrow[\text{HMPA}]{\text{Ni catalyst}} \underset{ArCCH_3}{\overset{O}{\|}}$$

Syn 47 (1981)

TL *28* 4745 (1987)

11. Organopalladium Compounds

$$RX \longrightarrow R\overset{\overset{\displaystyle O}{\|}}{C}H$$

CO, cat $PdCl_2$, H_2, py (R = aryl)   BCSJ *49* 1681 (1976)

CO, cat $PdX_2(PPh_3)_2$ (X = Br, I), H_2, R_3N JACS *96* 7761 (1974)
  (R = aryl, heterocyclic, vinylic)

CO, cat $Pd(PPh_3)_4$, NaO_2CH or   JOC *49* 4009 (1984)
  $[-SiMeHO-]_x$ (R = aryl)

CO, cat $Pd(PPh_3)_4$, n-Bu_3SnH JACS *105* 7175 (1983);
  (R = aryl, allylic, benzylic,   *108* 452 (1986)
  vinylic; X = halide, OTf)

$$RX + CO + HC \equiv CR' \xrightarrow[\text{Et}_3\text{N}]{\text{cat } PdCl_2L_2} R\overset{\overset{\displaystyle O}{\|}}{C}C \equiv CR'$$

CC 333 (1981)

$$RX + CO + R'SnR''_3 \xrightarrow{\text{Pd catalyst}} R\overset{\overset{\displaystyle O}{\|}}{C}R'$$

R = allylic, aryl, benzylic, vinylic; X = halide, OTf; R′ = alkyl, allylic, aryl, vinylic

TL 2601 (1979)
JOMC *205* C27 (1981)
JACS *105* 7173 (1983); *106* 4833, 6417, 7500 (1984)
Organomet *3* 1108 (1984)

$$ArNH_2 \longrightarrow (ArN_2)BF_4 \xrightarrow[\text{cat Pd(OAc)}_2]{\text{CO, Et}_3\text{SiH}} Ar\overset{\overset{\displaystyle O}{\|}}{C}H$$

JOMC *270* 283 (1984)

$$ArNH_2 \longrightarrow (ArN_2)X \xrightarrow[\text{cat Pd(OAc)}_2]{\text{CO, R}_4\text{Sn}} Ar\overset{\overset{\displaystyle O}{\|}}{C}R$$

$X = BF_4, PF_6$

CL 35 (1982)

12. Organoplatinum Compounds

$$RI + CO + H_2 \xrightarrow[K_2CO_3]{\text{cat } PtCl_2(PPh_3)_2} R\overset{\displaystyle O}{\overset{\|}{C}}H$$

$$R = 1°, 2° \text{ alkyl}$$

CC 351 (1986)

13. Hydroformylation and Hydroacylation

$$RCH = CH_2 + CO + H_2 \xrightarrow[\text{catalyst}]{\text{transition metal}} RCH_2CH_2\overset{\displaystyle O}{\overset{\|}{C}}H + R\overset{\displaystyle CHO}{\overset{|}{C}}HCH_3$$

Reviews:

J. Falbe, "Carbon Monoxide in Organic Synthesis," Springer, New York (1970)
Catal Rev 6 49 (1972)
"Organic Synthesis via Metal Carbonyls," Eds. I. Wender and P. Pino, Wiley, New York (1977)
J Mol Catal 4 243 (1978)
Topics Curr Chem 105 77 (1982) (chiral)
JACS 109 7714 (1987) (fluoroalkenes)

Co	JACS 99 1058 (1977); 103 7590, 7594 (1981)
Rh	JACS 95 6504 (1973); 99 1058 (1977)
	BCSJ 47 1698 (1974) (chiral)
	Org Syn 57 11 (1977)
	Angew Int 19 178 (1980) (review)
	JOC 46 4422 (1981) (chiral)
	JOMC 279 193 (1985) (chiral)
Pd	JOC 51 4189 (1986) (chiral)
Pt	JACS 97 3553 (1975); 109 7122 (1987) (chiral)
	J Catalysis 45 256 (1976)
	Helv 59 642 (1976) (chiral)
	CL 361 (1978) (chiral)
	JOMC 213 503 (1981); 279 193 (1985) (chiral);
	296 281 (1985) (chiral)
	Helv 64 1865 (1981)

$$RCH = CH_2 + (CH_2O)_n \xrightarrow{\text{cat } H_2Rh(O_2COH)(i\text{-}Pr_3P)_2} RCH_2CH_2\overset{\displaystyle O}{\overset{\|}{C}}H$$

TL 23 4967 (1982)

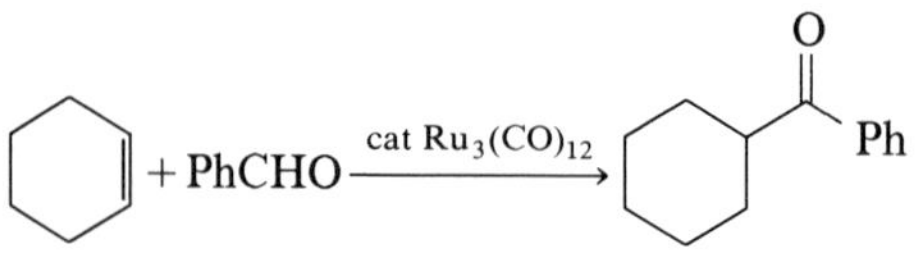

TL *28* 6229 (1987)

$$RCH{=}CH_2 + H\overset{O}{\overset{\|}{C}}R' \xrightarrow[\text{(AcO)}_2]{h\nu \text{ or}} RCH_2CH_2\overset{O}{\overset{\|}{C}}R'$$

JOC *14* 250 (1949)

$$XCH{=}CH_2 + H\overset{O}{\overset{\|}{C}}R \longrightarrow XCH_2CH_2\overset{O}{\overset{\|}{C}}R$$

$$X = COR, CO_2R, CN, \text{etc.}$$

See conjugate addition under the appropriate functional group.

$$R^1C{\equiv}CR^2 + H_2C{=}CH_2 + CO + H_2 \xrightarrow{\text{cat Rh}_4\text{(CO)}_{12}} \underset{H}{\overset{R^1}{>}}C{=}C\underset{COCH_2CH_3}{\overset{R^2}{<}}$$

CL 401 (1982)

8. ELECTROPHILIC ACYLATION

1. Synthesis of Aldehydes

Reviews:

Quart Rev *20* 179 (1966)
Heterocycles *6* 731 (1977)

$$RX + Li + \overset{\overset{\displaystyle O}{\|}}{HCNR'_2} \xrightarrow{\text{(ultrasound)}} \xrightarrow{H_2O} R\overset{\overset{\displaystyle O}{\|}}{CH}$$

R = 1°, 2°, 3° alkyl; benzylic; aryl

Syn 160 (1973)
TL *23* 3361 (1982); *27* 1791 (1986)

$$RM \longrightarrow R\overset{\overset{\displaystyle O}{\|}}{CH}$$

$\underline{M}$	Reagents	
Li	$CNC(CH_3)_2C(CH_3)_3/H_3O^+$	JACS *91* 7778 (1969); *92* 6675 (1970)
	![structure] $/H_3O^+$	TL 5151 (1969)
MgX	$Fe(CO)_5/H_3O^+$	BCSJ *55* 1663 (1982)
	![structure] OCH_3/H_3O^+	JACS *92* 584 (1970)
	$EtOCH{=}NPh/H_3O^+$	JOC *6* 489 (1941)
	HCO_2MgBr/H_3O^+	TL *21* 2869 (1980)
	$HC(OEt)_3/H_3O^+$	JOC *6* 437 (1941)

681

$$RM \longrightarrow \overset{\overset{O}{\|}}{RCH} \quad (Continued)$$

<u>M</u>	Reagents	

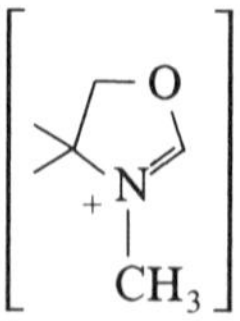

I⁻, HMPA/H₃O⁺

JACS *92* 6676 (1970)
Angew Int *15* 270 (1976)

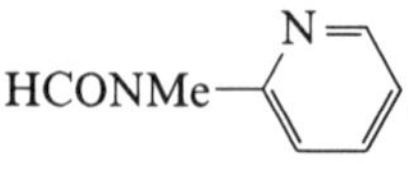

HCONMe—⟨N=⟩/H₃O⁺

Syn 403 (1978)
TL 5179 (1978)

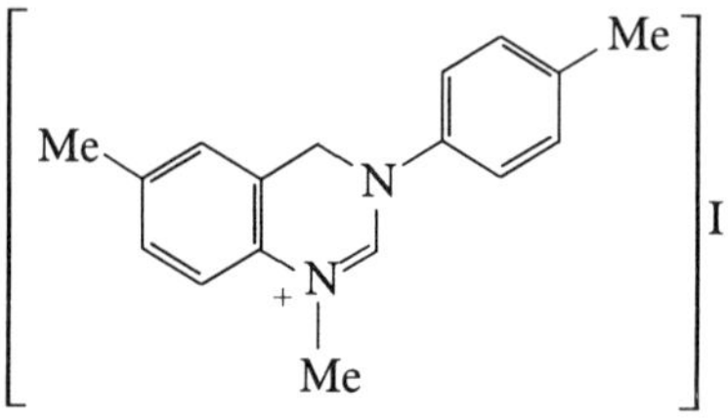

I⁻/   JACS *77* 5118 (1955)

H₃O⁺

(Me₂N⁺=CHN=CHNMe₂)Cl⁻/H₃O⁺   Syn Commun *11* 571 (1981)

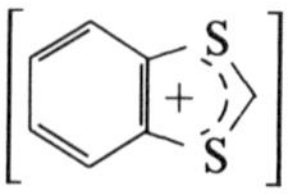

ClO₄⁻/hydrolysis

JCS Perkin I 1886 (1976)
TL *22* 1821 (1981)

CH₃SOCH₂SCH₃/SO₂Cl₂, SiO₂-H₂O   TL 3883 (1977)

Li or MgX   HCONMe₂/H₃O⁺

JOC *6* 437 (1941); *51* 3762,
  5106 (1986)
JCS 3334 (1955); 4691 (1956);
  1054 (1958)
Chem Ind 1596 (1957)
Compt Rend *275* 511 (1972)
Syn 160 (1973)

HCON⟨ ⟩/H₃O⁺

Angew Int *20* 878 (1981)
Org Syn *64* 114 (1985)
JOC *50* 2423, 2427 (1985);
  51 142 (1986); *52* 5560 (1987)

HCON⟨ O⟩/H₃O⁺   JOC *49* 3856 (1984)

Cl₂CHB(O-*i*-Pr)₂/H₂O₂   JOMC *122* 145 (1976)

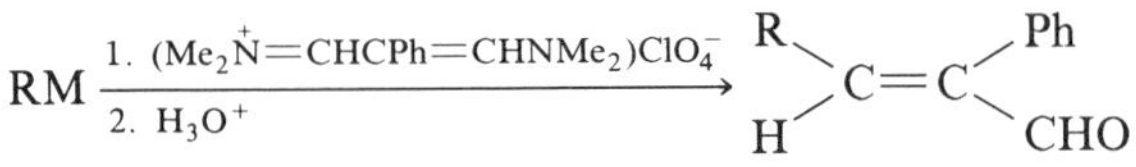

$$RM \xrightarrow[\text{2. } H_3O^+]{\text{1. } (Me_2\overset{+}{N}=CHPh=CHNMe_2)ClO_4^-} \overset{R}{\underset{H}{>}}C=C\overset{Ph}{\underset{CHO}{<}}$$

R = 1° alkyl, benzylic, aryl

M = Li, MgX

Syn Commun *11* 561 (1981)

$$RM + PhOCH(OR')_2 \longrightarrow RCH(OR')_2$$

RM	
RMgX (R = 1°, 2° alkyl; aryl; alkynyl; propargylic; vinylic)	Ber *103* 643 (1970) Syn 379 (1975) J Chem Res (S) 343 (1981)
$RCH=CHCH_2M$ (M = MgX, ZnX, $Al_{2/3}Br$)	J Chem Res (S) 343 (1981)
$RCH=C=CHCH_2Al_{2/3}Br$	J Chem Res (S) 343 (1981)

2. Synthesis of Ketones

2.1. From Carbonyl Cation Equivalents

$$2\,RLi + CO_2 \xrightarrow{H_2O} R\overset{O}{\overset{\|}{C}}R$$

JACS *55* 1258 (1933); *75* 1771 (1953); *77* 2806 (1955)

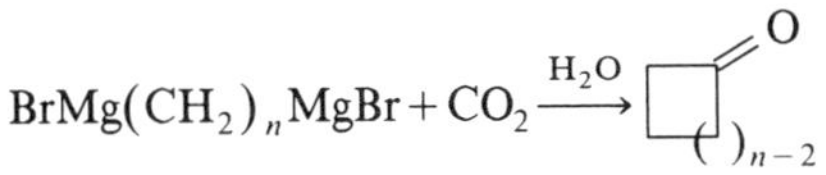

$$BrMg(CH_2)_n MgBr + CO_2 \xrightarrow{H_2O} \square\overset{O}{(\,)_{n-2}}$$

n	
3	Syn 721 (1983)
4	Ber *44* 1918 (1911)
5	Compt Rend *144* 1358 (1907)

$$2\,RMgX + MeOCH_2CO_2Me \xrightarrow[\text{oxidation}]{\text{anodic}} R\overset{O}{\overset{\|}{C}}R$$

TL 3625 (1977)

$$RMgX \xrightarrow[]{Fe(CO)_5} \xrightarrow{R'X} R\overset{O}{\overset{\|}{C}}R'$$

TL 761 (1978)

$$\left[\begin{array}{c} \text{benzene ring with} \\ \text{S}^+ \text{S} \end{array}\right] ClO_4^- \xrightarrow{RMgX} \begin{array}{c} \text{benzene ring} \\ \text{S} \quad H \\ \text{S} \quad R \end{array} \xrightarrow[\text{2. R'MgX}]{\text{1. }(Ph_3C)ClO_4} \begin{array}{c} \text{benzene ring} \\ \text{S} \quad R \\ \text{S} \quad R' \end{array}$$

$$\begin{array}{ccc} & O & & O \\ & \parallel & & \parallel \\ & RCH & & RCR' \end{array}$$

JCS Perkin I 1886 (1976)
TL *22* 1821 (1981)

$$\begin{array}{ccc} O & & O & & O \\ \parallel & RMgX & \parallel & R'MgX & \parallel \\ ClCSPh & \xrightarrow{\text{cat NiCl}_2\text{(dppe)}} & RCSPh & \xrightarrow{\text{cat Fe(acac)}_3} & RCR' \end{array}$$

TL *26* 3595 (1985)

2.2. From Aldehydes

See page 709, Section 9, and page 721, Section 10, for related reactions.

$$\begin{array}{ccc} O & & O \\ \parallel & \text{1. R'MgX} & \parallel \\ RCH & \xrightarrow{\text{2. PhCHO}} & RCR' \end{array}$$

TL *28* 769 (1987)

$$RM + VCl_3 \longrightarrow RVCl_2 \xrightarrow[\Delta]{R'CHO} \begin{array}{c} O \\ \parallel \\ RCR' \end{array}$$

M = Li, MgX; R = alkyl, alkynyl, aryl, vinylic

JACS *107* 7179 (1985)
TL *27* 933 (1986)

$$\begin{array}{ccc} O & & & O \\ \parallel & & \text{2 R'Li or} \quad H_3O^+ & \parallel \\ RCH & \longrightarrow RCH{=}NOR'' & \xrightarrow{\text{2 R'MgX}} & RCR' \end{array}$$

R = aryl > alkyl

TL *27* 3033 (1986)

$$\begin{array}{ccc} O & & NNH\text{-}t\text{-Bu} \\ \parallel & & \parallel \\ RCH & \longrightarrow & RCH \end{array} \xrightarrow{n\text{-BuLi}} \begin{cases} \xrightarrow{R'X} \quad \begin{array}{c} \text{1. }n\text{-BuLi} \\ \text{2. }H_2O \end{array} \xrightarrow{H^+} \begin{array}{c} O \\ \parallel \\ RCR' \end{array} \\ \xrightarrow{R'_2CO} \begin{array}{c} \text{1. }n\text{-BuLi} \\ \text{2. }H_2O \end{array} \xrightarrow{H^+} \begin{array}{cc} O & OH \\ \parallel & \mid \\ RC{-}CR'_2 \end{array} \end{cases}$$

CC 1040 (1983)

$$\begin{array}{ccc} O & & O \\ \parallel & \text{1. PhSOCXYLi} & \parallel \\ RCH & \xrightarrow{\text{2. }\Delta} & RCCHXY \end{array}$$

X, Y

H, F   TL *24* 725 (1983)

Cl, Cl   TL *24* 527 (1983)

$$\underset{RCH}{\overset{O}{\overset{\|}{}}} \longrightarrow \underset{\underset{\overset{\|}{N_2}}{RCHCCX}}{\overset{HO\ \ O}{\overset{|\ \ \ \|}{}}} \xrightarrow{\ cat\ Rh_2(OAc)_4\ } \underset{RCCH_2CX}{\overset{O\ \ \ O}{\overset{\|\ \ \ \|}{}}}$$

$$X = R,\ OR$$

CC 959 (1979)
JCS Perkin I 2566 (1981)
JACS *109* 6187 (1987)

See page 642, Section 7.16.

2.3. From Carboxylic Acids

See also page 702, Section 3.

$$2\ RCO_2H \longrightarrow (RCO_2)_2M \xrightarrow{\ \Delta\ } \underset{RCR}{\overset{O}{\overset{\|}{}}}$$

M	
Mg	Org Syn Coll Vol *4* 854 (1963)
Ba	Org Syn Coll Vol *1* 192 (1941)

$$RCO_2^-\ or\ RCO_2H + R'M \xrightarrow{\ reagent(s)\ } \xrightarrow{\ H_2O\ } \underset{RCR'}{\overset{O}{\overset{\|}{}}}$$

M	Reagent(s)	
Li	—	JCS 2012 (1950)
		JACS *91* 456 (1969); *92* 2590 (1970); *95* 4873 (1973); *103* 6157 (1981); *106* 1095, 2064, 6702 (1984); *107* 996, 5717, 7546 (1985); *108* 1617, 4484 (1986); *109* 3025 (1987)
		Org Rxs *18* 1 (1970)
		Org Syn Coll Vol *5* 775 (1973)
		TL 2877 (1974); *28* 2933 (intramolecular), 4965 (1987)
		JOC *40* 1768 (1975); *41* 1176 (1976); *48* 2260, 4789 (1983); *50* 325 (1985); *51* 2361 (1986); *52* 4171, 5501 (1987)
		Syn Commun *11* 7 (1981)
MgX	—	JACS *78* 2268 (1956) (perfluorinated acids); *106* 1095 (1984); *108* 7727 (1986) (LiO$_2$CCF$_3$)
		Austral J Chem *30* 427 (1977); *35* 1739 (1982) (R'M = H$_2$C=CHMgBr)
		JOC *48* 2260 (1983); *50* 325 (1985); *52* 3474 (1987)
	NiCl$_2$(dppe)	TL *24* 3677 (1983)
	[*p*-MeOC$_6$H$_4$CCl=$\overset{+}{N}$Ph$_2$]Cl$^-$	TL *23* 5059 (1982)
	Ph$_2$POCl, Et$_3$N	Syn Commun *8* 59 (1978)
		CC 412 (1986)

$$\underset{\text{Me}_3\text{SiC}=\text{CRCH}_2\text{CH}_2\text{CO}_2\text{H}}{\overset{\text{I}}{|}} \xrightarrow[\text{2. 2 } t\text{-BuLi}]{\text{1. MeLi}} \overset{\text{O}}{\overset{\|}{\text{SiMe}_3}}$$

JACS *105* 6761 (1983)

$$\text{R}^1\text{R}^2\text{CHCO}_2\text{H} \xrightarrow[\substack{\text{2. R}^3\text{COCl} \\ \text{3. HCl}}]{\text{1. 2 LDA}} \left[\underset{\text{R}^1\text{R}^2\text{CCO}_2^-}{\overset{\text{COR}^3}{\overset{|}{|}}} \right] \text{H}_2\overset{+}{\text{N}}(i\text{-Pr})_2 \xrightarrow{\Delta} \text{R}^1\text{R}^2\text{CHCR}^3$$

JOC *42* 1189 (1977)

$$\underset{\text{EtOCCH}_2\text{COH}}{\overset{\text{O} \quad \text{O}}{\overset{\| \quad \|}{}}} \xrightarrow[\text{2. RCOCl}]{\text{1. 2 } n\text{-BuLi} \quad \text{H}^+} \underset{\text{EtOCCH}_2\text{CR}}{\overset{\text{O} \quad \text{O}}{\overset{\| \quad \|}{}}}$$

JOC *44* 310 (1979)

$$\underset{\text{RCH}_2\text{CO}_2\text{H}}{} \xrightarrow[\substack{\text{2. (EtO}_2\text{C)}_2 \\ \text{3. H}_2\text{O, NaHCO}_3}]{\text{1. 2 LDA}} \underset{\text{RCH}_2\text{C}-\text{COEt}}{\overset{\text{O} \quad \text{O}}{\overset{\| \quad \|}{}}}$$

TL *22* 2439 (1981)

$$\underset{\text{RCOH}}{\overset{\text{O}}{\overset{\|}{}}} \xrightarrow[\text{2. H}_2\text{O}]{\text{1. LiCH(CO}_2\text{SiMe}_3)_2} \underset{\text{RCCH}_2\text{COH}}{\overset{\text{O} \quad \text{O}}{\overset{\| \quad \|}{}}}$$

Syn 787 (1979)

2.4. From Acid Halides

See also page 702, Section 3.

$$2 \underset{\text{RCCl}}{\overset{\text{O}}{\overset{\|}{}}} \longrightarrow \underset{\text{RCR}}{\overset{\text{O}}{\overset{\|}{}}}$$

Et$_3$N/H$_2$O, Δ JACS *69* 2444 (1947)
                                                             Org Syn Coll Vol *4* 555, 560 (1963)

Fe$_2$(CO)$_9$ TL 3861 (1977)

$$2 \underset{\text{RCCl}}{\overset{\text{O}}{\overset{\|}{}}} \xrightarrow{\text{SmI}_2} \underset{\text{RC}-\text{CR}}{\overset{\text{O} \quad \text{O}}{\overset{\| \quad \|}{}}}$$

TL *22* 3959 (1981); *25* 2869 (1984)

$$\text{RM} + \underset{\text{ClCR}'}{\overset{\text{O}}{\overset{\|}{}}} \longrightarrow \underset{\text{RCR}'}{\overset{\text{O}}{\overset{\|}{}}}$$

Reviews: Org Rxs *8* 28 (1954) (Mg, Zn, Cd); *22* 253 (1975) (Cu)

RM

o-LiC$_6$H$_4$CO$_2$Li	JOC 46 1057 (1981)
LiCH$_2$CO$_2$SiMe$_3$ (R = Me after H$_3$O$^+$, Δ)	Syn Commun 13 183 (1983)
RLi, cat CuX (X = Cl, I) (R =1°, 2°, 3° alkyl)	TL 829 (1971)
RMgX, THF, $-78°$C (R =1°, 2°, 3° alkyl; aryl)	TL 4303 (1979); 21 2303 (1980) Ber 114 2479 (1981) CL 13 (1986)
RMgX, HMPA (R =1°, 2° alkyl; aryl; alkynyl; vinylic)	Compt Rend C 275 511 (1972)
RMgX, FeCl$_3$ (R =1°, 3° alkyl)	JACS 75 3731 (1953) JOC 26 1768, 1772 (1961)
RMgX, cat Fe(acac)$_3$ (R =1°, 2°, 3° alkyl; aryl)	TL 25 4805 (1984); 26 1285 (1985); 28 2053 (1987)
RMgX, cat CuX (X = Cl, I) (R =1°, 3° alkyl)	JACS 71 4141 (1949); 75 3731 (1953) TL 829 (1971); 467 (1972) Tetr 29 3943 (1973); 37 4189 (1981) J Chem Res (S) 46 (1978)
RMgX, cat Cu-CuCl (R =1°, 3° alkyl)	TL 2523 (1970)
NaCo(I)Salen/MeMgI/H$_3$O$^+$	Tetr 37 863 (1981)
R$_4$BLi (R =1° alkyl, benzylic)	JOC 40 1676 (1975)
(R$_3$BMe)Li, CuCl·COD (R =1° alkyl)	TL 173 (1977)
RAlCl$_2$ (R =1° alkyl)	JACS 108 6036 (1986)
R$_3$Al, Cu(acac)$_2$, PPh$_3$ (R =1° alkyl)	BCSJ 54 1281 (1981)
(R$_4$Al)Li, cat CuCl (R =1° alkyl)	CL 623 (1979)
R$_4$Si, AlCl$_3$ (R =1° alkyl)	Syn 677 (1977)
RSiR$'_3$, ZnCl$_2$ (R = thienyl, furyl)	JOC 47 3219 (1982)
RSiR$'_3$, AlCl$_3$ or TiCl$_4$ (R = vinylic)	CC 633 (1975) JCS Perkin I 2485 (1980); 2527 (1981) TL 22 2985 (1981); 23 263 (1982); 27 4869 (1986) JOC 46 2400 (1981) (intramolecular)
R$_2$NC≡CSiR$'_3$	Ann 1907 (1981)
RSiR$'_3$; AlCl$_3$, ZnCl$_2$ or TiCl$_4$ (R = alkynyl)	Ber 96 3280 (1963) JOMC 37 45 (1972) Syn 438 (1979); 29 (1981) Helv 62 852 (1979); 64 1123 (1981); 65 13, 2413 (1982) CC 459 (1981) JOC 47 3219 (1982); 52 3662 (1987) TL 23 3203 (1982); 28 5543 (1987) Can J Chem 60 379 (1982) JACS 106 3548 (1984); 107 6046 (1985)

$\underline{RM}$ (*continued*)

RSiR$'_3$; AlCl$_3$, TiCl$_4$ or ZnCl$_2$ JOMC *85* 149 (1975)
   (R = allylic)   TL 1871 (1976); 1385 (1977); 429 (1979);
                                     21 831, 3783, 4369 (1980); *23* 723 (1982)
                                 JOC *44* 3397 (1979); *47* 3219 (1982)
                                 Syn 446 (1979)
                                 CL 961 (1982)
                                 JACS *104* 4962 (1982)

$$\overset{\overset{\displaystyle CH_2}{\|}}{H_2C=CHCCH_2SiMe_3},\ TiCl_4$$ Tetr *39* 883 (1983)

H$_2$C=CH(OSiR$_3$)SiR$_3$, TiCl$_4$ (product   JOC *43* 2551 (1978)
  after hydrolysis is R$'$COCH$_2$CH$_2$CHO)

R^{2}CH=CR^{1}CH(SPh)SiMe$_3$, AlCl$_3$ CC 377 (1981)

R$'$C≡CCH$_2$SiMe$_3$, AlCl$_3$ TL *22* 3401 (1981)
  (R = CR$'$=C=CH$_2$)

  , AlCl$_3$ TL *22* 2883 (1981)
  (R = CH$_2$CH=CHCH$_3$)

— CH$_2$SiMe$_3$, AlCl$_3$ Can J Chem *59* 802 (1981)
  (R = CH$_2$CH$_2$CH=CH$_2$)

ArSiMe$_3$, cat KO-t-Bu (on R$'$COF)   Angew Int *20* 265 (1981)

R$_4$Sn, Pd catalyst   CL 1423 (1977)
  (R =1° alkyl, aryl, vinylic)   JOC *51* 2405 (1986)

RSnR$'_3$, Pd catalyst   CL 1423 (1977)
  (R =1° alkyl, allylic, alkynyl, benzylic,   JACS *100* 3636 (1978);
  aryl, vinylic)   *105*, 6129 (1983)
                                   JOC *44* 1613 (1979); *47* 2549 (1982);
                                   48 4634 (1983)
                                   TL *24* 4283 (1983); *27* 2801 (1986);
                                   28 759 (1987)

R$_2$NCH$_2$Sn(n-Bu)$_3$ TL *27* 2361 (1986)

RSnR$'_3$ (R = alkynyl), Δ   Ann *716* 29 (1968)

ROC≡CSnR$'_3$ TL *22* 2637 (1981)
                                   Org Prep Proc Int *14* 189 (1982)

R$_2$NC≡CSnR$'_3$ Angew Int *18* 405 (1979)
                                   Ann 1907 (1981); 196 (1982)

RSnR$'_3$, AlCl$_3$ (R = vinylic)   BSCF 1251 (1977)

E-(n-Bu)$_3$SnCH=CHSn(n-Bu)$_3$, AlCl$_3$ TL *27* 5947 (1986)
                                   JOC *52* 1493 (1987)

RSnR$_3'$, cat ClRh(PPh$_3$)$_3$ (R = allylic)	JOMC *129* C36 (1977)
R″COSnR$_3'$, cat PdCl$_2$(PPh$_3$)$_2$ (R = R″CO)	TL *26* 6075 (1985)
R$_4$Pb, cat Pd(PPh$_3$)$_4$ (R =1° alkyl)	CC 1302 (1987)
RPbR$_3'$ (R = alkynyl)	TL 2265 (1967) JCS C 317 (1968)
Cp$_2$ZrClR, AlCl$_3$ (R =1°, 2° alkyl; vinyl)	JACS *96* 8115 (1974); *99* 638 (1977)
RMgBr, VCl$_3$ (R =1° alkyl, aryl)	TL *27* 929 (1986)

RMnI (R =1°, 2°, 3° alkyl; alkynyl;
   aryl; vinylic)
	TL 3155 (1976); *22* 1239 (1981);
	   26 1285 (1985); *27* 4869 (1986)
	 Syn 130 (1977); 37 (1984); 50 (1985)
	 Can J Chem *58* 287 (1980)
	 Tetr *40* 683 (1984)

Fe(CO)$_4^{2-}$, R′CH—CH$_2$ (O) (R = R′CH=CH)	CL 1067 (1979)
RRh(CO)L$_2$ (R =1° alkyl, aryl, allylic)	JACS *97* 5448 (1975) JOC *42* 1194 (1977)
RCu·COD (R =1° alkyl)	CC 1030 (1982)
RCu·n-Bu$_3$P (R =1° alkyl)	CC 1030 (1982)
RCu, LiI (R =1° alkyl)	CC 1030 (1982)
R$_f$CF=CFCu	JACS *108* 4229 (1986)
RC≡CCu, (LiI or Et$_3$N)	JOC USSR *3* 1298 (1967) JACS *91* 6464 (1969) TL 2659 (1970) JOC *40* 131 (1975); *46* 4107 (1981)
R$_2$CuLi (R =1° alkyl, aryl)	BSCF 797 (1970) TL 4647 (1970); 2113 (1971) JACS *94* 5106 (1972); *108* 3731, 4603 (1986) CC 1030 (1982) JOC *52* 4274 (1987)
(Z-RCH=CH)$_2$CuLi, [cat Pd(PPh$_3$)$_4$]	Tetr *42* 1369 (1986)
R$_2$C=CHCu·MgX$_2$, cat Pd(PPh$_3$)$_4$	Tetr *42* 1369 (1986)
R$_2$CuMgX (R =1°, 3° alkyl; allylic; aryl)	TL 2113 (1971); 3377 (1980) JOMC *188* 293 (1980)
(**R**CuMe)Li (R = 2°, 3° alkyl)	TL *21* 3151 (1980)
(**R**CuMe)MgX (R =1°, 2°, 3° alkyl; aryl)	JOC *41* 2750 (1976)
[RCuC≡CC(CH$_3$)$_3$]Li (R =1° alkyl)	CC 1030 (1982)
(RCuCN)Li (R =1° alkyl)	CC 1030 (1982)
(Me$_3$SiCH=CHCH$_2$CuCN)Li	TL *22* 2985 (1981)

<u>RM</u>   (*continued*)

$(R_2CuCN)Li_2$ (R = 1° alkyl)	CC 1030 (1982)
$(R'SO_2CH_2CuR)Li$ (R' = Me, Ph) (R = 1°, 2°, 3° alkyl; aryl)	CC 358, 1030 (1982)
$(CH_3SOCH_2CuR)Li$ (R = 1°, 2°, 3° alkyl; aryl)	JOC *52* 1885 (1987)
$[CH_3SOCH_2Cu(CN)R]Li_2$ (R = 2° alkyl)	JOC *52* 1885 (1987)
$(PhSCuR)Li$ (R = 2°, 3° alkyl)	JACS *95* 7788 (1973) Syn 662 (1974) Org Syn *55* 122 (1976) CC 1030 (1982)
$(t\text{-BuOCuR})Li$ (R = 1°, 2°, 3° alkyl)	TL 1815 (1973) JACS *95* 7788 (1973)
$(RCuNR'_2)Li$ (R' = Et, Ph; R = 1° alkyl)	CC 1030 (1982)
$(RCuNCy_2)Li$ (R = 1°, 3° alkyl)	JACS *104* 5824 (1982) CC 1030 (1982)
$(RCuPPh_2)Li$ (R = 1°, 3° alkyl)	JACS *104* 5824 (1982) CC 1030 (1982)
$(RCuPCy_2)Li$ (R = 1° alkyl)	CC 1030 (1982)
$RC{\equiv}CAg$	JACS *78* 1675 (1956) TL 1303 (1970)
RI, Zn-Cu (R = 1° alkyl)	Ber *80* 129 (1947)
RZnX or R_2Zn (R = 1°, 2° alkyl; aryl)	JACS *69* 2350 (1947); *71* 3804 (1949) Org Rxs *8* 28 (1954)
RZnX, cat $Pd(PPh_3)_4$ (R = 1°, 3° alkyl; benzyl; aryl; vinylic; alkynyl)	TL *24* 5181 (1983); *28* 1055 (1987)
EtO_2CCH_2ZnBr, cat Pd(O)	CL 1559 (1982)
$(EtO_2CCH_2CH_2)_2Zn$, Me_3SiCl, HMPA	JOC *50* 2802 (1985)
$(RO_2CCHRCH_2)_2Zn$, cat $PdCl_2(o\text{-Tol}_3P)_2$	TL *27* 83 (1986); *28* 337 (1987)
$(EtO_2CCH_2CH_2)_2Zn$, Me_3SiCl, CuI	JACS *106* 3368 (1984)
$(RO_2CCH_2CH_2)_2Zn$; HMPA or cat $PdCl_2(PPh_3)_2$ or cat $CuBr \cdot SMe_2$	JACS *109* 8056 (1987)
$ArCH_2Br$, cat Pd(O), Zn	CL 1135 (1981)
$RC{\equiv}CZnX$	JOC USSR *2* 1859 (1966)

RCdX or R_2Cd, MgX_2 or LiX   JACS *67* 740 (1945);
    ($R = 1° > 2° > 3°$ alkyl; aryl)   *72* 5333 (1950)
                                            Chem Rev *40* 15 (1947)
                                            Org Rxs *8* 28 (1954)
                                            Org Syn Coll Vol *3* 601 (1955)
                                            JCS A 453, 456 (1966)
                                            Can J Chem *58* 287 (1980)
                                            JOC *47* 2590 (1982)

RCdX or R_2Cd, $FeCl_3$ ($R = 1°$ alkyl)   JOC *26* 1772 (1961)

$RC\equiv CCdX$ JOC USSR *3* 210 (1967);
                                            4 2032 (1968)

R_2Hg, cat $Pd(PPh_3)_4$ CL 951 (1975)
    ($R = 1°$ alkyl, aryl)

$RCH=CHHgCl$, $AlCl_3$ JOC *43* 710 (1978)

$$2\ RMnI + COCl_2 \longrightarrow R\overset{\displaystyle O}{\overset{\|}{C}}R$$

$$R = 1°\ \text{alkyl}$$

BSCF 570 (1977)

$$ClCSPh \xrightarrow[\text{cat } NiCl_2(dppe)]{R^1MgX} \xrightarrow[\text{cat } Fe(acac)_3]{R^2MgX} R^1-\overset{\displaystyle O}{\overset{\|}{C}}-R^2$$

TL *26* 3595 (1985)

$$R\overset{\displaystyle O}{\overset{\|}{C}}Cl + 2\ R'CH=PPh_3 \xrightarrow{H_2O} R\overset{\displaystyle O}{\overset{\|}{C}}CH_2R'$$

Ber *95* 1513 (1962)

$$ArCCl + RCH_2\overset{\displaystyle Ph}{\overset{\|}{C}}=C=O\ \text{or}\ RCH_2CHPhCOCl \xrightarrow[\text{Et}_3N]{\text{Pd catalyst}} Ar\overset{\displaystyle O}{\overset{\|}{C}}C(Ph)=CHR$$

TL *26* 5143 (1985)
JOC *52* 3186 (1987)

$$RC\equiv CH + Cl\overset{\displaystyle O}{\overset{\|}{C}}R' \xrightarrow[\substack{\text{cat CuI} \\ \text{Et}_3N}]{\text{cat } PdCl_2(PPh_3)_2} RC\equiv C\overset{\displaystyle O}{\overset{\|}{C}}R'$$

Syn 777 (1977)

$$R\overset{\displaystyle O}{\overset{\|}{C}}Cl + R_2\overset{-}{C}CO_2^- \xrightarrow{H^+} R\overset{\displaystyle O}{\overset{\|}{C}}CHR_2$$

Compt Rend *286* 401 (1978)

$$\underset{\text{RCCl}}{\overset{\text{O}}{\|}} + \underset{\text{R'CClCO}_2\text{Li}}{\overset{\text{Li}}{|}} \longrightarrow \underset{\text{RCCHClR'}}{\overset{\text{O}}{\|}}$$

Syn 284 (1982)

$$\underset{\text{RCCl}}{\overset{\text{O}}{\|}} + \underset{\text{XCCH}_2\text{COEt}}{\overset{\text{O}\quad\text{O}}{\|\quad\|}} \xrightarrow[\text{2 Et}_3\text{N}]{\text{MgCl}_2} \underset{\overset{|}{\text{COX}}}{\underset{\text{RCCHCOEt}}{\overset{\text{O}\quad\text{O}}{\|\quad\|}}}$$

X = Me, OEt

JOC *50* 2622 (1987)

$$\underset{\text{RCCl}}{\overset{\text{O}}{\|}} + \underset{\text{R'C(CO}_2\text{R''})_2}{\overset{\text{Na}}{|}} \longrightarrow \underset{\text{RCCR'(CO}_2\text{R''})_2}{\overset{\text{O}}{\|}} \xrightarrow{\text{reagent(s)}} \underset{\text{RCCH}_2\text{R'}}{\overset{\text{O}}{\|}}$$

Reagent(s)

H$_2$, cat Pd-C (R'' = PhCH$_2$)   JCS 325 (1950)

H$^+$ JCS 322 (1950)
                                                    JACS *74* 831 (1952)

Syn Commun *10* 221 (1980)

JOC *43* 2087 (1978);
52 1889 (1987)
Syn 451 (1982)
Org Syn *63* 198 (1984)

$$\underset{\text{RCCl}}{\overset{\text{O}}{\|}} + {}^-\text{CH}_2\underset{\text{CR'}}{\overset{\text{O}}{\|}} \longrightarrow \underset{\text{RCCH}_2\text{CR'}}{\overset{\text{O}\qquad\text{O}}{\|\qquad\|}}$$

Org Rxs *8* 59 (1954) (review)
JACS *107* 1280 (1985)

$$\underset{\text{RCCl}}{\overset{\text{O}}{\|}} \longrightarrow \underset{\text{RCCH}_2\text{COR}}{\overset{\text{O}\qquad\text{O}}{\|\qquad\|}}$$

LiCH$_2$CO$_2$R   See page 873, Section 2.

NaCH(CO$_2$R)CO$_2$CH$_2$Ph/H$_2$, cat Pd-C   JCS 2758 (1951)

LiO$_2$CCHLiCO$_2$R/H$^+$ JOC *44* 310 (1979)
                                                  Org Syn *61* 5 (1983)

ROMgCH(CO$_2$R)$_2$, Δ   J Prakt Chem 322 (1980)

BrMgO$_2$CCH(MgBr)CO$_2$R/H$^+$ BSCF 945 (1964)
                                                    Syn 142 (1978)

$$\mathrm{R\overset{O}{\overset{\|}{C}}Cl + Me_3SiO\overset{O}{\overset{\|}{C}}CH_2\overset{O}{\overset{\|}{C}}NR_2 \xrightarrow{KO\text{-}\mathit{t}\text{-}Bu} R\overset{O}{\overset{\|}{C}}CH_2\overset{O}{\overset{\|}{C}}NR_2}$$

JACS *108* 5559 (1986)

$$\mathrm{R\overset{O}{\overset{\|}{C}}Cl + LiO_2C\overset{Li}{\overset{|}{C}}HCN \longrightarrow R\overset{O}{\overset{\|}{C}}CH_2CN}$$

Syn 308 (1983)

2.5. From Acid Anhydrides

$$\mathrm{R'\overset{O}{\overset{\|}{C}}O\overset{O}{\overset{\|}{C}}R' \longrightarrow R'\overset{O}{\overset{\|}{C}}R}$$

RX (R = allylic, benzylic), electrolysis	TL *27* 4175 (1986)
RC≡CLi	Ber *97* 1649 (1964) Monatsh *102* 214 (1971)
ArLi	JOC *41* 1268 (1976)
ArLi, TMEDA	Ann 2247 (1981)
RMgX (R = 1°, 2°, 3° alkyl; aryl)	JCS 1367, 1370 (1935) JACS *64* 2226 (1942); *67* 154 (1945); *80* 1225 (1958) JOC *13* 592 (1948); *25* 214 (1960); *51* 3502 (1986) CL 663, 687 (1974) Austral J Chem *30* 427 (1977) TL *27* 2001 (1986)
RMgX, HMPA, (5% $FeCl_3$) (R = 1°, 2° alkyl)	Compt Rend C *275* 511 (1972)
RC≡CMgX	JACS *58* 1861 (1936)
ArMgX, TMEDA	Angew *90* 1000 (1978) JOC *46* 2601 (1981) Ann 2247 (1981)
$LiAlR_4$, cat CuX or $CuCl_2$ (R = 1° alkyl)	CL 623 (1979)
$ArSiMe_3$, cat KO-*t*-Bu	Angew Int *20* 265 (1981)
$RC{\equiv}CSiMe_3$, $AlCl_3$	TL *28* 5543 (1987)
$RSnR'_3$, $AlCl_3$ (R = vinylic)	BSCF 1251 (1977)
RMnX (R = 1° alkyl, aryl, alkynyl, vinylic)	Syn Commun *9* 639 (1979) Tetr *40* 683 (1984) Syn 37 (1984); 50 (1985)
RC≡CLi, CuI	JACS *109* 6899 (1987)

R$_2$CuMgX (R = aryl)   JOMC *188* 293 (1980)

R$_2$Cd (R = 1° alkyl, aryl > 2°,   JOC *6* 462 (1941)
   3° alkyl)

$$\underset{\text{RCOH}}{\overset{\overset{\displaystyle O}{\parallel}}{}} \longrightarrow \underset{\text{RCOCX}}{\overset{\overset{\displaystyle O\;\;O}{\parallel\;\;\parallel}}{}} \xrightarrow{\text{reagent}} \underset{\text{RCR}'}{\overset{\overset{\displaystyle O}{\parallel}}{}}$$

X	Reagent	
OR	RCHLiCOR	JOMC *127* C65 (1977)
	RCHLiCO$_2$R	JOMC *127* C65 (1977)
	RCHLiCO$_2$SiMe$_3$	TL 3713 (1978)
	(decarboxylation also)	Tetr *37* 307 (1981)
	RCHLiCONR$_2$	Syn 954 (1979)
	RCHLiCN	TL 1585 (1979)
	$^-$CR(CO$_2$Et)$_2$	JOC *22* 245 (1957)
	R′MgX	CL 687 (1974)
	RMnI	Syn 50 (1985)
	Et$_2$Cd	JOC *22* 245 (1957)
t-Bu, Ar	R′MgX	CL 663, 687 (1974)

$$\underset{\text{R}'\text{COH}}{\overset{\overset{\displaystyle O}{\parallel}}{}} \longrightarrow \underset{\text{R}'\text{COPPh}_2}{\overset{\overset{\displaystyle O\;\;O}{\parallel\;\;\parallel}}{}} \xrightarrow{\text{RMgX}} \underset{\text{R}'\text{CR}}{\overset{\overset{\displaystyle O}{\parallel}}{}}$$

Syn Commun *8* 59 (1978)
CC 412 (1986)

JACS *102* 3848 (1980)

n = 1, 2

R = H, Me

TL *22* 3351 (1981)

Org Rxs *8* 59 (1954) (review)
JOC *52* 3986 (1987)

$$\underset{\text{RCOCR}}{\overset{\overset{O}{\parallel}\,\overset{O}{\parallel}}{}} \xrightarrow{\text{NaOH}} \quad \xrightarrow{\overset{R'OH}{\Delta}} \quad \underset{\text{RCCH}_2\text{COR}'}{\overset{\overset{O}{\parallel}\,\overset{O}{\parallel}}{}}$$

Syn 451 (1982)

2.6. From Esters and Lactones

$$\underset{R^1COR^2}{\overset{O}{\parallel}} \longrightarrow \underset{R^1CR^3}{\overset{O}{\parallel}}$$

R^3	Reagents	
CH$_3$	CH$_3$MgBr, $-78°$C/H$_2$O (R^2 = 2-pyridyl)	TL *28* 1603 (1987)
	$^-$CH$_2$SOCH$_3$/Al(Hg), H$_2$O	JACS *87* 1345 (1965) JOC *31* 2355 (1966); *46* 4825 (1981)
	$^-$CH$_2$SOCH$_3$/Zn, NaOH	JOC *52* 4477 (1987)
	$^-$CH$_2$SO$_2$CH$_3$/Al(Hg), H$_2$O	JOC *33* 61 (1968)
	$^-$CH$_2$SONAr/H$_2$O	JACS *90* 5548 (1968)
	R$_3$SnCH$_2$Li (R = Me, Ph)/ H$_2$O	TL *27* 4339 (1986)
CH$_2$R	PhSO$_2$CRLi$_2$/Al(Hg), H$_2$O	JOC *45* 4002 (1980); *47* 564 (1982)
	$R\bar{C}H[B(OR)_2]_2$	JACS *99* 3196 (1977) Organomet *1* 20 (1982)
CH$_2$CH$_2$R	RCH$_2$CH$\bar{B}$⟨ ⟩ with B group, Me (from RC≡CH, ⟨ ⟩BH, MeLi)/H$_2$O (R^1 = Ph)	CL 1193 (1981)
1° alkyl	1° RI, cat NiCl$_2$, Zn/H$_3$O$^+$	CL 531 (1981)
1° alkyl, aryl	RMgX (R^1 = CO$_2$R)	JOC *45* 2883 (1980) Syn Commun *11* 943 (1981)
1° alkyl, aryl	RMgX, Et$_3$N	Syn 877 (1980)
1° alkyl	RLi	JACS *109* 8071 (1987) (R^1 = RCF$_2$)
1°, 2°, 3° alkyl; 1° benzylic; aryl	RLi or RMgX/H$_2$O (R^1 = CF$_3$, CO$_2$Et)	JOC *52* 5026 (1987)
1°, 2° alkyl	RMgX, HMPA	Compt Rend C *273* 1543 (1971)
allylic	RCH=CRCH$_2$MgCl/KH	TL *23* 335 (1982)

$$\underset{\text{R}^1\text{COR}^2}{\overset{\text{O}}{\overset{\|}{}}} \longrightarrow \underset{\text{R}^1\text{CR}^3}{\overset{\text{O}}{\overset{\|}{}}} \quad (\textit{Continued})$$

R³	Reagents	
CHMeCH₂CMe=CH₂	H₂C=C(Me)MgBr	Austral J Chem *30* 427 (1977)
RCHX (X = Cl, Br)	RCHXLi	Syn 68 (1981)
		TL *25* 835 (1984)
CF₂CF₃	LiCF₂CF₃	JOC *52* 2481 (1987)
CH₂OH	Me₃SiCHLiOCO₂Li/H₃O⁺	TL *28* 1847 (1987)
RCH(OH)	RCHLiOCO₂Li/H₃O⁺	TL *28* 1847 (1987) (R = SiMe₃)
		Syn 415 (1987) (R = Ph)
PhCHNH₂	PhCHLiNLiCO₂Li/H₃O⁺	Syn 415 (1987)
CH₂SPh	LiCH(SPh)SiMe₃	TL *22* 2803 (1981)
CHRSPh	RC̄(SPh)[B(OR)₂]₂	Organomet *1* 280 (1982)
CClRCH=CH₂	RCCl=CHCH₂Li/H₃O⁺	JOMC *215* 1 (1981)
CCl=CHCH₃	ClCH=CHCH₂Li/H₃O⁺	JOMC *215* 1 (1981)
C≡CR	LiC≡CR	Syn 307 (1978)
		TL 937 (1978) (lactones);
		28 3651 (1987) (lactones)
CH₂COR	⁻CH₂COR	JACS *67* 1510 (1945)
		Org Rxs *8* 59 (1954) (review)
CH₂CO₂R	LiCH₂CO₂R	See page 873, Section 2.

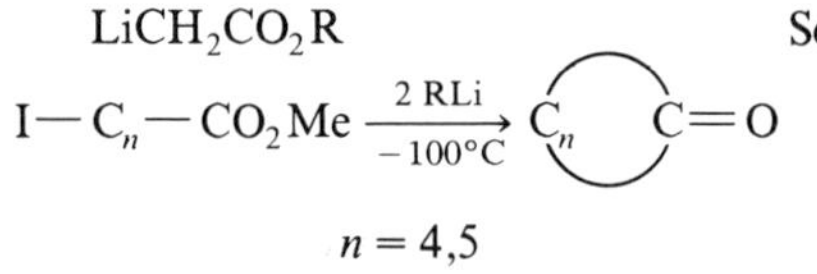

$$I{-}C_n{-}CO_2Me \xrightarrow[-100°C]{2\ RLi} C_n\ \ C{=}O$$

$$n = 4,5$$

TL *26* 4987 (1985); *27* 519 (1986)

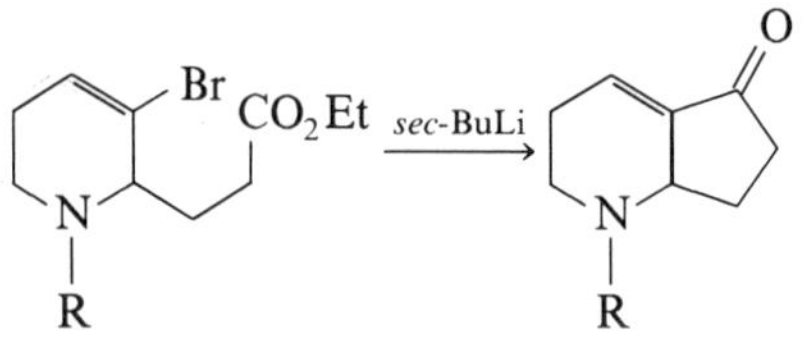

JACS *109* 6115 (1987)

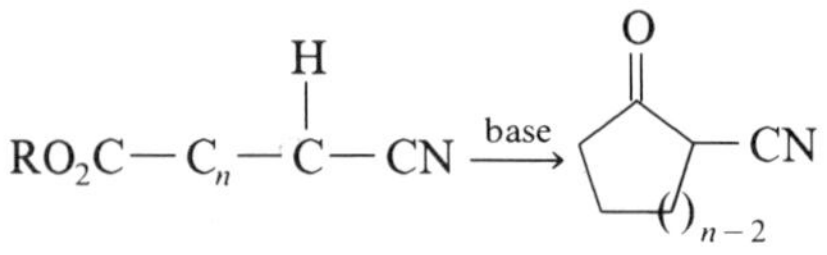

$$n = 3, 4$$

TL *28* 4641, 4645 (1987)

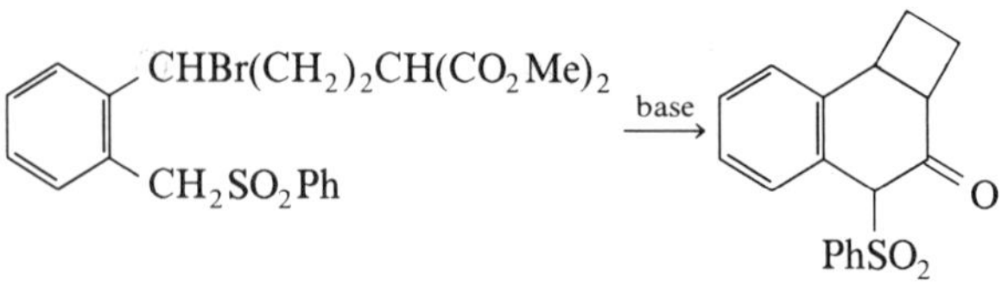

TL *28* 709 (1987)

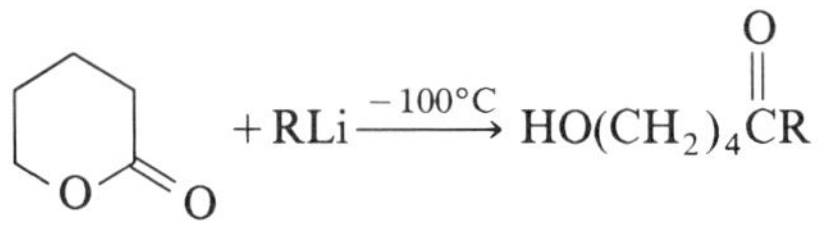

$$+ \text{RLi} \xrightarrow{-100°C} \text{HO(CH}_2)_4\text{CR}$$

TL *28* 651 (1987)

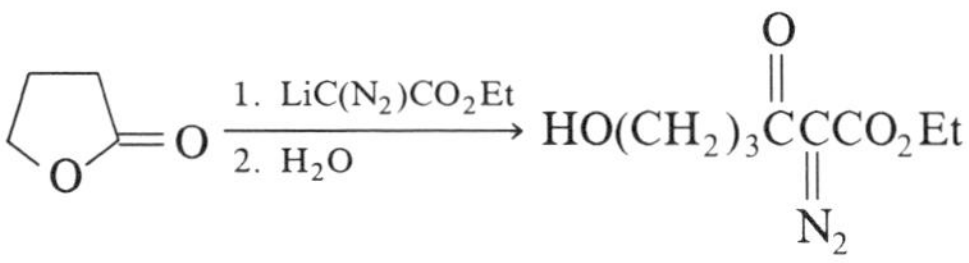

$$\xrightarrow[\text{2. H}_2\text{O}]{\text{1. LiC(N}_2)\text{CO}_2\text{Et}} \text{HO(CH}_2)_3\underset{\underset{\text{N}_2}{\|}}{\text{C}}\text{CCO}_2\text{Et}$$

TL *28* 5351 (1987)

$$\xrightarrow[\substack{\text{2. 1 HOAc} \\ \text{3. }\Delta}]{\text{1. 2 RCHLiPO(OMe)}_2}$$

JACS *107* 7967 (1985)

2.7. From Thioesters

$$\underset{\text{R'CSR''}}{\overset{\text{O}}{\|}} \longrightarrow \underset{\text{R'CR}}{\overset{\text{O}}{\|}}$$

RMgX (R″ = C$_5$H$_4$N)
  (R = 1°, 2° alkyl; aryl)

JACS *95* 4763 (1973)
BCSJ *47* 1777 (1974)
J Med Chem *23* 1392 (1980)
JOC *46* 3760 (1981); *51* 938 (1986)
TL *23* 2533 (1982)

RMgX, cat Fe(acac)$_3$
  (R″ = Ph; R = 1°, 2° alkyl; aryl)

TL *26* 3595 (1985)

R$_2$CuLi (R = 1°, 2°, 3° alkyl; aryl)

JACS *96* 3654 (1974);
  108 4943 (1986)
CC 1231 (1981)

2.8. From Ortho Esters

$$\text{RM} + \text{HC(OR')}_3 \longrightarrow \text{RCH(OR')}_2$$

RM

RMgX   TL *22* 545 (1981)

RC≡CMgX   JCS 4244 (1955)

RCH=CR'CH$_2$MBr (M = Mg, Zn)   JOMC *222* 1 (1981)

2.9. From Ketenes

$$R_2C=C=O \xrightarrow[\text{2. R''X}]{\text{1. R'Li}} R_2C\overset{\displaystyle O}{\overset{\|}{C}}CR'$$
$$\underset{R''}{|}$$

JOC *50* 2105 (1985)

2.10. From Amides

$$RBr + 2\,Li + R^1\overset{\displaystyle O}{\overset{\|}{C}}NR^2_2 \longrightarrow R\overset{\displaystyle O}{\overset{\|}{C}}R^1$$

R = Et, Ph

$$2\,ArBr + 4\,Li + Et_2N\overset{\displaystyle O}{\overset{\|}{C}}NEt_2 \longrightarrow Ar\overset{\displaystyle O}{\overset{\|}{C}}Ar$$

Syn 160 (1973)

$$RM + R_2N\overset{\displaystyle O}{\overset{\|}{C}}R' \xrightarrow{H_2O} R\overset{\displaystyle O}{\overset{\|}{C}}R'$$

RM

RLi (R = 1°, 2° alkyl)

Ber *88* 678 (1955)
JCS 4691 (1956)
JOC *24* 701 (1959); *51* 3290, 3566 (1986)
Can J Chem *54* 1098 (1976)
JOMC *177* 5 (1979)
Angew Int *20* 795 (1981)
TL *23* 109 (1982); *25* 811 (1984);
   28 651, 3573 (1987)

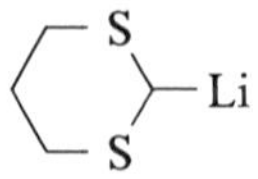

—Li

TL *25* 811 (1984)

ArLi

Ber *88* 678 (1955)
Chem Ind 1596 (1957)
Bull Acad Sci USSR, Div Chem Sci 769 (1961)
Tetr *27* 1221 (1971)
JOC *38* 901 (1973)
TL *25* 811 (1984)

RCH=CHLi

TL *27* 775 (1986) (intramolecular)

RC≡CLi

TL *25* 811 (1984)
CL 13 (1986)

RC≡CLi, BF₃·OEt₂

CL 35 (1983)
JOC *52* 1372 (1987)

RMgX (R = 1°, 2° alkyl)	JACS *61* 232 (1939)
	Ber *88* 678 (1955)
	Can J Chem *54* 1098 (1976)
	TL *25* 811 (1984); *28* 2999 (1987)
	JCS Perkin I 795 (1985)
	JOC *51* 3566 (1986)
ArMgX	Ber *88* 678 (1955)
	TL *25* 811 (1984)
RC≡CMgX	TL *25* 811 (1984)
RLa(OTf)$_2$ (R = 1°, 2° alkyl)	TL *28* 4391 (1987)

$$R^1CH_2\overset{O}{\overset{||}{C}}N Me\overset{*}{C}HMe\overset{*}{C}HOHPh \xrightarrow[\substack{2.\ R^2X \\ 3.\ MeLi}]{1.\ 2\ LDA} R^1R^2\overset{O}{\overset{||}{C}}HCCH_3$$

chiral

TL 3961 (1978)
JOMC *177* 5 (1979)

$$R\overset{O}{\overset{||}{C}}N(Me)OMe \longrightarrow R\overset{O}{\overset{||}{C}}R'$$

R'Li (R' = 1° alkyl)	TL *22* 3815 (1981); *27* 5467 (1986)
ArLi	TL *22* 3815 (1981); *28* 6331 (1987)
	JOC *51* 5106 (1986); *52* 2615 (1987)
RCH=CR'Li	JACS *107* 8066 (1985)
RC≡CLi	TL *22* 3815 (1981); *28* 1857 (1987)
	JOC *50* 2309, 3972 (1985)
PhSCH(OMe)Li, [structure]—Li, or [structure]—SCH$_2$Li	TL *28* 1857 (1987)
R'MgX (R' = 1° alkyl, allyl, benzyl, alkynyl, aryl, vinyl)	TL *22* 3815 (1981); *28* 1857, 6331 (1987)
R(C=CH)$_n$C=CH$_2$ (n = 0–3) with OLi, OLi	TL *24* 1851 (1983)

$$R\overset{O}{\overset{||}{C}}N\underset{}{\overset{O}{\diagup}} \xrightarrow{R'C≡CLi} R\overset{O}{\overset{||}{C}}CC≡CR'$$

JOC *51* 5320 (1986)

$$R\overset{O}{\overset{||}{C}}N\diagup{}^N \longrightarrow R\overset{O}{\overset{||}{C}}R'$$

R'MgBr (R' = Me, Et, Ph)	Ann *655* 90 (1962)
R'MgX (R' = aryl > alkyl; R = CO$_2$R)	JOC *46* 211 (1981)

Mg(O$_2$CCHRCOXR)$_2$ Angew Int *18* 72 (1979)
   (X = O, S; R' = CHRCOXR)   TL *28* 5249 (1987)

$$\underset{\text{EtO}}{\overset{\displaystyle O}{\underset{\Vert}{\text{C}}}} \quad \overset{O}{\underset{\Vert}{\overset{}{}}} \quad \overset{\text{O}}{\underset{\text{O}}{\text{Mg}}} /H_2O \ (R' = CH_2CO_2Et)$$ BSCF 945 (1964)

(Na)KCH$_2$NO$_2$ Syn 478 (1978)
LiCHRCO$_2$R   TL *27* 5281 (1986); *28* 2837 (1987)
                                                     JACS *109* 4717 (1987)
                                                     CC 1228 (1987)

$$\underset{RCNMe_2}{\overset{\displaystyle O}{\Vert}} \xrightarrow[\text{2. H}_3\text{O}^+]{\text{1. Me}_3\text{SiCHLiOCO}_2\text{Li}} \underset{RCCH_2OH}{\overset{\displaystyle O}{\Vert}}$$

TL *28* 1847 (1987)

$$Br(CH_2)_nCOCl \longrightarrow \longrightarrow \quad \text{(ring structure)} \xrightarrow[\text{2. Al(Hg)}]{\text{1. LDA}} O{=}C \quad (CH_2)_n$$

n = 7,11

TL 4487 (1979)

2.11. From Thioamides

$$\underset{RCNR_2}{\overset{\displaystyle S}{\Vert}} \xrightarrow[\text{2. H}_3\text{O}^+]{\text{1. R'Li}} \underset{RCR'}{\overset{\displaystyle O}{\Vert}}$$

TL *28* 1529 (1987)

2.12. From Acid Nitriles

$$\underset{RCCN}{\overset{\displaystyle O}{\Vert}} + \underset{LiCH_2CR'}{\overset{\displaystyle O}{\Vert}} \longrightarrow \underset{RCCH_2CR'}{\overset{\displaystyle O \quad\quad O}{\Vert \quad\quad \Vert}}$$

TL 1339 (1979)

$$\underset{RCCN}{\overset{\displaystyle O}{\Vert}} + \underset{RCHCO_2R}{\overset{\displaystyle Li}{\vert}} \longrightarrow \underset{\underset{R}{\vert}}{\underset{RCCHCOR}{\overset{\displaystyle O \quad\ O}{\Vert \ \ \Vert}}}$$

TL *28* 4011 (1987)

$$\underset{CH_3CCN}{\overset{\displaystyle O}{\Vert}} + H_2C{=}CHCH_2SiMe_3 \xrightarrow{\text{TiCl}_4} \underset{\underset{CN}{\vert}}{\underset{CH_3CCH_2CH{=}CH_2}{\overset{\displaystyle OH}{\vert}}}$$

TL *22* 1171 (1981)

2.13. From Nitriles

$$R'CN \longrightarrow R'\overset{\displaystyle O}{\overset{\displaystyle \|}{C}}R$$

RLi (R = 1°, 2° alkyl; aryl)/H$_3$O$^+$

Helv 729 (1962)
JACS 92 336 (1970); 107 1028 (1985)
TL 22 1509 (1981); 28 4329 (1987)
JOC 47 4347 (1982)

Me$_3$SiCHLiOCO$_2$Li/H$_3$O$^+$ (R = CH$_2$OH)

TL 28 1847 (1987)

RMgX (R = 1° alkyl, aryl)/H$_3$O$^+$

JACS 52 1267 (1930); 67 2059, 2197 (1945);
    70 426 (1948); 72 876 (1950); 73 3948 (1951);
    74 4607 (1952); 79 881 (1957);
    109 3378 (1978)
BSCF 867 (1935)
Bull Soc Chim Belg 44 523 (1935)
JOC 15 359 (1950); 45 237 (1980)
Org Syn Coll Vol 3 26 (1955)
JCS Perkin I 795 (1985)

R$_2$C=CHCH$_2$MgX /H$_3$O$^+$

Ann Chim 287 (1964)
JACS 108 1311 (1986)

RMgX (R = 2°, 3° alkyl; aryl),
    cat CuBr/H$_3$O$^+$

JOC 52 3901 (1987)

R$_3$P=CHR″ /H$_3$O$^+$ (R = CH$_2$R″)

JACS 89 7009 (1967)

R$_2''$C=CHCH$_2$Br (R″ = H, Me), Zn(Ag)/H$_3$O$^+$

TL 22 649 (1981)

R'CH=CHCH$_2$M (M = Li, MgX, ZnX)/H$_3$O$^+$

JOMC 69 1 (1974) (review)

R'CH=CHCH$_2$TiCp$_2$ /H$_3$O$^+$

CC 342 (1981)
JOMC 224 327 (1982)

ROĊN (with S=C and N-heterocycle) (R = 2° alkyl), Ph$_3$SnH,
    cat AIBN/H$_3$O$^+$

JOC 49 1313 (1984)
    (intramolecular)

HCl, MeOH/ (dimethyl malonate-type acetonide),
    Et$_3$N, Δ /HCl (R = Me)

Syn 130 (1981)

HCl, MeOH/ (dimethyl ketal-type acetonide),
    Et$_3$N, Δ /NaOEt/H$_3$O$^+$ (R = CH$_2$CO$_2$Et)

Syn 130 (1981)

R'CHBrCO$_2$R', Zn/H$_3$O$^+$ (R = R'CHCO$_2$R')

Org Syn Coll Vol 4 120 (1963)
JOC 48 3833 (1983)

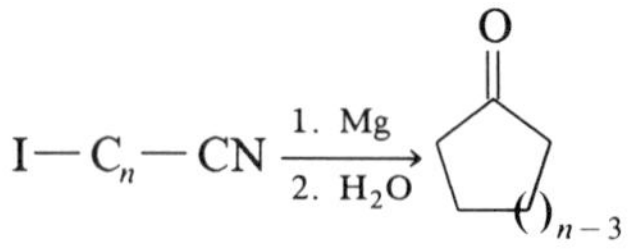

$$I-C_n-CN \xrightarrow[\text{2. H}_2\text{O}]{\text{1. Mg}}$$

$$n = 4, 5$$

JOMC *87* 25 (1975)

$$RCH \longrightarrow R\overset{\underset{\displaystyle |}{OSiMe_3}}{C}HCN \xrightarrow[\text{2. H}_3\text{O}^+]{\text{1. R'MgX}} R\overset{\underset{\displaystyle |}{HO}}{C}H\overset{\underset{\displaystyle \|}{O}}{C}R'$$

TL *27* 1933 (1986)

$$\xrightarrow[\substack{\text{3. RCN} \\ \text{4. H}_3\text{O}^+}]{\substack{\text{1. RLi} \\ \text{2. Cp}_2\text{Zr(Me)Cl}}}$$

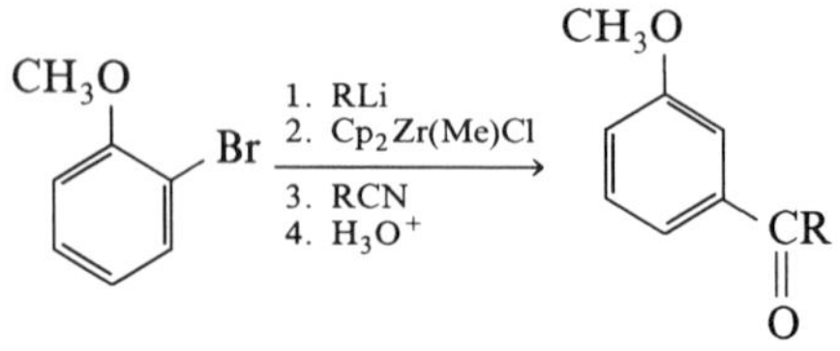

JACS *109* 7137 (1987)

$$2\ ArLi \xrightarrow[\substack{\text{2. RCN} \\ \text{3. H}_3\text{O}^+}]{\text{1. Cp}_2\text{ZrCl}_2} Ar\overset{\underset{\displaystyle \|}{O}}{C}R$$

TL *28* 3245 (1987)

3. Friedel–Crafts and Related Reactions

Reviews:

Chem Rev *55* 229 (1955)
"Friedel-Crafts and Related Reactions," Ed. G. A. Olah, J. Wiley and Sons, New York (1964), Vol 3, Pts 1 and 2
Syn 533 (1972)
G. A. Olah, "Friedel-Crafts Chemistry," Wiley Interscience, New York (1973)

$$ArH \longrightarrow Ar\overset{\underset{\displaystyle \|}{O}}{C}H$$

Review: "Friedel-Crafts and Related Reactions," Chpts 32 and 38

HCN, HCl, ZnCl₂ or AlCl₃/H₂O	Org Rxs *9* 37 (1957)
Zn(CN)₂, HCl/H₂O	TL *23* 4567 (1982)
Zn(CN)₂, AlCl₃, HCl/H₂O	Org Syn Coll Vol *3* 549 (1955)
	JOC *52* 336 (1987)
(CH₃)₂C(OH)CN, AlCl₃	Syn Commun *12* 485 (1982)
CO, HCl, AlCl₃	Org Rxs *5* 290 (1949)
CO, HCl, CuCl/AlCl₃	Org Syn Coll Vol *2* 583 (1943)

FCHO, BF_3	JACS *82* 2380 (1960)
Cl_2CHOCH_3, $AlCl_3$	Ber *93* 88 (1960)
Cl_2CHOR (R = Me, *n*-Bu), $SnCl_4$	Ber *93* 88 (1960)
Cl_2CHOCH_3, $TiCl_4$	Ber *93* 88 (1960); *96* 308 (1963) CC 214 (1972) JCS Perkin I 340 (1973) JOC *50* 3121 (1985); *52* 1972 (1987) TL *27* 6299 (1986)
Cl_2CHO-*n*-Bu, $TiCl_4$	Ber *93* 88 (1960)
$HC(OEt)_3$, $AlCl_3$	Ber *96* 308 (1963) (phenols)

$$\text{EtO}\overset{S}{\underset{S}{\overset{|}{C}}}\text{H}\Big],\ BF_3\cdot OEt_2/H_3O^+$$

BCSJ *54* 2120 (1981)

DMF, $POCl_3$	Org Syn Coll Vol *4* 331 (1963) "Friedel-Crafts and Related Reactions," Vol 3, Pt 2, p 1211
$HCCl_3$, NaOH (on ArOH)	Org Rxs *28* 1 (1982) (review)

$$NO_2\text{—}Ar\text{—}H \longrightarrow NO_2\text{—}Ar\text{—}CHO$$

$HCCl_3$, KO-*t*-Bu/H_2O, HCO_2H	TL *28* 3021 (1987)
$HC(SPh)_3$, NaOH, DMSO/hydrolysis	CL 1623 (1984)

$$ArH \longrightarrow Ar\overset{\displaystyle O}{\overset{\displaystyle \|}{C}}R$$

Review: "Friedel-Crafts and Related Reactions," Chpts 31 and 32

RCO_2H, $(CF_3CO)_2O$, H_3PO_4	Syn 303 (1979)
RCOCl, cat CF_3SO_3H	Angew Int *11* 300 (1972)
RCOCl, Nalfion-H	Syn 672 (1978)
RCOCl, $AlCl_3$	"Friedel-Crafts and Related Reactions," Vol 3, Pt 1, pp 1–382
RCOCl; cat $FeCl_3$, I_2, $ZnCl_2$ or Fe	Syn 533 (1972)
RCOCl, $(RCO)_2O$ or $CH_3C(=CH_2)O_2CR$; cat $(Ph_2B)SbCl_6$ (ArOR)	CL 165 (1986)
$(RCO)_2O$, H_3PO_4	Org Syn Coll Vol *3* 14 (1955)
$(RCO)_2O$, $AlCl_3$	Org Syn Coll Vol *1* 517 (1941) Ber *115* 3436 (1982) Tetr *38* 3555 (1982) JACS *109* 7122 (1987)
$(RCO)_2O$, $ZnCl_2$	Org Syn Coll Vol *2* 304 (1943)
$(RCO)SbF_6$	Syn 345 (1980)
$RCO_2SO_2CF_3$	Angew Int *11* 299 (1972)

RCO_2POX_2 (X = F, Cl)   Ber *114* 926 (1981)

RCO_2—⟨pyridyl⟩, CF_3CO_2H Syn 139 (1980)

RCOSMe, CuOTf   JACS *102* 860 (1980)

$RCONMe_2$, $POCl_3$ Helv *42* 1659 (1959)
                                    JOC *25* 2049 (1960)

$RCON$⟨imidazolyl⟩, CF_3CO_2H BCSJ *53* 1638 (1980)

RCN, HCl, $ZnCl_2/H_2O$ Org Syn Coll Vol *2* 522 (1943)
                                    Org Rxs *5* 387 (1949)
                                    "Friedel-Crafts and Related Reactions," Vol 3,
                                        Pt 1, p 383
                                    JOC *43* 4172 (1978)
                                        (intramolecular)

$(RC{\equiv}NH)OTf/H_2O$ JCS Perkin I 2894 (1980)

$(RC{\equiv}NMe)OTf/H_2O$ CC 1151 (1980)

$C_6H_5-(CH_2)_nCO_2H \longrightarrow$ indanone/tetralone $()_{n-1}$

Reviews: Org Rxs *2* 114 (1944)
                Angew *66* 435 (1954) (PPA)

n	Acid or Lewis Acid	
2	HF	JACS *61* 1272 (1939)
	CH_3SO_3H	JOC *46* 2974 (1981)
	CH_3SO_3H, P_2O_5	JOC *38* 4071 (1973)
	PPA	Helv *29* 859 (1946)
		JACS *72* 2965 (1950);
		108 1251 (1986)
		BSCF 810 (1957)
		Compt Rend *246* 779 (1958);
		249 2337, 2782 (1959)
		Syn Commun *11* 993 (1981)
		CC 185 (1982)
		JOC *51* 1402, 4250 (1986)
	$SnCl_4$	JCS 4306 (1954)
	$AlCl_3$, NaCl	JCS 2403 (1953)
3	HF	JACS *61* 1272 (1939)
		Can J Chem *49* 2712 (1971)
		JCS Perkin I 461 (1982)
		JOC *52* 1284 (1987)
	H_2SO_4	JCS 1125 (1932); 4306 (1954)
		Tetr *2* 271 (1958)
		JOC *28* 3571 (1963)
		TL *23* 2415 (1982)

	H_2SO_4, $B(OH)_3$	JACS *103* 4251 (1981)
	CH_3SO_3H	JOC *46* 2974 (1981); *52* 5574 (1987)
	$(CF_3CO)_2O$	JCS 1435 (1957)
	$POCl_3$	JCS 787 (1939)
	P_2O_5	Tetr *38* 3555 (1982)

	PPA	JCS 1995 (1938)
		JACS *69* 58 (1947); *72* 2965
		    (1950); *73* 1411 (1951);
		    74 5147 (1952);
		    78 450 (1956); *103* 4251
		    (1981)
		Org Syn Coll Vol *3* 798 (1955)
		JOC *43* 4172 (1978);
		    46 2547 (1981)

	$SnCl_4$	JCS 867 (1934); 790 (1939) JACS *69* 58 (1947)
	$AlCl_3$, NaCl	JCS 2403 (1953)
	$ZnCl_2$, HOAc, Ac_2O	JACS *69* 58 (1947)
4	PPA	JACS *73* 1411 (1951); *75* 720 (1953) BCSJ *35* 1380 (1962) Syn Commun *11* 993 (1981)
	$AlCl_3$, NaCl	JCS 2403 (1953)
5	PPA	JACS *77* 4596 (1955)

JOC *50* 5886 (1985)

n	Reagent	
1	HF	JOC *44* 2158 (1979); *52* 1355 (1987)
	PPA	JCS 4306 (1954)
2	HF	JACS *62* 1855 (1940)
2 or 3	H_2SO_4 or PPA or $SnCl_4$	JCS 4306 (1954)
3	P_2O_5 or $SnCl_4$	JCS 787 (1939)

Houben-Weyl, Vol VII/3C, p 31 (review)
JOC *52* 3205 (1987)

Acta Chem Scand B *36* 371 (1982)

n	
3	JCS 1435 (1957)
	JCS C 217 (1968)
4	JOC *35* 858 (1970)

n	Lewis Acid	
2	AlCl$_3$	Ann *376* 269 (1910);
	SnCl$_4$	*468* 277 (1929);
		586 52 (1954)
		Ber *46* 1700 (1913);
		55 1835 (1922)
		J Prakt Chem [2] *139* 95 (1934)
		JCS 1460, 3499 (1951)
		BSCF 810 (1957)
		Tetr *2* 271 (1958)
		JACS *80* 1243 (1958)
		Compt Rend *249* 2337, 2782
		(1959); *252* 1971 (1961)
		BCSJ *35* 1380 (1962)
		JOC *46* 2431 (1981);
		51 5252 (1986)
	SnCl$_4$	Compt Rend *246* 779 (1958);
		249 2782 (1959)
	FeCl$_3$	Ann *323* 246 (1902)
	AgClO$_4$	JCS 1718 (1957)

3	AlCl$_3$	JCS 75 144 (1899); 187 (1936); 844, 3499 (1951)
		Ber 56 1424 (1923); 61 441 (1928)
		Ann 531 129 (1937); 586 52 (1954)
		Org Syn Coll Vol 2 569 (1943)
		JOC 15 950 (1950); 28 357 (1963)
		Tetr 2 271 (1958)
	SnCl$_4$	JACS 57 782 (1935); 59 475 (1937); 62 2750 (1940); 63 1682 (1941); 69 58 (1947)
		JCS 4306 (1954)
		Ber 95 1786 (1962)
		JOC 28 3571 (1963)
		Org Syn Coll Vol 4 900 (1963)
	AgClO$_4$	JCS 1718 (1957)
4	AlCl$_3$	JCS 79 602 (1901)
		Helv 27 801 (1944)
		BSCF 769 (1953)
		Ann 586 52 (1954)
		Compt Rend 244 2513 (1957)
		Tetr 2 271 (1958)
		J Chem Res (S) 226 (1982)
	AgClO$_4$	JCS 1718 (1957)
5	AlCl$_3$	Ber 85 826 (1952); 90 1946 (1957)
		JACS 75 3744 (1953); 76 5462 (1954)
		Ann 586 52 (1954)
		Tetr 2 271 (1958)
		BCSJ 35 1380 (1962)
		J Chem Res (S) 226 (1982)
6, 7	AlBr$_3$	Ber 90 1946 (1957)

JACS 54 4373 (1932)

AlCl$_3$ Ber 55 1835 (1922)
                                                JCS 1991 (1938)

SnCl$_4$ JACS 72 733 (1950)

$$RCH{=}CH_2 \xrightarrow[\text{Lewis acid}]{\overset{\overset{\displaystyle O}{\|}}{R'CCl}} \underset{Cl}{R}CHCH_2\overset{\overset{\displaystyle O}{\|}}{C}R' \xrightarrow{\text{base}} RCH{=}CH\overset{\overset{\displaystyle O}{\|}}{C}R'$$

"Friedel-Crafts and Related Reactions," Vol 3, Pt 2, p 1033 (review)
JOC 35 858 (1970) (intramolecular)
CC 84 (1973) (intramolecular)
TL 2441 (1976); 27 2341 (1986) (intramolecular)
JACS 108 1265 (1986) (intramolecular)

$$\underset{\underset{\displaystyle CO_2Me}{|}}{\overset{}{N}}\text{-ring} \xrightarrow[\text{POCl}_3\text{-DMF (R = H)}]{\text{RCOCl-SnCl}_4 \text{ or}} \underset{\underset{\displaystyle CO_2Me}{|}}{\overset{\overset{\displaystyle CR}{\overset{\displaystyle \|}{O}}}{N}}\text{-ring}$$

TL 23 1201 (1982)

$$C{=}C{-}C_n{-}\overset{\overset{\displaystyle O}{\|}}{C}Cl \longrightarrow \square{\overset{O}{}}(\ )_n$$

	n	
AlCl$_3$	2	BCSJ 52 216 (1979)
SnCl$_4$	4	JOC 35 858 (1970)
	2	JACS 109 3025 (1987)

$$\overset{\overset{\displaystyle O}{\|}}{R}CCl + H_2C{=}CHR' \xrightarrow[\text{Et}_3N]{\text{Pd catalyst}} \overset{\overset{\displaystyle O}{\|}}{R}CCH{=}CHR'$$

Transition Met Chem 4 398 (1979)
TL 28 4215, 5883 (1987)

$$\overset{\overset{\displaystyle O}{\|}}{R}CCl \longrightarrow \overset{\overset{\displaystyle O}{\|}}{R}CCo(salophen)py \xrightarrow[\Delta]{R'CH{=}CH_2} \overset{\overset{\displaystyle O}{\|}}{R}CCH{=}CHR'$$

TL 28 5949 (1987)

$$\overset{\overset{\displaystyle O}{\|}}{R}CCl + HC{\equiv}CR' \xrightarrow{\text{Lewis acid}} \overset{\overset{\displaystyle O}{\|}}{R}CCH{=}\underset{Cl}{C}R'$$

JOC 1 163 (1936); 29 385 (1964); 50 2796 (1985)
"Friedel-Crafts and Related Reactions," Vol 3, Pt 2, p 1081 (review)
Chem Rev 66 161 (1966) (review)
TL 1821 (1970)
Tetr 29 4241 (1973); 31 177 (1975)

9. NUCLEOPHILIC ACYLATION

See also page 721, Section 10, and page 729, Section 11.

Reviews:

Angew Int *8* 639 (1969)
Chem Ind 687 (1974)
Tetr *32* 1943 (1976); *36* 2531 (1980)
Aldrichimica Acta *14* 73 (1981); *15* 35 (1982)

1. Cyanohydrin Derivatives

Review: Tetr *39* 3207 (1983)

$$\underset{R'CH}{\overset{\overset{O}{\|}}{}} \longrightarrow \underset{R'CHCN}{\overset{OCHMeOEt}{\overset{|}{}}} \xrightarrow[\substack{2.\ E^+ \\ 3.\ H_3O^+}]{1.\ \text{base}} \underset{R'CE}{\overset{\overset{O}{\|}}{}}$$

$\underline{E^+}$

RX

JACS *93* 5286 (1971); *96* 5272 (1974);
    103 5259 (1981) (intramolecular);
    108 4912 (1986)
Syn 358 (1973)
TL *22* 1359*, 1363*, 2651*, 3683 (1981);
    23 4361* (1982); *24* 4683 (1983)
    (*intramolecular)
JOC *51* 3393 (1986) (X = OTs, intramolecular)

RCHO, R₂CO

JOC *45* 395 (1980)
Org Syn *63* 79 (1984)

enone (1,4-addition)

TL *22* 2175 (1981)

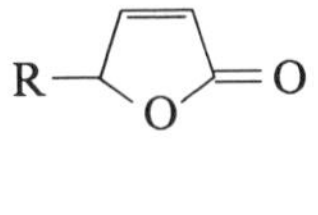

R—⟨ ⟩=O (1,4-addition)

TL *28* 3551 (1987)

PhCH=CHNO$_2$ (E = PhĊ=CH$_2$)   Syn Commun *10* 717 (1980)

H$_2$C=CMeCH=CHSO$_n$Ph   Syn Commun *11* 709 (1981)
  (n =1,2; 1,6-addition)

$$RCH \; (O) \xrightarrow{R_3SiCN} RCHCN \; (OSiR_3) \xrightarrow[\text{2. E}^+]{\text{1. LiNR}_2} \xrightarrow{H_3O^+} RCR' \; (O)$$

E$^+$ = RX; RCHO; R$_2$CO; RCOCl; enone; α,β-unsaturated esters; Michael acceptors

R
—

aryl   Syn 777 (1973); 180, 391 (1975)
                                              Ber *112* 2045, 2062 (1979); *113* 302, 324 (1980)
                                              Tetr *39* 841 (1983)
                                              JOC *52* 564 (1987)

vinylic   Syn 777 (1973)
                                              JOC *44* 462 (1979); *45* 395 (1980);
                                                52 4135 (1987)
                                              Ber *113* 3783 (1980); *114* 959 (1981)
                                              Syn Commun *11* 709 (1981)
                                              JACS *106* 718 (1984)
                                              TL *27* 5359 (1986)

$$R_2CHCHRCCl \; (O) \longrightarrow R_2CHCR=CCN \; (OSiMe_3) \xrightarrow[\substack{\text{2. R'X} \\ \text{3. hydrolysis}}]{\text{1. LDA}} R_2C=CRCR' \; (O)$$

Ber *119* 722, 1772 (1986)

$$R_2CO \xrightarrow[\substack{\text{2. } i\text{-Bu}_2\text{AlH} \\ \text{3. H}_3\text{O}^+}]{\text{1. Me}_3\text{SiCN}} R_2CCHO \; (OH(SiMe_3))$$

JACS *107* 4577 (1985)

$$RCH \; (O) \xrightarrow[\text{(R}_2\overset{+}{\text{N}}\text{H}_2)\text{Cl}^-]{\text{KCN}} RCHCN \; (NR_2) \xrightarrow[\text{2. E}^+]{\text{1. base}} RCE \; (NR_2, CN) \xrightarrow{H_3O^+} RCE \; (O)$$

R —	E$^+$ —	
H	RX enone (1,4-addition)	Heterocycles *19* 481 (1982) Heterocycles *19* 1395 (1982)
alkyl	RX	BSCF 1653 (1961) Syn 127 (1979) JCS Perkin II 1645 (1982) TL *23* 3369 (1982); *24* 4683 (1983); *27* 1569 (1986)

RX, RCH—CH$_2$, H$_2$C=CHCOCH$_3$ TL 5175 (1978)
(1,4-addition)

RCHO, R$_2$CO   CL 71 (1980)
     TL 23 639 (1982)

RCH=CRCO$_2$R (1,4-addition)   Z Chem 21 68 (1981)

aryl   D$_2$O   CC 218 (1967)
     JCS C 2049 (1970)

     RX   JACS 82 1786 (1960)
     JOC 26 4740 (1961);
       48 1909 (1983)
     Tetr 31 1219 (1975)
     Syn 127 (1979)
     TL 24 4683 (1983)

enone (1,4-addition)   TL 25 4641 (1984); 27 2985
       (1986) (added Lewis acids)
     JOC 51 1293 (1986)

H$_2$C=CHCN(CO$_2$R)   JOC 37 4465 (1972)
     JACS 98 6321 (1976)
     J Heterocyclic Chem 15 881
       (1978)
     CL 399 (1979)

RX, RCOCl, RCH—CH$_2$,   JOC 44 4597 (1979)
H$_2$C=CHCN(CO$_2$R)

vinylic   RX   CL 1263 (1982)
     R$_2$CO   TL 21 1205 (1980)

hydrolysis   H$_2$O, CuSO$_4$ TL 2763 (1978)

$$RLi \xrightarrow[\substack{2.\ R'X\ (or\ H_2O) \\ 3.\ HCl,\ H_2O}]{1.\ H_2C=C(CN)NMePh} RCH_2\overset{\displaystyle O}{\overset{\displaystyle \|}{C}}R'(H)$$

R

alkyl   Syn 897 (1978)

enolate or imine anion   Syn 413 (1980)

2. Isonitriles

$$ArSO_2CH_2N=C \xrightarrow[2.\ RX]{1.\ base} \xrightarrow[2.\ R'X]{1.\ base} \xrightarrow[2.\ base]{1.\ acid} R\overset{\displaystyle O}{\overset{\displaystyle \|}{C}}R'$$

TL 4229 (1977); 5335 (1982) (cyclophanes); 28 3825 (1987)
Syn 325 (1980)
JOC 46 5159 (1981); 51 1551 (1986)
Syn Commun 13 331, 379, 1067 (1983)
JACS 107 5238 (1985)

$$RM + t\text{-}C_4H_9\overset{\overset{\displaystyle CH_3}{|}}{\underset{\underset{\displaystyle CH_3}{|}}{C}}\!-\!NC \longrightarrow R\!-\!\overset{\overset{\displaystyle NR'}{\|}}{C}\!-\!M \xrightarrow[\text{2. }H_3O^+]{\text{1. }E^+} R\!-\!\overset{\overset{\displaystyle O}{\|}}{C}\!-\!E$$

M = Li, MgX

E^+ = H_2O, D_2O, $ClSiMe_3$, $RCHO$, $H_2C\!\!-\!\!\overset{\displaystyle O}{\overset{\diagup\;\diagdown}{C}}\!HR$, CO_2, $ClCO_2Et$, RX (R = 1° alkyl, aryl, alkynyl, vinylic), $H_2C\!=\!CHCO_2Et$ (1,4-addition)

JACS *91* 7778 (1969); *92* 6675 (1970)
JOC *39* 600 (1974); *46* 5405 (1981); *47* 52 (1982)

$$ArN\!\equiv\!C \longrightarrow ArN\!=\!C\!\!\overset{\textstyle \diagup SiMe_2(t\text{-}Bu)}{\diagdown SnMe_3} \xrightarrow[\substack{\text{3. R'X} \\ \text{4. }H_3O^+}]{\substack{\text{1. }n\text{-BuLi} \\ \text{2. }R_2CO}} R'\overset{\overset{\displaystyle O}{\|}}{C}\!-\!\overset{\overset{\displaystyle OH}{|}}{C}R_2$$

Ar = 2,6-xylyl

JACS *109* 7888 (1987)

3. Iron Carbonyl Reagents

Review: Acct Chem Res *8* 342 (1975)

$$1°\ RX \xrightarrow[\text{on polymer}]{HFe(CO)_4^-} R\overset{\overset{\displaystyle O}{\|}}{C}H$$

JOC *43* 1598 (1978)

$$RX \xrightarrow[NaOH,\ n\text{-}Bu_4NBr]{Fe(CO)_5} R\overset{\overset{\displaystyle O}{\|}}{C}R$$

R = 1° alkyl, benzylic

CL 321 (1979)

$$RX \xrightarrow[\substack{n\text{-}Bu_4N(HSO_4) \\ NaOH}]{\substack{CO \\ Fe(CO)_5}} \xrightarrow{R'X} R\overset{\overset{\displaystyle O}{\|}}{C}R'$$

CC 1754 (1986)

$$RX + Na_2Fe(CO)_4 \longrightarrow [RFe(CO)_4]^- \begin{cases} \xrightarrow[PPh_3]{HOAc} R\overset{\overset{\displaystyle O}{\|}}{C}H \\ \\ \xrightarrow[PPh_3]{R'X} R\overset{\overset{\displaystyle O}{\|}}{C}R' \end{cases}$$

$$\underset{RCX}{\overset{O}{\parallel}} + Na_2Fe(CO)_4 \longrightarrow \left[\underset{RCFe(CO)_4}{\overset{O}{\parallel}} \right]^- \xrightarrow[R'COCl]{R'X \text{ or}} \underset{RCR'}{\overset{O}{\parallel}}$$

JOC *35* 4183 (1970)
JACS *92* 6080 (1970); *94* 1788, 5905 (1972); *95* 249, 2689 (1973); *99* 2515 (1977); *100* 4766 (1978)
Org Syn *59* 102 (1980)

$$RMgBr \xrightarrow[2.\ R'I]{1.\ Fe(CO)_5} \underset{RCR'}{\overset{O}{\parallel}}$$

TL 761 (1978)

$$RLi \xrightarrow[2.\ Ar_2IX]{1.\ Fe(CO)_5} \underset{RCAr}{\overset{O}{\parallel}}$$

TL 1255 (1979)

$$H_2C=CH(CH_2)_nCH_2OTs \xrightarrow[2.\ HOAc]{1.\ Fe(CO)_4^{2-}}$$

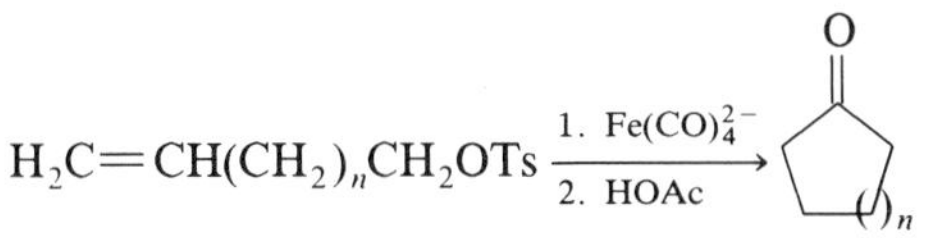

TL *21* 4687 (1980)

4. Allylic Amides

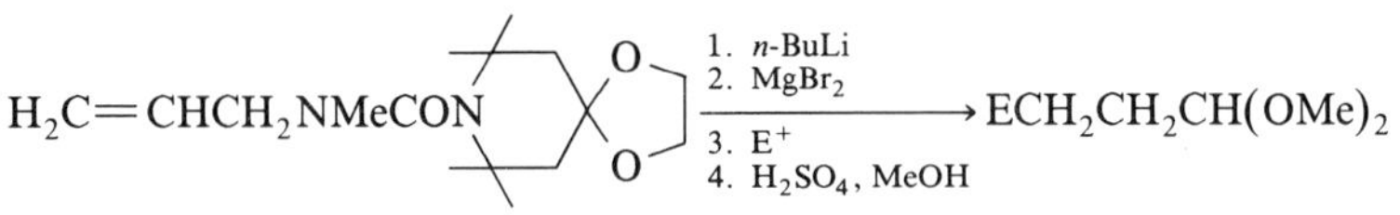

$$H_2C=CHCH_2NMeCON \underset{}{\overset{}{\diagup\diagdown}} \xrightarrow[\substack{3.\ E^+ \\ 4.\ H_2SO_4,\ MeOH}]{\substack{1.\ n\text{-BuLi} \\ 2.\ MgBr_2}} ECH_2CH_2CH(OMe)_2$$

$E^+ = RX, RCHO, R_2CO$

Angew *18* 399 (1979)

5. Allylic Ethers and Carbamates

$$H_2C=CHCH_2OR' \xrightarrow[2.\ E^+]{1.\ base} ECH_2CH=CHOR' \longrightarrow \underset{ECH_2CH_2CH}{\overset{O}{\parallel}}$$

R'	E⁺	
alkyl	RX	JACS *96* 5560 (1974)
		TL 833 (1979)
		CL 1637 (1982) (chiral)
	RX, epoxide (intramolecular)	TL 2115 (1976)

$$H_2C=CHCH_2OR' \xrightarrow[\text{2. E}^+]{\text{1. base}} ECH_2CH=CHOR' \longrightarrow ECH_2CH_2\overset{\displaystyle O}{\overset{\|}{C}}H \quad (\textit{Continued})$$

SiEt$_3$	RX	JACS 96 5561 (1974)
CONMe$_2$	(MeO)$_2$CO, RX	Angew Int 19 625 (1980)
CON(i-Pr)$_2$	RCHO, R$_2$CO	Angew Int 20 1024 (1981); 21 372 (1982); 25 160 (1986)
	(i-PrO)$_3$TiCl/RCHO	TL 27 4873 (1986)
	(i-PrO)$_3$TiCl, (Et$_2$N)$_3$TiCl or (i-PrO)$_4$Ti/RCHO	TL 28 5149 (1987)
	(Et$_2$N)$_3$TiCl/RCHO	Angew Int 25 160 (1986)
CONHR'	RX, Me$_3$SiCl, (MeS)$_2$, (MeO)$_2$CO	Angew Int 20 127 (1981)
CONR$_2$	review	Angew Int 23 932 (1984)

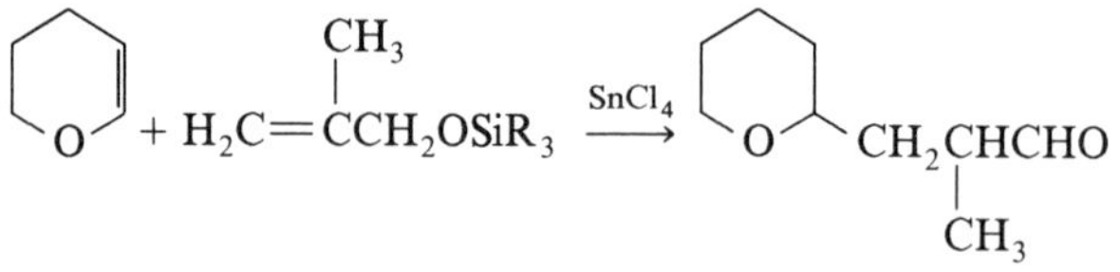

JOC 52 5691 (1987)

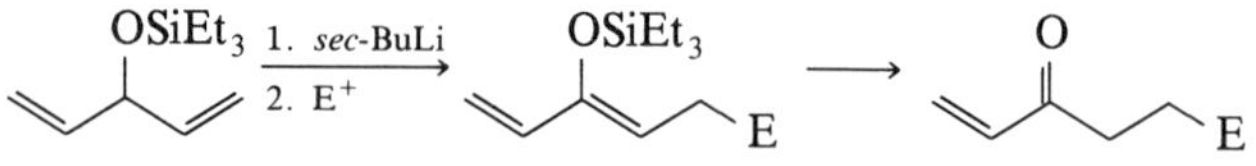

E$^+$ = H$_2$O, RX, RCHO, R$_2$CO, RCOCl, Me$_3$SiCl, (MeS)$_2$

TL 4187 (1976)
Helv 63 967 (1980); 64 2002, 2022, 2592 (1981)
JACS 104 5708 (1982)

6. Vinylic Ethers

$$CH_3CH=CHOC_6H_5 \xrightarrow[\text{2. CH}_3\text{I}]{\text{1. base}} CH_3CH_2CH=CHOC_6H_5$$

Helv 57 2261 (1974)

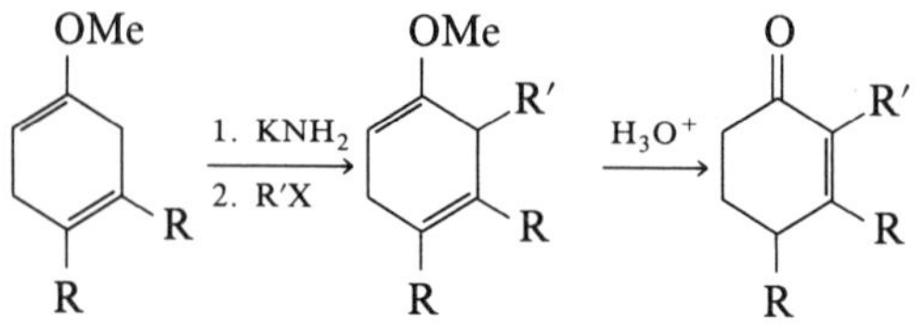

JCS Perkin I 7 (1983)
TL 28 4195 (1987)

JOC *42* 3755 (1977); *47* 1576 (1982); *51* 1407 (1986)
TL *28* 4163 (1987)

$$(RCHO)\ R_2C{=}O \xrightarrow{\text{reagent}} R_2C{=}CHOR' \xrightarrow{H_3O^+} R_2CHCH{\overset{\displaystyle O}{\overset{\displaystyle \|}{}}}$$

Review: Org Rxs *14* 270 (1965)

Reagent(s)

Ph$_3$P=CHOMe	JACS *80* 6150 (1958)
	Angew *71* 127 (1959)
	Ber *94* 1373 (1961)
	JOC *35* 1385 (1970); *50* 2668, 2676 (1985);
	51 2162 (1986)
	Syn 604 (1973)
	TL 3979 (1973); *22* 1063 (1981)
	Tetr *31* 89 (1975)
Ph$_3$P=CHO-*n*-Bu	Ber *95* 2526 (1962)
Ph$_3$P=CHOCH$_2$CH=CH$_2$	JACS *92* 5522 (1970)
Ph$_3$P=CHOAr	Ber *95* 2514 (1962)
Ph$_3$P=CHOCH$_2$OCH$_3$	Tetr *31* 89 (1975)
Ph$_3$P=CHOTHP	Tetr *31* 89 (1975)
(EtO)$_2$POC̄HOCH$_2$CH$_2$OCH$_3$	TL 3629 (1978)
	JOC *44* 4847 (1979)
(RO)$_2$POC̄HOTHP (R = Et, *n*-Bu)	TL 3629 (1978)
	JOC *44* 4847 (1979)
Ph$_2$POC̄HOCH$_3$	CC 314 (1977)
	JCS Perkin I 3099 (1979)
	CL 1143 (1982)
Ph$_2$PC̄HOCH$_3$	TL *21* 3535 (1980)
(MeO)$_2$POCHN$_2$, R'OH, KO-*t*-Bu	TL *21* 2041, 5003 (1980)
m-CF$_3$C$_6$H$_4$SeCHLiOCH$_3$/MsCl, Et$_3$N	JACS *101* 6638 (1979)
Me$_3$SiCH(Li)OCH$_3$	CC 822 (1979)
	Organomet *1* 553 (1982)

$$(RCHO) \quad R_2C{=}O \xrightarrow{\text{reagent}} R_2C{=}CROR \xrightarrow{H_3O^+} R_2CHC\overset{\displaystyle O}{\overset{\|}{C}}R$$

Reagent

$(n\text{-BuO})_2PO\overline{C}(CH_3)OTHP$ TL 3629 (1978)

$Ph_2PO\overline{C}(CH_3)OCH_3$ CC 314 (1977)
                    JCS Perkin I 3099 (1979)

$$Ph_3P{=}CR_2 + HCO_2R \longrightarrow R_2C{=}CHOR$$

CL 967 (1981)
JOC *49* 3595 (1984)

$$(RCHO) \quad R_2C{=}O \xrightarrow{Ph_3P{=}CHCH{=}CHOMe} R_2C{=}CHCH{=}CHOMe \xrightarrow{H_3O^+} R_2CHCH{=}CHC\overset{\displaystyle O}{\overset{\|}{}}H$$

TL 3875 (1977)

$$R_2CHCH(OR)_2 \xrightarrow[R_3N]{AlCl_3 \text{ or } MgBr_2} R_2C{=}CHOR$$

Helv *62* 1451 (1979)

$$RC{\equiv}CR(H) \longrightarrow \underset{H}{\overset{R}{\diagdown}}C{=}C\underset{SiF_5^{2-}}{\overset{R(H)}{\diagup}} \xrightarrow[R'OH]{\substack{O_2 \\ \text{cat } Cu(OAc)_2}} \underset{H}{\overset{R}{\diagdown}}C{=}C\underset{OR'}{\overset{R(H)}{\diagup}}$$

TL *21* 4105 (1980)

$$H_2C{=}CHOR' \xrightarrow[2.\ E^+]{1.\ RLi} H_2C{=}C\overset{\displaystyle OR'}{\overset{|}{E}} \xrightarrow{H_3O^+} CH_3C\overset{\displaystyle O}{\overset{\|}{}}E$$

R' = Me, Et, THP; E⁺ = RX, RCHO, R₂CO, RCO₂Me, RCN

See page 185, Section 1.

$$R_3B \xrightarrow{H_2C{=}C(OMe)Li} \xrightarrow[NaOH]{H_2O_2} RC\overset{\displaystyle O}{\overset{\|}{C}}CH_3$$

JOMC *156* 123 (1978)

$$(RCHO) \quad R_2C{=}O \xrightarrow[2.\ HCl]{1.\ \textit{cis-}LiCH{=}CHOSiMe_3} R_2C{=}CHC\overset{\displaystyle O}{\overset{\|}{}}H$$

JOC *46* 3741 (1981)

$$HC{\equiv}COEt \longrightarrow \underset{H}{\overset{EtO}{\diagdown}}C{=}C\underset{BR'_2}{\overset{H}{\diagup}} \xrightarrow[\substack{\text{cat } Pd(PPh_3)_4 \\ NaOH}]{RX} RCH{=}CHOEt$$

R = aryl, benzylic

JOC *47* 2117 (1982)

$$HC \equiv COEt \xrightarrow[\text{2. } n\text{-BuLi}]{\text{1. } n\text{-Bu}_3\text{SnH}} \underset{Li}{\overset{H}{\diagup}} C = C \underset{OEt}{\overset{H}{\diagdown}}$$

1. RX
2. H_3O^+ → $RCH_2\overset{O}{\overset{\|}{C}}H$

1. R_2CO (RCHO)
2. H_3O^+ → $R_2C = CH\overset{O}{\overset{\|}{C}}H$

CuI

cyclohexenone → cyclohexanone with $\diagdown C = C \diagup$ OEt substituent

JACS *99* 7365 (1977)
TL 3589 (1977)
JOC *43* 1595 (1978)

$$HC \equiv CCH = CHOEt \xrightarrow[\text{2. } n\text{-BuLi}]{\text{1. } n\text{-Bu}_3\text{SnH}} \xrightarrow{R_2CO \text{ (RCHO)}} \xrightarrow{H_3O^+} R_2C = CHCH = CH\overset{O}{\overset{\|}{C}}H$$

TL 717 (1978)

7. Allenic, Propargylic and Cyclopropyl Ethers

$$RCH = C = CHOR \xrightarrow[\text{2. } E^+]{\text{1. base}} RCH = C = \overset{OR}{\underset{|}{C}}E \xrightarrow{H_3O^+} RCH = CH\overset{O}{\overset{\|}{C}}E$$

$$E^+ = R'X, Me_2SO_4, RCHO, R_2CO, Me_3SiCl$$

Rec Trav Chim *87* 916 (1968)
TL 2585 (1973)
Angew Int *18* 875 (1979)
JACS *104* 1119 (1982); *106* 5360 (1984)

$$H_2C = C = CHOMe \xrightarrow[\text{2. RX}]{\text{1. RLi}} H_2C = C = \overset{OMe}{\underset{|}{C}}R \xrightarrow[\text{2. R'X}]{\text{1. RLi}} R'CH = C = \overset{OMe}{\underset{|}{C}}R \xrightarrow{H_3O^+} \underset{H}{\overset{R'}{\diagup}} C = C \underset{\overset{\|}{\underset{O}{CR}}}{\overset{H}{\diagdown}}$$

TL 1137 (1978)

$$H_2C = C = CHOMe \xrightarrow[\substack{\text{2. ArI,} \\ \text{Pd catalyst}}]{\text{1. } n\text{-BuLi}} H_2C = C = \overset{OMe}{\underset{|}{C}}Ar \xrightarrow{H_3O^+} H_2C = CH\overset{O}{\overset{\|}{C}}Ar$$

$$H_2C = C = \overset{OMe}{\underset{|}{C}}R \longrightarrow ArCH = C = \overset{OMe}{\underset{|}{C}}R \longrightarrow ArCH = CH\overset{O}{\overset{\|}{C}}R$$

Syn 738 (1982)

$$R^1C\equiv CCH_2OR^2 \xrightarrow[2.\ E^+]{1.\ base} \begin{array}{c} R^1 \\ E \end{array}\!\!C=C=C\!\!\begin{array}{c} OR^2 \\ H \end{array}$$

$$E^+ = H_2O,\ RX,\ R_2SO_4,\ Me_3SiCl$$

TL 2585 (1973)
Angew Int *18* 875 (1979)
JOMC *232* Cl (1982)

$$\begin{array}{c} OH \\ | \\ RCC\equiv CH \\ | \\ SiMe_3 \end{array} \xrightarrow[2.\ R'X]{1.\ n\text{-}BuLi} \begin{array}{c} OSiMe_3 \\ | \\ RC=C=CHR' \end{array} \xrightarrow{H_3O^+} \begin{array}{c} O \\ \| \\ RCCH=CHR' \end{array}$$

TL *28* 1299 (1987)

$$RX \xrightarrow[\substack{2.\ n\text{-}Bu_4NF \\ 3.\ H_3O^+}]{1.\ LiCH=C=C(OMe)SiMe_3} RCH=CHCH$$

(with O double bond on CH)

TL *21* 3987 (1980)

$$(RCHO)\ R_2C=O \xrightarrow[\substack{2.\ HSCH_2CH_2OH,\ BF_3\cdot OEt_2 \\ 3.\ H_2O,\ HgCl_2}]{1.\ Li\text{—}\triangle\text{—}OCH_3} R_2C=CHCH_2CH$$

(with O double bond)

TL 3685 (1975)

8. Miscellaneous Nucleophilic Acylation Reactions

$$RCH_2CH=CH_2 \longrightarrow R-C\begin{array}{c} H \\ | \\ C \\ \diagup\diagdown \\ \underset{\substack{| \\ Pd \\ Cl/_2}}{C} \end{array}\!\!C-H \xrightarrow{R'CONi(CO)_xLi} RCH=CHCH_2CR'$$

(with O double bond)

Organomet *1* 1188 (1982)

$$\begin{array}{c} O \\ \| \\ RCCH_2R' \end{array}$$

$$\uparrow [O]$$

$$base \uparrow R'CHO$$

$$CH_2\!\left(B\!\begin{array}{c} O\! \\ O\! \end{array}\!\right)_{\!2} \xrightarrow[RX]{base} RCH\!\left(B\!\begin{array}{c} O\! \\ O\! \end{array}\!\right)_{\!2} \xrightarrow[R'X]{base} RR'C\!\left(B\!\begin{array}{c} O\! \\ O\! \end{array}\!\right)_{\!2} \xrightarrow{[O]} RCR'$$

(final product with O double bond)

$$base \downarrow R'CO_2CH_3$$

$$H_2O \downarrow \qquad\qquad\qquad\qquad base = \begin{array}{c}\diagup\diagdown \\ \diagdown\diagup \end{array}\!NLi$$

$$\begin{array}{c} O \\ \| \\ RCH_2CR' \end{array}$$

JACS *97* 5608 (1975); *99* 3196 (1977)
Organomet *1* 20, 280 (1982)

$$\underset{RCH}{\overset{O}{\|}} \xrightarrow{Me_3SiOP(OEt)_2} \underset{RCHPO(OEt)_2}{\overset{OSiMe_3}{|}} \xrightarrow[\substack{2.\ R'X \\ 3.\ OH^-}]{1.\ LDA} \underset{RCR'}{\overset{O}{\|}}$$

BCSJ *55* 224 (1982)

$$H_2C{=}CRCH_2Br \xrightarrow[\substack{2.\ Zn(Cu) \\ 3.\ TiCl_3}]{1.\ R'C{\equiv}NO} H_2C{=}CRCH_2\overset{O}{\overset{\|}{C}}R'$$

Angew Int *18* 78 (1979)

$$CH_2(CO_2Et)_2 \xrightarrow[2.\ RX]{1.\ base} \xrightarrow[2.\ R'X]{1.\ base} \xrightarrow[2.\ H^+]{1.\ OH^-} RR'C(CO_2H)_2 \xrightarrow[2.\ KOH]{1.\ Pb(OAc)_4,\ py} \underset{RCR'}{\overset{O}{\|}}$$

TL 6145 (1966)
Org Rxs *19* 279 (1972) (decarboxylation review)

$$HC(OR)_2CO_2R \xrightarrow[2.\ Het-X]{1.\ LDA} Het-C(OR)_2CO_2R \xrightarrow{H_3O^+} Het-\overset{O}{\overset{\|}{C}}-CO_2R$$

TL *27* 6059 (1986)

$$RCH_2CO_2H \xrightarrow[2.\ ArSO_2NHNH_2]{1.\ SOCl_2} RCH_2\overset{O}{\overset{\|}{C}}NHNHSO_2Ar \xrightarrow[\substack{2.\ O\ \ \ NH}]{1.\ PCl_5} RCH_2\overset{NNHSO_2Ar}{\overset{\|}{C}}\underset{O}{\bigcirc}$$

$$\xrightarrow[\substack{2.\ E^+ \\ 3.\ H_2O}]{1.\ t\text{-BuLi}} RCH_2\overset{O}{\overset{\|}{C}}E$$

$$E^+ = MeI,\ R_2CO$$

CC 1121 (1981)

$$\underset{RCH}{\overset{O}{\|}} \longrightarrow \underset{RCH}{\overset{NNH\text{-}t\text{-}Bu}{\|}} \xrightarrow{n\text{-BuLi}}$$

$$\xrightarrow[\substack{2.\ n\text{-BuLi} \\ 3.\ H_2O \\ 4.\ H_3O^+}]{1.\ R'_2CO\ (R'CHO)} \underset{RC-CR'_2}{\overset{O\ \ \ \ OH}{\overset{\|}{}\ \ \ |}}$$

$$\xrightarrow[\substack{2.\ CF_3CO_2H \\ 3.\ H_3O^+}]{1.\ R'X} \underset{RCR'}{\overset{O}{\|}}$$

CC 1040 (1983)

10. SULFUR, SELENIUM AND TELLURIUM REAGENTS

See also page 729, Section 11.

Review: Syn 357 (1977)

1. Sulfur Acetals

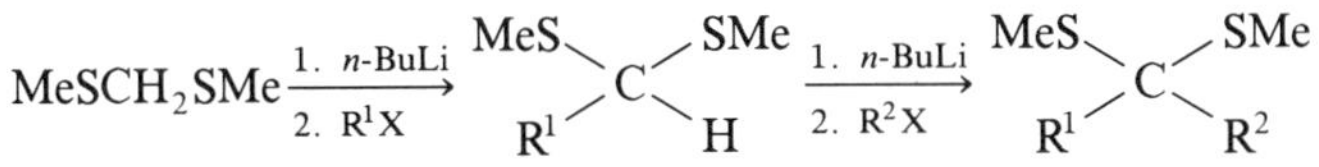

MeSCH₂SMe $\xrightarrow[\text{2. R}^1\text{X}]{\text{1. } n\text{-BuLi}}$

$\overset{\text{MeS}}{\underset{\text{R}^1}{\diagdown}}\text{C}\overset{\text{SMe}}{\underset{\text{H}}{\diagup}}$ $\xrightarrow[\text{2. R}^2\text{X}]{\text{1. } n\text{-BuLi}}$ $\overset{\text{MeS}}{\underset{\text{R}^1}{\diagdown}}\text{C}\overset{\text{SMe}}{\underset{\text{R}^2}{\diagup}}$

TL *23* 1047 (1982)

S S $\xrightarrow[\text{2. RX}]{\text{1. RLi}}$ S S $\overset{}{\underset{\text{R}\quad\text{H}}{}}$ $\xrightarrow[\text{2. R'X}]{\text{1. RLi}}$ S S $\overset{}{\underset{\text{R}\quad\text{R'}}{}}$

$\downarrow$

O‖RCH

$\downarrow$

O‖RCR'

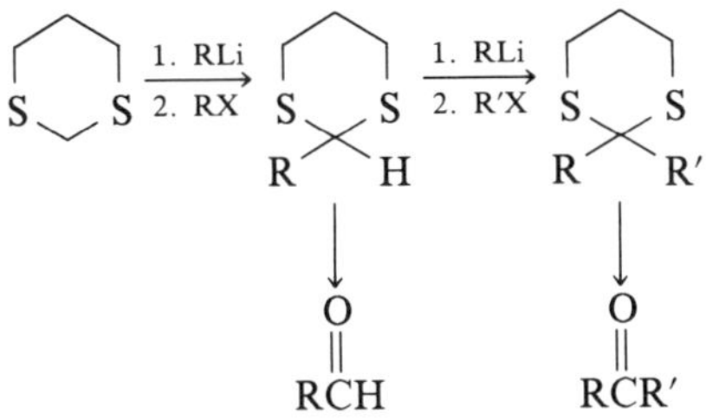

Syn *1* 17 (1969); 357 (1977); 561 (1980) (hydrolysis); 135 (1981) (hydrolysis); 580, 679 (1982) (both hydrolysis)
JOC *36* 366, 3553 (1971); *40* 231 (1975); *46* 1512 (1981); *47* 1145, 2212 (hydrolysis) (1982)
CC 529 (1972); 255 (1978) (hydrolysis)
Heterocycles *6* 731 (1977)
JCS Perkin I 1036 (1978)
Syn Commun *11* 343 (1981)
JACS *103* 3112 (1981)
Ann 1589 (1982)
TL *27* 2965 (hydrolysis), 6305 (alternate preparation) (1986)

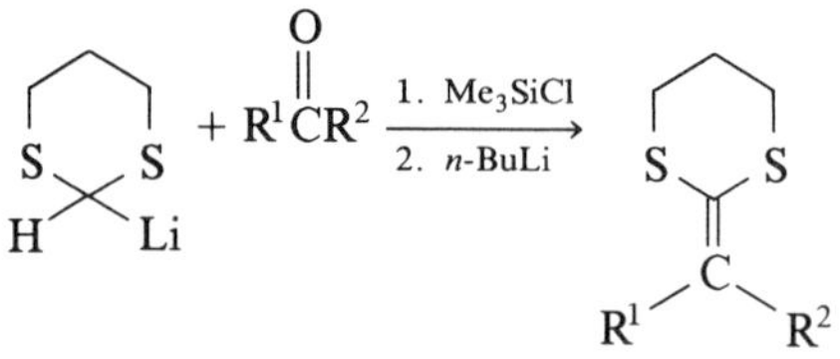

Syn 579 (1982)

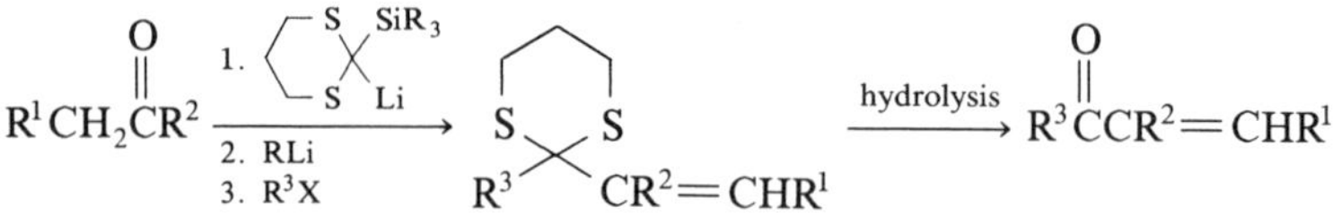

TL 3171 (1974)
Ann 811 (1977)
JCS Perkin I 2678 (1980)

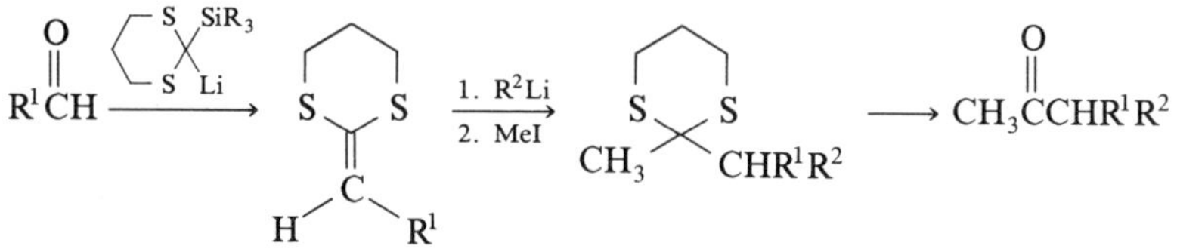

Ann 830 (1977)

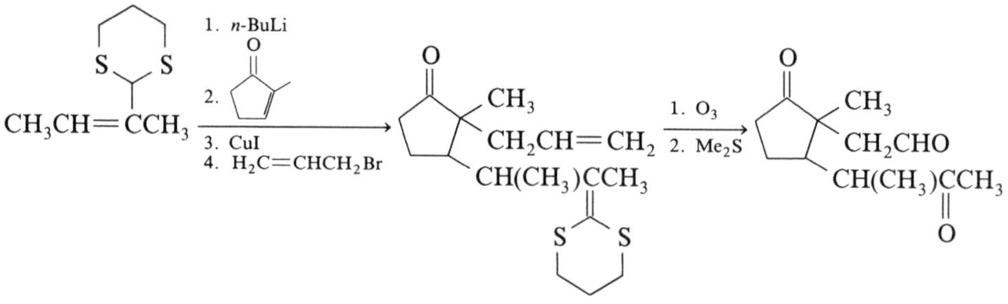

JOC 46 826 (1981)

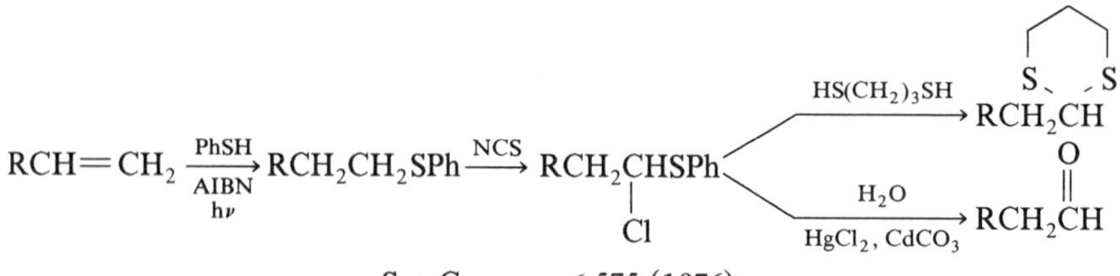

$RCH=CH_2 \xrightarrow[\substack{AIBN \\ h\nu}]{PhSH} RCH_2CH_2SPh \xrightarrow{NCS} RCH_2\underset{\underset{Cl}{|}}{C}HSPh$

$\xrightarrow{HS(CH_2)_3SH} RCH_2CH$ (dithiane)

$\xrightarrow[HgCl_2,\ CdCO_3]{H_2O} RCH_2\overset{O}{\overset{\|}{C}}H$

Syn Commun 6 575 (1976)

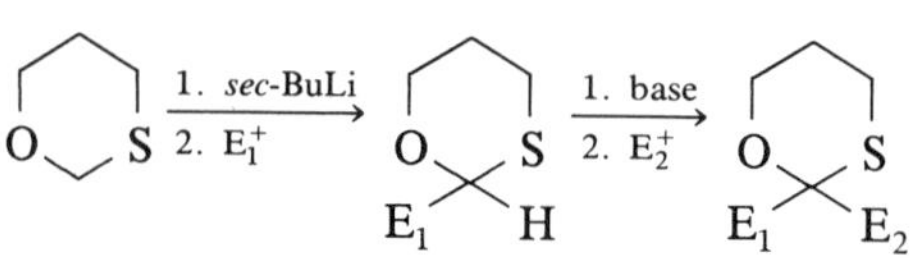

	$\underline{E_1^+}$	$\underline{E_2^+}$	
RX		—	CC 814 (1977)
Me$_3$MCl (M = Si, Ge, Sn), Me$_3$PbOAc or (MeX)$_2$ (X = S, Se)		—	TL 22 2005 (1981)

RCHO — JACS *106* 2937, 2943 (1984)

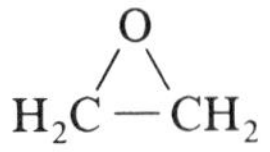

—

$$H_2C \overset{O}{\underset{\displaystyle \triangle}{-}} CH_2$$

JACS *103* 3112 (1981)
(E_1 = 2-furyl)

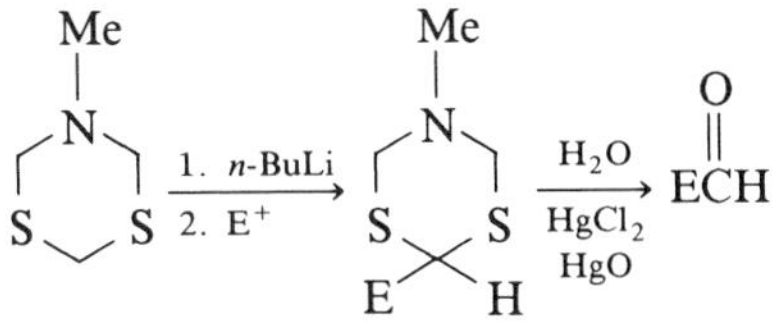

E^+ = RX, RCHO, R$_2$CO

JOC *42* 393 (1977)

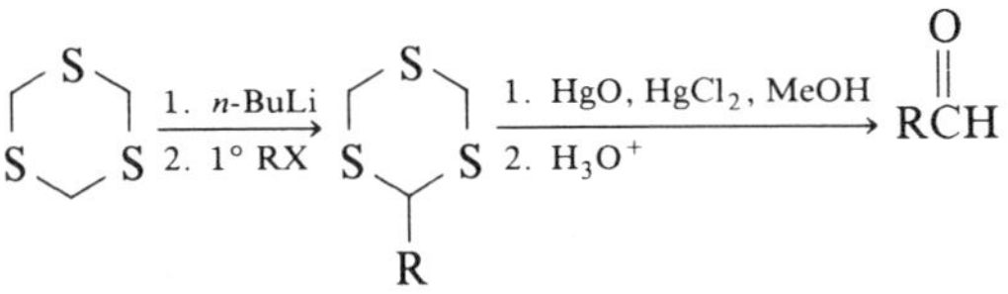

Org Syn *51* 39 (1971)

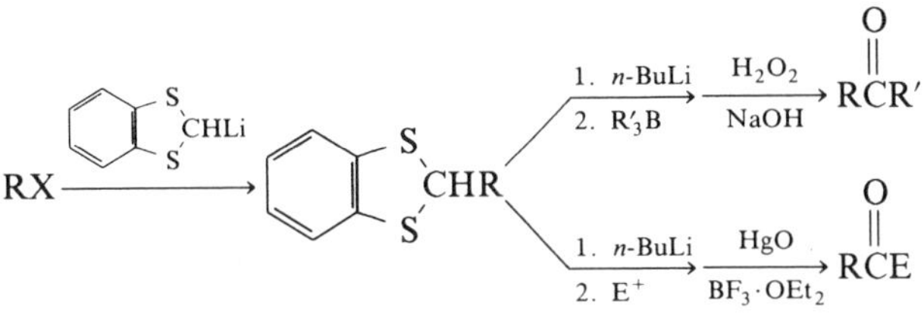

$\xrightarrow{\text{1. } n\text{-BuLi}}$ $\xrightarrow[\text{NaOH}]{\text{H}_2\text{O}_2}$ RCR′ TL 1893 (1979)

2. R′$_3$B

$\xrightarrow{\text{1. } n\text{-BuLi}}$ $\xrightarrow[\text{BF}_3 \cdot \text{OEt}_2]{\text{HgO}}$ RCE TL 2345 (1978)

2. E$^+$

$$E^+ = RX, R_2CO, H_2C \overset{O}{\underset{\displaystyle \triangle}{-}} CH_2$$

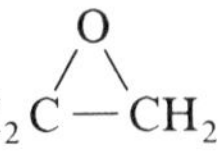

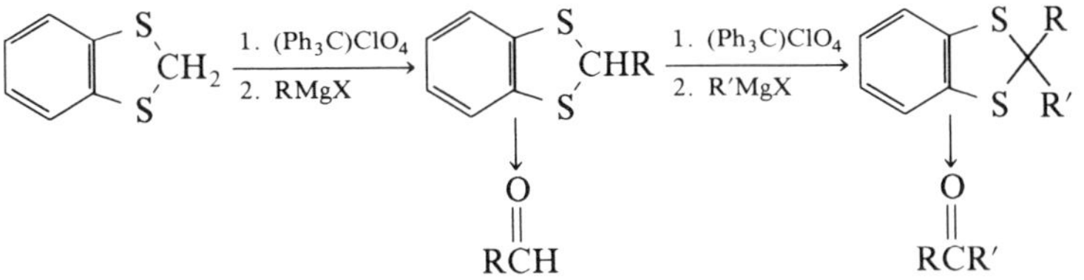

JCS Perkin I 1886 (1976)
TL *22* 1821 (1981)

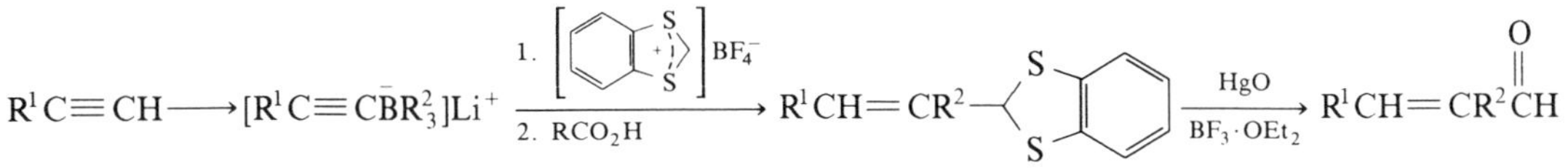

$R^1C\equiv CH \longrightarrow [R^1C\equiv C\bar{B}R_3^2]Li^+$

CC 164 (1981)

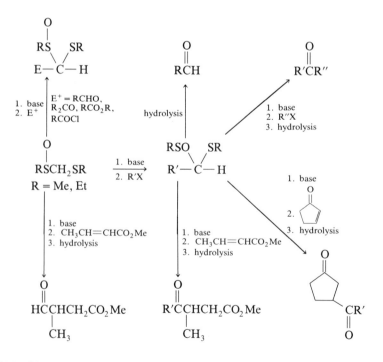

$E_1^+ = RX; E_2^+ = RX,$ epoxide

JCS C 1005 (1966)
TL 1837 (1972)
Syn 720 (1975)

$$RCH(SPh)_2 \xrightarrow[\text{2. R'X}]{\text{1. } n\text{-BuLi·TMEDA}} RR'C(SPh)_2 \longrightarrow R\overset{\overset{\displaystyle O}{\|}}{C}R'$$

TL *21* 4763 (1980)

$$RC(SPh)_2Li \text{ or } \overset{\text{S}}{\underset{\text{S}}{\bigcirc}}CRLi \xrightarrow[\text{2. [O]}]{\text{1. R'}_3B} R\overset{\overset{\displaystyle O}{\|}}{C}R'$$

CL 961 (1973)
TL 1893 (1979)

2. α-Thiosulfoxides

TL 3151 (1971); 3267, 3271, 3275, 4707 (1973); 3653 (1974); 2767 (1975); 1303 (1978); *27* 6381 (1986)
Syn 117 (1974)
JOC *52* 5466 (1987)

$$\underset{EtSCH_2SEt}{\overset{O}{|}} \xrightarrow[\substack{3.\ AcCl \\ 4.\ KOH}]{\substack{1.\ n\text{-BuLi} \\ 2.\ RCHO}} \underset{RCH=CSEt}{\overset{SOEt}{|}} \xrightarrow[\substack{also \\ LiCH_2CO_2R, \\ enamines \\ and\ \alpha\text{-thioketones}}]{NaCH(CO_2R)_2} \underset{(RO_2C)_2CHCHCHSEt}{\overset{SOEt}{\underset{R}{|}}}$$

TL 4707, 4711, 4715 (1973)
JACS *96* 3701 (1974)

3. α-Thiosulfones

$$MeSCH_2SO_2R \xrightarrow[\substack{3.\ base \\ 4.\ R^2X}]{\substack{1.\ base \\ 2.\ R^1X}} \underset{R^1R^2CSO_2Me}{\overset{SMe}{|}} \xrightarrow[H_2O]{HCl} \underset{R^1CR^2}{\overset{O}{\|}}$$

CL 813 (1982) (R = Me); 767 (1983) (R = *p*-Tol)

$$R^1CH_2SO_2Ph \xrightarrow[\substack{3.\ n\text{-BuLi} \\ 4.\ MeSSMe}]{\substack{1.\ n\text{-BuLi} \\ 2.\ R^2CH_2X}} \underset{R^1R^2CSO_2Ph}{\overset{SMe}{|}} \xrightarrow[SiO_2]{CuCl_2} \underset{R^1CR^2}{\overset{O}{\|}}$$

BCSJ *56* 2539 (1983)

4. Thiocarbamates

$$RX \xrightarrow[\substack{2.\ H_2O,\ HgCl_2}]{1.\ LiCH[SCN(CH_3)_2]_2} \underset{RCH}{\overset{S\ \ \ \ \ O}{\underset{}{\| \quad \ \|}}}$$

CL 731 (1974)

$$RX \xrightarrow[NaOH]{\overset{S}{\|}} \xrightarrow[NaOH]{R'X} \underset{R}{\overset{NC}{\diagdown}}\underset{R'}{\overset{SCN(CH_3)_2}{C}} \xrightarrow[NBS,\ NaOH]{NaOH\ or} \underset{RCR'}{\overset{O}{\|}}$$

TL 2967 (1976)

$$RX \xrightarrow[\substack{2.\ H_2O,\ HgX_2}]{1.\ MeS-\underset{Li}{\overset{R'}{C=C}}-SCSNMe_2} RCH=CR'CH \overset{O}{\|}$$

TL 3625 (1974)

5. Thioesters

$$\underset{\text{RCSEt}}{\overset{\overset{\displaystyle S}{\|}}{\text{RCSEt}}} \xrightarrow[\text{2. E}^+]{\text{1. EtMgI}} \underset{\underset{\displaystyle |}{\overset{\displaystyle \diagdown \diagup}{R-C-E}}}{\overset{\text{EtS}\quad\text{SEt}}{R-C-E}} \xrightarrow[\substack{\text{HgCl}_2 \\ \text{CaCO}_3}]{\text{H}_2\text{O}} \underset{\text{RCE}}{\overset{\overset{\displaystyle O}{\|}}{\text{RCE}}}$$

$$E^+ = D_2O, CO_2, ClCO_2Et, RCHO, R_2CO$$

TL 4657 (1978)

6. Vinylic and Allylic Sulfides and Selenides

Reviews:

TL 4437 (1975) (hydrolysis of vinyl sulfides)
Syn 357 (1977)

$$(\text{RCHO}) \; R_2C{=}O \xrightarrow{\text{reagent}} R_2C{=}CRSR \xrightarrow{\text{hydrolysis}} R_2\overset{\overset{\displaystyle O}{\|}}{\text{CHCR}} \; \text{or} \; R_2\overset{\overset{\displaystyle O}{\|}}{\text{CHCH}}$$

Reagent

$Ph_3P{=}CHSMe$	Ber *94* 1373 (1961)
$Ph_3P{=}CHS\text{-}n\text{-Bu}$	TL 3787 (1968)
$Ph_3P{=}CHSPh$	TL 3665 (1969); 4451 (1974)
$Ph_3P{=}CHSR$ (R = Ph, $PhCH_2$, n-Pr)	Ann 2085 (1974)
$Ph_3P{=}C(CH_3)SPh$	TL 3787 (1968)
$(MeO)_2PO\bar{C}HSMe$	Syn 170 (1969)
$(EtO)_2PO\bar{C}HSMe$	JCS 1324 (1963)
	Syn 278 (1975)
	JOC *40* 1979 (1975)
$(EtO)_2PO\bar{C}HSAr$	Syn 145 (1970); 278 (1975)
$(EtO)_2PO\bar{C}RSMe$	JOC *35* 777 (1970); *44* 2967 (1979)
	JCS Perkin I 1879 (1974)
$Ph_2PO\bar{C}RSR$	JCS Perkin I 2263 (1977)
$Me_3SiCH(Li)SPh$	TL *22* 2803 (1981)

$$(\text{RCHO}) \; R_2C{=}O \xrightarrow{(EtO)_2PO\bar{C}HSCH_2CH{=}CH_2} R_2C{=}CHSCH_2CH{=}CH_2 \xrightarrow[\Delta]{\text{HgO}} R_2\overset{\overset{\displaystyle \text{CHO}}{|}}{\text{C}}CH_2CH{=}CH_2$$

JACS *92* 5522 (1970)

$$\text{RX} \xrightarrow{R'CH{=}C(Li)SR''} R'CH{=}\overset{\overset{\displaystyle SR''}{|}}{\text{CR}} \xrightarrow[\text{HgCl}_2]{\text{H}_2\text{O}} R'CH_2\overset{\overset{\displaystyle O}{\|}}{\text{CR}}$$

JACS *95* 2694 (1973)

TL 1015 (1979)

$$H_2C=CHCH_2SH \xrightarrow[\substack{2.\ E^+ \\ 3.\ RX}]{1.\ 2\ n\text{-BuLi}\cdot TMEDA} Z\text{-ECH}_2CH=CHSR$$

$$E^+ = R'X, R_2CO, R_3SiCl, \text{epoxide}$$

Ber *110* 1833 (1977)

$$RCH=CHCH_2X \xrightarrow{CuCH_2CH=CHS\text{-}i\text{-Pr}} \overset{\displaystyle CH_2CH=CHS\text{-}i\text{-Pr}}{\underset{\displaystyle RCHCH=CH_2}{|}} \longrightarrow \overset{\displaystyle CH_2CH_2CHO}{\underset{\displaystyle RCHCH=CH_2}{|}}$$

JACS *95* 7926 (1973)

$$RX \longrightarrow RCH=\overset{\displaystyle O}{\overset{\displaystyle \|}{C}}CR'$$

R'	Reagents	
H	MeS$\bar{C}$HCH=CHSMe /H$_2$O, HgCl$_2$	TL 311 (1970)
		JACS *93* 1724 (1971)
		BCSJ *48* 1567 (1975)
		JOC *41* 2506 (1976)
	PhS$\bar{C}$HCH=CHOMe /H$_2$O, HCl,	CL 345 (1977)
	HgCl$_2$	
	PhSe$\bar{C}$HCH=CHSePh /H$_2$O$_2$	JOC *47* 1618 (1982)
Me	PhSe$\bar{C}$HCH=CClCH$_3$ /H$_2$O$_2$, py	JOC *40* 2570 (1975)

7. Miscellaneous Reactions

$$RX \longrightarrow \overset{\displaystyle O}{\overset{\displaystyle \|}{R}}CH$$

Me$_3$Si$\bar{C}$HSPh/m-ClC$_6$H$_4$CO$_3$H/H$_3$O$^+$ TL *21* 1559, 1677 (1980);
               22 2803 (1981)
               JCS Perkin I 1131 (1983)

Me$_3$Si$\bar{C}$HSePh/H$_2$O$_2$ TL 4223 (1976)

LiCH(TePh)$_2$/NaI, I$_2$ CL 1081 (1982)

$$RCH \overset{\displaystyle O}{\overset{\diagdown\diagup}{}} CH_2 \xrightarrow[\substack{2.\ \text{protect} \\ 3.\ m\text{-ClC}_6H_4CO_3H \\ 4.\ H_3O^+}]{1.\ LiCH(SPh)SiMe_3} RCHCH_2\overset{\displaystyle O}{\overset{\displaystyle \|}{C}}H$$
with OR' on the middle carbon

TL *21* 1559 (1980); *27* 4403 (1986)

$$\text{RX} \xrightarrow[\text{2. H}^+]{\text{1. LiO}_2\text{CCLi(SMe)R}'} \underset{\underset{\text{SMe}}{\overset{|}{}}}{\text{RR}'\text{CCO}_2\text{H}} \xrightarrow[\text{EtOH}]{\text{NCS}} \text{RR}'\text{C(OEt)}_2 \xrightarrow{\text{H}_3\text{O}^+} \overset{\text{O}}{\overset{\|}{\text{RCR}'}}$$

TL 3797 (1975)
JACS *97* 3529 (1975)
Chem Rev *78* 363 (1978) (review)
Acct Chem Res *11* 453 (1978) (review)

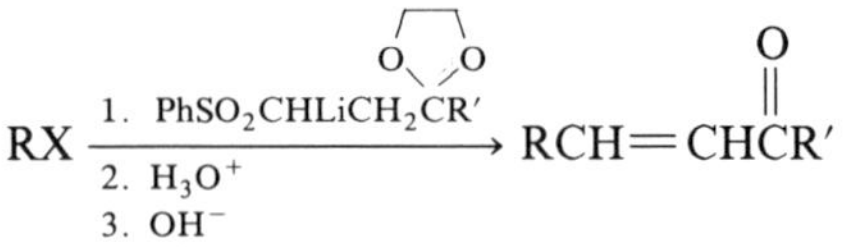

$$\text{RX} \xrightarrow[\substack{\text{2. H}_3\text{O}^+ \\ \text{3. OH}^-}]{\text{1. PhSO}_2\text{CHLiCH}_2\text{CR}'} \text{RCH}{=}\text{CHCR}'$$

TL 1007 (1975)

$$(\text{RCHO}) \ \text{R}_2\text{C}{=}\text{O} \xrightarrow[\text{2. CF}_3\text{CO}_2\text{H}]{\text{1. (PhS)}_2\text{CLiR}'} \overset{\text{O}}{\overset{\|}{\text{R}_2\text{CHCR}'}}$$

CC 547 (1976)

11. HETEROCYCLES

See also page 721, Section 10.

Review: Heterocycles *6* 731 (1977)

1. Dioxanes

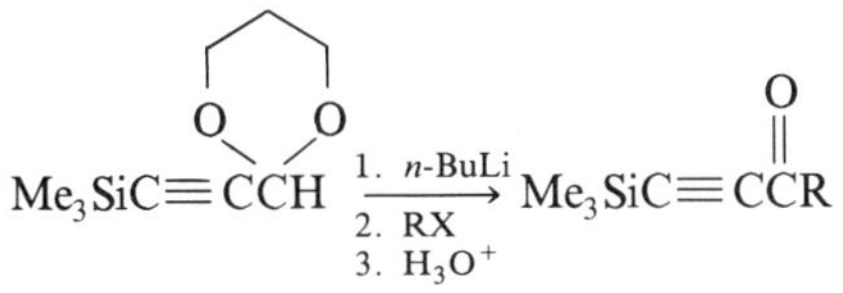

$$Me_3SiC\equiv CCH \xrightarrow[\substack{2.\ RX \\ 3.\ H_3O^+}]{1.\ n\text{-BuLi}} Me_3SiC\equiv CCR$$

Tetr *39* 3073 (1983)

2. Dihydro-1,3-Oxazines

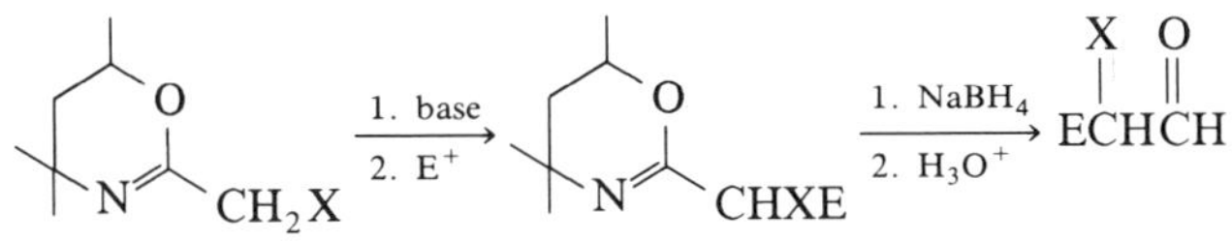

$$\xrightarrow[\substack{2.\ E^+}]{1.\ base} \xrightarrow[\substack{2.\ H_3O^+}]{1.\ NaBH_4} ECHCH$$

X = H, R, Cl

$$E^+ = RX,\ RCHO,\ R_2CO,\ H_2C\!-\!CH_2$$

CC 1163 (1967)
JACS *91* 763, 764, 765, 2155 (1969)
Org Prep Proc Int *1* 193 (1969)
JOC *38* 36 (1973); *39* 618 (1974); *43* 2907 (1978)

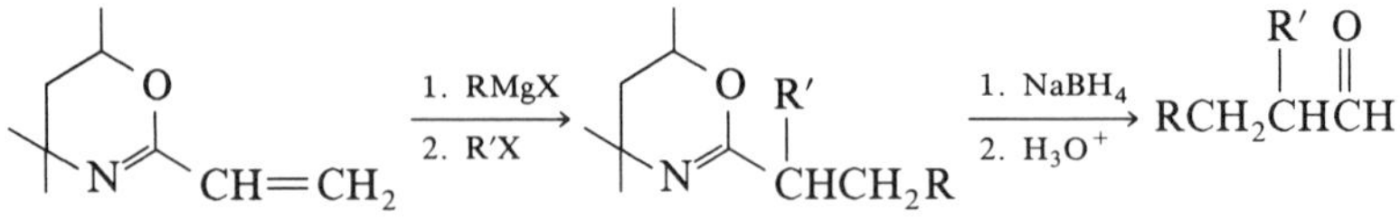

TL 1783, 4809 (1969)
Org Prep Proc Int *1* 213 (1969)
JOC *38* 36, 2136 (1973)

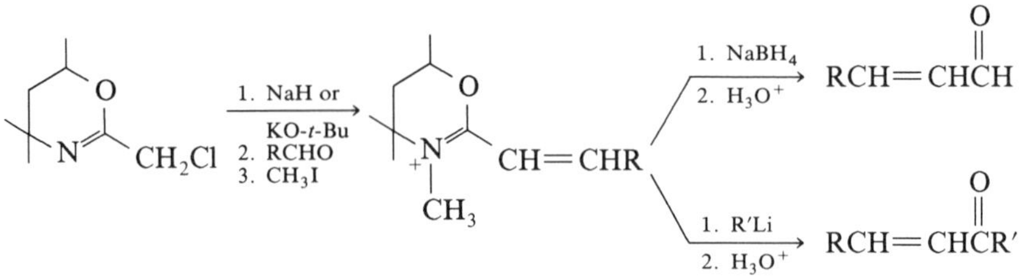

JOC *39* 623 (1974)

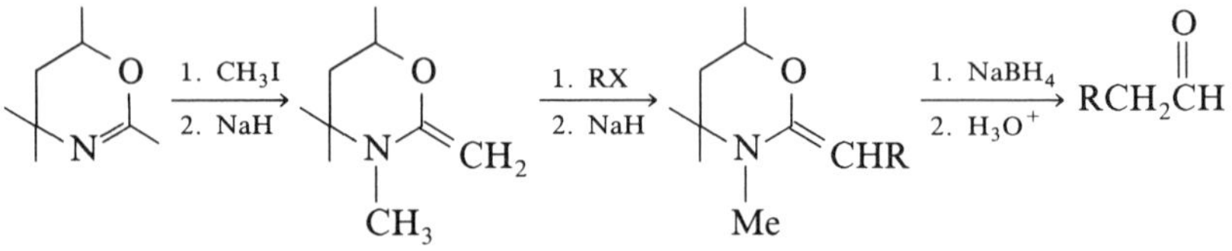

JACS *94* 3243 (1972)

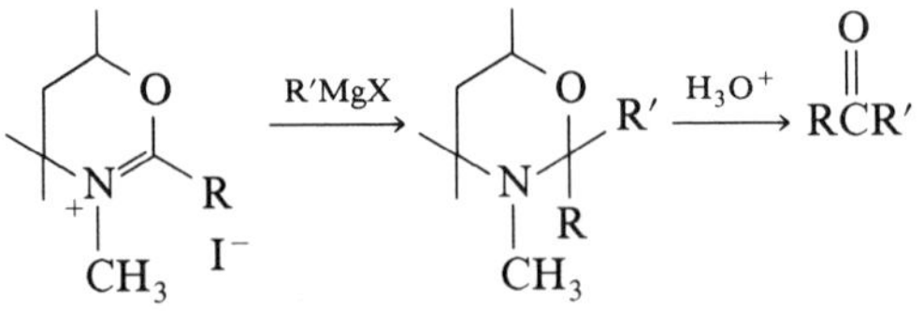

JACS *92* 1084 (1970)
JOC *37* 4289 (1972); *38* 175 (1973)

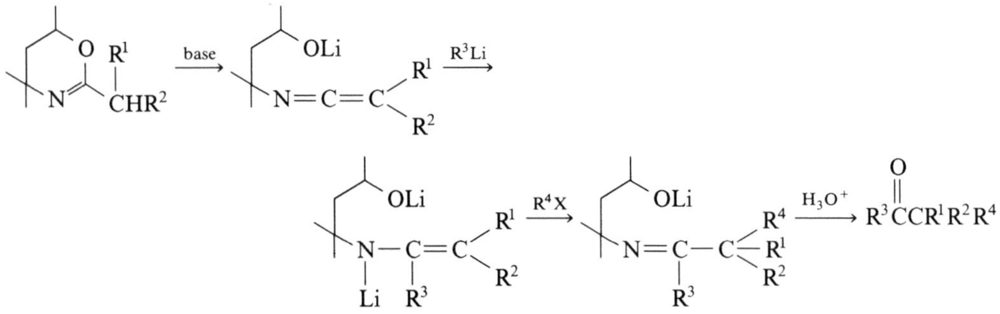

JACS *93* 2314 (1971)
JOC *38* 2129 (1973)

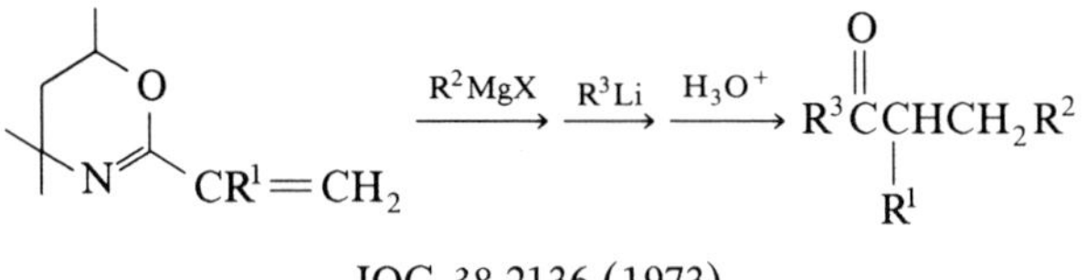

JOC *38* 2136 (1973)

3. Oxazolines

$$\text{(oxazoline)} \xrightarrow[\text{2. E}^+]{\text{1. } n\text{-BuLi}} \text{(oxazoline)}-CH_2E \xrightarrow[\substack{\text{2. NaBH}_4 \\ \text{3. H}_3\text{O}^+}]{\text{1. MeI}} ECH_2\overset{\overset{\displaystyle O}{\|}}{CH}$$

$$E^+ = RX, R_2CO, H_2C\overset{\displaystyle O}{-}CH_2$$

J Heterocyclic Chem *3* 531 (1966)
Tetr *29* 3417 (1973)
BSCF 2673 (1973)
Pure Appl Chem *51* 1255 (1979)

4. Imidazolines

$$\text{(imidazoline, NCH}_2\text{Ph, CH}_3) \xrightarrow[\text{2. } n\text{-BuLi/R}^2\text{X}]{\text{1. } n\text{-BuLi/R}^1\text{X}} \text{(imidazoline, NCH}_2\text{Ph, CHR}^1\text{R}^2) \xrightarrow[\substack{\text{2. R}^3\text{M} \\ (M = \text{Li, MgBr}) \\ \text{3. H}_3\text{O}^+}]{\text{1. MeI}} R^1R^2\overset{\overset{\displaystyle O}{\|}}{CH}CR^3$$

TL *22* 261 (1981)
CC 282 (1982)

5. 2-Thiazolines

$$\text{(thiazoline)}-CH_3 \xrightarrow[\text{2. E}^+]{\text{1. } n\text{-BuLi}} \text{(thiazoline)}-CH_2E$$

$$\xrightarrow[\text{2. H}_2\text{O, HgCl}_2]{\text{1. Al(Hg)}} ECH_2\overset{\overset{\displaystyle O}{\|}}{CH}$$

$$\xrightarrow[\substack{\text{2. E}^+ \\ \text{3. Al(Hg)} \\ \text{4. H}_2\text{O, HgCl}_2}]{\text{1. } n\text{-BuLi}} E_2\overset{\overset{\displaystyle O}{\|}}{CH}CH$$

$$E^+ = RX, RCHO, R_2CO$$

TL 3929 (1972)
JOC *40* 2021, 2025 (1975)

6. Thiazoles

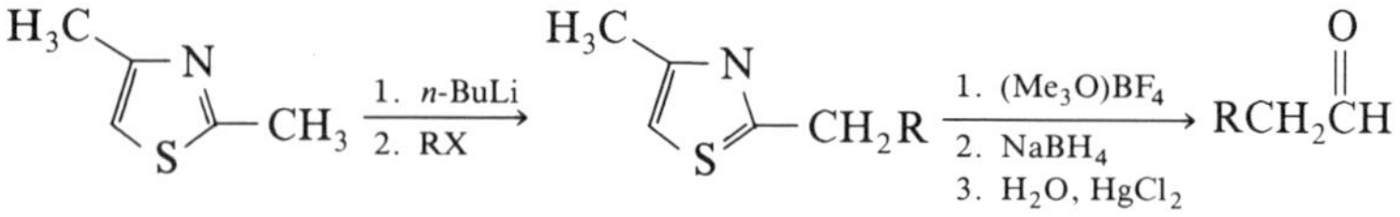

$$\text{(thiazole-CH}_3) \xrightarrow[\text{2. RX}]{\text{1. } n\text{-BuLi}} \text{(thiazole-CH}_2R) \xrightarrow[\substack{\text{2. NaBH}_4 \\ \text{3. H}_2\text{O, HgCl}_2}]{\text{1. (Me}_3\text{O)BF}_4} \text{RCH}_2\text{CHO}$$

TL 4709 (1971)
JACS *95* 3408 (1973)
JOC *39* 1189, 1192 (1974)

7. Benzothiazoles

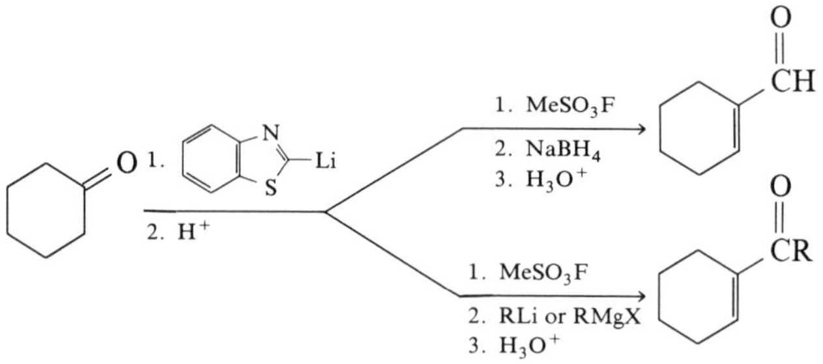

1. MeSO$_3$F
2. NaBH$_4$
3. H$_3$O$^+$

1. MeSO$_3$F
2. RLi or RMgX
3. H$_3$O$^+$

TL 5 (1978)

8. 1,3-Oxathiolane-3,3-Dioxides

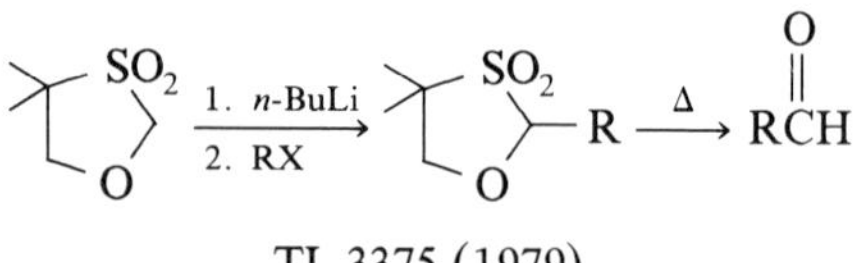

$$\xrightarrow[\text{2. RX}]{\text{1. } n\text{-BuLi}} \cdots -R \xrightarrow{\Delta} \text{RCHO}$$

TL 3375 (1979)

9. 1,3-Benzoxathiolium Salts

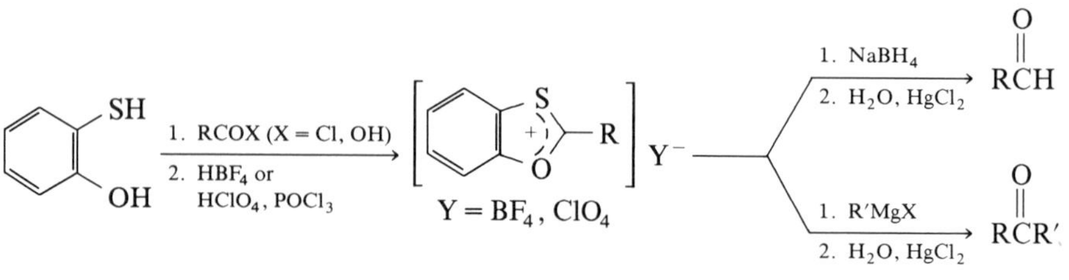

1. RCOX (X = Cl, OH)
2. HBF$_4$ or HClO$_4$, POCl$_3$

Y = BF$_4$, ClO$_4$

1. NaBH$_4$
2. H$_2$O, HgCl$_2$ → RCHO

1. R′MgX
2. H$_2$O, HgCl$_2$ → RCR′

J Heterocyclic Chem *11* 507, 943 (1974)
Gazz Chim Ital *105* 907 (1975)
JCS Perkin I 323 (1976)

12. CARBONYL HOMOLOGATION

See also page 627, Section 4; page 709, Section 9; page 721, Section 10; page 729, Section 11; and page 737, Section 13.

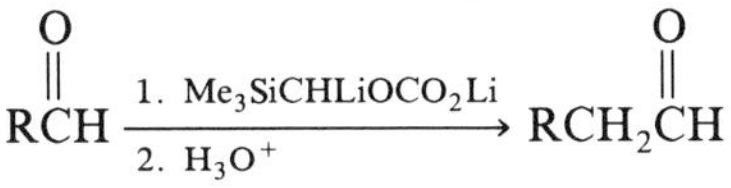

$$\underset{RCH}{\overset{O}{\parallel}} \xrightarrow[\text{2. } H_3O^+]{\text{1. } Me_3SiCHLiOCO_2Li} \underset{RCH_2CH}{\overset{O}{\parallel}}$$

TL *28* 1847 (1987)

$$\underset{RCR\ (RCH)}{\overset{O\quad O}{\parallel\quad\parallel}} \xrightarrow{LiCH\left[-B\overset{O}{\underset{O}{<}}\right]_2} R_2C=CHB\overset{O}{\underset{O}{<}} \xrightarrow{NaBO_3 \cdot 4\,H_2O} \underset{R_2CHCH}{\overset{O}{\parallel}}$$

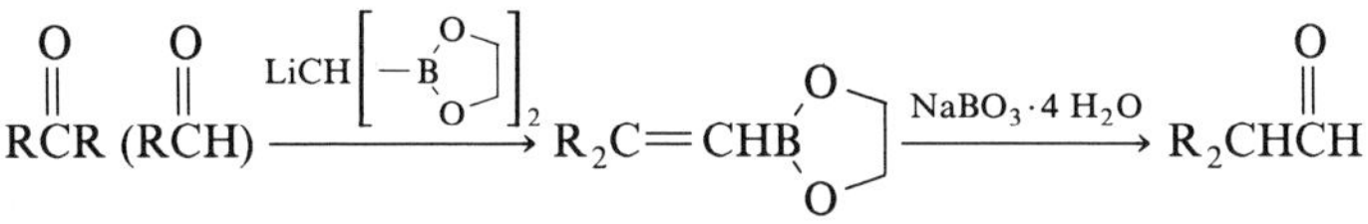

JACS *97* 5608 (1975); *99* 3196 (1977)
JOC *45* 1091 (1980)
Organomet *1* 20 (1982)

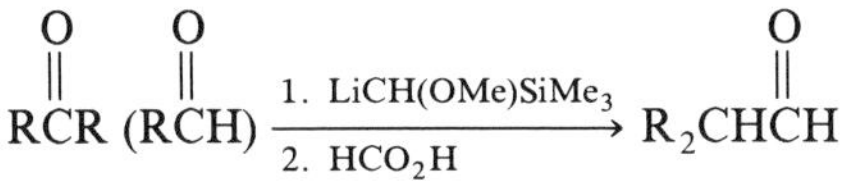

$$\underset{RCR\ (RCH)}{\overset{O\quad O}{\parallel\quad\parallel}} \xrightarrow[\text{2. } HCO_2H]{\text{1. } LiCH(OMe)SiMe_3} \underset{R_2CHCH}{\overset{O}{\parallel}}$$

Organomet *1* 553 (1982)
JACS *107* 5391 (1985)

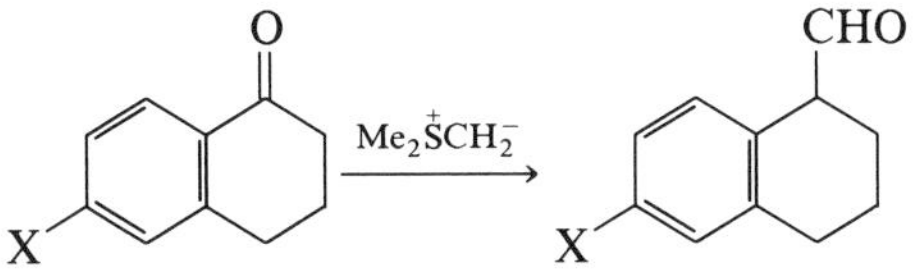

J Nat Prod *44* 557 (1981)
Syn Commun *12* 613 (1982)

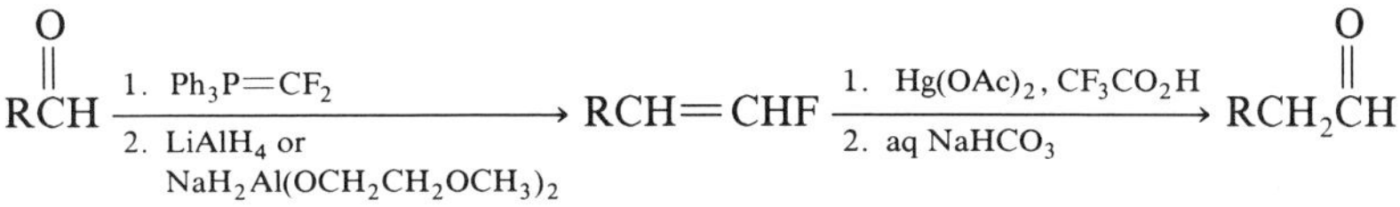

$$\underset{RCH}{\overset{O}{\parallel}} \xrightarrow[\substack{\text{2. } LiAlH_4 \text{ or}\\ NaH_2Al(OCH_2CH_2OCH_3)_2}]{\text{1. } Ph_3P=CF_2} RCH=CHF \xrightarrow[\text{2. aq } NaHCO_3]{\text{1. } Hg(OAc)_2,\ CF_3CO_2H} \underset{RCH_2CH}{\overset{O}{\parallel}}$$

CL 651 (1980)

$$\left(\begin{array}{c} O \\ \parallel \\ RCH \end{array}\right) \ \begin{array}{c} O \\ \parallel \\ RCR \end{array} \xrightarrow[]{\underset{(EtO)_2POCHN=CHPh}{}} R_2C=CHN=CHPh \xrightarrow[\substack{2.\ E^+ \\ 3.\ H_3O^+}]{1.\ n\text{-BuLi}} \begin{array}{c} E \\ \mid \\ R_2CCHO \end{array}$$

$$E^+ = RX, RCHO$$

See page 755, Section 10.

$$\begin{array}{cc} O & O \\ \parallel & \parallel \\ RCR & (RCH) \end{array} \xrightarrow{Cl\overline{C}R'SiMe_3} \begin{array}{c} O \\ \diagup \backslash \\ R_2C-CR'SiMe_3 \end{array} \xrightarrow{H_3O^+} \begin{array}{c} O \\ \parallel \\ R_2CHCR' \end{array}$$

$$R' = H, CH_3$$

CC 513 (1977)
JACS *99* 4536 (1977)
Organomet *1* 893 (1982)
Tetr *39* 867 (1983)

$$\begin{array}{c} O \\ \parallel \\ RCH \end{array} \xrightarrow{Ph_3P=CF_2} RCH=CF_2 \xrightarrow{R'Li} RCH=CFR' \xrightarrow[\substack{2.\ aq\ NaHCO_3}]{1.\ Hg(OAc)_2,\ CF_3CO_2H} \begin{array}{c} O \\ \parallel \\ RCH_2CR' \end{array}$$

CL 935 (1980)

$$\left(\begin{array}{c} O \\ \parallel \\ RCH \end{array}\right) \ \begin{array}{c} O \\ \parallel \\ R^1CR^2 \end{array} + R^3CHXCO_2Et \xrightarrow[NaNH_2]{NaOEt\ or} \begin{array}{c} O \\ \diagup \backslash \\ R^1R^2C-CR^3CO_2Et \end{array} \longrightarrow \longrightarrow \begin{array}{c} O \\ \parallel \\ R^1R^2CHCR^3 \end{array} \ or \ \begin{array}{c} O \\ \parallel \\ R^1R^2R^3CCH \end{array}$$

Org Rxs *5* 413 (1949)

$$\begin{array}{c} O \\ \parallel \\ R^1CR^2 \end{array}$$
aldehydes
or ketones
$$\xrightarrow{PhSOCHClCH_2CH_2SPh}$$

PhSeNa $\longrightarrow$ $\begin{array}{c} O \\ \parallel \\ R^1R^2CHCCH_2CH_2SPh \end{array} \xrightarrow[\substack{2.\ OH^-}]{1.\ m\text{-}ClC_6H_4CO_3H} \begin{array}{c} O \\ \parallel \\ R^1R^2CHCCH=C \end{array}$

PhSNa $\longrightarrow$ $\begin{array}{c} O \\ \parallel \\ R^1R^2CCCH_2CH_2SPh \\ \mid \\ PhS \end{array} \xrightarrow[\substack{2.\ \Delta \\ 3.\ OH^-}]{1.\ 2\ m\text{-}ClC_6H_4CO_3H} \begin{array}{c} O \\ \parallel \\ R^3CH=CR^2CCH \end{array}$

TL *27* 2471 (1986)

$$\begin{array}{c} O \\ \parallel \\ RCCH_2R' \end{array} \xrightarrow[DMF]{POCl_3} \begin{array}{c} O \\ \parallel \\ RCCl=CR'CH \end{array} \xrightarrow[cat\ Pd\text{-}C]{H_2} \begin{array}{c} O \\ \parallel \\ RCH_2CHR'CH \end{array}$$

Org Prep Proc Int *14* 9 (1982)

$$(RCHO) \quad \text{(cyclohexanone with O)} \longrightarrow \text{(cyclohexene with CHO)}$$

PhSOCHClLi/KOH/Δ

TL 1377 (1977)
JOC *44* 450 (1979)

LiCHCl$_2$/Δ/CaCO$_3$, LiClO$_4$, HMPA, Δ

TL 2465 (1973)
CC 1351 (1987)

CH$_3$NO$_2$, ⬡NH/TiCl$_3$ Syn 196 (1974)

TsNHNH$_2$/n-BuLi/DMF   TL 2287 (1976)

$$\text{cyclohexanone} \xrightarrow{\text{reagents}} \text{2-X-cyclohexenyl-CHO} \xrightarrow[\substack{\text{cat Pd} \\ \text{K}_2\text{CO}_3}]{\text{H}_2} \text{cyclohexenyl-CHO}$$

X	Reagents	
Cl	COCl$_2$, DMF	Coll Czech Chem Commun *24* 2385 (1959)
		Ber *93* 2743 (1960)
	POCl$_3$, DMF	Proc Chem Soc 227 (1958)
		Coll Czech Chem Commun *24* 2385 (1959)
		Ber *93* 2743 (1960)
		JOC *30* 1126 (1965); *48* 1921 (1983)
		(regioselectivity); *51* 2162 (1986)
Br	PBr$_3$ or POBr$_3$, DMF	Coll Czech Chem Commun *26* 3059 (1961)

$$\text{cyclohexanone} \xrightarrow[\text{2. NaOH}]{\substack{\text{1. RCHBrCO}_2\text{R}', \\ \text{NaOEt}}} \text{(epoxide)}\text{—CO}_2\text{Na} \xrightarrow[\Delta]{\text{Pb(OAc)}_4} \text{cyclohexenyl—CR(=O)}$$

TL 1321 (1979)

$$\overset{O}{\underset{\|}{\text{RCR}}} \xrightarrow[\text{2. LiAlH}_4]{\text{1. LiC}\equiv\text{CSiMe}_3} \overset{OH}{\underset{|}{\text{R}_2\text{CCH}}}=\text{CHSiMe}_3 \xrightarrow[\text{2. AgNO}_3, \text{H}_2\text{O}]{\text{1. PhSCl}} \text{R}_2\text{C}=\text{CHCH}(=O)$$

TL *22* 2021 (1981)

$$\left(\overset{O}{\underset{\|}{\text{RCH}}}\right)\overset{O}{\underset{\|}{\text{RCR}}} \longrightarrow \text{R}_2\text{C(OEt)}_2 \xrightarrow[\text{2. H}_3\text{O}^+]{\substack{\text{1. H}_2\text{C}=\text{CHOEt,} \\ \text{cat acidic clay}}} \text{R}_2\text{C}=\text{CHCH}(=O)$$

Syn 137 (1981)

See also page 173, Section 4, for analogous transformations.

$$\overset{O}{\underset{\|}{\text{RCR}}} \xrightarrow[\text{Zn}]{\text{BrCH}_2\text{C}\equiv\text{CR}'} \overset{OH}{\underset{|}{\text{R}_2\text{CCH}_2}}\text{C}\equiv\text{CR}' \xrightarrow{\text{H}^+} \text{R}_2\text{C}=\text{CHCCH}_2\text{R}'(=O)$$

Syn Commun *10* 637 (1980)

$$\overset{O}{\underset{\|}{\text{R}_2\text{CHCR}'}} \longrightarrow \overset{OSiMe_3}{\underset{|}{\text{R}_2\text{C}}}=\text{CR}' \xrightarrow[\text{2. }\Delta, (\text{Et}_3\text{N})]{\text{1. :CClCH}_3} \overset{O}{\underset{\underset{\text{CH}_3}{|}}{\text{R}_2\text{C}=\text{CCR}'}}$$

R' = H or alkyl

Syn 289, 291 (1981)
TL *22* 645 (1981)

$$RCH_2\overset{O}{\overset{\|}{C}}H \xrightarrow[\text{Et}_2\text{NH, (HOAc)}]{\text{PhSOCH}_2\text{COR}'} RCH\overset{OH}{\underset{|}{C}}HCH=CH\overset{O}{\overset{\|}{C}}R'$$

TL *28* 649 (1987)

$$R\overset{O}{\overset{\|}{C}}H \xrightarrow[\substack{\text{2. LiAlH}_4 \\ \text{3. H}_3\text{O}^+}]{\text{1. LiC}\equiv\text{CCH}=\text{CHOMe}} RCH=CHCH=CHC\overset{O}{\overset{\|}{C}}H$$

JCS 4082 (1956)
JOC *47* 4611 (1982)

$$R\overset{O}{\overset{\|}{C}}H \xrightarrow[\text{or KC(SEt)(SOEt)CH=CHPO(OEt)}_2]{\text{KC(SEt)}_2\text{CH}=\text{CHPO(OEt)}_2/m\text{-ClC}_6\text{H}_4\text{CO}_3\text{H}} \xrightarrow{\text{LiHBEt}_3} \xrightarrow[\text{HgCl}_2]{\text{HgO}} RCH_2CH=CHC\overset{O}{\overset{\|}{C}}H$$

Can J Chem *58* 2780 (1980)

$$R\overset{O}{\overset{\|}{C}}H\ (R\overset{O}{\overset{\|}{C}}R) \xrightarrow[\text{2. H}_2\text{O}_2]{\text{1. PhSe}\overline{\text{C}}\text{HCH}=\text{CHSePh}} RCH\overset{OH}{\underset{|}{C}}HCH=CHC\overset{O}{\overset{\|}{C}}H$$

JOC *47* 1618 (1982)

$$R\overset{O}{\overset{\|}{C}}OEt \xrightarrow[\substack{\text{3. 1,3-cyclohexadiene or LiH} \\ \text{4. ClSiMe}_3 \text{ (or Ac}_2\text{O)}}]{\substack{\text{1. LiCHBr}_2 \\ \text{2. } n\text{-BuLi}}} RCH=CHOSiMe_3(Ac) \xrightarrow{\text{H}_3\text{O}^+} RCH_2\overset{O}{\overset{\|}{C}}H$$

JACS *108* 1325 (1986)
TL *28* 2463 (1987)

13. ALKYLATION OF ALDEHYDES AND KETONES AND THEIR DERIVATIVES

1. Free Radical Alkylation

$$RCH_2CHO + R'CH=CH_2 \xrightarrow{Mn(OAc)_3} R'CH_2CH_2\overset{\displaystyle R}{\overset{|}{C}}HCHO$$

JOC USSR *8* 1422 (1972)

$$+ \; H_2C=CHR \xrightarrow[\text{MnO}_2 \text{ or CuO}]{PbO_2, Ag_2O,} $$

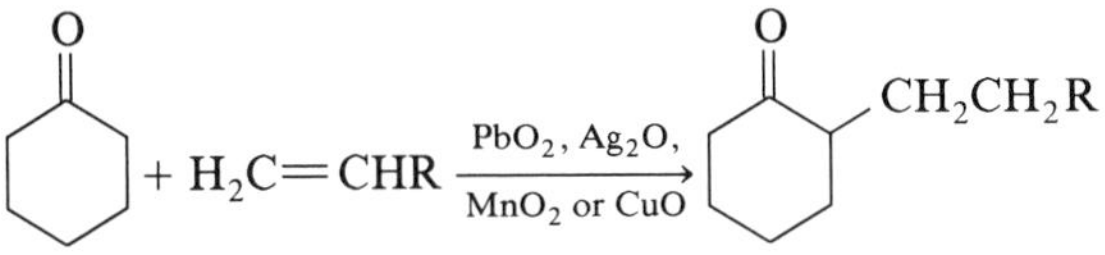

Syn 315 (1976)
Coll Czech Chem Commun *41* 746 (1976)

$$10 \; CH_3\overset{O}{\overset{||}{C}}CH_3 + H_2C=CHR \xrightarrow[\text{30\% Na}_2S_2O_8]{2\% \text{ AgNO}_3} CH_3\overset{O}{\overset{||}{C}}CH_2CH_2CH_2R$$

J Chem Res (S) 310 (1983)

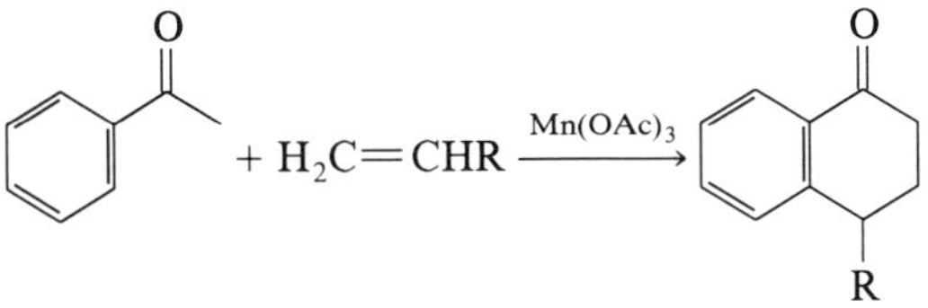

$$+ \; H_2C=CHR \xrightarrow{Mn(OAc)_3}$$

JACS *94* 2888 (1972)

$$\xleftarrow[\substack{(NH_4)_2Ce(NO_3)_6 \\ MeOH}]{H_2C=CHOAc} \qquad \xrightarrow[(NH_4)_2Ce(NO_3)_6]{H_2C=CCH_3}$$

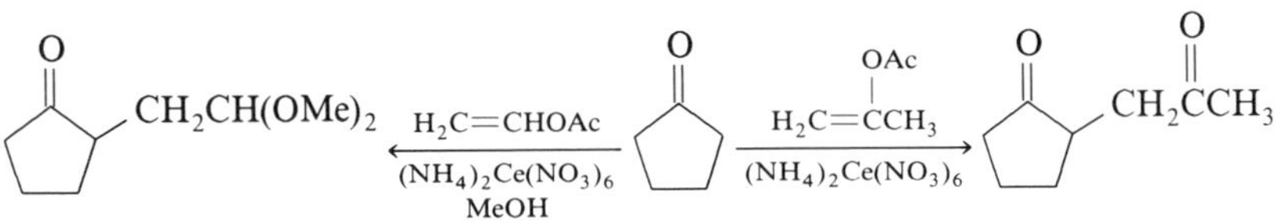

TL *28* 5357 (1987)

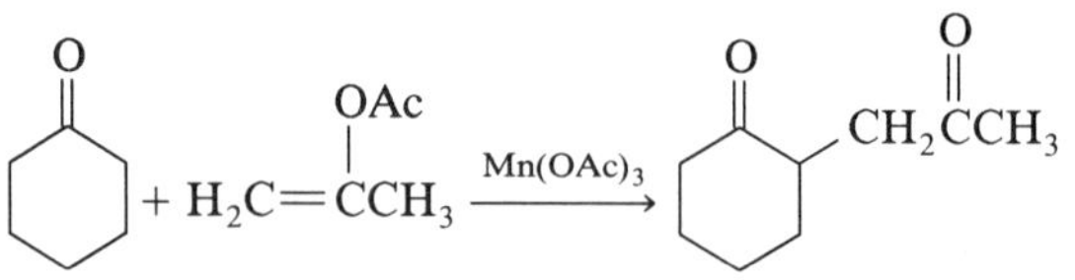

(low yields)

JOC *39* 3457 (1974)

2. Direct Enol or Enolate Alkylation

Reviews:

Rec Chem Prog *24* 43 (1963) (general); *28* 99 (1967) (general)
H. O. House, "Modern Synthetic Reactions," W. A. Benjamin, Menlo Park, California, 2nd ed
(1972), pp 546–70 (general)
Tetr *32* 2979 (1976) (regiospecific preparation of ketone enolates and synthetic uses); *33* 2737
(1977) (structure and reactivity of alkali metal enolates)
Ber *114* 2866 (1981) (substituent effects on O vs C alkylation)
JOC *48* 4789 (1983) (stereochemistry and O vs C alkylation)
D. A. Evans in "Asymmetric Synthesis," Ed. J. D. Morrison, Academic Press, New York
(1984), Vol 3, Chpt 1 (stereoselective alkylation reactions of chiral metal enolates)

Bases for kinetic enolate formation

$(C_6H_5)_3CLi$

JOC *30* 1341 (1965); *34* 3070 (1969);
37 3873 (1972)

$LiNR_2$

JOC *34* 2324 (1969); *36* 2361 (1971);
39 3459 (1974)
Org Syn *52* 39 (1972)
TL 965 (1973); *22* 4119 (1981)
JACS *99* 247 (1977); *101* 934 (1979)
JCS Perkin I 2306 (1981)
Tetr *37* 3981 (1981)

$LiN(i\text{-}Pr)CHArCH_2X$
$(X = H, OMe, NR_2)$

JACS *108* 543 (1986)
(enantioselective)

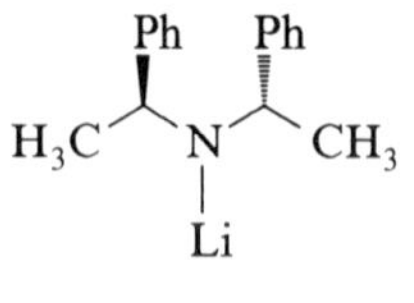

JOC *45* 755 (1980)
TL *23* 105 (1982); *27* 631, 2767 (1986)
(all enantioselective)

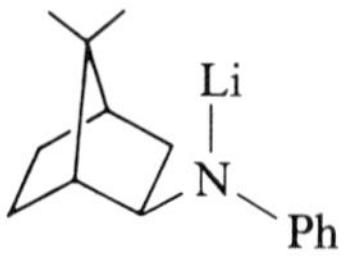

CC 88 (1986) (enantioselective)

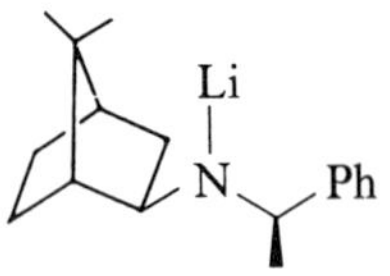

CC 88 (1986)
TL *28* 3723 (1987)
(both enantioselective)

$LiN[Si(CH_3)_3]_2$

JOMC *1* 476 (1964)
CC 1497, 1498 (1969)

$$R^1 R^2 CHCH + R^3 X \xrightarrow{\text{base}} R^1 R^2 R^3 CCH$$

(with carbonyl O above each CHCH and CCH)

R^3 = allylic, benzylic, Me > 1° alkyl > 2° alkyl

Base

KH	TL 491 (1978); *21* 4005 (1980)
	JOC *47* 2479 (1982) (R^1 = SPh)
NaOH, *n*-Bu$_4$NI	TL 1273 (1973)
	Chem Ind 731 (1978)
	Ann 1585 (1979)
KCPh$_3$	JOC *50* 2668, 2676 (1985)

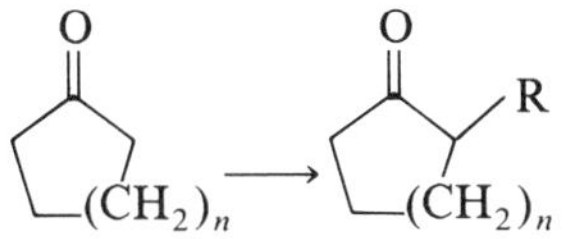

$n = 1, 2$

KO-*t*-Am, DME/RX (R = Me, PhCH$_2$)	Syn Commun 8 563 (1978)
K-Al$_2$O$_3$, hexane/*n*-BuBr	JOMC *204* 281 (1981)

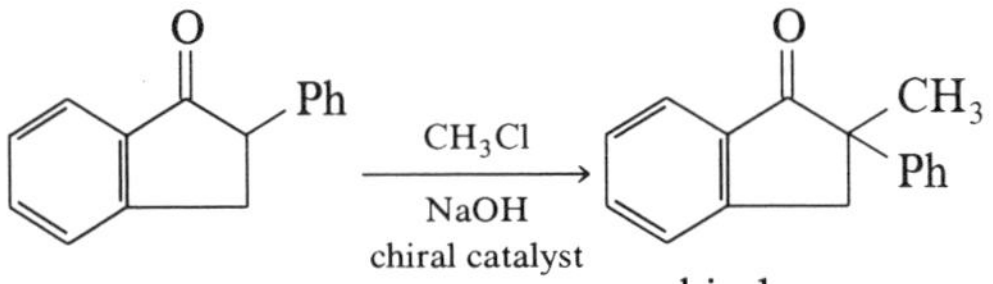

JACS *106* 446 (1987)
JOC *52* 4745 (1987)

$$RCCH_3 \xrightarrow[R'CHO]{KFe(CO)_4H} RCCH_2CH_2R'$$

$$RCH_2CR \longrightarrow R'CH_2CHRCR$$

TL 2491 (1973)
JCS Perkin I 1273 (1975)

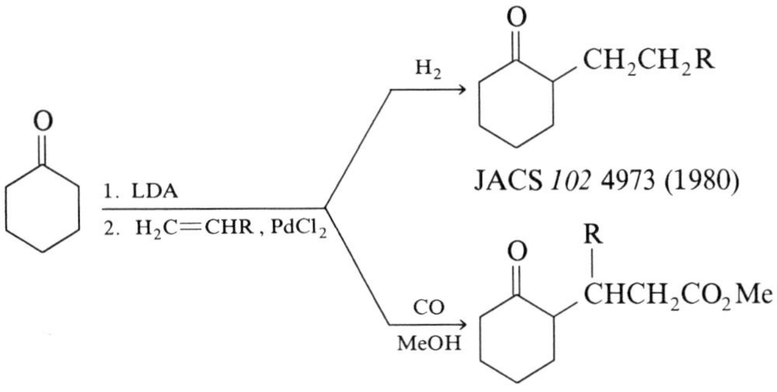

JACS *102* 4973 (1980)

JACS *102* 4980 (1980)

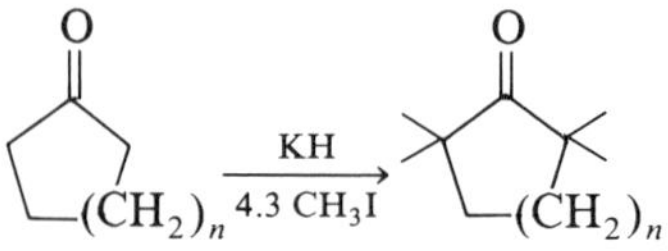

JOC *43* 1834 (1978)

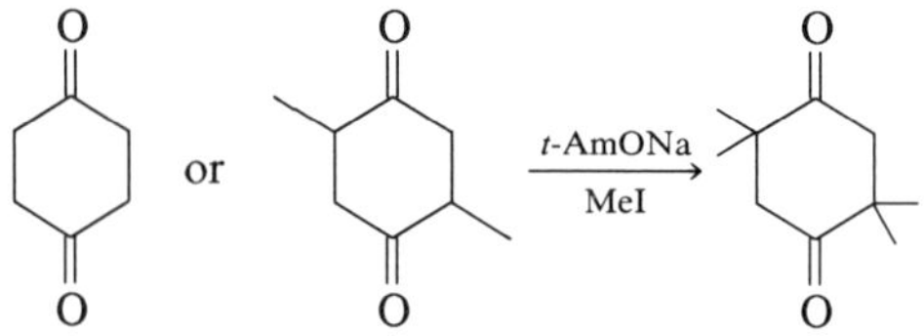

TL *24* 2104 (1983)

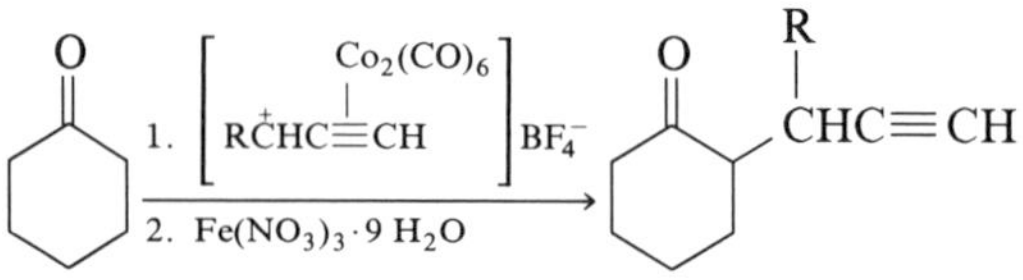

JACS *102* 2508 (1980)

$$\underset{RC=CH_2}{\overset{O^-}{|}} \longrightarrow \underset{RCCH_2R'}{\overset{O}{||}}$$

R′	Reagent(s)	
CH$_3$	CH$_3$I	CC 149 (1985) (stereochemistry)
RCH=CHCH$_2$	RCH=CHCH$_2$OAc, cat Pd(DBA)$_2$	CC 1158 (1981)
H$_2$C=C(SO$_2$Ph)CHR	RCH=C(SO$_2$Ph)CH$_2$Br	TL *27* 5095 (1986)
RCH=C(SO$_2$Ph)CH$_2$	H$_2$C=C(SO$_2$Ph)CH(OAc)R	TL *27* 5095 (1986)
CMe$_2$NO$_2$	ClCMe$_2$NO$_2$	JACS *101* 3378 (1979); *103* 4610 (1981) JOC *47* 1879 (1982)
Ph	Ph$_3$BiCO$_3$	CC 246 (1980)
aryl	R′X, Ni(COD)$_2$	TL 4519 (1973)
aryl, vinyl	R′X, hν	JACS *94* 683 (1972); *102* 7765 (1980); *107* 2183 (1985) (purines) TL 4519 (1973); *28* 91 (1987) JOC *38* 1407 (1973); *41* 1702, 1707 (1976); *42* 1481, 2481 (1977); *44* 2604 (1979); *46* 5022 (1981); *52* 3880 (1987) Acct Chem Res *11* 413 (1978) Org Prep Proc Int *10* 225 (1978) (review) Bull Soc Chim Belg *93* 547 (1984) (heterocycles)

$$\text{cyclohexanone} \xrightarrow[\text{KH}]{\text{Ph}_3\text{BiCO}_3} \text{2,2,6,6-tetraphenylcyclohexanone}$$

CC 246 (1980); 732 (1982)

$$\underset{\text{RCCH}_3}{\overset{\text{O}}{\|}} \longrightarrow \underset{\text{RCCH}_2\text{C}=\text{CHR}^2}{\overset{\text{O}\quad\text{R}^1}{\|\quad|}}$$

RLi (on enol silane)/ZnCl$_2$/PhSeCH$_2$COR^1/   CC 434 (1981)
MsCl, Et$_3$N (R^1 = H or Me, R^2 = H)   JOC *47* 1632 (1982)

MeNHCH$_2$CH$_2$NMe$_2$/*n*-BuLi/ZnCl$_2$/   CL 1007 (1987)
ICH=CHR^2, cat Pd(PPh$_3$)$_4$

LDA/PhSeCH$_2$CHO/MsCl, Et$_3$N   JACS *102* 7950 (1980)
(R^1 = R^2 = H)   JOC *52* 2760 (1987)

LDA/R^{2}CH(CHO)SiMe$_2$(*t*-Bu)/BF$_3$·OEt$_2$ JACS *103* 6251 (1981)
(R^1 = H)

LDA/[R^{2}CH=C(OEt)R^1—Fe$^+$(CO)$_2$C$_5$H$_5$]BF$_4^-$/   JACS *102* 5930 (1980);
HBF$_4$/NaI (R^1, R^2 = H, H; H, Me; Me, H)   *106* 7264 (1984)
     JOC *46* 4103 (1981)

$$\underset{-\text{C}-\text{CR}_2}{\overset{\text{O}\quad\text{H}}{\|\quad|}} \xrightarrow[\substack{\text{2. R'C}\equiv\text{CCl}\\ \text{R'}=\text{Cl, SPh, Ph}}]{\text{1. LDA}} \underset{-\text{C}-\text{CR}_2}{\overset{\text{O}\quad\text{C}\equiv\text{CR'}}{\|\quad|}}$$

$$\xrightarrow[\text{R'}=\text{Cl}]{\text{Cu, HOAc}} \underset{-\text{C}-\text{CR}_2}{\overset{\text{O}\quad\text{C}\equiv\text{CH}}{\|\quad|}}$$

$$\xrightarrow[\text{R'}=\text{Cl}]{\text{H}_2} \underset{-\text{C}-\text{CR}_2}{\overset{\text{O}\quad\text{CH}=\text{CH}_2}{\|\quad|}}$$

TL *23* 2373 (1982)
JACS *106* 3551 (1984)

$$\underset{\text{RCH}_2\text{CCH}_3}{\overset{\text{O}}{\|}} \xrightarrow[\text{2. R'CHO}]{\text{1. LiNR}_2} \underset{\text{RCH}_2\text{CCH}_2\text{CHR'}}{\overset{\text{O}\quad\quad\text{OH}}{\|\quad\quad|}}$$

JOC *39* 3459 (1974)

$$\underset{\text{RCCH}_3}{\overset{\text{O}}{\|}} \xrightarrow[\text{2. MX}_2]{\text{1. LiNR}_2} \underset{(\text{RCCH}_2)_2}{\overset{\text{O}}{\|}}$$

<u>MX$_2$</u>

FeCl$_2$ JOC *45* 5408 (1980)

CuCl$_2$ JACS *97* 2912 (1975); *99* 1487 (1977)

Cu(OTf)$_2$ TL 3741 (1977); 3555 (1978)
     Chem Pharm Bull *28* 262 (1980)
     (inter- and intramolecular)

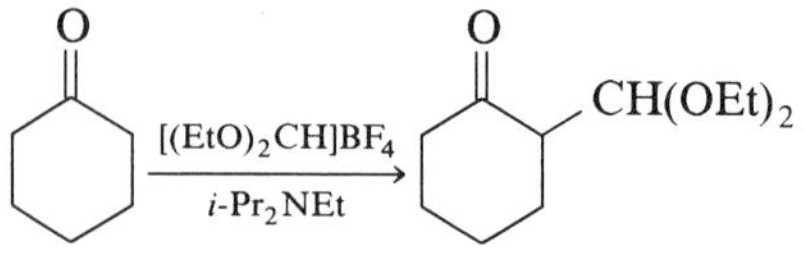

JOC *46* 2557 (1981)

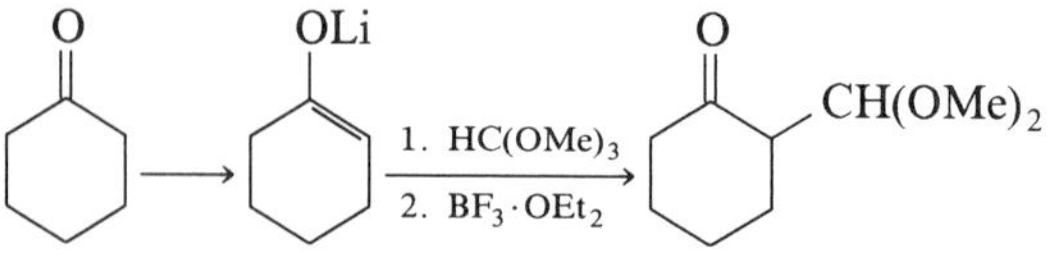

TL *23* 3595 (1982)

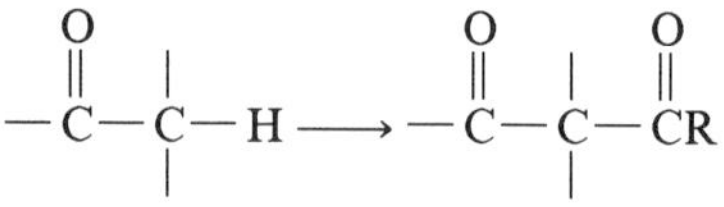

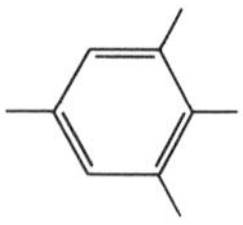 —Li /RCOCl	TL 1187 (1977) Helv *64* 716 (1981)
MeLi (on enol silane)/RCOCl	Helv *64* 716 (1981)
LDA/RCO$_2$CO$_2$Et	JOMC *127* C65 (1977)
LDA/ArCOCN (R = Ar)	TL 1339 (1979)
LiN(SiMe$_3$)$_2$/RCOPO(OEt)$_2$	CL 1087 (1981)
RCOCl (on enamine)/H$_3$O$^+$	JACS *76* 2029 (1954); *85* 207 (1963) Ber *92* 652 (1959); *93* 913 (1960); *99* 823 (1966)
(CH$_3$CO)$_2$O, BF$_3$	J Prakt Chem *141* 149 (1934) JACS *75* 5030 (1953) CC 85 (1979) TL *23* 1115 (1982)

3. Enol Ethers

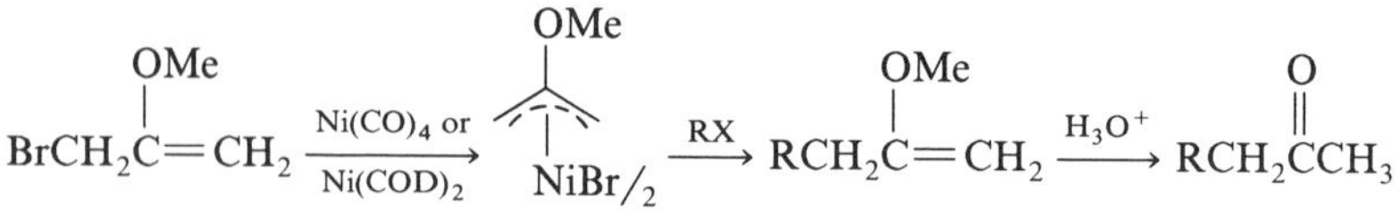

R =1°, 2° alkyl; allylic; benzylic; aryl; vinylic

JACS *96* 3250 (1974)

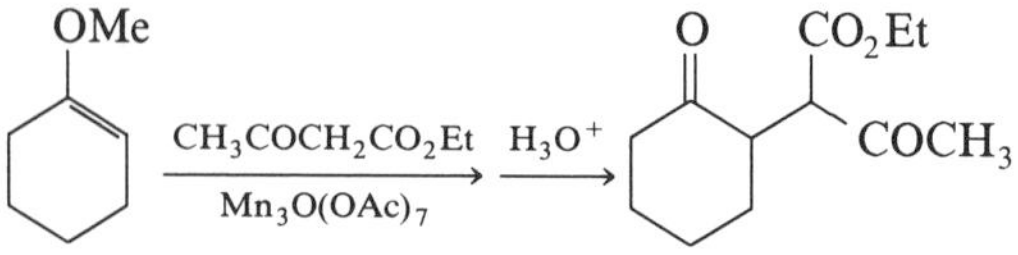

TL *28* 175 (1987)

4. Enol Esters

4.1. Synthesis of Enol Esters

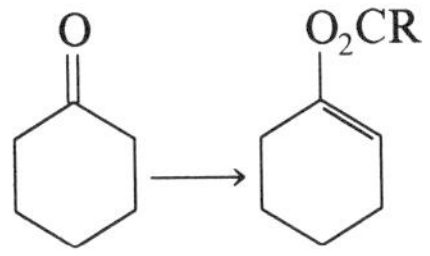

Review: JACS *102* 1966 (1980)

Ac$_2$O, NaOAc	Ber *42* 1161, 2014 (1909)
Ac$_2$O, KOAc	JACS *66* 1325 (1944); *74* 5381 (1952) BCSJ *59* 751 (1986)
Ac$_2$O, K$_2$CO$_3$	JOC *48* 2705 (1983)
Ac$_2$O, Et$_3$N, cat DMAP	Syn Commun *9* 157 (1979)
LiN(SiMe$_3$)$_2$/Ac$_2$O	JOC *47* 5088 (1982)
KH/AcCl, (DMAP)	Syn 504 (1979)
Ac$_2$O, cat TsOH	JACS *67* 1430 (1945) JOC *26* 3729 (1961); *28* 3362 (1963); *32* 1741 (1967); *48* 1921 (1983) (regioselectivity) JCS 2933 (1965)
Ac$_2$O, cat HClO$_4$	JOC *30* 2502 (1965); *32* 1741 (1967); *33* 943 (1968); *36* 2361 (1971) Tetr *23* 4143 (1967) Org Syn *52* 39 (1972)
(PhCO)$_2$O, cat HClO$_4$	JOC *34* 1962 (1969)

H$_2$C=C(Me)OAc, cat TsOH	Ind Eng Chem *41* 2920 (1949)
	JOC *28* 3362 (1963); *30* 1341, 2502 (1965);
	    33 935 (1968); *36* 2361 (1971); *48* 1921 (1983)
	    (regioselectivity)

H$_2$C=C(Me)OAc, cat TsOH, Cu(OAc)$_2$	BCSJ *59* 751 (1986)

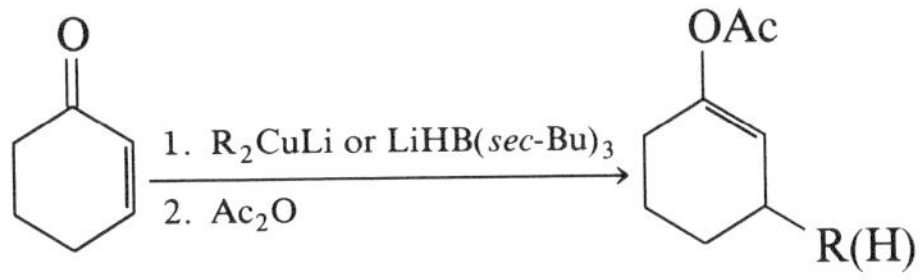

JOC *47* 5088 (1982)

RCO$_2$Et $\xrightarrow{\substack{\text{1. LiCHBr}_2 \\ \text{2. } n\text{-BuLi} \\ \text{3. 1,3-cyclohexadiene} \\ \text{4. Ac}_2\text{O}}}$ RCH=CHOAc

JACS *108* 1325 (1986)

4.2. Applications

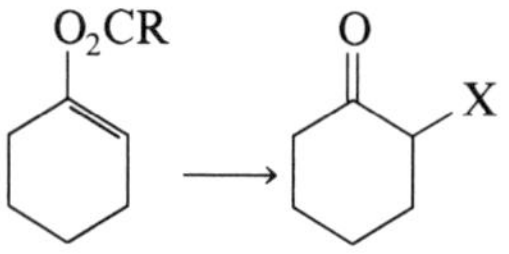

X	Reagent(s)	
R	MeLi/RX	JOC *30* 2502 (1965); *32* 1741 (1967); *33* 935, 943 (1968); *36* 2361 (1971); *51* 2408, 2416 (1986)
		Org Syn *52* 39 (1972)
		JACS *109* 3147 (1987)
	MeMgI/RX	JCS 2933 (1965)
	CH$_2$I$_2$, Zn-Cu/RX	JACS *102* 1966 (1980)
CH$_3$	MeLi/CH$_2$I$_2$, Zn-Cu	JOC *34* 1962 (1969)
CH$_2$CH=CH$_2$	H$_2$C=CHCH$_2$OCO$_2$Me, cat Pd$_2$(DBA)$_3$·HCCl$_3$, cat dppe, cat *n*-Bu$_3$SnOMe	TL *24* 4713 (1983)
CH$_2$C≡CR	$\overset{\overset{\displaystyle Co_2(CO)_6}{\displaystyle |}}{\overset{+}{C}H_2C{\equiv}CR}$	Syn Commun *10* 503 (1980)
COR	acylation or rearrangement	JACS *102* 1967 (1980) (review)
Br	Br$_2$	JACS *66* 1325 (1944); *67* 1430 (1945); *101* 2782 (1979)
		Org Syn Coll Vol *3* 127 (1955)
		JCS 907, 911 (1959); 1312, 3839 (1962); 2933 (1965)
		Tetr *21* 273 (1965)
		JCS Perkin I 126 (1978)
O$_2$CR	epoxidation	JACS *102* 1967 (1980) (review)
	electrolysis	JACS *97* 6144 (1975)
ArSO$_3$	(ArSO$_3$)$_2$	Syn 760 (1985)
		TL *27* 5811 (1986)

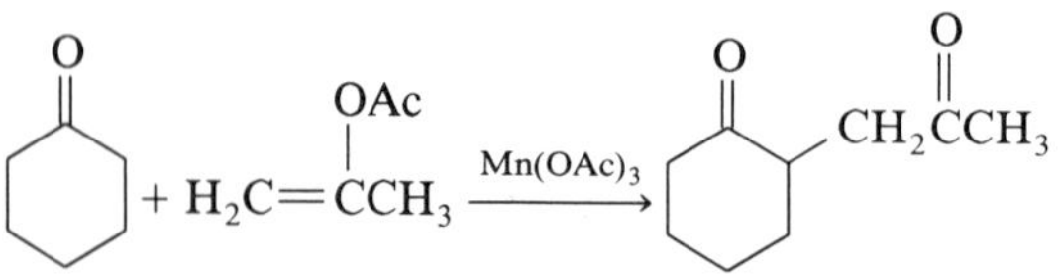

(low yields)

JOC *39* 3457 (1974)

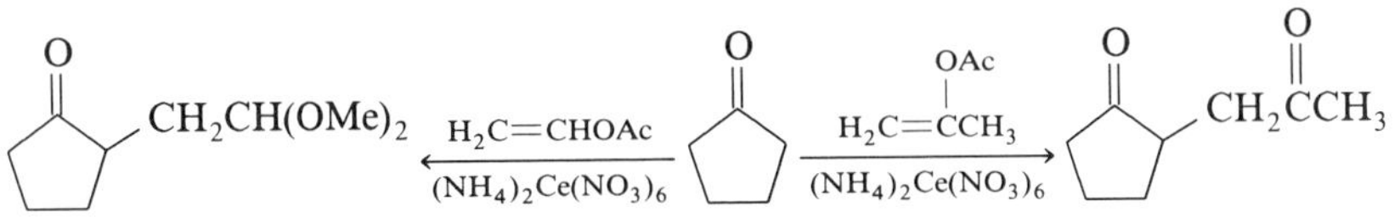
CH2CH(OMe)2
H2C=CHOAc
(NH4)2Ce(NO3)6
OAc
H2C=CCH3
(NH4)2Ce(NO3)6
CH2CCH3

TL *28* 5357 (1987)

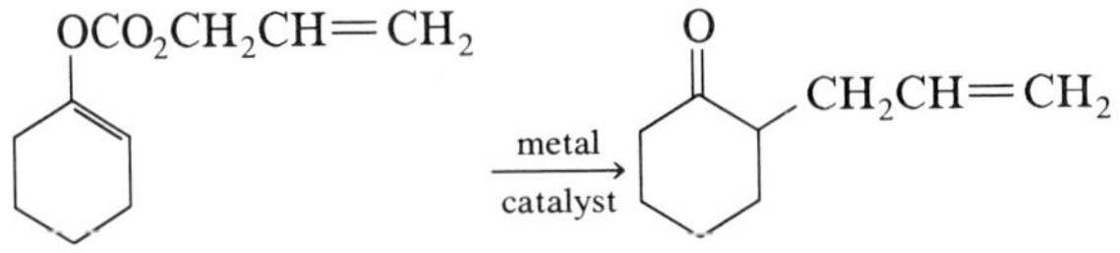
OCO2CH2CH=CH2
metal
catalyst
CH2CH=CH2

Metal	
Mo, Rh, Ni	CL 1721 (1984)
Pd	TL *24* 1793 (1983)

5. Enol Alanes

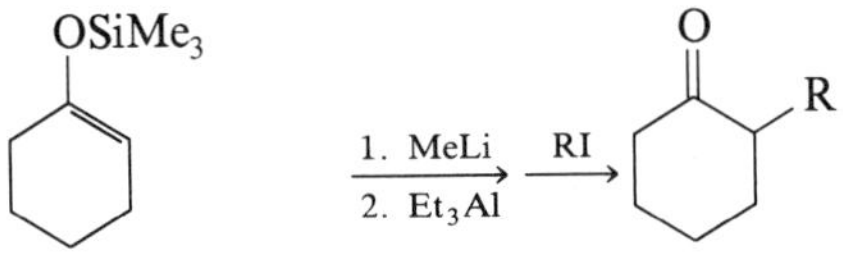
OSiMe3
1. MeLi
2. Et3Al
RI
R

R = Me, *n*-Bu

TL 1117 (1969)

6. Enol Boranes

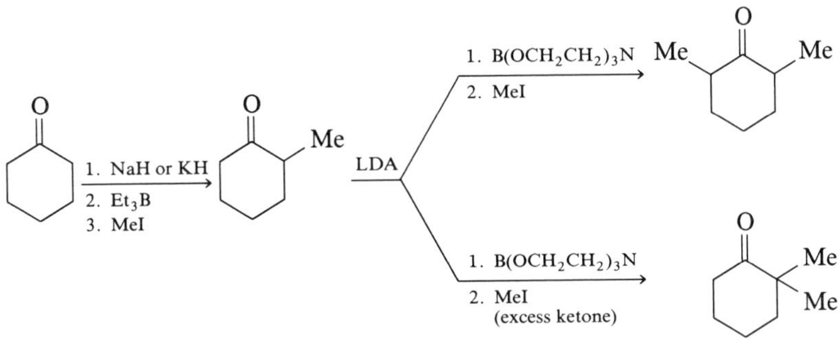
1. NaH or KH
2. Et3B
3. MeI
Me
LDA
1. B(OCH2CH2)3N
2. MeI
Me
Me
1. B(OCH2CH2)3N
2. MeI
(excess ketone)
Me
Me

Syn Commun *8* 9 (1979)
TL 845 (1979)

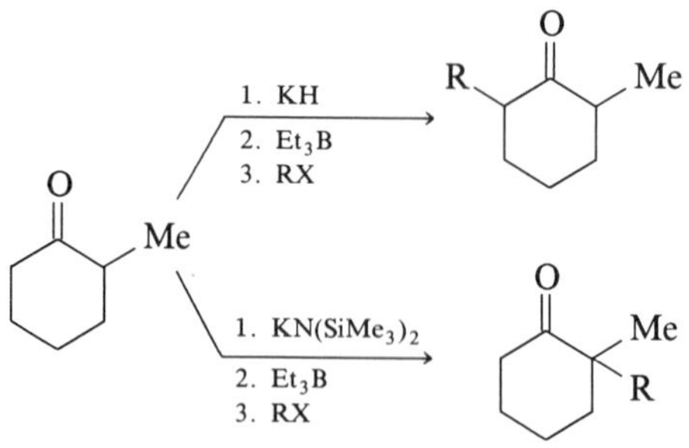

TL *24* 1341 (1983)
(RX = MeI)
JACS *106* 6690 (1984)
(RX = MeI); *107* 5391 (1985)
(RX = MeI); *108* 7864 (1986)
(RX = H₂C=CHCH₂Br)

TL *24* 1341 (1983)
(RX = MeI); *28* 1483 (1987)
(RX = BrCH₂CO₂Me)
JOC *52* 4810 (1987)
[RX = ICH₂C≡CPO(OEt)₂]

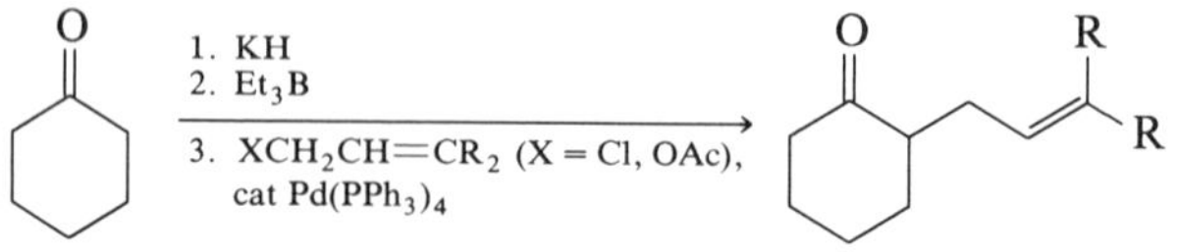

JOC *47* 3188 (1982); *48* 2427, 4098 (1983)
TL *24* 1341 (1983)

7. Enol Silanes

Reviews:

Syn 91 (1977); 1, 85 (1983)

7.1. Preparation

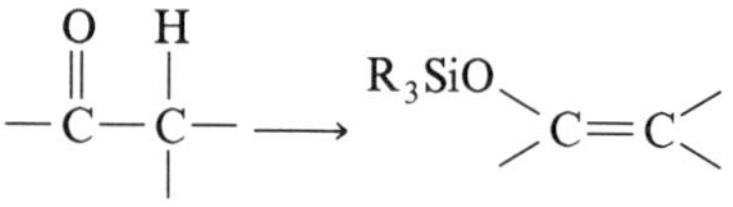

$$\underset{\text{O H}}{-\overset{\parallel}{C}-\overset{\mid}{C}-} \longrightarrow \underset{}{\text{R}_3\text{SiO}}\!\!\diagdown\!\!\underset{}{\text{C}=\text{C}}\!\!\diagup$$

Thermodynamic enol silane

Me₃SiCl, Et₃N, DMF	JOC *34* 2324 (1969); *45* 2307 (1980) TL 2671 (1978)
Me₃SiCl, NaI, Et₃N, CH₃CN	JOMC *201* C9 (1980) Org Syn *65* 1 (1987)
Me₃SiCl, LiI; (Me₃Si)₂NH	JACS *107* 268 (1985)
Me₃SiI, (Me₃Si)₂NH	Syn 730 (1979)
R₃SiOTf (R = Me, Et), Et₃N	Syn 259 (1976) Ann 1643 (1981) Syn 1 (1982) (review)
NaH/Me₃SiCl, Et₃N	JACS *90* 4462 (1968)
KH/Et₃B/Me₃SiCl	TL *24* 1341 (1983)

BrMgN(i-Pr)$_2$/Me$_3$SiCl	TL *24* 1345 (1983)
Na, anthracene/R$_3$SiCl (R = Me, Et)	BSCF 3552 (1967)

Kinetic enol silane

LiN(i-Pr)$_2$/Me$_3$SiCl	JOC *34* 2324 (1969); *45* 2307 (1980)
LiN(i-Pr)$_2$ or LiN(t-C$_4$H$_9$)(t-C$_8$H$_{17}$), Me$_3$SiCl, Et$_3$N	TL *25* 495 (1984)
LiN(i-Pr)CHArCH$_2$X (X = H, OMe, NR$_2$)/Me$_3$SiCl	JACS *108* 543 (1986) (enantioselective)

CC 88 (1986)
TL *28* 3723 (1987)
(both enantioselective)

JACS *102* 3959 (1980)

LiN(SiMe$_3$)$_2$/Me$_3$SiCl	JOC *45* 2307 (1980)
KN(SiMe$_3$)$_2$/Et$_3$B/Me$_3$SiCl	TL *24* 1341 (1983)
LiN(SiPhMe$_2$)$_2$/Me$_3$SiCl	JACS *104* 5526 (1982); *108* 3435, 3841 (1986)
LiN(i-Pr)$_2$/t-BuMe$_2$SiCl	JACS *98* 2868 (1976) Helv *65* 385 (1982)
LiN(SiMe$_3$)$_2$/t-BuMe$_2$SiCl (enones)	CC 564 (1973)
cat n-Bu$_4$NF, Me$_3$SiCH$_2$CO$_2$Et	JACS *98* 2346 (1976); *108* 3435, 3841 (1986)

$$\overset{\text{OSiMe}_3}{|}$$

cat n-Bu$_4$NF, CH$_3$CH=COMe	Syn 1089 (1982)
(Me$_3$Si)$_2$NH, imidazole (β-diketones)	Syn 722 (1976)

Unknown or mixed regiochemistry

Me$_3$SiCl, n-C$_4$H$_9$SO$_3$K, Et$_3$N	Syn 34 (1979)
Me$_3$SiCl, ZnCl$_2$, Et$_3$N	JACS *96* 7807 (1974)
t-BuMe$_2$SiOTf, Et$_3$N	TL *25* 5953 (1984) JACS *109* 7575 (1987)
t-BuMe$_2$SiOTf; 2,6-lutidine	TL *25* 5953 (1984)
i-Pr$_3$SiOTf, Et$_3$N	TL *22* 3455 (1981)
Me$_3$SiNMe$_2$	TL 3553 (1974)

TL *27* 631 (1986) (enantioselective)

KH/Me$_3$SiCl, cat Et$_3$N, DME   Syn 504 (1979)

$$\text{RCHCSiMe}_3 \xrightarrow{\text{R'M}} \text{RCH}=\text{CR'}$$

stereoselective

R'M = LiAlH$_4$, RLi, Me$_3$SnLi

JOC *52* 513 (1987)

R	Reagents	
H	Li, NH$_3$, *t*-BuOH/Me$_3$SiCl	JACS *96* 6181 (1974); *107* 1440 (1985)
	LiHB(*sec*-Bu)$_3$/Me$_3$SiCl	JACS *107* 1440 (1985) JOC *52* 1870 (1987)
	Et$_3$SiH, cat ClRh(PPh$_3$)$_3$	TL 5035 (1972) Tetr *37* 4515 (1981)
	Et$_3$SiH, EtMe$_2$SiH or MePh$_2$SiH; cat [Pt(μ-H)(SiR$_3$)PR$_3$]$_2$	JOMC *191* 39 (1980)
alkyl, aryl	RMgBr, cat CuBr·SMe$_2$, Me$_3$SiCl, HMPA (enals and enones)	TL *27* 4025 (1986)
	RCu, Me$_3$SiCl, HMPA or DMAP (enals and enones)	TL *27* 4029 (1986)
	RCu, Me$_3$SiCl, Me$_2$NCH$_2$CH$_2$NMe$_2$	TL *28* 27 (1987)
	R$_2$CuLi/Me$_3$SiCl	JACS *96* 6179 (1974) Tetr *37* 4027 (1981)
	R$_2$CuLi/Ph$_2$SiMeCl	JOC *52* 165 (1987)
	R$_2$CuLi, Me$_3$SiCl, HMPA or DMAP (enals or enones)	TL *27* 4029 (1986)
	R$_2$CuLi, Me$_3$SiCl	TL *26* 6015, 6019 (1985); *27* 1047 (1986); *28* 1973 (1987)
	(MeCuCN)Li, Me$_3$SiCl	TL *26* 6019 (1985)
	(MeCuCN)Li, *t*-BuMe$_2$SiCl	TL *28* 3589 (1987)
	R$_2$CuMgX/Me$_3$SiCl (enals)	JOMC *228* 321 (1982)

JACS *106* 7619 (1984)

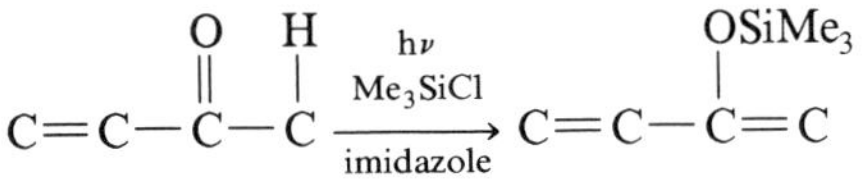

$$C=C-\overset{\overset{\displaystyle O}{\|}}{C}-\overset{\overset{\displaystyle H}{|}}{C} \xrightarrow[\text{imidazole}]{\overset{\displaystyle h\nu}{\text{Me}_3\text{SiCl}}} C=C-\overset{\overset{\displaystyle OSiMe_3}{|}}{C}=C$$

JOC *51* 3335 (1986)

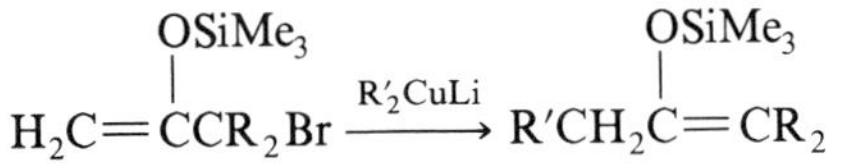

$$H_2C=\overset{\overset{\displaystyle OSiMe_3}{|}}{C}CR_2Br \xrightarrow{R'_2CuLi} R'CH_2\overset{\overset{\displaystyle OSiMe_3}{|}}{C}=CR_2$$

TL *21* 2325 (1980)

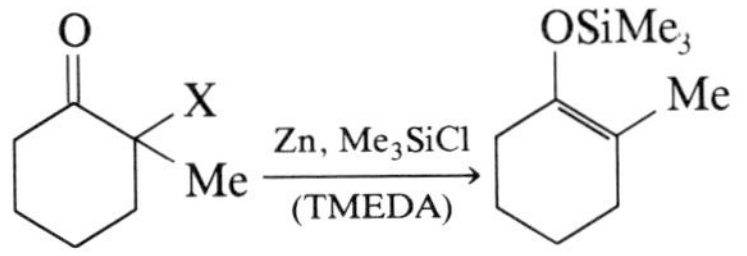

X = Cl, Br

Syn Commun *7* 327 (1977)
TL *24* 507 (1983)

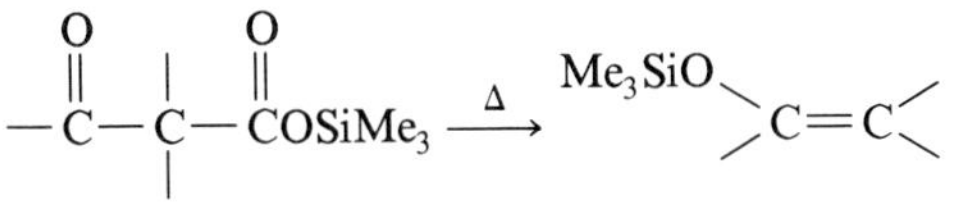

$$-\overset{\overset{\displaystyle O}{\|}}{C}-\overset{\overset{\displaystyle |}{\underset{\displaystyle |}{C}}}{}-\overset{\overset{\displaystyle O}{\|}}{C}OSiMe_3 \xrightarrow{\Delta} \underset{\displaystyle /}{\overset{\displaystyle Me_3SiO}{\diagdown}}C=C\overset{\diagup}{\diagdown}$$

JACS *97* 1619 (1975)

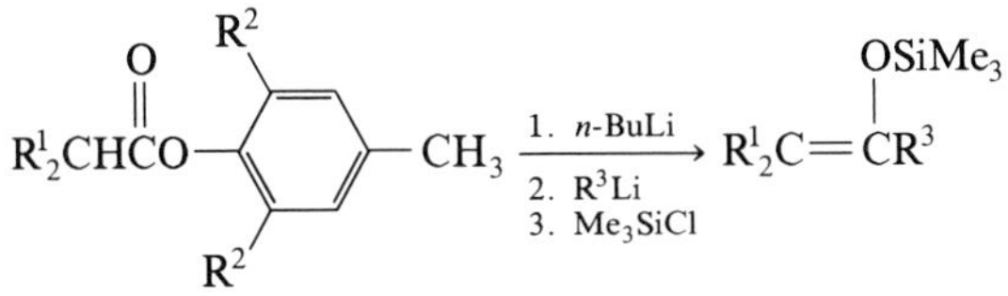

$R^2 = t\text{-Bu}$

JACS *107* 5396 (1985)

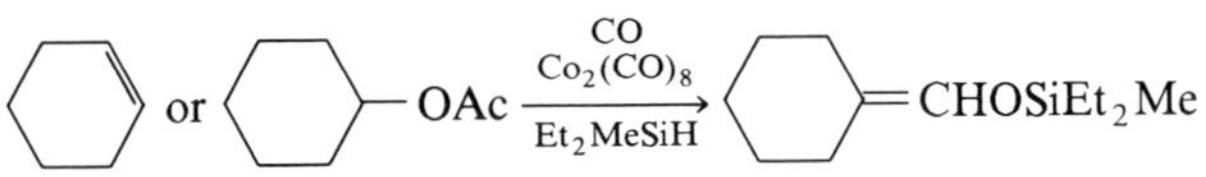

JACS *108* 7361 (1986)

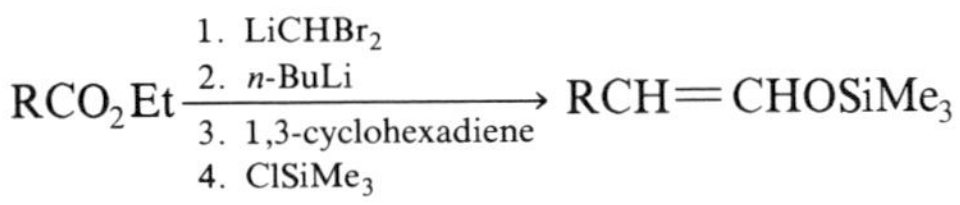

$$RCO_2Et \xrightarrow[\substack{\text{3. 1,3-cyclohexadiene} \\ \text{4. ClSiMe}_3}]{\substack{\text{1. LiCHBr}_2 \\ \text{2. } n\text{-BuLi}}} RCH=CHOSiMe_3$$

JACS *108* 1325 (1986)

$$R_2C=C=O+R'Li \xrightarrow{Me_3SiCl} R_2C=\overset{\overset{\displaystyle OSiMe_3}{|}}{C}R'$$

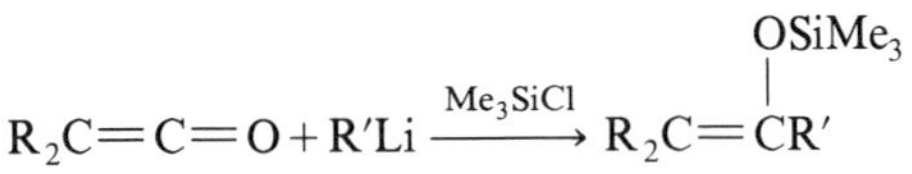

JACS *107* 5391 (1985)
JOC *50* 2105 (1985)

$$\overset{\overset{\text{OSiMe}_3}{|}}{\underset{}{\text{(bicyclic)}}} \quad R \quad \xrightarrow{\Delta} \quad \overset{\overset{\text{OSiMe}_3}{|}}{R C} = CH_2$$

BSCF 1122 (1976)

$$\text{(cyclopropyl)} \quad OSiMe_3 \quad \xrightarrow{\Delta} \quad \overset{\overset{\text{OSiMe}_3}{|}}{\text{(cyclopentenyl)}}$$

JACS *95* 5311 (1973)

$$RCH=CHSiMe_3 \quad \xrightarrow[\text{2. NaH}]{\text{1. cat OsO}_4, \text{Me}_3\text{NO} \cdot 3 \text{ H}_2\text{O}} \quad RCH=CHOSiMe_3$$

JACS *107* 4260 (1985)

$$-\overset{}{\underset{}{C}}\overset{O}{\overbrace{}}\overset{}{\underset{|}{C}}-SiMe_3 \quad \xrightarrow{\Delta} \quad \overset{Me_3SiO}{\underset{}{>}}C=C\overset{}{\underset{}{<}}$$

JOMC *94* C21 (1975)
TL 1449, 1453 (1976)

7.2. Reactions

See also page 776, Section 21.

$$\overset{\overset{\text{OSiR}'_3}{|}}{\text{(cyclohexenyl)}} \quad \longrightarrow \quad \overset{O}{\underset{}{\text{(cyclohexanone)}}} R$$

Reviews: Chimia 265 (1980)
Pure Appl Chem *55* 1749 (1983) (aldol review)

R	Reagents	
CH$_3$	CH$_2$I$_2$, Zn(Ag)/OH$^-$	TL 2767 (1973); 3327, 3333 (1974) JOC *39* 858 (1974)
1°, 2° alkyl; allylic; benzylic	MeLi/RX	JACS *90* 4464 (1968); *97* 1619 (1975); *101* 934 (1979) TL 1117 (1969) JOC *36* 2361 (1971); *51* 2408 (1986); *52* 165 (1987)
1° alkyl, allylic	LiNH$_2$, NH$_3$/RX	JOC *39* 2506 (1974); *40* 2156 (1975)
1° alkyl, benzylic, allylic	RX, [(R$'_2$N)$_3$S]F$_2$SiMe$_3$ (R$'$ = Et, *i*-Pr)	TL *21* 2085 (1980) JACS *102* 1223 (1980); *105* 1598 (1983)

1° alkyl, benzylic	RX, $(PhCH_2NMe_3)F$	JACS *104* 1025 (1982); *109* 1269 (1987) JOC *52* 3745 (1987)

2° and 3° alkyl,
  allylic, benzylic,
  RCHSR, RCHOR,
  CH_2NMeCO_2Me
	RX; $SnCl_4$, $TiCl_4$, $ZnCl_2$,
	  ZnI_2 or $FeCl_3$
		TL 4183 (1977); 1455, 4925
		  (1978); 995, 1427, 1519,
		  2179, 4971 (1979);
		  21 2010, 2033 (1980);
		  22 1101, 2321 (1981);
		  23 2399, 2601 (1982);
		  24 323, 327, 419, 2095 (1983)
		Angew Int *17* 48 (1978); *18* 72
		  (1979); *21* 96 (1982) (review)
		BCSJ *52* 1241 (1979)
		JACS *101* 984 (1979); *106* 7630
		  (1984)
		Ber *113* 3734, 3741 (1980)
		Syn 941 (1980); 1003 (1981)
		JOC *45* 3559 (1980); *47* 3219
		  (1982)
		Syn Commun *11* 217 (1981)
		Tetr *37* 319, 4027 (1981)
		JCS Perkin I 1099 (1982)

CH_2CH_2SAr	$ClCH_2CH_2SAr$, Lewis acid	TL *24* 961 (1983)
$C-C-XPh$ (X = S, Se)	$Cl-C-C-XPh$, $ZnBr_2$	TL *24* 5911 (1983)
allylic	allylic acetate, Lewis acid	CC 1180 (1981) JCS Perkin I 2079 (1982)

	$H_2C=CHCH_2OAc$ or
	  $H_2C=C(Me)CH_2OCO_2Me$,
	  cat $Pd_2(DBA)_3 \cdot HCCl_3$,
	  dppe
	n-BuLi/
	  $(R_2C=CHCH_2NEt_3)Br$,
	  cat $Pd(PPh_3)_4$
		CL 1325 (1983)
		TL *28* 2397 (1987)
		Acct Chem Res
		  20 140 (1987) (review)
		JOMC *236* 409 (1982)

	allylic nitro compound, $SnCl_4$	CC 1285 (1986)
R_f	$R_fI(Ph)O_3SCF_3$, py	TL *23* 1471 (1982)
Ar	ArX (X = Br, I), n-Bu_3SnF, 3% $PdCl_2(PAr_3)_2$	JACS *104* 6831 (1982)

NO_2-Ar	NO_2-Ar-H,
	  $[(Me_2N)_3S]Me_3SiF_2/Br_2$
	  or DDQ
		JACS *107* 5473 (1985)

$CR_2C \equiv CH$	$\left[\begin{array}{c} Co_2(CO)_6 \\ \| \\ R_2\overset{+}{C}C \equiv CH \end{array} \right] BF_4^- /$ $Fe(NO_3)_3 \cdot 9 H_2O$	JACS *102* 2508 (1980) CC 1353 (1987)
CH_2OH	$RLi/H_2CO/H_2O$	JOC *49* 3685 (1984)

R	Reagents	
CHROH or CR$_2$OH	RCHO or R$_2$CO, various reagents	See page 647, Section 7.
CH$_2$OCH$_2$Ph	H$_2$C(OCH$_2$Ph)$_2$, cat Me$_3$SiOTf	TL *28* 517 (1987)
	ClCH$_2$OCH$_2$Ph, TiCl$_4$	TL *28* 517 (1987)
CHROR (chiral)	RCH, TiCl$_4$	JACS *104* 7371, 7372 (1982); *106* 7588 (1984) JOC *52* 180 (1987)
CHROR	RCH(OR)$_2$, TiCl$_4$	CC 1691 (1986); 876 (1987) (both intramolecular) JACS *108* 3513 (1986) TL *28* 3747 (1987) (intramolecular)
	RCH(OR)$_2$, cat Me$_3$SiI	BCSJ *56* 3195 (1983)
	ClCHROR, cat Me$_3$SiX (X = I, OTf)	CL 405 (1983)
CR$_2$OR	(RO)$_2$CR$_2$, cat Me$_3$SiI	BCSJ *56* 3195 (1983)
	(RO)$_2$CR$_2$, cat Me$_3$SiO$_3$SCF$_3$	JACS *102* 3248 (1980)
	(RO)$_2$CR$_2$, TiCl$_4$	CL 15 (1974) JACS *109* 527 (1987) (chiral) TL *28* 4181, 4847 (1987) (both chiral)
CHROCHPhCO$_2$H	, BF$_3$·OEt$_2$	JOC *49* 2513 (1984)
	AcO— (n = 1, 2), Me$_3$SiOTf	TL *23* 2601 (1982)
	(n = 1, 2), AgOTf	JACS *107* 4289 (1985)
	, BF$_3$·OEt$_2$	CC 1245 (1987)
	OTf$^-$	TL *28* 6355 (1987)
CH$_2$SR	ClCH$_2$SR, cat Me$_3$SiX (X = I, OTf)	CL 405 (1983)
CHRSAr	RCH(SAr)$_2$, TiCl$_4$	JACS *109* 7199 (1987)
	NO$_2$CHRSPh, SnCl$_4$	CC 947 (1987)

R_2CSR'	$R_2C(SR')_2$, $FeCl_3$ or $SnCl_4$	TL 4971 (1979) Syn Commun *11* 315 (1981)
$C(SPh)R^1CHR^2R^3$	$R^1C(SPh){=}CR^2R^3$, $ROH \cdot 2\ TiCl_4$	TL *27* 3029 (1986)
CH_2NR_2	$R'OCH_2NR_2$, Me_3SiI or Me_3SiOTf $(R_2\overset{+}{N}{=}CH_2)Cl^-$ (R = Me, Et)	TL *23* 547 (1982) CL 405 (1983) CC 269 (1986) TL *28* 6355 (1987)
(β-lactam: R at top, =O, N–H)	*(AcO–β-lactam: R at top, =O, N–H)*, cat Me_3SiOTf	TL *28* 507 (1987)
	(AcO–β-lactam: R at top, =O, N–H), ZnI_2	CL 1343 (1985)
$CHArNHAr$	$ArCH{=}NAr$, cat Me_3SiOTf	CC 1053 (1987)
$RCHNRCOR$	$MeOCHRNRCOR$, $TiCl_4$ or $BF_3 \cdot OEt_2$	JACS *103* 1172 (1981)
$CH_3C(OH)CN$	CH_3COCN, $TiCl_4$	TL *22* 1171 (1981)
$CHRCH_2COR'$	$RCH{=}CHCOR'$, Lewis acid	See page 792, Section 2.
$CHR^1\overset{\overset{\text{O}}{\|}}{C}CH_2R^2$	$R^2CH_2C(NO_2){=}CHR^1$, $SnCl_4/H_2O$	JACS *98* 4679 (1976); *106* 2149 (1984)
$CH(OR)_2$	$HC(OMe)_3$, cat $MeSiOTf$ $HC(OR)_3$, cat Me_3SiI $HC(OMe)_3$, $SnCl_4$ $HC(OMe)_3$, $TiCl_4$ $MeLi/HC(OMe)_3/BF_3 \cdot OEt_2$	JACS *102* 3248 (1980) BCSJ *56* 3195 (1983) TL *25* 2813 (1984) CL 15 (1974) TL *23* 3595 (1982)
$CR(OMe)_2$	$RC(OMe)_3$, cat Me_3SiOTf	TL *27* 5099 (1986) (intramolecular)
CHO	*(1,3-dithiolan-2-ylium)* $+ BF_4^-/HgO$, $BF_3 \cdot OEt_2$	TL *22* 2829 (1981)
$COCH_3$	$(CH_3CO)BF_4$	JOC *46* 3771 (1981)
COR	$RCOCl$ $RCOCl$, $ZnBr_2$ or $TiCl_4$ $RCOCl$; $ZnCl_2$ or $SbCl_3$ $RCOCl$, $TiCl_4$ or $SnCl_4$ $MeLi/RCOCl$	CC 946 (1972) TL *28* 6355 (1987) Tetr *39* 841 (1983) JOC *47* 5099 (1982) JOC *47* 3219 (1982) TL 1187 (1977) Helv *64* 716 (1981)

R	Reagents	
CO$_2$CH$_3$	2 MeLi/chiral HNRR′/ CO$_2$/MeI	TL *27* 2767 (1986) (chiral)
CONHAr	ArNCO/H$_2$O	JOMC *164* 123 (1979)
COSMe	MeLi/COS/MeI	JOC *47* 3193 (1982)
halogen	See page 369, Section 1.	
OH	See page 488, Section 5.	
OAc	Pb(OAc)$_4$/H$_3$O$^+$ Pb(OAc)$_4$, KOAc, HOAc AgOAc, I$_2$	Syn Commun *6* 59 (1976) Tetr *39* 861 (1983) JOC *46* 2717 (1981)
O$_2$CPh	Pb(O$_2$CPh)$_4$/(Et$_3$NH)F	JOC *41* 1673 (1976)
O$_3$SAr	(ArSO$_3$)$_2$	JOC *50* 5148 (1985) TL *27* 5811 (1986)
SPh	PhSCl	CC 946 (1972) Tetr *37* 4027 (1981)
SOR (R = Me, Ph)	RSOCl	Syn 283 (1982)

$$\overset{\overset{OSiMe_3}{|}}{RC}=CH_2 \longrightarrow (\overset{\overset{O}{||}}{RC}CH_2)_2$$

Pb(OAc)$_4$	TL *28* 873 (1987)
Cu(OTf)$_2$	TL 3741 (1977); 3555 (1978) Chem Pharm Bull *28* 262 (1980)
Ag$_2$O	JACS *97* 649 (1975)
PhIO, BF$_3$·OEt$_2$	CC 420 (1985) (R = Ar)

8. Enol Stannanes and Related Compounds

$$CH_3CH_2\overset{\overset{O}{||}}{C}CMe_2OR \xrightarrow[\substack{2.\ \\ \text{R'}\diagup\text{OAc} \\ \text{O}\diagdown\text{NH}}]{1.\ Sn(OTf)_2,\ \bigcirc\!\!NEt} R'\diagup\overset{\overset{CH_3}{|}}{CH}COCMe_2OR$$

JOC *52* 5491 (1987)

$$\overset{\overset{OAc}{|}}{RC}=CH_2 \xrightarrow{n\text{-Bu}_3SnOMe} \overset{\overset{OSnBu_3}{|}}{RC}=CH_2 \xrightarrow{R'X} \overset{\overset{O}{||}}{RC}CH_2R'$$

JOMC *55* 273 (1973) (R′ = Me, Et, PhCH$_2$, CH$_2$CH=CH$_2$)
TL 3791 (1977) (R′ = 1° RI); *21* 2591 (1980) [R′X = allylic OAc, cat Pd(PPh$_3$)$_4$]
CL 939 (1982) [R′X = ArBr, cat PdCl$_2$(PAr$_3$)$_2$]
JACS *109* 7223 (1987) [R′X = heterocyclic iodide, cat Pd(OAc)$_2$-(o-Tol)$_3$P]

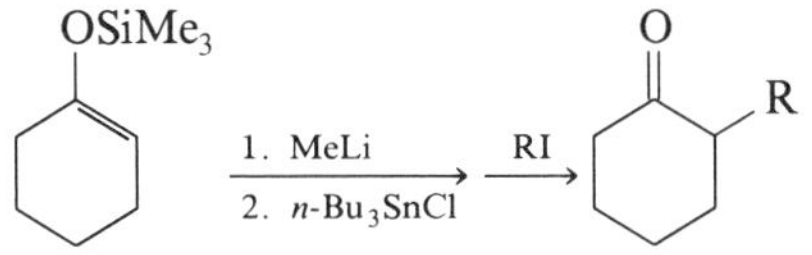

OSiMe$_3$

$$\xrightarrow[\text{2. } n\text{-Bu}_3\text{SnCl}]{\text{1. MeLi}} \xrightarrow{\text{RI}}$$

R = 1°, 2° alkyl

TL 1117 (1969)

$$n\text{-Bu}_3\text{SnCH}_2\text{CH(OEt)}_2 \xrightarrow[\text{2. PhCH}_2\text{Br}]{\text{1. } n\text{-BuLi}} \text{PhCH}_2\text{CH}_2\text{CH(OEt)}_2$$

JOMC *212* C31 (1981)

9. **Enol Phosphorus Compounds**

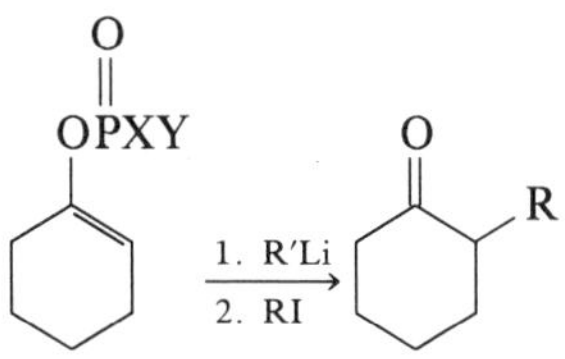

O
‖
OPXY

$$\xrightarrow[\text{2. RI}]{\text{1. R'Li}}$$

X	Y
Ph	Ph
Ph	OR
OR	OR

JOC *37* 3873 (1972)

10. **Enamines**

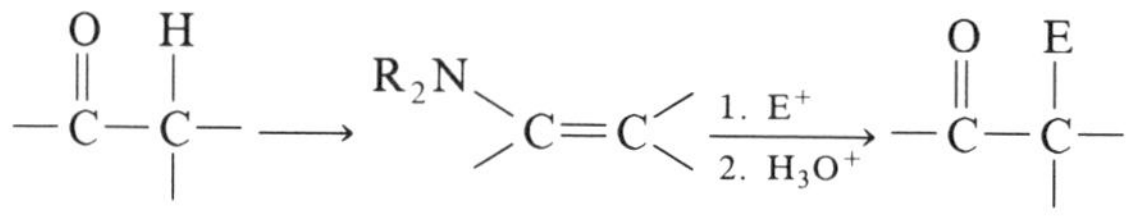

$$-\overset{\overset{\displaystyle O}{\|}}{C}-\overset{\overset{\displaystyle H}{|}}{C}- \longrightarrow \underset{}{\overset{R_2N}{\diagdown}}C{=}C{\diagup} \xrightarrow[\text{2. H}_3\text{O}^+]{\text{1. E}^+} -\overset{\overset{\displaystyle O}{\|}}{C}-\overset{\overset{\displaystyle E}{|}}{C}-$$

E$^+$ = X$_2$, RX, RCOCl, RO$_2$CCl, Michael acceptors, etc.

Reviews:

Adv Org Chem *4* 1 (1963)
"Enamines: Synthesis, Structure and Reactions," Ed. A. G. Cook, Marcel Dekker, New York (1969); 2nd ed. (1987)
Syn 510 (1970); 517 (1983)
S. F. Dyke, "The Chemistry of Enamines," Cambridge Univ. Press, New York (1973)
Org Syn *53* 48, 59 (1973)
Org Syn Coll Vol *5* 808 (1973)
Acta Chem Scand B *32* 335 (1978)
Tetr *38* 1975, 3363 (1982)

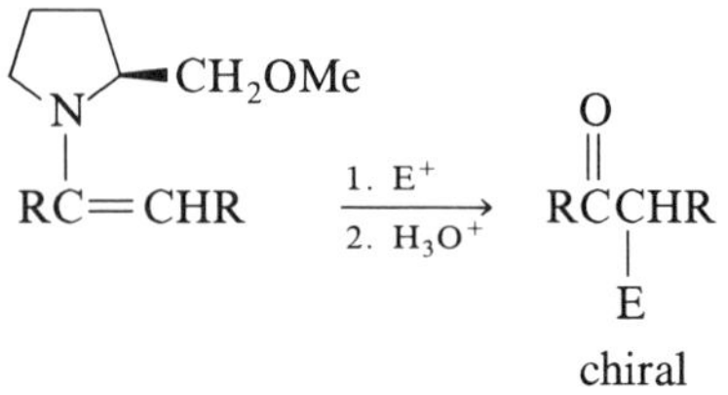

$$\underset{\substack{| \\ RC=CHR}}{\overset{\displaystyle \overset{\overset{\displaystyle \text{CH}_2\text{OMe}}{|}}{N}}{}} \quad \xrightarrow[\text{2. H}_3\text{O}^+]{\text{1. E}^+} \quad \underset{\substack{| \\ E \\ \text{chiral}}}{\overset{\overset{\displaystyle O}{\|}}{RCCHR}}$$

$$\underline{E^+}$$

R$_f$I, Zn, Cp$_2$TiCl$_2$, ultrasound (E = R$_f$)   JACS *107* 5186 (1985)

ArCH=CHNO$_2$ (conjugate addition)   Helv *65* 1637 (1982)

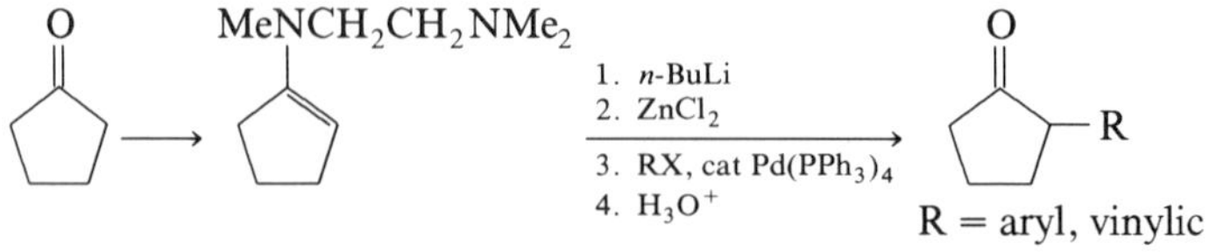

CL 1007 (1987)

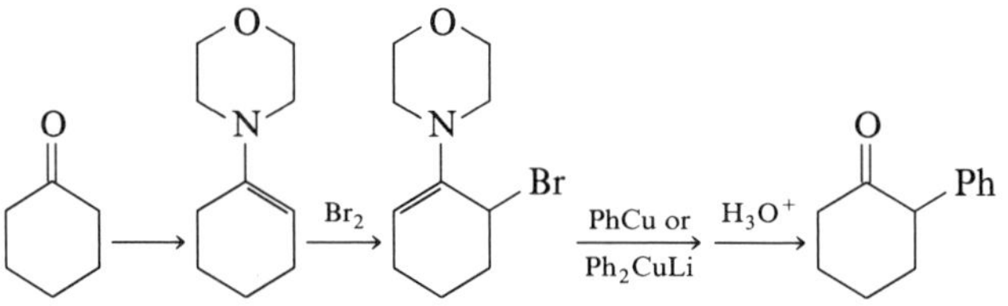

JOC *52* 3696 (1987)

Alternate syntheses of enamines and applications

$$\underset{\substack{\| \\ RCH}}{\overset{O}{}} \longrightarrow \underset{\substack{| \\ RCHNR_2}}{\overset{CN}{}} \xrightarrow[\text{2. R}_2'\text{CHX}]{\text{1. LDA}} \underset{\substack{| \\ R_2'CHCRNR_2}}{\overset{CN}{}} \xrightarrow[\Delta]{\text{KOH}} R_2'C{=}CRNR_2$$

Syn 127 (1979)

$$(RCHO)\ R_2C{=}O \longrightarrow R_2C{=}CRNR_2$$

Ph$_2$POCHLiNMePh   TL *21* 2671 (1980)

Ph$_2$POCHLiN⟨O⟩   TL 2433 (1979)

(EtO)$_2$POCHLiN⟨⟩   JOC *39* 2814 (1974); *42* 2520 (1977);
                                            44 3391 (1979)

(EtO)$_2$POCHLiN⟨O⟩   JOC *41* 3337 (1976); *42* 2520 (1977)

(EtO)$_2$POCLiArN⟨X⟩ (X = CH$_2$, O)   Arch Pharm *305* 88 (1972)

$(PhO)_2 POCLiArNHPh$

Ann *686* 107 (1965)

$(MeO)_2 POCHN_2$, HNR_2, KO-*t*-Bu

TL *21* 2041 (1980)

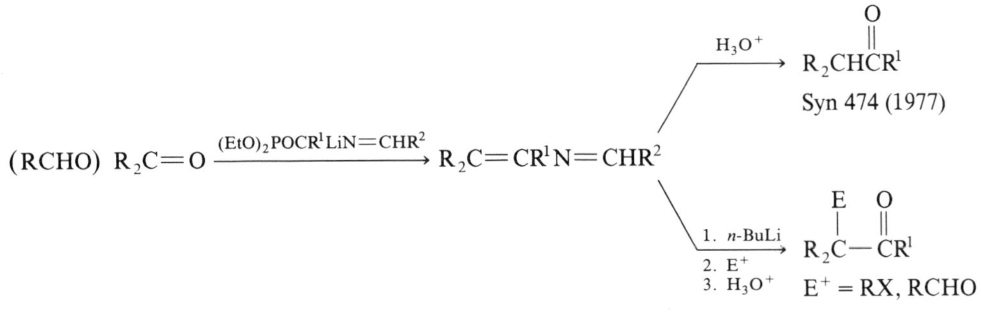

(RCHO) $R_2C=O$ $\xrightarrow{(EtO)_2POCR^1LiN=CHR^2}$ $R_2C=CR^1N=CHR^2$

$\xrightarrow{H_3O^+}$

$$\underset{\text{O}}{\overset{\text{O}}{R_2CHCR^1}}$$

Syn 474 (1977)

$\xrightarrow[\text{3. }H_3O^+]{\substack{\text{1. }n\text{-BuLi} \\ \text{2. }E^+}}$ $R_2\overset{E}{\underset{}{C}}-\overset{O}{\underset{}{C}}R^1$ $E^+ = RX$, RCHO

JOC *43* 782, 3792 (1978); *44* 3391 (1979);
46 3567 (1981); *47* 1513 (1982);
52 1962 (1987)
JACS *102* 5866 (1980); *106* 6431 (1984)
Syn Commun *11* 429 (1981)
Org Syn *65* 119 (1987)
TL *28* 503 (1987)
See also page 776, Section 21.

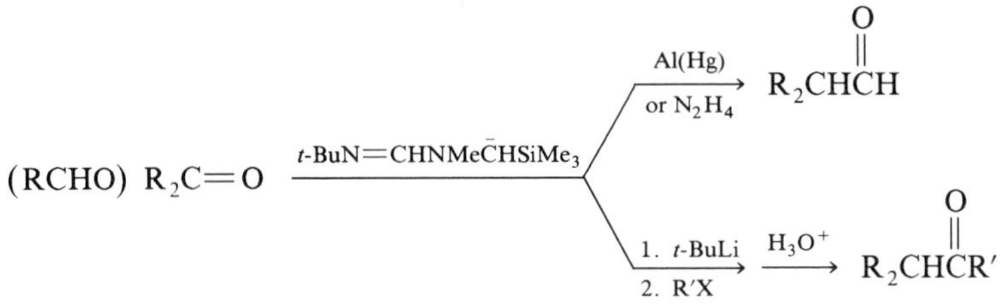

(RCHO) $R_2C=O$ $\xrightarrow{t\text{-BuN}=\text{CHNMeCHSiMe}_3}$

$\xrightarrow[\text{or }N_2H_4]{\text{Al(Hg)}}$ $\overset{O}{\underset{}{R_2CHCH}}$

$\xrightarrow[\text{2. R'X}]{\text{1. }t\text{-BuLi}}$ $\xrightarrow{H_3O^+}$ $\overset{O}{\underset{}{R_2CHCR'}}$

JACS *104* 877 (1982)

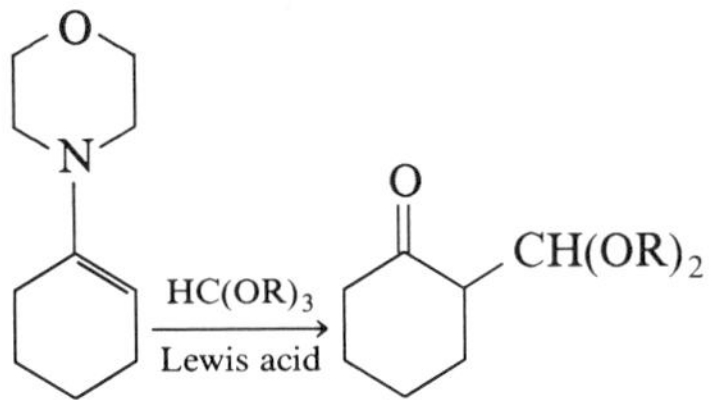

$\xrightarrow[\text{Lewis acid}]{\text{HC(OR)}_3}$

CL 1307 (1982)

$\xrightarrow{RNSnR_3}$ $\xrightarrow[\text{2. }H_2O]{\text{1. }H_2C=CHX}$

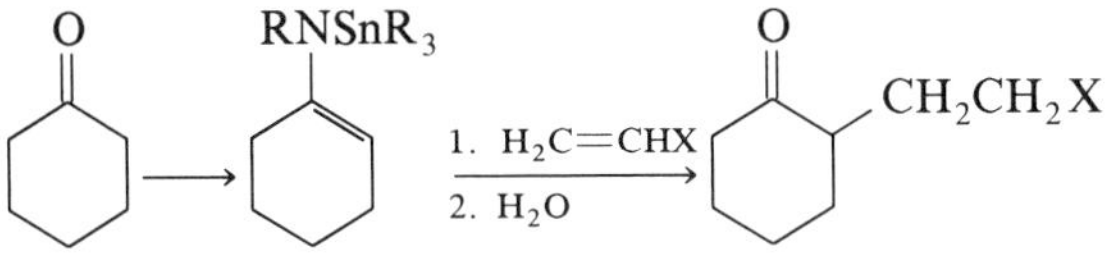

$X = CO_2R$, CN

JOMC *186* C9 (1980)
TL *21* 4511 (1980)

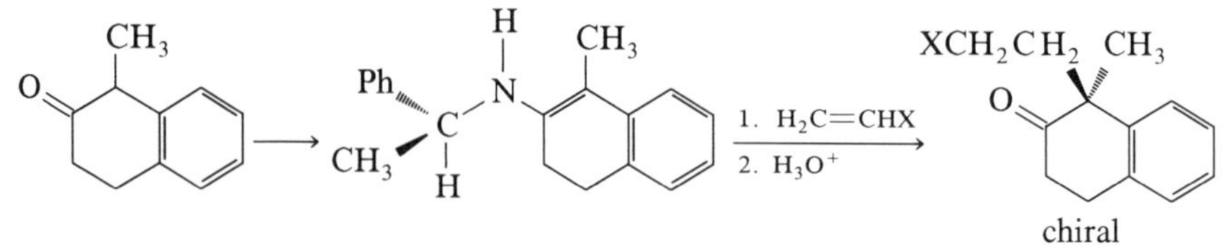

X = COCH₃, CO₂R

JACS *107* 273 (1985)
TL *28* 2367 (1987)

11. Imines

See also page 755, Section 10; and page 776, Section 21.

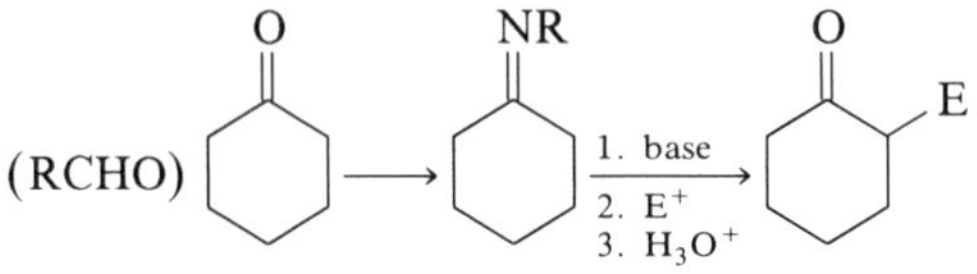

$$E^+ = RX, RCHO, R_2CO, H_2C\!-\!CH_2$$

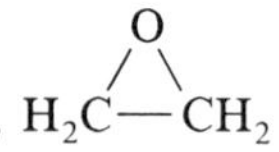

Review: Syn 517 (1983)

Alkylation	JACS *85* 2178 (1963); *92* 7593 (1970); *93* 5938 (1971); *100* 7999 (1978); *102* 1426, 5866 (1980); *104* 2081 (1982); *105* 4396 (1983) (regioselectivity)
	Helv *50* 2440 (1967)
	BSCF 3976 (1970)
	Organomet Chem Syn *1* 237 (1971)
	TL 1237 (1974); *24* 257, 511 (chiral), 3559 (regioselectivity) (1983)
	JOC *39* 3102 (1974); *42* 2545 (1977); *46* 3157 (regioselectivity), 4631 (1981)
	Ann 719 (1975)
	CC 47 (1979)
	Tetr *35* 1745 (1979)
	Can J Chem *60* 1836 (1982) (chiral); *61* 2466 (1983)
Epoxide opening	JOC *32* 1679 (1972)
	Ann 1075 (1973)
	Syn 256 (1975)
Directed aldol	Angew Int *2* 683 (1963); *7* 7 (1968)
	Ber *97* 3548 (1964)
	Helv *50* 2440 (1967)
	Rec Chem Prog *28* 45 (1967)
	JOC *34* 1122 (1969)
	JCS C 460 (1969)
	TL 381 (1970)
	JACS *102* 5866 (1980)

Regiochemistry

JACS *100* 292 (1978); *104* 2081 (1982)
JOC *43* 782, 3792 (1978)

Asymmetric induction

BSCF 4571 (1968)
Chem Pharm Bull *22* 459 (1974); *27* 2760 (1979)
JACS *98* 3032 (1976); *103* 3081, 3088 (1981)
JOC *42* 377 (1977); *43* 892, 3245 (1978)
TL 573 (1978); 3929 (1979); *24* 511 (1983)
Angew Int *18* 221 (1979)

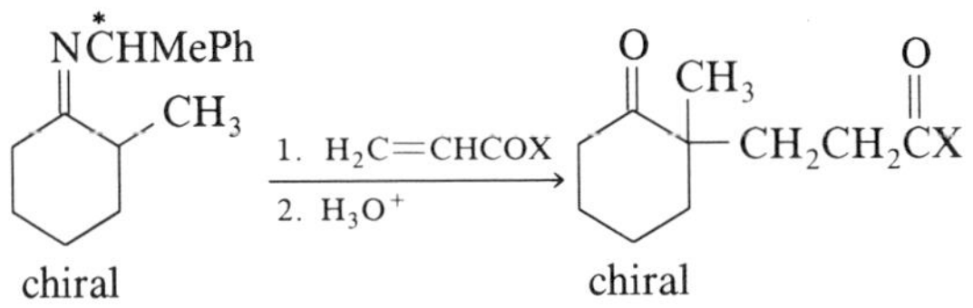

X = CH$_3$, OCH$_3$

JACS *107* 273 (1985)

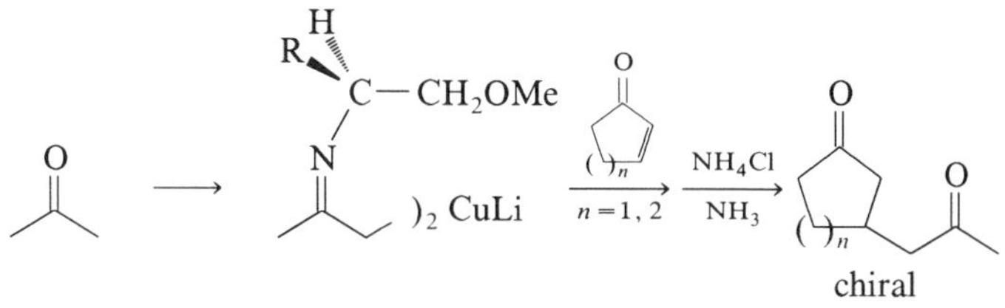

TL *23* 3711 (1982); *25* 2813 (1984)

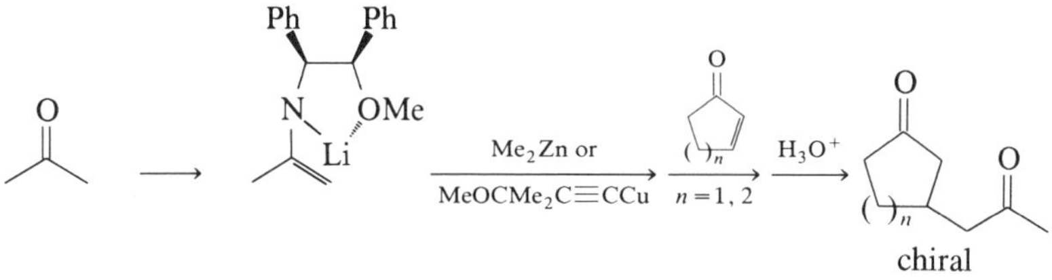

TL *28* 6347 (1987)

(RCHO) R$_2$C=O $\xrightarrow{\text{reagent}}$ R$_2$C=CHCH=NR $\xrightarrow{\text{H}_3\text{O}^+}$ R$_2$C=CHCHO

Reagent

(EtO)$_2$POCHCH=NR

JCS C 460 (1969)
JOC *43* 3788 (1978)

Me$_3$SiCHCH=NR

TL 7 (1976); *24* 2481 (1983); *26* 2391 (1985);
27 6177 (1986); *28* 259 (1987)
JOC *45* 2013 (1980); *50* 2798 (1985)

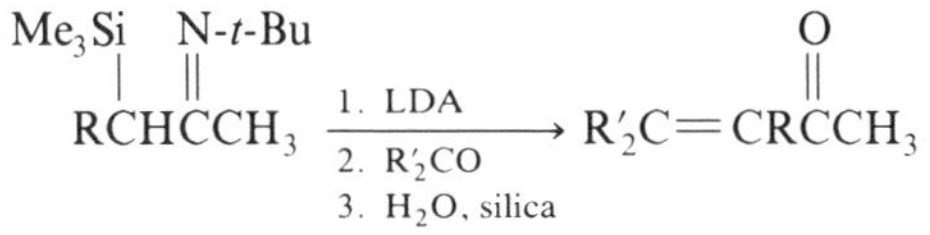

TL *24* 2481 (1983)

12. Hydrazone Anions

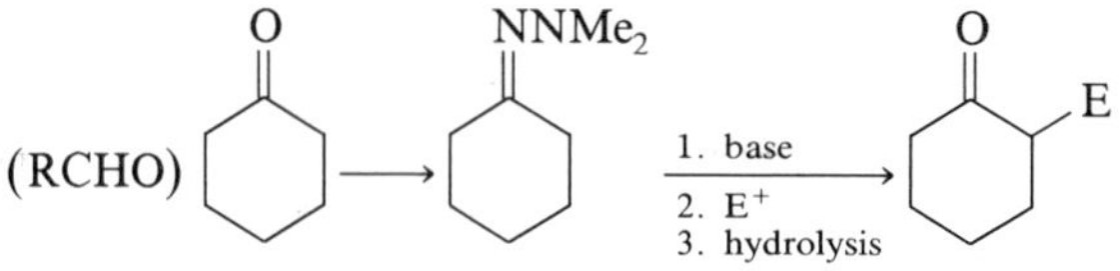

Review: Syn 517 (1983)

Preparation of dimethylhydrazones	Ber *111* 1337 (1978)
Deprotonation	TL 3691 (1978); 4145, 4149 (1979); *21* 3115 (1980) JACS *101* 5654 (1979)
Hydrolysis	TL 3 (1976)

Electrophile:

RX

JACS *93* 5938 (1971); *101* 5654 (1979); *106* 4865
(1984); *107* 2078 (stereoselectivity), 5303, (1985)
TL 3, 11, 4687 (1976); 3305 (1977); 4145, 4149
(1979); *27* 2595 (1986); *28* 813, 5275 (1987)
Ber *111* 1337 (1978)
JOC *46* 4631 (1981); *51* 3405, 4212 (1986);
52 5548 (1987)

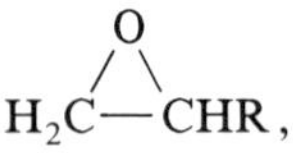

, I_2 (1,4-alkanedione)

TL 11 (1976)
Ber *111* 1362 (1978)

RCHO, R_2CO

TL 11 (1976)
Ber *111* 1362 (1978)
Angew Int *21* 864 (1982) (titanium species)
Ann 1439 (1983)

enone, CuX (1,4-addition)

TL 11 (1976); *21* 3115 (1980)
Ber *111* 1362 (1978)

RCOCl

TL 2853 (1978)

Me_3SiCl

TL 7 (1976)
Ber *111* 1362 (1978)

MeSSMe

TL 4687 (1976)

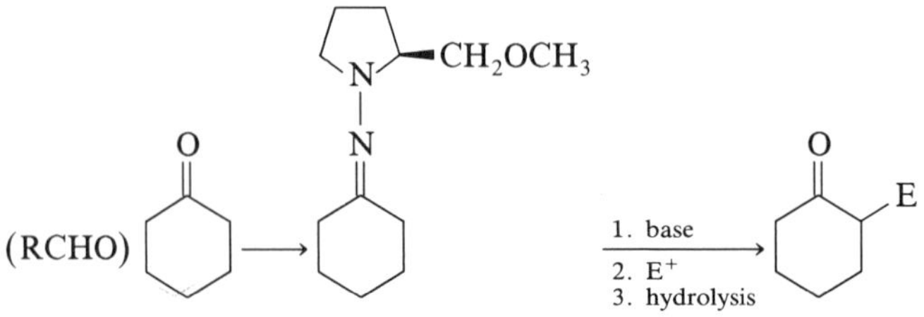

Review:

D. Enders in "Asymmetric Synthesis," Ed. J. D. Morrison, Academic Press, New York (1984), Vol 3, Chpt 4 (Alkylation of Chiral Hydrazones)

E^+	
RX	Angew Int *15* 549 (1976); *18* 397 (1979) TL 191 (1977) Ber *112* 2933 (1979) Acta Chem Scand B *35* 555 (1981) Tetr *38* 3705 (1982); *40* 1345 (1984) Ann 1439 (1983) Org Syn *65* 183 (1987) CC 358 (1987)
RCHO, R_2CO	Angew Int *17* 206 (1978) Ber *112* 3703 (1979)
$ArCH{=}CHNO_2$	Helv *65* 1637 (1982)
$RCH{=}CHCO_2R$ (1,4-addition)	TL *24* 4967 (1983); *27* 3491 (1986)

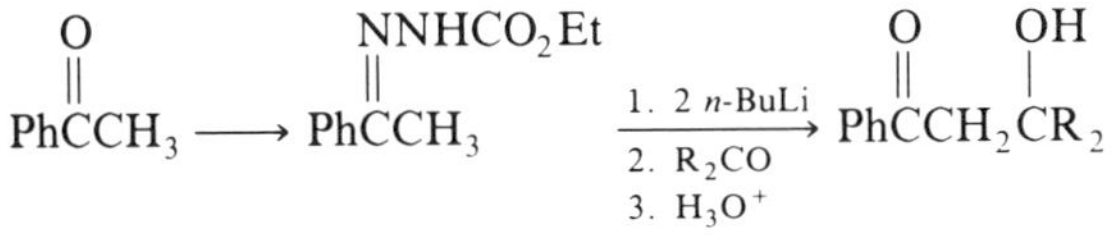

O
PhCCH3
NNHCO2Et
PhCCH3
1. 2 n-BuLi
2. R2CO
3. H3O+
O OH
PhCCH2CR2

TL *24* 3239 (1983)

13. Oxime Ether Anions

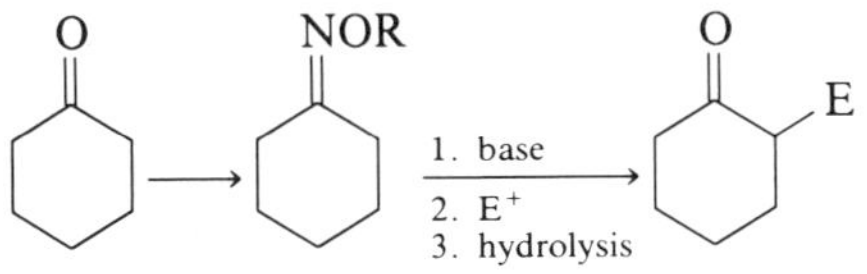

O
NOR
1. base
2. E+
3. hydrolysis
O
E

Preparation of oximes	TL 1415 (1978)
Deprotonation	TL 3889 (1975); 1415 (1978); *21* 3115 (1980)
Electrophile:	
RX	CC 674 (1976)
R_2CO	TL 1415 (1978)
enone, CuBr (1,4-addition)	TL *21* 3115 (1980)

14. Oxime Dianions

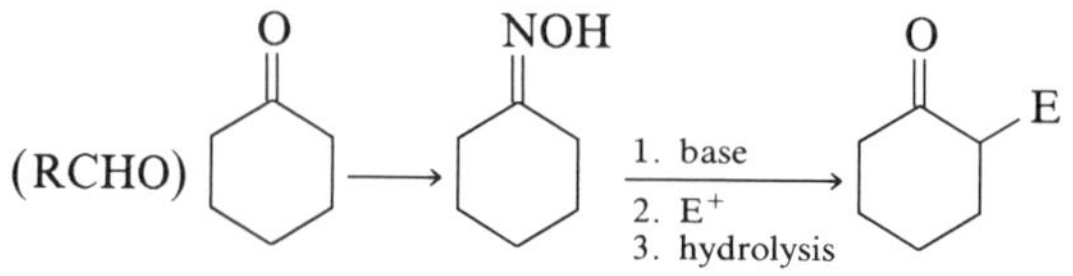

Deprotonation	TL 1439 (1976)
Electrophile:	
RX	JACS *91* 676 (1969)
	TL 1439, 4431 (1976)
	JOC *41* 439 (1976)
	JOMC *165* 1 (1979)

H_2C—CHR (with epoxide O)　　　JOMC *177* 35 (1979)

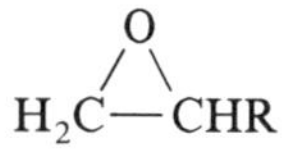

RCHO, R_2CO　　　JOC *41* 439 (1976)
　　　　　　　　　　　JOMC *165* 1 (1979)

RCO_2R (isoxazole)　　　JOC *35* 1806 (1970)

$RCONR_2$ (isoxazole)　　　JOC *43* 3015 (1978)

15. Blocking Groups

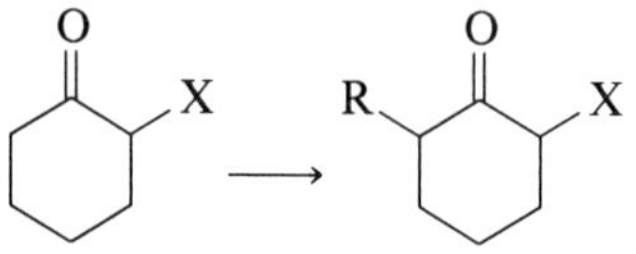

$\underline{X}$

=$CHNR_2$　　　JCS 501 (1944); 582 (1945)
　　　　　　　　JACS *74* 4223 (1952)

=CHOR　　　JACS *69* 1361 (1947)

=CHSR　　　JOC *27* 1615, 1620 (1962); *32* 1741 (1967)
　　　　　　　JACS *105* 6975 (1983)

$(SR)_2$　　　JCS 1131 (1957)

16. Alkylation and Acylation of Carbonyl Dianions

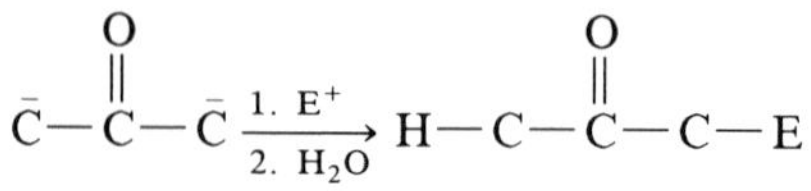

$$E^+ = RX, RCHO, RCH\overset{O}{\overbrace{\quad}}CH_2, RCO_2R$$

JACS *102* 2110 (1980)

$$ArCH_2\overset{O}{\overset{\|}{C}}CH_3 \longrightarrow ArCH\overset{O}{\overset{\|}{C}}CH_2$$

$\xrightarrow[\text{2. H}_2\text{O}]{\text{1. RX}}$ $Ar\overset{R}{\overset{|}{C}}H\overset{O}{\overset{\|}{C}}CH_3$

$\xrightarrow[\text{2. H}_2\text{O}]{\text{1. ArCHO}}$ $ArCH_2\overset{O}{\overset{\|}{C}}CH_2\overset{OH}{\overset{|}{C}}HAr$

JOC *48* 2957 (1983)

$$\overset{O}{\overset{\|}{C}}-\overset{O}{\overset{\|}{C}}-\overset{|}{\underset{R}{C}}-SPh \xrightarrow[\text{2. H}_2\text{O}]{\text{1. E}^+} E-\overset{O}{\overset{\|}{C}}-\overset{H}{\underset{R}{C}}-\overset{O}{\overset{\|}{C}}-SPh$$

$\xrightarrow{\text{Al(Hg)}}$ $E-C-\overset{O}{\overset{\|}{C}}-CH_2R$ JOC *39* 732 (1974) TL *28* 4629 (1987)

$\xrightarrow[\text{(R = Me)}]{\Delta}$ $E-C-\overset{O}{\overset{\|}{C}}-CH=CH_2$ TL 107 (1974) CC 497 (1974) JACS *106* 721 (1984)

$$E^+ = RX, H_2C\overset{O}{\overbrace{\quad}}CHR, RCHO, R_2CO, RCH=CHCO_2R \text{ (Michael addition)}$$

TL *28* 5677 (1987)

$$\overset{O}{\overset{\|}{C}}-\overset{O}{\overset{\|}{C}}-C-\overset{+}{P}Ph_3 \xrightarrow{E^+} E-\overset{O}{\overset{\|}{C}}-\overset{O}{\overset{\|}{C}}-C-\overset{+}{P}Ph_3$$

$\xrightarrow{\text{H}_2\text{O}}$ $E-C-\overset{O}{\overset{\|}{C}}-C$

$\xrightarrow{\text{R}_2\text{CO (RCHO)}}$ $E-C-\overset{O}{\overset{\|}{C}}-C=CR_2$

$$E^+ = RX, RCHO, R_2CO, RCO_2R$$

JOC *38* 4082 (1973); *41* 509 (1976)

$$CH_3\overset{O}{\overset{\|}{C}}CH_2P(OMe)_2 \xrightarrow[\text{2. }n\text{-BuLi}]{\text{1. NaH}} \overset{O}{\overset{\|}{C}}H_2\overset{O}{\overset{\|}{C}}\overset{-}{C}HP(OMe)_2 \xrightarrow[\text{2. R}_2\text{CO}]{\text{1. RX}} RCH_2\overset{O}{\overset{\|}{C}}CH=CR_2$$

JOC *38* 2909 (1973)

17. Alkylation and Acylation of β-Ketoaldehydes

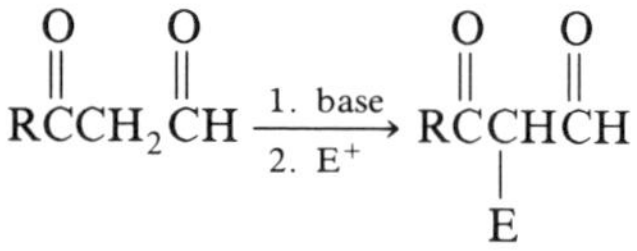

$$\underset{\text{RCCH}_2\text{CH}}{\overset{\text{O O}}{||\ ||}} \xrightarrow[\text{2. E}^+]{\text{1. base}} \underset{\text{RCCHCH}}{\overset{\text{O O}}{||\ ||}}$$
$$\underset{\text{E}}{|}$$

<u>E⁺</u>

RX   JACS *69* 1361 (1947); *79* 6313 (1957)
                                              JCS 1373 (1954)

(Me₃S)BF₄   JOC *48* 1362 (1983)

$$\underset{\text{RCH}_2\text{CCH}_2\text{CH}}{\overset{\text{O O}}{||\ ||}} \xrightarrow[\text{2. E}^+]{\text{1. 2 base}} \underset{\text{RCHCCH}_2\text{CH}}{\overset{\text{E O O}}{|\ ||\ ||}}$$

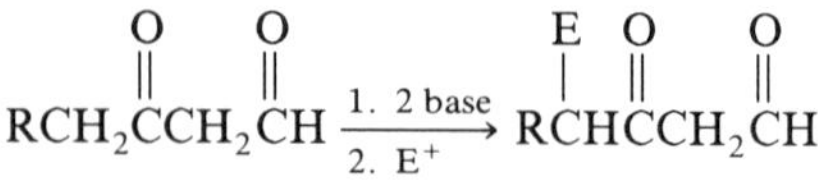

<u>E⁺</u>

RX   JACS *84* 1750 (1962); *85* 3273 (1963);
                                                  87 82, 3186 (1965)
                                              Org Syn *48* 40 (1968)
                                              Org Rxs *17* 155 (1969) (review)
                                              TL *24* 4769 (1983)

RCHO, R₂CO   JACS *84* 1750 (1962); *87* 3186 (1965)

enone (Michael addition)   JACS *87* 3186 (1965)

RCO₂R   JACS *87* 3186 (1965)

18. Alkylation and Acylation of β-Diketones

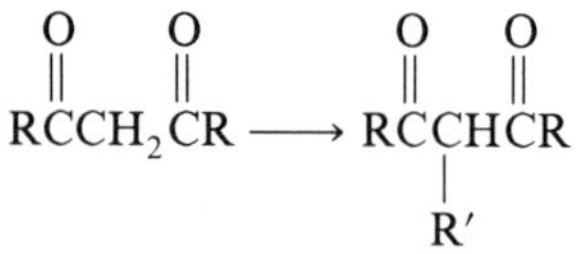

$$\underset{\text{RCCH}_2\text{CR}}{\overset{\text{O O}}{||\ ||}} \longrightarrow \underset{\text{RCCHCR}}{\overset{\text{O O}}{||\ ||}}$$
$$\underset{\text{R}'}{|}$$

Reviews:

    kinetics and thermodynamics   JACS *107* 2091 (1985)
    ion pairing and reactivity   JACS *100* 3514 (1978); *106* 6759 (1984)
    O vs C alkylation   Tetr *27* 4777 (1971)

Reagent(s):

    NaH/(Me₃S)BF₄   JOC *48* 1362 (1983)

    NaOH, R₄NX (chiral), phase transfer   BCSJ *52* 3119 (1979)

    KOH, R'X   JCS 803 (1953)
                                           Angew *67* 783 (1955)

    KOH, Cu, R'X (allylic)   Ber *85* 1061 (1952)

    K₂CO₃, MeI   Org Syn Coll Vol *5* 785 (1973)

K_2CO_3, R'X	Tetr *38* 1279 (1982)
NaOMe/R'X	Ber *85* 61 (1952) JCS 811 (1953) Tetr *2* 88 (1958)
NaOEt/R'X	TL 593 (1965)
MOEt (M = various metals)/R'X	JACS *100* 3514 (1978)
KO-*t*-Bu/R'X	JCS C 1973 (1966)
TlOEt/R'I	JACS *90* 2421 (1968)
DBU, R'X	BCSJ *52* 1716 (1979)
R_4NF, R'X	JCS Perkin I 1743 (1977) CC 64 (1977)
n-Bu$_4$NCl, KF·2 H$_2$O, MeI, CH$_3$CN	CC 514 (1979)
KF, Celite, R'X	CL 45 (1979) Syn Commun *11* 913 (1981)
KF, Al$_2$O$_3$, R'X	CL 755 (1979)
F$^-$ on basic anion exchange resin, R'X	Can J Chem *57* 2629 (1979)

base/π-allylpalladium compound
    (R' = allylic)
	J. Tsuji, "Organic Synthesis with Palladium
	    Compounds," Springer, New York (1980)
	Acct Chem Res *13* 385 (1980)
	CC 1162 (1982) (chiral)
	R. Heck, "Palladium Reagents in Organic
	    Syntheses," Academic Press, New York (1985)

R'OH (allylic, benzylic), cat Pd(acac)$_2$, cat PPh$_3$	Tetr *37* 3009 (1981)
i-PrOH, BF$_3$ (R' = *i*-Pr)	JOC *32* 2615 (1967)
H$_2$C=CHR ; PbO$_2$, AgO or Ag$_2$O (R' = CH$_2$CH$_2$R)	Syn 454 (1977)
H$_2$C=CHR, Mn(OAc)$_3$ or Mn(acac)$_3$ (R' = CH$_2$CH$_2$R)	JOC USSR *12* 1183 (1976)
(CH$_3$)$_2$C=CH$_2$, HClO$_4$ (R' = *t*-Bu)	TL 3599 (1966)

, BF$_3$·OEt$_2$ CC 68 (1983)

$$\left(R' = \ \ \right)$$

Ph$_4$BiX (X = O$_2$CR, O$_3$SR),
    N-*t*-butyl-N', N''-tetramethylquanidine
    (R' = Ph)
	TL *23* 3365 (1982)

ArPb(OAc)$_3$, py (R' = Ar)   Austral J Chem *32* 1561 (1979)

KO-*t*-Bu/NC-Ar-Br, NH$_3$, hν (R' = ArCN)   Tetr *38* 3479 (1982)

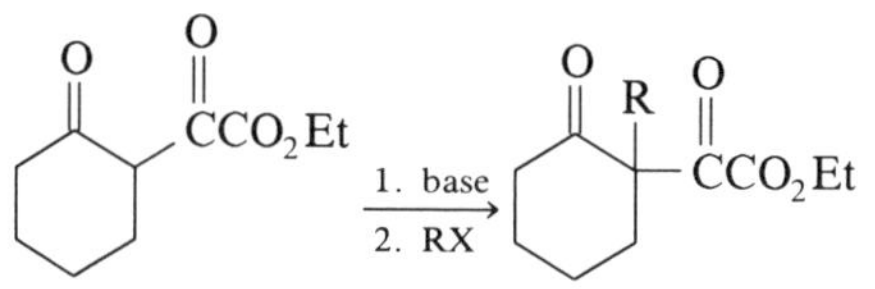

JACS *80* 5220 (1958)

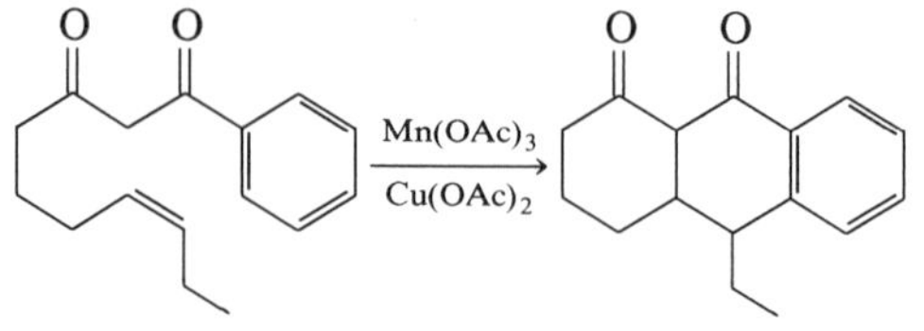

TL *28* 845 (1987)

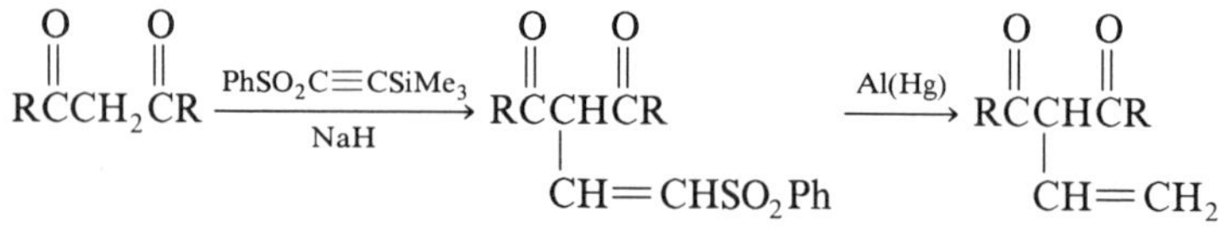

JOC *47* 4713 (1982)

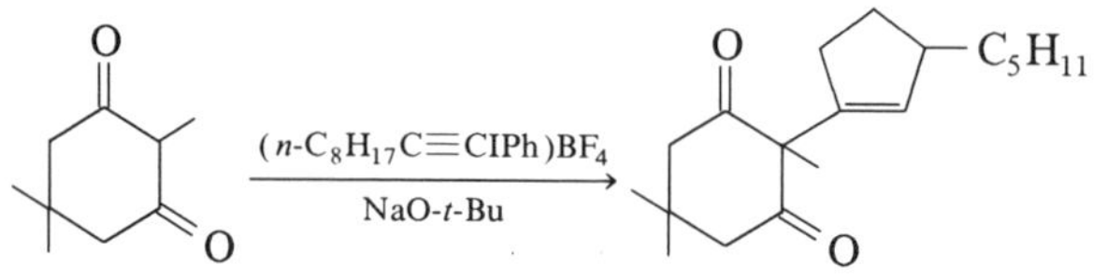

JACS *108* 8281 (1986)

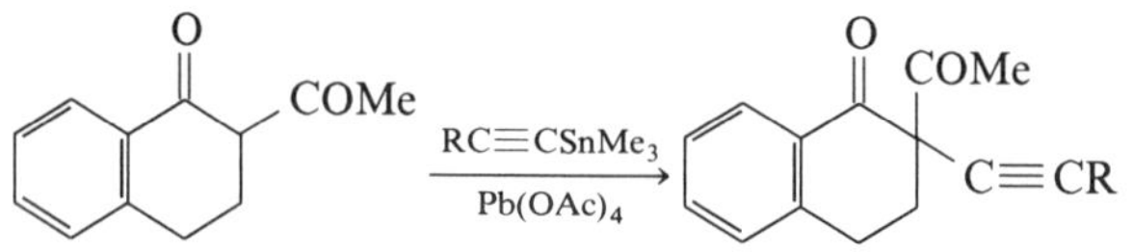

TL *27* 5025 (1986)

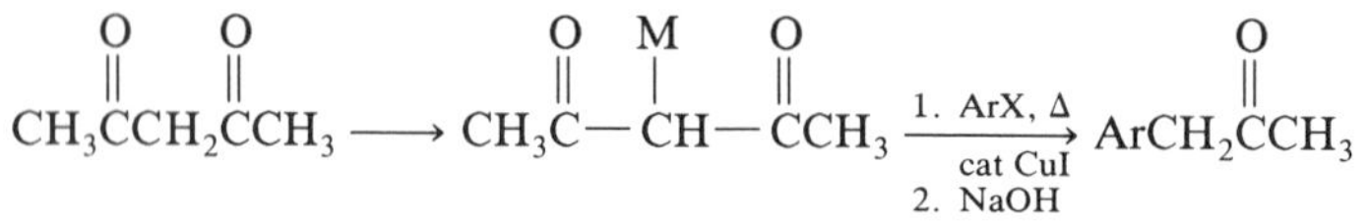

$$CH_3\overset{O}{\underset{\|}{C}}CH_2\overset{O}{\underset{\|}{C}}CH_3 \longrightarrow CH_3\overset{O}{\underset{\|}{C}}-\overset{M}{\underset{|}{C}}H-\overset{O}{\underset{\|}{C}}CH_3 \xrightarrow[\substack{cat\ CuI \\ 2.\ NaOH}]{1.\ ArX,\ \Delta} ArCH_2\overset{O}{\underset{\|}{C}}CH_3$$

M = Na, K

CL 597 (1982)

$$RCH_2\overset{O}{\underset{\|}{C}}CH_2\overset{O}{\underset{\|}{C}}R' \xrightarrow[2.\ E^+]{1.\ 2\ base} R\overset{E'}{\underset{|}{C}}H\overset{O}{\underset{\|}{C}}CH_2\overset{O}{\underset{\|}{C}}R'$$

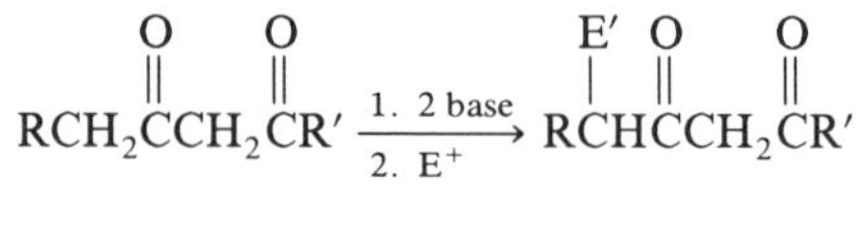

$\underline{E^+}$

RX	JACS *80* 6360 (1958); *81* 1160 (1959); 　　*96* 1082 (1974) JOC *25* 158, 1110 (1960); *30* 61 (1965); 　　*31* 663, 1035 (1966) BCSJ *40* 2698, 2909 (1967) Org Syn *47* 92 (1967) Org Rxs *17* 155 (1969) (review) Org Syn Coll Vol *5* 848 (1973) TL *28* 1997 (1987)
RCHO, R_2CO	JACS *80* 6360 (1958) JOC *26* 1344, 1716 (1961); *30* 1007 (1965) BCSJ *40* 2909 (1967) TL *28* 1997 (1987)
RCO_2R	JACS *80* 6360 (1958); *94* 8253 (1972); 　　*95* 6865 (1973); *99* 1631 (1977) JOC *25* 538, 1110 (1960); *28* 725, 2266 (1963); 　　*30* 1007, 4263 (1965); *46* 2260 (1981); 　　*51* 4254 (1986) BCSJ *40* 2909 (1967) Ber *114* 2786 (1981) Syn Commun *12* 621 (1982) TL *28* 5615 (1987) (lactone)
$CH_3CONMeOMe$ (E' = CH_3CO)	TL *24* 1851 (1983)
CO_2	JACS *80* 6360 (1958) JOC *25* 1110 (1960); *31* 1032 (1966)
RCH=CHCOX (X = R, OR) 　(Michael addition)	JOC *26* 1344 (1961); *28* 2266 (1963)

19. Alkylation and Acylation of β-Polyketones

$$\underset{CH_3C(CH_2C)_nR}{\overset{O\quad O}{\underset{\|\quad\|}{}}} \xrightarrow[\text{2. } E^+]{\text{1. base}} \underset{E'CH_2C(CH_2C)_nR}{\overset{O\quad O}{\underset{\|\quad\|}{}}}$$

$\underline{n}$	$\underline{E^+}$	
2	RX, RCHO, R_2CO	JOC *30* 4263 (1965)
	RCO_2R	JOC *30* 4263 (1965); *46* 2260 　　(1981) JACS *85* 3884 (1963); *93* 6708 　　(1971); *94* 8253 (1972); 　　*95* 6865 (1973) CC 442 (1979)
	CO_2	JOC *30* 4263 (1965) JACS *89* 6734 (1967); *91* 517 　　(1969); *98* 7733 (1976); 　　*99* 1631 (1977)
2, 3	$CH_3CONMeOMe$ (E' = $COCH_3$)	TL *24* 1851 (1983)
3	RCO_2R	JACS *93* 6708 (1971); 　　*95* 6865 (1973)
	CO_2	JACS *93* 6708 (1971)

20. Synthesis via β-Keto Acids, Esters, Amides and Imides

$$R'\overset{O}{\overset{\|}{C}}Cl + R_2\overset{-}{C}CO_2^- \longrightarrow R'\overset{O}{\overset{\|}{C}}CHR_2$$

Compt Rend *286* 401 (1978)
Syn 284 (1982)

$$R^1\overset{O}{\overset{\|}{C}}Cl + R^2\overset{Na}{\overset{|}{C}}(CO_2R^3)_2 \longrightarrow R^1\overset{O}{\overset{\|}{C}}CR^2(CO_2R^3)_2 \xrightarrow{\text{reagent(s)}} R^1\overset{O}{\overset{\|}{C}}CH_2R^2$$

Reagent(s)

H_2, cat Pd-C ($R^3 = PhCH_2$)	JCS 325 (1950)
H^+	JCS 322 (1950)
	JACS *74* 831 (1952)

$$RCCl \longrightarrow R\overset{O}{\overset{\|}{C}}CH_2\overset{O}{\overset{\|}{C}}OR$$

$LiCH_2CO_2R$	See page 873, Section 2.
$LiO_2CCHLiCO_2R/H^+$	JOC *44* 310 (1979)
	Org Syn *61* 5 (1983)
$NaCH(CO_2R)CO_2CH_2Ph/H_2$, cat Pd-C	JCS 2758 (1951)
$ROMgCH(CO_2R)_2$, Δ	J Prakt Chem 322 (1980)
$BrMgO_2CCH(MgBr)CO_2R/H^+$	Syn 143 (1978)

$$R\overset{O}{\overset{\|}{C}}CH_2\overset{O}{\overset{\|}{C}}OR \longrightarrow R\overset{O}{\overset{\|}{C}}CH\overset{O}{\overset{\|}{C}}OR$$
$$\underset{R'}{|}$$

$R'OH$, BF_3	JACS *62* 2389 (1940); *65* 552 (1943)
	Org Syn Coll Vol *3* 405 (1955)

$R_2'O$, BF_3 JACS *62* 2389 (1940); *65* 552 (1943)

t-BuBr, $AgClO_4$ Angew Int *5* 1044 (1966)

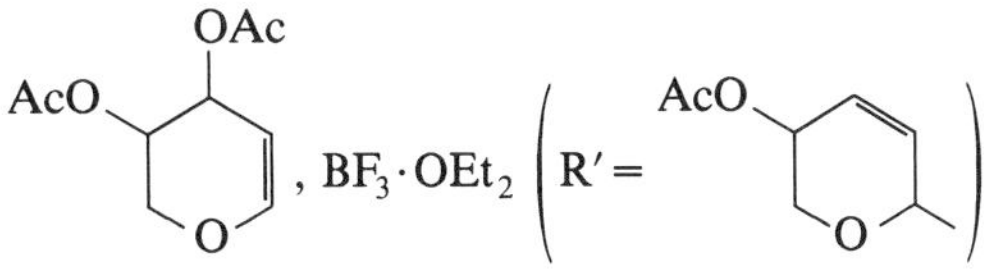

, $BF_3 \cdot OEt_2$ (R′ =)   CC 68 (1983)

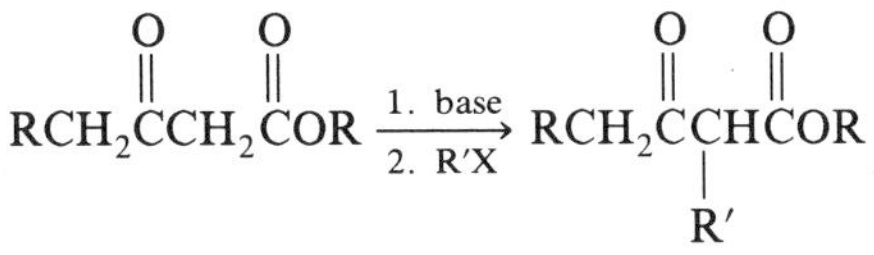

Newer Methods Prep Org Chem *2* 101 (1963)
Tetr *19* 1645 (1963) (rates); *27* 4777 (1971) (O vs C alkylation); *33* 2737 (1977) (review)
Org Syn *45* 7 (1965)
TL 593 (1965) (O vs C alkylation); 3679 (1968) (O vs C alkylation); 4903 (1972) (O vs C alkylation);
   23 1993 (1982) (copper enolate)
JOC *34* 1969 (1969) (O vs C alkylation); *35* 171 (1970) (stereochemistry); *48* 1362 (1983) (R′X =
   sulfonium salts)
Ann *736* 1 (1970) (R′X = Me_2SO_4)
BCSJ *49* 1126 (1976) (O vs C alkylation); *52* 1716, 3119 (1979)
CC 325 (1980)
Syn Commun *10* 279 (1980)
J Chem Res (S) 86 (1982)
JCS Perkin I 2293 (1983) (R′X = allylic halide, regiochemistry)

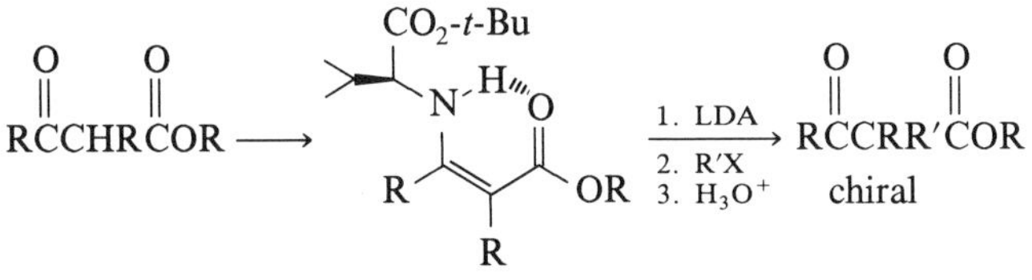

JACS *106* 2718 (1984)

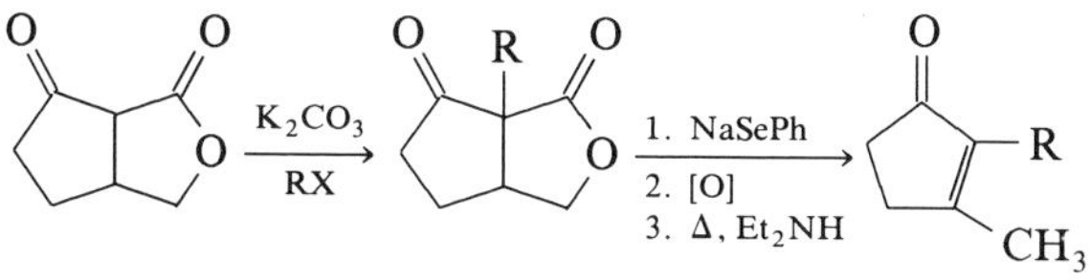

TL *22* 2447 (1981)

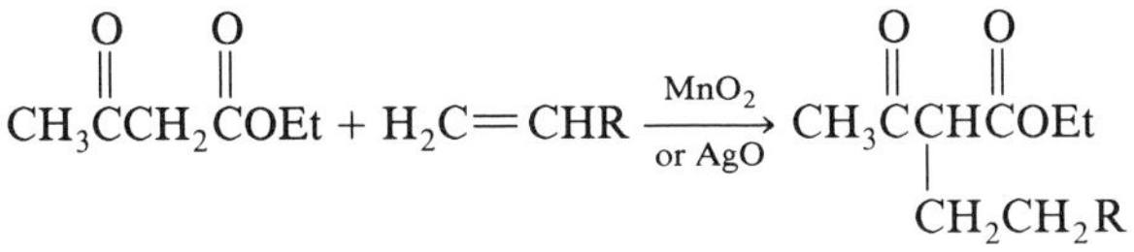

Syn 454 (1977)

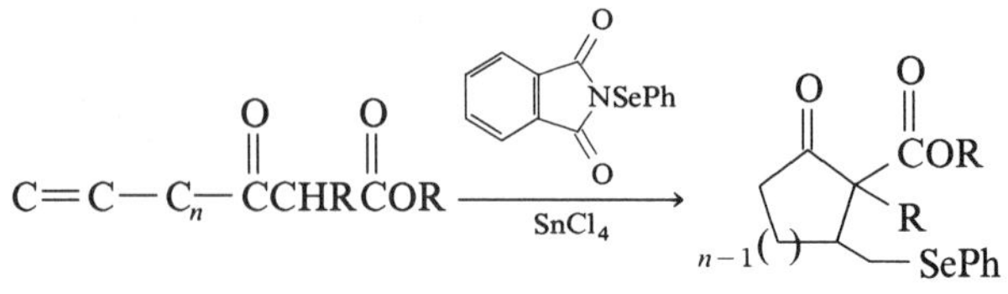

CC 1251, 1252 (1982)

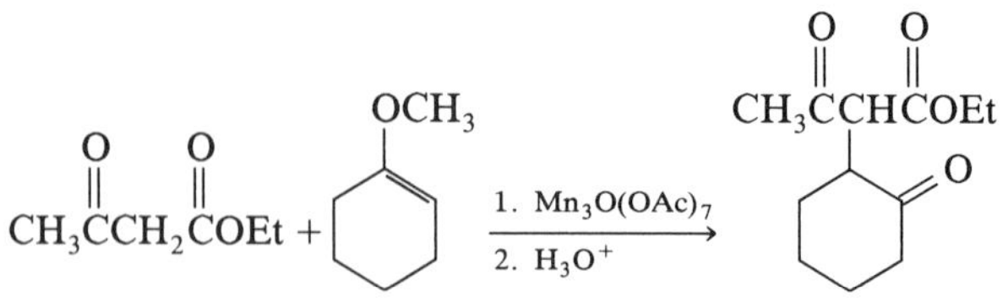

TL *28* 175 (1987)

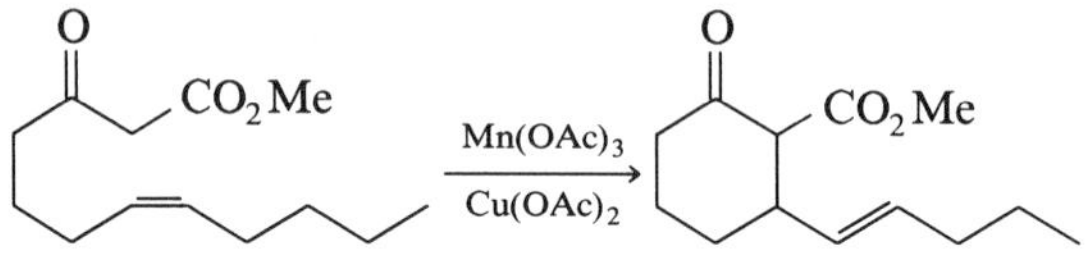

JOC *50* 3659 (1985)
TL *28* 845 (1987)

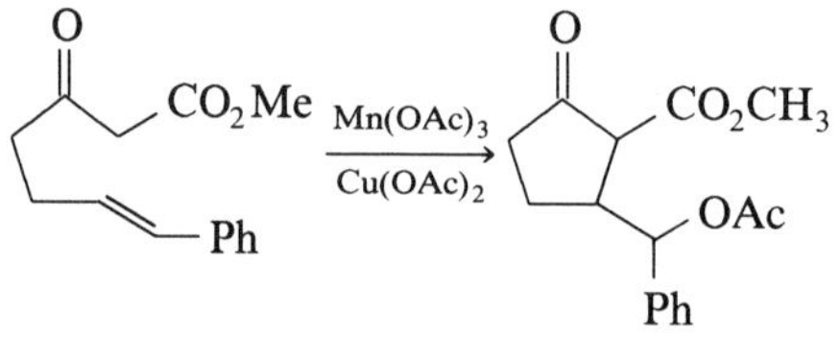

TL *28* 6109 (1987)

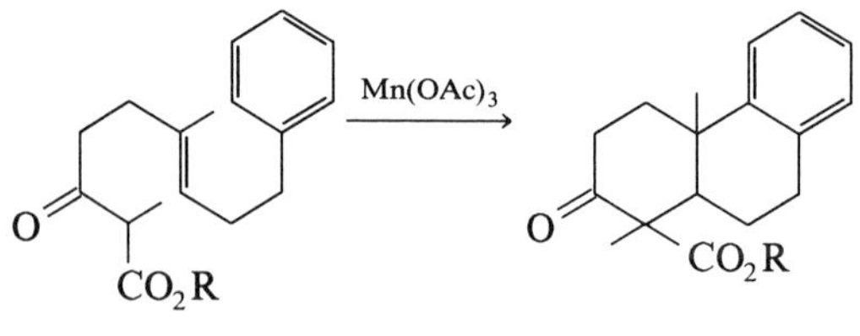

JOC *50* 3659 (1985)
TL *28* 845 (1987)

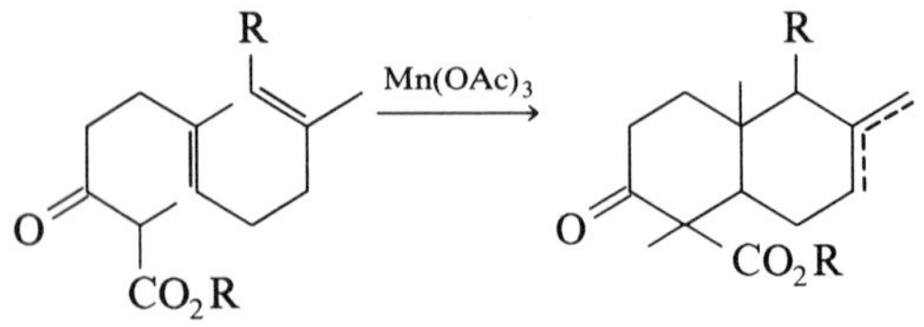

TL *28* 841 (1987)
JOC *52* 5487 (1987)

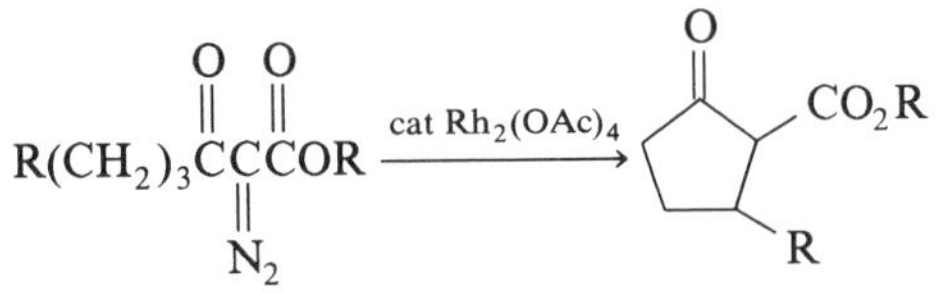

$$R(CH_2)_3\overset{\overset{\displaystyle O\ \ O}{|\ \ |}}{\underset{\underset{\displaystyle N_2}{||}}{C}}CCOR \xrightarrow{\text{cat } Rh_2(OAc)_4}$$

See page 653, Section 1.

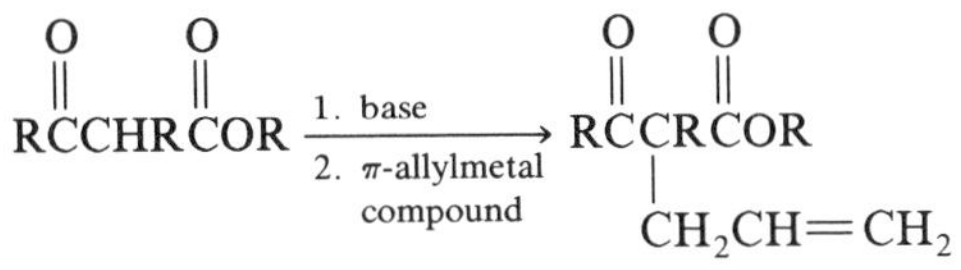

$$RC\overset{\displaystyle O}{\underset{}{C}}HRCOR \xrightarrow[\substack{2.\ \pi\text{-allylmetal} \\ \text{compound}}]{1.\ \text{base}} RCC\overset{\displaystyle O\ \ O}{RCOR}$$
$$\underset{CH_2CH=CH_2}{|}$$

Metal

Pd

Tetr *33* 2615 (1977) (review)
J. Tsuji, "Organic Synthesis with Palladium Compounds," Springer, New York (1980) (review)
Acct Chem Res *13* 385 (1980) (review)
R. Heck, "Palladium Reagents in Organic Syntheses," Academic Press, New York (1985) (review)

Mo

JACS *104* 5543 (1982); *105* 3343 (1983); *109* 1469 (1987)
Organomet *2* 1687 (1983)
JOMC *252* 105 (1983)

W

JACS *105* 7757 (1983)

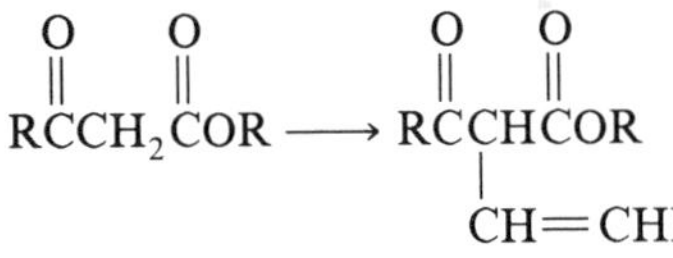

$$RCCH_2COR \longrightarrow RC\overset{\displaystyle O\ \ O}{\underset{\underset{CH=CHR'}{|}}{C}}HCOR$$

$PhSO_2C\equiv CSiMe_3$, KF, crown ether/Al(Hg)
($R' = H$)

JOC *47* 4713 (1982)

$R'CH=CHPb(OAc)_3$

CC 965 (1984)

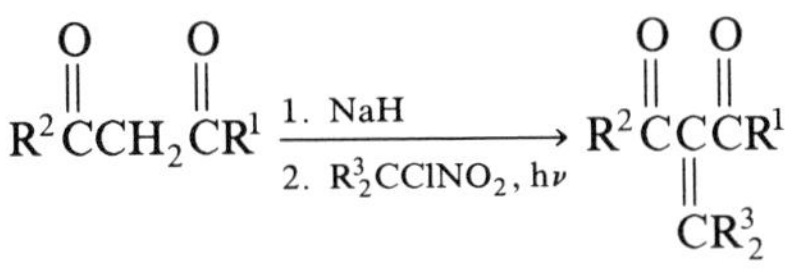

$$R^2CCH_2CR^1 \xrightarrow[2.\ R_2^3CClNO_2,\ h\nu]{1.\ \text{NaH}} R^2CCCR^1$$
with CR_2^3

$R^1 = $ alkyl, OEt

Syn 62 (1981)
TL *24* 1787 (1983)

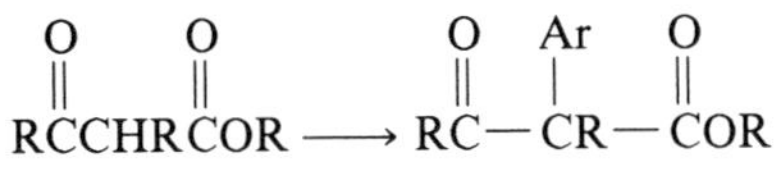

$$RCCHRCOR \longrightarrow RC-CR-COR$$
with Ar

$ArPb(OAc)_3$

Austral J Chem *33* 113 (1980); *38* 1155 (1985)
TL *23* 5365 (1982)

Ph$_4$BiX (X = O$_2$CR, O$_3$SR),   TL *23* 3365 (1982)
   N-t-butyl-*N'*,*N''*-tetramethylquanidine

KO-*t*-Bu/NC—Ar—Br, NH$_3$, hν Tetr *38* 3479 (1982)

TL *27* 5025 (1986)

TL *28* 3715 (1987)

JOC *50* 2622 (1985)

CC 240 (1985)

$$\underline{E^+}$$

RX   JOC *29* 3249 (1964); *38* 3428 (1973); *39* 2648
                                                                 (1974); *47* 381 (1982); *51* 4424 (1986); *52* 28,
                                                                 192 (1987)
                                                                 BCSJ *40* 2909 (1967)
                                                                 Org Rxs *17* 155 (1969) (review)
                                                                 JACS *92* 6702 (1970); *96* 1082 (1974); *106* 6006
                                                                 (1984) (RX = allylic phosphate); *107* 196, 2122,
                                                                 2712, 5732, 7967 (1985); *108* 8235 (1986)
                                                                 Can J Chem *60* 673 (1982)
                                                                 TL *24* 699 (1983); *26* 3059 (1985); *27* 131, 5555
                                                                 (1986); *28* 731 (1987)
                                                                 CC 265 (1986)

epoxide   JOC *38* 3428 (1973); *43* 788 (1978)
                                                                 JACS *100* 1938 (1978)
                                                                 Heterocycles *10* 111 (1978)
                                                                 Tetr *35* 1601 (1979)
                                                                 CL 337 (1985)
                                                                 TL *28* 3597 (1987)

RCHO, R$_2$CO

JOC *29* 3249 (1964); *47* 1779 (1982);
    52 4062 (1987)
BCSJ *40* 2909 (1967)
TL 4835 (1971); *27* 4713 (1986);
    28 5253, 5661 (1987)
CC 362 (1974)
JACS *104* 5528 (1982); *108* 2662 (1986)
Ann 1173 (1982)
JCS Perkin I 665 (1982)

RCO$_2$R

JOC *29* 3249 (1964)
TL 2405 (1972); *27* 1445, 2401, 3835 (1986);
    28 2017 (1987)
CC 362 (1974)
Can J Chem *52* 1343 (1974)
JACS *107* 7760 (1985) (lactone)

RCONR$_2$, BF$_3$·OEt$_2$

CL 1145 (1985)

RCON⟨N⟩

JOC *51* 268 (1986)

RCN

Can J Chem *52* 1343 (1974)

CO$_2$

JOC *31* 1032 (1966)

NCCO$_2$R

TL *28* 1051 (1987)

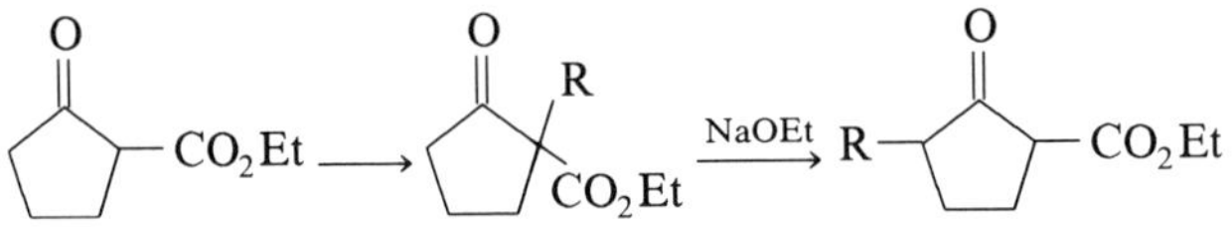

JOC *29* 2781 (1964); *52* 28 (1987)
JACS *106* 2064 (1984)

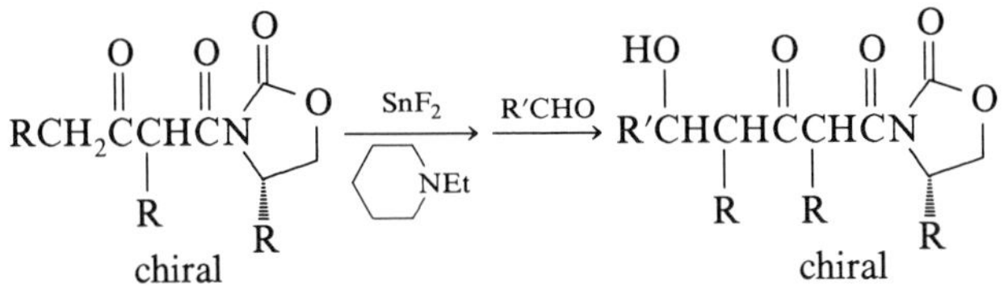

JACS *108* 2476 (1986)

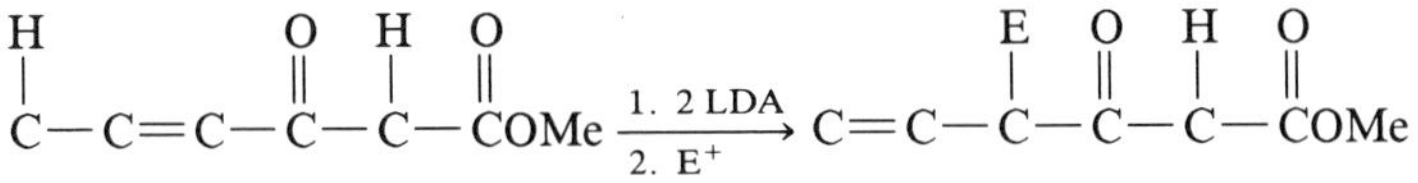

E$^+$ = RX, H$_2$CO, RCHO, R$_2$CO

Rec Trav Chim *102* 393 (1983)

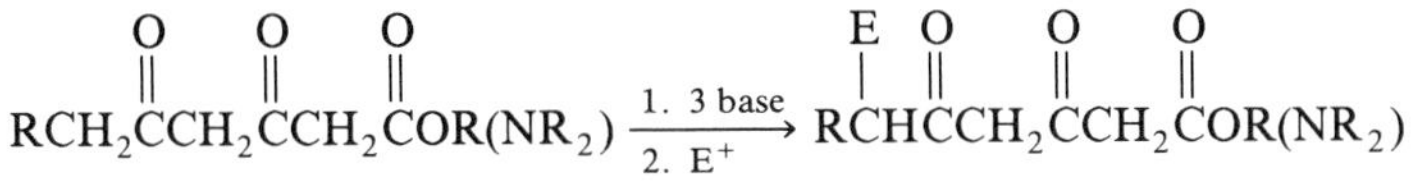

E$^+$ = RX, RCHO, RCO$_2$R

CC 362 (1974)
JOC *46* 2566 (1981)

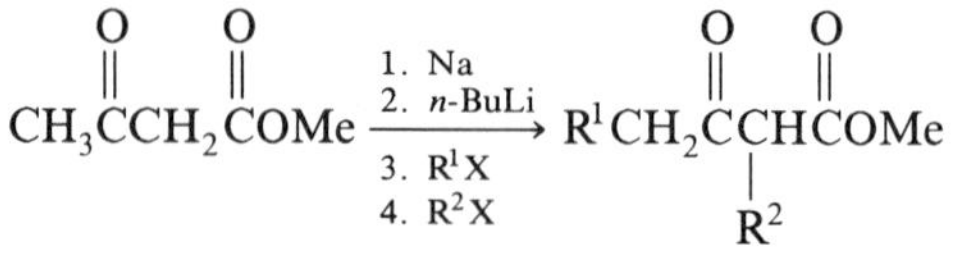

$$\underset{\substack{\| \\ O}}{CH_3C}\underset{\substack{\| \\ O}}{CH_2COMe} \xrightarrow[\substack{3.\ R^1X \\ 4.\ R^2X}]{\substack{1.\ Na \\ 2.\ n\text{-BuLi}}} R^1CH_2\underset{\substack{\| \\ O}}{C}\underset{R^2}{\overset{|}{CH}}\underset{\substack{\| \\ O}}{COMe}$$

Syn Commun *11* 7 (1981)

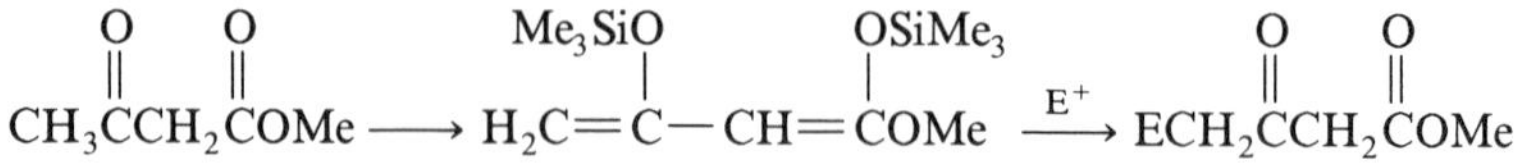

$$CH_3\underset{\substack{\| \\ O}}{C}CH_2\underset{\substack{\| \\ O}}{C}OMe \longrightarrow H_2C{=}\underset{\substack{| \\ Me_3SiO}}{C}{-}CH{=}\underset{\substack{| \\ OSiMe_3}}{COMe} \xrightarrow{E^+} E CH_2\underset{\substack{\| \\ O}}{C}CH_2\underset{\substack{\| \\ O}}{C}OMe$$

$E^+ = Br_2,\ RCOCl,\ RCHO\ (TiCl_4),\ RCH(OMe)_2\ (TiCl_4)$

CC 578 (1979); 860 (1986)
Can J Chem *61* 688 (1983)

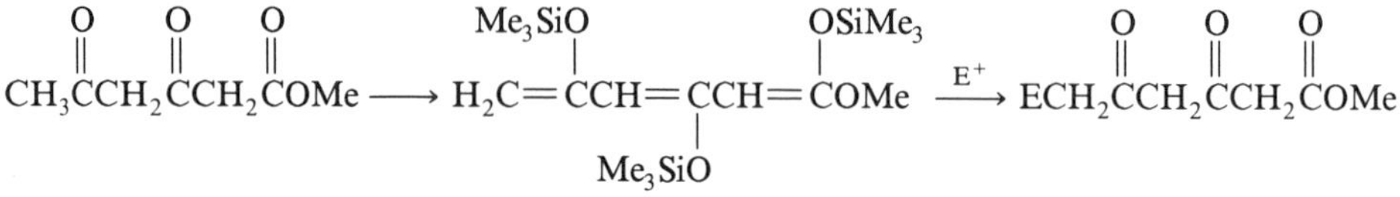

$$CH_3\underset{\substack{\| \\ O}}{C}CH_2\underset{\substack{\| \\ O}}{C}CH_2\underset{\substack{\| \\ O}}{C}OMe \longrightarrow H_2C{=}\underset{}{CCH}{=}\underset{\substack{| \\ Me_3SiO}}{CCH}{=}\underset{\substack{| \\ OSiMe_3}}{COMe} \xrightarrow{E^+} ECH_2\underset{\substack{\| \\ O}}{C}CH_2\underset{\substack{\| \\ O}}{C}CH_2\underset{\substack{\| \\ O}}{C}OMe$$

with Me₃SiO substituent shown below the chain.

$\underline{E^+}$

RCHO, TiCl₄   JOC *51* 2423 (1986)

ArCON⟨imidazole⟩, TiCl₄, Ti(O-*i*-Pr)₄   JOC *52* 2105 (1987) (biaryl product)

$$R\underset{\substack{\| \\ O}}{C}CH_2\underset{\substack{\| \\ O}}{C}OR' \longrightarrow R\underset{\substack{\| \\ O}}{C}CH_3$$

Review: Syn 893 (1982) (H₂O, salts and dipolar aprotic solvents)

H₂O, Δ	JOC *42* 459 (1977); *48* 2590 (1983)
B₂O₃	TL 3903 (1970) JOC *38* 3436 (1981) Syn Commun *11* 7 (1981)
Al₂O₃, aq dioxane	TL 2707 (1976)
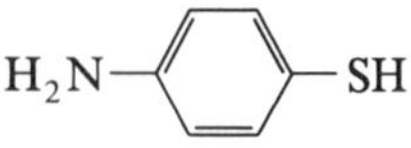 cat Cs₂CO₃, H₂N—⟨C₆H₄⟩—SH	JOC *51* 3165 (1986)
TsOH	JACS *66* 1286 (1944)
H₂SO₄	JACS *75* 3152 (1953)
CF₃CO₂H/Δ (R′ = *t*-Bu)	Syn 996 (1983)
H₂O, DMSO	TL 1091, 1095 (1974) JOC *43* 138 (1978); *52* 1880 (1987)
LiCl, H₂O, DMSO	JACS *107* 7967 (1985); *108* 3435 (1986) CC 265 (1986) TL *28* 1725 (1987) JOC *52* 1880 (1987)
LiCl, H₂O, DMF	TL *28* 5255 (1987)
LiCl, HMPA	CC 1351 (1987)

LiI·2 H$_2$O; 2,4,6-collidine

Org Syn *45* 7 (1965)

LiI·3 H$_2$O; 2,6-lutidine

JACS *107* 8066 (1985)

LiI·3 H$_2$O, DMF

JACS *106* 6690 (1984)

NaCl, H$_2$O, DMSO

TL 957 (1973); *28* 1439, 3131 (1987)
JOC *51* 5450 (1986); *52* 28 (1987)
JACS *109* 7477 (1987)

NaCl, H$_2$O, DMF

JACS *101* 7032 (1979)

CaCl$_2$·2 H$_2$O, DMSO

Syn 119 (1981)
JOC *48* 2590 (1983)

MgCl$_2$, (RSH), HMPA or DMSO, Δ

Syn 119 (1981)
TL *27* 3385 (1986)

Me$_3$SiI

Syn Commun *9* 233 (1979)

NaCN, HMPA

TL 3565 (1973)

DABCO

JOC *39* 2647 (1974)

DMAP, H$_2$O, Δ

TL *26* 3059 (1985)

~ 500°C

JOC *22* 1189 (1957)

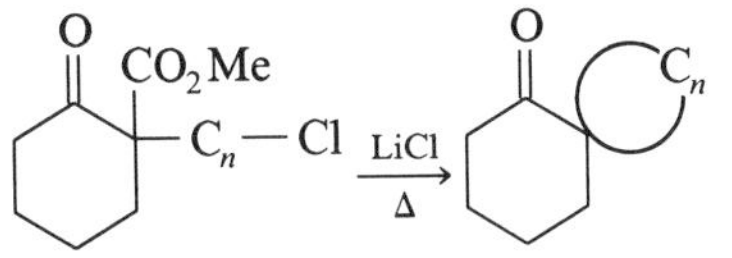

CC 30 (1981)

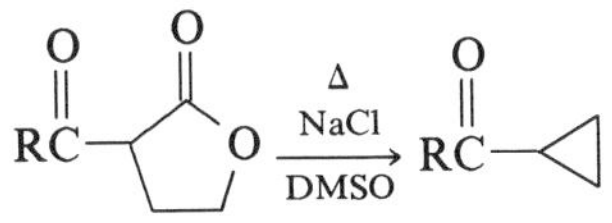

TL 4389 (1975)

Carroll reaction

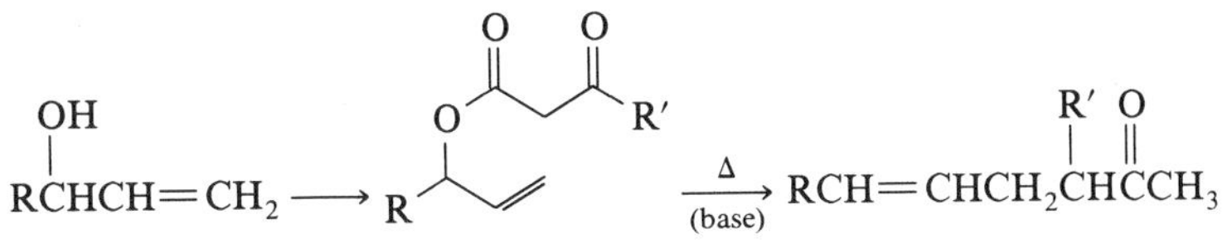

JCS 704, 1266 (1940); 507 (1941)
JACS *65* 1992 (1943); *102* 862 (1980)
JOC *23* 153 (1958) (propargylic alcohol); *33* 925 (1968); *49* 722 (1984)
TL 3253 (1969); *28* 4893 (1987) (on alumina)
Helv *57* 771 (1974)
CC 990 (1976)
Tetr *34* 2179 (1978)
Syn Commun *11* 237 (1981)
JCS Perkin I 2909 (1982)

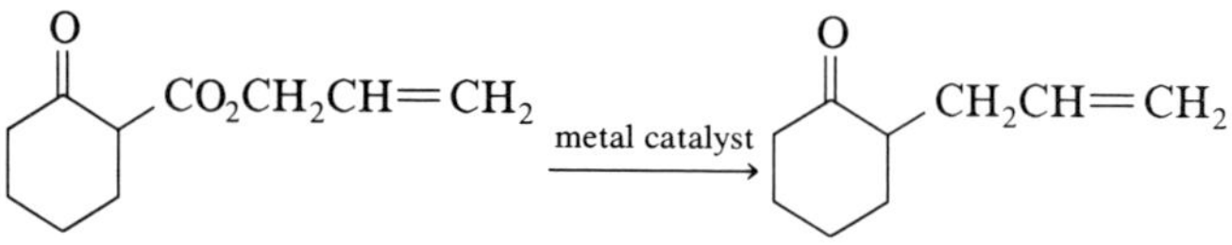

Metal

Mo, Rh, Ni   CL 1721 (1984)

Pd   TL *21* 3199 (1980); *23* 5279 (1982);
                                                 26 5575 (1985)
                                              JACS *102* 6381 (1980)
                                              JOC *52* 2988 (1987)

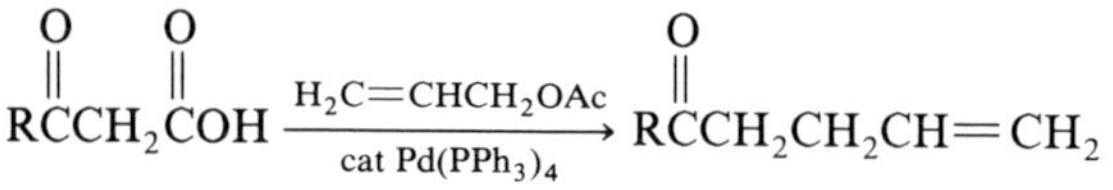

JOC *51* 421 (1986)

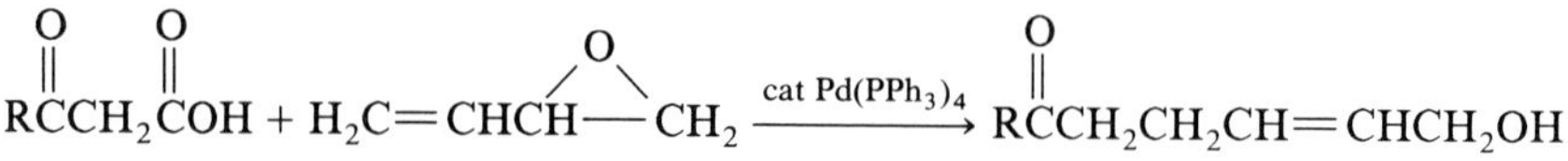

JOC *51* 5216 (1986)

21. Alkylation of Enones

See also page 587, Section 1, and page 791, Section 15.

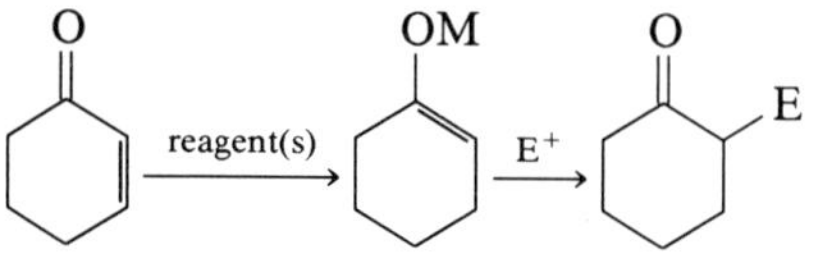

Reagent(s)	E$^+$	
Li, NH$_3$, *t*-BuOH	H$_2$O, RX, RCHO	JACS *87* 275 (1965); *89* 5464 (1967); *102* 1218, 1219 (1980) (enediones) JOC *32* 2851 (1967); *35* 1881 (1970); *40* 146 (1975) Ann Chim *5* 129 (1970)
i-Bu$_2$AlH, cat MeCu, HMPA/MeLi	RX	JOC *52* 439 (1987)
(Li)KHB(*sec*-Bu)$_3$	RX	JOC *41* 2194 (1976) JACS *107* 196 (1985)

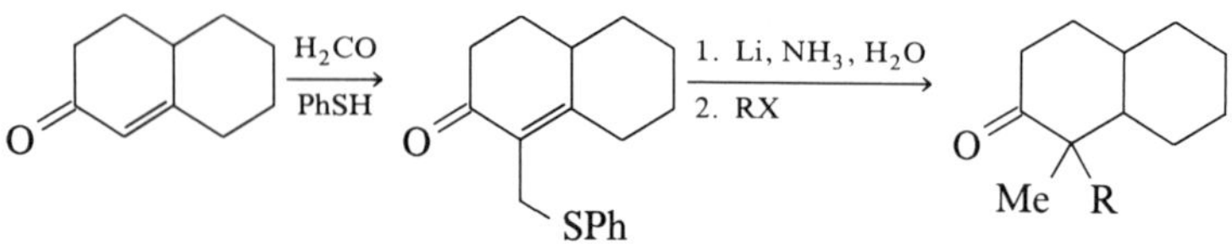

JOC *49* 3685 (1984)

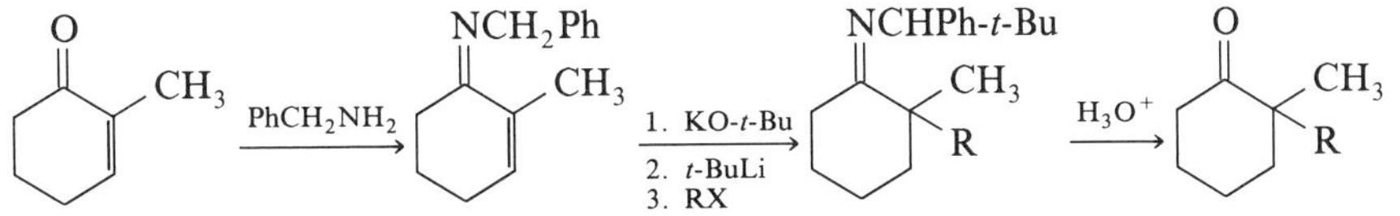

JACS *100* 292 (1978)

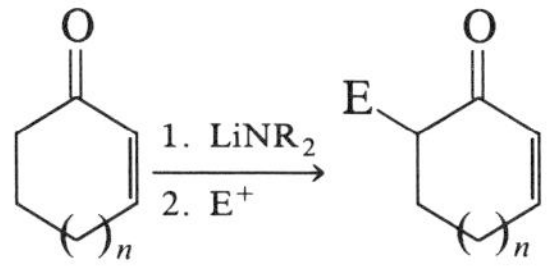

$n = 0, 1;$ E$^+$ = RX, RCHO, (MeO)$_2$CO

TL 965 (1973); *22* 15 (1981); *23* 1631 (1982); *27* 4461 (1986); *28* 3027, 5441 (1987)
CC 564 (1973); 1720 (1987)
JOC *38* 1775 (1973); *40* 862 (1975); *44* 2593 (1979); *46* 4103, 4643 (1981); *48* 2318 (1983)
   (intramolecular alkylation); *52* 3346 (1987)
JACS *98* 2351 (1976); *106* 1443, 3539 (1984); *108* 1106 (1986)
Syn Commun *7* 345 (1977)
Tetr *35* 961 (1979)

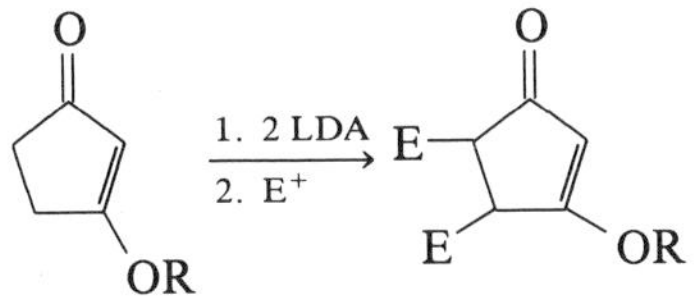

E$^+$ = RX, RCHO

CC 449 (1979)
JACS *105* 7203 (1983)

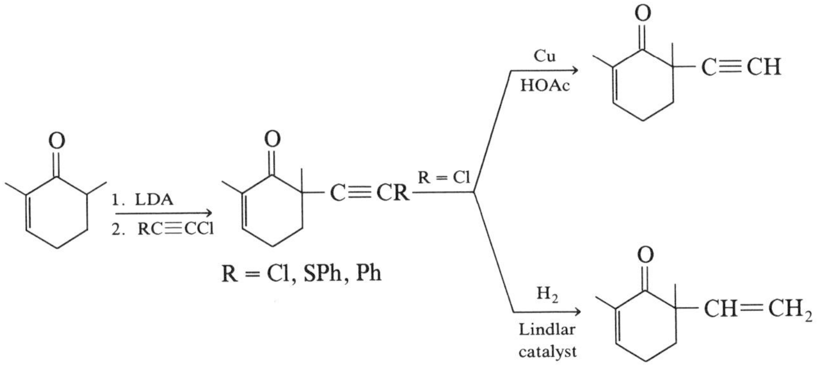

R = Cl, SPh, Ph

TL *23* 2373 (1982)

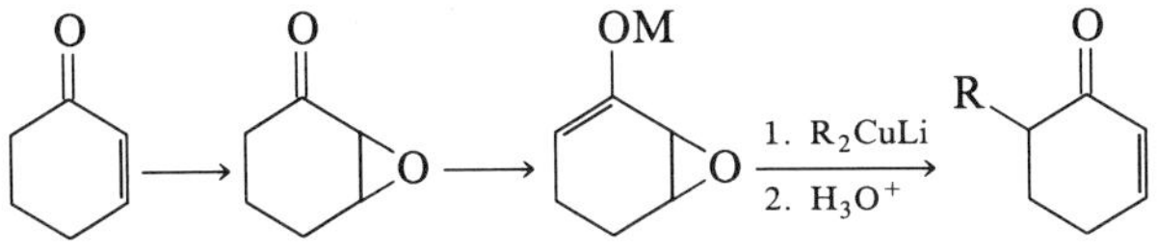

M = Li, SiMe$_3$

JACS *103* 2114 (1981)

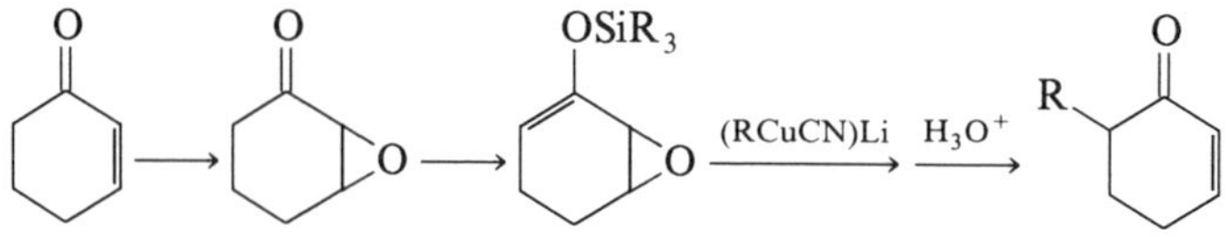

R = alkyl, aryl, not vinyl

JOC *44* 4467 (1979)
CC 1331 (1982)
JACS *104* 3165 (1982)

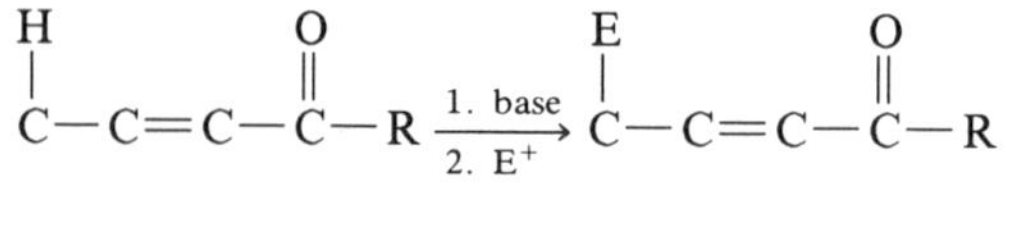

E$^+$

RCHO   TL *21* 711 (1980)

RCHO, ZnCl$_2$ TL *26* 4097 (1985)

Br$_2$, PhSO$_2$Cl, PhSeBr   TL 4103 (1978)

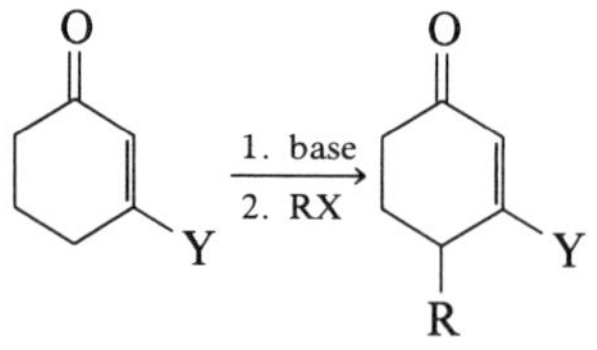

Y

OR   TL *22* 15 (1981)

NR$_2$ TL 39 (1973); 3963 (1974)
                                           Syn 401 (1976)
                                           JOC *46* 4643 (1981)

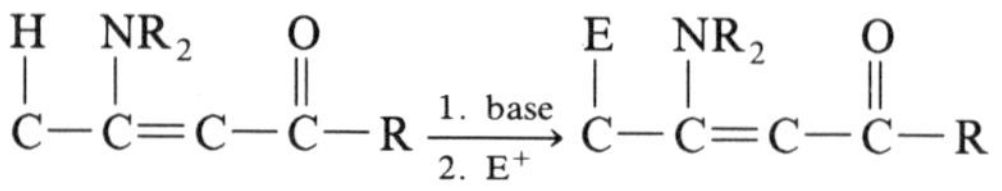

E$^+$

RX   TL 39 (1973); 3963 (1974)
                                           Syn 401 (1976)

RCHO, R$_2$CO   TL 39 (1973)

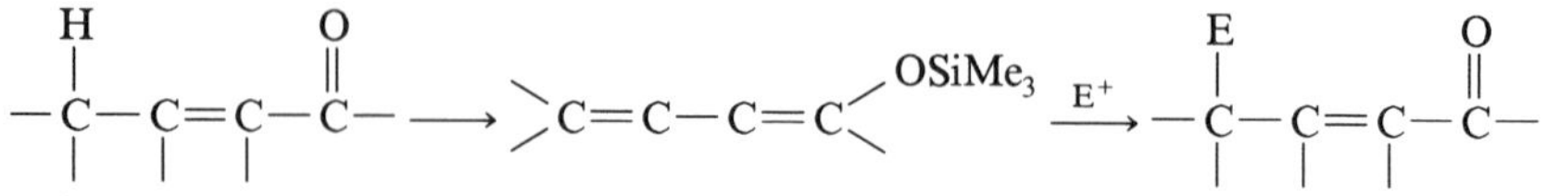

E$^+$

PhSCl   TL 3205 (1979)

PhSCHClR, ZnBr$_2$ TL 3209 (1979); *22* 705 (1981); *24* 2913 (1983)

ROCH$_2$Cl, ZnX$_2$	TL *22* 705 (1981) JOC *50* 4037 (1985)
RCH(OR)$_2$, cat Me$_3$SiOTf	TL *22* 705 (1981)
RCH(OR)$_2$, TiCl$_4$	CL 319 (1975) Tetr *39* 841 (1983)
RCH(OR)$_2$, TiCl$_4$, Ti(O-*i*-Pr)$_4$	CL 319, 1201 (1975); 467 (1977) TL *22* 705 (1981)
RCH(OR)$_2$, ZnCl$_2$	JOC USSR *18* 1001 (1982)
RCH(OR)$_2$, ZnBr$_2$	TL *22* 705 (1981)
R$_2$C(OR)$_2$, TiCl$_4$	TL *22* 705 (1981)
RCHO, TiCl$_4$ or BF$_3$·OEt$_2$	TL *22* 705 (1981); *27* 4533 (1986) BCSJ *55* 1907 (1982) JOC *51* 3400 (1986)
(RO)$_3$CH, BF$_3$·OEt$_2$	BCSJ *55* 1907 (1987)
(RO)$_3$CH, EtAlCl$_2$	TL *27* 2703 (1986)
(RO)$_3$CH, TiCl$_4$	TL *22* 705 (1981) BCSJ *55* 1907 (1982)
(RO)$_3$CH, ZnCl$_2$	JOC USSR *18* 834 (1982) Syn 227 (1984)

, ZnCl$_2$ Syn 227 (1984)

          TL *22* 2833 (1981)
JACS *106* 4862 (1984)

pyridinium salt   CC 1425 (1983)

JOC *50* 4037 (1985)

TL 2675 (1978)
JACS *102* 1602 (1980)

$$CH_3CH=CHCCH_3 \longrightarrow H-C\underset{\underset{Cl_{/2}}{\overset{H\ Pd\ H}{|\ \ |}}}{\overset{\overset{\overset{H}{|}}{C}}{\underset{=}{|}}}C-CCH_3 \xrightarrow{XYCH^-} XYCHCH_2CH=CHCCH_3$$

$$X = CO_2R;\ Y = CO_2R,\ CN,\ SO_2R$$

TL 1589 (1969); 2591 (1975)
Austral J Chem *30* 553 (1977)

$$-\overset{|}{\underset{|}{C}}-C=C-\overset{\overset{O}{\|}}{C}-R \xrightarrow[2.\ E^+]{1.\ base} \rangle C=C-\overset{\overset{E}{|}}{\underset{|}{C}}-\overset{\overset{O}{\|}}{C}-R$$

$\underline{R}$	$\underline{E^+}$	
H	RX	TL 1273 (1973); 1653 (1974)
alkyl	RX	JCS 1131 (1957); 4634, 4726 (1963)
		JACS *84* 284 (1962); *96* 7573 (1974); *108* 4556 (1986)
		Agric Biol Chem *28* 95 (1964)
		TL 1975 (1967); 459 (1968); 4103 (1978); *26* 4097 (1985)
		CC 611 (1969); 270 (1980); 75 (1986)
		JOC *35* 468 (1970); *51* 2408 (1986)
		CL 55 (1981)
		Syn Commun *12* 151 (1982)
	RCHO, ZnCl$_2$	TL *26* 4097 (1985)
	RCOCl, RO$_2$CCl	TL *26* 4097 (1985)

$$(CH_3)_2C=CHCCH_2SO_2Ph \xrightarrow[2.\ RX]{1.\ 2\ LDA} H_2C=C\overset{\overset{CH_3}{|}}{\underset{\underset{R}{|}}{C}}HCCH_2SO_2Ph$$

TL *27* 2725 (1986)

$$\xrightarrow[\substack{3.\ RCHO \\ 4.\ n\text{-}Bu_4NF\ or\ HF}]{\substack{1.\ (t\text{-}Bu)Me_2SiOTf,\ PPh_3 \\ 2.\ n\text{-}BuLi}}$$

JOC *51* 3400 (1986)

$$CH_3CH_2CH_2CH=\underset{\underset{Et}{|}}{\overset{\overset{O}{\|}}{C}}CH + H_2C=CHCN \xrightarrow{KOH} CH_3CH_2CH=CH\overset{\overset{Et}{|}}{\underset{\underset{CH_2CH_2CN}{|}}{C}}-\overset{\overset{O}{\|}}{C}H$$

JACS *66* 56 (1944)

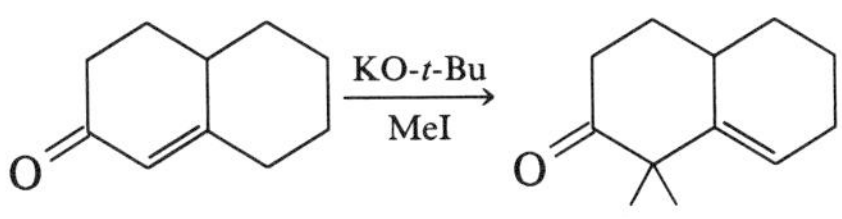

JACS 76 2852 (1954); 79 5542 (1957)
JCS 1131 (1957)
JOC 22 602 (1957); 23 841 (1958); 35 186 (1970)

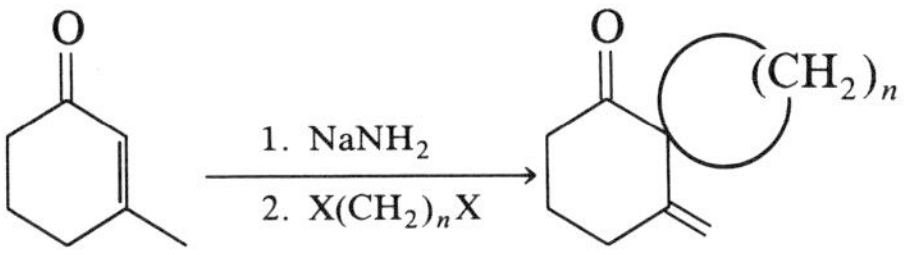

$n = 2, 4, 5$

JOC 31 2171 (1966)

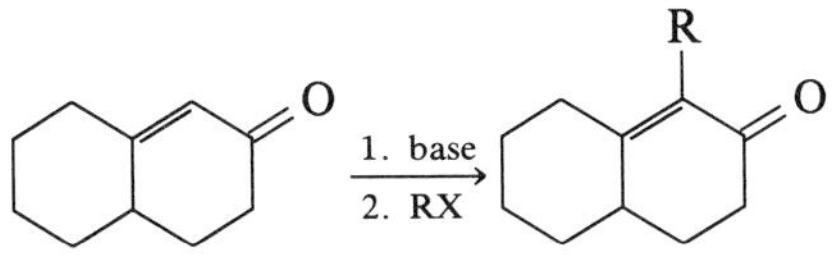

JACS 82 2847 (1960); 108 3841 (1986); 109 3017 (1987)
JOC 32 3008 (1967)

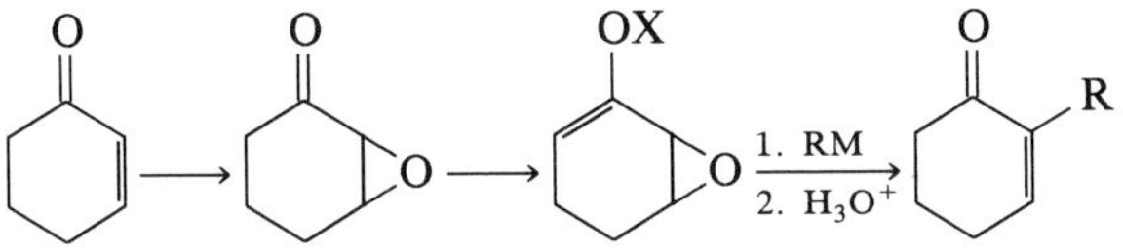

X	RM
Li	Ph$_2$CuLi, PhLi, PhMgBr
SiMe$_3$	Ph$_2$CuLi, (H$_2$C=CH)$_2$CuMgBr

JACS 103 2114 (1981)

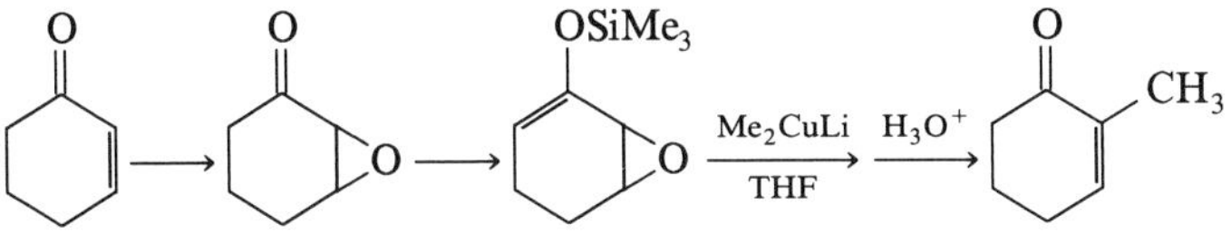

CC 1331 (1982)

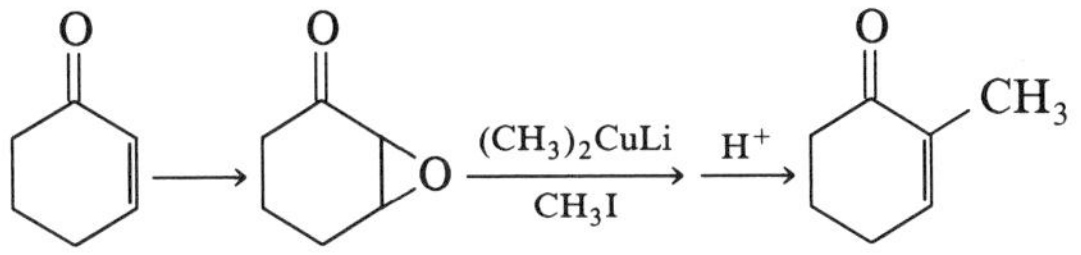

(not general for other R's)

JOC 43 1819 (1978)

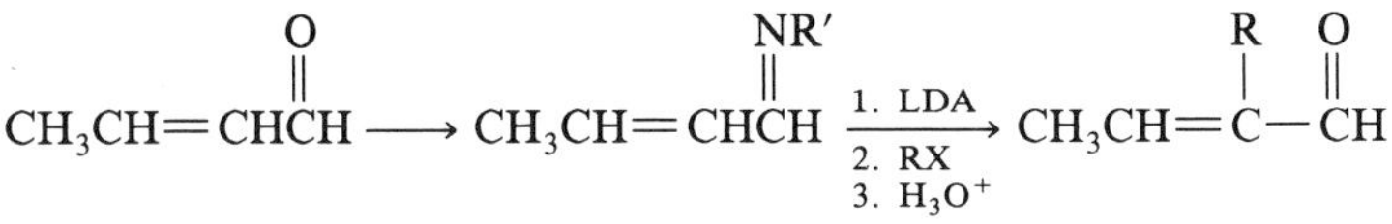

$R' = t\text{-Bu},$

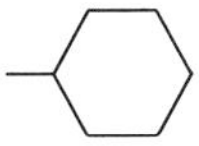

TL 1237 (1975); 597 (1976); 27 4533 (1986)

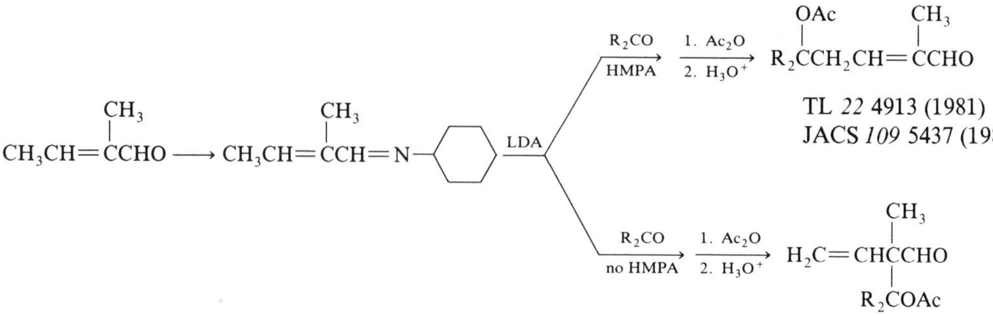

$$CH_3CH=CCHO \longrightarrow CH_3CH=CCH=N-\langle\;\rangle$$

CH₃ above both first structures.

LDA

R₂CO / HMPA → 1. Ac₂O / 2. H₃O⁺ →

OAc   CH₃
| |
R₂CCH₂CH=CCHO

TL *22* 4913 (1981)
JACS *109* 5437 (1987)

R₂CO / no HMPA → 1. Ac₂O / 2. H₃O⁺ →

CH₃
|
H₂C=CHCCHO
|
R₂COAc

TL *22* 4913 (1981)

1. LDA or NaH
2. RX
3. H₃O⁺

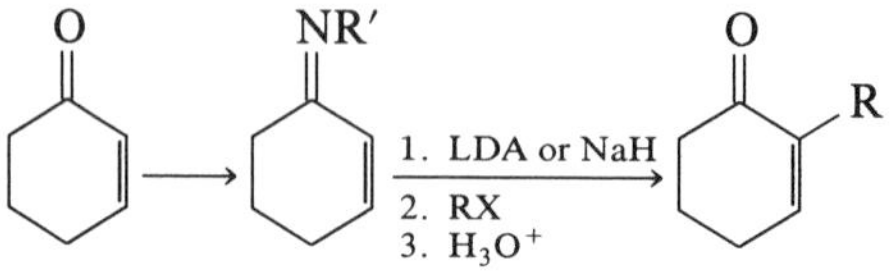

R'

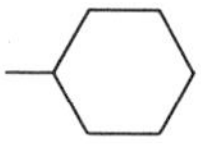

NMe₂

JACS *93* 5938 (1971)

TL *28* 3091, 3095 (1987)

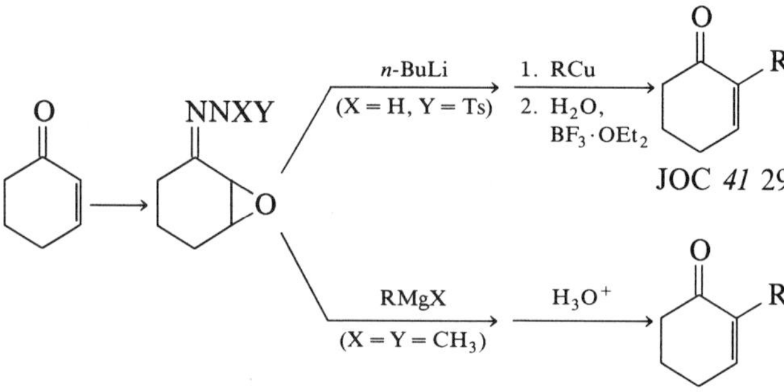

NNXY

n-BuLi (X = H, Y = Ts) → 1. RCu / 2. H₂O, BF₃·OEt₂ →

JOC *41* 2935 (1976)

RMgX (X = Y = CH₃) → H₃O⁺ →

JOC *41* 2937 (1976)

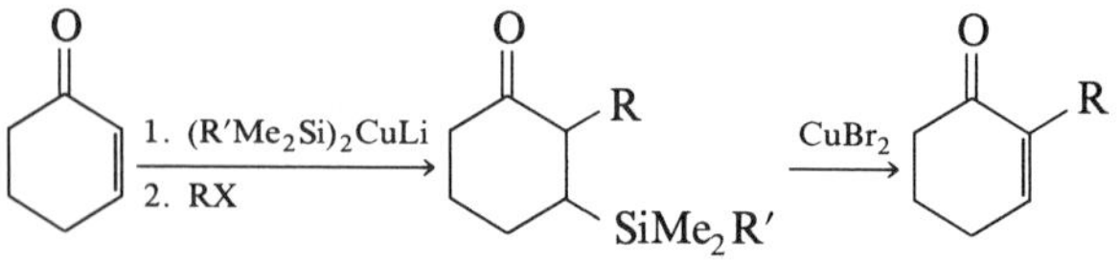

1. (R'Me₂Si)₂CuLi
2. RX →

SiMe₂R'

CuBr₂ →

JCS Perkin I 2520 (1981)

H₂C=C=CH₂ / h𝜈 → BF₃·OEt₂ →

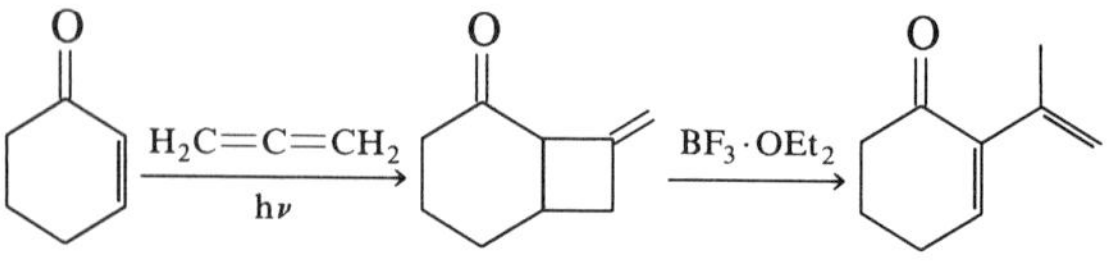

Syn 139 (1981)

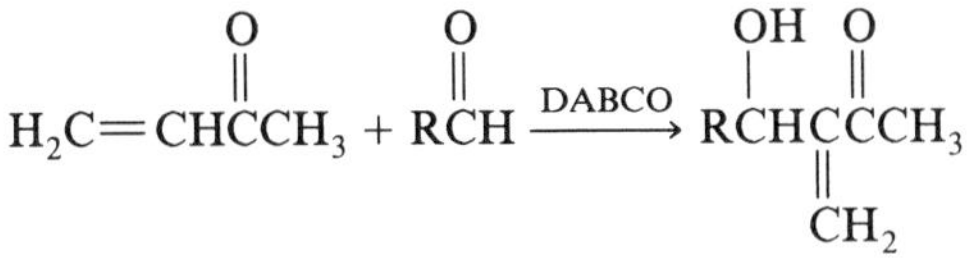

$$\underset{\text{H}_2\text{C}=\text{CHCCH}_3}{\overset{\text{O}}{\|}} + \underset{\text{RCH}}{\overset{\text{O}}{\|}} \xrightarrow{\text{DABCO}} \underset{\underset{\overset{|}{\text{CH}_2}}{\text{RCHCCCH}_3}}{\overset{\text{OH O}}{\overset{|}{\ }\overset{\|}{\ }}}$$

TL *27* 2031, 4307, 5007 (1986)

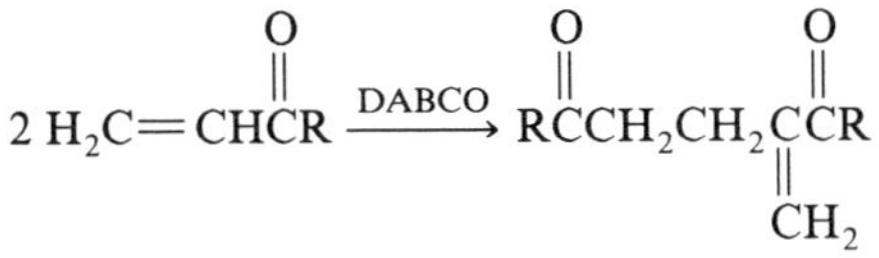

$$2\ \underset{\text{H}_2\text{C}=\text{CHCR}}{\overset{\text{O}}{\|}} \xrightarrow{\text{DABCO}} \underset{\underset{\overset{|}{\text{CH}_2}}{\text{RCCH}_2\text{CH}_2\text{CCR}}}{\overset{\text{O}\qquad\quad\text{O}}{\overset{\|}{\ }\qquad\overset{\|}{\ }}}$$

TL *28* 4591 (1987)

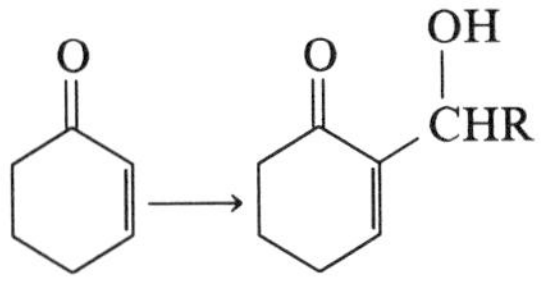

BSePh/RCHO/H$_2$O$_2$ JOC *50* 730 (1985)

Me$_2$AlX (X = SMe, SPh or SeMe)/   TL *21* 361 (1980)
  RCHO/NaIO$_4$ BCSJ *54* 274 (1981)
            JACS *109* 8017 (1987)

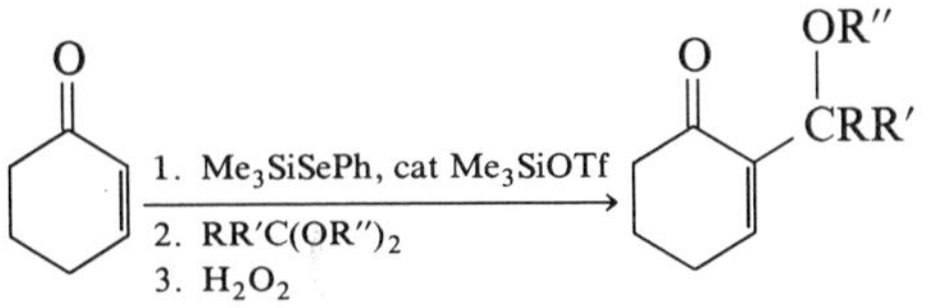

TL *22* 1809 (1981)

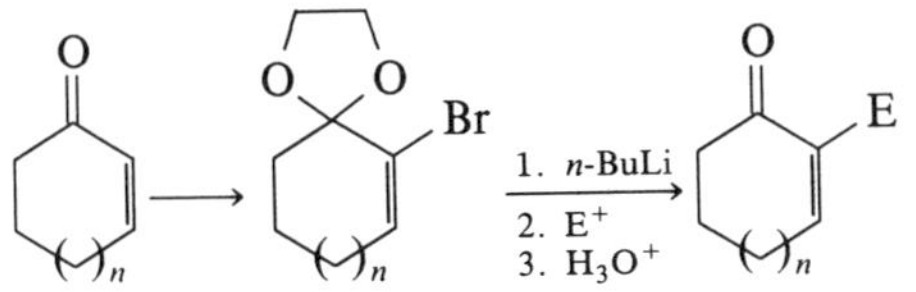

n = 0 or 1

E$^+$ = D$_2$O, RX, H$_2$CO, R$_2$CO, ClCO$_2$Et, Me$_2$SiCl, (MeS)$_2$

TL 4661 (1978)
Syn 389 (1982)
JOC *47* 1855 (1982); *51* 1490 (1986)
JACS *108* 3385 (1986)

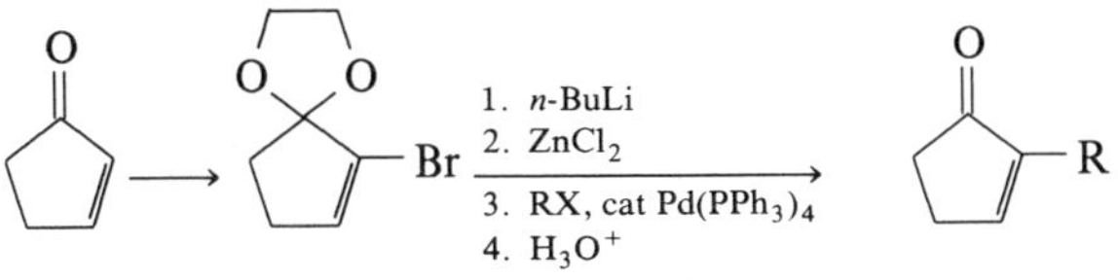

R = aryl, vinylic

CL 1007 (1987)

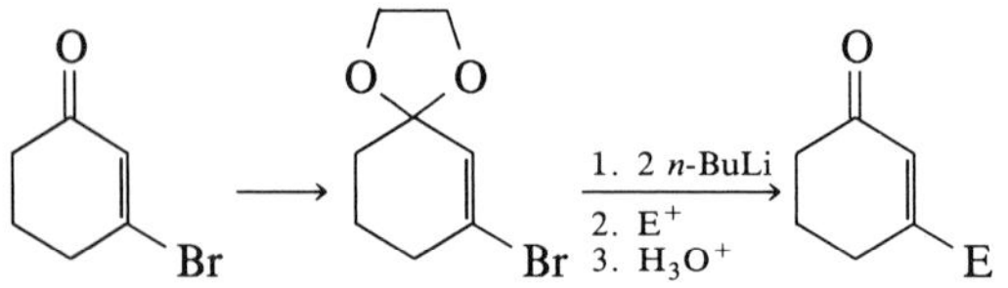

$E^+ = MeI, RCHO, R_2CO, CO_2, Me_3SiCl$

TL *22* 4217 (1981)
JOC *47* 2825 (1982)

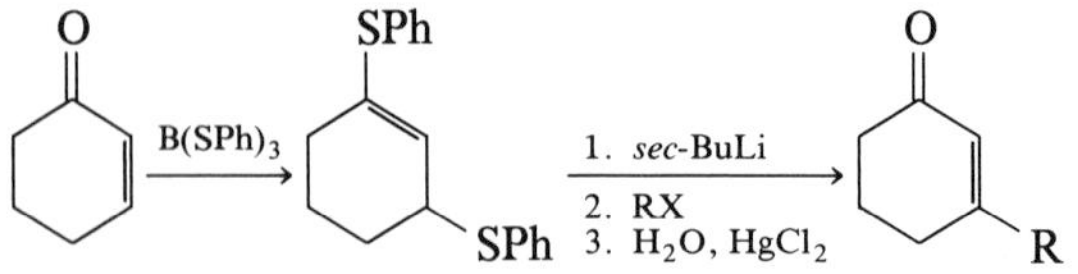

JOC *41* 2506 (1976)

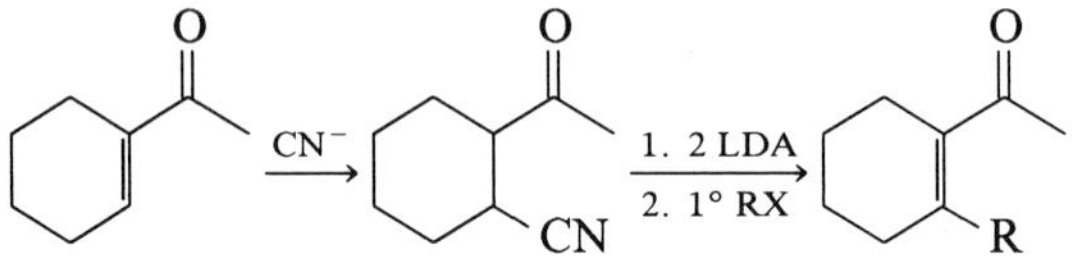

TL 3187 (1977)

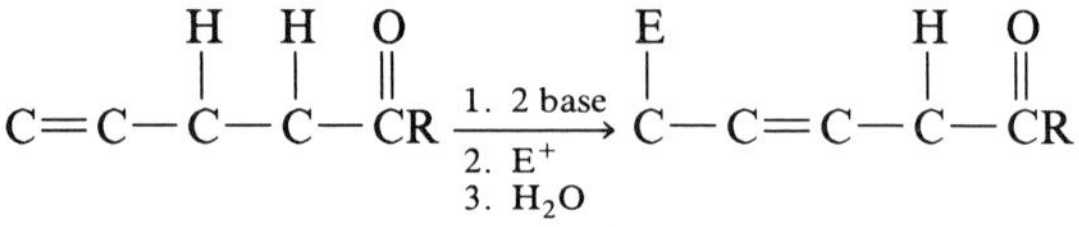

$E^+ = H_2O, D_2O, RX, RCH$—$CH_2, RCHO, R_2CO$

Angew Int *16* 320 (1977)
Tetr *37* 4047 (1981)
Ann 2272 (1981)
Helv *65* 419 (1982)

14. ALKYLATION OF α- AND β-SUBSTITUTED CARBONYL COMPOUNDS

1. Reduction of α-Substituted Ketones to Enolates

$$\underset{\substack{| \\ O}}{\overset{\substack{O \quad X \\ || \quad |}}{-C-C-}} \xrightarrow{\text{reagent(s)}} \overset{-O}{\underset{\diagup}{}}\!\!>\!\!C\!=\!C\!<\!\!\overset{\diagup}{} \xrightarrow{E^+} \underset{\substack{| \\ O}}{\overset{\substack{O \quad E \\ || \quad |}}{-C-C-}}$$

X	Reagent(s)	
Br	Li, NH$_3$	Chem Ind 118 (1963)
		Tetr 20 357 (1964)
	Zn	JACS 89 5727 (1967)
		Ann 459 (1982)
	CH$_3$MgBr	JCS 2933 (1965)
OH, OAc	Li, NH$_3$	Chem Ind 118 (1963)
		Tetr 20 357 (1964)
CN	Li, NH$_3$	JACS 104 2198 (1982)
HgCl	Li, NH$_3$	JOC 29 1868 (1964)

$$\underset{\substack{|| \\ O}}{\overset{\substack{O \quad O \\ || \quad ||}}{RCCH_2SPh}} \xrightarrow[\text{NH}_3]{\text{Li}} \overset{\substack{OLi \\ |}}{RC\!=\!CH_2} \xrightarrow[\text{2. R'Br}]{\text{1. } n\text{-Bu}_3\text{SnCl}} \overset{\substack{O \\ ||}}{RCCH_2R'}$$

R′ = allylic, benzylic, propargylic

JOC 50 3846 (1985)

2. Reduction and Alkylation of Cyclopropyl Ketones

$$\overset{C}{\underset{C\!-\!C}{\diagup\!\backslash}}\!\!-\!\!\overset{\substack{O \\ ||}}{C}\!-\!R \xrightarrow[\text{NH}_3]{\text{Li}} \overset{\substack{H \\ |}}{C}\!-\!C\!-\!\overset{\substack{OLi \\ |}}{C\!=\!C}\!-\!R$$

JOC 31 3794 (1966); 35 374, 2361 (1970); 52 3000, 4634 (1987)
TL 2879 (1974)
Austral J Chem 33 1061 (1980)
JACS 107 7724, 7732 (1985); 108 4149 (1986)

$$\text{cyclopropyl ketone} \longrightarrow R(CH_2)_3CCH_3$$

R_2CuLi	TL *24* 4543 (1983)
$R_2CuCNLi_2 \cdot BF_3$	TL *24* 5521 (1983)
$(\mathbf{RCH{=}CHCuC{\equiv}CC_3H_7)Li}$	JCS Perkin I 683 (1983)

$$\text{spiropentyl } \overset{COR}{\underset{CO_2R}{<}} \xrightarrow{Nuc} Nuc{-}C{-}C{-}\overset{\overset{\displaystyle CO_2R}{|}}{C}{-}COR$$

Review: Acct Chem Res *12* 66 (1979)

Nuc

R_2CuLi	TL 529 (1975); 3857 (1976) Can J Chem *60* 825 (1982)
$(RCuCH_2SOCH_3)Li, BF_3 \cdot OEt_2$	JOC *52* 1885 (1987)
$[RCu(CN)CH_2SOCH_3]Li_2$	JOC *52* 1885 (1987)
$[RCu(CN)CH_2SOCH_3]Li_2, BF_3 \cdot OEt_2$	JOC *52* 1885 (1987)
Et_2AlCN (Nuc = CN)	Can J Chem *60* 825 (1982)

3. Reduction and Alkylation of Ketones Containing Sulfur in the α-Position

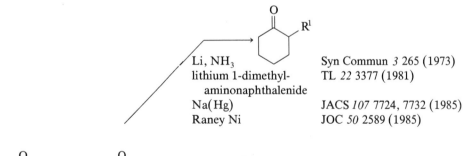

Li, NH$_3$	Syn Commun *3* 265 (1973)
lithium 1-dimethyl- aminonaphthalenide	TL *22* 3377 (1981)
Na(Hg)	JACS *107* 7724, 7732 (1985)
Raney Ni	JOC *50* 2589 (1985)

Angew Int *12* 586 (1973)
Syn Commun *3* 265 (1973)
TL 3955 (1973); *22* 3377 (1981)

JACS *108* 3385 (1986)

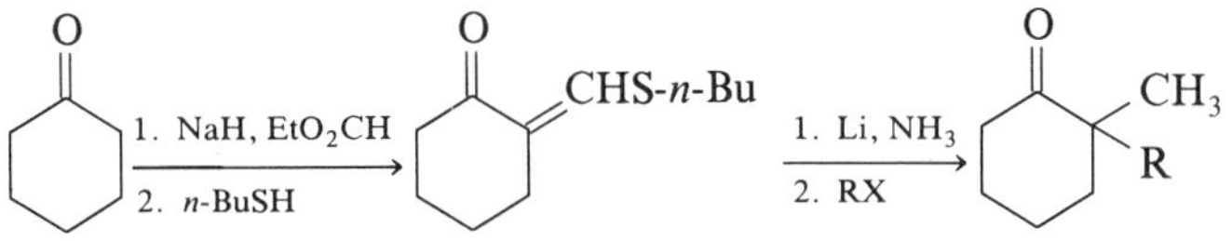

JACS *93* 1027 (1971); *104* 2198 (1982)

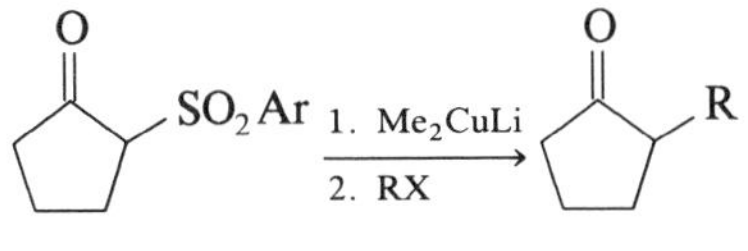

JACS *107* 1239 (1985)

4. Substitutive Alkylation of α-Halocarbonyl Compounds

4.1. Organomagnesium Compounds

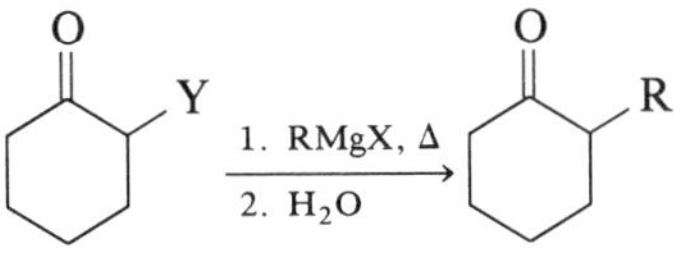

Y = Cl, Br

R	
alkyl (low yields)	JACS *56* 1990 (1934); *73* 1993 (1951) Compt Rend *198* 941 (1934) JOC *37* 4090 (1972)
aryl	JACS *66* 1550 (1944); *70* 495 (1948); *71* 3313 (1949); *72* 1995 (1950) JOC *12* 737 (1947); *19* 1363 (1954); *24* 843 (1959)
vinylic	TL *22* 2243 (1981) Tetr *37* 3967 (1981) JCS Perkin I 225 (1987)

4.2. Organoboron Compounds

Review: Organomet Chem Syn *1* 305 (1972)

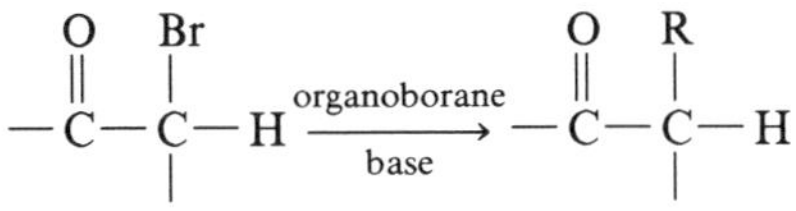

Organoborane	Base	
R₃B	KO-t-Bu	JACS *90* 6218 (1968)
R-9-BBN	KO-t-Bu 2,6-(t-Bu)₂C₆H₃OK	JACS *91* 2147, 4304 (1969) JACS *91* 6852 (1969) JOC *51* 3398 (1986) (R = vinylic)

4.3. Organocopper Compounds

$$-\overset{\overset{\displaystyle O}{\|}}{C}-\overset{\overset{\displaystyle Br}{|}}{C}-\ \longrightarrow\ -\overset{\overset{\displaystyle O}{\|}}{C}-\overset{\overset{\displaystyle R}{|}}{C}-$$

Me$_2$CuLi   TL 177 (1971)
                                                       JOMC *153* 259 (1983)

(H$_2$C$=$C)$_2$CuLi (with CH$_3$ on the vinyl)   J Ind Chem Soc *45* 1026 (1968)

(MeCuCN)Li   JOMC *153* 259 (1983)

$$-\overset{\overset{\displaystyle O}{\|}}{C}-\overset{\overset{\displaystyle Br}{|}}{C}-\ \xrightarrow[\text{2. CH}_3\text{I}]{\text{1. (CH}_3)_2\text{CuLi}}\ -\overset{\overset{\displaystyle O}{\|}}{C}-\overset{\overset{\displaystyle CH_3}{|}}{C}-$$

JACS *95* 3076, 7788 (1973)
Tetr *31* 1223, 1227 (1975)

cyclopentanone with Cl, Cl → 1. Me$_2$CuLi, 2. MeI → cyclopentanone with Cl, CH$_3$ → 1. Me$_2$CuLi, 2. MeI → cyclopentanone with CH$_3$, CH$_3$

JOC *45* 2036 (1980); *48* 4763 (1983)
JACS *105* 2435 (1983)

$$-\overset{\overset{\displaystyle Br}{|}}{C}-\overset{\overset{\displaystyle O}{\|}}{C}-\overset{\overset{\displaystyle Br}{|}}{C}-\ \xrightarrow[\text{2. H}_2\text{O}]{\text{1. R}_2\text{CuLi}}\ -\overset{\overset{\displaystyle H}{|}}{C}-\overset{\overset{\displaystyle O}{\|}}{C}-\overset{\overset{\displaystyle R}{|}}{C}-$$

JACS *95* 3076, 7788 (1973) (both R = Me)
Tetr *31* 1223, 1227 (1975) (both R = Me)
TL *27* 4671 (1986) (R = Ph)

$$-\overset{\overset{\displaystyle Br}{|}}{C}-\overset{\overset{\displaystyle O}{\|}}{C}-\overset{\overset{\displaystyle Br}{|}}{C}-\ \xrightarrow[\text{2. CH}_3\text{I}]{\text{1. (CH}_3)_2\text{CuLi}}\ -\overset{\overset{\displaystyle CH_3}{|}}{C}-\overset{\overset{\displaystyle O}{\|}}{C}-\overset{\overset{\displaystyle CH_3}{|}}{C}-$$

JACS *95* 3076, 7788 (1973)
Tetr *31* 1223, 1227 (1975)

cyclohexanone with Br → NNHTs cyclohexane with Br → 1. R$_2$CuLi, 2. H$_3$O$^+$ → cyclohexanone with R

JACS *97* 7372 (1975)

5. Substitutive Alkylation of β-Halocarbonyl Compounds

$$\text{BrCH}_2\text{CH}_2\overset{\overset{\displaystyle O}{\|}}{\text{CH}}\ \longrightarrow\ \text{BrMgCH}_2\text{CH}_2\text{—}\overset{\text{O}}{\underset{\text{O}}{\diagup}}\ \xrightarrow[\text{2. H}_3\text{O}^+]{\text{1. RX, cat Li}_2\text{CuCl}_4}\ \text{RCH}_2\text{CH}_2\overset{\overset{\displaystyle O}{\|}}{\text{CH}}$$

JOC *48* 1767 (1983)

6. Substitutive Alkylation of α-Diazocarbonyl Compounds

$$R_3B + N_2CHCH \xrightarrow[]{H_2O} RCH_2CH$$
(with carbonyl O on both CH groups)

Can J Chem *48* 868 (1970)

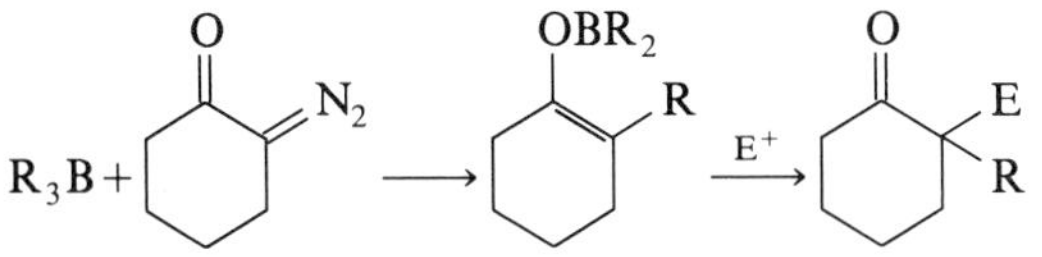

<u>E$^+$</u>

H$_2$O or D$_2$O	JACS *90* 5936 (1968); *91* 6195 (1969)
	CC 139 (1969)
	Can J Chem *49* 2371 (1971)
NCS or NBS	Can J Chem *50* 2387 (1972)
PhSeCl	Syn Commun *10* 667 (1980)
(Me$_2$N=CH$_2$)I	JACS *95* 602 (1973)
2 RLi/R′X (E = R′)	JOC *36* 1790 (1971)
Me$_2$NCH$_2$CH$_2$OLi/R′X (E = R′)	Syn Commun *10* 139 (1980)
R′CN/H$_3$O$^+$ (E = COR′)	Syn Commun *12* 189 (1982)

For other reactions of boron enolates see page 647, Section 7.

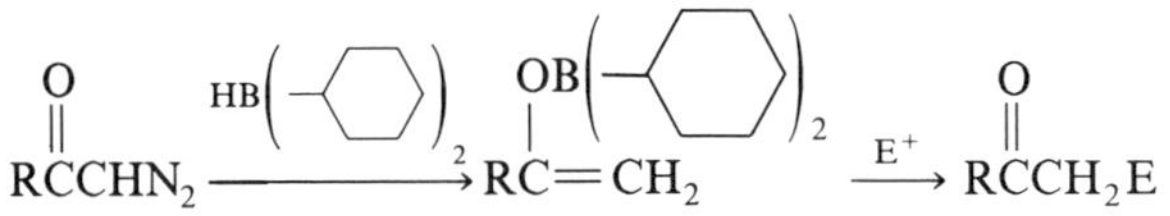

$$RCCHN_2 \xrightarrow[]{HB(\bigcirc)_2, \; OB(\bigcirc)_2} RC\!=\!CH_2 \xrightarrow[]{E^+} RCCH_2E$$

$$E^+ = (Me_2N\!=\!CH_2)I, RCHO, R_2CO$$

JOC *52* 1347 (1987)

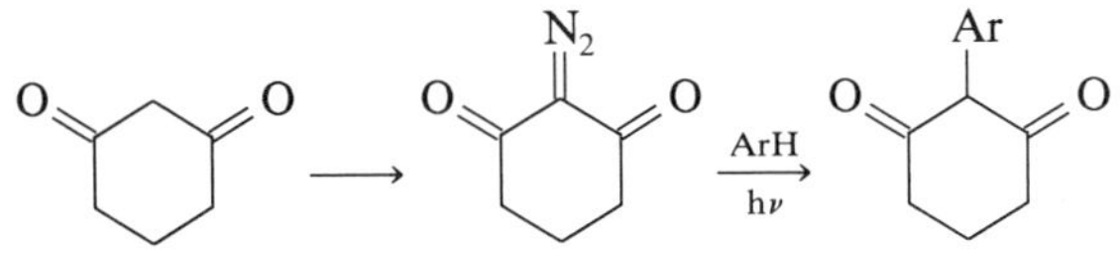

JOC *44* 4906 (1979)

7. Decarboxylation-Allylation of β-Keto Acids and Esters

See page 768, Section 20.

15. 1,4-ADDITION TO α,β-UNSATURATED CARBONYL COMPOUNDS

See also page 724, Section 2; and page 746, Section 7.1.

1. Free Radical and Related Addition Reactions

See also page 814, Section 16.

$$RX + H_2C = CHCCH_3 \longrightarrow RCH_2CH_2CCH_3$$

$$X = Br, I$$

$n\text{-Bu}_3\text{SnH}$	TL *28* 5853 (1987) (C-glycosides) Ann 427 (1987)
vitamin B_{12}, reductant	JACS *102* 3642 (1980) (intramolecular) Pure Appl Chem *55* 1791 (1983) (review) Chimia *39* 211 (1985) (C-glycosides)

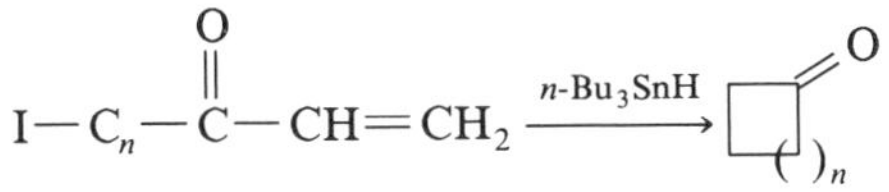

$$I - C_n - \overset{O}{\underset{\|}{C}} - CH = CH_2 \xrightarrow{n\text{-Bu}_3\text{SnH}} \boxed{}(\,)_n$$

$$n = 7, 11, 15$$

JACS *108* 2787 (1986)

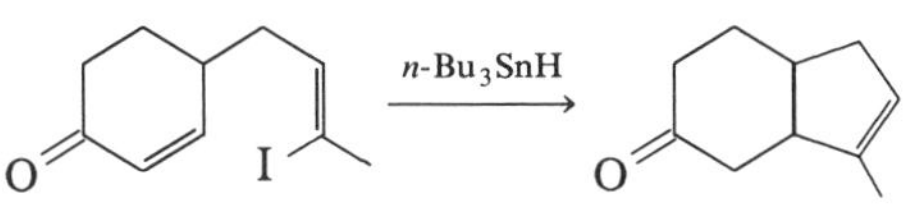

TL *24* 1871 (1983)

791

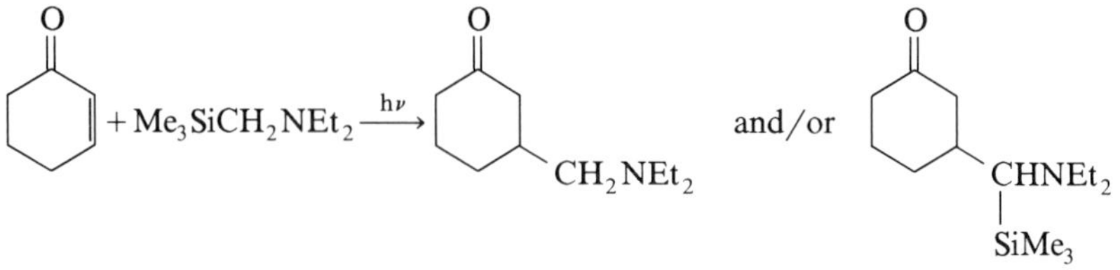

JACS *109* 4421 (1987)

2. Michael and Michael-like Addition Reactions

Reviews of and key references to the Michael reaction:

Org Rxs *10* 179 (1959) (review)
Syn 107 (1979) (phase transfer)
JOC *44* 2239 (1979) (asymmetric); *48* 4642 (1983) (annulation)
Angew Int *19* 1013 (1980) (high pressure); *20* 770 (1981) (high pressure, F⁻ catalyzed)
K. Tomioka and K. Koga in "Asymmetric Synthesis," Ed. J. D. Morrison, Academic Press, New York (1983), Vol 2, Part A, Chpt 7 (asymmetric additions, review)
TL *24* 3841 (1983) (high pressure)

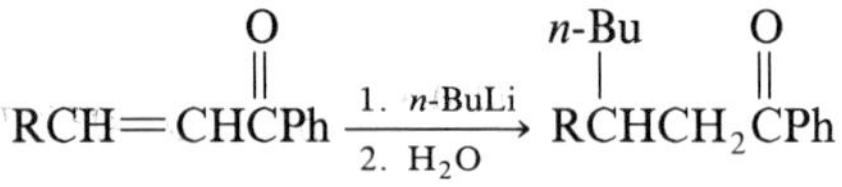

JOC *49* 1144 (1984)

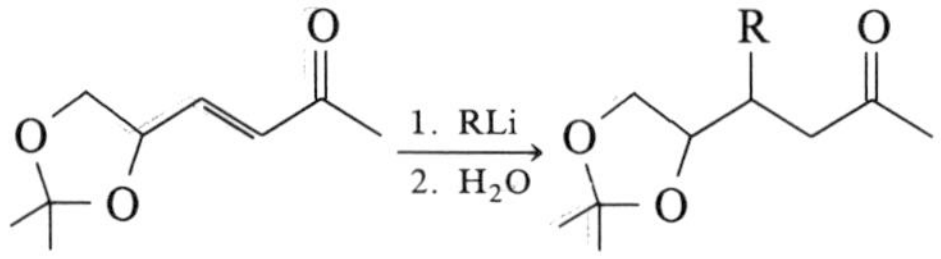

TL *28* 2525 (1987)

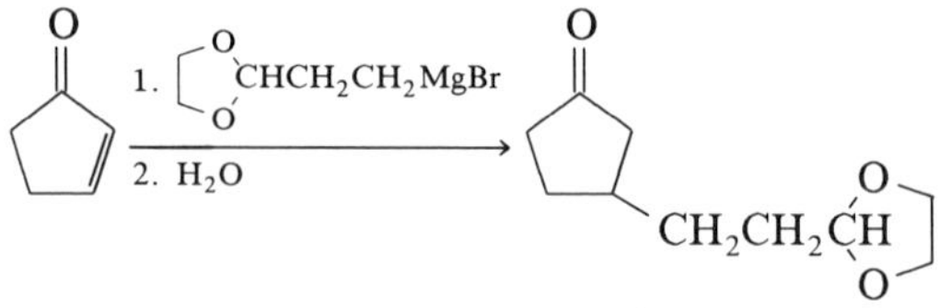

TL *28* 3217 (1987)

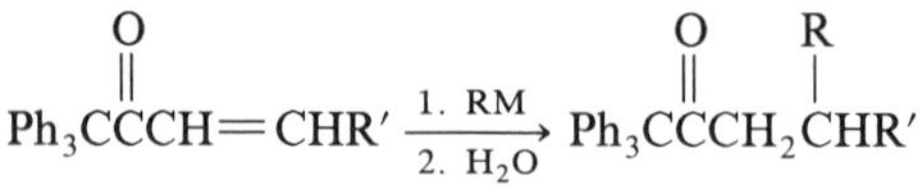

$$RM = RLi,\ RMgBr,\ RC{\equiv}CLi,\ ArLi,$$

Helv *68* 264 (1985)

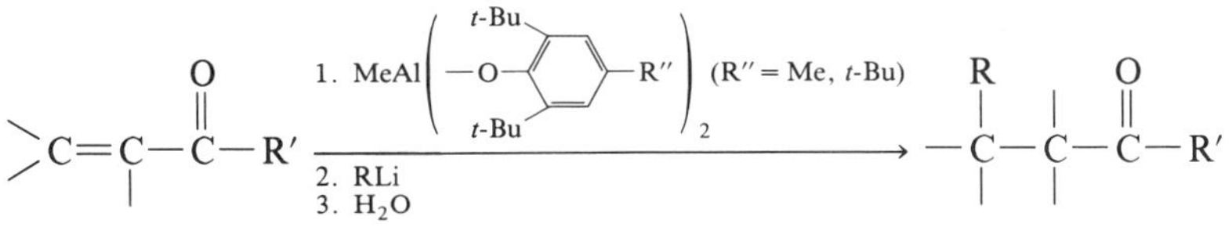

$$R = 1°, 2°, 3° \text{ alkyl; aryl; vinylic; ester enolate}$$

TL *28* 5723 (1987)

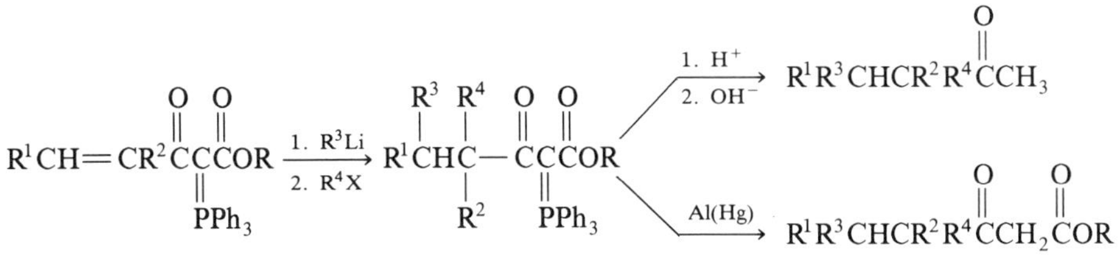

JOC *47* 4955, 4963 (1982); *52* 1381 (1987) (intramolecular)

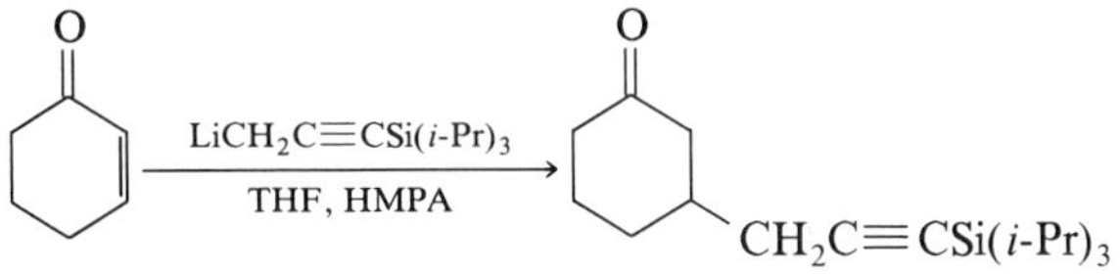

TL *23* 719 (1982)

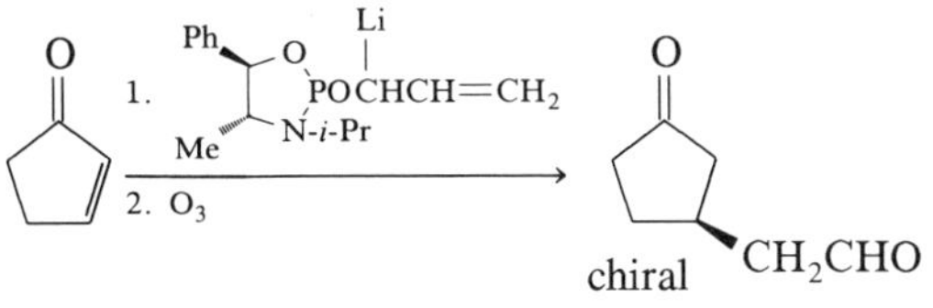

JACS *109* 5026 (1987)

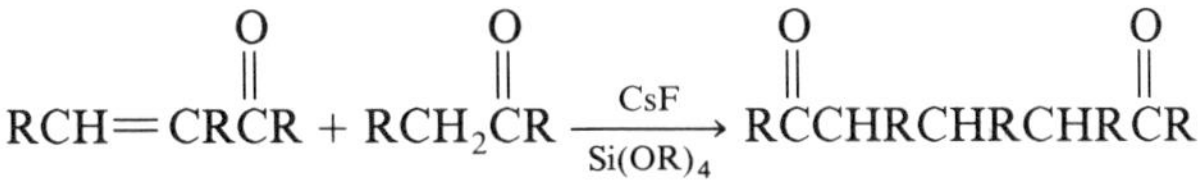

$$RCH=CRCR + RCH_2CR \xrightarrow[\text{Si(OR)}_4]{\text{CsF}} RCCHRCHRCHRCR$$

CC 122 (1981)
Tetr *39* 117, 999 (1983)

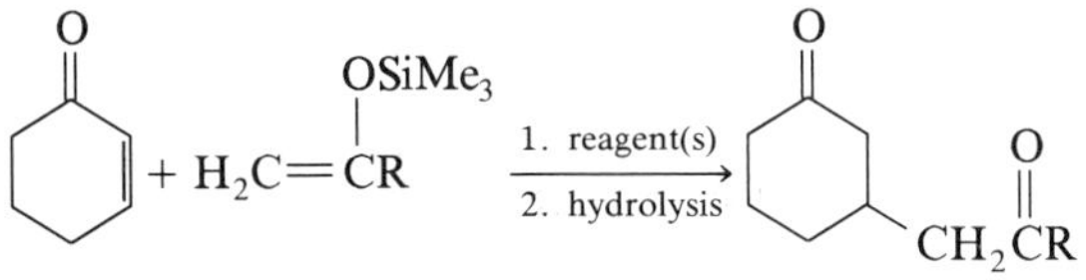

Reagent(s)	
montmorillonite clay	CC 1203 (1987)
cat (Ph₃C)ClO₄	CL 953 (1985); 1017 (1986)
CsF	JOMC *184* 157 (1980)
BF₃·OEt₂, ROH	TL *26* 6201 (1985) (also methyl enol ethers used)
SnCl₄	JACS *107* 2797 (1985)

$TiCl_4$	CL 1223 (1974) JOC *45* 607 (1980); *50* 2539, 4266 (1985); *51* 279 (1986) JACS *107* 2797 (1985) Org Syn *65* 12 (1987)
$TiCl_4$, $Ti(O\text{-}i\text{-}Pr)_4$	CL 1223 (1974) BCSJ *49* 779 (1976) JOC *43* 2720 (1978); *50* 4266 (1985)

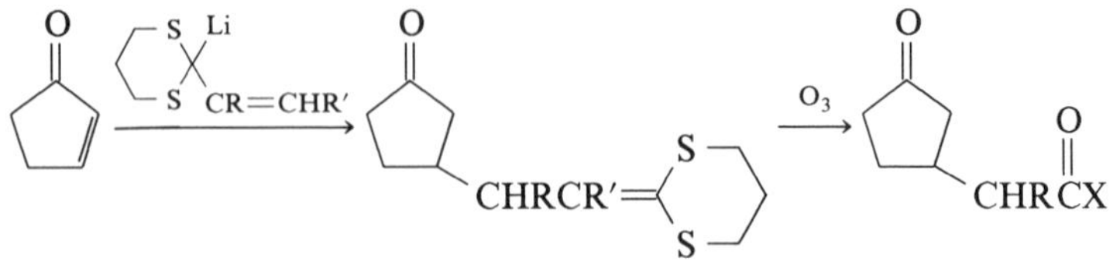

X = R′, OEt

TL *23* 3237 (1982)
JACS *104* 7174 (1982)

$$\underset{\displaystyle RCHX}{\overset{\displaystyle M}{|}} + C{=}C{-}\overset{\overset{\displaystyle O}{\|}}{C}{-}R \xrightarrow{H_2O} R\overset{\overset{\displaystyle X}{|}}{C}H{-}C{-}\overset{\overset{\displaystyle H}{|}}{C}{-}\overset{\overset{\displaystyle O}{\|}}{C}{-}R$$

M = alkali metal

X	
COR	BSCF 3543 (1966) JOC *42* 183 (1977) Tetr *40* 4127 (1984) TL *27* 6169 (1986)
CO_2R	JOC *41* 4044 (1976); *47* 3464 (1982); *50* 3022 (1985) J Polym Sci, Polym Chem Ed *17* 3509 (1979) TL *27* 3927 (1986) (intramolecular)
CS_2Me	TL *27* 1505 (1986)
$CONR_2$	JOC *50* 3019 (1985)

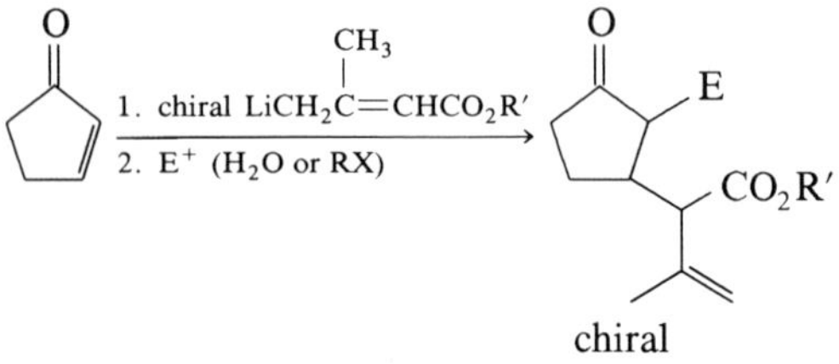

TL *24* 4975 (1983)

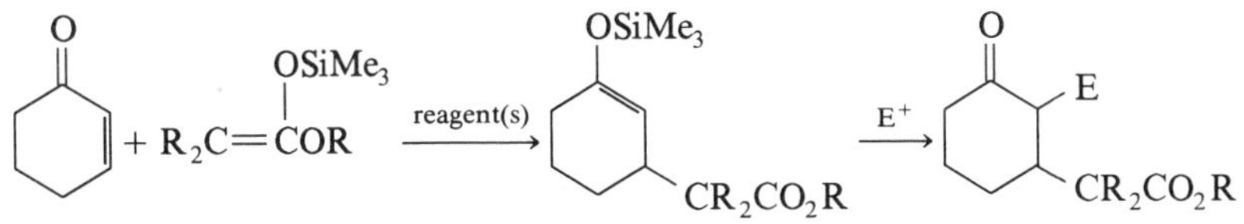

$E^+ = $ NBS; PhSeCl; H_3O^+; PhSCHClR, $ZnBr_2$; RX, $[(Me_2N)_3S]Me_3SiF_2$

Reagent(s)

high pressure	TL *24* 4943 (1983); *25* 1075 (1984)
CH_3NO_2	JOC *49* 2083 (1984)
Δ, CH_3CN	TL *21* 3779 (1980)
	JCS Perkin I 1099 (1982)
montmorillonite clay	CC 1203 (1987)
cat $(Ph_3C)ClO_4$	CL 1817 (1986)
cat $[(Me_2N)_3S]Me_3SiF_2$	JOC *49* 2083 (1984)
cat Me_3SiCl, CH_3NO_2	JOC *49* 2083 (1984)
$TiCl_4$	CL 163 (1976)
	JACS *102* 4262 (1980); *107* 2797 (1985)
	JOC *45* 237 (1980); *51* 279 (1986)
$TiCl_4$, $Ti(O-i-Pr)_4$	CL 163 (1976)
ZnI_2	JACS *109* 4390 (1987)

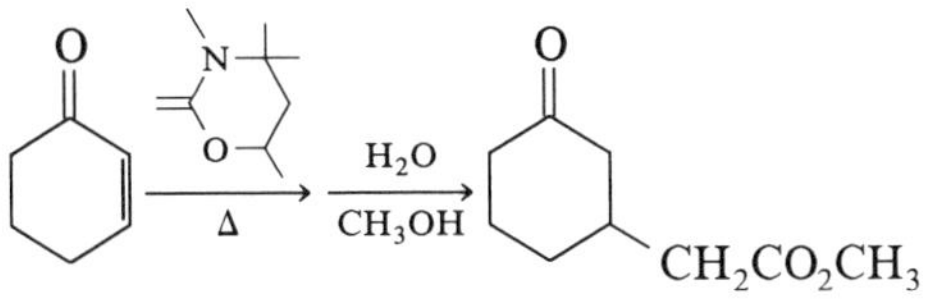

JOC *38* 175 (1973)

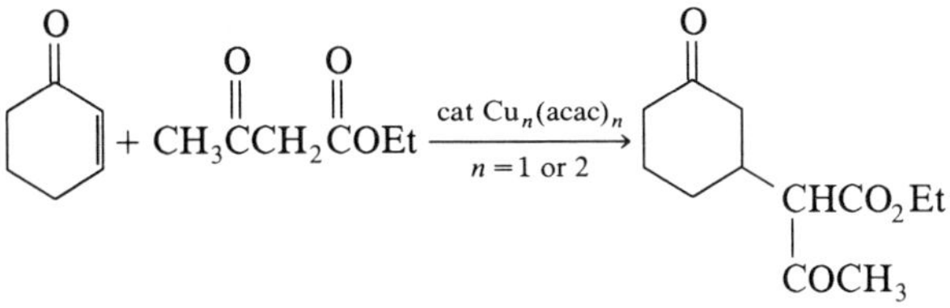

TL *27* 5015 (1986)

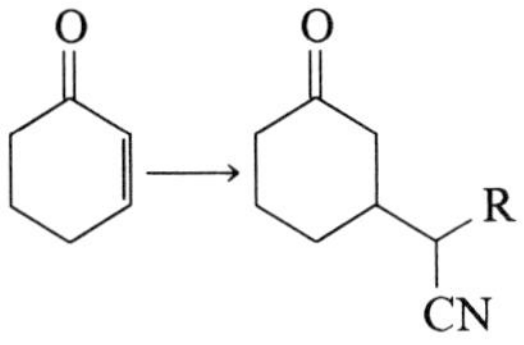

ArCHLiCN	Syn Commun *11* 85, 335 (1981)
	TL *22* 2171 (1981)

RCLi(CN)SPh/Raney Ni   JOC *47* 1131 (1982)

RCHLiCN $\left(R = N \langle \rangle \right)$ Heterocycles *19* 1395 (1982)

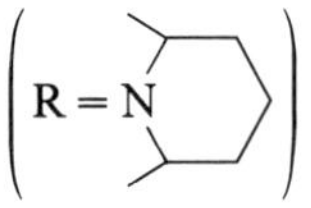

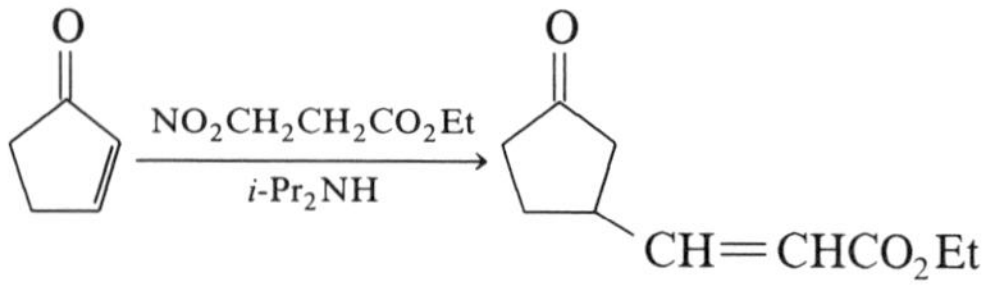

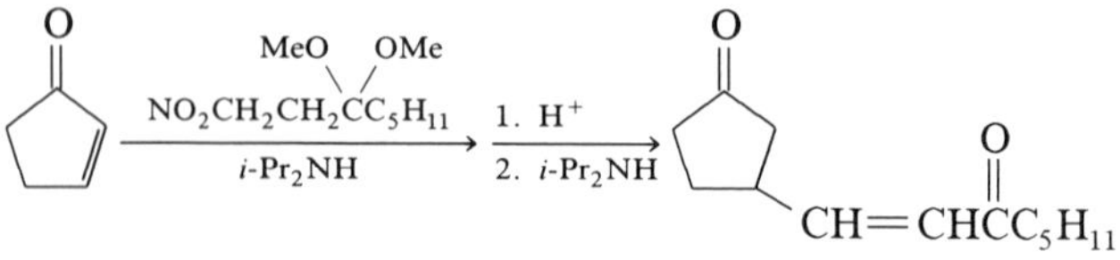

$$\xrightarrow[\text{i-Pr}_2\text{NH}]{\text{NO}_2\text{CH}_2\text{CH}_2\text{CO}_2\text{Et}}$$

CH=CHCO$_2$Et

$$\xrightarrow[\text{i-Pr}_2\text{NH}]{\text{NO}_2\text{CH}_2\text{CH}_2\overset{\text{MeO}\quad\text{OMe}}{\text{CC}_5\text{H}_{11}}} \xrightarrow[\text{2. i-Pr}_2\text{NH}]{\text{1. H}^+}$$

CH=CHCC$_5$H$_{11}$

TL 2371 (1978)

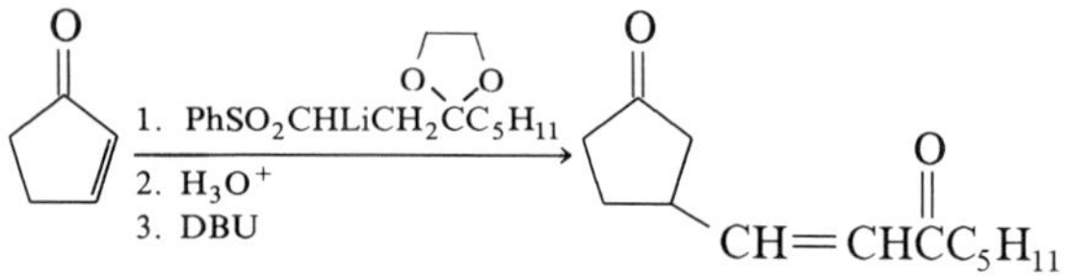

1. PhSO$_2$CHLiCH$_2$CC$_5$H$_{11}$
2. H$_3$O$^+$
3. DBU

CH=CHCC$_5$H$_{11}$

CL 607 (1982)

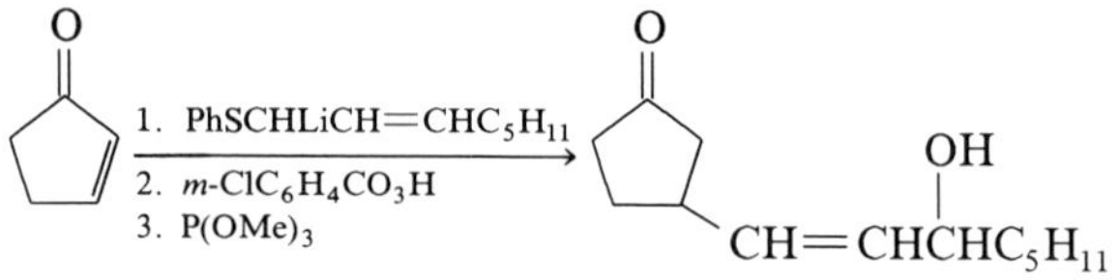

1. PhSCHLiCH=CHC$_5$H$_{11}$
2. *m*-ClC$_6$H$_4$CO$_3$H
3. P(OMe)$_3$

CH=CHCHC$_5$H$_{11}$ (OH)

BCSJ *55* 3043 (1982)

$$+ \text{R}_2\text{CHOH} \xrightarrow{h\nu}$$

CR$_2$OH

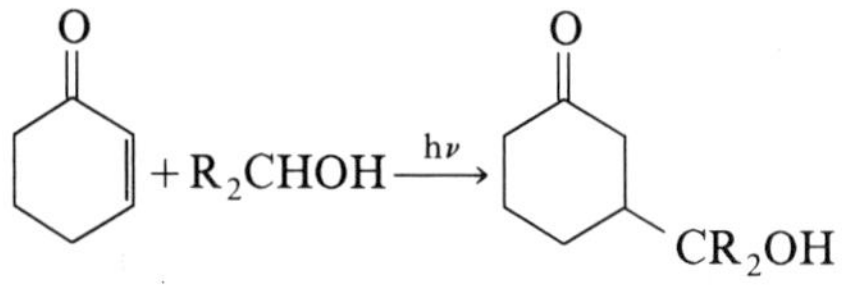

Compt Rend *255* 1817 (1962)
JCS C 2032 (1967)
CC 1286 (1972); 319 (1974)
TL 297 (1974); 1957 (1975)
Can J Chem *55* 3978, 3986 (1977)

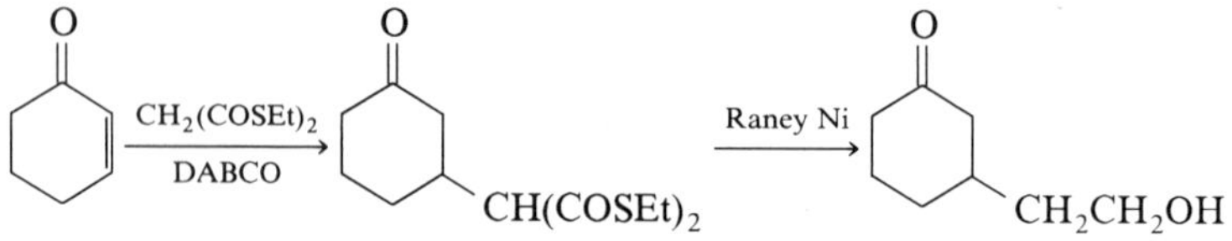

$$\xrightarrow[\text{DABCO}]{\text{CH}_2(\text{COSEt})_2} \qquad \xrightarrow{\text{Raney Ni}}$$

CH(COSEt)$_2$ CH$_2$CH$_2$OH

Can J Chem *60* 94 (1982)

$$\underset{}{>}\text{C=C}-\overset{\text{O}}{\overset{\|}{\text{C}}}- \longrightarrow \text{HC}-\overset{\text{O}}{\overset{\|}{\text{C}}}-\text{C}-\overset{\text{O}}{\overset{\|}{\text{C}}}-$$

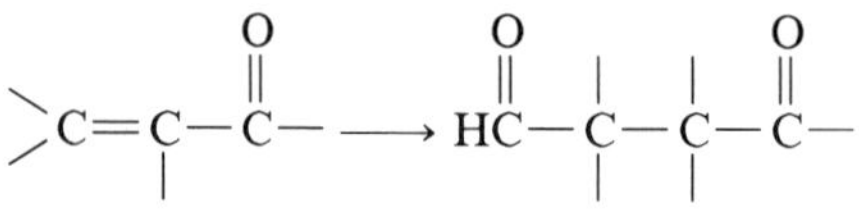

LiC(SPh)$_3$/CrCl$_2$, H$_2$O/hydrolysis   TL 3533 (1978)

ArSCHLiSOAr/I$_2$, KI/hydrolysis   JCS Perkin I 1284 (1981)
                                                  Syn 74 (1981)

TsCHLiSCH$_3$/H$_2$O, hν JOC *51* 508 (1986)

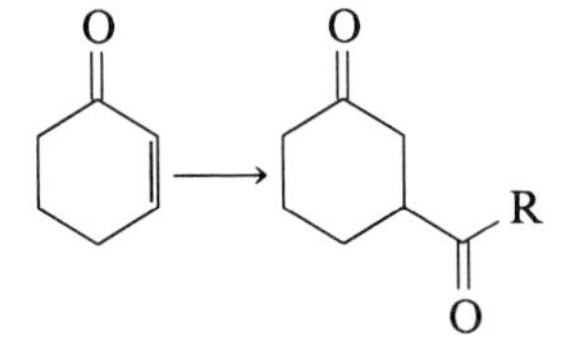

See subsequent sections on organometallic compounds and page 709, Section 9.

$$R_2CHNO_2 + H_2C{=}CHX \xrightarrow{\text{base}} R_2\overset{\displaystyle NO_2}{\underset{|}{C}}CH_2CH_2X$$

$$X = COR, CN, CHO, SOPh, SO_2Ph, CO_2Me$$

Org Syn Coll Vol *4* 652 (1963)
TL 2371 (1978); *23* 2957, 3521 (1982)
CL 1673 (1981) (chiral base)
Angew Int *19* 1013 (1980) (high pressure)
Syn 841 (1982)
JOC *47* 5017 (1982); *50* 3692 (1985); *52* 1601 (1987) (KF-alumina as base)
Syn Commun *12* 339 (1982)
CC 635 (1982)
BCSJ *56* 1885 (1983) (KF-alumina as base)

$$RCCH{=}CHR + R'CH_2NO_2 \xrightarrow[\text{18-crown-6}]{\text{KF}} RCCH_2CHRCHR' \xrightarrow[\text{silica gel}]{\text{KMnO}_4} RCCH_2CHRCR'$$

CC 635 (1982)

$$H_2C{=}CHCCH_3 + R{-}\text{(amidine)} \xrightarrow[\substack{(n\text{-Bu}_4\text{N})\text{BF}_4 \\ \text{Me}_3\text{SiCl}}]{\text{electrolysis}} \text{(product)}$$

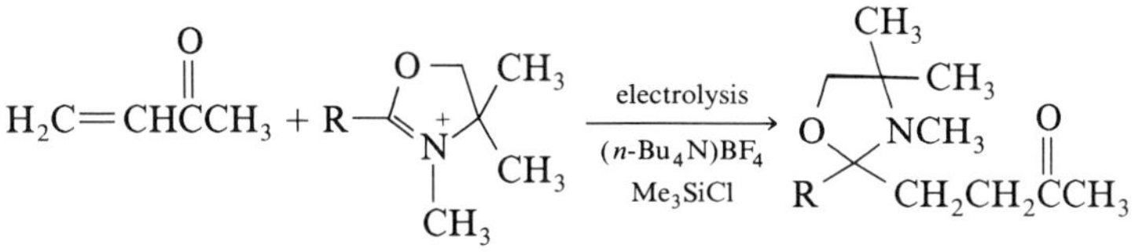

TL *28* 4411 (1987)

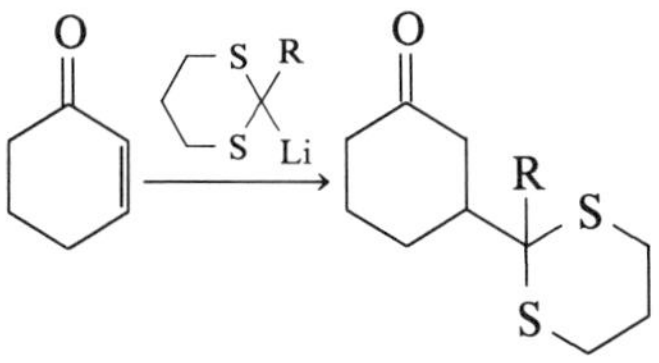

CC 100 (1979); 421 (1982)
Helv *65* 385 (1982)
Tetr *38* 3285 (1982)

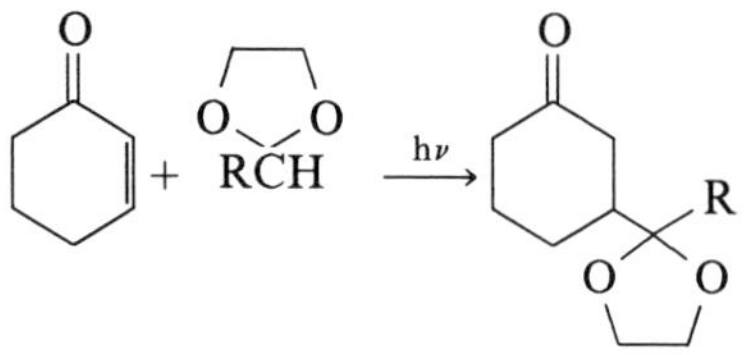

TL 297 (1974)
Can J Chem *55* 3986 (1977)

$$\underset{\substack{\|\\ O}}{RCH} + H_2C\!=\!CHX \longrightarrow \underset{\substack{\|\\ O}}{RCCH_2CH_2X}$$

$$X = COR,\ CN,\ CO_2R,\ SO_2Ph$$

Review: Angew Int *15* 639 (1976)

Reagent(s)

hν (R = Me, *n*-Pr, Ph)

Syn Commun *6* 417 (1976)
Can J Chem *55* 3986 (1977)

cat NaCN (R = aryl, heterocyclic)

TL 1461 (1973)
Angew Int *12* 81 (1973)
Ber *107* 210, 2453 (1974); *109* 534, 541 (1976)
Org Syn *59* 53 (1980)

cat [HOCH$_2$CH$_2$ ... NR′] X⁻ (R′ = Me,

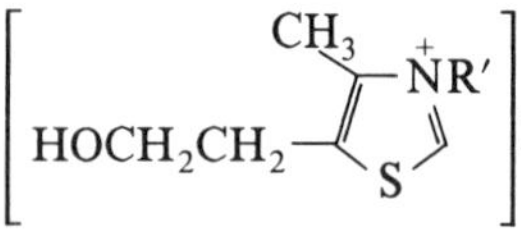

CH$_2$Ph; X = Cl, I), base
(R = alkyl, aryl, heterocyclic)

TL 4505 (1974)
Org Syn *65* 26 (1987)
Angew Int *13* 539 (1974);
    17 131 (1978)
Syn 379 (1975); 129, 626 (1981)
Ber *109* 2890, 3426 (1976); *112* 84, 1410, 2419
    (1979); *113* 690, 979, 1890, 2939 (1980); *114*
    564, 581, 1226, 2479 (1981)
Heterocycles *12* 369 (1979)
Ann 1550 (1981)
CC 1447 (1986)
JOC *51* 2712 (1986); *52* 2213 (1987)

$$(RC)_2O + H_2C\!=\!CHX \xrightarrow[\substack{\text{electrolysis}}]{\substack{h\nu\\ \text{vitamin B}_{12}}} \underset{\substack{\|\\ O}}{RCCH_2CH_2X}$$

$$X = CHO,\ COR,\ CO_2R,\ CN$$

JACS *105* 7200 (1983)

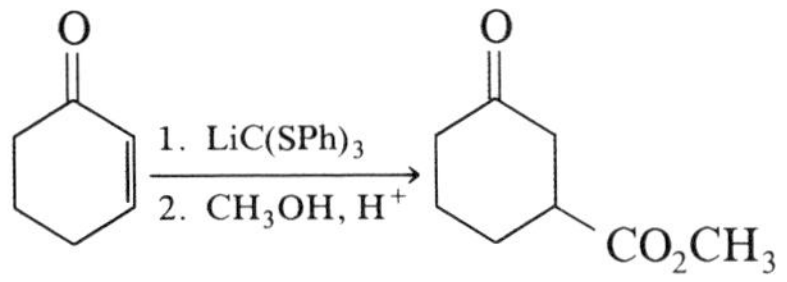

CC 216 (1975)

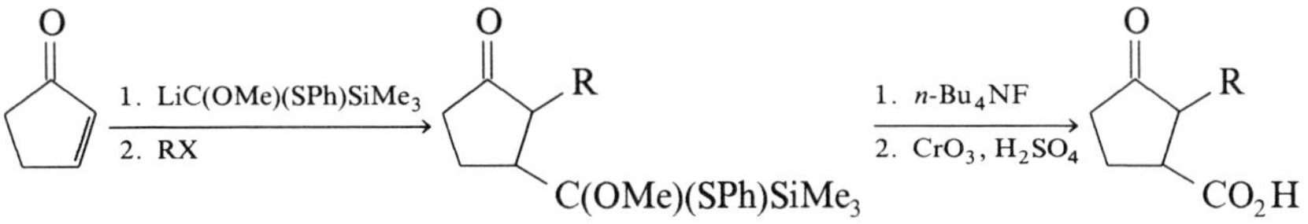

TL *28* 2147 (1987)

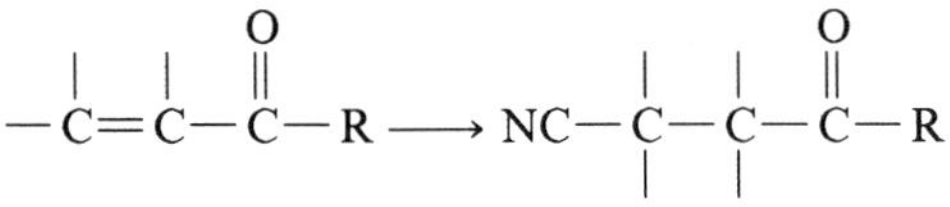

LiCN	TL *28* 4189 (1987)
NaCN, NH$_4$Cl	TL *28* 3061 (1987)
KCN	TL *28* 2537 (1987)
t-BuNC, TiCl$_4$ (R = alkyl)	JACS *104* 6449 (1982)
t-BuNC, EtAlCl$_2$ (R = H)	JACS *104* 6449 (1982)
HCN·py (R = alkyl)	JOC *51* 902 (1986)

See also page 800, Section 4.

3. Organoboron Compounds

Review: Angew Int *11* 692 (1972)

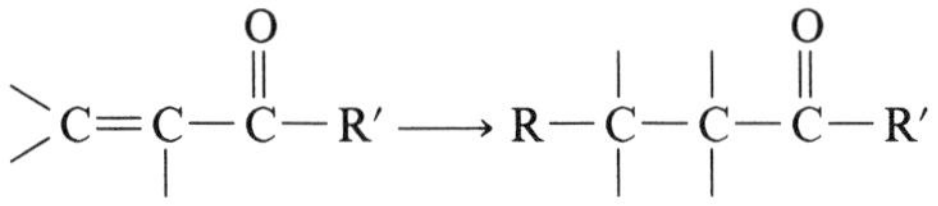

R′	Organoborane	
H	R$_3$B	JACS *89* 5709 (1967); *90* 4165 (1968); *92* 710, 712, 714 (1970)
alkyl	R$_3$B	JACS *89* 5708 (1967); *90* 4166 (1968); *92* 712, 714 (1970)
	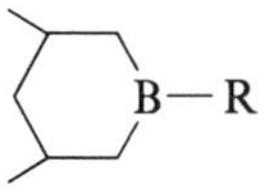 B—R	JACS *93* 3777 (1971); *95* 6757 (1973)
	Ph$_2$BR	JOMC *156* 101 (1978)
	E-RCH=CHB	JACS *98* 7832 (1976)
	Z-XCR=CHB, XB (X = Br, I)	JACS *107* 5225 (1985)
	(RC≡C)$_3$B	TL 2627 (1972); 235 (1976)
	RC≡CB	JACS *99* 954 (1977)

$$[R_3\bar{B}C\equiv CR']Li^+ \xrightarrow[\text{TiCl}_4]{H_2C=CHCOCH_3} \xrightarrow[\text{NaOH}]{H_2O_2} RCCHR'CH_2CH_2CCH_3$$

$$\text{(product with two C=O groups)}$$

CL 221 (1980)

$$R_3B + H_2C=CHCCH_3 \xrightarrow{D_2O} RCH_2CHDCCH_3$$

JACS *89* 5708 (1967); *91* 6195 (1969)

$$R_3B + H_2C=CHCCH_3 \xrightarrow[\text{2. R'I}]{\text{1. 2 }n\text{-BuLi}} RCH_2CHCCH_3 \ (R')$$

JOC *36* 1790 (1971)

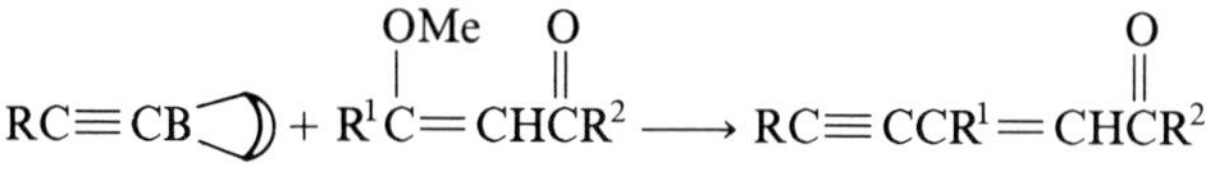

$$RC\equiv CB\!\!\left.\right) + R^1C=CHCR^2 \ (OMe) \longrightarrow RC\equiv CCR^1=CHCR^2$$

JOC *42* 3106 (1977)

$$R_3B + HC\equiv CCCH_3 \xrightarrow[\text{H}_2\text{O}]{O_2} RCH=CHCCH_3$$

JACS *92* 3503 (1970)

4. Organoaluminum and -gallium Compounds

$$\underset{\diagup}{\diagdown}C=C-C-R' \xrightarrow[\text{2. hydrolysis}]{\text{1. reagent(s)}} R-C-C-C-R'$$

R	Reagent(s)	
CH$_3$	(CH$_3$)$_3$Al or LiAl(CH$_3$)$_4$, Ni(acac)$_2$	JOC *39* 3297 (1974) Austral J Chem *28* 801, 817 (1975)
CH$_3$ or Ph	R$_2$AlI (R = Me, Ph)	JOC *44* 4792 (1979)
1°, 2°, 3° alkyl; aryl; vinylic; ester enolate	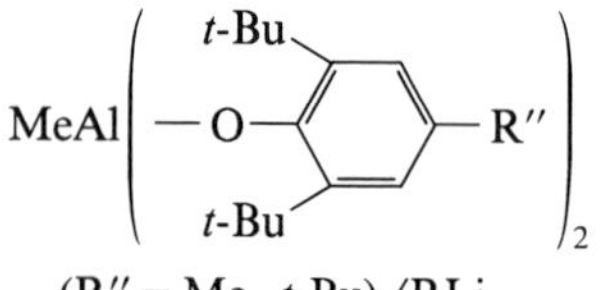 MeAl$\left(-O-\underset{t\text{-Bu}}{\overset{t\text{-Bu}}{\bigcirc}}-R''\right)_2$ (R'' = Me, t-Bu)/RLi	TL *28* 5723 (1987)
E-CH=CHR	[E-RCH=CHAlMe(i-Bu)$_2$]Li	TL 4083 (1972); 765 (1975) JOC *44* 71, 1438 (1979)

RC≡C	$Et_2AlC\equiv CR$	JACS *93* 7320 (1971)
	$Me_2AlC\equiv CR$, Ni catalyst	JACS *100* 2244 (1978)
		JOC *45* 3053 (1980)
	$(RC\equiv C)_3Al$ (Ga)	TL 2627 (1972)
CN	Et_2AlCN or Et_3Al-HCN	JACS *94* 4635, 4644, 4654, 4672 (1972)
		Org Syn *52* 100 (1972)
		TL 5113 (1973); *28* 1439 (1987)
		Org Rxs *25* 255 (1977)
		Syn Commun *8* 231 (1978)
		JOC *51* 902 (1986)

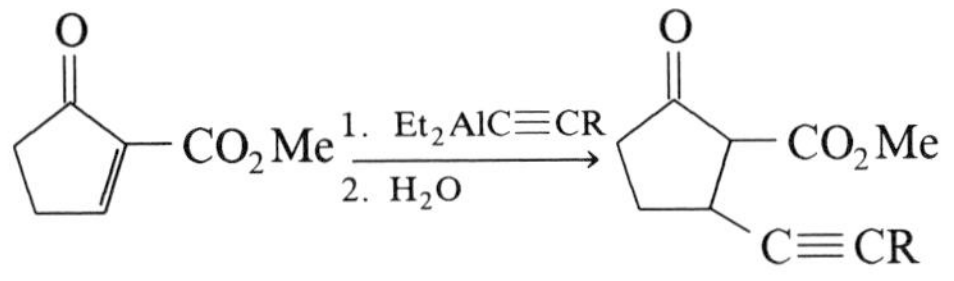

TL *27* 2885 (1986)

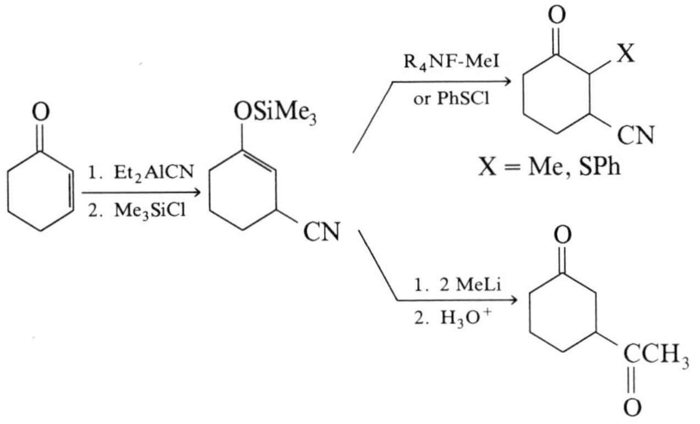

Syn Commun *8* 231 (1978)
See also TL *21* 3389 (1980)

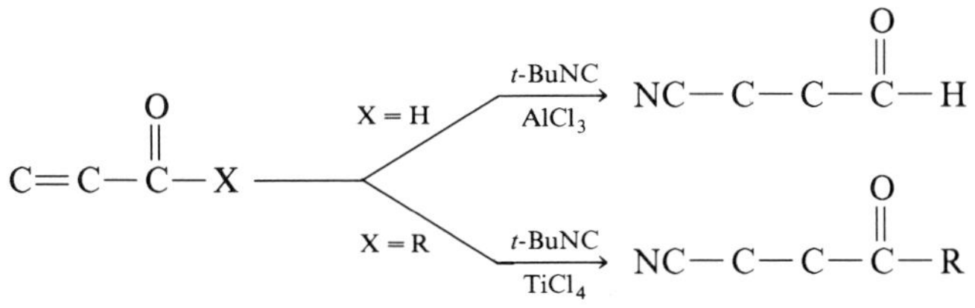

JACS *104* 6449 (1982)

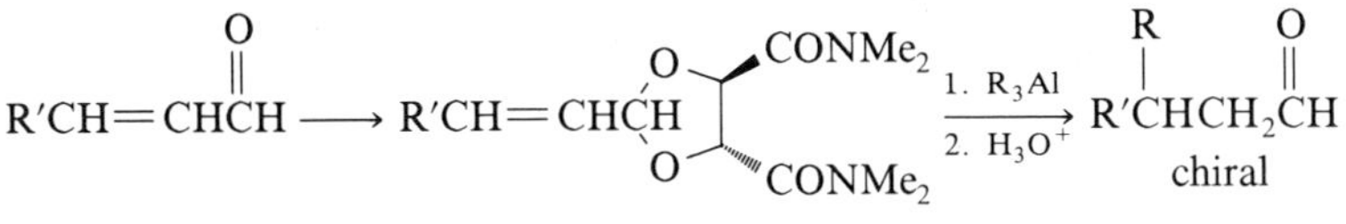

JACS *106* 5004 (1984)

5. Organosilicon Compounds

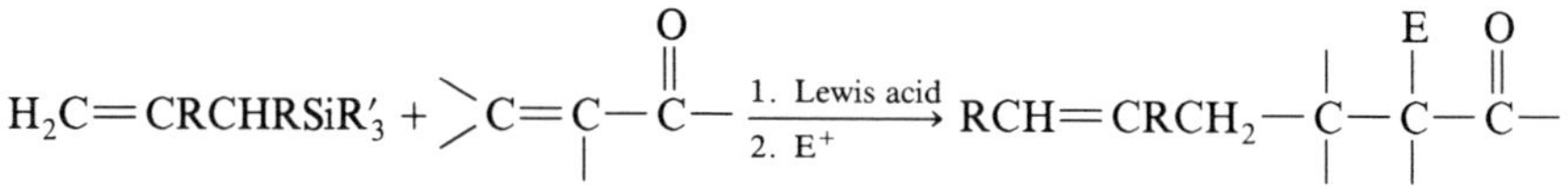

$$H_2C{=}CRCHRSiR'_3 + {>}C{=}C{-}\overset{\overset{O}{\|}}{C}{-} \xrightarrow[\text{2. E}^+]{\text{1. Lewis acid}} RCH{=}CRCH_2{-}\overset{\text{E}}{\underset{|}{C}}{-}\underset{|}{C}{-}\overset{\overset{O}{\|}}{C}{-}$$

$\underline{E^+}$

H_2O

JACS *99* 1673 (1977); *104* 1124 (1982)
    (intramolecular); *106* 721 (1984);
    107 2568 (1985); *108* 3835 (1986)
TL 1385 (1977); 4557 (1979); *21* 955, 4557 (1980);
    22 485 (1981); *28* 1483, 5441, 5793, 6413 (1987)
CC 525 (1979)
Syn 446 (1979)
CL 609, 961 (1982)
Pure Appl Chem *54* 1 (1982) (review)
JOC *49* 4214 (1984)
Org Syn *62* 86 (1984)

$RCHO, R_2CO, RCH(OR)_2, HC(OR)_3, RCOCl$ CL 245 (1979)

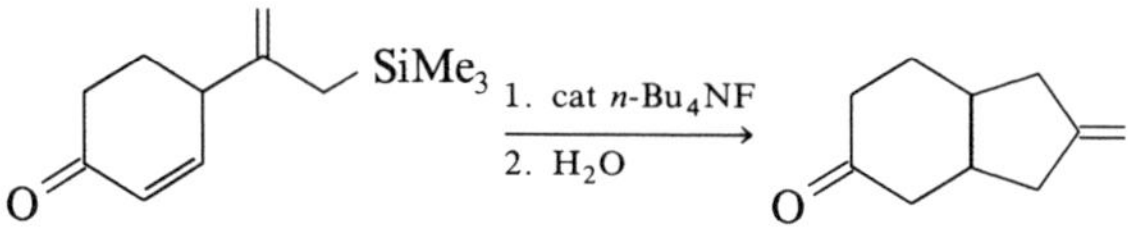

JOC *51* 1753 (1986)

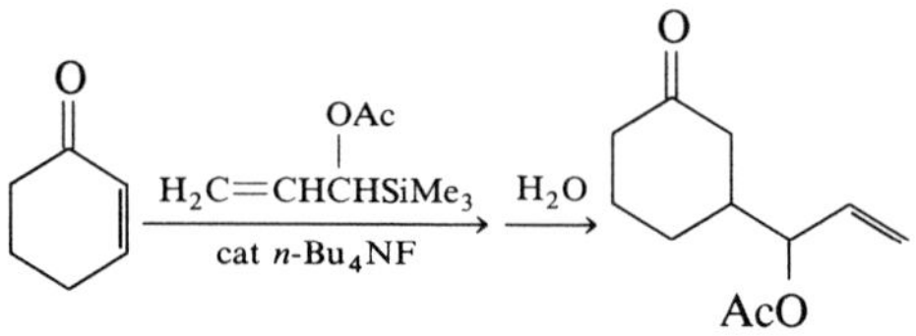

TL *28* 4649 (1987)

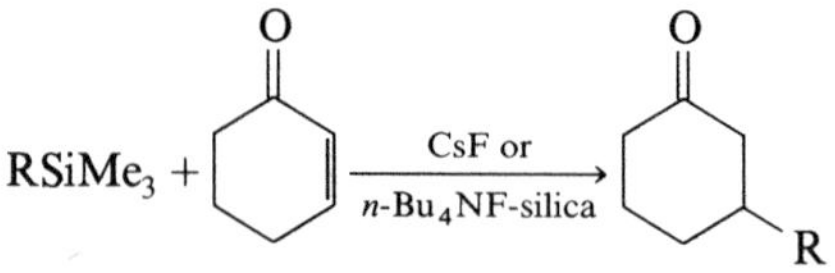

R = allyl, benzyl, heterocyclic

TL *23* 5079 (1982)

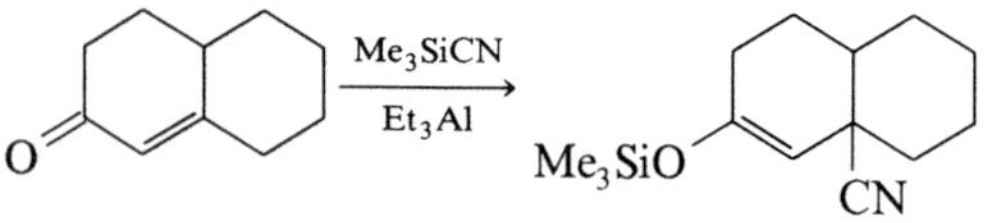

TL *21* 3389 (1980)
Tetr *39* 967 (1983)
See also Syn Commun *8* 231 (1978)

6. Organotin Compounds

See also page 805, Section 13.

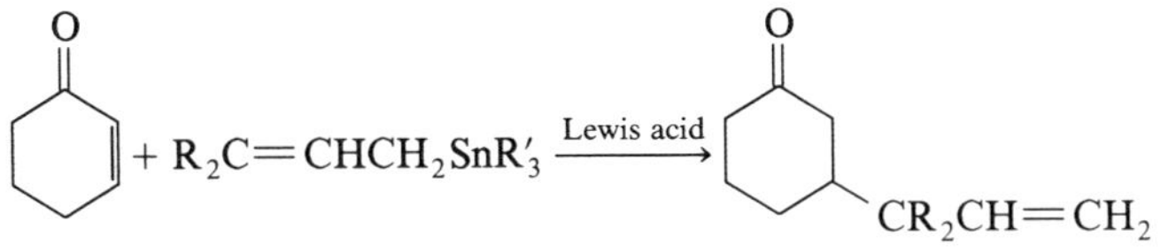

Lewis acid
$BF_3 \cdot OEt_2$	TL *28* 1483 (1987)
$AlCl_3$	TL *28* 1483 (1987)
$(Et_2Al)_2SO_4$	CL 977 (1979)
$TiCl_4$	JACS *102* 2112 (1980) (intramolecular)
	TL *28* 1483 (1987)

7. Sulfur and Selenium Compounds

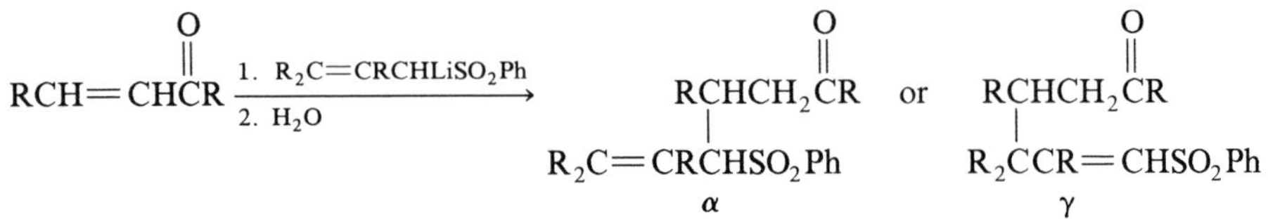

Syn Commun *8* 483 (1978) (α and γ)
TL *22* 1905 (1981) (α)

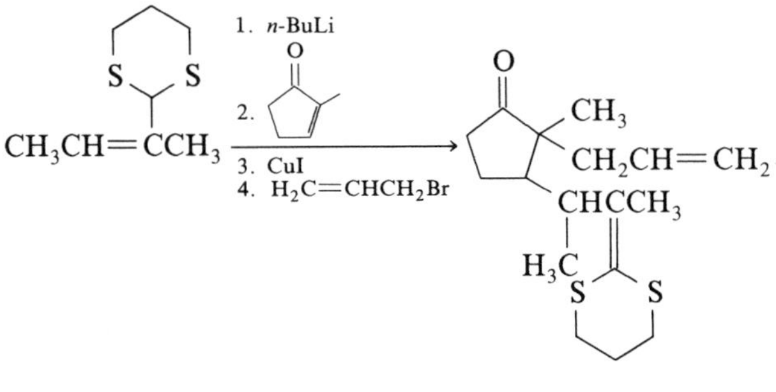

JOC *46* 825 (1981)

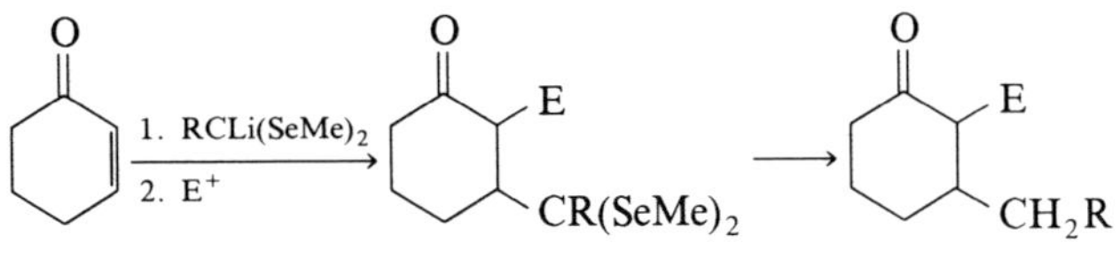

R = H, Me; $E^+ = H^+$, MeI

TL *22* 1623 (1981)

8. Organozirconium Compounds

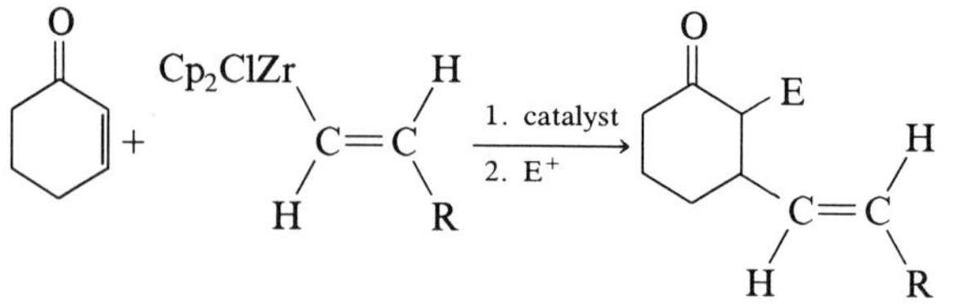

Catalyst	E^+	
CuO_3SCF_3, LiI	H^+	TL 1303 (1977)
$Ni(acac)_2$, $i\text{-}Bu_2AlH$	H^+	JACS *99* 8045 (1977); *102* 1333 (1980); *103* 4466 (1981) TL 4383 (1978)
	CH_2O	TL 4381 (1978)

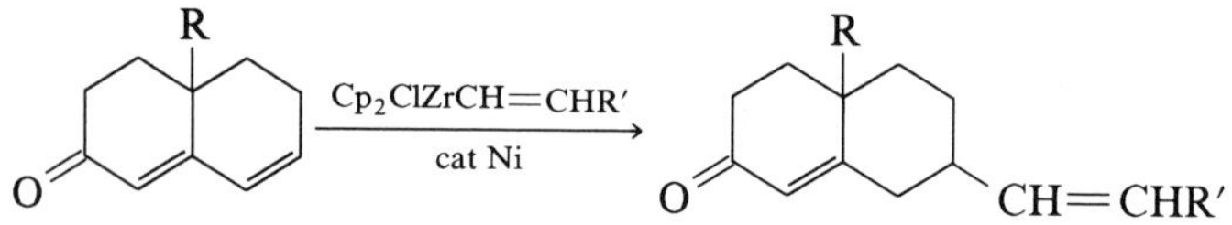

R = H, Me

JACS *103* 4466 (1981)

9. Organomanganese Compounds

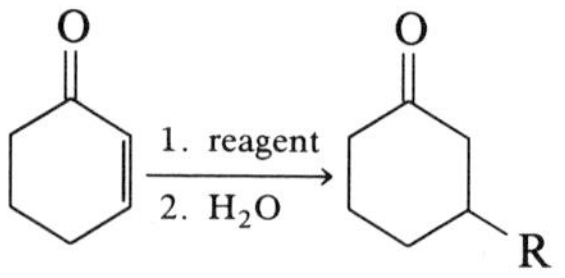

Reagent	
R_2Mn (R = 1° alkyl, aryl)	TL *27* 569 (1986)
$RMnCl$, R_2Mn or R_3MnLi (R = Me, *n*-Bu)	TL *25* 293 (1984)

10. Organoiron Compounds

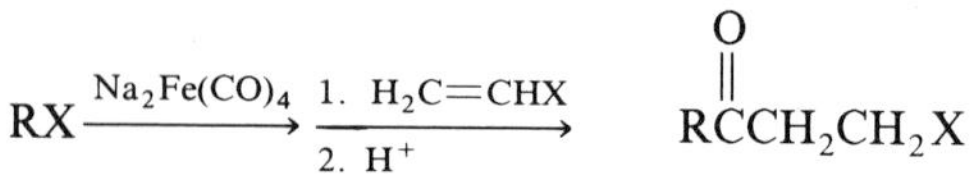

$$RX \xrightarrow{Na_2Fe(CO)_4} \xrightarrow[2.\ H^+]{1.\ H_2C=CHX} RCCH_2CH_2X$$

(with O double-bonded above the carbonyl carbon in $RCCH_2CH_2X$)

X = CO_2R, COR, CN

JACS *99* 5222 (1977)

$$\text{PhCH}{=}\text{CHCR}^1 \longrightarrow \text{Ph} \underset{\underset{\text{CC 226 (1987)}}{\text{Fe(CO)}_3}}{\overset{\overset{\text{R}^1}{\underset{\text{O}}{\big/\!\!\big/}}}{\bigg\langle}} \xrightarrow[\text{2. }t\text{-BuBr}]{\text{1. R}^2\text{Li(MgBr)}} \text{R}^2\text{CCHPhCH}_2\text{CR}^1$$

11. Organocobalt Compounds

$$\text{RLi} + \text{Co(NO)(CO)}_2\text{PPh}_3 \xrightarrow[\text{2. H}_2\text{O}]{\text{1. H}_2\text{C}{=}\text{CHCR}'} \text{RCCH}_2\text{CH}_2\text{CR}'$$

JOC *50* 4955 (1985)

12. Organonickel Compounds

$$\text{RLi} \xrightarrow[\substack{\text{2. H}_2\text{C}{=}\text{CHCOR}' \\ \text{3. E}^+}]{\text{1. Ni(CO)}_4} \text{RCCH}_2\text{CHCR}'$$

$$\underline{\text{E}^+}$$

$$\text{H}_2\text{O}$$

$$\text{H}_2\text{C}{=}\text{CHCH}_2\text{I}$$

JACS *91* 4926 (1969)

JOC *47* 4382 (1982)

13. Organopalladium Compounds

$$\text{RCH}{=}\text{CHCR} \xrightarrow[\text{PdCl}_2 \text{ (phase transfer)}]{\text{PhHgCl or Ph}_4\text{Sn}} \text{RCHCH}_2\text{CR}$$

TL 4591 (1979)

Tetr *37* 2941 (1981)

14. Organocopper Compounds

Reviews:

Org Rxs *19* 1 (1972)

G. H. Posner, "An Introduction to Synthesis Using Organocopper Reagents," J. Wiley, New
York (1980)

Tetr *40* 5005 (1984) (higher order cuprates)

$$\text{C}{=}\text{C}{-}\overset{\overset{\text{O}}{\|}}{\text{C}}{-} \xrightarrow[\text{2. H}_2\text{O}]{\text{1. copper reagent}} \text{R}{-}\text{C}{-}\text{C}{-}\overset{\overset{\text{O}}{\|}}{\text{C}}{-}$$

Copper reagent

RX, Li, CuI or $n\text{-}C_3H_7C\equiv CCu\cdot P(NMe_2)_3$,   JOC *47* 3805 (1982)
  ultrasound

Li, naphthalene/CuI·$(n\text{-}Bu)_3P$,   JOC *52* 5056 (1987)
  $n\text{-}Bu_3P/RBr$

$$\underset{\underset{CH_3}{\underset{|}{}}}{\overset{\overset{HO\ \ \ CH_3}{\overset{|\ \ \ \ |}{}}}{PhCHCHNCH_2CH_2N(CH_3)_2}}/RLi/CH_3I/$$ JACS *108* 7114 (1986)

  CuI/RLi/CH$_3$I (chiral)

RMgX, CuX   JACS *86* 269 (1964); *104* 6879 (1982)
                                                    JOC *33* 305 (1968); *42* 1709 (1977)
                                                    BSCF 568 (1969)
                                                    TL 3361 (1979); *28* 357, 3061 (1987)
                                                    CL 45 (1980) (chiral ligand)
                                                    JOMC *220* 295 (1981)
                                                    Helv *64* 1575, 2489 (1981)

$CH_3MgBr, CuBr/$ (pyrrolidine with CH$_2$OH and NCH$_3$)   CL 45 (1980) (enantioselective)

RMgX, CuBr·SMe$_2$ JACS *106* 1443 (1984)

RMgX, CuI, BF$_3$·OEt$_2$ JOC *51* 5311 (1986)

RMgX, cat CuBr   Org Syn *65* 203 (1987)

$$H_2C=\overset{\overset{CH_2}{\|}}{C}HCMgCl, \text{ cat CuBr, Me}_2S$$ TL *24* 1003 (1983)

RMgX, cat CuBr·SMe$_2$ JOC *47* 5045 (1982)

RMgX, cat CuBr·SMe$_2$, Me$_3$SiCl, (HMPA)   TL *27* 4025 (1986); *28* 5669 (1987)
                                                    JOC *51* 4323 (1986)

Zn(CH$_2$CH$_2$CO$_2$R)$_2$, cat CuBr·SMe$_2$,   JACS *106* 3368 (1984);
  Me$_3$SiCl   *109* 8056 (1987)

Zn(CH$_2$CH$_2$CO$_2$R)$_2$, cat CuBr·SMe$_2$,   JOC *51* 4323 (1986)
  BF$_3$·OEt$_2$, HMPA

RCu   JOC *33* 949 (1968)
                                                    TL *28* 2525 (1987)

RCu·n-Bu$_2$S (R = 2-pyridyl)   Tetr *38* 1509 (1982)

RCu·n-Bu$_3$P   JOC *31* 3128 (1966); *33* 949 (1968); *52* 3346
                                                      (1987) (R = vinyl)
                                                    Tetr *38* 1509 (1982)

RCu·2 n-Bu$_3$P   TL *21* 1247 (1980)

RCu[P(OR$'$)$_3$]$_3$ (R$'$ = Me, n-Bu)   JOC *33* 949 (1968)

RCu, n-Bu$_3$P, BF$_3$·OEt$_2$ TL *28* 1973 (1987)

RCu·BF$_3$	JACS *100* 3240 (1978), *104* 6081 (1982) JOC *47* 119 (1982) TL *28* 2525 (1987)
ROCH$_2$Cu, *i*-Pr$_2$S, BF$_3$·OEt$_2$	JACS *109* 4930 (1987)
RCu·AlCl$_3$	CC 1193 (1980); 703 (1982) (β-cyclopropyl-α,β- enones) TL *21* 4073 (1980)
RCu, Me$_3$SiCl, (HMPA)	TL *27* 4029 (1986); *28* 2525 (1987)
R$_2$CuMgX	JOMC *199* 9 (1980); *228* 321 (1982) (enals)

Me$_2$CuMgX, (structure: pyrrolidine ring with CH$_2$OH and NR)   TL *22* 3601 (1981) (enantioselective)

R$_2$CuMgX·COD	TL *21* 1311 (1980)
R$_2$CuLi	JOC *31* 3128 (1966); *33* 949 (1968); *35* 186 (1970); *41* 3629, 4031 (1976); *44* 4481 (1979); *48* 1404 (1983); *51* 4779 (1986) TL 2875, 3795 (1971); 3361, 3365 (1979); *22* 3585 (1981); *23* 3823 (1982) (chiral R); *28* 2525, 4251, 4943 (1987) Tetr *35* 2645 (1979) (α-haloenones); *36* 2305 (1980) (enals); *37* 1385 (enals), 3981, 4027 (1981); *38* 1509 (1982) CC 643 (1981) (enals) JACS *107* 2149 (1985); *109* 5731 (1987)

Me$_2$CuLi, (structure: pyrrolidine ring with CH$_2$OCH$_3$, NR, *t*-BuS)   TL *24* 3517 (1983) (enantioselective)

[Me$_3$Si(PhS)CH]$_2$CuLi	JOC *51* 3983 (1986)
(ROCH$_2$)$_2$CuLi, *i*-Pr$_2$S	JACS *109* 4930 (1987)
R$_2$CuLi, BF$_3$·OEt$_2$	Tetr *37* 3981 (1981) JACS *103* 194 (1981); *108* 3443 (1986) TL *28* 4061 (1987)
R$_2$CuLi, Me$_3$SiCl, (HMPA)	TL *26* 6015, 6019 (1985); *27* 1047, 4029 (1986)
(ROCH$_2$)$_2$CuLi, *i*-Pr$_2$S, Me$_3$SiCl	JACS *109* 4930 (1987)
(ROCH$_2$)$_2$CuLi, *i*-Pr$_2$S, BF$_3$·OEt$_2$	JACS *109* 4930 (1987)
(H$_2$C=CHCH$_2$)$_2$CuLi	JOC *34* 3615 (1969)
(RCH=CH)$_2$CuLi	Can J Chem *48* 1626 (1970) Helv *54* 1939 (1971) JACS *93* 7318 (1971); *94* 9256 (1972) TL 2455 (1971); 2627 (1972) JOC *42* 1709 (1977)
(H$_2$C=CH)$_2$CuLi, Me$_3$SiCl	JACS *109* 7575 (1987)

$(C_6H_5)_2CuLi$ TL 1579, 1583 (1970); *28* 3065 (1987)

polymer-PPh$_2$·R$_2$CuLi   JOC *44* 2705 (1979)

(MeCuR)Li   TL *21* 3151 (1980)

(PhCuR)Li (R = 2-pyridyl)   Tetr *38* 1509 (1982)

(ArCuR)Li   Tetr *34* 3023 (1978)
                                                      Acta Chem Scand B *32* 483 (1978)

TL *22* 192 (1981)
                                                      JOC *46* 192 (1981) (enals and enones)

(R = aryl)   TL *23* 3823 (1982)

TL *25* 5959 (1984) (R = vinyl); *28* 945 (1987)
                                                      JOMC *285* 437 (1985)

TL *25* 5959 (1984) (R = vinyl)

(n-PrC≡CCuR)Li (or MgX)   JACS *94* 7210 (1972); *106* 8296 (1984)
                                                      JOC *44* 3661 (1979)

(t-BuC≡CCuR)Li   JOC *38* 3893 (1973); *44* 1006 (1979)

[n-C$_5$H$_{11}$C≡CCuCH=CHSn(n-Bu)$_3$]Li   TL *28* 3065 (1987)

(MeOCMe$_2$C≡CCuR)Li   JOC *43* 3418 (1978)

(Me$_2$NCH$_2$C≡CCuR)Li   JOC *44* 1006 (1979)

(Me$_3$SiC≡CCuCH$_2$CH=CROSiMe$_3$)M   TL *25* 5307 (1984)
   (M = Li, MgBr)   JACS *107* 5495 (1985)

(Me$_3$SiC≡CCuCH$_2$CH=CROSiMe$_3$)Li,   TL *27* 4029 (1986)
   Me$_3$SiCl, HMPA

[Me$_3$SiC≡CCuCH=C=CROSiMe$_2$(t-Bu)]Li   TL *28* 1299 (1987)

(RCuCN)Li (or MgX)   CC 88 (1973)
                                                      TL 3879 (1977); *22* 2985 (1981)

(RCuCN)Li, Me$_3$SiCl   TL *26* 6019 (1985) (enals)

(R$_2$CuCN)Li$_2$ TL *23* 3755 (1982); *28* 2525 (1987)
                                                      JOC *49* 3938 (1984); *51* 1293 (1986)

(R$_2$CuCN)Li$_2$, BF$_3$·OEt$_2$ TL *25* 5959 (1984) (R = Ph)
                                                      JOC *52* 4647 (1987)

OR′
|
[(RCH)$_2$Cu(CN)]Li$_2$, Me$_3$SiCl   TL *28* 3911 (1987)

(RCuCH$_2$SO$_2$R′)Li (R′ = Me, Ph)   CC 358 (1982)

(RCuCH$_2$SOCH$_3$)Li   JOC *52* 1885 (1987)

[RCu(CN)CH$_2$SOCH$_3$]Li$_2$ JOC *52* 1885 (1987)

(RCuSR')Li (R' = Ph, chiral R')	JACS *95* 7788 (1973) Syn 662 (1974) Acta Chem Scand B *34* 443 (1978) TL 3849 (1980)
(RCuOR')Li (R' = *t*-Bu, chiral R')	JACS *95* 7788 (1973) TL 1815 (1973); *22* 1329 (1981) Acta Chem Scand B *34* 443 (1978)
(RCuNCy$_2$)Li	JACS *104* 5824 (1982)
(RCuNR'$_2$)Li (chiral amides)	JOC *51* 4953 (1986) JACS *109* 2041 (1987)
(MeCuX)Li (X = chiral alkoxide or amide)	Bull Soc Chim Belg *87* 369 (1978) (enantioselective)
(RCuPPh$_2$)Li	JACS *104* 5824 (1982)
Me$_3$CuLi$_2$	TL 2659 (1976)
Me$_3$CuLi$_2$, Me$_3$Cu$_2$Li, or Me$_5$Cu$_3$Li$_2$	JOC *42* 1099 (1977)
Me$_5$Cu$_3$Li$_2$	CC 643 (1981) (enals) JOC *47* 2572 (1982) (enals)
(PhMe$_2$Si)$_2$CuLi	TL *28* 965 (1987) (enal)

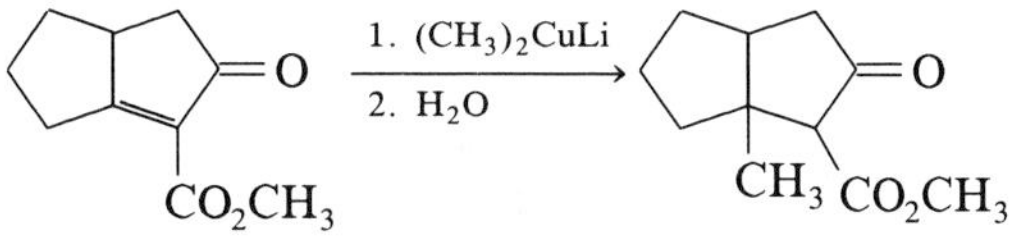

JACS *109* 6199 (1987)

$$RCH{=}CH_2 \xrightarrow[TiCl_4]{LiAlH_4} \xrightarrow[Cu(OAc)_2]{H_2C{=}CHCOR'} R(CH_2)_4\overset{\overset{\displaystyle O}{\|}}{C}R'$$

R' = H, Me

CL 167 (1979)

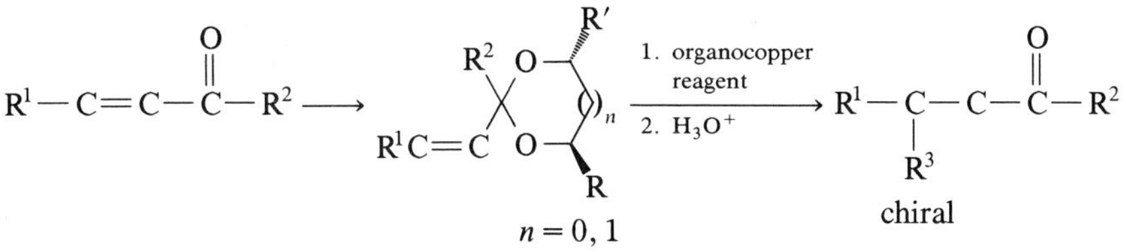

$n = 0, 1$

R^2 = H, alkyl

Organocopper Reagent

PhCu, BF$_3$·OEt$_2$	TL *27* 3143 (1986)
R^{3}Cu, CuBr, *n*-Bu$_3$P (R^3 = aryl, vinylic)	TL *28* 2363 (1987)
Me$_2$CuLi, BF$_3$·OEt$_2$	TL *25* 3083 (1984)

TL *24* 373, 585 (1983)

Review: Syn 364 (1985)

E^+	RM	
R'X	RMgX, cat CuI·n-Bu$_3$P	JACS *103* 4136 (1981); *108* 3435 (1986); *109* 6199 (1987)
	RMgX, CuBr·SMe$_2$	JOC *44* 3731 (1979)
	R$_2$CuLi	JOC *38* 4450 (1973); *29* 275 (1974)
		JACS *95* 6867 (1973); *97* 107 (1975); *101* 938 (1979)
		TL 2591 (1974); 4867 (1976); *27* 5103, 6295 (1986)
		CC 28 (1984)
	[(H$_2$C=CH)$_2$CuCN]Li$_2$	TL *27* 5103 (1986)
	(n-PrC≡CCu**CH**=**CHR**)Li	JCS Perkin I 1407 (1981)
		TL *23* 327 (1982)
R'CHO	RCu, PR$_3$	TL *23* 4057 (1982)
	RMgX, CuI	TL *28* 2489 (1987)
R'CHO (ZnCl$_2$)	R$_2$CuLi, (PR$_3$)	Helv *57* 1317 (1974)
		JACS *96* 7114 (1974); *97* 6260 (1975)
		Tetr *35* 425 (1979)
		JOC *46* 2932 (1981)
R'COCl	R$_2$CuLi	Helv *57* 1317 (1974)
		TL 1535 (1975); *21* 2337 (1980)
		Tetr *33* 1105 (1977)
		JOC *46* 3719 (1981)
		JCS Perkin I 1994, 2394 (1981)
ClCO$_2$Me	RMgX, CuI	JCS Perkin I 1516 (1981)
		TL *23* 703 (1982)
	R$_2$CuLi	JOC *40* 1488 (1975)
	(n-PrC≡CCu**CH**=**CHR**)Li	TL 4087 (1976)
CO$_2$/CH$_2$N$_2$	R$_2$CuLi	Tetr *35* 425 (1979)
		CL 57 (1982)
		JACS *106* 2954 (1984)
H$_2$C=C(SMe)SOMe	R$_2$CuLi	JOC *44* 3755 (1979)

Br$_2$	Ph$_2$CuLi	JACS *101* 934 (1979)
	R$_2$CuLi(MgX)	JOMC *228* 321 (1982) (on enals)
PhSSPh	(n-PrC≡CCuCH=CHR)Li	TL 4091 (1976)
PhSCl	(RCH=CHCuSPh)Li	TL 4091 (1976)
MeSOCl	R$_2$CuLi	CL 1159 (1981)
PhSeBr	RMgX, CuI	Tetr *38* 1959 (1982)
	(CH$_3$)$_2$C=CHMgBr, cat CuI	JOC *50* 2539 (1985)
	R$_2$CuLi	JOC *39* 2133 (1974)
		JACS *97* 5434 (1975)

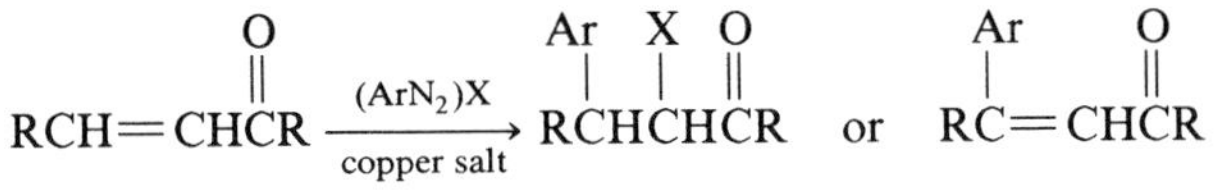

$$RCH=CHCR \xrightarrow[\text{copper salt}]{(ArN_2)X} RCHCHCR \quad or \quad RC=CHCR$$

X = halogen

Org Rxs *11* 189 (1960); *24* 225 (1976)

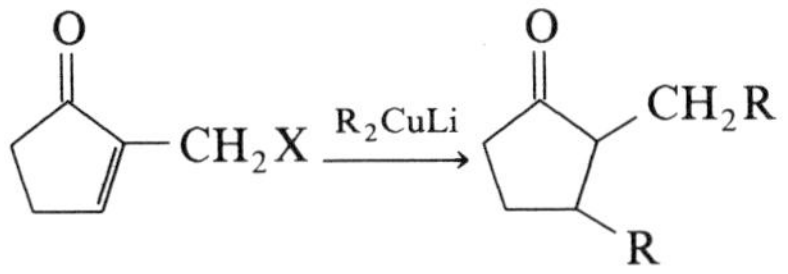

$$\xrightarrow{R_2CuLi}$$

X = Cl, OAc, SPh

TL *21* 3237 (1980)

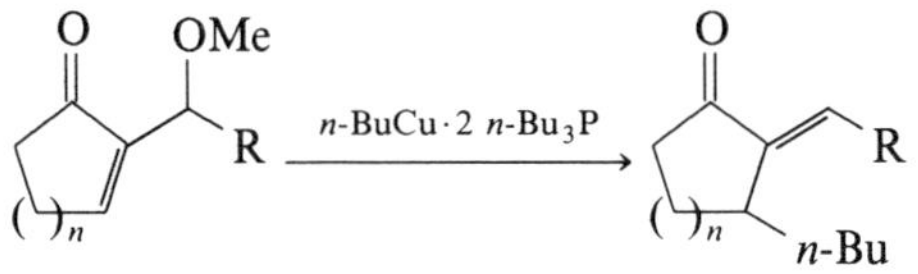

$$\xrightarrow{n\text{-BuCu}\cdot 2\ n\text{-Bu}_3P}$$

n = 1, 2

R = Ph, OMe

TL *22* 1809 (1981)

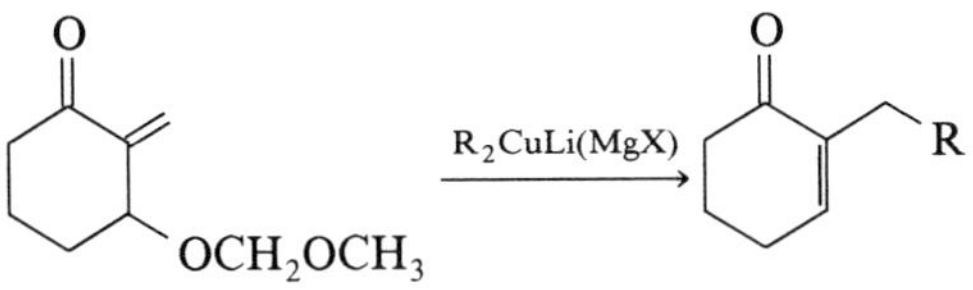

$$\xrightarrow{R_2CuLi(MgX)}$$

TL *22* 119 (1981)
CL 1189 (1981)

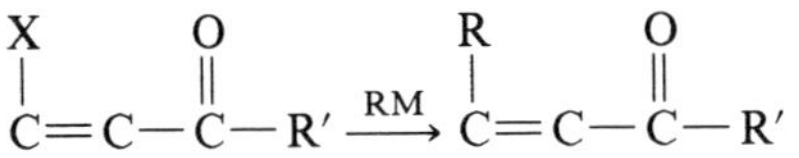

$$\underset{C=C-C-R'}{\overset{X \quad O}{|\quad\ \ ||}} \xrightarrow{RM} \underset{C=C-C-R'}{\overset{R \quad O}{|\quad\ \ ||}}$$

See page 185, Section 1; page 191, Section 2; page 192, Section 3; and page 206, Section 14.

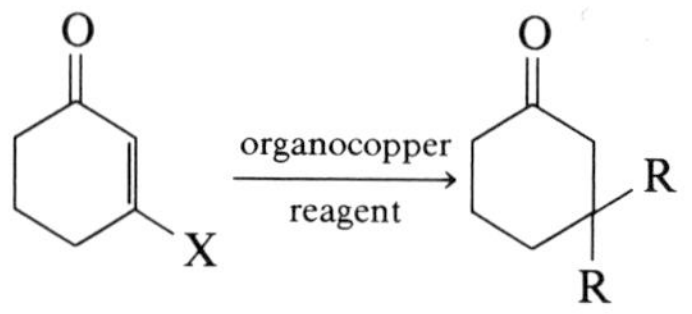

X	Organocopper Reagent	
Cl	R_2CuLi	JOC *41* 636 (1976)
Cl, Br	$[PhSCu(CH_2)_nCuSPh]Li_2$ ($n = 4, 5$)	TL 1245 (1977) (spiro product)
Cl, Br, I	R_2CuLi	JOC *40* 2694 (1975) Can J Chem *60* 1256 (1982)
OMs	R_2CuLi	JOC *46* 197 (1981)
OAc	R_2CuLi	TL 2071 (1973)
SR	R_2CuLi	CC 907 (1973) JOC *52* 110 (1987)

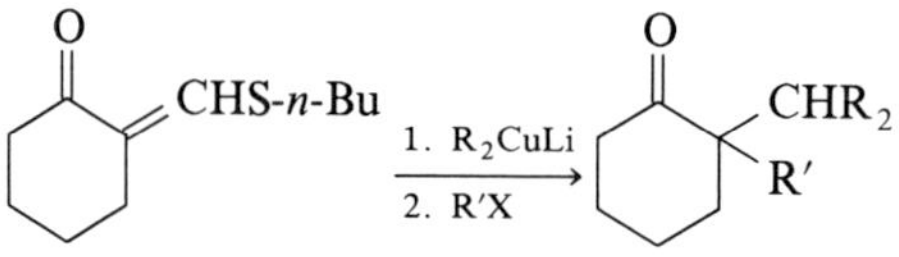

JOC *39* 275 (1974)

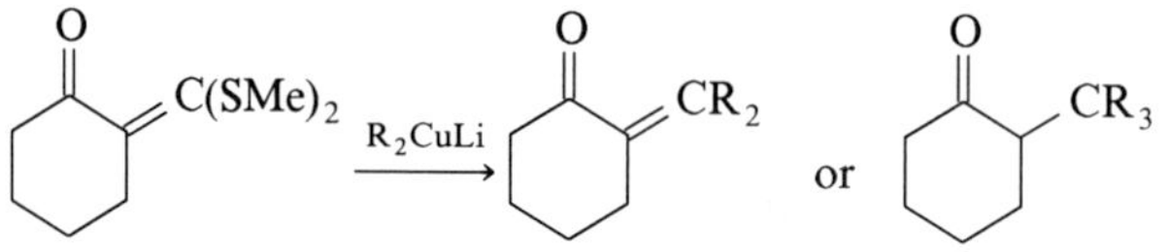

TL 3817 (1973)
Acta Chem Scand B *33* 460 (1979)

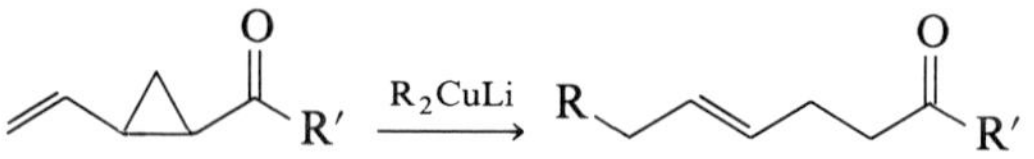

Syn 317 (1975)

$$C{=}C{-}C{=}C{-}\overset{\displaystyle O}{\overset{\|}{C}}{-}R' \longrightarrow R{-}C{-}C{-}C{=}C{-}\overset{\displaystyle O}{\overset{\|}{C}}{-}R'$$

RMgX, cat $CuCl/H_3O^+$	JACS *81* 4069 (1959)
RMgX, cat $CuBr \cdot SMe_2/H_3O^+$	JCS Perkin I 2840 (1979)
RMgX, cat $Cu(OAc)_2 \cdot H_2O/H_3O^+$	Proc Chem Soc 356 (1962) JOC *31* 3109 (1966) JACS *103* 4466 (1981)
R_2CuLi/H_3O^+	TL 3795 (1971)

$$RC\equiv CCH_3 \longrightarrow \underset{R'}{\overset{R}{\diagdown}}C=C\underset{H}{\overset{CCH_3}{\diagup}} \ \ (O)$$

See page 233, Section 2.17.

$$\text{(cyclohexenone)} \xrightarrow[\text{2. } H_3O^+]{\substack{OR' \\ 1. \ RCHLi\text{-}2 \ CuX \ (X=I, CN), \ Me_3SiCl}} \text{(product with R, OR')}$$

TL *27* 4553 (1986)

$$C=C-\overset{\overset{O}{\|}}{C}- \longrightarrow R\overset{\overset{O}{\|}}{C}-C-\overset{\overset{O}{\|}}{C}-$$

$(R_2CuCN)Li_2$, CO	JACS *107* 4551 (1985)
(RCuCN)Li, CO	TL *27* 1473 (1986)
n-Bu$_3$SnCH(OEt)$_2$, n-BuLi/CuI/H$_3$O$^+$	JOMC *212* C31 (1981)
$[RC(SPh)_2]_2$CuLi/H$_2$O, CuCl$_2$	JACS *94* 8641 (1972)
(H$_2$C=COR')$_2$CuLi/H$_3$O$^+$ (R' = Me, Et; R = Me)	CC 519 (1975) JACS *97* 3822 (1975) JOC *44* 4781 (1979)
(R'CH=CSiMe$_3$)$_2$CuLi/(HOCH$_2$)$_2$, TsOH/ m-ClC$_6$H$_4$CO$_3$H/aq HClO$_4$ or BF$_3$·OEt$_2$-CH$_3$OH (R = R'CH$_2$)	TL 3365 (1974) JOC *44* 4781 (1979)

15. Organozinc Compounds

See also page 805, Section 14.

$$ArX \xrightarrow[\substack{2. \ RCH=CHCHO, \ cat \ Ni(acac)_2 \\ 3. \ H_2O}]{1. \ Li, \ ZnBr_2, \ ultrasound} ArCHRCH_2CHO$$

TL *26* 829 (1985)

$$C=C-\overset{\overset{O}{\|}}{C}-R' \xrightarrow[\text{2. } H_3O^+]{\text{1. RM}} R-C-C-\overset{\overset{O}{\|}}{C}-R'$$

RM	
Et$_2$Zn	Polymer J *9* 595 (1977)
Ph$_2$Zn	JACS *63* 2046 (1941)
R$_2$Zn, cat Ni(acac)$_2$	JOC *50* 5761 (1985) JACS *109* 4752 (1987)

$$C = C - \overset{\overset{\displaystyle O}{\|}}{C} - R' \xrightarrow[\text{2. } H_3O^+]{\text{1. RM}} R - C - C - \overset{\overset{\displaystyle O}{\|}}{C} - R' \quad (\textit{Continued})$$

RM

(EtO$_2$CCH$_2$CH$_2$)$_2$Zn, Me$_3$SiCl,   JACS *106* 3368 (1984)
  cat CuBr·SMe$_2$

R$_3$ZnLi   CL 679 (1977)
                                               Helv *62* 1710 (1979)

Me$_2$RZnLi   TL *27* 1437 (1986)

3 RMgX, ZnCl$_2$·TMEDA   JOC *51* 3993 (1986)

JOC *50* 5761 (1985)

16. Organomercury Compounds

See also page 805, Section 13, and page 67, Section 2.16.

$$RHgX + H_2C = CHCR' \xrightarrow[\text{NaHB(OMe)}_3]{\text{NaBH}_4 \text{ or}} RCH_2CH_2CR'$$

Angew Int *20* 965 (1981)
JOC *47* 2231 (1982) (intramolecular)
TL *23* 931 (1982)
Syn 735 (1982)
Ber *117* 2132 (1984)
Tetr *41* 4025 (1985)
Ann 427 (1987)

17. Other Reagents

$$(ArN_2)Cl + RCH = CHCR \xrightarrow[\text{2. } H_2O]{\text{1. TiCl}_3} RCHCH_2CR$$

Syn 291 (1980)

$$RCH = CHCH \longrightarrow RCH = CHCH = NCH\text{-}t\text{-Bu} \xrightarrow[\substack{\text{2. } E^+ \\ \text{3. } H_3O^+}]{\text{1. R'M}} RCHCHCH$$

chiral   chiral

$$E^+ = H^+, RX$$

Review:

  K. Tomioka and K. Koga, "Asymmetric Synthesis," Ed. J. D. Morrison, Academic Press, New
  York (1983), Vol 2, Chpt 7

R′M	
R′MgX	Chem Pharm Bull *27* 771 (1979) TL 3009 (1979); *21* 4005 (1980) Tetr *37* 3951 (1981)
$KCH(CO_2Et)_2$	Chem Pharm Bull *27* 2437 (1979)

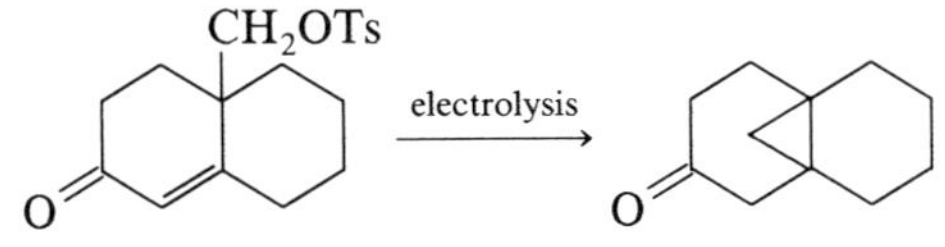
CH2OTs
electrolysis
O
O

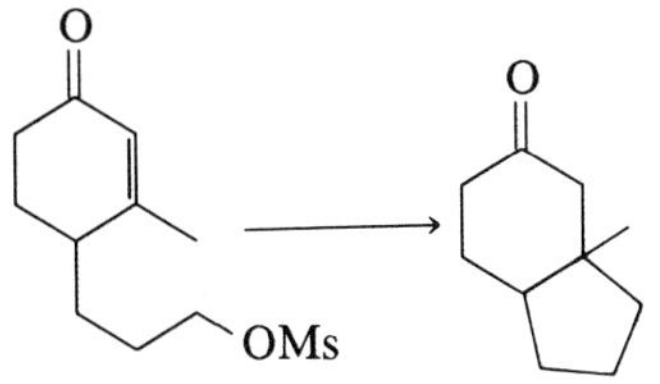
O
OMs
O

JOC *46* 5455 (1981)

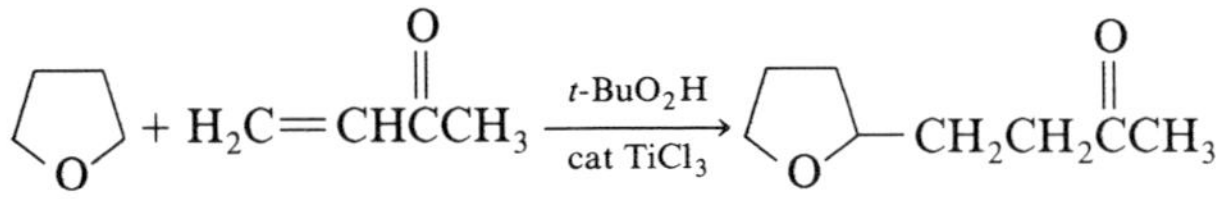
O
+ H2C=CHCCH3
t-BuO2H
cat TiCl3
O
CH2CH2CCH3

Tetr *38* 393 (1982)

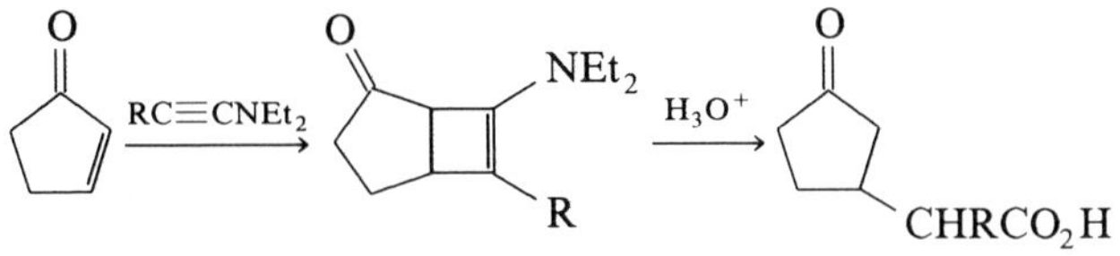
O
RC≡CNEt2
O
NEt2
R
H3O+
O
CHRCO2H

TL *23* 1821 (1982)

16. MISCELLANEOUS REACTIONS

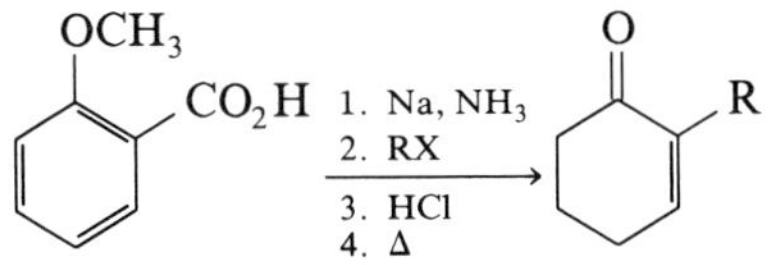

Org Syn *61* 59 (1983)

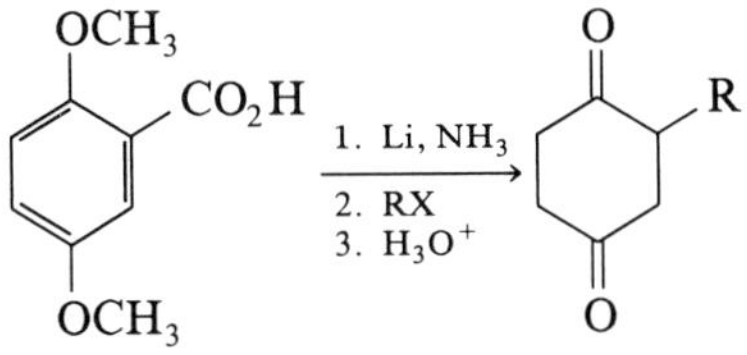

Syn 374 (1979)
JOC *45* 1722 (1980); *52* 2297 (1987)

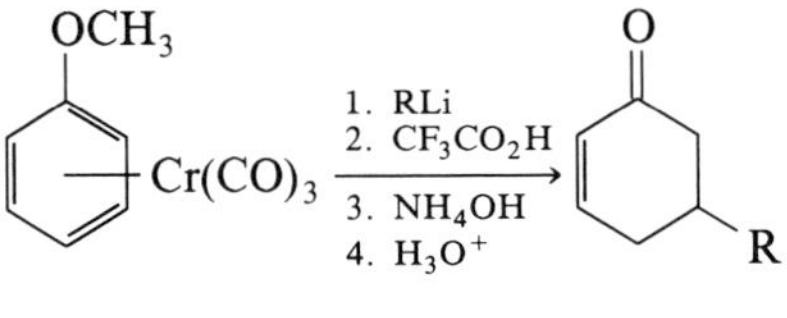

JOC *44* 3275 (1979)

$$\mathrm{RCHO + (CH_3CO)_2O \xrightarrow{\text{cat } CoCl_2} R\overset{\displaystyle O}{\overset{\|}{C}}{-}\overset{\displaystyle O}{\overset{\|}{C}}CH_3}$$

CC 692 (1987)

$$\mathrm{ArN_2^+ + CH_3\overset{\displaystyle O}{\overset{\|}{C}}{-}\overset{\displaystyle O}{\overset{\|}{C}}CH_3 \xrightarrow[\text{Cu or } FeSO_4]{h\nu,\ \Delta,} Ar\overset{\displaystyle O}{\overset{\|}{C}}CH_3}$$

TL *23* 1831 (1982)

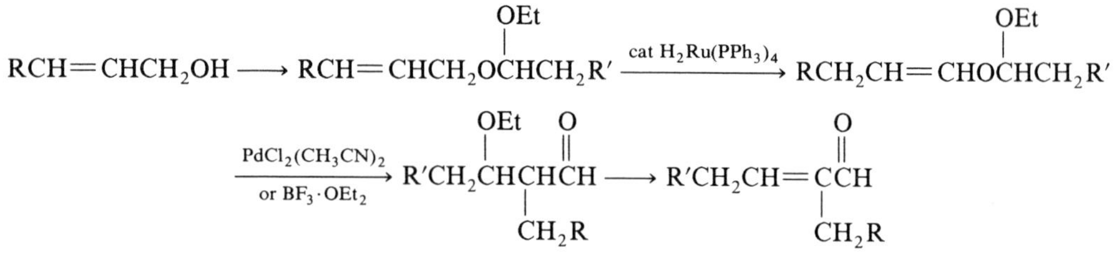

$$RCH{=}CHCH_2OH \longrightarrow RCH{=}CHCH_2O\overset{\overset{\displaystyle OEt}{|}}{C}HCH_2R' \xrightarrow{\text{cat } H_2Ru(PPh_3)_4} RCH_2CH{=}CHO\overset{\overset{\displaystyle OEt}{|}}{C}HCH_2R'$$

$$\xrightarrow[\text{or } BF_3 \cdot OEt_2]{PdCl_2(CH_3CN)_2} \quad R'CH_2\underset{\underset{\displaystyle CH_2R}{|}}{\overset{\overset{\displaystyle OEt}{|}}{C}}H\overset{\overset{\displaystyle O}{\|}}{C}H \longrightarrow R'CH_2CH{=}\underset{\underset{\displaystyle CH_2R}{|}}{\overset{\overset{\displaystyle O}{\|}}{C}}CH$$

CL 1361, 1435 (1981)

$$ArX + H_2C{=}CH\overset{\overset{\displaystyle OH}{|}}{C}HR \xrightarrow{\text{Pd catalyst}} ArCH_2CH_2\overset{\overset{\displaystyle O}{\|}}{C}R$$

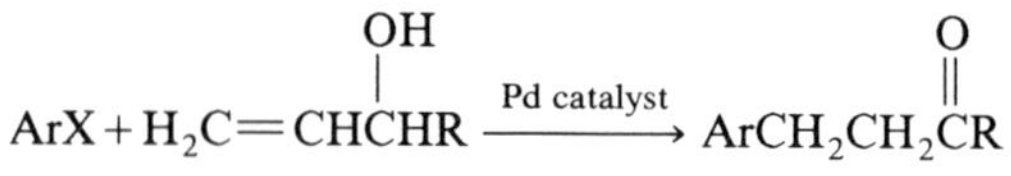

JOC *41* 265, 273, 1206 (1976)
Tetr *35* 329 (1979)
Syn Commun *11* 579 (1981)
TL *22* 2479 (1981)

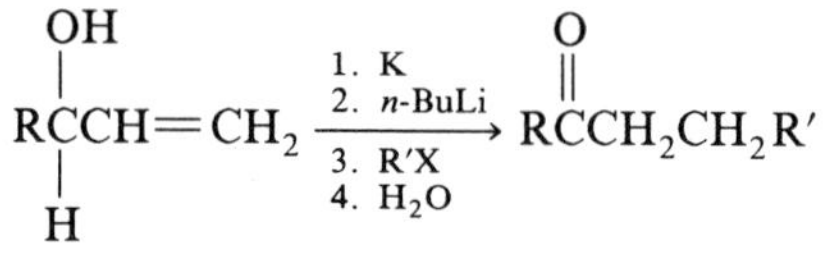

$$R\underset{\underset{\displaystyle H}{|}}{\overset{\overset{\displaystyle OH}{|}}{C}}CH{=}CH_2 \xrightarrow[\substack{3.\ R'X \\ 4.\ H_2O}]{\substack{1.\ K \\ 2.\ n\text{-BuLi}}} R\overset{\overset{\displaystyle O}{\|}}{C}CH_2CH_2R'$$

TL *28* 2587 (1987)

NITRILES, CARBOXYLIC ACIDS AND DERIVATIVES

GENERAL REFERENCES

"The Chemistry of Carboxylic Acids and Esters," Ed. S. Patai, Interscience, New York (1969)

"The Chemistry of Acyl Halides," Ed. S. Patai, Interscience, New York (1972)

"The Chemistry of Acid Derivatives," Ed. S. Patai, J. Wiley, New York (1979)

Houben-Weyl, "Methoden der Organischen Chemie," 4th ed, Vol E4 (acid derivatives), G. Thieme, Stuttgart-New York (1983)

Houben-Weyl, "Methoden der Organischen Chemie," 4th ed, Vol E5 (acids and derivatives), Part 1, G. Thieme, Stuttgart-New York (1985)

Houben-Weyl, "Methoden der Organischen Chemie," 4th ed, Vol E5 (acids and derivatives), Part 2, G. Thieme, Stuttgart-New York (1985)

1. OXIDATION, SUBSTITUTION AND ADDITION

1. Alkanes

$$ArCH_3 \longrightarrow ArCO_2H$$

O_2, KO-t-Bu, DMF	JOC *28* 410 (1963)
NaOCl, cat $RuCl_3 \cdot 3\ H_2O$, cat n-Bu_4NBr	JOC *51* 2880 (1986)
HNO_3	Org Syn Coll Vol *3* 820, 822 (1955)
PbO_2, KOH	Org Syn Coll Vol *5* 617 (1973)
CrO_3, HOAc	JACS *78* 1689 (1956)
$Na_2Cr_2O_7$, H_2SO_4	Org Syn Coll Vol *1* 392, 543 (1941) JOC *23* 1236 (1958)
$Na_2Cr_2O_7$/HCl	Org Syn Coll Vol *5* 810 (1973)
$KMnO_4$	Org Syn Coll Vol *1* 159 (1941); *2* 135 (1943); *3* 740 (1955)
(n-Bu_4N)MnO_4, py	CC 253 (1978)

$$ArCH_3 \longrightarrow ArCH_2OAc$$

$Na_2S_2O_8$, $Cu(OAc)_2$	Syn 477 (1980)
$K_2S_2O_8$, $Cu(OAc)_2$	JCS Perkin I 669 (1979)
$Pb(OAc)_4$	JCS 3943 (1954) JACS *91* 138 (1969) Syn 567 (1973) (review)
$Mn(OAc)_3$	JACS *91* 138 (1969)
$Co(OAc)_3$	JACS *91* 6830 (1969)
$Pd(OAc)_2$	Syn 567 (1973) (review)
$Hg(OAc)_2$	JOC *50* 3070 (1985)

Ce(NH$_4$)$_2$(NO$_3$)$_6$, HOAc   JOC *31* 2033 (1966); *42* 3682 (1977)

DDQ, HOAc   TL *28* 5403 (1987)

$$ArCH_3 \xrightarrow{\text{CrO}_2(\text{OAc})_2} ArCH(OAc)_2$$

Syn 567 (1973) (review)

$$R_3CH \longrightarrow R_3CO_2CR'$$

Pb(OAc)$_4$, LiCl, CF$_3$CO$_2$H   JCS Perkin I 2576 (1976)
                                       JOC *50* 2759 (1985)

CrO$_3$, Ac$_2$O, HOAc   JOC *50* 2759 (1985)

Pd(O$_2$CCF$_3$)$_2$ JACS *109* 8109 (1987)

$$R_3CH \xrightarrow[\text{CH}_3\text{CN}]{\text{Br}_2} \xrightarrow{\text{H}_2\text{O}} R_3CNHCCH_3 \overset{\overset{\displaystyle O}{\|}}{}$$

Syn 632 (1977)

$$\bigcirc + N_2CHCO_2Et \xrightarrow{\text{cat Rh}_2(\text{O}_2\text{CCF}_3)_4} \bigcirc\!\!-CH_2CO_2Et$$

CC 688 (1981)

2. Arenes

$$ArH \longrightarrow ArO_2CR$$

(RCO$_2$)$_2$; O$_2$, I$_2$, CuCl$_2$ Syn 1 (1972) (review)
  or Lewis acid

Pb(OAc)$_4$ Syn 567 (1973) (review)

K$_2$S$_2$O$_8$, cat Pd(O$_2$CCF$_3$)$_2$ JACS *109* 8109 (1987)

$$RC_6H_5 \longrightarrow RCO_2H$$

O$_3$/H$_2$O$_2$ JOC *42* 1254 (1977)
                                       JACS *98* 122 (1976); *108* 2343 (1986);
                                         109 5524 (1987)

NaIO$_4$, cat RuCl$_3$·H$_2$O   JOC *46* 3936 (1981)

3. Alkenes

For alkenol → lactone see page 941, Section 8.
For the addition of HCN to alkenes see page 860, Section 6.
See also page 325, Section 4; and page 832, Section 9.

$$RCH=CHCH_2R \longrightarrow RCH=CH\overset{\overset{\displaystyle O_2CR}{|}}{C}HR$$

$t\text{-}C_4H_8O_3CC_6H_5$, Cu(I) or Cu(II)

JACS *80* 756 (1958); *81* 5819 (1959); *84* 774, 4969 (1962); *86* 3753 (1964); *108* 8230 (1986)
Org Syn *48* 18 (1968)
Syn 1 (1972) (review)
TL *28* 1561 (1987)

cat $PdCl_2$, $CuCl_2$, HOAc,
  NaOAc, (PPh_3)

Angew Int *21* 366 (1982)

$Pb(OAc)_4$

Ann *481* 263 (1930)
JOC *29* 3353 (1964)
Syn 567 (1973) (review)

$Hg(OAc)_2$

JCS 2381 (1951)
JOC *29* 3353 (1964)
TL 3719 (1965); 4203 (1966); 3483 (1970)
JACS *94* 2320 (1972)
Syn 567 (1973) (review)
R. C. Larock, "Organomercury Compounds in Organic Synthesis," Springer, New York (1986), pp 190–196

$Mn(OAc)_3$, cat KBr

JCS C 2355 (1971)

$$RCH=CHCH_2CO_2R \xrightarrow[\textit{n-}AmONO,\ HOAc]{KOAc,\ PdCl_2} R\overset{\overset{\displaystyle OAc}{|}}{C}HCH=CHCO_2R$$

TL *22* 131 (1981)

$$RCH=CHCH_2R' \xrightarrow[\text{2. } H_2O_2]{\text{1. PhSeBr, HOAc}} R\overset{\overset{\displaystyle OAc}{|}}{C}HCH=CHR'$$

JACS *106* 1446 (1984)

$$RCH_2CH=CH_2 \longrightarrow R\overset{\overset{\displaystyle HNTs}{|}}{C}HCH=CH_2$$

TsN=S=NTs   JOC *41* 176 (1976)

TsN=Se=NTs   JACS *98* 269 (1976)

$$CH_3CH=C(CH_3)_2 \xrightarrow[\text{2. KOH, MeOH}]{\text{1. } MeO_2CN=S=NCO_2Me} CH_3CH=C\overset{\overset{\displaystyle CH_3}{|}}{C}H_2NHCO_2Me$$

JOC *48* 3561 (1983)
Org Syn *65* 159 (1987)

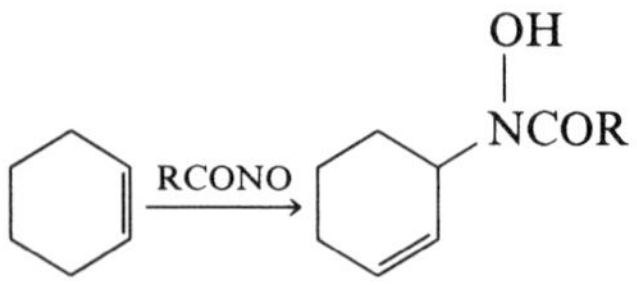

Tetr *37* 4007 (1981)

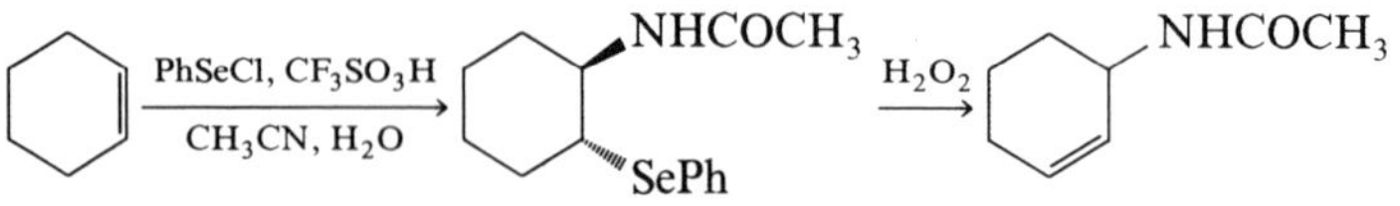

CC 546 (1981)
JOC *46* 4727 (1981)

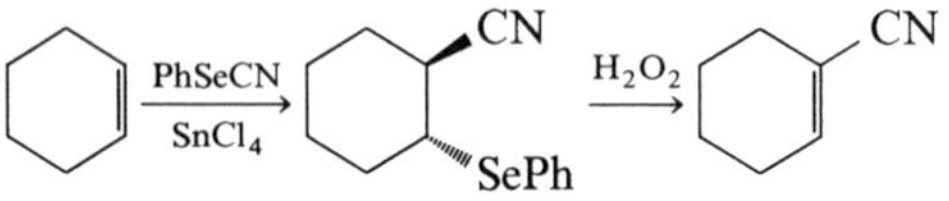

CC 871 (1982)

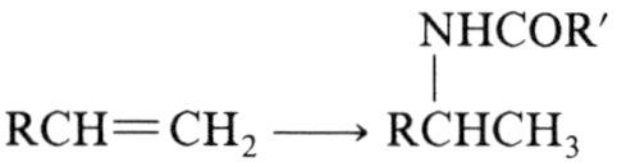

$$\text{RCH}=\text{CH}_2 \longrightarrow \overset{\overset{\textstyle \text{NHCOR}'}{\textstyle |}}{\text{RCHCH}_3}$$

H_2SO_4, $R'CN/H_2O$

JACS *70* 4045 (1948)
Org Rxs *17* 213 (1969) (review)

$Hg(NO_3)_2$, $R'CN/NaBH_4$, NaOH

JACS *91* 5647 (1969)
R. C. Larock, "Solvomercuration/Demercuration
  Reactions in Organic Synthesis," Springer
  Verlag, New York (1986), Chpt 7 (review)

$Hg(NO_3)_2$, $R'CONH_2/NaBH_4$, NaOH,
n-BuNH$_2$ (R' = Me, Ph, NH$_2$, OEt)

CC 670 (1981)
R. C. Larock, "Solvomercuration/Demercuration
  Reactions in Organic Synthesis," Springer
  Verlag, New York (1986), Chpt 7 (review)

$$\text{RCH}=\text{CH}_2 \xrightarrow[\text{Hg(NO}_3)_2]{\text{TsNH}_2} \xrightarrow[\text{NaOH}]{\text{NaBH}_4} \overset{\overset{\textstyle \text{NHTs}}{\textstyle |}}{\text{RCHCH}_3}$$

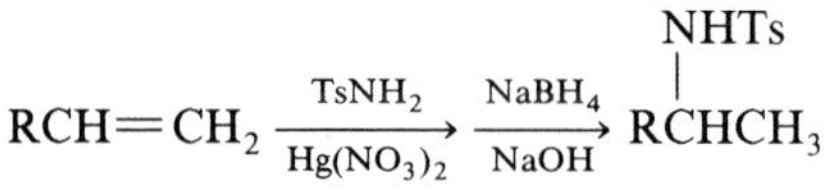

CC 1178 (1981)
R. C. Larock, "Solvomercuration/Demercuration Reactions in Organic Synthesis," Springer Verlag,
  New York (1986), Chpt 7 (review)

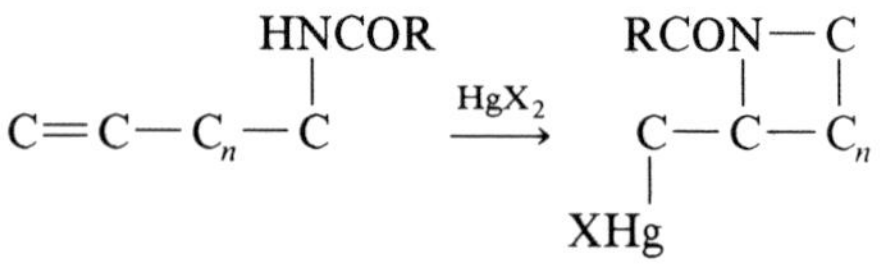

R. C. Larock, "Solvomercuration/Demercuration Reactions in Organic Synthesis," Springer Verlag,
  New York (1986), Chpt 7 (review)
JOC *52* 4717 (1987)

$$\underset{\overset{|}{C}}{C}=C-C_n-\overset{\overset{\displaystyle HNCOX}{|}}{\underset{|}{C}} \quad \xrightarrow{\text{PhSeY}} \quad \underset{\overset{|}{PhSe}}{\overset{\overset{\displaystyle O}{\overset{\|}{XCN-C}}}{C}}-\overset{|}{C}-C_n$$

X	Y	
R	Cl, Br	JOC *51* 1724 (1986)
OR	Cl	CC 379 (1978)
		JOC *45* 2120 (1980)

JACS *101* 3704 (1979)
Tetr *37* 4097 (1981)
TL *24* 1357 (1983)

$$C=C-C_n-\overset{\overset{\displaystyle OR}{|}}{C}-N=\overset{|}{C}-CH_3 \quad \xrightarrow[\text{X = Cl, Br}]{\text{PhSeX}} \quad \underset{\overset{|}{PhSe}}{\overset{\overset{\displaystyle O}{\overset{\|}{CH_3CN-C}}}{C}}-\overset{|}{C}-C_n$$

JCS Perkin I 1837 (1986)

$$RCH=CH_2 \quad \xrightarrow[\substack{\text{2. NaNClTs} \\ \text{3. NaOH}}]{\text{1. BH}_3} \quad RCH_2CH_2NHTs$$

TL 181 (1978)

$$RCH=CH_2 \quad \xrightarrow[\substack{\text{2. } p\text{-NO}_2C_6H_4SO_3NHCO_2Et, \\ (PhCH_2NEt_3)Cl, NaHCO_3, \\ H_2O}]{\text{1. BH}_3} \quad RCH_2CH_2NHCO_2Et$$

Syn Commun *11* 475 (1981)

$$RCH=CH_2 \longrightarrow RCH_2CH_2OAc$$

$BH_3/Hg(OAc)_2/I_2$	JOC *39* 834 (1974)
$BH_3/Pb(OAc)_4$ or $PhI(OAc)_2$	BCSJ *51* 901 (1978)
$BH_3/HOAc$, NaOAc, electrolysis	CL 1021 (1974)
$LiAlH_4$, $TiCl_4/Pb(OAc)_4$	TL 1405 (1979)

$$\underset{}{\overset{}{>}}C=C\overset{}{\underset{}{<}} \quad \xrightarrow[\text{HOAc, H}_2\text{O}]{\text{I}_2, \text{AgOAc}} \quad \overset{\overset{\displaystyle HO}{|}}{-\underset{|}{C}}-\overset{\overset{\displaystyle OAc}{|}}{\underset{|}{C}}-$$

JACS *109* 6403 (1987)

$$\diagup C = C \diagdown \xrightarrow[\text{I}_2]{\text{AgO}_2\text{CR}} \begin{array}{c} \text{RCO}_2 \\ | \\ -\text{C}-\text{C}- \\ | \\ \text{O}_2\text{CR} \end{array}$$

Org Rxs *9* 332 (1957) (review)
Adv Org Chem *1* 117 (1960) (review)
JCS C 1327 (1966)
JOC *50* 3070 (1985); *52* 2226, 5574, (1987)

$$\text{RCH}=\text{CH}_2 \xrightarrow[\text{2. CrO}_3]{\text{1. BH}_3} \text{RCH}_2\text{CO}_2\text{H}$$

CC 122 (1968)

$$\text{RCH}=\text{CHR} \longrightarrow 2\,\text{RCO}_2\text{H}$$

O_3/H_2O_2	Chem Rev *27* 437 (1940); *58* 925 (1958)
$O_3/K_2Cr_2O_7, H_2SO_4$	JACS *108* 4603 (1986)
O_3/H_2, Lindlar catalyst	Bull Acad Sci USSR, Div Chem Sci *25* 1790 (1976)
	JOC USSR *14* 48 (1978)
NaOCl, cat $RuCl_3$	CC 1420 (1970)
	JOC *46* 3936 (1981)
	JACS *103* 464 (1981)
	TL *28* 3061, 5441, 6331 (1987)
$NaIO_4$, cat RuO_4	JACS *85* 3419 (1963)
$NaIO_4$, cat $KMnO_4$	Can J Chem *33* 1701 (1955); *34* 1413 (1956)
	JACS *107* 7967 (1985)
$NaMnO_4$, *t*-BuOH, H_2O	JOC *26* 3734 (1961)
$KMnO_4$, H_2O	Org Syn Coll Vol *2* 53 (1943)
$KMnO_4$, Na_2CO_3, *t*-BuOH, acetone	JACS *108* 468 (1986)
$KMnO_4$, KNO_3, K_2CO_3	JOC *26* 3734 (1961)
$KMnO_4$, R_4NCl (phase transfer)	TL 1511 (1974)
	JOC *42* 3749 (1977)
	Org Syn *60* 11 (1981)
$KMnO_4$, dimethyl polyethylene glycol	JOC *43* 1532 (1978)
(*n*-Bu$_4$N)MnO$_4$, py	CC 253 (1978)

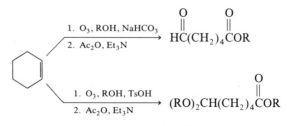

$$\begin{array}{c} \text{O} \qquad\qquad \text{O} \\ \parallel \qquad\qquad \parallel \\ \text{HC(CH}_2)_4\text{COR} \end{array}$$

1. O_3, ROH, NaHCO$_3$
2. Ac$_2$O, Et$_3$N

1. O_3, ROH, TsOH
2. Ac$_2$O, Et$_3$N

$$\begin{array}{c} \text{O} \\ \parallel \\ \text{(RO)}_2\text{CH(CH}_2)_4\text{COR} \end{array}$$

TL *23* 3867 (1982)

$$R^1R^2C=\overset{\overset{\displaystyle OSiMe_3}{|}}{CR^3} \xrightarrow[MoO_2(acac)_2]{t\text{-}BuO_2H} R^1R^2CO + R^3CO_2H$$

$$R^1CH=\overset{\overset{\displaystyle OSiMe_3}{|}}{CR^2} \xrightarrow{} R^1CO_2H + R^2CO_2H$$

TL *22* 2595 (1981)

$$RCH=CHR \xrightarrow[2.\ R'OH,\ HCl]{1.\ O_3} 2\ RCO_2R'$$

Angew Int *17* 939 (1978)

4. Dienes

$$C=C-C=C \longrightarrow X-C-C=C-C-O_2CR$$

X = halogen

See page 325, Section 4.

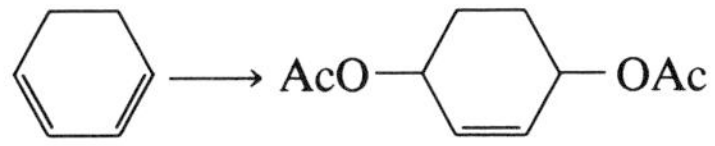

Stereochemistry	Reagents	
cis	HOAc, LiOAc, LiCl, cat Pd(OAc)$_2$, benzoquinone	JACS *103* 4959 (1981)
	HOAc, LiOAc, LiCl, MnO$_2$, cat Pd(OAc)$_2$, cat benzoquinone	JOC *49* 4619 (1984)
	HOAc, LiOAc, LiClO$_4$, LiCl, cat Pd(OAc)$_2$, cat hydroquinone, electrolysis	CC 1236 (1987)
trans	HOAc, LiOAc, cat Pd(OAc)$_2$, benzoquinone	JACS *103* 4959 (1981)
	HOAc, LiOAc, MnO$_2$, cat Pd(OAc)$_2$, cat benzoquinone	JOC *49* 4619 (1984)
	HOAc, LiOAc, LiClO$_4$, cat Pd(OAc)$_2$, cat hydroquinone, electrolysis	CC 1236 (1987)
	HOAc, O$_2$, cat Pd(OAc)$_2$, cat hydroquinone, cat Co(II) *meso*-tetraphenylporphyrin	JACS *109* 4750 (1987)

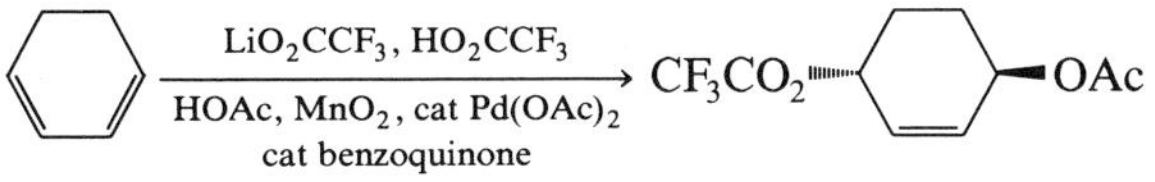

TL *25* 2717 (1984)

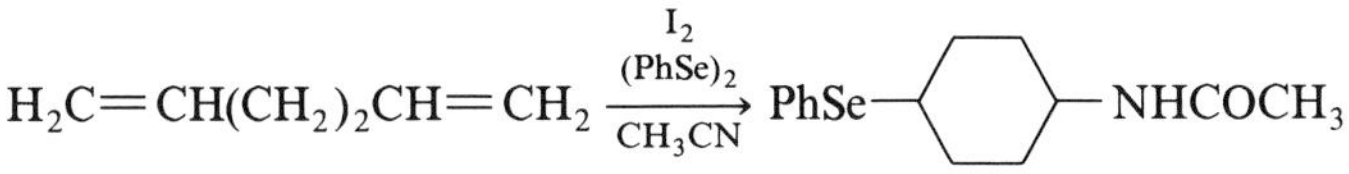

CC 87 (1982)

5. Alkynes

$$RC\equiv CH \longrightarrow RCO_2H$$

$Tl(NO_3)_3$	JACS *93* 7331 (1971); *95* 1296 (1973)
$NaIO_4$ or NaOCl, cat RuO_2	TL 2941 (1971)
PhIO, cat $RuCl_2(PPh_3)_3$	Helv *64* 2531 (1981)
$KMnO_4$, HOAc, H_2O, Aliquat 336	JOC *42* 3749 (1977)
$KMnO_4$, HOAc, H_2O, Adogen-464	Syn 462 (1978)

$$RC\equiv CR \longrightarrow 2\,RCO_2H$$

O_3	Chem Rev *27* 437 (1940) (review)
$KMnO_4$, H_2O	JOC *44* 2726 (1979)

$$RC\equiv CH \longrightarrow RCH_2CO_2H$$

$BH_3/m\text{-}ClC_6H_4CO_3H$	JACS *89* 291 (1967)
$n\text{-BuLi}/Me_3SiCl/\left(\langle\bigcirc\rangle-\right)_2 BH/$ H_2O_2, NaOH	JACS *99* 3184 (1977); *106* 6006 (1984)

$$RC\equiv CR' \xrightarrow[CH_3OH]{PhI(OH)OTs} RR'CHCO_2CH_3$$

TL *28* 2845 (1987)

$$RC\equiv CSiMe_3 \xrightarrow[R'OH]{\substack{cat\ OsO_4 \\ t\text{-BuO}_2H}} RC\overset{O}{\overset{\|}{-}}CO\overset{O}{\overset{\|}{}}R'$$

$R' = H$ or alkyl

TL *27* 1947 (1986)

$$RC\equiv CX \xrightarrow[cat\ RuCl_2(PPh_3)_3]{PhIO} RC\overset{O}{\overset{\|}{-}}CX\overset{O}{\overset{\|}{}}$$

$X = OR, NR_2$

TL *23* 3661 (1982)

$$RC\equiv CH \longrightarrow \underset{H}{\overset{R}{>}}C=C\underset{HgCl}{\overset{H}{<}} \xrightarrow[Hg(OAc)_2]{cat\ Pd(OAc)_2} \underset{H}{\overset{R}{>}}C=C\underset{OAc}{\overset{H}{<}}$$

JACS *102* 1966 (1980)

$$RC\equiv CH + HO_2CR' \longrightarrow RC\overset{O_2CR'}{\overset{|}{=}}CH_2$$

Review: JACS *102* 1966 (1980)

Reagent(s)

cat $[Ru(CO)_2(OAc)]_n$ Organomet *2* 1689 (1983)

cat $Ru_3(CO)_{12}$ Organomet *2* 1689 (1983)

cat $RuCl_3 \cdot 3\ H_2O$, (PR_3) TL *27* 6323 (1986)

cat $RuCl_2(PMe_3)(p\text{-cymene})$ TL *27* 6323 (1986)

cat bis(η^5-cyclooctadienyl)ruthenium,   JOC *50* 1566 (1985); *52* 2230 (1987)
  cat PR_3, (cat maleic anhydride)   TL *27* 2125, 5389 (1986)

cat $PdCl_2(PhCN)_2$, Et_3N TL *25* 5323 (1984) (intramolecular)

cat $Hg(OAc)_2$ JACS *77* 939 (1955); *103* 5459 (1981)
  (intramolecular)

cat $HgO\text{-}BF_3 \cdot OEt_2\text{-MeOH}$ JACS *56* 1802 (1934)

For intramolecular reactions see page 941, Section 8.

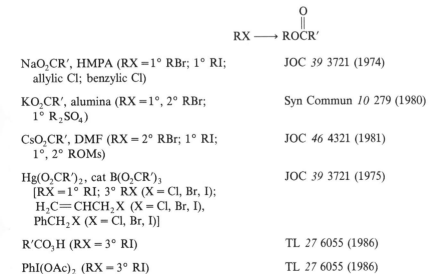

$$RC{\equiv}CH + Ac_2O \xrightarrow[\text{cat } BF_3 \cdot OEt_2]{\text{cat } Hg(OAc)_2} \overset{\overset{\displaystyle OAc}{|}}{RC}{=}CH_2$$

JOC *38* 4254 (1973)

$$RC{\equiv}CR \xrightarrow[\text{2. Zn}]{\text{1. } Hg(OAc)_2,\ HOAc} \underset{R}{\overset{H}{>}}C{=}C\underset{OAc}{\overset{R}{<}}$$

JOC *47* 3707 (1982)

6. Alkyl Halides

See also page 966, Section 3.

$$RX \longrightarrow RO\overset{\overset{\displaystyle O}{\|}}{C}R'$$

NaO_2CR', HMPA (RX $=1°$ RBr; $1°$ RI;   JOC *39* 3721 (1974)
  allylic Cl; benzylic Cl)

KO_2CR', alumina (RX $=1°, 2°$ RBr;   Syn Commun *10* 279 (1980)
  $1°\ R_2SO_4$)

CsO_2CR', DMF (RX $= 2°$ RBr; $1°$ RI;   JOC *46* 4321 (1981)
  $1°, 2°$ ROMs)

$Hg(O_2CR')_2$, cat $B(O_2CR')_3$ JOC *39* 3721 (1975)
  [RX $=1°$ RI; $3°$ RX (X $=$ Cl, Br, I);
  $H_2C{=}CHCH_2X$ (X $=$ Cl, Br, I),
  $PhCH_2X$ (X $=$ Cl, Br, I)]

$R'CO_3H$ (RX $= 3°$ RI)   TL *27* 6055 (1986)

$PhI(OAc)_2$ (RX $= 3°$ RI)   TL *27* 6055 (1986)

$$R_2C{=}CHI \xrightarrow{Cu(OAc)_2} R_2C{=}CHOAc$$

JOMC *93* 415 (1975)

$$R_3CX \xrightarrow[CH_3CN]{(NO_2)BF_4 \quad H_2O} R_3CNH\overset{\overset{\displaystyle O}{\|}}{C}CH_3$$

JOC *45* 165 (1980)

7. Amines

$$RCH_2NH_2 \longrightarrow RCH_2O\overset{\overset{\displaystyle O}{\|}}{C}R'$$

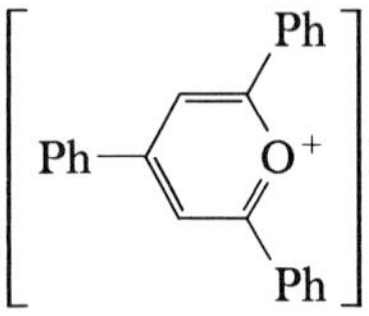

$$\left[\begin{array}{c} Ph \\ Ph{-}\underset{\underset{\displaystyle Ph}{}}{\bigcirc}{}^+ \end{array} \right] BF_4^-/R'CO_2Na \qquad\qquad CC\ 701\ (1977)$$

R'COCl/N$_2$O$_4$/Δ   Org Syn Coll Vol *5* 336 (1973)

8. Nitro Compounds

$$RCH_2NO_2 \xrightarrow[\text{2. MoO}_5\cdot py\cdot HMPA]{\text{1. LDA}} RCO_2H$$

TL *22* 5235 (1981)

9. Ethers

$$RCH_2OR \xrightarrow[Cu(I)\ or\ Cu(II)]{t\text{-BuO}_3CR'} RCHOR \overset{\displaystyle O_2CR'}{\underset{\displaystyle |}{}}$$

Syn 1 (1972) (review)

$$(RCH_2)_2O \longrightarrow R\overset{\overset{\displaystyle O}{\|}}{C}OCH_2R$$

CrO$_3$, HOAc   Helv *42* 1124 (1959)
                                 CC 752 (1966) (Me > RCH$_2$)
                                 Carbohydr Res *12* 147 (1970)
                                 JACS *107* 5289 (1985)

H$_2$CrO$_4$ JCS 221, 227 (1959)

(PhCH$_2$NEt$_3$)MnO$_4$ Angew Int *18* 69 (1979)

Zn(MnO$_4$)$_2$, silica gel   JACS *105* 7755 (1983)

RuO$_4$ JACS *80* 6682 (1958)
                                     JOC *28* 2729 (1963)
                                     TL *28* 435 (1987)

NaIO$_4$, cat RuO$_2$ Syn Commun *10* 205 (1980)
                                     JACS *102* 3904 (1980); *107* 1308 (1985)
                                     JOC *52* 83 (1987)

NaIO$_4$, cat RuCl$_3$·H$_2$O   JOC *46* 3936 (1981); *51* 1015 (1986)

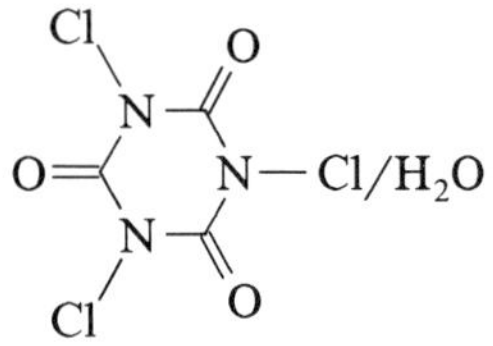

TL 5819 (1968)

Tetr *27* 2671 (1971)

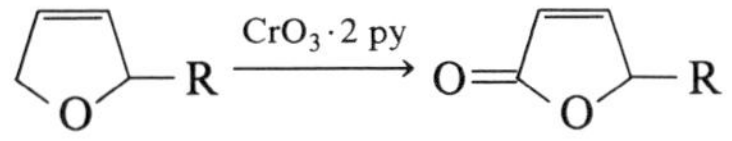

TL *28* 5819 (1987)

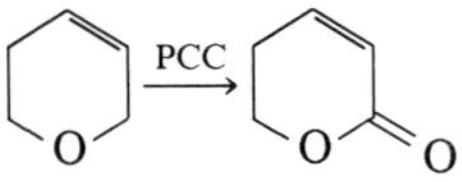

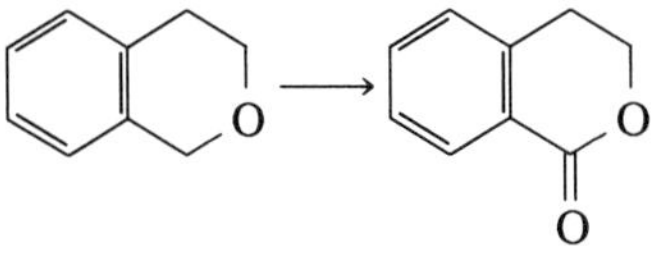

JOC *49* 1647 (1984)

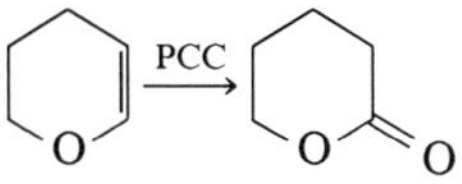

TL *27* 4011 (1986)

$$ROR' \longrightarrow ROCR''$$

Ac$_2$O, NaI   TL *28* 2537 (1987)

Ac$_2$O, Me$_3$SiCl (R' = CH$_3$, CH$_2$SCH$_3$)   TL *24* 1189 (1983)

Ac$_2$O, FeCl$_3$ JOC *39* 3728 (1974)

R''COCl, NaI   TL *23* 681 (1982); *28* 2537 (1987)

$$ROTHP \xrightarrow[\text{HOAc}]{\text{AcCl}} ROAc$$

Syn 567 (1972)

$$R_3COMe \xrightarrow[\text{CH}_3\text{CN}]{\text{(NO}_2)\text{BF}_4 \quad \text{H}_2\text{O}} R_3CNHCCH_3$$

JOC *45* 165 (1980)

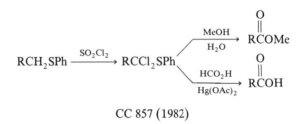

$$R = H, SiMe_3$$

$$n = 3\text{--}5$$

TL 4013 (1977)
JACS *107* 4230 (1985)

10. Sulfides

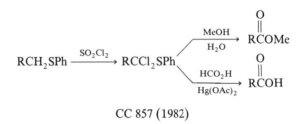

CC 857 (1982)

11. Alcohols

For alcohol → ester see also page 963, Section 9.
For alkenol → lactone see page 941, Section 8.

$$RCH_2OH \longrightarrow RCO_2H$$

CrO_3, HOAc	JACS *78* 2255 (1956)
CrO_3, H_2SO_4	JOC *48* 4404 (1983)
$K_2Cr_2O_7$, H_2SO_4	JACS *82* 2498 (1960)
$(C_5H_5NH)_2Cr_2O_7$ (PDC), DMF (non-allylic)	TL 399 (1979); *28* 5311, 6069 (1987) JACS *104* 1774 (1982); *109* 5437 (1987) JOC *50* 2607 (1985) CL 85 (1986)
$KMnO_4$	JCS 633 (1939); 2685 (1950) BCSJ *36* 1264 (1963) TL *28* 5263 (1987)
$KMnO_4$ (phase transfer)	TL 1511 (1974) JACS *109* 7280 (1987)
$NaMnO_4 \cdot H_2O$	TL *22* 1655 (1981)
$(n\text{-}Bu_4N)MnO_4$, py (benzylic)	CC 253 (1978)
$Zn(MnO_4)_2 \cdot 6\,H_2O$	J Biol Chem *241* 3970 (1966) JOC *50* 5480 (1985)
$Cu(MnO_4)_2 \cdot 8\,H_2O$	JOC *47* 2790 (1982)
$NaIO_4$, cat $RuCl_3 \cdot H_2O$, H_2O, CH_3CN, CCl_4	JOC *46* 3936 (1981); *50* 5696 (1985)

$NaIO_4$, cat RuO_2, H_2O, CH_3CN, CCl_4	TL *28* 6425 (1987)
H_5IO_6, cat $RuCl_3 \cdot H_2O$, H_2O, CH_3CN, CCl_4	JOC *50* 1560 (1985)
$RuCl_3$, $K_2S_2O_8$	TL *28* 4965 (1987)
RuO_4	JACS *80* 6682 (1958)
cat $K_2Ru_2O_4$, K_2SO_5	CC 58 (1979)
cat $RuO_2 \cdot 2\ H_2O$, electrolysis	JOC *51* 155 (1986)
nickel peroxide, NaOH	JOC *27* 1597 (1962)
O_2, cat PtO_2	Ber *89* 1648 (1956) Tetr *9* 67 (1960) JOC *52* 4898 (1987)
AgO	TL 5685 (1968)
HNO_3	Org Syn Coll Vol *1* 168 (1941)
H_2O_2	"Hydrogen Peroxide in Organic Chemistry," DuPont (1962), p 57

NaOCl, cat MeO—◯—NO· ,

  KBr, Aliquat 336

JOC *52* 2559 (1987)

electrolysis [$Ni(OH)_2$ anode]	Syn 513 (1979) Tetr *38* 3299 (1982)
Pseudomonas aeruginosa (R = allene, enantioselective)	TL *21* 1711 (1980) Appl Microbiol Biotechnol *21* 258 (1985)

$$\text{OH} \quad \bigcirc \longrightarrow HO_2C(CH_2)_4CO_2H$$

HNO_3	Rec Trav Chim *24* 19 (1905) JACS *52* 3235 (1930) Org Syn Coll Vol *1* 18 (1941)
$KMnO_4$	Ber *41* 575 (1908), *55B* 3526 (1922) J Chem Ed *10* 113 (1933)

$$RCH_2OH \longrightarrow R\overset{\overset{\displaystyle O}{\|}}{C}OCH_2R$$

$Ca(OCl)_2$	TL *23* 35 (1982)
KI, H_2O, electrolysis	TL 165 (1979)
$Na_2Cr_2O_7$, H_2SO_4	Org Syn Coll Vol *1* 138 (1941)
t-BuO_2H, cat $(PhCH_2NMe_3)OMoBr_4$	TL *25* 4417 (1984)
cat $H_2Ru(PPh_3)_4$	TL *22* 5327 (1981) JOC *52* 4319 (1987)

cat Ru$_3$(CO)$_{12}$, PhC≡CPh   TL *22* 1541 (1981)

cat PdCl$_2$, K$_2$CO$_3$, CCl$_4$ CL 1171 (1981)

$$2 \ RCH_2OMR_3' \xrightarrow{NBS} RCOCH_2R$$

$$RCH_2OMR_3' + R''CHO \xrightarrow{NBS} R''COCH_2R$$

R$_3'$M

Me$_3$Si   JOC *43* 371 (1978)

n-Bu$_3$Sn   JACS *98* 1629 (1976)

$$RCH_2OH \longrightarrow RCOR'$$

CrO$_3$·2 py, Ac$_2$O, *t*-BuOH (R′ = *t*-Bu)   JOC *49* 4735 (1984)
                                                 TL *27* 3049 (1986)

(COCl)$_2$-DMSO, Et$_3$N/(CH$_3$)$_2$C(OH)CN,   JOC *51* 3023 (1986)
   Et$_3$N/(COCl)$_2$-DMSO, Et$_3$N/MeOH
   (R′ = Me)

$$RCH_2OH \xrightarrow[\substack{MnO_2 \\ CH_3OH}]{NaCN \ or \ KCN} RCOCH_3$$

R

aryl   TL *28* 5157 (1987)

vinylic   JACS *90* 5616 (1968)
                                                 JOC *51* 253 (1986)

$$ROH \longrightarrow ROCR'$$

MsCl, Et$_3$N/CsO$_2$CEt (R′ = Et)   JOC *46* 4321 (1981)

Zn(O$_2$CR′)$_2$ (R′ = Me, Ph), PPh$_3$,   Syn Commun *16* 611 (1986)
   EtO$_2$CN=NCO$_2$Et

12. Diols

$$\underset{RCH-CHR}{\overset{OH \quad OH}{|\quad\quad|}} \longrightarrow RCO_2H$$

KMnO$_4$ JOC *50* 3007 (1985)

NaIO$_4$/AgNO$_3$, KOH   JACS *107* 3285 (1985)

NaIO$_4$, cat RuCl$_3$·H$_2$O   JOC *46* 3936 (1981)

NaOCl, cat $RuCl_3 \cdot 3\ H_2O$ CC 1420 (1970)

cat $RuO_2 \cdot 2\ H_2O$, electrolysis   JOC *51* 155 (1986)

H_2O_2, cat $Na_2WO_4 \cdot 2\ H_2O$-H_3PO_4 JOC *51* 1599 (1986)

t-BuO_2H, cat $VO(acac)_2$ TL *24* 5009 (1983)

t-BuO_2H, cat $Mo(CO)_6$, cat   Syn 59 (1986)
     $[C_5H_5N(CH_2)_{15}CH_3]Cl$, $MgSO_4$

$$HO(CH_2)_nOH \longrightarrow \overset{O}{\underset{O}{\overset{\|}{C}}} (CH_2)_{n-1}$$

Review: Heterocycles *14* 661 (1980) ($n = 4$)

O_2, cat Pt ($n = 4$)   TL 3899 (1973); 4045 (1976)
     JACS *98* 3379 (1976); *102* 3964 (1980)

$NaBrO_3$ ($n = 4$–6)   CL 1097 (1983)

NIS, AgOAc ($n = 4, 5$)   JOC *52* 5472 (1987)

$AgCO_3$, Celite ($n = 4$–6)   Tetr *31* 171, 987 (1975)
     TL 4045 (1976); *22* 3721, 3725 (1981)
     JOC *43* 1086 (1978); *50* 3017 (1985);
       52 325 (1987)
     JACS *104* 5523 (1982)
     JCS Perkin I 1579 (1983)

MnO_2 ($n = 4, 5$; allylic)   JACS *98* 4312 (1976)
     TL 111 (1978)
     Syn Commun *13* 237 (1983)

$BaMnO_4$ ($n = 4$)   JCS Perkin I 1579 (1983)
     JOC *52* 325 (1987)

$KMnO_4$ ($n = 4$)   JOC *28* 323 (1963); *42* 1623 (1977)

CrO_3, py ($n = 4, 6$)   JACS *76* 527 (1954)
     JOC *28* 323 (1963)
     TL *27* 1445 (1986)

$Na_2Cr_2O_7$, H_2SO_4 ($n = 4, 5$)   JOC *28* 323 (1963); *45* 1828 (1980)

copper chromite ($n = 4, 5$)   JACS *68* 1385 (1946); *69* 1545 (1947)

H_2O_2, $[C_5H_5N(CH_2)_{15}CH_3]_3PMo_{12}O_{40}$ JOC *52* 1868 (1987)
     ($n = 4$–6)

cat $Ru_3(CO)_{12}$, $PhC\equiv CPh$ ($n = 4, 5$)   JOMC *226* Cl (1982)

cat $RuO_2 \cdot 2\ H_2O$, electrolysis   JOC *51* 155 (1986)
     ($n = 4, 5$)

cat $RuCl_2L_3$ or $Ru_2Cl_4L_3$,   TL *24* 2677 (1983)
     $PhCH=CHCOCH_3$, Et_3N ($n = 4, 5$)

cat $H_2Ru(PPh_3)_4$, $PhCH=CHCOPh$ J Mol Catal *7* 337 (1980)
     ($n = 5$)

cat $H_2Ru(PPh_3)_4$, $PhCH{=}CHCOCH_3$ ($n = 4, 5$)	TL *27* 365 (1986) JOC *51* 2034 (1986)
cat $H_2Ru(PPh_3)_4$, CH_3COCH_3 ($n = 4, 5$)	TL *22* 5327 (1981) JOC *52* 4319 (1987)

, cat $RuCl_2(PPh_3)_3$/   TL *27* 3311 (1986)

Ag_2CO_3, Celite ($n = 5$)

($n = 4, 5$)   JOC *50* 3930 (1985)

$Ni(O_2CR)_2$, Br_2 ($n = 4$)	Syn Commun *10* 881 (1980) JOC *46* 4806 (1981)
Raney Ni ($n = 4$)	JOC *21* 1325 (1956)

horse liver alcohol dehydrogenase
   ($n = 4, 5$) (chiral)
	CC 469 (1977); 515 (1980)
	 JACS *104* 4659 (1982); *106* 1461 (1984);
	   107 2521 (1985)
	 Can J Chem *60* 2007 (1982)
	 Org Syn *63* 10 (1984)

13. Aldehydes

$Pb(OAc)_4$	JACS *79* 1938 (1957) Compt Rend *250* 4174 (1960)
$Pb(OAc)_4$, BF_3	BSCF 4083 (1968)

air	JOC *51* 449 (1986)
O_2, cat Mn(II) stearate	JOC *52* 287 (1987)
$KMnO_4$	Org Syn Coll Vol *2* 315, 538 (1943) JACS *106* 3297 (1984); *107* 199 (1985)
$KMnO_4$, 18-crown-6	JACS *94* 4024 (1972) TL *27* 4537 (1986)
$KMnO_4$, $MgSO_4$	JACS *109* 7122 (1987)
$KMnO_4$, *t*-BuOH, 5% (K)NaH_2PO_4	TL *27* 4537 (1986) JACS *109* 7575 (1987)

NaMnO$_4\cdot$H$_2$O	TL 22 1655 (1981)
(PhCH$_2$NEt$_3$)MnO$_4$, CH$_2$Cl$_2$, HOAc	Monatsh 110 1471 (1979)
(n-Bu$_4$N)MnO$_4$, py (R = Ar)	CC 253 (1978)
Cu(MnO$_4$)$_2\cdot$8 H$_2$O	JOC 47 2790 (1982)
cat (PhCH$_2$NMe$_3$)OMoBr$_4$, t-BuO$_2$H	TL 25 4417 (1984)
(NH$_4$)$_6$Mo$_7$O$_{24}\cdot$4 H$_2$O, H$_2$O$_2$, CeCl$_3\cdot$7 H$_2$O, n-Bu$_4$NCl, K$_2$CO$_3$	TL 25 173 (1984)
cat H$_2$Ru(PPh$_3$)$_4$, PhCH=CHCOCH$_3$, H$_2$O	JOC 52 4319 (1987)
cat RuCl$_2$(PPh$_3$)$_4$, PhIO	TL 22 2361 (1981)
NaIO$_4$, cat RuO$_2$	JOC 52 622 (1987)
NaIO$_4$, cat RuCl$_3\cdot$H$_2$O, CCl$_4$, CH$_3$CN, H$_2$O	JOC 46 3936 (1981)
cat RuO$_2\cdot$2 H$_2$O, electrolysis	JOC 51 155 (1986)
RuO$_4$	JACS 80 6682 (1958)
CrO$_3$, Ac$_2$O	JACS 86 2612 (1964)
K$_2$Cr$_2$O$_7$, H$_2$SO$_4$	JACS 80 3022, 3030 (1958); 107 199 (1985); 108 1039 (1986)
(C$_5$H$_5$NH)$_2$Cr$_2$O$_7$	TL 399 (1979) JOC 50 2095 (1985)
(bipy)H$_2$CrOCl$_5$	Syn Commun 10 951 (1980)
(phen)$_2$H$_2$CrOCl$_5$	Syn Commun 10 951 (1980)
(bipy)CrOCl$_5$	Syn Commun 10 951 (1980)
Ag$_2$O	JACS 73 2590 (1951); 92 336 (1970); 106 1029 (1984); 107 7978 (1985) Org Syn Coll Vol 4 919 (1963) Tetr 24 6583 (1968) JOC 44 1022 (1979); 50 2981 (1985); 51 956 (1986) Syn 74 (1981) JCS Perkin I 461 (1982)
AgO	TL 5685 (1968)
AgO, NaCN	JACS 90 5616 (1968)
AgNO$_3$, NaOH, H$_2$O$_2$, H$_2$O, O$_2$	Ber 93 2743 (1960)
NBS, H$_2$O	JACS 78 1689 (1956)
NaClO$_2$, CH$_3$CH=C(CH$_3$)$_2$	JOC 45 1175 (1980)
NaClO$_2$, (K)NaH$_2$PO$_4$, CH$_3$CH=C(CH$_3$)$_2$	Tetr 37 2091 (1981) JOC 50 470 (1985) JACS 109 6719 (1987) TL 28 5759, 5763 (1987)

$NaClO_2$, NaH_2PO_4, $(CH_3)_2C{=}CH_2$ — JACS *106* 7217 (1984)
TL *27* 3533 (1986)

$NaClO_2$, H_2O_2, NaH_2PO_4 — JOC *51* 567 (1986)

$NaClO_2$, H_2NSO_2OH — Acta Chem Scand *27* 888 (1973)
JOC *51* 3007 (1986)

$NaClO_2$, HOAc — Methods Carbohydr Chem *3* 182 (1963)

$NaClO_2$, resorcinol, acetate buffer — Acta Chem Scand *27* 888 (1973)

$Ca(OCl)_2$ — TL *23* 3131 (1982)

HNO_3 — Org Syn Coll Vol *1* 166 (1941)

H_2O_2 — "Hydrogen Peroxide in Organic Chemistry,"
DuPont (1962), p 56

nickel peroxide, NaOH (R = Ar) — Chem Pharm Bull *26* 299 (1978)

$CF_3C(OH)(O_2H)CF_3$ — TL *21* 685 (1980)

$NaHSO_3/Ac_2O$, $DMSO/H_2O$ — TL *27* 3995 (1986)

$$\overset{O}{\overset{\|}{RCH}} \longrightarrow \overset{O}{\overset{\|}{RCX}}$$

Cl_2 — Org Syn Coll Vol *1* 155 (1941)

$(PhCO_2)_2$, CCl_4 — JACS *69* 2916 (1947)

NBS, hν — TL 3809 (1979)

$$2\,\overset{O}{\overset{\|}{RCH}} \longrightarrow \overset{O}{\overset{\|}{RCOCH_2R}}$$

RCH_2ONa — Org Syn Coll Vol *1* 104 (1941)

$B(OH)_3$ — JOC *38* 1433 (1973)

$Al(OR)_3$ — Chem Zentr *77* 1309, 1554, 1556 (1906)
Org Syn Coll Vol *1* 104 (1941)
JACS *69* 2605 (1947); *74* 5133 (1952)
JOC *33* 3310 (1968)

cat $H_2Ru(PPh_3)_4$, H_2O — JOC *52* 4319 (1987)

cat $Ru_3(CO)_{12}$, H_2O, PhC≡CPh — TL *22* 1541 (1981)

$$\overset{O}{\overset{\|}{RCH}} \longrightarrow \overset{O}{\overset{\|}{RCOR'}}$$

NaOCl, HOAc, MeOH — TL *23* 4647 (1982)

HOCl, R'OH — TL *23* 4647 (1982)

t-BuOCl/R'OH, Et_3N or py — JOC *47* 1360 (1982)

t-BuO_2H, cat $(PhCH_2NMe_3)OMoBr_4$, R'OH — TL *25* 4417 (1984)

O_2, cat HCl, hν, R'OH (R = Ph) — CL 475 (1974)

O$_3$, KOH, R'OH	TL 1627 (1978)
NaHSO$_3$/Ac$_2$O, DMSO/R'OH	TL _27_ 3995 (1986)
H$_2$SO$_5$, R'OH	JOC _33_ 2525 (1968)
(NH$_4$)$_2$S$_2$O$_8$, H$_2$SO$_4$, MeOH	Tetr _37_ 2091 (1981)
KCN, PhNO$_2$, R'OH, phase transfer catalyst (R = aryl)	Tetr _38_ 337 (1982)
NaCN or KCN, MnO$_2$, CH$_3$OH (R = vinylic)	JACS _90_ 5616 (1968) Ber _103_ 3774 (1970) JOC _51_ 253, 3070 (1986)
Me$_3$SiCl, cat ZnI$_2$/H$_3$O$^+$/MnO$_2$, MeOH	Tetr _37_ 2091 (1981)
PDC, CH$_3$OH	TL _28_ 3235 (1987)
NBS, R'OSiMe$_3$	JOC _43_ 371 (1978)
NBS, R'OSn(_n_-Bu)$_3$	JACS _98_ 1629 (1976)
NBS, hν/R'OH	TL 3809 (1979)
I$_2$, CH$_3$OH, KOH	JCS C 1693 (1968) JOC _50_ 2707 (1985)
cat Ru$_3$(CO)$_{12}$, RCH$_2$OH (R' = R), PhC≡CPh	TL _22_ 1541 (1981)
cat H$_2$Ru(PPh$_3$)$_4$, R'OH	JOC _52_ 4319 (1987)
electrolysis, R'OH	JOC _50_ 4967 (1985)
electrolysis, NaCN, MeOH	BCSJ _55_ 335 (1982)

$$\overset{O}{\overset{\|}{ArCH}} \longrightarrow \overset{O}{\overset{\|}{ArOCH}}$$

_m_-ClC$_6$H$_4$CO$_3$H, KF	TL _22_ 3895 (1981)
H$_2$O$_2$, SeO$_2$	TL _27_ 6299 (1986)

$$\overset{O}{\overset{\|}{RCH}} \longrightarrow \overset{O}{\overset{\|}{RCNR_2}}$$

NBS, hν/RNH$_2$	TL 3809 (1979)
Me$_3$SiCN/LDA/Ph$_2$P(O)ONMe$_2$/H$_3$O$^+$	TL _23_ 3255 (1982)
NaHSO$_3$/Ac$_2$O, DMSO/HNR$_2$	TL _27_ 3995 (1986)

14. Ketones

See also page 929, Section 7.

Pb(OAc)$_4$

JCS 4426 (1955)
Austral J Chem *13* 121 (1960)
Syn 567 (1973) (review)
JACS *107* 5576 (1985)

Pb(OAc)$_4$, hν

JCS C 2484 (1969)

Pb(OAc)$_4$, cat BF$_3 \cdot$OEt$_2$

JCS 4472 (1961)
Syn 567 (1973) (review)

Hg(OAc)$_2$

Naturwiss *35* 125 (1948)
Ann *561* 165 (1949); *581* 59 (1953)
Helv *35* 1615 (1952)
JCS 3512 (1953); 705 (1954)
JACS *79* 4465 (1957)
Ber *93* 1374 (1960)
JOC *28* 1705 (1963)
Syn 567 (1973) (review)
R. C. Larock, "Organomercury Compounds in Organic Synthesis," Springer, New York (1985), p 205 (review)

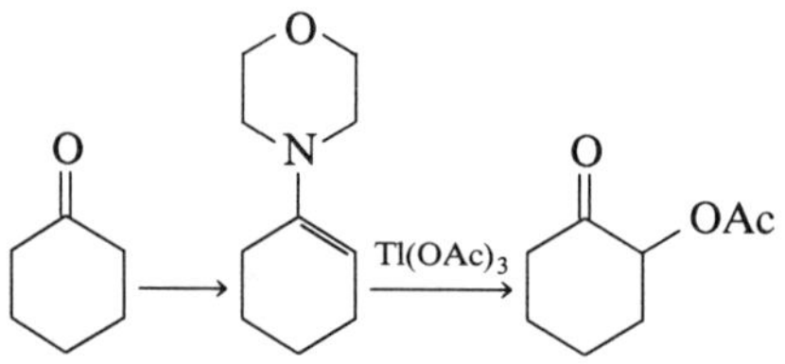

JOC *33* 3359 (1968)

$$\underset{\text{RCCH}_3}{\overset{\text{O}}{\overset{\|}{}}} \longrightarrow \text{RCO}_2\text{H}$$

HNO$_3$/KMnO$_4$ (R = aryl)

Org Syn Coll Vol *3* 791 (1955)

Na$_2$Cr$_2$O$_7$, HOAc (R = aryl)

Org Syn Coll Vol *3* 420 (1955)

NaOCl (R = aryl)

Org Syn Coll Vol *2* 428 (1943)

KOCl (R = vinylic)

Org Syn Coll Vol *3* 302 (1955)

Br$_2$, NaOH/HCl (R = 1°, 2° alkyl)

Org Syn Coll Vol *5* 8 (1973)
JACS *107* 2033 (1985)

I$_2$, py/NaOH (R = aryl)

JACS *66* 894 (1944)

C$_6$F$_5$I(O$_2$CCF$_3$)$_2$, H$_2$O, C$_6$H$_6$ (R = aryl)

CC 202 (1987)

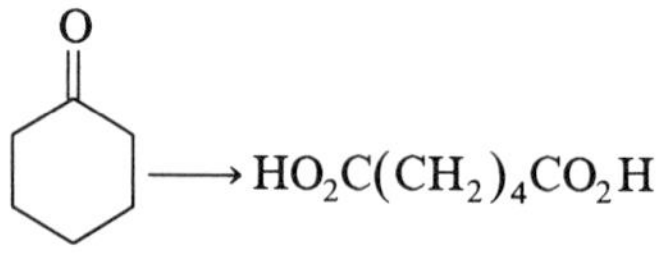

KO$_2$ (phase transfer)

TL 3689 (1978)

HNO$_3$

Rec Trav Chim *24* 19 (1905)

$KMnO_4$	Ber *39* 2202 (1906)
$Zn(MnO_4)_2$, silica gel	JACS *105* 7755 (1983)
$C_6H_5I(O_2CCF_3)_2$, H_2O, C_6H_6	CC 202 (1987)

$$\underset{RCR'}{\overset{\overset{\displaystyle O}{\|}}{}} \longrightarrow \underset{ROCR'}{\overset{\overset{\displaystyle O}{\|}}{}}$$

Reviews:

Chem Rev *45* 385 (1949)
Org Rxs *9* 73 (1957)
P. A. S. Smith, "Molecular Rearrangements," Ed. P. de Mayo, Interscience, New York (1963), Vol 1, pp 577–589
Quart Rev *21* 429 (1967)
B. Plesnicar, "Oxidation in Organic Chemistry," Ed. W. S. Trahanovsky, Academic Press, New York (1978), pp 254–262
Tetr *37* 2697 (1981)

RCO_3H	Org Rxs *9* 73 (1957) (review)
m-$ClC_6H_4CO_3H$	JOC *50* 2759 (1985)
m-$ClC_6H_4CO_3H$, $NaHCO_3$	JOC *52* 3560 (1987) TL *28* 4773 (1987)
m-$ClC_6H_4CO_3H$, cat Li_2CO_3	JOC *52* 5700 (1987)
m-$ClC_6H_4CO_3H$, cat TsOH	JOC *50* 2489 (1985)
CF_3CO_2H	TL *28* 3061 (1987)
H_2O_2, $(CF_3CO)_2O$	JACS *77* 188 (1955); *107* 5717 (1985) Tetr *6* 253 (1959)
98% H_2O_2, $(CF_3CO)_2O$, Na_2HPO_4	Tetr *37* 3981 (1981) JOC *47* 3871 (1982)
H_2O_2, $PhSeO_2H$	CC 870 (1977)
H_2O_2, CF_3COCF_3	TL 2741 (1970)
$Me_3SiOSO_3OSiMe_3$	JOC *44* 4969 (1979)

$$\underset{RCCH_3}{\overset{\overset{\displaystyle O}{\|}}{}} \xrightarrow[\text{CH}_3\text{OH}]{\text{aq KOCl}} \underset{RCOCH_3}{\overset{\overset{\displaystyle O}{\|}}{}}$$

JACS *107* 5570 (1985)

$$\underset{RC-CX}{\overset{\overset{\displaystyle O}{\|}\ \ \overset{\displaystyle O}{\|}}{}} \longrightarrow \underset{RCOH}{\overset{\overset{\displaystyle O}{\|}}{}}$$

X	Reagent(s)	
R, ONa	$Ca(OCl)_2$, H_2O, HOAc, CH_3CN	TL *23* 3135 (1982)
R	HIO_4	JCS 1467 (1935) JACS *81* 37 (1959)
R (R = Ph)	$C_6F_5I(O_2CCF_3)_2$, H_2O, C_6H_6	CC 202 (1987)

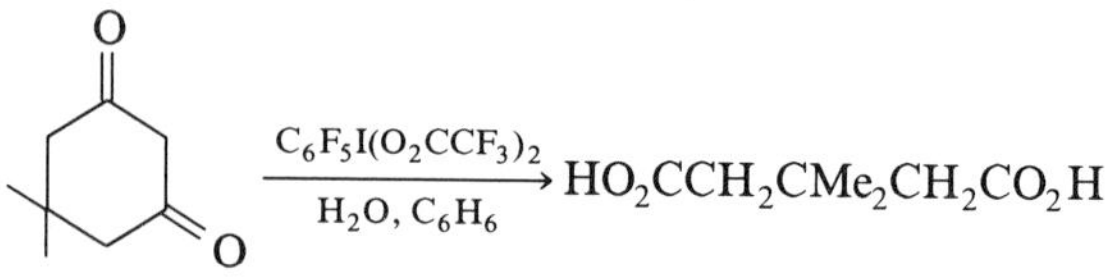

$$\xrightarrow[\text{H}_2\text{O, C}_6\text{H}_6]{\text{C}_6\text{F}_5\text{I(O}_2\text{CCF}_3)_2} \text{HO}_2\text{CCH}_2\text{CMe}_2\text{CH}_2\text{CO}_2\text{H}$$

CC 202 (1987)

15. Hemiacetals and Lactols

$$\underset{\text{OH}}{\overset{\text{OH}}{\text{RCHOR}'}} \xrightarrow[\text{py, Ac}_2\text{O}]{\text{CrO}_3} \underset{\text{O}}{\overset{\text{O}}{\text{RCOR}'}}$$

TL *27* 3049 (1986)

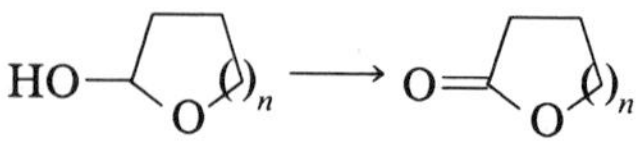

$$\text{HO}-\text{(O)}_n \longrightarrow \text{O}=\text{(O)}_n$$

$n = 1, 2$

CrO$_3$, HOAc	TL *27* 2757 (1986)
PCC	JACS *105* 3720 (1983)
	CC 1512 (1985)
	TL *26* 771 (1985); *27* 2757 (1986)
PCC, Al$_2$O$_3$	JOC *52* 3541 (1987)
PCC, NaOAc, Al$_2$O$_3$	TL *28* 6437 (1987)
PCC, NaOAc, molecular sieves	JACS *109* 7488 (1987)
PDC, molecular sieves	CC 1714 (1987)
	TL *28* 5457 (1987)
PDC, HOAc, molecular sieves	JACS *109* 7534 (1987)
Ag$_2$CO$_3$, Celite	CC 1720 (1987)
Br$_2$, Ba(O$_2$CPh)$_2$, H$_2$O	TL *28* 1073 (1987)

16. Acetals

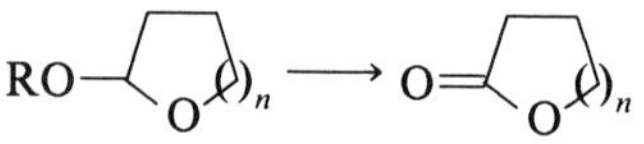

$$\text{RO}-\text{(O)}_n \longrightarrow \text{O}=\text{(O)}_n$$

n	Reagent(s)	
1	*m*-ClC$_6$H$_4$CO$_3$H, BF$_3$·OEt$_2$	TL 419 (1978)
		JACS *109* 7534 (1987)
2	H$_2$CrO$_4$	TL *27* 2703 (1986);
		28 671 (1987)

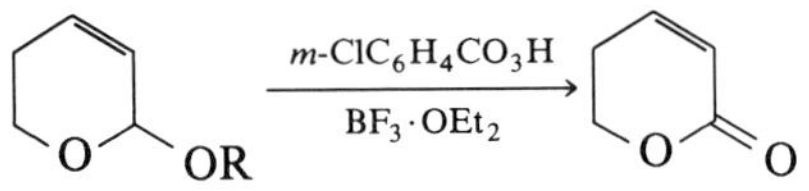

TL *23* 3781 (1982)
CC 1714 (1987)

$$R_2C(OR')_2 \xrightarrow{m\text{-}ClC_6H_4CO_3H} (RO)_2C(OR')_2$$

JACS *104* 1769 (1982)

17. Carboxylic Acids

$$RCO_2H \xrightarrow{Pb(OAc)_4} ROAc$$

Org Rxs *19* 279 (1972) (review)
TL 549 (1976); *27* 5319 (1986); *28* 781, 4143, 4147 (1987)
CC 1797 (1987)

$$R_2C(CO_2H)_2 \xrightarrow{Pb(OAc)_4} R_2C(OAc)_2$$

Org Rxs *19* 279 (1972) (review)

$$\overset{O}{\overset{\|}{RCCO_2H}} \longrightarrow RCO_2R'$$

R'	Reagents	
H	$Pb(OAc)_4/H_3O^+$	JACS *62* 1597 (1940)
Me, Et	NIS, R'OH	JOC *52* 3165 (1987)

$$CH_3CO_2H \xrightarrow{Pb(OAc)_4} AcOCH_2CO_2H$$

JOC *16* 533 (1951)

$$R_2CHCO_2H \xrightarrow{Tl(OAc)_3} \underset{\underset{O_2CCHR_2}{|}}{R_2CCO_2H}$$

TL 5285 (1970)

18. Esters

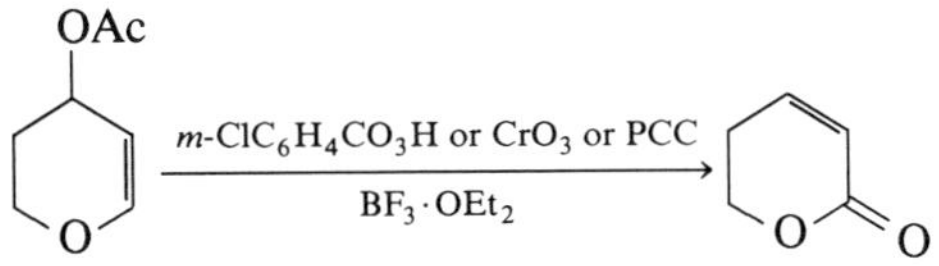

TL *23* 3781 (1982)

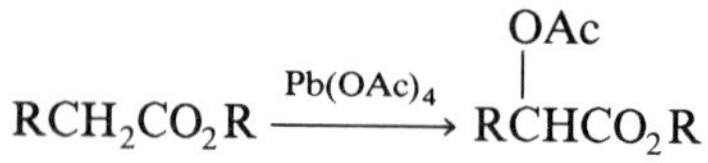

$$RCH_2CO_2R \xrightarrow{Pb(OAc)_4} \overset{\overset{\displaystyle OAc}{\displaystyle |}}{R}CHCO_2R$$

Proc Acad Sci USSR, Chem Sec *129* 995 (1959)
TL 2769 (1970) (lactones)
Syn 567 (1973) (review)

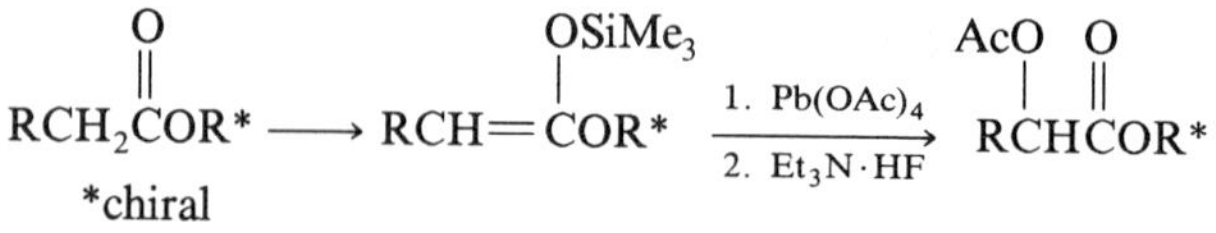

$$\overset{\overset{\displaystyle O}{\displaystyle \|}}{R}CH_2COR^* \longrightarrow RCH=\overset{\overset{\displaystyle OSiMe_3}{\displaystyle |}}{C}OR^* \xrightarrow[\text{2. Et}_3N\cdot HF]{\text{1. Pb(OAc)}_4} \overset{AcO\ \ \ O}{R CHCOR^*}$$

*chiral

Helv *68* 216 (1985)

19. Amides

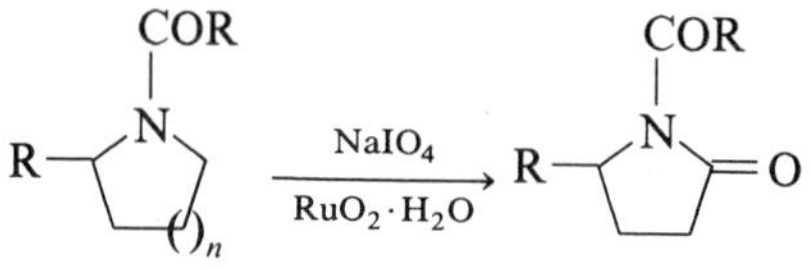

$n = 1, 2$

CC 71 (1975)

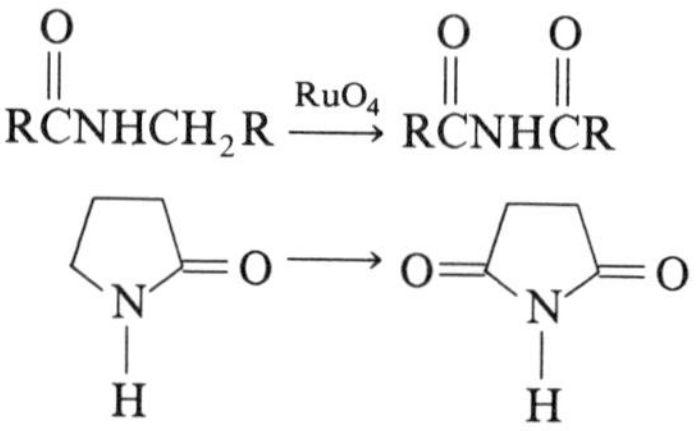

$$\overset{\overset{\displaystyle O}{\displaystyle \|}}{R}CNHCH_2R \xrightarrow{RuO_4} \overset{O\ \ \ O}{RCNHCR}$$

JACS *80* 6682 (1958)

$$PhCH_2NHCOCH_2Ph \xrightarrow[\text{silica gel}]{Zn(MnO_4)_2} \overset{O\ \ \ O}{PhCNHCOCH_2Ph}$$

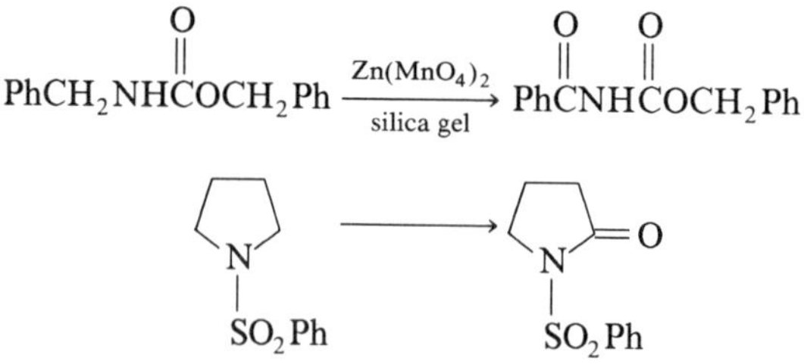

JACS *105* 7755 (1983)

20. Nitriles

$$RCH_2CN \xrightarrow[\substack{KO\text{-}t\text{-}Bu \\ \text{18-crown-6}}]{O_2} \xrightarrow{H^+} RCO_2H$$

JOC *45* 3630 (1980)

2. REDUCTION

$$R_2C=O \xrightarrow[\text{aq } H_2SO_4, \, CH_3CN]{R'_3SiH} R_2CHNHCOCH_3$$

$$ArCHO \xrightarrow{\hspace{3cm}} ArCH_2NHCOCH_3$$

JOC *39* 2740 (1974)

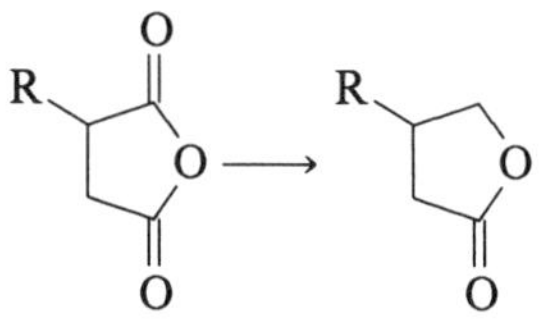

Na, C_2H_5OH	JCS 478 (1936)
NaBH$_4$	JOC *35* 3574 (1970)
	TL 4651 (1976); *28* 451, 455 (1987)
	Can J Chem *56* 1524 (1978) (review)
	CC 458 (1982)
	JACS *108* 4943 (1986)
KHB(*sec*-Bu)$_3$	CC 458 (1982)
LiAlH$_4$	JOC *32* 3919 (1967)
	TL 3589 (1968)
	Can J Chem *56* 1524 (1978) (review)

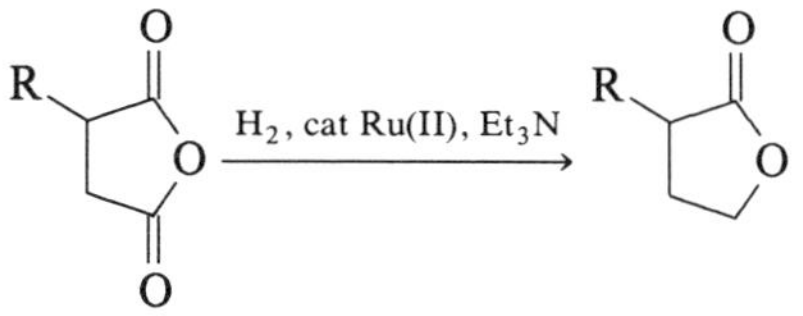

CC 412 (1975); 315 (1976)
BCSJ *57* 897 (1984)
TL *27* 365 (1986)

3. CARBONYLATION, CARBOXYLATION AND RELATED REACTIONS

See also page 211, Section 6, for the synthesis of α,β-unsaturated nitriles, carboxylic acids and derivatives.

1. Carboxylic Acids

$$ArH \longrightarrow ArCO_2H$$

CO, Pd(OAc)$_2$	CC 220 (1980)
CO, cat Pd(OAc)$_2$, t-BuO$_2$H, $\quad$ H$_2$C=CHCH$_2$Cl	JOMC 256 C35 (1983)
CO$_2$, cat Pd(OAc)$_2$, t-BuO$_2$H	JOMC 266 C44 (1984)
ClCOCOCl, (AlCl$_3$)/H$_2$O	Ber 41 3558 (1908); 44 202 (1911); $\quad$ 45 1186 (1912) $\quad$ JACS 70 1079 (1948) $\quad$ Org Syn 44 69 (1964)

(structure) CCl$_2$, AlCl$_3$/H$_2$O $\qquad$ Ber 96 1382 (1963)

$$\text{>C=C<} \longrightarrow \text{H-C-C-CO}_2\text{H}$$

CO, BF$_3$, H$_2$O	JACS 70 3793 (1948)
CO, NiCl$_2$ or Ni(CO)$_4$	Ann 582 53 (1953)
9-BBN/CO, KHB(O-i-Pr)$_3$/H$_2$O$_2$, $\quad$ NaOAc/AgNO$_3$, NaOH	CC 1273 (1982)
HCO$_2$H, H$_2$SO$_4$	JOC USSR 2 256 (1966)

$$\text{RCH=CHCH}_3 \text{ or } \text{RCH}_2\text{CH=CH}_2 \xrightarrow[\substack{\text{2. CO} \\ \text{3. NaOH, H}_2\text{O}_2}]{\text{1. Cp}_2\text{ZrHCl}} \text{R(CH}_2)_3\overset{\displaystyle O}{\overset{\displaystyle \|}{\text{C}}}\text{OH}$$

JACS 97 228 (1975)

$$H_2C=CHCH=CH_2 \xrightarrow[i\text{-PrMgBr}]{Cp_2TiCl_2} \xrightarrow[2.\ H^+]{1.\ CO_2} H_2C=CHCH(CH_3)CO_2H$$

CC 180, 342 (1981)
JOMC 224 327 (1982)

$$RCH_2CH=CH_2 \xrightarrow[2.\ IO_4^-\ \text{or Ce(IV)}]{1.\ R'COCOCOR'} RCH=CHCH_2CO_2H$$

<u>R'</u>

OEt   JACS 102 2473 (1980); 106 1092, 3797 (1984)
                                       JOC 47 4201 (1982)

o-C$_6$H$_4$ TL 23 1399 (1982)

$$\begin{array}{c}(CH_3)_3COH \\ \text{or} \\ (CH_3)_3CCl \\ \text{or} \\ (CH_3)_2C=CH_2\end{array} \xrightarrow[H_2SO_4]{H_2C=CCl_2} \xrightarrow{H_2O} (CH_3)_3CCH_2\overset{\displaystyle O}{\overset{\|}{C}}OH$$

Angew Int 4 956 (1965); 5 870 (1966); 19 171 (1980) (review)
CL 1107 (1980)
Acta Chem Scand B 34 621 (1980)

$$R_2C=CRCH_2OH + H_2NCOR + CO \xrightarrow[\substack{cat\ Co_2(CO)_8 \\ cat\ HRh(CO)(PPh_3)_3}]{H_2,\ CO} R_2CHCHRCHCO_2H \overset{\displaystyle HNCOR}{\overset{|}{}}$$

TL 23 2491 (1982)

$$RC\equiv CH \longrightarrow RC\equiv CCO_2H$$

CH$_3$—(ring with t-Bu top, OLi, CO$_2$, t-Bu bottom)   JOC 38 4086 (1973)

n-BuLi/CO$_2$/H$^+$ JACS 101 5364 (1979)

EtMgX/CO$_2$/H$^+$ JACS 65 2208 (1943)

$$RX \longrightarrow RCO_2H$$

RLi/CO$_2$/H$^+$ JACS 108 806, 1311 (1986)
                                         (both R = aryl)

Mg/CO$_2$/H$^+$ Org Syn Coll Vol 1 361 (2° RCl), 524 (3° RCl)
                                         (1941); 3 553 (1955) (ArBr)
                                         JOC 18 432 (1953) (2° RCl, benzylic Cl)
                                         Org Syn 59 85 (1980) (3° RCl)
                                         Syn 587 (1982) (1° RCl)
                                         TL 23 3901 (1982) (allylic Cl)

CO, CoCl$_2$, NaBH$_4$, NaOH/H$^+$ TL 28 2633 (1987) (R = benzylic)

CO, Co$_2$(CO)$_8$, NaOH, hν/H$^+$ TL 22 1013 (1981) (R = aryl, vinylic)

CO, NaOH, C_6H_6, (n-$C_{12}H_{25}NMe_3$)Cl,
  cat Co(CO)$_3$NO

JOMC *212* C23 (1981) (R = benzylic)

CO, KOH, n-BuOH, (PhNMe$_3$Br), cat
  Co$_2$(CO)$_8$

JOMC *232* 59 (1982) (R = 2° benzylic)

CO, NaOH, CH$_2$Cl$_2$ or C_6H_6, cat
  Pd(PPh$_3$)$_4$

Organomet *1* 775 (1982)
  (R =1° benzylic)

CO, NaOH, t-AmOH, (PhCH$_2$NEt$_3$)Cl, cat
  Pd(Ph$_2$PCH$_2$CH$_2$PPh$_2$)$_2$

Organomet *2* 801 (1983) (R = vinylic)

CO, Ca(OH)$_2$, MeOH, cat PdCl$_2$(PPh$_3$)$_2$

TL *28* 2721 (1987) (R = aryl)

CO, H$_2$O, n-Bu$_3$N, cat PdCl$_2$(PPh$_3$)$_2$, Δ

JOC *46* 4614 (1981) (R = aryl)

CO$_2$, cat Pd, electrolysis

CL 169 (1986) (R = aryl)

(RS)$_3$CLi/hydrolysis or esterification

Angew Int *6* 442 (1967) (R = alkyl)
Ber *105* 487 (1972) (R = alkyl)

$$ArI + CO \xrightarrow[\substack{cat\ PdCl_2(PR_3)_2 \\ i\text{-PrOH}}]{Ca(OH)_2} Ar\overset{\overset{\displaystyle OH}{|}}{C}HCO_2H$$

TL *28* 2721 (1987)

$$ArI + CO + H_2O \xrightarrow[base]{cat\ PdCl_2(PR_3)_2} Ar\overset{\overset{\displaystyle O}{\|}}{C}CO_2H$$

Base

Et$_3$N   CC 837 (1985)

Ca(OH)$_2$ J Mol Catal *41* 379 (1987)

$$R_3COH \xrightarrow[H_2SO_4]{HCO_2H} R_3\overset{\overset{\displaystyle O}{\|}}{C}COH$$

Org Syn Coll Vol *5* 739 (1973)

Org Syn Coll Vol *2* 557 (1943)
JOC *19* 510 (1954)
Chem Rev *57* 583 (1957) (review)
J Chem Eng Data *14* 388 (1969)
Org Prep Proc Int *13* 426 (1981)

$$RSPh \xrightarrow[\text{naphthalene}]{\text{Li}} \xrightarrow{\text{CO}_2} RCO_2H$$

JOC *43* 1064 (1978) (lithiation)
JACS *107* 196 (1985)
Tetr *42* 2803 (1986) (lithiation)

$$ArNH_2 \longrightarrow (ArN_2)BF_4 \xrightarrow[\text{NaOAc}]{\overset{\text{CO}}{\text{cat Pd(OAc)}_2}} \xrightarrow{\text{H}_2\text{O}} ArCO_2H$$

TL *21* 2877 (1980)
JOC *45* 2365 (1980)

$$RCH_2NO_2 \xrightarrow[\text{2. H}_3\text{O}^+]{\text{1. CO}_2, \text{Mg(OCH}_3)_2} R\overset{\overset{\text{NO}_2}{|}}{C}HCO_2H$$

JACS *85* 616 (1963)

$$ArSiMe_3 \xrightarrow[\text{KO-}t\text{-Bu}]{\text{CO}_2} Ar\overset{\overset{\text{O}}{\|}}{C}OH$$

Angew Int *20* 265 (1981)

$$Ar\overset{\overset{\text{O}}{\|}}{C}R\,(ArCHO) \xrightarrow[\text{electrolysis}]{\text{CO}_2} Ar\overset{\overset{\text{OH}}{|}}{\underset{\underset{\text{R}}{|}}{C}}CO_2H$$

Angew Int *22* 492 (1983)
CL 453 (1984)
BCSJ *58* 1723 (1985)
TL *27* 3429 (1986)

See also page 863, Section 4.

$$-\overset{\overset{\text{O}}{\|}}{C}-\overset{\overset{\text{H}}{|}}{\underset{\underset{|}{}}{C}}- \longrightarrow -\overset{\overset{\text{O}}{\|}}{C}-\overset{\overset{\text{CO}_2\text{H}}{|}}{\underset{\underset{|}{}}{C}}-$$

NaNH$_2$/CO$_2$/H$^+$ JACS *66* 1768 (1944); *106* 4186 (1984)

LiO—(aromatic ring with *t*-Bu, *t*-Bu)—CH$_3$, CO$_2$/H$^+$ JOC *38* 4086 (1973)

CO$_2$, DBU   CL 427 (1974)

CO$_2$, MgCl$_2$, NaI, CH$_3$CN, Et$_3$N   JOC *50* 4877 (1985)

CO$_2$, Mg(OCH$_3$)$_2$ JACS *81* 2598 (1959)
                                   JOC *31* 1747 (1966); *38* 2489 (1973) (lactone);
                                     52 1686 (1987)
                                     Tetr *38* 2797 (1982) (lactone)

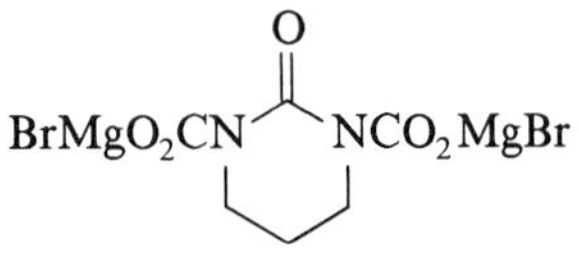

BrMgO$_2$CN—C(=O)—NCO$_2$MgBr   TL *21* 1967 (1980)

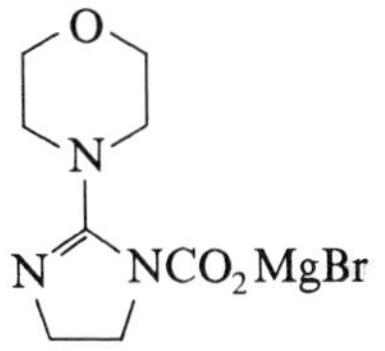

CC 326 (1980)

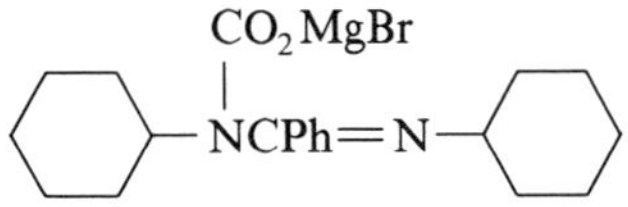

N—C(morpholine)—NCO$_2$MgBr

CO$_2$MgBr | cyclohexyl—NCPh=N—cyclohexyl   BCSJ *55* 3949 (1982)

HOCO$_2$CuL$_3$ (L = Et$_3$P, *t*-BuNC)   JACS *102* 431 (1980)

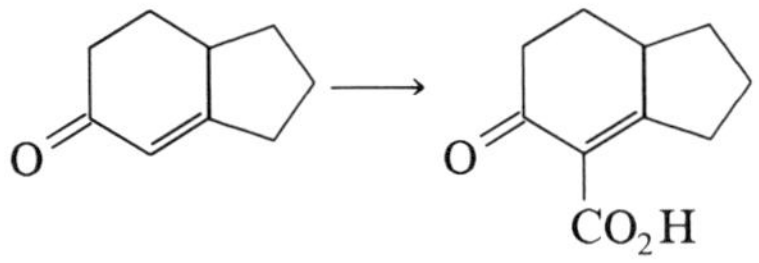

NaH, DMSO/CO$_2$ JOC *38* 3239 (1973)

Mg(OCH$_3$)$_2$, CO$_2$ JACS *107* 7184 (1985)

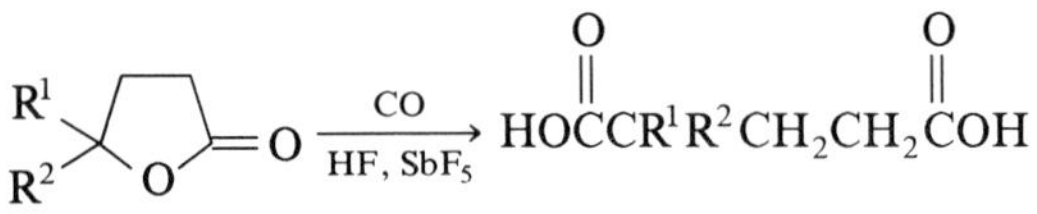

$$R^1R^2\text{-lactone}=O \xrightarrow[\text{HF, SbF}_5]{\text{CO}} HOCCR^1R^2CH_2CH_2COH$$

CC 767 (1981)

2. Esters

$$3° \text{ RH} \xrightarrow[\text{2. R'OH}]{\text{1. HCO}_2\text{H, H}_2\text{SO}_4\text{, R''OH}} RCO_2R'$$

J Gen Chem USSR *33* 3206 (1963)
JOC USSR *1* 1596 (1965); *2* 256 (1966)

$$\text{ArH} \longrightarrow \text{ArCO}_2\text{R}$$

n-BuLi/ClCO$_2$Et   JACS *106* 3286 (1984)

Tl(O$_2$CCF$_3$)$_3$/CO, cat Li$_2$PdCl$_4$, ROH   JOC *45* 363 (1980)
          JACS *104* 1900 (1982)

$$o\text{-X}-\text{Ar}-\text{H} \longrightarrow o\text{-X}-\text{Ar}-\text{PdOAc(Cl)} \xrightarrow[\text{ROH}]{\text{CO}} o\text{-X}-\text{Ar}-\text{CO}_2\text{R}$$

X

NHCOR   JOC *46* 4416 (1981)

CHRNR'$_2$ JOMC *182* 537 (1979)
          BCSJ *52* 142, 957 (1979)

CH$_2$CH$_2$NMe$_2$ TL *27* 1971 (1986)

$$\text{>C=C<} \xrightarrow[\text{2. ROH}]{\text{1. HCO}_2\text{H, H}_2\text{SO}_4} \text{H}-\overset{|}{\text{C}}-\overset{|}{\text{C}}-\text{CO}_2\text{R}$$

JOC USSR *1* 1814 (1965); *2* 256 (1966)

$$\text{RCH=CHCH}_3 \text{ or } \text{RCH}_2\text{CH=CH}_2 \xrightarrow[\text{3. Br}_2, \text{ MeOH}]{\substack{\text{1. Cp}_2\text{ZrHCl} \\ \text{2. CO}}} \text{R(CH}_2)_3\overset{\overset{\text{O}}{\|}}{\text{C}}\text{OMe}$$

JACS *97* 228 (1975)

$$\text{H}_2\text{C=CCH}_3 \longrightarrow \text{H}-\text{C}\overset{R}{\underset{H}{\overset{|}{\underset{\text{Pd}}{C}}}}\text{C}-\text{H} \xrightarrow[\text{ROH}]{\substack{\text{CO} \\ \text{NaO}_2\text{CR}}} \text{H}_2\text{C=CCH}_2\overset{\overset{\text{O}}{\|}}{\text{C}}\text{OR}$$

Organomet *1* 888 (1982)

$$\text{RCH=CHCH}_2\text{OC}\overset{\overset{\text{O}}{\|}}{}\text{OEt or } \text{RCHCH=CH}_2 \xrightarrow[\text{cat PPh}_3]{\substack{\text{CO} \\ \text{cat Pd(OAc)}_2}} \text{RCH=CHCH}_2\text{C}\overset{\overset{\text{O}}{\|}}{}\text{OEt}$$

TL *23* 5189 (1982)

$$\text{CH}_3\text{CH=CH(CH}_2)_2\overset{\overset{\text{OH}}{|}}{\text{C}}\text{HR} \xrightarrow[\text{CH}_3\text{OH}]{\substack{\text{CO} \\ \text{PdCl}_2 \\ \text{CuCl}_2}}$$

JACS *106* 1496 (1984)

$$\text{RC≡CR'} \longrightarrow \overset{R}{\underset{H}{>}}\text{C=C}\overset{R'}{\underset{\text{BX}_2}{<}} \xrightarrow[\substack{\text{NaOAc} \\ \text{MeOH}}]{\substack{\text{CO} \\ \text{cat PdCl}_2}} \overset{R}{\underset{H}{>}}\text{C=C}\overset{R'}{\underset{\text{CO}_2\text{Me}}{<}}$$

$$\text{X}_2 = \text{, (OSia)}_2$$

CL 879 (1981)

$$\text{R}^1\text{C≡CR}^2 \longrightarrow \overset{R^3}{\underset{R^2}{>}}\text{C=C}\overset{R^1}{\underset{\text{FeCp(CO)[P(OPh)}_3]}{<}} \xrightarrow[\text{ROH}]{\substack{\text{CO} \\ \text{(NH}_4)_2\text{Ce(NO}_3)_6}} \overset{R^3}{\underset{R^2}{>}}\text{C=C}\overset{R^1}{\underset{\text{CO}_2\text{R}}{<}}$$

JACS *108* 1940 (1986)

$$\text{RC≡CH} \xrightarrow[\text{NaBH}_4]{\text{(PhTe)}_2} \xrightarrow[\substack{\text{CuCl}_2 \\ \text{R'OH}}]{\substack{\text{CO} \\ \text{cat PdCl}_2}} \overset{R}{\underset{H}{>}}\text{C=C}\overset{\text{CO}_2\text{R'}}{\underset{H}{<}}$$

JOC *52* 4859 (1987)

$$RC\equiv CH \longrightarrow RC\equiv C\overset{\overset{\displaystyle O}{\|}}{C}OR'$$

CO, R'OH, NaOAc, CuCl$_2$, cat PdCl$_2$	TL 21 849 (1980); 28 1857 (1987) JACS 108 800 (1986)
LDA/ClCO$_2$Me	JOC 52 2378 (1987)
RLi/ClCO$_2$R'	CC 1290 (1982) JOC 47 4611 (1982) Org Syn 64 108 (1985); 65 47 (1987) TL 28 2941 (1987)
RMgX/ClCO$_2$Me	Org Syn 60 81 (1981) JOC 52 1889 (1987)

$$RC\equiv CH \longrightarrow RCH=CHCO_2R$$

See page 215, Section 2.

$$RX \longrightarrow R\overset{\overset{\displaystyle O}{\|}}{C}OR'$$

LiC(OMe)$_2$CN/H$_3$O$^+$ (R =1° alkyl)	TL 22 4279 (1981)
Mg/NCC(OEt)$_3$/H$_3$O$^+$ (R =1°, 2° alkyl; aryl)	TL 22 1509 (1981)
Mg/OC(OEt)$_2$ (R = aryl)	Org Syn Coll Vol 2 282 (1943)
Mg/ClCO$_2$Et (R = aryl)	Ber 36 3087 (1903)
Mg or Li/MnI$_2$/ClCO$_2$Et (R =1° alkyl, aryl, vinylic)	BSCF 570 (1977)
Na$_2$Fe(CO)$_4$/(CO)/I$_2$, R'OH (R =1° alkyl)	JACS 95 249 (1973)
Fe(CO)$_5$, NaOR' (R =1° alkyl)	CL 1355 (1977)
Ni(CO)$_4$, NaOR' (R =1° alkyl, aryl, vinylic)	JACS 91 1233 (1969); 92 6314 (1970); 104 6646 (1982)
(Ph$_3$P)$_2$Ni(CO)$_2$, Et$_3$N (R = vinylic; lactone)	JACS 104 6879 (1982)
CO, R'OH, Pd catalyst (R = benzylic, aryl, vinylic, heterocyclic)	JOC 39 3318 (1974); 40 532 (1975); 47 3630 (1982) (lactone) BCSJ 48 2075, 2091 (1975) JACS 98 5832 (1976); 102 4193 (1980) TL 133 (1979) (lactone); 21 3885 (1980); 23 4361 (1982) Heterocycles 12 921 (1979) (lactone) CL 369 (1980)
CO, Ti(OR')$_4$, cat Pd(PPh$_3$)$_4$ (R =1° alkyl, aryl, vinylic)	JOC 50 2134 (1985)
CO, Zr(OR')$_4$, cat Pd(PPh$_3$)$_4$, cat [ClRh(H$_2$C=CHCH$_2$CH$_2$CH=CH$_2$)]$_2$ (R = aryl, vinylic)	JOC 50 2134 (1985)

CO, MeOH, cat $PtCl_2(PPh_3)_2$, K_2CO_3 CC 351 (1986)
  (R =1°, 2° alkyl; aryl)

CO, $M(OR')_4$ (M = Ti, Zr), cat   JOC *50* 2134 (1985)
  $[ClRh(H_2C{=}CHCH_2CH_2CH{=}CH_2)]_2$
  (R = benzylic)

CO, HCO_2R', cat   CC 167 (1986)
  $[ClRh(H_2C{=}CHCH_2CH_2CH{=}CH_2)]_2$, cat
  KI (R =1° alkyl; 1°, 2° benzylic)

CO, HCO_2R', cat   CC 167 (1986)
  $[ClRh(H_2C{=}CHCH_2CH_2CH{=}CH_2)]_2$, cat
  $Pd(PPh_3)_4$, cat KI (R = aryl)

$Mg/Fe(CO)_5/I_2$, R'OH   TL 1477 (1978)
  (R =1° alkyl, aryl)

NaOH, $MeSCH_2SO_2Me/SOCl_2/MeOH$ TL *22* 4499 (1981)
  (R =1° alkyl, benzylic, allylic)

NaOH, $MeSCH_2SO_2Tol$ (phase transfer)/   CL 767 (1983)
  H_2O_2, HOAc/MeOH, H^+
  (R =1° alkyl, benzylic, allylic)

$\overset{O_2}{\underset{O_2}{\overset{S}{\underset{S}{\bigcirc\!\!\!\!\bigcirc}}}}$ —$OCH_2CH_2SiMe_3$, Cs_2CO_3/   JACS *106* 2469 (1984)

$BCl_3/R'OH$ (R =1° alkyl)

$$ArCH_2X \xrightarrow[\substack{cat\ Pd(dppe)_2 \\ (R_4N)HSO_4}]{CO} ArCH_2\overset{\overset{O}{\|}}{C}OCH_2Ar$$

Organomet *1* 775 (1982)

$$ArX + CO + ROH \xrightarrow[NEt_3]{cat\ PdCl_2(PR_3)_2} Ar\overset{\overset{O}{\|}}{C}\!-\!\overset{\overset{O}{\|}}{C}OR$$

CL 567 (1985)
J Mol Catal *32* 115 (1985); *34* 317 (1986)
JOC *52* 5733 (1987)

$$RX \longrightarrow R(CH_2)_n CO_2R$$

$$n = 1, 2$$

See page 873, Section 2.

$$ArOTf + CO + ROH \xrightarrow{Pd\ catalyst} ArCO_2R$$

TL *27* 3931 (1986)
CC 904 (1987)

$$\underset{\overset{|}{RCH=CR}}{\overset{OTf}{}} + CO + MeOH \xrightarrow{\text{Pd catalyst}} \underset{\overset{|}{RCH=CR}}{\overset{CO_2Me}{}}$$

TL *26* 1109 (1985)
CC 1002 (1987)

$$ArCH_2OH + CO + ROH \xrightarrow[\text{Co}_2(\text{CO})_8,\ n\text{-Bu}_3\text{N}]{\text{NaI, ethyl polyphosphate}} ArCH_2CO_2R$$

BCSJ *55* 643 (1982)

$$RCHO \xrightarrow[\text{2. MeOH, HgO, HgCl}_2]{\text{1. LiC(SMe)}_3} \underset{\overset{|}{RCHCO_2Me}}{\overset{OH}{}}$$

TL *28* 475 (1987)

$$\underset{RCCHR_2}{\overset{O\ \|}{}} \longrightarrow \underset{RCCR_2COR'}{\overset{O\ \ \ \ O\ \|\ \ \ \|}{}}$$

LDA/CO$_2$/CH$_2$N$_2$	JACS *108* 800 (1986)
KN(SiMe$_3$)$_2$/CO$_2$/CH$_2$N$_2$	JACS *102* 889 (1980)
$\underset{\overset{|}{Li}}{\overset{Ph\ \ \ \ \ Ph}{CH_3\diagup\overset{\diagdown}{N}\diagdown CH_3}}$/CO$_2$/MeI	TL *27* 2767 (1987) (enantioselective)
LDA/NCCO$_2$R'	TL *24* 5425 (1983); *26* 2291 (1985); *28* 1051 (1987) JACS *109* 2850 (1987)
(Li)NaN(SiMe$_3$)$_2$/NCCO$_2$Me	JACS *107* 2730 (1985) TL *28* 3985 (1987)
MeLi/NCCO$_2$R (on enol silane)	TL *26* 5433 (1985)
NaOMe, OC(OMe)$_2$	JACS *107* 7967 (1985)
NaH/OC(OEt)$_2$	JACS *69* 2677 (1947); *70* 2287 (1948); *75* 4287 (1953) Chem Ind 576 (1956) Org Syn Coll Vol *5* 198 (1973)
NaNH$_2$/OC(OEt)$_2$	JACS *66* 1768 (1944); *67* 2197 (1945)
KO-*t*-Bu/OC(OEt)$_2$	BSCF 543 (1959)
NaO-*t*-Am, CS$_2$/MeI/MeOH, BF$_3$·OEt$_2$, HgO	Syn Commun *12* 829 (1982)

$$RCH_2CO_2R \longrightarrow \underset{\overset{|}{RCHCO_2R}}{\overset{CO_2R'}{}}$$

ClCO$_2$R', DMAP	JOC *52* 5425 (1987)
LDA/NCCO$_2$R'	TL *27* 1221, 1225, 1229 (1986) (all lactones); *28* 1051, 1725 (1987)

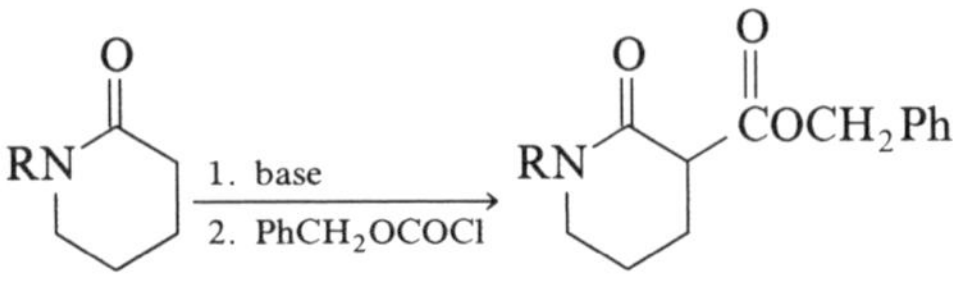

TL *28* 4901 (1987)

3. Lactones

See page 941, Section 8.

4. Acid Anhydrides

$$2\ \mathrm{ArH} \xrightarrow[\mathrm{Pd(OAc)_2}]{\mathrm{CO,\ BrCH_2CH_2Br}} \mathrm{ArC(=O)OC(=O)Ar}$$

CC 132 (1982)

$$\mathrm{(ArN_2)BF_4 + CO + RCO_2Na} \xrightarrow{\text{Pd catalyst}} \mathrm{ArC(=O)OC(=O)R}$$

TL *21* 2877 (1980)
JOC *46* 4413 (1981)

5. Amides

$$\mathrm{RNH_2 + CO} \xrightarrow{\text{cat } \mathrm{Ru_3(CO)_{12}}} \mathrm{RNHC(=O)H}$$

JOMC *309* 333 (1986)

$$\mathrm{R_2NH + CO} \xrightarrow{\mathrm{NiBr_2}} \mathrm{R_2NC(=O){-}C(=O)NR_2}$$

JOMC *236* C53 (1982)

$$\mathrm{ArH} \longrightarrow \mathrm{ArC(=O)NH_2}$$

$ClCONH_2$, $AlCl_3$ — Ann *244* 47 (1888); Houben-Weyl VIII (1952), p 380 (review)

$EtOCONH_2$, $AlCl_3$ — Syn 977 (1981)

$$\mathrm{RCH{=}CH_2 + CO + H_2NR} \xrightarrow{\text{cat } \mathrm{Ru_3(CO)_{12}}} \mathrm{RCH_2CH_2C(=O)NHR}$$

JOMC *309* 333 (1986)

$$RCH{=}CH_2 + H\overset{\overset{\displaystyle O}{\displaystyle \|}}{C}NR'_2 \xrightarrow[\text{initiator}]{\text{free radical}} RCH_2CH_2\overset{\overset{\displaystyle O}{\displaystyle \|}}{C}NR'_2$$

R' = H, Me

Angew *73* 621 (1961)
Chem Ind 362 (1962)
JOC *29* 1855 (1964); *30* 3361 (1965); *31* 3829 (1966)
Z Chem *4* 177 (1964)
Syn 99 (1970) (review)

$$H_2C{=}CHCH{=}CH_2 \xrightarrow[i\text{-PrMgBr}]{\overset{\text{Cp}_2\text{TiCl}_2}{} \; \overset{\text{1. PhNCO}}{\underset{\text{2. H}_2\text{O}}{}}} H_2C{=}CHCH(CH_3)\overset{\overset{\displaystyle O}{\displaystyle \|}}{C}NHPh$$

CC 342 (1981)
JOMC *224* 327 (1982)

$$RC{\equiv}CH \longrightarrow RC{\equiv}C\overset{\overset{\displaystyle O}{\displaystyle \|}}{C}NR'_2$$

RLi/R'NCO/H$_2$O   Ann 1844 (1981)

ClCONMe$_2$, cat PdCl$_2$(PPh$_3$)$_2$,   Syn 777 (1977)
  cat CuI, Et$_3$N

$$RX \longrightarrow R\overset{\overset{\displaystyle O}{\displaystyle \|}}{C}NR'_2$$

n-BuLi/ClCONEt$_2$ (R = aryl)   JACS *109* 3402 (1987)

Mg/PhNCO/H$_2$O   JOC *18* 441 (1953)
  (R = 2° alkyl, 2° benzylic)

Mg/Cp$_2$TiCl$_2$, PhNCO/H$_2$O   TL *28* 3815 (1987)
  (R =1°, 2° alkyl in RX;
  R =1° alkyl in RCONR'$_2$)

Mg, *t*-BuNCO, ultrasound/H$_2$O (R = aryl)   TL *27* 501 (1986)

Ni(CO)$_4$, HNR'$_2$ (R = vinylic)   JACS *91* 1233 (1969)

Li[(CO)$_3$NiCONMe$_2$]   JOC *36* 2721 (1971)
  (R = aryl, benzylic, allylic, vinylic)

LiCu(CONR'$_2$)$_2$ (R = aryl, allylic)   JOC *44* 3734 (1979)

CO, HNR'$_2$, Pd catalyst   JOC *39* 3327 (1974); *43* 1684 (1978) (lactam)
  (R = aryl, vinylic)   BCSJ *48* 2091 (1975)
                         Heterocycles *6* 1711, 1841 (1977) (lactam)
                         CC 741 (1982) (lactam)
                         Chem Pharm Bull *32* 3840 (1984) (lactam)
                         JACS *107* 4577 (1985)

CO, R'$_3$N, cat PhPdI(PPh$_3$)$_2$ JOMC *231* C12 (1982)
  (R = aryl, vinylic)

$$\text{(benzodithiole-1,1,3,3-tetraoxide)}-OCH_2CH_2SiMe_3,$$

JACS *106* 2469 (1984)

$Cs_2CO_3 / BCl_3 / HNR'_2 \ (R = 1° \ \text{alkyl})$

$$RX + CO + HNR'_2 \xrightarrow{\text{Pd catalyst}} RC(=O)-C(=O)NR'_2$$

R = aryl, vinylic

JOMC *233* C64 (1982)
TL *23* 3383 (1982)
CL 865 (1982)
Organomet *3* 683, 692 (1984)
JACS *106* 1506 (1984); *107* 3235 (1985)
JOC *51* 415 (1986)

$$ArCH_2OH + CO + HNR_2 \xrightarrow[\text{ethyl polyphosphate}]{\text{NaI, } Co_2(CO)_8} ArCH_2C(=O)NR_2$$

BCSJ *55* 643 (1982)

$$ArO_3SCF_3 \xrightarrow[R_2NH, Ar_3P]{\substack{CO \\ \text{cat } Pd(OAc)_2}} ArC(=O)NR_2$$

TL *27* 3931 (1986)

$$\underset{\substack{| \\ OTf}}{RCH=CR} \xrightarrow[\text{Pd catalyst}]{CO, R_2NH} \underset{\substack{| \\ CONR_2}}{RCH=CR}$$

TL *26* 1109 (1985); *28* 6117 (1987)

$$RCH_2CO_2R \longrightarrow \underset{\substack{| \\ OSiMe_3}}{RCH=COR} \xrightarrow[2.\ H_2O]{1.\ ArNCO} \underset{\substack{| \\ CONHAr}}{RCHCO_2R}$$

JOMC *164* 123 (1979)

6. Nitriles

$$ArH \xrightarrow[\text{2. DMF}]{\text{1. } ClSO_2NCO} ArCN$$

Org Syn *50* 52 (1970)

$$\underset{}{>}C=C\underset{}{<} \longrightarrow H-\overset{|}{\underset{|}{C}}-\overset{|}{\underset{|}{C}}-CN$$

Review:

    E. S. Brown in "Organic Syntheses via Metal Carbonyls," Eds. I. Wender and P. Pino, Wiley Interscience, New York (1977), Vol 2, p 655

HCN, cat $Co_2(CO)_8$ JACS *76* 5364 (1954)

HCN, Ni catalyst

J Catalysis *26* 254 (1972)
CC 1098 (1981)
Austral J Chem *35* 2041 (1982)
Adv Catalysis *33* 1 (1985) (review)

HCN, Pd catalyst

CC 112 (1969)
JACS *101* 6128 (1979)
TL *23* 1621 (1982)
Austral J Chem *35* 2041 (1982)

$Cp_2ZrHCl/RN=C$ (R = *t*-Bu, $SiMe_3$)/I_2 TL *28* 295 (1987)

$$RX + CN^- \longrightarrow RCN$$
$$R = 1°, 2° \text{ alkyl}$$

Chem Rev *42* 189 (1948)
"The Chemistry of the Cyano Group," Ed. S. Patai, Interscience, New York (1970), p 67

$$R_3CX \longrightarrow R_3CCN$$

Me_3SiCN, $SnCl_4$

Angew Int *20* 1017 (1981)
Tetr *39* 961 (1983)

t-BuLi/$(CN)_2$

JOC *52* 2674 (1987)

$$RCH=CHX \xrightarrow{\text{reagent(s)}} RCH=CHCN$$

$\underline{X}$	Reagent(s)	
Cl, Br	KCN, $NiBr_2(PPh_3)_2$, Zn, PPh_3	CL 1565 (1982)
	$Co(CN)_4^{3-}$	JACS *104* 1560 (1982)
Br	$K_4Ni_2(CN)_6$, KCN	JACS *91* 1233 (1969)
	KCN, cat $Pd(PPh_3)_4$	TL 4429 (1977)
	$Co(CN)_5^{3-}$	JOC *50* 3934 (1985)
	CuCN	BSCF 720 (1973)
		JOC *43* 2839 (1978);
		52 2674 (1987)
	$NaCu(CN)_2$	JOC *34* 3626 (1969)
I	CuCN	JOMC *93* 415 (1975)
		TL *28* 6351 (1987)

$$RCH=CHSO_2Ar \xrightarrow[\text{crown ether}]{KCN} \overset{\displaystyle CN}{\underset{\displaystyle |}{R}}C=CH_2$$

JOC *46* 4817 (1981)

$$ArX \longrightarrow ArCN$$
Review: Chem Rev *42* 189 (1948)

Reagent(s)	$\underline{X}$	
CuCN	Cl, Br, I	Org Syn Coll Vol *3* 212, 631 (1955)
		JACS *81* 3667 (1959)
		JOC *26* 2522, 2525 (1961);
		34 3626 (1969);
		50 2128 (1985); *51* 4169 (1986);
		52 3196 (1987)
		JCS 1097 (1964)
		Quart Rev *19* 95 (1965)

Reagent(s)	X	
CuCN-charcoal	Br	CC 877 (1986)
NaCu(CN)$_2$	I	JOC *34* 3626 (1969)
Me$_3$SiCN, cat Pd(PPh$_3$)$_4$	I	JOC *51* 4714 (1986)
NaCN, cat Ni(PPh$_3$)$_4$	Cl, Br, I	JOMC *54* C57 (1973)

(Na)KCN, cat Ni(PPh$_3$)$_4$
   or ArNiCl(PPh$_3$)$_2$,
   phase transfer
	Cl, Br	JOMC *173* 335 (1979)

KCN, cat Pd(OAc)$_2$, NaOEt	Br, I	BCSJ *48* 3298 (1975)
KCN, cat Pd(PPh$_3$)$_4$	I	CL 277 (1975)
NaCN-alumina, cat Pd(PPh$_3$)$_4$	Br, I	JOC *44* 4443 (1979)

$$RM \longrightarrow RC{\equiv}N$$

RM	Reagent(s)	
RLi (R = vinylic, alkynyl)	PhOCN	Syn 150 (1980)
RC≡CLi	ClCN	Rec Trav Chim *92* 667 (1973)
RC≡CMgX	ClCN	BSCF IV *17* 228 (1915)
RCH=CHCu·MgX$_2$	ClCN, ArSO$_2$CN	Syn 784 (1977)
E-RCH=CHAlMe(*i*-Bu)$_2$	(CN)$_2$	JACS *90* 7139 (1968)

$$ArNH_2 \longrightarrow ArN_2{}^+ \xrightarrow{\ CuCN\ } ArCN$$

Org Syn Coll Vol *1* 514 (1941)

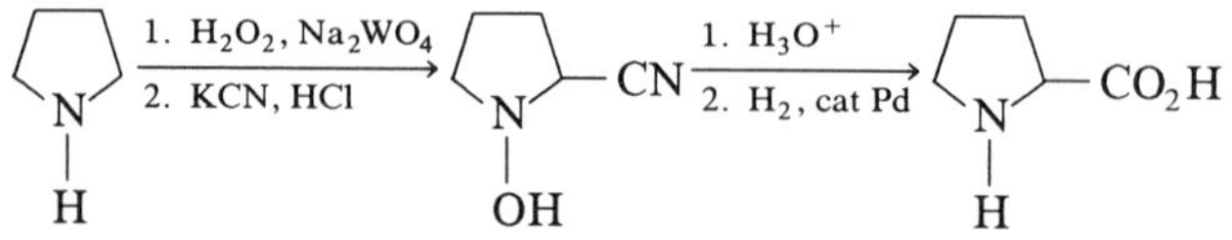

TL *28* 6469 (1987)

$$ROH \xrightarrow[\text{cat NaI}]{\text{Me}_3\text{SiCl, NaCN}} RCN$$

R = 1°, 2°, 3° alkyl

JOC *46* 2985 (1981)

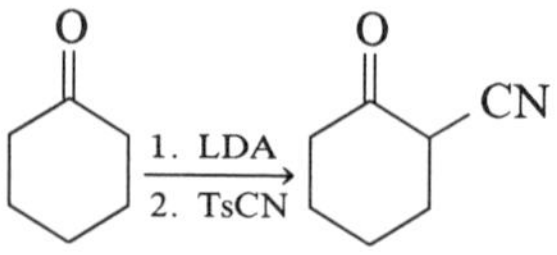

TL *22* 5011 (1981)

4. CARBONYL HOMOLOGATION AND DEGRADATION

See also page 941, Section 8.

1. Homologation

$$\underset{\text{RCH}}{\overset{\text{O}}{\|}} \longrightarrow \underset{\text{RCH}_2\text{COR}'}{\overset{\text{O}}{\|}}$$

R	Reagents	
aryl	NaC(NMe$_2$)[PO(OEt)$_2$]$_2$/H$_3$O$^+$ (R' = H)	Angew Int 7 391 (1968)
	n-Bu$_3$SnCH(OEt)$_2$, n-BuLi/H$_2$O or CH$_3$OH?	JOMC 212 C13 (1981)
	MeSOCH$_2$SMe, base/H$^+$, R'OH	TL 1383 (1972) BCSJ 52 2013 (1979)
	ArCONMeCH$_2$CN, NaOH (phase transfer)/H$_3$O$^+$	Syn 1043 (1983)
	CBr$_4$, PPh$_3$/OH$^-$, cat Pd(dppe)$_2$, PEG-400 (R' = H)	JOC 51 4354 (1986)
vinylic	CBr$_4$, PPh$_3$/OH$^-$, cat Pd(dppe)$_2$, PEG-400 (R' = H)	JOC 51 4354 (1986)

$$\underset{\text{ArCR}}{\overset{\text{O}}{\|}} \xrightarrow{\text{Me}_3\text{SiCN}} \underset{\text{ArCR}}{\overset{\text{Me}_3\text{SiO}\quad\text{CN}}{\diagdown\;\diagup}} \xrightarrow[\substack{2.\ \text{HOAc} \\ 3.\ \text{HCl}}]{1.\ \text{SnCl}_2} \underset{\text{ArCHRCOH}}{\overset{\text{O}}{\|}}$$

Syn Commun 12 763 (1982)

$$\text{(RCHO) } R_2C{=}O \longrightarrow R_2CH\overset{\overset{\displaystyle O}{\|}}{C}OR'$$

/H_2O, $HgCl_2$ (R' = H)   CC 526 (1972)

$NCCH_2Cl$, NaO-t-Am/HCl/Ac_2O, Et_3N,   CC 988 (1974)
   Δ/OH^-/H^+ (R' = H)

$NCCH_2Cl$, NaO-t-Am/$LiClO_4$/OH^-/H^+ CC 988 (1974)
   (R' = H)

$PhNMeCH_2CN$, KH/H_3O^+ (R' = H)   JOC 48 3566 (1983)

$(EtO)_2PO\overset{-}{C}H(CN)O$-$t$-Bu/$Ac_2O$, $ZnCl_2$,   JACS 99 182 (1977)
   Δ/ OH^- or $R'O^-$

$Ph_3P{=}CF_2$ / R'OH-H_2SO_4 or   CL 651 (1980)
   R'OH-$Hg(OAc)_2$-$(CF_3CO)_2O$

CBr_4, PPh_3/OH^-, cat   JOC 51 4354 (1986)
   $Pd(dppe)_2$, PEG-400 (R' = H)

$$\text{(RCHO) } R_2C{=}O \longrightarrow R_2\overset{\overset{\displaystyle OH}{|}}{C}CO_2R'$$

/$Hg(ClO_4)_2$,   Syn Commun 11 209 (1981)
MeOH (R' = Me, ketones only)

$(RS)_3CLi$/$HgCl_2$, H_2O Angew Int 6 442 (1967)
   (R' = H, aldehydes and ketones)   Ber 105 487 (1972)
                                       TL 22 4009 (1981)

See also page 849, Section 1.

$$\text{(RCHO) } R_2C{=}O \longrightarrow R_2CH\overset{\overset{\displaystyle O}{\|}}{C}NR'_2$$

$Ph_3P{=}CF_2$/$LiNR'_2$/H_2O (aldehydes only)   CL 935 (1980)

$(EtO)_2PO\overset{-}{C}H(CN)O$-$t$-Bu/$Ac_2O$, $ZnCl_2$,   JACS 99 182 (1977)
   Δ/R'_2NH (ketones only)

$$\text{(RCHO) } R_2C{=}O \longrightarrow R_2CHC{\equiv}N$$

$TsCH_2NC$, KO-t-Bu   TL 1357 (1973)
   (ketones only)

2,4,6-(i-Pr)$_3C_6H_2SO_2NHNH_2$/KCN, Δ CC 280 (1977)
   (aldehydes and ketones)

$$RCO_2R \longrightarrow RCH_2CO_2R'$$

$MeSOCHNaSMe$/$NaBH_4$/Ac_2O/R'OH, H^+ CL 659 (1974)

$LiCHBr_2$/n- or t-BuLi/EtOH (R' = Et)   JACS 104 321 (1982); 107 1429 (1985)

$$R_2CHCOR' \longrightarrow R_2C = \overset{\overset{\displaystyle OSiMe_3}{|}}{C}OR' \xrightarrow[\text{2. MeOH, Et}_3\text{N, }\Delta]{\text{1. }n\text{-BuLi, CH}_3\text{CHCl}_2} \underset{R}{\overset{R}{>}}C = C \underset{CO_2R'}{\overset{CH_3}{<}}$$

Syn 58 (1982)

$$\overset{O}{\overset{||}{RCOR'}} \xrightarrow[\text{2. Cu(OAc)}_2\text{, EtOH}]{\text{1. PhNMeCH}_2\text{CN, KH}} \overset{O}{\overset{||}{RC}} - \overset{O}{\overset{||}{CNPhMe}}$$

CL 859 (1983)

$$\overset{O}{\overset{||}{ClC}}(CH_2)_n\overset{O}{\overset{||}{CCl}} \xrightarrow[Me_2N\text{—}\langle N\rangle]{} \quad \xrightarrow[\text{2. HCl}]{\text{1. NaBH}_3\text{CN, HOAc}} \overset{O}{\overset{||}{HOC}}(CH_2)_{n+4}\overset{O}{\overset{||}{COH}}$$

Syn Commun 12 19 (1982)

$$\overset{O}{\overset{||}{HCNR_2}} \xrightarrow{t\text{-BuLi}} \xrightarrow{R_2CO \text{ (RCHO)}} \underset{R_2}{\overset{\overset{\displaystyle HO}{|}}{C}} - \overset{O}{\overset{||}{CNR_2}}$$

Angew Int 12 836 (1973); 15 293 (1976)
Can J Chem 52 185 (1974)
CC 387 (1976)
JCS Perkin I 1881 (1977)

$$R_2C = O \xrightarrow[\substack{\text{3. KOH} \\ \text{4. H}^+}]{\substack{\text{1. LiC} \equiv \text{COEt} \\ \text{2. H}_2\text{SO}_4}} R_2C = CH\overset{O}{\overset{||}{C}}OH$$

JACS 108 2691 (1986)

$$R_2C = O \longrightarrow R_2C = CH\overset{O}{\overset{||}{C}}OR$$
See page 173, Section 4.

$$RCH_2CHR'\overset{O}{\overset{||}{CH}} \xrightarrow[\langle NH\rangle]{PhSOCH_2CO_2Et} RCH_2CHR'CH = C(SOPh)CO_2Et \xrightarrow[\Delta]{K_2CO_3} RCH = CR'CH = CH\overset{O}{\overset{||}{C}}OEt$$

CL 781 (1980)

$$R_2CH\overset{O}{\overset{||}{CH}} \xrightarrow[\langle NH\rangle]{ArSOCH_2CO_2Me} E\text{-}R_2\overset{\overset{\displaystyle HO}{|}}{C}CH = CH\overset{O}{\overset{||}{C}}OMe$$

Syn 134 (1983)

$$(RCH_2\overset{O}{\overset{\|}{C}}CH_3) \quad RCH_2\overset{O}{\overset{\|}{C}}H \xrightarrow[\text{base}]{ArSOCH_2CN} RCH\overset{OH}{\overset{|}{C}}HCH=CHCN$$

TL *22* 4489 (1981); *27* 5109 (1986) (enantioselective)
JACS *106* 7890 (1984)

2. Degradation

$$R\overset{O}{\overset{\|}{C}}R' \longrightarrow R\overset{O}{\overset{\|}{C}}NH_2 + R'H$$

MNH$_2$ (M = Li, Na, K)   JACS *109* 6858 (1987)

NaNH$_2$ Org Rxs *9* 1 (1957) (review)

NaNH$_2$, DABCO   Syn 395 (1975)

$$RCH_2\overset{O}{\overset{\|}{C}}OH \xrightarrow[\text{2. }O_2]{\text{1. 2 LDA}} RCH\overset{HO}{\overset{|}{C}}\overset{O}{\overset{\|}{C}}OH \xrightarrow[\text{CrO}_3]{NaIO_4} R\overset{O}{\overset{\|}{C}}OH$$

Syn Commun *9* 63 (1979)

5. ALKYLATION, ACYLATION AND SUBSTITUTION OF NITRILES, CARBOXYLIC ACIDS AND DERIVATIVES

See also page 119, Section 8, for ene reactions leading to unsaturated carboxylic acids, esters and amides.

1. Carboxylic Acids

See also page 873, Section 2.

$$RCH{=}CH_2 + HCR_2X \xrightarrow[\text{initiator}]{\text{free radical}} RCH_2CH_2CR_2X$$

$$X = CO_2H, CO_2R, CONR_2, CN$$

Syn 99 (1970) (review)
JCS Perkin II 1655 (1973)

$$R_2CHCO_2H \xrightarrow[\substack{2.\ E^+ \\ 3.\ H_2O}]{1.\ \text{base}} R_2\overset{\overset{\displaystyle E}{|}}{C}CO_2H$$

Reviews:

Syn 509 (1977); 521 (1982)
P. L. Creger, "Annual Reports in Medicinal Chemistry," Vol 12, Academic Press, New York (1977), Chpt 12

$\underline{E^+}$

D_2O \qquad JOC *37* 451 (1972)

R'X \qquad JCS 1551 (1950)
JACS *78* 4942 (1956); *89* 2500 (1967); *92* 1397 (1970); *103* 5459 (1981); *104* 5523 (1982); *107* 7776 (1985); *109* 1186, 7816 (1987)

JOC *26* 3696 (1961); *32* 2797 (1967); *35* 262 (1970); *37* 451 (1972); *45* 3236 (1980) (stereochemistry); *46* 1616 (1981); *47* 893 (1982); *50* 2719 (1985); *51* 1541 (1986); *52* 1309, 2549 (1987)

Org Syn *50* 58 (1970)

Syn 517 (1975); 710 (1980)

Can J Chem *58* 716 (1980); *60* 1238 (1982)

Tetr *36* 775 (1980)

TL *27* 4545 (1986); *28* 2941 (1987)

CC 656 (1987) (chiral base)

$PhCH_2SeCH_2Br$ JOC *51* 2981 (1986)

epoxide

JACS *89* 2500 (1967)

Can J Chem *58* 716 (1980)

JOC *52* 34 (1987)

H_2CO JOC *37* 1256 (1972)

$RCHO, R_2CO$

JACS *74* 1730 (1952); *94* 2000 (1972); *102* 2841 (1980)

Compt Rend C *270* 1471 (1970)

BSCF 1848 (1970)

Israel J Chem *8* 731 (1970)

JOC *36* 1149 (1971); *40* 8 (1975); *46* 3359 (1981); *51* 1478 (1986); *52* 3143, 4124 (1987)

CC 52 (1979); 98 (1986)

Tetr *40* 2211 (1984) (enantioselective with chiral bases)

TL *28* 2753, 4787 (1987)

HCO_2R

TL 699 (1970); 603 (1974)

JOC *52* 4303 (1987)

RCO_2R JACS *93* 6321 (1971); *106* 1811 (1984)

$RCOCl$

JOC *42* 1189 (1977)

J Chem Res (S) 44 (1980)

EtO_2CCO_2Et TL *22* 2459 (1981)

CO_2

Israel J Chem *8* 731 (1970)

JOC *37* 451 (1972)

Syn 587 (1982)

$RCH{=}CHX$ (X = CO_2R, CN; 1,4-addition)   JACS *93* 6321 (1971)

$RCH{=}CHNO_2$ (1,4-addition)   CL 1505 (1982)

O_2

Syn 647 (1971) (hydroxy acid)

JOC *40* 3253 (1975) (hydroxy or hydroperoxy acid); *41* 370 (1976) (hydroperoxy acid)

MeSSMe   JACS *107* 8066 (1985)

RNCO, RNCS   JOC *45* 1106 (1980)

$$\text{RCH}_2\text{CO}_2\text{H} \xrightarrow[\text{2. I}_2]{\text{1. 2 LDA}} \overset{\overset{\displaystyle \text{HO}_2\text{C} \quad \text{CO}_2\text{H}}{|\qquad\quad|}}{\text{RCHCHR}}$$

TL *25* 5969 (1984); *27* 127 (1986); *28* 4441 (1987)
JOC *52* 2549 (1987)

$$\text{RCH}_2\text{CO}_2\text{H} \xrightarrow[\text{2. ICR}_2\text{CO}_2^-]{\text{1. 2 LDA}} \overset{\overset{\displaystyle \text{HO}_2\text{C} \quad \text{CO}_2\text{H}}{|\qquad\quad|}}{\text{RCHCR}_2}$$

TL *25* 5969 (1984)

$$\overset{\overset{\displaystyle \text{H}}{|}}{\underset{\underset{\displaystyle \text{X}}{|}}{\text{RCCO}_2\text{H}}} \xrightarrow[\text{2. E}^+]{\text{1. base}} \overset{\overset{\displaystyle \text{E}}{|}}{\underset{\underset{\displaystyle \text{X}}{|}}{\text{RCCO}_2\text{H}}}$$

X	R	E$^+$	
Cl	Me	RX, RCOCl (decarboxylation), α,β-unsaturated esters (cyclopropanes)	Syn 284 (1982)
PhO	H	D$_2$O, RX	Syn 828 (1978)
		D$_2$O, RX, RCHO, R$_2$CO	JOC *43* 772 (1978)
		$\overset{\displaystyle \text{O}}{\overset{\displaystyle /\backslash}{\text{H}_2\text{C}\!-\!\text{CHCH}_3}}$	TL 2835 (1977)
PhO	PhO	D$_2$O, RX	Syn 828 (1978)
MeO	Me	R$_2$CO	Can J Chem *51* 981 (1973)
MeS	CH$_2$CH=CH$_2$	RX, RCHO	CL 351 (1982)
PhS	H	$\overset{\displaystyle \text{O}}{\overset{\displaystyle /\backslash}{\text{RCH}\!-\!\text{CH}_2}}$	CL 385 (1974) BCSJ *50* 242 (1977) TL *26* 5623, 5627, 5631 (1985) JACS *108* 5352 (1986)
PhS	H, Me	RCH=CRNO$_2$ (1,4-addition)	CL 1505 (1982)
PhSe	H	$\overset{\displaystyle \text{O}}{\overset{\displaystyle /\backslash}{\text{H}_2\text{C}\!-\!\text{CHR}}}$	CC 754 (1986) TL *28* 1147 (1987)
CO$_2$R	H	RCOCl	JOC *44* 310 (1979) Org Syn *61* 5 (1983)
PhCONH	H	D$_2$O, RX	TL 2205 (1976)
CN	H	RCOCl	Syn 308 (1983)

$$\overset{O}{\underset{\|}{RCH_2COH}} \longrightarrow RCH=C(OSiMe_3)_2 \xrightarrow[2.\ H_3O^+]{1.\ E^+} \overset{E\ O}{\underset{|\ \|}{RCHCOH}}$$

$\underline{E^+}$

RCHClSPh, ZnBr$_2$ TL *23* 5083 (1982)

HC≡CCO$_2$Me,   JACS *107* 3879 (1985)
  TiCl$_4$ (E = CH=CHCO$_2$Me)

RCHO, TiCl$_4$ TL *25* 2143 (1984)

RCH=NR, TiCl$_4$ (lactam formation)   TL *25* 2143 (1984)

Pb(OAc)$_4$ (E = OAc)   TL *28* 3971 (1987)

$$\overset{H\ O}{\underset{|\ \|}{C=C-C-COH}} \text{ or } \overset{H}{\underset{|}{C-C=C-COH}} \xrightarrow[2.\ E^+]{1.\ \text{base}} \overset{E\ O}{\underset{|\ \|}{C=C-C-COH}}$$

$\underline{E^+}$

H$_2$O or D$_2$O   JOC *36* 3290 (1971); *52* 4471 (1987)
                                           TL *26* 1939 (1985)

R'X   JCS 1551 (1950)
                                           JACS *93* 4330 (1971); *98* 4925 (1976)
                                           JOC *36* 3290 (1971); *42* 260 (1977);
                                             46 239 (1981); *52* 4471 (1987)
                                           Gazz Chim Ital *104* 625 (1974)
                                           Tetr *32* 1347 (1976)
                                           TL 335 (1979); *22* 1691 (1981)

RCHO, R$_2$CO   TL 1163 (1973); *22* 4913 (1981); *23* 4773 (1982);
                                             28 3853 (1987)
                                           JCS Perkin I 400 (1973)
                                           Tetr *32* 107 (1976)
                                           JOC *40* 8 (1975); *46* 239 (1981); *49* 4424 (1984)
                                           JCS Perkin I 1651 (1978)
                                           Syn 802 (1985)

$$\overset{R\ NC}{\underset{|\ |}{CH_3C=CCO_2Et}} \xrightarrow[\substack{2.\ R'X \\ 3.\ H_3O^+}]{1.\ LDA} \overset{R\ NH_2}{\underset{|\ |}{H_2C=C-\underset{|}{C}-CO_2H}} \atop R'$$

Syn 646 (1981)

$$\overset{H\ O}{\underset{|\ \|}{C=C-C-COH}} \text{ or } \overset{H}{\underset{|}{C-C=C-COH}} \xrightarrow[2.\ E^+]{1.\ \text{base}} \overset{E}{\underset{|}{C-C=C-COH}}$$

$\underline{E^+}$

R'X, CuI   JACS *98* 4925 (1976)
                                           Tetr *32* 1347 (1976)
                                           CC 500 (1977)
                                           JOC *46* 239 (1981)
                                           TL *22* 1691 (1981)

RCHO, R_2CO

JOC 29 3161 (1964); 46 2439 (1981);
   49 4424 (1984)
JACS 90 3282 (1968)
Compt Rend C 270 1471 (1970)
BSCF 1848 (1970)
Chem Ind 80 (1972)
Austral J Chem 25 2393 (1973)
JCS Perkin I 400 (1973); 1651 (1978)
TL 1163 (1973); 3851 (1974); 22 4913 (1981);
   23 4773 (1982); 28 3853 (1987)
Gazz Chim Ital 103 117 (1973)
Tetr 32 107 (1976)
Syn 802 (1985)

H_2C=CHCOR (1,4-addition)

JCS Perkin I 1651 (1978)

H_2C=CHCH$\overset{O}{\overset{\diagup\diagdown}{—}}CH_2$, CuI (1,4-addition)

JOC 46 239 (1981)

$$CH_3CH=CHCH=CHCO_2H \xrightarrow[\text{2. }E^+]{\text{1. 2 LiNR}_2} H_2C=CHCH=CHCHCO_2H$$
$$\overset{E}{\overset{|}{}}$$

$\underline{E^+}$

RX

RCHO, R_2CO

JOC 48 3003 (1983)
TL 26 3625 (1985)

$$CH_3CH=CHCH=CHCO_2H \xrightarrow[\substack{\text{2. RCHO (R}_2\text{CO)} \\ \text{3. H}_2\text{O}}]{\text{1. 2 LiNEt}_2} R\overset{OH}{\overset{|}{C}}HCH_2CH=CHCH=CHCO_2H$$

TL 26 3625 (1985)

$$CH_3C\equiv CCO_2H \xrightarrow[\text{2. }E^+]{\text{1. 2 LiNR}_2} ECH_2C\equiv CCO_2H > H_2C=C=\overset{E}{\overset{|}{C}}CO_2H$$

$\underline{E^+}$

R′X

JOC 40 269 (1975)
Tetr 32 1347 (1976)

RCHO

TL 3221 (1975)

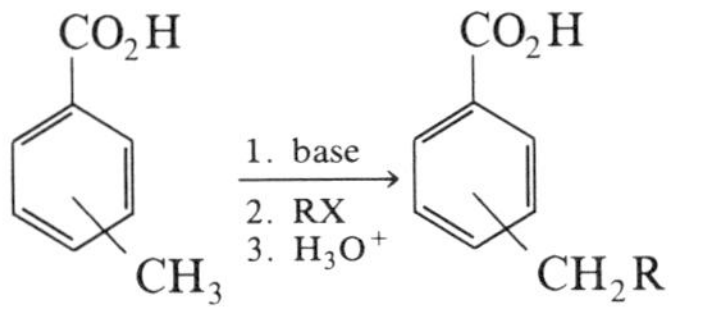

JACS 92 1396 (1970)

$$CH_3CO_2H \longrightarrow \underset{\substack{\text{(oxazine ring with N and O,} \\ CH_3)}}{} \xrightarrow[\text{2. RX}]{\text{1. }n\text{-BuLi}} \xrightarrow[\text{H}_2\text{O}]{\text{HBr}} RCH_2\overset{\displaystyle O}{\overset{\|}{C}}OH$$

JACS *91* 5886 (1969)

$$Ph_3COH + H_2C(CO_2H)_2 \xrightarrow[\Delta]{H^+} Ph_3CCH_2CO_2H$$

JACS *49* 1735 (1927)
JCS 716 (1962)

$$PhOCH_2CO_2H \xrightarrow{2\,LDA} PhO\bar{C}HCO_2^- \xrightarrow[\text{2. H}^+]{\text{1. R}_3\text{B}} RCH_2\overset{\displaystyle O}{\overset{\|}{C}}OH$$

TL 2891 (1978)

$$XCH_2CH_2CO_2H \xrightarrow[\text{2. E}^+]{\text{1. 2 LDA}} \overset{\displaystyle E}{\overset{|}{X}}CHCH_2CO_2H$$

$\underline{X}$	$\underline{E^+}$
SOPh	RX, R_2CO (lactone)
SO_2Ph	R_2CO (lactone)

Syn Commun *6* 357 (1976)

$$\underset{\substack{\text{HO} \\ |}}{}CH_3CHCH_2CO_2H \longrightarrow \underset{\substack{\text{BO}\quad\text{OB} \\ | \quad\ \ |}}{}CH_3CH=CSPh \xrightarrow[\substack{\text{2. HCl} \\ \text{3. KOH}}]{\text{1. RCH=NR}'} \underset{\substack{\text{OH} \\ | \\ \\ RCHNHR'}}{}CH_3CHCHCO_2H$$

TL *26* 1523 (1985); *27* 2149, 2153 (1986)

$$\underset{\substack{CH_2-C=O \\ | \qquad\ | \\ CH_2-O}}{} \xrightarrow{RM} RCH_2CH_2\overset{\displaystyle O}{\overset{\|}{C}}OH$$

$\underline{RM}$	
RMgX, cat CuX	TL *21* 935, 3377 (1980); *23* 3193, 3587 (1982)
	CL 571 (1980)
	JACS *109* 4649 (1987)
RMgX, cat Li_2CuCl_4	CL 569 (1982)
	BCSJ *56* 345 (1983)
R_2CuLi	TL *21* 935, 2181 (1980)
	JACS *109* 4649 (1987)
$(RCuC\equiv CR')Li$	TL *21* 935 (1980)
R_2CuMgX	TL *21* 935, 2181, 3377 (1980)
	CL 1123 (1980)
$R_2Cu(CN)Li_2$	JACS *109* 4649 (1987)
$RC\equiv CAlMe_2$	TL *27* 87 (1986)
RCdX	JOC *28* 2362 (1963)

$$\underset{\substack{CH_2-C=O \\ | \qquad\ | \\ H_2C=C\qquad\ \ O}}{} \xrightarrow[\text{CoI}_2]{1°\ RMgX} \underset{\substack{R\quad O \\ | \quad\ \| \\ H_2C=CCH_2COH}}{}$$

BCSJ *55* 3555 (1982)

$$\underset{\underset{\text{H}_2\text{C}=\text{CHCH}-\text{O}}{\overset{\text{CH}_2-\text{C}=\text{O}}{\mid\qquad\mid}}}{}\xrightarrow[\text{2. H}_3\text{O}^+]{\text{1. reagent}}\ \text{RCH}_2\text{CH}=\text{CHCH}_2\overset{\overset{\text{O}}{\parallel}}{\text{C}}\text{OH}$$

Reagent

R_2CuLi or R_2CuMgX or
  RMgX-cat CuI

TL *22* 1817 (1981)
CL 1307 (1981); 71, 219, 1521 (1982)

LiI (R = I)

CL 71 (1982)

$$\underset{\underset{\text{HC}\equiv\text{CCH}-\text{O}}{\overset{\text{CH}_2-\text{C}=\text{O}}{\mid\qquad\mid}}}{}\xrightarrow[\text{2. H}_3\text{O}^+]{\overset{\text{1. organocopper}}{\text{reagent}}}\ \text{RCH}=\text{C}=\text{CHCH}_2\text{CO}_2\text{H}$$

TL *22* 2375 (1981)

$$\overset{\text{O}}{\underset{(\ )_n}{\diagdown}}\!\!=\!\!\text{O}\ \xrightarrow[\text{or R}_2\text{CuMgX}]{\text{RMgX, cat CuI}}\ \text{RCH}_2\text{CH}=\text{CH(CH}_2)_{n+1}\text{CO}_2\text{H}$$

$n = 1, 2$

R = 1°, 2°, 3° alkyl; allyl; aryl; vinylic

TL *23* 3583, 3587 (1982)

2. Esters

$$\text{R}_2\text{CHCO}_2\text{R}\xrightarrow[\text{2. E}^+]{\text{1. base}}\ \underset{\underset{\text{R}_2\text{CCO}_2\text{R}}{}}{\overset{\overset{\text{E}}{\mid}}{}}$$

Reviews:

Org Rxs *9* 107 (1957)
Syn 521 (1982)

E^+

H_2O

CC 149 (1985) (stereochemistry)
TL *27* 2405 (1986) (lactone, stereochemistry);
  28 517 (1987)

D_2O

JACS *93* 2318 (1971)
CC 892 (1972) (lactone)

Me_2SO_4

JACS *109* 2426 (1987)

RX (X = halogen or sulfonate ester)

Organomet Chem Syn *1* 237 (1971)
JACS *93* 2318 (1971); *102* 3620 (1980); *104* 1735
  (1982); *107* 1435, 2512, 5570 (lactone) (1985);
  108 800, 2451 (lactone) (1986); *109* 6858 (1987)
TL 2425 (1973); *21* 1137 (1980); *23* 5271 (1982);
  24 1235, 3213 (1983) (both chiral); *26* 397,
  5623 (lactone) (1985); *27* 1781, 1785, 3247,
  3685, 3719, 5335, 5769, 5951 (1986) (all lactones
  except 3685 and 5951); *28* 183, 1623 (lactone),
  1685, 1933 (lactone), 2045, 2087, 2849 (chiral),
  4629, 5075 (intramolecular), 5161
  (intramolecular), 5205, 5659 (1987)

CC 892 (1972) (lactone); 711 (1973); 616 (1980); 904 (1984); 149, 1662 (lactone) (1985); 288 (1986); 992, 1721 (lactone), 1786 (lactone) (1987)

JCS Perkin I 694 (1977)

Syn 112 (1977); 710 (1980); 305 (1982)

J Polym Sci, Polym Chem Ed
17 3499, 3509 (1979)

JOC 44 2165 (1979); 45 891, 3236 (stereochemistry) (1980); 46 3756 (polymer), 4795 (inter- and intramolecular), 5364 (polymer) (1981); 47 180, 598 (1982); 50 2128 (intramolecular), 2668 (lactone) (1985); 51 4828 (lactone), 5492 (intramolecular) (1986); 52 569, 1309, 2563, 4601 (lactones), 4633 (intramolecular), 4641 (1987)

Rec Trav Chim 99 141, 311 (1980)

CL 1621 (1981); 81 (1986)

Angew Int 20 207, 574 (1981)

Tetr 37 3981 (1981); 40 2211 (1984)

Can J Chem 60 2007 (1982)

Org Prep Proc Int 15 149 (1983)

Ann 531 (1984)

$RC \equiv CCl$ (R = Cl, Ph, SPh)

TL 23 2373 (1982)
JACS 106 3551 (1984)

ArX, hν

JACS 102 7765 (1980)

RX (R = aryl, vinylic), Ni catalyst

JACS 99 4833 (1977)

epoxide

J Polym Sci, Polym Chem Ed 17 3509 (1979)
JOC 51 4840 (1986)

thiirane

J Polym Sci, Polym Chem Ed 17 3509 (1979)

▷—COX (X = OR, NR$_2$, Ph)
(cyclopropane opening)

Ber 114 32 (1981)

RCHO, R$_2$CO

JACS 92 3222 (1970); 95 3050 (1973); 101 2501 (1979); 105 1667 (1983) (diastereoselectivity); 106 3252 (1984); 107 1379, 2138 (lactone), 5541 (1985); 108 1019 (1986)

Org Syn 53 66 (1973); 63 99 (1984)

Syn 719 (1974); 112 (1977); 297 (1983)

TL 1745 (1975); 3975 (1980); 23 4285 (lactone), 5271 (1982); 24 1311 (1983); 27 2489, 3577, 4601, 4873, 6341, 6345 (1986); 28 655, 1761, 1925, 3059, 3723 (intramolecular), 5661, 5921 (1987)

JOC 45 1066, 1726, 3549, 3846 (1980); 50 3022 (1985) (enones); 51 1402, 5492 (1986); 52 2378 (lactone), 3541 (lactone) (1987)

Tetr 37 4087 (1981)

Acta Chem Scand B 35 273 (1981)

Helv *64* 2592 (1981); *69* 1699 (1986)
CL 57, 929 (1982)
Angew Int *21* 777 (1982) (lactone); *24* 874 (1985)
  (diastereoselective)
Syn Commun *12* 225 (1982) (silyl ester)
CC 305, 878, 1199, 1237 (1986)

$(Me_2\overset{+}{N}{=}CH_2)X^-$ CC 305 (1986)

$RCH{=}NOCH_2Ph$ (lactam formation)   CL 369 (1984)

$RCH{=}NSPh$ (lactam formation)   JOC *51* 1929 (1986)

$RCH{=}NSiMe_3$ (lactam formation)   JOC *48* 289 (1983); *50* 5120 (1985)
                                         JACS *106* 4819 (1984)
                                         TL *27* 1695 (1986); *28* 5369 (1987)

$RCH{=}NAr$ (lactam formation)   JOC *45* 3413 (1980); *50* 5120 (1985)
                                         JACS *106* 4819 (1984); *108* 6054 (1986)

$RCH_2CH{=}NR$, Me_2AlCl (lactam formation)   TL *28* 3377 (1987)

$RCH{=}C{=}NR$ (lactam formation)   TL *28* 4347 (1987)

enal (1,4-addition)   J Polym Sci, Polym Chem Ed *17* 3509 (1979)

enone (1,2- or 1,4-addition)   JOC *41* 4044 (1976); *47* 3464 (1982);
         50 3022 (1985)
         J Polym Sci, Polym Chem Ed *17* 3509 (1979)
         CC 305 (1986)
         TL *27* 3927 (1986) (intramolecular)

enone, $MeAl(-O-\underset{t\text{-Bu}}{\overset{t\text{-Bu}}{\bigcirc}}-Me)_2$ TL *28* 5723 (1987)

(1,4-addition)

$RCH{=}CHNO_2$ Chimia *38* 255 (1984)

HCO_2Me TL *27* 5397 (1986)

RCO_2R Syn 715 (1983)
       TL *27* 5177 (1986); *28* 2017, 2753 (1987)

lactone   TL 4323 (1978); *27* 6345 (1986); *28* 5661 (1987)
       JACS *107* 1691 (1985); *109* 1564 (1987)

$RCH_2CH{=}CHCO_2R(NR_2)$ (1,4-addition)   See page 915, Section 6.

$RO_2C-\overset{OR'}{\underset{()_n}{\bigcirc}}$ ($n=1, 2$; $R' = Ac$, $SiMe_3$;   TL *28* 5521 (1987)

S_N2' substitution of OR')

RCOCl   TL 2953 (1971); *28* 3551 (1987) (lactone)
       JOC *39* 3455 (1974); *46* 3756 (polymer), 4795
         (inter- and intramolecular), 5364 (polymer)
         (1981); *51* 4813 (1986); *52* 4531, 4601 (lactones)
         (1987)

	Compt Rend *286* 401 (1978)
	J Chem Res (S) 44 (1980)
	JACS *102* 3620 (1980); *107* 5289 (1985);
	108 7686 (1986); *109* 7488 (1987)
	JCS Perkin I 2885 (1982)
	Syn Commun *13* 183 (1983) (silyl ester)
RCO_2CO_2R'	JOMC *127* C65 (1977)
	TL 3713 (1978)
	Tetr *37* 307 (1981)
	JOC *52* 4531 (1987)

RCO_2—[pyridin-2-yl]

JOC *52* 4531 (1987)

RCOCN	TL *28* 4011 (1987)
RCN	TL *23* 1597 (1982)

$HCON$—[imidazolyl]

JOC *52* 4303 (1987)

$RCON$—[imidazolyl]

Bull Soc Chim Belg *91* 871 (1982)
TL *27* 5281 (1986); *28* 2837 (1987)
JACS *109* 4717 (1987)
CC 1228 (1987)

$\left(\text{[imidazol-1-yl]}N-\right)_2 CO$

Bull Soc Chim Belg *91* 871 (1982)

[imidazolyl]NCO_2R

Bull Soc Chim Belg *91* 871 (1982)

[imidazolyl]NCO_3R

Bull Soc Chim Belg *91* 871 (1982)

[imidazolyl]$NCONR_2$

Bull Soc Chim Belg *91* 871 (1982)

$ClCO_2R$

JOC *39* 2114 (1974); *46* 3151 (1981)
J Polym Sci, Polym Chem Ed *17* 3509 (1979)
JACS *102* 3620 (1980)
JCS Perkin I 2885 (1982)

CO_2	TL 3001 (1971)
$NCCO_2R$	TL *27* 1221, 1225, 1229 (1986)
	(all lactones); *28* 1051 (1987)

$p\text{-}MeOC_6H_4N{=}C{=}X$ (X = O, S)	TL *28* 3593 (1987)

[pyridinium with R] (1,4-dihydropyridine)

TL *28* 4457 (1987)

O_2 (hydroxy ester)	TL 1731 (1975)
	JOC *52* 3323 (1987)

$MoO_5 \cdot py \cdot HMPA$ (E = OH)	JOC *52* 3346 (1987)
	TL *28* 221 (1987)

$RSO_2N\!\!-\!\!CR_2$ (oxaziridine, chiral) (E = OH)

JOC *51* 2402 (1986)

RSSR (R = Me, Ph) JACS *95* 6840 (1973); *98* 4887 (1976)
TL 2429 (1973)
JOC *39* 2114 (1974)

PhSeX (X = Cl, Br) JACS *95* 5813 (1973)
JOC *39* 2114 (1974); *52* 2563 (1987)
Tetr *34* 1049 (1978)

PhSeSePh JOC *52* 2563, 2639 (1987)

Ph$_2$MeSiCl JACS *103* 2418 (1981)
Syn Commun *13* 833 (1983)

$$RCH_2CO_2R \xrightarrow[\text{2. reagent}]{\text{1. LDA}} \underset{\substack{| \ \ | \\ R \ \ R}}{RO_2CCHCHCO_2R}$$

I$_2$ Syn 396 (1975)
TL *28* 4441 (1987)

CuBr$_2$ or $(n\text{-}C_4H_9CO_2)_2$Cu JACS *93* 4605 (1971)

Cl$_3$CCCl$_3$ or BrCMe$_2$CMe$_2$Br JOC *35* 2085 (1970)

$$CH_3CO_2R \xrightarrow[\substack{\text{2. } t\text{-BuMe}_2\text{SiCH}_2\text{CHO} \\ \text{3. BF}_3\cdot\text{OEt}_2}]{\text{1. LDA}} H_2C{=}CHCH_2CO_2R$$

JACS *103* 6251 (1981)

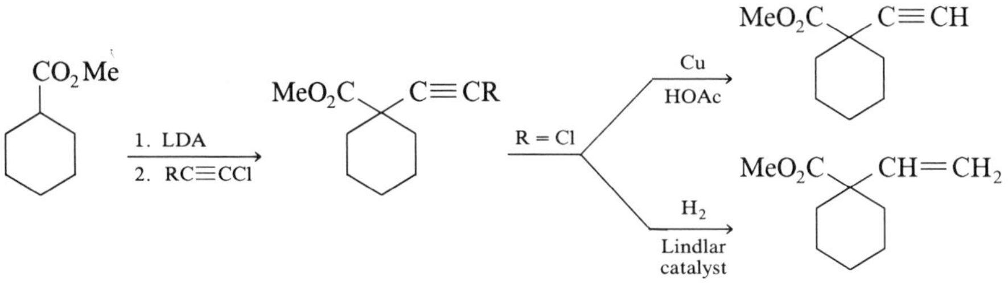

R = Cl, Ph, SPh

TL *23* 2373 (1982)

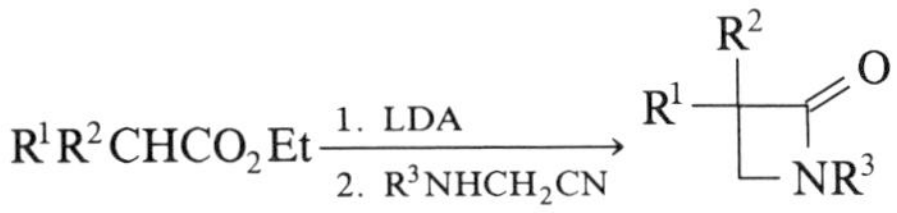

$$R^1R^2CHCO_2Et \xrightarrow[\text{2. } R^3NHCH_2CN]{\text{1. LDA}}$$

R^1 = alkyl, SPh, NHCOR, N(SiMe$_2$...SiMe$_2$)

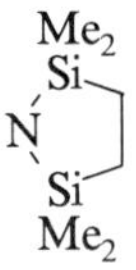

JACS *107* 1698 (1985)

$$\underset{X}{\overset{H}{RCCO_2R'}} \xrightarrow[\text{2. E}^+]{\text{1. base}} \underset{X}{\overset{E}{RCCO_2R'}}$$

X	E$^+$	
F	R$_2$CO	Syn 322 (1983)
OH	D$_2$O, RX, RCHO, R$_2$CO, H$_2$C—CH$_2$ (O) 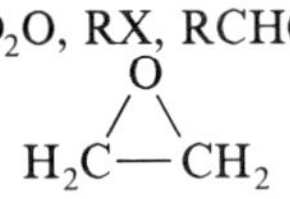	JOC *42* 2948 (1977)

$$\overset{\overset{\displaystyle H}{|}}{\underset{\underset{\displaystyle X}{|}}{R C C O_2 R'}} \xrightarrow[\text{2. E}^+]{\text{1. base}} \overset{\overset{\displaystyle E}{|}}{\underset{\underset{\displaystyle X}{|}}{R C C O_2 R'}} \quad (\textit{Continued})$$

$\underline{X}$	$\underline{E}^+$	
OR	RX	CC 951 (1971)
		Helv *64* 2704 (1981);
		65 385 (1982)
		TL *22* 4221 (1981);
		28 221 (1987)
		Angew Int *20* 1030 (1981);
		21 449 (1982)
		Ann 1930 (1983)
		Tetr *40* 1313 (1984) (lactone)
		JOC *51* 3746 (1986);
		52 3777 (1987)
	ArX (heterocyclic)	TL *27* 6059 (1986)
	RCHO, R_2CO	TL 1477 (1975); 2835 (1977);
		24 5869 (1983)
		JACS *103* 4972 (1981);
		106 8161 (1984)
		Helv *64* 2704 (1981)
		Angew Int *20* 1030 (1981)
		Tetr *40* 1313 (1984) (lactone)
		JOC *50* 2095 (1985); *52* 3176
		(lactone), 3777 (1987)
	enone (1,2- or 1,4-addition)	JOC *41* 4044 (1976)
	lactone	TL 4323, 4327 (1978)
	Ac_2O, AcCl, RCO_2R,	JOC *52* 3777 (1987)
	$\quad H_2C{=}CHCO_2R$	
	$\quad$ (1,4-addition),	
	$\quad CO_2$, $(CN)_2$	
SR	RX	Syn Commun *3* 265 (1973)
		BCSJ *50* 242 (1977) (lactone)
		TL *24* 3391 (1983)
	RCHO	TL *24* 523 (1983)
		JOC *48* 2705 (1983);
		52 4631 (1987) (lactone)
		CC 717 (1985)
	enone (1,2- or 1,4-addition)	JOC *41* 4044 (1976)
	$PhCH{=}NSiMe_3$	JACS *106* 4819 (1984)
	$\quad$ (lactam formation)	
SePh	RX	JACS *95* 6137 (1973)
		TL *27* 3297 (1986)
	RCHO	JACS *103* 4114 (1981)
NO_2	RX	BCSJ *43* 2277 (1970);
		46 337 (1973)
$PhCH_2OCONH$	RCHO, R_2CO	JOC *44* 3967 (1979)

RCONR	D_2O	TL 2205 (1976)
	RX	TL 2205 (1976); *28* 2243 (1987)
		Pol J Chem *53* 2397 (1979)
		JACS *109* 4649, 6537 (1987)
	RCHO	CC 1738 (1987)
	RCHO, R_2CO, RCOCl	JOC *46* 2809 (1981)
	RCOCl, $(RCO)_2O$	CC 753 (1978)
	RCOCl	CC 1283 (1987)
RO_2CNR	RCHO	JOC *52* 2881, 4804 (1987)
HCONH	RX	Angew Int *20* 971 (1981)
NMe_2	RCHO	TL 1477 (1975)
NR_2	RX	Helv *64* 2704 (1981)
		CC 1329 (1987)
	RCHO, R_2CO	CC 1329 (1987)
	D^+, RX, RCHO, R_2CO,	JACS *105* 5390 (1983)
	$C_6H_6 \cdot Cr(CO)_3$ (E = Ph),	
	PhSSPh	
$N(SiMe_3)_2$	RX	Angew Int *7* 809 (1968)
		Z Chem *10* 392 (1970)
	RCHO, R_2CO	Angew Int *7* 809 (1968);
		21 210 (1982)
		Z Chem *10* 393 (1970)
		JOC *44* 3967 (1979)
		($R' = SiMe_3$)
		TL *28* 2849 (1987)

$\begin{array}{c}Me_2\\Si\\N\quad\rceil\\Si\\Me_2\end{array}$	RX, RCHO	TL *22* 1787 (1981)
	$RCH=NSiMe_3$	TL *28* 5369 (1987)
	(lactam formation)	
$N=C:$	RX	Angew Int *10* 331 (1971)
	$RCH=CHCO_2R$	Ann 1571 (1973)
	(1,4-addition)	
$N=CHR$	RX	JOC *41* 3491 (1976)
		TL 1455 (1977); *23* 4255, 4259
		(1982); *27* 4435 (1986)
		BCSJ *55* 961 (1982)
		(enantioselective)
		Syn 789 (1983)
	$PhCH_2SeCH_2Br$	JOC *51* 2981 (1986)
	CH_2O	Syn 445 (1981)
	$RCH=CHCOR$,	JOC *41* 3491 (1976)
	$RCH=CHCO_2R$,	TL 1455 (1977)
	$H_2C=CHCN$	
	(all 1,4-addition)	
$N=CHNMe_2$	RX, epoxide (lactone),	JOC *42* 2639 (1977)
	$ArCH=C(CO_2R)_2{}^*$,	
	$ArCH=CHNO_2{}^*$,	
	$H_2C=CHCO_2R^*$	
	(*1,4-addition)	

$$\underset{X}{\overset{H}{R\overset{|}{\underset{|}{C}}CO_2R'}} \xrightarrow[\text{2. } E^+]{\text{1. base}} \underset{X}{\overset{E}{R\overset{|}{\underset{|}{C}}CO_2R'}} \quad (\textit{Continued})$$

X	E^+	
	PhCH$_2$SeCH$_2$Br	JOC *51* 2981 (1986)
N=CHN (pyrrolidine) CH$_3$OCH$_2$	RX	Ann 1668 (1983) (chiral)
N= (bicyclic) HO— CH$_3$	RX	TL *24* 3721 (1983) (chiral)
N=C(Me)C(OH)(Me)Ph	RCHO	CL 279 (1981) (chiral)
N=CR$_2$	RX	CC 136 (1976) TL *23* 2863, 4255 (1982); *27* 3839 (1986) (chiral); *28* 3801 (1987)
	PhCH$_2$SeCH$_2$Br H$_2$C=CRCH$_2$O$_2$CX (X = Me, OEt), cat PdL$_2$ (E = H$_2$C=CRCH$_2$)	JOC *51* 2981 (1986) TL *27* 23, 4573 (1986)
	RCHO	CL 145 (1982)
N=C(SR)$_2$	RX	Angew Int *14* 426 (1975)

$$\underset{}{\overset{O\qquad OMe}{R\overset{\|}{C}NH\overset{|}{C}HCO_2Me}} \xrightarrow[\text{2. ArCH=NAr}]{\text{1. 2 LDA}} \underset{ArH\overset{|}{C}-N-Ar}{\overset{O\qquad OMe}{R\overset{\|}{C}NH\overset{|}{C}-\overset{|}{C}=O}}$$

Syn 407 (1982)

$$\text{(thiazolidine-oxazolone)} \xrightarrow[\text{2. RCHO}]{\text{1. LDA}} \text{(hydroxy product)}$$

with C(CH$_3$)$_3$ substituent

TL *24* 3315 (1983)

$$\underset{O}{\overset{N}{Ph}} \text{(oxazoline ring) } \overset{CO_2CH_3}{\underset{CH_3}{}} \xrightarrow[\text{2. } E^+]{\text{1. LDA}} \underset{O}{\overset{N}{Ph}} \text{(ring) } \overset{E}{\underset{CH_3}{CO_2CH_3}}$$

$E^+ = D^+$, RX, RCHO, R$_2$CO

TL *24* 3311 (1983)
JOC *52* 3326 (1987)

$E^+ = RX, RCOCl$, Michael acceptors

Ber *102* 883 (1969); *113* 3706 (1980)
Angew Int *10* 653 (1971)

$$HO_2CCH_2CO_2R \xrightarrow{2 \, n\text{-BuLi}} \xrightarrow[\text{2. } H_3O^+]{\text{1. } R'COCl} R'\overset{O}{\underset{\|}{C}}CH_2\overset{O}{\underset{\|}{C}}OR$$

JOC *44* 310 (1979)
Org Syn *61* 5 (1983)

$E^+ = RX, R_2CO$

Ber *116* 3413 (1983)

$$RCNH-\overset{O}{\underset{\|}{C}}-\overset{O}{\underset{\|}{C}}-\overset{O}{\underset{\|}{C}}OR' \xrightarrow[\text{2. } E^+]{\text{1. 2 LDA}} RCNH-\overset{O}{\underset{\|}{C}}-\overset{E}{\underset{|}{C}}-\overset{O}{\underset{\|}{C}}OR'$$

R	E^+	
H	RX	Angew Int *20* 971 (1981)
Ph	RX	TL *24* 2733 (1983) (lactone); *28* 3103 (1987) JACS *108* 4943 (1986) (lactone)
	RCHO	TL *28* 3103 (1987)
PhCH$_2$O	RX	TL *24* 2733 (1983) (lactone)
	R$_2$CO	JACS *105* 1659 (1983) (lactone)

$$Me_2NCH_2CH_2\overset{O}{\underset{\|}{C}}OMe \xrightarrow[\text{2. } E^+]{\text{1. LDA}} Me_2NCH_2\overset{E}{\underset{|}{C}}HCO_2Me \xrightarrow[\text{2. DBN or DBU}]{\text{1. MeI}} H_2C=\overset{E}{\underset{|}{C}}-\overset{O}{\underset{\|}{C}}OMe$$

E^+	
RX	TL 3423 (1978) JOC *46* 4536 (1981) Syn Commun *11* 591 (1981)
RCHO	CC 1112 (1983) JOC *49* 3784 (1984); *50* 157 (1985); *52* 5452 (1987)
PhCH=NSiMe$_3$ (lactam formation)	JACS *106* 4819 (1984)

X	E⁺	
O	RX	TL *24* 2733 (1983)
S	RX	JACS *107* 1435 (1985); *108* 4943 (1986)
	RCHO	JOC *51* 3742 (1986)

$$NO_2CH_2CH_2COR \xrightarrow[\text{2. E}^+]{\text{1. 2 LDA}} NO_2CH_2\overset{E}{\underset{}{C}}HCOR \xrightarrow{\text{DBU}} H_2C{=}\overset{E}{\underset{}{C}}{-}COR$$

E⁺ = RX, RCHO, enone (1,4-addition)

Ber *115* 1705 (1982)
Helv *65* 385 (1982)

E⁺ = MeI, I₂, O₂, (RS)₂, PhSeBr

Can J Chem *64* 1781, 1788 (1986)

E⁺	
RX	TL 2429 (1973); *22* 425 (1981); *23* 3055, 4991 (1982); *27* 5281 (1986); *28* 35, 3189, 5033 (lactone) (1987) Helv *62* 2825, 2829 (1979); *63* 197, 1383 (1980); *65* 293, 344 (1982) JOC *46* 4319 (1981); *48* 1114 (1983); *49* 2168 (1984) (lactone); *52* 1780 (1987) Ber *114* 2786, 2802 (1981) Ann 939, 2114 (1983) Org Syn *63* 109 (1984) JACS *107* 5292 (1985); *108* 2105 (1986) CC 1368 (1987)
RCHO (ZnCl₂)	JOC *46* 4319 (1981); *49* 2168 (1984) (lactone)
R₂CO	Helv *63* 197 (1980)
R²CH=CR¹NO₂	Helv *63* 2005 (1980)

RCH=NAr (lactam formation)

TL *25* 3779 (1984); *26* 3903 (1985); *28* 4489 (1987)
CC 1433 (1985)
JACS *109* 1129 (1987)

RCH=NSiMe$_3$ (lactam formation)

JACS *106* 4819 (1984)
TL *26* 937, 5493 (1985)
CL 1927 (1984); 651 (1985)

RCH=NCO$_2$R

TL *28* 69, 83 (1987)

(MeS)$_2$

TL 2429 (1973)

I$_2$ (epoxide formed)

TL 4575 (1977)
Helv *63* 197 (1980)

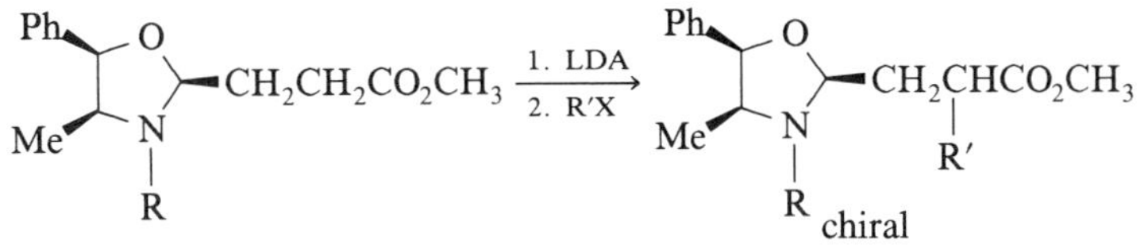

Syn Commun *17* 241 (1987)

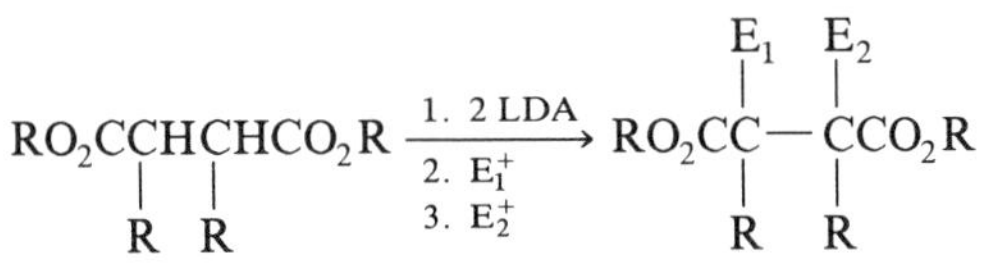

E_1^+	E_2^+	
D$_2$O	D$_2$O	JACS *100* 7753 (1978)
Me$_3$SiCl	Me$_3$SiCl	Syn Commun *11* 687 (1981) (O silylation)
RX	RX or R'X	TL 73 (1979); *23* 3683 (1982) Syn Commun *11* 687 (1981) CL 687 (1982)
RCHO, R$_2$CO	HCl	CL 687 (1982)
RCHO	RCHO	TL 73 (1979) JOC *47* 4731 (1982)
	X—C$_n$—X (n=1–5)	TL 1815 (1979); *27* 5951 (1986); *28* 589 (1987) Syn 389 (1980) Syn Commun *14* 227 (1984) JOC *49* 1412 (1984); *50* 5727 (1985) JACS *107* 3343 (1985) (chiral)

$$RO_2CCHCHCO_2R \xrightarrow[\substack{2.\ E_1^+ \\ 3.\ E_2^+}]{1.\ 2\ LDA} RO_2C\overset{E_1}{\underset{R}{C}}-\overset{E_2}{\underset{R}{C}}CO_2R \quad (\textit{Continued})$$

$\underline{E_1^+}$ $\underline{E_2^+}$

$Br(CH_2)_2\overset{O}{\overset{\triangle}{CH}}{-}CH_2$ TL *28* 351 (1987)

$BrCH_2CH_2COCH_3$ JOC *50* 5727 (1985)

$Br\!-\!C_n\!-\!CO_2Et \ (n = 2, 3)$ TL *22* 1755 (1981)
                                                     JOC *47* 4731 (1982);
                                                        51 5450 (1986)
                                                     Syn Commun *14* 227 (1984)

$RCH{=}C(CHRBr)CO_2Me$ TL *25* 669 (1984);
                                                        27 5951 (1986)

$RC{\equiv}CCO_2Ph$ JACS *109* 7534 (1987)

$o\text{-}BrCH_2ArCO_2Me$ TL *23* 1031 (1982)
                                                     JOC *49* 2785 (1984)

$o\text{-}EtO_2CC_6H_4CO_2Et$ TL 73 (1979); *23* 1031 (1982)

$H_2C{=}CHCO_2Ar$ TL *27* 5951 (1986);
                                                        28 5241 (1987)

$TsOCH{=}CHCOC(CH_3)_3$ TL *27* 5951 (1986)

$$RO_2C\!-\!C_n\!-\!CO_2R \xrightarrow[\substack{2.\ reagent}]{1.\ 2\ LDA} \begin{array}{c} RO_2C\!-\!C \\ | \quad \rangle C_{n-2} \\ RO_2C\!-\!C \end{array}$$

Reagent	$\underline{n}$	
I_2	2, 5	Syn Commun *14* 227 (1984) TL *28* 1831 (1987)
$Cu(OTf)_2$	3, 6	Chem Pharm Bull *28* 262 (1980)
$CuCl_2$ or $CuBr_2$	3–6	JOC *48* 1125 (1983); *52* 3462 (1987)

$$RO_2CCH_2CH{=}CHCH_2CO_2R \xrightarrow[\substack{2.\ BrCH_2CH_2CO_2Et}]{1.\ 2\ LDA} \text{(cyclopentanone ring with } {=}CHCH_2CO_2R \text{ and } CO_2R)$$

TL *25* 671 (1984)

$$R_2CHCO_2R \xrightarrow[\text{2. ClSiMe}_3]{\text{1. LiNR}_2} R_2C\overset{\overset{\displaystyle OSiMe_3}{|}}{=}COR \xrightarrow{E^+} R_2\overset{\overset{\displaystyle E'}{|}}{C}CO_2R$$

$\underline{E^+}$

NO_2-Ar-H, $[(Et_2N)_3S]Me_3SiF_2/Br_2$ or DDQ ($E' = NO_2-Ar$)	JACS *107* 5473 (1985)
$R_2C=CHCH_2X$ (X = Cl, Br), high pressure	TL *25* 1075 (1984)
R_3CCl, $ZnCl_2$	TL 1455 (1978) Tetr *37* 319 (1981)
CH_3OCH_2Cl, $ZnBr_2$	CC 305 (1986)
Cl_2CHOCH_3, $ZnCl_2$ ($E' = CHO$, $MeOCHCR_2CO_2R$)	Syn 723 (1982)
$H_2C=CHCH_2OCO_2Me$, cat $Pd_2(DBA)_3 \cdot HCCl_3$, cat dppe ($E' = H_2C=CHCH_2$)	TL *25* 4783 (1984)
$HC\equiv CCO_2H$, $TiCl_4$ ($E' = CH=CHCO_2H$)	JACS *107* 3879 (1985)
$RCH(OMe)ONR_2$, $TiCl_4$ or $Zn(OTf)_2$ ($E' = RCHOMe$)	JACS *107* 2569 (1985)
	TL *26* 2535 (1985)
$R_2C(OR)_2$, $TiCl_4$	TL *28* 1313 (1987) (intramolecular)
($n = 1, 2$), AgOTf	JACS *107* 4289 (1985)
, Me_3SiOTf	TL *28* 1035 (1987) (intramolecular)
RCHO or R_2CO, high pressure	TL *25* 1075 (1984)
RCHO, $BF_3 \cdot OEt_2$	JACS *105* 1667 (1983) JOC *51* 3027 (1986) (diastereoselection) TL *28* 5615 (1987)
RCHO, Ph—⟨B⟩—Ph (Cl)	TL *27* 4721 (1986) (enantioselective)
RCHO, $MgBr_2$	JOC *52* 888 (1987)
RCHO, $SnCl_4$	TL *25* 5973 (1984) JOC *51* 3027 (1986) (diastereoselection)
RCHO, $TiCl_4$	TL 4029 (1979); *25* 4655 (1984) CL 531 (1983) JACS *106* 5305 (1984); *107* 5812 (1985) (enantioselective)

	Angew Int *24* 874 (1985) (diastereoselective)
	JOC *51* 5032 (1986); *52* 2754 (1987)
	(enantioselective)
	Helv *69* 1699 (1986) (diastereoselective)
RCHO, TiCl$_4$, R$_3$P	TL *27* 1735 (1986)
RCHO, Rh catalyst	TL *28* 793 (1987)
RCHO; cat SmCl$_3$, CeCl$_3$, LaCl$_3$ or Eu(fod)$_3$	TL *28* 5513 (1987)
R$_2$CO, TiCl$_4$ or FeCl$_3$ or AlCl$_3$	Syn Commun *13* 449 (1983)
PhSCH$_2$Cl, ZnBr$_2$	TL 993 (1979)
	CC 305 (1986)
PhSCHClR, TiCl$_4$ or ZnBr$_2$	TL 2179 (1979); *23* 2399 (1982)
	CC 1472 (1987)

	CC 305 (1986)
PhSCH$_2$CH$_2$Cl, ZnBr$_2$	TL *24* 1315 (1983)

	TL *24* 913 (1983)

cat CF$_3$SO$_3$H (E′ = CH$_2$NHR)

, cat Me$_3$SiOTf	TL *28* 507 (1987)

, ZnI$_2$	CL 1343 (1985)

(Me$_3$Si)$_2$NCH$_2$OCH$_3$, cat Me$_3$SiOTf [E′ = CH$_2$N(SiMe$_3$)$_2$]	CC 883 (1984)
R$_2$NCH$_2$OR, ZnCl$_2$ (E′ = CH$_2$NR$_2$)	BCSJ *56* 645 (1983)
Me$_2$NCH(OMe)$_2$, ZnCl$_2$ (3-dimethylamino-2,4-dialkylpentanedioic esters)	Syn 787 (1983)
HC(OMe)$_3$, TiCl$_4$ [E′ = CH(OMe)$_2$]	TL *26* 4129 (1985) (chiral)
RCH=NR′, cat Me$_3$SiOTf (E′ = CHRNHR′)	CL 1371 (1985)
	CC 119 (1985)
	TL *28* 4331, 4335 (1987)
RCH=NSiMe$_3$, ZnI$_2$ (E′ = CHRNHSiMe$_3$)	CC 539 (1985)
RCH=NR′, TiCl$_4$/H$_2$O (E′ = CHRNHR′)	TL 3643 (1977); *21* 2077, 2081 (1980);
	28 227 (1987)
	Syn 545 (1981)

$RCH{=}NOCH_2Ph$, cat Me_3SiOTf
  $(E' = CHRNHOCH_2Ph)$

        TL *24* 4707 (1983)

$t{-}BuO_2CN{=}NCO_2{-}t{-}Bu$, $TiCl_4$,
  $[Ti(O{-}i{-}Pr)_4]$
  $[E' = N(CO_2{-}t{-}Bu)NHCO_2{-}t{-}Bu]$

        JACS *108* 6394 (1986)
        Helv *69* 1923 (1986)

$R^1CH{=}\overset{\underset{|}{O^-}}{\underset{+}{N}}R^2$ $(E' = CHR^1NR^2OSiR_3)$

        CL 1787 (1982)
        TL *28* 1431 (1987)

nitro-olefins, $TiCl_4$, $Ti(O{-}i{-}Pr)_4$
  (γ-keto esters)

        CL 1043 (1980)
        JACS *106* 2149 (1984)

enone, $TiCl_4$ (1,4-addition)

        JACS *107* 2797 (1985) (stereoselection)
        JOC *51* 279 (1986); *52* 2754 (1987)
          (enantioselective)

2-acyl-2-alkenoate ester
  (1,4-addition)

        JACS *102* 4262 (1980)

$Pb(OAc)_4$ $(E' = OAc)$

        Helv *68* 216 (1985)

$$R_2CHCO_2R \longrightarrow R_2C{=}\overset{\underset{|}{OSiMe_3}}{C}OR \xrightarrow[\text{2. }\Delta]{\text{1. MeLi, PhCHCl}_2} R_2C{=}\overset{\underset{|}{Ph}}{C}CO_2R$$

        Syn Commun *12* 401 (1982)

$$R\overset{\underset{|}{X}}{C}HCO_2R \longrightarrow R\overset{\underset{|}{X}}{C}{=}\overset{\underset{|}{OSiR_3}}{C}OR \xrightarrow{E^+} R\overset{\underset{|}{X}}{C}\overset{\underset{|}{E}}{X}CO_2R$$

$\underline{X}$	$\underline{E^+}$	
OR	$RCHO$ or R_2CO, $ZnCl_2$	CC 20 (1985)
SMe	$RCHO$, Lewis acid	TL *26* 6509 (1985)

$$RCH_2\overset{\underset{\|}{O}}{C}OH \longrightarrow \overset{O}{\underset{N}{\diagup}}{\diagdown}CH_2R \xrightarrow[\text{2. E}^+]{\text{1. }n\text{-BuLi}} \xrightarrow[\text{H}_2\text{SO}_4]{\text{EtOH}} RCH\overset{\underset{\|}{O}}{C}\overset{\underset{|}{E}}{O}Et$$

$\underline{E^+}$

RX

        JACS *92* 6644 (1970); *96* 6508 (1974)
        JOC *39* 2778, 2787 (1974); *46* 3097 (1981)
        Angew Int *15* 270 (1976)
        Pure Appl Chem *51* 1255 (1979)
          (asymmetric induction)
        TL *28* 5509 (1987)

RCHO

        JACS *103* 4278 (1981) (B azaenolate)

$$RX + CuCH_2CO_2Et \longrightarrow RCH_2CO_2Et$$

R = allylic, benzylic

TL 1163 (1972)

$$RX + BrZnCH_2CO_2{-}t{-}Bu \longrightarrow RCH_2CO_2{-}t{-}Bu$$

R = allylic or α-halo ester

Syn Commun *13* 523 (1983)

$$H_2C{=}CHCH_2Br + BrZnCH_2CO_2Et \xrightarrow{cat\ Cu(acac)_2} H_2C{=}CHCH_2CH_2CO_2Et$$

TL *24* 2749 (1983)

$$\triangleright\!\!<\!\!\begin{array}{l}OSiMe_3\\OEt\end{array} \xrightarrow{ZnCl_2} Zn(CH_2CH_2CO_2Et)_2 \xrightarrow[cat\ PdCl_2[(o\text{-}Tol)_3P]_2]{RX} RCH_2CH_2CO_2Et$$

R = aryl, vinylic, acyl

TL *27* 83 (1986)

$$RX + ClZnCH_2\overset{\overset{\displaystyle CH_3}{|}}{C}HCO_2Me \xrightarrow{cat\ CuBr\cdot SMe_2} RCH_2\overset{\overset{\displaystyle CH_3}{|}}{C}HCO_2Me$$

R = allylic, aryl, vinylic, acyl

TL *28* 337 (1987)

$$RCH{=}CHCH_2Y + XZn(CH_2)_nCO_2Et \xrightarrow{cat\ CuCN} H_2C{=}CH\overset{\overset{\displaystyle R}{|}}{C}H(CH_2)_nCO_2Et$$

Y = Cl, Br, OTs;   n = 2–4

JOC *52* 4418 (1987)

$$ArO_3SR_f + IZnCH_2CH_2CO_2Et \xrightarrow[LiCl]{cat\ Pd(PPh_3)_4} ArCH_2CH_2CO_2Et$$

TL *28* 2387 (1987)

$$\overset{H\quad O}{\underset{\quad\;\|}{C{=}C{-}C{-}COR}} \text{ or } \overset{H\qquad O}{\underset{\qquad\;\|}{C{-}C{=}C{-}COR}} \xrightarrow[2.\ E^+]{1.\ base} \overset{E\quad O}{\underset{\;\;|\;\;\|}{C{=}C{-}C{-}COR}}$$

$\underline{E^+}$

H$_2$O or D$_2$O   TL 4249 (1972); *25* 1333, 5177
(See also page 110, Section 2.2)   (*Z* alkene), 5181 (*Z* alkene) (1984);
                                                         28 5075 (1987)
                                                     BCSJ *51* 2970 (1978)
                                                     JACS *102* 3964 (1980); *107* 1293 (1985);
                                                         108 2776 (1986); *109* 6389, 8117 (1987)
                                                     J Med Chem *23* 525 (1980)
                                                     Helv *64* 1023 (1981) (stereochemistry)
                                                     JOC *47* 163 (1982) (stereochemistry)
                                                     Syn 129 (1982)

R'X   Agric Biol Chem *24* 685 (1960); *26* 705 (1962);
                                                         36 793 (1972)
                                                     JACS *94* 1790 (1972); *96* 5662 (1974);
                                                         107 1285, 7184 (1985)
                                                     TL 4249 (1972); 2433 (1973); 4135, 4171 (1975);
                                                         4485 (1977); *21* 2509 (1980); *22* 1691 (1981);
                                                         28 6253 (1987)
                                                     JOC *39* 2323 (1974); *44* 300 (1979); *47* 163
                                                         (1982); *49* 3278 (1984); *51* 123, 561 (1986);
                                                         52 353, 4471, 4517 (1987)
                                                     Syn Commun *7* 483 (1977)
                                                     CC 799 (1980); 502 (1982); 1717 (1986)
                                                     Syn 129 (1982)

RCHO, R$_2$CO

TL 4249 (1972); 4135 (1975)
JOC *41* 4065 (1976); *45* 1181 (1980); *47* 163 (1982); *52* 3956 (1987)
JACS *108* 3755 (1986)

PhCH=NSiMe$_3$ (lactam formation)

JACS *106* 4819 (1984)

O$_2$, SnCl$_2$ (E = OH)

TL 4215 (1976)

(PhCO$_2$)$_2$ (E = PhCO$_2$)

TL 4215 (1976)

PhSSO$_2$Ph (E = SPh)

JOC *51* 4594 (1986)

MeSSO$_3$Me (E = SMe)

TL 4215 (1976)

$$RCH_2CH=CHCO_2R \xrightarrow[\substack{2.\ Me_3SiCl \\ 3.\ HCl}]{1.\ LDA} RCH=CHCH_2CO_2R$$

JACS *109* 7063 (1987)

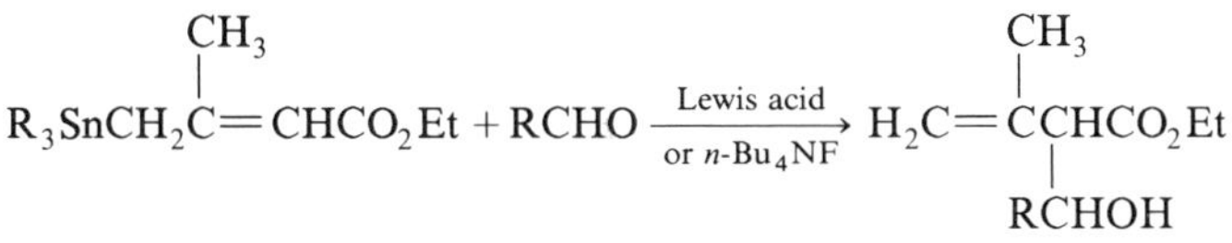

$$R_3SnCH_2\overset{\overset{\displaystyle CH_3}{|}}{C}=CHCO_2Et + RCHO \xrightarrow[\text{or } n\text{-Bu}_4NF]{\text{Lewis acid}} H_2C=\overset{\overset{\displaystyle CH_3}{|}}{\underset{\underset{\displaystyle RCHOH}{|}}{C}}CHCO_2Et$$

CC 561 (1987)

$$RY-\overset{\overset{\displaystyle H}{|}}{C}-C=C-\overset{\overset{\displaystyle O}{||}}{C}OR \xrightarrow[2.\ R'X]{1.\ base} RY-C=C-\overset{\overset{\displaystyle R'}{|}}{C}-\overset{\overset{\displaystyle O}{||}}{C}OR$$

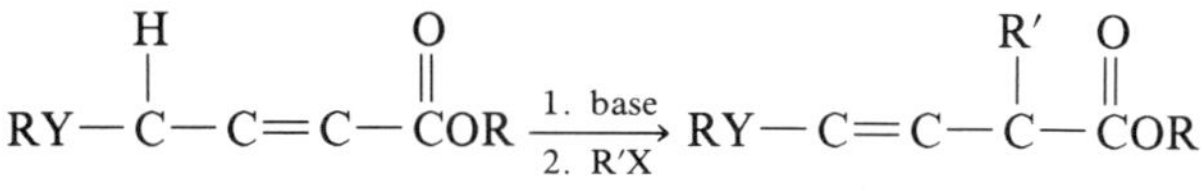

Y

O

Syn Commun 7 189 (1977)

S

TL 405 (1975); *24* 3391 (1983)

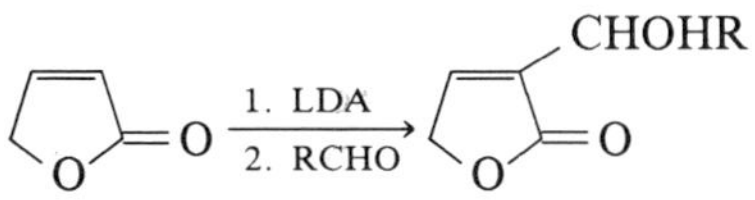

TL *28* 985 (1987)

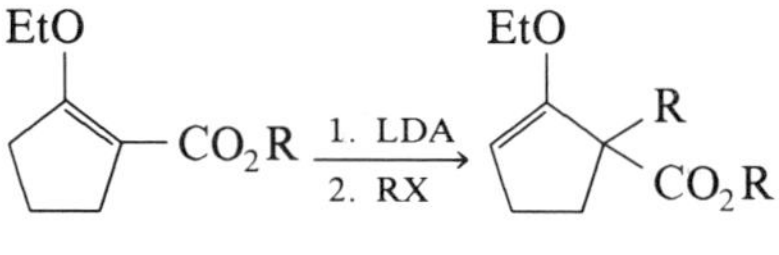

JOC *52* 5745 (1987)

$$RCH_2CR{=}CCO_2R \xrightarrow[\text{2. R'X}]{\text{1. KO-}t\text{-Bu}} RCH{=}CRCCO_2R$$

with SPh substituents and R' substituent

TL *24* 2113, 3391 (1983)

$$RCH_2CH{=}C(CO_2R)_2 \xrightarrow[\text{2. R'X}]{\text{1. base}} \underset{H}{\overset{R}{>}}C{=}C\underset{C(CO_2R)_2}{\overset{H}{<}}$$

with R' substituent

JACS *62* 314 (1940)
CL 1909 (1982)
JOC *51* 4944 (1986)

$$\xrightarrow[\text{2. RX}]{\text{1. LDA}}$$

cyclohexadiene with CO_2Et → with R and CO_2Et

JACS *109* 6187 (1987)

$$\underset{}{C{=}C{-}\overset{H}{\underset{}{C}}{-}\overset{O}{\underset{}{C}}OR} \text{ or } \overset{H}{\underset{}{C}}{-}C{=}C{-}\overset{O}{\underset{}{C}}OR \xrightarrow[\text{2. E}^+]{\text{1. base}} \overset{E}{\underset{}{C}}{-}C{=}C{-}\overset{O}{\underset{}{C}}OR$$

$\underline{E^+}$

H_2O	BCSJ *51* 2970 (1978) JOC *50* 3526 (1985)
R'X	JACS *94* 1790 (1972); *96* 5662 (1974); *98* 1204 (1976); *107* 7184 (1985) JOC *39* 669 (1974)

RCHO, R_2CO	JACS *74* 5529 (1952); *77* 4111 (1955)
	J Vitaminol (Osaka) *4* 178, 190 (1958)
	    [CA *53* 5329b,d (1959)]
	Helv *45* 528 (1962)
	Can J Chem *46* 3115 (1968)
	TL 1507 (1972); 4135 (1975);
	    23 4773 (1982); *28* 5249 (1987)
	Ber *106* 2643 (1973); *110* 1594 (1977)
	Syn 343 (1974)
	JOC *45* 1181 (1980)

$CdCl_2$/RCHO	TL *27* 5193 (1986)
PhSeCl	JOC *51* 5243 (1986)

$$CH_3(CH{=}CH)_nCOEt + ArCH \xrightarrow[\substack{\text{transfer} \\ \text{catalyst}}]{\text{phase}} ArCH{=}CH(CH{=}CH)_nCOEt$$

$n = 0, 1$

J Chem Res (S) 106 (1981)

$$\underset{\substack{|\\X}}{\overset{\substack{H\\|}}{C}}-C=C-\overset{\substack{O\\||}}{C}OR \xrightarrow[\text{2. }E^+]{\text{1. base}} X-\underset{\substack{|\\}}{\overset{\substack{E\\|}}{C}}-C=C-\overset{\substack{O\\||}}{C}OR$$

X	E⁺	
Br	RCHO	TL 1507 (1972)
PhSO₂	R'X	BSCF 743, 746 (1973)
Ph₃P⁺	RCHO, base (2,4-alkadienoate ester)	JOC *39* 821 (1974)

$$\text{(furanone)}=O \xrightarrow[\text{2. }E^+]{\text{1. base}} E-\text{(furanone)}=O$$

E⁺	
D₂O	TL 1627 (1979)
R'X	CC 81 (1979)
RCHO, R₂CO	JOC *39* 669 (1974)
	JCS Perkin I 70 (1979)
	CC 81 (1979)
	TL 1627 (1979); *28* 985 (1987) (lactone)
Michael acceptors (RCH=CHCOR, RCH=CHCO₂R, H₂C=CHCN)	TL 3129 (1977)

$$\text{MeO-(furanone)}=O \xrightarrow[\text{2. }E^+]{\text{1. base}} \text{MeO-(furanone)}-E=O$$

E⁺	
D₂O, NBS, RCHO	TL 1627 (1979)
BF₃·OEt₂ + RCH(OMe)₂, R(OMe)₃ or C(OMe)₄	TL *23* 5229 (1982)

$$\underset{\substack{|\\}}{\overset{\substack{H\\|}}{C}}-\underset{\substack{|\\X}}{\overset{\substack{X\\|}}{C}}=C-\overset{\substack{O\\||}}{C}OR \xrightarrow[\text{2. }E^+]{\text{1. base}} \underset{\substack{|\\}}{\overset{\substack{E\\|}}{C}}-\underset{\substack{|\\X}}{\overset{\substack{X\\|}}{C}}=C-\overset{\substack{O\\||}}{C}OR$$

X	E⁺	
OR	RCHO, R₂CO	JOC *45* 1181 (1980)
NR₂	RX	TL 39 (1973); 3963 (1974)
		Syn 401 (1976)
	RCHO, R₂CO	JOC *45* 1181 (1980)
		JACS *104* 357 (1982);
		107 1777 (1985);
		108 3112 (1986)
		TL *27* 6173 (1986) (ZnCl₂ added)
	RCOCl	JOC *52* 708 (1987)
	Me₃SiCl	TL *23* 3011 (1982)

$$\underset{\text{C}}{\overset{\text{H}}{|}}-\underset{\text{C}}{\overset{\text{NR}_2}{|}}=\text{C}-\overset{\text{O}}{\overset{||}{\text{COR}}} \xrightarrow[\text{2. Me}_3\text{SiCl}]{\text{1. LDA}} \underset{\text{C}}{\overset{\text{Me}_3\text{Si}}{|}}-\underset{\text{C}}{\overset{\text{NR}_2}{|}}=\text{C}-\overset{\text{O}}{\overset{||}{\text{COR}}} \xrightarrow{\text{E}^+} \underset{\text{C}}{\overset{\text{E}}{|}}-\underset{\text{C}}{\overset{\text{NR}_2}{|}}=\text{C}-\overset{\text{O}}{\overset{||}{\text{COR}}}$$

$\underline{\text{E}^+}$

RCHO, R$_2$CO, HC(OMe)$_3$	TL *23* 3011 (1982)
RCOCl	TL *23* 3011 (1982);
	28 5423 (1987)

$$\underset{\text{C}}{\overset{\text{H}}{|}}-\underset{\text{C}}{\overset{\text{NR}_2}{|}}=\text{C}-\text{CO}_2\text{R} \xrightarrow[\text{2. E}^+]{\text{1. LDA}} \underset{\text{C}}{\overset{\text{E}}{|}}-\underset{\text{C}}{\overset{\text{NR}_2}{|}}=\text{C}-\text{CO}_2\text{R}$$

$\underline{\text{E}^+}$

RX	Syn 58 (1983)
RCHO, R$_2$CO	Syn 61 (1983)
RCHO, ZnCl$_2$	JACS *109* 1587 (1987)

$$\text{H}_2\text{C}=\underset{\underset{\text{SiMe}_3}{|}}{\overset{\overset{\text{R}}{|}}{\text{C}}}\text{CHCOR}' \xrightarrow[\text{Lewis acid}]{\text{E}^+} \text{ECH}_2\text{C}=\underset{}{\overset{\overset{\text{R}}{|}}{\text{C}}}\overset{\text{O}}{\overset{||}{\text{HCOR}'}}$$

E$^+$ = RX (R = allylic, benzylic), RCHO, R$_2$C(OMe)$_2$, R$_2$CO, RCOCl, PhSCHClR

TL *23* 723, 2953 (1982)

CL 961 (1982)

$$\underset{\text{C}}{\overset{\text{H}}{|}}-\text{C}=\text{C}-\overset{\text{O}}{\overset{||}{\text{COR}}} \longrightarrow \text{C}=\text{C}-\underset{}{\overset{\overset{\text{OSiMe}_3}{|}}{\text{C}}}=\overset{\text{O}}{\overset{||}{\text{COR}}} \xrightarrow{\text{E}^+} \underset{\text{C}}{\overset{\text{E}}{|}}-\text{C}=\text{C}-\overset{\text{O}}{\overset{||}{\text{COR}}}$$

$\underline{\text{E}^+}$

PhSCl, PhSCHClR, RCOCl,	TL 3205, 3209 (1979)
RBr (R = allylic, benzylic),	
MeOCH$_2$Cl, MeCH(OEt)$_2$,	
or (MeO)$_3$CH; ZnBr$_2$	

XPh
, ZnBr$_2$ TL *24* 5911 (1983)
O Cl
(X = S, Se)

(MeO)$_3$CH, ZnBr$_2$	TL *26* 397 (1985)
RCOCl, ZnBr$_2$	JOC *52* 708 (1987)
RCH=NR, TiCl$_4$	TL *28* 613 (1987)
	(5,6-dihydro-2-pyridones or
	5-amino-2-alkenoates)
nitro-olefins, ZnBr$_2$	CL 1043 (1980)

$\underline{E^+}$

allylic halide, AgO$_2$CCF$_3$	TL *28* 949 (1987)
RCH(OMe)$_2$ or R$_2$C(OMe)$_2$, SnCl$_4$	TL *22* 4269 (1981)
RCHO, cat *n*-Bu$_4$NF-silica	Heterocycles *19* 2327 (1982)
RCHO or R$_2$CO, SnCl$_4$	TL *21* 4611 (1980); *22* 4269 (1981); *28* 985, 4037, 4041 (1987)
RCHO or R$_2$CO; BF$_3$·OEt$_2$, ZnCl$_2$, ZnBr$_2$, CsF, *n*-Bu$_4$NF, (Ph$_3$C)ClO$_4$, or R$_3$SiOTf (R = Me, Et)	TL *28* 4037, 4041 (1987)
HC(OR)$_3$, cat SnCl$_4$	BCSJ *52* 1953 (1979)
RC(OR)$_3$, cat Lewis acid	BCSJ *52* 1953 (1979)
RCHCl(OAc), cat SnCl$_4$	BCSJ *52* 1953 (1979)
Pb(OAc)$_4$ (E = OAc)	BCSJ *53* 1061 (1980)

$_{}^+$ = ZnBr$_2$ or TiCl$_4$ plus RCH(OMe)$_2$, RC(OMe)$_3$, C(OMe)$_4$, RCHO or R$_2$CO; Br$_2$; Pb(OAc)$_4$ (E = OAc)

TL *23* 353 (1982)

$$R_2C{=}CHCO_2R \xrightarrow[\text{2. } E^+]{\text{1. base}} R_2C{=}\overset{\overset{\displaystyle E}{|}}{C}CO_2R$$

See page 185, Section 1.

$$H_2C{=}CHCO_2R \xrightarrow[\text{cat DABCO}]{\text{RCHO}} H_2C{=}\overset{\overset{\displaystyle RCHOH}{|}}{C}CO_2R$$

Angew Int *22* 795, 796 (1983)
JCS Perkin I 2293 (1983)
Helv *67* 413 (1984)
JOC *50* 3849 (1985)
TL *27* 5007 (1986)

E$^+$ = MeOD, MeI, RCHO, (RCO)$_2$O, RCO$_2$R, RSSR, (RCO$_2$)$_2$

Angew Int *21* 637 (1982)
TL *23* 581, 585, 1793 (1982)

$$XCH\!=\!CRCO_2R \xrightarrow[\text{2. E}^+]{\text{1. base}} \overset{\overset{\displaystyle E}{\displaystyle |}}{X}C\!=\!CRCO_2R$$

$$X = Ph, OR, SR, NR_2, CO_2R$$

See page 185, Section 1.

$$CH_3CH\!=\!CHCH\!=\!CHCO_2CH_3 \xrightarrow[\text{2. RX}]{\text{1. LDA}} CH_3CH\!=\!CHCH\!=\!\overset{\overset{\displaystyle R}{\displaystyle |}}{C}CO_2CH_3 \xrightarrow[\text{2. HOAc}]{\text{1. LDA}}$$

$$H_2C\!=\!CHCH\!=\!CH\overset{\overset{\displaystyle R}{\displaystyle |}}{C}HCO_2CH_3$$

TL *24* 261 (1983)

$$CH_3CH\!=\!CHCH\!=\!CHCO_2CH_3 \xrightarrow[\text{2. RX}]{\text{1. LDA}} H_2C\!=\!CHCH\!=\!CH\overset{\overset{\displaystyle R}{\displaystyle |}}{C}HCO_2CH_3$$

JOC *48* 3003 (1983)

$$RCH_2C\!\equiv\!CCO_2R \xrightarrow[\substack{n\text{-Bu}_4\text{NI} \\ \text{electrolysis}}]{R'X} RC\!\equiv\!CCR'_2CO_2R$$

CC 188 (1980)

CC 1265 (1982)

$$CH_2(CO_2R)_2 \longrightarrow R^1R^2C(CO_2R)_2 \longrightarrow R^1R^2CHCO_2H(R)$$

Review: Org Rxs *9* 107 (1957)

Alkylation:

  base/R'X

                                       Org Syn Coll Vol *1* 250 (1941); *2* 474 (1943);
                                           3 495, 705 (1955); *4* 288 (1963)
                                       JACS *80* 622 (1958); *106* 1051 (1984)
                                           (intramolecular); *109* 7477 (1987)
                                        JCS B 67 (1968) (intramolecular)
                                        JOC *43* 4682 (1978); *46* 3127 (1981)
                                           (intramolecular); *47* 3769 (1982)
                                        CC 522 (1979)
                                        Rec Trav Chim *99* 96 (1980)
                                        J Chem Res (S) 283 (1980)
                                        Syn 452, 805, 893 (1982)

  BF$_3$/R'Cl   TL 3599 (1966)

  alkene, free radical initiator   Syn 99 (1970) (review)

Allylation:

  H$_2$C=CHCH$_2$X (X = OAc, O$_2$SPh),   JOMC *250* C21 (1983)
    cat NiCl$_2$L$_2$

via π-allyl Pd compounds	Tetr *33* 2615 (1977)
	J. Tsuji, "Organic Synthesis with Palladium Compounds," Springer, New York (1980) (review)
	Acct Chem Res *13* 385 (1980)
	Pure Appl Chem *54* 189 (1982) (chiral)
	R. Heck, "Palladium Reagents in Organic Syntheses," Academic Press, New York (1985) (review)
	JOC *52* 5430 (1987)
via π-allyl Mo compounds	JACS *104* 5543 (1982); *105* 3343 (1983); *109* 1469 (1987)
	Organomet *2* 1687 (1983)
via π-allyl W compounds	JACS *105* 7757 (1983); *109* 2176 (1987)

Arylation:

ArH, $(NH_4)_2Ce(NO_3)_6$	TL *27* 2763 (1986)
$ArPb(OAc)_3$	TL *21* 965 (1980)
	Austral J Chem *37* 1245 (1984)
KO-*t*-Bu/Ar_2IX (X = Cl, Br, I)	JOC *52* 4115 (1987) (Meldrum's acid)
KO-*t*-Bu/NC—Ar—Br, NH_3, hν	Tetr *38* 3479 (1982)

Acylation:

RCOCl, $MgCl_2$, 2 Et_3N	JOC *50* 2622 (1987)
base, RCOCl	Syn 451 (1982)
	Syn Commun *12* 19 (1982) (both Meldrum's acid)
base/$RCOPO(OEt)_2$	CL 1087 (1981)
ArI, CO, Pd catalyst	TL *27* 4745 (1986)

Decarboxylation:

$\sim$ 500°C	JOC *29* 1249 (1964)
Δ, CuCl, chiral alkaloid (enantioselective)	TL *28* 539 (1987)
hydrolysis/Δ, Cu_2O	Syn 1029 (1986)
DABCO, Me_2S, Celite, Δ	JACS *109* 3010 (1987)
$(Me_4N)OAc$, DMSO	JACS *98* 6188 (1976)
$(Me_4N)OAc$, HMPA	JACS *98* 630 (1976); *100* 3426 (1978)
$OH^-/H^+/\Delta$	Org Syn Coll Vol *2* 416, 474 (1943); *3* 495 (1955); *4* 630 (1963)
	Org Rxs *9* 107 (1957) (review)
B_2O_3, 170–190°C	Syn Commun *9* 609 (1979)
water, salts, dipolar aprotic solvent	Syn 805 (1982) (review)

H$_2$O, DMSO	TL 1091, 1095 (1974) JOC *43* 138 (1978)
LiCl, H$_2$O, DMSO	JOC *43* 138 (1978)
NaCl, H$_2$O, DMSO	TL 957 (1973); 1091 (1974); *28* 171 (1987) JOC *43* 138 (1978); *52* 3205 (1987)
NaCN, LiI, H$_2$O, DMF	JACS *100* 3426 (1978); *107* 2033 (1985)
KCN or NaCN, DMSO, (H$_2$O)	TL 215 (1967) Tetr *26* 5437 (1970) JOC *43* 138 (1978) JACS *107* 1421 (1985)
Me$_3$SiI	Syn Commun *9* 233 (1979)

$$NCCH_2CO_2R \longrightarrow \underset{\underset{R}{|}}{NCCHCO_2R} \longrightarrow RCH_2CN$$

Alkylation:

base/RX	Org Rxs *9* 107 (1957) (review)
diol, EtO$_2$CN=NCO$_2$Et, PPh$_3$ (dialkylation)	TL 2455 (1976)

Decarbalkoxylation:

H$_2$O, DMSO	TL 1095 (1974)
water, salts, DMSO	Syn 893 (1982) (review)
H$_2$O, NaCl, DMSO, Δ	Syn 893 (1982) (review) TL 957 (1973); *27* 3353 (1986)

$$RCH=CH_2 + XCH_2CO_2Et \longrightarrow RCH_2CH_2CHXCO_2Et$$

X	Reagent(s)	
CO$_2$Et	Mn(OAc)$_3$·2 H$_2$O	CC 693 (1973)
CO$_2$Et, CN	CuO, Cu$_2$O, NiO$_2$, PbO$_2$, MnO$_2$, AgO or Ag$_2$O	Syn 454 (1977)

$$RCH_2CH=CH_2 + H_2C(CO_2Et)_2 \xrightarrow[\text{cat Cu(OAc)}_2]{\text{Mn(OAc)}_3 \cdot 2 \text{ H}_2\text{O}} RCH=CHCH_2CH(CO_2Et)_2$$

CC 693 (1973)

$$R^1(R^1O)\overset{O}{\overset{||}{C}}CH_2\overset{O}{\overset{||}{C}}OR^2 + H_2C=CHR^3 \xrightarrow{PdCl_2} \begin{cases} \xrightarrow{H_2} R^1(R^1O)\overset{O}{\overset{||}{C}}\underset{\underset{R^3CHCH_3}{|}}{CH}\overset{O}{\overset{||}{C}}OR^2 \\ \\ \xrightarrow{25°C} R^1(R^1O)\overset{O}{\overset{||}{C}}\underset{\underset{R^3C=CH_2}{|}}{CH}\overset{O}{\overset{||}{C}}OR^2 \end{cases}$$

JACS *99* 7093 (1977)

$$RCH(CO_2Et)_2 \xrightarrow[\substack{\text{2. R'CHClSPh} \\ \text{3. NaBr or LiCl, } \Delta}]{\text{1. NaH}} R'CH = CRCO_2Et$$

Syn 131 (1982)

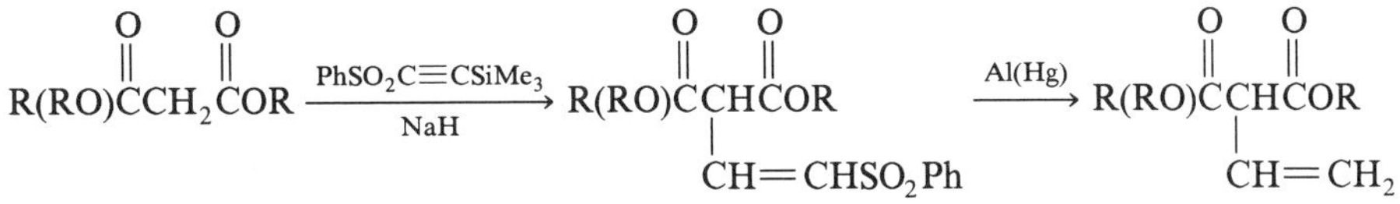

$$R(RO)\overset{O}{\overset{\|}{C}}CH_2\overset{O}{\overset{\|}{C}}OR \xrightarrow[\text{NaH}]{PhSO_2C\equiv CSiMe_3} R(RO)\overset{O}{\overset{\|}{C}}CH\overset{O}{\overset{\|}{C}}OR \xrightarrow{Al(Hg)} R(RO)\overset{O}{\overset{\|}{C}}CH\overset{O}{\overset{\|}{C}}OR$$

$$\underset{CH=CHSO_2Ph}{} \qquad \underset{CH=CH_2}{}$$

JOC 47 4713 (1982)

$$R^1(R^1O)\overset{O}{\overset{\|}{C}}CH_2\overset{O}{\overset{\|}{C}}OEt \xrightarrow[\text{2. } R_2^2CClNO_2, h\nu]{\text{1. NaH}} R^1(R^1O)\overset{O}{\overset{\|}{C}}\overset{}{C}\overset{O}{\overset{\|}{C}}OEt$$

$$\underset{CR_2^2}{\overset{\|}{}}$$

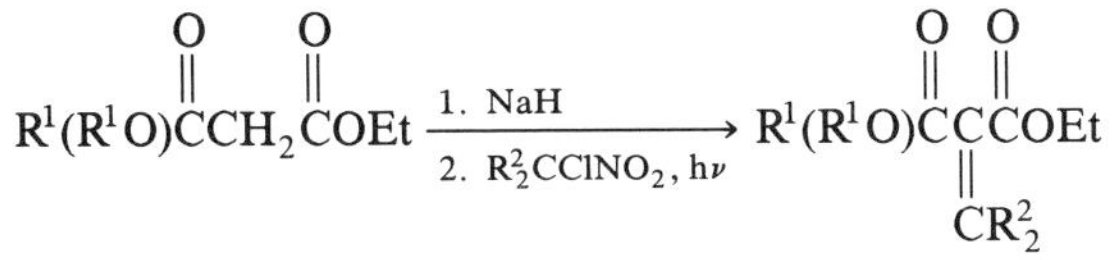

Syn 62 (1981)

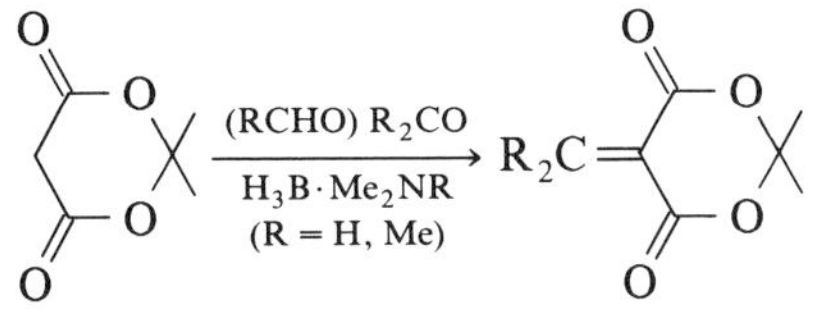

TL 24 4951 (1983)

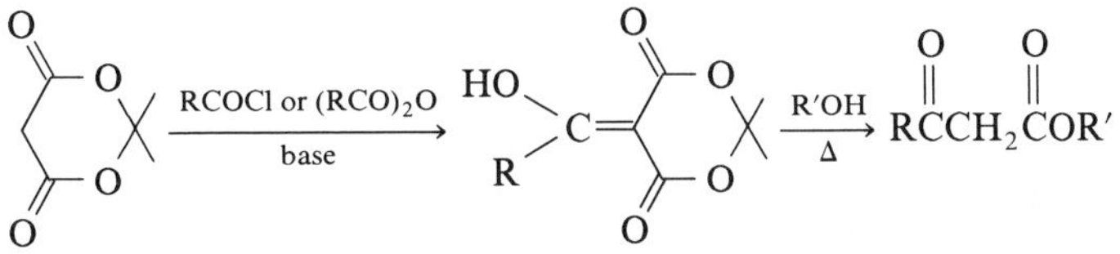

Syn 451 (1982)

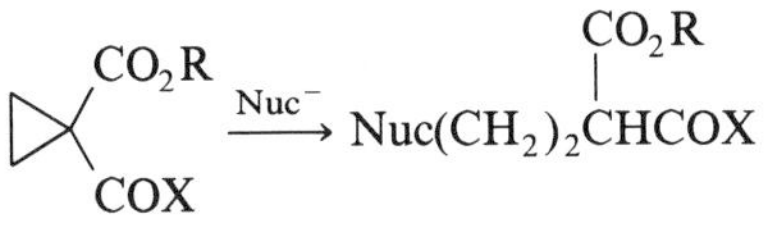

X = R, OR

Review: Acct Chem Res 12 66 (1979)

Nuc⁻

R_2CuLi

JACS 94 4014 (1972)
TL 997 (1972); 529 (1975); 3857 (1976)
Can J Chem 60 825 (1982)

Et_2AlCN (Nuc = CN)

Can J Chem 60 825 (1982)

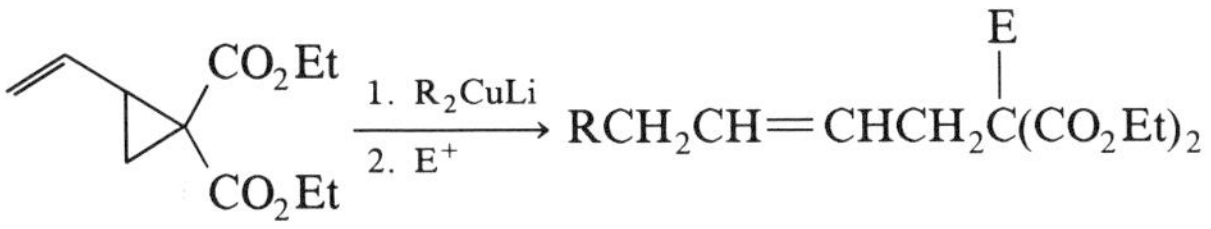

$$\xrightarrow[\text{2. E}^+]{\text{1. } R_2CuLi} RCH_2CH=CHCH_2\overset{E}{\overset{|}{C}}(CO_2Et)_2$$

E⁺

H_2O

TL 997 (1972)

R'X

JOC 38 2100 (1973)

$$\text{ArH or ArM} \xrightarrow{\text{OC(CO}_2\text{Et)}_2} \underset{\overset{|}{\text{OH}}}{\text{ArC(CO}_2\text{Et)}_2} \xrightarrow[\text{2. Li, NH}_3]{\text{1. Ac}_2\text{O, Et}_3\text{N}} \text{ArCH(CO}_2\text{Et)}_2$$

$$M = Li, MgX$$

JOC *47* 4692 (1982)

$$t\text{-BuSCR}_2\text{CO}_2\text{R} \xrightarrow[\text{2. E}^+]{\text{1. Li, NH}_3} \text{ECR}_2\text{CO}_2\text{R}$$

$$E^+ = H_2O, RX, RCHO, RCOCl$$

Syn Commun *3* 265 (1973)

$$p\text{-CH}_3\text{C}_6\text{H}_4\text{SOCHRCO}_2\text{-}t\text{-Bu} \xrightarrow[\substack{\text{2. RCHO or R}_2\text{CO} \\ \text{3. H}_2\text{O}}]{\text{1. }t\text{-BuMgBr}} \xrightarrow{\text{Al(Hg)}} \underset{\overset{|}{\text{OH}}}{\text{RR'CCHRCO}_2\text{-}t\text{-Bu}}$$

chiral

Tetr *36* 227 (1980)
Helv *65* 1602 (1982)

$$\underset{\overset{\|}{\text{O}}}{\text{MeSCH}_2\text{CO}_2\text{Et}} \xrightarrow[\text{TsOH}]{\text{ArH}} \underset{\overset{|}{\text{SMe}}}{\text{ArCHCO}_2\text{Et}}$$

TL *22* 81 (1981)

$$t\text{-BuSCH}_2\text{CO}_2\text{R} \xrightarrow[\text{2. R}^1\text{X}]{\text{1. LiNH}_2} \xrightarrow[\text{2. R}^2\text{X}]{\text{1. LiNH}_2} \underset{\overset{|}{\text{R}^2}}{\overset{\overset{|}{\text{R}^1}}{t\text{-BuSCCO}_2\text{R}}} \xrightarrow[\text{2. R}^3\text{X}]{\text{1. Li, NH}_3} \underset{\overset{|}{\text{R}^3}}{\overset{\overset{|}{\text{R}^1}}{\text{R}^2\text{—C—CO}_2\text{R}}}$$

Syn Commun *3* 265 (1973)

$$\text{RCH}{=}\text{CH}_2 + \text{Cl}_3\text{CCO}_2\text{CH}_3 \xrightarrow{\text{catalyst}} \underset{\overset{|}{\text{Cl}}}{\text{RCHCCl}_2\text{CO}_2\text{CH}_3}$$

Review: Syn 145 (1977)

Catalyst

$\text{RuCl}_2(\text{PPh}_3)_3$	JOC *41* 396 (1976)
CuCl	JOC *29* 2104 (1964)

$$\underset{\overset{|}{\text{Br}}}{\text{R}^1\text{CHCO}_2\text{R}^2} \longrightarrow \underset{\overset{|}{\text{R}}}{\text{R}^1\text{CHCO}_2\text{R}^2}$$

R_3B or RB⟨ ⟩, KO-*t*-Bu   JACS *90* 818, 1911 (1968);
 (R^1 = H, Br, alkyl; R = alkyl, aryl)   *91* 2146, 4304 (1969)

R_3B or RB⟨ ⟩, 2,6-(t-Bu)$_2$C$_6$H$_3$OK   JACS *91* 6855 (1969)
                                                    (R^1 = H, Br; R = alkyl)
                                                    Organomet Chem Syn *1* 95 (1970)
                                                    (R^1 = CN, R = alkyl)

$RCH\!=\!CHB\!\!\bigcirc$, 2,6-$(t$-Bu$)_2C_6H_3OK$ JOC *51* 3398 (1986)

Zn/ArX, Ni or Pd catalyst   JOMC *132* C17 (1977); *177* 273 (1979)
  (R = Ar, R^l = H)

Zn/RCH$=$CHCH(OAc)R, cat Pd(PPh$_3$)$_4$ TL *27* 4223 (1986)

PhCdCl (R^l = H)   JOC *33* 1675 (1968)

$$RMgBr + R'NH\overset{\underset{|}{Cl}}{C}HCO_2Me \longrightarrow R'NH\overset{\underset{|}{R}}{C}HCO_2Me$$

TL *27* 2435 (1986)

$$R_2Zn + Et_2N\overset{\underset{|}{OMe}}{C}HCO_2Me \longrightarrow Et_2N\overset{\underset{|}{R}}{C}HCO_2Me$$

JOMC *256* 193 (1983)

$$BrCH_2CO_2R' \xrightarrow[\substack{\text{2. }R_2CO \\ \text{3. }H_2O}]{\text{1. metal}} R_2\overset{\underset{|}{OH}}{C}CH_2\overset{\overset{O}{\|}}{C}OR'$$

See page 553, Section 2.1.

$$RX + ICF_2CO_2Me + Cu \longrightarrow RCF_2CO_2Me$$

R = 1° alkyl, allylic, benzylic, aryl, alkynyl, vinylic, heterocyclic

TL *27* 6103 (1986)

$$BrCH_2CH\!=\!CHCO_2R' \xrightarrow[\text{2,6-}(t\text{-Bu})_2C_6H_3OK]{R_3B} RCH\!=\!CHCH_2CO_2R'$$

JACS *92* 1761 (1970)

$$R'H + N_2CRCO_2R \xrightarrow{\text{Rh catalyst}} R'CHRCO_2R$$

CC 688 (1981)
TL *23* 4321 (1982)

$$N_2CHCO_2R' \xrightarrow{\text{organometallic}} \xrightarrow{E^+} R\overset{\underset{|}{E}}{C}H\overset{\overset{O}{\|}}{C}OR'$$

E^+	Organometallic	
H$_2$O	R$_3$B	JACS *90* 6891 (1968)
	R$_2$BCl	JACS *94* 3662 (1972)
	RBCl$_2$	JOC *38* 2574 (1973)
	(RC$\equiv$C)$_3$B	Can J Chem *50* 1105 (1972)
	PhCu	JACS *90* 2186 (1968)
		CC 515 (1969)
D$_2$O	R$_3$B	JACS *91* 6195 (1969)
NCS, NBS	R$_3$B	Can J Chem *50* 2387 (1972)

$$Me_2\overset{+}{S}\overset{-}{C}HCO_2R' \xrightarrow[\text{2. H}_2\text{O}]{\text{1. R}_3\text{B}} RCH_2CO_2R'$$

JACS *89* 6804 (1967)

$$i\text{-Pr} \overset{i\text{-Pr}}{\underset{i\text{-Pr}}{\bigodot}} CO_2CH_2R \xrightarrow[\text{2. E}^+]{\text{1. RLi}} i\text{-Pr} \overset{i\text{-Pr}}{\underset{i\text{-Pr}}{\bigodot}} CO_2\overset{E}{C}HR$$

$E^+ = RX, RCHO, R_2CO, Me_3SiCl, n\text{-Bu}_3SnCl$

JACS *99* 5213 (1977)
JOC *43* 4255 (1978); *46* 2363 (1981)
Chem Rev *78* 275 (1978) (review)

$$ArCHR\overset{O}{\overset{\|}{C}}NR'_2 \xrightarrow[\text{2. E}^+]{\text{1. } n\text{-BuLi}} Ar\overset{E}{\overset{|}{C}}R\overset{O}{\overset{\|}{C}}NR'_2$$

$E^+ = RX, RCHO, R_2CO, (MeO)_2CO, Me_3SiCl$

Syn 1045 (1982)

3. Anhydrides

$$(CH_3\overset{O}{\overset{\|}{C}})_2O + Ph_3CBF_4 \longrightarrow Ph_3CCH_2\overset{O}{\overset{\|}{C}}OH$$

JOC *35* 278 (1970)

$$RCH_2\overset{O}{\overset{\|}{C}}O\overset{O}{\overset{\|}{P}}X_2 \xrightarrow[\text{Et}_3\text{N}]{\text{RCH=NR}} \begin{matrix} RCH-C=O \\ | \quad\quad | \\ RCH-NR \end{matrix}$$

$$\underset{Cl_3CCH_2O}{\overset{\overset{X}{|}}{\underset{O}{\overset{\|}{\bigcirc}}} \overset{O}{\underset{}{\diagdown}}N-}$$

Syn 63 (1982)

Syn 1053 (1982)

4. Amides

$$H\overset{O}{\overset{\|}{C}}NR_2 \xrightarrow[\substack{\text{2. R}_2\text{CO (RCHO)} \\ \text{3. H}_2\text{O}}]{\text{1. base}} R_2\overset{HO}{\overset{|}{C}}-\overset{O}{\overset{\|}{C}}NR_2$$

Angew Int *12* 836 (1973); *15* 293 (1976); *20* 795 (1981)
Can J Chem *52* 185 (1974)
CC 387 (1976)
JCS Perkin I 1881 (1977)
Chem Rev *78* 275 (1978) (review)

$$Ph_3COH + CH_2(CONH_2)_2 \xrightarrow{H^+} Ph_3CCH(CONH_2)_2$$

JCS 716 (1962)

$$NCCH_2CONR_2 \xrightarrow[\text{2. RX}]{\text{1. base}} NCC\overset{\overset{\displaystyle R}{|}}{\text{H}}CONR_2$$

chiral   chiral

TL 27 2463 (1986)

$$R_2CHCONR_2 \xrightarrow[\text{2. E}^+]{\text{1. base}} R_2\overset{\overset{\displaystyle E}{|}}{\text{C}}CONR_2$$

Review: Syn 509 (1977) (polyalkali metal amides)

$\underline{E^+}$

RX	JOC 26 3696 (1961); 31 982, 989, 3873 (1966); 39 2475 (1974); 42 1688 (1977); 51 1541, 1936 (lactam), 3140 (lactam), 4080 (intramolecular) (1986) Ber 101 3113 (1968) Organomet Chem Syn 1 237 (1971) Can J Chem 54 1098 (1976) TL 3961 (1978); 21 4233 (1980); 25 857 (1984) (chiral); 28 651 (chiral), 2041, 6389 (1987) JOMC 177 5 (1979) (chiral) Pure Appl Chem 53 1109 (1981) (chiral) JACS 107 3915, 7776 (chiral lactam) (1985); 108 306 (1986) (chiral lactam); 109 4405 (1987) (lactam) CC 992 (1987)
ArX, hν	JOC 45 1239 (1980) JACS 107 435 (1985) (intramolecular)
ArBr, LiNR$_2$	JOC 52 2110 (1987)
epoxide	JOC 37 1907 (1972); 42 1688 (1977); 46 2831 (1981) Ber 105 1621 (1972) Can J Chem 55 266 (1977) JACS 107 7776 (1985) (chiral lactam)
RCHO, R$_2$CO	Ber 101 3113 (1968) JOC 39 2475 (1974); 42 1688 (1977); 45 1068 (1980); 50 3019 (1985) (enones) Can J Chem 55 266 (1977) Syn 954 (1979); 247 (1980) TL 21 3975 (1980); 24 3883 (1983); 26 5807 (1985) (chiral Li, Zr enolates); 28 651 (Zr enolate), 2037 (Ti enolate), 6137 (1987) Pure Appl Chem 53 1109 (1981) (chiral Li, B, Zr enolates) JACS 103 2876 (1981) (chiral Li, Zr enolates); 105 1667 (1983) (diastereoselectivity) Angew Int 22 788 (1983) CC 992 (1987)

enones (1,4-addition)   JOC *50* 3019 (1985)

RCOCl   TL *25* 6015 (1984)

ROCOCl   TL *28* 4901 (1987)

diketene   TL *28* 4901 (1987)

$$\underset{RCN}{\overset{O}{\parallel}}\diagup\diagdown N$$ CC 1447 (1986)

RCO_2CO_2Et Syn 954 (1979)

$$RSO_2N\overset{O}{\overbrace{}}CR_2$$ (chiral) (E = OH)   JOC *51* 2402 (1986)

MeSSMe   JOC *42* 3236 (1977)

PhSeSePh   JACS *108* 306 (1986) (chiral lactam)

$$\underset{RCHCONR_2}{\overset{X}{|}} \xrightarrow[\text{2. E}^+]{\text{1. base}} \underset{\underset{X}{|}}{\overset{\overset{E}{|}}{RCCONR_2}}$$

$\underline{X}$	$\underline{E^+}$	
F	RX	TL *27* 6103 (1986)
Cl	RCHO	JACS *108* 4598 (1986) (chiral)
RO	RX	TL *26* 1343 (1985) (chiral); 27 2731 (1986) (chiral)
	RCHO	JACS *106* 8161 (1984)
R_3SiO	RX	TL *27* 2731 (1986)
RCONR	RX	JOC *52* 4044 (1987)
	RCHO	TL *27* 6361 (1986)
ArCH=N	RX, RCHO, RCOCl, $ClCO_2R$	Helv *51* 1905 (1968) JACS *95* 4324 (1971) CC 1138 (1972) TL 375 (1972)
$Ar_2C=N$	RX, $ClCO_2R$	Helv *64* 1145 (1981)
$(MeS)_2C=N$	RX	TL *27* 3403 (1986)

$$Ph\overset{O}{\diagdown}\!\!\!\diagup\!\!\!\diagdown CO_2H \longrightarrow \underset{\underset{i\text{-Pr}}{}}{\overset{Ph}{\diagdown N \diagdown O}} \xrightarrow[\text{2. LDA/R}^2X]{\text{1. LDA/R}^1X} \underset{\underset{i\text{-Pr}}{}}{\overset{Ph}{\diagdown N \diagdown O}} \overset{R^1}{\underset{R^2}{}}$$

JACS *106* 1146 (1984)
TL *26* 2047 (1985)
JOC *51* 1541 (1986)

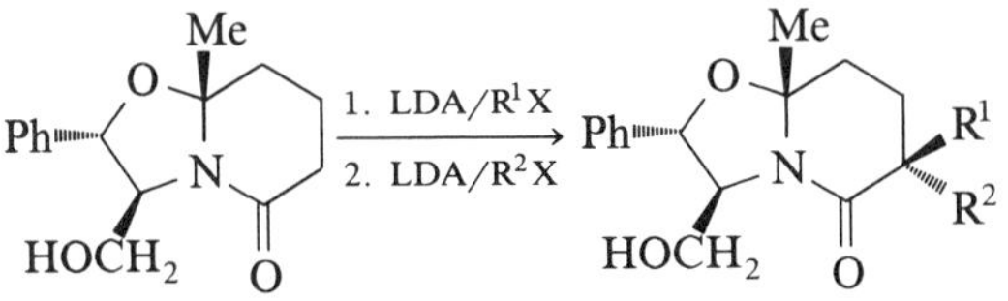

JOC *51* 1936 (1986)

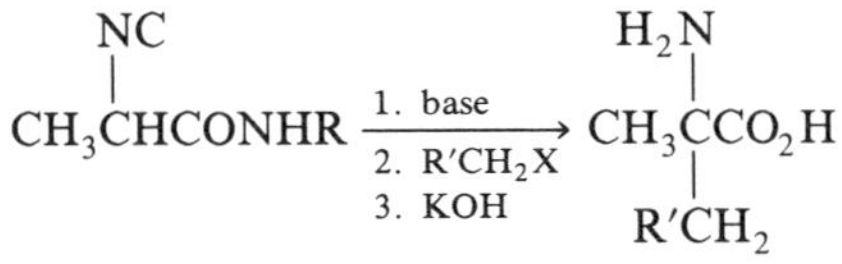

Ann 439 (1981)

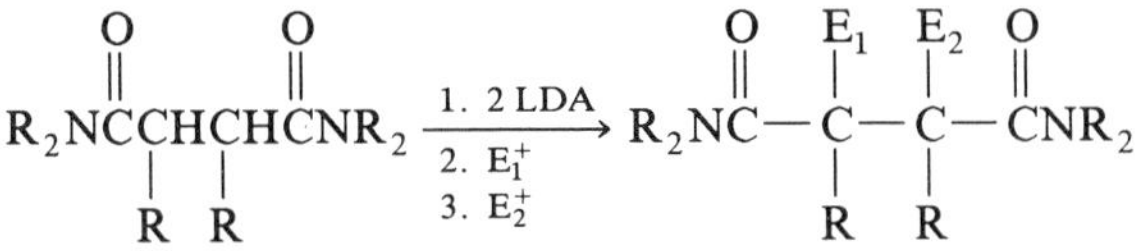

$E^+ = D_2O$, RX, RCHO, $Br(CH_2)_4Br$, $Br(CH_2)_3CO_2Et$, RCO_2Et, $o\text{-}C_6H_4(CO_2Et)_2$, $(PhS)_2$, $(PhSe)_2$

TL *23* 3971, 3975 (1982)

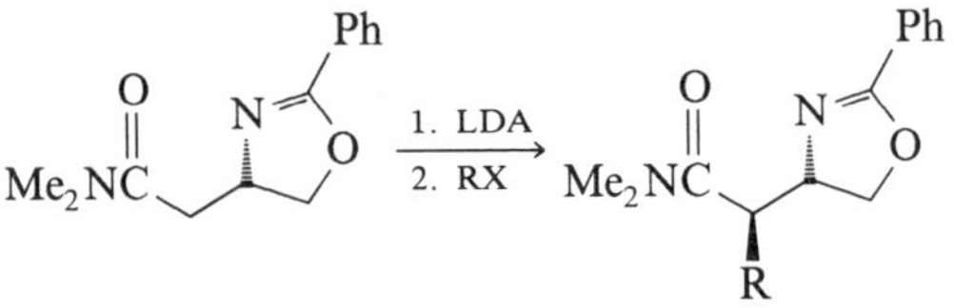

TL *24* 2733 (1983)
JACS *108* 4943 (1986)

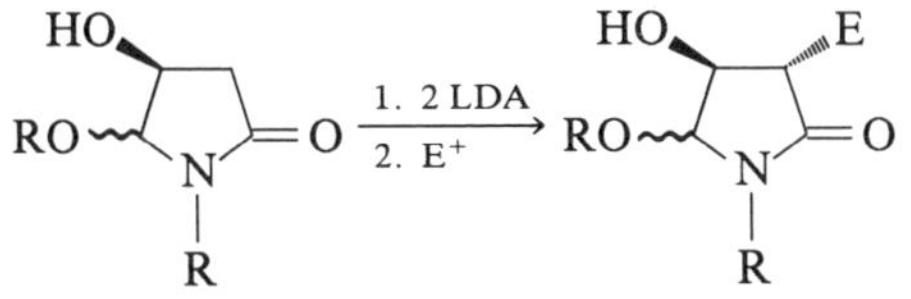

$E^+ = RX$, R_2CO

TL *28* 1581 (1987)

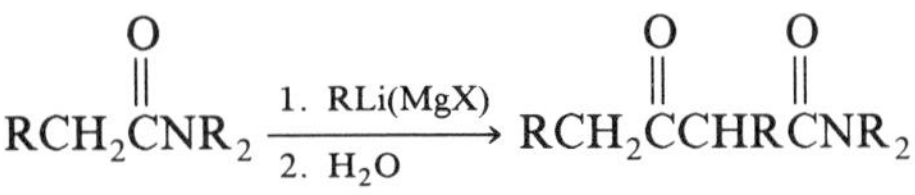

Tetr *38* 557 (1982)

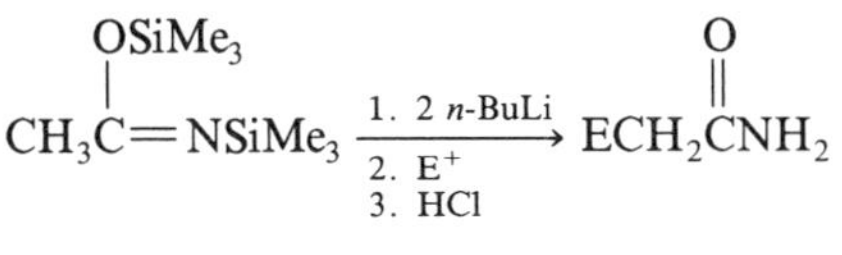

E^+

RCHO, R_2CO TL *21* 3227 (1980)

RCO_2R JACS *107* 7760 (1985)

$$CH_3(CH=CH)_nCNR_2 \xrightarrow[\substack{\text{phase} \\ \text{transfer} \\ \text{catalyst}}]{\text{RCHO}} RCH=CH(CH=CH)_nCNR_2$$

(with C=O double bonds on each C adjacent to NR$_2$)

$$n = 0-2$$

J Chem Res (S) 106 (1981)
Ann 1725 (1981)

$$\underset{\text{H O}}{C=C-\overset{|}{C}-\overset{\|}{C}NR_2} \text{ or } \underset{\text{H}}{C-C=C-\overset{\|}{C}NR_2} \xrightarrow[\text{2. } E^+]{\text{1. base}} \underset{\text{E O}}{C=C-\overset{|}{C}-\overset{\|}{C}NR_2}$$

$\underline{E^+}$

H_2O	JOC *34* 3263 (1969)
R'X	JOC *34* 3263 (1969); *46* 2029 (1981)
	TL 1645 (1978); *27* 531 (1986)
RCHO, R_2CO	TL 1645 (1978)
	JOC *46* 2029 (1981)
Me_3SiCl/Δ /RCHO, n-Bu$_4$NF	TL *27* 535 (1986)
or TiCl$_4$ (E = RCHOH)	
$PhCO_2Me$	TL 1645 (1978)
PhSeBr	TL 1645 (1978)

$$\underset{\text{H O}}{C=C-\overset{|}{C}-\overset{\|}{C}NR_2} \text{ or } \underset{\text{H}}{C-C=C-\overset{\|}{C}NR_2} \xrightarrow[\text{2. } E^+]{\text{1. base}} \underset{\text{E}}{C-C=C-\overset{\|}{C}NR_2}$$

$\underline{E^+}$

H_2O	JOC *50* 3526 (1985)
R'X	JOC *34* 3263 (1969)
	TL 2057 (1975)
R'X, CuI	TL 1645 (1978)
	JOC *46* 2029 (1981)
RCHO, R_2CO	JOC *34* 3263 (1969); *46* 2029 (1981)
CO_2	JOC *34* 3263 (1969)
Me_3SiCl, Δ	TL *27* 535 (1986)

$$\underset{CH_3}{H_2C=CCONHAr} \xrightarrow[\substack{\text{2. } R_2CO \text{ (RCHO)} \\ \text{3. } H_2O}]{\text{1. KO-}t\text{-Bu, } n\text{-BuLi}} \underset{CH_2CR_2OH}{H_2C=CCONHAr}$$

CL 1567 (1980)

$$
\begin{array}{c}
\text{C}-\text{H} \\
| \\
\text{C}=\text{C}-\text{CONR}_2
\end{array}
\xrightarrow[\text{2. E}^+]{\text{1. RLi}}
\begin{array}{c}
\text{C}-\text{E} \\
| \\
\text{C}=\text{C}-\text{CONR}_2
\end{array}
$$

$\text{E}^+ = \text{D}_2\text{O, RX, RCHO, R}_2\text{CO, H}_2\text{C}\!-\!\!\overset{\displaystyle O}{\triangle}\!\!-\!\text{CH}_2, \text{RCONR}_2, \text{CO}_2, \text{H}_2\text{C}=\text{CHCONR}_2$ (1,4-addition), $\text{B(OMe)}_3, \text{Me}_3\text{SiCl, (PhS)}_2, \text{RN}=\text{NR}$

Angew Int *13* 468 (1974)
JACS *102* 4550 (1980); *107* 4745 (1985)
JOC *45* 4257 (1980); *47* 1610 (1982); *51* 3921, 4627 (1986); *52* 218 (1987)

$$
\begin{array}{c}
\text{H} \quad\quad \text{O} \\
| \quad\quad\quad || \\
\text{C}-\text{C}=\text{C}-\text{C}-\text{C}-\text{CNR}_2
\end{array}
\xrightarrow[\substack{\text{2. MgBr}_2 \\ \text{3. E}^+}]{\text{1. sec-BuLi, TMEDA}}
\begin{array}{c}
\text{E} \quad\quad \text{O} \\
| \quad\quad\quad || \\
\text{C}-\text{C}-\text{C}=\text{C}-\text{C}-\text{CNR}_2
\end{array}
$$

$\text{E}^+ = \text{D}_2\text{O, RX, H}_2\text{CO, R}_2\text{CO, Me}_3\text{SiCl, Ph}_3\text{SnCl}$

JACS *105* 6350 (1983); *109* 5403 (1987)

$$
\begin{array}{c}
\text{O} \\
|| \\
\text{XCH}=\text{CRCNR}_2
\end{array}
\xrightarrow[\text{2. E}^+]{\text{1. base}}
\begin{array}{c}
\text{E} \quad\ \text{O} \\
| \quad\quad || \\
\text{XC}=\text{CRCNR}_2
\end{array}
$$

$\text{X} = \text{H, OR, SR, NR}_2$

See page 185, Section 1.

$$
\begin{array}{c}
\text{O} \\
|| \\
n\text{-Bu}_3\text{SnCH}_2\text{CH}_2\text{CNHR}
\end{array}
\xrightarrow[\substack{\text{2. E}^+ \\ \text{3. HCl}}]{\text{1. 2 } n\text{-BuLi, DABCO}}
\begin{array}{c}
\text{O} \\
|| \\
\text{ECH}_2\text{CH}_2\text{CNHR}
\end{array}
$$

$\text{E}^+ = \text{MeOD, RX, R}_2\text{CO, Me}_3\text{SiCl}$

TL *23* 1463 (1982)
JACS *105* 7182 (1983)

$$
\begin{array}{c}
\text{H} \quad\quad \text{O} \\
| \quad\quad\quad || \\
\text{X}-\text{C}-\text{C}-\text{CNR}_2
\end{array}
\xrightarrow[\text{2. E}^+]{\text{1. sec-BuLi, TMEDA}}
\begin{array}{c}
\text{E} \quad\quad \text{O} \\
| \quad\quad\quad || \\
\text{X}-\text{C}-\text{C}-\text{CNR}_2
\end{array}
$$

$\text{X} = \text{Ph, SPh, C}=\text{C}; \text{E}^+ = \text{D}_2\text{O, RX, R}_2\text{CO, Me}_3\text{SiCl, Ph}_3\text{SnCl, PhSSPh}$

JACS *109* 5403 (1987)

$$
\begin{array}{c}
\text{O} \\
|| \\
\text{PhSO}_2\text{CH}_2\text{CH}_2\text{CNHPh}
\end{array}
\xrightarrow[\substack{\text{2. RCHO (R}_2\text{CO)} \\ \text{3. H}_2\text{O}}]{\text{1. 2 } n\text{-BuLi}}
\begin{array}{c}
\text{R}_2\text{COH} \quad \text{O} \\
| \quad\quad\quad || \\
\text{PhSO}_2\text{CHCH}_2\text{CNHPh}
\end{array}
$$

CL 1359 (1984)

$$\underset{\substack{\| \\ O}}{X\text{C}}\text{NMe}_2 \xrightarrow[\text{2. E}^+]{\text{1. RLi}} \underset{\substack{\| \\ O}}{X\text{C}}\text{NMeCH}_2\text{E}$$

$$E^+ = RX, RCHO, R_2CO$$

Review: Chem Rev *78* 275 (1978)

<u>X</u>

2,4,6-(i-Pr)$_3$C$_6$H$_2$ TL 1839 (1977)
                                                     Helv *61* 512 (1978)

R$_2$N   Helv *61* 2239 (1978)

2,4,6-(t-Bu)$_3$C$_6$H$_2$O   Angew Int *17* 274 (1978)

$$\underset{\substack{| \\ X}}{\underset{N}{\bigcirc}} \xrightarrow[\text{2. E}^+]{\text{1. RLi}} \underset{\substack{| \\ X}}{\underset{N}{\bigcirc}}\text{E}$$

<u>X</u>	<u>E</u>$^+$	
Me$_3$CCO	RX, R$_2$CO, Me$_3$SiCl, n-Bu$_3$SnCl	Tetr *39* 1963 (1983)
Et$_3$CCO	MeOD, 1° RI, RCHO, R$_2$CO	JOC *46* 4316 (1981)
		JACS *106* 1010 (1984)
Ph$_3$CCO	RX, RCHO, R$_2$CO	Helv *64* 1337 (1981)
2,4,6-(i-Pr)$_3$C$_6$H$_2$CO	MeOD, RCHO, Me$_3$SiCl	JOC *46* 4108 (1981)
		JACS *106* 1010 (1984)
MeOCO	RX	JOC *45* 193 (1980)
		(allylic anion)
PO(NMe$_2$)$_2$	RX, RCHO, R$_2$CO, epoxide, ArH·Cr(CO)$_3$, I$_2$, Me$_3$SiCl, n-Bu$_3$SnCl	Helv *64* 643 (1981)
		Tetr *39* 1963 (1983)

$$\text{RCH}=\text{CH}_2 + \underset{\substack{| \\ H}}{\underset{N}{\bigcirc}}=\text{O} \xrightarrow{t\text{-butyl peroxide}} \text{RCH}_2\text{CH}_2-\underset{\substack{| \\ H}}{\underset{N}{\bigcirc}}=\text{O}$$

Bull Acad Sci USSR, Div Chem Sci 1745 (1964)
Proc Acad Sci USSR, Chem Sec *158* 1069 (1964)

$$\underset{\substack{\| \quad | \\ O \quad X}}{\text{RC}\text{NR}\text{CHR}} + \text{H}_2\text{C}=\text{CHCH}_2\text{SiMe}_3 \xrightarrow[\text{BF}_3\cdot\text{OEt}_2]{\text{SnCl}_4 \text{ or}} \underset{\substack{\| \quad | \\ O \quad \text{CH}_2\text{CH}=\text{CH}_2}}{\text{RC}\text{NR}\text{CHR}}$$

X = OAc, OEt

CC 134 (1982)

5. Imides

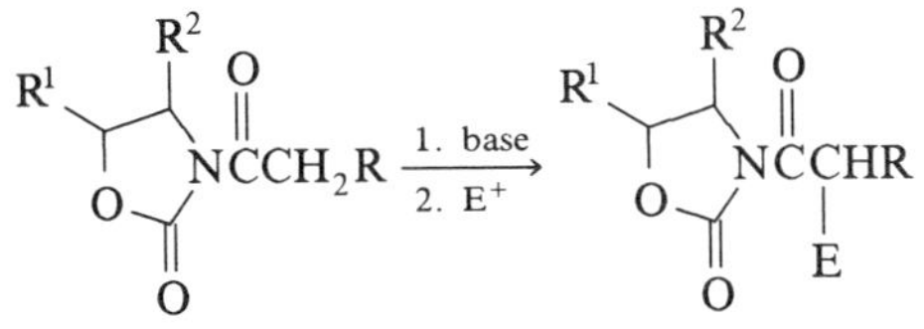

chiral   chiral

$\underline{E^+}$

NBS	TL *28* 1123 (1987) (B enolate)

RX	Pure Appl Chem *53* 1109 (1981)
	JACS *104* 1737 (1982); *107* 196 (1985);
	   109 1269 (1987)
	TL *23* 807 (1982); *27* 3059 (1986);
	   28 3651 (1987)
	JOC *52* 3168 (intramolecular), 3759 (1987)

RCHO	Pure Appl Chem *53* 1109 (1981)
	   (Li, B enolates)
	JACS *103* 2127 (1981) (B enolate);
	   107 5292 (1985) (B enolate);
	   108 2476, 4595 (Li, B, Sn, Zn enolates),
	   6757 (Sn enolate) (1986);
	   109 7151 (1987) (Sn enolate)
	TL *23* 807 (1982) (Li, B enolates);
	   24 3395 (1983) (B enolate); *27* 897
	   (Ti enolate), 3311 (B enolate), 5683
	   (B enolate) (1986); *28* 39
	   (B enolate), 5921 (B enolate), 6001
	   (B enolate) (1987)
	CC 1237 (1986) (B enolate)
	JOC *51* 4322 (1986) (B enolate);
	   52 5588 (1987) (B enolate)

RCHO, ZnI$_2$	JACS *108* 4675 (1986) (Si enolate)
RCOCl	JACS *106* 1154 (1984) CC 1102 (1987)
(RCO)$_2$O	TL *24* 4805 (1983)
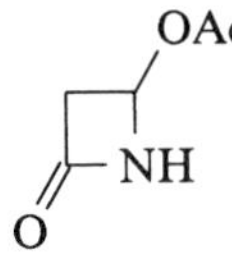 , ZnBr$_2$	JACS *108* 4675 (1986)
(PhCH$_2$OCO)$_2$	JOC *51* 3700 (1986)
2,4,6-(*i*-Pr)$_3$C$_6$H$_2$SO$_2$N$_3$/HOAc	JACS *109* 6881 (1987)
RO$_2$CN=NCO$_2$R	JACS *108* 6395, 6397 (1986)

$$R^1 \underset{O}{\overset{R^2}{\underset{\|}{\text{NCCH}}}}=\text{CHCHR}_2 \longrightarrow R^1 \underset{O}{\overset{R^2}{\underset{\|}{\text{NCCHCH}}}}=\text{CR}_2$$

X	Reagents	
RCHOH	$n\text{-Bu}_2\text{BOTf, Et}_3\text{N/RCHO}$	TL *27* 4957, 4961 (1986)
OH	$\text{NaN(SiMe}_3)_2 / \text{PhHC}\overset{O}{\overset{\triangle}{-}}\text{NSO}_2\text{Ph}$	JACS *107* 4346 (1985)

$$\underset{\text{chiral}}{\overset{\text{CO}_2\text{CH}_3}{X \underset{S}{\overset{}{\bigvee}} \text{NCOCH}_2\text{R}}} \xrightarrow[\text{2. R'CHO}]{\text{1. }i\text{-Pr}_2\text{NEt, }n\text{-Bu}_2\text{BOTf or Sn(OTf)}_2} \underset{\text{chiral}}{\overset{\text{CO}_2\text{CH}_3}{X \underset{S}{\overset{}{\bigvee}} \underset{\text{R'CHOH}}{\text{NCOCHR}}}}$$

X = O, S

JOC *52* 2201 (1987)

$$\text{RCH}_2\overset{O}{\overset{\|}{C}}\text{N}\underset{R'}{\overset{S}{\bigvee}}\text{S} \xrightarrow[\underset{\text{NEt}}{\bigcirc}]{\text{Sn(OTf)}_2} \overset{E^+}{\longrightarrow} \text{ECHR}\overset{O}{\overset{\|}{C}}\text{N}\underset{R'}{\overset{S}{\bigvee}}\text{S}$$

$$\xrightarrow{\text{R''OH}} \text{ECHR}\overset{O}{\overset{\|}{C}}\text{OR''}$$

$$\xrightarrow{\text{R''NH}_2} \text{ECHR}\overset{O}{\overset{\|}{C}}\text{NHR''}$$

E^+		
RCHO		CL 1903 (1982); 297 (chiral), 1799 (chiral) (1983)

$$\underset{O}{\overset{\text{OAc}}{\overbrace{}}}\text{NH}$$

JACS *108* 4673 (1986) (chiral)

$$\text{Et}\overset{O}{\underset{O}{\bigcirc}}\text{NH}\overset{}{=}O \xrightarrow{\text{RCH}=\text{CHPb(OAc)}_3} \text{RCH}=\text{CH}\underset{\text{Et}}{\overset{O}{\bigcirc}}\text{NH}\overset{}{=}O$$

CC 965 (1984)

$$\underset{\text{OCO-}t\text{-Bu}}{\overset{}{\bigvee_N}}=O \xrightarrow[\text{2. RCOCl}]{\text{1. LDA}} \underset{\text{OCO-}t\text{-Bu}}{\overset{\text{COR}}{\bigvee_N}}=O$$

TL *27* 2691 (1986)

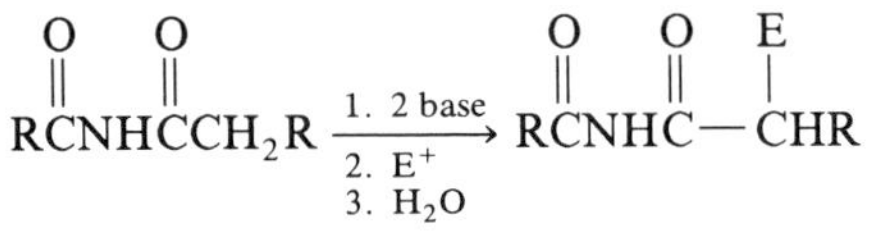

$$\underset{\substack{\| \quad \| \\ RCNHCCH_2R}}{O \quad O} \xrightarrow[\substack{2.\ E^+ \\ 3.\ H_2O}]{1.\ 2\ base} \underset{\substack{\| \quad \| \quad | \\ RCNHC-CHR}}{O \quad O \quad E}$$

$$E^+ = RX,\ RCHO,\ R_2CO,\ RCO_2R,\ PhSSPh$$

JACS *83* 3468 (1961)
Can J Chem *46* 2561 (1968)
JOC *51* 495 (1986)

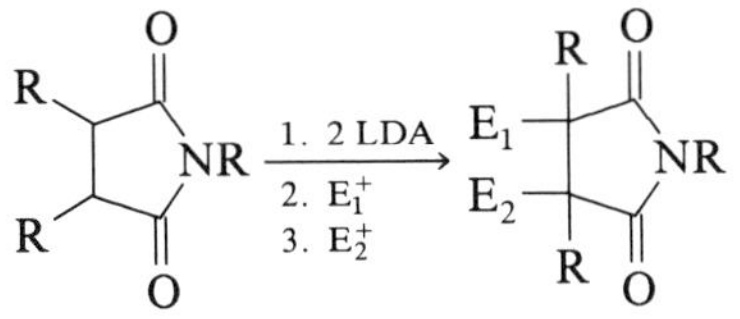

$$\xrightarrow[\substack{2.\ E_1^+ \\ 3.\ E_2^+}]{1.\ 2\ LDA}$$

$$\underline{E_1^+}$$

$$\underline{E_2^+}$$

RX

RX

JOC *47* 68 (1982)

$$X-C_n-X$$
$$(X = Br,\ OTs;\ n = 3-5)$$

JOC *47* 68, 4731 (1982)

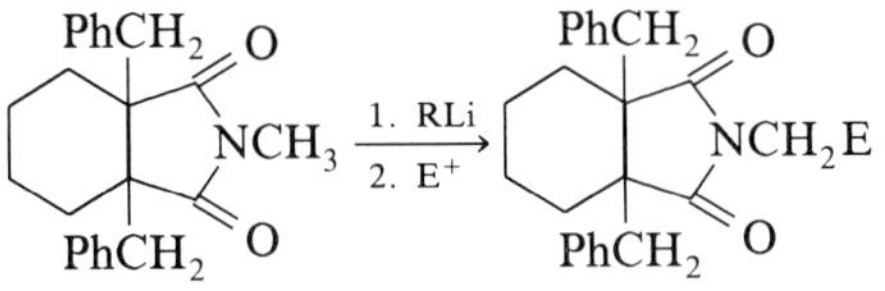

$$\xrightarrow[\substack{2.\ E^+}]{1.\ RLi}$$

$$E^+ = RX,\ RCHO,\ R_2CO,\ RCO_2Me,\ Me_3SiCl$$

Helv *60* 1459 (1977)

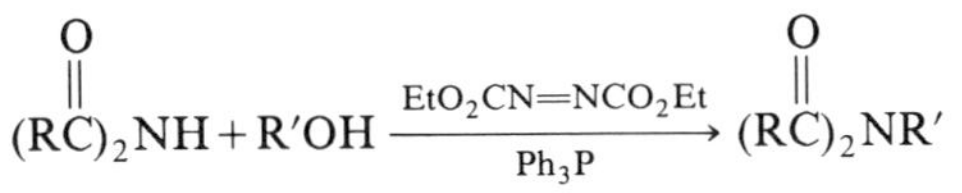

$$\underset{\substack{\| \\ (RC)_2NH}}{O} + R'OH \xrightarrow[Ph_3P]{EtO_2CN=NCO_2Et} \underset{\substack{\| \\ (RC)_2NR'}}{O}$$

JACS *94* 679 (1972)
JOC *52* 5127 (1987)

For further alkylations of imides see page 397, Section 3.

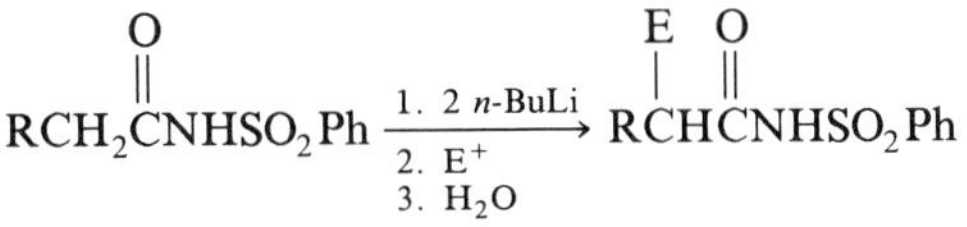

$$\underset{\substack{\| \\ RCH_2CNHSO_2Ph}}{O} \xrightarrow[\substack{2.\ E^+ \\ 3.\ H_2O}]{1.\ 2\ n\text{-BuLi}} \underset{\substack{| \quad \| \\ RCHCNHSO_2Ph}}{E \quad O}$$

$$E^+ = RX,\ RCH{=}CHCO_2R\ (1,4\text{-addition}),\ I_2\ (dimerization)$$

TL *27* 131 (1986)

6. Nitriles

$$RCH(CN)_2 \xrightarrow[\substack{2.\ R'X,\ h\nu \\ (R' = \text{allylic})}]{1.\ \text{Na base}} \overset{\overset{\displaystyle R'}{|}}{RC(CN)_2}$$

Austral J Chem *36* 527 (1983)

$$RBr + CH_2(CN)_2 \xrightarrow{AlBr_3} RCH(CN)_2$$

Angew Int *5* 1044 (1966)

$$Ph_3COH + CH_2(CN)_2 \xrightarrow{\Delta} Ph_3CCH(CN)_2$$

JCS 716 (1962)

$$R_2CHCN \xrightarrow[2.\ E^+]{1.\ \text{base}} \overset{\overset{\displaystyle E}{|}}{R_2CCN}$$

Reviews:

Chem Rev *12* 135 (1933); *20* 451 (1937)
Org Rxs *9* 107 (1957); *31* 1 (1984)

$\underline{E^+}$

D$_2$O   JACS *88* 2348 (1966)

RX   J Prakt Chem (2) *39* 233 (1889)
                                             Compt Rend *182* 1226 (1926)
                                             Ann *495* 84 (1932); *652* 99 (1962)
                                             JACS *67* 2152 (1945); *88* 2348 (1966)
                                             BSCF 1881 (1965)
                                             Tetr *24* 175 (1968); *31* 153 (1975); *36* 775 (1980)
                                             TL 707 (1974); *27* 3685 (1986);
                                               28 4329 (1987)
                                             J Polym Sci, Polym Sci Ed *17* 3499 (1979)
                                             JOMC *204* 281 (1981)
                                             JOC *46* 4600 (1981); *51* 3007, 4080
                                               (intramolecular) (1986); *52* 4142 (1987)
                                             Syn 305 (1982)
                                             Can J Chem *61* 2006 (1983)
                                             CC 279 (1986); 1342 (1987)

ArX   JOC *48* 4397 (1983); *50* 1334 (1985);
                                               51 5157 (1986); *52* 1333, 2619 (1987)

RCH──CHR (with O bridge)   JOC *51* 2230 (1986)

RCHO, R$_2$CO   JOMC *9* 125 (1967); *57* C36 (1973)
                                             Ber *101* 3113 (1968)
                                             JOC *33* 3402 (1968)
                                             Gazz Chim Ital *103* 117 (1973)
                                             Coll Czech Chem Commun *46* 1682 (1981)

		Syn 297 (1983) JACS *108* 1311 (1986) TL *28* 1611 (1987)
RCO_2R		JACS *67* 2152 (1945); *79* 725, 728 (1975) TL *24* 2059 (1983); *28* 4641, 4645 (1987) (both intramolecular)
RCO_2CO_2Et		TL 1585 (1979)
CO_2		JOC *31* 3873 (1966)
RCN		JOMC *9* 125 (1967) TL *24* 3509 (1983)
Me_3SiCl		JOMC *9* 125 (1967)

$$\overset{\text{X}}{\underset{|}{\text{ArCH}_2\text{CN} \longrightarrow \text{ArCHCN}}}$$

X	Reagents	
R	Na, ROH	TL 1509 (1966) Chem Pharm Bull *15* 1811 (1967); *18* 550 (1970) JOC *36* 2948 (1971)

	1° ROH, cat $RhCl_3\cdot$
	  3 H_2O-PPh_3-Na_2CO_3
	  or cat $H_2Ru(PPh_3)_4$
		TL *22* 4107 (1981)

	$NaNH_2$, RX	Org Syn Coll Vol *3* 219 (1955)
	n-BuLi/RO_3SPh	Syn Commun *13* 35 (1983)
COR	Na, RCO_2R	JOC *36* 2948 (1971)

n-BuLi/ (cyclohexenone)   Tetr *37* 1927 (1981)

$$\text{PhCH}_2\text{CN} \xrightarrow[\text{2. BrCMe}_2\text{CMe}_2\text{Br}]{\text{1. KNH}_2} \overset{\text{NC \quad CN}}{\underset{|\quad\;|}{\text{PhCHCHPh}}}$$

JOC *35* 2085 (1970)

$$\text{XCH}_2\text{CN} \xrightarrow[\text{2. E}^+]{\text{1. base}} \overset{\text{E}}{\underset{|}{\text{XCHCN}}}$$

X	E^+	
R_2N	RX R_2CO	TL *28* 547 (1987) JOC *52* 2427 (1987)
t-BuS	RX	Syn Commun *3* 265 (1973)

$$RX + n\text{-}Bu_3SnCH_2CN \xrightarrow{\text{Pd catalyst}} RCH_2CN$$

CL 1511 (1984) (RX = aryl bromide)
JACS *109* 7223 (1987) (RX = heterocyclic iodide)

$$RX + CuCH_2CN \longrightarrow RCH_2CN$$

TL 487 (1972)

$$Ar^1CH_2CN + Ar^2\overset{\overset{\textstyle O}{\|}}{CH} \xrightarrow[\text{transfer catalyst}]{\text{phase}} Ar^1\overset{\overset{\textstyle CHAr^2}{}}{\underset{}{C}}CN$$

Syn 913 (1981)

$$RCH_2CN + R'\overset{\overset{\textstyle O}{\|}}{CH} \xrightarrow[i\text{-}Pr_2NEt]{R_2BOTf} R'\overset{\overset{\textstyle OH}{|}}{CH}CHRCN$$

CL 1401 (1982)

$$H_2C{=}CHCN + RCHO\,(R_2CO) \xrightarrow{\text{DABCO}} R\overset{\overset{\textstyle OH}{|}}{CH}\underset{\underset{\textstyle CH_2}{\|}}{C}CN$$

TL *27* 4307, 5007 (1986)
Syn Commun *17* 587 (1987)

$$XCR{=}CHCN \xrightarrow[\text{2. } E^+]{\text{1. base}} XCR{=}\overset{\overset{\textstyle E}{|}}{C}CN$$

X = Ar, OR, NR$_2$, SR

See page 185, Section 1.

$$XCH{=}CRCN \xrightarrow[\text{2. } E^+]{\text{1. base}} X\overset{\overset{\textstyle E}{|}}{C}{=}CRCN$$

X = OR, NR$_2$

See page 185, Section 1.

$$C{=}C{-}\overset{\overset{\textstyle H}{|}}{C}{-}CN \text{ or } \overset{\overset{\textstyle H}{|}}{C}{-}C{=}C{-}CN \xrightarrow[\text{2. } E^+]{\text{1. base}} C{=}C{-}\overset{\overset{\textstyle E}{|}}{C}{-}CN$$

$\underline{E^+}$

RX

BSCF 951 (1965)
JOC *36* 877 (1971); *40* 1162 (1975);
  44 300 (1979)
TL 1377 (1974); 4647 (1975)
Tetr *31* 153 (1975)

RCH=CHCO$_2$R (1,4-addition)

CL 1085 (1986)
TL *28* 1785 (1987)

H$_2$C=CHCN (1,4-addition)

JACS *65* 18 (1943)

$$\underset{\substack{| \quad | \\ \text{H} \quad \text{NR}_2}}{\text{C}-\text{C}=\text{C}-\text{CN}} \xrightarrow[\text{2. RX}]{\text{1. base}} \underset{\substack{| \quad | \\ \text{R} \quad \text{NR}_2}}{\text{C}-\text{C}=\text{C}-\text{CN}}$$

TL 3963 (1974)

$$\xrightarrow[\text{2. R'X}]{\text{1. LDA}}$$

TL *28* 6179 (1987)

$$\underset{\text{R'CHCN}}{\overset{\text{X}}{|}} \xrightarrow[\text{2,6-}(t\text{-Bu})_2\text{C}_6\text{H}_3\text{OK}]{\text{R}_3\text{B or RB}} \underset{\text{R'CHCN}}{\overset{\text{R}}{|}}$$

X = Cl, Br; R' = H, Cl, alkyl, CO_2R; R = alkyl, vinylic

JACS *91* 6854 (1969); *92* 5790 (1970)
Organomet Chem Syn *1* 95 (1970)
JOC *46* 229 (1981); *51* 3398 (1986) (R = vinylic)
TL *23* 2077 (1982)

$$\text{ArZnCl} + \text{BrCH}_2\text{CN} \xrightarrow[\text{PPh}_2]{\text{cat Ni(acac)}_2} \text{ArCH}_2\text{CN}$$

Syn 40 (1987)

$$t\text{-BuSCR}_2\text{CN} \xrightarrow[\text{2. E}^+]{\text{1. Li, naphthalene}} \text{ECR}_2\text{CN}$$

$E^+ = H_2O$, RX, RCHO, R_2CO, RCOCl, HCO_2R

Syn Commun *3* 265 (1973)

$$\text{N}_2\text{CHCN} \xrightarrow[\text{2. H}_2\text{O}]{\text{1. R}_3\text{B}} \text{RCH}_2\text{CN}$$

JACS *90* 6891 (1968)

$$\text{HO}_2\text{CCH}_2\text{CN} \xrightarrow[\substack{\text{2. RCOCl} \\ \text{3. H}_3\text{O}^+}]{\text{1. 2 } n\text{-BuLi}} \underset{\overset{\|}{\text{O}}}{\text{RCCH}_2\text{CN}}$$

Syn 308 (1983)

6. CONJUGATE ADDITION TO α,β-UNSATURATED NITRILES, CARBOXYLIC ACIDS AND DERIVATIVES

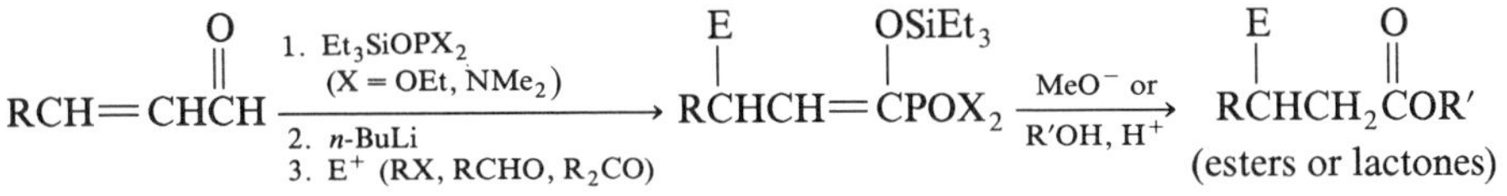

$$RCH=CHCH\overset{O}{\overset{\|}{}} \xrightarrow[\substack{2.\ n\text{-BuLi}\\3.\ E^+\ (RX,\ RCHO,\ R_2CO)}]{\substack{1.\ Et_3SiOPX_2\\(X=OEt,\ NMe_2)}} RCHCH=CPOX_2\overset{E\quad OSiEt_3}{\overset{|\qquad|}{}} \xrightarrow[R'OH,\ H^+]{MeO^-\ or} RCHCH_2COR'\overset{E\quad O}{\overset{|\qquad\|}{}}$$
(esters or lactones)

TL 2047 (1979)
JACS *101* 371 (1979)
BCSJ *55* 224 (1982)

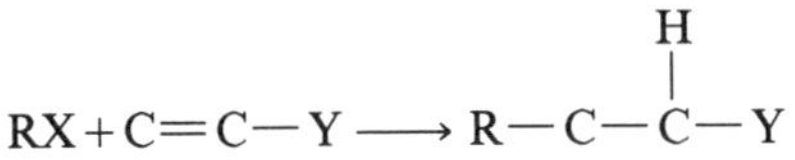

$$RX+C=C-Y \longrightarrow R-C-\overset{\overset{H}{|}}{C}-Y$$

Reviews:

Angew Int *24* 553 (1985)
Chimia *39* 203 (1985)

Y	X	Reagent	
CO₂H	halogen	*n*-Bu₃SnH	CC 944 (1983)
CO₂R	halogen	*n*-Bu₃SnH	JOC *47* 3348 (1982);
			48 1782 (1983)
			(intramolecular);
			50 546 (1985)
			(intramolecular)
			TL *27* 5071 (1986)
			(intramolecular,
			R = vinylic);
			28 5853 (1987)
			JACS *108* 3102 (1986)
			(intramolecular);
			109 4976 (1987)
			(intramolecular)
			Angew Int *25* 450 (1986)
			CC 1006, 1438
			(intramolecular) (1987)
			Ann 231 (1987)

$\underline{Y}$	$\underline{X}$	Reagent	
		Ph_3SnH	CC 944 (1983)
		cat $Pd(PPh_3)_4$	TL *28* 3179 (1987) (R = Ar)
	SePh	n-Bu_3SnH	JOC *47* 3348 (1982)
			TL *28* 5853 (1987)
		Ph_3SnH	CC 944 (1983)
CO_2R, CN	halogen	n-Bu_3SnH	Angew Int *23* 69 (1984)
			Israel J Chem *26* 387 (1985)
			JOC *52* 3659 (1987)
		cat n-Bu_3SnCl, $NaBH_4$, $h\nu$	Angew Int *23* 69 (1984) JOC *51* 3726 (1986)
		vitamin B_{12}, reductant	Pure Appl Chem *55* 1791 (1983)
	NO_2	n-Bu_3SnH	CL 635 (1985)
	$\overset{\overset{\text{S}}{\|\|}}{OCSCH_3}$	cat n-Bu_3SnCl, $NaBH_4$, $h\nu$	Angew Int *23* 69 (1984)
$CONR_2$	halogen	n-Bu_3SnH	TL *28* 5853 (1987)
CN	halogen	n-Bu_3SnH	Angew Int *22* 622 (1983) CC 1006 (1987) Org Syn *65* 236 (1987)
		vitamin B_{12}, reductant	Chimia *39* 211 (1985)
	NO_2	n-Bu_3SnH	JACS *107* 4332 (1985)
	$\overset{\overset{\text{S}}{\|\|}}{OCSCH_3}$	n-Bu_3SnH	Angew Int *23* 69 (1984)
	$\overset{\overset{\text{S}}{\|\|}}{OCN}$⟨N⟩	n-Bu_3SnH	Angew Int *23* 69 (1984) TL *28* 4645 (1987)
OAc	halogen	n-Bu_3SnH	JOC *52* 3659 (1987)

$$RNO_2 + H_2C{=}CHX \xrightarrow[\text{2. } n\text{-Bu}_3\text{SnH, AIBN}]{\text{1. base}} RCH_2CH_2X$$

R = 2° alkyl, 1° allylic or benzylic

X = CN, CO_2R, SOPh, SO_2Ph, CHO, COR

JOC *50* 3692 (1985); *52* 1601 (1987) (KF-alumina as base, no reduction)

$$I{-}C_n{-}C{\equiv}C{-}CO_2R \xrightarrow{n\text{-BuLi}} RO_2C{-}C{-}\overset{\frown}{C}\underset{\smile}{{}_n}$$
$$n = 3, 4$$
JOC *49* 1144 (1984)

$$R'CH{=}CHX \xrightarrow[\text{2. } H_2O \text{ or } H_3O^+]{\text{1. RM}} R'CH\overset{\overset{\text{R}}{\|}}{C}H_2X$$

Reviews: Org Rxs *10* 179 (1959) (Michael addition); *19* 1 (1972) (organocopper reagents)

X	RM	
CO_2H	RLi or RMgX	JOC *52* 5729 (1987) (α-silyl acids)
	RMgX	JACS *75* 6342 (1953)
	$RCu \cdot BF_3$	JACS *100* 3240 (1978) JOC *47* 119 (1982)
CO_2R'	RMgX	BCSJ *39* 910 (1966) Org Syn Coll Vol *5* 762 (1973)

	RMgX, cat CuX	JOC *27* 2706 (1962);
		     42 3209 (1977);
		     45 4117 (1980)
		 BCSJ *39* 910 (1966)
		 Syn Commun *9* 325 (1979)
		 CC 1472 (1987)

	RMgBr, cat $CuBr \cdot SMe_2$, Me_3SiCl, HMPA	TL *27* 4025 (1986)
	R_2CuMgX	JOMC *199* 9 (1980) TL *28* 3791 (1987)

	R_2CuLi	JACS *94* 5495 (1972);
		     103 1222 (1981)
		 JOC *38* 3893 (1973); *46* 3874
		     (1981); *51* 3376, 5041 (1986);
		     52 3541 (lactone), 4603
		     (lactone) (1987)
		 CC 907 (1973)
		 Acta Chem Scand B *31* 667
		     (1977); *34* 113 (1980)
		 Syn Commun *9* 325 (1979)
		 Tetr *37* 3981 (1981)
		 TL *27* 2519 (1986) (lactone); *28*
		     949, 3791 (1987)
		     (both lactones)

	$(py)_2CuLi$, LiI	Tetr *38* 1509 (1982)
	R_2CuLi, Me_3SiCl	TL *27* 1047 (1986) JOC *51* 5041 (1986)
	(RCuAr)Li	Acta Chem Scand B *31* 667 (1977); *32* 483 (1978) Tetr *34* 3023 (1978)
	(**py**CuPh)Li, LiI	Tetr *38* 1509 (1982)
	$[(\mathbf{H_2C\!=\!CR})CuC\!\equiv\!C(CH_2)_2CH_3]^-$	TL *28* 3551 (1987)
	RLi, $CuC\!\equiv\!CSiMe_3$, Me_3SiCl	TL *28* 5719 (1987)
	(RCuCN)Li	TL *22* 2985 (1981)
	$(R_2CuCN)Li_2$	TL *24* 127 (1983)
	$[(H_2C\!=\!CH)_2CuCN]Li_2$	JACS *109* 7495 (1987)
	$\left[\underset{S}{\boxed{}}\!\!-CuR(CN) \right]Li_2$	JOMC *285* 437 (1985)
	$RCu \cdot BF_3$	JACS *100* 3240 (1978); *109* 5820 (1987) Helv *64* 2808 (1981) JOC *47* 119 (1982)

<u>X</u>	<u>RM</u>	
		CC 904 (1984); 464, 1572 (1987)
		TL *27* 943 (1986)
	RLi, n-Bu$_3$P·CuI	TL *27* 4713 (1986)
	RCu·n-Bu$_3$P·BF$_3$	TL *24* 4971 (1983);
		27 1139 (1986)
		Helv *68* 212 (1985)
	R$_2$CuLi, BF$_3$·OEt$_2$	Tetr *37* 3981 (1981)
		JACS *104* 1774 (1982);
		109 5820 (1987)
		CC 1572 (1987)
	R$_3$CuLi$_2$, BF$_3$·OEt$_2$	CC 1572 (1987)
	RCu[P(OMe)$_3$]$_3$	JOC *45* 4117 (1980)
	[(PhMe$_2$Si)$_2$CuCN]Li$_2$	CC 1472 (1987) (lactones)
	H$_2$C=CHCH$_2$SiMe$_3$, TiCl$_4$	JOC *51* 1745 (1986)

t-Bu, CO$_2$ ... OMe, t-Bu (structure) RLi or LiCH$_2$CO$_2$R' — JOC *51* 1637 (1986)

X	RM	
CO$_2$R', CONR'$_2$, CN	H$_2$C=CHCH$_2$SiMe$_3$, cat n-Bu$_4$NF	JOC *51* 1745, 1753
		(intramolecular) (1986)
		CC 1472 (1987) (lactone)
CO$_2$R', CN	(H$_2$C=CHCH$_2$)$_2$CuLi	JOC *51* 1745 (1986)
CONHR	RMgBr, cat CuCN	JOC *51* 3921 (1986)
	[Ph$_2$Cu(CN)$_2$]Li$_2$	JOC *51* 3921 (1986)
CONR'$_2$	RLi	JOC *22* 1013 (1957)
		TL *21* 1881 (1980)
	RMgX	BSCF I 1087 (1934)
		JACS *77* 4413 (1955)
		JOC *22* 1013 (1957)
		TL 3251 (1971)
	R$_2$CuLi, Me$_3$SiCl	TL *27* 1047 (1986)
	[(PhMe$_2$Si)$_2$CuCN]Li$_2$	CC 1472 (1987) (lactam)
CSNR'$_2$	RLi or RMgX	JACS *100* 5221 (1978)
CONHCOCH$_3$	R$_2$CuLi, Me$_3$SiCl	TL *27* 1047 (1986)
CONRCOR	RMgCl, 0.5 CuBr·SMe$_2$/HCl, MeOH/KOH (R =1° or 2° alkyl, aryl, vinylic)	TL *27* 369 (1986)
CN	(R$_3$BMe)Li, CuBr	TL 255 (1976)
	Me$_2$CuLi, Me$_3$SiCl (product is R'CHMeCH$_2$COMe)	TL *27* 1047 (1986)

COCN

$H_2C=CR'CH_2SiMe_3$, $TiCl_4/R'OH$ TL *21* 4487 (1980)
($R = H_2C=CRCH_2$; $X = CO_2R'$
in product)

$H_2C=C=CHSiMe_3$, $TiCl_4/R'OH$ TL *21* 4487 (1980)
($R = HC≡CCH_2$; $X = CO_2R'$
in product)

$RSiMe_3$, $TiCl_4$ JOC *51* 1199 (1986)
(R = allylic, allenic,
propargylic, alkynyl)

$$RHgX + C=C-Y \xrightarrow[\text{NaHB(OMe)}_3]{\text{NaBH}_4 \text{ or}} R-\overset{\overset{\displaystyle H}{|}}{C}-\overset{|}{C}-Y$$

$Y = CO_2R$, CO_2COR, CN

$R = 1°, 2°, 3°$ alkyl; benzyl; $MeO-C-C-$; $MeO-C-C-C$; $C=C-C(OMe)-C-$;
HCO—C—C— ; $RCOCHRCH_2$; RCONR—C—C—

Ber *110* 2588 (1977); *112* 3759 (1979); *113* 1192, 2787 (1980); *114* 1572 (1981); *115* 2526 (1982); *117*
859, 2132, 3160, 3175 (1984); *118* 1289, 1345, 1616 (1985); *119* 1291 (1986)
TL *21* 1829, 3569 (1980); *22* 2155 (1981); *23* 931, 2765 (1982); *24* 11, 15, 2051, 3221 (1983); *25* 2743
(1984); *27* 4841 (1986)
Angew Int *20* 965 (1981); *21* 130 (1982); *24* 553 (1985) (review); *26* 479 (1987)
JOC *47* 3348 (1982); *49* 1313 (1984)
Syn 735 (1982)
Organomet *1* 675 (1985)
Tetr *41* 4025 (1985)

$$H_2C=C=\overset{\overset{\displaystyle R}{|}}{C}CO_2H \xrightarrow[\text{2. H}_2O]{\text{1. EtMgBr}} H_2C=\overset{\overset{\displaystyle Et}{|}}{C}CHRCO_2H$$

JACS *74* 2559 (1952)

$$CH_3CH=\overset{\overset{\displaystyle Br}{|}}{C}CO_2R \xrightarrow[\text{2. H}_2O]{\text{1. PhMgBr}} CH_3\overset{\overset{\displaystyle Ph}{|}}{C}H-\overset{\overset{\displaystyle Br}{|}}{C}HCO_2R$$

R = H, CH_3

JOC *35* 666 (1970)

RM

$Et_2AlC≡CR$ TL *27* 2885 (1986)

Me_2CuLi JACS *109* 6199 (1987)

$$ArCH=\overset{\overset{\displaystyle CN}{|}}{C}CO_2Me \xrightarrow[\text{2. H}_2O]{\text{1. Ar}_2\text{CuLi}} Ar_2CH\overset{\overset{\displaystyle CN}{|}}{C}HCO_2Me$$

TL *27* 5319 (1986)

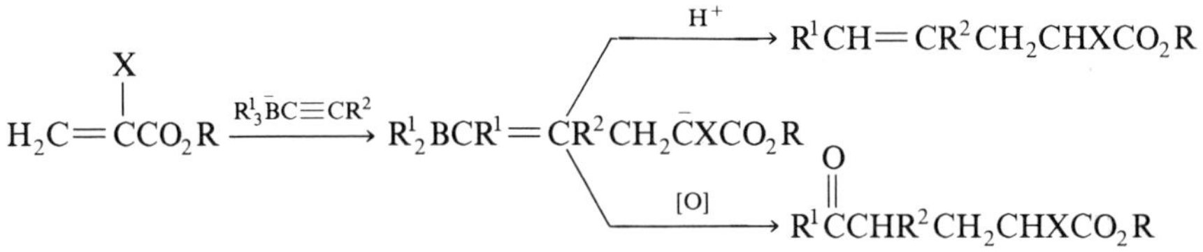

$$H_2C=\underset{\underset{X}{|}}{C}CO_2R \xrightarrow{R_3^1\bar{B}C\equiv CR^2} R_2^1BCR^1=CR^2CH_2\bar{C}XCO_2R$$

$$\xrightarrow{H^+} R^1CH=CR^2CH_2CHXCO_2R$$

$$\xrightarrow{[O]} R^1\overset{\overset{O}{\|}}{C}CHR^2CH_2CHXCO_2R$$

$$X = COMe, CO_2Et$$

JCS Perkin I 719 (1982)

$$RCH=C(CO_2R)_2 \xrightarrow[\text{2. H}_2\text{O}]{\text{1. R'M}} R\underset{\underset{R'}{|}}{CH}CH(CO_2R)_2$$

R'M	
R'Li	CC 464 (1987)
R'MgX	CC 464, 1572 (1987)
ArMgX	TL *27* 2235 (1986)
RC≡CMgX	BSCF 2542 (1964)
R'MgX, cat CuCl	Org Syn *50* 38 (1970) TL *23* 75 (1982)
R'Cu	CC 464 (1987)
R'Cu, BF$_3$·OEt$_2$	CC 464, 1572 (1987)
R$_2'$CuLi	CC 464, 1572 (1987)
R$_2'$CuLi, BF$_3$·OEt$_2$	CC 464, 1572 (1987)
R$_3'$CuLi$_2$, BF$_3$·OEt$_2$	CC 1572 (1987)
(ArCuCN)Li	TL *27* 2235 (1986)
LiAlR$_4'$	CC 1572 (1987)
H$_2$C=CHCH$_2$Sn(n-Bu)$_3$, TiCl$_4$	CC 1572 (1987)
RCH=CHCH$_2$ZnBr (R' = H$_2$C=CHCHR)	TL 997 (1972)
R'NLi | RC=CRCO$_2$R (chiral)	CC 1345 (1987) (enantioselective)

$$RCH=C(CO_2R)_2 \xrightarrow[\text{2. R}^2\text{X}]{\text{1. R}_2^1\text{CuLi}} R\underset{\underset{R^1}{|}}{CH}-\underset{\underset{R^2}{|}}{C}(CO_2R)_2$$

JOC *38* 2100 (1973)

$$R'CH=CH-\underset{N}{\overset{O}{\diagdown}}\underset{CH_2OMe}{\overset{Ph}{\diagup}} \xrightarrow[RMgX]{RLi \text{ or}} \xrightarrow{H_3O^+} R'\overset{R}{\underset{*}{CHCH_2}}\overset{O}{\overset{\|}{COH}}$$

$$\text{chiral}$$

JOC *44* 2247, 2250 (1979); *46* 3874 (1981)

$$H_2C=CHC(OEt)_3 \xrightarrow{RM} \xrightarrow{H_3O^+} RCH_2CH_2\overset{O}{\overset{\|}{C}}OEt$$

RM

RMgX, cat CuBr	BSCF II 305 (1979)
E-RCMe=CCHAlMe$_2$, cat Pd(PPh$_3$)$_4$, ZnCl$_2$	JOC *50* 3406 (1985)
PhZnCl, cat Pd(PPh$_3$)$_4$, ZnCl$_2$	JOC *50* 3406 (1985)

$$RCH=CH\overset{O}{\overset{\|}{C}}X \longrightarrow RCH_2\overset{R'}{\underset{|}{CH}}COX$$

X	Reagents	
OR	LiHB(sec-Bu)$_3$/R'X	JOC *40* 2846 (1975)
	i-Bu$_2$AlH, cat MeCu/MeLi/R'X	JOC *52* 439 (1987)
NR$_2$	LiHB(sec-Bu)$_3$/R'X	TL *27* 4717 (1986)

$$(RCHO)\ R_2CO+H_2C=\overset{CH_3}{\underset{|}{C}}CO_2Me+HSiMe_3 \xrightarrow{cat\ RhCl_3\cdot3\ H_2O} R_2\overset{Me_3SiO}{\underset{|}{C}}C(CH_3)_2\overset{O}{\overset{\|}{C}}OMe$$

TL *28* 4809 (1987)

$$Br(CH_2)_{n+1}CH=C(CO_2Me)_2 \xrightarrow{reagent} \underset{(\ )_n}{\overset{MeO_2C\diagdown\diagup CO_2Me}{\diagdown\diagup}}X$$

X	Reagent	n	
H, CN, CH(CO$_2$Me)$_2$ or t-BuS	LiHB(sec-Bu)$_3$, KCN, NaCH(CO$_2$Me)$_2$, or t-BuSLi(Na)	1, 2	JOC *47* 362 (1982)
H	electrolysis	0, 1, 2	TL *23* 1339 (1982)

$$R'X+RCH=CHY+R_2CO\ (RCHO) \xrightarrow[2.\ H_2O]{1.\ Zn} R'\overset{R_2COH}{\underset{|}{R}}CHCHY$$

$$Y=CO_2R,\ CN$$

JACS *100* 4314 (1978)

$$R'CH{=}CHC(O)X \xrightarrow{RM} \xrightarrow{E^+} R'CH(R)CH(E)C(O)X$$

$E^+ = H^+$, RX, RCHO, R_2CO, $ClCO_2Et$, RSSR, MeSOCl

X	RM	
OR′	RMgX, cat CuX	CL 1159 (1981)
	ArMgX or Me_2CuLi	TL *23* 3287 (1982)
	RCu, $BF_3 \cdot OEt_2$	JACS *109* 5820 (1987)
	R_2CuLi	JOC *51* 3376 (1986) (intramolecular acylation to 2-carboalkoxycycloalkanone)
		TL *28* 3791 (1987)
	$[(PhMe_2Si)_2CuCN]Li_2$	CC 1472 (1987) (lactone)
	Me_3SnLi	TL *27* 5417 (1986)
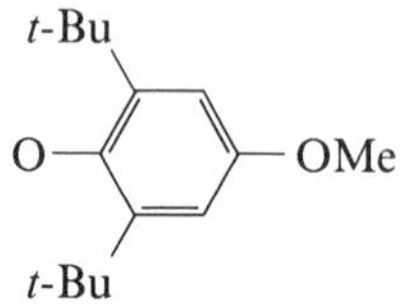	RLi, $LiCH_2CO_2R$	JOC *51* 1637 (1986)
NHPh	RLi	TL *21* 1881 (1980)
NR_2	RLi, RMgX or $(RS)_2CHLi$	TL *21* 4823 (1980)
	$ArC(SR)_2Li$	TL *21* 4827 (1980)
	RMgX	CL 913 (1981)
	R_2CuLi, Me_3SiCl	TL *27* 1047 (1986)
$NMeNMe_2$	RLi	Syn Commun *10* 837 (1980)

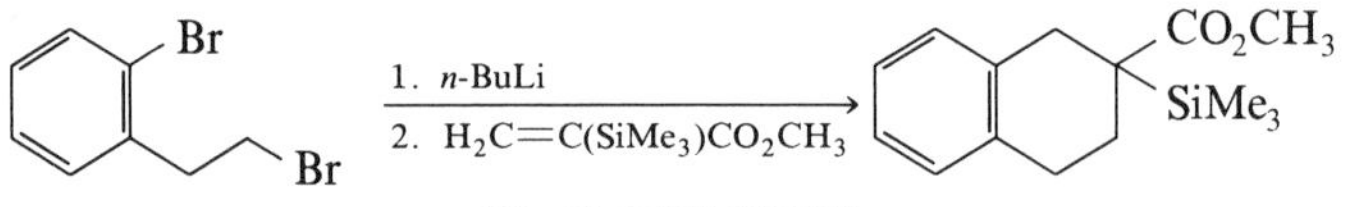

TL *22* 3707 (1981)

$$H_2C{=}C(SiMe_3)CO_2CH_3 \xrightarrow[2.\ R'CHO\ (R'_2CO)]{1.\ RM} RCH_2C(=CHR')CO_2CH_3$$

RM = PhMgBr (cat CuCl), $H_2C{=}C(Li)CH{=}CH_2$, MeSCHLiSOMe

CL 1993 (1984)

$$H_2C{=}CHC(O)Cl \xrightarrow{Ph_3P{=}CHCO_2Et} H_2C{=}CHC(O)C({=}PPh_3)C(O)OEt \xrightarrow[3.\ H^+,\ MeOH]{1.\ R^1Li\quad 2.\ R^2X} R^1CH_2CH(R^2)C(O)OMe$$

JACS *99* 642 (1977)

$$H_2C{=}CHCO_2R'(CN) + RX \longrightarrow RCH_2CH(X)CO_2R'(CN)$$

X = halogen

various initiators	Syn 145 (1977) (review)
CuCl, hν	TL *21* 4457 (1980)

$$(ArN_2)X + RCH{=}CHCOR \xrightarrow[\text{salt}]{\text{copper}} \overset{Ar}{\underset{}{R}}\overset{X}{\underset{}{CH}}\overset{O}{\underset{}{CHCOR}} \quad \text{or} \quad \overset{Ar}{\underset{}{RC}}{=}\overset{O}{\underset{}{CHCOR}}$$

X = halogen

Org Rxs *11* 189 (1960); *24* 225 (1976) (both reviews)
JOC *52* 2997 (1987)

$$X{-}C{-}H + C{=}C{-}Y \xrightarrow{\text{base}} X{-}C{-}\overset{H}{\underset{}{C}}{-}C{-}Y$$

X = COR, CO$_2$R, CONR$_2$, CN; Y = CO$_2$R, CONR$_2$, CN

Org Rx *10* 179 (1959) (review)

$$NO_2{-}C{-}H + C{=}C{-}CO_2R \xrightarrow{\text{base}} NO_2{-}C{-}C{-}\overset{H}{\underset{}{C}}{-}CO_2R$$

Angew *19* 1013 (1980) (high pressure); *20* 770 (1981) (high pressure, F$^-$ catalyzed)

$$X{-}C{-}H + C{=}C{-}Y \xrightarrow[\text{Si(OR)}_4]{\text{CsF}} X{-}C{-}C{-}\overset{H}{\underset{}{C}}{-}Y$$

X = COR, CN; Y = CO$_2$R, CN

CC 122 (1981)
Tetr *39* 117, 999 (1983)

$$RCH{=}CH\overset{O}{\underset{}{C}}NR_2 \xrightarrow[\text{CsF, Si(OMe)}_4]{\text{Nuc-H}} RCHCH_2\overset{O}{\underset{}{C}}NR_2$$

Nuc-H = PhCH$_2$CN, CH$_2$(CO$_2$Et)$_2$, NCCH$_2$CO$_2$Et, RCH$_2$NO$_2$,

TL *23* 5531 (1982)
Tetr *39* 999 (1983)

$$R\overset{O}{\underset{}{C}}CH_2R + H_2C{=}CH\overset{O}{\underset{}{C}}OR \xrightarrow{\text{base}} R\overset{O}{\underset{}{C}}CHRCH_2CH_2\overset{O}{\underset{}{C}}OR$$

BSCF 823 (1962)

TL *27* 4611 (1986)

$$(ROC)_2CH_2 + H_2C=CHCOR \xrightarrow{\text{NaOEt}} (ROC)_2CHCH_2CH_2COR$$

Org Syn Coll Vol *4* 630 (1963)

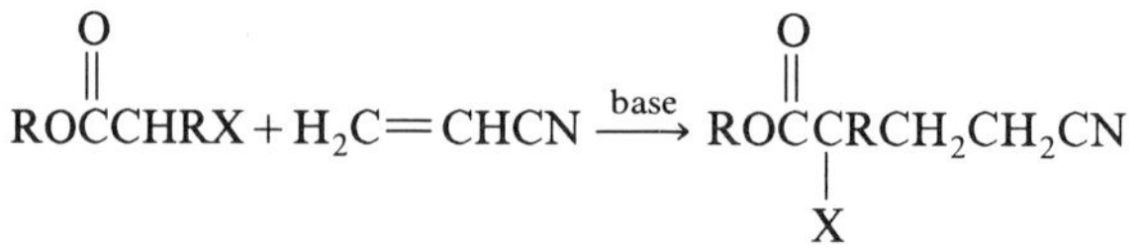

$$ROCCHRX + H_2C=CHCN \xrightarrow{\text{base}} ROCCRCH_2CH_2CN$$
$$\qquad\qquad\qquad\qquad\qquad\qquad X$$

X	Base	
CO$_2$R	NaOEt	JACS *67* 2044 (1945)
CN	KOH	Org Syn Coll Vol *4* 776 (1963)

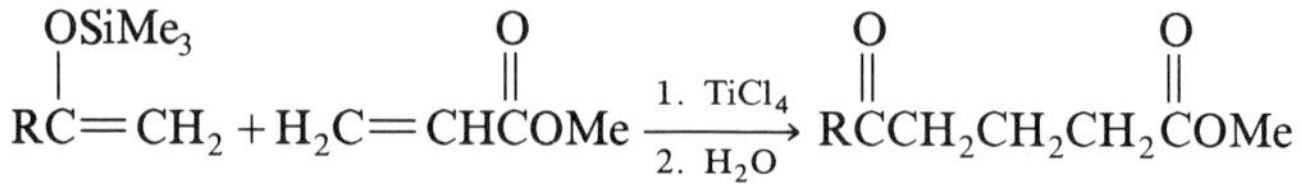

$$RC=CH_2 + H_2C=CHCOMe \xrightarrow[\text{2. H}_2\text{O}]{\text{1. TiCl}_4} RCCH_2CH_2CH_2COMe$$
with OSiMe$_3$

BCSJ *49* 779 (1976)

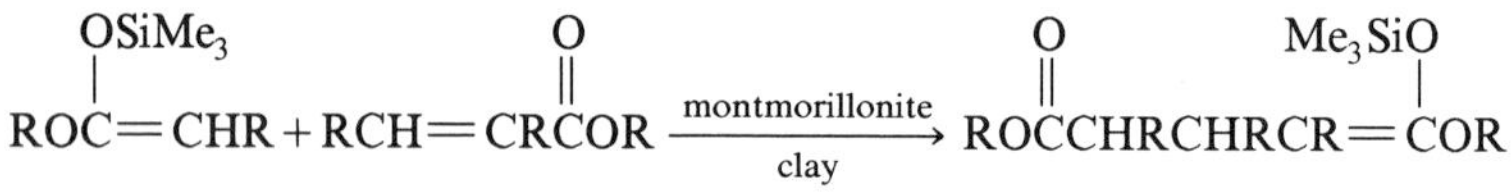

$$ROC=CHR + RCH=CRCOR \xrightarrow[\text{clay}]{\text{montmorillonite}} ROCCHRCHRCR=COR$$
with OSiMe$_3$ ... Me$_3$SiO

CC 1203 (1987)

$$RCCHR + RCH=CHCOMe \xrightarrow{\text{O}_3} RCCHRCHRCH_2COMe$$

chiral

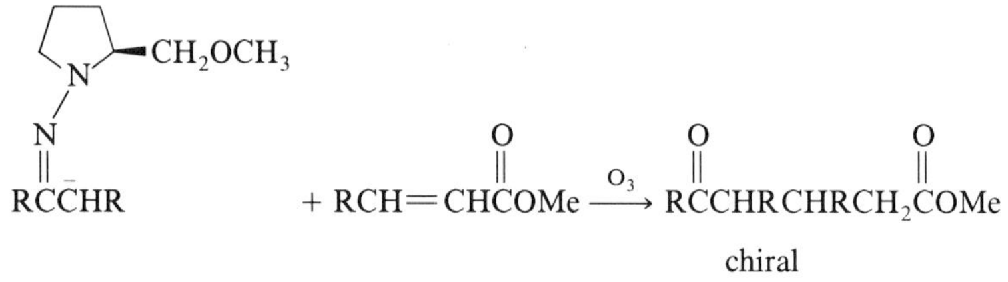

TL *24* 4967 (1983); *27* 3491 (1986)

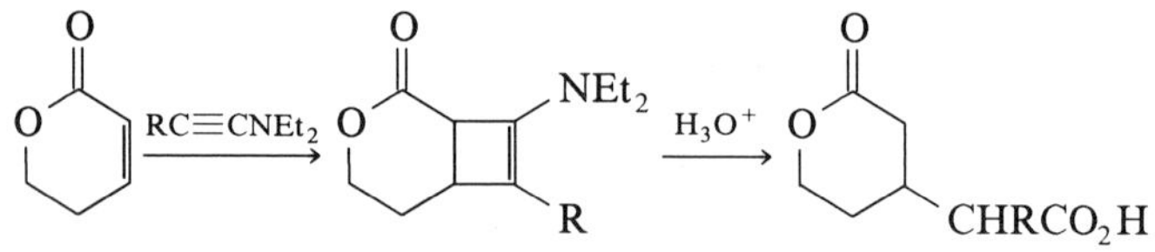

$$\text{(pyranone)} + RC\equiv CNEt_2 \longrightarrow \text{(bicyclic NEt}_2\text{)} \xrightarrow{\text{H}_3\text{O}^+} \text{(pyranone)} CHRCO_2H$$

TL *23* 1821 (1982)

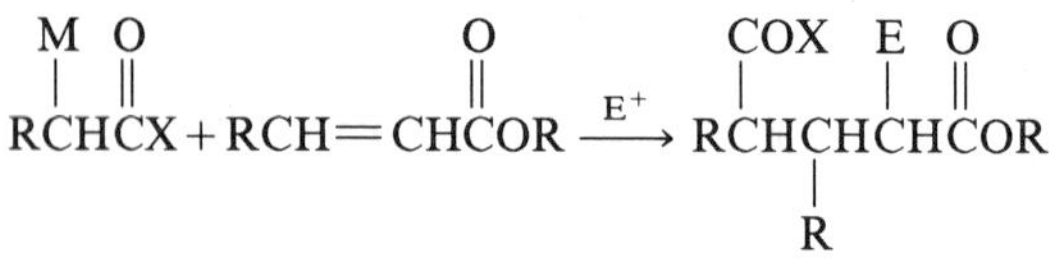

$$RCHCX + RCH=CHCOR \xrightarrow{E^+} RCHCHCHCOR$$
with M O ... COX E O ... R ; M = Li, Na

X	E$^+$	
OR	H$^+$	JACS *62* 1763 (1940); *101* 1544 (1979); *109* 287 (1987) TL 4087 (1974); *25* 5661 (1984); *26* 5025 (1985) (chiral); *27* 3551 (1986) (chiral catalysis)

J Polym Sci, Polym Chem Ed
17 3509 (1979)
Gazz Chim Ital *109* 95 (1979)
CC 869 (1983)
Syn 715 (1983)
CL 375 (1984)
JACS *101* 1544 (1979)
TL *26* 1723 (1985)
TL *27* 959 (1986)

$$\text{NR}_2$$

$$\begin{array}{c} \text{I}_2 \\ \text{R'X} \\ \text{H}^+ \end{array}$$

$$+ \text{RCH}{=}\text{CHCOR} \xrightarrow{\text{H}_2\text{O}}$$

CL 1085 (1986)
TL *28* 1785 (1987)

$$\overset{O}{\overset{\|}{\text{XC}}}\text{CR}_2^- + \text{R}_2\text{C}{=}\text{C(Y)CN} \xrightarrow{\text{H}_2\text{O}} \overset{O}{\overset{\|}{\text{XC}}}\text{CR}_2\text{CR}_2\text{CH(Y)CN}$$

$$X = R, OR; Y = CO_2R, CN$$

JOC *51* 5480 (1986)

$$\text{R}_2\text{CHOH} + \text{C}{=}\text{C}{-}\overset{O}{\overset{\|}{\text{C}}}\text{OR} \xrightarrow[\text{PhCOPh}]{h\nu} \text{R}_2\overset{OH}{\overset{|}{\text{C}}}{-}\overset{H}{\overset{|}{\text{C}}}{-}\overset{O}{\overset{\|}{\text{C}}}\text{OR}$$

Angew *69* 177 (1957)

$$\text{R}\overset{O}{\overset{\|}{\text{C}}}(\text{CH}_2)_n\text{CH}{=}\text{CH}\overset{O}{\overset{\|}{\text{C}}}\text{OEt} \xrightarrow{\text{electrolysis}}$$

$$n = 3, 4$$

JOC *50* 2202 (1985)

$$+ \text{C}{=}\text{C}{-}\text{CO}_2\text{R} \xrightarrow{\text{H}_2\text{O}}$$

R	
H	JOC *49* 2682 (1984)
Ar	JOC *51* 1637 (1986)

$$\text{RC(SMe)}_2\text{Li} + \text{RCH}{=}\text{CH}\overset{O}{\overset{\|}{\text{C}}}\text{OR} \xrightarrow{\text{E}^+} \overset{\text{MeS}}{\underset{\text{MeS}}{\overset{|}{\text{RC}}{-}\overset{|}{\text{CH}}}}\overset{\text{E}}{\underset{\text{R}}{\overset{|}{\text{CH}}}}\overset{O}{\overset{\|}{\text{C}}}\text{OR}$$

$$E^+ = HOAc, MeI$$

TL *26* 3031 (1985)

$$\underset{\text{RCH}=\text{CHCOR}}{\overset{\text{O}}{\overset{\|}{}}} \xrightarrow{p\text{-TolSO}_2\text{CHLiSMe}} \underset{p\text{-TolSO}_2\text{CHCHRCH}_2\text{COR}}{\overset{\text{SMe}}{\overset{|}{}}\overset{\text{O}}{\overset{\|}{}}} \xrightarrow[h\nu]{\text{H}_2\text{O}} \underset{\text{HCCHRCH}_2\text{COR}}{\overset{\text{O}}{\overset{\|}{}}\overset{\text{O}}{\overset{\|}{}}}$$

JOC *51* 508 (1986)

TL *28* 3551 (1987)

TL *28* 4411 (1987)

$$\underset{\text{RCH}}{\overset{\text{O}}{\overset{\|}{}}} + \text{H}_2\text{C}=\text{CHX} \longrightarrow \underset{\text{RCCH}_2\text{CH}_2\text{X}}{\overset{\text{O}}{\overset{\|}{}}}$$

X	Promoter	
CO$_2$H	hν	Syn 490 (1980)
CO$_2$R	hν	Syn 490 (1980) JOC *52* 3323 (1987)
CO$_2$R' or CN	thiazolium catalyst	Ber *113* 690 (1980); *114* 564 (1981) Syn 129, 626 (1981)

See also page 792, Section 2.

$$\underset{(\text{RC})_2\text{O}}{\overset{\text{O}}{\overset{\|}{}}} + \text{H}_2\text{C}=\text{CHCO}_2\text{R}'(\text{CN}) \longrightarrow \underset{\text{RCCH}_2\text{CH}_2\text{CO}_2\text{R}'(\text{CN})}{\overset{\text{O}}{\overset{\|}{}}}$$

electrolysis   JACS *99* 7396 (1977)

vitamin B$_{12}$, electrolysis, hν JACS *105* 7200 (1983)

$$\underset{\text{RCH}=\text{CHCOR}'}{\overset{\text{O}}{\overset{\|}{}}} \xrightarrow[\text{AlCl}_3]{t\text{-BuNC}} \underset{\text{RCHCH}_2\text{COR}'}{\overset{\text{CN}}{\overset{|}{}}\overset{\text{O}}{\overset{\|}{}}}$$

JACS *104* 6449 (1982)

$$\underset{\text{RCH}=\text{CHCOR}}{\overset{\text{O}}{\overset{\|}{}}} \xrightarrow[\text{2. H}_2\text{O}]{\text{1. LiC(SMe)}_3} \underset{(\text{MeS})_3\text{CCHRCH}_2\text{COR}}{\overset{\text{O}}{\overset{\|}{}}}$$

TL *28* 1147 (1987)

$$\begin{array}{ccc} \overset{\text{X}}{\underset{|}{\text{C}}}=\overset{|}{\underset{|}{\text{C}}}-\overset{\text{O}}{\overset{||}{\text{COR}'}} \longrightarrow \overset{\text{R}}{\underset{|}{\text{C}}}=\overset{|}{\underset{|}{\text{C}}}-\overset{\text{O}}{\overset{||}{\text{COR}'}} \end{array}$$

See page 198, Section 12; page 201, Section 13; and page 206, Section 4.

$$\overset{n\text{-BuS}}{\underset{|}{\text{CH}_3\text{C}}}=\text{CHCOEt} \xrightarrow[\text{2. H}_2\text{O}]{\text{1. R}_2\text{CuLi}} \overset{\text{CH}_3}{\underset{|}{\text{R}_2\text{C}}}\text{CH}_2\overset{\text{O}}{\overset{||}{\text{COEt}}}$$

CC 907 (1973)

$$\text{HC}\equiv\text{CCO}_2\text{R} \longrightarrow \text{RCH}=\text{CHCO}_2\text{R}$$

See page 215, Section 2.

7. REARRANGEMENTS

1. Willgerodt and Related Reactions

Review: Angew Int *23* 413 (1984)

$$ArCCH_2CH_3 \xrightarrow[\substack{O \quad NH, S}]{(NH_4)_2S \text{ or}} ArCH_2CH_2CNH_2$$

JACS *64* 3051 (1942); *68* 2029 (1946)
Org Rxs *3* 83 (1946)
Angew Int *3* 19 (1964)
Newer Methods Prep Org Chem *3* 1 (1964)

$$ArCCH_3 \longrightarrow ArCH_2COCH_3$$

$BF_3 \cdot OEt_2$, $Pb(OAc)_4$, CH_3OH	Syn 126 (1981)
$Tl(NO_3)_3$, CH_3OH	JACS *93* 4919 (1971)
	JCS Perkin I 235 (1982)
	JOC *51* 1607 (1986)
$Pb(OAc)_4$, $HC(OMe)_3$, $HClO_4$	Syn 456 (1982)
$AgNO_3$, I_2, CH_3OH	JCS Perkin I 235 (1982)

$$ArC-CHR \longrightarrow ArCHRCOR'$$

X	Reagents	
Cl, Br	$Tl(NO_3)_3 \cdot 3\ H_2O$, $HC(OR')_3$	Syn 444 (1983)
Br	Ag^+, R'OH	TL *23* 1385 (1982)

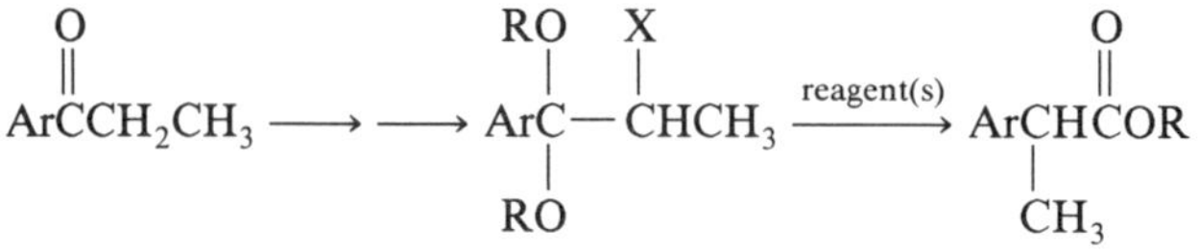

$$\underset{\substack{\| \\ O}}{ArCCH_2CH_3} \longrightarrow \longrightarrow \underset{\substack{| \\ RO}}{ArC} \overset{RO \ X}{\underset{}{-}} \underset{}{CHCH_3} \xrightarrow{reagent(s)} \underset{\substack{\| \\ O}}{ArCHCOR} \underset{CH_3}{}$$

Review: Angew Int *23* 413 (1984)

X	Reagent(s)	R	
Cl	ZnCl$_2$	Me	JOC *52* 10 (1987)
Cl, Br	ZnX$_2$ (X = Cl, Br), SnCl$_2$, CoCl$_2$, Hg$_2$Cl$_2$, PdCl$_2$ or Cu$_2$Br$_2$	Me, Et	JOC *48* 4658 (1983)
Br	Δ, ROH	Me, $-C_n-$ (*n* = 2, 3)	Syn 505 (1985)
	AgBF$_4$, H$_2$O	H	JOC *52* 3018 (1987) (chiral)
I	RCO$_3$H	Me	JCS Perkin I 1483 (1983)
RSO$_2$	CaCO$_3$, H$_2$O, CH$_3$OH	Me	TL *22* 4305 (1981); *23* 5427 (1982)

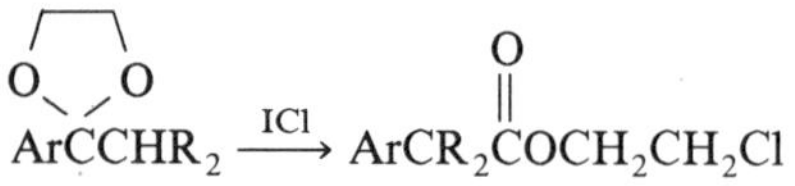

$$\underset{ArCCHR_2}{\overset{\overset{O \frown O}{\diagdown \diagup}}{}} \xrightarrow{ICl} \underset{\substack{\| \\ O}}{ArCR_2COCH_2CH_2Cl}$$

CC 1311 (1982)

$$\underset{\substack{\| \\ O}}{ArCCH_2R} \longrightarrow \underset{}{ArC} \overset{OSiR_3}{\underset{}{=}} CHR \xrightarrow[\substack{NaN_3 \\ (PhCH_2NEt_3)Cl}]{NCS} \underset{\substack{| \\ N_3}}{ArC} \overset{OSiR_3}{\underset{}{}} CHClR \xrightarrow{\Delta} \underset{\substack{\| \\ O}}{RCHClCNHAr}$$

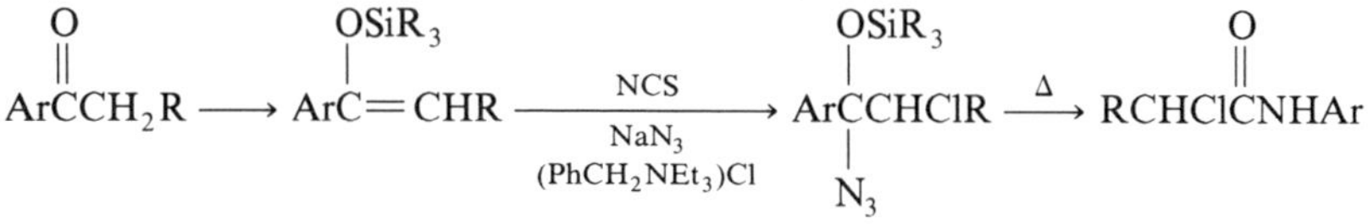

CC 1727 (1987)

2. Baeyer–Villager

See page 841, Section 14.

3. Beckmann

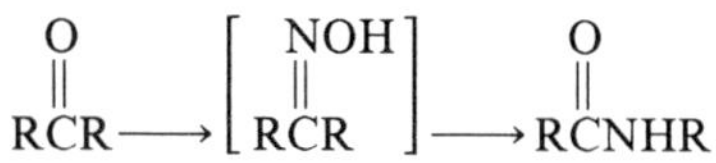

$$\underset{\substack{\| \\ O}}{RCR} \longrightarrow \left[\underset{\substack{\| \\ O}}{RCR} \overset{NOH}{} \right] \longrightarrow \underset{}{RCNHR}$$

Reviews:

Chem Rev *12* 215 (1933); *35* 335 (1944)
Org Syn Coll Vol *2* 76 (1943)
Org Rxs *11* 1 (1960)
P. A. S. Smith, "Molecular Rearrangements," Ed. P. de Mayo, Interscience, New York (1963), Vol 1, p 483
C. G. McCarty, "Chemistry of the Carbon-Nitrogen Double Bond," Ed. S. Patai, Interscience, New York (1970), pp 408–439
R. T. Conley and S. Ghosh, "Mechanisms of Molecular Migrations," Ed. B. S. Thyagarajan, J. Wiley, New York (1971), pp 203–250
Tetr *37* 1283 (1981)

H_2NOSO_3H	JACS *67* 1941 (1945)
H_2NOSO_3H, HCO_2H, Δ	Syn 537 (1979) Org Syn *63* 188 (1984)
H_2NOH/P_2O_5, $MeSO_3H$	JOC *38* 4071 (1973); *47* 3876 (1982)
H_2NOH/H_3PO_4, Ac_2O	JOC *38* 4073 (1973)
$H_2NOH/TsCl$, py, cat DMAP	JACS *106* 6414 (1984)
$H_2NOH/TsCl$, py/Et_3N, Δ	JACS *106* 6702 (1984)

JOC *47* 3876, 3881 (1982)

4. Schmidt

$$\underset{RCR}{\overset{O}{\|}} \xrightarrow{HN_3} \underset{RCNHR}{\overset{O}{\|}}$$

JACS *67* 1941 (1945)
Org Rxs *3* 307 (1946)
P. A. S. Smith, "Molecular Rearrangements," Ed. P. de Mayo, Interscience, New York (1963), Vol 1, pp 507–527
D. V. Banthorpe, "Chemistry of the Azido Group," Ed. S. Patai, Interscience, New York (1971), p 397
Russ Chem Rev *40* 835 (1971)
Tetr *37* 1283 (1981)

5. Curtius

See page 431, Section 1.

6. Oxaziridine → Amide

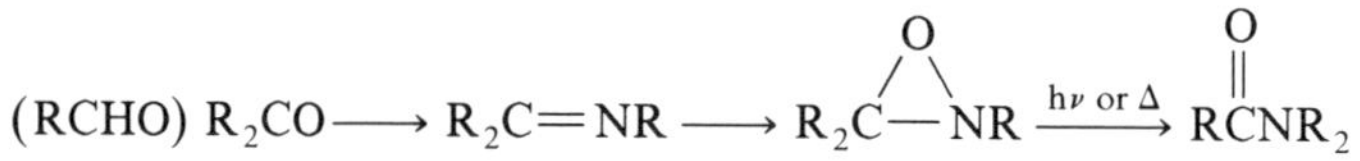

$$(RCHO) \ R_2CO \longrightarrow R_2C=NR \longrightarrow R_2\overset{O}{\overset{\diagup \diagdown}{C-NR}} \xrightarrow{h\nu \ or \ \Delta} R\overset{O}{\overset{\|}{C}}NR_2$$

JACS *79* 5739 (1957)
Adv Heterocyclic Chem *24* 63 (1979) (review)
TL *28* 2595 (1987)

7. Favorski and Related Reactions

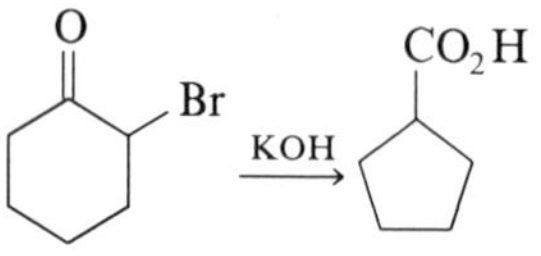

Org Rxs *11* 261 (1960) (review)
Org Syn Coll Vol *4* 594 (1963)
Russ Chem Rev 732 (1970) (review)
JOC *47* 4485 (1982); *52* 4885 (1987)
JACS *109* 7230 (1987)

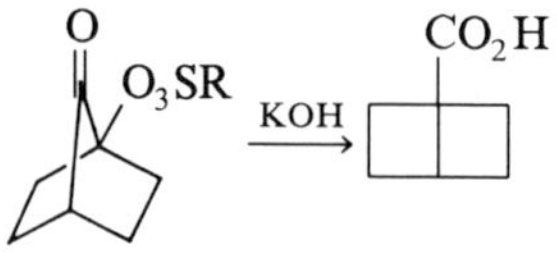

Tetr *42* 1621 (1986)

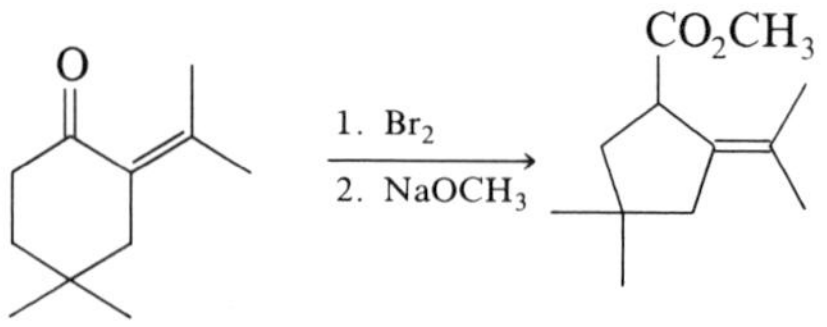

JOC *46* 5434 (1981)

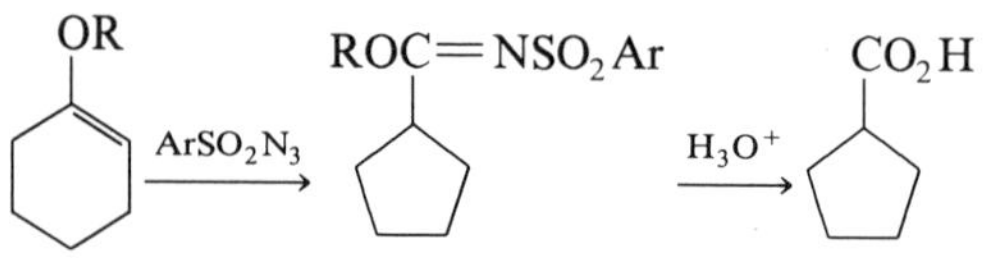

$\underline{R}$	
Me	TL 3111 (1973)
Me$_3$Si	Helv *56* 1826 (1973)
	JOC *47* 5042 (1982)

8. Arndt–Eistert and Wolff

$$\underset{RC-CR}{\overset{O \quad N_2}{\|\quad\|}} \longrightarrow \underset{R_2CHCX}{\overset{O}{\|}}$$

$$X = OH, OR, NR_2$$

Reviews:

Newer Methods Prep Org Chem *1* 513 (1948); *3* 451 (1964)
Angew *72* 535 (1960)
Fortschr Chem Forsch *5* 1 (1965)
Russ Chem Rev *36* 260 (1967)
Adv Alicyclic Chem *3* 125 (1971) (ring contraction)
Angew Int *14* 32 (1975)
JACS *101* 7675 (1979) (ring contraction)

$$\underset{RCCl}{\overset{O}{\|}} \xrightarrow{CH_2N_2} \underset{RCCHN_2}{\overset{O}{\|}} \xrightarrow{reagents} \underset{RCH_2CX}{\overset{O}{\|}}$$

Review: Org Rxs *1* 38 (1942)

X	Reagents	
OH	hν, H$_2$O	Ber *91* 430 (1958)
	Ag$_2$O or hν, H$_2$O	Org Rxs *1* 38 (1942)
	Ag$_2$O, Na$_2$CO$_3$, Na$_2$S$_2$O$_3$, H$_2$O	TL 2667 (1979)
OR	hν, ROH	JACS *107* 5732 (1985); *109* 4626 (1987)
	AgO$_2$CPh, Et$_3$N, ROH (R = Me, Et)	Org Syn *50* 77 (1970) TL 2667 (1979) JACS *109* 5432 (1987)
NH$_2$	hν/NH$_4$OH	JACS *106* 6437 (1984)

$$\underset{RCCl}{\overset{O}{\|}} \xrightarrow[Et_3N]{Me_3SiCHN_2} \xrightarrow[\Delta]{R'OH} \underset{RCH_2COR'}{\overset{O}{\|}}$$

TL *21* 4461 (1980)

$$\underset{RCCl}{\overset{O}{\|}} \longrightarrow \underset{RCCHN_2}{\overset{O}{\|}} \xrightarrow[NH_4OH]{AgNO_3} \underset{RCH_2CNH_2}{\overset{O}{\|}}$$

JOC *5* 606 (1940)

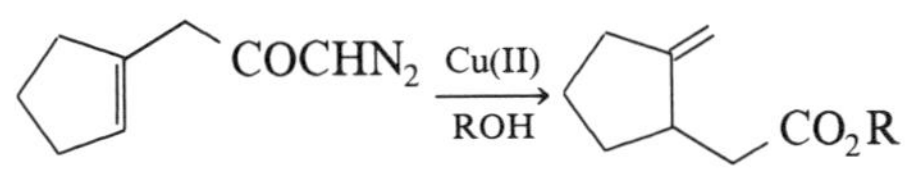

JACS *106* 3995, 4001 (1984)

9. Cope

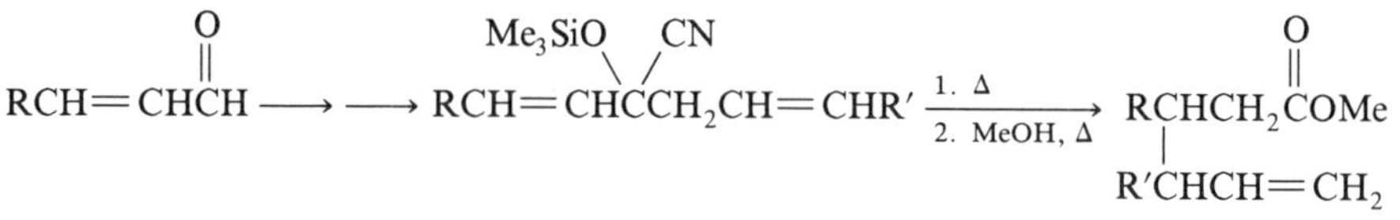

$$RCH=CHCH \xrightarrow{\quad} \xrightarrow{\quad} RCH=CHCCH_2CH=CHR' \xrightarrow[2.\ MeOH,\ \Delta]{1.\ \Delta} RCHCH_2COMe$$

(with $\overset{O}{\overset{\|}{}}$ on starting material; Me_3SiO and CN groups on intermediate; $\overset{O}{\overset{\|}{}}$ on product and $R'CHCH=CH_2$ branch)

TL *21* 2125 (1980); *22* 1179 (1981)
JACS *106* 718 (1984)

10. Claisen

Reviews:

Syn 589 (1977)
Acct Chem Res *10* 227 (1977)
R. K. Hill, "Asymmetric Synthesis," Ed. J. D. Morrison, Academic Press, New York (1984), Vol 3, Part B, Chpt 8 (chirality transfer)

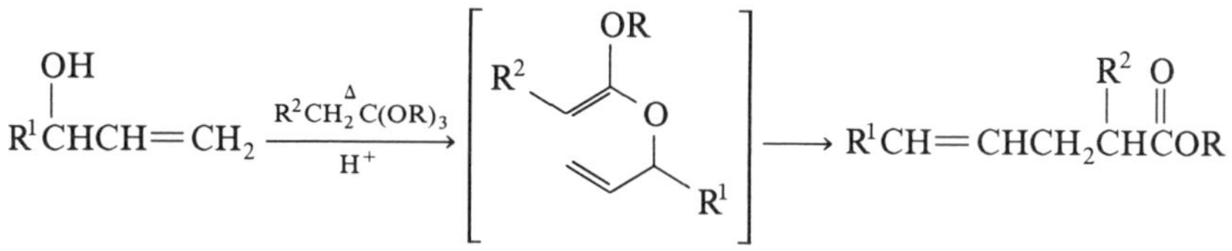

$$R^1CHCH=CH_2 \xrightarrow[H^+]{R^2CH_2C(OR)_3\ \Delta} \left[\begin{array}{c} OR \\ R^2 \diagdown \diagup O \\ \diagup \diagdown R^1 \end{array} \right] \longrightarrow R^1CH=CHCH_2CHCOR$$

(starting material with OH; product with R^2 and $\overset{O}{\overset{\|}{}}$)

JACS *92* 741 (1970); *98* 1583 (1976); *100* 8272 (1978); *101* 3066 (1979); *102* 6891 (1980); *103* 2419, 5259 (1981); *108* 1039, 4603 (1986); *109* 3025 (1987)
CC 1512, 1513 (1970); 858 (1979); 123 (1982)
TL 1281 (1973); 691 (1975); 2543 (1977); 2575 (1978) (allenic alcohol); 3057 (1979); *21* 1285, 4335 (1980); *23* 947, 3531 (1982); *28* 2597 (1987)
JOC *39* 3315 (1974); *41* 3497 (1976); *43* 3435 (1978); *44* 3374 (1979); *45* 891, 2080 (1980); *46* 1485, 3896 (1981); *47* 337, 620, 2420 (1982); *48* 1829 (1983); *51* 3402, 5429 (1986); *52* 1201, 3541 (1987)
BSCF 2040 (1974) (allenic alcohols)
CL 1721 (1981); 1113 (1982)
JCS Perkin I 2909 (1982)
Syn Commun *12* 395 (1982)
Org Syn *64* 175 (1985)

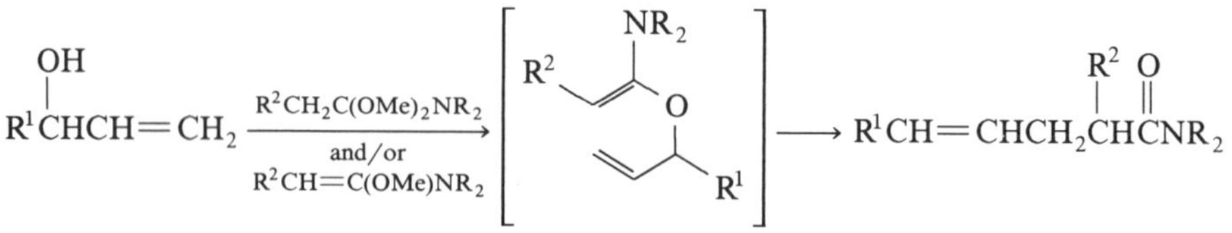

$$R^1CHCH=CH_2 \xrightarrow[and/or]{R^2CH_2C(OMe)_2NR_2} \left[\begin{array}{c} NR_2 \\ R^2 \diagdown \diagup O \\ \diagup \diagdown R^1 \end{array} \right] \longrightarrow R^1CH=CHCH_2CHCNR_2$$

with $R^2CH=C(OMe)NR_2$

(starting material with OH; product with R^2 and $\overset{O}{\overset{\|}{}}$)

Helv *47* 2425 (1964); *52* 1030 (1969); *62* 1922 (1979)
Ber *104* 3679 (1971)
BSCF 2040 (1974) (allenic alcohols)
JOC *42* 3828 (1977); *49* 2682 (1984); *52* 1372 (1987)
TL *23* 109 (1982); *27* 3053 (1986); *28* 789 (allenic alcohol), 2041 (1987)
Angew Int *24* 700 (1985) (allenic alcohol)
CC 1759 (1986)

$$R^1CH=CHCH_2OH \xrightarrow[\Delta]{R^2C\equiv CNR_2^3} \left[\begin{array}{c} R^2 \overset{NR_2^3}{\diagup} \\ \diagdown O \\ R^1 \diagdown \end{array} \right] \longrightarrow H_2C=CHCH\overset{R^1}{\underset{}{C}}H\overset{R^2}{\underset{}{C}}H\overset{O}{\underset{}{C}}NR_2^3$$

TL 6425 (1966)
Compt Rend C *268* 1446 (1969) (allenic alcohol)
JOC *44* 882 (1979); *46* 3896 (1981)

$$CH_3CH=CHCH_2OLi + RCH_2\overset{+OCH_3}{\underset{}{C}}NR_2' \longrightarrow H_2C=CHCH\overset{CH_3}{\underset{R}{C}}H\overset{O}{\underset{}{C}}NR_2'$$

JOC *50* 5909 (1985)
JACS *109* 6716 (1987)

$$CH_3CH_2\overset{S}{\underset{}{C}}NR_2 \xrightarrow[Et_3N]{CH_3CH=CHCH_2Br} CH_3\overset{S}{\underset{CH_3CHCH=CH_2}{C}}HCNR_2 \xrightarrow[K_2CO_3]{MeI} CH_3\overset{O}{\underset{CH_3CHCH=CH_2}{C}}HCNR_2$$

JACS *109* 6716 (1987)

$$R^1CH=CHCHR^2 \atop \overset{OH}{|} \xrightarrow[\Delta]{ArSC\equiv CNR_2} \left[\begin{array}{c} ArS \overset{NR_2}{\diagup} \\ \diagdown O \\ R^1 \diagdown R^2 \end{array} \right] \longrightarrow R^2CH=CHCH\overset{R^1}{\underset{}{C}}H\overset{SAr}{\underset{}{C}}HCONR_2$$

$$\xrightarrow[\Delta]{NaIO_4} R^2CH=CHC\overset{R^1}{\underset{}{C}}=CH\overset{O}{\underset{}{C}}NR_2$$

TL *22* 4097 (1981)

$$R^1CH=CHCHR^2 \atop \overset{OH}{|} \xrightarrow[2\,KH]{CF_3CH_2SOPh} \left[\begin{array}{c} O \\ PhS \overset{\overset{\|}{}}{\diagup} F \\ \diagdown O \\ R^1 \diagdown R^2 \end{array} \right] \longrightarrow R^2CH=CHCH\overset{R^1}{\underset{}{C}}H\overset{SOPh}{\underset{}{C}}HCO_2H$$

$$\xrightarrow{\Delta} R^2CH=CHC\overset{R^1}{\underset{}{C}}=CH\overset{O}{\underset{}{C}}OH$$

CL 1289 (1981)

$$(CH_3)_2\overset{Br}{\underset{}{C}}-\overset{O}{\underset{}{C}}OCH_2CH=CH_2 \xrightarrow[2.\ H_2O]{1.\ Zn,\ \Delta} H_2C=CHCH_2\overset{CH_3}{\underset{CH_3}{C}}CO_2H$$

CC 117 (1973)

$$RCH{=}CHCH_2OH \longrightarrow RCH{=}CHCH_2O\overset{\overset{O}{\|}}{C}CH_2R' \xrightarrow[\substack{2.\ (R_3SiCl) \\ 3.\ H^+}]{1.\ LDA} \underset{RCHCH{=}CH_2}{\overset{R'CHCO_2H}{|}}$$

JACS *71* 1150, 2439 (1949); *94* 5897 (1972); *98* 2868 (1976); *102* 1155 (R' = OR), 6889, 6891 (1980); *103* 3205 (1981); *104* 1124, 4030 (1982); *106* 3668, 5002 (1984); *107* 1448, 3271, 3279, 3285 (1985); *108* 2105, 2662 (1986)

Syn Commun *2* 27 (1972)

JOC *39* 3315 (1974); *41* 986 (1976); *43* 784 (1978) (R' = OR); *45* 48, 4259 (1980); *46* 479, 3896 (1981); *47* 337, 3933 (R' = RCONH), 3941 (R' = OH, OR) (1982); *48* 1829, 5221 (R' = OR) (1983); *50* 1128 (1985) (R' = OR); *51* 503, 1152 (R' = OR), 3247, 4023, 4322, 4485, 5019 (1986); *52* 3889 (1987) (R' = OR)

TL 3975 (1975); 2839 (1977); *23* 619 (R' = RCONH), 623 (R' = RCONH), 2825, 3419 (R' = OH, PhS), 3799, 4309, 5455 (1982); *24* 729 (R' = OH), 5177 (R' = OR) (1983); *25* 1543, 5155 (R' = OH) (1984); *27* 3345 (1986); *28* 1035, 1439, 1925, 3031, 4143 (R' = OR), 4147 (R' = OR), 4629 (1987)

CL 1721 (1981)

BCSJ *55* 3555 (1982)

CC 1220 (1987)

$$\underset{CH_3\overset{\overset{CH_3}{|}}{C}{=}CH\overset{\overset{O}{\|}}{C}OCH_2CH{=}CHR}{} \xrightarrow[2.\ H_2O]{1.\ base} \underset{H_2C{=}CHCHR\overset{\overset{H_2C{=}CCH_3}{|}}{C}HCO_2H}{}$$

JOC *40* 3309 (1975)

Chimia *29* 528 (1975)

$$RCH{=}CHCH_2OH \longrightarrow RCH{=}CHCH_2O\overset{\overset{O}{\|}}{C}CH_2OMe \xrightarrow[\substack{3.\ O_2 \\ 4.\ H^+}]{\substack{1.\ LDA \\ 2.\ LDA}} \underset{RCHCH{=}CH_2}{\overset{CO_2CH_3}{|}}$$

CC 594 (1977)

JOC *43* 784 (1978); *45* 4135 (1980)

11. Aza-Claisen

$$R_2C{=}CRCH_2X \xrightarrow[2.\ n\text{-BuLi}]{1.} R_2C{=}CRCH_2N \xrightarrow[\substack{2.\ Me_2SO_4 \\ 3.\ KOH}]{1.\ \Delta} \underset{\text{diastereoselective}}{H_2C{=}CRCR_2CH_2CO_2H}$$

JACS *107* 443 (1985)

JOC *50* 5769 (1985); *51* 1377 (1986)

TL *28* 1031 (1987)

$$R_2C{=}CHCH_2NH_2 \longrightarrow \longrightarrow \quad \xrightarrow{\Delta} \quad$$

JOC *39* 421 (1974)

12. Allylic Selenides

$$\underset{\underset{\displaystyle RCHCH=CHR'}{|}}{PhSe} \xrightarrow[\text{2. NCS}]{\text{1. } XCNH_2, HC(OR)_3} \underset{\underset{\displaystyle RCH=CHCHR'}{|}}{HNCOX}$$

$\underline{X}$

OR   JOC *51* 5243 (1986); *52* 3759 (1987)

CH(OMe)Ph   JOC *51* 5243 (1986)

JOC *51* 5243 (1986)

13. Imidic Esters

$$RCH=CHCH_2OH \xrightarrow[Cl_3CCN]{\text{(base)}} \underset{\underset{\displaystyle RCH=CHCH_2OCCCl_3}{||}}{NH} \longrightarrow \underset{\underset{\displaystyle RCHCH=CH_2}{|}}{NHCOCCl_3}$$

Review: Acct Chem Res *13* 218 (1980)

Δ   JACS *96* 597 (1974); *98* 2901 (1976)
                                            BCSJ *49* 3247 (1976)
                                            Org Syn *58* 4 (1978)
                                            CC 770 (1984)
                                            JOC *52* 1487, 5127 (1987)

cat Hg(O$_2$CCF$_3$)$_2$ JACS *96* 597 (1974); *98* 2901 (1976)

cat PdCl$_2$(PhCN)$_2$ Angew Int *23* 579 (1984)

14. 2,3-Sigmatropic Rearrangement

$$\underset{\underset{\displaystyle R^1CHCH=CH_2}{|}}{OH} \xrightarrow[\Delta]{(RO)_2CHNR_2^2} \underset{\underset{\displaystyle R^1CH=CHCH_2CNR_2^2}{||}}{O}$$

JACS *96* 5563 (1974)
JOC *42* 3828 (1977)

$$RCH{=}CHCH_2OCH_2\overset{\overset{\displaystyle O}{\|}}{C}X \xrightarrow{\text{base}} H_2C{=}CHCH\overset{\overset{\displaystyle HO}{|}}{C}H\overset{\overset{\displaystyle O}{\|}}{C}X$$

$$\underset{R}{|}$$

$$X = OH, OR, NR_2$$

See page 521, Section 5.

$$\overset{\overset{\displaystyle OH}{|}}{R^1CH}CR^2{=}CHR^3 \xrightarrow[\substack{2.\ BrCH_2CO_2H \\ 3.\ LDA \\ 4.\ TsCl}]{1.\ NaH} R^1CH{=}CR^2CHR^3\overset{\overset{\displaystyle OTs}{|}}{C}HCO_2H \xrightarrow{\text{KO-}t\text{-Bu}} R^1CH{=}CR^2CR^3{=}CH\overset{\overset{\displaystyle O}{\|}}{C}OH$$

TL *22* 69 (1981)

15. Ene Reaction—Oxidation

$$RCH_2CH{=}CH_2 \xrightarrow[\Delta\ \text{or Lewis acid}]{EtO_2C\overset{\overset{\displaystyle O}{\|}}{C}CO_2Et} \xrightarrow[\text{Ce(IV)}]{\text{NaIO}_4\ \text{or}} RCH{=}CHCH_2CO_2H$$

JACS *102* 2473 (1980); *106* 1092, 3797 (1984)
JOC *45* 1228 (1980); *47* 4201 (1982)

For related reactions see page 119, Section 8.

16. Ring Opening

$$\xrightarrow[2.\ H^+]{1.\ NaOH} R\overset{\overset{\displaystyle O}{\|}}{C}(CH_2)_x\overset{\overset{\displaystyle O}{\|}}{C}OH$$

$$n = 1, 2; \ x = 4, 5$$

JACS *70* 4023 (1948); *75* 5030 (1953)

$$\xrightarrow[2.\ H^+]{1.\ Ba(OH)_2,\ \Delta} RCH_2\overset{\overset{\displaystyle O}{\|}}{C}(CH_2)_3\overset{\overset{\displaystyle O}{\|}}{C}OH$$

Ber *85* 1061 (1952)

$$\xrightarrow[2.\ HCl]{1.\ NaOH} HO\overset{\overset{\displaystyle O}{\|}}{C}(CH_2)_5\overset{\overset{\displaystyle O}{\|}}{C}OH$$

Org Syn Coll Vol *2* 531 (1943)

17. Decarboxylation—Allylation

$$\underset{\text{XCHRCOCH}_2\text{CH}=\text{CH}_2}{\overset{\overset{\displaystyle O}{\|}}{}} \xrightarrow{\text{Pd catalyst}} \text{XCHRCH}_2\text{CH}=\text{CH}_2$$

$$X = CN, CO_2R$$

JOC *52* 2988 (1987)

18. Other Reactions

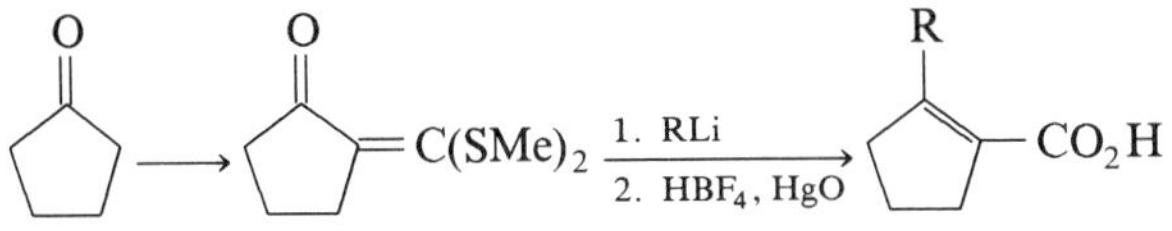

TL *23* 3747 (1982)

8. LACTONE AND LACTAM FORMATION

See also page 836, Section 12.

Reviews:

Syn 67 (1975) (α-methylene lactones)
Tetr *33* 683, 3041 (1977)
Angew Int *16* 585 (1977)
Org Prep Proc Int *13* 59 (1981) (saturated γ-lactones)

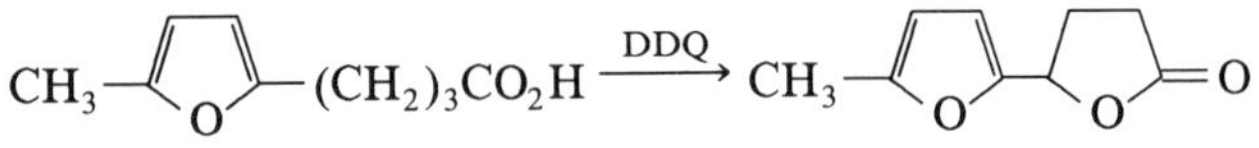

$$CH_3 \underset{O}{\text{—}} (CH_2)_3 CO_2 H \xrightarrow{\text{DDQ}} CH_3 \underset{O}{\text{—}}\underset{O}{\text{—}} O$$

TL *28* 5175 (1987)

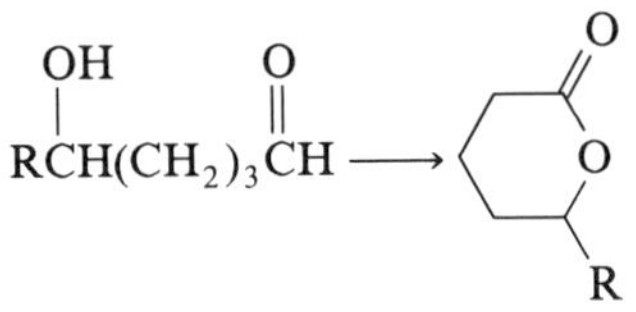

$$\underset{RCH(CH_2)_3CH}{\overset{OH \qquad O}{|\qquad\quad ||}} \longrightarrow \underset{R}{\text{—}} O$$

See page 844, Section 15.

$$HO-C_n-\overset{O}{\overset{||}{C}}OH \longrightarrow \underset{O}{\overset{C_n \qquad C}{\bigcirc}} O$$

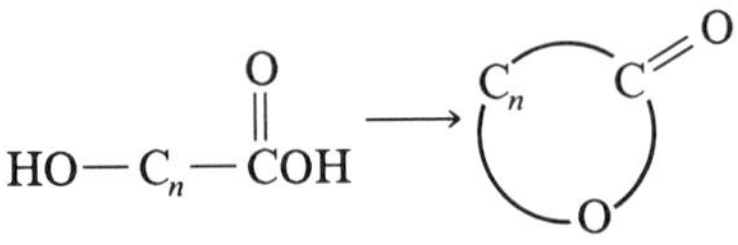

Reviews:

Houben-Weyl, Vol VI/2 (1963), p 511
Tetr *33* 683, 3041 (1977)
Angew Int *18* 707 (1979)

cat TsOH, Δ	Helv *17* 1283 (1934); *18* 1087 (1935)
TsOH, microemulsion, Δ	JOC *46* 2594 (1981)
n-Bu$_2$SnO	JACS *102* 7578 (1980); *105* 7130 (1983)
n-Bu$_2$SnO $-$ n-Bu$_2$SnCl$_2$	TL *27* 4501 (1986)
2,6-Cl$_2$C$_6$H$_3$COCl, Et$_3$N/DMAP, Δ	TL *28* 2409 (1987)

2,4,6-Cl$_3$C$_6$H$_2$COCl, Et$_3$N/DMAP, Δ

BCSJ *52* 1989 (1979)
TL *25* 2163 (1984); *28* 4569 (1987)

Me$_2$NCH(OCH$_2$CMe$_3$)$_2$

Angew Int *16* 876 (1977)

EtO$_2$CN=NCO$_2$Et, Ph$_3$P

TL 2455 (1976); 2371 (1978); *25* 2163 (1984)
Syn 1 (1981) (review)
JOC *51* 4840 (1986)
JACS *109* 6176 (1987)

[structure: 1-phenyl-tetrazoline-5-thione with NH], *t*-BuNC

Angew Int *20* 771 (1981)

[structure: 2,4,6-trichloro-1,3,5-triazine], Et$_3$N

TL *21* 1893 (1980); *27* 2369 (1986)

[Me$_2$NC≡$\overset{+}{N}$Me$_2$]Cl$^-$, collidine
(*n* = 11, 12, 14)

CL 1891 (1982)

[structure: 2-halo-1-methylpyridinium iodide] I$^-$, Et$_3$N   X = Cl

CL 49 (1976); 441, 763, 959 (1977)
Angew Int *20* 286 (1981)
JOC *51* 3247 (1986)
JACS *108* 2105 (1986); *109* 8117 (1987)
TL *28* 3031, 5759, 5763 (1987)
CC 1220 (1987)

X = Br

JACS *109* 5437 (1987)

[structure: 2-chloro-6-methyl-1,3-diphenylpyridinium] BF$_4^-$,

CL 885 (1978)

pyridine base, (PhCH$_2$NEt$_3$)Cl

[structure: 2,2'-dipyridyl disulfide], Ph$_3$P

JACS *96* 5614 (1974); *97* 653, 654, 2287 (1975);
98 222 (1976); *103* 3213 (1981); *104* 1774
(1982); *108* 6800 (1986)
TL 3405 (1976); *21* 2791 (1980); *27* 1815 (1986);
28 4993 (1987)

[structure: R-substituted imidazoline-2-thione disulfide with NR'], S)$_2$, Ph$_3$P/Δ

TL 3409 (1976)
JACS *100* 4621 (1978); *101* 7131 (1979);
106 2735 (1984)

[structure: dicyclohexylcarbodiimide] —N=C=N—, DMAP

JACS *108* 2776 (1986)

$$\left\langle \bigcirc \right\rangle - N=C=N - \left\langle \bigcirc \right\rangle,$$ JOC *50* 2394 (1985)

DMAP, DMAP·HCl

$$\underset{\text{OH}}{\text{NaO}_2\text{C(CH}_2)_2\overset{|}{\text{CH}}(\text{CH}_2)_2\text{CO}_2\text{Na}} \xrightarrow{\text{(+)-10-camphorsulfonic acid}} O{=}\left\langle \underset{O}{} \right\rangle{-}(\text{CH}_2)_2\text{CO}_2\text{H}$$

                                                                     chiral

JACS *107* 6404 (1985)
TL *27* 5381 (1986)

$$\text{HO}-\text{C}_n-\text{CO}_2\text{R} \longrightarrow \overset{}{\underset{\text{C}_n\diagdown_O}{\diagup}} \text{C}{=}\text{O}$$

Lipase P	TL *28* 805 (1987)
porcine pancreatic lipase	TL *28* 805, 3861, 5367 (1987)
imidazole	JACS *109* 2208 (1987)
EtO$_2$CN=NCO$_2$Et, Ph$_3$P	JACS *99* 646 (1977)
KH	JACS *108* 2662 (1986)
n-BuLi (R = CH$_2$SO$_2$CH$_3$)	CL 455 (1982) JACS *106* 2954 (1984)
H$^+$	Houben-Weyl, Vol VI/2 (1963), p 511
ZnCl$_2$, molecular sieves	CC 1797 (1987)

$$\text{HO}-\text{C}_n-\overset{\overset{\text{O}}{\|}}{\text{COH}} \longrightarrow \text{HO}-\text{C}_n-\overset{\overset{\text{O}}{\|}}{\text{CSR}} \xrightarrow{\text{reagent}} \overset{}{\underset{\text{C}_n\diagdown_O}{\diagup}} \text{C}{=}\text{O}$$

<u>Reagent</u>

KO-t-Bu	TL 1469 (1979) JOC *45* 1535 (1980); *46* 3209 (1981) CC 251 (1982)
CuO$_2$CCF$_3$	CC 1805 (1985) (dimerization)
AgX	Helv *57* 2661 (1974); *58* 2036 (1975); *59* 755 (1976) JACS *99* 6756 (1977); *109* 6205 (1987) Org Syn *63* 192 (1984)
HgX$_2$	JACS *97* 3513, 3515 (1975); *98* 7874 (1976)

$$\text{X}-\text{C}_n-\overset{\overset{\text{O}}{\|}}{\text{COH}} \longrightarrow \overset{}{\underset{\text{C}_n\diagdown_O}{\diagup}} \text{C}{=}\text{O}$$

KHCO$_3$, (polymer-PBu$_3$)OMs, H$_2$O, toluene (X = OMs)	JACS *104* 2064 (1982)
K$_2$CO$_3$ (X = Br, I)	Ber *80* 129 (1947)

Cs$_2$CO$_3$ (X = Br, I)   CC 286 (1979); 251 (1982)
                                                            JACS 103 5183 (1981)

Cs$_2$CO$_3$ (X = OMs)   CC 668 (1982)

NaOH (X = Br)   JACS 95 8374 (1973); 99 2591 (1977)

KOH (X = Br)   CC 251 (1982)

KOH, microemulsion (X = Br)   JOC 46 2594 (1981)

KOH, n-Bu$_4$NBr (phase transfer)   JOC 48 1533 (1983)
  (X = Br)

(Et$_4$N)N⬠(X = Br)   JOC 51 546 (1986)

$$X-C_n-\overset{\displaystyle O}{\overset{\|}{C}}OR \longrightarrow C_n \overset{\displaystyle C=O}{\underset{O}{\big)}}$$

X = halogen

Δ (n = 3)   JOC 50 2128 (1985) (phthalide)

Δ, silica gel (n = 3)   CL 1909 (1982)
                                                            JOC 51 4944 (1986)

TsOH, Hg(OAc)$_2$ or AgClO$_4$ (n = 3)   Heterocycles 12 699 (1979)

$$I-C_n-O-\overset{\displaystyle O}{\overset{\|}{C}}-C=C \xrightarrow{n\text{-Bu}_3\text{SnH}} C_{n+2}\overset{\displaystyle C=O}{\underset{O}{\big)}}$$

n = 7–9, 11, 12, 16

JACS 109 4976 (1987)

$$H_2N-C_n-\overset{\displaystyle O}{\overset{\|}{C}}OH \longrightarrow C_n \overset{\displaystyle C=O}{\underset{NH}{\big)}}$$

alumina or silica gel (n = 3–5)   TL 21 2443 (1980)

n-Bu$_2$SnO (n = 3–5)   JACS 102 7578 (1980); 105 7130 (1983)

o-NO$_2$C$_6$H$_4$SCN, n-Bu$_3$P (n = 3–5)   JOC 44 2945 (1979)

(pyridyl–S)$_2$, Ph$_3$P (n = 2)   JACS 103 2406 (1981)
                                                            TL 28 4335 (1987)

(PhO)$_2$PON$_3$, Et$_3$N (n = 7)   CC 344 (1986)

(Me$_3$Si)$_2$NH, Δ (n = 3–5)   Syn 614 (1978)

(catechol)BH, py (n = 3, 5)   JOC 43 4393 (1978)

$$H_2N-C_n-\overset{\overset{\displaystyle O}{\|}}{C}OMe \longrightarrow C_n \overset{C=O}{\underset{NH}{\overbrace{}}}$$

NaOH/$(PhO)_2PON_3$, Et_3N CC 344 (1986)

t-BuMgCl   TL *28* 5481 (1987)

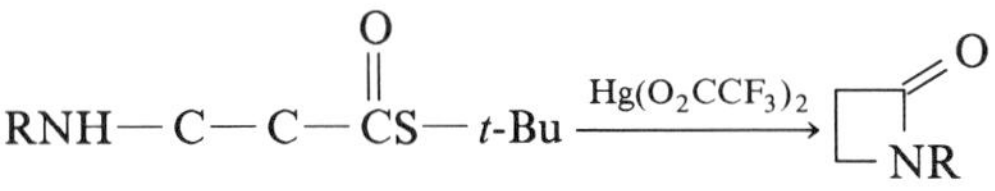

$$RNH-C-\overset{\overset{\displaystyle O}{\|}}{C}-CS-t\text{-}Bu \xrightarrow{Hg(O_2CCF_3)_2} \overset{O}{\underset{NR}{\overbrace{}}}$$

CL 915 (1986)

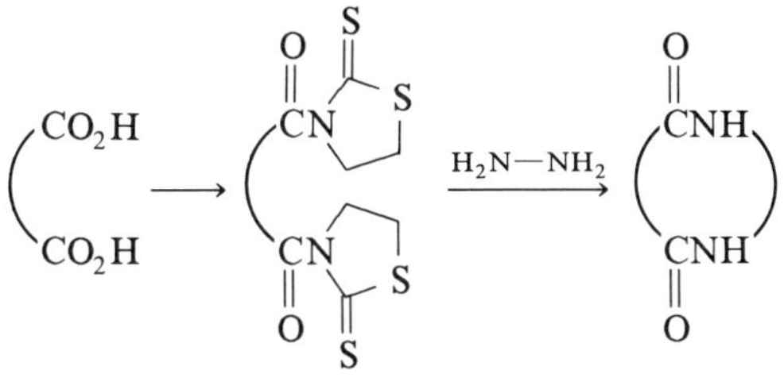

$$\overset{CO_2H}{\underset{CO_2H}{\big(}} \longrightarrow \overset{\overset{S}{\|}}{\underset{\overset{\|}{S}}{\big(}} \xrightarrow{H_2N-NH_2} \overset{\overset{\displaystyle O}{\|}}{\underset{\overset{\|}{O}}{\big(}}$$

CL 159 (1980)

$$\underset{PhC=CHCO_2H}{\overset{\overset{\displaystyle HCR_2}{|}}{}} \xrightarrow{H_2SO_4} \underset{R}{\overset{Ph}{\underset{R}{\big\langle}}}O=O$$

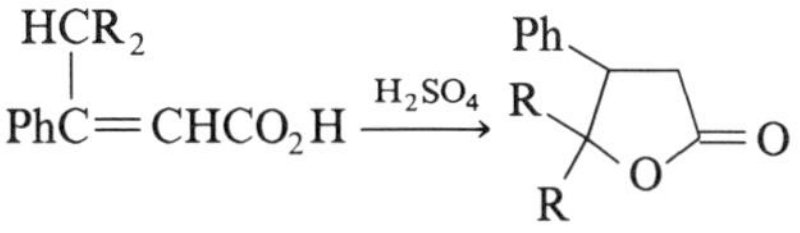

Acta Chem Scand B *36* 371 (1982)

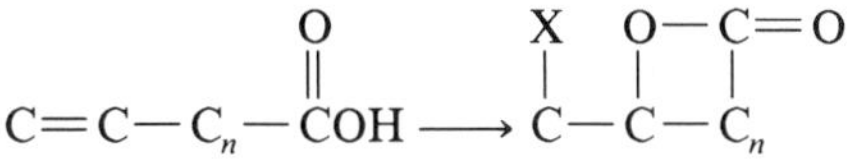

$$C=C-C_n-\overset{\overset{\displaystyle O}{\|}}{C}OH \longrightarrow \underset{C-C-C_n}{\overset{X\quad O-C=O}{|\quad|\quad|}}$$

Reviews:

Russ Chem Rev *40* 272 (1971) (X = halogen)
Chem Soc Rev *8* 171 (1979) (X = halogen)
Tetr *37* 4097 (1981) (X = SePh)
P. A. Bartlett, "Asymmetric Synthesis," Ed. J. D. Morrison, Academic Press, New York (1984), Vol 3, Part B, Chpt 6

$\underline{X}$	Reagent(s)	
H	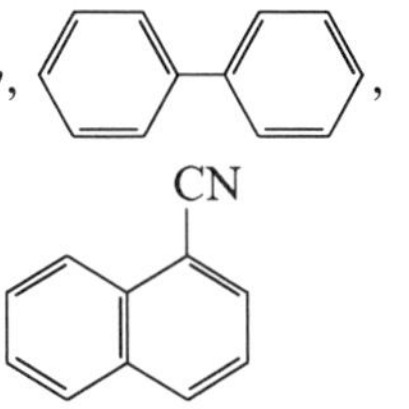 $h\nu$, [biphenyl],	JACS *109* 7547 (1987)
	[naphthalene-CN]	
Cl	Cl_2	Tetr *4* 393 (1958)
	TsNClNa	JOC *46* 3552 (1981)
Br	Br_2	Tetr *4* 393 (1958)
		Tl 2595 (1971)
	Br_2 (on TlO_2CR)	TL 335 (1979)
	Br_2, TlO_2CCMe_3	CC 1045 (1987)

	NBS	TL 3983 (1967); 1005 (1977) (chiral); *21* 2733 (1980); *27* 6079 (1986); *28* 2801 (1987)
		Tetr *30* 819 (1974); *35* 2337, 2345 (1979)
		CL 1109 (1977) (chiral)
		JOC *52* 1372 (1987)
I	I_2	JOC *38* 800 (1973); *44* 1625 (1979); *51* 2505, 4840 (1986); *52* 3346 (1987)
		TL 2543 (1977); 335 (1979); *22* 4611 (1981); *27* 6079 (1987)
		JACS *100* 3950 (1978); *102* 2118 (1980); *103* 4114 (1981); *105* 5819 (1983); *108* 5559 (1986); *109* 6844 (1987)
		Tetr *40* 2317 (1984)
		Org Syn *64* 175 (1985)
	I_2 (on TlO_2CR)	JCS Perkin I 1864 (1974)
	I_2, TlOAc	Austral J Chem *32* 2793 (1979)
	I_2, KI	JACS *76* 2315 (1954); *91* 5675 (1969); *92* 397 (1970); *106* 7854 (1984); *109* 6389 (1987)
		TL 1777 (1972); 2543 (1977); *27* 3297, 5467 (1986)
		CC 472 (1972)
		Tetr *30* 819 (1974)
		CL 351 (1982)
		JOC *51* 4600, 4944 (1986); *52* 4399 (1987)
OTs	PhI(OH)OTs	TL *27* 4557 (1986)
MeS	$(MeSSMe_2)BF_4$, *i*-Pr_2NEt	TL *26* 6159 (1985)
PhS	PhSCl, R_3N	CC 293 (1977)
		JACS *101* 3884 (1979)
		TL *28* 523 (1987)
ArSe	PhSeCl	CC 484 (1977)
		JACS *99* 3185 (1977); *101* 3884 (1979); *103* 4114 (1981); *106* 6060 (1984)
		TL 4801 (1979); *24* 4769 (1983)
		Tetr *36* 1399 (1980); *37* 4097 (1981) (review); *40* 2317 (1984)
		JOC *51* 3023 (1986)
	ArSeX (X = Cl, Br)	Ber *93* 317 (1960)
	PhSeOTf	CL 849 (1987)

PhSeN (phthalimide structure) — Tetr *40* 2317 (1984)

(PhSe)$_2$, electrolysis — TL *28* 6511 (1987)

ArTe	ArTeI	Ber *93* 317 (1960)
ArTeCl$_2$	ArTeCl$_3$	TL (6) 11 (1959) Ber *93* 317 (1960) Tetr *18* 521 (1962) Syn Commun *13* 889 (1983)
HgX	HgX$_2$	Tetr *40* 2317 (1984) R. C. Larock, "Solvomercuration/Demercuration Reactions in Organic Synthesis," Springer, New York (1986), Chpt 5 (review)

$$RCH{=}CHCH{=}CH(CH_2)_3CO_2H \xrightarrow[X = Se, S]{PhXCl}$$

(product: R–CH(PhX)–CH=CH–(tetrahydropyranone ring with O and C=O))

TL *27* 5919 (1986)

$$RCH{=}CHHgCl + H_2C{=}CH(CH_2)_nCO_2H \xrightarrow[2.\ K_2CO_3]{1.\ (CH_3CN)_2PdCl_2} RCH{=}CH{-}\overset{()_n}{\underset{O}{\bigcirc}}{=}O$$

$$n = 1, 2$$

TL *28* 4977 (1987)

$$C{=}C{-}C_n{-}\overset{O}{\overset{\|}{C}}Y \longrightarrow \overset{X}{\underset{}{C}}{-}\overset{O-C=O}{\underset{}{C}}{-}C_n$$

X	Y	Reagent	
Cl, Br, I	OCH$_3$	X$_2$	JOC *45* 839 (1980)
PhS	OCH$_3$	PhSCl	JOC *45* 839 (1980)
PhSe	OSiMe$_2$(*t*-Bu)	PhSeCl	JACS *107* 1448 (1985)
	OCH$_3$	PhSeCl	JOC *45* 839 (1980)
		PhSeOTf	CC 849 (1987)
HgX	OR	HgX$_2$	R. C. Larock, "Solvomercuration/Demercuration Reactions in Organic Synthesis," Springer Verlag, New York (1986), Chpt 6 (review)

$$C{=}C{-}C_n{-}\overset{\overset{\displaystyle O}{\|}}{C}NR_2 \xrightarrow[\text{2. H}_2\text{O}]{\text{1. reagent}} \overset{X\ \ O{-}C{=}O}{\underset{\displaystyle C{-}C{-}C_n}{|\ \ \ |}}$$

X	Reagent	
Cl	NCS	JACS *106* 1079 (1984)
Br	NBS	JACS *106* 1079 (1984)
I	I$_2$	TL 1625 (1977)
		JACS *106* 1079 (1984);
		109 6716 (1987)
		CC 156 (1986)

$$Ph_2C{=}CHCH_2\overset{\overset{\displaystyle O}{\|}}{C}NH_2 \xrightarrow[\text{CH}_3\text{SO}_3\text{H}]{\text{P}_2\text{O}_5} \begin{matrix} Ph \\ Ph \end{matrix}\!\!\!\underset{\underset{\displaystyle H}{|}}{\overset{}{\underset{N}{}}}{=}O$$

JOC *50* 2220 (1985)

$$C{=}C{-}C_n{-}\overset{\overset{\displaystyle O}{\|}}{C}NHR \longrightarrow \overset{X\ RN{-}C{=}O}{\underset{\displaystyle C{-}C{-}C_n}{|\ \ \ |}}$$

X	Reagent	n	
Br	Br$_2$	1	JOC *52* 4471 (1987)
			TL *28* 6257 (1987)
			(both R = OCO$_2$CH$_2$Ph)
	Br$_2$, K$_2$CO$_3$	1	TL *26* 5385 (1985)
			(R = O$_2$CR′)
Br, I	X$_2$, HCO$_3^-$	1	JACS *104* 3233 (1982)
			(R = SO$_2$R′)
I	Me$_3$SiOTf, Et$_3$N/I$_2$	2, 3	TL *26* 1803 (1985)
PhS	PhSCl/base	0 (β-lactam formation)	JACS *105* 7345 (1983)
PhSe	PhSeX	2, 3	JOC *51* 1724 (1986);
			52 2018 (1987)

$$C{=}C{-}C{-}C{-}\overset{\overset{\displaystyle SMe}{|}}{C}{=}NR \longrightarrow X\!\!\diagdown\!\!\underset{\underset{\displaystyle R}{|}}{\overset{}{\underset{N}{}}}{=}O$$

X	Reagent	
Br	BrClO$_4$(collidine)$_2$	Heterocycles *26* 359 (1987)
I	I$_2$	CC 1627 (1987)

$$C{=}C{-}C_n{-}NH{-}Y \xrightarrow[n=3,4]{X^+\ (X = Br,\ I)} \overset{X\ \ \ \ \ \ Y}{\underset{\displaystyle C{-}C{-}C_n}{|\ \ \ \diagup N}}$$

Y = CO$_2$Me, *p*-MeC$_6$H$_4$SO$_2$

Heterocycles *23* 192 (1985)

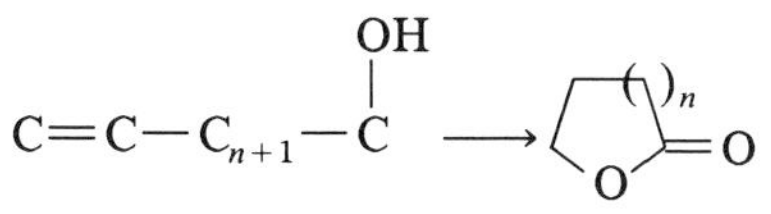

$$C=C-C_{n+1}-\overset{\overset{\displaystyle OH}{|}}{C} \longrightarrow \text{(lactone)}_n$$

n	Reagent(s)	
1, 2	$(n\text{-}C_{16}H_{33}NMe_3)MnO_4$	TL *27* 4079 (1986)
1, 2	$(bipyH_2)CrOCl_5$	CL 551 (1985)
1, 2	CrO_3, HOAc, Ac_2O	TL *26* 127 (1985)

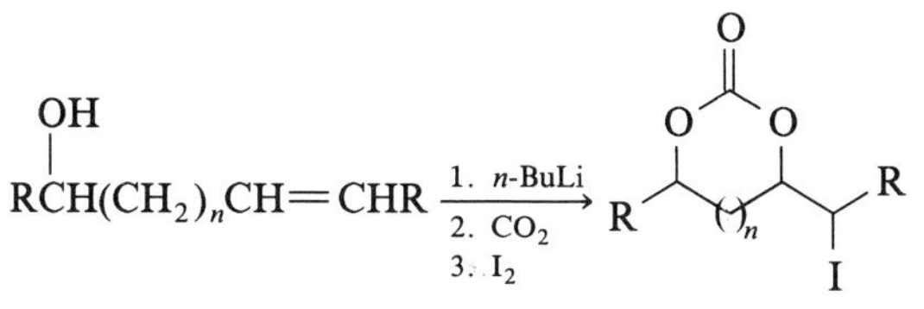

$$RCH(CH_2)_n CH=CHR \xrightarrow[\substack{2.\ CO_2 \\ 3.\ I_2}]{1.\ n\text{-BuLi}} \text{(product)}$$

with $\overset{\overset{\displaystyle OH}{|}}{}$ on the left reactant

$$n = 0, 1$$

CC 465 (1981)

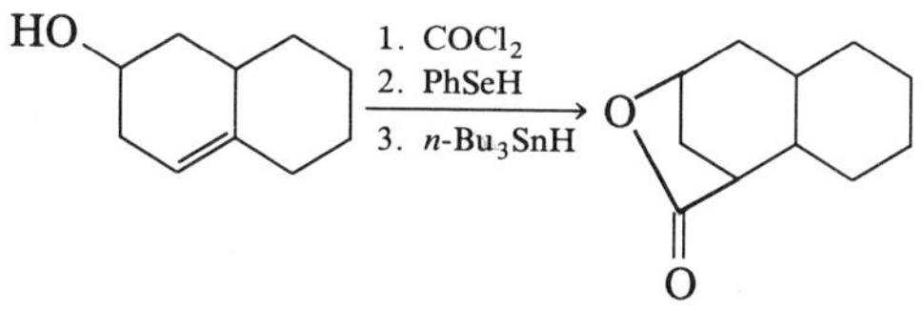

$$\xrightarrow[\substack{2.\ PhSeH \\ 3.\ n\text{-Bu}_3SnH}]{1.\ COCl_2}$$

JACS *109* 6187 (1987)

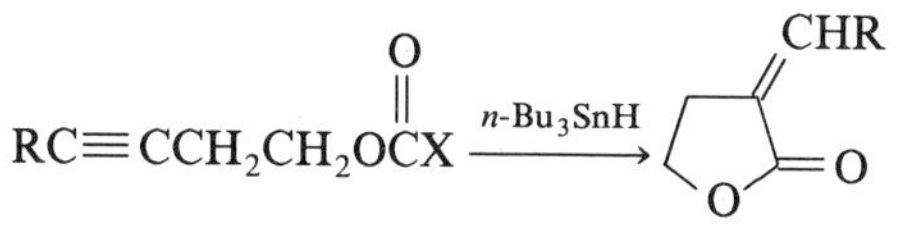

$$RC{\equiv}CCH_2CH_2O\overset{\overset{\displaystyle O}{\|}}{C}X \xrightarrow{n\text{-Bu}_3SnH} \text{(lactone with =CHR)}$$

$$X = Cl, SePh$$

TL *27* 641 (1986)

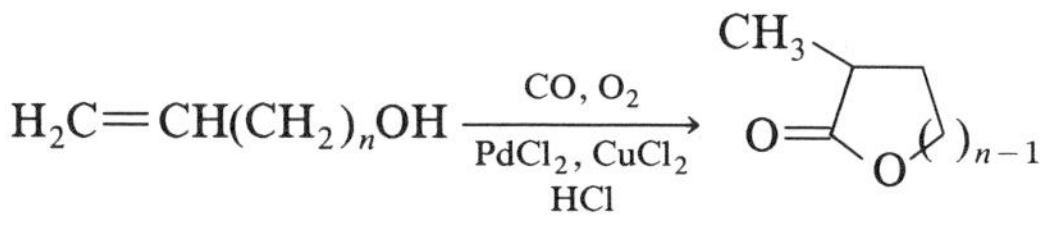

$$H_2C=CH(CH_2)_nOH \xrightarrow[\substack{PdCl_2,\ CuCl_2 \\ HCl}]{CO,\ O_2} \text{(lactone with CH}_3)_{n-1}$$

$$n = 2, 3$$

CC 511 (1985)

$$H_2C=CHCH_2CH_2OH \xrightarrow[\substack{cat\ PdCl_2,\ CuCl_2 \\ CH_3OH}]{CO} \text{(lactone with CO}_2CH_3)$$

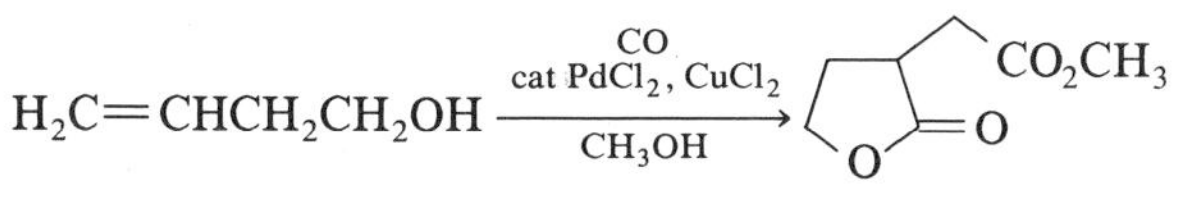

TL *26* 3207 (1985); *28* 325 (1987)

$$Cl_3C-\overset{O}{\underset{||}{C}}-O-X-\text{(cyclohexene)} \xrightarrow{\text{catalyst}} O=\text{(bicyclic, } Cl, Cl, Cl\text{)}-X$$

X	Catalyst	
O	CuCl	TL *24* 2395 (1983)
		Heterocycles *22* 1779 (1984)
NR	CuCl or RuCl$_2$(PPh$_3$)$_3$	CC 652 (1984); 518 (1985)

$$R_2C=C=CHCMe_2CONH_2 \xrightarrow{AgBF_4} \text{(dihydropyridinone ring with Me, Me, NH, R, R)}$$

TL *27* 5089 (1986)

$$H_2C=C=CHCH_2CH_2CO_2H \xrightarrow{\text{cat HgO}} \text{(pyranone ring with CH}_3\text{, O, O)}$$

TL *25* 3179 (1984)

$$RC\equiv CCH_2CH_2\overset{O}{\underset{||}{C}}OH \longrightarrow RCH=\text{(furanone ring)}=O$$

Catalyst	
$\left(\text{cyclohexyl}\right)_2 PCH_2CH_2P\left(\text{cyclohexyl}\right)_2 RhCl$	JACS *109* 6385 (1987)
Pd(PPh$_3$)$_4$	JACS *109* 6385 (1987)
PdCl$_2$(PhCN)$_2$	TL *25* 5323 (1984)
Ag$_2$CO$_3$	TL *28* 6448 (1987)
AgNO$_3$	JCS 3962 (1958)
	JOC *49* 736 (1984)
HgO	JCS Perkin I 582 (1981)
	TL *25* 3179 (1984)
Hg(OAc)$_2$	JOC *43* 560 (1978)
	JACS *103* 5459 (1981)
Hg(O$_2$CCF$_3$)$_2$	JACS *103* 4114 (1981); *108* 5589 (1986)
	JOC *50* 2331 (1985)
HCO$_3^-$	Acta Chem Scand *11* 582 (1957)
	JCS 1313 (1958)
	JACS *103* 4114 (1981)

For intermolecular reactions see page 830, Section 5.

$$HC{\equiv}C{-}C_n{-}CO_2H \longrightarrow ICH{=}\overset{(\,)_{n-1}}{\underset{O}{\diagup}}{=}O$$

Reagent(s)	n	
I_2	3	JOC *50* 2331 (1985)
NIS, NaHCO$_3$	2	JACS *108* 5589 (1986)

$$RC{\equiv}C(CH_2)_nCO_2Li \xrightarrow[PdCl_2(CH_3CN)_2]{H_2C{=}CHCH_2Cl} \underset{R}{\overset{H_2C{=}CHCH_2}{\diagdown}}C{=}\overset{(\,)_{n-1}}{\underset{O}{\diagup}}{=}O$$

$$n = 2\text{--}4$$

JACS *108* 2753 (1986)

$$X{-}C_n{-}OH \xrightarrow[Pd\ catalyst]{CO} C_n\overset{C{=}O}{\underset{O}{\diagup}}$$

JACS *102* 4193 (1980)

$$\overset{OH}{\underset{\mid}{Z\text{-}RCHCH_2CH}}{=}CHBr \xrightarrow[Et_3N]{Ni(CO)_4}$$

TL *24* 3209 (1983)

$$\overset{OH}{\underset{\mid}{RC{\equiv}CCHR'}} \longrightarrow \underset{I}{\overset{R}{\diagdown}}C{=}C\underset{HO}{\overset{H}{\diagup}}CHR' \xrightarrow[Pd\ catalyst]{CO} O{=}\overset{R}{\diagup}{-}R'$$

TL 133 (1979); *28* 723 (1987)
JACS *102* 4193 (1980)
JOC *52* 3860, 3883 (1987)

$$\overset{HO}{\underset{\mid}{R_2CC{\equiv}CH}} \xrightarrow[NaBH_4]{(PhTe)_2} \xrightarrow[CuCl_2]{CO\ cat\ PdCl_2} O{=}\overset{R}{\underset{R}{\diagup}}$$

JOC *52* 4859 (1981)

$$\overset{X}{\underset{\mid}{C{=}C{-}C_n{-}YH}} \xrightarrow{reagent(s)} \overset{O{=}C{-}Y}{\underset{\mid\quad\mid}{C{=}C{-}C_n}}$$

Reagent(s)

Ni(CO)$_4$, KOAc CL 773 (1978) ($n = 2$, X = Br, Y = O)

Ni(CO)$_4$ or Ni(CO)$_2$(PPh$_3$)$_2$ JOC *46* 1723 (1981)
($n = 2, 3$; X = Br; Y = O)

Ni(CO)$_2$PPh$_3$ JACS *104* 6879 (1982)
($n = 2$; X = Br; Y = O)

CO, Pd catalyst JOC *47* 3630 (1982)
($n = 2$, X = Br, Y = O);
48 4058 (1983)
($n = 2$–4; X = Br, I; Y = O, NH)

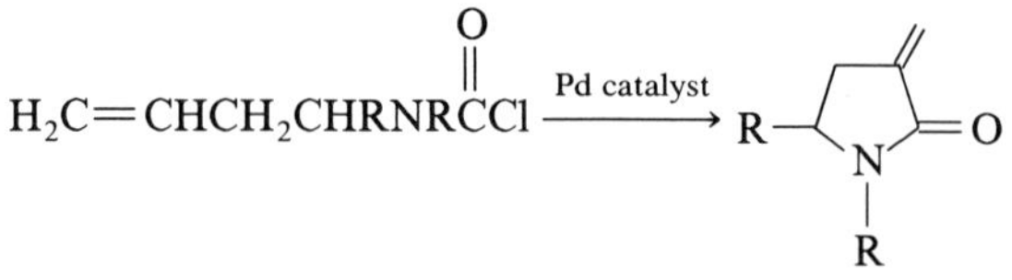

$$H_2C=CHCH_2CHRNRCCl \xrightarrow{\text{Pd catalyst}}$$

(with carbonyl O above CCl)

TL *27* 6339 (1986)

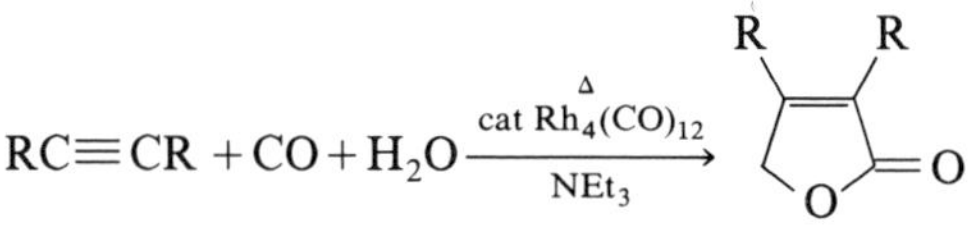

$$RC{\equiv}CR + CO + H_2O \xrightarrow[\text{NEt}_3]{\overset{\Delta}{\text{cat Rh}_4(\text{CO})_{12}}}$$

CC 649 (1987)

$$\xrightarrow[\text{PdCl}_2]{\text{CO}}$$

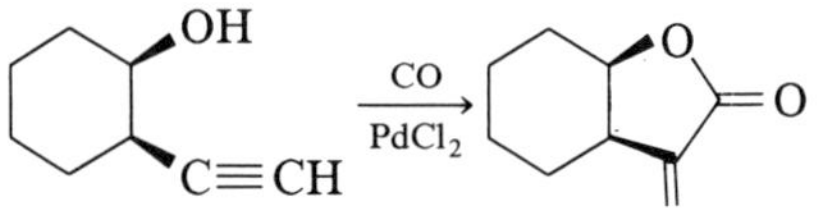

TL 51 (1975)
JACS *101* 4107 (1979); *103* 7520 (1981); *106* 5505 (1984)

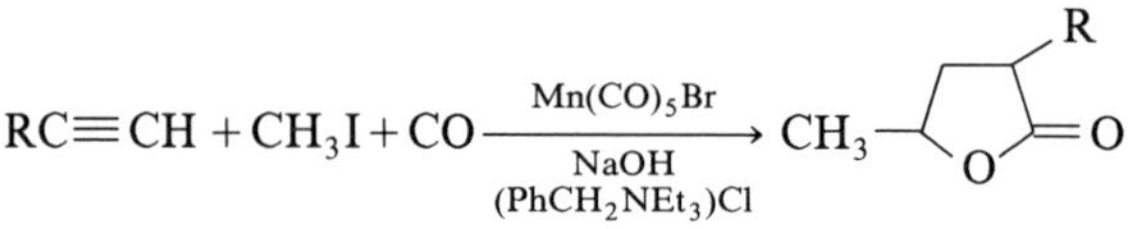

$$RC{\equiv}CH + CH_3I + CO \xrightarrow[\substack{\text{NaOH} \\ (\text{PhCH}_2\text{NEt}_3)\text{Cl}}]{\text{Mn(CO)}_5\text{Br}}$$

JOC *51* 273 (1986)

$$\xrightarrow[\text{cat Li}_2\text{PdCl}_4]{\text{Tl(O}_2\text{CCF}_3)_3 \quad \text{CO}}$$

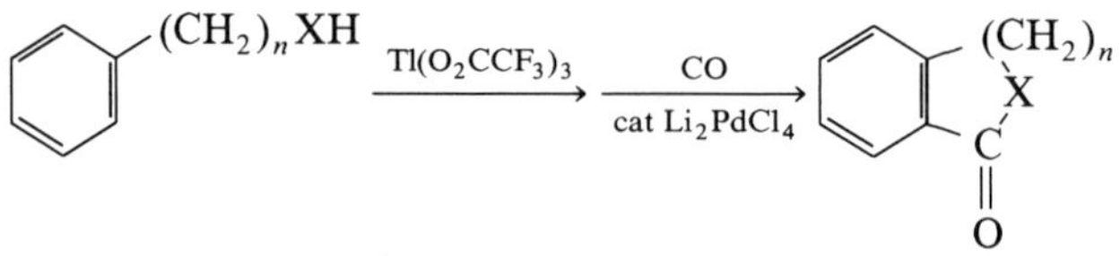

X = O, CO$_2$, CONH

n = 0–2

JOC *45* 363 (1980)
JACS *104* 1900 (1982)

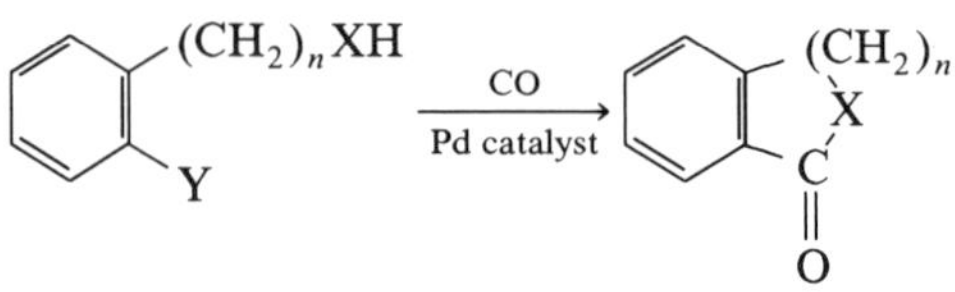

$$\xrightarrow[\text{Pd catalyst}]{\text{CO}}$$

X = O, NR; Y = Br, I; n = 1–3

Heterocycles *6* 1711, 1841 (1977); *12* 921 (1979); *16* 1491 (1981)
JOC *43* 1684 (1978)
JACS *102* 4193 (1980)
CC 841 (1986)

X	Reagents	
CO$_2$H	2 n-BuLi/R$_2$CO or RCHO/H$_3$O$^+$	JOC *41* 2628 (1976)
CO$_2$-i-Pr	n-BuLi/R$_2$CO or RCHO/H$_3$O$^+$	JOC *41* 2704 (1976)
(oxazoline)	Mg/R$_2$CO or RCHO/H$_3$O$^+$	JOC *39* 2787 (1974)

CC 764 (1983) (chiral); 520 (1987)

JOC *52* 183 (1987)

JOC *29* 853 (1964); *51* 3566 (1986)

For other ortho lithiation reactions see page 49, Section 2.1.

JOC *29* 3514 (1964); *51* 3566 (1986)
J Heterocyclic Chem *6* 83 (1969)
TL 633 (1970)

Heterocycles *23* 825 (1985)
JOC *52* 5378 (1987)

JOC *47* 3787 (1982)

JOC *45* 1835 (1980)

$$\underset{\substack{Br \quad O \\ | \quad \| \\ RCHCO-C_n-CR'}}{} \xrightarrow[\text{2. H}_2\text{O or Ac}_2\text{O}]{\text{1. SmI}_2} \underset{\substack{(Ac)HO \quad R \quad O \\ | \quad | \quad \| \\ CR'CHCO}}{\overset{}{\underset{C_n}{}}}$$

$\underline{n}$	
2	JACS *109* 6556 (1987)
8–14	TL *27* 3889 (1986)

$$R_2C=O \ (RCHO) \longrightarrow$$

$\underline{n}$	Reagents	
0	$H_2C=CHMgBr/H_2O/EtMgBr$, cat $Cp_2TiCl_2/CO_2/H^+$ ($R^1 = R^2 = H$)	JOMC *160* C8 (1978)
	$Li(CH_2)_3OCH(OEt)CH_3/H_3O^+/CrO_3$, H^+ or py ($R^1 = R^2 = H$)	JOC *37* 1947 (1972) TL 4651 (1976)
	$Li(CH_2)_2CH(OEt)_2/NH_4Cl, H_2O/$ $m\text{-}ClC_6H_5CO_3H, BF_3\cdot OEt_2$	CC 1534 (1987)
	$BrMg(CH_2)_2\underset{O}{\overset{O}{CH}}/Ac_2O/NaOH/H_3O^+/$	Syn Commun *7* 27 (1977)
	H_2CrO_4 ($R^1 = R^2 = H$)	
	$Li(CH_2)_2CO_2Li/TsOH$ ($R^1 = R^2 = H$)	TL 883 (1978)
	$H_2C=CHCH_2MgBr/H_2O/BH_3/H_2O_2$, OH^-/H_2CrO_4 ($R^1 = R^2 = H$)	BSCF 3377 (1973) JOC *42* 1623 (1977)
	$LiCH_2CH=CHSiMe_3/CH_3CO_3H, H_2SO_4$ ($R^1 = R^2 = H$)	CC 772 (1977)
	$LiCH_2CH=CHSiMe_3/t\text{-}BuO_2H$, cat $VO(acac)_2/MeOH, BF_3\cdot OEt_2/CrO_3$, H_2SO_4 ($R^1 = R^2 = H$)	TL *21* 11 (1980) JACS *102* 5004 (1980)

$LiCH_2CH=C(OLi)PO(NMe_2)_2/H_2O$
$(R^1 = R^2 = H)$

TL 47 (1976)

Li or $NaC\equiv CCH(OEt)_2/H_2$, cat Pd-C/
H_2CrO_4 $(R^1 = R^2 = H)$

JOC 22 570 (1957)
JACS 86 485 (1964)

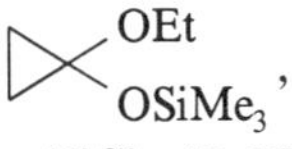

$TiCl_4/TsOH$ $(R^1 = R^2 = H)$

JACS 99 7360 (1977);
108 3745 (1986)

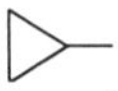

$\overset{+}{S}Ph_2/H^+/NaOH, H_2O_2$ or NaOBr
or HOCl $(R^1 = R^2 = H)$

TL 887 (1972); 923 (1973)
JACS 93 3773 (1971);
94 4777 (1972);
95 5321 (1973)

$CH_3CHLiCH=C(CN)NMe_2/H_3O^+$
$(R^1 = H, R^2 = Me)$

TL 21 1205 (1980)

$CH_3CHLiCH=C(CN)OR/H_3O^+$
$(R^1 = H, R^2 = Me)$

JOC 45 395 (1980)

$H_2C=CHCO_2Et$, electrolysis

Rec Trav Chim 93 47 (1974)

$R^2CH=CR^1CO_2CH_3$, Me_3SiCl,
electrolysis

TL 21 5029 (1980)

$R^2CH=CR^1CO_2Et$, SmI_2, ROH, (HMPA)

CC 624 (1986); 920 (1987)
(intramolecular)
TL 27 5763 (1986)

$MeO_2CCH_2CH_2Sn(n\text{-}Bu)_3$, $TiCl_4$
$RO_2CCH_2CH_2TiCl_3$
$MeO_2CCH_2CH_2Br$; Ce, La, Nd or Sm

JOC 50 5907 (1985)
JACS 108 3745 (1986)
CC 475 (1986)

$H_2C=CHCH_2MgBr/H_2O/EtMgBr$, cat
$Cp_2TiCl_2/CO_2/H^+$ $(R^1 = R^2 = H)$

JOMC 160 C8 (1978)

$H_2C=CH(CH_2)_2MgBr/H_2O/BH_3/H_2O_2$,
$NaOH/H_2CrO_4$ $(R^1 = R^2 = H)$

BSCF 3377 (1973)

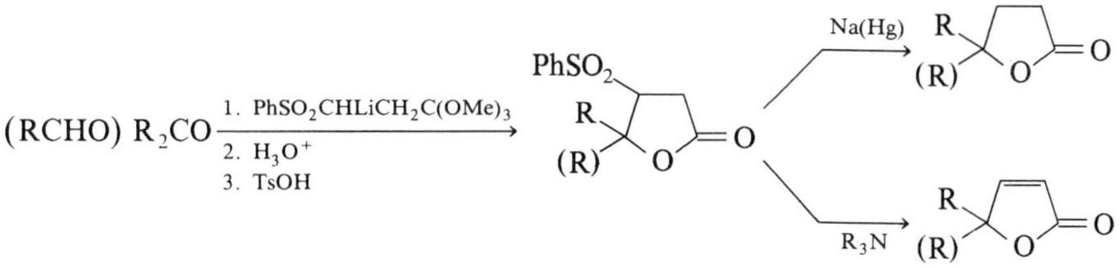

TL 28 2135 (1987)

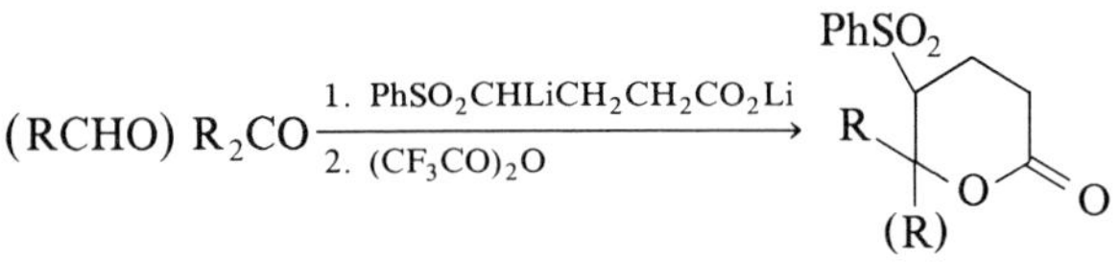

TL 28 4243 (1987)

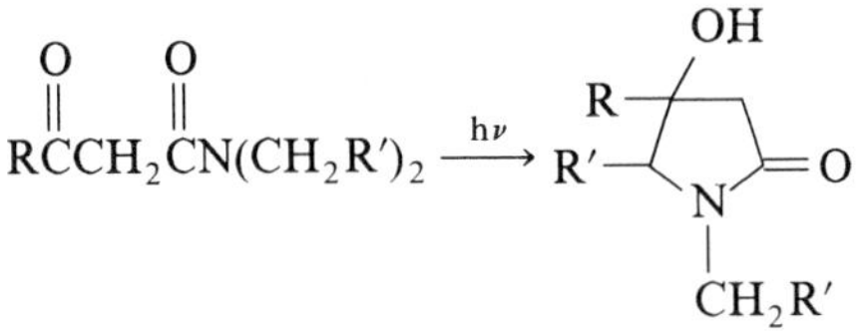

CC 743 (1974)
JOC 43 1005 (1978)

$$\text{RCHO (R}_2\text{CO)} \longrightarrow \quad \overset{(R')}{\underset{(R)}{R}}\text{lactone}$$

H$_2$C=CBrSiMe$_3$, TiCl$_4$/CO, cat   JACS *104* 6879 (1982)
   Ni(CO)$_2$(PPh$_3$)$_2$

n-Bu$_3$SnCH$_2$CCONHR, BF$_3$·OEt$_2$ JOC *51* 1856 (1986)
   or TiCl$_4$/H$^+$ (with CH$_2$= group)

R'CH=CCO$_2$Et, Sn(Al)/H$^+$ CL 541 (1986)
  (with CH$_2$Br group)

H$_2$C=CCO$_2$R, Zn/H$^+$ Angew *9* 457 (1970)
  (with CH$_2$Br group)   J Med Chem *17* 672 (1974);
                              18 812 (1975); *19* 309 (1976);
                              23 1031 (1980); *25* 650 (1985)
                     J Pharm Sci *70* 84 (1981)
                     TL *26* 5693 (1985)

H$_2$C=CCO$_2$SiMe$_3$, Zn graphite   JOC *48* 4108 (1983)
  (with CH$_2$Br group)

H$_2$C=CCO$_2$H, Et$_3$N/Zn/H$^+$ TL *26* 5693 (1985)
  (with CH$_2$Br group)

$$\overset{CH_2Br}{H_2C=CCO_2Et} \xrightarrow[\text{2. } R_2C=NR \text{ (RCH=NR)}]{\text{1. Zn}} \text{pyrrolidinone}$$

Syn Commun *15* 1233 (1985)
TL *28* 59 (1987)

$$\text{cyclohexanone} \xrightarrow[\text{MeC≡CCO}_2\text{Me}]{\text{FpBF}_4(\text{H}_2\text{C=CMe}_2)} \text{product}$$

Fp = CpFe(CO)$_2$

Organomet *1* 397 (1982)

$$\text{cyclohexyl-CHO} \xrightarrow[\text{2. PhSO}_2\text{Cl}]{\text{1. RCHLiCO}_2\text{Li}} \text{β-lactone} \xrightarrow{\text{MgBr}_2} \text{spirolactone}$$

TL *28* 4787 (1987)

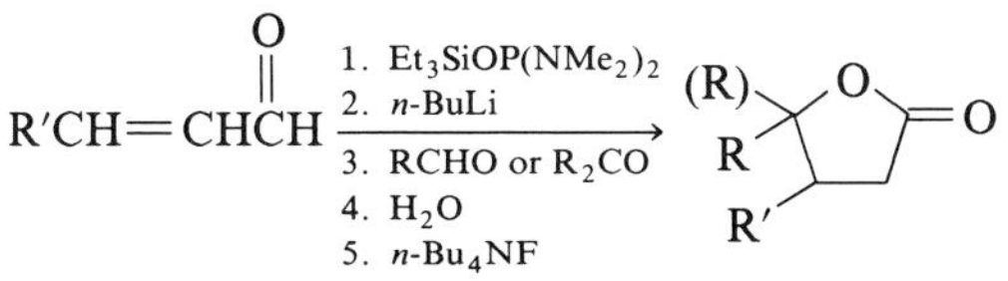

$$R'CH=CHCH \xrightarrow[\substack{1.\ Et_3SiOP(NMe_2)_2 \\ 2.\ n\text{-BuLi} \\ 3.\ RCHO\ or\ R_2CO \\ 4.\ H_2O \\ 5.\ n\text{-Bu}_4NF}]{}$$

JACS *101* 371 (1979)

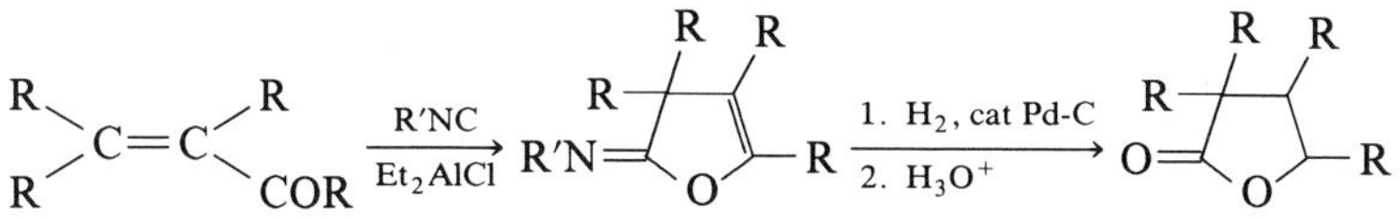

$$\xrightarrow[Et_2AlCl]{R'NC} R'N \xrightarrow[\substack{1.\ H_2,\ cat\ Pd\text{-}C \\ 2.\ H_3O^+}]{}$$

JOC *47* 741 (1982)

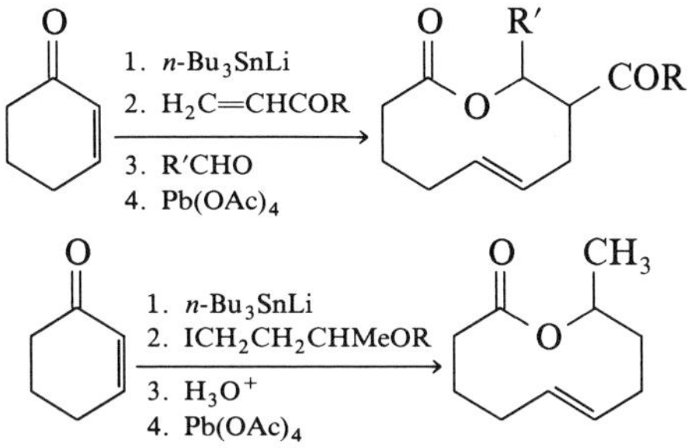

1. *n*-Bu$_3$SnLi
2. H$_2$C=CHCOR
3. R'CHO
4. Pb(OAc)$_4$

1. *n*-Bu$_3$SnLi
2. ICH$_2$CH$_2$CHMeOR
3. H$_3$O$^+$
4. Pb(OAc)$_4$

TL *28* 5071 (1987)

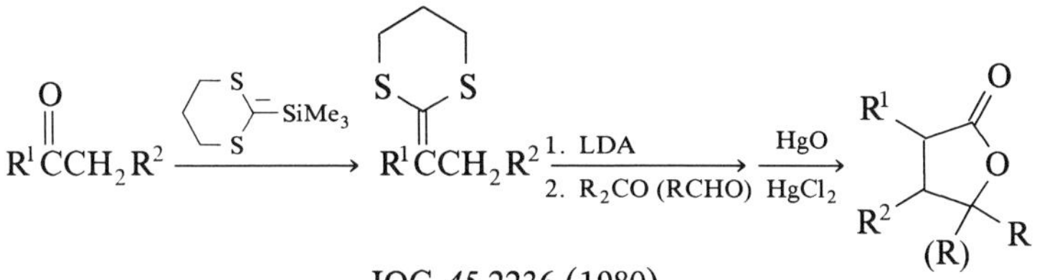

$$R^1CCH_2R^2 \longrightarrow R^1CCH_2R^2 \xrightarrow[\substack{1.\ LDA \\ 2.\ R_2CO\ (RCHO)}]{HgO \atop HgCl_2}$$

JOC *45* 2236 (1980)

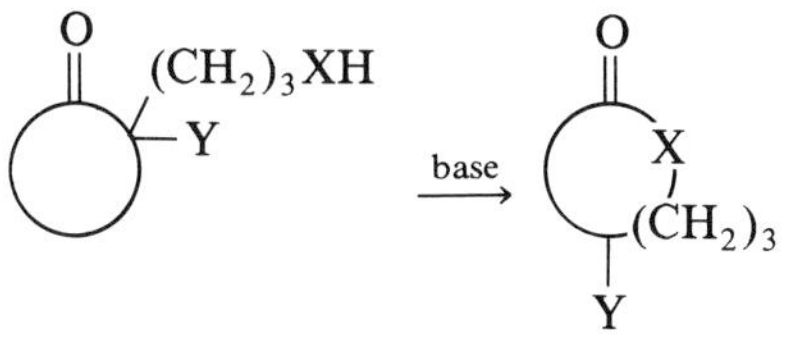

$$\xrightarrow{base}$$

X	Y	
O	NO$_2$	TL *23* 3521 (1982)
		Helv *67* 1713 (1984)
	CN	Helv *68* 2115 (1985)
NH	NO$_2$	Helv *68* 484 (1985)

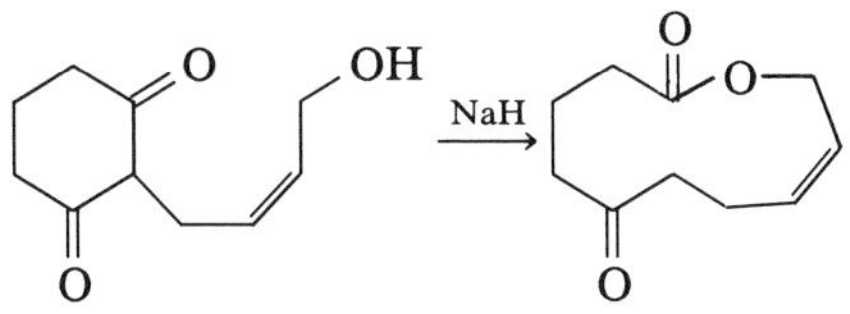

$$\xrightarrow{NaH}$$

Syn 110 (1976)
JOC *52* 5296 (1987)

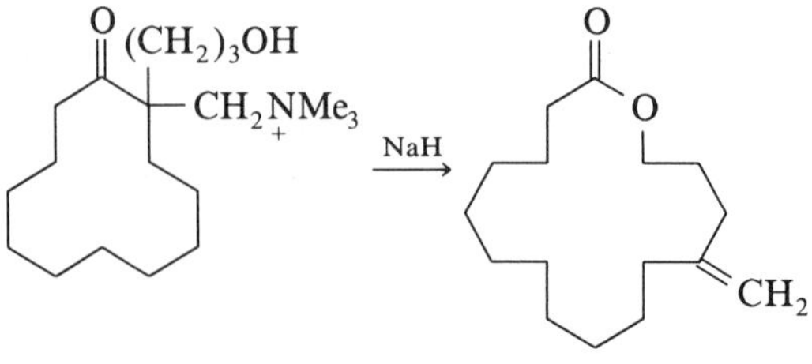

TL *28* 315 (1987)

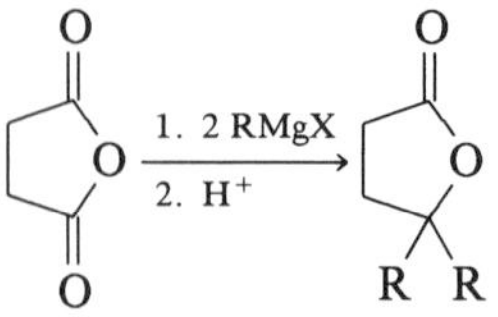

TL *22* 2611 (1981)

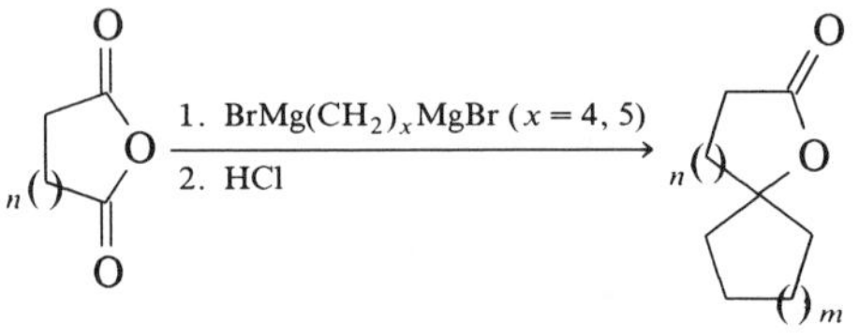

n = 1, 2; *m* = 1, 2

CC 125 (1980)
TL *21* 4167 (1980); *22* 2611 (1981); *28* 4997 (1987)
JOC *46* 3091 (1981); *47* 3953 (1982)

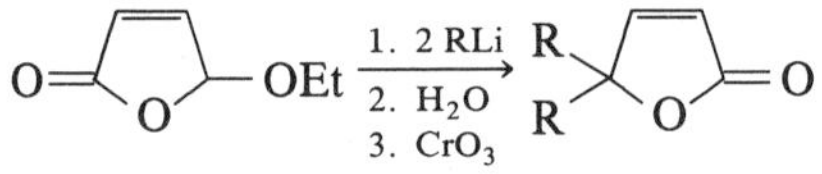

Tetr *38* 2897 (1982)

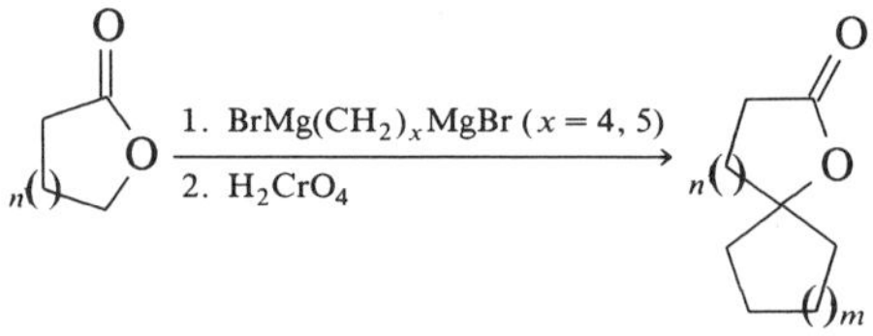

n = 0–3; *m* = 1, 2
JOC *45* 1828 (1980)

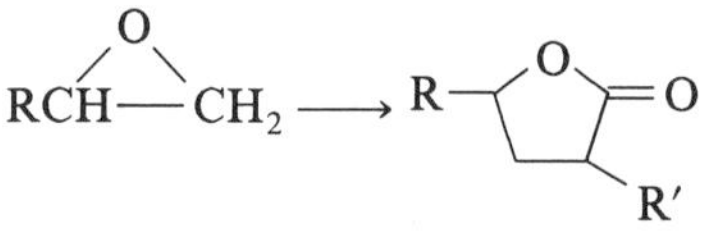

$$RCH{-}CH_2 \longrightarrow R{-}\text{(lactone)}{=}O$$

R′CHLiNCONMe$_2$/acid ion exchange resin

JOC *39* 2783 (1974)

Ber *108* 48 (1975)

See also the reaction of epoxides with alpha metallated carboxylic acids and derivatives under page 867, Section 5.

$$\underset{RCH-CH_2}{\overset{O}{\triangle}} \xrightarrow[\substack{2.\ H^+ \\ 3.\ \text{oxidation}}]{1.\ PhXCHLiCO_2Li} R-\underset{O}{\overset{}{\langle}}\!\!=\!O$$

X	
S	CL 385 (1974)
	BCSJ *50* 242 (1977)
	TL *26* 5627 (1985)
Se	CC 754 (1986)
	TL *28* 1147 (1987)

$$\underset{RCH-CH_2}{\overset{O}{\triangle}} \xrightarrow[\substack{2.\ H_2O}]{1.\ LiCH_2CH=C(OLi)PO(NMe_2)_2} R\underset{}{\overset{O}{\langle}}\!=\!O$$

TL 47 (1976)

$$RCH_2COH \longrightarrow RCH=C(OSiMe_3)_2 \xrightarrow[\substack{2.\ H_3O^+}]{\substack{1.\ Me_3SiO\quad Cl \\ R'CH(CH_2)_nCHSPh \\ (n=0,1),\ ZnBr_2}}$$

1. *m*-ClC$_6$H$_4$CO$_3$H
2. DBU

Raney Ni

TL *23* 5083 (1982)

$$RCH=CH_2 \longrightarrow \underset{R}{\overset{R'}{\langle}}\!\!\overset{O}{\langle}\!=\!O$$

R'CHBrCO$_2$H, (PhCO$_2$)$_2$ CL 415 (1981)

R'CH$_2$CO$_2$H, Mn(OAc)$_3$·2 H$_2$O

JACS *90* 5903, 5905 (1968); *96* 7977 (1974);
    106 5384 (1984) (intramolecular)
BCSJ *49* 1041 (1976)
Org Syn *61* 22 (1983)
TL *26* 3761 (intramolecular), 4921
    (R' = CN, CO$_2$Et) (1985)
JOC *50* 10, 1026
    (R' = CO$_2$H, spiro lactones),
    3143 (1985)
Tetr *42* 3429 (1986) (mechanism)

$$\underset{O}{\langle}\!\!\bigcirc + ICH_2CO_2SnBu_3 \xrightarrow{AIBN} \underset{O\quad O}{\overset{}{\langle\langle}}\!=\!O$$

TL *25* 3939 (1984); *27* 3715, 5927 (1986)
Tetr *41* 4039 (1985)

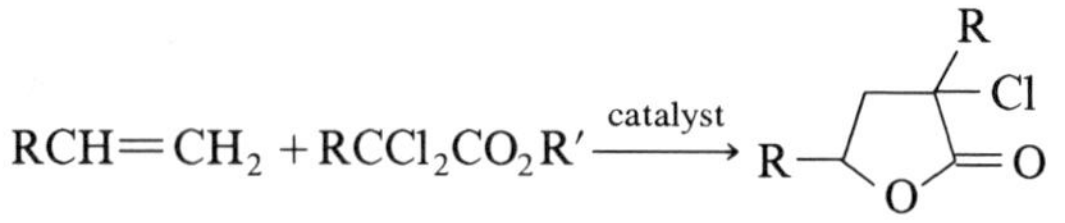

$$RCH=CH_2 + RCCl_2CO_2R' \xrightarrow{\text{catalyst}}$$

R'	Catalyst	
H	$RuCl_2(PPh_3)_3$	CC 363 (1978)
H, alkyl	$RuCl_2(PPh_3)_3$	JOC 51 5501 (1986) (intramolecular)
H, alkyl	$FeCl_2[P(OEt)_3]_3$	JOC 51 5501 (1986) (intramolecular)
Me	$[CpMo(CO)_3]_2$ or $[CpFe(CO)_2]_2$	Tetr 28 29 (1972)
SiMe₃	$RuCl_2(PPh_3)_3$	CC 1011 (1979)

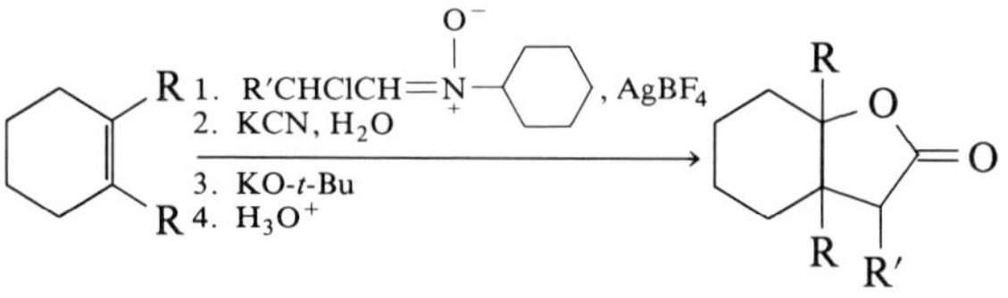

Helv 55 2198 (1972)

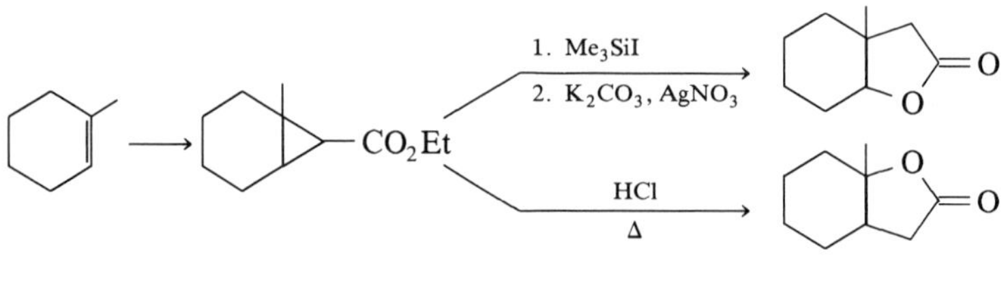

TL 22 4891 (1981)

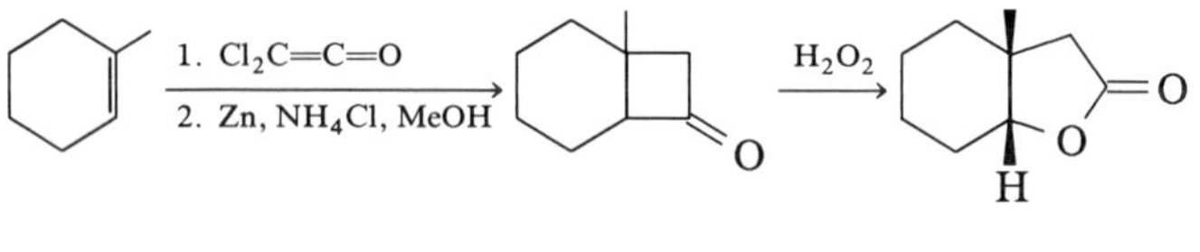

JOC 47 3871 (1982)

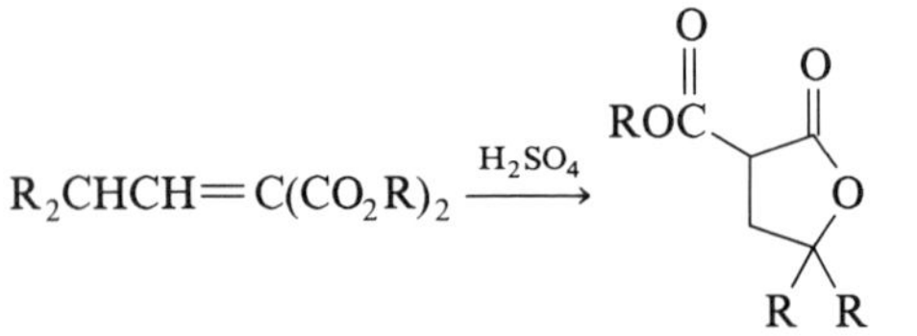

$$R_2CHCH=C(CO_2R)_2 \xrightarrow{H_2SO_4}$$

Syn Commun 11 35 (1981)

$$(CH_3)_2C=CH_2 \xrightarrow[RC\equiv CCO_2Me]{FpBF_4(H_2C=CMe_2)}$$

R = H, Me

Fp = CpFe(CO)$_2$

Organomet *1* 397 (1982)

$$(RCH=NR)\ R_2C=NR \longrightarrow \underset{R_2C-NR}{\overset{R_2C-C=O}{|\qquad|}}$$

Reviews:

Org Rxs *9* 388 (1957)
Syn 327 (1973)
Chem Soc Rev *5* 181 (1976)
Tetr *34* 1731 (1978)

$R_2C=C=O$	Ann *356* 51 (1907); *401* 292 (1913); 760 (1977) Ber *50* 1035 (1917); *90* 2460 (1957) JACS *73* 3172 (1951) Tetr *24* 1011 (1968) CC 302 (1977) Syn 989 (1982)
$R_2CHCOCl$, Et_3N	JACS *73* 1204, 4367 (1951); *109* 1798 (1987) Tetr *23* 4769 (1967); *36* 3427 (1980) BSCF 2450 (1968) JOC *34* 2846 (1969); *38* 3437 (1973); *39* 115, 312, 2877 (1974); *41* 1112 (1976) TL 2633, 3135 (1974) Heterocycles *12* 405 (1979) Syn 933 (1980); 989 (1982)
R_2CHCO_2H, CBr_4, PPh_3	Syn 689 (1976)
R_2CHCO_2H, $ClPO(OR)_2$, Et_3N	Syn Commun *6* 435 (1976) Heterocycles *12* 405 (1979)
R_2CHCO_2H, Cl_2PONMe_2, Et_3N	TL *28* 1945 (1987)

$$R_2C\underset{O=\overset{}{}\diagdown COC \diagup^{O}}{\overset{O\diagdown COC\diagup^{O}}{<}}CPh_2,\ \Delta$$

Helv *6* 291, 304 (1923)

RCOC(N$_2$)R, hν	Ber *89* 2759 (1956) Tetr *23* 957 (1967) BSCF 2450 (1968)
RCOCHN$_2$, Ag$_2$O	Ber *90* 2460 (1957)
RCH$_2$CO$_2$R, LDA	See page 873, Section 2.

$RC\equiv COEt, \Delta$

Rec Trav Chim *78* 551 (1959)
BSCF 2450 (1968)

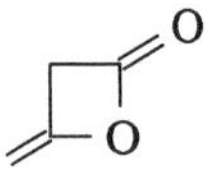

($CH_3COCH\!=\!C\!=\!O$ equivalent)

TL *27* 6241 (1986)

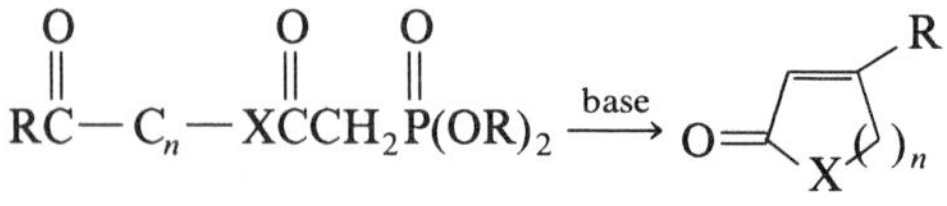

$$\underset{RC}{\overset{O}{\underset{\|}{}}}\!-\!C_n\!-\!\underset{XC}{\overset{O}{\underset{\|}{}}}CH_2\underset{}{\overset{O}{\underset{\|}{}}}P(OR)_2 \xrightarrow{\text{base}} O\!=\!\langle\underset{X}{\overset{R}{}}\rangle_n$$

$\underline{X}$

O

Angew Int *7* 300 (1968)
JACS *100* 7069 (1978)
JOC *44* 4010, 4011 (1979)
TL *28* 2717 (1987)

NH

CC 445 (1970)

9. INTERCONVERSION OF NITRILES, CARBOXYLIC ACIDS AND DERIVATIVES

1. Carboxylic Acids to Acid Halides

$$\underset{\text{RCOH}}{\overset{\overset{\displaystyle O}{\|}}{}} \longrightarrow \underset{\text{RCX}}{\overset{\overset{\displaystyle O}{\|}}{}}$$

X	Reagent(s)	
F, Cl, Br, I	$Me_2C{=}CXNMe_2$	CC 1180 (1979)
F	SeF_4, py	JACS *96* 925 (1974)
	$CF_3CF\overset{\displaystyle O}{\overbrace{}}CF_2$, Et_3N	CL 483 (1977)
	$FClCHCF_2NEt_2$	J Gen Chem USSR *29* 2125 (1959)
	![triazine structure with F], py	Syn 487 (1973)
	$\left[\begin{array}{c}\text{Me pyridinium with F}\end{array}\right]$ OTs^-, Et_3N	CL 303 (1976)
Cl	$SOCl_2$	Org Syn Coll Vol *1* 12, 147 (1941); *3* 169, 490, 547, 555, 712 (1955); *4* 154, 263, 339, 715, 739 (1963) JOC *50* 2719 (1985)
	$SOCl_2$, DMF	Helv *42* 1653 (1959)

$$\underset{\text{RCOH}}{\overset{\overset{\displaystyle O}{\|}}{}} \longrightarrow \underset{\text{RCX}}{\overset{\overset{\displaystyle O}{\|}}{}} \quad (\textit{Continued})$$

X	Reagent(s)	
	SOCl$_2$, py	JACS *75* 2347 (1953)
		JCS 491 (1963)
		Can J Chem *46* 2549 (1968)
	SOCl$_2$, py	JACS *85* 643 (1963)
	(on KO$_2$CR)	
	PCl$_3$	Org Syn Coll Vol *2* 156 (1943)
	PCl$_5$	Org Syn Coll Vol *1* 394 (1941)
		JACS *67* 2239 (1945)

		Ber *96* 1387 (1963)
		Z Chem *22* 126 (1982) (review)
	CCl$_4$, Ph$_3$P	JACS *88* 3440 (1966)
	CCl$_4$, polymer-PPh$_2$	CC 622 (1975)
	polymer-PPh$_2$Cl$_2$	JACS *96* 6469 (1974)
	PhCOCl	JACS *60* 1325 (1938)
	ClCOCOCl	JACS *42* 599 (1920),
		69 2568 (1947)
		JCS 3490 (1953)
		Can J Chem *33* 1515 (1955)
		JOC *29* 843 (1964)
	OH$^-$/ClCOCOCl	JACS *42* 599 (1920); *70* 2427
		(1948); *87* 3958 (1965)
		Helv *37* 45 (1954)
		Can J Chem *33* 1515 (1955)
		TL 3379 (1977)
	Cl$_2$CHOMe	Syn 163 (1975)
		Org Syn *61* 1 (1983)
	XCH$_2$CCl$_2$OEt	Rec Trav Chim *76* 969 (1957)
	(X = H, Cl)	

Ann *694* 78 (1966)

TL 3037 (1979)

Br	PBr$_3$	JCS 1406 (1934)
	PBr$_5$	JACS *42* 599 (1920)
	Ph$_3$PBr$_2$	Ann *693* 132 (1966)
		Syn 684 (1982)
	BrCOCOBr	JACS *42* 599 (1920)

2. Carboxylic Acids to Acid Anhydrides

$$\underset{RCOH}{\overset{O}{\parallel}} \longrightarrow \underset{(RC)_2O}{\overset{O}{\parallel}}$$

Ac$_2$O	JACS *63* 699 (1941); *72* 3294 (1950) Org Syn Coll Vol *1* 91, 410 (1941); *3* 449 (1955); *4* 630, 790 (1963)
H$_2$C=C=O	Org Syn Coll Vol *3* 164 (1955)
H$_2$C=C(OAc)CH$_3$, cat H$_2$SO$_4$	Ind Eng Chem *41* 2920 (1949)
RCOCl, 4-vinylpyridine copolymer	TL *27* 4933 (1986)
CH$_3$COCl	Org Syn Coll Vol *2* 560 (1943)
COCl$_2$, Et$_3$N	Helv *47* 162 (1964) Org Syn Coll Vol *5* 822 (1973)
ClCOCOCl	JACS *42* 599 (1920)
NaOH/ClCOCOCl	JACS *42* 599 (1920); *76* 5803 (1954)
ClCO$_2$Et, Et$_3$N/(Et$_3$NH)O$_2$CR	JOC *28* 1905 (1963)
TsCl, py	JACS *77* 6214 (1955)
SOCl$_2$, py	JACS *70* 2964 (1948) JCS 741 (1952); 2117 (1953)
SOCl$_2$, 4-vinylpyridine copolymer	TL *27* 4937 (1986)
ClSO$_2$NCO, Et$_3$N	Syn 506 (1982)
POCl$_3$	Org Syn Coll Vol *2* 560 (1943)
(PhO)$_2$POCl, Et$_3$N	Syn 218 (1981)
(PhO)(PhNH)POCl, R$_3$N	Syn 218 (1981)
$\left(\underset{O}{\overset{O}{\parallel}} \!\!\! N- \right)_2$ POCl, R$_3$N	Syn 616 (1981)
Cl$_2$, (Me$_2$N)$_3$P/Et$_3$N	BSCF 3034 (1971)
(Me$_2$N)$_3$P, CCl$_4$	BSCF 3034 (1971)
(Me$_2$NC=NMe$_2$)Cl, Et$_3$N, with Cl substituent	BCSJ *56* 3529 (1983)
⬡—N=C=N—⬡	J lipid Res *7* 174 (1966) JOC *50* 2323 (1985)
polymer-CH$_2$N=C=N-*i*-Pr	TL 3281 (1972)
PhCOCH=CHCOPh, *n*-Bu$_3$P	JOC *29* 1385 (1964)

BCSJ *54* 1470 (1981)

HC≡COMe

JCS 1860 (1954)

HC≡COEt

JOC *33* 3808 (1968)

Me₃SiC≡COEt

TL *25* 6027 (1984); *27* 3689 (1986); *28* 3971 (1987)
JOC *51* 4150 (1986)
CC 1474 (1987)

Ber *95* 2073 (1962)

$$\underset{\text{RCOH}}{\overset{O}{||}} + \underset{\text{XCR}'}{\overset{O}{||}} \xrightarrow{R_3N} \underset{\text{RCOCR}'}{\overset{O\ \ O}{||\ \ ||}}$$

Org Syn *26* 1 (1946)
JCS 2117 (1979)
CL 145 (1979)
TL *28* 4711 (1987)

$$\underset{\text{RCOH}}{\overset{O}{||}} + H_2C{=}C{=}O \xrightarrow{\text{cat } H^+} \underset{\text{RCOCCH}_3}{\overset{O\ \ O}{||\ \ ||}}$$

JACS *54* 3427 (1932)

$$\underset{\text{RCOH}}{\overset{O}{||}} + \underset{\text{H}_2C{=}CCH_3}{\overset{OAc}{|}} \longrightarrow \underset{\text{RCOCCH}_3}{\overset{O\ \ O}{||\ \ ||}}$$

Ind Eng Chem *41* 2920 (1949)

3. Carboxylic Acids to Esters

See also page 941, Section 8.

$$\underset{\text{R}'\text{COH}}{\overset{O}{||}} \longrightarrow \underset{\text{R}'\text{COR}}{\overset{O}{||}}$$

Review: Tetr *36* 2409 (1980)

alkene, H₂SO₄

Org Syn Coll Vol *4* 261, 417 (1963)

ROH, lipase (enantioselective)

Proc Natl Acad Sci USA *82* 3192 (1985)
JACS *107* 7072 (1985); *109* 2812 (1987)
TL *26* 1857 (1985); *27* 29 (1986); *28* 1647 (1987)
JOC *52* 3477 (1987)

ROH, HCl

Org Syn Coll Vol *1* 237, 451 (1941);
2 261, 276, 292 (1943)
JOC *52* 4689 (1987)

ROH, H$_2$SO$_4$	Org Syn Coll Vol *1* 241, 254 (1941); *2* 264, 365, 414 (1943); *3* 381 (1955); *4* 329, 532, 635 (1963); *5* 762 (1973) JOC *50* 2128 (1985)
100% H$_2$SO$_4$/ROH	JACS *63* 2431 (1941)
ROH, TsOH	Org Syn Coll Vol *3* 610 (1955)
ROH, graphite bisulfate	JACS *96* 8113 (1974)
ArOH, B(OH)$_3$, H$_2$SO$_4$	TL 3455 (1971)
ROH, BF$_3$·OEt$_2$	TL 4011 (1970) Syn 628 (1972) JACS *108* 468 (1986)
ROH, BF$_3$·2 MeOH (R' = H)	Org Prep Proc Int *14* 177 (1982)
ROH, AlCl$_3$-polymer	TL 1823 (1973)
ROH, cat R'$_2$SnCl$_2$ (R' = Me, Ph)	TL *28* 3713 (1987)

Sn$\left(\begin{array}{c} \text{Me} \\ \bigcirc \end{array} \right)_2$/ROH   CL 683 (1983)

R$_3$N/[pyridine with NO$_2$ and SCl], Ph$_3$P/ROH   CL 979 (1978)

[triazine: Cl, N, N, Cl, N, Cl], Et$_3$N/ROH   TL 3029, 3037 (1979)

NaOH/*n*-BuSO$_3$Cl/Δ (R = *n*-Bu)	JACS *69* 1046 (1947)
ClSO$_2$NCO, Et$_3$N/ROH	Syn 506 (1982)
SO$_2$ClF, Et$_3$N/ROH, Et$_3$N	Syn 790 (1981)
MsCl, Et$_3$N/ROH	TL *23* 3799 (1982) Syn Commun *12* 727 (1982)
TsCl, py/ROH	JACS *77* 6214 (1955)
ROH, Me$_3$SiCl	BCSJ *54* 1267 (1981)
MeOH, CH$_3$COCl (R = Me)	JACS *108* 1039 (1986)
2,4,6-Cl$_3$C$_6$H$_2$COCl, NEt$_3$/ROH, DMAP	BCSJ *52* 1989 (1979) JACS *108* 4645 (1986)
o-BrCH$_2$C$_6$H$_4$COBr, Et$_3$N/ROH, Δ	CL 145 (1979)
o-BrCH$_2$C$_6$H$_4$COBr, Et$_3$N/AgBF$_4$, ROH	TL *28* 4711 (1987)
RO$_2$CCl, py/ROH	TL *28* 1665 (1987)

RO$_2$CCl, Et$_3$N/cat DMAP | TL *24* 3365 (1983)
JOC *50* 560 (1985)

$$\text{CH}_3$$

ROH, H$_2$C=COCOCl, Et$_3$N or py,   TL *28* 1661, 1665 (1987)
(cat DMAP)

(CF$_3$CO)$_2$O/ROH   Chem Rev *55* 787 (1955)
TL 1285 (1964)
Tetr *21* 3531 (1965)
JOC *30* 927 (1965)
Can J Chem *59* 2617 (1981)

(CF$_3$CO)$_2$O, NaF/ROH   Can J Chem *59* 2617 (1981)

ROH, PPh$_3$, CCl$_4$ (on KO$_2$CR′)   JOC *50* 4991 (1985)

ROH (R = 2°, 3° alkyl),   Austral J Chem *35* 517 (1982)
   n-Bu$_3$P·I$_2$, HMPA

Ph$_3$P(OTf)$_2$/ROH   TL 277 (1975)

py/Me$_2$NPOCl$_2$ or PhOPOCl$_2$/ROH   TL 4461 (1978)

PhOPOCl$_2$, DMF/ROH/py   Syn Commun *12* 681 (1982)

n-BuLi/(EtO)$_2$POCl/ROH   JOC *52* 3937 (1987)

$$\left(\substack{O \\ \| \\ O\text{—}\overset{}{C}\text{—}N\text{—}} \right)_2 POCl, \; Et_3N/ROH$$ Syn 547 (1980)

Cl$_3$CPO(OEt)$_2$ (R = Et)   Tetr *38* 1457 (1982)

ROP(OEt)$_2$, EtO$_2$CN=NCO$_2$Et   BCSJ *40* 2380 (1967)
   (R = Et, allyl)

ROP(NEt$_2$)$_2$, EtO$_2$CN=NCO$_2$Et   BCSJ *44* 3427 (1971)

ROH, EtO$_2$CN=NCO$_2$Et, Ph$_3$P   BCSJ *40* 2380 (1967); *44* 3427 (1971)
TL 1619 (1973); *27* 5813 (1986)
   (on pyranose hemiacetals)
Syn 1 (1981) (review)
JOC *52* 3468, 3784 (1987)
JACS *109* 3017 (1987)

ROH, EtO$_2$CN=NCO$_2$Et, PPh$_3$/NaO$_2$CPh   JOC *52* 4235 (1987)

ROH, EtO$_2$CN=NCO$_2$Et, polymer-PPh$_2$ JOC *48* 3598 (1983)

ROH, (PhO)$_2$PO—[benzisoxazole], Et$_3$N   JOC *50* 760 (1985)

(EtO)$_2$PN[benzosultam], Et$_3$N/ROH   CL 123 (1985)

$[(RO)_3\overset{+}{P}CH_3]BF_4^-$ (on $R'CO_2H$ or $R'CO_2Na$)	JOC *49* 4877 (1984)
P_2O_5/ArOH (R = Ar)	Chem Ind 2102 (1964)
ArOH, polyphosphate ester, DMF (R = Ar)	Syn 429 (1979)

ROH, (pyridyl)O)$_2$CO, cat DMAP — CC 473 (1985)

(imidazolyl)NCON(imidazolyl), ROH, Δ or NaOR — Angew Int *1* 351 (1962)

(imidazolyl)NCON(imidazolyl)/t-BuOH, DBU (R = t-Bu) — Syn 833 (1982)

(azepine)OCH$_3$ (R = Me) — Syn Commun *12* 453 (1982)

$R'NHC(OR)=NR'$ (R' = i-Pr, c-C$_6$H$_{11}$) — Syn 561 (1979)

ROH, (c-hexyl)N=C=N(c-hexyl)	Tetr *21* 3531 (1965) Ber *100* 16 (1967) Tetr *37* 233 (1981) (review) Chem Rev *81* 589 (1981) (review)
ArOH, (c-hexyl)N=C=N(c-hexyl)	Compt Rend *256* 1804 (1963) Tetr *21* 3531 (1965)
ROH, (c-hexyl)N=C=N(c-hexyl), TsOH	Acta Chem Scand B *33* 410 (1979)
ROH, (c-hexyl)N=C=N(c-hexyl), (pyridyl)NR'$_2$ [R'$_2$ = Me$_2$, (CH$_2$)$_5$]	TL 4475 (1978); *28* 4019 (1987) Angew Int *17* 522 (1978) Syn Commun *9* 539 (1979) BCSJ *54* 631 (1981) CC 1132 (1982) Org Syn *63* 183 (1984) JACS *108* 3112 (1986)
(c-hexyl)N=C=N(c-hexyl), DMAP, DMAP·HCl/ROH	JOC *50* 2394 (1985)
ROH, (EtN=C=C(CH$_2$)$_3$NMe$_2$)·HCl, DMAP	JOC *47* 1962 (1982)

$(R_3O)BF_4$ (R = Me, Et), i-Pr$_2$NEt	TL 4741 (1971) JOC 44 1149 (1979)
Me$_3$SOH (R = Me)	JOC 44 638 (1979)
K$_2$CO$_3$/R$_3$SX or RSPh$_2$X (X = BF$_4$, ClO$_4$) (phase transfer)	Syn 926 (1980)
Me$_3$SeOH (R = Me)	TL 1787 (1979)
Me$_4$NOH/Δ (R = Me)	JACS 61 1290 (1939)
(R$_3$NCH$_2$Ar)OH, Δ (R = CH$_2$Ar)	Syn 727 (1974) Austral J Chem 28 2065 (1975)

$$\text{ROH, } [Me_2N\overset{\underset{|}{Cl}}{C}=\overset{+}{N}Me_2]Cl^-, \text{ py} \qquad \text{CL 1891 (1982)}$$

ROH, [2-fluoro-1-ethylpyridinium] BF$_4^-$, CsF   CL 391 (1980)

[2-chloro-1-ethylpyridinium] BF$_4^-$,   CL 563 (1980)

[6-(2-pyridyl)-2(1H)-pyridone]/ROH, CsF

ROH, [2-halo-1-alkylpyridinium] Y$^-$, n-Bu$_3$N   BCSJ 50 1863 (1977)

(R' = Me, Et; X = Cl, Br; Y = I, BF$_4$)

2,4,6-(NO$_2$)$_3$C$_6$H$_2$F, Et$_3$N/ROH	BCSJ 51 1866 (1978)
ROH, 2,4,6-(NO$_2$)$_3$C$_6$H$_2$Cl, DMAP	Syn Commun 11 121 (1981)
ROH, 2,4,6-(NO$_2$)$_3$C$_6$H$_2$Cl, py	BCSJ 54 1470 (1981)
NaOH or K$_2$CO$_3$/Me$_2$SO$_4$ (R = Me)	TL 757 (1972)
Et$_2$SO$_4$, DBN (R = Et)	Syn Commun 6 89 (1976)
MeI, KOH, HMPA, EtOH (R = Me)	TL 4063 (1972)
MeI, KOH, DMSO (R = Me)	Tetr 31 2169 (1979)
MeI, CaO, DMSO (R = Me)	Syn 262 (1972)
MeI, Cs$_2$CO$_3$, DMF (R = Me)	JOC 46 4321 (1981)
EtI, anion exchange resin (R = Et)	JOC 44 2425 (1979)

RX, NaHCO$_3$, DMF (R = 1°, 2° alkyl)	Syn 961 (1979)
RX, NaHCO$_3$, [(n-C$_8$H$_{17}$)$_3$NCH$_3$]Cl, CH$_2$Cl$_2$ (R = 1° alkyl, benzyl)	Syn 957 (1979)
CsCO$_3$ or Cs(HCO$_3$)$_2$/RX, DMF (R = 1°, 2°, 3° alkyl; benzylic; X = halogen or sulfonate ester)	Helv 56 1476 (1973) JOC 42 1286 (1977); 46 4321 (1981); 52 3777, 4230 (1987) Syn Commun 13 553 (1983) CL 1555 (1984) TL 26 5257 (1985); 28 1873 (1987)
1°, 2° RI; K$_2$CO$_3$ or KOH; acetone	JOC 44 2425 (1979)
MeI or EtI, K$_2$CO$_3$	JOC 50 2668 (1985)
K$_2$CO$_3$ or NaOH/1°, 2° RX, HMPA	TL 689 (1973) JOC 39 1968 (1974)
Me$_4$NOH/RX, DMF	Syn Commun 2 215 (1972)

(Et$_4$N)N— (O) /RX, DMF   JOC 51 546 (1986)
(R = 1°, 2° alkyl; benzylic; allylic)

1°, 2° RX; DBU (X = Br, I)	BCSJ 51 2401 (1978) Org Prep Proc Int 12 225 (1980)
Cu$_2$O; 1°, 2° RX (X = Br, I); ⟨ ⟩—N≡C	Syn Commun 2 1 (1972)
(MeO)$_3$CH, MeOH, cat TsOH (R = Me)	JOC 50 2607 (1985)
(RO)$_2$CHNMe$_2$	Angew 75 296 (1963) (2 publications) Helv 48 1746 (1965) Ann 821 (1974) Tetr 35 1675 (1979) Syn 135 (1983)

ROH, Et$_3$N, (structure with NO$_2$, O$_2$S, N)   CL 1161 (1980)

(benzothiazole structure) —S—(structure SO$_2$/ROH,   Syn 933 (1982)

Et$_3$N (X = O, S)

CH$_2$N$_2$ (R = Me)	Org Syn Coll Vol 2 165 (1943) JACS 76 4481 (1954) Ber 89 933 (1956) Org Syn 41 16 (1961) J Chem Ed 47 710 (1970)

CH_2N_2 (R = Me)   (*continued*)

TL 1397 (1973)
JOC *47* 578, 1962 (1982);
50 2323, 2607 (1985)

$$\underset{HOC(CH_2)_n COH}{\overset{O\quad\quad O}{\|\quad\quad\|}} \xrightarrow[\text{2. } CH_2N_2]{\text{1. alumina}} \underset{CH_3OC(CH_2)_n COH}{\overset{O\quad\quad O}{\|\quad\quad\|}}$$

JACS *107* 1365 (1985)

$$\underset{RCOH}{\overset{O}{\|}} + HC\equiv CR' \longrightarrow \underset{RCOCH=CHR'}{\overset{O}{\|}}$$

See page 830, Section 5.

4. Carboxylic Acids to Amides

Reviews:

Syn 453 (1972); 549 (1974) (both peptide coupling)

$$\underset{RCOH}{\overset{O}{\|}} + HNR'_2 \longrightarrow \underset{RCNR'_2}{\overset{O}{\|}}$$

Δ	JACS *53* 1879 (1931); *59* 401 (1937); *71* 2215 (1949)
	Org Syn Coll Vol *1* 3, 83, 111 (1941); *3* 590 (1955)
	JOC *8* 473 (1943)
$NaBH_4$	JOC *40* 3453 (1975)
	Syn 766 (1978)
$BH_3 \cdot Me_3N$	Syn 1013 (1983)
$BH_3 \cdot n\text{-}Bu_3N$	Tetr *26* 1539 (1970)
$HB(OR)_2$ (R = *i*-Pr, *t*-Am)	Tetr *26* 1539 (1970)
benzodioxaborole	JOC *43* 4393 (1978)
$ClB(OMe)_2$	Tetr *26* 1539 (1970)
BX_3 (X = $n\text{-}C_8H_{17}$, OMe)	Tetr *26* 1539 (1970)
$Sn(\text{methylcyclopentadienyl})_2$	CL 683 (1983)
$TiCl_4$	Can J Chem *48* 983 (1970)
$MeSO_2Cl$, $(n\text{-}Bu_4N)HSO_4$, H_2O, CCl_4	CL 443 (1981) (*β*-lactams)
TsCl, py	JACS *77* 6214 (1955)

SO_2ClF, Et_3N (1° amines)	Syn 661 (1980)
$ClSO_2NCO$, Et_3N	Syn 506 (1982)

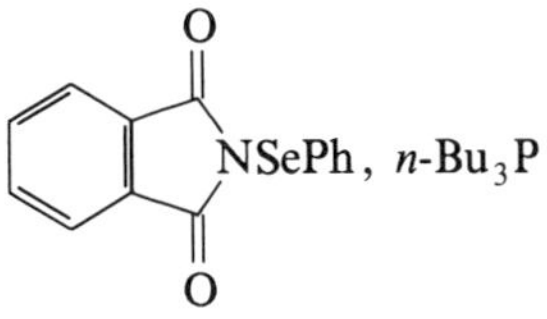

NSePh, n-Bu$_3$P — JOC *46* 1215 (1981)

P_2I_4	CL 449 (1983)
Ph_3P, CCl_4 or $BrCCl_3$	JOC *36* 1305 (1971)
R_3P, CX_4 (R = n-Bu, Ph, NR_2; X = Cl, Br)	TL 3595 (1971)
$P(NMe_2)_3$, $COCl_2$	Ber *116* 2037 (1983)
$P(NMe_2)_3$, CCl_4	BSCF 3034 (1971)
o-$NO_2C_6H_4SCN$, n-Bu$_3$P	JOC *44* 2945 (1979)
Ph_3P, $(ArS)_2$, metal reagents	BCSJ *44* 1373 (1971)

Ph_3P, — TL 1901 (1970)

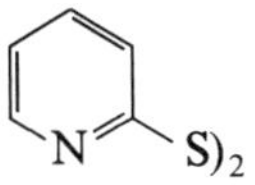

Ph_3P, , Et_3N — Syn 287 (1981)

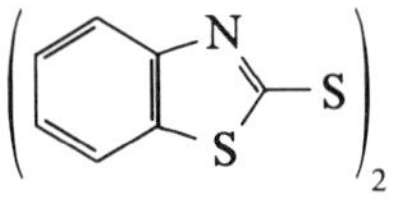

$Ph_3P(OTf)_2$	TL 277 (1975)

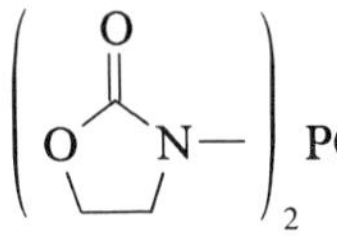

POCl, Et_3N — Syn 547 (1980); 413 (1984) / JACS *107* 1421, 4342 (1985)

(PhO)(PhNH)POCl, Et_3N	Syn 288 (1982) / TL *28* 4875 (1987)
Ph_2POCl	Syn 385 (1980)

POCl — TL *25* 4825 (1984)

(n-Bu$_4$N)OH/R'_2NH, — CL 1367 (1981)

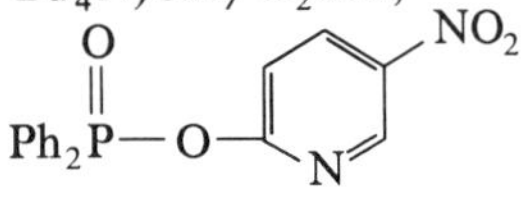

KOH, n-Bu$_4$NX (X = HSO_4, Br)/ (o-$NO_2C_6H_4O)_2POPh$, R'_2NH	CL 285 (1981)

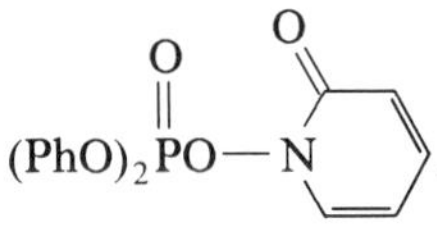

, Et_3N — CC 719 (1986)

(PhO)$_2$PO—[benzisoxazole], Et$_3$N	JOC *50* 760 (1985)
(PhO)$_2$PON[succinimide], Et$_3$N	TL *21* 1467 (1980)
(PhO)$_2$PON[norbornene imide], Et$_3$N	CC 1029 (1980)
(EtO)$_2$PN[benzotriazole], Et$_3$N	TL *26* 1341 (1985)
(EtO)$_2$PN[benzisothiazolone dioxide], Et$_3$N	CL 123 (1985)
(PhO)$_2$PN[oxazolone]	TL *22* 1257 (1981)
(PhO)$_2$PON$_3$, Et$_3$N	JACS *94* 6203 (1972)
(EtO)$_2$POCN, Et$_3$N	TL 1595 (1973)
Me$_3$SiC≡COEt, cat HgO	JOC *51* 4150 (1986)
CF$_3$CF—[epoxide]—CF$_2$, Et$_3$N	CL 483 (1977)
ROCOCl (R = Me, *i*-Bu)	Syn 385 (1980) JOC *50* 2323 (1985) TL *28* 939 (1987)
MeO$_2$CCl, [piperidine]NOH	TL 2697 (1971)
EtO$_2$CCl, Et$_3$N	Ann *673* 186 (1964) JCS C 3540 (1971) JCS Perkin I 2909 (1982)
t-BuCOCl, Et$_3$N	Tetr *11* 39 (1960)
t-BuCOCl, DMAP, py	TL *28* 1131 (1987)
N[imidazole]NCON[imidazole]N	Angew Int *1* 351 (1962)

1-cyclohexyl-3-(2-morpholinoethyl)-
  carbodiimide metho-4-*p*-toluenesulfonate

TL *28* 3163 (1987)

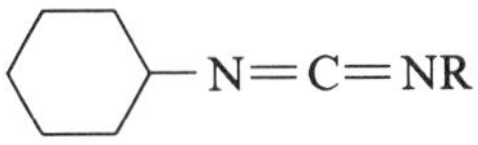

—N=C=NR

JOC *21* 439 (1956)

—N=C=N—

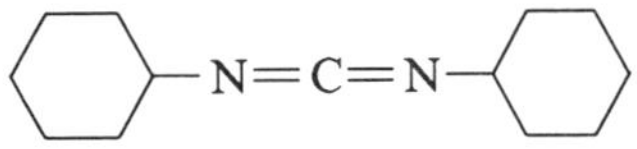

Chem Ind 1087 (1955)
JACS *77* 1067 (1955); *78* 1367 (1956)
Ber *92* 2813 (1959)
Tetr *37* 233 (1981) (review)
Chem Rev *81* 589 (1981) (review)
JOC *52* 5717 (1987)

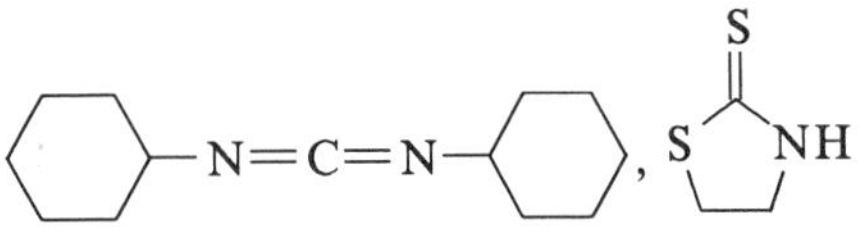

TL *21* 841 (1980)
Pure Appl Chem *53* 1141 (1981)

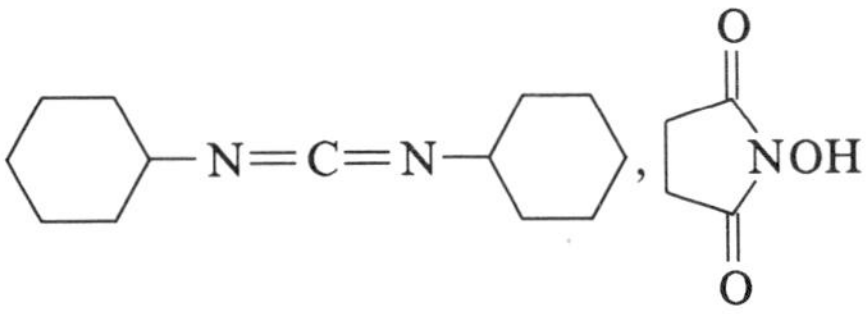

JCS Perkin I 2909 (1982)

PhCCl=NNMePh

JACS *102* 4537 (1980)

2,4,6-(NO$_2$)$_3$C$_6$H$_2$F

CL 647 (1977)
BCSJ *51* 1866 (1978)

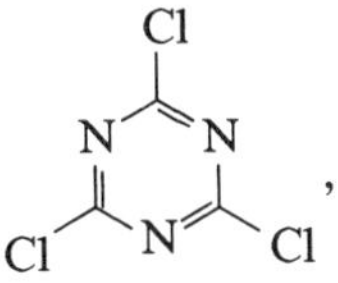

, Et$_3$N

TL 3037 (1979)

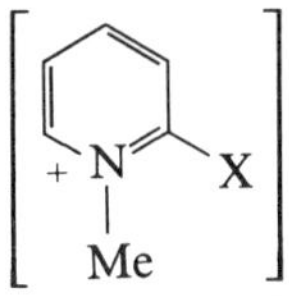

I$^-$ (X = Cl, Br, I), *n*-Bu$_3$N

JCS 4650 (1964)
CL 1163 (1975)

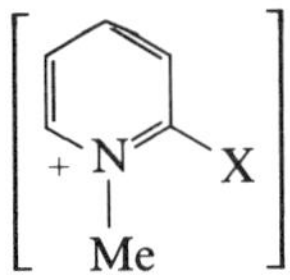

OTs$^-$ (X = F, Cl), *n*-Bu$_3$N

CL 57 (1976)

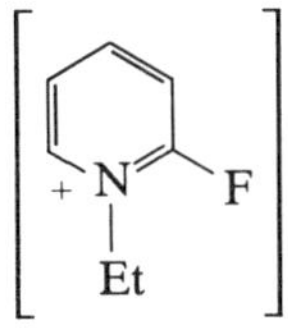

BF$_4^-$, CsF

CL 391 (1980)

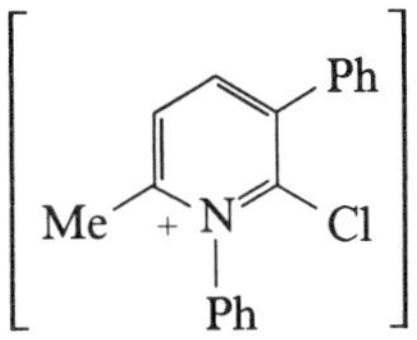

BF$_4^-$,

CL 1551 (1981)

proton sponge, cat DMAP

Syn 933 (1982)

$$\underset{\text{RCOH}}{\overset{\overset{\displaystyle O}{\|}}{}} \longrightarrow \underset{\text{RCNR'R''}}{\overset{\overset{\displaystyle O}{\|}}{}}$$

HCONH$_2$, Δ (R' = R'' = H)	JACS *71* 2215 (1949)
NH$_2$CONH$_2$, Δ (R' = R'' = H)	JACS *71* 2215 (1949)
	Org Syn Coll Vol *4* 513 (1963)
CH$_3$CONHCONH$_2$, Δ (R' = R'' = H)	JACS *71* 2215 (1949)
R'NCO (R'' = H)	TL *27* 1251 (1986)
R'R''NPOCl$_2$ (R' = R'' = Me or Et)	Syn Commun *9* 31 (1979)
(Me$_2$N)$_3$P·I$_2$, HMPA (R' = R'' = Me)	Austral J Chem *35* 517 (1982)
R'N$_3$, Ph$_2$PX (X = Ph, OEt; R'' = H)	JOC *50* 2601 (1985)

$$\underset{\text{HCONa}}{\overset{\overset{\displaystyle O}{\|}}{}} \xrightarrow[\text{2. RNH}_2 \text{ or (RNH}_3)\text{Cl}]{\text{1. } t\text{-BuCOCl, 18-crown-6}} \underset{\text{HCNHR}}{\overset{\overset{\displaystyle O}{\|}}{}}$$

Recl J R Neth Chem Soc *101* 460 (1982)

5. Carboxylic Acids to Nitriles

$$\underset{\text{RCOH}}{\overset{\overset{\displaystyle O}{\|}}{}} \longrightarrow \text{RCN}$$

NH$_3$, silica gel, Δ	Org Syn Coll Vol *4* 62 (1963)
NH$_3$, ethyl polyphosphate, Δ	Syn 142 (1983)
NH$_2$CONH$_2$/Δ	Org Syn Coll Vol *3* 768 (1955)
ArCN, Δ (R = aryl)	JOC *23* 1350 (1958)
NCCH$_2$CH$_2$CHMeCN, cat H$_3$PO$_4$, Δ	JOC *36* 3050 (1971)
CH$_3$SO$_2$Cl, py/NH$_3$/CH$_3$SO$_2$Cl	Org Prep Proc Int *14* 396 (1982)
ClSO$_2$NCO/DMF	Ber *100* 2719 (1967)
	Tetr *24* 1063 (1968)
	Org Syn *50* 18 (1970)
	Syn Commun *12* 25 (1982)
ClSO$_2$NCO/Et$_3$N	TL 1631 (1968)

PhSO$_2$NH$_2$, Δ   JCS 763 (1946)

TsNH$_2$, TsOH, Δ   JCS 763 (1946)

TsNH$_2$, PCl$_5$ Org Syn Coll Vol 3 646 (1955)

6. Acid Halides to Other Acid Halides

$$\underset{\text{RCCl}}{\overset{\overset{\displaystyle O}{\|}}{}} \longrightarrow \underset{\text{RCX}}{\overset{\overset{\displaystyle O}{\|}}{}}$$

X   Reagent(s)

F   KF, CH$_3$CN   CL 761 (1981)
                         KF, cat 18-crown-6   JACS 96 2250 (1974)
                                                          JOC 44 1016 (1979)
                         KF, cat (Et$_3$NCH$_2$Ph)Cl   Syn Commun 12 513 (1982)
                         KF, CF$_3$COCF$_3$ JOC 31 2316 (1966)
                         KF, CaF$_2$, sulpholane   CC 791 (1986)
                         n-Bu$_4$NF   JOC 49 3217 (1984)
                         ZnF$_2$, py   BCSJ 51 1267 (1978)
                         KHF$_2$ Ber 89 862 (1956)

$$\left[\underset{}{\left\langle\bigcirc\right\rangle} \text{NH}\right](\text{HF})_x\text{F}$$ JOC 44 3872 (1979)

                         Na$_2$SiF$_6$, Δ   Angew 71 274 (1959)
                         KSO$_2$F   Ber 91 2553 (1958)
                         R$_2$NSF$_3$ Syn 801 (1975)

Br   HBr   Ber 46 1417 (1913)
                         Me$_3$SiBr   Syn 216 (1981)

I   HI   Ber 46 1417 (1913)
                                                          JACS 55 374 (1933)
                         NaI, CH$_3$CN   Syn 715 (1981); 237 (1982)
                         Me$_3$SiI   Syn 216 (1981)

$$\underset{\text{XCCl}}{\overset{\overset{\displaystyle O}{\|}}{}} \xrightarrow[\text{18-crown-6}]{\text{KF}} \underset{\text{XCF}}{\overset{\overset{\displaystyle O}{\|}}{}}$$

X = OR, NR$_2$

JOC 44 1016 (1979)

7. Acid Halides to Acid Anhydrides

$$\underset{RCX}{\overset{\overset{\textstyle O}{\|}}{}} \longrightarrow \underset{(RC)_2O}{\overset{\overset{\textstyle O}{\|}}{}}$$

H_2O, py	JACS *71* 2242 (1949)
H_2O, py, $NaHCO_3$ (R = Ar)	Angew Int *7* 465 (1968)
H_2O, Cl_3CCOCF_3/py	JOC *51* 3390 (1986)
py/H_2O	Org Syn Coll Vol *3* 28 (1955)
Ac_2O	Helv *9* 177 (1926)
	J Am Oil Chem Soc *31* 151 (1954)
	JOC *24* 388 (1959)

$$\underset{RCX}{\overset{\overset{\textstyle O}{\|}}{}} + \underset{HOCR'}{\overset{\overset{\textstyle O}{\|}}{}} \xrightarrow{R_3N} \underset{RCOCR'}{\overset{\overset{\textstyle O\ \ O}{\|\ \|}}{}}$$

Org Syn *26* 1 (1946)
JCS 2117 (1953)
Org Syn Coll Vol *3* 28 (1955)
CL 145 (1979)
TL *27* 4933 (1986); *28* 4711 (1987)

$$\underset{RCCl}{\overset{\overset{\textstyle O}{\|}}{}} + \underset{MOCR'}{\overset{\overset{\textstyle O}{\|}}{}} \longrightarrow \underset{RCOCR'}{\overset{\overset{\textstyle O\ \ O}{\|\ \|}}{}}$$

$\underline{M}$	
Na	JACS *73* 4911 (1951)
	JCS 755 (1964)
	Rec Trav Chim *85* 627 (1966)
	Org Syn *50* 1 (1970)
K	JACS *61* 684 (1939); *69* 2231 (1947)
Tl	JACS *90* 2422 (1968)
Ag	JACS *75* 232 (1953)

8. Acid Halides to Esters

$$\underset{R'CCl}{\overset{\overset{\textstyle O}{\|}}{}} \longrightarrow \underset{R'COR}{\overset{\overset{\textstyle O}{\|}}{}}$$

ROH	Org Syn Coll Vol *5* 1 (1973)
	JACS *108* 468 (1986)
ROH, Et_3N	Org Syn Coll Vol *5* 258 (1973)
ROH, py	JACS *73* 5487 (1951); *109* 6726, 7838 (1987)
	JOC *21* 1362 (1956)

ROH, PhNMe$_2$

JACS *54* 2088 (1932); *65* 986 (1943)
Org Syn Coll Vol *3* 142 (1955); *5* 171 (1973)

ROH, N◯—NR$_2$

Angew Int *17* 569 (1978)
Tetr *34* 2069 (1978)
JACS *107* 3279 (1985); *109* 6726 (1987)
JOC *50* 2390 (1985)

ROH, Mg

Org Syn Coll Vol *3* 144 (1955)

LiOR

JOC *35* 1198 (1970); *52* 2927 (1987)
Org Syn *51* 96 (1974)

TlS-*t*-Bu/ROH, Hg(O$_2$CCF$_3$)$_2$

JACS *97* 3515 (1975)

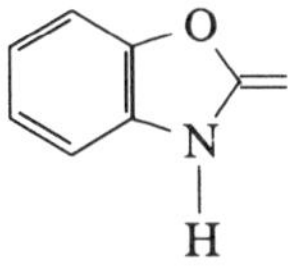

◯=S, Et$_3$N/ROH

Syn 991 (1981)

◯(n =1, 2), NaI [R = I(CH$_2$)$_{n+3}$]

TL *23* 681 (1982)

$$\underset{RCX}{} \xrightarrow{2\ Me_2\overset{+}{S}O\overline{C}H_2} RCO\overline{C}H\overset{+}{S}OMe_2 \xrightarrow[R'OH]{h\nu} RCH_2COR'$$

X = halogen, OPh

JACS *86* 1640 (1964)

9. Acid Halides to Amides

$$\underset{RCCl}{} \longrightarrow \underset{RCNR'_2}{}$$

NH$_3$ (R' = H)

JACS *71* 2215 (1949)
Org Syn Coll Vol *3* 490 (1955)

R'$_2$NH

JACS *71* 2215 (1949); *108* 1039 (1986)
Org Syn Coll Vol *4* 339 (1963); *5* 387 (1973)

R'$_2$NH, NaOH

Org Syn Coll Vol *1* 99 (1941)

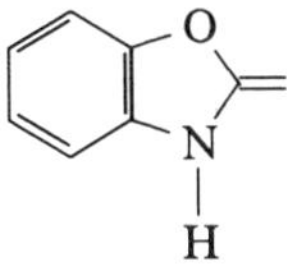

◯=S, Et$_3$N/R'$_2$NH

Syn 991 (1981)

polymer-ArOH, py/R'$_2$NH

Syn Commun *12* 709 (1982)

$$\underset{RCF}{} + Me_3SiNR_2 \xrightarrow{cat\ n\text{-}Bu_4NF} \underset{RCNR_2}{}$$

TL *28* 5099 (1987)

10. Acid Halides to Nitriles

$$\underset{\text{RCCl}}{\overset{\displaystyle O \atop \|}{}} \longrightarrow \text{RCN}$$

H_2NCONH_2, Δ	Org Syn Coll Vol *3* 768 (1955)
H_2NCONH_2, H_2NSO_3H, Δ	Chimia *25* 94 (1971)
$H_2NSO_2NH_2$, Δ	TL *23* 1505 (1982)
$TsNH_2$, PCl_5, Δ	Org Syn Coll Vol *3* 646 (1955)
$(PNCl_2)_n$ ($n = 3$, 4), Δ	TL 3825 (1973)

11. Acid Anhydrides to Acid Halides

$$\underset{(RC)_2O}{\overset{\displaystyle O \atop \|}{}} \longrightarrow \underset{RCCl}{\overset{\displaystyle O \atop \|}{}}$$

PCl_5 Org Syn Coll Vol *2* 528 (1943)

benzo-PCl_3 structure   Ber *96* 1387 (1963)
Z Chem *22* 126 (1982) (review)

12. Acid Anhydrides to Esters

$$\underset{(RC)_2O}{\overset{\displaystyle O \atop \|}{}} + R'OH \longrightarrow \underset{RCOR'}{\overset{\displaystyle O \atop \|}{}}$$

Δ	Org Syn Coll Vol *3* 169 (1955)
TsOH	Org Syn Coll Vol *4* 304 (1963)
TsOH, RCO_2H	JACS *75* 3489 (1953)
Zn, Mg or $ZnCl_2$	JACS *54* 2088 (1932)
$ZnCl_2$	Org Syn Coll Vol *3* 141 (1955)
cat $COCl_2$	CC 114 (1987)
Me_3SiCl	TL *24* 1189 (1983) JOC *52* 5034 (1987)
NaOAc	Org Syn Coll Vol *1* 285 (1941)
py	JACS *109* 7838 (1987)
pyridine–NR_2 structure	Angew Int *8* 981 (1969); *17* 569 (1978) Syn 619 (1972) Tetr *34* 2069 (1978) JACS *108* 4603 (1986) JOC *52* 3784, 4495 (1987)

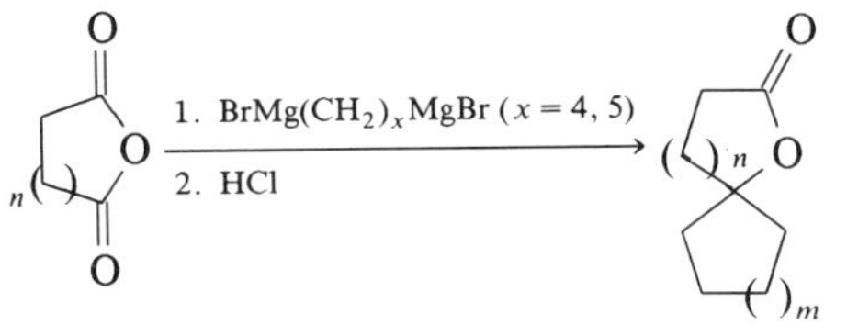

$$n = 1, 2; \; m = 1, 2$$

CC 125 (1980)
TL *21* 4167 (1980); *22* 2611 (1981); *28* 4997 (1987)
JOC *46* 309 (1981); *47* 3953 (1982)

13. Acid Anhydrides to Amides

$$(R\overset{O}{\overset{\|}{C}})_2O + HNR'_2 \longrightarrow R\overset{O}{\overset{\|}{C}}NR'_2$$

Δ	Org Syn Coll Vol *2* 11 (1943)
py	Org Syn Coll Vol *4* 5 (1963)
RCO_2H, H_2O	Org Syn Coll Vol *3* 661 (1955)
polymer-ArOH, py	Syn Commun *12* 709 (1982)
cat $CoCl_2$	CC 114 (1987)

14. Esters to Carboxylic Acids

$$R'\overset{O}{\overset{\|}{C}}OR \longrightarrow R'\overset{O}{\overset{\|}{C}}OH$$

Reviews:

Org Rxs *24* 187 (1976) (S_N2-type)
Tetr *36* 2409 (1980)
Ann Rep Med Chem *19* 263 (1984) (enzymes)
Angew Int *24* 617 (1985) (enzymes)

cholesterol esterase (enantioselective)	JOC *52* 1765, 2608 (1987)
pig liver esterase (enantioselective)	JACS *104* 7294 (1982); *106* 3695 (1984); *108* 4603 (1986)
	Helv *66* 2501 (1983)
	Angew Int *23* 64, 66, 67 (1984)
	CC 236 (1984); 808, 1545 (allenic esters) (1986); 1041 (1987) (diester)
	TL *26* 2073, 4957, 5831 (1985); *27* 2543, 4639, (1986); *28* 781, 1887, 2767, 3103, 4661 (1987)
	JOC *51* 1003, 2047 (1986); *52* 4565 (1987)
	Bioorg Chem *14* 176 (1986)
	Chimia *40* 314 (1986)
	Ann 687 (1986)

protease subtilisin (enantioselective)

JACS *108* 2767 (1986)
TL *28* 5169 (1987)

various cellulases (enantioselective)

JOC *51* 1003 (1986)

porcine pancreatic lipase
  (enantioselective)

JACS *106* 3695, 7250 (epoxy esters) (1984);
  109 2845 (1987)
CC 1563 (1985); 1298 (1986); 1080 (1987)
TL *26* 2073 (1985); *27* 5707 (1986); *28* 531, 1973,
  2767, 2989, 3471, 4661, 4669 (1987)
JOC *52* 1765 (1987)

Candida cylindracea lipase
  (enantioselective)

JOC *51* 1003 (1986); *52* 1765 (1987)
JACS *108* 6421 (1986); *109* 2845 (1987)
TL *27* 2843 (1986); *28* 2767, 2989, 4661 (1987)

Mucor meihei lipase
  (enantioselective)

CC 1298 (1986)
TL *27* 5203 (1986)
JACS *109* 2845 (1987)

Pseudomonas fluorescens lipase
  (enantioselective)

CC 838 (1987)

lipase-MY (enantioselective)

JOC *52* 3211 (1987)
  (trifluoromethyl carbinol esters)

lipase Amano A or Amano A-6
  (enantioselective)

TL *27* 5241 (1986)

lipase Amano P (enantioselective)

JOC *52* 5079 (1987)

α-chymotrypsin (enantioselective)

JACS *83* 4228 (1961); *90* 3495 (1968);
  108 2767 (1986)
JOC *49* 3657 (1984); *51* 1003 (1986)
TL *26* 4957 (1985); *28* 4661, 4935 (1987)

baker's yeast (enantioselective)

TL *27* 4293 (1986)

pancreatin (enantioselective)

Agric Biol Chem *46* 1593 (1982)
JOC *52* 5079 (1987)

steapsin (enantioselective)

Agric Biol Chem *46* 1593 (1982)
JOC *52* 5079 (1987)

Trichoderma S (enantioselective)

Agric Biol Chem *37* 1687, 1691, 1695 (1973);
  38 1961, 1965 (1974); *39* 89 (1975)

Saccharomyces sp. (enantioselective)

Biochim Biophys Acta *316* 363 (1973)

Bacillus subtilis var. *Niger*
  (enantioselective)

Agric Biol Chem *37* 1687, 1691, 1695 (1973);
  38 1961, 1965 (1974); *39* 89 (1975)
Tetr *36* 91 (1980)
JACS *108* 4603 (1986)

bacterium *Corynebacterium equi*
  (IFO 3730) (enantioselective)

TL *25* 5235 (1984); *28* 1303 (1987)
  (α-benzyloxy esters)
CL 217 (1986) (arylsulfinyl esters)

mold *Rhizopus nigricans*
  (enantioselective)

JOC *48* 3017 (1983); *49* 675 (1984);
  52 2400 (1987)

other microbiological reagents (enantioselective)	Agric Biol Chem *37* 1687 (1973) TL *25* 5235 (1984); *27* 5203 (1986); *28* 2767 (1987) JACS *108* 4603 (1986); *109* 2845 (1987) CC 1298 (1986) JOC *51* 1003 (1986)
H_2, cat Pd (R = benzylic)	Org Rxs *7* 263 (1953)
cat Pd-C, 1,4-cyclohexadiene, CH_3OH (R = CH_2Ph)	TL *28* 3225 (1987)
$(NH_4)O_2CH$, cat Pd-C (R = CH_2Ph)	Syn 929 (1980)
$KO_2CCHEt(CH_2)_3CH_3$, cat $Pd(PPh_3)_4$-PPh_3 (R = allylic)	JOC *47* 587 (1982)
cat $Pd(PPh_3)_4$, R_2NH (R = allylic)	Angew Int *23* 71 (1984) TL *28* 4371 (1987)
Li, NH_3 (R = Me)	JACS *80* 217 (1958)
Zn (R = CH_2CCl_3)	Syn 457 (1976) JOC *52* 1790 (1987)
Zn, H_2O, HOAc (R = CH_2CCl_3)	JACS *88* 852 (1966)
Zn, MeOH (R = aryl)	TL *22* 335 (1981)
$LiOH/H_3O^+$	JACS *108* 4603 (1986)
$NaOH/H_3O^+$	JACS *58* 1014 (1936); *109* 6726 (1987) Org Syn Coll Vol *1* 379, 391 (1941); *3* 526, 531, 652 (1955); *4* 582, 616, 628 (1963) JOC *50* 2128 (1985)
KOH/H_3O^+	Org Syn Coll Vol *3* 267 (1955); *4* 608, 633 (1963) JOC *52* 4647 (1987)
KOH, Al_2O_3/H_3O^+	Syn Commun *11* 413 (1981)
$Ba(OH)_2/H_3O^+$	Org Syn Coll Vol *4* 635 (1963)
NaO_2, DMSO (R = Me)	JOC *44* 4727 (1979)
KO-*t*-Bu, DMSO (R = Me)	TL 2969 (1964)
2 KO-*t*-Bu, 1 H_2O (R = 1°, 3° alkyl)	JOC *42* 918 (1977) TL *28* 1131 (1987)
Na_2S, H_2O (R = p-$NO_2C_6H_4CH_2$)	JOC *43* 1243 (1982)
Na_2S, CH_3CN [R = $(CH_2)_nCl$; n = 4, 5]	Syn Commun *4* 307 (1974)
KSCN, DMF (R = Me, $PhCH_2$)	JOC *37* 744 (1972) Syn Commun *5* 305 (1975)
EtSH, AlX_3 (X = Cl, Br) (R = Me, $PhCH_2$)	TL 5211 (1978) JOC *46* 1991 (1981)
Me_2S or ⬠S, AlX_3 (X = Cl, Br) (R = Me, Et, *n*-Pr, *i*-Pr)	JOC *46* 1991 (1981); *50* 3957 (1985)

LiSMe (R = Me or lactone)	TL 3859 (1977) JACS *102* 3904 (1980)
NaSMe (R = 9-anthrylmethyl, lactone)	Ber *83* 265 (1950) JACS *96* 590 (1974)
NaSEt, DMF (R = Me)	Austral J Chem *25* 1731 (1972)
LiS-*n*-Pr, HMPA (R = Me)	TL 4459 (1970) JACS *106* 5304 (1984) JOC *52* 4634 (1987)
NaS-*n*-Pr, DMF (R = Me, Et, PhCH$_2$)	JOC *27* 739 (1962)
NaSPh, DMF (R = Me, benzylic, phenacyl)	JOC *29* 2006 (1964)
(NaSCH$_2$)$_2$, CH$_3$CN [R = CH$_2$CH$_2$Cl(Br)]	Syn 510 (1975)
Na$_2$CS$_3$, CH$_3$CN [R = CH$_2$CH$_2$Cl(Br)]	Syn 715 (1974)
KSeO$_3$K (lactone)	JOC *31* 1202 (1966)
LiSeCH$_3$, DMF	Syn Commun *13* 617 (1983)
NaSeCH$_3$, DMF	TL *28* 4225 (1987)
NaSeCH$_2$Ph (lactones)	Arkiv Kemi *24* 415, 573 (1965)
NaSePh (R = 1°, 2° alkyl; benzylic; lactone)	Coll Czech Chem Commun *34* 3801 (1969) TL 4365, 4369 (1977) JACS *102* 3904 (1980) JOC *46* 2605 (1981)
Me$_3$SiSePh, KF, 18-crown-6 (R = Et, lactone)	TL 5087 (1978)
NaCN, HMPA (R = Me)	TL 3565 (1973)
LiI, DMF, (NaOAc or NaCN) (R = Me)	JCS 6655 (1965) Syn Commun *2* 389 (1972)
LiI, various pyridines (R = Me, Et)	Rocz *30* 323 (1956) Helv *43* 113 (1960) Org Syn *45* 7 (1965)
BCl$_3$, CH$_2$Cl$_2$ (R = Me)	CC 667 (1971)
MeSiCl$_3$, NaI (R = Me, *t*-Bu, PhCH$_2$)	JOC *48* 3667 (1983)
Me$_3$SiOTf, Et$_3$N/H$_2$O (R = *t*-Bu)	Syn 545 (1980); 1 (1982) (review)
Me$_3$SiI (R = 1° alkyl, *t*-Bu, benzylic)/ H$_2$O	Angew Int *15* 774 (1976) JACS *99* 968 (1977); *106* 5335 (1984) CC 495 (1979)
Me$_3$SiCl, NaI/H$_2$O (R = 1°, 2°, 3° alkyl; benzylic)	CC 874 (1978) JOC *44* 1247 (1979)
Me$_3$SiSiMe$_3$, I$_2$/H$_2$O (R = Me, Et, PhCH$_2$)	Angew Int *18* 612 (1979)
PhSiMe$_3$, I$_2$/H$_2$O (R = 1°, 2° alkyl; benzylic)	Syn 417 (1977)

ZnCl$_2$ (R = Et)	Helv *36* 1203 (1953)
CF$_3$CO$_2$H (R = Me)	Syn Commun *12* 855 (1982)
HCl (R = *t*-Bu)	JACS *73* 4752 (1951); *81* 3089 (1959)
HCl, HBr or H$_2$SO$_4$ (phase transfer) (R =1° alkyl, *t*-Bu)	JOC *47* 154 (1982)
Na$_2$S$_2$O$_4$, H$_2$O, Na$_2$CO$_3$ (R = *p*-NO$_2$C$_6$H$_4$CH$_2$)	Syn Commun *12* 219 (1982)
Ar$_3$N$^{+\cdot}$ (R = benzylic)	Angew Int *21* 780 (1982)
NOPF$_6$/H$_2$O (R = benzylic)	Syn 418 (1977)
Me$_2$CuLi (R = allyl)	Syn Commun *8* 15 (1978) TL *28* 5921 (1987)
Hg(OAc)$_2$, CH$_3$OH/KSCN (R = cinnamyl)	TL 2081 (1977)

15. Esters to Acid Halides

$$\underset{RCOR'}{\overset{\displaystyle O \atop \displaystyle \|}{}} \longrightarrow \underset{RCX}{\overset{\displaystyle O \atop \displaystyle \|}{}}$$

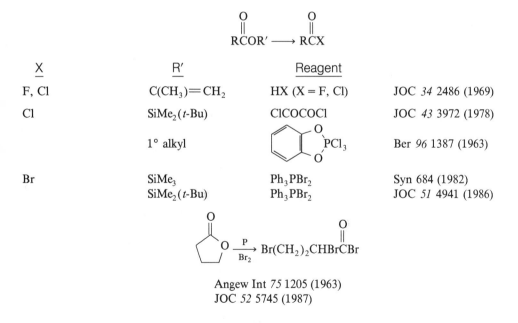

X	R'	Reagent	
F, Cl	C(CH$_3$)=CH$_2$	HX (X = F, Cl)	JOC *34* 2486 (1969)
Cl	SiMe$_2$(*t*-Bu)	ClCOCOCl	JOC *43* 3972 (1978)
	1° alkyl	PCl$_3$	Ber *96* 1387 (1963)
Br	SiMe$_3$	Ph$_3$PBr$_2$	Syn 684 (1982)
	SiMe$_2$(*t*-Bu)	Ph$_3$PBr$_2$	JOC *51* 4941 (1986)

Angew Int *75* 1205 (1963)
JOC *52* 5745 (1987)

16. Esters to Other Esters

$$\underset{R^1COR^2}{\overset{\displaystyle O \atop \displaystyle \|}{}} \longrightarrow \underset{R^1COR}{\overset{\displaystyle O \atop \displaystyle \|}{}}$$

BBr$_3$/ROH	TL 3995 (1974)
Me$_3$SiI, I$_2$/ROH	Syn 142 (1981)

ROH, CsF $\left(R^2 = \right.$ $\left. \right)$ CL 563 (1980)

ROH, TsOH   Org Syn Coll Vol *3* 146 (1955)

ROH, cat KCN   JCS Perkin I 1186 (1972)
                              Syn 790 (1973)

ROH, Al_2O_3 (R^1 = Me, R^2 = Et;   Syn 789 (1981)
   R =1° > 2° alkyl)   TL *22* 5003, 5007 (1981)

ROH (R =1°, 2° alkyl), cat   TL *27* 2383 (1986)
   $XSn(n\text{-Bu})_2OSn(n\text{-Bu})_2OH$
   (X = Cl, NCS)

ROH, $Ti(OR')_4$ (R' = Et, i-Pr)   Syn 138, 826 (1982)
                              Org Syn *65* 230 (1987)

CH_3OH, cat $(EtO)_3TiOCH_2CH_2OTi(OEt)_3$ Helv *65* 1197 (1982)
   (R = Me)

$CH_3CH_2CO_2CH_3$, cat $Ti(OEt)_4$ (R = Me)   Helv *65* 1197 (1982)

MeOH, MeO_2CEt, cat $Ti(OEt)_4$,   Helv *65* 1197 (1982)
   $(HOCH_2CH_2OH)$ (R = Me)

ROH, cat DMAP   JOC *50* 3618 (1985) (*β*-keto esters)
                              TL *28* 2713, 2717 (1987)
                              JACS *109* 7488 (1987)

ROH, n-BuLi (R^1 = aryl, vinylic)   CC 695 (1986)

t-BuOH, KO-t-Bu, molecular sieves   Syn 49 (1972)

KOR, NH_3 JOC *39* 855 (1974)

BrMgOR (R^2 = CH_2SOCH_3)   JACS *106* 2954 (1984)

ROH, electrolysis (R^2 = p-C_6H_4OH)   TL *22* 3715 (1981)

ROH, porcine pancreatic lipase   Science *224* 1249 (1984)
   (enantioselective)   Proc Natl Acad Sci USA *82* 3192 (1985)
                              JACS *107* 7072 (1985); *109* 3977 (1987)
                              TL *27* 5707 (1986); *28* 1607, 2091, 3471 (1987)

ROH, porcine pancreatic lipase, py   JACS *108* 5638 (1986)
   (enantioselective)

ROH, porcine pancreatic lipase,   TL *28* 953 (1987)
   molecular sieves
   (R^2 = CH=CH_2, enantioselective)

ROH, porcine pancreatic lipase, solid   TL *28* 3563 (1987)
   support (enantioselective)

ROH, pig liver esterase, solid support   JACS *106* 2687 (1984)
   (enantioselective)

ROH, horse liver esterase, solid   TL *28* 3563 (1987)
   support (enantioselective)

ROH, α-chymotrypsin (enantioselective)	JACS *108* 2767 (1986)
ROH, α-chymotrypsin, solid support (enantioselective)	TL *28* 3563 (1987)
ROH, subtilisin (enantioselective)	JACS *108* 2767 (1986)
ROH, *Candida cylindracea* lipase (enantioselective)	JACS *107* 7072 (1985); *109* 3977 (1987) TL *27* 29 (1986)
ROH, *Candida cylindracea* lipase, solid support (enantioselective)	JACS *106* 2687 (1984) TL *28* 3563 (1987)
ROH, other lipases (enantioselective)	TL *28* 1607 (1987) JOC *52* 5079 (1987) JACS *109* 3977 (1987)

$$\underset{\text{RCSR}^1}{\overset{\text{O}}{\|}} \longrightarrow \underset{\text{RCOR}^2}{\overset{\text{O}}{\|}}$$

R^2OH, AgO_2CCF_3 ($R^1 = t$-Bu)	TL *24* 5143 (1983) JCS Perkin I 121, 131 (1987)
R^2OH; $Hg(O_2CCF_3)_2$, $HgCl_2(CdCO_3)$ or $Tl(O_2CCF_3)_3$	JACS *97* 3515 (1975)
R^2_2CuLi, O_2 ($R^1 = 2$-pyridyl)	CC 1231 (1981)

17. Esters to Amides

$$\underset{\text{R'COR}}{\overset{\text{O}}{\|}} \longrightarrow \underset{\text{R'CNR}_2}{\overset{\text{O}}{\|}}$$

NH_3	Org Syn Coll Vol *1* 153, 179 (1941); *3* 516, 536 (1955) JOC *52* 4379 (1987)
n-BuNH$_2$, C$_6$H$_6$ (R' = aryl > alkyl)	TL *27* 2263 (1986)
HNR$_2$, Δ	JACS *71* 2215 (1949) Org Syn Coll Vol *3* 108, 765 (1955); *4* 80 (1963)
HNR$_2$, cat NaCN	JOC *52* 2033 (1987)
HNR$_2$, electrolysis	TL *28* 441 (1987)
HNR$_2$, porcine pancreatic lipase	Proc Natl Acad Sci USA *82* 3192 (1985)
HNR$_2$, enzymes	TL *28* 1629 (1987)
NH$_3$, NH$_4$X	JACS *60* 579 (1938)
NH$_4$OH, NH$_4$Cl	Org Syn Coll Vol *4* 486 (1963)
HNR$_2$, [pyridine-N-OH structure]	JCS C 89 (1969)

HNR$_2$, NaOCH$_3$	JOC 28 2915 (1963)
LiNR$_2$	TL 1791 (1970)
	JOC 52 5745 (1987)
NaNR$_2$	JACS 77 469 (1955)
	TL 321 (1971)
	Chem Ind 277 (1956)
XMgNR$_2$	JCS 1188 (1954)
R$_2$AlNR$_2$	TL 4171 (1977)
	Org Syn 59 49 (1980)
	JACS 107 1695, 5732 (1985)
	JOC 52 4665 (1987)
R$_3$SnNR$_2$	JCS A 992 (1969)
	JCS C 2565 (1969)
Ti(NR$_2$)$_4$	JCS C 2565 (1969)
BBr$_3$/R$_2$NH	TL 3995 (1974)
KOH, n-Bu$_4$NBr/R$_2$NH, (o-NO$_2$C$_6$H$_4$O)$_2$POPh, molecular sieves	CL 285 (1981)

$$\underset{\text{RCSR}^1}{\overset{O}{\|}} + H_2NR^2 \longrightarrow \underset{\text{RCNHR}^2}{\overset{O}{\|}}$$

CuI	JOC 52 4531 (1987)
AgO$_2$CCF$_3$	TL 28 3019 (1987)
Hg(O$_2$CCF$_3$)$_2$	CL 915 (1986); 293 (1987) (both intramolecular)

18. Esters to Nitriles

$$\underset{\text{RCOR}'}{\overset{O}{\|}} \xrightarrow{\text{Me}_2\text{AlNH}_2} \text{RCN}$$

TL 4907 (1979)
JOC 52 1309 (1987)

19. Amides to Carboxylic Acids

$$\underset{\text{RCNR}'_2}{\overset{O}{\|}} \longrightarrow \underset{\text{RCOH}}{\overset{O}{\|}}$$

H$_2$SO$_4$, H$_2$O	JACS 70 3091 (1948)
H$_3$PO$_4$, H$_2$O	Rec Trav Chim 46 600 (1927)

HCl, H$_2$O	Org Syn Coll Vol *3* 66, 88 (1955); *4* 39, 496 (1963)
HCl, HOAc, H$_2$O	JACS *63* 2494 (1941)
acidic resin, H$_2$O	Chem Ind 736 (1957) JOC *46* 5351 (1981)
KOH	JOC *52* 4689 (1987)
NaOH, ethylene glycol	JACS *78* 450 (1956); *79* 2530 (1957)
KOH, ethylene glycol	TL *27* 4941 (1986)
KOH, diethylene glycol	JOC *15* 617 (1950)
NaOH/HCl	Org Syn Coll Vol *4* 58 (1963)
Na$_2$O$_2$, H$_2$O/H$^+$	JOC *40* 1187 (1975)
N$_2$O$_4$, HOAc (RCONHR′)	JACS *60* 235 (1938)
(NO)BF$_4$, CH$_3$CN (RCONHR′)	JOC *30* 2386 (1965) TL *27* 4941 (1986)
n-BuONO, HCl, HOAc (RCONHR′)	JACS *70* 3091 (1948) J Med Chem *9* 603 (1966)
NOCl, Ac$_2$O, HOAc, py/KOH (RCONHR′)	JACS *83* 1492 (1961)
NaNO$_2$, H$_2$SO$_4$	JACS *78* 5416 (1956); *79* 2530 (1957); *107* 2111 (1985)
HOSO$_2$ONO, H$_2$O (RCONH$_2$)	Org Prep Proc Int *14* 357 (1982)

PCl$_3$ (RCONR′$_2$)   Ber *96* 1387 (1963)

Ce(NH$_4$)$_2$(NO$_3$)$_6$ (NR′$_2$ = ...)   CL 1551 (1981)

20. Amides to Esters

$$\underset{\text{RCNR′}_2}{\overset{O}{\parallel}} \longrightarrow \underset{\text{RCOR″}}{\overset{O}{\parallel}}$$

MeOH, BF$_3$·OEt$_2$	CC 414 (1969)
MeOH, acidic resin	JOC *46* 5351 (1981)
EtOH, HCl (R″ = Et)	Arch Pharm *290* 218 (1957) JOC *25* 560 (1960); *35* 125 (1970)
R″OH, DBU (NR′$_2$ = ...)	JOC *50* 3224 (1985)
R″Br, H$_2$O	JOC *34* 3204 (1969)

N_2O_4 or $NaNO_2$-Ac_2O, Δ JACS 76 4497 (1954); 77 6008, 6011 (1955)
  (on RCONHR′)   TL 2627 (1965)
                             JOC 34 3834 (1969)

21. Amides to Other Amides

$$\underset{\displaystyle R^1CNHR^2}{\overset{\displaystyle O}{\overset{\displaystyle \|}{}}} \longrightarrow \underset{\displaystyle R^1CNRR^2}{\overset{\displaystyle O}{\overset{\displaystyle \|}{}}}$$

Na, Me_2SO_4 (R = Me)	Ber 68 751 (1935)
Na, RX (R =1° alkyl, $PhCH_2$; X = Cl, Br, I)	Ber 10 327 (1877) JACS 74 1010 (1952)
NaH, RX (R =1°, 2° alkyl; allylic; X = Br, I)	JOC 14 1099 (1949) JACS 74 1010 (1952)
NaOH, K_2CO_3, C_6H_6, cat (n-Bu_4N)HSO_4, RX (R =1° alkyl, benzyl; X = Cl, Br)	Syn 527, 549 (1979); 1005 (1981)

NaOH, (Et_3NCH_2Ph)Cl, aq C_6H_6, RX
  [RX = Me_2SO_4, Et_2SO_4; 1° alkyl and
  benzylic Cl, Br, I; R^2 = aryl]
	Syn 113 (1976)

KOH, acetone, MeI (R^2 = Ar; R = Me)	JACS 74 1321 (1952)
KOH, EtI (R = Et)	Ber 20 3422 (1887)
KOH, EtOH, RBr (R =1° alkyl)	JACS 74 1010 (1952)
KOH, DMSO, RX (R =1°, 2° alkyl, $PhCH_2$; X = Cl, Br, I)	Syn 266 (1971) Tetr 35 2169 (1979)
KO-t-Bu, crown ether/RX (RX = MeI, allyl bromide; R^2 = aryl)	Syn Commun 9 757 (1979)
n-BuLi/Me_2SO_4 (R = Me)	JOC 38 1677 (1973)
KF, alumina, RX (RX = MeI, $PhCH_2Cl$)	CL 1143 (1981)
ROH, $EtO_2CN{=}NCO_2Et$, PPh_3 (intramolecular, β-lactam)	JOC 46 1229 (1981)
$R_2'C(OR'')_2$ (R″ = Me, Et), H_2, cat Pd-C, HOAc, H_2SO_4 (R^2 = H, R = CHR_2')	JOC 27 2205 (1962)

$$\underset{\displaystyle RCNRCHRAr}{\overset{\displaystyle O}{\overset{\displaystyle \|}{}}} \xrightarrow[\displaystyle NH_3,\ H_2O]{\displaystyle Li} \underset{\displaystyle RCNHR}{\overset{\displaystyle O}{\overset{\displaystyle \|}{}}}$$

TL 27 4941 (1986)
JOC 51 5226 (1986)

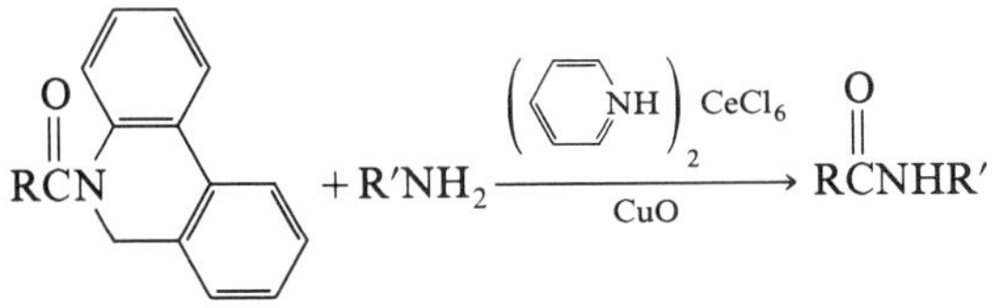

$$RCN + R'NH_2 \xrightarrow[\text{CuO}]{\left(\left\langle\!\!\!\!\bigcirc\!\!\!\!\right\rangle NH\right)_2 CeCl_6} RCNHR'$$

CL 991 (1982)

$$RCNHR' \xrightarrow[\text{2. } R''NH_2]{\text{1. } N_2O_4, NaOAc} RCNHR''$$

TL *23* 1127 (1982)

$$HCN(CH_3)_2 + H_2NAr \xrightarrow{NaOCH_3} HCNHAr$$

JOC *26* 2563 (1961)

22. Amides to Nitriles

$$RCNH_2 \longrightarrow RC\equiv N$$

Review: Chem Rev *42* 189 (1948)

SOCl$_2$	JACS *69* 2663 (1947); *82* 2498 (1960); *83* 2354, 2363 (1961) JOC *27* 4608 (1962) Org Syn Coll Vol *4* 436 (1963)
SOCl$_2$, DMF	JOC *24* 26 (1959); *36* 3960 (1971); *50* 2323 (1985) Tetr *21* 2239 (1965) JACS *88* 2025 (1966)
ClSO$_2$NCO, Et$_3$N	CC 227 (1979)
PhSO$_2$Cl	JCS 763 (1946)
PhSO$_2$Cl, py	JACS *77* 1701 (1955)
TsCl, py	JACS *77* 1701 (1955) BSCF 2262 (1965)
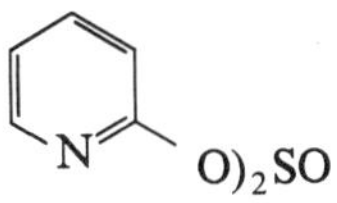 (R = aryl) $O)_2SO$	TL *27* 1925 (1986)
(CF$_3$CO)$_2$O, py	TL 1813 (1977)
P$_2$O$_5$	JOC *27* 4608 (1962) Org Syn Coll Vol *4* 144, 486 (1963)
P$_2$O$_5$, Me$_3$SiOSiMe$_3$	Syn 591 (1982)
(Ph$_3$PO$_3$SCF$_3$)O$_3$SCF$_3$	TL 277 (1975)

$(EtO)_2POP(OEt)_2$	JACS *88* 2025 (1966)
$(EtO)_3PI_2$	TL 1725 (1979)

PCl, py   JACS *88* 2025 (1966)

PCl$_3$ Ber *96* 1387 (1963)
Z Chem *22* 126 (1982) (review)

$POCl_3$	JACS *70* 3315 (1948) JOC *27* 4608 (1962); *50* 5451 (1985)
$POCl_3$, DMF/py	Syn Commun *10* 479 (1980)
CCl_4, PPh_3	TL 4383 (1970) Ber *104* 1030 (1971)
CCl_4, Ph_2P-polymer	Syn 41 (1977)
$(PNCl_2)_3$	Can J Chem *50* 3857 (1972)
$P(NEt_2)_3$	CL 577 (1973)
$COCl_2$, py	JCS 3730 (1954)
$COCl_2$, DMF/py	Syn Commun *10* 479 (1980)
ClCOCOCl, DMF/py	Syn Commun *10* 479 (1980)
$ClCO_2Me$	Bull Acad Polon Sci, Ser Sci Chem *10* 227 (1962) [CA *59* 4031g (1963)]
Cl_3CCOCl, Et_3N	Syn 184 (1985)
$Cl_3COCOCl$, $OP(OMe)_3$	TL *27* 2203 (1986)
$Cl_3CN{=}CCl_2$	Syn 599 (1972)
$NaCl \cdot AlCl_3$, Δ	JACS *62* 1432 (1940)
cat $ClRh(PPh_3)_3$ (R = aryl)	TL 1963 (1970)
$TiCl_4/R_3N$	TL 1501 (1971)
$HCCl_3$, NaOH, $(PhCH_2NEt_3)Cl$	TL 2121 (1973)
$HN(SiMe_2)_n$ ($n = 3, 4$)	JOC *35* 3253 (1970)

, DMF   Syn 657 (1980)

, py   JOC *26* 3356 (1961); *36* 3960 (1971)
JACS *88* 2025 (1966)

$LiAlH_4$	Can J Chem *44* 2113 (1966)

$$\underset{\substack{\| \\ R^1CNHR^2}}{O} \longrightarrow R^1C{\equiv}N$$

R¹	R²	Reagent	
aryl	benzylic	cat ClRh(PPh₃)₃	TL 1963 (1970)
alkyl, aryl, vinylic	t-Bu	POCl₃	Org Prep Proc Int 15 297 (1983)

23. Nitriles to Carboxylic Acids

$$RCN \longrightarrow \underset{\substack{\| \\ RCOH}}{O}$$

HBr, H₂O	Org Syn Coll Vol 1 131 (1941)
HCl, H₂O	Org Syn Coll Vol 1 21, 289, 336, 451 (1941); 3 84, 114, 591, 851 (1955); 4 496, 804 (1963)
HCl, H₂O, HOAc	Org Syn Coll Vol 4 790 (1963)
H₂SO₄, H₂O	Org Syn Coll Vol 1 406, 436 (1941); 2 25, 588 (1943); 3 557 (1955)
H₃PO₄	Rec Trav Chim 46 600 (1927)
NaOH, H₂O/H⁺	Org Syn Coll Vol 1 321 (1941); 2 376 (1943)
Ba(OH)₂, H₂O/H⁺	Org Syn Coll Vol 1 298 (1941); 3 34 (1955)
NaOH, H₂O, MeOH/H⁺	JACS 107 7967 (1985)
KOH, H₂O, EtOH/H⁺	Org Syn Coll Vol 2 292 (1943)
NaOH, ethylene glycol/H⁺	JACS 78 450 (1956) JOC 51 4169 (1986)
NaOH, diethylene glycol/H⁺	JOC 50 2128 (1985)
KOH, ethylene glycol/H⁺	JACS 78 5413 (1956) Org Syn Coll Vol 4 93 (1963)
enzyme	TL 28 4057 (1987) (α-amino acids)

24. Nitriles to Esters

$$RCN \longrightarrow RCO_2R'$$

R′OH, H₂SO₄	JOC 50 2128 (1985)
R′OH, H₂SO₄, H₂O	Org Syn Coll Vol 1 270 (1941)
R′OH, HCl	Ber 105 1778 (1972) JOC 50 2128 (1985) TL 27 2103 (1986) JOC 51 5463 (1986)

R'OH, TsOH

JOC *23* 1225 (1958)

NaOH/MeI (R' = Me)

JACS *108* 1039 (1986)

25. Nitriles to Amides

$$RC{\equiv}N \longrightarrow RC\overset{\displaystyle O}{\overset{\|}{}}NR_2$$

KF, Al$_2$O$_3$ (RCONH$_2$)	Syn Commun *12* 177 (1982)
H$_2$O, cat Cu (RCONH$_2$)	JOC *47* 4812 (1982)
H$_2$SO$_4$ (RCONH$_2$)	JACS *78* 5416 (1956); *79* 2530 (1957)
H$_2$SO$_4$, R'OH (RCONHR')	Org Syn Coll Vol *5* 73 (1973)
PPA (RCONH$_2$)	JACS *76* 3039 (1954); *79* 725 (1957)
H$_2$O, HCl (RCONH$_2$)	Org Syn Coll Vol *3* 66, 88 (1955); *4* 58, 496, 760 (1963)
BF$_3$, HOAc (RCONH$_2$)	JOC *20* 1448 (1955) JACS *79* 725 (1957)
BF$_3$, HOAc, H$_2$O (RCONH$_2$)	JOC *20* 1448 (1955)
NaO$_2$, DMSO (RCONH$_2$)	JOC *44* 4727 (1979)
H$_2$O$_2$, base (RCONH$_2$)	Ber *18* 355 (1885) Org Syn Coll Vol *2* 44, 586 (1943) JACS *75* 3961 (1953), *77* 2519 (1955) JOC *36* 3048 (1971)
H$_2$O$_2$, NH$_4$OH (RCONH$_2$)	JOC *51* 1065 (1986)
H$_2$O$_2$, NaOH, (*n*-Bu$_4$N)HSO$_4$ (phase transfer) (RCONH$_2$)	Syn 243 (1980)
KOH, diethylene glycol (RCONH$_2$)	JOC *15* 617 (1950)
KOH, *t*-BuOH (RCONH$_2$)	JOC *41* 3769 (1976)
KOH, *t*-BuOH/1° R'X (RCONHR')	Syn 303 (1978)
NOPF$_6$, R'X (R' =1°, 2°, 3° alkyl; benzylic) (RCONHR')	Syn 274 (1979)
R$_2$NH, H$_2$O, cat H$_2$Ru(PPh$_3$)$_4$	JACS *108* 7846 (1986)

26. Other Interconversions

$$RC\overset{\displaystyle O}{\overset{\|}{}}{-}\overset{\displaystyle O}{\overset{\|}{P}}(OEt)_2 + R'OH \xrightarrow{\text{DBU}} RC\overset{\displaystyle O}{\overset{\|}{}}OR'$$

TL *22* 3617 (1981)

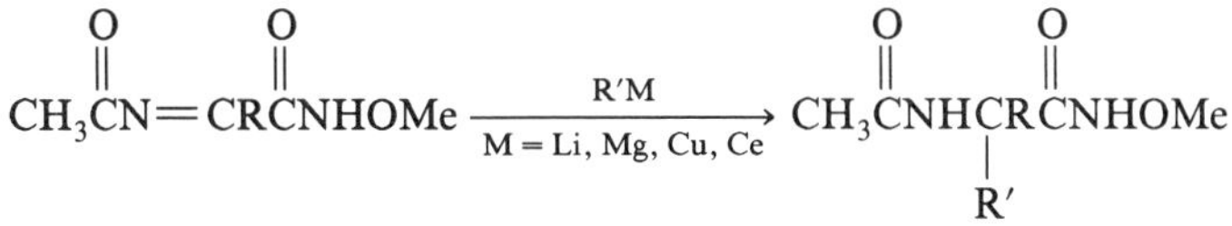

TL *27* 4241 (1986)

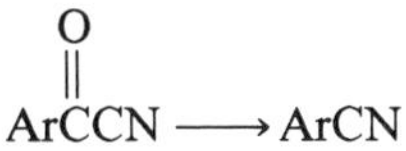

cat $ClRh(PPh_3)_3$ JACS *89* 2338 (1967)

cat $Pd(PPh_3)_4$ JOC *51* 898 (1986)

TRANSFORMATION INDEX

The following index tabulates all organic transformations covered in this book. It describes in the first column the organic product of the transformation, in the second column the organic starting material, and in the last column the page on which that transformation occurs. Inorganic reagents have been ignored. Transformations of the type A → B → C have been indexed as follows: to C from B, to C from A, and to B from A. In equations with many variables, such as various electrophiles, nucleophiles and substituents, only those transformations actually reported in the literature have been indexed.

Each transformation has been indexed in its most general form. The reader is cautioned that for reactions whose scope is ill-defined, the author has often had to make assumptions as to a reaction's generality in order to index it. Not only the presence of key functional groups, but also the presence of carbon-carbon double or triple bonds and aryl groups have been considered carefully in the indexing of all transformations. The reader should keep in mind that not all reactions general for both aldehydes and ketones have been illustrated as such in the text. It is, therefore, advisable for the reader to look under both alkanal (aldehyde) and alkanone (ketone) for a given reaction and to check the original literature.

To shorten an already lengthy index and to increase its utility, the names of key functional groups have often been simplified. For example, trimethylsilyl has been shortened to silyl, methylthio or arylthio to thio, etc. The halogens have not been differentiated from one another. Alkyl or aryl groups that are not central to the transformation in question have been ignored.

Current IUPAC and Chemical Abstracts nomenclature has been employed in this index, with the former given preference. Substitutive nomenclature has been given preference over radicofunctional, additive, subtractive, conjunctive or replacement nomenclature except where this becomes unwieldy.

Prefixes for all names are listed in alphabetical order, not in order of complexity. The following less obvious prefixes have been employed for the group or metal indicated; lithio (Li), boryl (B), sodio (Na), magnesio (Mg), aluminio (Al), silyl (Si), thio (SR), mercapto (SH), halo (F, Cl, Br, I), potassio (K), calcio (Ca), titanio (Ti), vanadio (V), chromio (Cr), manganio (Mn), ferrio (Fe), nickelio (Ni), cuprio (Cu), zincio (Zn), germyl (Ge), seleno (Se), zirconio (Zr), niobio (Nb), rhodio (Rh), palladio (Pd), argentio (Ag), cadmio (Cd), stannyl (Sn), telluro (Te), hafnio (Hf), tantalio (Ta), mercurio (Hg), thallio (Tl), plumbyl (Pb), bismuthio (Bi), lanthanio (La), cerio (Ce), ytterbio (Yb).

The suffix is assigned according to the principal functional group, keeping in mind the following priorities: cation > alkanoic acid > alkanoate ester > acyl halide > alkanamide > alkanenitrile > alkanal > alkanone > alkanol.

The principle chain in acyclic compounds has been assigned according to the following criteria applied successively: (1) maximum number of principal groups, (2) maximum number of carbon-carbon double or triple bonds considered together, (3) maximum number of double bonds, (4) highest locant for multiple bonds, (5) lowest locant for all substituents cited as prefixes and (7) lowest locant for substituents cited first as prefix in alphabetical order. Note that this does not follow IUPAC or Chemical Abstracts rules exactly since chain length has been ignored.

To	From	Page
1,4-alkadiene	π-allylnickel halide	62, 198, 255
1,4-alkadiene	π-allylpalladium halide	230, 254, 256
1,4-alkadiene	1-aluminio-1-alkene	193, 225, 250, 251
1,4-alkadiene	1-boryl-1-alkene	220
1,4-alkadiene	1-cuprio-1-alkene	234, 236, 258
1,4-alkadiene	cycloalkene	260
1,4-alkadiene	enol triflate	202
1,4-alkadiene	5-halo-1,3-alkadiene	253
1,4-alkadiene	1-halo-1-alkene	62, 198, 255
1,4-alkadiene	3-halo-1-alkene	62, 198, 203, 208, 220, 224, 225, 229, 230, 234, 250–252, 254, 258, 260
1,4-alkadiene	1-halo-1-alkyne	220
1,4-alkadiene	4-halo-1-alkyne	239
1,4-alkadiene	1-mercurio-1-alkene	260
1,4-alkadiene	1-silyl-1-alkene	227, 251
1,4-alkadiene	1-stannyl-1-alkene	229, 252
1,4-alkadiene	3-sulfonyl-1-alkene	193, 251
1,4-alkadiene	1-zirconio-1-alkene	230, 254, 256
1,5-alkadiene	1,3-alkadiene	226
1,5-alkadiene	1,5-alkadiene (Cope)	241, 242
1,5-alkadiene	alkanoate 2-alkenyl ester	194, 252, 257
1,5-alkadiene	alkene	250, 257
1,5-alkadiene	2-alken-1-ol	48, 49, 262
1,5-alkadiene	2-alkenylidenetriphenylphosphorane	253
1,5-alkadiene	3-alkoxy-1-alkene	194, 252
1,5-alkadiene	π-allylpalladium halide	257
1,5-alkadiene	3-boryl-1-alkene	250
1,5-alkadiene	3-halo-1-alkene	47, 48, 194, 226, 250, 252–255, 257, 259, 260
1,5-alkadiene	3-magnesio-1-alkene	191, 257
1,5-alkadiene	3-nitro-1-alkene	194
1,5-alkadiene	phosphate 2-alkenyl ester	191
1,5-alkadiene	3-silyl-1-alkene	194, 251
1,5-alkadiene	3-stannyl-1-alkene	252
1,5-alkadiene	3-sulfonyl-1,5-alkadiene	253
1,5-alkadiene	3-sulfonyl-1-alkene	253
1,5-alkadiene	3-thio-1,5-alkadiene	253
1,5-alkadiene	3-thio-1-alkene	253
1,5-alkadiene	titanium di(2-alken-1-olate)	48
1,5-alkadiene	3-zincio-1-alkene	260
2,4-alkadienenitrile	2-alkenenitrile	251
2,4-alkadienenitrile	1-silyl-1-alkene	251
2,4-alkadienoate ester	2,4-alkadienoate ester (alkylation)	894
2,4-alkadienoate ester	alkanal	865, 891
2,4-alkadienoate ester	alkene	245
2,4-alkadienoate ester	2-alkenoate ester	251
2,4-alkadienoate ester	alkyne	227, 236
2,4-alkadienoate ester	2-alkynoate ester	236, 237, 258
2,4-alkadienoate ester	1-cuprio-1-alkene	236, 258
2,4-alkadienoate ester	haloalkane	894
2,4-alkadienoate ester	4-halo-2-alkenoate ester	145
2,4-alkadienoate ester	phosphorus ylid	891
2,4-alkadienoate ester	1-silyl-1-alkene	227, 251
2,4-alkadienoate ester	2-sulfinylalkanoate ester	245, 865

To	From	Page
alkanal	alkanal (homologation)	631, 715, 726, 733, 734, 757
alkanal	alkanal dithioacetal	721, 723, 724
alkanal	alkanal hydrazone	760, 761
alkanal	alkanal imine	627, 758, 759
alkanal	alkanal oxime	762
alkanal	alkanamide	623, 624
alkanal	alkane	673
alkanal	1,2-alkanediol	615, 616
alkanal	alkanenitrile	624, 625
alkanal	alkanethioate S-alkyl ester	622
alkanal	alkanoate 1-alkenyl ester	736
alkanal	alkanoate ester (homologation)	736
alkanal	alkanoate ester (reduction)	621, 622
alkanal	alkanoic acid	619, 620, 732
alkanal	alkanol (1°)	604, 605, 607–613
alkanal	alkanone	629, 715, 726, 733, 734, 757
alkanal	2-alkenal (conjugate addition)	799, 801, 807–810, 814, 815, 916
alkanal	2-alkenal (reduction)	8, 9
alkanal	2-alkenal acetal	801, 809
alkanal	2-alkenal N,O-acetal	810
alkanal	2-alkenal imine	814, 815
alkanal	alkene	593–596, 630, 674, 676, 679, 722, 737, 809
alkanal	2-alkoxyalkanoate ester	629
alkanal	2-alkoxy-1-alkanol	629
alkanal	2-alkoxy-1-alkanone	629
alkanal	1-alkoxy-1-alkene	713, 715
alkanal	3-alkoxy-1-alkene	713
alkanal	1-alkoxy-1-alkyne	717
alkanal	1-alkoxy-2-lithio-1-alkene	717
alkanal	alkyl nitrate	599
alkanal	alkyl sulfate anion	599
alkanal	aluminioalkane	801
alkanal	amine (1°)	601, 602
alkanal	2-aminoalkanenitrile	710
alkanal	1-amino-1-alkene	627
alkanal	3-amino-1-alkene	627
alkanal	2-amino-2-alkenenitrile	711
alkanal	arenesulfonate alkyl ester	599
alkanal	1,3-benzodithiole	684
alkanal	1,3-benzodithiolylium salt	684
alkanal	1,3-benzoxathiolium salt	732
alkanal	borylalkane	674, 799
alkanal	1-boryl-1-alkene	733
alkanal	carbamate O-(1-alkenyl)	714
alkanal	carbamate O-(2-alkenyl)	714
alkanal	cuprioalkane	807–810
alkanal	dihydro-1,3-oxazine	729, 730
alkanal	epoxide	600, 628
alkanal	2,3-epoxyalkanoate ester	734
alkanal	formaldehyde	710
alkanal	formamide	681, 682
alkanal	2-formylalkanoate anion	620
alkanal	2-haloalkanal	23

To	From	Page
alkane	alkanoate ester (reduction)	41
alkane	alkanoate *N*-hydroxypyridine-2-thione ester	42
alkane	alkanoate perester	42
alkane	alkanoate selenoester	42
alkane	alkanoic acid (alkylation)	57
alkane	alkanoic acid (decarboxylation)	40, 42
alkane	alkanoic acid (decarboxylative dimerization)	49, 67
alkane	alkanol (3°, alkylation)	57, 60
alkane	alkanol (reduction)	27, 28, 30, 31
alkane	alkanone (alkylation)	39, 40, 57, 59, 60, 67
alkane	alkanone (cleavage)	39, 866
alkane	alkanone (reduction)	35–38
alkane	alkanone *O*-acyl oxime	42
alkane	alkanone diselenoacetal	35, 59
alkane	alkanone dithioacetal	34
alkane	alkanone hydrazone	37, 38, 67
alkane	alkanone tosylhydrazone	37, 38
alkane	alkene (alkylation)	58, 67
alkane	alkene (dimerization)	46
alkane	alkene (reduction)	6–8
alkane	alkyl methyl oxalate	41
alkane	alkyne	17
alkane	amine (1°)	24, 25
alkane	arenesulfonate alkyl ester (alkylation)	57
alkane	arenesulfonate alkyl ester (reduction)	28, 29
alkane	azoalkane	67
alkane	benzothiazolylsulfide	31
alkane	cuprioalkane	48
alkane	cyclopropane	5
alkane	1,1-dihaloalkane (alkylation)	60
alkane	1,2-dihaloalkane (alkylation)	60
alkane	1,ω-dihaloalkane (alkylation)	65
alkane	dithiocarbonate *O*-alkyl *S*-alkyl ester	30
alkane	dithiocarbonate *O*-alkyl *S*-aryl ester	30
alkane	enol phosphate	38
alkane	enol triflate	38
alkane	ether alkyl silyl	26
alkane	ether dialkyl (alkylation)	60
alkane	ether dialkyl (reduction)	26
alkane	haloalkane (alkylation)	57, 59, 60, 63, 65, 66
alkane	haloalkane (dimerization)	47, 48
alkane	haloalkane (reduction)	18–20
alkane	1-halo-1-alkene	24
alkane	haloformate ester	30
alkane	isourea *O*-alkyl	31
alkane	magnesioalkane	48
alkane	mercurioalkane (alkylation)	67
alkane	nitroalkane	26
alkane	phosphonium salt	60
alkane	phosphorus ylid	60
alkane	selenide alkyl aryl	35
alkane	selenide dialkyl	35, 39
alkane	sulfide alkyl aryl	32
alkane	sulfide dialkyl	31
alkane	sulfone alkyl aryl	33, 40, 59
alkane	1-sulfonyl-1-alkene	40

To	From	Page
1,2-alkanediol	hafnioalkane	565
1,2-alkanediol	2-halo-1-alkanol	482
1,2-alkanediol	2-halo-1-alkanone	578
1,2-alkanediol	2-hydroxyalkanoate ester	550
1,2-alkanediol	2-hydroxy-1-alkanone	530, 541, 545–547, 565
1,2-alkanediol	(hydroxymethyl)lithium lithium salt	561
1,2-alkanediol	2-mercurio-1-alkanol	493
1,2-alkanediol	2-silyl-1-alkanol	516
1,3-alkanediol	alkanal	482, 493, 578
1,3-alkanediol	1,3-alkanedione	529–533, 541
1,3-alkanediol	alkanoate 3-alkenyl ester	494
1,3-alkanediol	alkanone	482, 493, 578
1,3-alkanediol	alkene	493
1,3-alkanediol	3-alken-1-ol	494
1,3-alkanediol	chromioalkane	565
1,3-alkanediol	cuprioalkane	516, 517
1,3-alkanediol	1,3-dioxane	525
1,3-alkanediol	2,3-epoxy-1-alkanol	506, 507, 516, 517
1,3-alkanediol	3,4-epoxy-1-alkanol	517
1,3-alkanediol	2-halo-1-alkanol	482
1,3-alkanediol	2-halo-1-alkanone	578
1,3-alkanediol	3-hydroxy-1-alkanone	529–532, 534, 565
1,3-alkanediol	2-mercurio-1-alkanol	493
1,3-alkanediol	vanadioalkane	565
1,4-alkanediol	alkanal	561
1,4-alkanediol	1,4-alkanedione	532
1,4-alkanediol	4-alkanolide	550
1,4-alkanediol	alkanone	561
1,4-alkanediol	(3-hydroxypropyl)lithium lithium salt	561
1,4-alkanediol	2-hydroxytetrahydrofuran	578
1,5-alkanediol	1,5-alkanedione	532
1,5-alkanediol	5-alkanolide	550, 579
1,5-alkanediol	2-hydroxytetrahydropyran	578
1,5-alkanediol	magnesioalkane	579
1,ω-alkanediol	ω-alkanolide	550, 551
1,2-alkanedione	acid anhydride	817
1,2-alkanedione	acyl halide	621, 686, 689
1,2-alkanedione	alkanal	817
1,2-alkanedione	1,2-alkanediol	612
1,2-alkanedione	alkanone	616, 617
1,2-alkanedione	ω-alkyne	597, 598
1,2-alkanedione	2-halo-1-alkanone	600
1,2-alkanedione	2-hydroxy-1-alkanone	607
1,2-alkanedione	1-stannyl-1-alkanone	689
1,3-alkanedione	acid anhydride	651, 694, 742
1,3-alkanedione	acid anhydride, mixed	694, 742
1,3-alkanedione	acyl halide	651, 692, 742, 753, 755, 760, 810
1,3-alkanedione	acyl tetrafluoroborate	753
1,3-alkanedione	alkanal	685
1,3-alkanedione	1,3-alkanedione (2-alkylation)	65, 764–766
1,3-alkanedione	1,3-alkanedione (4-alkylation)	766, 767
1,3-alkanedione	alkanenitrile	652, 789
1,3-alkanedione	alkanoate 1-alkenyl ester	744
1,3-alkanedione	alkanoate ester	651, 696, 762, 763
1,3-alkanedione	alkanoic acid	694

To	From	Page
alkanoate ester	3-thio-2-alkenoate ester	927
alkanoate ester	trialkoxymethylphosphonium tetrafluoroborate	969
alkanoate ester	trialkyloxonium tetrafluoroborate	970
alkanoate ester	trialkylsulfonium salt	970
alkanoate ester	trihalomethylphosphonate diester	968
alkanoate ester	trimethylselenonium hydroxide	970
alkanoate 1-ethenyl-4-oxoalkyl ester	alkanoate 1-silyl-2-alkenyl ester	802
alkanoate 1-ethenyl-4-oxoalkyl ester	2-alken-1-one	802
alkanoate 2-formyl-3-alkenyl ester	alkanone	782
alkanoate 2-formyl-3-alkenyl ester	2-alkenal	782
alkanoate 2-formyl-3-alkenyl ester	2-alkenal imine	782
alkanoate 4-formyl-3-alkenyl ester	alkanone	782
alkanoate 4-formyl-3-alkenyl ester	2-alkenal	782
alkanoate 4-formyl-3-alkenyl ester	2-alkenal imine	782
alkanoate 1-formyl alkyl ester	alkanal	838
alkanoate 2-halo-1-alkenyl ester	alkyne	240
alkanoate 4-halo-2-alkenyl ester	1,3-alkadiene	329
alkanoate 1-haloalkyl-2-alkenyl ester	1,3-alkadiene	329
alkanoate 2-haloalkyl ester	alkanoate 2-mercurioalkyl ester	331
alkanoate 2-haloalkyl ester	alkene	328, 329, 331
alkanoate 2-haloalkyl ester	ether, cyclic	349, 350
alkanoate 3-haloalkyl ester	ether, cyclic	350
alkanoate 4-haloalkyl ester	acyl halide	979
alkanoate 4-haloalkyl ester	ether, cyclic	350
alkanoate 4-haloalkyl ester	tetrahydrofuran	979
alkanoate 5-haloalkyl ester	acyl halide	979
alkanoate 5-haloalkyl ester	ether, cyclic	350
alkanoate 5-haloalkyl ester	tetrahydropyran	979
alkanoate 2-hydroxyalkyl ester	alkanoate 2-oxoalkyl ester	541, 546
alkanoate 2-hydroxyalkyl ester	alkene	496, 827
alkanoate 2-hydroxyalkyl ester	1,3-dioxolane	526
alkanoate 2-hydroxyalkyl ester	epoxide	509
alkanoate 3-hydroxyalkyl ester	1,3-dioxane	525, 526
alkanoate 3-hydroxy-1-methylenealkyl ester	alkanal	554
alkanoate 3-hydroxy-1-methylenealkyl ester	alkanoate 1-(halomethyl)ethenyl ester	554
alkanoate 3-hydroxy-1-methylenealkyl ester	alkanone	554
alkanoate N-hydroxypyridine-2-thione ester	acyl halide	42
alkanoate N-hydroxypyridine-2-thione ester	alkanoic acid	42
alkanoate 2-mercurioalkyl ester	alkene	331
alkanoate 2-oxoalkyl ester	alkanoate 1-alkenyl ester	744
alkanoate 2-oxoalkyl ester	alkanone	841, 842
alkanoate 2-oxoalkyl ester	1-amino-1-alkene	842
alkanoate 2-oxoalkyl ester	1-silyloxy-1-alkene	754
alkanoate 3-oxoalkyl ester	alkanoate 2-alkenyl ester	595
alkanoate perester	acyl halide	42
alkanoate 2-selenoalkyl ester	alkene	149
alkanoate selenoester	alkanoic acid	42
alkanoate sodium salt	2-alkenoate sodium salt (reduction)	13
alkanoate 1-succinimidyl ester	alkanoic acid	552
alkanoate 2-sulfonylalkyl ester	alkanal	294
alkanoate 2-sulfonylalkyl ester	1-lithio-1-sulfonylalkane	294
alkanoate thioester	see alkanethioate S-alkyl or O-alkyl ester or alkanedithioate ester	
alkanohydrazide	alkanoic acid	719
alkanohydrazide	2-alkenohydrazide	922

To	From	Page
alkanohydrazide	lithioalkane	922
alkanoic acid	acyl halide	933
alkanoic acid	alkanal	838–840, 864
alkanoic acid	alkanamide	988, 989
alkanoic acid	1,2-alkanediol	836, 837
alkanoic acid	1,2-alkanedione	843
alkanoic acid	alkanenitrile	846, 993
alkanoic acid	alkanimidate ester	932
alkanoic acid	alkanoate 2-alkenyl ester	983, 985
alkanoic acid	alkanoate 1-arylalkyl ester	983–985
alkanoic acid	alkanoate aryl ester	983
alkanoic acid	alkanoate *t*-butyl ester	984, 985
alkanoic acid	alkanoate ω-epoxyalkyl ester	982
alkanoic acid	alkanoate ester	981–985
alkanoic acid	alkanoate ω-haloalkyl ester	983, 984
alkanoic acid	alkanoate methyl ester	983–985
alkanoic acid	alkanoate 2,2,2-trihaloethyl ester	983
alkanoic acid	alkanoate trihalomethyl ester	982
alkanoic acid	alkanoic acid (alkylation)	867, 868, 872
alkanoic acid	alkanoic acid (degradation)	866
alkanoic acid	alkanol (homologation)	850, 851
alkanoic acid	alkanol (1°, oxidation)	834, 835
alkanoic acid	3-alkanolide	872
alkanoic acid	alkanone	842, 864
alkanoic acid	alkene	596, 828, 849, 850, 867
alkanoic acid	2-alkenoic acid (conjugate addition)	915–917
alkanoic acid	2-alkenoic acid (reduction)	7, 12, 13
alkanoic acid	3-alkenoic acid	7
alkanoic acid	1-alkenyl-2-oxazoline	921
alkanoic acid	1-alkoxy-1-alkene	932
alkanoic acid	alkyne	830
alkanoic acid	arylalkane	824
alkanoic acid	2-(aryloxy)alkanoic acid	872
alkanoic acid	borylalkane	872
alkanoic acid	cadmioalkane	872
alkanoic acid	cuprioalkane	872, 916, 917
alkanoic acid	2-diazo-1-alkanone	933
alkanoic acid	1,1-dihalo-1-alkene	850
alkanoic acid	1,3-dihydrooxazine	872
alkanoic acid	haloalkane	850, 851, 867, 868, 872, 915
alkanoic acid	2-haloalkanoic acid	23
alkanoic acid	2-halo-1-alkanone	932
alkanoic acid	2-hydroxyalkanoic acid	866
alkanoic acid	lithioalkane	921
alkanoic acid	magnesioalkane	917, 921
alkanoic acid	malonate diester	894–896
alkanoic acid	nitroalkane	832
alkanoic acid	2-oxoalkanoate sodium salt	843
alkanoic acid	2-oxoalkanoic acid	845
alkanoic acid	1-silyloxy-1-alkene	829, 932
alkanoic acid	sulfide alkyl aryl	852
alkanoic acid	2-sulfinylalkanoate ester	982
alkanoic acid	2-sulfonyloxy-1-alkanone	932
alkanol	alkanoic acid	490
alkanol	2-halo-1-alkanol (alkylation)	65, 482
alkanol	silylalkane	491

To	From	Page
alkanone	2-thio-1-alkanone	34, 786
alkanone	1-thio-1-alkene	726
alkanone	3-thio-2-alken-1-one	787, 812
alkanone	2-(1-thioalkyl)-2-alken-1-one	811
alkanone	2-thiomethyl-2-alken-1-one	776
alkanone	2-tosyloxy-1-alkanol	630
alkanone	2-tosyloxy-1-alkanone	27, 29
alkanone	vanadioalkane	684, 689
alkanone	zincioalkane	686, 690, 813, 814
alkanone	zirconioalkane	689
alkanone acetal	alkanone	463–465, 735
alkanone acetal	haloalkane	728
alkanone acetal	2-thioalkanoic acid	728
alkanone *O*-acyl oxime	acyl halide	42
alkanone *O*-acyl oxime	alkanoic acid	42
alkanone *O*-acyl oxime	alkanone	424
alkanone *O*-alkyl oxime	alkanal	427
alkanone *O*-alkyl oxime	alkanone	424, 761
alkanone arylsulfonylhydrazone	alkanone	158, 864
alkanone *N*-diphenylphosphinyl amine	alkanone *N*-diphenylphosphinyl imine	424
alkanone *N*-diphenylphosphinyl amine	alkanone oxime	424
alkanone *N*-diphenylphosphinyl imine	alkanone	424
alkanone *N*-diphenylphosphinyl imine	alkanone oxime	424
alkanone diselenoacetal	alkanone	59, 372
alkanone dithioacetal	alkanal	722
alkanone dithioacetal	alkanal dithioacetal	721, 724
alkanone dithioacetal	1,1-dithio-1-alkene	722
alkanone dithioacetal	haloalkane	719, 722, 724
alkanone hydrazone	alkanone	67, 428, 759, 760
alkanone imine	alkanone	421–423, 758, 932
alkanone imine	2-alken-1-one	777
alkanone imine	2-alken-1-one imine	777
alkanone imine	amine (1°)	421–423
alkanone imine	haloalkane	777
alkanone lithium enolate	1-cyclopropyl-1-alkanone	785
alkanone *O*-mesyl oxime	alkanone	428
alkanone oxime	alkanone	424, 762, 930, 931
alkanone *O*-silyl oxime	alkanone	424
alkanone tosylhydrazone	alkanone	158, 587
alkanone *O*-tosyl oxime	alkanone	424, 428
alkanoyl	*see* acyl *except for substituted acyl compounds*	
alkanoyl halide	*see* acyl halide *except for substituted alkanoyl groups*	
2,4,6,8-alkatetraenoate ester	alkyne	236
2,4,6,8-alkatetraenoate ester	2-alkynoate ester	236
2,4,6,8-alkatetraenoate ester	1-cuprio-1-alkene	236
2,4,6-alkatrienamide	2,4-alkadienamide	904
2,4,6-alkatrienamide	alkanal	904
1,3,5-alkatriene	2-alkenal	245
1,3,5-alkatriene	2-alken-1-one	161
1,3,5-alkatriene	3-alkoxy-4-sulfonyl-1-alkene	245
1,3,5-alkatriene	alkyne	236
1,3,5-alkatriene	1-cuprio-1-alkene	236
1,3,5-alkatriene	haloalkane	236
1,3,5-alkatriene	3-halo-4-sulfonyl-1-alkene	245
1,3,5-alkatriene	1-lithio-1-sulfonylalkane	245

To	From	Page
alkene	alkanone	146, 157, 158, 160–163, 173–184, 196, 197, 293, 633
alkene	alkanone arylsulfonylhydrazone	158
alkene	alkanone dithioacetal	157, 160, 201
alkene	alkanone tosylhydrazone	158
alkene	2-alkenal (decarbonylation)	38
alkene	alkene (alkylation)	197, 204, 206, 210
alkene	alkene (allylic substitution)	189, 190
alkene	alkene (isomerization)	109–115
alkene	alkene (vinylic substitution)	186–188
alkene	2-alken-1-ol (alkylation)	56, 200, 201, 203
alkene	2-alken-1-ol (alkylative transposition)	120–122
alkene	2-alken-1-ol (reduction)	27, 28, 31
alkene	2-alken-1-ol (reductive transposition)	116
alkene	2-alken-1-one	36, 158, 159
alkene	3-alken-1-one	36
alkene	2-alken-1-one arylsulfonylhydrazone	159
alkene	2-(2-alkenyloxy)benzothiazole (alkylation)	209
alkene	2-(2-alkenyloxy)benzothiazole (alkylative transposition)	121
alkene	2-alkenyl phosphonium salt (reductive transposition)	115
alkene	2-alkenyl sulfonium salt (alkylation)	209
alkene	2-(2-alkenylthio)benzothiazole (alkylative transposition)	121
alkene	(2-alkenyl)trialkylammonium salt (alkylation)	188, 191, 209
alkene	1-alkoxy-1-alkene (alkylation)	200
alkene	1-alkoxy-1-alkene (reduction)	160
alkene	3-alkoxy-1-alkene (alkylation)	191, 201, 203
alkene	3-alkoxy-1-alkene (alkylative transposition)	120, 121
alkene	3-alkoxy-1-alkene (reduction)	26
alkene	2-alkoxy-1,3-dioxolane	155
alkene	1-alkoxy-2-haloalkane	136, 137
alkene	1-alkoxy-2-selenoalkane	148
alkene	N-alkyl-N,N-disulfonimide	140
alkene	alkylpyridinium salt	140
alkene	alkyl triflate	194
alkene	alkyltrimethylammonium hydroxide	139
alkene	alkyne (alkylation and functionalization)	215–240
alkene	alkyne (hydrogen addition)	212–214
alkene	π-allylnickel halide	62, 198
alkene	π-allylpalladium compound	206
alkene	1-aluminio-1-alkene	223, 225
alkene	amine (1°)	140
alkene	amine oxide	139
alkene	1-amino-1-alkene (reduction)	160
alkene	arenesulfonate 2-alkenyl ester (alkylation)	209
alkene	arenesulfonate alkyl ester	153, 154, 192
alkene	3-aryloxy-1-alkene (alkylative transposition)	121
alkene	3-aryloxy-1-alkene (reduction)	26
alkene	1,2-bis(mesyloxy)alkane	156
alkene	borylalkane	57, 114, 192
alkene	1-boryl-1-alkene	219
alkene	3-boryl-1-alkene	192

To	From	Page
2-alkoxyalkanoate ester	2-alkoxyalkanoate ester (alkylation)	878
2-alkoxyalkanoate ester	haloalkane	878
2-alkoxyalkanoate ester	2-haloalkanoate ester	446
3-alkoxyalkanoate ester	alkanal acetal	885
3-alkoxyalkanoate ester	alkanoate ester	885
3-alkoxyalkanoate ester	alkanone acetal	885
3-alkoxyalkanoate ester	3-alkoxyalkanoate ester (alkylation)	882
3-alkoxyalkanoate ester	1-alkoxy-1-haloalkane	885
3-alkoxyalkanoate ester	1-alkoxy-1-silyloxy-1-alkene	885
3-alkoxyalkanoate ester	haloalkane	882
5-alkoxyalkanoate ester	2-alkenoate ester	919
5-alkoxyalkanoate ester	1-alkoxy-2-mercurioalkane	919
6-alkoxyalkanoate ester	2-alkenoate ester	919
6-alkoxyalkanoate ester	1-alkoxy-3-mercurioalkane	919
2-alkoxy-1-alkanol	alkanal	468, 557, 560
2-alkoxy-1-alkanol	alkanone	468, 557, 560, 629
2-alkoxy-1-alkanol	2-alkoxyalkanal	562
2-alkoxy-1-alkanol	2-alkoxyalkanoate ester	629
2-alkoxy-1-alkanol	2-alkoxy-1-alkanone	559, 562, 564, 565, 629
2-alkoxy-1-alkanol	1-alkoxy-2,3-epoxyalkane	517
2-alkoxy-1-alkanol	1-alkoxy-1-haloalkane	557
2-alkoxy-1-alkanol	1-alkoxy-1-lithioalkane	560
2-alkoxy-1-alkanol	1-alkoxy-1-stannylalkane	468
2-alkoxy-1-alkanol	aluminioalkane	517
2-alkoxy-1-alkanol	cuprioalkane	517
2-alkoxy-1-alkanol	1,3-dioxolane	524, 525
2-alkoxy-1-alkanol	epoxide	509
2-alkoxy-1-alkanol	haloalkane	468
2-alkoxy-1-alkanol	lithioalkane	559
2-alkoxy-1-alkanol	magnesioalkane	562, 629
2-alkoxy-1-alkanol	titanioalkane	564
3-alkoxy-1-alkanol	3-alkoxy-1-alkanone	530
3-alkoxy-1-alkanol	1-alkoxy-3,4-epoxyalkane	517
3-alkoxy-1-alkanol	aluminioalkane	517
3-alkoxy-1-alkanol	3,5-dioxa-1-alkene	467, 524
3-alkoxy-1-alkanol	1,3-dioxane	523–525
4-alkoxy-1-alkanol	2-alkoxytetrahydrofuran	525
4-alkoxy-1-alkanol	1,3-dioxepane	525
5-alkoxy-1-alkanol	2-alkoxytetrahydropyran	525
2-alkoxy-ω-alkanolide	2-alkoxy-ω-alkanolide (alkylation)	878
2-alkoxy-ω-alkanolide	haloalkane	878
2-alkoxy-1-alkanone	alkanol	447
2-alkoxy-1-alkanone	2-diazo-1-alkanone	447
3-alkoxy-1-alkanone	alkanal acetal	752
3-alkoxy-1-alkanone	alkanol	450
3-alkoxy-1-alkanone	alkanone acetal	752
3-alkoxy-1-alkanone	2-alken-1-one	450
3-alkoxy-1-alkanone	3-alkoxy-1-alkene	595
3-alkoxy-1-alkanone	1-alkoxy-1-haloalkane	751, 752
3-alkoxy-1-alkanone	1-silyloxy-1-alkene	751, 752
4-alkoxy-1-alkanone	2-alken-1-one	807, 808, 813
4-alkoxy-1-alkanone	1-alkoxy-1-cuprioalkane	807, 813
2-alkoxy-2-alkenamide	2-alkoxy-2-alkenamide (substitution)	188
2-alkoxy-2-alkenamide	haloalkane	188
3-alkoxy-2-alkenamide	3-alkoxy-2-alkenamide (substitution)	188
1-alkoxy-1-alkene	alkanal	177, 178, 715, 716
1-alkoxy-1-alkene	alkanal acetal	157, 716

To	From	Page
2-alkoxy-3-silyl-2-alkenamide	2-alkoxy-2-alkenamide	188
1-alkoxy-1-silyloxy-1,3-alkadiene	2-alkenoate ester	892
1-alkoxy-1-silyloxy-1-alkene	alkanoate ester	80, 131, 846, 860, 865, 885–887
5-alkoxy-5-silyloxy-4-alkenoate ester	2-alkenoate ester	924
5-alkoxy-5-silyloxy-4-alkenoate ester	1-alkoxy-1-silyloxy-1-alkene	924
4-alkoxy-2-(silyloxy)furan	3-alkoxy-2-alken-4-olide	893
4-alkoxy-4-silyl-4-thio-1-alkanone	2-alken-1-one	799
1-alkoxy-1-stannylalkane	alkanal	468
1-alkoxy-1-stannylalkane	haloalkane	468
1-alkoxy-2-sulfonylalkane	alkanal	245
1-alkoxy-2-sulfonylalkane	1-lithio-1-sulfonylalkane	245
1-alkoxy-2-sulfonyl-1-alkene	1-alkoxy-2-sulfonyl-1-alkene (substitution)	186
1-alkoxy-2-sulfonyl-1-alkene	haloalkane	186
3-alkoxy-4-sulfonyl-1-alkene	2-alkenal	245
3-alkoxy-4-sulfonyl-1-alkene	1-lithio-1-sulfonylalkane	245
3-alkoxy-2-sulfonyl-2-alken-1-ol	alkanal	186
3-alkoxy-2-sulfonyl-2-alken-1-ol	alkanone	186
3-alkoxy-2-sulfonyl-2-alken-1-ol	1-alkoxy-2-sulfonyl-1-alkene	186
1-alkoxy-3-sulfonyloxy-2-alkanol	1,2-epoxy-3-(sulfonyloxy)alkane	511
5-alkoxy-3-thia-1,5-alkadiene	haloalkane	638
3-alkoxy-2-thioalkanoate ester	3-alkoxyalkanoate ester	882
2-alkoxy-3-thio-2-alkenamide	2-alkoxy-2-alkenamide	188
2-alkoxy-3-thio-2-alkenamide	2-alkoxy-3-thio-2-alkenamide (substitution)	188
2-alkoxy-3-thio-2-alkene-1,4-dioate diester	2-alkoxy-2-alkene-1,4-dioate diester	187, 188
3-alkoxy-2-thio-2-alkenoate ester	3-alkoxy-2-alkenoate ester	187
3-alkoxy-3-thio-2-alkenoate ester	3-alkoxy-2-alkenoate ester	187
3-alkoxy-2-thio-2-alken-4-olide	3-alkoxy-2-alken-4-olide	893
1-alkoxy-1,3,5-tris(silyloxy)-1,3,5-alkatriene	3,5-dioxoalkanoate ester	774
alkylammonium salt	alkene	392
alkylammonium salt	dialkyl N-(alkoxycarbonyl)-phosphoramidate	399
alkylammonium salt	haloalkane	399
alkylammonium salt	phosphoramide	392
alkylarene	see arylalkane	
N-alkylarenesulfonamide	N-tosyl aziridine	435
alkylbenzoic acid	haloalkane	871
alkylbenzoic acid	toluic acid	871
N-alkyl bis(benzenesulfenimide)	haloalkane	399
alkylcycloalkane	see cycloalkylalkane or the specific cyclo-alkyl ring desired	
5-alkyl-2,3-dihydrofuran	4-alkanolide	166
5-alkyl-2,3-dihydrofuran	2-silyl-4-alkanolide	166, 180
1-alkyl-1,4-dihydro-4-pyridineacetate ester	alkanoate ester	876
1-alkyl-1,4-dihydro-4-pyridineacetate ester	pyridinium salt	876
N-alkyl-N,N-disulfonimide	amine (1°)	140
alkyl halide	see haloalkane	
2-alkyl-3-halotetrahydrofuran	2,3-dihalotetrahydrofuran	137
2-alkyl-3-halotetrahydrofuran	2,3-dihydrofuran	137
2-alkyl-3-halotetrahydropyran	2,3-dihalotetrahydropyran	137
2-alkyl-3-halotetrahydropyran	3,4-dihydro-2H-pyran	137
2-alkylidene-1,6-alkanediamide	2-alkenamide	905
2-alkylidene-1,4-alkanedioate diester	alkanal	170
2-alkylidene-1,4-alkanedioate diester	1,4-alkanedioate diester	170
2-alkylidene-1,5-alkanedioate diester	2-[1-(acyloxy)alkyl]-2-alkenoate ester	875
2-alkylidene-1,5-alkanedioate diester	alkanoate ester	875

To	From	Page
arenecarboxamide	haloalkane	906
arenecarboxamide	haloarene	56, 859
arenecarboxamide	isocyanate	56
arenecarboxamide	lithioarene	859
arenecarboxamide	magnesioarene	859
arenecarboxamide (2°, 3°)	arenecarboxamide (2°, 3°; substitution)	54
arenecarboxylate ester	1,3-alkanedione	97
arenecarboxylate ester	alkanol	853, 855, 856
arenecarboxylate ester	alkenylketenimine	103
arenecarboxylate ester	alkyne	103
arenecarboxylate ester	arene	61, 853
arenecarboxylate ester	arenecarboxaldehyde	840, 841
arenecarboxylate ester	arenecarboxylate ester (substitution)	50, 900
arenecarboxylate ester	arenecarboxylate ester (transesterification)	985
arenecarboxylate ester	arene chromium tricarbonyl	61
arenecarboxylate ester	1-aryl-1-alkanol	836
arenecarboxylate ester	aryl oxazoline	56
arenecarboxylate ester	aryl triflate	856
arenecarboxylate ester	carbonate dialkyl ester	855
arenecarboxylate ester	cyclohexanone	102
arenecarboxylate ester	4,6-dioxo-2-alkenoate ester	97
arenecarboxylate ester	formate ester	856
arenecarboxylate ester	haloalkane	900
arenecarboxylate ester	haloarene	855, 856
arenecarboxylate ester	haloformate ester	61, 855
arenecarboxylate ester	lithioarene	855
arenecarboxylate ester	magnesioarene	855, 856
arenecarboxylate ester	metal alkoxide	855
arenecarboxylate ester	3-oxo-6-alkenoate ester	97
arenecarboxylate ester	palladioarene	853
arenecarboxylate ester	pyrone	102
arenecarboxylate ester	2-[(silyloxy)methylene]cyclohexanone	100
arenecarboxylate ester chromium tricarbonyl	haloformate ester	61
arenecarboxylate 2-hydroxyalkyl ester	alkanal	900
arenecarboxylate 2-hydroxyalkyl ester	alkanone	900
arenecarboxylate 2-hydroxyalkyl ester	arenecarboxylate ester	900
arenecarboxylate 1-silylalkyl ester	arenecarboxylate ester	900
arenecarboxylate 1-stannylalkyl ester	arenecarboxylate ester	900
arenecarboxylic acid	aminoarene (1°)	852
arenecarboxylic acid	arene	50–55, 61, 849
arenecarboxylic acid	arenecarboxaldehyde	839, 840
arenecarboxylic acid	arenecarboxylic acid (substitution)	50, 54
arenecarboxylic acid	arene chromium tricarbonyl	61
arenecarboxylic acid	arenediazonium salt	852
arenecarboxylic acid	arylalkane	823
arenecarboxylic acid	1-aryl-1,2-alkanedione	843
arenecarboxylic acid	1-aryl-1-alkanol	834
arenecarboxylic acid	1-aryl-1-alkanone	842
arenecarboxylic acid	benzil	843
arenecarboxylic acid	haloarene	56, 850, 851
arenecarboxylic acid	silylarene	852
arenecarboxylic acid halide	*see* aroyl halide	
arene chromium tricarbonyl	arene	61
arene chromium tricarbonyl	arene chromium tricarbonyl (substitution)	61, 64
arenediazonium salt	aminoarene (1°)	25, 678, 852, 862
arenesulfonamide	alkene	826, 827

To	From	Page
4-aryl-4-oxoalkanenitrile	arenecarboxaldehyde	711, 798
1-aryl-4-oxoalkanoate ester	arenecarboxaldehyde	710
2-aryl-2-oxoalkanoate ester	alkanol	856
2-aryl-2-oxoalkanoate ester	haloarene	856
2-aryl-2-oxoalkanoate ester	imidazole-1-glyoxylate ester	699
2-aryl-2-oxoalkanoate ester	magnesioarene	695, 699
2-aryl-2-oxoalkanoate ester	oxalate diester	695
2-aryl-3-oxoalkanoate ester	arene	58
2-aryl-3-oxoalkanoate ester	bismuthioarene	772
2-aryl-3-oxoalkanoate ester	haloarene	772
2-aryl-3-oxoalkanoate ester	mercurioarene	58
2-aryl-3-oxoalkanoate ester	3-oxoalkanoate ester	58, 771, 772
2-aryl-3-oxoalkanoate ester	plumbylarene	58, 771
3-aryl-2-oxoalkanoate ester	lithioarene	695
3-aryl-2-oxoalkanoate ester	magnesioarene	695
3-aryl-3-oxoalkanoate ester	aroylmalonate diester	895
3-aryl-3-oxoalkanoate ester	haloarene	895
3-aryl-3-oxoalkanoate ester	malonate diester	895
4-aryl-4-oxoalkanoate ester	2-alkenoate ester	710, 711, 798
4-aryl-4-oxoalkanoate ester	2-amino-2-arylalkanenitrile	711
4-aryl-4-oxoalkanoate ester	4-amino-4-aryl-4-cyanoalkanoate ester	711
4-aryl-4-oxoalkanoate ester	arenecarboxaldehyde	711, 798
4-aryl-4-oxoalkanoate ester	2-aryl-2-(silyloxy)alkanenitrile	710
6-aryl-6-oxoalkanoate ester	alkanoate ester	874
6-aryl-6-oxoalkanoate ester	aryl cyclopropyl ketone	874
2-aryl-2-oxoalkanoic acid	haloarene	851
2-aryl-2-oxoalkanoic acid	isonitrile	712
2-aryl-2-oxoalkanoic acid	lithioarene	712
3-aryl-3-oxoalkanoic acid	aroylmalonate diester	895
3-aryl-3-oxoalkanoic acid	haloarene	895
3-aryl-3-oxoalkanoic acid	malonate diester	895
3-aryl-2-(3-oxoalkyl)-2-alkenenitrile	2-alken-1-one	187
3-aryl-2-(3-oxoalkyl)-2-alkenenitrile	3-aryl-2-alkenenitrile	187
2-aryl-3-oxooxacycloheptane	5-sulfinyl-5,6-epoxy-1-alkanol	453
2-aryl-3-oxotetrahydropyran	4,5-epoxy-4-sulfinyl-1-alkanol	453
3-aryloxy-1,2-alkanediol	2,3-epoxy-1-alkanol	511
2-(aryloxy)alkanoic acid	2-(aryloxy)alkanoic acid (alkylation)	869
2-(aryloxy)alkanoic acid	haloalkane	869
1-aryloxy-1-alkene	1-aryloxy-1-alkene (alkylation)	714
1-aryloxy-1-alkene	haloalkane	714
3-aryloxy-1-alkene	acetate aryl ester	472
3-aryloxy-1-alkene	3-halo-1-alkene	448, 472
2-aryloxy-2-deuteroalkanoic acid	2-(aryloxy)alkanoic acid	869
2-aryloxy-3-hydroxyalkanoic acid	alkanal	869
2-aryloxy-3-hydroxyalkanoic acid	alkanone	869
2-aryloxy-3-hydroxyalkanoic acid	2-(aryloxy)alkanoic acid	869
2-aryloxy-4-hydroxyalkanoic acid	2-(aryloxy)alkanoic acid	869
2-aryloxy-4-hydroxyalkanoic acid	epoxide	869
1-aryloxy-1-silylalkane	phenol	521
aryl pyrazole	aryl pyrazole (substitution)	53
2-arylpyridine	2-arylpyridine (substitution)	55
aryl quinone	quinone	204
1-aryl-1-selenoalkane	arenecarboxaldehyde diselenoacetal	40, 59
1-aryl-1-selenoalkane	1-aryl-1-alkanone	40, 59
1-aryl-1-selenoalkane	1-aryl-1-alkanone diselenoacetal	40, 59
1-aryl-1-selenoalkane	haloalkane	59
3-aryl-3-silylalkanamide	3-arylalkanamide	905

To	From	Page
cyclohexane	arene	5, 6
cyclohexane	1,6-dihalohexane	88
1,2-cyclohexanedicarboxamide	1,4-dihaloalkane	903
1,2-cyclohexanedicarboxamide	succinamide	903
1,1-cyclohexanedicarboxylate diester	(5-haloalkyl)malonate diester	88
1,2-cyclohexanedicarboxylate diester	1,4-dihaloalkane	89
1,2-cyclohexanedicarboxylate diester	succinate diester	89
1,3-cyclohexanedione	1,5-dialkoxy-1,4-cyclohexadiene	715
1,4-cyclohexanedione	1,4-cyclohexanedione (alkylation)	740
1,4-cyclohexanedione	2,5-dialkoxybenzoic acid	817
1,4-cyclohexanedione	haloalkane	817
cyclohexanol	alkanoate ester	579
cyclohexanol	1,5-dihaloalkane	579
cyclohexanol	1,5-dimagnesioalkane	579
cyclohexene	1,3-alkadiene	263–272
cyclohexene	alkene	263–272
cyclohexene	6-thio-1-alkyne	216
3-cyclohexenol	2-alken-1-one	279
3-cyclohexenol	2-(1-alkenyl)cyclobutanone	279
2-cyclohexenone	alkanone	668–670
2-cyclohexenone	5-alkenal	655
2-cyclohexenone	5-alken-1-ol	655
2-cyclohexenone	2-alken-1-one	668–670
2-cyclohexenone	2-alkoxybenzoic acid	817
2-cyclohexenone	1-alkoxy-1,4-cyclohexadiene	714
2-cyclohexenone	2-alkyn-1-one	217
2-cyclohexenone	anisole chromium tricarbonyl	817
2-cyclohexenone	haloalkane	817
2-cyclohexenone	lithioalkane	817
3-cyclohexenone	1,3-alkadiene	270, 271
3-cyclohexenone	ketene	270, 271
2-(3-cyclohexenyl)-1-alkanone	bicycloalkenol	630
1-cyclohexyl-1-alkanone	7-halo-1-alkanone	88
1,5-cyclooctadiene	1,3-alkadiene	256, 280
1,5-cyclooctadiene	1,2-divinylcyclobutane	280
2-cyclopentadienylirondicarbonyl-2-alken-5-olide	alkanone	956
2-cyclopentadienylirondicarbonyl-2-alken-5-olide	alkene	961
2-cyclopentadienylirondicarbonyl-2-alken-5-olide	2-alkynoate ester	956, 961
cyclopentane	alkene	86
cyclopentane	1,5-dihaloalkane	48, 88
cyclopentane	1,5-dimagnesioalkane	48
cyclopentane	3-ferrio-1-alkene	86
cyclopentane	ferriocyclopentane	86
cyclopentane	6-halo-1-hexene	85
cyclopentane	ω-nitro-1-alkene	91
cyclopentane	6-seleno-1-hexene	85
cyclopentanecarboxylate ester	alkene	86
cyclopentanecarboxylate ester	3-ferrio-1-alkene	86
cyclopentanecarboxylate ester	ferriocyclopentane	86
1,1-cyclopentanedicarboxylate diester	alkene	86
1,1-cyclopentanedicarboxylate diester	1,1-cyclopropanedicarboxylate diester	86
1,1-cyclopentanedicarboxylate diester	(4-haloalkyl)malonate diester	88
1,2-cyclopentanedicarboxylate diester	1,3-dihaloalkane	89
1,2-cyclopentanedicarboxylate diester	succinate diester	89

To	From	Page
3-halo-1-alkene	haloalkane	189, 342
3-halo-1-alkene	3-halo-1-alkene (rearrangement)	343
3-halo-1-alkene	3-halo-1-alkene (substitution)	189
3-halo-1-alkene	3-silyloxy-1-alkene	350
3-halo-1-alkene	3-(2-tetrahydropyranyloxy)-1-alkene	351
4-halo-1-alkene	1,3-alkadiene	329
4-halo-1-alkene	alkene	342
4-halo-1-alkene	3-alken-1-ol	341
4-halo-1-alkene	1-cyclopropyl-1-alkanol	153
4-halo-1-alkene	1-halo-1-alkene	342
4-halo-1-alkene	3-halo-1-alkene	341
5-halo-1-alkene	alkene	342
5-halo-1-alkene	3-halo-1-alkene	342
5-halo-1-alkene	halobicycloalkane	343
2-halo-2-alkenenitrile	2-halo-2-thioalkenenitrile	143
2-halo-2-alkenoate ester	alkanal	182
2-halo-2-alkenoate ester	alkanone	182
2-halo-2-alkenoate ester	1,1-diseleno-1-alkene	197
2-halo-2-alkenoate ester	haloformate ester	197
2-halo-2-alkenoate ester	2-halo-2-thioalkanoate ester	143
2-halo-2-alkenoate ester	2-seleno-2-alkenoate ester	197
2-halo-2-alkenoate ester	2,2,2-trihaloacetate ester	182
3-halo-2-alkenoate ester	2-alkenoate ester	325
2-halo-2-alkenoic acid	1,1-diseleno-1-alkene	197
2-halo-2-alkenoic acid	2-seleno-2-alkenoic acid	197
3-halo-2-alkenoic acid	3-halo-2-alkenoic acid (substitution)	187
5-halo-3-alkenoic acid	diketene	873
2-halo-2-alken-1-ol	alkanal	186, 197
2-halo-2-alken-1-ol	alkanone	186, 197
2-halo-2-alken-1-ol	2-alkyn-1-ol	224
2-halo-2-alken-1-ol	1,1-diseleno-1-alkene	197
2-halo-2-alken-1-ol	1-halo-1-alkene	186
2-halo-2-alken-1-ol	2-seleno-2-alken-1-ol	197
3-halo-2-alken-1-ol	2-alkyn-1-ol	218, 224, 951
3-halo-3-alken-1-ol	alkanal	554, 570
3-halo-3-alken-1-ol	alkanone	554, 570
3-halo-3-alken-1-ol	2,3-dihalo-1-alkene	554
3-halo-3-alken-1-ol	2-halo-3-silyl-1-alkene	569
4-halo-3-alken-1-ol	alkanal	568
4-halo-3-alken-1-ol	alkanone	568
4-halo-3-alken-1-ol	3-boryl-3-halo-1-alkene	568
4-halo-3-alken-1-ol	1-halo-3-lithio-1-alkene	568
5-halo-4-alken-4-olide	4-alkynoic acid	951
6-halo-5-alken-5-olide	5-alkynoic acid	951
2-halo-2-alken-1-one	alkanoate ester	696
2-halo-2-alken-1-one	2-alken-1-one	325, 372
2-halo-2-alken-1-one	1-halo-3-lithio-1-alkene	696
2-halo-2-alken-1-one	2-halo-2-thio-1-alkanone	143
2-halo-3-alken-1-one	alkanoate ester	696
2-halo-3-alken-1-one	2-alken-1-one	372
2-halo-3-alken-1-one	1-halo-3-lithio-1-alkene	696
3-halo-2-alken-1-one	acyl halide	708
3-halo-2-alken-1-one	1,3-alkanedione	375, 663
3-halo-2-alken-1-one	alkyne	708
3-halo-2-alken-1-one	2-alkyn-1-one	228, 335
3-halo-2-alken-1-one	3-mesyloxy-2-alken-1-one	375

To	From	Page
4-hydroxy-1-alkanone	alkanol	796
4-hydroxy-1-alkanone	alkanone	758–760, 762
4-hydroxy-1-alkanone	alkanone dianion	762, 763
4-hydroxy-1-alkanone	alkanone hydrazone	760
4-hydroxy-1-alkanone	alkanone imine	758, 759
4-hydroxy-1-alkanone	alkanone oxime	762
4-hydroxy-1-alkanone	2-alken-1-one	796
4-hydroxy-1-alkanone	epoxide	758–760, 762, 763
4-hydroxy-1-alkanone	5-hydroxy-1-sulfinyl-2-alkanone	763
4-hydroxy-1-alkanone	2-sulfinyl-1-alkanone	763
5-hydroxy-1-alkanone	5-alkanolide	697
5-hydroxy-1-alkanone	2-alken-1-one	796
5-hydroxy-1-alkanone	lithioalkane	696
2-hydroxy-1-alkanone acetal	2,3-epoxy-1-alkanone acetal	507
3-hydroxy-1-alkanone acetal	2,3-epoxy-1-alkanone acetal	507
2-hydroxy-1-alkanone dithioacetal	alkanal	726
2-hydroxy-1-alkanone dithioacetal	1,2-alkanedione dithiomonoacetal	546, 547
2-hydroxy-1-alkanone dithioacetal	alkanedithioate ester	726
2-hydroxy-1-alkanone dithioacetal	alkanone	726
3-hydroxy-1-alkanone dithioacetal	alkanal dithioacetal	724
3-hydroxy-1-alkanone dithioacetal	epoxide	724
3-hydroxy-1-alkanone dithioacetal	haloalkane	724
N-(3-hydroxyalkanoyl)oxazolidinethione	N-acyloxazolidinethione	908
N-(3-hydroxyalkanoyl)oxazolidinethione	alkanal	908
N-(2-hydroxyalkanoyl)oxazolidinone	N-acyloxazolidinone	489
N-(3-hydroxyalkanoyl)oxazolidinone	N-(3-oxoalkanoyl)oxazolidinone	530
N-(3-hydroxyalkanoyl)oxazolidone	N-acyloxazolidone	907
N-(3-hydroxyalkanoyl)oxazolidone	alkanal	907
N-(3-hydroxyalkanoyl)thiazolidinethione	N-acylthiazolidinethione	908
N-(3-hydroxyalkanoyl)thiazolidinethione	alkanal	908
4-hydroxy-2-alkenal	alkanal	736
4-hydroxy-2-alkenal	alkanone	736
2-hydroxy-4-alkenamide	3-oxa-5-alkenamide	522
3-hydroxy-4-alkenamide	alkanamide	901
3-hydroxy-4-alkenamide	2-alken-1-one	901
5-hydroxy-2-alkenamide	alkanal	904
5-hydroxy-2-alkenamide	alkanone	904
5-hydroxy-2-alkenamide	2-alkenamide	904
5-hydroxy-2-alkenamide	3-alkenamide	904
6-hydroxy-3-alkenamide	alkanal	905
6-hydroxy-3-alkenamide	alkanone	905
6-hydroxy-3-alkenamide	4-alkenamide	905
2-hydroxy-4-alkenenitrile	3-oxa-5-alkenenitrile	522
2-hydroxy-4-alkenenitrile	2-oxoalkanenitrile	700
2-hydroxy-4-alkenenitrile	3-silyl-1-alkene	700
4-hydroxy-2-alkenenitrile	alkanal	866
4-hydroxy-2-alkenenitrile	alkanone	866
4-hydroxy-2-alkenenitrile	2-sulfinylalkanenitrile	866
2-hydroxy-3-alkenoate ester	2-alkenoate ester	889
2-hydroxy-3-alkenoate ester	3-alkenoate ester	889
2-hydroxy-4-alkenoate ester	3-oxa-5-alkenoate ester	522
3-hydroxy-4-alkenoate ester	alkanoate ester	874, 875
3-hydroxy-4-alkenoate ester	2-alken-1-one	874, 875
4-hydroxy-2-alkenoate ester	alkanal	865
4-hydroxy-2-alkenoate ester	2-hydroxy-1-alkanone	175
4-hydroxy-2-alkenoate ester	2-sulfinylalkanoate ester	865
5-hydroxy-2-alkenoate ester	alkanal	555, 556, 566, 890, 892

To	From	Page
4-(1-hydroxyalkyl)-1,3-dioxolane-4-carboxylate ester	alkanone	881
4-(1-hydroxyalkyl)-1,3-dioxolane-4-carboxylate ester	1,3-dioxolane-4-carboxylate ester	881
(1-hydroxyalkyl)malonate diester	alkanal	575
(1-hydroxyalkyl)malonate diester	halomalonate diester	575
(1-hydroxyalkyl)malonate diester	malonate diester	575
2-(1-hydroxyalkyl)oxacycloalkane	ω-alken-1-ol	451
4-(1-hydroxyalkyl)-3-oxo-5-alkenoate ester	alkanal	773
4-(1-hydroxyalkyl)-3-oxo-5-alkenoate ester	alkanone	773
4-(1-hydroxyalkyl)-3-oxo-5-alkenoate ester	3-oxo-4-alkenoate ester	773
N-(2-hydroxyalkyl)phosphorylamide	alkanal	906
N-(2-hydroxyalkyl)phosphorylamide	alkanone	906
N-(2-hydroxyalkyl)phosphorylamide	phosphorylamide	906
N-(3-hydroxyalkyl)phosphorylamide	epoxide	906
N-(3-hydroxyalkyl)phosphorylamide	phosphorylamide	906
2-(1-hydroxyalkyl)-3-stannyl-2-alkenoate ester	alkanone	229
2-(1-hydroxyalkyl)-3-stannyl-2-alkenoate ester	2,3-distannyl-2-alkenoate ester	229
2-(1-hydroxyalkyl)tetrahydrofuran	alkanal	575
2-(1-hydroxyalkyl)tetrahydrofuran	4-alken-1-ol	452, 453
2-(1-hydroxyalkyl)tetrahydrofuran	tetrahydrofuran	574
2-(1-hydroxyalkyl)thiazole	1-(2-thiazolyl)-1-alkanone	546
2-(1-hydroxyalkyl)-3-thio-2-alkenenitrile	alkanal	187
2-(1-hydroxyalkyl)-3-thio-2-alkenenitrile	3-thio-2-alkenenitrile	187
2-hydroxy-4-alkynoate ester	1-aluminio-1-alkyne	518
2-hydroxy-4-alkynoate ester	2,3-epoxyalkanoate ester	518
4-hydroxy-2-alkynoate ester	alkanal	561
4-hydroxy-2-alkynoate ester	alkanone	561
4-hydroxy-2-alkynoate ester	3-lithio-2-alkynoate ester	561
4-hydroxy-2-alkynoic acid	alkanal	561
4-hydroxy-2-alkynoic acid	alkanone	561
4-hydroxy-2-alkynoic acid	3-lithio-2-alkynoate lithium salt	561
5-hydroxy-2-alkynoic acid	alkanal	871
5-hydroxy-2-alkynoic acid	2-alkynoic acid	871
5-hydroxy-2-alkynoic acid	epoxide	513
5-hydroxy-2-alkynoic acid	3-lithio-2-alkynoate lithium salt	513
ω-hydroxy-ω-alkynoic acid	ω-alkyn-1-ol	297
ω-hydroxy-ω-alkynoic acid	ω-haloalkanoic acid	297
ω-hydroxy-1-alkyn-3-one	ω-alkanolide	696
ω-hydroxy-1-alkyn-3-one	1-lithio-1-alkyne	696
1-(3-hydroxy-1-alkynyl)cycloalkanol	cycloalkanone	667
N-hydroxy-2-aminoalkanenitrile	amine	862
2-hydroxyarenecarboxylate ester	3-oxo-6-alkenoate ester	97
4-hydroxyarenecarboxylate ester	4,6-dioxo-2-alkenoate ester	97
hydroxybicycloalkanone	3-(2-oxocycloalkyl)alkanenitrile	577
3-hydroxycycloalkanecarbonitrile	ω,ω-epoxyalkanenitrile	518, 519
2-hydroxycycloalkanecarboxylate ester	ω-oxoalkanoate ester	874
2-hydroxycycloalkanone	1,ω-alkanedioate diester	645, 646
2-hydroxycycloalkanone	1,2-bis(silyloxy)cycloalkene	646
3-hydroxycycloalkanone	1,3-cycloalkanedione	546
1-(3-hydroxycycloalkyl)-1-alkanone	ω,ω-epoxy-1-alkanone	518, 519
1-(1-hydroxycycloalkyl)-2-alken-1-one	cycloalkanone	667
2-hydroxycyclobutanecarbonitrile	alkene	671
3-hydroxycyclohexanecarboxamide	6,7-epoxyalkanamide	518, 519